The Birds of
Shropshire

The Birds of
Shropshire

Leo Smith
on behalf of Shropshire Ornithological Society

Leo Smith was supported by an Avifauna Working Party of six other members,
all of whom have made a major contribution to the finished publication.

Jim Almond   **Geoff Holmes**
John Arnfield   **John Tucker**
Allan Dawes   **Graham Walker**

Short biographies of all seven can be found on p. 513.

Liverpool University Press

First published 2019 by
Liverpool University Press
4 Cambridge Street
Liverpool
L69 7ZU
www.liverpooluniversitypress.co.uk

British Library Cataloguing-in-Publication data
A British Library CIP record is available

ISBN 978-1-78138-259-2 cased

Printed in Poland by Opolgraf.

Front cover photograph by Jim Almond
Back cover photograph by John Fielding

Contents

Foreword

Amanda Craig (Operations Director, Natural England)

Publication of *The Birds of Shropshire* is particularly special because it draws on local research which complements the national BTO Bird Atlas 2007–11. The combination of species accounts alongside detailed local maps provides the most detailed and comprehensive record of Shropshire's avifauna ever produced.

The book is an impressive outcome of collective effort, based on six years of intensive survey, accompanied by records of change and distribution from several surveys over a much longer time period – all carried out by volunteers. Each individual species account is a labour of love and the result is an easily readable, top-quality publication, of which the Society should be proud.

I am particularly enthused by the contents of this book as it makes understanding our birds accessible to ordinary people, and a copy of this book will be placed in every state secondary school in Shropshire, and in Telford and Wrekin, thereby introducing the wonderful world of birds to future generations.

This Avifauna takes our knowledge and understanding of birds in Shropshire to a new level. It shows current distribution and relative abundance of bird species that currently occur locally and how the distribution has changed since 1990. I find the sections on climate change particularly poignant, showing earlier arrival dates of some migrants – or even increased overwintering of some species such as Blackcap. The Pied Flycatcher, however, is now under pressure as its chicks miss out on peak food availability because caterpillars are emerging much earlier, but the birds have yet to adapt their arrival dates, so their breeding success has declined.

Natural England is the UK Government's statutory advisor for the natural environment in England. Environmental conservation in the UK has achieved much over the past 60 years, and we continue to build up a vast body of knowledge about habitats, species and landscapes, through the combined efforts of many people, particularly volunteers. Having this knowledge at our fingertips is vital if we are to understand more about what's happening to our birds and it will helps us decide where best we need to focus our conservation effort.

The Government has set out in its 25 year Environment Plan – published in 2018 – its goals for improving the environment within a generation and leaving it in a better state.

A significant proportion of Natural England's role is responsibility for protected sites, the nationally important Sites of Special Scientific Interest and National Nature Reserves, and for European and internationally important sites for habitats, birds and protected species. Part of Natural England's role is providing advice on how land can be managed for nature conservation. Having the necessary evidence available to us is essential and the analysis of detailed local maps set out in this book will contribute significantly to our local knowledge.

Some of the mechanisms we currently have available to help deliver nature conservation outcomes are the various agri-environment schemes. In the recent past, Shropshire Ornithological Society has provided us with maps at the tetrad level showing the distribution of target species, and the Community Wildlife Groups have provided us with more detailed information on nest sites and foraging areas for Curlew and Lapwing. This enabled us to secure better management options in the agri-environment agreements that we recently put in place in the Shropshire Hills and the River Clun catchment.

The timing of this book couldn't be better to help us and landowners understand more about the needs of those species we are working on as part of delivering the 25 Year Environment Plan locally. This, and initiatives such as the Big Farmland Bird Count, will particularly help us identify target areas and options for where potential future action is needed to help farmland birds, such as the Tree Sparrow, Corn Bunting, Yellowhammer, Lapwing and Curlew, and the habitats that they need to feed, breed and raise their young, and guide all of us in how we might stem the declines and secure better environmental outcomes.

This book will most definitely shape conservation priorities in the County for at least the next 20 years, and I would like to say a tremendous thank you to everyone who has contributed.

Shropshire Ornithological Society gratefully acknowledges a grant of £5,000 from Natural England towards the cost of producing *The Birds of Shropshire*.

Foreword

Dawn Balmer (Head of Surveys at the British Trust for Ornithology, and Co-ordinator and Lead Author of the National Bird Atlas 2007–11)

Shropshire is a special county. I grew up watching Yellowhammers in the hedges around Vennington, listening to the Curlews bubbling in the summertime out the back of our house in Horsebridge and when my family moved to Pontesbury, spent endless hours exploring Earl's Hill. I was lucky to have Wood Warblers, Redstarts and Pied Flycatchers on my doorstep and to be able to watch Buzzards and Ravens overhead, and even find breeding Peregrines not too far from home. I adopted Polemere as my local patch in the mid-1980s and used to cycle there through the lanes from Pontesbury, or get my parents to stop off if we were driving past, and look over the water and fields from the roadside fence. It gave me great pleasure to open the bird hide in June 2007 and I still like to call in whenever I'm over visiting and see what's around – and enjoy much closer views of the birds!

I was fortunate to be part of the Young Ornithologists' Club local group in Pontesbury in the mid-1980s, ably run by the late June North, which encouraged my enthusiasm for local bird-watching, and trips outside the county opened my eyes to new and exciting species and habitats. School work experience placements, and later, summer holiday volunteering, at the Shropshire Wildlife Trust (SWT) gave me a fantastic opportunity to meet and work alongside some great people and learn more about the county and its wonderful wildlife. John Tucker and the late Colin Wright have played an important role in my life; always encouraging and providing many opportunities for a young person to get involved. They got me involved in local BTO-led surveys and perhaps most memorably, fieldwork for the local atlas in the late 1980s. I used to get around by bicycle and remember the excitement of finding breeding Sand Martins out at Stapleton, and best of all, hearing my first Quail at Farley.

One of my tasks whilst volunteering at the SWT was to enter data from atlas cards and to help prepare the dot distribution maps for *An Atlas of the Breeding Birds of Shropshire*. All good experience for things to come later in my life! The publication of this local atlas was instrumental in providing baseline information on the distribution of birds in the county, and provided an important tool for local conservationists.

I must have caught the 'atlas bug' in Shropshire and was very fortunate go on to co-ordinate and be lead author for the *Bird Atlas 2007–11* project for Britain and Ireland; a once in a lifetime opportunity. One of the highlights of working on *Bird Atlas 2007–11* was working with bird clubs and local groups across Britain who were undertaking their own intensive atlas work at the same time. It's been so exciting to see the publication of county atlases, avifaunas, websites and DVDs across Britain; more than 20 in total. An enormous effort by so many people. Of course, all these local projects ensured that coverage for the Britain and Ireland atlas was excellent; a wonderful example of partnerships in action.

At the core of *The Birds of Shropshire* is the atlas fieldwork undertaken between 2007 and 2013, co-ordinated by the then BTO Regional Representative Allan Dawes, and supported by the local Atlas team. Underlying the maps that form such an important part of this book are thousands of hours of fieldwork covering every corner of Shropshire, many hundreds of hours checking the records to ensure they are correct and the skilful manipulation and presentation of the records. The local organisers of the atlas work are to be congratulated for achieving such excellent coverage through motivating volunteers. These new maps of distribution, and more importantly, the change maps, will be important for identifying priority species requiring urgent conservation action and more detailed monitoring.

Shropshire can be immensely proud of its long history of volunteer survey work, monitoring of the scarcer species by specialist groups and long-term ringing studies, all of which makes this such a data-rich county. Bringing all this together in an accessible way to a wide range of readers is no easy task, and the authors are to be congratulated on their perseverance, leadership and teamwork. This is a wonderful legacy for the next generation of ornithologists to aspire to.

Foreword

Frank Gribble and Peter Deans (SOS Presidents)

This Avifauna has been published to celebrate the sixtieth anniversary of the Society. We trust it will be appreciated and used by current and future members as a record of what has happened to our birds in the last 60 years. We hope it can highlight where future conservation effort is needed.

We think our founder Chairman E.M. Rutter, would be delighted at the effort that has been put in by so many observers and pleased with the result. Thank you all who have contributed.

In our early days there was little information of substance published on bird populations and distribution since H.E. Forrest's *The Fauna of Shropshire* in 1899. It was clear to our founders that so much had changed over the previous 60 years. This led to the Society publishing *A Handlist of the Birds of Shropshire* in 1964 as a baseline for further studies. Since then our members have participated in various national surveys, mostly organised by the British Trust for Ornithology, as well as contributing records and articles to our annual County Bird Reports (SBRs). Our annual reports were complemented by *An Atlas of the Breeding Birds of Shropshire*, published in 1992.

However this did not deal with passage migrants and winter birds, hence when the BTO decided on a new national Bird Atlas 2007–11 the committee decided to incorporate our local results into a new Avifauna. To get complete coverage, the work of collecting records was extended by two years.

Much has changed since the early days; we now have more observers; computers have enabled us to record so much more easily; and identification aids are so much better.

As a generally agricultural county we have also seen major changes to our countryside and its bird life. Dairy farming is no longer such a prominent feature, a lot of marginal land has been brought into cultivation, and, especially in the uplands, stock rearing and sheep in particular have led to short grassland over much of the area. More intensive cereal growing, along with sugar beet and potatoes, has occurred on the lower ground in the east, although intensive sugar beet cultivation ceased with the closure of Allscott Sugar Factory in 2007.

Much of the former mosses and wet ground at the head of our smaller rivers has been deeply drained to bring the land into more intensive production. This in turn has led to quicker run-off of water, so, coupled with similar effects in the Welsh Hills to the west of the Severn–Vyrnwy confluence, flooding is often worse than previously.

This has contributed to large declines in our population of Lapwing, Curlew, Snipe, Skylark and Yellow Wagtails.

Agriculture has become highly mechanised and old buildings have been replaced with metal barns, not conducive to nesting Barn Owls and Swallows. Hedgerows are mechanically trimmed and old trees are no longer tolerated. Herbicides have led to less weed infested crops so even if stubbles are left in the winter they contain few seeds to sustain Grey Partridge, Skylarks, finches, buntings and sparrows in their former numbers.

Insecticides have reduced insect populations, the good and the bad, and this in turn has led to reductions among many species, especially those dependent on insects in their early stage of life such as waders and game chicks, Yellow Wagtails, probably some warblers and birds like Spotted Flycatchers. As evidence of this change, those of us who drove at night 40–50 years ago had to clean the dead moths, beetles, flies and midges from our windscreens and headlights next morning before we drove again.

The growth of urbanisation, industry and major road improvements, around all our villages and towns, but especially Telford New Town, have led to more loss of land, much of it marginal or left abandoned from former industrial use. This too has had a big effect, some good and some bad. On the plus side, new quarries for stone, sand and gravel extraction to support this work have given opportunities to create new habitats.

The Society's acquisition of Venus Pool, a site of former gravel extraction, has provided an opportunity to create a reserve of differing habitats not just for water birds. It has also encouraged public access for all and we have been able to demonstrate how effective some of the agricultural support schemes can be in providing winter food and cover for our birds in an intensively farmed area.

The Shropshire Wildlife Trust reserve at Wood Lane in the north is another good example of a former quarry being turned into excellent bird habitat. However these successes have been offset by the loss of the Sugar Beet Factory site at Allscott in recent years. It is sad that the bird-ringing effort there for more than 50 years will not continue.

All is not doom and gloom! Against these losses are the notable gains. Sixty years ago the small population of Buzzards and Ravens were confined to the south-west and western hills. Buzzards were just recovering from the death of most rabbits due to myxomatosis. Both species are now widespread and have been joined by Peregrine, Red Kite and a few Goshawks. The growth of nest box schemes for Barn Owls and Pied Flycatchers are positive examples of the ways that a few keen volunteers can make a difference for species that primarily lack suitable nesting habitat. Goosanders are found nesting, along with Shelduck and Oystercatcher, along rivers cleaned of most pollution, and it may not be long before Little Egrets breed in the County.

Better bird protection laws have reduced egg collecting, catching of wild birds for caging and improved gamekeeping. There are still the odd rogues out there but no longer do we see keepers' gibbets or pole traps set on our moors as used to be the case.

Much has been lost, but much has been gained, over the 60 years since the Society was formed, and there is no sign that change will be any slower in future. Climate change may well be something we can do little about as individuals, but we can all try and make sure our birds are recorded, conserved where possible, and cherished by future generations.

Frank C. Gribble
President 1995–2016, Vice-President 1965–1995
Honorary Secretary 1958–1963

Peter G. Deans
President 2017–, Chairman 1979–2016

Preface

Leo Smith

The Birds of Shropshire is the most comprehensive record of the County's avifauna ever published. It has been produced by Shropshire Ornithological Society (SOS, also referred to as the Society), and is based largely on records collected since the early nineteenth century, most recently the results of six years of Atlas fieldwork between November 2007 and July 2013, winter and summer, by over 650 different observers who submitted over 333,400 records. These records have been used to produce maps showing the distribution of almost 200 different species.

The book includes an account for each of these species, describing its distribution and relative abundance, and the breeding status where relevant. The current breeding and winter maps have been compared with those shown in *An Atlas of the Breeding Birds of Shropshire* (1992) and with the Shropshire part of the national *An Atlas of Wintering Birds in Britain and Ireland* (1986). Historical data, results of specific local studies, and population estimates have been incorporated.

All 301 species on the 'Shropshire List' have been covered, including migrants, County rarities, and a few that have only been seen here prior to 1950. Full references have been given in each account. The standard works, and abbreviations used for them, are listed in Table 3 on p. 71, and a complete list of references is given in Appendix 4.

The book shows that massive changes have occurred in the population and distribution of many species. It will shape conservation priorities in the County for the next 20 years.

Other chapters describe the history of ornithology and bird recording, the changing arrival dates for migrants, the wide variety of habitats, and the factors that affect them, particularly agricultural and climate change.

Forewords have been contributed by Amanda Craig, Operations Director of Natural England, confirming the importance of the results and research for setting conservation priorities and targeting scarce resources; by Dawn Balmer, co-ordinator of the acclaimed *Bird Atlas 2007–11* for the whole of Britain and Ireland, published in 2013 by the British Trust for Ornithology, putting the local book in its national context; and by the previous and current Presidents of SOS, Frank Gribble MBE and Peter Deans.

An Avifauna working party has helped steer the production, and *The Birds of Shropshire* has been written by 27 of the County's most experienced and knowledgeable birders. Stunning images of over 220 species have been contributed by 21 local photographers, both amateur and professional, and nine photographers have provided the images in the *Habitats* chapter. The book was planned as part of the sixtieth anniversary celebrations of the Society.

Each account or article has been written by a named author. Views expressed are not necessarily the views of SOS.

Area Covered

The Birds of Shropshire covers the whole of the current county of Shropshire, an area of 3,519sq.km, including the borough of Telford and Wrekin, which was established as a unitary authority through local government reorganisation in 1998. SOS has covered this area since its formation in 1956, and continues to do so. The area covered by the local bird atlases is almost identical, but it does not follow the County boundary precisely (see p. 484).

The boundary was largely defined in the sixteenth century. There were a few small changes in the first half of the nineteenth century, but they have remained unchanged since the publication of the Watsonian vice-county system for botanical recording in 1852. Bird recording has followed the same boundary.

Why?

A Bird Atlas for Shropshire is especially important. The County lies at the meeting point of the Welsh hills to the west and the Midland plain to the east, with their contrasting terrain, land use, habitats and wildlife. This geographical position also determines the climate, which is intermediate between that of the warm, dry south-east of England and the cooler, wetter regions to the north and west, with considerable variation between the uplands and lowlands. Thus Shropshire is on the edge of the range of many British birds and any changes in the future should manifest themselves early here.

The Birds of Shropshire therefore provides a benchmark against which future trends can be monitored, so it will become a major reference book. Local authorities, voluntary bodies and landowners, and anyone else concerned with planning, nature conservation and protection of important wildlife habitats, will use it for the foreseeable future. Habitat protection is especially important, as birds can only flourish if their breeding and feeding areas are safeguarded. It has highlighted which areas and habitats are particularly important, and has helped identify sites, species and land uses which require care in the future.

Publication has also opened up new horizons and opportunities. Further fieldwork is already underway to survey in detail the important sites that have now been identified, and the results and distribution maps pose as many questions as they resolve. Several issues worthy of additional research are listed in the summary and conclusions, and it is hoped that *The Birds of Shropshire* will inspire further inquiries into their fascinating world.

Thank You

The publication is only possible thanks to the efforts of the field-workers for several separate local and national Atlas projects, mostly members of the SOS, many of whom spent a considerable amount of time over several years searching for the more elusive species; the Atlas Organiser and Area Co-ordinators, who organised the fieldwork and checked the results; all the many thousands of individuals who have submitted records over two centuries to the early authors, the Caradoc and Severn Valley Field Club (CSVFC) and SOS County Bird Recorders; and Liverpool University Press and BBR Design.

In addition to anticipated sales income, production costs have been funded by SOS, Natural England's Regional Innovation Fund, the Shropshire Ecological Data Network, the William Dean Trust, and bequests from Molly Donoghue and Jack Sankey. The Walker Trust has agreed to fund provision of a copy to every state secondary school in Shropshire, and Telford and Wrekin.

Contributors are named and thanked in Appendix 5. All these people can be justly proud of their contribution to *The Birds of Shropshire*, a major reference book for many years to come.

Conventions Used in the Text

The full name of each chapter appears in the Contents. In each chapter, there are cross references to the content of other chapters. Rather than use the full title, chapters are referred to as follows:

* in the *History* chapter
* in the *Arrival Dates* chapter
* in the *Habitats* chapter
* in the introduction to the *Species Accounts*
* in the *Breeding Status* chapter ...

The abbreviations below are used for the named organisations throughout the book.

A list of other frequently used abbreviations, mainly for standard references, regular publications and monitoring surveys, referred to in *Species Accounts*, can be found in Table 3 on p. 71, and a list of references cited in the text can be found in Appendix 4.

BOU	The British Ornithologists Union, which maintains the British List, the official list of wild birds recorded in Great Britain
BTO	The British Trust for Ornithology, which has organised bird surveys and ringing since 1933
CSVFC	The Caradoc and Severn Valley Field Club, a voluntary organisation set up to further the knowledge and study of the natural and built environments, and which published the *Record of Bare Facts* and *Transactions*, the forerunners of Shropshire Bird Reports (SBRs), from 1892 onwards
Defra	The Department for the Environment, Food and Rural Affairs, a UK government department formed in 2001, and responsible for Natural England, the Forestry Commission and the Environment Agency
EA	The Environment Agency is an executive non-departmental public body, sponsored by Defra, responsible for river management, water quality and pollution control, and flood defences
EN	English Nature, the government agency responsible for the conservation of wildlife and wild places from 1990 to 2006
FC	The Forestry Commission, a non-ministerial government department which owns many forests and woodlands, is responsible for forestry policy
FE	Forest Enterprise, an agency of the Forestry Commission responsible for the direct management of its estate
JNCC	Led by a Joint Committee, JNCC is the successor body to the UK-wide Nature Conservancy Council, which was split into three national agencies (English Nature, the Countryside Commission for Wales and Scottish National Heritage) in 1990. The JNCC supports the devolved agencies, and co-ordinates UK government policy
MAFF	The Ministry of Agriculture, Fisheries and Food, a UK Government department which became part of Defra in 2001
NE	Natural England replaced English Nature, and incorporated parts of the Rural Development Service and the Countryside Agency, in 2006. It is funded by Defra, and is responsible for the condition of Sites of Special Scientific Interest (SSSIs), the management of National Nature Reserves (NNRs) and the implementation of agri-environment schemes (e.g. payments to owners and tenant farmers for managing their land to benefit the environment, including bird habitats)
RSPB	The Royal Society for the Protection of Birds
RSPCA	The Royal Society for the Protection of Cruelty to Animals
SOS	The Shropshire Ornithological Society, also known as the Society
SWT	The Shropshire Wildlife Trust, formerly STNC, the Shropshire Trust for Nature Conservation
WWT	The Wildfowl and Wetland Trust

These organisations have their own websites, which provide further information about their work.

History of Bird Recording and Ornithology in Shropshire

John Tucker (Part 1) and Leo Smith (Part 2), with both contributing to Part 3 and the table

Part 1. The Early Period (Up to 1950)

The Earliest Records

Identified bird bones from the excavations of the baths basilica at Roman Viroconium (Wroxeter), dating from the late third to the seventh century, provide the earliest bird list (Meddens 1987; Hammon 2011). The two contributions give 32 species, the identification of birds from bone fragments being a matter of experience and judgement. Alphabetically the species are Barn Owl, Barnacle Goose, Black Grouse, Blackbird, Buzzard, Carrion Crow, Common Crane, Corncrake, Fieldfare, Garganey, Grey Heron, Grey Partridge, Golden Plover, Goldeneye, House Sparrow, Jackdaw, Jay, Lapwing, Long-eared Owl, Magpie, Mallard, Moorhen, Mute Swan, Quail, Raven, Rook, Song Thrush, Teal, Water Rail, Whooper Swan, Woodcock and Woodpigeon. Woodcock bones were the most abundant in the assemblage, and the Common Crane, with two of the three bones showing signs of butchery, might also be counted among the food species (Yalden 2002). The other species were likely either prey species taken for food or incidentals, although Raven, the second most abundant, was probably resident. Perhaps the Barnacle Goose was traded from the wintering grounds on the Solway Firth via the then Roman town of Luguvalium (Carlisle).

Long before formal recording, people were observing birds closely. Old local names, such as 'Brand-Tail' (firebrand) for Redstart, 'Nettle-Creeper' for Whitethroat and 'Water-Weasel' for Water Rail, are evidence of a keen appreciation of the field-points and behaviour of the birds which were part of everyday life.

Several place names derive from associations with birds: Kitesnest, Raven's Nest and Cronkhill, a name mimicking the Raven's call. Other names are based on goose, heron and Woodcock. Intriguingly, crane has two, Cranmoor Gorse on what was Knockin Heath and Cranmere, near Worfield, which may also have been heath, but as the name 'crane' was often used for Grey Heron we cannot be certain which was meant. Place names also offer clues to past habitats such as heaths and commons, moors, mosses and marshes, especially in the north, before enclosure and farming. The map in the Black Grouse species account (p. 122) shows the location of 331 place names which include 'heath', 'common', 'moor' or 'moss', which suggests that these habitats were once widespread. The village of Cockshutt encapsulates both landscape history and a bird in its name: the suffix -shutt or -shot (found in 10 place names) derives from woodland clearings created, or at least used, for netting Woodcock.

The first written reference appears in parish accounts. White-tailed Eagles may have been patrolling the skies here in the early sixteenth century (Holloway 1996), perhaps even breeding around the meres, for in 1514 a payment was made 'to William Hichecox for an yron [White-tailed Eagle] iij [3 pennies]' in the parish of Worfield, near Bridgnorth (Lovegrove 2007). Nationally, two Acts were later passed 'for the preservation of grayne' in 1532 and 1556 under which rewards were offered and birds of prey were a target. Listed in the Acts as vermin were 40 species of bird, including Kingfisher and Dipper for their supposed depredations upon fish, and Bullfinch for its known damage to orchards. Locally, the surviving accounts of churchwardens, who continued making these payments, have been analysed by Collingwood (2012); they include, for example, a total of 596,374 House Sparrows killed before the Acts were repealed in 1861, although only a small proportion of parishes made payments for sparrows.

The Itinerary of John Leland in or About the Years 1538–43, describing his tour of England and Wales at the behest of Henry VIII, written in Latin and translated into Old English, tells 'Ther is another [hill] cawllyd Caderton's Cle, and ther be many hethe cokks' (Toulmin Smith 1910). The hill is Catherton Common and the 'hethe cokks' Black Grouse. At that time much of the County would have been uncultivated, rather wild, lightly wooded open heath, ideal country for Black Grouse.

From the seventeenth century, first is a list of food and accommodation items, probably resulting from a visit to Shrewsbury by the Lord President of the Council of Wales and the Marches on 7 October 1634 (Shropshire Archives ref. 3365/579/48). The list includes Woodcock, wild Mallard, Snipe, lark, Quail, Teal and Widgeon (an early generic term for 'ducks'), along with chicken, turkey, ducks, geese and pigeon.

Second is the description of what is clearly a Nightjar, from a duck decoy somewhere in the south, on 4 May 1668, reported ultimately to Francis Willoughby, the very early ornithologist (Wood 1958). It

A wood engraving, *Haymaker*, by Thomas Bewick (1753–1828), reproduced by kind permission of the Natural History Society of Northumbria, Great North Museum: Hancock. It depicts the accidental exposure of a Corncrake's nest and eggs by a reaper who killed the sitting bird. A local example of this occurred when a mechanical harvester killed a female on her nest at Morda in July 1867. Of Bewick's work Charlotte Bronte wrote, in Jane Eyre (1847): 'Each picture told a story: mysterious often to my undeveloped understanding and imperfect feelings, yet ever profoundly interesting … with Bewick on my knee, I was happy'.

was 'catched the night before in a tunnell nett … It was almost as big as a Cukcow, long wings as a Martin, speakled like a Woodcock, a sharp little bill or beak, the eyes standing backward as big as an Owles, with long hairs on each side of the beak like a ratt, with some white feathers on each wing'.

The revelation of a landscape of extensive wetlands in the north-west, on the Baggy (Boggy) Moors bordering the River Perry, appears in the chapter on Waste Land by Archdeacon Plymley (1803). He wrote:

> About 20 years ago there were large tracts of lands [Baggymoor], and other moors from Boreatton to St Martins in winter usually covered with water, but which are now, in consequence of enclosures and drainage, at no great expense, rendered of considerable value … Hither wild-fowl of all sorts usually resorted, and astonishing quantities were annually taken at the decoy near Whittington, the property of Mr. Lloyd, of Aston, but which, from the above improvement, has been suffered to go out of repair, never again, probably, to be appropriated to its former use.

The extent of the tracts of land that were drained in this short period is illustrated by the distance from Boreatton to St Martins, about 15km.

The decoy at Whittington has been traced to the precise location SJ352295, close to Decoy Farm, though no sign of it remains.

There were four other known decoys, at Hawkstone Park in the north, Onslow Hall west of Shrewsbury, Sundorne Castle east of the town, and at Oakley Park near Bromfield (Payne-Gallwey 1886; Heaton 2001; *in litt.*). Brown (2018) discusses them in detail and adds a fifth, just beyond the County boundary at Lymore near Montgomery. Two others have been cited, at Attingham Park and on the Long Mynd, though both were duck flight pools rather than traps. In addition is the small four-sided apparent trap illustrated on an estate map showing Burlington Pool dated to about 1738 (Shropshire Archives ref. SA-972-7-1-33). Decoys generally went out of use in the nineteenth or early twentieth century, the final record being a female Mallard

trapped at the Oakley decoy in November 1950; it had been ringed at Peter Scott's Severn Wildfowl Trust site two months earlier.

The extent of the toll of waterfowl being trapped by decoymen or shot by wildfowlers near Shrewsbury is not recorded, though several nineteenth-century authors refer to the 'widgeon' being on sale in the town's markets.

From the same early period there are three intriguing references, commencing with two from Thomas Pennant. Writing in 1768, he referred to the first British Little Bittern being shot in the Quarry, Shrewsbury (no date given), while in 1776 he wrote of receiving 'a male and a female [Parrot Crossbill] out of Shropshire', but no further information was given, and this latter record was not considered to be acceptable in a recent review (see p. 463). Next comes a note about Hawkstone (Rodenhurst 1783) describing the croaking calls of Ravens there over the imposing crags on which they were nesting.

Shropshire also holds the first British records for four other species; the Grasshopper Warbler, described by Pennant (1768) using a County specimen, the White-rumped Sandpiper shot at Stoke Heath before 1839, the Short-toed Lark described below and the Magnificent Frigatebird found dying near Whitchurch in 2005.

Thomas Bewick produced his two-volume *History of British Birds* (1797 and 1804) with his fine wood-engravings, the first popular book illustrating birds well.

Both Bewick and Gilbert White in Hampshire, who produced his *Natural History of Selborne* in 1789, encouraged and enabled interest in nature. Entirely new species were still being discovered, and Gilbert White had only recently separated Willow and Wood Warblers as distinct species, by their songs. In the early nineteenth century, there was a limited network of perhaps a dozen bird-watchers and recorders here. The handful of naturalists we know of were educated men, often clergymen, in contact with each other through the gentlemanly family and social networks of the time, and by post. All had the books and journals necessary for identification, and time to devote to the collection, contemplation of and correspondence about their specimens and records.

Collections

In those days collecting specimens was essential when dealing with difficult species, by shooting or otherwise 'obtaining' them to be sure of their identity. It used to be said 'what's hit is history, what's missed is mystery'. Thomas Campbell Eyton, then curator of ornithology at Shrewsbury Museum, in 1835 appealed to gentlemen to shoot 'any rare birds which they may meet with on their shooting excursions' in order to increase the collection; their gamekeepers were to be similarly instructed.

It was not only for identification that birds were killed. The big houses of the time often had collections of stuffed mounted birds in cases as items of interest and as statements of their owners' prestige. Notable among these was that of John Rocke in Clungunford, who in the late 1860s was believed to have the most complete collection in Britain. Rocke had on display, in a special wing built onto Clungunford House, examples of both sexes and the young of every British bird. On 8 September 1898 the CSVFC made a train excursion to Clungunford Hall and were 'glad of the claret cup which was hospitably offered. By kind invitation of J.C.L. Rocke, Esq., the club [viewed] the splendid collection of British birds and eggs made by his late father ... displayed in a series of glass cases entirely covering the walls ... even the extinct Great Auk and its egg'. The collection was subsequently sold and dispersed.

Lord Hill started another fine collection in 1848, at Hawkstone, and many of the specimens were collected on the estate, but others came from elsewhere. It was sold to Peplow Hall for its museum in 1904, and catalogued by Forrest (1907). This collection too no longer exists.

The County Museum, and its annexe at Ludlow, also had, and still maintain, their scientific collections, most of which survive as study skins. Some of Rocke's and Hill's collections, and those of other smaller museums, may survive there.

Henry 'Harry' Shaw (1812–1887), a taxidermist with a shop on the High Street in Shrewsbury, was one of the leading suppliers of mounted stuffed birds, famous for the naturalistic settings and backgrounds in his cases. The Earl of Powis was one of Harry's well-to-do patrons and some impressive examples of Shaw's work can still be seen in Powis Castle. Together with his brother John (?–1888), also a taxidermist in the town,

and others in the same trade, Harry kept an eye on the larks trapped for the pot and brought to market, along with the occasional by-catch of other species. The markets provided the taxidermists with occasional unusual and valuable specimens to mount and sell to wealthy patrons competing to build the choicest collections. Such specimens often became the subjects of short notes in the local newspapers, a valuable source of early records which is still not exhausted.

Shaw had a keen interest in birds and sent significant contributions to the *Field*. He recognised a lark brought to him in 1841, caught by a lark-catcher at the Isle, west of Shrewsbury, on 25 October, as the first Short-toed Lark to be recorded in Britain. He passed the bird to William Yarrell, who was at the time writing his *History of British Birds* (1843), and the specimen was used for the illustration in the book. Also, one of only three records of Lapland Bunting came from a lark-catcher, near Shrewsbury in 1852, and the sole Richard's Pipit was brought in alive to Shaw on 2 October 1866 having been caught in a lark-net at Shrawardine.

Henry Shaw's shop and home at 43 High Street, Shrewsbury, in about 1898. The gentleman in the doorway is believed to be his son. Photo: Joseph Lewis della-Porta. Acks. Shropshire Archives.

The sole record of White's Thrush, *Zoothera dauma*, shot near Moreton Corbet on 14 January 1892. Preserved by the taxidermist G. Cooke of Dogpole, Shrewsbury. First in the possession of William E. Beckwith, subsequently with his sister and later passed to Shropshire Museums in Ludlow Museum Resource Centre (specimen SHYMS: Z/2006/123) by his brother Capt. Henry Beckwith in 1925. Photo: Gareth Thomas FRPS. Courtesy Shropshire Museums in Ludlow Museum Resource Centre.

The enigmatic Pallas's Sandgrouse, *Syrrhaptes paradoxus*, labelled 'Edgmond 1880'. Currently with the Shropshire Museums in Ludlow Museum Resource Centre (specimen SHYMS: Z/2006/141) it was initially in the possession of William E. Beckwith, and he was proud of it, but it is not known to have been reported in the literature. 1880 was not, nationally, an 'invasion' year for this species and its existence is rather a mystery. Photo: Gareth Thomas FRPS. Courtesy Shropshire Museums in Ludlow Museum Resource Centre.

Two lark lures, manual left, clockwork right. The circular mother-of-pearl or mirror insets are made into triangular 'Toblerone'-shaped pieces of wood. When slowly turned these sparkle and, seen by larks passing overhead, attract the birds down to be shot or trapped in nets. Photo: Simon Holloway, from his collection.

The Short-toed Lark as it appears in Yarrell (1843)

This female Little Bustard, *Tetrax tetrax*, the only County record, is in the Shropshire Museums in Ludlow Museum Resource Centre (specimen SHYMS: Z/2006/159), labelled 'Edgmond spring 1883'. Unlike the Sandgrouse, from the same locality three years earlier, this bird was widely reported at the time. Photo: John Tucker. Courtesy Shropshire Museums in Ludlow Museum Resource Centre.

The Rise of Ornithology as a Subject for Study

John Freeman Milward Dovaston (1782–1854), of West Felton, was the first inquisitive observer, an early scientific ornithologist. For example he experimented by putting loops of violin wire around the necks of Swallows nesting at his house and found that the same individuals returned the following year; he was among the first to test such ideas. In his early days some people believed that Swallows spent the winter under water in ponds or hibernating in roofs (perhaps a confusion with bats). One investigator of the time went so far as testing Swallows under water, as if confirmation were needed that they would drown! Dovaston was also among the first observers to note that Robins held their own patches of ground, which we now call territories.

Putting quill to paper for the first substantial published work specific to the wildlife of the area, Thomas Campbell Eyton (1809–90) wrote his two-part 'An attempt to ascertain the Fauna of Shropshire and North Wales', published in the *Annals and Magazine of Natural History* in 1838 and 1839. Eyton was a contemporary, friend and correspondent of Charles Darwin, who lived in Shrewsbury until 1831, though the latter's interests lay then in invertebrates rather than birds. Eyton was able to report that the Black Grouse was still to be found on most of the extensive heaths 'but appears to be decreasing'. Red Kites were then rare as a result of the depredations of gamekeepers, but the 'Wrekin Dove' (Turtle Dove) was common. He also noted that Red-backed Shrikes were easily tamed and made mischievous but amusing pets.

The Zoologist was launched in 1843 in response to the increasing popularity of natural history. This publication was posted to subscribers and included short notes, some about birds, contributed by specialists

The Rev H.O. Wilson's Case of Falcons

This story is an example of the contentious and sometimes imprecise and confusing material contained within the historical record.

In 1909, Forrest wrote in *British Birds*:

A case of Falcons has just been presented to the Shrewsbury Museum by Mrs. H.O. Wilson, who states that all the specimens in it were collected on or near the Longmynd, between 1848 and 1857, when her husband was rector of Church Stretton. Besides examples of the commoner species, the series includes all three of the British Harriers, a pair of Kites, an adult Red-footed Falcon (*F. vespertinus*), Goshawk (*Astur palumbarius*), and an immature Iceland Falcon (*F. islandus*). This last had been recorded by Rocke, and was one of two examples obtained at Leebotwood, the other being placed in the Hawkstone collection. The date was not given by Rocke, but I now learn that in the late Rev. H.O. Wilson's diary there is an entry on 5th April, 1853, of payment to Millington (keeper) 'for the Jer-Falcon'; so that the bird was probably obtained before that date. The Red-footed Falcon has been obtained on three other occasions in Shropshire, but the Goshawk never; the species is, therefore, new to the county fauna (Forrest 1909a).

Goshawk was already scarce at the start of the nineteenth century

but then suffered relentless persecution and was exterminated in Britain by the end of that century. Given the absence of suitable habitat and its rarity, and the long time lag between the specimen being collected and the second-hand account of its provenance, there must be some doubt on the validity of this record.

The doubt is reinforced by recent accounts from 'the heyday of taxidermy' of dealers sourcing wild birds from the continent and selling them through bird auctions to collectors and taxidermists: 'If sales could be guaranteed merely by falsifying data labels so collectors could satisfy their desire for large national collections, we should not be surprised that deception was widespread' (Harrop *et al.* 2012). Rev. Wilson paid for at least one specimen and he was dependent on the integrity of the source, and whereas the normal practice was for specimens to be labelled with the locality and date taken, it is likely that the contents of this case were not, otherwise Forrest would presumably have quoted the details, rather than relying on the testimony of Mrs Wilson.

The case also included one of only six historic records of Montagu's Harrier and one of only four historic records of Red-footed Falcon, as well as the only historic record of Goshawk, all supposedly collected in a small area over a very short period of time. While not impossible, this seems very unlikely and casts doubt on the provenance of all the specimens in the case.

throughout Britain, including some from this County. Likewise it was possible to add snippets of bird news to the Nature Notes column of the weekly *The Field* from 1853. These two publications, with a few other journals and magazines which came and went, were a mainstay of ornithological communication until the launch of *British Birds* in 1907. This helped seal the demise of the *Zoologist*, which had become increasingly bird-oriented; and it ceased publication in 1916. *The Field* is still published, though with a different emphasis.

William Beckwith

Late nineteenth-century ornithology was dominated locally by two leading authorities, William Edmund Beckwith (1844–92) and later by Herbert Edward Forrest (1859–1942). Beckwith began his well-informed and expansive outputs, mostly specific to birds but two on botany, aged 23, with a short note in the *Zoologist* on 'Fire-crested Wren', Richard's Pipit and Velvet Scoter, in 1867. He followed that next by a *List of Birds found near Shrewsbury*, published in a local guide book in 1878. Forrest, who was also a reader of, and contributor to, both *The Field* and *Zoologist*, began publishing on his astonishingly diverse interests with a note on 'Ornithological visitors to Shropshire' – about seabirds occurring around Shrewsbury – in the obscure journal *Hardwicke's Science-Gossip* in 1876, when he was aged just 17.

Among the accounts by Beckwith were two fascinating snippets relating to the Great Bustard, though neither allows the species onto the County list. A letter written to Beckwith by Rev. Gledowe of Frodesley in May 1879 stated that his father-in-law saw a pair of Great Bustards on the Longmynd many years previously: 'He was riding from Church Stretton to Ratlinghope, about July in 1826, and saw two birds, apparently a pair. ... They only remained a moment or two and then took flight ... [they were] of a chestnut colour above and ashy grey beneath, but the neck was more of the same thickness throughout than a turkey's'. Beckwith added to the intrigue in 1885: 'I am inclined to think that about that time [ca. 1825] there was a nest on one of the heaths, between Shrewsbury and Oswestry, as a gentleman, now living at Fitz, remembers seeing some tame bustards, probably reared from eggs found in the neighbourhood, at a timber yard belonging to Lord Powis'. Beckwith's description of the observer as a 'gentleman' lends his credence to the account, suggesting that the informant was a cultured, educated and trustworthy witness. However, Forrest (1899) quoted the first of these accounts in full, but concluded that '[t]he occurrence of this fine bird ... is somewhat doubtful'. Modern research supports Forrest's scepticism. Great Bustard used to breed in only 11–13 British counties, and they last nested in 1832 (national BTO Atlas 2013), so they were very rare at this time, and unlikely to occur. The second record suggests captive-bred birds.

William Beckwith was the first to write a full series of species accounts, initially in the two-part 'Birds of Shropshire' published in the *Transactions of the Shropshire Natural History & Philosophical Society* (SNHPS) in 1879 and 1881. The son of the Rector of Eaton Constantine, Beckwith spent much of his time observing, corresponding about and documenting birds and he was famous for carrying field glasses when most of his fellow bird enthusiasts were still carrying guns. A prototype County Bird Recorder, he was evidently corresponding with many contacts as his accounts are, for their time,

---

The Historical Ornithology of Shropshire

The *Historical Ornithology of Shropshire*, the 'Histo' website, is a continually updated record containing everything published about birds in the County. It contains copies of everything known to be formally published in book or journal form, with an expanding range of additional material, together comprising more than 7,400 pages. The emphasis is on material relating to the distribution and abundance of species and specific records. For example, it includes every issue of the CSVFC *Record of Bare Facts* and *Transactions*, from 1892 onwards.

A species index of over 14,400 entries is a powerful addition, allowing easy and rapid access to virtually every species entry. A citations table lists the full bibliographic data for every work, there is a gazetteer of 7,550 entries with grid references and there are other map resources.

Histo is unique in Britain and has been created for the SOS by brothers John and Peter Tucker; Peter designed and maintains the website while John continues the research and development. It can be accessed from the menu on the home page on the SOS website by clicking on 'Historical Ornithology of Shropshire'.

strikingly comprehensive; regrettably his papers have not survived. In 1885–86, in two issues of *The Field*, he produced his 'Notes on Shropshire Birds', in effect appealing to gentleman sportsmen, and their gamekeepers, to send him interesting specimens. In the following year, 1887, he began an expanded set of notes, also in the *Transactions* of the SNHPS; nine more or less annual contributions which ceased in 1892 due to his sudden illness, and, within days, death, on 1 July 1892, aged only 48. He is buried at Eaton Constantine. According to Forrest, it had been Beckwith's intention to write a book on the birds of the County. A book about Beckwith, his story and including all his previously scattered published work, has recently been published (Tucker & Tucker 2018).

The Caradoc and Severn Valley Field Club and H.E. Forrest

The CSVFC was formed in 1892, largely on the initiative of Herbert Edward Forrest (1858–1942); he was to remain the club's leading light and a mainstay of natural and historical enquiry until his death in 1942. From its inception the club published the invaluable annual *Record of Bare Facts* (the *Record*) listing the most significant records of mammals, fish and birds. Beckwith was the compiler of the first annual birds section of the *Record* but did not live to see it published.

The *Record*, together with the *Transactions*, remained the mainstay of the CSVFC's activities for more than half a century. The latter comprised notes of the club's indoor and field meetings together with transcripts of papers presented at its evening meetings. These gatherings were often accompanied by exhibits of recently collected specimens, the talks sometimes illustrated with lantern slides.

This initiative was followed in 1899 by Forrest's *Fauna of Shropshire*, which listed birds and included useful notes on many of the naturalists of the time. He took the opportunity to update some of the information in this work in his account of birds in the *Victoria*

County History (1908). Forrest's works are the only comprehensive account of the birds of Shropshire in single volumes to appear in the early period.

The CSVFC *Transactions* provided an opportunity for the few ornithologists, notably John Hugh Owen (1877–1959) who retired to Llanymynech, and Llewelyn Cyril Lloyd (1905–1968), to publish some of their studies. Owen was an accomplished nest-finder, egg-collector and field naturalist specialising in breeding behaviour and often published in *British Birds*. A comment by John Owen, on a postcard written in his later years – 'Only 177 Robins' and 143 Spotted Flycatchers' [nests found] this year ... I must be getting old' – speaks volumes for both his nest-finding abilities and the sheer numbers of birds at that time. Lloyd is notable particularly for his 1938 analysis of migrant arrival dates using all the data collected up to that time. In 1939 he published a detailed study of rookeries in the Shrewsbury area, and in 1943 a review of 'Changes in the bird population of Shropshire in the present century'.

Every issue of the *Record* and *Transactions* is published on the Histo website (see box). Many of the Species Accounts include specific sightings from the *Record*. In most cases they are not attributed (see *Historic Data* on p. 63), but can be found in the *Record* for the particular year on the Histo website. A few accounts do include a specific reference, cited as RBF (year), and/or a summary of the studies published in *Transactions*, cited as Transactions (year).

Other Authors and Publications

Books about the birds of the County have been few and far between. Paddock, of Wellington, tried twice, first with his modest *Notes on Some of the Birds Found in the Neighbourhood of Newport, Salop* (1890), followed by the more comprehensive *Catalogue of Shropshire Birds* (1897). Both were published privately in limited numbers and are now very rare, but can be consulted in Shropshire Archives.

Conclusion

Interest in birds increased in the twentieth century and, concurrently, the recording of other wildlife declined. The editorship of the *Record* had passed to Lloyd after Forrest's death in 1942, and throughout the 1940s almost all the notes submitted to it were accounts of birds. It was this growing concentration upon birds which led to the formation of the Shropshire Ornithological Society in 1955.

Between them, the gamekeepers, sportsmen, lark netters, taxidermists, gentlemen, collectors, early authors, and members and editors of the CSVFC *Record* and *Transactions*, compiled records of 19 species with no modern (post-1950) records: Squacco Heron, Little Crake, Little Bustard, Great Snipe, Puffin, Razorbill, Roseate Tern, Pallas's Sandgrouse, Tengmalm's Owl, Gyr Falcon, Chough, Short-toed Lark, Icterine Warbler, Great Reed Warbler, White's Thrush, American Robin, Alpine Accentor, Richard's Pipit and Cirl Bunting. Details of the records are included in the Species Accounts. Many other Species Accounts include interesting historical records.

Species with No Modern Records Which Failed to Make the County List

Several species crept into the early accounts but cannot now be accepted to the County list.

- Great Bustard has been discussed above.
- American Bittern. A record published in *The Field* in 1871 has been judged to be an error.
- Marsh Sandpiper. An 1855 list of the birds of the north-west (based on the collection of Mr Cross of Oswestry) included Marsh Sandpiper among 'specimens of recent date' but without a precise location; even if correctly identified it may have been from elsewhere.
- Black Guillemot. Rocke (1866) included Black Guillemot (along with Common Guillemot), as 'very rare indeed so far inland, though an occasional specimen of each ... has been obtained'. At least three reputable authors subsequently repeated the tale but no such specimen has been recorded.
- A 'Purple Gallinule' was caught by boys at Overton in 1901 and two years later another was shot nearby by a keeper – both perhaps accounting for an introduced pair. Even if they were not escapes, it is not clear which of the three possible species was involved. Purple Swamphen (*Porphyrio porphyrio*) from Mediterranean countries is rather sedentary and not on the British list. Purple Gallinule *P. martinicus* is from the Americas, is on the British list and is perhaps the most likely candidate. However, these two species had not been separated by that time. Allen's Gallinule *P. alleni*, an African species, is also on the British list and a possibility.

Of these five, Great Bustard and Black Guillemot were noted in the *Handlist* (1964) as species with records prior to 1950. These records are no longer considered acceptable. The other three were not noted in the *Handlist*, and were therefore already considered doubtful at that time.

An American Robin record from 1927 was originally judged an escape, but modern research suggests it may have been a wild vagrant. This illustrates the difficulty of judging historic records by modern standards, but the identification is not in doubt, so it is included in the Species Accounts.

Part 2. The Modern Period (1950–2014)

Shropshire Ornithological Society

Shropshire Ornithological Society was formed in March 1955, with the Objects:

- To encourage the study and protection of bird life in Shropshire, and to co-operate in such matters with national and local bodies, education authorities and others.
- To organise field meetings, lectures, exhibitions, etc., in all parts of the county, with a view to furthering the Society's aims.
- To publish an annual Bird Report, and any other matter relevant to the Society's interests.

A typescript Bird Report for 1955 was prepared, but not published at the time, with 27 individuals contributing records. There were 142 members by December 1955. In common with many subsequent Shropshire Bird Reports (SBRs), the 1955 report included a summary of the County results collected as part of a national survey organised by the British Trust for Ornithology, in this case on Mute Swan.

The first printed SBR was published for 1956, with records from 17 individuals. It repeated the results of the 1955 Mute Swan survey, and there were 160 members listed in the report.

It soon became apparent to the editors that there was no up-to-date account of the general status of each species to provide a context for SBR accounts or indicate what type of reports observers should be asked for.

Previously, the most recent full accounts were in *The Fauna of Shropshire* (Forrest 1899), and a list, with notes, in Volume I of the *Victoria County History* (Forrest 1908). CSVFC had published annual reports, but to obtain a more complete picture of the status of any species it was necessary to go through these many reports. It also became apparent that there had been many changes since Forrest's day, but the CSVFC reports were by no means a comprehensive summary of changes in status.

The *Handlist*

SOS therefore compiled and published an up-to-date account of the status of all species that had occurred here in *A Handlist of the Birds of Shropshire* (Rutter, Gribble and Pemberton, 1964, abbreviated to the *Handlist*). It reviewed all records available from 1950 to 1963, and the CSVFC reports, and also included a description of the geography, in eight 'divisions', thereby providing the first attempt at correlating distribution of each species with local habitat.

It also aimed to present a balanced picture of what had happened during the previous 50 years, and included a summary overview:

Apart from several of the wading birds which were proved to be regular passage migrants, and probably always were, the number of changes since the turn of the century is quite considerable. On balance we have probably gained as much as we have lost. The Wryneck and Red-backed Shrike have probably disappeared and possibly the Cirl Bunting and Black Grouse also. The Sparrowhawk, Snipe, Corncrake, Stonechat, Hawfinch and Corn Bunting have all decreased in number over this period. On the credit side the Little Owl, Black-headed

Gull and Redshank are now established as breeding birds and probably the Collared Dove. The Shoveler, Tufted Duck, Canada Goose, Mute Swan, Buzzard, Curlew, Raven and Pied Flycatcher have all increased as breeding species. The Cormorant, White-fronted Goose and Goosander are more common than formerly and the wintering flocks of gulls are a relatively new feature in our countryside. (*Handlist* 1964)

The Species Accounts show considerable further changes in the status of some of these species.

County Atlases, and BTO Atlases and Surveys

The distribution of breeding species described in the *Handlist* was confirmed, with more systematic detail, in the first national BTO *Atlas of Breeding Birds in Britain and Ireland* (Sharrock 1976). SOS members helped with the fieldwork between 1968 and 1972, and it was completed within 10 years of the end of the period reviewed by the *Handlist*. The presence, absence and status of every breeding species can be determined for each of the 33 10km squares wholly (19) or mainly (14) within the County.

Similarly, the distribution of wintering species 1981–84 can be determined at the 10km square level from the first national BTO winter Atlas, *An Atlas of the Wintering Birds of Britain and Ireland* (Lack 1986).

Updated distribution maps, at finer resolution, were published by SOS in *An Atlas of the Breeding Birds of Shropshire* (Deans *et al.* 1992). The whole project was organised by SOS, with fieldwork carried out between 1985 and 1990. The survey unit was the tetrad (a 2x2 kilometre square on the Ordnance Survey map national grid). There are 25 tetrads in each 10km square, and 870 wholly or mainly in the County (see Appendix 1, p. 484). All 870 squares were surveyed, and additional records were submitted, by a total of 452 observers who were acknowledged in that publication.

The final two years of the local Atlas project overlapped with the 1988–91 fieldwork period for the second national BTO Atlas, and the results were fed in, so again SOS members contributed to *The New Atlas of Breeding Birds in Britain and Ireland* (Gibbons *et al.* 1993).

When BTO announced plans to produce a combined breeding and winter season Atlas, with fieldwork between 2007 and 2011, SOS decided to repeat the local breeding Atlas, and undertake a first winter Atlas, again at the tetrad level, but it was necessary to extend the fieldwork period to 2013, to obtain acceptable coverage of all 870 tetrads. The results, and a comparison of changes in breeding distribution between the two Atlases, are included in the Species Accounts in this book. Over 650 people contributed fieldwork records, acknowledged in Appendix 5.

In addition to the national Atlases, BTO surveys have been well supported, with good local coverage for individual species surveys, monthly non-breeding waterbird (WeBS) counts, primarily in winter, the Common Bird Census (CBC) up until 2000, and the Breeding Bird Survey (BBS) which replaced it from 1994, with an overlap period to reconcile the trend data. One CBC plot, at the Old Racecourse, Oswestry, was surveyed every year from 1964 to 2000, providing fascinating detail of species' increases, losses and fluctuations over 36 years. An average of more than 50 BBS squares have been surveyed every year since 1997, allowing the calculation of population trends for 34 species 1997–2014, which are included in the relevant Species Accounts.

Conservation Sub-committee

By the 1980s, ornithology became more applied, as national research and monitoring was increasingly identifying large declines in common species, particularly on farmland, and highlighting the need for conservation action to reverse the declines. Locally, to ensure that monitoring and survey work was effective and co-ordinated, and the results were used for conservation purposes, the SOS Conservation Sub-committee was set up in 1987. The sub-committee is responsible for the collection and management of records, and publication of SBRs. It oversaw the computerisation of records, starting in 1992, and ensures that systems are upgraded as necessary. A review of modern (post-1950) records was carried out, to ensure they were all acceptable by current standards. Published records for two species, Melodious Warbler and Aquatic Warbler, were considered not proven, and they are not now on the County list. The results were published in SBR 2011. The County list at the end of 2018 is included in Appendix 6.

Local surveys of Lapwing, Yellow Wagtail and Corn Bunting were organised in 2003–04, and the Lapwing survey was repeated in 2008. To ensure effective co-ordination, the sub-committee includes the BTO Regional Representative and the local RSPB staff Conservation Officer.

The sub-committee is also responsible for the management of the Society's reserves, including maintaining and improving the habitats, at Venus Pool (see p. 53), Bushmoor Coppice and Ward's Coppice. Although Venus Pool is mainly a facility for recreational bird-watchers, it provides breeding sites for Lapwing, Little Ringed Plover, Oystercatcher, Black-headed Gull and Yellow Wagtail, as well as some other less common species. The wild bird crops are locally important for farmland birds, especially in winter when large flocks of Yellowhammer, Reed Bunting and Corn Bunting have been found. It is well-watched and many records are received, and a log book kept in each hide provides additional records, so it is a valuable source of ornithological data, particularly of passage migrants. For example, the rapid decline in the number of Curlews passing through each year is powerful evidence for the overall decline of the local breeding population.

Both Bushmoor Coppice and Ward's Coppice are woodland reserves with a range of typical species, which have included Pied Flycatcher before their recent decline. Both receive only a fraction of the visitors to Venus Pool.

An access agreement with South Staffordshire Water Board for a hide and management of the scrape at Chelmarsh has also been negotiated, though at the time of writing this needs to be updated.

SOS has also partly fulfilled its conservation role through representation on other partnerships. It was represented on the Shropshire Biodiversity Action Plan (BAP) Partnership, which included all the leading conservation organisations. Through the UN Convention on Biological Diversity, each country has the responsibility to conserve and enhance biodiversity within its own jurisdiction. The national BAP was underpinned by local (County level) Action Plans which included the national target species, plus some locally distinctive species. This was the main government-led process for planning nature conservation, and therefore had a major influence on how money and other resources were spent. The Shropshire BAP was published by Shropshire Council in 1996, and revised in 2002; the subsequent review of bird species in 2005 was co-ordinated by SOS. The Action Plans for Farmland seed-eating birds (Tree Sparrow, Corn Bunting, Reed Bunting, Linnet, House Sparrow, Skylark, Yellowhammer and Bullfinch), Lapwing, Ring Ouzel, Snipe and Song Thrush were updated, by assessing progress towards each of the targets, and adding new initiatives and targets. New species were added: Curlew, Barn Owl, Dipper, Nightjar and Spotted Flycatcher, mostly adding local plans for national BAP species.

Unfortunately, in 2012 the government and other partners removed financial support for biodiversity action planning, and monitoring progress towards targets to reverse declines. SWT took over the co-ordinating role from Shropshire Council, but the partnership only meets rarely and there are no resources for plan development or monitoring.

The government promoted Local Nature Partnerships (LNPs) to replace BAP partnerships in England, to work more at the strategic level, and they include representation from business and the Health and Wellbeing Board. The Shropshire LNP was formed in 2012, but it has been hampered by a lack of core funding. It primarily acts as a communication channel and in an influencing capacity, but useful links have been made with the Marches Local Enterprise Partnership and it is hoped that this could lead to greater influence and access to large funding streams, such as the EU Life £2.5m project led by the Shropshire Wildlife Trust on wetland creation and habitat restoration at Whixall Moss.

The UK BAP still aims to reverse the decline of many bird species in the government's 'Biodiversity 2020' targets. The local BAPs are now out of date, but still reflect species and habitats 'of importance', and are taken into account in the planning process. They can be found under 'Biodiversity Action Plan' on the Shropshire Council website (www.shropshirecouncil.co.uk).

SOS has also been represented on the Shropshire Hills AONB Partnership, and ensured that addressing the plight of Lapwing and Curlew was included in the statutory AONB Management Plans for 2009–14 and 2014–19.

Such arrangements for promoting conservation action locally have been in place for much less than half the time since the formation of SOS, and will no doubt change periodically to reflect government, and conservation, priorities.

Other Monitoring and Conservation Work

A major contribution to local ornithology is being made by several groups or projects undertaking monitoring and conservation of individual species.

- Peregrine
- Red Grouse
- Snipe
- Dipper
- Barn Owl
- Raven
- Rook

The Shropshire Raptor Study Group was founded in 2010, to monitor in particular Hobby, Merlin and Goshawk, and support the Welsh Kite Trust in monitoring Red Kite.

Most of these surveys, groups and projects contribute annual reports to SBRs, and their results are summarised in the appropriate Species Accounts. More information about each of them can be found on the SOS website, www.shropshirebirds.com.

Several community wildlife groups are surveying Lapwing and Curlew, and monitoring their population trends. The longest running are in the Upper Onny area, from 2004, and Upper Clun, from 2007, while others started more recently. The Upper Clun group is also monitoring species that inhabit the mires and bogs, and has helped get such habitat adopted as Local (County) Wildlife Sites for their bird assemblage (www.ShropsCWGs.org.uk).

There are a number of individuals and groups with nest box schemes, primarily for Pied Flycatcher, as described in that species account.

In addition, there are a number of research surveys that cover more than one species, usually as part of multi-annual systematic monitoring of particular areas or sites. The Long Mynd and the Stiperstones each have two such surveys, the former in 1994–98 and 2006–09, and the latter in 1995–96 and 2004–07, so population estimates and trends of the upland specialists have been established. The surveys were commissioned by the owners, National Trust and Natural England respectively, to inform site management.

Other similar surveys include:

- The RSPB Shropshire Moors Breeding Wader Survey on Baggy Moor (near Oswestry) and the Weald Moors (north of Telford). It started in 2009, and led to the Lapwing Meadows project. Surveyors also found key farmland birds such as Corn Buntings, Tree Sparrows and Grey Partridges. The survey is part of landscape scale conservation work in this area over several years.
- Periodic surveys of the Severn Gorge Countryside Trust (SGCT) woodlands around Ironbridge Gorge, and other woodlands owned by Telford and Wrekin Council, listed in Bishton, SBR 2003. Reports of later annual surveys for SGCT up until 2014 in the same woods can be found on the Histo website. These surveys, and nest box schemes in some of the same woods, have informed the Pied Flycatcher and Wood Warbler accounts.
- Birds of Harper Adams University College farm (Bishton 2012).
- Bird population change on the National Trust estate at Dudmaston, over the course of a Countryside Stewardship Scheme 2002–11 (Tucker 2011).

Surveys or research studies spread over several years may only appear in SBR at the end, in a multi-annual report in the first section, not in the systematic list of Species Accounts. The *Histo* website lists all such articles in SBRs, while the species index lists references to all the individual species covered by them. Many of the full reports can also be found on *Histo* (see box on p. 6).

Ringing

Ringers have to be trained in catching and handling birds over several years. The process is regulated by the BTO, which keeps a record of each metal ring fitted to the leg, with a unique number on it. If it is found again, usually dead but sometimes alive, the details and ring number are also sent to BTO. The data builds up into a picture of how long each species lives, whether it is sedentary or migratory, and its pattern of movements or migration route.

The first ringing schemes started in 1909, and became the responsibility of BTO in the 1930s. Most of what we know about longevity and migration comes from ringing. The increase in ringers across the country, and in Europe and the rest of the world, means that an increasing number of ringed birds are caught alive by other ringers.

The main focus of ringing has changed over the years, and now aims at understanding and monitoring population change. Retrapping Adults for Survival (RAS) schemes now capture, and subsequently recapture, as many adults of a species as possible in a predefined location annually. Increasingly, such projects use colour-rings so each individual has a unique combination of colours, or code, so it can be identified without the need to recapture it. Constant Effort Site (CES) ringing involves spending the same amount of time at the same site each year, so variation in results reflect changes in the wider populations. Ringing complements other BTO monitoring schemes, particularly BBS and Nest Recording, and more detail is available on the BTO website.

Some ringing and recoveries occurred in the pre-war period, but in the modern period there have been three main local ringing groups, all of which contribute to the training of new ringers.

Allscott ringing station (also known as Walcott) was founded in 1961, in the grounds around the settling pools at Allscott Sugar Factory (see p. 47). It was not a formal ringing group, but around 40 ringers have been trained there. It operated until April 2013, when the new owners withdrew permission, and in its 52 years of operation 109,710 birds of 123 species were ringed. The total site list was 191 species recorded, including many rarities, some from America and Russia, and 83 species were proved to breed. Swallows used the reed bed for roosting on passage, and a total of 16,888 were ringed, together with 6,828 House Martins and 6,038 Sand Martins. Breeding and passage Reed Warbler (6,255), Sedge Warbler (4,833), Blackcap (4,586), Willow Warbler (5,685), Chiffchaff (3,364) and Blue Tit (6,293) were ringed in large numbers, and Greenfinch (5,204) and Redwing (4,778) were ringed during the winter months. Many of these birds were re-caught, providing information on site faithfulness and longevity, and birds ringed in other places were also re-trapped. The Allscott ringers were also active at other winter roost sites, at Hinstock where over 25,000 birds were ringed between 1966 and 1999, Pell Wall and Attingham, but catches declined from 2000 onwards, probably due to milder winters, and is no longer worthwhile (Julian Langford *pers. comm.*).

The Chelmarsh Ringing Group operates at the reservoir and scrape (see p. 53). Ringing began in 1978, and the Group was formally set up in 2002: 25,473 birds from 95 different species had been ringed by the end of 2014, with Reed Bunting (3,800), Blue Tit (2,500) and Reed Warbler (1,500) the most frequent. Rarities ringed include Spotted Crake, Grey Phalarope and Cetti's Warbler. Regular monitoring of Reed Warbler and Sedge Warbler has been carried out through a BTO RAS scheme since the early 2000s, and a CES ringing scheme started in 2014. The Group also colour-rings Wheatear and Stonechat on Titterstone Clee hill. There are currently five ringers (Dave Fulton, *pers. comm.*)

The Shropshire Ringing Group in its present form was established in 2010, although several of its members have been ringing for many years, and has steadily grown to a membership of 11 keen and active ringers. Most of its activity is in the north, reflecting its current membership and ringing sites, although, as some species are sought to the south of Shrewsbury, BTO were insistent on the Group calling itself the Shropshire Ringing Group.

It works at a variety of sites which it rotates throughout the year in order to target various species to fulfil scientific objectives. With the permission of Natural England, Whixall Moss is visited to monitor three Amber-listed species: in the breeding season Willow Warbler and Reed Bunting, and later in the year Common (and Lesser) Redpoll on migration. A private woodland site is used to monitor several species, in particular, red-listed Marsh Tit.

RAS schemes are underway for red-listed House Sparrows at both Bicton Heath and Whixall, for Marsh Tits (site confidential) and for Sand Martins at Wood Lane. In addition two Pied Flycatcher (red-listed) studies are undertaken, one in the Clun Valley (started in 1986 and taken over by the Group in 2015) of about 800 nest boxes, and the other a long-running scheme at Hawkstone Park of about 200 boxes.

The Group also undertakes colour-ringing studies on Peregrine, in collaboration with the Shropshire Peregrine Group, Goshawk (with the Shropshire Raptor Group), and Kestrel, and works with ringers in other parts of the UK, and abroad, to share knowledge, experience and ideas (Gerry Thomas, Bob Harris *pers. comm.*).

In addition there are many active individual ringers, ringing at specific sites or looking for particular species. Raven, Golden Plover and Dipper have all been the subject of local colour-ringing projects, Red Kites have been wing-tagged, and Barn Owls in Shropshire Barn Owl Group nest boxes have been ringed.

The respective Species Accounts include many detailed results of this valuable local ringing work (see p. 70).

The Shropshire List

Fifty-five species were recorded for the first time since 1950, shown at the end of this chapter. The total list, including pre-1950 records but excluding escapes, stood at 301 species at the end of 2017, and all are included in the Species Accounts. There were no new species recorded in 2018, but the first modern record of Little Bittern was accepted.

Breeding Species

Of the total list, 120 species were confirmed breeding during the recent Atlas project, compared with 116 in 1985–90. The *Breeding Status* chapter includes a list of all those that bred for the first time since 1950 and bred in 2008–13, a separate list of occasional breeders that were not found breeding in the recent Atlas period, and the species whose breeding status changed between the two Atlas periods.

Growth of Recording Effort

SBRs have been published every year from 1956 to 2017, and the 2018 report is in preparation. As well as a systematic list of Species Accounts, and summaries of BTO and other surveys, SBRs have included articles on many other ornithological topics, and the results of surveys organised by the Conservation Sub-committee.

The number of individuals contributing to local ornithology through SOS, and submitting records, has grown considerably. In 1956, there were records from 17 individuals. Just after the end of the first Atlas period, in 1991, 106 observers submitted records. In 2006, the last year in which a list of observers was published in SBR without including Atlas fieldworkers, but including those submitting records indirectly via BirdTrack or surveys, the number reached 252. No list of observers has been published since 2009.

However the number of observers submitting Atlas records has also grown substantially, from 485 in 1985–90 to 650 in 2007–13.

On-going Development of SOS

SOS was registered as a charity in 1983. The Objects clause in the SOS Rules and Constitution was brought up to date in April 2000, with the addition of a fourth clause:

> *To acquire and manage areas of land as nature reserves primarily of ornithological value.*

The responsibilities inherent in owning and managing land led to SOS becoming a company limited by guarantee in 2005. The number of SOS members (including joint and life members) had grown to 746 at the end of 2014. *The Birds of Shropshire* was planned to summarise the County avifauna immediately following the sixtieth anniversary.

Part 3. Period or Date of First Record

There are acceptable records of 301 species that comprise the 'Shropshire List'. Most of them were known to the late nineteenth-century authors, Beckwith and/or Forrest.

The *Handlist* (1964) referred to species recorded between 1950 and 1963, but noted, in Appendix A, several species that had not been recorded since 1950. The records for three of these species have not been considered acceptable for including in the 'Shropshire List' in this Avifauna, but there were pre-existing records for several other species recorded for the first time after Forrest's deadline, but not since 1950 (i.e. they should have been included in Appendix A).

The following table lists all 301 species, with columns showing whether they were referred to by Beckwith, Forrest or the *Handlist*, and another pair of columns showing the date of the first record if the species was not included in Forrest (1899). These columns split the date of the first record between the early period (first record for 10 species), and the modern period since 1950, with first records for 55 species.

The current status of each of these species is shown in the heading to the species account, and they are all summarised in Appendix 6.

The table below makes no reference to breeding status. Changes in breeding status since 1950 are described in the *Breeding Status* chapter.

British (English) Vernacular Name	Species Included in			Date of First Record		Note
	Beckwith 1881	Forrest 1899	Handlist 1964	Between Forrest (1899) and 1949	First Record 1950–2018	
Brent Goose	X	X	X			
Canada Goose	X	X	X			
Barnacle Goose	X	X	X			
Greylag Goose	X	X	X			
Pink-footed Goose	X	X	X			
Tundra Bean Goose	X	X	X			1
White-fronted Goose	X	X	X			
Mute Swan	X	X	X			
Bewick's Swan	X	X	X			
Whooper Swan	X	X	X			
Egyptian Goose	X	X				2
Shelduck	X	X	X			
Mandarin Duck				1966		
Garganey	X	X	X			
Shoveler	X	X	X			
Gadwall	X	X	X			
Wigeon	X	X	X			
Mallard	X	X	X			
Pintail	X	X	X			
Teal	X	X	X			
Green-winged Teal					1996	
Red-crested Pochard					1977	
Pochard	X	X	X			
Ferruginous Duck				1937		2
Ring-necked Duck					1995	
Tufted Duck	X	X	X			
Scaup	X	X	X			

British (English) Vernacular Name	Species Included in			Date of First Record		Note
	Beckwith 1881	Forrest 1899	Handlist 1964	Between Forrest (1899) and 1949	First Record 1950–2018	
Lesser Scaup					2005	
Eider					1979	
Velvet Scoter	X	X	X			
Common Scoter	X	X	X			
Long-tailed Duck		X	X			
Goldeneye	X	X	X			
Smew	X	X	X			
Goosander	X	X	X			
Red-breasted Merganser	X	X	X			
Ruddy Duck					1965	
Black Grouse	X	X	X			
Red Grouse	X	X	X			
Red-legged Partridge	X	X	X			
Grey Partridge	X	X	X			
Quail	X	X	X			
Pheasant	X	X	X			3
Red-throated Diver	X	X	X			
Black-throated Diver	X	X	X			
Great Northern Diver	X	X	X			
Storm Petrel	X	X	X			
Leach's Petrel	X	X	X			
Fulmar			X	1909		
Manx Shearwater	X	X	X			
Little Grebe	X	X	X			
Red-necked Grebe	X	X	X			
Great Crested Grebe	X	X	X			
Slavonian Grebe	X	X	X			

British (English) Vernacular Name	Species Included in			Date of First Record		Note
	Beckwith 1881	Forrest 1899	Handlist 1964	Between Forrest (1899) and 1949	First Record 1950–2018	
Black-necked Grebe	X	X	X			
Black Stork			X		1956	
White Stork					1978	
Glossy Ibis	X	X	X			
Spoonbill					1965	
Bittern	X	X	X			
Little Bittern		X	X			
Night-heron	X	X	X			
Squacco Heron	X	X	X			
Cattle Egret					1987	
Grey Heron	X	X	X			
Purple Heron					1995	
Great White Egret					1995	
Little Egret					1992	
Magnificent Frigatebird					2005	
Gannet	X	X	X			
Shag	X	X	X			
Cormorant	X	X	X			
Osprey	X	X	X			
Honey-buzzard	X	X	X			
Sparrowhawk	X	X	X			
Goshawk			X		1951	
Marsh Harrier	X	X	X			
Hen Harrier	X	X	X			
Montagu's Harrier	X	X	X			
Red Kite	X	X	X			
White-tailed Eagle	X	X	X			
Rough-legged Buzzard	X	X	X			
Buzzard	X	X	X			
Little Bustard		X	X			
Water Rail	X	X	X			
Corncrake	X	X	X			
Little Crake			X	1898		
Spotted Crake	X	X	X			
Moorhen	X	X	X			
Coot	X	X	X			
Crane			X		1962	3
Stone-curlew	X	X	X			

British (English) Vernacular Name	Species Included in			Date of First Record		Note
	Beckwith 1881	Forrest 1899	Handlist 1964	Between Forrest (1899) and 1949	First Record 1950–2018	
Oystercatcher	X	X	X			
Black-winged Stilt					1965	
Avocet		X				2
Lapwing	X	X	X			
White-tailed Plover					1984	
Golden Plover	X	X	X			
Grey Plover			X	1913		
Ringed Plover	X	X	X			
Little Ringed Plover			X		1957	
Dotterel	X	X	X			
Whimbrel	X	X	X			
Curlew	X	X	X			
Bar-tailed Godwit	X	X	X			
Black-tailed Godwit	X	X	X			
Turnstone	X	X	X			
Knot	X	X	X			
Ruff	X	X	X			
Curlew Sandpiper	X	X	X			
Temminck's Stint					1986	
Sanderling			X		1957	
Dunlin	X	X	X			
Purple Sandpiper		X	X			
Little Stint		X	X			
White-rumped Sandpiper	X	X	X			
Buff-breasted Sandpiper					1986	
Pectoral Sandpiper					1979	
Woodcock	X	X	X			
Jack Snipe	X	X	X			
Great Snipe	X	X	X			
Snipe	X	X	X			
Red-necked Phalarope	X	X	X			
Grey Phalarope	X	X	X			
Common Sandpiper	X	X	X			
Green Sandpiper	X	X	X			
Lesser Yellowlegs					1995	
Redshank	X	X	X			
Wood Sandpiper			X		1952	
Spotted Redshank			X	1923		

British (English) Vernacular Name	Species Included in			Date of First Record		Note
	Beckwith 1881	Forrest 1899	Handlist 1964	Between Forrest (1899) and 1949	First Record 1950–2018	
Greenshank	X	X	X			
Kittiwake	X	X	X			
Sabine's Gull	X	X	X			
Black-headed Gull	X	X	X			
Little Gull	X	X	X			
Mediterranean Gull					1982	
Common Gull	X	X	X			
Ring-billed Gull					1992	
Great Black-backed Gull	X	X	X			
Glaucous Gull	X	X	X			
Iceland Gull					1982	
Herring Gull	X	X	X			
Caspian Gull					2006	1
Yellow-legged Gull					1989	1
Lesser Black-backed Gull	X	X	X			
Gull-billed Tern					1997	
Sandwich Tern		X	X			
Little Tern	X	X	X			
Roseate Tern	X	X	X			
Common Tern	X	X	X			
Arctic Tern	X	X	X			
Whiskered Tern					2010	
White-winged Black Tern					1967	
Black Tern	X	X	X			
Great Skua	X	X	X			
Pomarine Skua	X	X	X			
Arctic Skua	X	X	X			
Long-tailed Skua	X	X	X			
Little Auk	X	X	X			
Common Guillemot	X	X	X			
Razorbill	X	X	X			
Puffin	X	X	X			
Pallas's Sandgrouse	X	X	X			
Feral Pigeon/Rock Dove		X	X			
Stock Dove	X	X	X			
Woodpigeon	X	X	X			
Turtle Dove	X	X	X			
Collared Dove			X		1961	

British (English) Vernacular Name	Species Included in			Date of First Record		Note
	Beckwith 1881	Forrest 1899	Handlist 1964	Between Forrest (1899) and 1949	First Record 1950–2018	
Cuckoo	X	X	X			
Barn Owl	X	X	X			
Tawny Owl	X	X	X			
Little Owl			X	1899		
Tengmalm's Owl	X	X	X			
Long-eared Owl	X	X	X			
Short-eared Owl	X	X	X			
Nightjar	X	X	X			
Swift	X	X	X			
Kingfisher	X	X	X			
Bee-eater					1992	3
Hoopoe	X	X	X			
Wryneck	X	X	X			
Lesser Spotted Woodpecker	X	X	X			
Great Spotted Woodpecker	X	X	X			
Green Woodpecker	X	X	X			
Kestrel	X	X	X			
Red-footed Falcon	X	X	X			
Merlin	X	X	X			
Hobby	X	X	X			
Gyr Falcon	X	X	X			
Peregrine	X	X	X			
Ring-necked Parakeet				1975		
Red-backed Shrike	X	X	X			
Great Grey Shrike	X	X	X			
Steppe Grey Shrike					2011	1
Woodchat Shrike				1977		
Golden Oriole		X	X			
Jay	X	X	X			
Magpie	X	X	X			
Nutcracker				1968		
Chough		X	X			
Jackdaw	X	X	X			
Rook	X	X	X			
Carrion Crow	X	X	X			
Hooded Crow	X	X	X			
Raven	X	X	X			
Waxwing	X	X	X			

British (English) Vernacular Name	Species Included in			Date of First Record		Note
	Beckwith 1881	Forrest 1899	Handlist 1964	Between Forrest (1899) and 1949	First Record 1950–2018	
Coal Tit	X	X	X			
Marsh Tit	X	X	X			4
Willow Tit	n/a	n/a	X	1937		1, 4
Blue Tit	X	X	X			
Great Tit	X	X	X			
Bearded Tit			X		1960	3
Woodlark	X	X	X			
Skylark	X	X	X			
Shore Lark					2017	
Short-toed Lark	X	X	X			
Sand Martin	X	X	X			
Swallow	X	X	X			
House Martin	X	X	X			
Red-rumped Swallow					1978	
Cetti's Warbler					1975	
Long-tailed Tit	X	X	X			
Willow Warbler	X	X	X			
Chiffchaff	X	X	X			
Iberian Chiffchaff					2016	
Wood Warbler	X	X	X			
Pallas's Warbler					1987	
Yellow-browed Warbler					2008	
Great Reed Warbler		X	X			
Sedge Warbler	X	X	X			
Reed Warbler	X	X	X			
Marsh Warbler					2016	
Icterine Warbler				1941		2
Grasshopper Warbler	X	X	X			
Blackcap	X	X	X			
Garden Warbler	X	X	X			
Lesser Whitethroat	X	X	X			
Whitethroat	X	X	X			
Dartford Warbler				1903		2
Firecrest	X	X	X			
Goldcrest	X	X	X			
Wren	X	X	X			
Nuthatch	X	X	X			
Treecreeper	X	X	X			

British (English) Vernacular Name	Species Included in			Date of First Record		Note
	Beckwith 1881	Forrest 1899	Handlist 1964	Between Forrest (1899) and 1949	First Record 1950–2018	
Rose-coloured Starling	X	X	X			
Starling	X	X	X			
White's Thrush		X	X			
Ring Ouzel	X	X	X			
Blackbird	X	X	X			
Black-throated Thrush					2007	
Fieldfare	X	X	X			
Redwing	X	X	X			
Song Thrush	X	X	X			
Mistle Thrush	X	X	X			
American Robin				1927		2
Spotted Flycatcher	X	X	X			
Robin	X	X	X			
Bluethroat					1987	
Nightingale	X	X	X			
Pied Flycatcher	X	X	X			
Black Redstart	X	X	X			
Redstart	X	X	X			
Whinchat	X	X	X			
Stonechat	X	X	X			
Wheatear	X	X	X			
Desert Wheatear					2011	
Dipper	X	X	X			
House Sparrow	X	X	X			
Tree Sparrow	X	X	X			
Alpine Accentor		X				2
Dunnock	X	X	X			
Yellow Wagtail	X	X	X			
Grey Wagtail	X	X	X			
Pied Wagtail	X	X	X			3
Richard's Pipit	X	X	X			
Meadow Pipit	X	X	X			
Tree Pipit	X	X	X			
Water Pipit	n/a	n/a	n/a		1957	1, 4
Rock Pipit		X	X			4
Chaffinch	X	X	X			
Brambling	X	X	X			
Hawfinch	X	X	X			

British (English) Vernacular Name	Species Included in			Date of First Record		Note
	Beckwith 1881	Forrest 1899	Handlist 1964	Between Forrest (1899) and 1949	First Record 1950–2018	
Bullfinch	X	X	X			
Greenfinch	X	X	X			
Twite	X	X	X			
Linnet	X	X	X			
Common Redpoll	X	X				2, 5
Lesser Redpoll	X	X	X			2, 5
Arctic Redpoll					2013	6
Parrot Crossbill	X	X				2
Crossbill	X	X	X			
Two-barred Crossbill					1973	

British (English) Vernacular Name	Species Included in			Date of First Record		Note
	Beckwith 1881	Forrest 1899	Handlist 1964	Between Forrest (1899) and 1949	First Record 1950–2018	
Goldfinch	X	X	X			
Siskin	X	X	X			
Corn Bunting	X	X	X			
Yellowhammer	X	X	X			
Pine Bunting					2017	
Cirl Bunting	X	X	X			
Reed Bunting	X	X	X			
Lapland Bunting	X	X	X			
Snow Bunting	X	X	X			
Total	222	236	245	10	55	

NOTES

1 In these cases, the date of the first record reflects the positive identification of the sub-species (race) that was subsequently given the status of a species.

2 In addition to the species in the *Handlist* (1964) column, there were pre-existing records of Egyptian Goose, Ferruginous Duck, Avocet, Dartford Warbler, Icterine Warbler, American Robin, Alpine Accentor and Parrot Crossbill, which should have been included. 'Redpoll' (including Common (Mealy) Redpoll and Lesser Redpoll) was considered to be one species at that time.

3 In addition to the species in the Forrest column in the table, he also listed as species White Wagtail (a race of Pied/White Wagtail) and Ring-necked Pheasant (a sub-species of Pheasant). He listed three records of Eagle Owl, which may have escaped from aviaries, and Virginia Colin, an American Quail now known as Northern Bobwhite, shot on the Hawkstone estate. Both these species are viewed as 'Escapes' and not given species status in this Avifauna. Records of Crane shot at Trippleton on the Teme and Bee-eater near Tenbury are not included, as the locations are in Herefordshire, and a Shore Lark shot near Enville and a dubious account of Bearded Tit at Aqualate have also been excluded, as the locations are in Staffordshire. Forrest also included Black Guillemot, but listed no records, and Great Bustard, with dubious records (see box 'Species with No Modern Records Which Failed to Make the County List' on p. 7), so these are also not included in the table.

4 Two species were not known in the nineteenth century. Willow Tit was not separated from Marsh Tit until 1897, and Water Pipit was not separated from Rock Pipit until 1987. Therefore both were unknown to Beckwith or Forrest, and Water Pipit was unknown to the authors of the *Handlist*.

5 There are several races of Redpoll, but the way in which they have been grouped into species has changed several times. In Forrest's day, 'Lesser Redpoll' was common in winter, and 'Mealy Redpoll' was rare. They were considered to be separate species. From the early twentieth century until 2001, they were considered to be different races of the same species, 'Redpoll'. From 2001, the Redpoll species was split into two separate species by the BOU, the keeper of the Official British List: 'Common Redpoll', including the 'Mealy Redpoll' race, and 'Lesser Redpoll'.

6 The above excludes 'Arctic Redpoll', which was not known in Forrest's day, or to the authors of the *Handlist*. It was a race of 'Redpoll' up until 2001, and has been a separate species since then.

Appendix A in the *Handlist* noted species recorded prior to 1950, but not since. The list included Great Bustard, Black Guillemot and Eagle Owl, taken from Forrest (1899), which are referred to in note 3 above. However, these records are not considered acceptable, so the three species concerned are not listed in the table, nor are they included in the 'Shropshire List' (Appendix 6).

Changes in Migrant Arrival Dates

John Arnfield and John Tucker

In the Beginning

On 17 March 1886, by the River Severn in Ironbridge, Mr F. Rawdon Smith saw his first House Martin of the year and made a note of it. That day was notable for Shropshire's ornithology for two reasons: first, it remains the earliest ever House Martin arrival date and, second, as far as we know, it was the first time anyone in the County had begun to compile a list of migrant arrival dates. Over the next eight years, Mr Smith noted the first dates of 10 summer migrants including, unsurprisingly, Cuckoo and Swallow, but also Corncrake and Nightingale, the latter's stronghold being then around Ironbridge.

The publication of CSVFC's *Record* began in 1892, providing an outlet for the publication of Mr Smith's observations (in 1894), thus beginning the recording of migrant arrival dates.

First arrival records, where observers note the first migrants they see each year, often result from chance encounters with aberrant individuals unrepresentative of the general population: hence the popular saying 'one swallow doesn't make a summer'. For this reason, the relative values of first arrival and main arrival dates vexed Lloyd in the 1930s (see box below) and the subject remains a live issue today.

Simple first dates are useful because they do not require observers to follow any methodology beyond the obvious. Hence, they are comparable across long periods (a few date back more than two centuries) and wide areas (e.g. across Europe). Nowadays these statistics are more robust because of the larger number of better informed and equipped observers, better transport and communications, more sophisticated and convenient reporting systems (e.g. BTO's BirdTrack), and other factors. The number of people recording bird observations in the County in the nineteenth century was barely a dozen, especially before the start of the CSVFC; between 1900 and 1940 it rose to some 30–40 but, after World War II, fell below 20 before rising steeply through the 1960s and 1970s following the formation of SOS. Since the 1980s, there have been well over 100 each year.

Shropshire's Data

Shropshire's Migrant Arrival Database (SMAD, Tucker 2016) holds the first dates for 32 species from 1886 to 2014, plus a single Cuckoo record from 1871 by Rocke, described by Beckwith (1892). Apart from this case, they have all come from the CSVFC's *Record* and *Transactions* and latterly from SBRs, and now number over 3,000 records. These dates appeared in the published record within individual species accounts or in summary tables. In SMAD, first dates are held for each species as day, month and year, along with the equivalent day-number (count of days into the year) used to calculate, for example, average arrival dates. A few of the early records come from two locations just outside the County boundary, Churchstoke and Knighton. The SMAD is among the most comprehensive such data sets in Britain. A few other series extend further back in time (for example, Rutland's series from the early 1700s: Kington 1988) but cover fewer species. Table 1 shows the 32 species included, ranked by column in order of the number of years (not necessarily continuous) for which first arrival dates are known. Dates after 1974 have been excluded for Chiffchaff and Blackcap, because of the difficulty of separating early arrivals from late over-wintering birds that started to appear around that time.

Table 1. Number of Early Arrival Records (N) by Species in SMAD, 1886–2014. Records began with a House Martin on 17 March 1886 and are ongoing: total 129 years.

Species	N	Species	N	Species	N	Species	N
Swift	125	Redwing	107	Lesser Whitethroat	98	Nightingale	77
Swallow	124	Spotted Flycatcher	107	Garden Warbler	97	Brambling	76
Cuckoo	121	Tree Pipit	107	Pied Flycatcher	97	Corncrake	73
House Martin	120	Common Sandpiper	106	Grasshopper Warbler	96	Blackcap	65
Sand Martin	116	Wood Warbler	106	Yellow Wagtail	95	Reed Warbler	54
Willow Warbler	111	Whitethroat	105	Sedge Warbler	88	Nightjar	42
Redstart	109	Turtle Dove	103	Whinchat	88	Wryneck	36
Fieldfare	108	Wheatear	102	Chiffchaff	77	Little Ringed Plover	17

Arrival Date Trends in SMAD

To identify potential long-term changes in migrant arrival dates over the period covered by the SMAD, and to identify differences among species, dates were extracted for all years having records for three 29-year sub-periods centred on the years 1900, 1950 and 2000, extending 14 years on each side of the nominal date and averaged for each species.

Table 2 presents a summary of the average arrival dates obtained in this way (columns 2–4). The differences in arrival dates (in days) between the sub-periods are shown following the average first arrival dates, in columns 5–6, and the difference over the whole century is shown in column 7.

A full statistical analysis of the SMAD data shown in columns 2–7 has been carried out and can be found on the Avifauna section of the SOS website www.shropshirebirds.com/AvifaunaSupplement. Table 2 columns 8–10 present the results of this analysis. In cases where the statistical analysis found that the difference between averages was not large enough, given the number of data points and the amount of variability around the averages, to permit evaluation of an arrival date change with a reliability of 95% or more, the difference is shown in parentheses in columns 5–7.

Table 2 (column 7) shows that, for 15 of the 29 springtime arrivals, there has been an apparent shift in arrival dates to earlier in the year. Species Group 1 (Table 2) contains a single species – Yellow Wagtail – which shows the clearest change in arrival date over time, with advances in both sub-periods and over the whole century. Group 2 is similar, with advances of at least five days between 1950 and 2000 and over the whole century, but with small or unreliable arrival date advances in the period 1900–50.

Species Group 3 exhibits more ambiguous results, with unreliable changes over the century and in one or both of the 50-year sub-periods for most species. The data for Whitethroat and Cuckoo suggest earlier arrival by about a week over the century but not in the sub-periods, but those for the other species in the group are not clear cut. It should be noted, however, that in the case of Reed Warbler, there were very few records available until regular access to suitable habitat at Allscott Sugar Factory was gained in 1960, and the period after that date shows a well-marked advance in arrival date. Arrival dates for Spotted Flycatcher are particularly unreliable, as it has declined by 86% since 1970, and 44% since 1995, greatly reducing the likelihood of observers encountering the first-arriving individual.

Group 4 represents species for which data deficiencies weaken conclusions that might be drawn from the available records. The first five have shown large population declines over the period: three are on the Red list of *Birds of Conservation Concern* (50% decline since 1970), one is on the Amber list (25% decline since 1970) and one is a former breeder, now only an uncommon passage visitor. SMAD data for Nightingale and Wryneck end in 1990 and Corncrake and Nightjar in 2001. Dates for Turtle Dove exhibit great variability in recent decades. Arrival dates for this group show no clear trend, but any change that may have occurred has almost certainly been obscured by the decreasing likelihood of the first recorded date being close to the 'true' first date with so few recent observations. Chiffchaff and Blackcap, while not scarce, are subject to erroneous first arrival dates owing to an increasing propensity for over-wintering and/or winter in-migration in recent years, which is why results from 1975 onwards have been excluded from this analysis. Little Ringed Plover, in contrast, has only been recorded as a spring migrant in recent years and so lacks data from the early and mid-1900s with which to compare.

Group 5 includes only the autumn migrant species. Redwing shows a marked advance in arrival date over the twentieth century. Fieldfare and Brambling, in contrast, show *later* arrival in the first sub-period and the opposite trend in the second, leading to no meaningful change over the whole century. The ambiguity in these results may perhaps reflect the differences in the driving forces on spring and autumn migrants. The former need to arrive at their breeding sites to claim the best territories (and maybe the fittest mates), as well as to time their breeding cycle to synchronise with the availability of food items for their young, factors that do not resonate to the same extent with immigrants from Scandinavia and other northern European locations in autumn.

First Arrivals Date Change 1900–2014

SMAD data end in 2014, the cut-off date for observations contributing to this *Avifauna*. To estimate 'generalised' first arrival dates for 2014 and the advance of arrival date to that year, the following procedure was adopted. Both the data of Table 2 (columns 5–6) and visual inspection of graphs of arrival dates versus year suggest that, for many species, the pattern of change before and after 1950 was different. Hence, the statistical analysis determined a 'best fit' linear

Table 2. First Arrival Date Calculations

Species	Average First Arrival Date			Average First Arrival Date Change (dy) Over Periods ...			Estimated First Arrival Date and Change to 2014		
(1)	(2)	(3)	(4)	(5)	(6)	(7)	(8)	(9)	(10)
	1900	1950	2000	1900–1950	1950–2000	1900–2000	Date in 2014	Change (dy) 1900–2014	Change (dy) 1950–2014
Species Group 1									
Yellow Wagtail	22 Apr	17 Apr	8 Apr	5	9	14	5 Apr	17	12
Species Group 2									
Wheatear	4 Apr	8 Apr	19 Mar	(−4)	20	17	13 Mar	22	26
Sand Martin	2 Apr	4 Apr	17 Mar	(−2)	18	16	14 Mar	19	21
Sedge Warbler	1 May	25 Apr	16 Apr	(6)	9	15	15 Apr	16	11
Swallow	9 Apr	7 Apr	25 Mar	(2)	13	14	24 Mar	16	14
Garden Warbler	29 Apr	24 Apr	15 Apr	(5)	9	14	11 Apr	18	12
Pied Flycatcher	28 Apr	27 Apr	15 Apr	(2)	12	13	10 Apr	18	17
Common Sandpiper	15 Apr	18 Apr	4 Apr	(−3)	14	11	31 Mar	15	17
Willow Warbler	10 Apr	6 Apr	31 Mar	(4)	5	9	29 Mar	11	7
Redstart	18 Apr	18 Apr	10 Apr	(0)	8	9	6 Apr	12	12
House Martin	10 Apr	13 Apr	2 Apr	(−3)	12	8	30 Mar	11	14
Swift	30 Apr	28 Apr	23 Apr	(2)	5	7	22 Apr	8	6
Wood Warbler	26 Apr	29 Apr	21 Apr	(−3)	8	5	17 Apr	9	12
Species Group 3									
Whitethroat	22 Apr	18 Apr	15 Apr	(4)	(3)	7	15 Apr#	7	3
Cuckoo	15 Apr	12 Apr	9 Apr	(4)	(3)	6	9 Apr#	6	3
Lesser Whitethroat	27 Apr	21 Apr	23 Apr	6	(−2)	(5)	21 Apr	6	0
Grasshopper Warbler	26 Apr	26 Apr	22 Apr	(0)	(4)	(4)	22 Apr#	4	4
Tree Pipit	14 Apr	19 Apr	10 Apr	(−5)	9	(4)	10 Apr#	4	9
Whinchat	26 Apr	25 Apr	24 Apr	(1)	(1)	(2)	24 Apr#	2	1
Reed Warbler	20 Apr	7 May	24 Apr	(−17)	13	(−4)	18 Apr	2	19
Spotted Flycatcher	3 May	8 May	8 May	(−6)	0	−5	8 May#	−5	0
Species Group 4									
Nightjar	14 May	10 May	8 May	(4)	(2)	(6)	n/a	n/a	n/a
Turtle Dove	5 May	5 May	3 May	(0)	(2)	(2)	3 May#	2	2
Wryneck	20 Apr	21 May	23 Apr	(−31)	(28)	(−4)	n/a	n/a	n/a
Nightingale	25 Apr	26 Apr	1 May	(−2)	(−5)	(−6)	n/a	n/a	n/a
Corncrake	28 Apr	10 May	13 May	−12	(−3)	−15	n/a	n/a	n/a
Blackcap	21 Apr	18 Apr	x	(3)	x	x	n/a	n/a	n/a
Chiffchaff	31 Mar	22 Mar	x	10	x	x	n/a	n/a	n/a
Little Ringed Plover	x	x	22 Mar	x	x	x	18 Mar	n/a	4
Species Group 5									
Redwing	20 Oct	14 Oct	3 Oct	(6)	11	17	28 Sep	21	15
Brambling	25 Oct	9 Nov	19 Oct	−15	21	(6)	12 Oct	14	28
Fieldfare	9 Oct	22 Oct	5 Oct	−13	17	(4)	5 Oct#	4	17

NOTES

- Within each group, entries are ordered by decreasing arrival date change from 1900 to 2000 (column 7).
- Differences are calculated as (early date)–(later date). Hence, advances in arrival dates are positive numbers.
- Discrepancies between dates and the number of days shown for differences are due to rounding.
- Parentheses around differences in columns 5, 6 and 7 denote differences that are not large enough to be reliable at the 95% confidence level.
- The symbol # in column 8 denotes a 1950–2014 trend line showing neither an increase nor a decrease, implying indistinguishable arrival dates for 2000 and 2014.

trend line for the arrival dates, using all data that occur in SMAD for the period 1950–2014. This trend line was extrapolated to the year 2014 to estimate the arrival date in that year, except when there was no trend to earlier or later dates, when the date for 2000 was used also for 2014. These cases are identified by the symbol '#' against the estimated 2014 date (column 8 in Table 2). Estimated arrival date changes for both 1900–2014 and 1950–2014 were calculated (where possible), and are shown in columns 9 and 10 in Table 2.

For Species Groups 1 and 2, changes in first arrival dates over the 1900–2014 period were notable (all in excess of one week), with Wheatear showing a remarkable 22-day change. The species in Group 3, which were characterised by less pronounced arrival date change over time, naturally exhibited smaller advances (generally a week or less) and are likely to be unreliable.

Autumn-arriving migrants (Group 5) exhibited little consistency in arrival date change to 2014 other than the fact that all were advances over both the 64-and 114-year periods.

Overall, there is no obvious correlation between the size of the arrival date change to 2014 and bird taxonomic group, behaviour, habitat needs or over-wintering location.

Discussion

How credible are these results and do they resemble those from other assessments of first arrival date advance? Cotton (2003) determined advances for a variety of spring migrants in Oxfordshire that include 17 species in the first three groups in Table 2. For these species, the average advance was eight days over the period 1971–2000. This compares with 10 days for the same species in Table 2 for the longer 1950–2000 period. The ranking of the 17 species by the size of the advance showed little relationship between the two studies. Sparks *et al.* (2007) found that 50% of the cases they examined for six locations in England showed significantly earlier dates over a 50-year period, with a springtime average advance of 0.25 days/year. The reliable data of column 6 in Table 2 for those species in Groups 1–3 show a rather larger proportion of advances and an average rate of change of 0.20 days/year. More recently, Newson *et al.* (2016), using UK data from three citizen science sources (Inland Observation Points, Migration Watch and BirdTrack) reported an earlier arrival of 9 days for 11 of the species in the first three groups in Table 2, from the mid-1960s to 2002–11 (equivalent to a rate of change of 0.22 days/year). Notably, the date shifts recorded by Newson *et al.* were significantly correlated with those found from SMAD, with Sand Martin showing the largest and Whinchat the smallest advance.

It seems likely, therefore, that the results shown in Table 2 are consistent with data reported by others (using different methods of analysis), which is both reassuring and lends support to the methodology employed here.

A Note of Caution

As with all generalisations drawn from observed data, our assessment of their reliability should be moderated by an understanding of the uncertainties in the original observations and in their subsequent manipulations. Uncertainties in the case of calculating trends from recorded first arrival dates reflect both ornithological judgements (how likely is it that the recorded date is close to the 'true' date), and uncertainty inherent in inferences from statistical techniques.

Arrival date observations are straightforward and, short of misidentification or recording error, can only be erroneously late, not early. However, there may be little or no correlation between the date of the first record and the 'true' first date, especially for scarce species or in areas with few observers or poor accessibility. Many of the species listed have niche habitats, which are unlikely to have been visited by observers at the right time in every year. The increase in observers over time noted above may have affected the detectability of early arrivals, potentially reducing the internal consistency of the dataset over time. Furthermore, 'population size may influence the first observation … particularly pronounced for bird species which are difficult to observe and/or are at low population numbers' so 'caution needs to be used in interpreting arrival dates of species whose population is low and changing' (Sparks *et al.* 2001). This is an important factor in interpreting the results for many species, particularly those in Group 4.

Naturally, like all statistics, averaged dates of arrival and the arrival date advance have levels of reliability associated with them. Hence, a calculated advance of (say) 10 days in arrival for a species is associated with a given probability (say 95%) that the true advance is between (say) 7 and 13 days. This aspect of the analysis should be explored in more detail in subsequent work with SMAD. The practical implication of this is that we should be careful about attributing biological meaning to arrival date differences that are not large.

Despite these caveats, the results of the SMAD analysis display sufficient internal consistency and adequate correspondence to results from other studies to suggest that they are reflecting real changes in bird behaviour over the past 114 years.

Implications

Notwithstanding these reservations, and accepting that causality cannot be inferred from this analysis, the changes observed are consistent with the hypothesis that a warming climate (see p. 44) is driving earlier migration into the County. This position is lent further support in that Table 2 suggests greater changes in first arrival dates occurred in the second half of the twentieth century than in the first half, which is consistent with the fact that global and regional temperature series show a much more modest warming prior to 1950 than since (Met Office 2017). Sparks *et al.* (2001) reviewed numerous studies and stated that 'in many incidences there is sufficient correlative evidence with temperature to suggest that temperature has an influence on bird migration timing', and called for further discussion and research. Analysis of the extensive SMAD records provides an important contribution to this, and further work is encouraged.

Changes in Egg-laying Dates

Nationally, advances in first egg-laying dates during the period 1966–2014, have been found using BTO Nest Record Card data (Walker *et al.* 2014). An attempt was made to correlate this advance with the advance in arrival dates in Shropshire over the same period for the 18 species that feature in both studies, but no relationship was found.

It should be noted that use of observations from very different geographical contexts and different provenances, and the regional bias associated with the non-random nature of the egg-laying data, may not be appropriate for such a comparison.

Coda

None of this analysis would be possible without the combined effort of many observers over the years. It is easy, when making a single observation, to assume it has no wider importance. We hope this chapter will encourage observers to continue to submit records, however commonplace and trivial they seem at the time. The BTO's national BirdTrack system is probably the best place for such records and its use is highly recommended; such records are automatically passed to the SOS and are available for conservation use locally.

Where to Find More Information

The more statistically detailed version of the arrival date advance analysis, together with graphs of SMAD first arrival dates shown as a function of year of observation, are available by visiting the SOS website home page and clicking on 'Avifauna Supplement'.

Shropshire and Its Bird Habitats

Leo Smith

Shropshire

Ours is a county of contrasting and dramatic landscapes. It is the largest landlocked county in Britain, with an area of 3,487 square kilometres (sq.km) or 1,346 square miles.

The administrative county was split in two when Telford and Wrekin became an independent unitary local authority in 1998. Its largely urban area is 290sq.km, while the much larger and more rural Shropshire Council area is 3,197sq.km.

This Avifauna covers the whole of the County as shown in Map 1. It coincides with Watsonian vice-county 40 (see p. xi).

Map 1. Topography and Major Towns

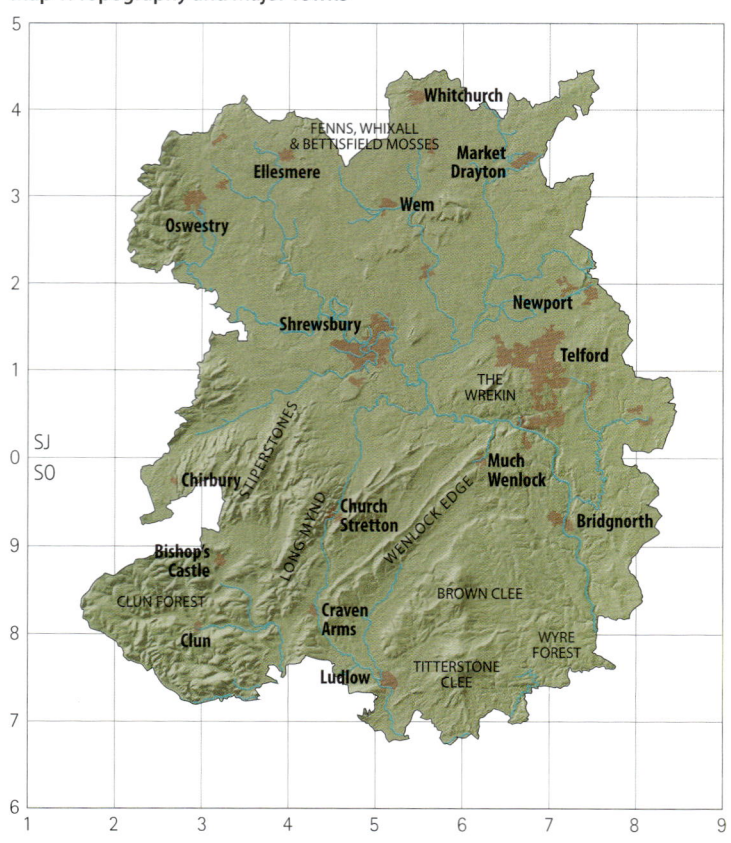

This chapter is intended to provide a brief overview of the natural foundations, climate, land management and other factors that have shaped bird habitats, and the distribution of species. It does not go into great depth on any aspect of this, but more detailed publications are cited, and specific references are included in Appendix 4 for readers that wish to explore further.

Natural Foundations

The complex geology is fully described in *Geology of Shropshire* (Toghill 2006), with 10 of the 12 recognised periods of geological time represented.

The Shropshire Hills, designated as an Area of Outstanding Natural Beauty, dominate the south and south-west. The west is bounded by the hills of the Welsh borders, while the north and north-east comprise the southern part of the Shropshire, Cheshire and Staffordshire Plain, an expanse of flat or gently undulating, lush, pastoral and arable farmland, which includes the internationally important meres and mosses, and a series of small but prominent sandstone ridges. The east and south-east are part of the larger mid-Severn Sandstone Plateau. A major river, the Severn, and its tributaries, drains virtually the whole County.

More detail can be found in the six National Character Area profiles, which between them cover the whole County, at 'National Character Area Profiles: Data for Local Decision Making' at www.gov.uk.

This natural foundation is overlain by human land use: agriculture and forestry, mining and quarrying, drainage and water management, and urban development.

The resulting interactions have created very varied habitats, reflected in the diversity of breeding birds, passage migrants and winter visitors.

The natural regions, adapted from *The Shropshire Landscape* (Rowley 1972), are shown in Map 2.

Natural Regions

The north-west uplands, now more commonly referred to as the Oswestry uplands, are an extension of the Clwyd hills. Up to 330m high, the Carboniferous limestone and sandstone scarps have steep

Map 2. The Natural Regions (adapted from *The Shropshire Landscape*, Rowley 1972)

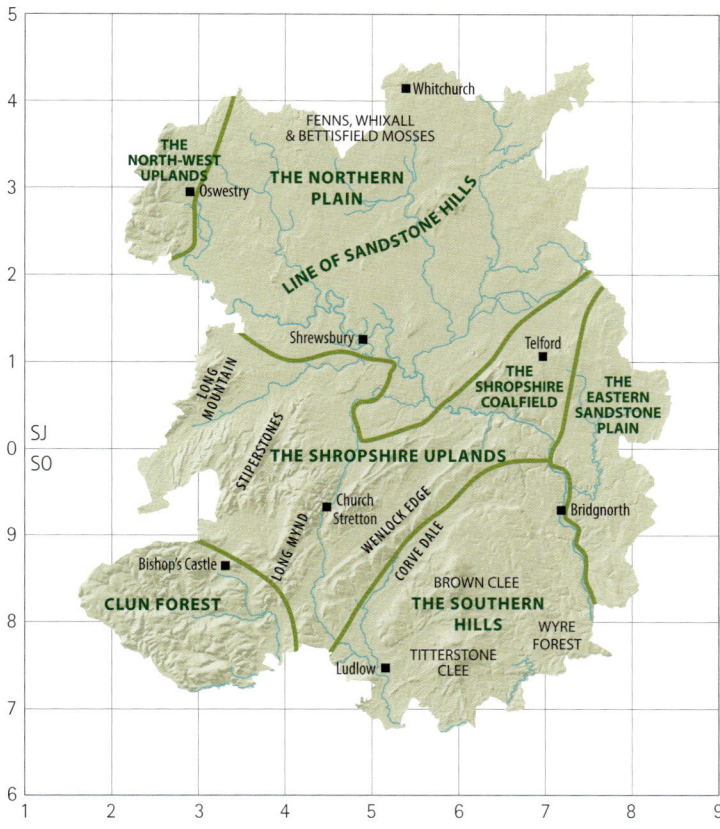

drift deposits, cover much of the northern area, with many ponds and meres in the associated hollows. Many became vegetated, forming peat mosses and marsh land, but most have now been drained and reclaimed for agriculture. The gently rolling landscape at around 80m is also broken by a line of higher sandstone hills – Nesscliffe, Pim Hill, Grinshill and Hawkstone – and in some areas sandy soils come to the surface. Moving south-east, the plain is slightly higher at about 120m, and here the countryside is more steeply rolling with deeper river valleys and sandstone outcrops.

The most important part of the Shropshire coalfield was around Coalbrookdale, as it included the last deep mine, and the proximity of coal and ironstone was a major factor in Ironbridge becoming the birthplace of the Industrial Revolution. It, along with Wellington, Oakengates, Madeley and Dawley were subsequently incorporated into the site of the sprawling Telford New Town, with major new development occurring from the 1960s onwards.

Brown Clee, the highest summit, from Titterstone Clee, the third highest. Stephanie Hayes, Shropshire Hills AONB, 30 July 2012

west-facing and gentle east-facing slopes, deep valleys and occasional limestone cliffs and quarries.

Clun Forest in the south-west is also an extension of the Welsh hills. A rolling plateau of Silurian sandstones and shales, and Old Red Sandstone, it is around 460m high, and is cut deeply by the major valleys of the rivers Clun and Teme and their tributaries.

Offa's Dyke, the historic earthwork that defined the Welsh border many centuries ago, traverses both these hilly western regions, and in the miles between them passes over Long Mountain, an outlying hill of the Shropshire uplands. These uplands rise to over 500m and are based on Precambrian rocks, some of the oldest in Britain. The four long parallel ridges of the uplands – Stapeley Hill, the Stiperstones, the Long Mynd and the Stretton hills/Caer Caradoc/the Lawley/the Wrekin – together with the Silurian limestone escarpment of Wenlock Edge, all run south-west to north-east, and are separated by the valleys of the rivers West Onny and East Onny, the Stretton valley and Ape Dale.

Corvedale separates Wenlock Edge from the southern hills where Brown Clee and Titterstone Clee rise from an undulating plateau at around 230m. Brown Clee (540m) is the highest point in Shropshire, and Titterstone Clee is only 7m lower.

Although the hills are the major landscape feature in west and south Shropshire, the dales and valleys that separate them, especially Corvedale, are often wide and fertile. They are fed by countless streams which frequently cut steep narrow valleys of their own.

All the upland areas, except the Oswestry uplands, lie south and west of the River Severn. Plains lie on the northern and eastern side, based on New Red Sandstone and largely overlain by extensive glacial

The River Severn flows from west to east then south-east across the southern parts of the plain, while its major tributaries from the north, the rivers Perry, Roden, Tern, and Worfe, drain the plain itself. The rivers and main streams are shown on Maps 1 and 2 (topography and natural regions respectively), and the background map to each species account (p. 66).

Relief

These natural foundations are reflected in the variation in the height of the land above sea level, shown in Map 3. However, the steep-sided hills and deep valleys create very considerable height differences over short distances. The Atlas is based on survey areas of 2km squares, known as tetrads (see Appendix 1), and within the area of a single tetrad it is not uncommon to find a height difference of more than 150m. Some contain height differences of over 200m, and the maximum appears to be 290m where the whole western slope of Caer Caradoc rises from the Stretton valley to the summit at 459m. Birds typical of upland and lowland habitats may therefore be found in the same tetrad.

Between the hills and the plains there is some low-lying land, and most of that below 30m shown in Map 3 is part of the river valley system of the Severn and its tributaries, the only major exception being the area to the east of the Tern valley and north of Telford, the ancient marshland of the Weald Moors.

Climate (Temperature and Rainfall)

Shropshire possesses a fairly typical example of the marine west coast climate (Arnfield, 2015a; 2015b) which covers all of the UK and much of Western Europe. Such climates are characterised by warm (not hot) summers, cool (not cold) winters and, as a result, a moderate annual temperature range. Precipitation is evenly distributed throughout the year, with no dry season, and is mostly in the form of rain. These characteristics originate from the annual persistence of the prevailing mid-latitude westerlies, with associated cyclonic

Looking northwards from the Long Mynd (foreground), with Wrekin, Lawley, Caer Caradoc and Hope Bowdler on the skyline west to east, and the Stretton valley and Church Stretton. Peter and Jane Howsam, 23 August 2014

and frontal activity, and ready access to humid air masses from the north Atlantic, moderated by the warming influence of the North Atlantic Drift ocean current.

The climate is transitional between the mild and dry conditions of the English Midlands and the cool, wet upland environments of mid-Wales. The remoteness of the moderating influence of the ocean leads to an annual range of temperature slightly greater than that of more westerly locations closer to the sea. Variations of both temperature and precipitation are largely topographic in origin, and not surprisingly there is a high correlation between these maps and Map 3, the Relief map.

The Met Office produces temperature and rainfall data averaged over 30-year periods, most recently for 1971–2000. Map 4 (Met Office 2015a) shows that July daily mean temperatures vary between 12.4 and 16.8°C, with the mildest conditions on the plains north of Shrewsbury, down the Severn Valley south of Ironbridge and in the valleys in the south. The coolest conditions are found in the uplands

Map 3. Relief

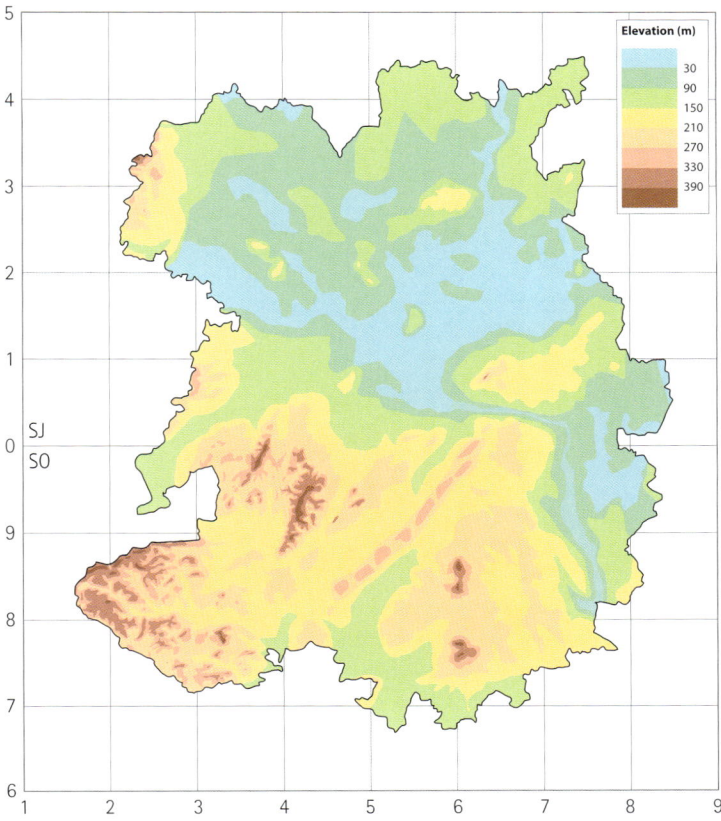

Map 4. July Daily Mean Temperature 1971–2000
© Crown copyright, adapted from map provided by Met Office

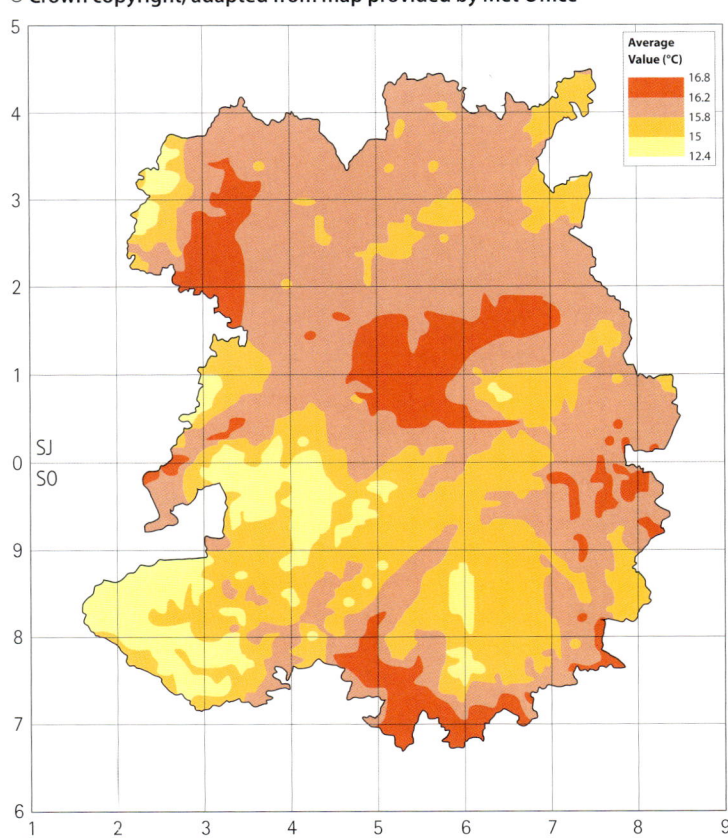

Map 5. January Daily Mean Temperature 1971–2000
© Crown copyright, adapted from map provided by Met Office

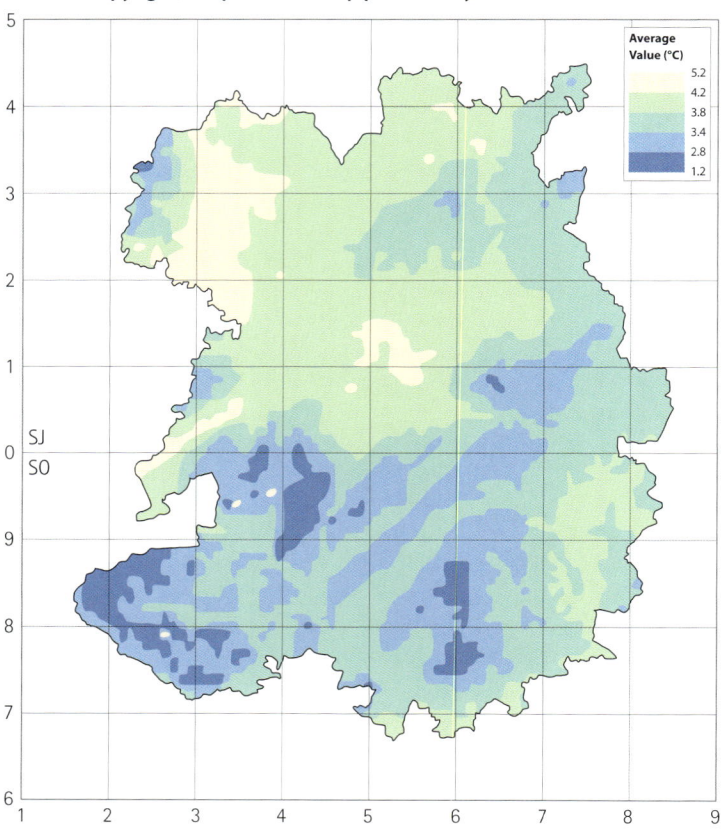

Map 6. Annual Average Rainfall 1971–2000
© Crown copyright, adapted from map provided by Met Office

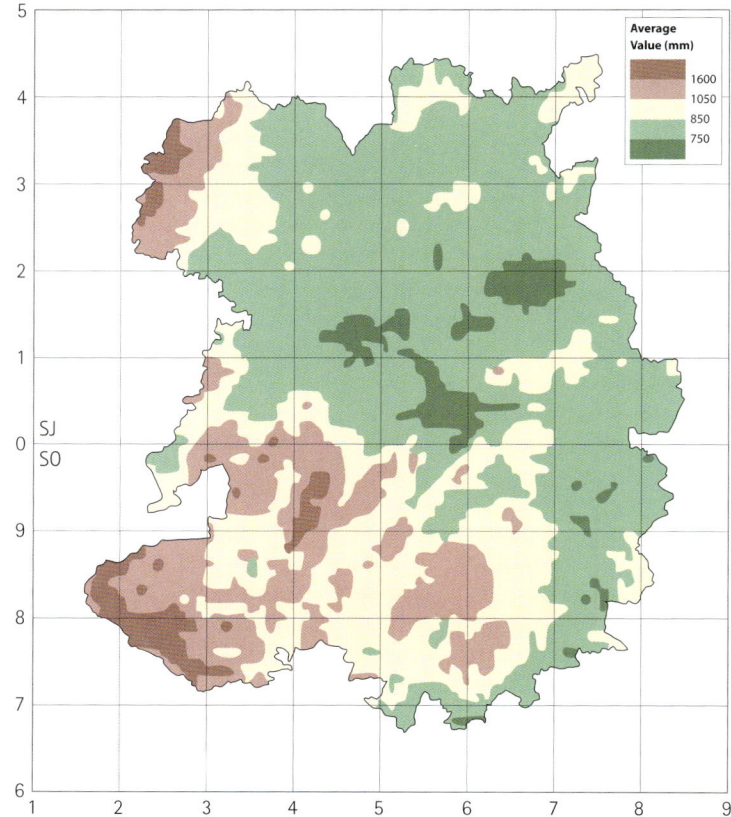

Winter conditions on the highest hills are hard; only Red Grouse remain on the Long Mynd. Peter and Jane Howsam, 1 February 2015

of the north-west, the Stiperstones and Long Mynd, the Clun Forest, Wenlock Edge, the Wrekin and the Clee hills. The pattern is similar for the average July maximum temperature but temperatures are 4–5°C higher.

January mean temperature (Map 5) again is highest in the lowland areas (3.4–5.2°C) and lowest over the upland areas of the south and south-west (1.2–3.4°C). Night-time minima at the highest elevations average –1.0–0.6°C. A measure of the severity of the climate in the hills of the south and west is that they experience air frost between 50 and 98 days per year. Ground frost is more widespread throughout the west and south and may occur on more than 118 days in the Clun Forest, the Long Mynd and Stiperstones, the Clee hills and in the Severn Valley between Shrewsbury and the Welsh border. While much of this frost is short-lived, longer spells may limit feeding by species that probe moist ground in search of invertebrate prey (e.g. Woodcock, Snipe, Lapwing, Blackbird and Starling).

Rainfall distribution (Map 6) again reflects topography with the lowlands of the north and Severn Valley showing annual totals of 500–750mm. The wettest areas, in contrast, may receive up to 1,600mm due to uplift of moist air streams over high ground. On up to 58 days per year, precipitation may occur as sleet or snow in the Clee hills, south Shropshire hills and Clun Forest. Lying snow – another winter hazard for ground-feeding species – may occur from 10 to 38 days in the Oswestry uplands, and south-west of a line connecting Llanymynech, Shrewsbury, Ironbridge and Cleobury Mortimer.

Land Use

The wide range of rocks and drift deposits has produced an equally wide range of minerals and soil types, which, coupled with the relief, climate, rainfall and drainage, has determined the land use.

Agriculture

Farming dominates the landscape, using around 84% of the land. The national agricultural land classification grades land according to the physical suitability of the area in which it lies for agricultural production. It takes into account climate (temperature and rainfall, aspect, exposure and frost risk), site (gradient, micro-relief and flood risk) and soil (texture, structure, depth and stoniness, and chemical properties which cannot be corrected). A map showing the classification in the West Midlands, updated in 2010, can be found on the Natural England website. The best agricultural land is grade 1, but there is none in Shropshire. Grade 2 land occurs where there are well-drained soils, and no climatic and few relief considerations affecting agricultural use. Grade 3 land is widespread below 200m, and grade 4 land has severe limitations for agriculture; it is generally above 200m with high annual rainfall, or on the poorly drained soils of the coalfields and river valleys. Grade 5 is generally on the highest hills over 300m. Therefore the distribution of the grades broadly matches the temperature, rainfall and relief maps above.

The quality of land determines the type of farming, which in turn influences the landscape that we see today. The predominant farm types in 1986 are shown in Map 7. However, the cultivated land on the mixed farms in the fertile valleys of the south and west, and scattered across the northern plain, does not show through on the map as it is swamped by the predominant livestock farming. Land in use for cereal production at that time is therefore under-represented on this map.

It represents farming at the time of fieldwork in 1985–90 for the *Atlas* (1992), but there have been several major changes in farming land use since then, which are described on p. 40. Unfortunately, Defra no longer collects statistics at the parish level, so no equivalent revised map has been published since. Arable land in the north, and in the river valleys in the south, is more extensive now, and therefore even more under-represented on Map 7. Some of these changes have influenced the population and distribution of many birds, reflected in the species accounts.

Woodland

Shropshire is well wooded, with 8.5% of its area still covered by woodlands of at least 0.1ha, an increase from 7.3% since 1980. Almost 58% is broadleaved, 29% is conifer, and 9% mixed, with small amounts of open space, felled and coppice with standards making up the remainder. The Forestry Commission owns or leases 16%, including 62% of the total coniferous woodland, with 84% in other ownership. Oak is the most common broadleaved species (31% of all broadleaved), and larch is the most common conifer (22% of all conifer species). In addition, there are over five million trees not in woodland, almost all broadleaved. Map 8 shows all 1,644 woods of two hectares or more (Forestry Commission 2002).

Map 7. Predominant Farm Types 1986
© Crown copyright, adapted from MAFF 1988

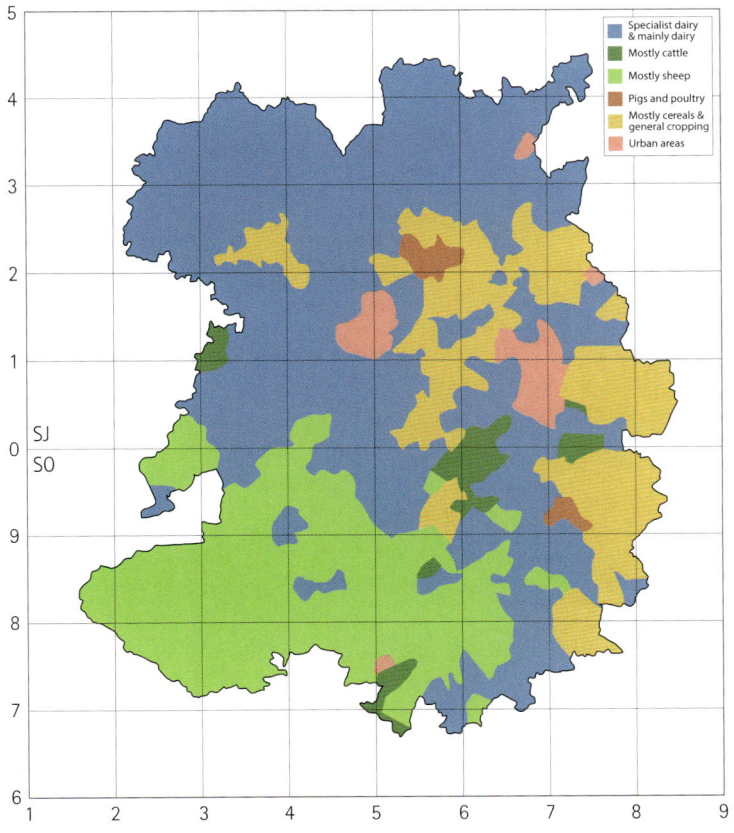

Map 8. Woodlands of 0.2ha or More
Contains Forestry Commission information licensed under the Open Government License v3.0

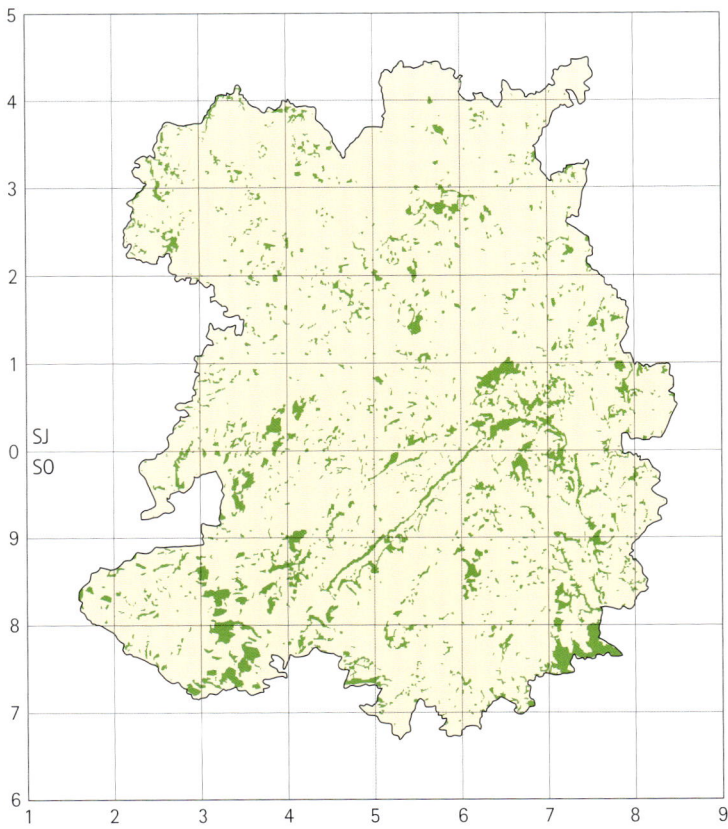

Old quarry building and spoil near the summit of Titterstone Clee hill. Jim Almond, 2 April 2011

Most of the large conifer blocks owned by the Forestry Commission are in the south-west, and parts of the Wyre Forest, and almost all have been planted since 1941 (1945 in practice; data is presented in 10-year periods), with 52% planted between 1951 and 1970. Just over half of the privately owned coniferous woodland was planted in the same 20 year period (*ibid.*).

Much of the remainder is ancient semi-natural woodland, with major remnants in other parts of the Wyre Forest, in the Ironbridge Gorge, and along Wenlock Edge, the latter owned by the National Trust. Smaller open woods remain on other hillsides and riverbanks where the steepness of the slope has hindered agricultural clearance, and small woods, coppices and game coverts are still widespread.

Industrial and Other Human Influences

Apart from agriculture many other human activities have affected the landscape.

Mining used to be very important. The area west of the Stiperstones, around Shelve, used to be one of the richest sources of lead in Britain, and coal, iron, limestone, dhustone (dolerite) and copper were all intensively mined in the Clee hills, and dhustone still is. New habitats have been created as a result; shafts, overgrown spoil tips, cliff faces and boulder-strewn slopes. More recently, here and elsewhere, quarrying for building and roadstone has created most of the cliff faces.

Similarly, interesting habitats have been created around Ironbridge and elsewhere in the Telford area as a result of industrial development based on the coalfield. The early iron industry's need to maintain a continuous supply of charcoal, through coppicing, is one of the main reasons why the extensive ancient semi-natural woodlands of the Ironbridge Gorge have survived.

Apart from the natural meres, many open still-water habitats are also man-made. These include ornamental lakes around the old manor houses, pools resulting from gravel or sand extraction or subsidence of old mines, the construction of reservoirs and canals, and the extensive pools resulting from sugar beet processing at the Allscott factory. Ponds, drainage ditches and small reservoirs for livestock and irrigation are a frequent by-product of farming, and trout fisheries are increasing.

The total population counted in the 2011 census was just over 472,000. Only Telford (population of the 'built up area' around 142,700) and Shrewsbury (71,700) can be considered large towns, and they make up about 44% of the total. Both are predominantly suburban, with gardens and parks providing breeding habitats for the more common birds. The former was largely created by Telford Development Corporation, which was charged with reclaiming the derelict land from the coalfield, and aimed to invest sufficient funds in landscaping and new habitats to create an abundance of wildlife, although ironically some of the 'derelict' sites were preferable to the subsequent development from the wildlife point of view. The Development Corporation was wound up in 1991, and Telford and Wrekin became a unitary local authority in 1998, becoming independent of Shropshire County Council, which was renamed Shropshire Council after the subsequent abolition of the six District Councils in 2009.

Apart from Telford and Shrewsbury, five more 'built up areas' had populations of more than 10,000 in 2011: Oswestry (18,700), Newport (12,700), Bridgnorth (12,300), Market Drayton (11,800) and Ludlow (10,500); Whitchurch, Shifnal, Wem and Broseley have more than 5,000, and Albrighton, Church Stretton, Cleobury Mortimer, Craven Arms, Ellesmere, Highley, Minsterley and Pontesbury, and Much Wenlock have more than 2,000. The Shropshire Council area is predominantly rural: 54% of the population lives in these settlements, but they cover only 4% of the area. Telford and Wrekin is more built up, with large industrial estates, but just over 60% of the land is still farmed.

Telford had a population growth of around 17% between the 1991 and 2011 censuses, and Shropshire around 15%. However, the average household size decreased, so the number of households increased by almost 25,000 to just under 130,000 in Shropshire (23.6% increase), while Telford and Wrekin increased by just over 14,000 to around 66,600 (27.0% increase). Both councils as planning authorities have housebuilding targets for the next 10 years or so: another 19,300 homes between 2014–26 in Shropshire, and about 7,500 in Telford and Wrekin up to 2031. Building on such a large scale has already incurred a substantial wildlife habitat loss, particularly around Telford, and wildlife organisations need to be vigilant to oppose future development of important sites, particularly those inhabited by scarce species, or those that occupy niche habitats. However, paradoxically the conversion of monoculture farmed fields that support little wildlife into suburban gardens with trees, shrubs, lawns and bird feeders probably represents a biodiversity gain, and will certainly be beneficial to the more common 'garden birds', provided that the increasing use of slug pellets, weedkillers, insecticides, artificial grass 'lawns' and the modern fetish for tidiness can all be reversed by householders, with the encouragement of conservation and wildlife organisations.

Bird Habitats

The interaction of the landscape and land use has created a wide range of bird habitats, many of which have changed for the worse in the last 25 years.

Open Moorland and Upland Heaths

Large areas of the highest tops in the Shropshire uplands and southern hills are open moorland, corresponding to the highest hills (Map 2) and areas of low temperature and high rainfall. They are mainly

Heathland on the Long Mynd – Pole Bank from behind Pole Cottage. Leo Smith, 26 July 2008

heathland dominated by heather and bilberry, or extensive areas of bracken with a few scattered stunted trees. On the Long Mynd and Stiperstones, the heather is cut or burnt periodically to promote new growth to benefit Red Grouse, but many adjacent areas have been cleared to create improved grassland. Small pools and bogs are frequent on the flat tops. There are some rock outcrops, with extensive boulder-fields on the Stiperstones and Titterstone Clee. As the hillsides are cut by fast-flowing streams in steep-sided valleys, single trees and small open woods can grow in very sheltered positions close to the summits. Gorse and other shrubs are also prevalent in some of these upland valleys.

Several different upland habitats and the birds associated with them exist side by side in a small area: Meadow Pipit, Skylark, and perhaps Curlew and Red Grouse on the heather moor, Reed Bunting and Snipe in the wet flushes, Wheatear among the rocks and rabbit burrows, Whinchat and Stonechat in the bracken and gorse, Dipper and Grey Wagtail along the streams, and Tree Pipit, Redstart, Pied Flycatcher, Wood Warbler, Buzzard and Raven among the scattered trees and woods.

Upland Grassland

Elsewhere on the high hills in the south and west, land has been 'improved' for agriculture, mainly sheep grazing, and most is well drained with short, well-cropped, grass. Rock outcrops and rabbit burrows provide homes for the occasional Wheatear, but in general there are few ground-nesting birds, apart from Meadow Pipit and Skylark in the less uniform swards. Small numbers of the more common species such as crows and tits inhabit the few trees and unimproved land that might remain.

Upland field boundaries are usually fences rather than walls, providing little cover for nesting birds. Traditionally-laid hedgerows enclosed many fields on the lower hillsides, but most have been neglected, with the resulting mature gnarled hawthorn trees providing numerous sites for hole-nesting birds, especially Redstart. The hollows between the hills are sometimes marshy, and the drainage ditches overgrown, so patches of rushes or longer grass may give some cover for Reed Bunting, and the occasional Lapwing and Snipe. Clumps of gorse, occasionally extensive, can also be found, providing homes for Linnet and Yellowhammer.

Upland grassland – close cropped sheep pasture at Nordy Bank Fort, with Titterstone Clee in the distance. Stephanie Hayes, Shropshire Hills AONB, 4 January 2006

**Potato crop near Bishops Castle.
Leo Smith, 13 June 2004**

The banks of the main streams and rivers draining the uplands are usually lined with deciduous trees rich in common woodland birds and species such as Nuthatch, Pied Flycatcher, Redstart, Treecreeper and woodpeckers, and mature hedgerows often enclose the fields alongside these streams. Many of them have steep-sided banks that cannot be worked by machinery, so the trees and scrub are safe from clearance. Most of the species found in tetrads dominated by upland grassland occur almost entirely in these other incidental habitats, which usually occupy only a small proportion of the tetrad.

Lowland Grass and Pasture

Descending the northern slopes of the uplands onto the plain and down into the Severn Valley around Shrewsbury, the land is richer and was used mainly for dairy farming, although increasingly cereals have been grown. This land use continues northwards, and then extends eastwards to the north of the line of sandstone hills (Maps 2 and 7).

Many species have declined substantially in these areas in the last 25 years, and the grassland and pasture itself supports few birds, but many scattered small woods, coppices and game coverts, and mature hedgerow trees, are still left and provide cover for many hedgerow, woodland and scrub birds.

Cultivated Farmland

Most of east and north-east Shropshire is grade 2 or grade 3 agricultural land, and is heavily cultivated. The main crops are barley and wheat, but these are usually rotated with potatoes, cabbage, peas and, increasingly, Oilseed Rape and Maize (see p. 40). Yellow Wagtail and Corn Bunting nest on the land here, although both are declining.

Though many hedgerows have been removed to create larger fields, Shropshire appears to have suffered less in this respect than many counties. Unfortunately, there are no records to quantify the loss. However, fields are still often enclosed by mature hedges, and small woods and mature hedgerow trees also remain, so there is usually some diversity of bird life even in tetrads where the land is predominantly cultivated.

Woodland

The deciduous and mixed woods support a wealth of wildlife and have high numbers of breeding birds. However, the typical woodland species, such as Great Spotted Woodpecker, Nuthatch and Treecreeper, also use small coppices and scattered trees, and the distribution maps for these species indirectly demonstrate the widespread occurrence of trees as well as the large areas of woodland (see Map 8).

Trees in many of the coniferous plantations are close together, so the mature woodland is dark and impenetrable, devoid of all except the most adaptable or catholic birds: Woodpigeon, Goldcrest, Coal Tit, Blue Tit and Chaffinch. However, when the trees are first planted, and again after thinning, more open and interesting habitats are created, offering opportunities to Tree Pipit, and occasionally Nightjar and Lesser Redpoll.

Oak woodland at Whitcliffe, Ludlow, home of Pied Flycatcher, Redstart and many other woodland specialists, including the increasingly rare Marsh Tit and Treecreeper. Gareth Thomas, 1 May 2009

Most of the coniferous plantations were planted in the early post-war period, and are now producing crops of seed-bearing cones, providing food for the increasing Siskin and Crossbill populations.

Rivers and Streams

The major rivers and tributaries are included on the background map for every species. They, and all significant streams, are shown on Map 1, Topography.

The Severn is far and away the largest river. On the western border its confluence with the River Vyrnwy has created an extensive lowland floodplain, which floods in most winters, and provides habitat for Whooper Swans, other wildfowl and the main wintering Curlew flock. This plain narrows as the river approaches Shrewsbury, but it widens considerably again to the east of the town at the confluence with the River Tern, and remains so as the river meanders to Buildwas. After the 15km of the steep-sided Ironbridge Gorge, the valley again opens out at the confluence with the River Worfe. These river systems have created most of the land below 30m (Map 3), and large sections flood every winter. The floods mean the land is largely pasture and some spring-planted crops, but while some remains damp into the breeding season, providing habitat for Curlews, in the main it is now well drained and no longer provides habitat for breeding waders.

The Severn and its main tributaries are currently the responsibility of the Environment Agency, which seeks to maintain flows, protect the built environment from flooding and manage fishing. All have been engineered to varying extents over many years to achieve these objectives and facilitate land drainage, but also to raise water levels or to prevent natural processes such as a river changing course. These modifications to river channels and changing weather patterns have resulted in more rapid fluctuations in both water flows and water levels. The Severn, especially, is prone to sudden changes in water level following heavy rain upstream in Wales, and this can be particularly disastrous in the spring when birds nesting on or close to the river can be washed out.

At the beginning of the breeding season the banks seem rather bare, but as the water level falls, a rich margin of emergent and bankside vegetation appears, with large expanses of water crowfoot in the shallower riffles between the deeper reaches. On some stretches willow and alder add to the diversity, and the vegetation attracts Sedge Warbler and Reed Bunting, while Kingfisher and Sand Martin make nest holes in the steep banks.

In contrast, the rivers and streams that rise in the hills and flow south of the watershed into the Clun/Onny/Corve/Teme system have been less intensively managed. The Teme itself is one of the UK's fastest flowing rivers because of its gradient, falling almost 500m in its total length of 130km. Many of the waters in this catchment are generally boulder-strewn, fast flowing and shallow, with rapids and some relatively deep pools where the gradient temporarily slackens off. The banks are often lined with trees and many stretches are overgrown and inaccessible, but in many places they are grazed right to the water's edge, so there is little cover, and there is pollution in the water from sediment. However, various projects have begun recently to fence the waterway and minimise access for cattle.

Elsewhere, the same characteristics are shown by the streams flowing from the Oswestry uplands, those streams that descend steep valley-sides into the Severn around Ironbridge, and the upper reaches of the rivers Tern and Worfe, although the latter has been affected by abstraction. Many of these rivers and streams are rich in bird life, and it is not uncommon for the same stretch of river to have both pools that support Kingfisher, and rapids occupied by Dipper and Grey Wagtail.

All three of these riverine species have declined since 1985–90, and disappeared altogether from some watercourses. They are dependent on food, such as aquatic invertebrates and small fish that require sufficient flows and good water quality. There has also been an apparent substantial decline over recent years in a number of fish species across the Severn basin, though this is very hard to quantify as it is largely anecdotal (Mike Morris, Severn Rivers Trust *pers. comm.*)

Changes in land use, increased drainage of agricultural land, modifications to the river channels themselves and changing weather patterns have resulted in more rapid changes in water flows, while water quality continues to deteriorate as a result of domestic, industrial and agricultural pollution. The Environment Agency is monitoring the quality of all water bodies, to assess progress towards meeting the requirement of the European Union Water Framework Directive (WFD), which requires that all water bodies are brought into good ecological condition by 2027. It carries out a technical assessment of every water body every few years, including biological monitoring, most recently in 2013. There are five classifications: High, Good, Moderate, Poor and Bad. The last three classifications all reflect a failure to meet the required standards for good ecological condition.

The 2013 results, for all 123 water bodies in Shropshire (including Telford and Wrekin, but excluding still or standing waters), are shown in Table 1. In summary, it shows that over half (56%) of the water bodies are failing. Detailed results for each water body can be found on the Environment Agency website.

River Teme looking east (downstream) from Ludford Bridge, Ludlow. The river is mostly fast-flowing, but the flow is controlled by weirs in Ludlow, so there are breeding Mute Swans on its more tranquil sections, but Dipper and Grey Wagtail on the rapids, above the bridge and below the weir. Gareth Thomas, 7 June 2009

Table 1. Condition of Water Bodies in Shropshire 2013

Condition	Number
High	19
Good	35
Moderate	60
Poor	7
Bad	2
Total	123

Apart from the Severn itself, most of the rivers and streams are upstream from large settlements, so 'agriculture and rural land management' is the most frequently cited reason for the failure of these water bodies, although the waste water from even small settlements can add a considerable amount of phosphate pollution. Enhanced nutrient levels result in excessive algal growth while an increase in sediments smother the bed of the watercourse and can lead to its cementation; both considerably increase turbidity. This results in a reduction in the aquatic flora and a severe loss of aquatic invertebrates, the main food supply for riverine birds. While the source of domestic and industrial pollution is often clear cut (it comes out of a pipe), agricultural sources are often more obscure and referred to as 'diffuse pollution'.

Farming methods such as ploughing and livestock management, the increasing size of farm machinery (including the increased run-off from damaged roadside verges), drainage systems on the land that rainwater has to flow over before it reaches the streams, fertilisers applied to nearby fields, run-off from recently cultivated land and erosion of banks by elevated livestock levels are among the most obvious sources of diffuse pollution.

Criteria for assessing the condition of SSSIs perhaps provide a better basis for assessing the value of rivers as wildlife habitat, although they focus on the reasons for the original designation, which might not relate to wildlife. The whole of the River Teme, and the lower reaches of the River Clun, is an SSSI. The latter is included because of its freshwater pearl mussel population, which is internationally important, so it is also designated a Special Area of Conservation, protected under the EU Habitats Directive. The WFD assessment divides the SSSI into five water bodies, two of high quality, one good and two moderate, but the assessment of its condition as an SSSI, in six 'river units', classes one as 'unfavourable, declining' and the other five as 'unfavourable, no change'. Natural England has produced River Restoration Plans for this SSSI, and several other freshwater SSSIs, but there are no similar remedial action plans for the vast majority of watercourses, and improvement of the stretches upstream of the SSSIs is necessary before there is any chance of returning the SSSIs themselves to favourable condition.

Chelmarsh Reservoir, the largest open water, frozen over. Jim Almond, 26 December 2010

Open water is often surrounded by reedbeds, for example at Venus Pool. Can you seen the Bittern? Jim Almond, 26 November 2008

Open Water

The natural meres, mainly around Ellesmere, but also Fenemere, both Marton Pools and stretching to Berrington Pool south of Shrewsbury, are the largest areas of open water. These are supplemented by the network of canals in the north, man-made ornamental lakes around the widely scattered old manor houses and parks, such as Shavington, and balancing lakes and landscape features around Telford New Town.

There is only one large reservoir, at Chelmarsh, with a surface area of about 39ha, which opened in 1966 to provide water to the Black Country and parts of South Staffordshire.

Sand and gravel removal has also created several interesting pools and wetlands, some of which have subsequently become important bird reserves, such as Wood Lane.

There are few large waters in the south and west, but of particular note are Shelve Pool at over 300m, where the surrounding reed beds support Sedge Warbler and Reed Bunting, and Boyne Water at 455m near the top of Brown Clee, with breeding Moorhen, Coot and Tufted Duck.

Small pools and ponds are widespread on the plain, and in the valleys between the uplands, as reflected in the large number of tetrads where both Mallard and Moorhen breed.

The larger meres, lakes and pools are widespread on the northern plain, and some have good quality water, cover to the water's edge, gently sloping banks and marshy areas with reeds or rushes, emergent vegetation and overhanging trees providing breeding habitat for Little Grebe, Great Crested Grebe, Mute Swan, Canada Goose, Tufted Duck and Coot among other water birds. Reed and Sedge Warbler, and Reed Bunting, nest in the bankside vegetation and reeds.

However, many of these waters also suffer from diffuse pollution. The Environment Agency's assessment of still waters shows that of 11 natural water bodies assessed in 2016, all but one of which are SSSIs, five were poor, four were moderate and only two were in good ecological condition. Of the 10 SSSIs, only one is in 'Favourable' condition (and in good ecological condition), six are 'Unfavourable but recovering' and three are 'Unfavourable and not improving'. Elevated nutrient levels are the major threat to the quality of the still waters and while there is no specific evidence that diffuse pollution affects particular bird species, it might account for the unexplained decline of many waterbirds on the north Shropshire plain, and it is a further indication of the poor state of the natural environment.

Aerial view of Fenn's, Whixall and Bettisfield mosses, an oasis in an agricultural landscape. © Natural England, 1 March 2006

Lowland Mosses and Bogs

Originally the north Shropshire plain contained many meres and 'mosses', meres which have become fully overgrown through the natural encroachment of successional vegetation, forming peat. Rowley (1972) commented that 'about four-fifths of the peatland acreage of north Shropshire is now wholly reclaimed and under pasture' and more has been drained since. The only extensive remnants are the SSSIs that include Fenn's, Whixall and Bettisfield Mosses National Nature Reserve, mostly in Wales, and Wem Moss SWT reserve.

The Weald (i.e. Wild) Moors north of Telford are extensive mosses lying lower than the surrounding plain. Map 3 shows them as the large area below 30m to the east of the River Tern and north of Telford.

They proved difficult to drain completely and still retained some of the original moss wetland character in 1985–90, although they have suffered further agricultural 'improvement' since then. Most of the best habitat now is the reeds, emergent vegetation, scrub and small trees in the extensive drainage ditches. Map A1.4 in Appendix 1 showing the change in the number of breeding species between the two Atlas periods (see p. 482) highlights the Weald Moors as the area that has lost the highest number of species.

The re-created rush pasture at Wall Farm is one of the few good bird-friendly sites in this area now.

Snipe, Redshank, Whinchat, and Tree and Meadow Pipit have been lost from all three of these areas of extensive mosses since 1985–90, and Oystercatcher is the only less common species that has increased.

The scrape at Wall Farm. Jim Almond, 25 August 2017

Hawkstone Park: Grotto Rock and Raven shelf, part of one of the very few natural cliffs. Jim Almond, 13 June 2010

Lowland Heath

This once covered the larger part of the light sandy soils of the north Shropshire plain, but most has been cleared for agriculture and only a few relics, such as Prees Heath, Steel Heath and Hodnet Heath, remain. The extent to which this habitat has been lost is shown in the map in the Black Grouse species account on p. 122.

The old industrial pit mounds from the intensive mining around what is now Telford frequently reverted to heathland once disturbance ceased, though it has largely been lost again through urban development in recent years. These areas used to provide much of the lowland habitat of Meadow Pipit, but they too have now disappeared.

Cliffs and Quarries

While there are many cliff faces in quarries, some of which provide nest sites for Peregrine and Raven, and the softer ones cater for Sand Martin, there are very few natural cliffs, apart from Hawkstone Park.

The Changing Face of Shropshire

Reference has already been made to the extent to which agricultural improvement has created the Shropshire landscape. Enclosure of the land, followed by drainage and other human influences, has been well documented (Rowley 1972).

Even so, it is hard to imagine the environment that the early nineteenth-century ornithological authors lived in. There were no cars, metalled roads or tractors, horses provided human transport and motive power on farms, and pastoral farming predominated. The human population was only half of what it is now, and lived in overcrowded conditions, so dwellings occupied a lot less land. The House Sparrow, one of the main beneficiaries of the horse-powered economy, was frequently described as far *too* numerous.

Figure 1 shows the tripling of the human population since 1801, Figure 2 shows the almost total disappearance of horses from the farm labour force between 1905 and 1975 (the statistics were not collected after that date), and Figure 3 shows a permanent switch of around 50,000 hectares (about 15% of the land area) from pasture to arable during the Second World War. Much of the arable land was devoted to growing oats to feed the horses, so their disappearance allowed more land to be used to grow other cereals, as shown by the increase in barley after 1955.

Loss of horses, accompanied by mechanisation, is reflected in a decline in the number of people employed in agriculture: the number of agricultural labourers fell from 18,450 in 1861 to 12,056 in 1931, a decrease of 35%, with a further fall to just over 2% of the population (about 8,100, but including farmers as well as labourers) by 1991.

These changes were largely driven by increased investment to improve productivity and yields. MAFF (1988) documented a strong trend to increased farm sizes between 1978 and 1986, coupled with increased purchase of farms by financial investment institutions. Only 65% of farmland in the West Midlands was owner-occupied in 1986. Investment in machinery has increased while the workforce has decreased, and larger fields and destruction of hedgerows have been an inevitable consequence.

Agricultural Change by 1990

Much agricultural change had occurred by 1985–90, when the fieldwork for the *Atlas* (1992) was undertaken. Changing farming practices led to habitat loss and also affected bird populations in other ways. Sometimes the impact on their food supplies only occurred over a limited period during the year, but as birds need to feed all year round, short-term impacts were just as deadly as prolonged ones.

These changes included:

- **The switch from spring to autumn planting of wheat and barley,** which had been largely completed by 1985. Winter wheat accounted for 95% of the wheat grown in the West Midlands region, except in years when the weather was unfavourable for autumn sowing, and the proportion of autumn barley increased by more than 50% between 1980 and 1986, 'mainly attributable to the high yield potential and lower disease risk associated with modern winter varieties' (MAFF 1988). Apart from an inherently lower yield, a successful spring crop is also much more dependent on suitable weather conditions in spring, to get it established, and again at a later period in the autumn for harvesting, and so the risk associated with spring sowing is far higher.

Figure 4 shows that wheat overtook barley as the main cereal crop about 1992, but spring barley made up less than one-third of the barley crop, and only about 15% of the total cereal crop, by the mid-1990s. Apart from weather-related fluctuations in recent years, when heavy rainfall prevented access to the land at the appropriate time, the balance of cereal cropping has not changed since.

This switch has substantially reduced the area of stubble available to sustain several seed-eating species through the winter, and also the bare ground that Lapwings in particular need at the start of the breeding season.

Comparison of Figures 3 and 4 shows that cereal production occupies around half the arable land.

- **Use of herbicides and pesticides** seriously affected many more species, either through eliminating their food, such as 'weed' seeds

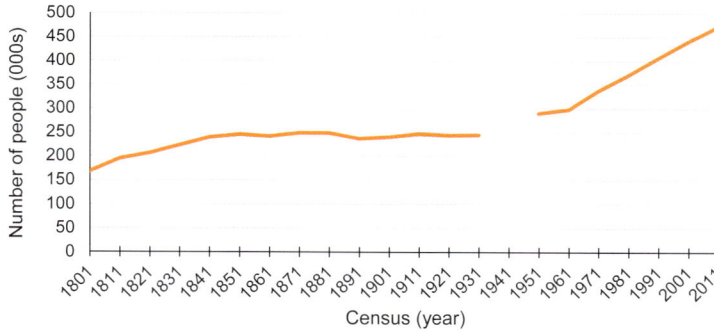

Figure 1. Human Population Increase 1801–2011

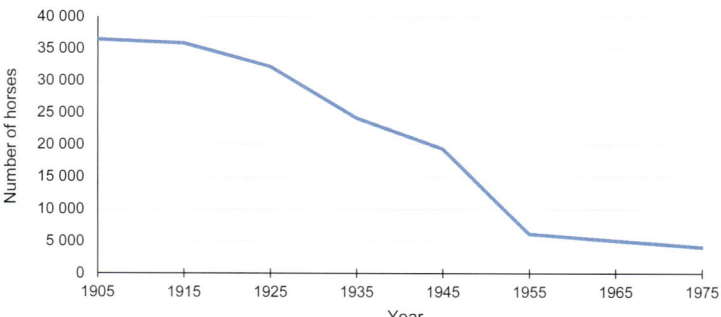

Figure 2. Disappearance of Horses from Farms

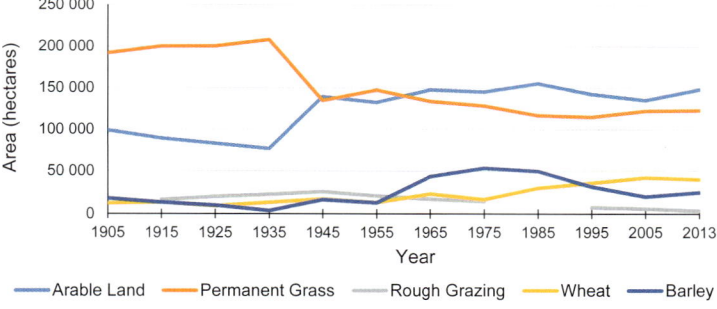

Figure 3. Farm Land Use 1905–2013

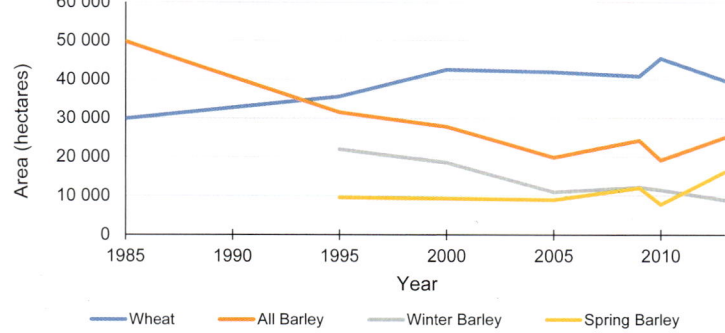

Figure 4. Farm Land Use (Cereals) 1985–2013

or insect prey, or poisoning the birds themselves as chemicals accumulated in their body tissue. Where over-winter stubbles did remain they were devoid of weed seed, and more efficient harvesting reduced spillage.

- **The management of grassland** also changed substantially. The major changes included a doubling in the use of inorganic nitrogen, and increased stocking densities, particularly of sheep. Structurally diverse and species-rich swards had been largely replaced by relatively dense, fast-growing and structurally uniform swards. The most important direct effects were the deterioration of the sward as nesting and wintering habitat, and loss of seed resources as food (Vickery *et al.* 2001). The earlier cutting of hay, and increasing silage production, which cuts the grass two or three times annually, destroyed many nests each year. Corncrake disappeared early, and other species such as Curlew and Skylark are still being affected.

- **Introduction of avermectins** in the 1980s to 'worm' cattle and sheep. This was a big improvement for farm animal welfare, but it makes their dung harmful to the beetles and other invertebrates that used to break it down. Invertebrate populations on farm grassland have plummeted, and a further source of bird food has been lost.

- **Ponds,** another important habitat, were also being lost as they were infilled, largely for agricultural intensification. For example around Shrewsbury there were 490 ponds at the turn of the century, 360 in 1963, and only 50 by 1990 (Jepson 1991).

A field of the commercially important and increasingly widespread oilseed rape crop, with the Wrekin in the background. Jim Almond, 22 April 2011

There was overwhelming evidence that the breeding numbers of many farmland species were declining rapidly by 1990, and the scientific research which produced it is fully documented in *Population Trends in British Breeding Birds* (Marchant *et al.*) published by BTO in 1990.

Map 7, Predominant Farm Types (p. 27) shows the position based on agricultural statistics in 1985 and other data collected up until 1986. It therefore provides a good assessment of the position during the 1985–90 fieldwork period for the Atlas (1992), although, as noted above, arable land is under-represented.

Agricultural Intensification and Change Since 1990

The drive to increase the efficiency of farming by increasing production and productivity has intensified since 1990, a process largely driven by increasing demand for food on world markets, various government and EU policies, and the purchasing policy of the increasingly powerful supermarkets and their wish to provide consumers with low-cost food.

Many species accounts refer to the effect of this agricultural intensification, which in a variety of ways has indirectly further removed nest sites and reduced or eliminated food supplies. This has continued since then, assisted by a continuing trend towards larger farms, and an accelerating increase, over the last few years, in holdings purchased by rich business people, encouraged by a favourable tax regime, who then let the land on share farming or contract farming arrangements.

In addition, several other major changes have also occurred in agriculture that have had dramatic effects on bird habitats and populations.

- **Sugar beet production has been replaced.** A substantial amount of beet was grown locally for the Allscott Sugar Factory. The future of the factory was in doubt for some years, and it eventually closed in 2007, so the demand for this cash crop disappeared, and two new crops, maize and oilseed rape, were planted to replace it. The demand for both has been substantially enhanced by the EU Renewable Energy Directive in 2009, which requires that 10% of the energy from transport fuels should come from renewable sources by 2020. Rape is a major source of biodiesel, and maize can be fermented for bioethanol.

  The change is shown in Figure 5, a 580% increase from 1995 to 2013 in the land devoted to rape, to more than 15,000ha, 4.3% of the County area. This has led to a substantial increase in the Woodpigeon population, and has also provided new habitat for Linnet and Reed Bunting. One of the only three confirmed breeding records for Grasshopper Warbler in the recent Atlas period also came from a rape crop. Maize production was rare in 1985, but has increased to 10,000ha, 2.9% of the area. Maize needs a frost-free growing season, so it is planted late. The bare ground prior to planting is attractive to Lapwing, so many have already laid, and maybe even hatched chicks, before the fields are ploughed and seeded, and the eggs or chicks are destroyed. Skylark are also likely to be affected.

- **'Set-aside'** required arable farmers to keep fallow around 10% of their productive land. It was introduced by the EU in 1992, to reduce overproduction, and there was over 10,000ha of 'set-aside' between 1995 and 2005. This land provided good habitat for Lapwing, Skylark and other farmland birds. Unfortunately, the

Figure 5. Farm Land Use 1985–2015: Increase in Rape and Maize

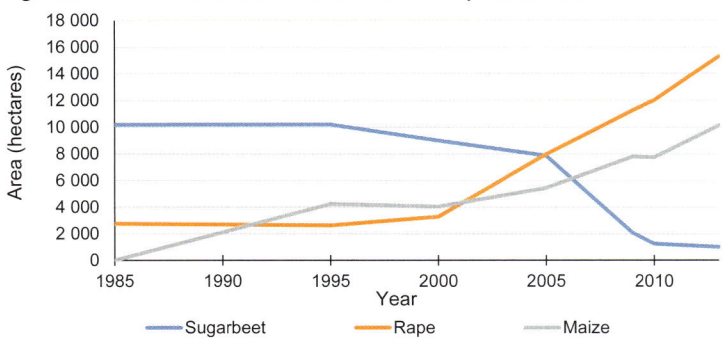

Figure 6. Sheep Population 1950–2013

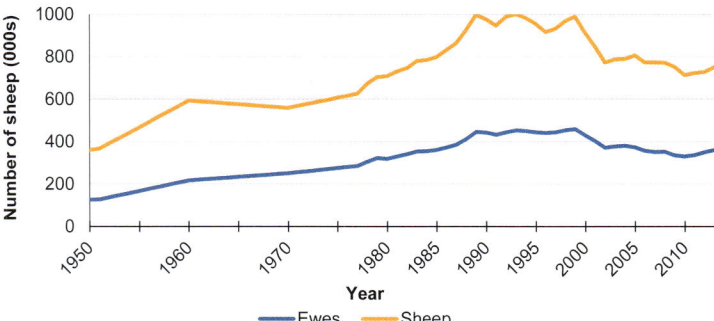

percentage was reduced to zero in 2007, and the requirement was abolished altogether in 2008 to encourage an increase in food production and biofuels.

- **A decline in the number of mixed farms** has continued. Many livestock farms grew various arable and root crops for winter fodder. There has been an increasing trend to buy it in, rather than grow it, reducing the stubbles, bare earth in spring, and seeds and arable weeds that supported farmland bird populations in the uplands. Unfortunately there are no statistics to quantify this, but it has accounted for the total loss of Lapwings in the Upper Clun Community Wildlife Group area, and at sites in the Oswestry uplands (Allan Dawes *pers. comm.*).

- **Milk quotas** limiting the overproduction of dairy products were introduced by the EU in 1984 and substantially revised in 1994. Quotas were raised steadily from 2008 onwards in preparation for being lifted altogether in 2015, exposing smaller farmers to more competition from more 'efficient' or cheaper producers, from inside and outside the EU. Many local dairy farms have therefore reduced their stock, or moved out of diary into beef or arable production, although some moved into larger units. Dairy cows decreased from 106,000 to 63,000 (40% decline), beef cattle increased from 21,000 to 27,000 (28% increase) and the land used for dairy farming decreased from 82,000ha (23% of the County area) to 46,000ha (44% decrease), all between 1985 and 2013. The number of cattle has also been reduced by the slaughter of those with bovine tuberculosis. This is a high risk area for the disease, with a steadily increasing number of slaughtered cattle until 1,000/year was passed for the first time in 2007, and a total of 14,113 have been slaughtered from that year to the end of 2014, with more than 2,000 in each of the three years, 2010, 2011 and 2013. While some farmers restock, others do not.

- **The impact of sheep on the uplands** has been mitigated. The 'headage payment' – a per head subsidy paid to farmers, which encouraged overstocking – was abolished in 2005.

Numbers had been steady at just under 500,000 up until the start of the Second World War, fell to 333,000 by 1945, then increased threefold in the next 40 years, peaking at over 1,000,000 in 1993, as shown in Figure 6.

In anticipation of the abolition of the subsidy, which also influenced restocking levels following foot and mouth disease in 2001 (when 49,000 sheep were culled, nine times the number of cattle), the number fell from 990,000 in 1999 to 749,000 in 2013, a

drop of 18%. This ensured the remaining sheep were more healthy, reducing mortality and hence the amount of carrion available to feed Ravens, Crows, Buzzards and foxes. BBS shows a comparable reduction in the Carrion Crow population (see p. 329).

Agri-environment Agreements

Subsidies for farming have been substantial since the Second World War, initially through price guarantees, and grants to increase production through draining wetlands, ploughing grassland, and increasing field size by removing hedgerows. When Britain joined the EEC in 1973 it also joined the Common Agricultural Policy (CAP), which used its considerable funds to increase food production, a policy which led to food mountains. Partly to reduce them, and also to ensure that farming contributed to safeguarding the environment, Environmentally Sensitive Areas (ESAs) were introduced in 1986, to offer incentives to encourage farmers to adopt agricultural practices to safeguard and enhance parts of the country which had particularly high landscape, wildlife or historic value. The Clun ESA was one of the first to be launched and the Shropshire Hills ESA started in 1993. These two ESAs made targeted payments to farmers for landscape management, and also mitigated the effect of sheep on the uplands, as farmers received payments, theoretically to compensate for loss of earnings by farming less intensively, if they maintained rough grazing, damp and rush pasture, and wet areas.

The other main benefit of the ESAs was encouraging hay production, rather than silage, with delayed grass cutting for hay until 15 July. These measures were designed to help Lapwing and Curlew, and a number of other bird species such as Skylark, Meadow Pipit and Reed Bunting.

On the Long Mynd, a specific additional ESA agreement, starting in 2000, between English Nature, National Trust and the commoners, reduced the number of sheep on the hill, particularly in winter, which facilitated regeneration of the heather and grassland.

Another consequence of the attempt to reduce food mountains was the introduction of 'set-aside', which removed land from production. It was introduced on a voluntary basis in 1988 and became compulsory in 1992, and had benefits for birds such as Skylark, Grey Partridge and Lapwing, and other wildlife. The scheme lasted until 2008 when it was ended by pressure to increase food production and grow biofuels.

The policy of using agri-environment schemes to promote countryside management continued when MAFF was incorporated into Defra in 2001. It operated against the backdrop of the Single Farm Payment Scheme (SFPS), introduced by the EU in 2003, which decoupled agricultural support from food production. It, and its successor the Basic Payment Scheme, provides area-based payments to almost all farmers, proportionate to the amount of land they owned, providing that certain 'cross-compliance' rules are met. These have varied over time but have generally required the land to be farmed in an 'environmentally friendly' way by buffering watercourses and other boundary features, minimising soil erosion, using pesticides and ferti-lisers to specified standards, and complying with food safety, animal health and welfare law. Land has had to be kept in good agricultural and environmental condition. These conditions have provided only limited benefits for wildlife, and some of the rules restricted options to modify fields to benefit species such as Lapwing.

In addition, through its Rural Development Programme, the UK government has been able to develop a sequence of agri-environment schemes which make payments over and above the basic payment for environmental benefits. Environmental Stewardship was introduced in 2005. Within that, the Entry Level Scheme (ELS) made payments over and above SFPS for specific but simple environmental benefits such as planting and restoration of hedgerows, maintaining field margins and Skylark plots (unplanted ground in the middle of winter cereal crops). The Higher Level Scheme (HLS) paid for specific environmental benefits, including wildlife habitats for priority species and biodiversity, but also including historic environment, landscape and resource protection. However, scheme agreements were voluntary. The farmer had to choose to apply, and could then decide which options to apply for. Natural England frequently tried to influence the applicant, and as a last resort could decide to not accept an application, but it was unable to insist that the most important wildlife habitats or species received any protection from an agreement, and in some cases they did not do so. Thus, although some HLS agreements deliver high-quality bird habitats, often targeted at specific Red list species, such as wet scrapes to provide feeding areas for Lapwing chicks, restrictions on grass cutting for Curlew, and spring-planted crops to provide over-winter stubbles for seed-eating farmland birds, only a very small proportion of farmland is under agreement to benefit birds. HLS agreements run for 10 years (with a five-year break option for both parties), so most still have several years to run.

ESA agreements ended not later than 2014, but farmers could then apply to ELS and HLS to provide continued support. However, the money available was less, so many of the benefits paid for by the ESAs for many years could not be included in the new agreements, and they have been lost.

In 2012, the last year for which figures are readily available, 33.9% of the land area was covered by the 1,304 different ELS agreements with individual farmers, landowners or tenants, 18.7% was covered by ELS with additional HLS options targeted on specific fields (494 agreements), 5.9% was in 330 ESA agreements, and over 40% of the land was not covered by any agreement at all. In total, there were 2,284 local agreements, at a cost of £13.9M over their lifetime (NE 2012).

The Countryside Stewardship Scheme replaced Environmental Stewardship in 2015, and the latter closed to new applicants at the end of 2014. Although the new scheme includes options to implement the recommendations of the Lawton *et al.* report, 'Making Space for Nature', to join up habitats on a landscape scale, and the targeting and eligibility of schemes is becoming increasingly focused and applications not meeting the criteria can be rejected, it is more

Set-aside with a clutch of Lapwing eggs, between Bishop's Castle and Norbury. Leo Smith, 13 May 2005

bureaucratic and less generous than its predecessor, and it is not clear if it will deliver the intended benefits. SFPS has been renamed the Basic Payments System, but continues to spend most of the budget in return for minimal wildlife benefit.

The agri-environment schemes have rarely delivered the wildlife benefits that could have been expected for the money spent. An applicant-led approach to the distribution of agreements and the options selected, with no overall strategy, has often meant that consid-erable opportunities have been lost. Therefore questions are being increasingly asked whether sufficient wildlife benefit is being gained in return for the grants being paid out through agri-environment agreements, and the much larger amount paid to almost all farmers through SFPS. In 2013, the total paid out to farmers through all agri-environment schemes and SFPS together in the UK was £3.26 billion but only £525 million (16%) was for the targeted agri-environment schemes that might produce specific environmental benefits (ONS 2018). It is hard to see £525 million worth of gain on the ground, and the conservation and wildlife charities could produce far more benefits with an annual income of £3.26 billion.

Reflecting the reduced provision for Countryside Stewardship, by 2016 the expenditure on the UK agri-environment schemes had fallen by 17%, and made up only 14% of the total payments to farmers and landowners (ONS 2018).

As Countryside Stewardship is part of the EU CAP until 2020, its future is in doubt, with debate now starting on the nature and scope of the farm payment system that might replace the current CAP.

The Future of Farming and Wildlife

The population loss of farmland birds has arisen as an indirect consequence of large-scale changes in farm management, decisions of the UK government and the EU, and pressures from consumers, including measures to improve animal welfare, increase yields and productivity, increase competition and reduce prices. As detailed above, most agricultural subsidy is paid according to the size of the land holding (the more you own, the more you get), and only a small proportion of the total budget is spent on schemes to enhance wildlife habitat. These are not direct subsidy schemes, as they are intended to compensate farmers for 'income foregone' as a result of using less intensive regimes. They only influence farming on a small proportion of the land. Several of the species accounts show that these schemes have a positive effect where they exist, but there are not enough of them to reverse the overall declines. It is unrealistic, and unfair, of conservationists and wildlife enthusiasts to expect farmers to pay the cost of maintaining habitats on farmland, in spite of all the other increasing pressures they face. We all need to contribute, through more extensive agri-environment schemes to help farmers meet these costs, support farmers in resisting pressures that damage wildlife habitats, and paying more for food that is produced through less intensive regimes. The vast majority of land is farmed (84% here, over 74% nationally), so we need to work with farmers if we are to have any chance of safeguarding the habitats that remain, and gradually restore them.

Botanical Change and its Significance for Bird Populations

The Flora and Vegetation of Shropshire (Lockton & Whild 2015) and *The Ecological Flora of the Shropshire Region* (Sinker *et al.* 1985) are both based on tetrad (2km x 2km) and monad (1km x 1km) recording. This has allowed a quantitative comparison of the flora between the two recording periods 1970–84 and 1985–2013. This comparison was included as a chapter in the recent Flora (Trueman 2015).

It shows that the flora is still overwhelmingly native, although there was a drop in native species from 78% to 74% of tetrad records. Introduced plant species are conventionally divided into the anciently introduced 'archaeophytes' and the more recent 'neophytes'. There has been a massive increase in neophyte species, although they still only account for 14% of tetrad records. The increase may, however, partly mark the trend towards urbanisation, since many of these plants are escapes from gardens. A few of them, such as the Butterfly Bush (Buddleia), might have some positive attributes for some fauna, but it is highly invasive and causes damage to some sites in the wider countryside. For some species the rate of increase is quite alarming. Probably most significant for the avifauna is the 12% decrease in archaeophyte tetrad records. Many of these are weeds of cultivation and are important for farmland birds.

Species extinctions have continued, although not at an obviously accelerating pace. Most species lost were long known from a single, or a very small number of, lowland sites, and their loss reflects the apparently inexorable nutrient pollution and inappropriate management of our choicest lowland, and especially wetland and open water habitats. However, in addition, many commoner native species have shown a considerable change in tetrad numbers. A total of 689 native species were recorded in the period of the new *Flora*. Of these, 39 species (0.6%) were found in over 100 less tetrads and a further 52 (0.8%) were found in 50 less. Many, particularly non-wooded, habitat types are represented in these losses, with obvious consequences for avifauna.

This new *Flora* (2015) identified 376 vascular plant species considered to be of the highest nature conservation significance, and the number of each of these 'axiophytes' in each tetrad was identified. When mapped, these showed very similar distributions and areas of concentration in the two survey periods. Numbers of species in the most biodiverse areas, with over 40 axiophytes per tetrad, seem to be fairly stable. However, the big changes were in the less diverse areas, with a large increase in the numbers of the poorest tetrads: 643 tetrads in the new Flora had fewer than 20 axiophytes, compared with 541 in the old Flora, while 165 and 361 tetrads respectively had 20–39 axiophytes. These differences are significant statistically and ecologically. The maps show a widespread loss of axiophytes in tetrads with 20–39 axiophytes which previously tended to surround and link the richer areas.

There are wider ecological implications of these trends. In part due to the creation of nature reserves and the spread of management for nature conservation since the 1970s, protection of the richer botanical areas has been modestly successful, although the threat remains. The zones between these core areas of biodiversity are however continuing to deteriorate relentlessly. This process of isolating and fragmenting the core areas must make them more unstable in the long run and render the avifauna and other wildlife populations which depend on them more vulnerable. The solution proposed (for example by Sir John Lawton in his 2010 report to government) is conservation in the whole landscape, rather than just the core sites, creating a more continuous and hence much more resilient 'ecological network'. The botanical data shows how far we have to go to even stop the deterioration of the existing ecological network.

Changes to Woodland

Although relatively well wooded, a high proportion of the woods (43%) are now planted with conifers and other non-native trees which are largely colonised by a few specialist bird species. The woods that are best for birds are ancient semi-natural woods with native trees and shrubs that have been present since at least the year 1600. Of the 9,267ha of ancient woodland in Shropshire, only 3,869ha (41%) is still considered to be semi-natural with the remainder having been replanted with non-native species, usually conifers (Natural England website). Therefore, only 13% of the woodland area is now habitat of the highest quality for birds.

Urbanisation

Farming is not the only cause of habitat loss. The population of the area covered by Telford New Town was intended to double following the creation of the Development Corporation in the 1960s, and many of the SWT Prime Sites lost in the 10 years prior to publication in

1989 were in the New Town. They totalled about 10% of the area lost, with woodland, heath and flower-rich grassland/tall herb categories suffering particularly. Other towns and villages are continuing to expand, and while careful planning may increase some habitats, such as gardens, parkland and ponds, this will mainly benefit the more common birds that adapt to suburban life, and will not help the less common and more specialised species. Even here, the increasing use of herbicides and pesticides by gardeners is likely to suppress bird populations.

Growing suburbanisation was also indirectly responsible for damage to Whixall and Fenn's Mosses, where the peat was being cut for sale in garden centres. Fortunately, total destruction of this site was averted, as a strong local campaign led to purchase of the land and management as a National Nature Reserve by Natural England, but this is an example of the kind of threat that remains elsewhere.

New roads to cater for the increasing levels of traffic destroy much valuable habitat, though in some cases the large embankment verges and woodland planting are better and provide more variety than the agricultural land lost. The Shrewsbury bypass, constructed in the mid- to late 1980s, cut through woodlands very close to a heronry, which has since been abandoned, presumably because the Herons were unable to cope with increased disturbance. Many birds are killed by traffic collisions, and, as the species account demonstrates, the Barn Owl suffers substantial losses from this cause, especially along the Shrewsbury and Oswestry bypasses.

Leisure Activities

Many species are vulnerable to disturbance from increased human use of their living quarters. If scared from their nest, many birds will abandon it altogether, while others may return later, only to find the eggs or nestlings were predated in the meantime. Nests at the water's edge are prone to flooding by the wash from boats. If eggs or young are deserted or lost too late in the season for the adults to lay again, reproduction is lost for that year, and the population will suffer.

Cars allow an ever increasing number of people to enjoy the countryside, and a whole range of activities – fishing, rambling, climbing, orienteering, riding, pony trekking, mountain biking, motocross, school field trips, Duke of Edinburgh awards, canoeing, water-skiing, raft-racing, and even bird-watching – all increase disturbance of nesting birds. New development for leisure activities also threatens important sites.

Changes in Predator-Prey Relationships

Reference has been made to the reduction in population of Carrion Crows as a result of reducing their food supply, sheep carrion. The other main source of carrion, and live prey for many predators, is the increasing number of artificially reared Pheasants that are released into the countryside by shoots each year, described on p. 131. This is likely to have enabled the fox population to increase above naturally sustainable levels, and the early results from the Stiperstones-Corndon Landscape Partnership Scheme Curlew Recovery Project suggest that high levels of Curlew nest predation by foxes is the main reason for their continuing decline. Dwindling populations of other ground-nesting birds may also be affected by excessive fox predation.

Climate Change

Climate change too is having a major effect on bird populations and habitats, globally, nationally and locally.

Weather Versus Climate Impacts

It is important to distinguish between the biological effects of weather fluctuations and those of climate change. The former bring about short-term impacts such as changes in distributions or higher rates of mortality that will generally be reversed in subsequent seasons and years. The *Handlist* (1964) was written shortly after the very severe winter of 1962–63, and several species accounts refer to its impact, often a big decline followed by a subsequent recovery, taking longer for some species than others. In the recent Atlas fieldwork period, one of the largest impacts was the effect of the very cold and frosty winter of 2009–10 which brought species into gardens that are not generally thought of as garden birds (e.g. Fieldfare) and which had a major impact on smaller species, such as a 45% decline in Goldcrest in the West Midlands (BTO Garden BirdWatch) and a crash in Wren population figures (Pearce-Higgins 2011b). Such impacts are reflected in the fluctuations in BBS graphs in the accounts for the more common species. Many other local examples of such effects are included in the accounts, and the weather during the recent Atlas period is summarised in Appendix 1. The local climate (temperature and rainfall) for the period 1970–2000 is described in the *Habitats* chapter on p. 24.

In contrast to the ephemeral effects of weather, climate change impacts are long term, generally unidirectional, largely persistent, and already evident.

Climate Change

Global climate change is considered to be a major conservation threat to avian populations. The Earth's average temperature has increased by 0.8°C over the past century, and it is now possible to detect the effects of climatic variability on population numbers, the timing of biological events, geographic range shifts and changes in both the species mix and functional interrelationships in ecosystems. Detecting climate change at the local level is notoriously difficult as the statistical 'noise' associated with year-to-year weather events is hard to disentangle from long-term trends when dealing with small amounts of data. The only local UK Met Office climate station (at Shawbury) was founded in 1946 but air temperature monitoring did not begin until 1957. The dependence on observation year of monthly means of daily maximum, minimum and average temperatures for all years 1957–2014 (Met Office 2015b) was determined and all three temperature measures were found to have increased over time at a rate of about 0.2°C per decade. This increase is of a similar order to that found for several stations in Central England over the same period (Met Office 2015c) and suggests that Shropshire is in step with broader measures of air temperature increase that have known ecological impacts elsewhere in the country.

Evidence for, and modelling of, the biological effects of climate change is largely available at the regional and national scales. However, there is no basis for assuming that Shropshire is immune from all or some of the observed larger-scale trends, which are enumerated below.

Both computer modelling and meteorological observations support the hypothesis that planetary temperatures have been increasing since the early part of the twentieth century (and particularly after 1940) driven by human-induced changes in atmospheric composition and land surface cover.

Effects of Climate Change

These temperature increases have already had a major impact. For example, breeding ranges have moved northwards in response to warming. Some evidence suggests that the southern edges of species ranges have remained static or shifted less than the northern edge, making the range area larger (Massimino *et al.* 2015). Other evidence suggests that northerly limits have shifted north-westerly or north-easterly rather than simply latitudinally (Gillings *et al.* 2015; MacLean 2008). The centre of winter distribution of several waders has moved over 100km (MacLean 2008). This trend increases the likelihood of some species colonising the UK, or deserting it. Locally, Cetti's Warbler appears to be becoming established, observations of Little Egret have increased considerably, and Great White and Cattle Egrets are occurring more frequently. Likewise, currently existing species may show population increases or decreases dependent on their thermal preferences (Pearce-Higgins, 2011c).

An important mechanism by which climate change affects species abundance and distribution is via inter-species relationships (especially food availability) rather than direct thermal effects. Ockendon *et al.* (2014) found it was changes to the populations or activity of these other species that were responsible for many of the impacts observed. Although birds are now breeding earlier, the timing of insect prey emergence has changed more rapidly, and the timings of plant biological stages have advanced even faster. This could be problematic for species that rely on seasonal peaks in food abundance (Pearce-Higgins, 2011a), and is a likely contributory factor in Pied Flycatcher decline here. Additionally, advances in Cuckoo host species breeding timetables may have outstripped the advanced arrival date of Cuckoo so that Reed Warbler (a late-nesting species), which was almost unknown as a host species until the 1970s, is now the fourth most common host here.

Another important mechanism is through change to habitats. Upland areas are being affected by an increase in bracken, which is spreading and moving further uphill in response to climate change. This may contribute to population changes on local sites such as the Long Mynd (Andrew Perry *pers. comm.*).

Migratory timings have responded to global warming with Swallows now arriving in the UK about 20 days earlier than they did during the 1970s, while the first Sand Martins arrive 25 days earlier (Pearce-Higgins 2011a). Several species also leave for their wintering grounds later than they used to which, when combined with earlier arrivals, suggests multi-brooded birds may be able to incorporate an extra clutch in their breeding period (Newson *et al.* 2016), a possible cause of the local increase in the Reed Warbler population.

The *Arrival Dates* chapter shows that, over the century following 1900, advances of two weeks or more in migratory arrival date here were detectable for Chiffchaff, Garden Warbler, Pied Flycatcher, Sand Martin, Sedge Warbler, Swallow, Wheatear and Yellow Wagtail, with changes of over a week for many other species. Sparks *et al.* (2001) – reviewing a number of papers on first arrival dates in the country as

A Swallow carrying nesting material at Dudleston Heath. Swallows are arriving two weeks earlier than they did in 1950, and nesting earlier too, as a result of climate change. John Hawkins, 24 April 2016

a whole – conclude that 'there is so much correlative evidence that it is difficult to believe that there is no influence on bird migration of temperature'.

For summer visitors, such changes have led to earlier egg-laying dates, but also have other implications on breeding success resulting from asynchrony between migration, breeding dates and prey species availability. The breeding success of Curlew Sandpipers (p. 218) is closely tied to both the relationship between Arctic Fox and lemming cycles, with good years occurring only when lemmings are abundant and when weather conditions are benign (Barshep *et al.* 2011; Summers *et al.* 1998). These sandpipers visit only when breeding is successful but there is evidence that lemming cycles are breaking down due to climate amelioration, which potentially could reduce the occurrence of Curlew Sandpiper here.

Milder winters have affected the numbers and distribution of winter swans and other waterfowl, as the east coast of Britain and the near Continent are no longer subject to the severe winter temperatures which encouraged migration to the milder west coast, and several of the species accounts refer to declines due to 'short-stopping'. Small songbirds can remain later in the year on Long Mynd providing food

Very low water levels at Venus Pool, at the end of a very dry summer (compare with p. 53). Such conditions are likely to become more frequent as a result of climate change, as summer temperatures are expected to increase, and rainfall to decrease. Jim Almond, 25 September 2011

for Merlin and Hen Harrier, and improved over-winter survival has increased the Stonechat population (Leo Smith *pers. comm.*).

Blackcaps from central Europe have been wintering in gardens and other habitats in Britain with increasing frequency over the past several decades, rather than migrating south to the Mediterranean. Plummer *et al.* (2015) have shown that the genetic propensity for central European Blackcaps to migrate to the UK in winter, while once perilous, is now a successful strategy due to climatic amelioration and the provision of supplementary foods in our gardens. Blackcap abundance in the UK has increased consistently since the late 1970s, with an extraordinary acceleration in the upward trend over the last five years. Since those wintering in the UK arrive back at their breeding grounds earlier than those that adopt the traditional migration route, they secure favourable territories and breed more productively, effectively enhancing the genetic tendency for a northern migration route in the population. Hence, supplementary feeding and climatic change, in tandem, are directly affecting the distribution and evolutionary direction of this warbler species.

Climate changes elsewhere in the world can also produce impacts on our avifauna. Many summer visitors from sub-Saharan Africa (such as Swallow, Sand Martin, Swift, Turtle Dove, Cuckoo, Whitethroat and several other warblers) have shown dramatic declines in population,

part of which has been accounted for by increased drought incidence and desertification in the Sahel. At high latitudes, according to temperature estimates based on computer modelled temperatures with tripled CO_2 (equivalent) concentrations (Lindstrom and Agrell 1999), warming in the northern hemisphere has the potential to reduce by 65% the availability of tundra breeding grounds for waders and other birds (e.g. see Knot, p. 215) and to disrupt predator-prey relations that put Curlew Sandpiper chicks at risk (p. 218). The emissions scenario on which such predictions are based may be excessively pessimistic but even more modest accumulations of greenhouse gases, such as the 2°C targets negotiated (but not yet fully implemented) at the 2015 United Nations Conference on Climate Change in Paris, are likely to shrink suitable breeding habitats for Arctic waders in future.

Global climate models are generally regarded as most successful in simulating temperatures, and predictions of other climatic elements can be expected to be less reliable. Precipitation is unlikely to change by more than 10% in annual terms but seasonal analyses suggest dryer summers and wetter winters (Natural England 2011). The former can be expected, in combination with enhanced evaporation, to lower discharge in the region's rivers, posing threats to riparian birds through nest site and food (plant and animal) availability (Royan *et al.* 2013). It is not possible to accurately determine how such features as frontal systems, storm tracks and pressure and wind patterns might change in detail, but most climatologists expect an increase in storminess in the mid-latitudes, an implication that might be expected to affect established migratory patterns and the frequency and nature of vagrant records resulting from storm-disrupted migratory behaviour and 'overshooting'.

Future Effects of Climate Change

Climate change has already occurred, and led to dramatic differences in population, distribution and behaviour of many species here in Shropshire. There is no rational basis for believing that it will cease in the future. Global climate simulation models offer the most reliable means of assessing likely climate scenarios, including those based on human modification of the atmosphere in conjunction with natural processes. Such models predict that, even with a low greenhouse gas emissions scenario, by 2050 the western section of the West Midlands, including Shropshire (Sustainability West Midlands 2003) can expect temperatures 1.0–1.5°C higher, with up to 10% less precipitation and soil moisture. Changes elsewhere in the world will be even more dramatic and the impact on both resident and migratory birds can be expected to be profound.

Wildlife Protection

The changing face of Shropshire, changes in human activity and climate change are all obviously affecting bird habitats and populations. The *State of the UK's Birds 2012*, produced by several organisations including BTO, RSPB and Natural England, estimated a net loss of 44 million birds between 1966 and 2011. Assuming that the loss here was proportionate to the total land area, Shropshire has lost almost 700,000 birds in 50 years.

Against this background, wildlife protection is essential. It takes two main forms, site protection, either through designation or land ownership, and species protection. Working with landowners, and the necessary land-use policies to support it, is also essential. Ownership of important sites by conservation organisations such as Natural England, National Trust, SOS or SWT provides the highest level of protection, although even these sites can be affected, either directly or indirectly, through, for example, national infrastructure schemes such as roads or railways, or diffuse and atmospheric pollution.

Sites of Special Scientific Interest

SSSIs are the only wildlife sites that have statutory protection through UK legislation. These sites were originally set up by the National Parks and Access to the Countryside Act 1949, but the current legal framework is provided in England and Wales by the Wildlife and Countryside Act 1981, amended in 1985 and further substantially amended in 2000 (by the Countryside and Rights of Way Act 2000). These sites are 'notified' (the legal process of designating a site as an SSSI) because they are the best and most important wildlife sites nationally for their fauna, flora or geology. Most are privately owned, and the owner is expected to protect the important features mentioned in the 'citation' (the reason why the site was designated).

Although implementation of the Wildlife and Countryside Act (1981) has led to protection of both sites and species, it is unfortunately still the case that sites are damaged or lost, and protected species killed. Indeed, only 29% of the total designated area (8,438ha, only 2.4% of the County area) is in 'favourable condition', while 67% is 'Unfavourable Recovering', which means they have deteriorated, but are being managed to restore their condition, and 4% are either destroyed or unfavourable and declining. One site given statutory protection by designation as an SSSI for its bird interest, and hence deemed to be nationally important, has been damaged here, probably irretrievably.

There are 111 SSSIs wholly or partly within Shropshire, and five of them included birds as an interest feature when they were 'notified'. The following SSSIs, therefore, are nationally important for their specified bird interest (apart from Allscott Settling Ponds, the other four sites also have other features for which they are nationally important):

- Allscott Settling Ponds: assemblage of breeding birds associated with lowland open waters and their margins;
- Fenn's, Whixall, Bettisfield, Wem and Cadney Mosses: rare bird species or feature (wet meadow wader) – Curlew;
- Long Mynd: assemblage of breeding birds associated with upland moorland and grassland without water bodies;
- The Stiperstones and the Hollies: assemblage of breeding birds associated with upland moorland and grassland without water bodies; and
- Wyre Forest: assemblage of breeding birds associated with woodland.

Citations for other SSSIs sometimes mention bird interest, such as that for Marton Pool (Chirbury), but if it is not a 'designated feature' the bird interest is not directly protected by the SSSI status.

Further legal protection for some of these sites as Special Areas of Conservation through European legislation, or Ramsar sites through an international convention, is outlined in the next section (see p. 52).

Although most SSSIs are privately owned, some are owned by conservation agencies and managed specifically for their wildlife interest. In particular, the greater parts of Fenn's, Whixall, Bettisfield, Wem and Cadney Mosses, the Stiperstones and the Hollies, and the Wyre Forest, are managed by Natural England as National Nature Reserves, and the Long Mynd is managed by the National Trust.

Natural England monitors the condition of all SSSIs. In summary, Allscott is 'Unfavourable, Declining', the Mosses and Long Mynd are 'Unfavourable, Recovering', the Stiperstones is 'Favourable' and Wyre Forest is partly 'Favourable' and partly 'Unfavourable, Recovering'. More detail can be found on its website, including an assessment of each of the component units that make up the SSSI.

ALLSCOTT SETTLING PONDS

The Sugar Factory, which opened in 1927, included up to 17 settling ponds that were used to both separate the silt from the water used to wash the beet crop, and to reduce the biological oxygen demand of this water before its discharge back into the River Tern. The sequential filling of the settling ponds during the winter beet campaign resulted in a series of shallow pools of various depths which, along with associated areas of swamp, marsh and scrub, attracted a typical breeding bird community of County importance, including Little Grebe, Shoveler, Little Ringed Plover, six species of warbler and

Mirelake: shallow water and exposed mud in the autumn, providing an ideal stopover site for rare passage migrants such as Knot. Jim Almond, 30 October 2008

Yellow Wagtail. As a consequence, it was first designated as an SSSI in 1963 and re-notified in 1986.

The Mirelake, the largest pool in the complex, was filled over the winter and spring before being drawn down in the late summer in preparation for the forthcoming winter processing season. These anoxic conditions were ideal for bloodworm, which, as the water levels fell, acted as a magnet for migrating waders on autumn passage. As a consequence, the Mirelake, was the premier wader site in the County and was often specifically named in records.

The scrub around the ponds provided breeding habitat for warblers, and together with the large phragmites reed bed, provided breeding sites for Reed and Sedge Warblers. Hirundines were numerous on autumn passage, and sometimes up to 10,000 roosted overnight. The grounds around the pools were also the foremost ringing site (see p. 10).

Prior to the factory closing in 2007, the process had already changed, in part to meet tighter environmental requirements, and the value of the pools to breeding birds had already been considerably reduced. The factory closed in 2007, and housing is being built on the factory site, but there are no plans to develop the SSSI. The pools themselves now have a new private owner, but their value as bird habitat has largely been lost now that the pools are no longer managed in a way that maintains their bird interest. They have almost all dried out and Natural England, with the encouragement of SOS, has been working to try and restore the SSSI to good condition. Unfortunately, however, this will not be possible unless Natural England uses its powers to enforce the reintroduction of a water level management regime that will restore the conditions that were once so attractive to birds. As there is no apparent economic reason to manage the water levels, this is unlikely.

This case does highlight the vulnerability of artificial sites based on an industrial process with complex management practices; even the best sites with (theoretically) the highest level of protection are at risk.

FENN'S, WHIXALL, BETTISFIELD, WEM AND CADNEY MOSSES SSSI

Straddling the border with Wales, parts of this SSSI were first notified in 1955, with a final revision covering the whole site in 1994. It includes the vast majority of the original lowland raised bog peat deposit, and is one of the biggest and best raised bogs in Britain. It was designated a Wetland of International Importance under the Ramsar Convention in 1997, and is also a European Special Area of Conservation. The whole SSSI covers 948ha, of which 258ha is in Shropshire.

The greater part of the site is managed as Fenn's, Whixall and Bettisfield Mosses NNR, which covers approximately 700ha. A separate part of the site, Wem Moss, is also an NNR, managed by the Shropshire Wildlife Trust.

Since acquiring the NNR in 1989, Natural England and its predecessors, in conjunction with Natural Resources Wales, have steadily carried out management to reverse the effects of commercial peat cutting and afforestation of the lowland raised bog, particularly by blocking drainage ditches and installing water control structures to reinstate the natural hydrological regime, and by removing conifer plantations. Significant portions of the site have already returned to a more favourable condition, and further restoration works will be carried out from October 2016 through a five-year EU funded Marches

Mosses BogLIFE Project, including the acquisition of surrounding more marginal land to convert it back to active raised bog habitat, further ditch blocking, and experimental works to improve the hydrology. Some scrub will be removed, but only where it is necessary for raised bog restoration.

Several habitats are important for breeding, passage and wintering birds. The expanse of the typical lowland raised bog habitat, which mainly comprises Sphagnum mosses, sedges and patches of heather, provides nesting habitat for Meadow Pipit, Reed Bunting and particularly the locally important Curlew population, a feature of interest in the SSSI designation. The drier parts of the raised bog, with heather as a major constituent along with sporadic birch scrub, provide nesting habitat for Stonechat. Significant pools or flashes occur across the moss, often as a result of flooded hand cuttings; these acidic pools are feeding and nesting habitat for Black-headed Gull, Shoveler and Teal.

Some degraded parts of the bog offer little habitat to birds at any time, for example some drier areas are dominated by Purple Moor Grass and are typically of low botanical diversity. Natural England is aiming to restore these areas through the LIFE project. There are several other important habitats in marginal areas of the reserve. Fen habitats exist in areas of thin peat, as the peat/soil is less acidic and minerals are available for uptake by plants. In these areas willow scrub is present, providing feeding and nesting habitat for summer migrants such as Willow Warblers and Chiffchaffs. Some scrub will be removed to allow old drainage channels to be blocked up, to prevent the drying of these habitats. Tree Pipit are scarce visitors; although Natural England would like to see them breeding on site, this would probably require the creation of a mosaic of short vegetation, bare ground and tall scrub, which at present would be difficult to achieve on a large scale.

Some of the fen habitats also provide improved feeding habitat for Snipe and appear to be extremely valuable to breeding and wintering Reed Bunting. Marginal wet woodland on the site is important for breeding Willow Tit, several pairs of which have been found nesting in these habitats. The coverage of wet woodland will increase on some newly acquired areas of land during the BogLIFE project, but it may take several years before the habitat is mature and therefore suitable for Willow Tit.

These habitat improvements should increase the populations of several of the more scarce breeding species, and Natural England will monitor the changes.

Further information can be found on the Natural England and the Marches Mosses BogLIFE Project websites.

Restored raised bog at Fenn's, Whixall and Bettisfield mosses NNR.
© Natural England, 8 January 2019

THE LONG MYND

The Long Mynd is the largest upland area, a long whaleback hill 6km wide in the north and extending 10km southwards. The highest point is the top of Pole Bank at 516m, the fourth highest summit behind Brown Clee (540m), Manstone Rock on the Stiperstones (536m) and Titterstone Clee (533m).

The National Trust owns and manages 2,322ha. Just over half of that is a heather moorland plateau, accessed by a few roads, and up several steep-sided valleys. Although Carding Mill Valley can be full of visitors in summer, most stay in the lower parts of the valley. The other valleys are quieter, and might be deserted on weekdays, but increasing recreational disturbance is becoming more of a problem.

The SSSI, first notified in 1953 and re-notified in 1990, is slightly larger (2,729ha) and includes some privately owned heathland to the west and south of the Trust land. In addition to the upland bird communities, it is of special interest for its geology and upland plant communities. The most recent SSSI citation, in 1990, stated:

> The Long Mynd is the most important site in Shropshire for upland birds. Species which regularly breed here include Merlin, Red Grouse, Ring Ouzel, Wheatear, Curlew and Snipe. In addition, the site is also important for Dipper and Grey Wagtail which breed in the numerous stream valleys.

It will be seen from the species accounts that Merlin still breeds, Wheatear, Curlew, Snipe and Grey Wagtail have declined but are hanging on, Ring Ouzel and Dipper have been lost, and only Red Grouse has increased (from a very low ebb) since designation. More than 99% of the SSSI is in 'unfavourable but recovering' condition.

In addition to the species mentioned in the SSSI citation, Teal has been lost, Whinchat has declined, Stonechat has increased, but Buzzard, Peregrine and Raven have been gained, reflecting their wider population increase.

The 2000 ESA agreement between Natural England, National Trust and the commoners was followed in 2009 by an HLS agreement, which continued to regulate the number of sheep on the hill, particularly in winter, allowing for regeneration of the heather and grassland. These two agreements allowed the restoration of heather management. The heather is burnt in small blocks to optimise the edge effect, the mix of tall older heather with shorter new regrowth. This management provides feeding habitat adjacent to shelter and nesting habitat, and has been crucial for the Red Grouse recovery. Winter visitors such as Hen Harrier and Short-eared Owl have also benefited from the heather management. The HLS agreement will continue until 2019.

However, while the reduction in sheep grazing may have allowed regeneration of some vegetation, comparison of the habitat surveys in 1995 and 2015 shows that the proportion of the area covered by

The Long Mynd, showing the various habitats in the upper reaches of Ashes Hollow, with the heather moorland and the trees at Pole Cottage on the skyline. Peter and Jane Howsam, 13 May 2019

The Stiperstones summit ridge, with tors, heather and scree. Simon Cooter, 4 May 2010

bracken is essentially unchanged, but in some places it is getting thicker and moving further uphill in response to reduced grazing and trampling, increased nitrogen deposition and climate change (Andrew Perry, National Trust *pers. comm.*), perhaps affecting the habitat quality for Whinchat and the pipits.

Other management designed to help the breeding bird population includes cutting rushes on Wild Moor, to improve Snipe habitat, and routes for large events have to be approved to avoid the 'Ground-nesting Birds Recovery Area' during the nesting season.

The main upland species were surveyed for the Trust in 1994–98 (Smith SBR 2003) and 2006–09 by the Long Mynd Breeding Bird project, and this survey is being repeated, using volunteers, starting in 2017. The Red Grouse population has been surveyed annually since 2011. The results of these surveys are included in the relevant species accounts.

Wild Mynd: Birds and Wildlife of the Long Mynd was published by the National Trust (Smith *et al.* 2007), and lists Skylark, Tree Pipit, Meadow Pipit, Redstart and Reed Bunting as other numerous upland species. Kestrel nests too, and Red Kite and Hobby are seen increasingly frequently. Hen Harrier, Dotterel and Ring Ouzel might be seen on passage, but numbers vary from year to year.

THE STIPERSTONES AND THE HOLLIES SSSI

Like the Long Mynd, this site of 588ha was first notified in 1953 (re-notified in 1989) for its geology, upland habitats and assemblage of upland breeding birds. The greater part of the site is owned and managed by Natural England as an NNR of 488ha, and it is a Special Area of Conservation.

The Stiperstones is one of only two sites with Red Grouse, and also has breeding Snipe, Wheatear and Whinchat and, in the fringing woodland, Redstart and Pied Flycatcher.

In addition to managing the NNR, the NE staff have focused in recent years on the ambitious and highly successful project, 'Back to Purple: Conserving and Restoring the Stiperstones', largely on land outside the SSSI and NNR. This project has removed four large conifer plantations, totalling 85ha and dating mostly from the 1960s, from the ridge and flanks of the hill. Within a few years 79ha had been replaced by the natural regeneration of drifts of flowering heather, with the remaining 6ha replanted as deciduous woodland. The project was initiated in 1998 through a partnership between English Nature, SWT and Forest Enterprise, and supported by a local landowner, and is now jointly led by Natural England and SWT. More remains

Back to purple on the Stiperstones, with newly restored heather in bloom, the stumps of the felled confers still visible, and the Rock in the background. Leo Smith, 4 June 2010

to be done, but thanks to support from the Heritage Lottery Fund, charitable trusts, landfill tax credits and, crucially, from hundreds of SWT supporters, much has been achieved, and each year brings further recovery and greater opportunities for nesting Stonechat, Skylark, and Tree and Meadow Pipits, and Red Grouse have now started using the restored heathland.

The first coniferous plantation to be felled, Gatten Plantation, was on privately owned land leased to Forest Enterprise, who removed the timber and sold the lease to Natural England, who now manage it as an extension to the NNR. Conifers were then felled around Nipstone Rock, where SWT acquired the land as a new reserve, and on the southern part of the Stiperstones ridge, owned by the Linley Estate, all for the restoration of upland heathland. Thus, in effect, Back to Purple work has complemented and extended the NNR. During this period SWT acquired two other reserves, Brook Vessons, and the Hollies, both within the SSSI.

Other management on the Stiperstones involves heather management by burning, cutting and grazing, and legal predator control, to benefit Red Grouse and other ground-nesting birds; the restoration of wet flushes through scrub clearance and drainage blocking, including on part of Brook Vessons SWT reserve, which hopefully will help the nearby Snipe population to expand; a nest box scheme for Pied Flycatcher and Redstart on Resting Hill, in cooperation with the Rea Valley Community Wildlife Group; and acquisition of other small sites nearby for heathland or woodland restoration.

When the Nature Conservancy Council bought the Stiperstones to create the NNR, the previous owner retained and exercised the shooting rights. When the decline in the Grouse population made that pointless, NE subsequently leased these rights, but they were then acquired by shooting interests with the intention of increasing Grouse numbers, but that did not materialise, and in early 2018 NE acquired the full shooting rights.

WYRE FOREST

Although the majority of the site is in Worcestershire, there is still a significant portion in Shropshire, part of which is an NNR. First notified in 1955, with the latest revision in 1998, this site is designated for its woodland, heathland, grassland and mire communities and their associated invertebrate, reptile, vascular plant and woodland bird assemblages. The latter includes species such as Wood Warbler, Pied Flycatcher, Redstart, Tree Pipit, Tawny Owl, all three woodpeckers and, until recently, Hawfinch.

ELLESMERE MERES

Cole, Oss, Sweat and Crose, and White Meres are all SSSIs in their own right, but they were not notified for their bird interest. They are more fully described below.

Other Legal Protection for Important Sites

Special Areas of Conservation (SACs) are designated by the British Government using criteria in the EU Directive 92/43/EEC on the conservation of natural habitats and of wild fauna and flora, known as the Habitats Directive.

Special Protection Areas (SPAs) are designated by the British Government using criteria in the EU Directive 2009/147/EC on the conservation of wild birds, known as the Birds Directive.

SACs and SPAs are internationally important at the European level and, together, form a network of protected sites across the EU called Natura 2000. There are five SACs: Brown Moss; Downton Gorge; Fenn's, Whixall, Bettisfield, Wem and Cadney Mosses; River Clun; and the Stiperstones and the Hollies. There are no SPAs here.

The Ramsar Convention is an international treaty for the conservation and sustainable use of wetlands. It is named after the city in Iran where it was signed in 1971. There are only two Ramsar sites, the Midland Meres and Mosses Phase 1 and Phase 2 which comprise a number of SSSIs, including Fenn's, Whixall, Bettisfield, Wem and Cadney Mosses.

Apart from the sites named above (including the SSSIs), no sites listed in this chapter have any legal protection.

Other Important Sites for Birds

VENUS POOL SOS RESERVE

The site was originally a natural wet hollow that had been drained. When a culvert collapsed in the 1950s, the pool re-formed and it became well known to local birdwatchers. The land to the east became the Coldharbour Lane sand and gravel quarry, and a bund was created to protect the pool from being drained. Waste water was pumped from the gravel-washing plant in the quarry into the pool. When it ceased to operate, the main workings were re-profiled to form the present day fishing pool to the south-east, and Venus Pool gradually returned to a more natural condition. It was acquired by SOS in 1986, and, following extensions in 1999 and 2003, now covers almost 27ha.

In addition to the pool itself, with its islands and areas of open shoreline, other habitats include stands of willow scrub, extensive marginal vegetation, flower-rich grassland, hedgerows and woodland. Management work is undertaken regularly by members to maintain and improve the habitat for wildfowl and wading birds, and an arable field to the south of the pool has been planted with bird-friendly crops. There are five hides, and the reserve is undoubtedly the most well-watched birding site. It boasts a wide range of breeding waterfowl and wading birds, including some of the more unusual species such as Oystercatcher and Little Ringed Plover, but Redshank and Black-headed Gull have been lost, although the latter still makes occasional breeding attempts. It is the best site now for passage waders, and the occasional Osprey and rarity turns up. A total of 206 species had been recorded by the end of 2014, and two more (Glossy Ibis and Pine Bunting) since.

Managing the grassland and arable field has been supported by an agri-environment agreement with Natural England, and the Heritage Lottery Fund has supported the purchase of land, creation of the car park and the erection of hides. Further information can be found on the SOS website.

CHELMARSH

Chelmarsh reservoir attracts many species of wintering wildfowl and passage migrants, including waders, terns and the occasional Osprey. It provides a good feeding site year round for Grey Heron, and since Common Tern first bred there in 2011 this is one of two sites where

Venus Pool at first light. Jim Almond, 27 December 2011

The Mere (Ellesmere). Jim Almond, 13 November 2016

they are likely to be found (Priorslee Lake is the other). There is a large winter gull roost, which sometimes includes rarer gulls.

A reed bed was created at the infall (north-western) end by planting just 100 phragmites roots (from Allscott) in 1978 (Dave Fulton *pers. comm.*), and SWT created a scape within it in 1987 to attract waders and make the reed bed more attractive to breeding Reed Warbler and Reed Bunting. The former have increased from nil to at least 30–40 pairs, and the reed bed is used as a roost by hirundines on passage, and as a winter roost by Starling, Reed Bunting and Wagtails. This is also the only place where Water Rail is known to breed regularly. The scape is overlooked by a bird hide provided by SWT, and access is restricted to members of the SWT and SOS. The Chelmarsh ringing group operate here (see p. 10).

THE ELLESMERE GROUP OF MERES

The cluster of meres near Ellesmere, locally known as the Ellesmere Group of Meres (EGM), are natural water bodies. They include Cole, Sweat and Crose, and White Meres, which are notified as SSSIs for their aquatic plant communities and included within a Ramsar site, but they were not notified for their bird interest. The whole cluster hosts several species of breeding waterfowl and are known for wintering species such as Goldeneye and Goosander, as well as a regular site for water bird rarities. They have been regularly counted as part of the WeBS programme since 1960–61, providing useful information on population trends for the species accounts. The Mere itself at Ellesmere hosts a notable winter gull roost with large numbers of Black-headed and Lesser Black-backed Gulls, along with a number of rare and scarce winter visitors including Glaucous, Iceland and Ring-billed Gulls.

Wood Lane Nature Reserve, with wintering Lapwings. John Hawkins, 7 January 2007

WOOD LANE

Near Ellesmere, this is one of the top wetland sites, and more than 180 species have been recorded. It was created from worked-out sand and gravel pits, designed to attract passage and breeding waders, and became an SWT reserve in 1999. It has been managed for wildlife since then, particularly as a haven for birds. There is a large Black-headed Gull breeding colony, which hosted the first confirmed breeding of Mediterranean Gull, Little Ringed Plover also breed here, and Lapwing, Greenshank, Godwits and Whimbrel regularly drop in to feed up on their migration journeys in spring and autumn.

There are two hides (accessible through a permit system) and bird feeding stations, and the patches of Common Reed planted in the adjoining fishery host breeding Reed Warbler. The quarry workings themselves and the associated sand storage mounds are often used by Sand Martins forming, at times, often quite substantial colonies.

OTHER SWT RESERVES

SWT has 40 nature reserves, each of which is valuable to birds as a result of conservation management. Some woodlands are traditionally coppiced, which brings a diversity of vegetation at different heights, from the tree canopy, through a bushier understorey to ground flora. Along with piles of dead, rotting wood this creates a good range of nesting and feeding habitat. Several woods, including the Ercall, Craig Sychtyn and Earl's Hill have nest box schemes to attract Pied Flycatchers.

An SWT project to help Lapwing and Curlew in the Severn–Vyrnwy confluence area resulted in several farmers entering Environmental Stewardship schemes to manage their land in ways that would benefit wetland birds. With grant funding and support from members the Trust succeeded in buying Holly Banks, one of the lowest lying parts of the area. Several new ponds have been dug, which regularly attract wildfowl. Winter floods bring Whooper Swans and Pintail to these fields, but for the bird-watcher, access can be a problem then as roads are under flood water.

Three reserves are within the Stiperstones SSSI, and two others were acquired as part of the Back to Purple project, which has increased heathland habitat and been jointly led by SWT (see the section 'The Stiperstones and the Hollies SSSI' above).

Three reserves in the Clun Forest (Rhos Fiddle, Lower Short Ditch Turbary and Mason's Bank) include most of the remnant of heathland, and provide important foraging areas for Curlew.

In the north, the Meres & Mosses Landscape Partnership Scheme, led by SWT, has worked with farmers and landowners to restore wetlands and reduce pollution. Seventeen hectares of flood-prone fields at Whixall were bought by the Esmée Fairbairn Foundation

Rhos Fiddle SWT reserve, Clun Forest. Leo Smith, 18 May 2004

on behalf of SWT and will be handed over to Natural England on lease, following receipt of funding from the Heritage Lottery Fund to match the EU BogLIFE grant. This work complements that of the Fenn's, Whixall and Bettisfield Mosses NNR, and Natural England will manage the fields and install a bird hide (see the section 'Fenn's, Whixall, Bettisfield, Wem and Cadney Mosses SSSI' above).

Also among SWT nature reserves are several quarries, two of which are intermittently used by breeding Peregrines. Scarce species on the Red list of *Birds of Conservation Concern* have been recorded at 12 of these reserves.

POLEMERE

Polemere was well known as a birding site for many years, but infrequently visited. In 2007 an access agreement was negotiated with the landowner, the pool was re-profiled to improve habitat to attract wildfowl and waders, and a hide erected. Several rarities have also been found there.

SHREWSBURY SEWAGE FARM AND MONKMOOR POOL

Shrewsbury Sewage 'Farm', the treatment works at Monkmoor in a loop of the River Severn, has changed substantially since the formation of SOS. At that time there were open settling pools and, from the late 1950s onwards, it was the premier site for observing passage waders, with several County 'firsts', but records declined after access permission was withdrawn in 1962. The pools were replaced by open filter beds, which were particularly attractive to pipits, wagtails and other passerines, but they were later enclosed. The pools that remain on the site have become overgrown, although they still provide valuable habitat, and Monkmoor Pool, which is also part of the complex, has attracted a number of interesting species, including

the only record for Lesser Scaup. It is managed as a conservation area alongside a recently formed scrape, together with some of the former settling ponds and scrub, but access is restricted and now it is not well-watched.

OTHER GOOD BIRD-WATCHING SITES

Location, access and ownership details of more than a dozen of the best bird-watching sites, including more information about Venus Pool and some of the other sites referred to above, and other places mentioned in the species accounts, can be found on the SOS website, www.shropshirebirds.com, including:

- Chetwynd Pool
- Eardington Nature Reserve
- Haughmond Hill
- Howle Pool
- Priorslee Balancing Lake
- Priorslee Flash
- Severn Valley Country Park
- Wall Farm
- Whitcliffe Common

It is intended to expand this part of the website.

A description of local sites can also be found in *Where to Watch Birds in the West Midlands* (Third edition, Gribble *et al.* 2007).

Local (County) Wildlife Sites

In 1978–79, over 5,000 sites were surveyed and the best 766 were identified as 'Prime Sites for Nature Conservation'. *Losing Ground in Shropshire* (SWT 1989) described the process, but, by 1989, 95 (12%) had been destroyed, leaving 671, of which 133 (17%) had been damaged. Of this destruction, 64% was due to agriculture, 21% to forestry, and 15% to development.

Between 1989 and 2010, a number of new sites were identified, but a higher number were lost, leaving 555, a net loss of 28% since 1978–79 and 17% since 1989, with a total area of 10,405ha (3% of the County total). During the same period the status of these sites was clarified in a national framework, and their name changed, initially to County Wildlife Sites, and they are now known as Local Wildlife Sites (LWS). They have to meet strict locally defined scientific criteria, and are approved by a 'Local Sites Partnership' of statutory and voluntary organisations, co-ordinated by SWT. Although they are the second tier of important wildlife sites, after SSSIs, they have no legal protection. Their survival depends almost entirely on the attitude of the landowner.

The landowner has to be consulted before adoption. Although consent is technically not necessary, in practice goodwill is needed if the site is to survive, so the Partnership does not normally adopt a site if the landowner is unwilling. Management plans are provided for many sites. Most sites have been selected for their botanical importance, but many have specific bird habitats too.

The council's planning policy includes a 'Shropshire Environmental Network', which consists of areas of high biodiversity value and the areas that act as connective 'corridors and stepping stones' between them. The Wildlife Sites are 'core areas' in the network; they are normally included in Local Plan maps, and they are 'material considerations' that have to be taken into account when planning applications are considered. They also highlight key areas for wildlife to inform Natural England when they are considering applications for agri-environment scheme funding approval.

In 2010, SWT started a new project, which aimed to resurvey all the 555 sites that had LWS status at that time. By 2015, 310 had been surveyed and 29 (9.4%) of these were found to be destroyed, of which 7% had disappeared due to neglect, 68% to agriculture, 9% to forestry, 13% to development and 3% to drainage. Of the remainder, 43% were in a declining condition.

In October 2015 there were 634 Local Wildlife Sites covering 11,888ha. The total area is perhaps more meaningful than the number of sites, as the latter can be divided, merged, remapped with revised boundaries, or redefined by 'type', all of which can affect the figures. Despite the losses, around 15 new sites are adopted each year, and between 2010 and 2015 there has been a net increase of 79 Local Wildlife Sites, with an accompanying increase of 1,484ha (14%). However, 43% of those surveyed in the original number were in declining condition, and it is known that the new sites being adopted are of poorer quality than those defined in the 1970s and 80s. The best wildlife sites are still being lost at an alarming rate.

There were a further 130 sites (2,014ha) which had been accepted in principle by the Partnership, but which had not received landowner approval, and which therefore had not been approved for LWS status.

In 2010, 37 LWS (762ha) were within SWT reserves, and by 2015 that had risen to 43 (883ha).

Several sites in the Clun uplands that hold a suite of wetland birds have been adopted as LWS, and SOS intends to use the recent Atlas results to identify the best bird habitat sites across the County and propose their adoption as LWS.

Working with Landowners

The vast majority of birds, even the less common ones, nest outside the reserves and SSSIs managed as wildlife habitat, so the attitude and support of landowners is crucial for maintaining their nest and feeding sites. Some landowners are making real efforts to improve the habitats on their land by more sympathetic management of hedgerows and field margins, planting up headlands and establishing pools, but the protection of most of our bird populations is in the hands of individual landowners and managers. We need to support and work with as many of them as possible to safeguard remaining habitats, and create new ones.

For example, hedgerow and scrub clearance during the breeding season can destroy large numbers of nests, but if such works are discovered and discussed, most landowners are supportive and will wait until the young have flown. It is much harder to persuade them that similar actions at other times of the year, or draining or ploughing a meadow, are just as deadly in the long run because the parents will be homeless when looking for a nest site the following year. The effect of herbicides and pesticides is even more insidious.

Commons

Common land is owned by one or more people where other people, known as 'commoners' are entitled to use the land or take resources from it. The commoners' rights are included in their land deeds. All common land is owned by somebody, either privately or by a public body such as a local council or the National Trust. There are 86 commons, which include all the different types of habitat described above. Some, like the Long Mynd and the Stiperstones, are large and important bird habitats, but most consist of good habitat because the number of people involved, and their different interests, has made development or 'improvement' difficult. Conversely, management and conservation work on commons is usually complex.

Preventing Persecution of Birds of Prey

The Wildlife and Countryside Act (1981) provides legal protection for most wild birds and their nests, but enforcement is difficult. The *Atlas* (1992) made reference to an RSPB publication, *Death by Design* (1991), which looked at persecution in the UK between 1979 and 1989. During that period, this County had the second highest number of pesticide abuse incidents in England, after North Yorkshire. The Buzzard species account, referring to this publication, commented that 'Shropshire is a black spot for the poisoning of raptors, and the Buzzard is the main victim'. It referred to a 'particularly distressing incident' of deliberate poisoning being investigated by the RSPCA and added that 'several injured by shooting have been taken in for treatment in recent years, but those killed outright probably remain undiscovered'.

Since that publication the RSPB has produced an annual *Birdcrime* report. This remains the only centralised source of incident data for birds of prey persecution offences in the UK and includes details of related prosecutions. However, only a small of fraction of incidents are ever discovered and recorded, and only a very small percentage of reported cases ever result in court proceedings. In recent years a number of peer-reviewed scientific papers have been published which clearly spell out the impact that persecution continues to have on a

Table 2. Raptor Persecution Data for the UK and Shropshire

Report type	UK	Shropshire (% UK) total)
Total reports of shooting and destruction involving raptors	4,768	58 (1.2%)
Confirmed shooting and destruction incidents	1,293	24 (1.9%)
Totals reports of poison related incidents involving raptors	1,802	30 (1.6%)
Confirmed poison related incidents	1,133	15 (1.3%)

range of species, particularly on upland areas of England and Scotland managed for driven grouse shooting.

The raptor persecution data for the 25 years from 1991 to 2014 is shown in Table 2.

With about 1.4% of the total land area, levels of raptor persecution appear about average for the UK. However, the highest recorded levels are typically associated with driven grouse shooting. Consequently, Shropshire figures will be above average when compared with other counties which also have no driven grouse shooting interests. Persecution incidents during this period include:

- the shooting of numerous Buzzards, five Peregrines, and singles of Osprey, Red Kite and Goshawk;
- the poisoning of nine Buzzards, four Peregrines and a Red Kite in 2014;
- a few incidents relating to the use of illegal traps to take raptors.

In these figures, robbery of nests is not classified as persecution. In addition to the previous incidents, at least 13 Peregrine nests have been robbed of eggs or chicks (mainly believed to be connected to falconry interests), and at least three Goshawk sites have been robbed of eggs. The taking of one of the Goshawk's clutches in 1994 led to the prosecution of a falconer and his partner.

To maintain these long-term datasets, the RSPB relies heavily on the monitoring work of many raptor study groups around the UK, and has worked closely with the Shropshire Peregrine Group.

While North Yorkshire still retains its place as the worst county in England for raptor persecution, it does appear there has been a welcome reduction in confirmed pesticide offences here since the *Death by Design* report. Encouragingly, the situation for raptors across much of lowland UK has significantly improved since that time, with dramatic increases in the range and population of Buzzard and Red Kite.

The profile of the 172 people convicted of persecution offences nationally since 1990 shows that around three-quarters had game shooting interests, the majority being gamekeepers. There have been three local prosecutions:

- In 2007, a gamekeeper was convicted for the possession and use of pesticides following an incident involving the poisoning of two Buzzards. He received a conditional discharge;
- In 2008, two gamekeepers from a large commercial pheasant shooting estate pleaded guilty to a range of offences. This included the shooting of Buzzards, use of illegal traps and killing badgers. Both men got suspended jail sentences, an indication of how seriously the court took this case, which attracted national media attention. During the West Mercia Police and RSPB investigation,

a 'vermin diary' for 2007 was seized from one of the men. In addition to birds and animals that they were lawfully controlling, the records indicated that during part of 2007 a total of 102 Buzzards, 40 Ravens and 37 badgers had been illegally killed. This provides a graphic and disturbing insight into the levels of illegal persecution that are still taking place on some game shooting estates in the UK;

- During 2014, a cage trap illegally baited with two live domestic quail was found close to an active pheasant release pen near the Stiperstones. This was considered to be a typical 'hawk trap' and believed to be intended for accipiters and Buzzards. A local gamekeeper was covertly filmed by the RSPB attending to the trap on a number of occasions. He was charged, and appeared at court in 2015, but was acquitted in relation to these offences after the court deemed the surveillance footage to be inadmissible as evidence.

In conclusion, in common with the rest of the UK, it is clear that raptor persecution problems remain. However, levels of pesticide abuse appear to have reduced. The national BTO *Bird Atlas 2007–11* shows that Shropshire is important for Goshawk. Unfortunately, this species appears to have made little national progress since 1990, and remains absent from large areas of suitable habitat across the UK. It is hoped the County can help act as a stepping stone for the future expansion of the Goshawk population into other parts of the UK.

Habitat Quality

Several sections of this chapter have highlighted the deterioration in quality of a variety of habitats for birds in the Shropshire landscape.

Only 29% of SSSIs, nominally the best habitats for wildlife, and the only ones with statutory protection, are in 'favourable' condition. Over half (56%) of the water bodies are failing to meet the standards set by the WFD.

All our rivers and streams have been engineered to a varying extent over many years, principally to facilitate land drainage, but also to raise water levels or to prevent natural river processes such as a river changing course. These modifications to the river channel, and changing weather patterns, have resulted in more rapid changes in both water flows and water levels and this can be particularly disastrous in the spring when birds nesting on or close to the river can be washed out, or lose access to their food supplies.

The vast majority of the land is farmland, but, during the recent Atlas period, less than 60% of the land area was covered by agri-environment ELS agreements with individual farmers, landowners or tenants, with less than a fifth targeted on specific fields to provide specific benefits, not necessarily for wildlife, and over 40% of the land was not covered by any agreement at all. Where the agreements are targeted at providing habitat for farmland birds, they do provide some benefits, but the proportion of such agreements is small, and they have made little impact on arresting the declines. The total cost of the agreements that did exist was £13.9M over their lifetime. In comparison, the budgets of the nature conservation organisations, statutory and voluntary, is miniscule, although they provide much higher benefits for birds.

Farmland bird numbers continue to decline because of the deterioration in habitat quality, or its total loss, largely through the impact of the different aspects of agricultural intensification summarised on pp. 38–41, resulting in the loss of 44 million birds nationally between 1966 and 2011.

The less common but more important habitats – moorland, upland grassland, heathland, mosses and bogs – support a number of our key species, but can be difficult to manage. They are often marginal in agricultural terms and are either overgrazed or not grazed at all, both of which are detrimental to the bird interest. These habitats are usually nutrient poor and acid-sensitive habitats and an insidious threat to them comes in the form of atmospheric deposition. Nitrous oxide and sulphur dioxide from burning fossil fuels (including petrol and diesel by cars), ammonia from farm animals and ammonia and nitrous oxide from high fertiliser use all contribute to this form of pollution which affects these habitats, and the SSSIs and other important sites. For example, the Stiperstones includes a range of upland habitats, including heathland where the critical load for nitrogen is 10–15kg/ha/year, but it actually receives an average 23.2kg/ha/year (www.apis.ac.uk), which, over a period of time will undoubtedly affect the composition of the vegetation and impact on species like Meadow Pipit, Red Grouse and Skylark.

The quality of woodland for birds has also declined considerably in the post-war period, and only 13% of the woodland area is now habitat of the highest quality for birds.

The proportion of the land taken over for direct human use (urbanisation, including more housing, offices, retail, business and industrial parks, and leisure activities) is also growing. In addition to all these changes for the worse in habitat quality, the impact of climate change has yet to assert itself to any great extent.

This chapter has shown that even the habitats that support our characteristic bird species are degraded and, in many cases, continuing to decline, despite the efforts of our statutory and non-statutory agencies. Without a change of direction, it is inevitable that many of the specialist birds the habitats support will become increasingly rare. There are few examples of the necessary level of monitoring the quality of specific habitats. Indeed, the assessment of water bodies by the Environment Agency, and publication of site-by-site results (see p. 33), is the only example of a statutory agency monitoring the condition of a specific wildlife habitat. Most of the monitoring that is done, such as the monitoring of SSSIs by Natural England (see p. 47), and Special Areas of Conservation, is site based and covers only a tiny proportion of the respective habitats.

Although reversing the decline in habitat quality will require a considerable amount of work, and perhaps additional legislation, the budgets of both the Environment Agency and Natural England are insufficient to address the issues, and are likely to be cut even further, rather than increased, over the next few years. The result of the referendum on membership of the European Union throws even progress towards the objectives of the WFD into doubt, and potentially removes legal protection provided through Special Areas of Conservation, and the Birds and Habitats directives. Whatever the future relationship between the UK and the EU, the loss of habitat quality highlights the need for legislative protection for the environment in the UK to be no less than that in the EU.

The Birds of Shropshire

The primary purpose of the *Atlas* (1992) was to provide a benchmark against which future trends in distribution and population of breeding species could be monitored. This Avifauna has utilised that benchmark, updated it, and provided a new one for wintering species and passage migrants.

The commentary in each species account, with the associated maps, relates distribution to the various habitats described above, estimates the current population levels, and wherever possible compares the current position with previous assessments, particularly that in the Atlas (1992). However, the results must be interpreted with care, and the limitations of Atlas survey work are described more fully in Appendix 1.

Even so, judging from the results, it is clear that some species are increasing, many more are declining, and a few have disappeared as breeding birds in Shropshire during the last half-century, as shown in the *Breeding Status* chapter. The current status of Shropshire's breeding birds is summarised in Appendix 6. Although some species are known to be increasing, their growth in numbers is no compensation for the decline or loss of a much larger number.

The SOS aims to 'encourage the study and protection of birds of Shropshire'. This Avifauna, and the Atlas before it, has provided invaluable base data for this task. Work is now advancing to ensure that all the best sites found by Atlas fieldwork are included in the SWT's Local Wildlife Sites system.

It is in this context that the SOS, along with the SWT, will continue to monitor both the population level and distribution of our birds. We will assess the ongoing effects of the changes described above, particularly those of land management upon habitats, and, wherever necessary, seek to influence conservation policy both locally and nationally.

By building upon the records and trends presented in this Avifauna, farmers, landowners, planners, conservationists, ornithologists and amateur bird watchers can all help to safeguard the future for birds in Shropshire.

The Species Accounts

Leo Smith

Introduction

The Species Accounts review the status, occurrence and distribution of all species that have been recorded in Shropshire, mainly up until the end of December 2014. The data cut-off dates are set out in detail below (p. 62). The full list of 301 species is shown in Appendix 6. There are dubious old records for several other species that have not been included, but which are listed in the *History* chapter in the box on p. 7).

The review includes historical references dating from the first half of the nineteenth century, occasionally earlier, through to the formation of SOS in 1956; annual County Bird Reports (SBRs) from that date, and an unpublished report for 1955; *A Handlist of the Birds of Shropshire*, published by SOS in 1964; the Shropshire part of maps published in national BTO Bird Atlases; the maps from 1985–90 fieldwork published in *An Atlas of the Breeding Birds of Shropshire* (1992); Breeding Bird Survey (BBS) population trends since 1997 for the 34 most common and widespread species; the results of fieldwork for the recent 2007–13 Atlas project, published here for the first time; recoveries of birds ringed or recovered here; and, in a few cases, the local results of national surveys organised by BTO, or of local research projects.

Background information published in 'Bird Facts' and 'Bird Trends' on the BTO website on habitat preferences, typical lifespan, and the age of the oldest recorded individual nationally, has also been utilised.

Further detail is provided on all these data sources, and what the maps, graphs and charts show, later in this chapter. This information is not duplicated in the individual species accounts.

For just over half the species (96 residents, 29 summer visitors and 27 winter visitors), the main focus of the account is the distribution and relative abundance maps from the 2007–13 Atlas project. The distribution of breeding species has been compared with that published in the 1992 Atlas, which used the same methodology as the recent Atlas. In most cases, distribution is also related to the landscapes and land use described in the *Habitats* chapter.

A further 58 species are primarily passage migrants, and these accounts review the number, pattern and timing of records.

Fifty-two species are vagrants. Of these, accounts of rarities with less than 10 records since 1950 list all the records, while the accounts

of species that have between 11 and 20 records summarise all the records, without listing them individually. Another 19 species have not been recorded since 1950.

Several species have established feral breeding populations as a result of deliberate releases or escapes, in Shropshire or elsewhere, and where this had led to local breeding populations, or an increase in passage or winter records, accounts have been included. Examples are mainly wildfowl. Other escapes and releases which are not considered to be self-sustaining in the wild, are summarised at the end of the Species Accounts, but no detailed accounts are included.

Apart from a statement about the species range, usually in the first paragraph, the accounts are specifically about birds in Shropshire, and all statements in them apply to, and perhaps only to, this County, unless explicitly stated otherwise. Usually any commentary about the species status elsewhere is included only to explain changes in distribution or occurrence here. The terms 'in Shropshire' or 'in the County' are therefore usually unnecessary. They should be taken for granted. Commentary may or may not apply elsewhere, though the context should make that clear.

Place names are those shown on the current Ordnance Survey (OS) map. All such names and geographical features mentioned are listed in the Gazetteer (Appendix 2), along with the 4-figure OS map grid reference that identifies the 1km square, so all locations can be related to the distribution map if required. Wider areas (for example, the north Shropshire plain or Clee hills), are not specific locations named on the OS map, and such names do not appear in the gazetteer.

However, many place names have changed over time, and in quotations, and comments attributed to particular authors, the place names are those used at the time. Some of the changes are relatively recent – for example, the *Handlist* (1964) refers to 'Longmynds', and old names for the Ellesmere meres.

The full scientific names of any plants or animals referred to are listed in Appendix 3.

Descriptions of the birds themselves, their behaviour and how to identify them, are omitted as they can be found in any of several excellent field guides.

The accounts have been written by 27 different authors. The author(s) of each account is credited at its end.

Species Names and Order

The British Ornithologists Union is the keeper of the official List of British Birds. The names of species, classification into species and races, and the taxonomic order are kept under constant review. This Avifauna uses the list published in January 2018, which includes both the vernacular name used by British birders and the International English name now agreed by the IOC International Ornithological Congress, together with the scientific name (see www.bou.org.uk/british-list/).

Some of the names are different from those used in recent years. The taxonomic order has also been revised, so it is now different from that used in the current edition of many reference works and field guides, and may be unfamiliar. An index at the end, listing all the page numbers where each bird name appears, with the main species account in bold, will help readers navigate the new order.

Status and Abundance

As part of the heading to each account, an indication is given of the species status and abundance in this County. Similar headings have been given in SBRs since 1991, but they have often changed and appear to have reflected the opinion of individual editors. This was rectified in 2007, when a list of criteria was published. These criteria were reviewed during the preparation of this Avifauna, and the criteria below will be used in SBRs in future. It is intended to review them regularly in future, every five years or so.

Assessing the abundance is relatively easy for most breeding species, which are tied to territories or nest sites for most of the season. This is based on the population estimate given in the account, but numbers may vary from year to year depending on breeding success in the previous year, over-winter survival, and weather at the start of the breeding season, which might depend on conditions in southern Europe and parts of Africa in the case of summer visitors. Therefore these estimates are not always sufficiently accurate for the status to be listed with certainty, and it should be taken as a guide rather than a statement of fact.

Defining the status of wintering species is much more problematic, as detailed in many accounts. For most resident species, there will be an increase in population, the size of which will depend on the success of the breeding season, and natural mortality will see a fall throughout the winter. Other species usually thought of as resident, such as Woodpigeon and Blackbird, are joined by an unknown number of winter visitors from northern Europe, some of which stay, but others pass through. A number of these species are partial migrants, with some breeders moving south and others staying here, but even they may be supplemented with arrivals from further north, as for example happens with Goldfinch and Linnet.

For species only present in winter, numbers are highly dependent on weather conditions, both on the continent and here. Mild winters and climate change have increased the numbers of northern European breeding birds that exhibit 'short stopping' and now over-winter on the continent, especially in the Netherlands, rather than come to Britain as they used to. In these cases there has been a distinct falling trend, but hard weather may drive them over here from time to time. Others attempting to winter here may be driven further south by freezing conditions as well. Such conditions vary considerably from year to year, with a trend to milder winters here too. Other species depend on specific food supplies, and only come here, often in large numbers, when the supply is restricted. Some may be more accurately described as nomadic, visiting several locations during a winter, rather than being migratory (moving between specific wintering areas and breeding grounds). In many cases, numbers regularly reach an annual peak then fall again.

Therefore any attempt at including an assessment of the winter population is likely to be meaningless, and, while an estimate and a general classification can be attempted for the rare and scarce species, in most cases the status is given only as 'Winter Visitor', without any indication of numbers. Instead, the proportion of tetrads with winter records has been used as an indicator of winter abundance.

Given the inherent difficulties of defining the status outlined above, the headings provide only a general guide, and the number of categories is deliberately not extensive.

The status and abundance of each species is based on the data collected during the recent Atlas period (2007–13). There are two exceptions to this:

- where there have been significant changes since the population estimate in the account was made, the status has been reconsidered. Turtle Dove appears to have declined to less than 10 pairs, and the status has been revised from scarce to rare. Snipe, Willow Tit and Lesser Spotted Woodpecker are believed to have declined as well, but not sufficiently to affect their defined status category;

- where a species occurs infrequently and the Atlas data does not provide a meaningful indication of its status and abundance. These cases have been considered in terms of the pattern of numbers involved in their individual occurrences.

In all cases, the status and abundance is based on numbers in a typical year (although it is recognised that some years are atypical), and it is ordered with the predominant status first. The conspicuousness of a species has not been taken into account when defining status and abundance.

Taking this into account, each species has been allocated to at least one of the categories in Table 1. The table includes a code for each definition, which has been used as part of the 'Shropshire Status' definition in the Shropshire List summary table in Appendix 6.

Each species has also been allocated an abundance score based on the definitions below.

'Resident' species may have seasonal, sometimes substantial, movements of individuals in and out of the County, and, particularly in the case of species that breed in the uplands, may occupy different areas and habitats in the winter. Many 'Summer' and 'Winter Visitors' pass through in larger numbers as they arrive or depart; in neither case do the status definitions cover these movements, and they usually refer to the predominant status unless there is a significant discrepancy between the two, in which case both are mentioned (e.g. a scarce summer visitor but common passage migrant). In other cases, the 'Resident' population is largely stable, but there is a significant influx of 'Winter Visitors', and they determine the primary status. Occasional 'out of season' records have been discounted, but may be referred to in the account. Some passage migrants are still present into the winter season, but have usually departed by mid-November, and others occasionally pass through during the winter period; in neither

Table 1. Status Definitions

Status	Code	Definition
Resident	R	Breeds and present all year
Summer Visitor	S	Mainly occurs as a breeding visitor in summer
Winter Visitor	W	Mainly occurs as a visitor throughout the winter
Visitor	V	Can appear at any time during the year, with no obvious pattern as to when it occurs
Passage Migrant	P	Mainly passes through in the spring and/or autumn; this category is only used when it is the predominant status
Vagrant	Va	A very rare bird well outside its usual range
Naturalised	N	Introduced deliberately or accidentally by man and now breeds in the wild
Irruptive	I	Occurs sporadically, or in very variable numbers, in response to severe food shortages within its home range
No Modern Record	NMR	A species that has only occurred before 1 January 1950
Has Bred	HB	Species that have bred before 2008, either regularly or infrequently, but no longer breed; they may have another status now
Breeding Species	BS	Species that breed in low numbers and not always annually. This includes those that are predominantly winter visitors or passage migrants, where the breeding population is considerably lower; and those that have bred for the first time since 2008, but have not yet become established as a regular breeding species

Table 2. Abundance Definitions

Abundance	Code	Definition
Residents and Summer Visitors (i.e. breeding species)		
Very rare	1	Less than annual
Rare	2	Annual and up to 10 Breeding Pairs
Scarce	3	11–100 Breeding Pairs
Uncommon	4	101–1,500 Breeding Pairs
Fairly Common	5	1,501–3,500 Breeding Pairs
Common	6	3,501–15,000 Breeding Pairs
Very Common	7	Over 15,000 Breeding Pairs
Winter Visitors, Passage Migrants and Rarities		
Very Rare	1	Less than annual
Rare	2	Annual or nearly annual and up to 20 individuals each year
Scarce	3	21–200 individuals each year
Uncommon	4	Found in 10–24% of tetrads (2007–13)
Fairly Common	5	Found in 25–39% of tetrads (2007–13)
Common	6	Found in 40–89% of tetrads (2007–13)
Very Common	7	Found in 90–100% of tetrads (2007–13)

case are they considered to be winter visitors. Occasionally, some regular visitors are also irruptive and arrive in very large numbers. The 'Has Bred' category makes no distinction between previous occasional or regular breeding, but the status of these species is described more fully on p. 481. Two species, Cormorant and Cetti's Warbler, appear to be resident, but breeding has yet to be proven and they are defined as 'Non-breeding Residents'.

The definitions in Table 2 have been used in conjunction with the status category to give some indication of the abundance of an individual species. Where a population is supplemented by releases this is also noted in the heading to the account, and denoted by the code 'SR' in the Shropshire List summary table in Appendix 6. This table also incorporates the code in Table 2 as part of the definition of the 'Shropshire Status'.

Where the local estimate for the number of Breeding Pairs is different from the estimate based on TTV counts, the former has been used. Where a range for the number of Breeding Pairs has been given, and the range straddles a boundary in the definition, the lower band of abundance has been used.

It should be noted that some species can be highly visible, highly mobile or in mobile flocks during the winter and have been recorded in a higher proportion of tetrads than reflects their actual abundance. In such cases, their abundance rating has been downgraded to give a better picture of their actual winter population by taking into account their breeding population, if relevant, and known movements. Conversely, some species are numerous, but occur in a restricted number of tetrads that reflect their habitat requirements and in these cases the abundance rating has been increased.

Cut-off Date

All records and data have been reviewed up until 31 December 2014, and no subsequent records have been included, except:

- Four species recorded for the first time, up to the finalisation of the manuscript for publication: three (Iberian Chiffchaff, Pine Bunting and Shore Lark) not previously recorded, and a fourth, Marsh Warbler, identified with certainty for the first time.

- Little Bittern: the first modern record (accepted by British Birds Rarities Committee [BBRC]).

- Rarities with 10 or fewer records, and near rarities with 11–20 records: (in alphabetical order) Bearded Tit, Cattle Egret, Cetti's Warbler, Glossy Ibis, Great Northern Diver, Great White Egret, Green-winged Teal, Little Tern, Long-tailed Duck, Manx Shearwater, Montagu's Harrier, Night-heron, Pectoral Sandpiper, Purple Heron, Ring-billed Gull, Ring-necked Duck, Rough-legged Buzzard, Sandwich Tern, Spoonbill, Temminck's Stint, Tundra Bean Goose, Velvet Scoter, White-winged Black Tern and Yellow-browed Warbler.

- Species with rare races (Chiffchaff), previously more common species that have become rare in the twenty-first century (White-fronted Goose, Corncrake) and one with many records of a few individuals (Crane).

There are 30 of these species altogether with post-2014 records.

A few ringing recoveries, of birds ringed before the data cut-off, have been included if they add important information to the account.

The Narrative

The text explains the history, distribution, status and breeding habitat of each species. Where possible this is related to the habitat maps and other content of the *Habitats* chapter.

For most species the account is written in chronological order, starting with a very brief description of its breeding and wintering quarters, but more emphasis is put on the recent past, and the main focus of most accounts is the results of the recent Atlas project, covering November 2007–February 2013 for wintering species, and 2008–13 for breeding species.

Apart from the 30 species listed above, the Species Accounts include records up to 31 December 2014. Comments noting the most recent, or last, record, or using similar phraseology, relate to that date, and do not exclude the possibility of records being received subsequently. These can be found in SBRs for the years from 2015 onwards.

Historic Data

Historical nineteenth-century data includes the works of William Beckwith and H.E. Forrest, and many other authors. SOS is fortunate in that all the known historical data can be found on a website, *The Historical Ornithology of Shropshire* (Tucker & Tucker 2012). Known colloquially as '*Histo*', it can be accessed from the menu on the home page on the SOS website by clicking on 'Historical Ornithology of Shropshire'. Some relevant information can also be found in the national *Historical Atlas of Breeding Birds 1875–1900* (Holloway 1996).

Early records often refer to a 'specimen' being 'obtained' or 'collected'. These are euphemisms for birds being shot or otherwise killed, taken to a taxidermist for preservation and mounting, and incorporating into a collection (see p. 2 in the *History* chapter).

The main source of twentieth-century records prior to the formation of SOS in 1955 are the annual CSVFC *Record of Bare Facts*, and the *Transactions*. All these publications can also be found on *Histo*. The accounts provide a brief overview of records and status, rather than a detailed summary, and full references for each record have not been provided. CSVFC covered a large area, so individual records have only been included if they were clearly from within this County.

As noted above, place names in quotations, or used by the early authors, are those used at the time.

SOS Records

SOS encouraged the submission of records to the County Bird Recorder, and annual County Bird Reports (SBRs) have been published since 1956. The Systematic List in each SBR has been an important source for every account. SBRs were not available for the later years covered by this Avifauna, so authors were provided with a list of all records of their species for those years.

The validity of several individual records was queried during the preparation of the accounts, and where possible they were checked against the original record cards up until 1992, or the computerised data base from that year, and suspect records have been omitted. There are other examples where avifauna accounts include different dates or details from those published in SBRs, and, where the discrepancy has been identified, the original record has been checked and the information published in the Avifauna is correct, and supersedes that published in SBRs.

The Handlist

The first modern benchmark publication, *A Handlist of the Birds of Shropshire* (Rutter *et al.*), referred to as the *Handlist*, was published in 1964. It included an assessment of the local status of all species at that date, a summary of the records of the less common species from 1950 and a review of historic records.

National and Local Atlases

Some accounts draw on the Shropshire part of the maps published in the first national breeding BTO Atlas (Sharrock 1976) and the first winter Atlas (Lack 1986). These national Atlases were based on surveys of 10km squares, rather than the more thorough and detailed coverage at the 2x2 kilometre square ('tetrad') level of the *Atlas* (1992), and the recent 2007–13 Atlas project. Where the species accounts refer to 10km squares they are comparing local Atlas results with those from these earlier national Atlases. There are 19 10km squares wholly in Shropshire, and a further 14 which have more than half their 25 tetrads in it. These 33 10km squares that are wholly or mainly in Shropshire are described in the accounts as the 'County's 33 10km squares'. The definition of 10km squares and tetrads, and the relationship between them, is described in Appendix 1.

An Atlas of the Breeding Birds of Shropshire (1992), based on fieldwork in 1985–90 in all 870 local tetrads, was the first publication to map the distribution of breeding birds, and review their status up until 1990. No comparable previous exercise has been carried out for winter visitors or passage migrants, and this has been published for the first time in this Avifauna.

All the County Bird Reports, and the *Handlist*, can be found on the *Histo* website.

Sites

The best bird-watching sites are described briefly in the *Habitats* chapter. SOS Records from these sites have steadily increased, partly due to an increase in observers (see *Growth in Recording Effort* on p. 11), and, in the case of the more uncommon species, an increasing use of pagers and bird lines, and, more recently, the internet and social media, to alert birders to make special journeys to see them.

The most watched sites have been the SOS Reserve at Venus Pool, Allscott Sugar Factory, and, more recently, Wood Lane and Polemere. These sites have been visited regularly, and feature consistently in the species accounts. Mirelake, the largest pool at Allscott, had a hide and was often specifically named in records. Log books in each hide at Venus Pool, and at Polemere and Wood Lane, each produce many records in addition to the normal submissions to the County Bird Recorder.

The grounds around the settling pools at the Sugar Factory were also the foremost ringing site. Unfortunately this site has been lost. The reed bed and scrub around the settling pools became unsuitable after the factory closed in 2007, and permission for access was withdrawn by the new owners at the end of 2009 (see p. 47).

Before access was gained to the Sugar Factory, Shrewsbury Sewage 'Farm', the treatment works at Monkmoor, was the foremost wader site, and the first record for several species was obtained there. However, the works have been modernised several times since, and records from the site declined from the early 1970s onwards (see p. 56).

The effective loss of these two important sites has affected the trend graphs and charts for some species.

Ringing is also carried out at Chelmarsh reservoir, and at various other sites, as described in the *History* chapter on p. 10.

Review of Records of Rarities

Records submitted to the County Bird Recorder were largely accepted at face value until 1996, when SOS established a Rarities Committee to validate records before acceptance. Subsequently it was decided to review all modern records received between 1950 and 1996 for certain rare species, and a number of such records, many previously published in SBRs, were rejected. These rejected records are not included in the Species Accounts.

A full description of the process, the species considered, and a list of rejected records, has been published (Holmes & Walker SBR 2011).

Other records of the less common species held by SOS, whether they have been published in SBRs or not, have been reviewed during the preparation of the species accounts, and a few suspect records have also not been included.

Escapes and Feral Populations

Initial records of several species, particularly wildfowl, were considered to be of escapes, and were noted as such in SBRs. There was no consistency of presentation, which appears to have reflected the judgement of individual SBR editors. As the number of records received, and the knowledge and experience of the County Bird Recorder and SBR editor in judging them, increased, interpretation evolved, and recently all records have been treated as being of wild birds (having originated from the feral population in the UK or near continent) unless there is overwhelming evidence suggesting captive origin. The review of records did not include an assessment of whether birds were wild or escapes and the records were accepted as published. The species accounts have reviewed all such records, whether or not any of them were initially reported as escapes.

Charts

Bar charts, compiled from SOS records, showing when species have occurred here, are included in the accounts for many of the passage migrants, and for some rarities and scarce species. Most charts are based on data for one of two standard time periods, either 1950–2014, or from 1992, when records were computerised, until 2014.

So far as possible, the annual occurrence charts start at 1950, and show an estimate of the number of individual birds present in each year, not the number of records. Data for the period 1950–55, before the foundation of SOS, has been taken from CSVFC records. In the case of rare species covered by the recent 'Review of Records of Rarities' (Holmes and Walker, SBR 2011, covering records before 1996), a spreadsheet including every known record from 1950 onwards was prepared. This includes the relevant CSVFC records 1950–55 and SOS records 1956–2014. This spreadsheet can be found on the Avifauna website (see p. 73).

Rarity records have usually been compiled on a calendar year basis, so two winter visitors in the same year, one arriving in January and the other in December, both appear in the same year even though they arrived in different winter periods. However, if the latter stayed into or beyond January, it has only been counted once, on the first date it was recorded. To give a clearer picture, where appropriate, some charts have been compiled for winter periods, rather than calendar years.

In many cases, several records referred to what appears to be the same individual bird, so an assessment has been made of the records to estimate the number of different individuals present each year or winter season, and the species account is based on this assessment. For some scarce species, SBRs have included an assessment of the number of individuals involved in some years, but in the main this assessment has been undertaken and published here for the first time.

However, in the case of three species not covered by the review, Hen Harrier, Curlew Sandpiper and Sanderling, CSVFC records from 1950–55 and the results of a manual search of SBRs and SOS records from 1956 to 2014 have been used to compile the chart.

As far as possible, the bottom axis on these annual occurrence charts starts at 1950, even if the first modern record was somewhat later, so all charts are to the same scale. However, different time scales have been used for a few species, and if so the reason for the difference is explained in the account.

Several other annual occurrence charts were produced, but they showed no apparent pattern or trend, and contained only a handful of birds in half or less of the 65 years covered. These charts are published on the Avifauna website (see p. 73).

Charts for seven wildfowl species (Goldeneye, Goosander, Mallard, Pochard, Shoveler, Tufted Duck and Wigeon) have been compiled from WeBS counts, usually from the Ellesmere meres and/or Shavington. These charts cover winter periods, not calendar years, and cover the whole period from the start of WeBS in 1960–61. Two other species (Coot and Great Crested Grebe) have charts starting in later winters, when they were added to the WeBS target species and counted for the first time. Pintail has a chart starting in 1995–96, when a WeBS count started at its main site, the Severn–Vyrnwy confluence. Ruddy Duck has a chart from 1965–66, when it was first recorded on a WeBS, until 2008–09, when it was last found.

Seven other winter visitors also have charts showing numbers split by winter periods, rather than calendar years. Apart from Ruddy Duck, the data cut-off at the end of December 2014 means that the figure for the final period (labelled 2014) is not comparable with the other winter periods shown on the chart, particularly for species that usually peak early in the new year.

The title on each chart makes it clear which timescale has been used.

For more common species, where numbers present each year have shown a trend, but an assessment of the number of individuals is not possible, the computerised records from 1992 onwards have been analysed. All duplicate records (i.e. the same species seen at the same site on the same day) have been removed, leaving the record with the maximum number of individuals. These have been added to the other records, so the number of individuals present in each period can be estimated. If a single record covers more than one day, all days after the first have been discounted. Some records state a range of dates, and state a maximum number seen. In these cases this maximum has been used as the first day count. In this way each bird listed on the records is counted once. However, if records of what was presumably the same bird were submitted by different people on 10 successive dates, it will appear on the chart as 10 separate entries. There is no easy way round this difficulty, but the records utilised start in 1992, so the bias of more recorders in the later period is largely avoided, and the pattern of the records is valid, even if the numbers are only fairly accurate approximations.

Depending on the source and the number of records available for analysis, three different titles have been used for these annual occurrence charts. For rarities, the number is almost certain, so they are titled 'Number per year'. For less rare, the scarce species, the number can be estimated reasonably accurately, and they are titled 'Estimated number per year'. For the more common species, the number calculated from records, as outlined above, is less accurate, and they are titled 'Approx. number per year'.

The analysis of computerised SOS records described above has also been used to calculate a seasonal profile of the estimated number of birds recorded, in 10-day intervals. Even if all the records had noted the length of stay, splitting them into the different 10-day periods would have been extremely time-consuming, and would probably not have improved the picture of the pattern of occurrence for most species. Starting this analysis in 1992 has the advantage that any shifts in migration periods due to climate change over a long period are avoided. However, in a few cases with few records to analyse, the seasonal occurrence charts have been compiled from the rarities spreadsheet. Again, there are different levels of uncertainty in the number of birds covered by the records, and only the first date on the record has been taken into account. These charts are intended to show the periods when each species is most likely to be found, but in a few cases they might be occasionally found later than the chart suggests.

As outlined above, the data cut-off for 30 rarity species was extended until the end of 2017. These include six species (Little Tern, Long-tailed Duck, Pectoral Sandpiper, Ring-necked Duck, Sandwich Tern and Temminck's Stint) where the accounts include annual and/or seasonal occurrence charts, and these charts include the post-2014 records. The chart title reflects the end date of 2017, irrespective of which years the additional records came from.

The charts therefore have some limitations, arising from the limitations of the data itself. The number of records may be very different from the number of individual birds, particularly in more recent times when modern communications mean the same bird(s) may be seen by many different observers on successive days, especially at well-watched sites. Ideally, counts of bird-days would have been used in the charts, but unfortunately such counts are not readily or reliably available from SBRs or from the records themselves.

Making an assessment of whether different reports relate to the same or different individuals is often difficult, especially in years when there are multiple sightings. Also, because the sites most likely to attract wildfowl or waders are relatively close together, there is almost certainly some movement from one site to another, evidenced by the sequential dates of some sightings. Other individuals may stay for many days or weeks, with only sporadic records to indicate their continued presence. Particularly in earlier years, when there were fewer birders and no modern communications, some species were recorded at a single site on several successive weekends, but on none of the days in between.

Nationally, in articles on 'scarce migrant birds', published in *British Birds*, 'only new birds have been included in the totals, and those believed to have been returning from previous years or seen elsewhere in the same year (mostly in the same recording area) have been omitted'. This convention has not been followed in this Avifauna, as our much lower numbers would lead to charts which showed none at all being present in years when it was believed that the one(s) seen were the same individual(s) present in previous years. Also, it is impossible to make such an assessment in most cases. However, if it is believed that the same rare individual returned in more than one year, that is stated in the account. Obviously this does not apply to more common species, where many individuals may return in successive years.

All the caveats, particularly recognising that the charts only show the first date from records of birds that may have been present for extended periods, must be taken into account when interpreting what they show. Even so, species charts showing the number of individuals per year or per winter season, and the seasonal variation in the number of individuals in 10-day intervals, give a good indication of the patterns of occurrence.

Weather Reports

Many continental, coastal and pelagic species only occur here after very hard or stormy weather. Many accounts correlate the records with such weather conditions, sometimes sourced from the records themselves, but more frequently from the Monthly Weather Report that can be found on the Met Office website. Any note of weather conditions in the accounts comes from one or other of these sources, unless a different reference is provided.

The Recent Atlas Project

The results of the recent Atlas project are published here for the first time, and provide the main focus of many accounts.

Coverage

Fieldwork started in November 2007 for the winter period, and April 2008 for the breeding season, to coincide with that for the national BTO Atlas. The methodology and survey Instructions are described in Chapter 2 and Appendix 2 respectively of the national BTO Atlas (2013).

The Atlas recording unit is a tetrad, a 2x2km square on the OS national grid. Locally, the fieldwork period was extended to 2013, to ensure adequate coverage of every tetrad, and, in the case of breeding species, ensure equal levels of coverage to those achieved for the 1985–90 Atlas.

The County boundary passes through many tetrads. If half or more of the tetrad area is in Shropshire it is included in the Atlas area, but if there is less than half it is not. The same criterion was applied in both 1985–90 and 2007–13. This means that, on the border, there are a few records included on Atlas maps of observations just outside the boundary, and conversely there are records from some sites that are not shown on the Atlas maps. Perhaps the most notable example is Chetwynd Pool, near Newport, which is well within the County boundary, but the tetrad is more than 50% in Staffordshire so the site is not shown on the maps, even though it now has a heronry, and is an important wintering site for wildfowl.

There are 870 tetrads in the Atlas area, and they were all surveyed, in both the breeding season and in winter.

An analysis of the coverage achieved, and comparison with that in 1985–90, is given in Appendix 1.

Atlas Maps

There are up to five standard maps from the recent Atlas project for each species. Each map has the same background (Map 1), which highlights the County boundary, the main rivers, towns and upland areas, and the OS National Grid 10 kilometre squares, so distribution and relative abundance can be related to local landmarks and OS maps.

The breeding distribution maps are based on detailed surveys over six years from 2008 to 2013. They show only what was found by Atlas fieldworkers, or reported to the County Bird Recorder, during this period. Some species breed outside the main breeding season, so the maps include all records with validated breeding evidence, irrespective of the month of the record.

Fieldworkers submitted records of evidence of breeding using standard Atlas definitions. On the breeding distribution map, the large dots represent confirmed breeding, the medium dots probable breeding, and the small dots possible breeding, somewhere in the relevant tetrad. The definitions were specified by BTO, and included in the Atlas instructions.

All recent Atlas records were subject to validation checks by the Atlas organisers, as required by BTO. In particular, observers were advised to distinguish between recently fledged young which are almost certain to have come from nests in the tetrad, and more mobile, slightly older independent juveniles, that might have moved considerable distances to feeding areas. The latter should not have been recorded as confirmed breeding. The breeding distribution maps have been compiled from the submitted records, as amended by the validation process. In addition, for many of the more scarce species, the SOS records were inspected, and added to the Atlas database if there were no equivalent records, provided a breeding code and the tetrad could be determined. The account includes any necessary interpretation or explanation. For example, some species are skulking, nocturnal or favour inaccessible habitat, and are therefore hard to find and under-recorded. Others have large territories, or move round farms following the crop rotation, and may change their nest sites from year to year, so the same pair may have been recorded in two or more tetrads in the Atlas period, and the species is therefore over-represented on the map.

For a few species, particularly those that forage over long distances from the nest, a special breeding distribution map has been produced, usually showing tetrads with confirmed and probable breeding, and the range. The content of each special map is described in the account.

The breeding season relative abundance maps are based only on counts made during timed tetrad visits (TTVs) in the breeding season. Two two-hour visits were made to the vast majority of tetrads, the first in April or May, and the second in June or July. Such maps have only been produced for species recorded on more than five TTVs.

The colour scheme highlights the areas of highest density. The darkest colours represent around 25% of squares where the highest numbers were counted on TTVs. Then, going from deepest to palest colours, the next band also includes about 25%, then the lowest 50%. The number of squares in each of the three bands varies, because the total reflects the number of squares where the species was found during TTVs. However, the boundary between bands has been adjusted, so the same count does not appear in two bands (i.e. after counting up from the lowest figure, if the band at the top of the lowest 50% has the same number above and below the band boundary, the boundary has been moved upwards until the next number is reached, and similarly for the band at 75%). The palest colour shows squares where the species was found during the relevant Atlas period, but not during a TTV. The numbers at each band boundary are not included in this book, but can be found on the Avifauna website (see p. 73).

It must be stressed that the relative abundance maps only give a general impression of densities. Only a small proportion of a tetrad can be visited in two hours. Observers were asked to visit all the habitats in the square, but in many cases this was not possible, and in others it meant that roughly the same amount of time was spent in each habitat, when one habitat may be predominant in the square. Luck plays a part as well, as widespread but scarce species like Sparrowhawk, nocturnal species such as the owls, or skulking species like some warblers, may be overlooked, and detectability for most species varies according to the prevailing weather conditions and time of day. TTVs should not have been carried out during periods of inclement weather likely to affect bird detectability, and they should have been done in the morning during the breeding season, but these expectations were occasionally not complied with. The time within the two-month period when the count was made also had an effect. In particular, many counts in the early period were carried out before

Map 1. Atlas Background Map

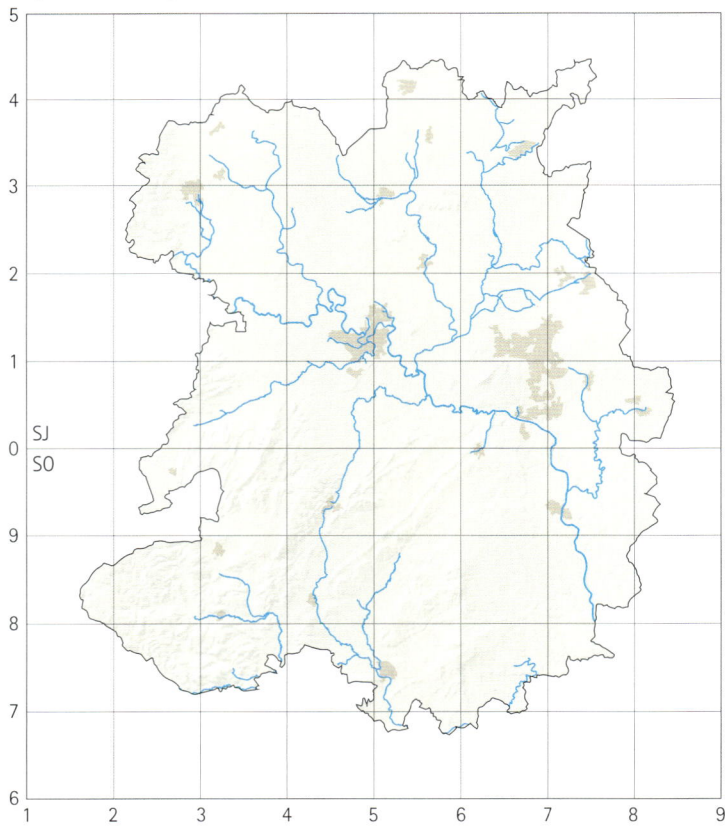

the bulk of the summer visitors arrived, and, in the late period, many species, but not all, would still be in song in early June, but not in late July. Even so, these maps are the first indication we have of the relative abundance of different species in different habitats. Most accounts relate the pattern to the maps in the *Habitats* chapter.

The winter season distribution maps only show records from the months November–February collected between November 2007 and February 2013. There is only one size of dot, indicating that the species was seen at least once in the Atlas period. Fieldworkers were asked to record only those species that were actively using the square and to avoid any that were flying high overhead, such as migrating geese or gulls and Starlings flying to or from roosts. Raptors actively hunting were to be included. Records therefore involved an element of judgement over and above identification of the species, with inevitably some inconsistencies, but these do not affect the maps unduly.

The number of tetrads where each species was found in the breeding season, and the percentage that represents of the total 870 tetrads, is listed in a tetrad occupancy table, and compared with the results from the 1985–90 Atlas (see below). Similar figures listing the occurrence in winter are also given, either under a winter map if there is one, or otherwise in a combined tetrad occupancy table including the breeding season.

Again, the relative abundance maps are based on two two-hour TTVs in almost every tetrad, the first in November or December, the second in January or February. The same qualifications apply as outlined for the breeding season abundance maps.

For less than 10% of tetrads (163 in the breeding season, 173 in winter), one or both of the TTVs were limited to one hour. In these cases the counts have been adjusted, and standardised to the equivalent of a two-hour visit.

The fifth standard map, showing the change in breeding distribution over a 23-year period, compares the distribution map from the recent (2008–13) Atlas fieldwork with the one for the same species obtained from 1985–90 fieldwork, published in *An Atlas of Birding Birds of Shropshire*, published by SOS in 1992. Both Atlases used the same methodology, and the level of survey effort was comparable (see Appendix 1). Every tetrad with any level of breeding evidence – possible, probable or confirmed – is shown as having 'breeding' in the appropriate period. Tetrads shaded grey had breeding evidence in both periods (no change). In tetrads with a green triangle pointing upwards, there was breeding evidence in the recent period only (a gain). A red triangle pointing downwards indicates that no breeding evidence was found in the recent period, and the species has apparently disappeared from that square as it was found in the earlier period.

However, less rigorous scrutiny and validation was applied to 1985–90 Atlas records, and new non-breeding codes were introduced for the recent Atlas, so in a few cases a direct comparison of the Atlas maps for the two periods may give a misleading impression. Two specific cases occur several times.

Firstly, some observations in April and early May of scarce breeding species, some of which have more substantial wintering populations, were submitted as possible breeding records, but in 2008–13 they were considered to be late-departing winter visitors, and downgraded during validation. Many similar records were shown as possible breeding in 1985–90, so the breeding distribution change map for

e.g. Water Rail excludes possible breeding records from both periods to provide a realistic comparison.

Secondly, birds seen only in flight, with no immediate evidence that they were breeding in the tetrad, were largely recorded as such during the recent Atlas, but no similar code existed in the earlier one, so many similar observations would have been recorded as possible breeding in 1985–90. Even in the recent Atlas, some birds in flight might have been recorded as possible breeding if the observer made the judgement that there was suitable breeding habitat in the square. To try and remove this inconsistency, on maps of some species, such as Red Kite, Hobby, Grey Heron, Black-headed Gull and Swift, the 'F' and 'H' records have been combined to show their range away from breeding sites in the recent Atlas, and, if a breeding distribution change map has been included, it has taken into account only probable and confirmed breeding.

Conversely, some observations in April of what are believed to have been passage birds were submitted as possible and probable breeding records and accepted as such because they complied with the definitions, and are recorded as such in the table of tetrad occupancy. However, for e.g. Redshank and Ring Ouzel, the sites where these declining rare and scarce species were seen were revisited, and no further evidence of breeding was found. These species are believed to be extinct now, and the special breeding distribution change map in the accounts reflects this.

The species accounts do not include all the available maps. Some species occur everywhere, and the content of the distribution map can be summed up in a single sentence. Conversely, some species are very scarce, and maps with five or less dots have usually been omitted. In these cases the account lists the tetrads with records.

Several species were recorded in the breeding season which were late-returning winter visitors about to depart, or passage migrants, and it would be misleading to produce maps of such occurrences. Relative abundance maps are only reliable for fairly common species, found in some numbers in most tetrads, and only those which show a pattern are published in this book. If the abundance map is similar for both seasons, only one is included. Breeding distribution change maps may reflect different levels of fieldwork effort in the same square for the two different Atlases, especially for the more scarce species in the less-visited tetrads, so maps showing an approximately equal number of up and down triangles may reflect no real measurable change, and have been omitted. Although they have not been included in this book, all Atlas maps, and all those published in the Atlas (1992), can be viewed on the Avifauna website (see p. 73).

Advice has been sought, and followed, from the Rare Breeding Bird Panel (RBBP) on the mapping of rare species. As a result, distribution maps are published at the 10km square level, rather than the tetrad level, for three species which are vulnerable to disturbance or illegal persecution, and whose breeding sites must be kept secret – Goshawk, Merlin and Peregrine. Anyone chancing across breeding sites of any of these species, or any others shown in this book to be rare or scarce, should report the information in confidence to the County Bird Recorder or appropriate Raptor group, but otherwise keep it to themselves.

All cases where special maps have been produced using non-standard criteria are clearly identified in the account, and the criteria are described.

Some gaps in the distribution maps are inevitable in covering such a large county with a relatively low human population. Several species breed all over the County, and there appears to be suitable habitat in every tetrad. The gaps that appear in the maps for these species – Woodpigeon, Wren, Dunnock, Robin, Blackbird, Blue Tit, Magpie, Carrion Crow and Chaffinch – are believed to be due to absence of observers at the right time, rather than the absence of the species themselves. Most other species will be similarly under-recorded to some extent.

Some of the maps must be interpreted with caution, particularly for skulking, nocturnal and crepuscular species, which are likely to be under-recorded. They are based on actual fieldwork results, and other factors may also influence them, as described in Appendix 1.

Comparison Between Recent Atlas Maps and the Atlas (1992)

For many breeding species, the change in distribution over the last 25 years or so has been dramatic. The two Atlases were carried out using very similar methodologies, and involved similar levels of fieldwork effort, so the results are directly comparable (see Appendix 1).

A 'tetrad occupancy table' is included in every account where evidence of probable or confirmed breeding was found in at least one of the two Atlases. The table shows the number of tetrads with possible, probable and confirmed breeding evidence, then the total of the three categories, and the percentage of the 870 tetrads where that level of breeding evidence was found, for the 1985–90 and 2008–13 Atlases respectively. The final two columns show the change between the two periods. The first shows the difference in the number of tetrads where that level of breeding evidence was found. The final column is the percentage change between the two numbers of tetrads, so a species found initially in say 600 tetrads (69%) in the earlier Atlas, but in 300 tetrads (34%) in the later one, has disappeared from 300 tetrads, a 50% decline in the number of tetrads occupied.

There are a few species for which a complete tetrad occupancy table cannot be provided, because either the Atlas (1992) did not include the relevant data, or records of birds in flight and on passage, or late-departing winter visitors, have been treated differently in the two Atlases (see p. 485).

Change in Range or Population

In addition to the comparison between the results of the two Atlases, the distribution of every breeding species found in one or both of them has been compared with that described in the earlier sources referred to above. Any apparent change in the range or population at any time since the early nineteenth century is explicitly discussed. If no such statement is made, there is no significant evidence of any change in status, though of course this may be due to a lack of research or historical data, rather than stability in the distribution.

Overview

A summary of the status of all species, including their breeding status where appropriate, is given in Appendix 6. Changes in the numbers and distribution of the latter are described in the *Breeding Status* chapter.

Non-breeding Species

Many species are not resident here, and some only pass through occasionally. Accounts provide a brief summary of their breeding and wintering areas, and passage routes, where they spent the parts of their lives when they are not here. Most accounts refer to countries or parts of continents, but two important regions are often referred to:

- The Sahel region is a belt of semi-arid country which runs across Africa south of the Sahara, spanning many countries. Variable rainfall, and drought, in the Sahel is implicated in the fluctuating population levels of breeding species such as Sand Martin and Whitethroat.

- The Arctic is circumpolar, and includes Greenland and the most northerly parts of North America, Scandinavia and Russia.

Passage Migrants

The Atlas fieldwork periods, April–July and November–February, ensured that few spring migrants, and virtually no autumn migrants, were recorded. Validation checks ensured that any such records were classed as migrants, and were not included on Atlas maps. These accounts have therefore been compiled from historic and SOS records, rather than Atlas data.

As well as noting the breeding and wintering grounds, and summarising the records, each account provides a commentary on any charts showing annual and/or seasonal occurrence, and indicates whether the migrants generally pass through quickly, or whether some linger. The charts might highlight different sub-species or populations moving through at different times, reflecting their starting point and destination, or different migration routes used in spring and autumn (more birds in one season compared to the other) and the timing of migration, usually dependent on how far north (and west or east) the breeding grounds are.

Data Provided by BTO

BTO staff have been very helpful in extracting local data from national surveys and schemes. The data is for the geographic County, and not the 'BTO Shropshire Region'.

Breeding Bird Survey

The national BBS is the main scheme for monitoring the population change of the UK's common breeding birds. It is run by the BTO for a partnership between BTO, the JNCC and RSPB. It started in 1994, so it spans most of the period between the two breeding Atlases. Observers survey 1km squares on the OS national grid, chosen at random by BTO. They walk two 1km transects, and record all the birds they see or hear. Coverage in Shropshire has been good, and over 50 squares have been surveyed in almost every year from 1997 onwards, except 2001 (when the survey was suspended because access to the countryside was severely limited to help prevent the spread of Foot and Mouth disease in farm livestock).

Statistically valid trends can be calculated for any species that has been found, on average, in 30 or more squares each year. Thirty is considered a reasonable sample size. Similar trends, but slightly less reliable because of the smaller sample size, can be calculated for species found on average in 20 or more squares.

Trends have been calculated by BTO for *The Birds of Shropshire* for the period 1997–2014 for 25 species found in an average of 30 squares or more: Blackbird, Blackcap, Blue Tit, Buzzard, Carrion Crow, Chaffinch, Chiffchaff, Dunnock, Goldfinch, Great Spotted Woodpecker, Great Tit, Greenfinch, House Sparrow, Jackdaw, Magpie, Pheasant, Robin, Rook, Skylark, Song Thrush, Swallow, Willow Warbler, Woodpigeon, Wren and Yellowhammer.

Trends have also been calculated for another nine species found in an average of 20 squares or more: Collared Dove, House Martin, Long-tailed Tit, Mallard, Mistle Thrush, Pied/White Wagtail, Starling, Stock Dove and Whitethroat.

BBS measures population change, not the actual number of birds, so the number counted in 1997, divided by the number of squares surveyed in that year, is the baseline. That number is given an index value of 1, by definition. The number counted in each subsequent year is converted to a relative index, again by dividing the number counted by the number of squares surveyed, and then dividing the result by the 1997 figure. The annual change in the index shows the proportionate change in the population.

A BBS graph in a species account shows the estimated annual fluctuation in the number of birds counted. Some annual fluctuations are considerable, often reflecting the variable effect of preceding good or poor breeding seasons, and hard or mild winters for resident species, or weather conditions on the wintering grounds or migration routes for summer visitors. The success of the current breeding season may have an effect too, as most species are more detectable when there are territories to defend and young to feed. These factors affect different species in different ways. However, it must also be recognised that, as with any other sample survey, the sample may not be representative of the whole. No statistical analysis has been carried out, but the trendline on each graph takes the count from every year into account, so it shows the underlying trend, and is a more reliable indicator of change than year-on-year comparisons.

Some species show a steady increase or decrease over the period. Some others, particularly the smaller resident species, show a similar pattern – an increase up until 2006 or thereabouts, as a result of a run of mild winters allowing a relatively good over-winter survival rate, followed by good weather in the breeding seasons, but a decline or marked fluctuation since then because of a run of wet or cold breeding seasons, followed by inhospitable winters. Other effects are also apparent, for example the impact of disease on Chaffinch and Greenfinch, and the benefit of providing appropriate seeds in gardens for Goldfinches. The factors influencing the fortunes of each species reflected in its BBS graph are more fully described in its individual species account.

BBS complements the Atlas maps. The distribution map shows the range of a species, and the breeding distribution change map shows only if the range has expanded or contracted. It is sometimes possible to infer a population change, but not always. For example, many farmland bird species have declined considerably, but they are still sufficiently numerous to have occurred in the same squares in the recent Atlas period as they did in 1985–90, so the change map shows a similar distribution in both periods. Therefore the BBS results showing the population trends for our more common birds

are a vital supplement to the distribution and relative abundance maps for planning local conservation measures to protect wildlife from the dual pressures of development, and agricultural change and intensification. SOS will continue to encourage members, and other people interested in birds and conservation, to participate.

Most species are not found in sufficient local BBS survey squares for a reliable population trend to be calculated, and in these cases accounts may quote BBS figures for the West Midlands region, which includes Shropshire, or for adjacent Wales, or national figures for England or the UK. The full national BBS report for 2015 includes indices for population change between 1995 and 2014 for these wider areas (Harris *et al.* 2016).

Waterways Breeding Bird Survey

Waterways Breeding Bird Survey (WBBS) is similar to BBS, but only one transect, alongside the river or canal, is walked. It replaced the Waterbird Survey in 2007. Results complement BBS for the species that use this specialist habitat, and provide more statistically valid results nationally. However, there are few WBBS transects here, so no local trends can be produced, but some species accounts refer to results on specific waterways.

Wetland Bird Survey

The first scheme to assess the numbers, distribution and trends of wildfowl wintering in the UK started in 1947. It has evolved and expanded into its current form, when the National Wildfowl Count was integrated with the Birds of Estuaries Enquiry giving rise to the Wetland Bird Survey (WeBS), and the BTO took on responsibility for its management in 1993. Initially it concentrated on the main estuaries, but then included the larger inland waters, and now seeks to cover all waters. A full 'History of WeBS' can be found on the BTO website.

Local coverage started with the Ellesmere group of meres, Shavington and Venus Pool in 1960–61. The Ellesmere group contains six large meres and Shavington three. Due to the large numbers that congregate at the first two of these sites and the long running data set, WeBS trends from them feature prominently in some species accounts. However, the acquisition of Venus Pool by SOS and its subsequent management has improved the habitat quality, and consequently the number of wildfowl using this site, so the WeBS results indicate the effects of reserve management rather than wider population trends. As a result, counts from Venus Pool have not been included.

By 1972–73, seven sites were counted. However only one of these, Oerley Reservoir, has been covered every year since it started in 1966–67. The important Severn–Vyrnwy confluence was added in 1995, and four other stretches of the Severn are also now included. Other sites with over 100 counts are Allscott Sugar Factory pools, Berthpool, Chelmarsh Reservoir, Chetwynd Pool, Dudmaston Pools, Fenemere, Howle Pool, Marton Pool (Baschurch), Middle Pool, Polemere, Sambrook Mill Ponds, Shrewsbury Sewage Farm (treatment works), Trench Pool, Walcot Hall Lakes and Willey Estate (Broseley). WeBS trends from some of these sites have been used when the species is not well represented at the Ellesmere meres or Shavington.

Up until 2014, counts had been submitted for 87 sites, and 116 species had been recorded. However, there are few sites that have had regular coverage from the first count, and, although coverage in 2014 was the best to date, only 51 sites were counted that year (including several different Ellesmere meres and sections of the Severn). Including sites other than the main Ellesmere meres and Shavington in the WeBS charts would therefore introduce fluctuations that are due to variable monitoring, rather than changes in the bird populations.

Garden BirdWatch

Recording for BTO Garden BirdWatch (GBW) began in January 1995. Participants are asked to spend about the same amount of time recording birds in the same area (usually their whole garden) each week, and record the highest number of each species seen together at any one time during the week. Further details can be found on the BTO website.

GBW has been well supported, with up to just over 30 gardens included in 1995, 100 for the first time in 2000, and more than 100 in over 400 weeks from 2000 onwards.

The basic results are presented as a 'reporting rate', a measure of the proportion of gardens where a particular species was observed during a particular week. They show the periods each year when the species makes use of the garden, highlighting abundance or shortage of food in the wider countryside, influxes in hard weather, and the underlying population trend. The results complement those of BBS.

BTO has supplied GBW graphs showing the reporting rate for each of the 52 or 53 weeks each year for several species, and examples are included in the Greenfinch, Goldfinch and Siskin accounts, demonstrating the importance of gardens for those species, and population change since 1995. The other graphs and data supplied by BTO are on the Avifauna website (see p. 73).

Reference is made to GBW data in several other accounts, and the rank of the reporting rate for the top 36 species is shown in Appendix 6.

Ringing Recoveries

Bird ringing activities are described on p. 10. All records of birds when ringed and recovered are sent to BTO, and summaries are published in an annual Online Ringing Report on the BTO website.

Most records in the species accounts have been taken from the 2014 report (Robinson *et al.* 2015). However, until all the records were computerised, and ringers could submit their records online, the report did not record retraps (birds caught alive) or resightings (colour-rings read in the field) at, or within 5km of, the original ringing site. Thus many of the local longevity records could not be found on the BTO website, including the colour-ringed Ring Ouzel which apparently held the national longevity record for six years from 2003 to 2009. Such records are only available if they have been published in SBR, or made available for the species account by the individual ringer.

The totals in the 2014 ringing report apparently included the retraps and resightings that had been excluded from earlier reports, and showed large increases for many species in the numbers ringed and recovered in Shropshire, compared with the 2013 report. It appears that the BTO subsequently reverted to the earlier definition, because total numbers in the 2015 Online Ringing Report are slightly above the 2013 totals, but well below the 2014 totals. This means that no reliable numbers are available for 2014, the cut-off date for the species accounts. The accounts therefore generalise about the proportions of

recoveries here and elsewhere, which unfortunately limits the value of the ringing recovery information.

The Shropshire part of the Online Ringing Reports for 2013, 2014 and 2015 are posted on the Avifauna website (see p. 73).

Population Estimates

An estimate of the number of breeding pairs of each species was published in the *Atlas* (1992), but it must be stressed that no attempt was made to establish it during the fieldwork. In many cases the number of breeding pairs per tetrad was estimated, often using national data included in the BTO Atlas (1976), or in *Population Trends in British Breeding Birds* (Marchant *et al.* 1990), or from local results from national BTO surveys, or impressions gained from Atlas fieldwork. This was multiplied by the number of tetrads in which the species was recorded at the appropriate level. For many species only tetrads with probable or confirmed breeding were taken into account, particularly if territories are easy to locate but breeding is difficult to confirm, and possible breeding records might refer to migrants. For other more elusive species all records were likely to relate to breeding pairs, and the estimate may also have been adjusted to take into account whether the particular species was likely to have been under- or over-recorded, for example because it is very elusive or nocturnal, or each pair has a large territory likely to cover several tetrads. Often there was little information on regional variation in abundance patterns, so some of these estimates were very crude.

The national BTO Atlas (1993) included relative abundance maps for the first time, but these were published after the *Atlas*. Data from the Common Bird Census (CBC) informed the *Population Trends* book, but results were known to be biased, as observers picked the survey plots, usually in the best habitats, and it covered mainly farmland and woodland, so BTO replaced it with BBS, surveying randomly selected 1km squares, in 1994.

As a result, better estimates are available now, but they still contain much uncertainty. The Avian Population Estimates Panel produced a national estimate for most species, published in *British Birds* in February 2013 (Musgrove *et al.* 2013). Different methods and baselines were used to produce these estimates, and each estimate was classified as of good, moderate or poor reliability. BBS only contributed to these estimates by applying population change trends to the original estimates, so for many species the 2013 figures still incorporated earlier inaccuracies. Recording the detection method, added to BBS in 2014, is designed to produce more accurate population estimates in the future.

BTO has supplied an estimate of the Shropshire proportion of this national figure, based on analysis of national Atlas TTV counts across the country between 2008 and 2011. The resulting County estimate is often quoted in species accounts, using the formulation 'The population estimate, based on TTV counts, is ...'. However, even the most accurate assessment of the Shropshire proportion of an unreliable estimate will produce an unreliable result.

There are potential shortcomings in the TTV data itself used to calculate the proportion of the total population in this County, as the proportion of tetrads in each 10km square with TTV counts varied considerably across the country, depending particularly on whether or not the square was included in a County tetrad Atlas project, and the estimates for summer visitors, which may not have arrived when a high proportion of the early TTVs were carried out, are potentially even less reliable. While theoretically these factors should apply equally across the whole country, and have a negligible effect on these calculations, in practice this is unlikely.

For example, the national estimate for Rook is of moderate reliability, and the estimate based on TTV counts is 16,100–16,750. All Rookeries were counted in 2008, giving an estimate of 20,950 pairs, considerably higher (by 28%) than the estimate based on TTV counts. Therefore, wherever possible, a population estimate has also been made using local data. In some accounts the two estimates are both quoted, with appropriate commentary.

For species that are not widespread and conspicuous, or which have only small populations, the numbers counted on TTVs were small, and the estimate based on TTV counts is unreliable. A threshold has been applied: the population estimate based on TTV counts should be at least 1,000 pairs, with counts from TTVs in 10% of tetrads, to be considered reasonably reliable. However, the table in Appendix 6 includes the calculated estimate for all species for which BTO provided a Shropshire proportion of the national estimate.

Population Change and Agricultural Intensification

If the account includes evidence of population change, it will so far as possible explain it, based on available research and scientific evidence. Reference has often been made to habitat change due to one or more aspects of 'agricultural intensification'. Rather than describe this in detail in each account, a cross-reference is provided to the relevant section of the *Habitats* chapter, which describes the many aspects of agricultural change that have occurred, particularly since the Second World War, and their respective impacts on the populations of different species.

References

To minimise interruptions to the flow of the narrative, in most accounts the quoting of references has been kept to the minimum necessary. Therefore, material from the *Handlist* (1964), *An Atlas of the Breeding Birds of Shropshire* (1992), the annual Shropshire Bird Reports from 1956 (SBRs), the national BTO Atlases, 'Bird Facts' and Ringing Data on the BTO website, and records on the *Histo* website (including various publications and the works of Beckwith and Forrest, but particularly the annual reports produced by CSVFC from 1892 onwards, the 'Record of Bare Facts' and the 'Transactions') has largely been quoted or reproduced without reference to the source. In the case of SBR and CSVFC records, this only applies if the record appears in the publication for the same year, so if, for example, an SBR article which summarises survey work and records over several years has been cited in a species account, it has been referenced as (SBR year).

A number of publications or activities are referred to in many accounts, using an abbreviated form. The abbreviations, and the full title or name of the publication or activity, are listed in Table 3. Some are described in some detail earlier in this chapter.

Abbreviations for the names of organisations are listed in the *Preface*.

All these publications include extensive references to original research and, in the main, only references not listed there are included

Table 3. Abbreviations Used in the Species Accounts

Standard Abbreviation Used in Text	Publication or Activity Referred To
SOS Publications	
the *Handlist* (1964)	*A Handlist of the Birds of Shropshire* (1964)
the *Atlas* (1992)	*An Atlas of the Breeding Birds of Shropshire* (1992)
Atlas fieldwork 1985–90	fieldwork resulting in the map in the *Atlas* (1992)
SBR (year)	the *Shropshire County Bird Report* for that year
National Atlases	
national *Historical Atlas* (1996)	*The Historical Atlas of Breeding Birds in Britain and Ireland 1875–1900* (Holloway 1996)
national BTO Atlas (1976)	*Atlas of Breeding Birds in Britain and Ireland* (Sharrock 1976)
national BTO Winter Atlas (1986)	*An Atlas of the Wintering Birds of Britain and Ireland* (Lack 1986)
national BTO Atlas (1993)	*The New Atlas of Breeding Birds in Britain and Ireland* (Gibbons et al. 1993)
national BTO Migration Atlas	*The Migration Atlas: movements of the birds of Britain and Ireland* (Wernham et al. 2002)
national BTO Atlas (2013)	*Bird Atlas 2007–11: The Breeding and Wintering Birds of Britain and Ireland* (Balmer et al. 2013)

Standard Abbreviation Used in Text	Publication or Activity Referred To
Atlas Fieldwork 2007–13	
recent Atlas fieldwork	fieldwork in one or other, or both (according to context), of the two periods referred to on the next two rows
breeding season fieldwork (2008–13)	breeding season Atlas fieldwork in Shropshire 2008–13
winter fieldwork (2007–13)	winter Atlas fieldwork in Shropshire 2007–13
TTV	Timed Tetrad Visit, a two-hour visit to a tetrad to count all birds seen or heard. Four visits were made to each tetrad, two in the breeding season (April or May, and June or July) and two in Winter (November or December, and January or February). If a visit was curtailed to one hour, the count was adjusted to create a standardised four-hour count for the appropriate season
Atlas Maps	
Atlas map 2007–13 or Atlas map 2008–13	maps showing results of fieldwork described in the previous section
recent Atlas map(s)	one or other, or both (according to context), of the two maps referred to on the previous row
breeding distribution map (2008–13)	tetrads with confirmed, probable and possible breeding, the results of fieldwork 2008–13 and records sent to the County Bird Recorder in the same period

(ctd.)

Table 3 (ctd.)

Standard Abbreviation Used in Text	Publication or Activity Referred To
breeding season relative abundance map (2008–13)	The result of the breeding season TTVs, with tetrads shaded to show three density bands calculated from the counts, with the darkest colour representing the highest density (the 'hotspots'). (Tetrads where the species was found, but not during a TTV, are shaded in the palest colour)
breeding distribution change map	distribution change between the 1985–90 and 2008–13 breeding season fieldwork (all levels of breeding evidence included)
winter distribution map (2007–13)	tetrads where the species was recorded in the four winter months (November–February) between November 2007 and February 2013
winter relative abundance map (2007–13)	the result of the winter season TTVs, with tetrads shaded to show three density bands calculated from the counts, with the darkest colour representing the highest density (the 'hotspots'). (Tetrads where the species was found, but not during a TTV, are shaded in the palest colour)
Tetrad Occupancy Tables	
tetrad occupancy table (breeding season)	table comparing the results of 1985–90 and 2008–13 Atlas fieldwork for a particular species, in terms of number and proportion of tetrads occupied
tetrad occupancy table (winter season)	the number of tetrads where the species was recorded at least once in the four winter months (November–February) between November 2007 and February 2013
National Surveys and Censuses	
BBS	Breeding Bird Survey, organised annually by BTO since 1994 to monitor population changes of common species
CBC	The Common Birds Census (CBC) ran from 1962 to 2000 and was the first of the BTO's schemes for monitoring population trends among widespread breeding birds. It has now been superseded for this purpose by BBS
GBW	Garden BirdWatch survey, organised by BTO to monitor monthly changes in garden use since January 1995
WeBS	The Wetland Bird Survey, organised annually by BTO since 1960 to assess the numbers, distribution and trends of wildfowl wintering in the UK
Local Surveys and Censuses	
BBP	The Long Mynd Breeding Bird Project, which carried out surveys in 1994–98 and 2006–09
British Birds	
BB	*British Birds* is a monthly journal for all keen bird-watchers which publishes articles on a wide variety of topics, including behaviour, conservation, distribution, identification, status and taxonomy. Contributors include both professional and amateur ornithologists. It is regarded as THE journal of record in Britain, and publishes the annual reports of the Rarities Committee and the Rare Breeding Birds Panel

Standard Abbreviation Used in Text	Publication or Activity Referred To
BBRC	British Birds Rarities Committee is the official adjudicator of rare bird records in Britain. It publishes its annual report in the monthly journal *British Birds* (BB). Records for the relevant species are only accepted in Shropshire if they have been accepted by BBRC
RBBP	Rare Breeding Birds Panel is an independent body of experts that monitors rare species in the UK, and supports their conservation and study. Its annual report is published in BB. It has provided advice on the mapping of rare species by national and local Atlases, which has been followed in this Avifauna
Charts and Histograms	
BBS graph	BBS results for Shropshire 1997–2014, showing the population trend for 34 common species
Birds/year 1950 (or 1992)–2014 or Birds/winter period	histogram showing the estimated number of different birds present in each year, or each winter period, of occurrence
chart of birds present in ten-day intervals (1950–2014) or (1992–2014)	histogram showing the estimated number of birds present on the first calendar date of all records, in ten-day bands, for the period stated
WeBS chart	WeBS results for Shropshire, usually from 1960–61 onwards, showing the maximum count for the species in each winter period at selected sites
UK Conservation Status	
Red list and Amber list	All regularly occurring species are allocated to Red, Amber and Green lists in the *Birds of Conservation Concern* in the UK, mainly according to their rate of decline in the last 25 years, or since the first list was produced in 1969 (Red list species have declined most). Revisions to the lists were made in 2002, 2009 and, most recently, in 2015 (Eaton et al. 2015)
International Conservation Status	
IUCN	International Union for Conservation of Nature is the global authority on the status of the natural world. Species of conservation concern are placed on the Red list, with classifications that include 'threatened' (vulnerable, endangered or critically endangered) or 'near threatened'
Websites	
Avifauna website	Part of the SOS website www.shropshirebirds.com showing additional information not published in this book, including all Atlas maps. It can be found via the website homepage under the menu item 'Avifauna Supplement'.

in the species accounts in this Shropshire Avifauna. Anyone wishing to study any species further, or verify the statements made, should refer to these publications for the relevant source.

All references in the accounts are listed in full in Appendix 4. In a few cases, references in the accounts are referred to as *in prep*. In these cases, the work has not been completed ready for publication, so there can be no listing in Appendix 4, but readers may wish to look out for (or enquire about) these future publications.

Comments attributed to these publications and references usually apply to Britain as a whole, or named research areas. While they may not explicitly apply to Shropshire, they are only included if it is believed that they probably do so.

SOS Avifauna Website

The 'Avifauna website' has been referred to several times in this chapter in relation to material used in the production of this publication, but which is not included in it. The Avifauna website has been created to publish all this material, and is available to all readers.

All maps published in the Atlas (1992), and five different types of standard map compiled from records collected during recent Atlas fieldwork, have been prepared. None of the former, and only 218 of the latter (less than one-quarter of the total) have been published in this Avifauna.

Records from early authors, and from the *Record of Bare Facts* and the *Transactions* of the Caradoc and Severn Valley Field Club, which have not been specifically referenced, have been published on the *Histo* website. An index to *Histo*, in the form of an excel spreadsheet with all records listed in species and then in chronological order, which can be pasted into the search facility, has been very helpful in searching *Histo*.

Data supplied by BTO has been used to calculate the BBS charts.

Ringing Recoveries have been taken from the Online Ringing Report 2014 on the BTO website. This report is updated annually, so the totals for each species change, and the size of the database means that only the more interesting records can be published. To make space for newer, more interesting records, older ones are often removed. It is therefore no longer possible to find many of the records and figures quoted in the species accounts on the BTO website. The 2014 report has therefore been saved. In view of the problems with the 2014 report, the Online Ringing Reports for 2013 and 2015 have also been posted on the Avifauna website.

The Government Agricultural Statistics have been used to create the graphs in the *Habitats* chapter.

Some of the material drafted for the species accounts was too detailed for inclusion, and this too has been published.

TTV counts have been used to calculate the numbers of birds in each band on the relative abundance maps for each species.

All the Atlas maps, the index to *Histo* as the website stood at April 2015, when authors were consulting it for their accounts, the BBS data and BTO Online Ringing Reports for Shropshire 2013–15, other material referred to in other chapters of *The Birds of Shropshire*, and additional detail for some species, can all be found on the SOS website www.shropshirebirds.com under menu item 'Avifauna Supplement'.

Conclusion

The accounts have been compiled from all sources with records, and fieldwork in all of the 870 tetrads was carried out for both the 1985–90 and 2007–13 Atlases. They, together with the maps, give an excellent indication of the status and distribution of *The Birds of Shropshire*.

Brent Goose (Brant Goose)
Branta bernicla
Very rare winter visitor

Jim Almond, Venus Pool, 16 April 2011

The dark-bellied form which occurs locally derives from the arctic Russian breeding population.

Described by early authors as very rare, there are just eight historic records. The first was of one killed in January 1861 near Shrewsbury, and apart from 'one killed out of a small flock on the hills near Church Stretton' in October 1883 and 'one shot out of four on stubble near Shrewsbury' in February 1917, the other records appear to be of single birds.

Early SBRs up until 1968 included sightings from Leighton Flats and the Camlad Valley. Although the county border follows the River Camlad in places, Leighton Flats are several kilometres further west. At that time many White-fronted Geese wintered in this area and other species were often identified among the flocks, including Brent Geese on four occasions. A long-staying single at Leighton Flats from 31 January to 2 March 1958 was followed by one at Hem Farm

in February 1961, in March 1964 'some' were with White-fronts at Chirbury, and finally four adults and an immature were at Leighton Flats in January 1968. It is not known with certainty whether any of them actually came into Shropshire, but those near Chirbury probably did. These records have been included to give an indication of the species' status in the area at that time. The attraction of these sites was diminished by agricultural improvements, which resulted in the loss of the White-fronts and no further Brent Geese were reported from this area until the 1990s, when occasional individuals were seen at the then-new Montgomeryshire Wildlife Trust reserve at Dolydd Hafren.

Apart from those, the recent rarities review accepted records of 23 modern sightings (concerning 25 individuals), but six of these records (concerning eight individuals) reported them as probably 'escapes'. Judging the origin of wildfowl is always difficult, but the recent increase in records casts possible doubt on this assumption. Therefore all these sightings are included in the totals to provide a clear picture of recent trends.

Just four singles were seen prior to 1990; at Fenemere in November 1965 and in January 1973, Newton Mere in March 1976 and Leighton, near Buildwas, in January 1986. Two pairs and a singleton in 1990 and two singles in 1992 were originally considered to be escapes, but they were also the first of a series of more regular sightings as the national population increased. Since then there have been a further 14 records, all of singles, in 11 different years, which included only one believed to be an escape or feral bird, at Wood Lane in August 2005. Excluding the latter, and June records from 1990 and 1992, the above records show two clear peaks in occurrence, in January (six) and April (seven), with all the remainder occurring in the autumn passage or early winter periods. National populations are at their highest in February, indicating continued arrival of wild winter visitors from September onwards, and they drop off sharply in April as flocks depart to their Russian breeding grounds. The records, and the peaks in January and April, fit well with this pattern, but there may be some doubt on the origin of a few of them. This issue is an ever-present dilemma with wildfowl.

There has been a single record from most of the regularly watched wetlands, but four from the Ellesmere area, and seven (eight individuals) from Venus Pool and Cound Fishery. All the accepted records have been of the dark-bellied sub-species *B.b. bernicla*.

During the recent Atlas period, a single was seen at Venus Pool and the adjacent fishing pool in January, April and September 2011, another was at Priorslee Lake in April 2012 and one was seen at two different locations near Ponthen in January 2013. Some of these records probably refer to the same individual.

The national population has risen steadily since the early 1970s, reaching a peak in 1993–94 before stabilising at a slightly lower level. Formerly occurring mainly on the south and east coasts of England, they have now expanded their range and they winter annually both in south-west and north-west Wales. As a result of the increase in both numbers and range more sightings are expected in the future.

Allan Dawes

Canada Goose
Branta canadensis
Uncommon naturalised resident

Jonathan Cartwright, Venus Pool, February 2011

Canada Goose is native to North America. It was first introduced to the UK and parts of Europe in the seventeenth century, although its colonisation was subsequently consolidated by further releases.

Rocke (1866) wrote 'This bird has been for so many years a constant occupant of our various ornamental meres and ponds ... Instances of its capture ... have of course been very numerous'. Beckwith (1886a) attributed their source to young from these ornamental lakes that had not been pinioned. By the end of the nineteenth century, breeding had been taking place at many of these lakes for some time. In addition to recent escapes, Forrest (1899) noted 'a certain number of these Geese constantly resort to Hawkstone, Combermere [Cheshire] and Ellesmere, and there are generally seven or eight nests on the mere at the place named last'.

The feral population was slow to become established and it was not until 1951 that a substantial flock of 99 was recorded at the Mere (Ellesmere). Just five years later, between 40 and 200 were said to be there regularly, and breeding occurred frequently at many lakes and pools in northern and central areas.

In addition to the increasing numbers commented upon in many early SBRs, they were also expanding their range. Breeding in the Clun area was noted in 1969, where they were virtually unheard of only 10 years previously. In 1968, a count of moulting birds was made in July and this was repeated in 1974. A comparison of sites covered in both periods showed an increase of 29.95% and in total 1,439 were

found in the later year. Decreases were noted at a few places, and at one of them egg pricking was being used as a control measure.

In 1979 the first flock to number 1,000 was counted at Shavington, in October, and 1,600 were found there in September two years later, the largest flock recorded by then. Between 1983 and 1992, the River Severn at Leighton was one of the more regular wintering sites, and around 1,000 were present in most years.

During the breeding season Canada Geese are widely distributed, having increased their range considerably since 1985–90, mainly in the south-west. Altitude is no barrier, with goslings fledged on Wildmoor Pool (425m) in the recent Atlas period, but a shortage of pools in the south-west is a limiting factor. They have been lost from over 100 tetrads in the north and east, which is difficult to explain; habitat loss and local persecution may account for some losses, but not on the scale observed. For every loss, however, there have been two gains.

Tetrads with a large water body, or a number of small waters, are likely to hold several breeding pairs, but many others will only have a single pair. Taking an average of two to two-and-a-half pairs per tetrad gives an estimated breeding population of 800–1,000 pairs. This figure is consistent with the estimate of 900–1,000 pairs based on TTV counts.

Wood Lane has become a regular moulting refuge since 2000, and around 1,000 have been counted there in late summer in subsequent years, with a maximum of 1,500 in August 2004. Prior to this, many

Occupied Tetrads

Atlas period (breeding)	1985–90		2008–13		Change	
	Number	%	Number	%	Number	%
Confirmed	195	22	204	23	9	5
Probable	78	9	120	14	42	54
Possible	33	4	85	10	52	158
TOTAL	306	35	409	47	103	34
Tetrads with Winter Records (2007–13): 274 (31%)						

Breeding Distribution Change (1985–90 to 2008–13)

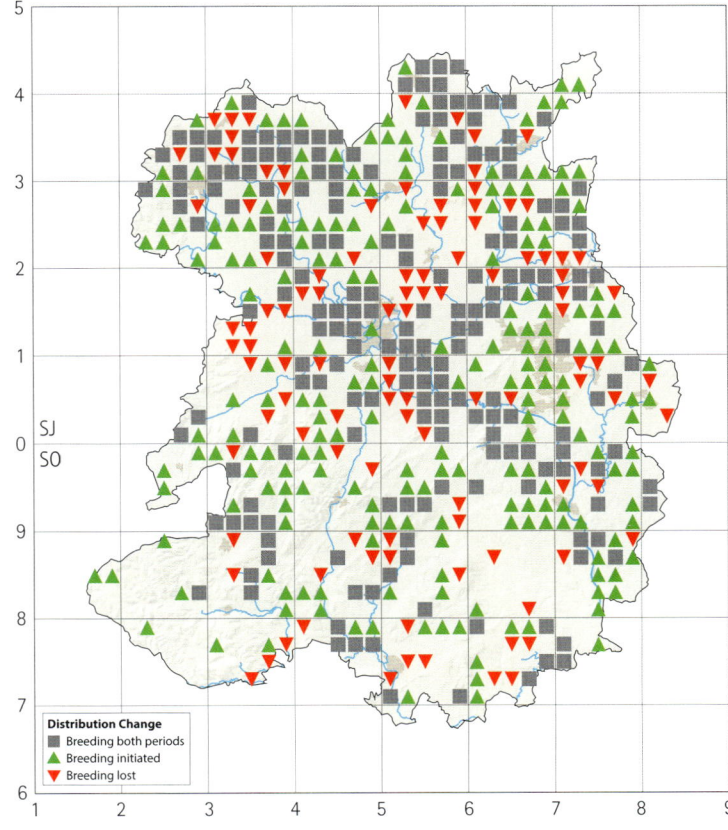

Distribution Change
- Breeding both periods
- Breeding initiated
- Breeding lost

Winter Relative Abundance (2007–13)

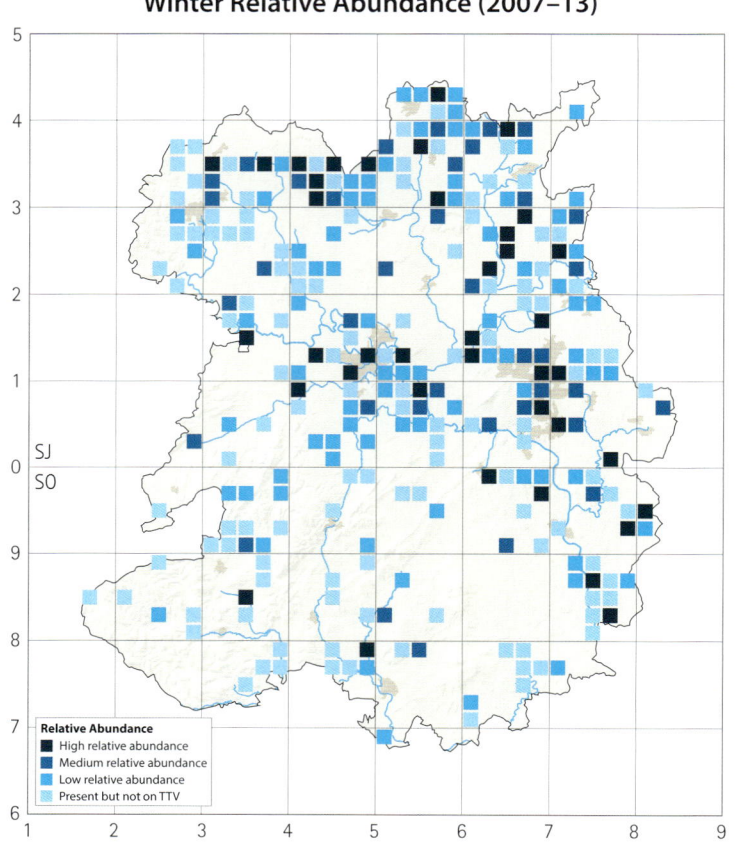

Relative Abundance
- High relative abundance
- Medium relative abundance
- Low relative abundance
- Present but not on TTV

travelled to the Beauly or Moray Firths in north-east Scotland, to moult on the estuary in late summer. There have been nine ringing recoveries involving this moulting area. These are the longest movements recorded by ringing, but most reflect the sedentary nature of this goose, with 73% of all 331 recoveries from this or adjacent counties.

The ring from a nestling at Worfield in 1984 was found less than four miles away just over 19 years later, but there were no other remains. The national longevity record is nearly 32 years so this was a mere youngster.

In winter, the range is more restricted when small water bodies used for breeding are abandoned and flocks gather on larger waters. Non-breeding and failed breeding birds summer on these larger waters too, so the abundance maps for both seasons highlight areas such as the meres at Ellesmere, those north of Whitchurch, Allscott Sugar Factory, and the lakes around Telford, where summer flocks of up to 200 can reach 1,000 during the winter.

A winter population estimated at between 2,650 and 3,000 individuals, based on TTV counts, seems reasonable when compared to numbers counted by WeBS, bearing in mind that flocks are mobile and form part of a larger Midlands population. This tendency to wander widely, both within and between winters, means that numbers at any one site can vary enormously, making it difficult to assess local trends. The national steady increase which has continued since monitoring began seems to have levelled off, particularly in the core areas. WeBS counts suggest that this is the case locally too, and there are probably few suitable breeding pools left unoccupied.

Allan Dawes

Barnacle Goose
Branta leucopsis
Scarce naturalised resident, has bred, possible vagrant

Jim Almond, Polemere, 21 October 2010

The natural populations breed mainly in arctic Greenland and Svalbard. These two populations remain separate in winter, the former mainly in Ireland and western Scotland, and the latter on the Solway Firth and the east coast.

Rocke (1866) began the tale of the then 'Barnicle' Goose, stating that it was 'very rare, but occasionally met with in winter', without giving further evidence, a statement reiterated by Beckwith (1879) but changed to it 'now appears not to visit us' (Beckwith 1886a). Forrest (1908) summed up the whole situation with 'no recent occurrence is known'.

It now occurs as a naturalised escape with no evidence of truly wild birds from natural populations. Records have increased from a handful each year in the 1960s and 1970s, from eight different sites by the end of 1982, up to around 10 or 12 since 1983, since when the number of sites monitored by WeBS has increased, and these geese have now been counted at 28 sites, with annual fluctuations in the number of records and maximum counts. The extraordinary count of 52 at Blake Mere, Whitchurch, in January 2010 is probably accounted for by a wandering flock from elsewhere.

Out of a total of 221 records since 1961, only nine have been of 10 birds or more, from Blake Mere (52 in January 2010), Chelmarsh Reservoir and Cole Mere (17, in December 1992 and October 2006

Occupied Tetrads

Atlas period (breeding)	1985–90		2008–13		Change	
	Number	%	Number	%	Number	%
Confirmed	1	0	0	0	-1	-100
Probable	2	0	0	0	-2	-100
Possible	2	0	0	0	-2	-100
TOTAL	**5**	**1**	**0**	**0**	**-5**	**-100**
Tetrads with Winter Records (2007–13): 14 (2%)						

Winter Distribution (2007–13)

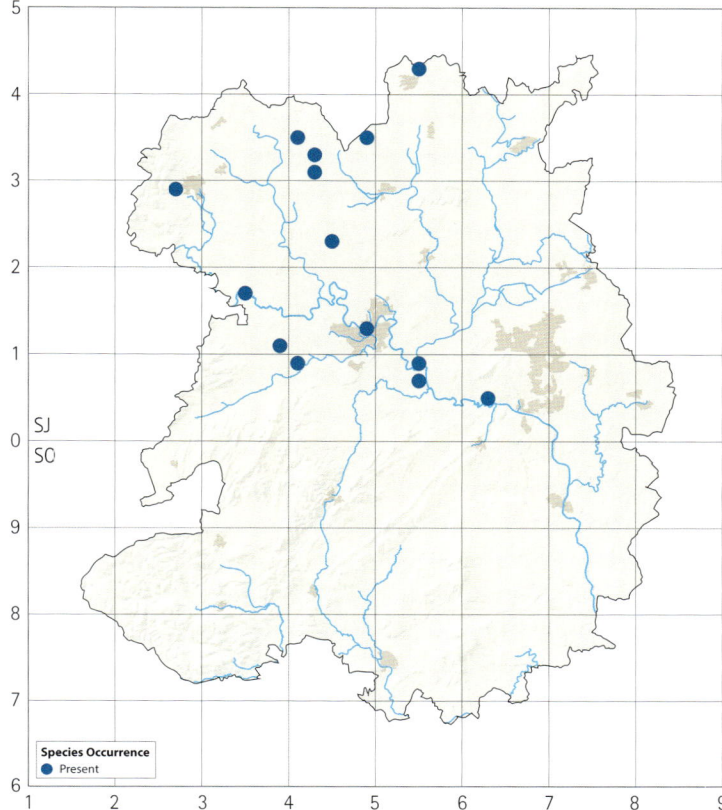

respectively), Dudmaston Pools (three records of 13–14, all in the first three months of 1993, and probably the same flock as those at Chelmarsh in the previous month), Venus Pool (16 in March 1978, and 11 in January 1979) and the Ellesmere meres (10 in March 1976). Only two of these double figure counts have occurred since 1993.

The waters where Barnacle Geese have been found during WeBS counts are almost exclusively in the north. They have been found at only four sites in the south, Walcot Hall Lakes, on the Teme between Ludlow and Knighton, Chelmarsh Reservoir and Dudmaston Pools, but not at any of them since 1993.

The recent winter Atlas map shows records from 14 tetrads, all near the Severn or north of it, and west of Telford.

There was a total lack of breeding evidence. Thus the small breeding population found in 1985–90 (five tetrads with breeding evidence: one confirmed, and two each of probable and possible) has not become established, and no SBRs since 1990 contain any suggestion of breeding. Thus two young raised at Cranmere Bog, Worfield, in 1985 is the only confirmed breeding record.

Both of the wild populations, and the naturalised one, have increased substantially in recent years, and a flock from Greenland began wintering on the Dyfi Estuary in mid Wales in the 1990s. This is the nearest regular wintering site, and the flock now numbers over 300. Although there is little habitat here for winter visitors, the increases elsewhere may result in occasional visits from flocks of the Greenland race, which may perhaps account for the two double figure flocks since 1993. There may be more such visitors in the future, but distinguishing them from naturalised individuals will be difficult.

John Tucker

Greylag Goose
Anser anser
Uncommon naturalised resident

The nominate race *anser* occurs in north-west Europe from Iceland to the Baltic and south to Britain.

The only native breeding goose, it formerly bred in many parts of the UK. Hunting and land drainage caused widespread decline and it was largely restricted to north-west Scotland by the end of the nineteenth century. Early local commentators considered this the rarest of the geese, and only four records were documented before 1950. The first was a single at Bromfield in 1885, and 32 wintered at this site after arriving in December 1890. One of a party of five was shot near the River Severn at Bridgnorth in February 1929 and 35 flew north over Bridgnorth in December 1939.

One Greylag among White-fronted Geese on the Camlad in February 1957 was the first for almost 20 years and heralded a change in status. From 1961 to 1980 mainly singles or small groups, up to a maximum of 10, were to be seen annually and breeding was noted just over the Cheshire border at Bickley in 1969. At this time they were being released in various parts of the UK by the Wildfowlers Association of Great Britain and Ireland. Local releases started in 1981 at Nib Heath, where in June a count of 46 included young of the year, after which numbers rose steadily. By the end of the 1980s over 100 were to be found at Nib Heath in most years, and they probably fuelled the local population growth, with Greylags being noted more frequently at other locations. By 1990 the breeding population was estimated at 50–75 pairs.

This trend has continued and is clearly shown by the tetrad occupancy table and the breeding distribution change map.

Breeding is easily confirmed at large wetlands where several pairs may be present and goslings can be seen accompanying their parents. At smaller sites some breeding attempts that fail in the early stages may go unnoticed.

Dawn Micklewright, Venus Pool, 16 March 2014

Occupied Tetrads

Atlas period (breeding)	1985–90		2008–13		Change	
	Number	%	Number	%	Number	%
Confirmed	11	1	48	6	37	336
Probable	14	2	78	9	64	457
Possible	16	2	40	5	24	150
TOTAL	41	5	166	19	125	305
Tetrads with Winter Records (2007–13): 116 (13%)						

As would be expected for a resident species, there has also been a corresponding increase during the winter period; and the number of occupied 10km squares has doubled since the national BTO Winter Atlas (1986). There is little difference between the breeding and winter distribution maps, but most small pools used for breeding are abandoned. Winter flocks will take advantage of temporary flooding, and can also be found feeding in wet fields but usually within a short flight of water. Winter distribution is concentrated in the Severn Valley and the areas to the north which correspond with the main wetland

Breeding Distribution Change (1985–90 to 2008–13)

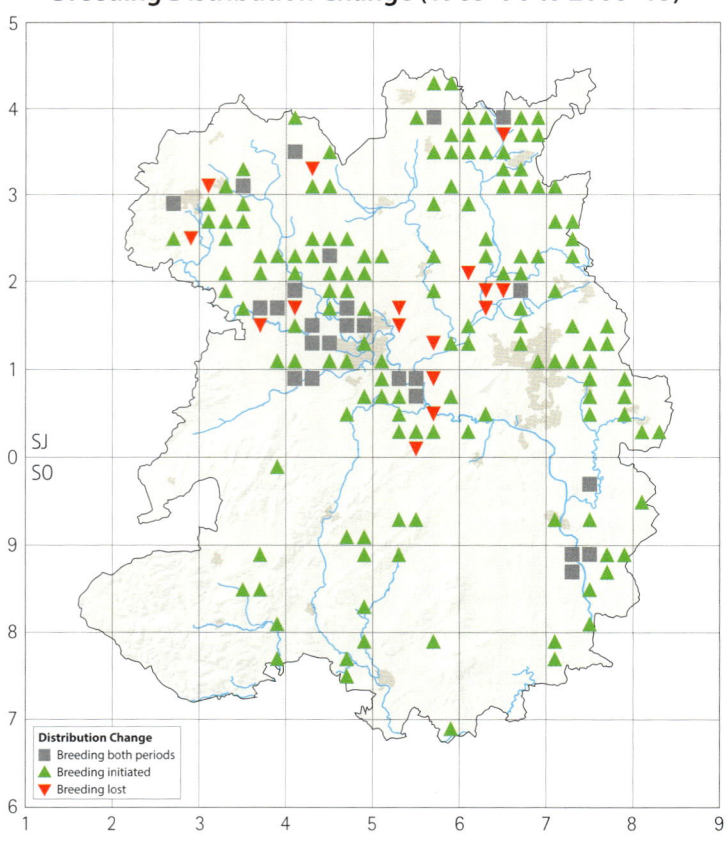

Distribution Change
- ■ Breeding both periods
- ▲ Breeding initiated
- ▼ Breeding lost

Winter Relative Abundance (2007–13)

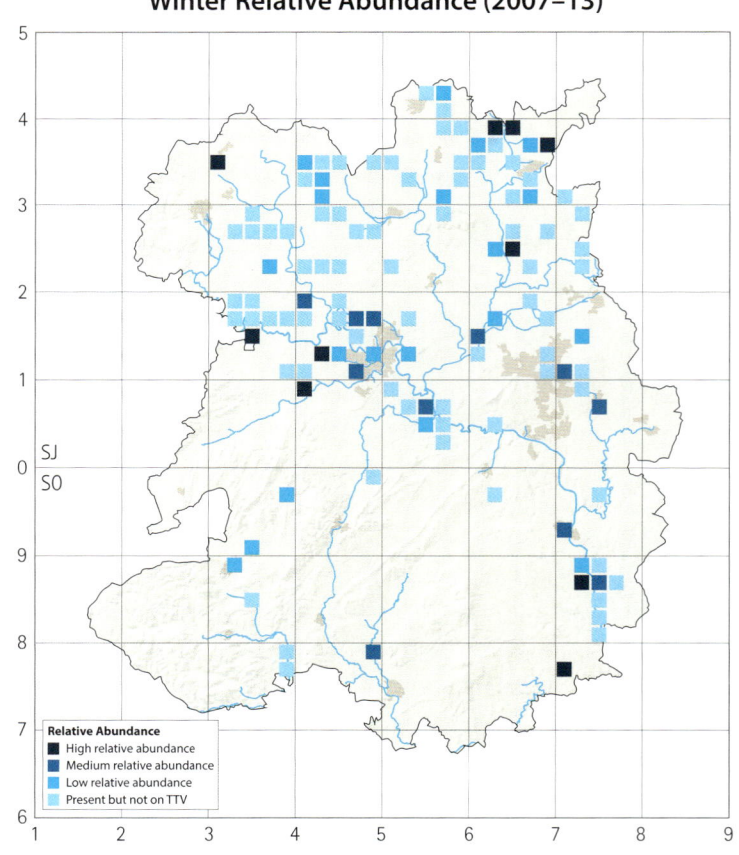

Relative Abundance
- ■ High relative abundance
- ■ Medium relative abundance
- ■ Low relative abundance
- ■ Present but not on TTV

habitats. The largest counts during recent Atlas fieldwork, all of over 200, were at Polemere (three counts of over 200, maximum 250 in December 2007), Cound Lane (246 in January 2012) and Venus Pool (208 in January 2013).

Moulting flocks have become an annual event, starting in the late summer of 1999, when 52 geese remained at Venus Pool, with numbers rapidly rising to a peak of 435 in 2006, before falling back and levelling out at about 300. Another moulting flock has become established at Wood Lane, beginning with 80 in 2003 and increasing to about 300 each year. Large flocks are occasionally reported elsewhere at this time but there do not appear to be any other regular moulting sites.

The breeding population is estimated at 180–220 pairs, based on 166 occupied tetrads, with some of the larger waters having several pairs. Moulting flocks at Venus Pool and Wood Lane and those present elsewhere during this period would suggest a total of between 700 and 800 individuals at the end of the breeding season. This figure is slightly higher than would be expected from the estimated breeding population but may include some attracted to safe moulting areas from further afield.

National ringing data shows that re-established Greylag Geese are largely sedentary, although odd individuals make longer movements.

There are only six recoveries, three of which were short distance movements to or from Cheshire and the furthest was a colour-ringed individual from North Yorkshire re-sighted at Venus Pool in August 2006. Two ringed at Slimbridge in September 1989 were found the following year, one shot in March, the other identified from a colour ring at Ellesmere in June. This timing could indicate a moult migration but it may be a coincidence, a reflection of the volume of ringing carried out at Slimbridge.

Because of the size and extent of the current population it would now be impossible to detect migrants from Greenland or from the native population in the north-west of Scotland unless they were ringed. The small flocks in the historic period and an unusual record of 300 flying south-west over Waters Upton in January 1992, when local numbers were still small, would suggest that they may occur.

Both the timing and scale of the local increase mirrors that of the national re-established population which shows no sign of levelling off. The BBS trend in England shows an increase of 283% since 1995. Comparison of the distribution map with that of Canada Goose suggests there is scope for further growth.

Allan Dawes

Pink-footed Goose

Anser brachyrhynchus

Scarce winter visitor

There are two distinct populations of Pink-footed Geese. Those from Iceland and east Greenland migrate to the British Isles, while those from Svalbard winter mainly in the Netherlands, Denmark and Belgium.

A specimen was taken from the River Tern near Hodnet in 1842. Two (one shot) at Eyton in 1879, and another shot near Oswestry (at 'Bryntanat') in 1894 or 1895, were the only other nineteenth century records (Forrest 1899).

There was one further occurrence before 1950, when eight visited Walcot Hall Lake in the last week of February 1942.

Pink-footed geese remained very scarce, with only a few records, mainly of singles, received up to 1979, and the national BTO Winter Atlas (1986) showed them as present in only four of the County's 10km squares. After that date there was a marked change which coincided with a four-fold increase in the national population. Flocks from Iceland and eastern Greenland usually make landfall in north-east Scotland before moving south, to either Lancashire or East Anglia. There is interchange between these sites during the winter, and skeins moving between them are responsible for the majority of sightings. In general their number and size has continued to increase in line with the national trend, but with two notable large occurrences. Firstly, on the morning of 10 January 1993, numerous flocks, ranging from 10 to 400, moved west across the Staffordshire border, passing over Oswestry. Records collected by SWT after an appeal on local radio suggested a total of about 2,000 geese. Secondly, although they rarely make landfall, 300 were seen on floods at Melverley on 5 February 2004.

Records on the recent winter Atlas map show a clear correlation with the River Severn and its major tributaries, the only exceptions being five at Shavington in January 2008, two at Whixall in January

Jim Almond, Venus Pool, 31 December 2007

2011 and one at Priorslee Flash in November 2012. All the other sightings were of singles, including one with a damaged wing which frequented the Leighton area for several years. Another thought to be injured was found at Cranmere Bog in March 2010 and was still

Occupied Tetrads

Tetrads with Winter Records (2007–13): 19 (2%)

present in April. These latter two individuals were the only ones found outside the winter period.

Observations of flocks passing overhead are not usually included on the Atlas maps, but an exception has to be made in this case, as many are of the interchange between East Anglia and Lancashire. The majority are seen flying across the north-east, although some follow the River Severn corridor. Observations of skeins are mapped with pale blue dots, but they give only a general impression. Firstly, each dot represents a single sighting of a flock which will have flown over many other tetrads too. Secondly, some dots represent skeins seen on more than one occasion: 400 were noted over Market Drayton in December 2009, and 'hundreds' at the same place during the night on 10 October 2010 would have involved several skeins; a flock was seen flying over Whixall, and another over Shavington, but these two tetrads are marked in dark blue because geese had previously been found using the squares; while 200 passed over Prees in foggy conditions on 6 February 2010, but many more were thought to have gone unseen. Apart from these larger flocks, the remainder ranged from 11 to 77.

One hundred flying south over Trefonen in the evening of 4 November 2014 is the only post-Atlas record. Did they continue south or follow the Severn Valley unseen during the night?

Allan Dawes

Winter Distribution (2007–13)

Species Occurrence
- Present
- Flying over

Tundra Bean Goose

Anser serrirostris
Very rare winter visitor

Jim Almond, Venus Pool, 18 October 2013

In 2018 the two former races of Bean Goose were upgraded to individual species. The Tundra race *Anser fabalis rossicus* is now Tundra Bean Goose *Anser serrirostris* while the Taiga race *A.f. fabalis* has become the Taiga Bean Goose *Anser fabilis*. The former breeds in northern Siberia and overflies the main European wintering range of the Taiga Bean Goose to winter further south; it is the more likely to occur locally. The latter breeds in northern Fennoscandia and northern Russia; two flocks, thought to originate from Lapland, winter annually in Britain, one in the Yare Valley in Norfolk and the other on the Slamannan Plateau in Central Scotland.

All of the records prior to 1997 were of 'Bean Goose' and none were assigned to the sub-species level. Since then, all records have been positively identified as Tundra Bean Goose and it is thought likely that all previous records were also of this sub-species.

Early writers thought this was the commonest of the geese, but highlighted the difficulty of identification, especially of flying birds. In most cases, identification was confirmed by examining shot specimens and Forrest (1899) detailed five such occurrences between 1861 and 1891. Their appearance in hard weather or after prolonged periods of frost was regularly cited.

The start of the twentieth century marked a change in fortune for the Bean Goose. After a flock of over 100, the largest ever noted, was seen near Wenlock in December 1901 it became remarkably scarce. A single at Ellesmere in January 1904 and another near Ludlow in December of the same year were the last positive sightings for many years. Thirty flying south over Dowles in September 1921 were thought to be this species, but there is a question mark by the original record and the date is rather early when compared to other sightings. There were no further records until 1983.

There are eight modern records, all except one of single birds:

Cole Mere	18 December 1983–15 January 1984
Venus Pool	11 April 1990
Venus Pool	25 January–2 February 1997
Priorslee Lake and Priorslee Flash	
	2 December 2011–12 February 2012
Chelmarsh	10 January–11 April 2012
Whixall Canal Floods	12 May 2012
Venus Pool	1 October–29 November 2013
Chelmarsh Reservoir/Cranmere Bog	
	six, 18 November 2016–2 February 2017

Arrivals during the mid-winter period suggest a wild origin, rather than an escape, and the April and May reports are likely to relate to late passage migrants.

A small influx of the Tundra Bean Goose occurred nationally in the late winter of 1996–97 and the sighting at Venus Pool coincided with this. It was the first record to refer specifically to this race of Bean Goose. Another influx occurred in late 2011 and provided the only records during the recent Atlas period. The dates suggest that the same individual moved between Priorslee and Chelmarsh and was then seen briefly at Whixall as it moved north. However, some observers considered them to be different.

Since the Atlas period, one arrived at Venus Pool on the unusually early date of 1 October 2013 and stayed until the end of November. It arrived with Greylag Geese and there was some speculation that it may have been one from the spring of 2012 which frequently accompanied the Greylag flocks.

The six in 2016, three adults and three juveniles, were first seen at Chelmarsh Reservoir, but they relocated to the fields around Cranmere Bog after about three weeks, although they did occasionally return to Chelmarsh during their 11 week stay.

Allan Dawes

White-fronted Goose (Greater White-fronted Goose)
Anser albifrons
Rare winter visitor

The European race *A. albifrons* breeds in Russia west of the Taimyr peninsular, and this is the most frequently encountered race in England. The Greenland race *A.a. flavirostris* breeds, as the name suggests, in Greenland, mainly on the west coast, and most winter in the west of Scotland and Ireland.

It was described in 1866 as not uncommon but becoming less numerous, except on the lower River Severn. The first recorded specimens were a pair obtained near Ludlow in February 1855, one taken on the River Teme in December 1871 and one shot out of a flock of eight at Ruyton-XI-Towns in January 1891. There were no further reports in the nineteenth century and only nine in the first half of the twentieth, the largest flock being 30 flying north at Atcham in December 1927.

In the late 1940s a flock began wintering on the border with Montgomeryshire, moving between Leighton Flats (in Wales) and the Camlad Valley. They were thought to have been displaced from the Mersey Estuary by increased disturbance. Each mid-winter, 200–300 were usually present, increasing to 1,500–1,700 at the end of February, before leaving in mid-March. A record count of about 4,000 was made in February 1960. Such large influxes were often a result of severe weather conditions in the Netherlands.

Increased land drainage in the 1970s probably reduced feeding opportunities, and these geese ceased to visit the Camlad Valley on a regular basis. Dwindling numbers continued to winter at Leighton Flats, but none have been recorded there since 1985. This wintering population was probably responsible for the large number of records received during that period, which often referred to flocks flying overhead. While it was in terminal decline, small numbers became a regular feature at the Ellesmere meres, peaking

Jim Almond, Venus Pool, 28 January 2013

at 24 in 1983, but their presence was short-lived and they were last seen there in 1986.

Sightings continued almost annually until 2002. Most were of singles or pairs, but some larger flocks put in an appearance, the largest being 56 in flight over Chelmarsh in November 1988, thought to be on their way to Slimbridge.

Since 2002 there have been only six records, all during the recent winter Atlas period: two at Polemere in January in both 2009 and 2013, three at Chetwynd Park in November 2011, two at Cressage in January 2012, and one at Venus Pool in January 2011, with two at the same site in February 2013. The latter were thought to have come from Polemere.

This lack of recent records reflects milder conditions, which have enabled the geese to remain in the Netherlands throughout the winter.

The flock on the Montgomeryshire border was of the European race, and it is likely that most other records were of this subspecies, although race was seldom noted. Only two records specifically refer to the Greenland race, a single at Wollaston Pond on 29 December 1998 and another seen at Venus Pool on 1 January 2015, which was subsequently seen at Polemere on 3 February and 26 March. This population is declining and is classed as endangered. The nearest regular wintering site is the Dovey Estuary, but numbers there are falling and the future of this flock is uncertain.

Thus both races seem destined to become even rarer in future years.

Allan Dawes

Mute Swan
Cygnus olor
Uncommon resident

Dawn Micklewright, Venus Pool, 10 November 2013

Mute Swan is resident in temperate Western Europe with migratory populations breeding in southern Scandinavia and across central Asia.

Here, historical records show the species bred in the sixteenth century at Waters Upton and Moreton Corbet. Prior to the 1530s, when turkeys were introduced into Britain, Mute Swan was the traditional Christmas meal of royalty and lords of the manor. This privilege was extended in 1482 to wealthy landowners, provided the birds were marked and pinioned. However if such birds nested on land belonging to the lord of the manor he was entitled to one cygnet as compensation for the food eaten, and this custom was known as land or ground bird. This happened in 1532, when Richard Chorleton and Thomas Eyton of Waters Upton were ordered by the courts to compensate the lord of the manor of Waters Upton, and again in 1596, when Roland Hill and John Ironmonger were ordered to compensate the same lord. (The same reference covers the two incidents, Stafford Record Office D593/J/20a/1–3.) The lord had to pay the Crown 12d for the privilege of having the land bird. Another early reference to swans occurs in 1588 when the court roll for Moreton Corbet notes there were three ponds in the demesne (land surrounding the manor house) for the fattening of swans (Shropshire Archives 322/2/1).

Rocke, writing in 1866, dismissed these swans as 'far too common and domesticated to require any remark', while Beckwith went on to say they were very numerous on the Ellesmere meres where they lived in a semi-wild state, and suggested in his 1887 work that they

Breeding Distribution Change (1985–90 to 2008–13)

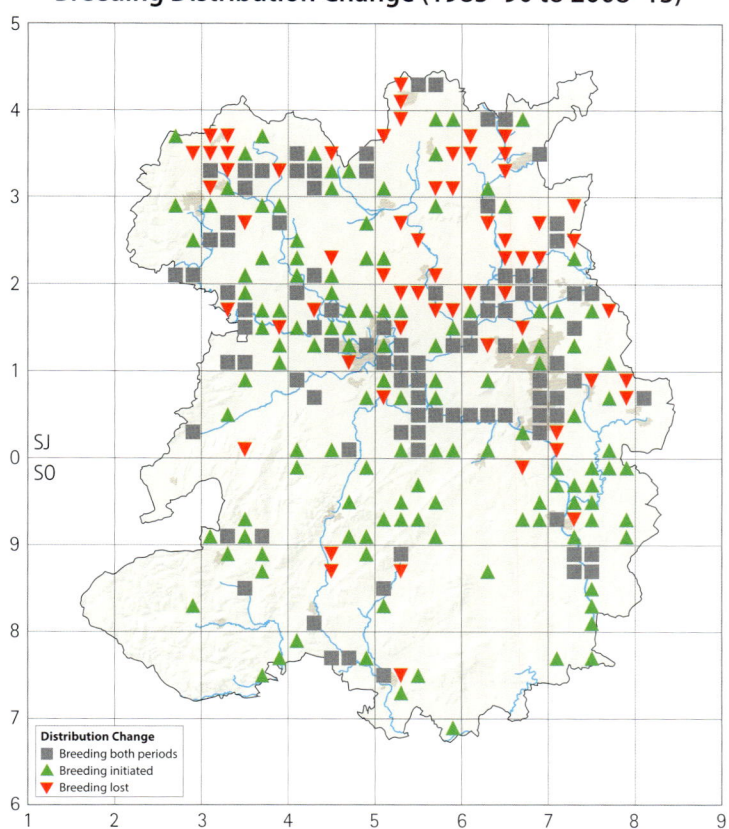

Distribution Change
- ■ Breeding both periods
- ▲ Breeding initiated
- ▼ Breeding lost

Winter Relative Abundance (2007–13)

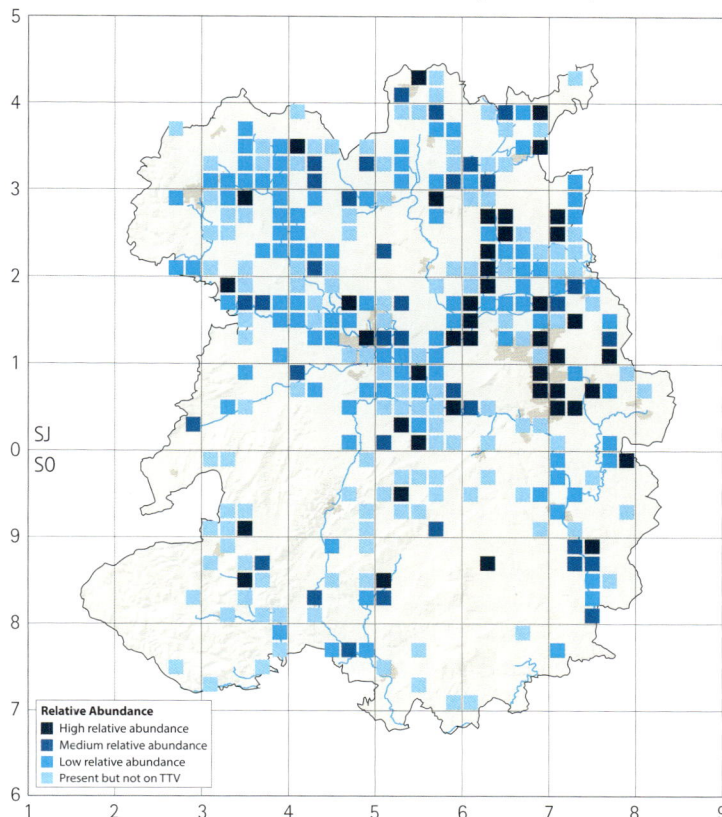

Relative Abundance
- ■ High relative abundance
- ■ Medium relative abundance
- ■ Low relative abundance
- ■ Present but not on TTV

Occupied Tetrads

Atlas period (breeding)	1985–90		2008–13		Change	
	Number	%	Number	%	Number	%
Confirmed	88	10	130	15	42	48
Probable	41	5	58	7	17	41
Possible	28	3	34	4	6	21
TOTAL	157	18	222	26	65	41
Tetrads with Winter Records (2007–13): 307 (35%)						

should be deleted from the Shropshire list as they are a domestic species. Most writers of the period appear to have dismissed them due to their apparent semi-wild state, while Forrest (1899) listed them as 'Tame, or Mute Swan', and noted 'A pair used to nest under the English Bridge, in Shrewsbury, but never reared the nestlings; the eggs were generally destroyed by floods or rats, and at length the birds went away altogether'. They had returned by 1912, when they raised three cygnets. The following year the nest was washed away by a high flood on 22 April, despite the pair making 'strenuous efforts to save the eggs by building up the nest using straw let down to them by ropes from the bridge'. They laid again in May and hatched eight cygnets on the rather late date of 5 July, which were successfully reared. They were successful again in 1914 and 1916 and still nest at this site, but they still usually lose out to flooding.

One of the earliest records of a non-breeding flock was on the River Severn at Atcham in 1915 when 35 were counted. In 1936, this time on the River Severn in Shrewsbury, a flock of 30 to 40 was

counted, and in the following year 40 flew over Shrewsbury School. An exceptional flock of nearly 100, all adults apart from six cygnets, was on the River Severn at Shrewsbury in December 1940. The following year, at the same site, only 40 were counted, while there were 60 on 22 July 1942 and 67 in the first week of September 1943.

A pair with 14 cygnets walking across a lawn at Coleham Head Shrewsbury on 9 June 1955 had an exceptionally large brood. The average brood size is six or seven, and double figures are exceptional. However, two pairs nesting on Newport Canal in 2003, one with nine cygnets the other with seven, were involved in a territorial dispute and the pair with nine cygnets ended up with the additional seven and successfully raised 15 to fledging.

The *Handlist* (1964) stated it was 'a common resident except in the south', but a considerable decline in breeding pairs was evident when a census in 1978 recorded only 27, and in 1983 only 21 pairs were found. Although the latter count could have been affected by high water levels on the River Severn, due to the wet cold spring of that year, it was also noted that the non-breeding birds were as few as ever and non-breeding flocks such as the 46 on the River Severn in Shrewsbury in 1955 were no longer to be found.

The decline up to the 1980s was attributed to the ingestion of lead weights and fishing lines. The sale of lead weights was made illegal in 1987 and their continued use was banned on many waters. The improved survival of young birds resulted in the re-establishment of non-breeding flocks, and then an increase in breeding pairs as they matured, bringing numbers in 1990 back up to those found in 1961.

The size of the non-breeding flocks increased through the 1990s. By 1994 they included 44 at Nesscliffe in February, 35 on the River

Severn at Shrewsbury and 76 at Ellesmere in August, and 48 at White Mere in November. The following year the largest flock was 81 at the Mere (Ellesmere) in July, while in October 1996 there were 105 on the River Severn at Montford Bridge. Flocks were also developing in new areas, and in 2001 there were 82 on the Weald Moors at Kynnersley in February, with 80 there again the following year at the end of March 2002.

The population continued to grow, and the recent Atlas found a considerable expansion of range.

Much of it was in the south where the *Atlas* (1992) had mapped confirmed breeding in only eight tetrads and probable breeding in a further six, compared to 37 and 16 respectively in 2008–13. It had apparently disappeared from 62 tetrads, all except six in the north, but most of these losses were due to possible breeding records in the earlier Atlas, perhaps of non-breeding individuals, not being repeated, while most of the gains are confirmed breeding records. However, there have been permanent losses: some pools have been filled in, while disturbance has become unacceptable at others.

Some nest sites are not used every year, and some of the probable breeding records relate to non-breeding pairs. Conversely, some tetrads have more than one breeding pair, for example four pairs on nests in April 2011 at Walcot Lake. It is probably safe to say that the total breeding population is 130 to 150 pairs. This is a considerable improvement on the 52 to 74 breeding pairs estimated in 1990.

Notable flocks found during the recent Atlas period include one at Walcot Lake, where monthly counts in 2008 produced a maximum of 64 in August, with 63 still present in November. The same site held between 30 and 60 every month in 2009, with a peak of 69 in November, and had annual maxima of 59 in October 2010, 54 in October 2011, 41 in September 2012 and 49 in February 2013. At Cound and/or Venus Pool, the annual maxima were 61 in October 2008, 79 in February 2009, 59 in January 2010, 93 in January 2011 and 58 in May 2012. Other notable flocks were at the Mere (Ellesmere: 74 in March) at Priorslee Lake (65 in October 2009), Allscott Sugar Factory (66 in April 2011 and 82 in March 2012) and Trench Pool (32 in January 2010, and 30–36 from October to December 2012).

In addition, winter flocks, such as the one associated with Venus Pool, are commonly found early in the year feeding on the young shoots of oilseed rape, producing records away from the rivers and lakes favoured in the breeding season. The winter relative abundance map highlights both the wintering flocks, and the rivers and lakes.

The table shows the results of the five censuses since 1955, and an estimate of the total population in 2011. The latter was derived by adding the non-breeding flocks recorded in the first three months of the year to the breeding pairs that would be holding territory at that time. Allowance has also been made for additional pairs that go unrecorded, particularly along the River Severn. Maximum WeBS counts at 24 lakes and meres in January–April 2011 appear to corroborate the estimate; they totalled 267 birds, and adding casual records from non-WeBS sites in the same period gives a total of 468.

Given the regular flocks detailed above, and the counts summarised in the table, the non-breeding birds (almost 240) numbered considerably more than the 152 recorded in the 1990 census, and together with 130+ breeding pairs, represents a new population peak.

Colour-ringing, which enables individual birds to be identified without the need for recapture, has facilitated a study of the local

Year of survey	Breeding pairs	Non-breeding pairs	Non-breeding birds	Total number of birds
1955	78	(19)	181	341
1961	72	(11)	154	298
1978	43+	–	34+	120+
1983	21+	–	122	164
1990	52	22	152	300
2011	130+	16	206	466+

breeding population. Mute Swans, mainly cygnets, have been ringed with a unique blue Darvic and BTO metal ring, every September since 1987, at sites in Oswestry, Shrewsbury, Telford, Newport, Sambrook and Ellerton. Many of the breeding swans at nest sites often have orange Darvics from Worcestershire or yellow from Staffordshire. The colour-ringing is controlled by WWT, but recoveries appear in the ringing reports on the BTO website in the usual way.

Two 1991 cygnets, a cob ringed at Tamworth (Staffordshire) and a pen from the River Severn at Bewdley, made a failed attempt to breed on the banks of the Severn above the Shrewsbury weir in 1994. They bred successfully the following year at Ellerton Mill, and every subsequent year until 2008, when the female flew into overhead wires. The male then paired with one of his own (2002) cygnets in 2009. He failed to breed in 2010 and 2011, but successfully raised cygnets at Ellerton Mill in 2012 and 2013. He finally died early in 2014 at 23 years old having raised 71 cygnets.

It is by no means unusual for a bird to pair with one of its own young, and this has been found at three further sites. At the Hem Shifnal a pen bred successfully with her 2003 cygnet for five years 2005–09. At Priorslee Lake in 2011, the resident male raised five young with his own 2008 cygnet, and this pair bred again in each of the following two years, while at Sambrook Mill in 2012 a pen paired with her 2001 cygnet.

Colour-ringing also allows longevity and movements to be monitored. A cob ringed at Ellesmere in August 1990 (a cygnet from the previous year) was found first breeding at Springfield Mere Shrewsbury in 1992 and was still breeding in 2014, at 25 years old. After an unsuccessful breeding attempt in 2015, he was found 'long dead' on 22 March 2016, 25y 7m 17d after ringing. This is close to the longevity record of just over 29 years.

The vast majority of swans were ringed and recovered here, or involve immediately adjacent counties, confirming their highly sedentary nature. However, a cygnet ringed at Ellerton Mill in September 2001 was found freshly dead (oiled) on 4 February 2009 in Irvine Harbour, Strathclyde, a distance of 343km, and a second year bird ringed in July 1984 in Christchurch, Dorset was alive and well at the Mere (Ellesmere) in September 1987, a distance of 253km. There are no known instances of Swans ringed here being found on the continent, or vice versa, although a cygnet ringed on 26 December 1989 at Stirchley Grange, Telford was reported in a flock of 18 in Kilcoole Estuary just south of Dublin on 5 January 1992, and an adult female ringed on the River Severn at Bridgnorth on 28 March 1999 was reported at Castletown, Isle of Man in June 2000, a distance of 226km.

Martin Grant

Bewick's Swan (Tundra Swan)
Cygnus columbianus
Rare winter visitor

Jim Almond, Cound, 2 January 2011

Bewick's Swan breeds across the Russian Tundra and winters in western Europe, with 30–40% of the total population currently coming to Britain and Ireland.

The first historical reference was in 1841 by Ogier Ward, the location given as 'Drayton', presumably Market Drayton. In 1861 a male was shot on the River Severn at Melverley. In 1866, Rocke described it as 'by no means rare in the neighbourhood of Shrewsbury in hard winters', and Beckwith wrote in 1879 'seldom obtained now, although in former years it appears to have been frequent and most collections contain a specimen'.

At the turn of the nineteenth century it was still described as very rare, although 20 adults and three cygnets stayed for a week on the Mere (Ellesmere) from 20 December 1899, and nine were still on Newton Mere on 4 January 1900.

There were records in only eight of the years between 1900 and 1950, and there were only two with more than five birds, a flock of 25 at the Mere (Ellesmere) on 26 October 1921 that flew off the next morning, and 70 that flew north-east over Ellesmere on 6 January 1938.

There were no reports in the 1950s until the founding of SOS in 1956 encouraged an increase in records, with 13 at Crose Mere on 7 April 1956 and three at the Mere on 19–20 February 1957, then probably the same three at Crose Mere on 8–11 April the same year. Although none were recorded the following year (1958), the *Handlist* (1964) described it as a 'non-breeding visitor, occurs annually, sometimes in fairly large numbers' and with many more records than Whooper Swan. It listed eight sites with recent records; Allscott Sugar Factory, Camlad Valley, Criftins (Dudleston Heath), Crose Mere,

Ellesmere, Prees, Venus Pool and the River Severn in severe weather, with the largest number being 27 at Ellesmere on 17 January 1960.

An exceptionally large passage flock of 33 was seen at Cole Mere in the first half of March 1976, while 37 were at Crickheath on 26 February 1979 with 26 still there the following day. In 1980 another flock built up to 23 on 7 December and 31 by the turn of the year. They spent most of the time feeding on a potato field at Rowe Lane, near Bettisfield, before flying to roost at Hanmer Mere (Clwyd), or Cole Mere, remaining until 6 March before dispersing.

The peak counts in 1979–80 and 1984–85 were of flocks flying east, possibly to East Anglia, and probably not stopping here at all. The former involved 77 over Shrewsbury on 20 February, and the latter, at least 80, passed over Bayston Hill on 2 March after a particularly cold start to the year. It was mapped as present in only four of the County's 33 10km squares in the 1981–84 national BTO Winter Atlas.

Individuals ringed and fitted with coloured neck collars by several Dutch expeditions to Russian breeding sites at Korovinskaya Bay near the mouth of the Pechora River at about 65N 55E in August were seen here in the winters of 1991, 1993 and 1997. One of two at Maesbrook on 31 December 1991 had been ringed earlier that year. It was also seen, in the same winter, at Martin Mere (Lancashire), the Wirral (Cheshire) and Kingsbury Water Park in the West Midlands. It subsequently returned to the Russian breeding grounds as part of a non-breeding flock in the summer 1992. In 1993, a first year at Dyffryd on 14 November had been ringed earlier that year (unpublished SOS records). In 1997, an adult male, ringed a few months earlier in 1996, was seen in a flock of 21 at Allscott Sugar Factory on 4–7 January. It was subsequently seen on 19 February 1997 in the Netherlands at Lauwersmeer.

In the severe winter 2005–06 a flock of five flew north over Shavington on 22 January, one was recorded in a flock of Mute Swans at Tern Hill from 4–19 February and 10 were at Wall Farm Kynnersley from 26–30 March.

During the recent winter Atlas period, this swan was found in only four tetrads, and in only two winter periods. At the end of 2010, a flock of seven (five adults and two cygnets) arrived at Coundlane, increasing to 32 on New Year's Day 2011.This flock was seen mainly in the same vicinity (tetrad SJ50S), except on 2 January when 25 were recorded at Acton Pigott (SJ50L) before they moved to Noneley

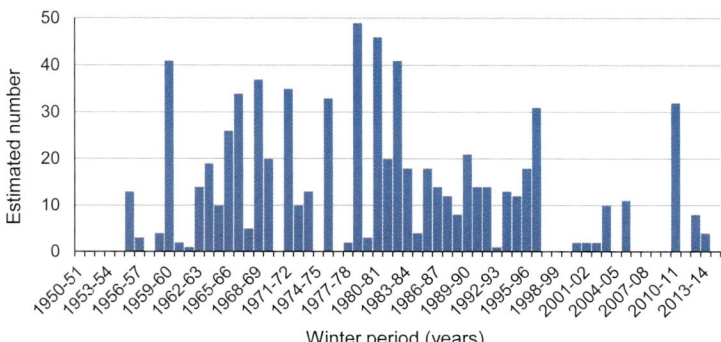

Estimated Winter Period Totals (1950–51 to 2014)

Dawn Micklewright, Venus Pool, 19 January 2014

(SJ42T), where 32 were recorded 19–25 February 2011. Early in 2013, between 5 January and 7 March, eight (six adults and two juveniles) were seen regularly coming in to roost at Venus Pool.

After the Atlas period, the winter of 2013–14 saw a family group of two adults and two cygnets arrive on 27 November 2013 at Coundlane. They were seen regularly, both there and in the evenings, coming in to roost at Venus Pool, until 23 February 2014.

Nationally, the number of Bewick's increased from the mid-twentieth century, and reached a peak in the early 1990s. The subsequent decline has been most noticeable in western Britain, attributed partly to milder winters, enabling a higher proportion to 'short-stop' on the continent in most years. However, since 1995, the number making the migration from arctic Russia to northern Europe has plummeted by nearly a half, from 29,000 to just 18,000 in 2010. Threats to the population are habitat loss, including loss of wetlands, illegal hunting, collisions with windfarms and power pylons on their flight path, and climate change (WWT website).

Reflecting this trend, smaller numbers have been wintering here from 1997, with annual totals of 10 or less since then, except in the severe winters of 2005–06 and 2010–11. The number of Bewick's was usually greater than Whoopers until the late 1990s, since when they have been relatively much lower, reflecting both the decrease in Bewick's and the continued increase in Whoopers.

All SBR records have been assessed to estimate the total number occurring each winter (October–April). Individuals or flocks seen only in flight, with no evidence that they were using local waters or feeding areas, and those identified only as 'wild swans', have been excluded. Most do not appear to stay for any length of time, so it has generally been assumed that different records are of different individuals, unless they come from the same (or nearby) locations within a short period of time, or there is evidence that a flock stayed in an area for a lengthy period. The chart shows the result, the estimated total number recorded each winter from 1955–56 to the end of 2014.

Martin Grant

Whooper Swan

Cygnus cygnus

Scarce winter visitor

Whooper Swan breeds in sub-arctic Eurasia, including Iceland, and migrates south for the winter. A handful breed in northern Scotland and Ireland.

In 1879, Beckwith described it as very rare, though it had occurred frequently in severe winters such as 1837, when Henry Shaw of Shrewsbury had no less than 25, mostly adults, sent to him for preservation.

Early in the twentieth century, Forrest also described it as a rare winter visitor, although several were on the Mere (Ellesmere) in

the winter of 1892–93, and a flock of 10 were over Ellesmere on 4 December 1899.

It continued to be a rare winter visitor in the first half of the twentieth century with only three records: two at Betton Pool on 25–27 December 1927, three on flood water at Bayston Hill on 26 February 1928, and one on a newly made pool at Betchcott on 3 November 1935.

The *Handlist* (1964) described it as occurring most winters in small numbers, and in smaller numbers than Bewick's Swan. Between 1950

Occupied Tetrads

Tetrads with Winter Records (2007–13):	23	(3%)

and 1980 this was a fair description, as it was mapped in only three of the County's 33 10km squares in the 1981–84 national BTO Winter Atlas, it was not recorded in nine of the 25 years and reported in double figures in only four winters, with 30 at Ellesmere on 4 January 1975 being an exceptional count.

The entire Icelandic population, estimated at around 30,000 in the 2010 International Swan Census, winters exclusively in Britain & Ireland. This population has increased almost three-fold since 1984, and by 11% since 2005, resulting in a large extension of winter range, including areas in central and eastern England. This population growth is likely to be the main source of the increasing numbers recorded here. Whoopers outnumbered Bewick's in only three winters from 1956–79, and then in seven of the winters from 1980 to 2000. During this 20-year period they were not recorded in only two, with counts in 1983 of 22 at Sleap in November and 17 at Cound in December. In January and February 1995 there were also good numbers reported from the River Severn flood meadows around Melverley, with maxima of 29 at Cae Howell and 23 at Alberbury, probably involving the same birds.

The increase has been maintained in this century, with estimated annual totals in double figures every year to date, and a steady increase year on year, and they now easily outnumber Bewick's. There

Winter Distribution (2007–13)

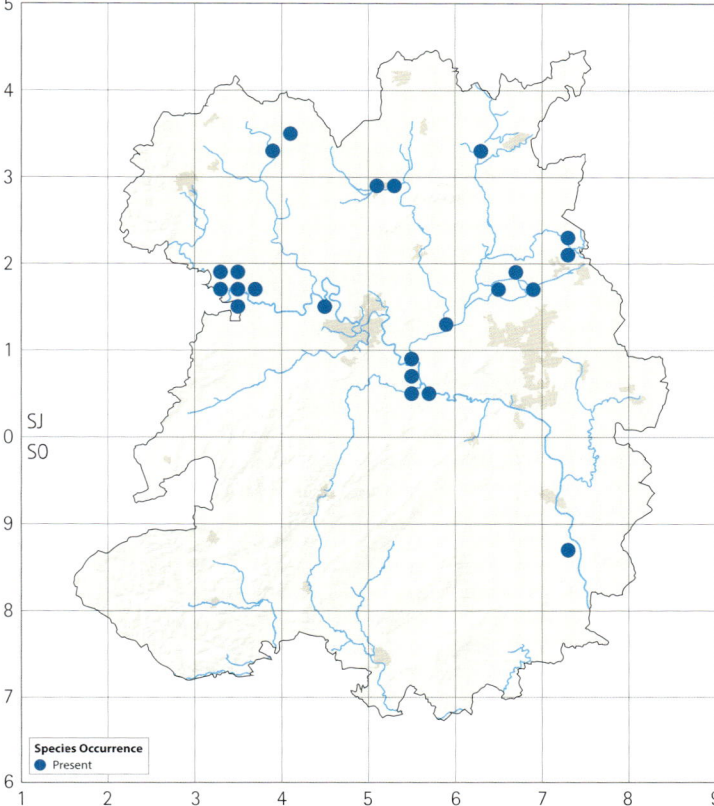

Species Occurrence
● Present

Jim Almond, Edgerley Floods, 9 January 2016

have been over 20 in every winter, except one, since 2004–05, and between 36 and 44 every winter since 2010–11.

A single Whooper Swan that was with up to 42 Mute Swans, at Wall Farm Kynnersley from 8 November to 18 December 2000, had been ringed in Iceland on 19 August 1998, and was subsequently seen in the winter of 1998–99 at Welney (Norfolk), Upwell Fen (Cambridgeshire) and at two sites in Yorkshire. The following winter it was back at Welney from 2 November 1999 to 25 March 2000, before visiting Wall Farm eight months later, presumably after its third trip from Iceland.

A flock of 12 Whooper Swans at Melverley on 18 November 2007 was seen to include three adults (two males and a female) ringed on a lake in northern Iceland on 9 August 2007.

In the recent Atlas period it was found in 23 tetrads. The River Severn flood plain around Melverley is the favoured wintering area, and the maximum single count in each of the six winters, ranging between 19 and 39, average 28, were all in the Melverley area. There were several other double figure counts in this cluster of six tetrads, perhaps involving the same birds, with the only other double figure count being 13 at Venus pool from 28 January to 7 March 2013.

Other occupied tetrads were four in the Severn Valley (Atcham, Cound, Cressage and Venus Pool), two (three sites) near Newport, two near Ellesmere and two near Wem, and one at Bicton, Buttery, Chelmarsh, Crudgington, Mirelake, Tern Hill and Wall Farm. Most of these records were counts of one or two and none were more than seven. Outside the Atlas months, there were five at Polemere on 20 October 2010.

After the end of the Atlas period, in the winter 2013–14 there were

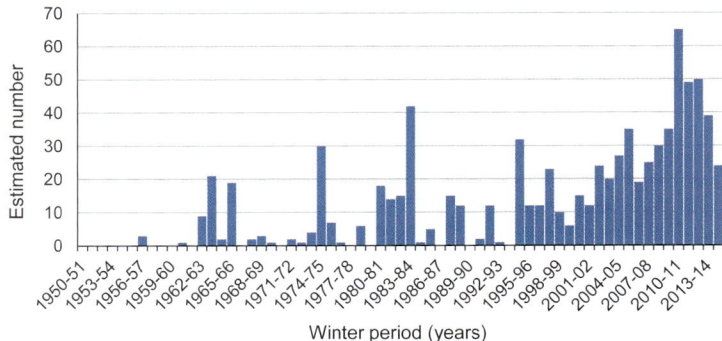

Estimated Winter Period Totals (1950–51 to 2014)

maxima of 24 (18 adults and six juveniles) at Melverley in January 2014 and 15 at Coundlane in February 2014, and, at the end of the year, 15 at Melverley and 12 (including seven juveniles) at Coundlane and Venus Pool in December 2014. All records in these periods were from sites on the winter Atlas map, apart from a sub-adult on Whixall Canal Floods between 13–27 July 2014.

This was only the third record from the summer months. Previously, a late spring sighting at Chelmarsh on 19 May 1991 was almost certainly the same bird as the one at Priorslee Flash on 4 June 1991.

All SBR records have been assessed, to estimate the total number occurring each winter (October–April). The same criteria have been applied as noted in the Bewick's account. The chart shows the result, the estimated total number recorded each winter from 1955–56 to the end of 2014.

Martin Grant

Egyptian Goose
Alopochen aegyptiaca
Rare naturalised visitor, very rare breeding species

Jim Almond, Polemere, 11 August 2008

Native to Africa mainly south of the Sahara, and introduced into the UK from the seventeenth century onwards, a small feral population became established in Norfolk in the mid-twentieth century, and is now spreading.

These geese were a popular addition to ornamental lakes in the second half of the nineteenth century and escapes, particularly young that had not been pinioned, were thought responsible for sightings elsewhere. However, some observers noted them to be wary and difficult to approach so they may have survived in the wild for some time. According to Rocke, 'a few were obtained' and Beckwith mentions 'several', but only two were cited; one near Shrewsbury with no date and one at Hatton in the winter of 1878–79. Then in 1884, there were three at Halston and one near Ludlow, after which there were no further reports for over a century.

A spate of sightings in 1988–89, which may have involved a single wandering individual, were the first modern records. They began at Chelmarsh on 10 February, followed two days later at Dudmaston, a spring sighting at Halfway House, at Apley Park in May and finally back at Dudmaston the following March. Further singletons were at Venus Pool in April 1997 and Ponthen in March 2000, and a pair were at Bicton Heath in April 2001. After a five year gap, a pair at Venus Pool in May 2006 heralded the start of annual records.

They have been found at 18 locations, both large well-watched waters and small ones that may not receive many visits from birders. Most reports are for a single day, but at Polemere one stayed for 15 days in December 2007 and 16 days in August 2008. Because of the short stays and the likelihood that some return in subsequent years, determining the number of individuals involved is difficult, but between 21 and 38 are estimated to have occurred.

Twenty-five were first reported between January and May, suggesting dispersal to potential breeding areas. A downy chick at Chetwynd Park in 2009 was the first confirmed breeding and a juvenile was also found there the following year. A pair at Radbrook Pool in June 2012 and another pair at Priorslee Flash in May 2013 are the only other potential breeding records.

Winter Atlas records were received from five tetrads: singles in two years at Polemere, referred to above; one at Onslow; two at Allscott Sugar Factory and at Marsh Green were probably the same birds; and three at Knighton Reservoir. All bar the Polemere records were from 2012.

The first recent records were often considered to be of local escapes, but the feral population in Norfolk has expanded and they are now regularly seen in the Midlands. The national BTO Atlas (2013) suggested that a new population is becoming established in the East Midlands. This range expansion has almost certainly been responsible for the local increase. It is likely that the Egyptian Goose will soon become a regular year-round feature, with many dots predicted in the next atlas.

Allan Dawes

Shelduck (Common Shelduck)

Tadorna tadorna

Scarce summer and winter visitor

Jonathan Cartwright, Venus Pool, 11 June 2011

Resident on British coasts, Shelduck breeds across temperate Eurasia, but northern populations are migratory. Historical records from the nineteenth century are limited to just three, all of single birds shot at Corfton Manor, Lutwyche and Minsterley. However Beckwith (1879) cast doubt on whether these birds were truly wild, as 'they are often kept on ornamental pools and were probably escapes not truly wild'. The only other report, prior to the middle of the twentieth century, was of three at Betton Pool on 26 December 1934.

There were no further reports until 1956 when single birds were at Crose Mere in April, Cole Mere in September and Trench Pool in December.

Two or three records a year were then the norm until the first breeding was confirmed at Allscott Sugar Factory in 1963, when a pair hatched eight young on 14 July and raised three to free flying stage.

The following year heralded a sudden increase in numbers, with four at Allscott Sugar Factory in January, five at Ellesmere in April, then 10 at Venus Pool in September and five at Crose Mere in December.

This reflected an expansion of range nationally. Previously, breeding was largely restricted to coastal estuaries; but the 40 years from 1968 saw a colonisation of central and northern England, utilising shallow lakes and river valleys, preferably with little vegetation, and nesting in rabbit burrows.

Despite this expansion, there were no Shelduck mapped in the County's 33 10km squares in the 1968–72 national BTO Atlas, and it was not until 1983 that breeding was confirmed for a second time, when a pair raised six young near Cole Mere. Shelduck remained a rare breeding species, and Atlas work in 1985–90 found only three instances of confirmed breeding, with no more than one in any one year, at Crose Mere (1985), Wood Lane (1986) and Allscott Sugar Factory (1988). Probable breeding was found in a further nine tetrads at Baggy Moor, Chelmarsh, Crudgington Moor, Cloverley Pool, Isombridge, Newton Mere, Shavington, Shrewsbury Sewage Farm and Venus Pool.

After the next successful breeding at Venus Pool in 1992, colonisation continued in the years leading up to the recent Atlas, when 15–20 pairs were found, with confirmed breeding from 15 sites, all in the north. As well as the traditional sites at Allscott Sugar Factory, Crose Mere and Wood Lane, there was also confirmed breeding at Baggy Moor, Cloverley Pool, Knighton Reservoir, Mirelake, Onslow Park, Quina Brook, Shavington, Tittenley, Venus Pool, Wall Farm, Walford and White Mere. Breeding was confirmed in 14 tetrads (15 sites) but the sites at Allscott Sugar Factory and Mirelake are no longer suitable habitat. Probable breeding records also came from an additional 22 tetrads, but some of these will relate to breeding pairs on pools that span tetrad boundaries, and others will be non-breeding pairs, as they do not breed until their third calendar year. However, the vast majority of confirmed breeding records were observations

Breeding Distribution Change (1985–90 to 2008–13)

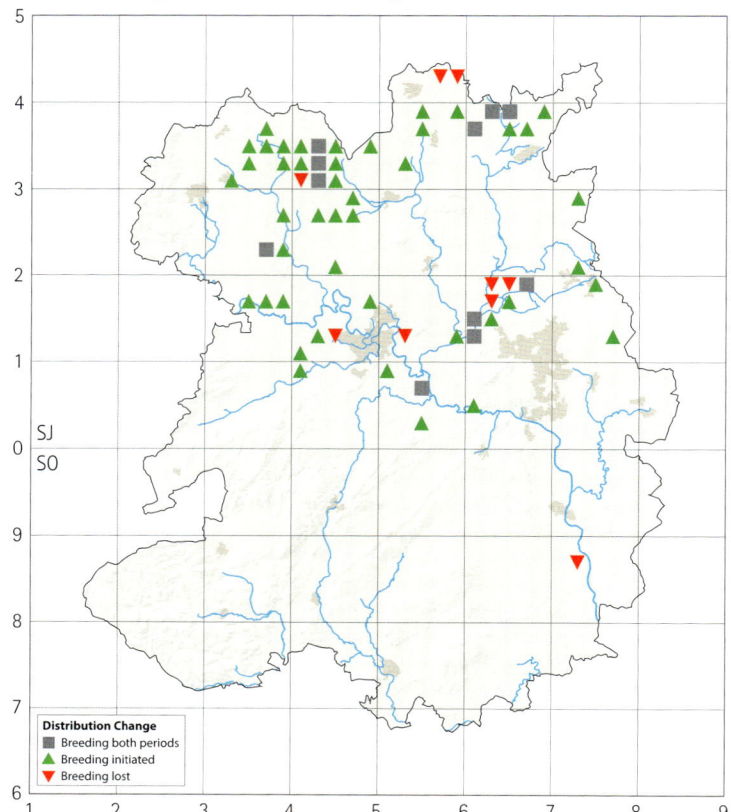

Winter Relative Abundance (2007–13)

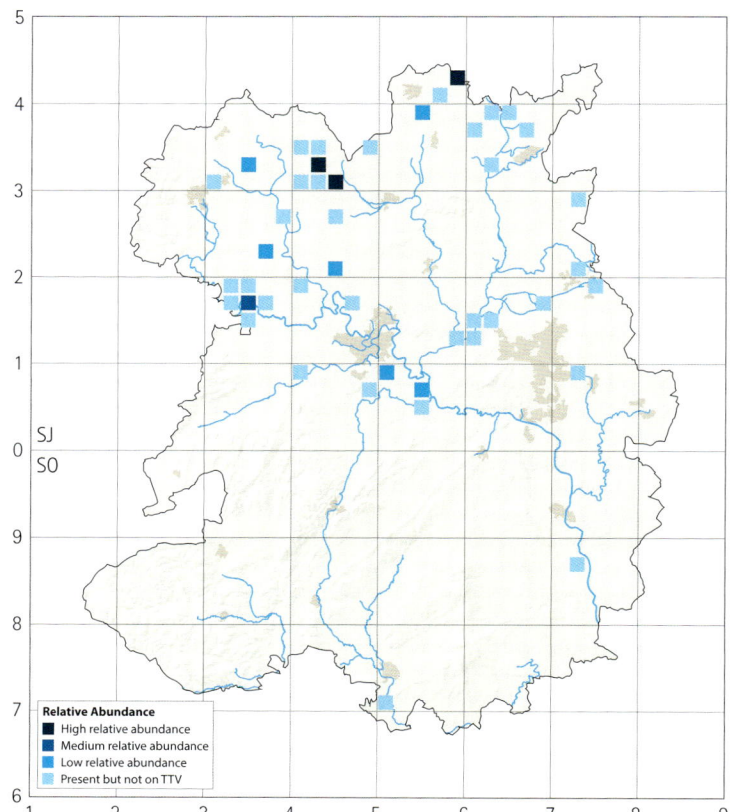

Occupied Tetrads

Atlas period (breeding)	1985–90		2008–13		Change	
	Number	%	Number	%	Number	%
Confirmed	3	0	14	2	11	367
Probable	9	1	22	3	13	144
Possible	8	1	19	2	11	138
TOTAL	20	2	55	6	35	175
Tetrads with Winter Records (2007–13): 45 (5%)						

Maximum Annual Count (Individuals) at Main Sites (1980–2014)

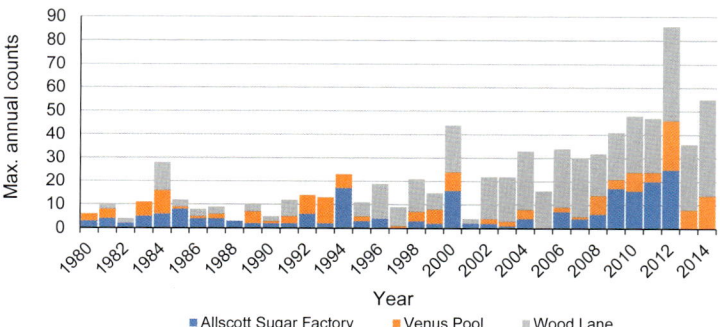

of ducklings, so unsuccessful nests of some of these pairs will have gone unrecorded, and the breeding population is probably around 10–15 pairs.

The breeding distribution change map illustrates the colonisation that has taken place since 1990, and shows that the only site in the south occupied in 1985–90, Chelmarsh, is no longer utilised.

A pair at Wall Farm, Kynnersley, fledged eight young in 2014, the first successful breeding at that site, and breeding was confirmed for the first time at three other sites that year, Cound Moor, Knighton Reservoir and Wilcott Marsh. Breeding was also confirmed at Venus Pool.

The two charts show the increase in population, and the number of fledged young, each year since 1980 at the three main sites, Allscott Sugar Factory, Venus Pool and Wood Lane. The former shows the maximum annual count, usually in January, February or early March. Twenty-five at Wood Lane on 1 March 2004 was the highest ever single site count up until that date, but it was equalled

Fledged Young at Main Sites (1980–2014)

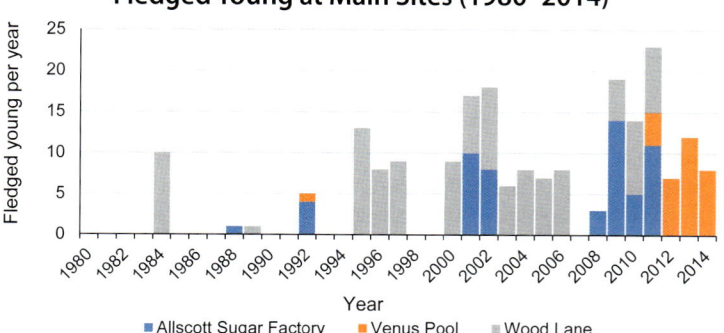

in February 2006 and in the same month in 2007, and exceeded by counts of 40 in February 2012 and 41 in January 2014. Numbers remaining through the breeding season have also increased during the twenty-first century.

At the end of July adults migrate to moult, traditionally to the Heligoland Bight area of the Wadden Sea, but in more recent years moulting sites in the UK have developed, including the Dee and Mersey estuaries, which are probably the sites favoured by local Shelducks. Most juveniles leave with their parents, only the odd immature remaining. Post-moulting, they start to return at the end of October, with numbers building up to an annual peak in the first three months of the following calendar year.

The winter population and distribution has steadily increased in the last 30 years. During the 1981–84 national Winter Atlas, Shelduck was found in eight of the County's 33 10km squares, compared to 16 in the recent winter Atlas, when it was found in 45 tetrads, all except two, Chelmarsh and Overton, in the north.

The relative abundance map highlights the regular site at Wood Lane and the River Severn floods at Cae Howell, together with Pikes End Farm (SJ43K) and Combermere Park (SJ54W) that held good numbers during the TTVs.

There were record counts in early 2012 at the three main sites, probably including some passage migrants, with 40 at Wood Lane on 2 February, 21 at Venus Pool on 10 March and 25 at Allscott Sugar Factory on 19 March. There were also 25 on flooded farmland at both Long Lane (Sleapford) on 27–28 February 2013 and Wall Farm Kynnersley on 4 March 2014.

Given the recent increases in the breeding, passage and wintering numbers, it is likely that the population will continue to grow.

Martin Grant

Mandarin Duck
Aix galericulata
Scarce naturalised resident

Mandarin Ducks were brought to England from East Asia by 1745 but they were not recorded breeding prior to 1834, and it was not until 1866 that one was reported in the wild; it was shot on the Thames in Berkshire (Savage 1952; Lever 2013). It seems likely that the pair at Coedway on 27 March 1901, when the male was shot, were on the Welsh side of the border, so it was 100 years before the first record of an apparently wild bird here, one seen at Chelmarsh in December 1966 on a Wildfowl Count (WeBS online records downloaded from BTO website 29 November 2013). Yet Mandarins had been around for a long time. In 1935, the Stevens brothers, owners of Walcot Hall (Lydbury North), had acquired a number of 'full winged' Mandarin Ducks for their wildfowl collection, although it is not known how many. For the following two years 'swarms of Jackdaws' are said to have occupied all possible nest sites, but then 'thousands' were trapped and Mandarin broods were 'numerous' the following year; by 1939 there was a free-flying flock of some 100. Walcot was requisitioned by the army during the Second World War and, apparently as a consequence, the flock scattered, but it is said that as many as 40 reappeared after 1945 (Savage 1952). It is not known how long they survived, nor what happened at a second site, reported in 1964 by the Shrewsbury School Natural History Society, 'feral birds [are] now free ... which originated from the school collection'. In view of the lack of subsequent records, the Mandarins of Walcot Hall and Shrewsbury School presumably died out.

From 1977 onwards sightings were reported annually in SBRs, although all records up until a pair at Polemere in 1984 were of singles, generally males. Most of these early records came from Chelmarsh, but from 1986 the Severn at Quatford came to the fore, with up to a dozen reported, and this area has generated more reports than any other, including, in 1988, the first confirmed breeding record. It is thought that this flock may have originated as 'escapes' from the former West Midland Bird Gardens which lay less than a mile from the river. By this time some were occurring on the Dowles Brook; they were first recorded there in 1983 and a pair bred in a nest-box in

John Robinson, Dowles Brook, 2008

1988. This seems to have been on the Worcestershire side of the brook (Harrison & Harrison 2005), but the record appears in the *Atlas* (1992) because more than half of the tetrad in question lies in Shropshire. These small local beginnings contrasted with the national population estimate at that time of 7,000 birds (Davies 1988); most were to be found in Buckinghamshire, Berkshire and Surrey. After 1988 sightings increased markedly; typically they were of singletons, pairs, or parties of less than five. But despite this increase, *Atlas* fieldwork 1985–90 generated probable or confirmed breeding in only four tetrads, and, prior to the recent Atlas, there had been, in all, only eight confirmed breeding records. This is, however, a furtive bird. Nests are rarely found and females keep their broods close to dense marginal vegetation, so the breeding distribution map (2008–13), showing occupation of 46 tetrads, is likely to understate the recent spread.

The map shows a markedly southerly distribution, indicating

Breeding Distribution (2008–13)

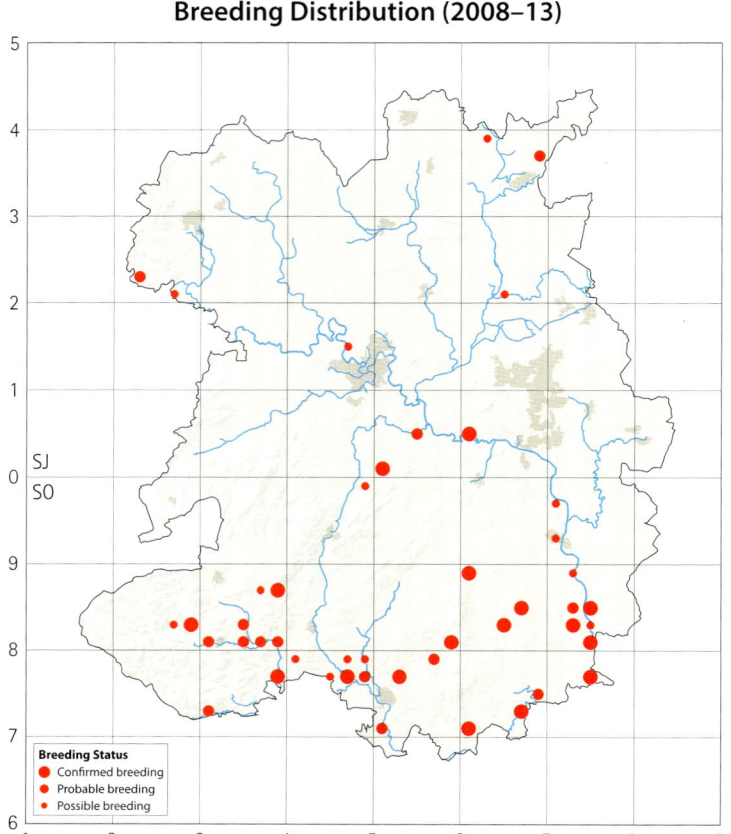

Breeding Status
● Confirmed breeding
● Probable breeding
● Possible breeding

Winter Distribution (2007–13)

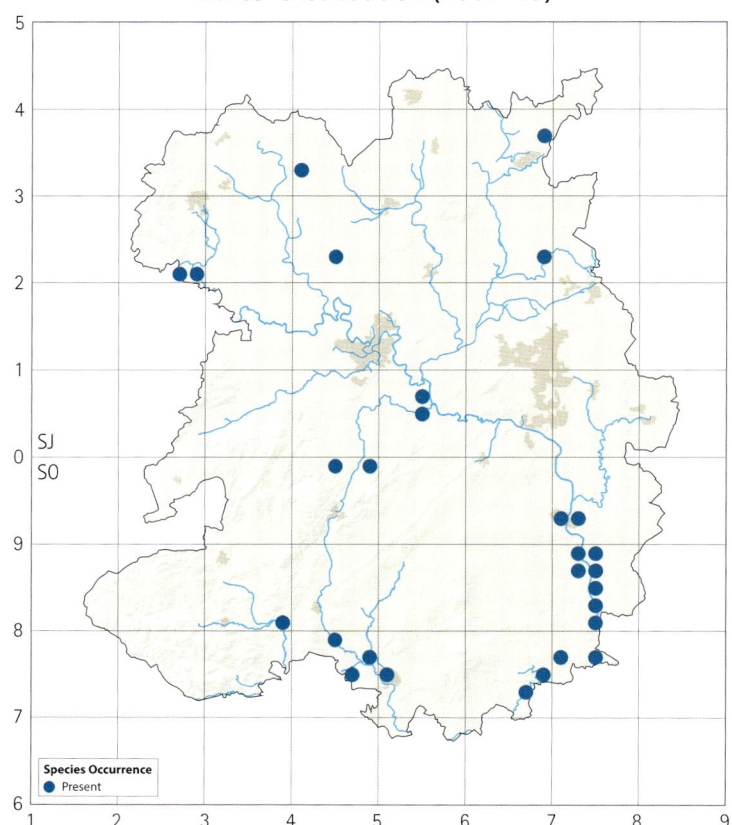

Species Occurrence
● Present

Occupied Tetrads

Atlas period (breeding)	1985–90		2008–13		Change	
	Number	%	Number	%	Number	%
Confirmed	3	0	17	2	14	467
Probable	1	0	14	2	13	1300
Possible	0	0	15	2	15	x
TOTAL	4	0	46	5	42	1050
Tetrads with Winter Records (2007–13): 28 (3%)						

perhaps that colonisers come from the south, possibly from the well-established population in the Forest of Dean. Most records are from running water, notably the Severn, and the Teme and its tributaries, particularly the Clun, but there are records from standing water too, including north-east of Ludlow where broods were found on small, well-vegetated, tree-fringed pools well away from significant rivers. The lack of records from the larger waters in the Ellesmere area is notable; perhaps colonisers have yet to reach this far north, or it may reflect the species' preference throughout the year for more secluded habitats.

Survey work in two tetrads along the River Kemp in 2013, and in four tetrads along the River Teme just over the boundary in northern Herefordshire, in 2013 and 2014, revealed up to five pairs per tetrad, with a mean of two pairs (Wall *in prep.*). This would suggest a population of some 60–90 pairs, a major increase since the estimate of three to six for the *Atlas* (1992). Significant further population growth appears certain.

The national BTO Winter Atlas (1986) shows records from just four of the County's 33 10km squares, amounting in all to fewer than 10 birds. Many subsequent winter records have continued to be of just one or two, and the only counts exceeding 10 were 14 at Quatford in February 1991, 25 on the Borle Brook in December 2008, 14 at Bromfield in January 2010 and 17 at Onibury in January 2011.

The rule of thumb used by Musgrove *et al.* (2013), that winter numbers are 1.5 times greater than the tally of breeding adults, would indicate a wintering population of some 180–270, but the winter distribution map shows sightings from only 28 tetrads, again with few of them in the north, suggesting that many move away. WeBS counts reinforce this view as, despite 87 sites having been surveyed over the years, generally on a regular basis, Mandarins have only ever been seen at nine of them, and all 24 sightings have been of singletons. So it may well be that counts of 64 at Trimpley Reservoir (Worcestershire) in December 2009, and of 101 on the adjacent River Severn, a location just 6km downstream of Highley, in December 2010 (WeBS), included Mandarins that had bred here.

It may seem perverse to regard this stunningly beautiful bird as an example of environmental pollution – 'a form of litter … carelessly tossed into … natural ecosystems' (Fox 2009) – but such it is. And those who survey wetlands, counting Canada Geese, introduced Greylags, Mallard (often released) and now Mandarin Ducks, may well agree with the late Colin Bibby (2000): 'the spread of exotic species somehow seems to rob further wildness from a country where this is already scarce'.

Tom Wall

Garganey
Spatula querquedula
Rare passage migrant, very rare summer visitor

Jim Almond, Venus Pool, 8 May 2014

Garganey is strictly migratory, breeding across much of Europe and western Asia, with the western European population wintering in southern Africa. Britain is at the north-western edge of the range, and has a small widely scattered breeding population.

Beckwith (1879) described it as 'very rare; but has, I believe, been obtained on the Severn, though not recently'. This was probably the same as the specimen in the collection at Clungunford House (Rocke 1866). Forrest (1899) stated 'a nest was found near Shrewsbury, about 1888 and identified by H Shaw, from the eggs and lining of down, as belonging to Garganey'. A pair was also recorded at Ellesmere in April 1906.

There were no further records until 1957, and the *Handlist* (1964) described it as a passage migrant in small numbers, and noted records in recent years from only five sites, all in or near the Severn Valley.

The *Atlas* (1992) noted only three records, all of single males, at Chelmarsh (31 August 1986), Tong Lake (25 April 1987) and Venus Pool (29 May 1989).

Since then the species has been seen every year, at a number of different sites, but most frequently at Venus Pool, in 15 out of the 25 years 1989–2014, and at Allscott Sugar Factory, in 13 out of the 20 years 1989–2009, after which the site became unsuitable. In 2003 a pair was found at Pen-yr-Estyn. A female first seen 15 April was joined by a male, and the pair were seen from 13 to 24 May, with the male last seen on 26 May, but no evidence was seen to suspect nesting.

During recent Atlas field work, 2008 was probably the best year, producing records of 10 birds (all males) from four sites: Eyton Moor (three on 31 May); Venus Pool (two on 3 June); Allscott Sugar Factory (five in total, the first on 18/19 May; two eclipse males on 3 August; and two at Mirelake on 30 August). A female or immature male

at Venus Pool on 14 and 15 January 2012 was an unusual winter record, while in 2013 a summer plumage drake was at Polemere from 30 March to 2 May, singles were at Wood Lane on 1 May and 23 September, and at Venus Pool there were four records: a pair on 11 June, a juvenile from 15 September to 1 October, two on 6 October and one on 8 October.

These records reflect a national pattern of individuals or pairs being seen over a prolonged period in the spring, often at sites with apparently suitable breeding habitat, but they usually move on without any obvious breeding attempt. Juveniles and adults in eclipse return over an even longer period in the autumn, extending into the winter months.

In 2014, a female arrived on the floods at Wall Farm on 6 May, with a pair regularly present from 19 to 30 May. Only the male was recorded from 31 May to 7 June; presumably the female was incubating, as on the 7 June she was seen with nine ducklings. She was recorded next day and also on 16 and 17 June but sadly the ducklings were only seen the once. Neither of the pair was seen after 17 June, by which time the floods had dried up.

This represents the first confirmed breeding since the historical record of a nest near Shrewsbury towards the end of the nineteenth century.

Also in 2014, there was a drake at Wood Lane for a week from 18 April, a single at Venus Pool on 8 May, a juvenile (probably not raised locally) at Whixall floods on 26 July, and one at Chelmarsh Reservoir on 22 August.

The chart shows the number of different individuals seen each year. Records came from no more than two sites per year, until 1992, when there were four. There have been four in three further years, but only in 2001 and 2014 have there been as many as five. Although numbers remain small and fluctuate from year to year, there is a steady underlying upward trend. The breeding habitat (shallow freshwater and marshes) is uncommon, so their scarcity is unsurprising. However, the increase in observers and their ability to identify females and eclipse, assisted by better optics, may have also contributed to the apparent increase.

Martin Grant

Estimated Number per Year (1950–2014)

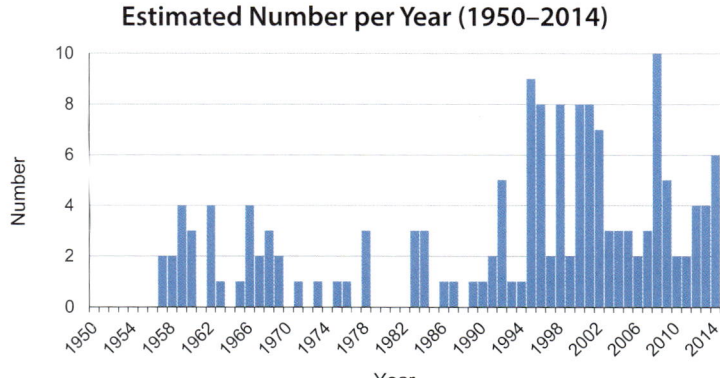

Shoveler (Northern Shoveler)

Spatula clypeata

Uncommon winter visitor, very rare breeding species

Dawn Micklewright, Venus Pool, 16 March 2014

Shoveler breed across north and central Europe and Asia, close to shallow lakes and small pools in marshy ground, particularly where rank vegetation provides adequate cover for the nest. It moves south for the winter.

In the early part of the nineteenth century it was said to be 'not uncommon in the winter', but at the end of that century it was 'rare, more often found in spring and autumn' on passage, frequenting large lakes and reservoirs. It was more numerous in some winters than others, a status similar to that in the present day.

The first reference to confirmed breeding was at Hencott when a very young bird shot in August 1869 was 'evidently reared there'. Beckwith (1886) wrote that 'within the last few years there have been five or six instances of its breeding in north Shropshire and the Staffordshire border'. A few pairs also nested annually between 1895 and 1897 on Burlington Pool and nearby at Weston Park.

In the early part of the twentieth century, at least one pair nested on marshy ground at Minsterley and raised young in both 1907 and 1908. Most winter reports were of pairs, with a flock of 35 at Shrawardine Pool on 21 February 1937 being the exception.

A considerable increase in winter numbers was noted in the 10 years before publication of the *Handlist* (1964), with a record count of 102 at Venus Pool on 21 February 1959. Other maxima in this period were at Ellesmere, 48 in December 1957, and Shavington, 46 in November 1961.

Breeding at Venus Pool was described as 'fairly regular' at that time, including a pair with young in 1956 and a female with young in May 1960. Two pairs also bred at 'Worfield Bog' (now known as Cranmere Bog) in May 1963, with probable breeding at Shrawardine,

Breeding Distribution Change (Probable & Confirmed Breeding Only) (1985–90 to 2008–13)

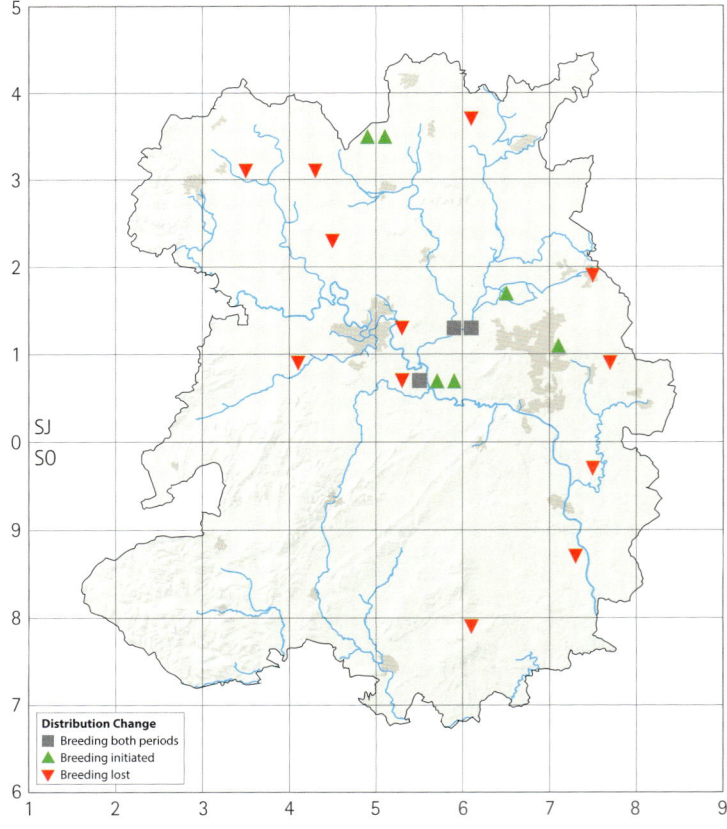

Distribution Change
- ■ Breeding both periods
- ▲ Breeding initiated
- ▼ Breeding lost

Occupied Tetrads

Atlas period (breeding)	1985–90		2008–13		Change	
	Number	%	Number	%	Number	%
Confirmed	4	0	3	0	-1	-25
Probable	11	1	6	1	-5	-45
Possible	8	1	7	1	-1	-13
TOTAL	23	3	16	2	-7	-30
Tetrads with Winter Records (2007–13): 54 (6%)						

and possibly at sites in the lowlands to the north-west and north-east of Shrewsbury, and further south-east down the Severn Valley.

There was a marked increase in breeding season records in the late 1960s, suggesting more than the odd pair at the meres around Ellesmere. A peak occurred in 1968, with pairs at nine sites, but breeding was unconfirmed at most, although there were about 24 juveniles reported at Shavington on 5 August 1968. Breeding evidence was mapped in the 1968–72 BTO national Atlas in eight of the County's 33 10km squares, all except two in the north, one with confirmed, two with probable and five with possible, breeding.

There then followed a rapid decline, for no apparent reason, with no confirmed breeding for 10 years until 1977, when six juveniles were recorded in August at both Cranmere Bog and Allscott Sugar Factory. Probable or confirmed breeding occurred in 1979 at five sites: Allscott Sugar Factory, Shrewsbury Sewage Farm, Baggy Moor, Oss Mere and Venus Pool. In 1983 there were three broods at Baggy Moor with a total of 21 young. Breeding was confirmed again the following year at Baggy Moor, and also at Fenemere.

The *Atlas* (1992) showed confirmed breeding in four tetrads (Allscott Sugar Factory, Cranmere Bog, Venus Pool and a pool near Shifnal), but only at Venus Pool did it occur in more than one year of the six, and probable breeding in a further 11 tetrads, including Fenemere, but not Baggy Moor, where the river had been canalised and the water level lowered.

There appears to have been a slight reduction in the breeding population since then. In the recent Atlas period, breeding was confirmed in only three tetrads with probable breeding in only six more.

Whixall Moss (two tetrads) has emerged as an important breeding site with 18 juveniles in June 2008, and a pair and three downy young were seen at Allscott Sugar Factory on 28 July 2008, where breeding was also confirmed in the 1985–90 Atlas, although the latter site has since become unsuitable. Pairs were seen in suitable habitat (probable breeding) in six tetrads, at Mirelake in May and June 2009, and May 2010 and May 2011; Priorslee Flash on 11 April 2013; Venus Pool in May, June and July 2013; and Eyton on the Weald Moors in May and June 2008. This later site was only suitable in 2008, when at least two pairs were present on a flooded potato field. The probable breeding at Beckmoor spans two tetrads, where at least one pair were seen on both sides of the boundary on 22 April 2013. Although they were in suitable breeding habitat – a series of shallow weedy pools – they may have been late departing winter migrants. This could also apply to some of the other pairs, which were only

Winter Relative Abundance (2007–13)

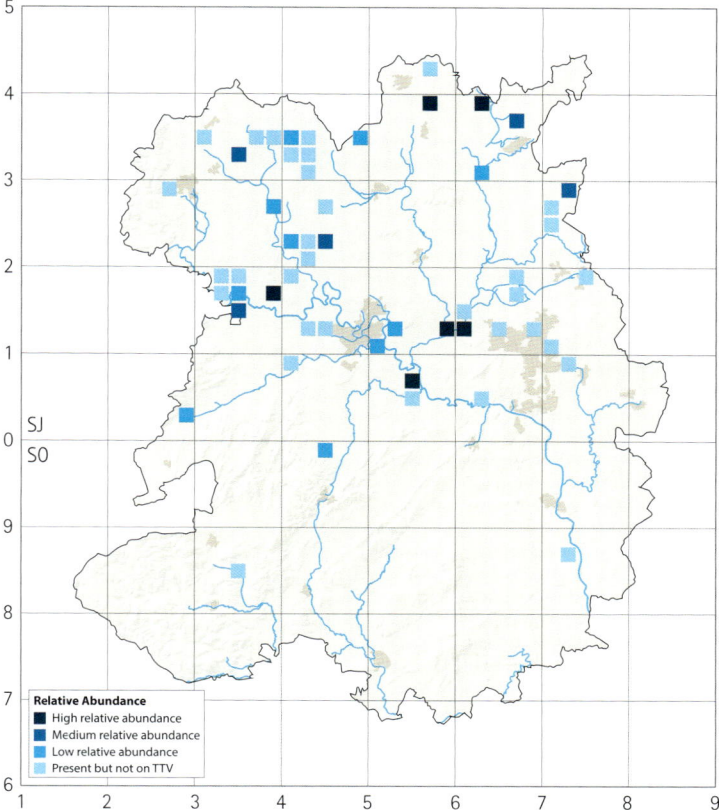

Relative Abundance
- High relative abundance
- Medium relative abundance
- Low relative abundance
- Present but not on TTV

recorded early in the breeding season, and to the records of possible breeding in both Atlas periods, which have been omitted from the breeding distribution change map.

Although present in 16 tetrads between 2008 and 2013, breeding does not occur every year, but in a good year such as 2008 there may be up to five breeding pairs. There were also occasional records of non-breeding birds.

In winter, Shoveler were found in 10 of the County's 33 10km squares in the 1981–84 national BTO Atlas, and have been found at 51 of the 87 WeBS sites since 1960. The two main sites are Chetwynd Pool, where a site and County record of 275 were counted on 19 October 2008, and Shavington Pool, with 200 on 1 November 1973. There was also a site record of 130 on the canal floods at Whixall on 19 March 2007. Apart from these three sites, there have only been two WeBS counts of more than 60, both at Allscott Sugar Factory, 64 in September 1999 and 62 in February 2012.

WeBS counts at the two main wintering sites show Shoveler is more numerous in some winters than others. In line with national trends, peak counts occur in October or November and include birds on passage to wintering areas in France and Spain. However, a greater proportion linger in milder winters. The winter of 2008–09 saw the highest wintering population of around 200, with the lowest in 2010–11 of around 70. Mean temperature in November 2008 was slightly above average, while November 2010 was the coldest since 1993 and December the coldest in over 100 years, with considerable snow and most areas of water frozen over.

The species is uncommon in winter, and was recorded in 50 tetrads during the recent Atlas, with the highest counts at the

WeBS Counts at the Main Sites (1960–61 to 2014)

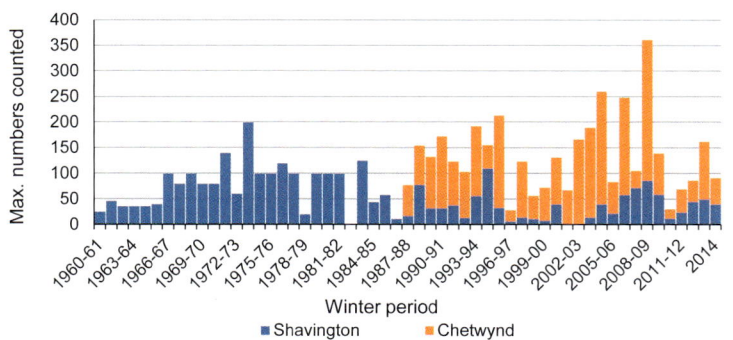

Winter period

■ Shavington   ■ Chetwynd

favoured sites of Mirelake (93) and Allscott Sugar Factory (80) in November 2010 (probably including some of the same birds), Shavington (75 in October 2008), Brown Moss (55 in February 2008), Venus Pool (100 in November 2012) and on the canal floods at Whixall Moss (80 in November 2013). Chetwynd Pool (275 in October 2008) does not appear on the map, as more than half the tetrad is in Staffordshire.

The ringing recoveries illustrate the large migratory distances travelled by Shoveler. Two nestlings were ringed at Venus Pool on 8 July 1962. One was reported dead (shot) on 4 September 1963 in Boguchar (Russia) a distance of 3,025km (the second furthest recorded distance travelled). The other was shot on 15 March 1966 at St Etienne-De-Mont Luc (France), a distance travelled of 609km.

A female nestling, ringed near Stockholm (Sweden) on 19 June 1959, was shot at Edgerley on 19 December 1960, a distance of 1,565km, while, travelling in the opposite direction, an adult male ringed at Amager Sjaelland (Denmark) on 3 June 1970 was shot only four and a half months later on 29 October 1970 near Whitchurch, a distance of 1,032km.

Martin Grant

Gadwall

Mareca strepera

Scarce winter visitor, rare summer visitor

Gadwall breeds across central and northern Europe and Asia, with most of the population migratory. At the end of the nineteenth century, it was described as a very rare winter visitor. Beckwith (1886) stated 'the Gadwall has always been rare, about thirty years ago a young male was killed on the Severn [he does not say where], but I have no note of its occurrence since'. Forrest (1899) noted that it had occurred only twice, and added a record of one shot at Winsley on 1 November 1889, which 'when first seen had a mate'. A male at Hawkstone in December 1907 was said to be 'only the fourth record'.

The *Handlist* (1964) described it as a non-breeding visitor and passage migrant, with records of 'odd birds ... possibly some are escapes'. It noted that five males at Atcham Bridge on 14 January 1963 was the largest number recorded to date, but went on to say that 'an increase in reports in future might occur, owing to release of considerable numbers by the Wildfowlers Association of Britain and Ireland (in other parts of the country), commencing in 1964'.

In the 1981–84 national Winter Atlas, it was mapped as occurring in very small numbers in only four of the County's 33 10km squares, and in 1985, five at White Mere at the end of October and at Chelmarsh in December were the largest flocks recorded by then. The first count in double figures was a flock of 12 at Allscott Sugar Factory on 15 August 1989.

Breeding habitat is shallow lowland lakes and pools with surrounding vegetation to hide the nest. None were mapped here in the 1968–72 national BTO Atlas, and the first confirmed breeding was at Venus Pool, where three young were raised in 1980, followed by Shrewsbury Sewage Farm, with seven ducklings on 7 May 1981, and Priorslee, with five young on 25 July 1984.

The *Atlas* (1992) showed confirmed breeding only at Cranmere Bog (in 1988), but refers to it also occurring at Chetwynd Pool in 1986 ('although the lake at Chetwynd is wholly in the county, more than half the tetrad is in Staffordshire, so the record does not appear on the map'). Probable breeding was shown at four other sites: Allscott Sugar Factory, Chelmarsh, Knighton Reservoir and Venus Pool, and possible breeding at Shavington and the Weald Moors near Tibberton.

The population has continued to grow, and in 2008–13 there was confirmed breeding at Allscott Sugar Factory, and probable breeding at Cranmere Bog, Chelmarsh Scrape and Venus Pool, with additional confirmed breeding sites at Eyton Moor, where a male, two females

Jim Almond, Venus Pool, 15 January 2015

and five small juveniles were seen on 3 July 2008; at Ellerton, with juveniles on 15 August 2010 and at Mirelake/Allscott Sugar Factory, where three pairs with five downy young were seen on 25 May 2011. Of the sites with probable or confirmed breeding in 1985–90, only Knighton Reservoir had no records, and the breeding distribution change map also shows that pairs relocated from Chelmarsh Reservoir to the Scrape, and, on the Weald Moors, from Tibberton to the adjacent tetrad at Wall Farm. There were five additional probable breeding sites, with pairs at Priorslee Lake, Sambrook, Shavington, Wood Lane and near by at Lyneal. Chetwynd Pool also had a pair again, on 20 April 2012.

Although breeding was confirmed or probable in 11 tetrads, some sites are not occupied every year, while Allscott Sugar Factory and Mirelake are no longer suitable, and the Eyton Moor site was only suitable in 2008. No year had more than one confirmed breeding record, and some of the pairs were only seen in April and were probably late departing over-wintering birds, so the breeding population may be as low as one pair, and is unlikely to be as many as five.

The early years of the twenty-first century saw wintering flocks increase dramatically, in line with national trends. A flock of 17 at Dudmaston in October 2000 was followed by 23 at Ellerton in

Breeding Distribution Change (1985–90 to 2008–13)

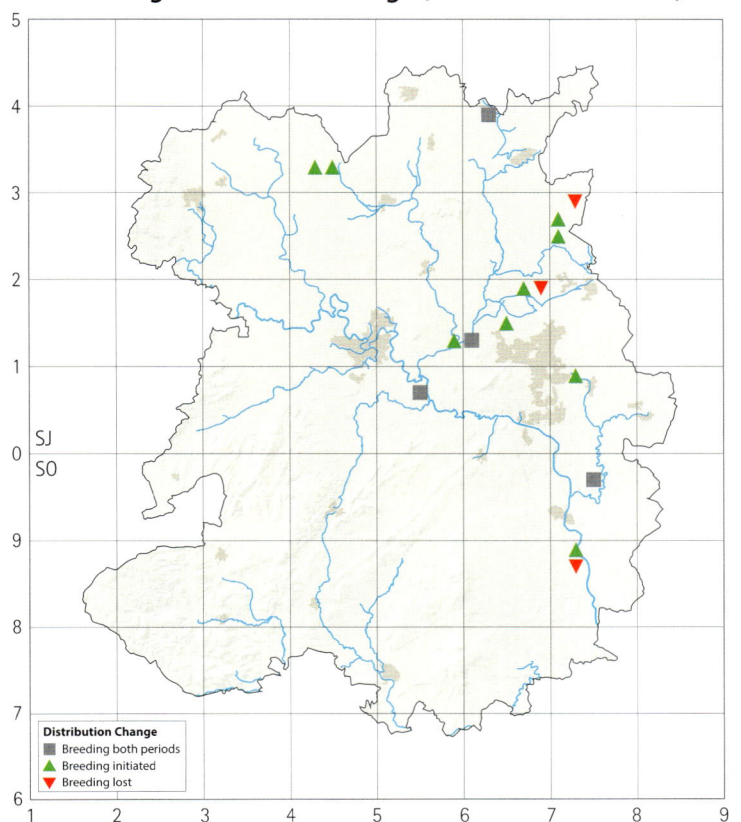

Distribution Change
- ■ Breeding both periods
- ▲ Breeding initiated
- ▼ Breeding lost

Winter Distribution (2007–13)

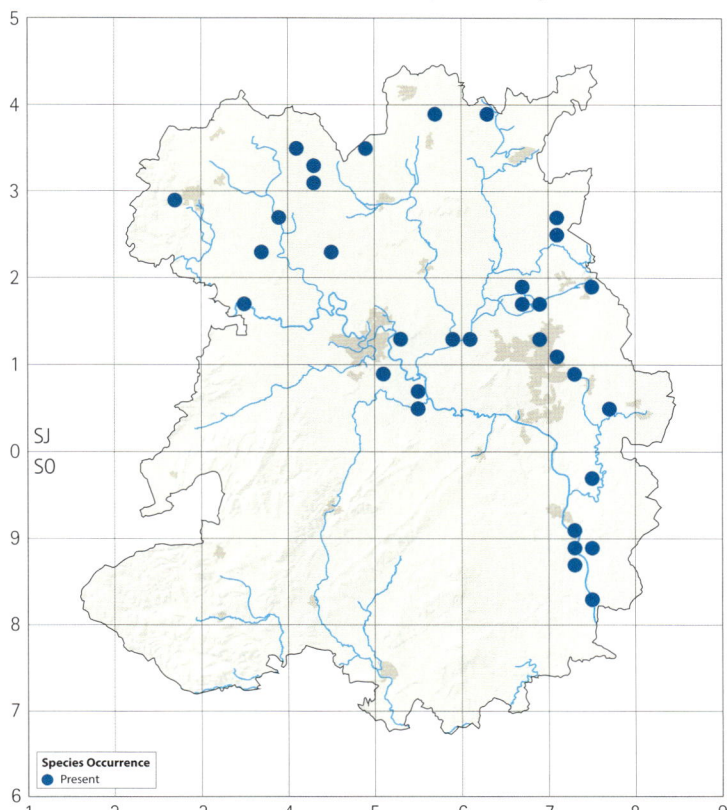

Species Occurrence
- ● Present

Occupied Tetrads

Atlas period (breeding)	1985–90		2008–13		Change	
	Number	%	Number	%	Number	%
Confirmed	1	0	4	0	3	300
Probable	4	0	7	1	3	75
Possible	2	0	2	0	0	0
TOTAL	7	1	13	1	6	86
Tetrads with Winter Records (2007–13): 33 (4%)						

November 2003. This was eclipsed by an incredible flock of 42 at Sambrook Mill on 24 October 2011.

Gadwall is more widespread in winter, when dispersal from natal areas is supplemented by migrants from Iceland and the near Continent, and here it is now found sporadically on many northern lakes and meres and along the Severn Valley in the south east. A few of these sites are favoured: there have been only 16 WeBS counts of more than 10, from only four sites, maximum 29 at Ellerton, also in October 2011. It has occurred at 30 different WeBS sites, but on more than 10 occasions at only seven of them, and is less than annual at the Ellesmere meres. In 2012, for example, it was found at only seven WeBS sites, and the total of the maximum count at each of them was 51. During recent winter Atlas fieldwork, 10 or more were counted in only three tetrads.

Most of the national breeding population is derived from captured, pinioned and then released wild birds in Norfolk around 1850, and the population has increased and expanded since then, rapidly in the last 20–30 years, and the continental populations have also increased, so it is likely that the local breeding and wintering populations will continue to slowly increase.

Martin Grant

Wigeon (Eurasian Wigeon)

Mareca penelope

Uncommon winter visitor

Wigeon come to the UK for winter from breeding grounds in Scandinavia and across European Russia, with smaller numbers from Iceland.

In the mid-nineteenth century, Wigeon were common on the pools and meres in the north, and in freezing weather they moved to the River Severn and its tributaries. By 1886, according to Beckwith:

'The Widgeon from some cause or other that is impossible to explain, has of late years, become rare ... now only a few small lots visit the meres, and killed birds are scarcely ever exposed for sale by the local game dealers'. In 1908 Forrest wrote 'sometimes numerous in severe weather, though it rarely stays long in one place'.

By the middle of the twentieth century numbers had recovered, and they could be found on all suitable waters between October and May, with peak numbers occurring from January to March. The highest number recorded was about 600 at the Mere (Ellesmere) in March 1955, although between 250 and 350 were more commonplace. Similar numbers could be found at Venus Pool, but during cold spells when pools were frozen, they moved to the River Severn at Leighton. Regular flocks of up to 100 could also be found at Allscott Sugar Factory and Shavington (*Handlist* 1964).

Numbers at Shavington increased and reached a peak of 1,000 in 1974, and along with the Ellesmere meres continued to attract 200–300 in most winters. Both sites showed a dip during the 1990s, at which time Leighton became a regular wintering location, with up to 300 in some years, but regular wintering ceased there after 2001, and in 2007, 78 at Leighton was described as unusual.

Regular WeBS counts at the Severn–Vyrnwy confluence began in 1995. The number present there depends very much on the amount of flooding, but high counts of 1,370 in December 2000, 1,203 in February 2002 and 1,211 in January 2011 show the value of this area.

The chart shows maximum WeBS counts for each winter period at Ellesmere meres and Shavington from 1960, and the Severn–Vyrnwy confluence from 1995.

The winter relative abundance map clearly illustrates the areas along the River Severn which are prone to flooding: the Severn–Vyrnwy confluence, around Leighton, and the stretch south of

Jim Almond, Venus Pool, 15 January 2015

Bridgnorth. In addition to the main meres in the north, small pools are regularly used, as are wetlands such as Whixall Canal Floods, Baggy Moor and the Weald Moors. There are fewer suitable sites in the south-west, but some were found along the River Onny around Bromfield, at Walcot Park lakes, and on several smaller pools. Wigeon numbers vary from year to year, and many sites will not be occupied annually, but they can also be found on very small ponds, so some may have been overlooked.

The maximum monthly WeBS totals for each of the winter periods during recent Atlas fieldwork varied between 1,689 in January 2011 and 480 in January 2012. The main Wigeon sites will have been counted but small numbers at isolated pools will not be included in these totals.

Nationally, the wintering population increased steadily from the mid-1960s, peaking in 2005–06, then, after falling slightly the following year, it has remained stable. Locally, the winter population fluctuates more, depending both on the severity of the weather and the amount of flood water present, and based on WeBS results it probably varies between 500 and 2,000. This is a more accurate guide than the winter population estimate based on TTV counts of 1,200–1,450 individuals, which does not adequately reflect the annual fluctuations.

A pair was believed to have nested at Peplow Hall in 1906, as young birds were shot there in the early autumn, but no dates were given. Wintering birds usually return in September, but late August

Occupied Tetrads

Tetrads with Winter Records (2007–13):   103   (12%)

Winter Relative Abundance (2007–13)

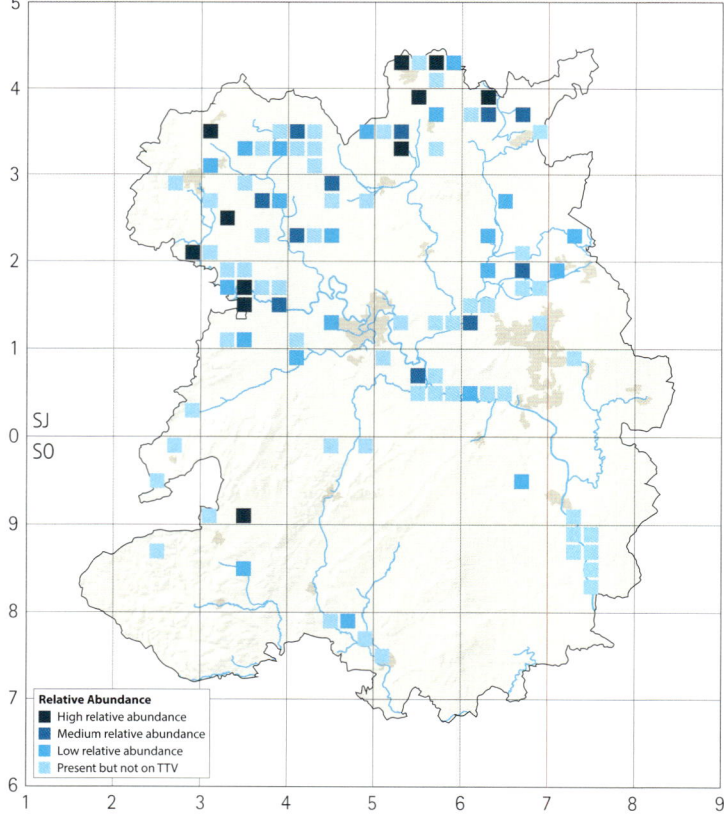

Relative Abundance
- High relative abundance
- Medium relative abundance
- Low relative abundance
- Present but not on TTV

WeBS Counts at the Main Sites (1960–61 to 2014)

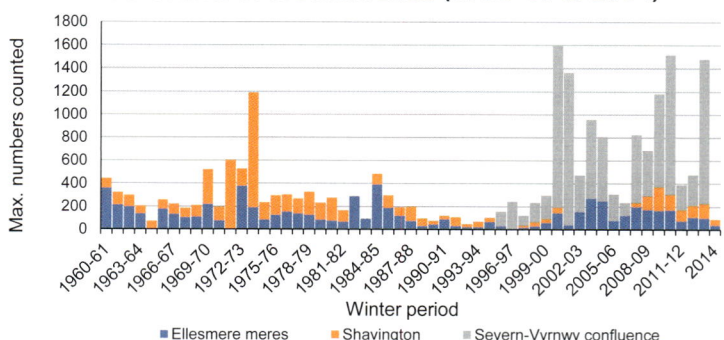

Max. numbers counted

Winter period

- Ellesmere meres
- Shavington
- Severn-Vyrnwy confluence

records are not infrequent, so there must be some doubt attached to this claim. However, it seems likely that breeding has been attempted occasionally. Sometimes individuals remain throughout the summer and females can disappear for some time, presumably incubating eggs, but a successful outcome has never been recorded.

Recent Atlas fieldwork in the breeding season found late wintering birds at six locations during April. Additional June records came from Whixall Moss in 2008 and 2010, Shavington in 2008 and Walcot Park in 2009, and there was a late July report from Venus Pool in 2008. This is similar to the 1985–90 Atlas results, with seven breeding

season records, including a drake at Chelmarsh in May 1987 and a duck at Venus Pool through most of May and June 1990.

There have only been three ringing recoveries. One shot at Shrewsbury in December 1959 had been ringed in Belgium in February 1958. This is the only overseas movement noted and only two have come from within the UK; one ringed in Norfolk in December 1972 was shot just six weeks later near Shrewsbury, and one from Leicestershire in January 1989 was shot at a confidential site the following December.

Allan Dawes

Mallard

Anas platyrhynchos

Fairly common resident; population supplemented by annual releases

Mallard is the most widely distributed of our ducks, occurring across North America, temperate Eurasia, parts of North Africa and Australia, and New Zealand.

Generally referred to as the Wild Duck, it was 'sparingly distributed over the various streams and brooks. The greatest bulk of them are to be found in the decoys or upon those lakes and ponds where they are fed and protected' (Rocke 1866). At this time, many were killed during the breeding season, but the Wild Birds Protection Act of 1872 provided a close season between 1 April and 1 August and put a stop to this practice. By 1879 the status of Mallard was changing, and it was 'becoming more numerous every year, many now breeding about our large pools, and even the small wet bogs, so common throughout the county, are usually frequented in summer by a pair of ducks' (Beckwith 1879).

By the end of the nineteenth century it was mapped as 'common' (Holloway 1996).

There was no apparent change in the first half of the twentieth century, other than occasional weather-related movements, such as that caused by severe weather in early 1939. This resulted in a large influx, but they 'nested in normal numbers, though rather late'.

This status was repeated (*Handlist* 1964), and 'from September to March between 500 and 600 could be found at both the Ellesmere meres and Shavington, occasionally rising to a peak of 800 to 1,000'.

The chart shows the maximum winter numbers at these two sites, where WeBS counts have been completed almost annually since 1960. Both sites show a decline starting in the late 1980s, in

Jim Almond, Chelmarsh, 11 September 2010

accord with the national trend, which is thought to be due to fewer crossing the channel since the advent of milder winters. WeBS has demonstrated a 36% decline nationally over the past 25 years, and 17% in the last 10.

The winter population includes residents and the decreasing autumn arrivals from north-west Europe, and numbers are boosted still further by releases on shooting estates. In 1985, an estimated 400,000 were released nationally, and about 600,000 were shot annually. These numbers have since fallen, but recent figures are not available. However, two estates in the north-west released 1,000 and 500 annually during the recent Atlas period. Such releases usually take place on small waters, and many are shot in September, so the survivors may not be included in WeBS counts on the larger waters, but many will have been found during the early winter TTV counts.

The effect of severe winter weather can be seen by the peaks at the Ellesmere meres in 1980–81 and 2012–13, which are caused

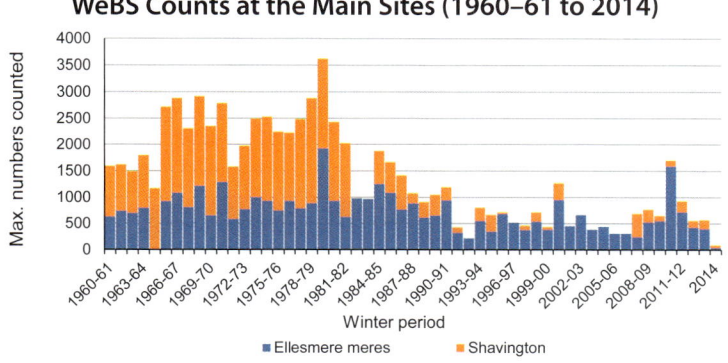

WeBS Counts at the Main Sites (1960–61 to 2014)

Max. numbers counted (y-axis: 0 to 4000)

Winter period (x-axis: 1960-61 to 2014)

■ Ellesmere meres   ■ Shavington

Breeding Distribution Change (1985–90 to 2008–13)

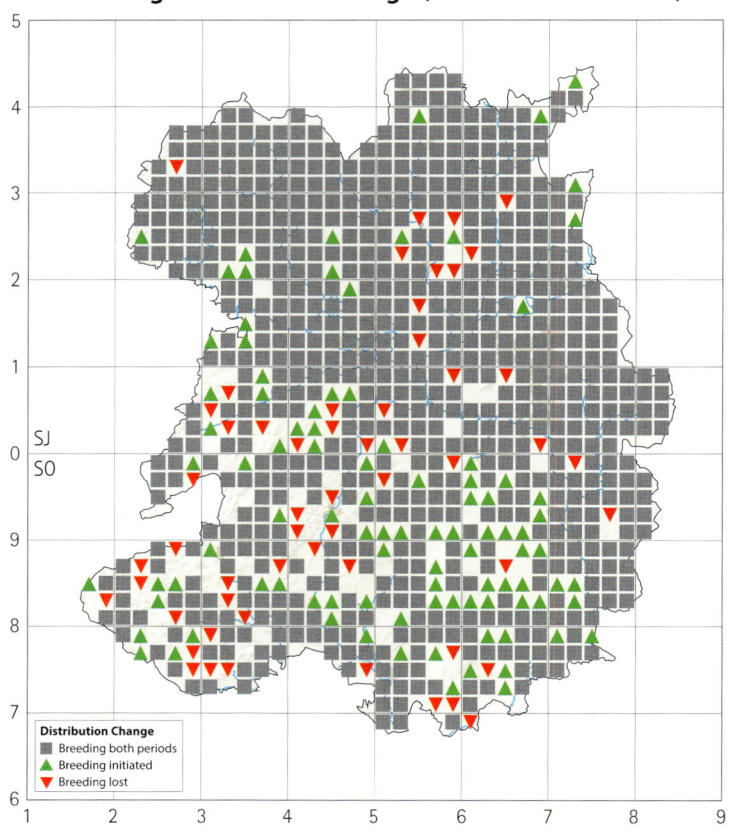

Distribution Change
- ▪ Breeding both periods
- ▲ Breeding initiated
- ▼ Breeding lost

Occupied Tetrads

Atlas period (breeding)	1985–90		2008–13		Change	
	Number	%	Number	%	Number	%
Confirmed	519	60	440	51	-79	-15
Probable	147	17	242	28	95	65
Possible	56	6	82	9	26	46
TOTAL	722	83	764	88	42	6
Tetrads with Winter Records (2007–13): 702 (81%)						

both by an influx of visitors arriving from the continent and local movements from small ponds and marshes that have iced over. The large numbers that gather on the meres at these times usually manage to keep small patches of water free from ice but once these become frozen they have to move on, probably to the coast.

Mallard are widespread during the winter, though less so than in the breeding season, because some of the less hospitable sites, particularly on the higher ground in the south, are vacated. Highest abundance occurs along the Severn Valley, and at the numerous meres and pools in the north.

The highest WeBS total during the recent winter Atlas period was 3,010, counted at 42 sites in December 2011, but Mallards were found in 699 tetrads and most of these have no WeBS site, so the total number will be far higher. Based on TTV counts, the estimated winter population is between 9,000 and 10,500, but this will fluctuate in accordance with both local and European weather conditions.

Mallard have an extended breeding season. In 1894, a nest with eggs was found on 24 December at Woolstaston, and during breeding season fieldwork for the recent Atlas, young could be found throughout the recording period. There was a 6% increase in the number of tetrads where they were found, compared with the *Atlas* (1992).

The areas of high relative abundance are broadly the same in both seasons, mainly in the north where most waters are to be found. However, the breeding distribution change map shows that the increase in range has occurred mainly in the south. The recent increase in population recorded by BBS, about 60% between 1995 and 2014, may be forcing some to move into sub-optimal habitats found in the southern uplands.

Based on TTV counts, the estimated breeding population is between 850 and 2,000 pairs, equivalent to just one or two pairs per tetrad. The *Atlas* (1992) used an average of around three to six pairs per tetrad, and this seems more appropriate. Given the increase in range,

Winter Relative Abundance (2007–13)

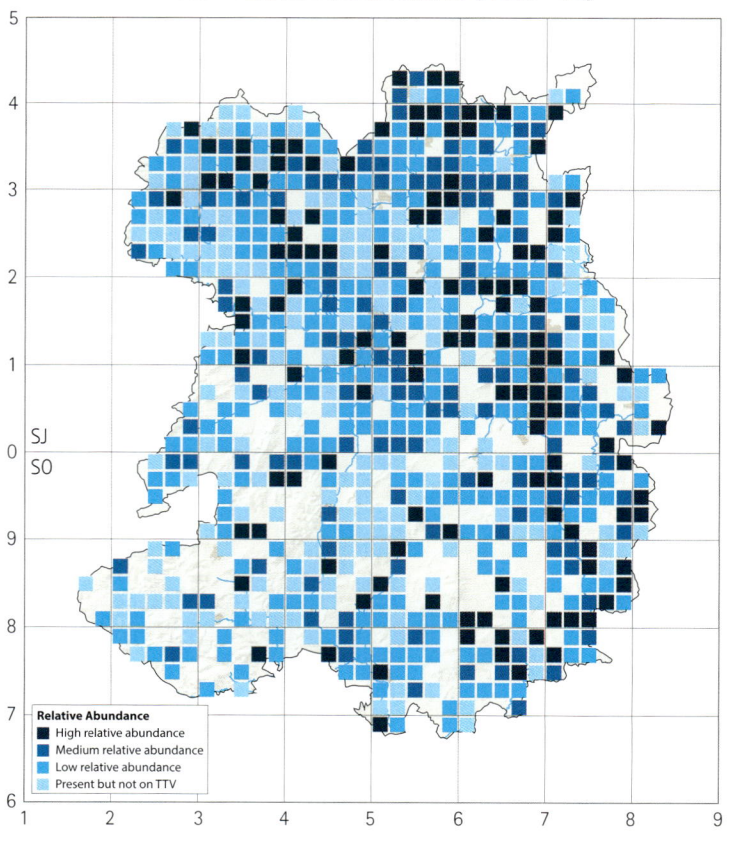

Relative Abundance
- ▪ High relative abundance
- ▪ Medium relative abundance
- ▪ Low relative abundance
- ▪ Present but not on TTV

BBS Trend (1997–2014)

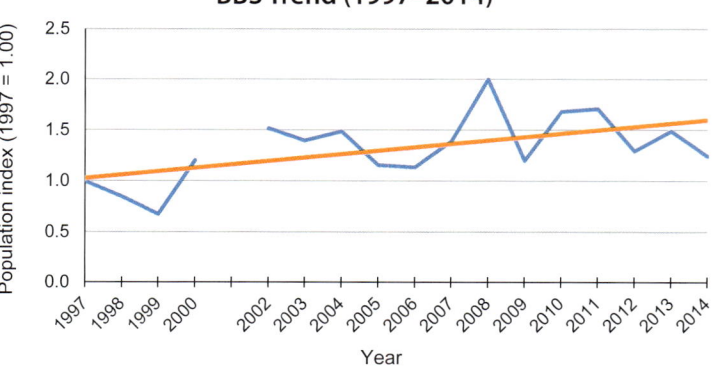

the population is estimated at between 2,300 and 4,600 pairs.

Most of the 215 ringing recoveries have been of shot birds; 117 have been from this or adjacent counties. A further 54 have originated from Slimbridge, reflecting the large amount of ringing carried out there. Two from Germany have been taken here, and seven movements to or from Denmark have been recorded, the furthest travelled being

a full-grown male ringed in September 1971 in Viborg and shot near Bedstone, 3y 0m 26d later, and 929km distant.

Although winter flocks are falling, the Mallard is still widespread, especially during the breeding season, and should remain the familiar 'Wild Duck' for the foreseeable future.

Allan Dawes

Pintail (Northern Pintail)

Anas acuta

Scarce winter visitor

This species breeds widely over northern temperate and arctic zones, and winters as far south as the tropics. Those wintering in Britain come mainly from the north-west European population, including Iceland.

One obtained on the Ruyton Brook in the winter of 1832–33 was the only inland record known to Eyton (1839). Later, in the 1860s and 1870s, a few were being taken annually in decoys, but by 1908, although described as 'not very uncommon' they were not found every year. A remarkable influx of several species, including Pintail, was noted in 1903 and 1908, but no details were provided, and only four further reports were received prior to 1950; under recording must be suspected.

The *Handlist* (1964) described it as a 'non-breeding visitor. Small numbers at the Ellesmere meres in most years' and 'a few records from Venus Pool ... including 11 in March 1963'.

Singles, pairs and the occasional small party then occurred almost annually up to 1980, but although sightings were widespread, over half came from Polemere, the Ellesmere meres and Hem Flash. After 1980, the number of sightings began to increase, but 16 over the River Severn at Cross Houses in January 1986 was considered unusual enough to be described as 'astonishing'.

In January 1993, a flock of 40 was the first report from the Severn–Vyrnwy confluence at Melverley. Pintail have since been recorded annually at this site (with the exception of 1997) and after a period of flooding they often arrive in good numbers, exceeding 300 in November 2000, February 2002 and January 2014. Since 2000, they have also become a regular feature at Whixall Canal Floods, with a maximum count of 100 in February 2007.

A steady growth in the national population began in the early 1970s and the increase in local sightings coincides with this, but

regular WeBS along the Severn–Vyrnwy confluence since 1995–96 may also be locating flocks which were previously overlooked.

The recent winter Atlas map shows the main sites along the Severn–Vyrnwy confluence, Whixall and the Ellesmere meres. Drainage schemes and arable crops at Hem Flash have reduced the suitability of the site although flooding still occurs periodically. The remaining locations are mainly well-watched water bodies that Pintail visit occasionally. The latest spring sighting recorded for the recent Atlas was 18 April, in 2010, and the first returning bird was on 18 July in the same year.

Based on WeBS from the main sites, the winter population during

Occupied Tetrads

Tetrads with Winter Records (2007–13):   26   (3%)

Winter Distribution (2007–13)

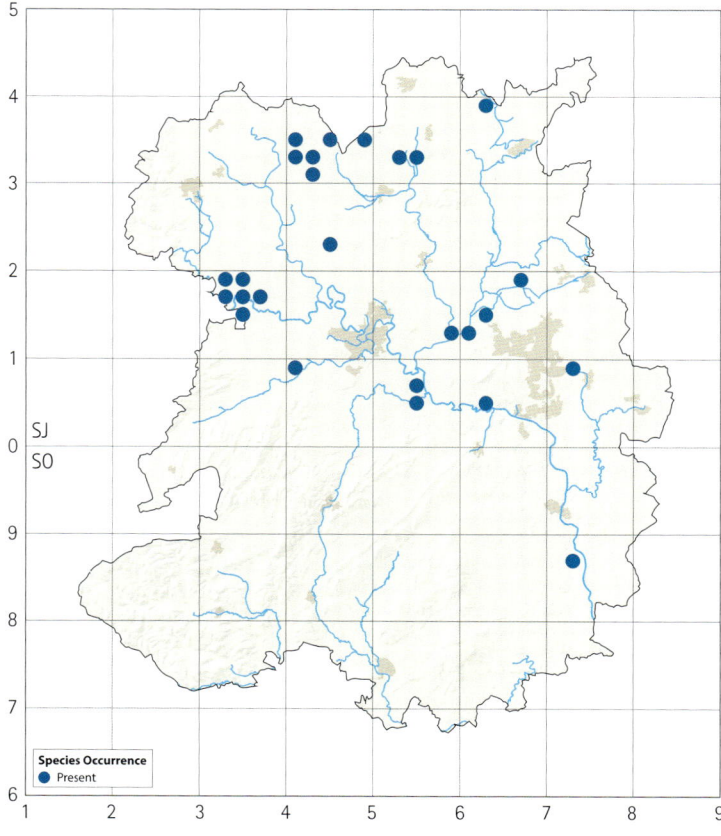

Species Occurrence
● Present

WeBS Counts at the Severn–Vyrnwy Confluence (1995–96 to 2014)

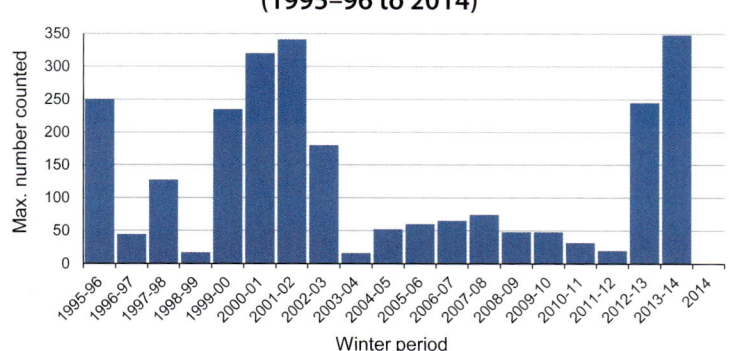

the recent Atlas period has varied from about 30 in 2011–12 to about 285 in 2012–13. The variation is due to the amount of flood water present, with the 2011–12 winter being particularly dry. It is always surprising how quickly Pintail appear after flooding. Large numbers winter on the Dee Estuary, from where high flying birds will be able to see the floodwater, and this is the likely origin of these visitors.

Although the wintering population in the UK has fallen by 27% in the past 10 years, variation in local flood conditions probably plays a greater part in determining the numbers present.

There have been no confirmed breeding records, and only two summer records since 1950. A male at Great Hay Golf Course in June 2000 was said to be of dubious origin, and in the same year a pair with clipped wings were seen mating at Priorslee Flash; they were present for the following two years but no progeny were ever seen.

A first year male ringed at Slimbridge on 4 January 2009 was shot near Preston Montford, 114km distant, just nine days later. This unfortunate individual is the only ringing recovery.

If national numbers continue to fall there may be none to take advantage of local conditions when they are suitable.

Allan Dawes

Jim Almond, Venus Pool, 10 October 2007

Teal (Eurasian Teal)

Anas crecca

Fairly common winter visitor, rare breeding species

Teal breed in temperate Eurasia and migrate south in winter, when large numbers arrive to add to the British breeding population.

Historical records from the nineteenth century describe Teal as abundant in winter on rivers, meres and large pools, with many staying to breed on the open boggy ground around Whixall and Ellesmere, and on wet marshy fields elsewhere.

Early in the twentieth century, breeding was confirmed at Sandford Pool Whitchurch when a nest was found in May 1903, and again at the same site the following year.

There was a long gap until the next breeding records: two pairs nested in 1934 at Betton Pool; a female with young was seen on the peat bog at Whixall Moss on 18 May 1939, and a pair nested at Nobold in May 1942.

The *Handlist* (1964) described it as mainly a winter visitor with peak numbers at Venus Pool of 400, and in most winters the Ellesmere meres held 150–200, Allscott Sugar Factory 100 and Venus Pool 200. Breeding was reported only occasionally from the lowlands of the north-west and north-east, and from the hills around Church Stretton.

The favoured breeding habitat is shallow acidic pools with emergent vegetation, where nests and young can be hidden away in dense vegetation on boggy ground surrounding the small pools.

A nest was reported from 'near the Staffordshire border' on 23 May 1964. More recently, breeding was confirmed on the Long Mynd, when broods of ducklings were reported in five of the 10 years from 1975 to 1984. Breeding was suspected here in the intervening years when several pairs were also present. These 10 years also had breeding season records from Allscott Sugar Factory, Baggy Moor, Brown Clee and Polemere, with confirmed breeding at Venus Pool in 1980 (a pair

with seven ducklings on 11 June) and again in 1983, at Alkmund Park, also in 1983, and at Shifnal in 1984.

The *Atlas* (1992) showed confirmed breeding in 12 tetrads between 1985 and 1990, but only the Long Mynd returned such records in more than one year (1985, 1986 and 1989). Probable breeding was recorded in 28 tetrads, but as many of these were probably non-breeding pairs or passage birds, it was estimated the breeding population at the time was only around five pairs in any one year.

Intensive fieldwork was carried out by the Long Mynd BBP between 1994 and 1998. In most years the population was estimated at four breeding pairs, with two pairs at the favoured site of Wildmoor Pool and the pools immediately south, another on the upper reaches of Wild Moor and another in the upper reaches of Callow Hollow. In some years the population was higher and in other years lower than four pairs. The lower populations were thought to be due to relatively dry weather in the first half of the year or hard conditions the previous winter.

Between the two Atlas periods, apart from the Long Mynd, breeding was confirmed by sightings of ducklings at Venus Pool in 1995, and at Whixall Moss in three of the first seven years of the twenty-first century: 2003, 2004 (two broods), and 2006 (two broods, and a third agitated female was thought to also have young hidden close by).

The breeding population on the Long Mynd was down to only one pair by 2004, and none have been found there since. A pair seen on Wild Moor on 6 June 2008 were not thought to have bred. The pools on the Long Mynd that were used for breeding have become overgrown, and are apparently no longer suitable habitat. Human disturbance was unlikely to have been a factor, but increased predation may have been.

The pair seen on Wild Moor in 2008 was the only breeding season record from the south-west during Atlas fieldwork 2008–13, and the breeding range appears to have shifted to the north, with the only regular site being Whixall Moss, which had two or three pairs in each year and confirmed breeding in three of the six years (2008, 2010 and 2012). Breeding was confirmed in only four other tetrads, all in the north. A pair at Frodesley was reported by the landowner to have been injured in the previous shooting season and were unable to leave at the end of the winter, so had stayed and raised a brood in 2009. Also, in June 2009, ducklings were seen at Onslow. In June 2010, ducklings were recorded at Cold Hatton and a pair were found on a nest at Woolaston.

Most of the sites with probable breeding only had pairs in one year, the exception being Venus Pool with pairs present in each year and two pairs throughout the 2010 breeding season. Most of the others would have been late wintering birds recorded in April or early May, or wintering birds returning early, in July.

It would therefore be optimistic to think that the present breeding population is more than three or four pairs.

Large winter flocks have continued to grace our water bodies. These flocks begin to build up from late August onwards, and increase steadily until November or December. These numbers are maintained until early March when they start the return journey to their breeding grounds in northern Europe. Only small numbers remain into April, and even fewer in May.

The settling pools at Allscott Sugar Factory regularly held wintering flocks numbering over 200 with flocks of 500 in September 1967, January 1985 and December 1988, with a maximum count of 750 in January 1997. However, the greatest number reported was a large flock of 6–7,000 migrating south over Allscott Sugar Factory on 17 December 1968.

They also took advantage of winter floods around the river Severn–Vyrnwy confluence at Crew Green (680 in February 1983) and Hayes Farm Alberbury (700 in February 1996).

Large wintering flocks early in the twenty-first century included 500 on the canal floods at Whixall Moss in November 2001, 608 close to the Severn–Vyrnwy confluence at Ponthen in January 2003 and 600 at Pen-yr-Estyn in November 2004. Wall Farm also attracted good numbers with 400 in November 2002 and January 2005, and 320 in January 2007.

Wintering Teal will use almost any pool, large or small, reflected in the relative abundance map. Most are in the north, the main concentrations being on the Meres and Mosses around Ellesmere and around the Severn–Vyrnwy confluence near Melverley. The flood meadows close to the confluence attracted flocks of 466 at Ponthen in January 2008, 660 at Hayes Farm Alberbury in December 2009

Jim Almond, Venus Pool, 10 December 2011

Breeding Distribution Change (1985–90 to 2008–13)

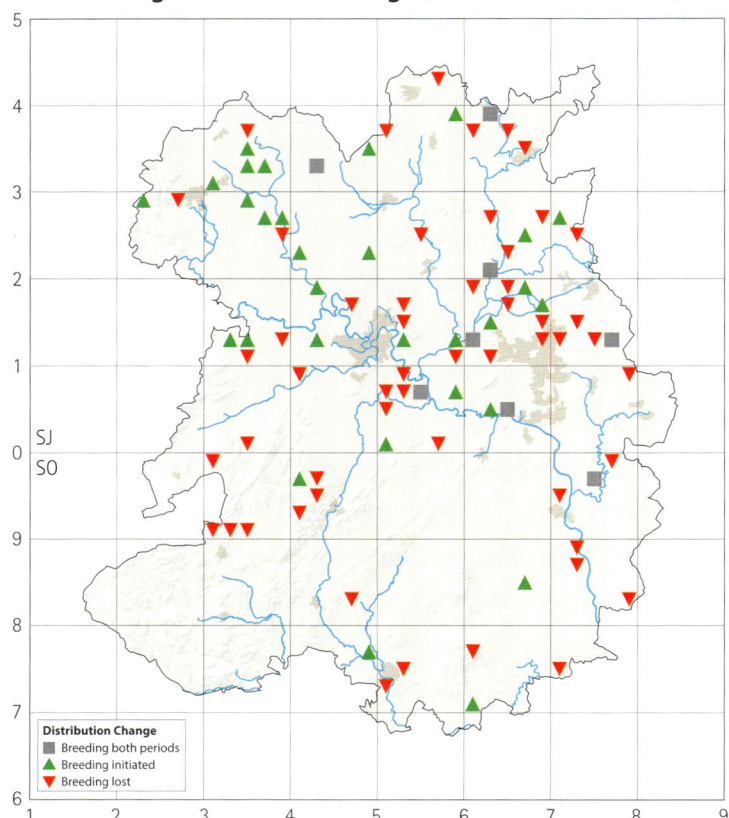

Distribution Change
- ■ Breeding both periods
- ▲ Breeding initiated
- ▼ Breeding lost

Winter Relative Abundance (2007–13)

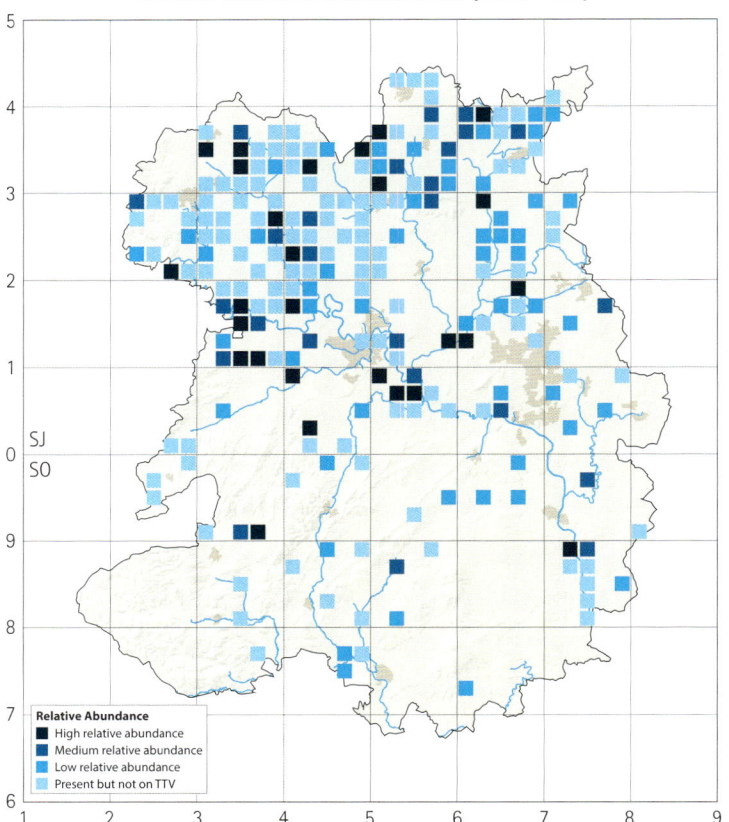

Relative Abundance
- ■ High relative abundance
- ■ Medium relative abundance
- ■ Low relative abundance
- ■ Present but not on TTV

Occupied Tetrads

Atlas period (breeding)	1985–90		2008–13		Change	
	Number	%	Number	%	Number	%
Confirmed	12	1	5	1	-7	-58
Probable	28	3	23	3	-5	-18
Possible	21	2	10	1	-11	-52
TOTAL	61	7	38	4	-23	-38
Tetrads with Winter Records (2007–13): 225 (26%)						

and 996 at the same site in February 2013, and 700+ at Cae Howell in January 2012.

Wall Farm (300 in February 2009), Allscott Sugar Factory (465 in February 2011) and Whixall (about 1,200 on the Canal Floods and another 250 on the Moss on 5 October 2013) also held exceptionally good numbers.

Apart from these sites, there were only 31 TTV counts from 24 tetrads of 10 or more, only four of which were more than 100, so most pools hold only small numbers, and there are large numbers on the Ellesmere meres only when the smaller pools are frozen. The monthly maxima at selected sites in winter (SBRs) suggest that numbers have remained fairly constant since 1990, only fluctuating in response to suitable habitat created by flooding in the upper Severn around Melverley and the canal floods at Whixall. Based on TTV counts the winter population is estimated at around 2,000, which, given the monthly maxima in SBRs, is likely to be the upper limit of the population in most winters.

Few Teal are ringed here, and only nine have been recovered. They include an adult female ringed at Roddington on 1 January 1964 and shot on 19 February 1965 at Fraserburgh, Aberdeenshire, 553km distant, and an adult female ringed at Allscott Sugar Factory on 12 October 1980 that was shot on 11 January 1986 in Galway, Ireland, a distance of 405km. Two were recovered here, but there is no obvious pattern to the remainder. None have been recovered outside Britain and Ireland.

More interestingly, 127 have been ringed elsewhere and recovered here. Most of the recoveries have been shot, but the locations demonstrate the migration routes to and from breeding areas in northern Europe, and known mobility in search of mild winter weather. Most are movements from eastern England (52 from Essex, Norfolk, Suffolk, Cambridgeshire or Lincolnshire) and the Netherlands (28), while 16 have come from Scandinavia, and one from Iceland. Only two other counties, Dyfed and Gloucestershire, were the origin of 10 or more.

An adult ringed in Finland on 7 August 1969 which was shot less than three months later on 2 November at Eyton-on Severn (1,879km), was the furthest travelled, but most of the 16 from Scandinavia travelled over 1,000km. From the other direction, the nestling female ringed at Mývatn, Iceland in August 1979 was shot near Broseley only three and a half months later, having travelled 1,658km.

An adult male ringed at Slimbridge Gloucestershire on 22 December 1986 and found freshly dead (shot) 17y 2m 21d later near Marchamley on 14 March 2004 was not far short of the national longevity record of 18y 20d. Four others were more than 5 years old.

Martin Grant

Green-winged Teal
Anas carolinensis
Vagrant

Jim Almond, Polemere, 21 January 2013

This transatlantic vagrant breeds in the northern United States and Canada, and winters in the southern states and central America.

Although a scarce migrant to Britain, it is nevertheless regularly seen, with over 50 new birds in some years; this is in addition to those that have remained on this side of the Atlantic from a previous year.

There are six records:

Venus Pool	27 February–10 March 1996
Allscott Sugar Factory	8–12 December 2001
Whixall Canal Floods	19 January–16 March 2002
Polemere	2–31 January 2013
Venus Pool	24 November–10 December 2016
Venus Pool	8 January 2017

Typically, all were wintering birds and all were drakes. However, the individual at Whixall was often seen away from the main Teal flock in the company of what was assumed to be a female Teal, although two Green-winged Teal cannot be ruled out given the difficulties of identifying females. The drake at Allscott in 2001 may possibly have been the one found at Whixall a month later, although 2002 was an exceptional year with nearly 60 individuals located across the country. It is not possible to say whether the individuals at Venus Pool in 2016 and 2017 were one and the same, but it is a distinct possibility.

Graham Walker

Red-crested Pochard
Netta rufina
Very rare naturalised visitor

Widely scattered populations are present in central and southern Europe.

The national population was founded by escapes, mainly from the 1960s onwards, and is now centred on Cotswold Water Park. This is assumed to be the main source of local visitors. Recent escapes from elsewhere or vagrants from populations in Europe may also occur, but unless they are ringed their origins are difficult to assess. Some SBR entries have been listed in the escapes section, while others in the systematic list were noted as escapes. The treatment has not been consistent, and may have depended on the opinion of the individual observer or SBR editor, but this account summarises all 26 published records, all of singletons.

A female on the river by Greyfriars Bridge, Shrewsbury in May 1977, listed at the time as an escape, was the first reported. A male at Chelmarsh in November 1983 heralded the start of almost annual records, which continued until 1996. During this period, 10 males, five females and a juvenile were found at widespread locations. After a gap, a single female was seen in 2001, 2002 and 2005, after which there were no further sightings until two males were found towards the end of the recent Atlas period, one on the river near Bridgnorth in December 2010 and one at Polemere in December 2012, the last recorded.

Jim Almond, Hampton Loade, 28 December 2010

Estimated Number per Year (1950–2014)

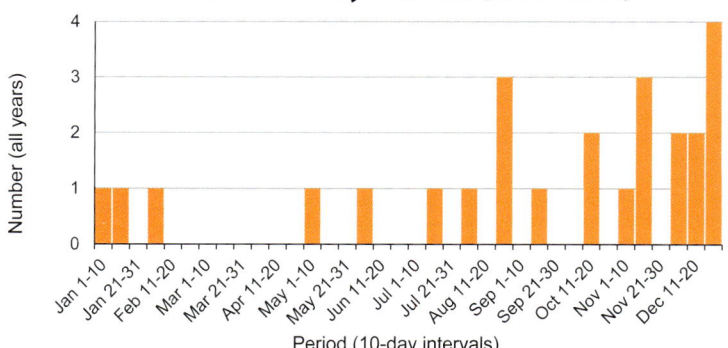

Arrival Date in 10-day Intervals (1950–2014)

One that arrived at the Mere (Ellesmere) in October 1994, before relocating to White Mere, stayed for 41 days, and single arrivals at Venus Pool in August, in 1992 and 2005, remained for 29 and 21 days respectively. The remainder all moved on within one week, with many only being seen on a single day.

The first chart shows the year in which each was first recorded, and the second shows the time of year each was first recorded.

They have been found in every month except March and April but half were first seen in the period from November to January. At this time, hard weather on the continent often precipitates movements to milder climes, and influxes have been recorded in eastern England, so it seems highly likely that some have originated from mainland Europe.

Despite the recent downturn in records, the population is gradually increasing in both the UK and Europe. They now breed regularly in the East Midlands and lower Thames Valley, so this duck is destined to become a more familiar sight in the future.

Allan Dawes

Pochard (Common Pochard)

Aythya ferina

Scarce winter visitor, has bred

Pochard breed throughout the temperate zone of Eurasia and move south and west for the winter. The small British resident population is boosted in winter by visitors, mainly from the Baltic countries.

During the second half of the nineteenth century, the Pochard was considered to be common during the winter. It frequented most of the larger pools, and flocks of 40–50 could be found at the Ellesmere meres. Their preference for large waters offered some protection from the guns, which was just as well since the flesh was described as excellent. In contrast to other wintering ducks, adult males were said to predominate and flocks of 20–30 often contained no females or juvenile males.

Although rarely found on rivers, severe weather in February 1929 forced many onto the River Severn at Atcham, and three were found dead at Bridgnorth. The winter of 1939–40 was also harsh, and they were reported from running water near Oswestry and the River Severn at Shrewsbury.

There were very few historic records of breeding. A nest with eggs was found at Tong Mere in May 1885, a female and four young were on a pool near Shrewsbury in the 1880s, and two nests containing eggs at Whixall Moss in 1916, were the only ones prior to 1950.

After that, breeding took place near Shrewsbury in 1961 and 1963, and also near Worfield in the latter year. The 1968–72 BTO Atlas shows confirmed breeding in SJ43, which the *Atlas* (1992) credited to the Ellesmere meres. However SJ43 is a border square and, as the record is not documented, it might have been from Clwyd, although breeding was later confirmed at Crose/Sweat Mere during the period 1985–90. Summering became a regular feature at Venus Pool from 1971, but breeding was not confirmed until 1980. After that, a pair bred there in nine different years, with two pairs raising broods in 1982. This run, albeit intermittent, ended in 1995, after which no further breeding activity has been noted, from this or any other location. Nesting attempts were probably more frequent, but unless young were seen they could have easily been overlooked, and many nests fail before reaching this stage. Throughout this period

Jim Almond, Priorslee Flash, 10 December 2011

Winter Distribution (2007–13)

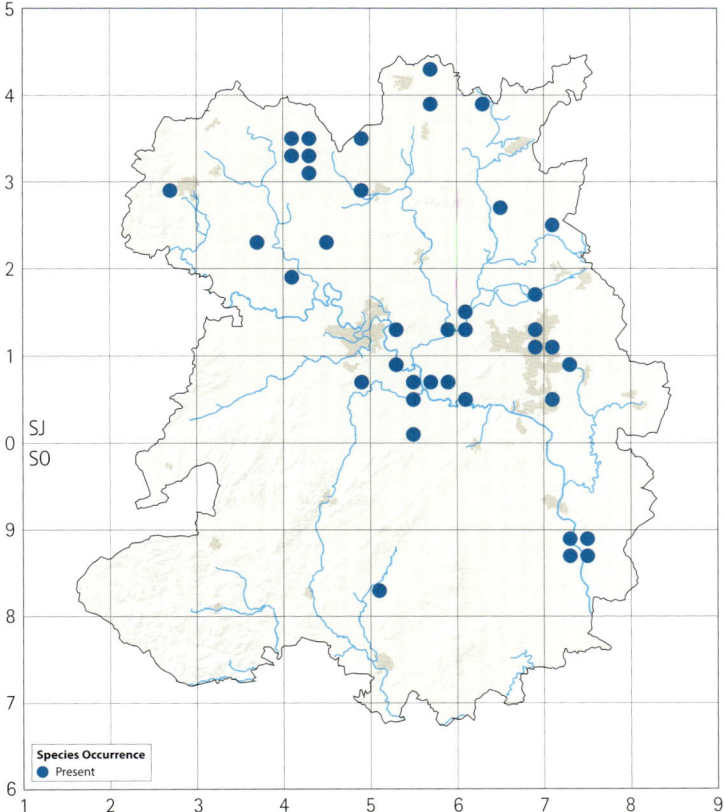

Occupied Tetrads

Atlas period (breeding)	1985–90		2008–13		Change	
	Number	%	Number	%	Number	%
Confirmed	5	1	0	0	-5	-100
Probable	6	1	2	0	-4	-67
Possible	16	2	3	0	-13	-81
TOTAL	27	3	5	1	-22	-81
Tetrads with Winter Records (2007–13): 39 (4%)						

summering occurred at several locations but, in addition to Venus Pool, only Brompton in 1986 and Madeley Court in 1988, together with SJ43F (Crose/Sweat Mere) and SJ63D (Cloverley Pool), both undated records, had breeding confirmed during fieldwork for the *Atlas* (1992).

During fieldwork for the recent Atlas, April sightings in four tetrads were thought to have been late wintering or passage birds. May records came from Allscott Sugar Factory, including Mirelake, where a pair was seen in both 2008 and 2009, and Kinnersley Moor in 2010, while in July singles, which may have been failed breeders returning early, were at Longdon on Tern in 2010 and Shavington Park in 2011.

The maximum number of possible breeding pairs recorded in the *Atlas* (1992) was between two and eight per year. Many of these would have involved late wintering or passage birds, and the decline in the winter population is reflected in the lower number of April records during fieldwork for the recent Atlas.

Although five tetrads had confirmed breeding records in 1985–90, there were never more than two pairs in any one year. Recent Atlas fieldwork suggested a maximum of one pair at Allscott Sugar Factory (in two tetrads) but only in the early years of the survey. This species has ceased to be an annual breeder and occasional breeding is becoming more unlikely as the winter population continues to decline. The increase in breeding pairs from the early 1970s seems to have been a temporary phenomenon. The BTO Atlas (2013) also recorded a 39% decline nationally in range since 1968–72, mainly since 1988–91.

Outside the breeding season, the *Handlist* (1964) stated:

> Small numbers are present with flocks of Tufted Ducks on passage and in winter, particularly at Ellesmere and Shavington, and occasionally at Venus Pool, if the water conditions are good. Highest numbers recorded are: 162 at Shavington in December 1958, 110 at Venus Pool in February 1959 and 133 at Ellesmere in 1963. Usually numbers are about half the above figures for the waters mentioned and smaller numbers elsewhere.

It also echoed the earlier statement that males are predominant in the flocks.

WeBS results from the Ellesmere meres and Shavington are shown in the chart. The virtual disappearance from Shavington at the start of this century, and the more recent decline at the Ellesmere meres – a reversal of the increasing numbers at both sites in the 1970s – are clearly visible. Similar changes have occurred throughout the country, and the national WeBS index has fallen by 47% over the previous 10 years, with numbers in 2013–14 at a record low. This local decline may be partly explained by milder winters enabling Pochards to remain closer to their breeding grounds, but data from the International Waterbird Census has shown a significant long-term decline in both the north-west European population and those from further east.

The occasional peaks are difficult to explain, but they appear to be local and not related to national influxes. The highest counts from Shavington; 160 in December 1961 and 200 in January 1981, may have been due to cold weather movements, as these winters were quite severe. However, at the Ellesmere meres, the large counts of 177 in 1978 and 276 in 1987 both occurred in November and were not associated with unusually cold conditions.

In 1983–84, the Wildfowl Trust asked counters to include separate totals for males and females. Out of the 213 present at the Ellesmere meres during that winter, 59% were male and 41% female or immature males (unpublished data). Male and female Pochard show differential

WeBS Counts at the Main Sites (1960–61 to 2014)

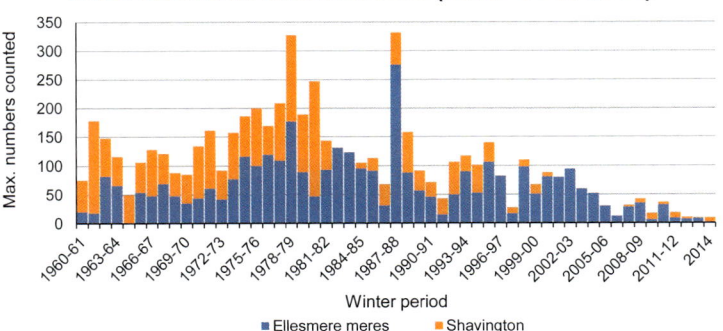

migration strategies, males remaining closer to the breeding grounds and females wintering further south. This change in the sex ratio from that noted in the *Handlist* and earlier accounts may have been the first indication of a shift in wintering areas caused by milder conditions.

Wintering Pochard prefer larger waters and the recent Atlas distribution map reflects this. They were found in 39 tetrads, and were almost annual at the Mere (Ellesmere), Priorslee Lake, Shavington, Shrewsbury Sewage Farm, Trench Pool, Venus Pool and Chelmarsh. All these sites occasionally had flocks which exceeded the threshold for the highest TTV relative abundance category, so the winter abundance map is not representative. Elsewhere, 44 were at Polemere in March 2011, but they are not regular at this site, and double figures were only occasionally noted from other large pools. The single record from the south, other than the cluster around Chelmarsh, was from a recently constructed pool – part of a new wetland habitat in

Corve Dale – and a few were also seen on the River Severn between Shrewsbury and Buildwas.

During the recent Atlas period, the maximum annual WeBS totals varied between 94 in February 2010 and 25 in February 2012. Most of the main sites are covered by WeBS so few will have been overlooked. The winter population would now appear to be 100 at most.

An adult ringed near Wellington in September 1969 was shot in Armagh in January 1974 and a nestling male ringed in June 1982 in Latvia was shot near Shifnal in the following December, 1,934km distant. Three other recoveries of shot males were two from Slimbridge and one from Norfolk. The latter, an adult when ringed in March 1994, was shot at Morville 6y 9m 24d later.

If milder winters and population decline continue, Pochard sightings seem destined to become a rare event. Fewer overwintering may also reduce the chances of future breeding.

Allan Dawes

Ferruginous Duck

Aythya nyroca

Vagrant

Ferruginous Duck, also known as White-eyed Pochard, breed from Central and Eastern Europe, and North Africa, through India to China. Its wintering range overlaps with the breeding range, but extends southwards.

In the late summer of 1937, one was seen on a pond at Bomere near Bayston Hill by an observer who was confident of the identification, being familiar with the species in India from where she had recently returned (NB this record was published in the CSVFC Recorder's Report for 1947–50 and 1947 may have been the actual year of the sighting, given that a lot of colonial and military staff returned to

the UK from the subcontinent around that time; the reference to 1937 may have been typographic error).

There is one accepted modern record (Rogers *et al.* 2005; Butter 2003), a drake on a typical mid-winter date:

Cole Mere   4 December 2002

In 2002, perhaps 17 individuals were seen in Great Britain during a period when the national average was closer to 12 (Fraser 2013a).

Graham Walker

Ring-necked Duck

Aythya collaris

Very rare passage migrant and winter visitor

This duck is a native of North America, with seasonal migration between the northern and southern parts of the sub-continent.

A male found at Oss Mere in March 1995, which remained into April, was the first accepted record.

The twenty-first century then saw an increase, with almost annual appearances after the second record in December 2000, when two females arrived at Venus Pool, and remained for the winter period, leaving in mid-March. Individuals often return to a wintering or passage site used previously, and a male at Chelmarsh for 75 days from 12 October 2002 was thought to be the same one that had been present for 45 days from 28 December 2001 to 10 February 2002, and, true to form, it often commuted to nearby Dudmaston, as had the one in the previous winter. A female spent 15 days at Venus Pool in April 2005 during the passage period.

During the recent Atlas period, a female moved between Monkmoor, Venus Pool and Allscott Sugar Factory during the winter of 2007–08.

Jim Almond, Priorslee Lake, 22 March 2015

Estimated Number per Year (1950–2014)

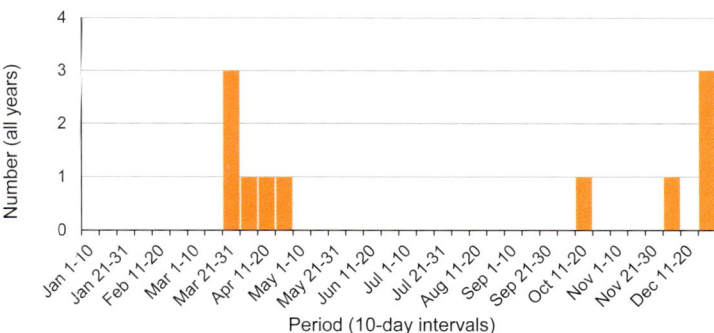

Arrival Date in 10-day Intervals (1950–2017)

Allscott Sugar Factory also hosted a male in March 2008, a female in April 2009 and a male in March 2010. It is possible that both the latter reports were of returning individuals from previous winters, but they are treated as separate individuals in the totals.

In 2014, a drake on spring passage was found at Wood Lane on 27 April. It remained in the area until 5 May, during which time it was also seen at the Mere (Ellesmere) and White Mere. The adult male at Priorslee Lake on 21 March 2015 was only present until the next day before it moved on.

Estimating numbers is difficult due to the wandering nature of some individuals, but the 12 accepted records shown on the chart probably relate to 11 individuals, with a roughly even split between males and females. It shows the year that each was first recorded, and does not take into account either the second calendar year when over-wintering individuals were still present, or the possibility that the same individual has occurred in more than one year.

Having crossed the Atlantic, there is some evidence that they maintain their normal migration instincts in Europe, and regular transatlantic migration by part of the adult population cannot be ruled out (White & Kehoe 2014).

The records fall into two clear categories; winter arrivals in December staying until the spring, and brief passage during March and April. The male that returned to Chelmarsh in October 2002 is an exception, but it is surprising that autumn passage has not been recorded.

A large increase in the North American population is likely to be responsible for the recent increase in sightings shown in the annual occurrence chart.

Allan Dawes

Tufted Duck

Aythya fuligula

Uncommon resident

Tufted Duck is found throughout northern and temperate Eurasia during the summer. Northern populations depart before the winter, so local residents are joined by visitors from Scandinavia and European Russia.

In 1866, Rocke wrote: 'One of the commonest of our winter stragglers, though I have seldom seen more than one at a time', and, referring to the adult male, 'in that handsome dress they are certainly rare'. This contrasts with the findings of Beckwith in 1879: 'Young birds and females are very common in winter, and the handsome adult males are by no means rare'.

The status of the Tufted Duck was changing rapidly during this period, with a marked range expansion in Western Europe. This was driven by an increase in artificial waters such as gravel pits and reservoirs and the spread of the introduced zebra mussel, a staple food, in these same waters. Whether this expansion was responsible for the different observations of Rocke and Beckwith, just 13 years apart, is unclear.

A few years after British breeding was first documented in 1849, Hatton Grange, near Shifnal, became a regular breeding location for several pairs, with the first nest found there around 1855. Birds from Hatton were thought to have colonised Weston Park by 1871, where

Dawn Micklewright, Wood Lane, 16 February 2014

up to 20 pairs became established, although most were just across the border in Staffordshire. From 1891, four pairs bred annually at Sandford, near Whitchurch, but despite its proximity to the Ellesmere meres, which hosted large numbers each winter, the Ellesmere flock always departed before the summer.

Breeding Distribution Change (1985–90 to 2008–13)

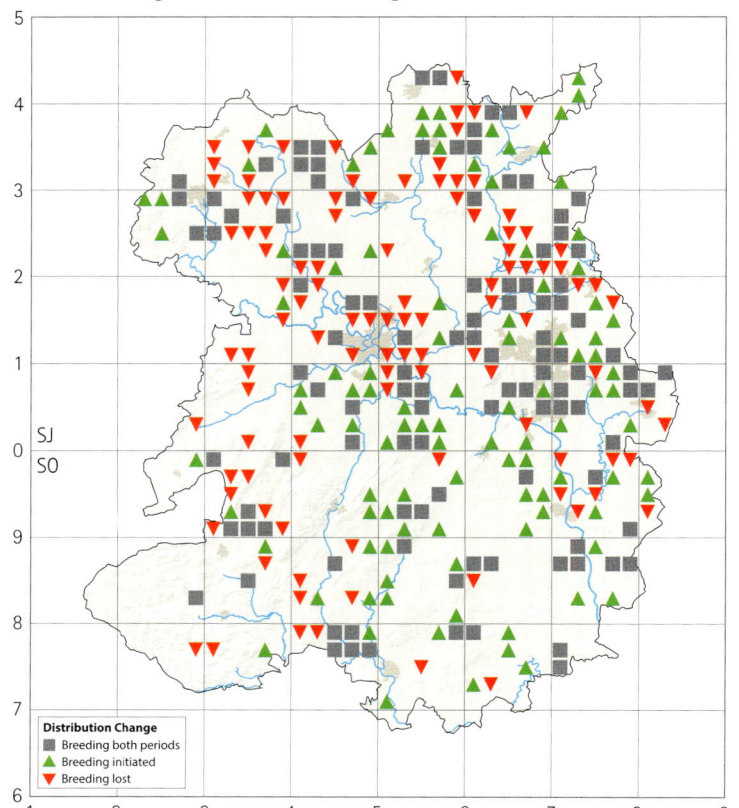

Distribution Change
- ■ Breeding both periods
- ▲ Breeding initiated
- ▼ Breeding lost

Winter Relative Abundance (2007–13)

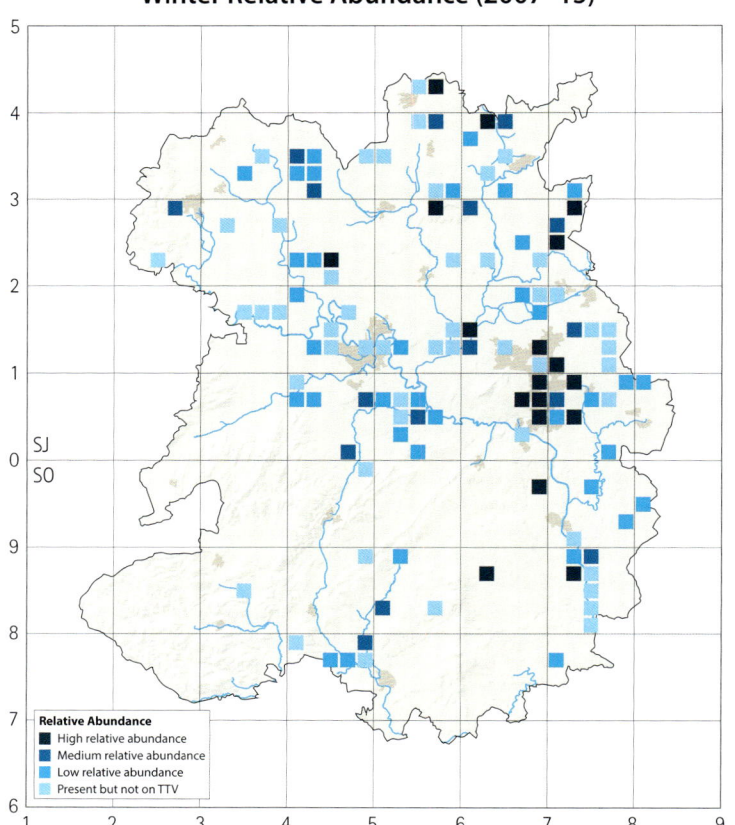

Relative Abundance
- ■ High relative abundance
- ■ Medium relative abundance
- ■ Low relative abundance
- ■ Present but not on TTV

Occupied Tetrads

Atlas period (breeding)	1985–90		2008–13		Change	
	Number	%	Number	%	Number	%
Confirmed	83	10	55	6	-28	-34
Probable	107	12	118	14	11	10
Possible	27	3	43	5	16	59
TOTAL	217	25	216	25	–1	0
Tetrads with Winter Records (2007–13): 120 (14%)						

WeBS Counts at the Main Sites (1960–61 to 2014)

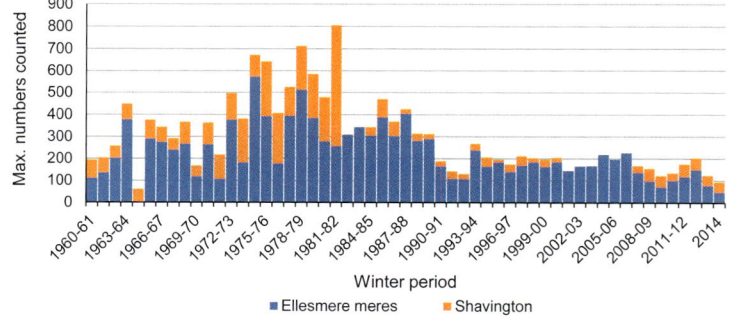

Winter period
- ■ Ellesmere meres ■ Shavington

A pair nested near Bridgnorth in 1901, and by 1943 many pools in the north had been colonised. Even so, numbers remained low and the *Handlist* estimated 30–40 breeding pairs in 1964.

The *Atlas* (1992) estimated a population of between 160 and 240 pairs and commented upon the increased use of both small pools and those at altitude, such as Boyne Water at 455m on Brown Clee. Even greater heights were achieved in the recent Atlas period, with young seen at 510m near Abdon Burf in July 2013, a record which may be impossible to break. Despite this, the overall picture is mixed, with losses from many tetrads in the north and west just outnumbering gains, mostly in the south-east. The pattern of losses in the north is shared by Canada Goose and Reed Bunting, but close inspection reveals that losses have often occurred in different tetrads. This rules out widespread habitat loss but local change will affect species independently as they all have different requirements.

The breeding distribution change map and tetrad occupancy table

suggest little variation in the population in recent years. However, an estimate of one pair per tetrad with probable breeding, and two to three pairs in tetrads with confirmed breeding, is reasonable now, and gives a population of between 230 and 380 pairs. This is higher than the estimate given in the *Atlas* (1992), but an increase is evidenced by the BBS trend for England, which increased by 22% between 1995 and 2014. This calculation is based on annual breeding in all tetrads with probable and confirmed records, rather than the assumption made in the *Atlas* (1992) that only a third of the tetrads had breeding birds in any one year.

The *Handlist* noted autumn passage of between 100 and 150 at the Ellesmere meres, with a high of 300 at Newton Mere in November 1962. At the onset of bad weather they moved away, returning in mid-April before departing in early May, leaving only a few residents behind.

Peak wintering numbers at the Ellesmere meres and Shavington, two of the main sites, showed a steady increase from the 1960s, when regular monitoring began, until the beginning of the 1980s. Following a period of stability, they then fell sharply at the end of the decade and have since remained low. This is in contrast to the WeBS trend for England which increased steadily until 2010–11, after which it dropped very slightly.

The winter distribution mainly reflects the availability of large pools, the majority of which are in the north and east. Small pools are often abandoned at this time. Occasionally found on flowing water, numbers increase there when the large pools are frozen.

During the recent Atlas period, maximum winter numbers found by WeBS ranged from 262 in January 2012 to 468 in December 2012. Because Tufted Ducks are also found at sites which are not covered by WeBS, many are not included in these counts. Based on TTV counts,

the estimated population is between 800 and 920.

A juvenile ringed at Shrewsbury in July 1988 was found dead at Markermeer in the Netherlands in September 1992. This is the only recovery outside the UK and shows that not all juveniles fledged locally remain in the area. Out of the total of 15 recoveries, two were found in Northern Ireland, while one from Loch Leven, Tayside, was found in the Ellesmere area. The remainder were scattered throughout England with three ringed and recovered here, and two involving adjacent counties.

The Tufted Duck may be near breeding capacity, and despite the decline at two of the main sites noted above they are not, as yet, showing any signs of changing their wintering distribution as many other wildfowl appear to be doing.

Allan Dawes

Scaup (Greater Scaup)

Aythya marila

Rare winter visitor

Scaup have a circumpolar breeding distribution, and north-west European populations winter mainly in the Baltic and North Sea. Primarily a sea duck, there are regular winter flocks at several places around the Welsh coast. Records during passage periods suggest some overland migration takes place, and occasional reports mention that they have appeared following gales.

In the nineteenth century, they were occasionally found on large pools and rivers, both in small groups and individually, and Forrest (1899) cited five instances of them being 'obtained' or 'shot', including a nearly adult male on the River Severn near Kinnerley in January 1899. This was taken from a flock of about 15 which is the largest group ever recorded.

A minimum of 17 were found at eight locations between 1901 and 1910, but they were only sighted once in each of the following decades up until 1950.

They often move between adjacent waters, and sometimes further afield, and 'quite probably the same bird' moved from Newton Mere to Venus Pool and on to Crose Mere in December 2005. Such movements make calculating numbers difficult, so a review of both published records and the original record cards was undertaken to obtain a more accurate estimate of annual numbers in the modern period. As a result, 31 previously published records were no longer

considered acceptable and six previously unpublished ones were found.

The chart shows the estimated number of individuals in each winter period. Eight occurred in the 1950s, followed by three and five in the next two decades. Two in 1982 marked the beginning of an unprecedented run of records which is still ongoing. Since then they have occurred annually, with the exception of 1998. There were 35 new arrivals in the 1980s, 50 in the 1990s, 36 in the 2000s and a further 19 up to the end of 2014, a total of 156 since 1950.

The total for each winter shown in the chart includes those recorded on passage, with those in spring added to the previous winter total and autumn ones to the later one.

Scaup have been found in every month except June and there has only been one in July. Most arrivals have been during the winter months from November to February and these often remain for some weeks. In contrast, those on passage during March to May and August to September normally depart quickly, although on two occasions September arrivals appear to have remained for the winter.

The local trend closely matches that reported by the WeBS for Wales where numbers peaked sharply in the mid-1980s, fell almost as quickly towards the end of the 1990s and have continued to decline slowly since, although they have yet to fall back to their former level.

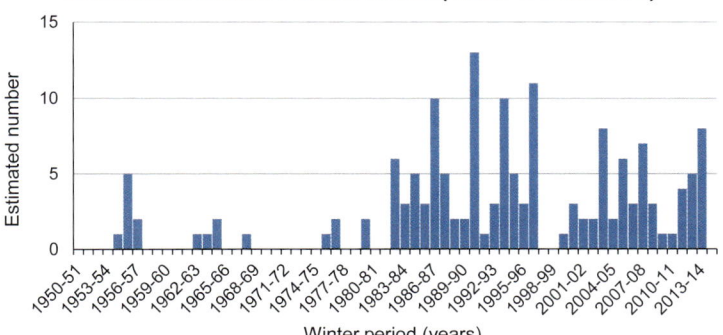

Estimated Winter Period Totals (1950–51 to 2014)

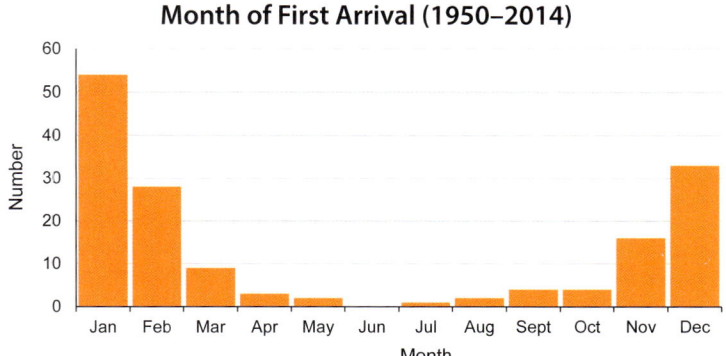

Month of First Arrival (1950–2014)

Winter Distribution (2007–13)

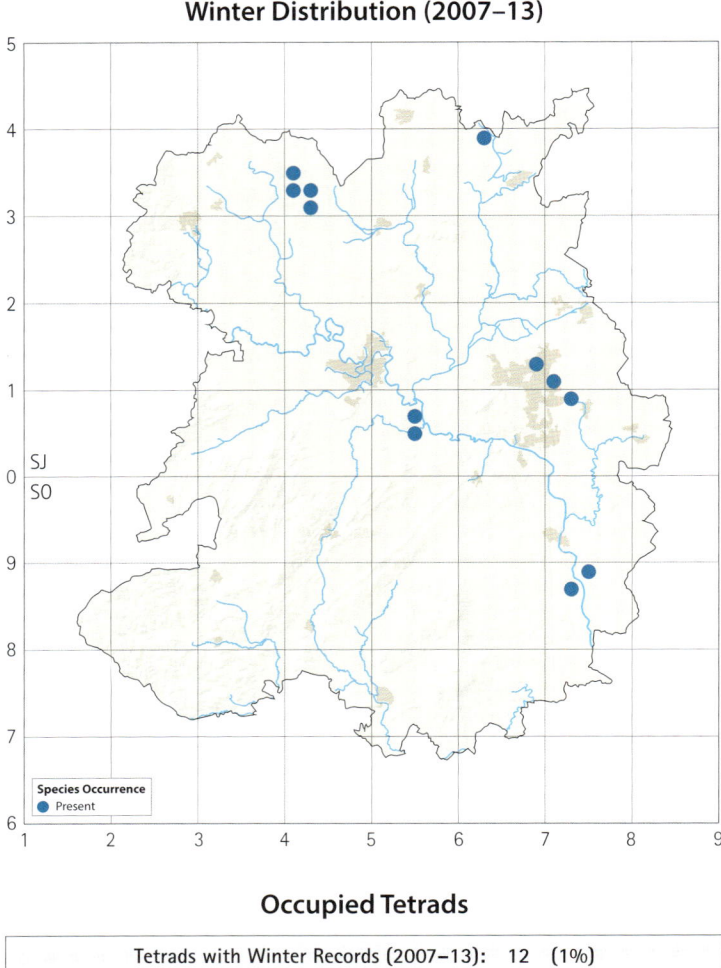

Species Occurrence
● Present

Occupied Tetrads

Tetrads with Winter Records (2007–13):	12	(1%)

Jim Almond, Priorslee Flash, 17 January 2014

but reduced grain discharges from breweries and distilleries, and new sewage treatment at several sites in the mid-1970s, particularly the Forth Estuary where 25,000 formerly wintered, will have forced them to move further afield.

Seven at Cole Mere in December 2007, with six remaining into the New Year, was a good start for the recent winter Atlas, but three at Venus Pool in November 2011 was the only other multiple sighting. The remainder were all singletons. The clusters on the map show the Ellesmere meres, Venus Pool and Cound Lake, Priorslee Lake, Priorslee Flash and Trench Pool, and Chelmarsh Reservoir and Dudmaston, with a single record in the north-east from Shavington. These are some of the largest water bodies and they are also some of the most frequently watched. Occasional visitors to other sites may have been missed but large waters seem to be preferred.

The WeBS indicates a national decline of 31% over the past 10 years. The reasons behind this have yet to be determined but the Scaup seems destined to become much scarcer in future years.

Allan Dawes

The reasons behind these fluctuations are not fully understood, but a recovery of the Icelandic breeding population, which winters mainly in the west of the British Isles, was thought to be responsible for the increase. Sewer outfalls in Scotland used to attract large numbers,

Ian Butler (ianbutlerphotography.co.uk), Monkmoor, 11 June 2005

Lesser Scaup
Aythya affinis
Vagrant

Lesser Scaup breed in the north of North America and winter in the southern states through to northern Columbia and the West Indies. There is one record of this national rarity (Fraser *et al.* 2007a; Holmes 2005a):

Monkmoor Pool, Shrewsbury   6–26 June 2005

This long-staying male attracted much attention, including a live outside broadcast shown on BBC Midlands Today. Originally identified as a Scaup (*A. marila*), it may have been present from at least 7 May because the site's logbook recorded a Scaup from that date onwards. It was one of perhaps 17 records nationally for the year, a typical number at that time.

Graham Walker

Eider (Common Eider)

Somateria mollissima

Vagrant

This truly marine species has a circumpolar distribution and is only occasionally recorded inland. It is Britain's commonest sea-duck, breeding along the coast from North Wales to Northumberland, but with the bulk of the population in Scotland. In winter, it can be found all around the British coast, albeit in smaller numbers where it is away from its breeding grounds.

There are no historic reports and just two modern records, both males:

Kempton	30 March 1979
River Severn, Leighton	6 November 1993

Although the first was found dead, it had been ringed as an adult on the Ythan Estuary, Aberdeenshire, on 16 May 1973, 5y 10m 14d earlier. It had travelled 546km, the greatest movement recorded in Britain and Ireland (BTO online ringing reports, summary for Shropshire 2014).

Although only recorded on one day, the male at Leighton may also have been present on both 5 and 7 November according to local fishermen. This sighting coincided with an extraordinary influx of some 200 Eiders in to the Midlands at that time, probably as a result of a period of very cold weather in the North Sea (Harrison & Harrison 2005).

Graham Walker

Velvet Scoter

Melanitta fusca

Vagrant

The nearest breeding grounds for this sea duck are in Fennoscandia. It is a scarce winter visitor to Britain, mainly along the North Sea and south coasts, and is rarely recorded inland.

In 1841, 'Velvet Duck' was included on a list of rarer birds seen in the vicinity of Shrewsbury, while in 1866, Rocke stated that it was 'very rarely met with so far from the sea coast, though I believe a few instances have occurred'. An adult male was found exhausted near Whitchurch on 23 November 1866 while a second, or conceivably the same individual, was discovered stuffed in a collection having been being reported as either shot or found moribund near Shrewsbury in the autumn/winter, possibly November, also in 1866. Later, an immature female was shot at Clungunford on 12 December 1890. There were no records after that until the flock at Venus Pool in 1983, the first of only six modern records:

Venus Pool	six, 24–28 October 1983
River Severn, Cressage	10 December 1991
Venus Pool	30 December 1991–4 April 1992
Newton Mere	27 April 1994
Priorslee Lake/Priorslee Flash	15 January–15 February 2014
Trench Pool	28 November–8 December 2016

Jim Almond, Priorslee Lake, 22 January 2014

The party of six, consisting of two adult females and four juveniles, also thought to be females, were 'quite untroubled by heavy plant working nearby, and periodically approaching to within 20–30m of edge of pool'. They dived regularly, often synchronously, and for up to 73 seconds.

There is a strong possibility that the individual on the River Severn in 1991 was the same as the one seen at Venus Pool some 20 days later. The individual at Priorslee was specifically identified as a first winter as was the one at Trench Pool two years later.

Graham Walker

Common Scoter

Melanitta nigra

Rare passage migrant and winter visitor

The Common Scoter breeds mainly near north European and Russian lakes and rivers, and is mainly a winter visitor around the British coast with only small numbers found inland, particularly after storms. The small British breeding population is found only in northern Scotland.

Historical records show that only occasional single birds were obtained, but five were shot from a flock of 20 at Acton Burnell Park on 24 November 1892 and three on the River Teme at Ludlow on 21 November 1907. This suggests little change in the status of Common Scoter from then up to the present day.

During the 1950s it was said to be 'a rare visitor but possibly

Jim Almond, Priorslee Lake, 17 September 2010

an exceptional flock of 10 males at Trench Pool on 25 June 1973, and a pair at Chelmarsh on 13 April 1975.

Records have increased since then, with only four out of the 35 years from 1980 to 2014 having no sightings. Most of the counts were of no more than three. The exceptions were at Venus Pool (11 on 2 June 1984), Chelmarsh (eight on 17 September 1989), Ellesmere (nine on 30 June 1997), Priorslee Lake (19 on 25 July 2000) and, the largest flock to date, 28 at Ellesmere on 29 October 2004.

During recent Atlas field work, Common Scoter was found, on passage, in the breeding season at Cole Mere (a flock of 20, including 18 males, on 11 April 2009, and a single male on 23 April 2010) and at Priorslee Lake (two males on 11 May 2012). In the winter period it was found only at Ellesmere (a female on 15 December 2008), Mirelake (a male on 11 November 2009), Cound Fishery (a female on 27 November 2012), and Chelmarsh Reservoir (a single drake on 20 January 2013).

After the Atlas period, it was seen at Chelmarsh (a drake again on 11 November 2013), at Cole Mere (a female on 16–20 November 2013) and at Priorslee Lake (a female on 20 July 2014).

The international population trend is uncertain, but it appears that the winter range in Britain is increasing, with fewer remaining in the Baltic.

Martin Grant

overlooked', with only three records, all at Ellesmere: a male on 17 November 1955; a flock of five on 31 July 1957; and a single bird on 20 December 1959. The 1960s only returned two records, both of single females in 1966, at Shavington on 11 August and Chelmarsh on 24 September, while in the 1970s it was recorded in five years, with no more than two records a year, all of single birds, apart from

Long-tailed Duck
Clangula hyemalis
Very rare winter visitor

Jim Almond, Trench Pool, 12 November 2014

Most of the European population breed in north-west Russia but this duck has a circumpolar distribution, mainly within the Arctic Circle. It winters at sea, often out of sight of land.

There are just four historic records. A male at Tong Mere in November 1881 met with the usual fate at that time and ended up in the collection at Weston Park. Four at the Mere (Ellesmere) in November 1881 was the largest group ever recorded and, along with

one at White Mere in November 1885, they appeared to escape the gun. One in the winter of 1887–88, when the species was said to be 'numerous in England', was not so lucky: it was reported as shot, but no details were given.

Since 1950, there have been records of a maximum of 23 individuals. The first was at White Mere in January 1970 and the second followed just two years later when a female was nearby at the Mere (Ellesmere). The bulk of the occurrences took place between 1976 and 1993 when sightings were almost annual. Both the single in 1983–84 and the two in 1988–89 frequented two different meres in the Ellesmere area.

Since 1993, Long-tailed Duck has become much scarcer, with only three appearances over the next 21 years: a female at White Mere in

Estimated Number per Year (1950–2014)

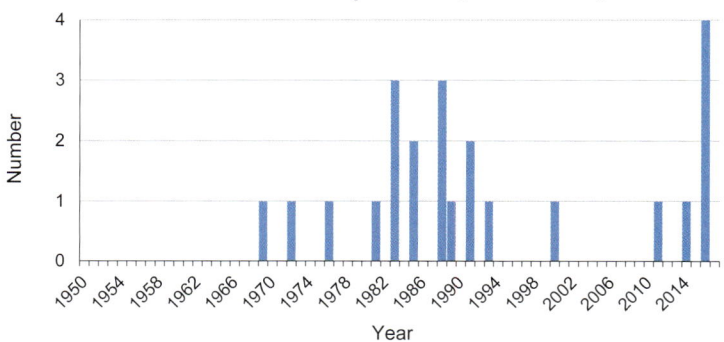

Arrival Date in 10-day Intervals (1950–2017)

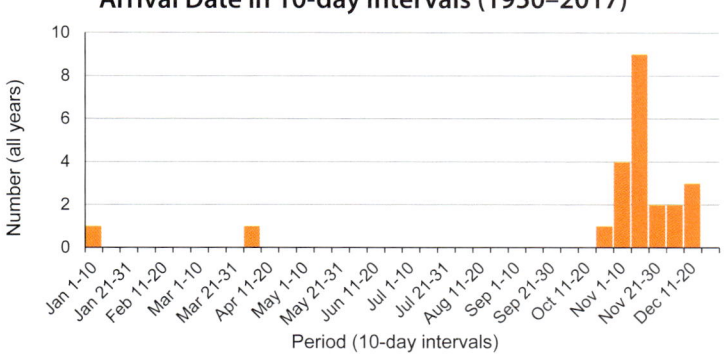

October 2000, a single Atlas record of an immature male at Adeney in November 2011, and a first winter drake at Trench Pool in November 2014. However, in 2016 there were three records: a first winter male, together with a female, which visited several of the Meres around Ellesmere between November 2016 and January 2017, one at Trench Pool in December 2016, and a first winter female at Chetwynd Pool from December 2016 until January 2017.

Sixteen were first seen in November, and once arrived they usually remained for some time, often into the New Year; others were discovered in January, April, October and December. The Ellesmere meres have always been the most reliable location to search for these ducks, 12 having been recorded there. Venus Pool has had three and Chelmarsh and Trench Pool two each, while only single appearances have been made at the remaining locations.

The lack of recent records shown in the annual occurrence chart was thought to be caused by migrants 'short-stopping' further east in response to milder winters, although this doesn't explain the records for 2016. However, the population wintering in the Baltic has also declined considerably in recent years and, because of this, the Long-tailed Duck has now been listed as 'vulnerable' by IUCN. It seems unlikely that the totals seen in the 1980s will be repeated in the foreseeable future, and 2016 may be an exception.

Allan Dawes

Goldeneye (Common Goldeneye)
Bucephala clangula
Scarce winter visitor

John Hawkins, Ellesmere, 6 January 2009

The nominate race *clangula* extends from Scotland, where a small breeding population has been encouraged by the provision of nest boxes, through Fennoscandia to Siberia. In the autumn a south-westerly passage brings some, mainly from Scandinavia, to the British Isles.

In the second half of the nineteenth century, small groups of three or four could be found annually on meres and pools, and when these waters were frozen they moved onto rivers and streams. Adult males were seldom seen: they occurred mainly during frosty weather and, even then, they were rare.

A rather imprecise 'Many Ellesmere in winter' was noted in 1902, and in December 1933 an adult male was shot out of 'a considerable number' at Eyton on Severn. Apart from these vague estimates, all other accounts were of single figures until 1945 when a combined total of 10 came from the Mere (Ellesmere) and Cole Mere.

'Unprecedented' numbers were reported in January 1950 when 49 were at Cole Mere on 12 January and 50 were at the Mere (Ellesmere) on 15 January, although some duplication may have been involved. The following winter only two were noted at the Ellesmere meres.

Numbers had tended to increase over the previous 10 years, according to the *Handlist* (1964). Between 30 and 40 were regular at the Ellesmere meres at this time, and small numbers were said to be present at Shavington, Apley Park (Castle), Dudmaston, Sundorne, Bomere, Marton Pool (Chirbury) and Venus Pool, and also on the River Severn during icy conditions.

The Ellesmere meres have continued to attract Goldeneye, and the chart shows maximum annual numbers from the WeBS.

The steady increase from the early 1970s followed by a recent decline mirrors the national trend for England, although the national decline began a little earlier. Falling numbers are thought to be caused by milder winters enabling them to remain closer to their breeding grounds in Fennoscandia and western Russia. An increase in those spending the winter in Sweden from 18,800 in 1971 to 75,000 in 2004 (Holt *et al.* WeBS 2012) adds weight to this theory.

WeBS Counts at the Ellesmere Meres (1960–61 to 2014)

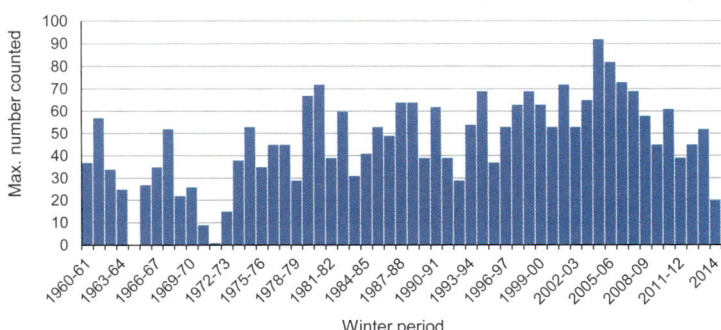

Winter Distribution (2007–13)

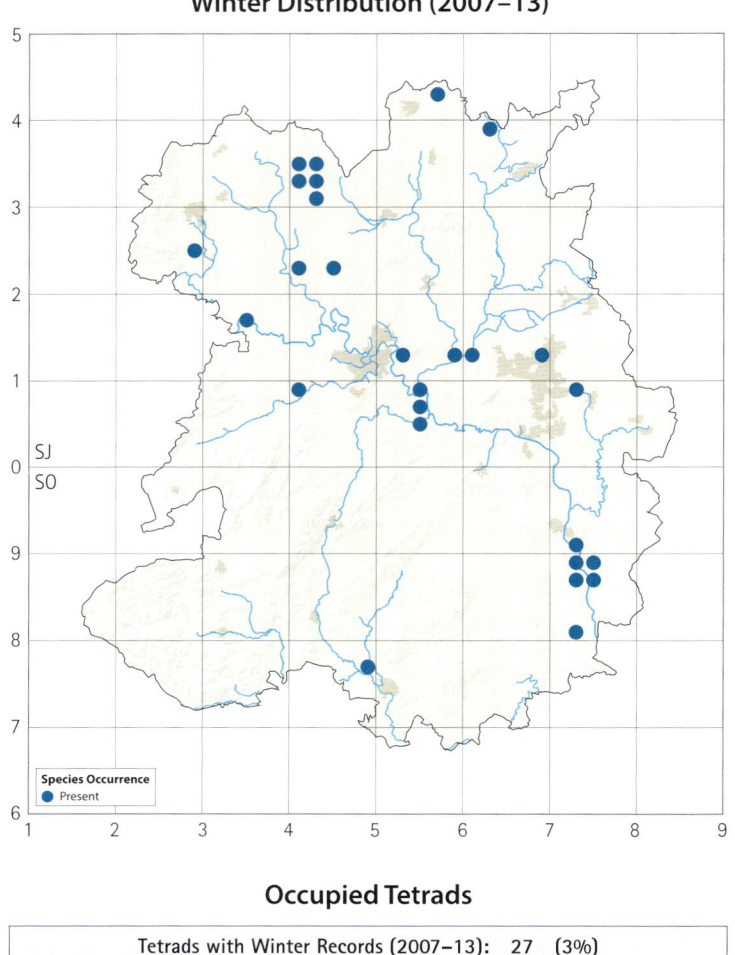

Occupied Tetrads

Tetrads with Winter Records (2007–13): 27 (3%)

Males can often be seen and heard displaying to small groups of females on fine days in the spring, and a visit to the Ellesmere meres to witness this is recommended. Pair formation at this time enables breeding to commence without further delay once they have completed their spring migration. Some linger into April and occasionally May. On a few occasions individuals have remained throughout the summer, most recently a male at the Ellesmere meres in both 1994 and 1995. Six at Oss Mere on 20 April 2009 were the latest recorded during the recent Atlas period. Goldeneye have never been known to nest here, but confirmed breeding in Avon in 2008, and possible breeding there in 2010, suggests the possibility of them breeding here should not be ruled out.

Mid-October usually marks the arrival of the first winter visitors but numbers build up slowly and generally peak in the New Year.

The only site where Goldeneye occur annually is the Ellesmere meres, but even here they tend to favour the Mere (Ellesmere) with smaller numbers at Cole Mere and White Mere. Some were present at most of the main wetland sites at some point during the recent Atlas period, but numbers were always low, never reaching double figures. A few were noted on the River Severn at Ponthen, Atcham and Hampton Loade. Bromfield, with singles in two years, was the only location in the south-west.

The maximum combined WeBS total during the recent Atlas period varied between 76 in February 2008 and 39 in January 2012. Only a few, if any, are likely to have been missed, suggesting an annual population of between 40–100.

Goldeneye have never been numerous and with current trends they appear to be becoming even less so. The peak of 92 at the Ellesmere meres in March 2005 seems unlikely to be repeated.

Allan Dawes

Smew

Mergellus albellus

Very rare winter visitor

The Smew is a short-distance migrant breeding in boreal zones from northern Fennoscandia east through Russia.

The first recorded specimen, a male, was shot at Wroxeter in 1861, although Smew had previously been described as 'not uncommon' (Ogier Ward 1841). A further 15 were documented during the nineteenth and early twentieth centuries. One shot near Shrewsbury in January 1909 was to be the last for over 40 years, after which two were at the Ellesmere meres in January 1951. In contrast to the modern era, records came from a wide spread of localities, including the Rivers Severn and Onny, and surprisingly only two were noted at the Ellesmere meres.

Between 1956 and 2004 wintering Smew were almost annual, but numbers increased markedly after 1980 when the average doubled from 2.1 to 4.2 a year. However, some are very mobile, frequently moving between adjacent waters and generating multiple records, making exact numbers difficult to calculate. In such cases the records are presumed to relate to the same individual, and the chart shows the estimated total number of 137 by winter periods.

During this time, redheads outnumbered adult males by more than

Jim Almond, Venus Pool, 3 January 2009

two to one and over half of the sightings came from the Ellesmere meres. Others frequented large water bodies such as Oss Mere, Chelmarsh, Venus Pool and Dudmaston, but none were reported from rivers.

Since 2004, Smew have become extremely rare. Three females visited Cound Fishing Pool briefly in January 2010, and two females were at White Mere the following month. These were the only recent Atlas records.

Smew have always been associated with hard winters, and the peaks in the chart correspond with especially cold periods. The majority of new arrivals are found between December and February and this is determined by the onset of harsh weather. However, fewer are now crossing the channel, and numbers remaining to winter in Sweden have increased substantially as conditions there have become milder. Birders needing this species for their local list may find it much harder to find in the future.

Allan Dawes

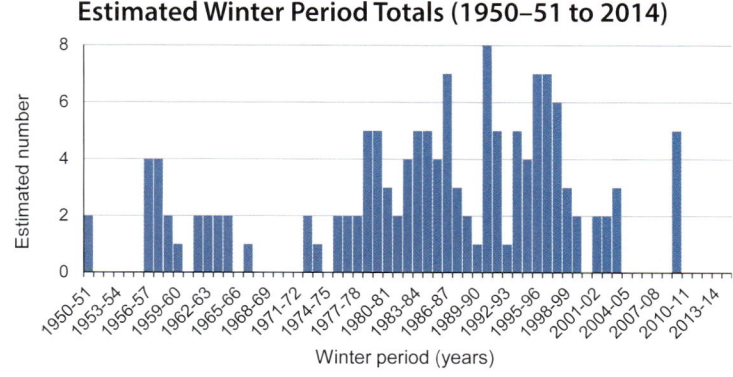

Estimated Winter Period Totals (1950–51 to 2014)

Goosander (Common Merganser)
Mergus merganser
Scarce resident, uncommon winter visitor

Jim Almond, Shrewsbury, 28 January 2011

Mainly found in the temperate and boreal zones of northern Europe, small numbers are resident in the British Isles and parts of southern Europe. In winter they are joined by migrants from the north.

Frosty weather often preceded sightings of small groups on the River Severn in the latter half of the nineteenth century. A flock of 16, which spent two months at Hawkstone in 1876, was described as being 'unusually large', but just 10 years later the same author, Beckwith, said these numbers occurred 'often' on the River Severn. The winter of 1885–86, when two groups, each of between 20 and 30, could be found regularly between Atcham and Cressage may have prompted this change. In the early twentieth century, apart from 1907, when 12–15 were on the river at Cressage and six were

at Ellesmere, all the remaining accounts were of single figures. The *Handlist* (1964) reported a changing status: 'Prior to 1948 numbers were small, mainly odd birds from scattered locations. It has since become a regular visitor in winter months to the Ellesmere group of meres, particularly Hanmer (Flintshire), and Shavington. Numbers vary from about 10–25 birds'.

Despite large annual variations at Shavington, there has been no underlying trend since the WeBS began there in 1960, the blank years on the chart indicating no survey rather than a lack of birds. In contrast, the Ellesmere meres ceased to be a regular wintering site in 1970–71, and only occasional individuals were seen until 1993–94, since when these visitors have returned in good numbers. Why the Ellesmere meres were abandoned for this period is unknown. There are five main waters and it is unlikely that a single factor would have affected all of them. The steady increase in the national wintering population throughout this period just adds to the mystery.

National numbers peaked in 1996–97, which coincided with the highest ever count of 257 at Chelmarsh in January. In Wales, the increase continued and peaked in 2010–11, while 87, the highest total at the Ellesmere meres, was in February 2012.

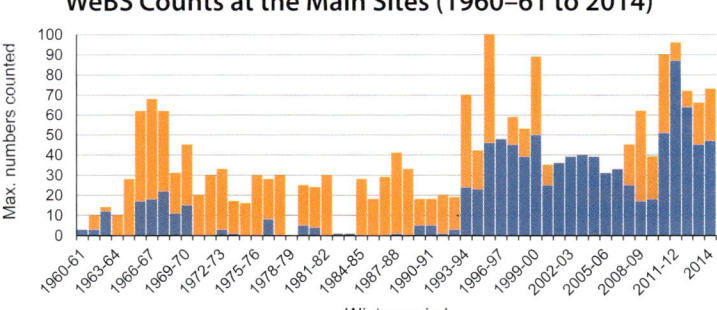

WeBS Counts at the Main Sites (1960–61 to 2014)

Breeding Distribution Change (1985–90 to 2008–13)

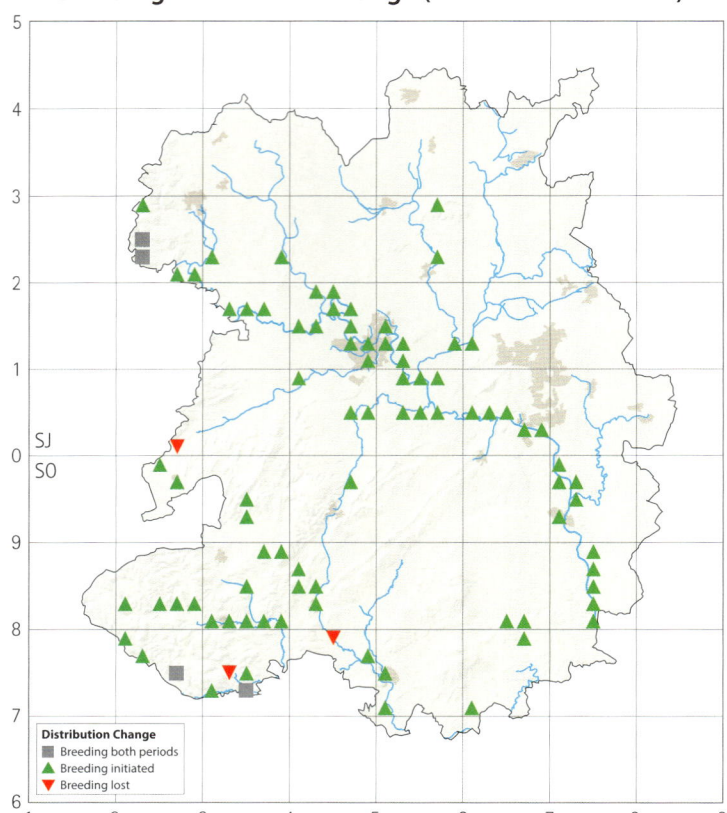

Distribution Change
■ Breeding both periods
▲ Breeding initiated
▼ Breeding lost

Winter Relative Abundance (2007–13)

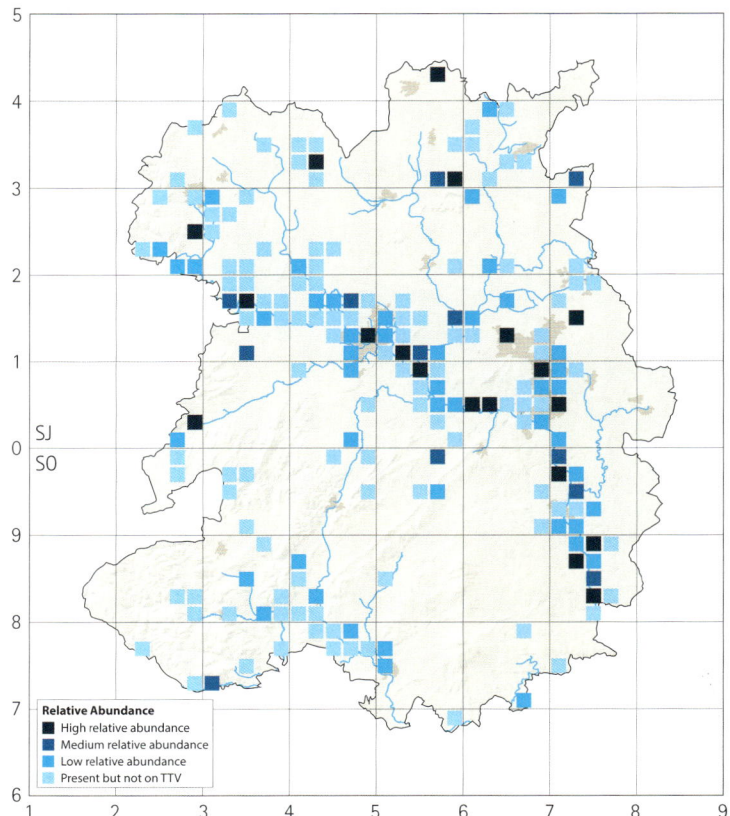

Relative Abundance
■ High relative abundance
■ Medium relative abundance
■ Low relative abundance
■ Present but not on TTV

Occupied Tetrads

Atlas period (breeding)	1985–90		2008–13		Change	
	Number	%	Number	%	Number	%
Confirmed	2	0	36	4	34	1700
Probable	3	0	26	3	23	767
Possible	2	0	24	3	22	1100
TOTAL	7	1	86	10	79	1129
Tetrads with Winter Records (2007–13): 188 (22%)						

After first breeding in Scotland in 1871, Goosanders spread slowly, and the first recorded Welsh nest was not until 1970. Young broods were regularly encountered on the small streams and rivers in Montgomery in the mid-1980s (*pers. obs.*), so it was not unexpected when breeding was confirmed close to the border on the River Tanat in 1987. This coincided with fieldwork for the *Atlas* (1992). Further breeding was noted on the River Teme in 1989 and summer sightings began to feature at other sites, but only seven tetrads had any level of breeding evidence between 1985 and 1990. Prior to 1987 they were only recorded during the winter months, although four at Betton Pool on 6 April 1899 were described as 'being so late' and four at Fenemere on 30 April 1962 were 'exceptionally late'.

Colonisation has continued at a pace, as can be seen from the breeding distribution change map. It would be logical to assume that Goosander continued to spread down the rivers from the border. Although observations are patchy, the few records received appear to support this. On the River Severn, possible breeding at Leighton in

1996 was followed by confirmed breeding at Coalport the following year and at Bridgnorth in 2000. In the south, confirmed breeding was noted at Purslow in 1994, Craven Arms in 2001 and Ashford Carbonell in 2002.

Their absence from the River Corve, the upper reaches of the Perry, and most of the Roden and Tern, stands out. These rivers flow through large arable areas, which may provide fewer nesting opportunities, but although tree cavities are generally used for the nest site, holes in the ground are also used, and nests can be up to 1km from the river, so suitable breeding sites should be present in many places along the vacant rivers. None of these rivers originate

Jim Almond, Radbrook, 28 January 2011

at the Welsh border, the original source of the breeding population. Recent records from the lower Perry, and at the confluence of the Tern and Roden, suggest that Goosanders are beginning to move up these water courses from the Severn.

In almost all tetrads where breeding was confirmed, it was by sighting broods. Goosander will nest by the smallest of streams and then the female takes the newly hatched ducklings downstream onto larger rivers. This may have exaggerated the number recorded on the larger rivers, but breeding near some of the smaller streams would probably have gone unrecorded if these movements were not witnessed.

Although Goosander were only found on a small number of TTVs, based on these counts the estimated population is 80–90 breeding pairs. This is close to one pair per occupied tetrad and therefore seems reasonable.

While the females are left with the parental responsibilities, the males depart in mid-summer. They travel to Tana Fjord in the north of Norway where most of the western European drakes, about 35,000, undergo their annual moult. They return from November onwards, along with others from northern Fennoscandia and Western Europe.

During the winter Goosander are much more widespread, although they abandon some of the smaller rivers. In addition to the larger rivers, which stand out clearly on the map, they also frequent many pools both large and small. During a cold spell in January 2011 they were also found on the Montgomery canal at both Maesbury and Pen-yr-estyn, where small patches of water remained unfrozen.

They often gather together in the late afternoon to roost on large waters, dispersing at first light to suitable feeding areas. This was probably the case at Chelmarsh, where 108 in December 2009 and 161 in November 2010, the largest counts during the recent winter Atlas period, were made at 16.00 and 16.30 respectively.

Those breeding at the south of their range winter within 150km, so the increased breeding population will have boosted numbers throughout the year. The national Winter Atlas (1986) showed them as present in just eight of the County's 33 10km squares, but this increased to 31 in the recent Atlas period. Based on TTV counts, the estimated winter population is between 320 and 400. The highest combined WeBS total during this period was 176 in January 2011, but because Goosander are dispersed over a large area. a great many will not have been counted, and the estimate from TTVs will be the better guide.

The breeding population still has room to expand into new areas, so the number that winter locally is likely to grow. However, numbers arriving from the continent will depend on future climate trends so winter numbers could go up or down.

Goosanders are perceived as a threat to fish stocks and licences are issued so that applicants can destroy them legally. As numbers grow they will come under increasing fire. Accurate information on trends is essential to counter wild accusations made by those who do not wish to share the countryside with native wildlife.

Allan Dawes

Red-breasted Merganser

Mergus serrator

Very rare winter visitor

This sawbill breeds throughout the northern latitudes across three continents, with small numbers in the north-western parts of the four home countries. It winters mainly in coastal waters.

In the nineteenth century it was said to be seldom obtained, except during hard winters. One or two were on the River Severn during the winter of 1838–39 and several were taken near Clungunford but not detailed. A male at Minsterley in November 1889 and a female on the Burway near Ludlow in February 1895 complete the historic records.

After an apparent gap of more than 50 years, there were two at White Mere in November 1954 and one at the Mere (Ellesmere) in February two years later, but it was another 10 years before the next record, two at Shavington in March 1965. A further 13 years were to pass before the next, one at Oerley reservoir in October 1978, which marked the beginning of an unprecedented run of 21 records in 18 years. Four were on the river at Cressage in February 1979 and seven at Chelmarsh reservoir in January 1987, but elsewhere singles predominated. In 1996, singles at the Ellesmere meres and Priorslee Flash, and two at Chelmarsh, all in the first two months, gave no indication that this sequence of records was about to cease abruptly. During this period sightings were almost annual, and 35 out of a total of 45 individuals since 1950 occurred at this time.

Since 1996 there have been just four records: a single at Priorslee Lake in January 2003, two at Chelmarsh reservoir in March 2004, a

single at the Ellesmere meres from February 2010 into March – which was the only one during the recent Atlas period – and the final one was at Chelmarsh in November 2013. The chart shows the estimated number of individuals seen each year, rather than by winter period, as none have remained from one year to the next, and the May and passage records listed below do not fit into winter periods.

The national WeBS shows a steady nine-fold increase from 1966–67 until 1994–95. Since then numbers have declined but they remain well above the original level. Most of the local occurrences were during this period of growth but the lack of recent records is in contrast to the national situation.

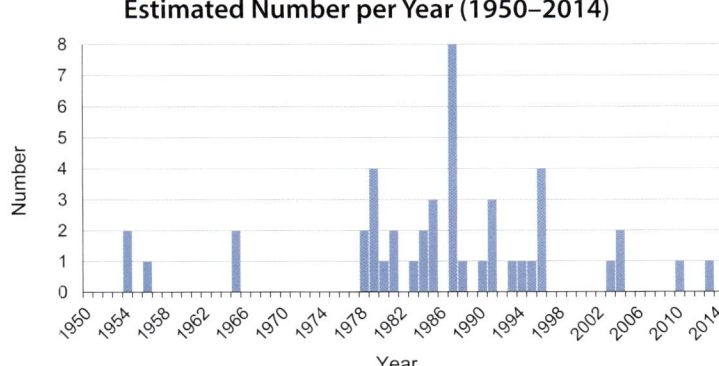

Estimated Number per Year (1950–2014)

Arrival Date in 10-day Intervals (1950–2014)

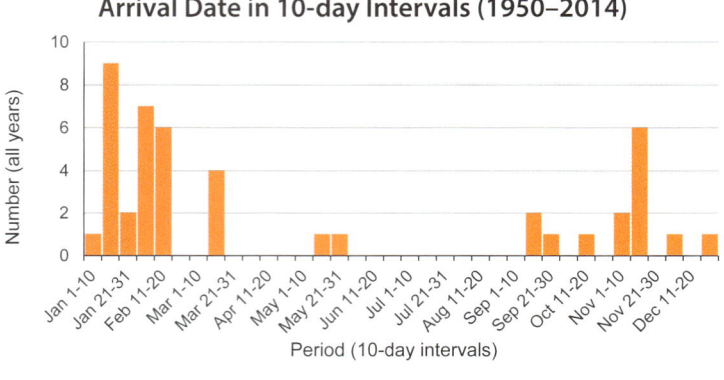

There are only two breeding season records, both of singles at Venus Pool in May, in 1983 and 1984. Nationally, after first breeding in north Wales in 1953, the population grew considerably during the 1960s and 1970s. both at coastal sites and on upland lakes and streams. Since then numbers breeding inland have fallen and they seem unlikely to follow the example set by their close relative the

Goosander. Three on passage in September and one in October did not linger more than a few days. The remainder occurred during the mid-winter period, with peaks in January and February indicative of cold weather movements. The first winter male that visited several of the Ellesmere meres in February 2010 was present for at least 41 days; other January arrivals at Chelmarsh in 1987 and 1996, and Cole Mere in 1996, also stayed for several weeks. The remaining reports were for a single date.

Differentiating first year males from females can be difficult and not all records specified gender, but of those that did, 13 were male and 11 female. Some females may also have been overlooked since Goosanders have become more widespread.

The reduction in annual occurrence since 1996 is thought to reflect fewer wintering on this side of the channel because of milder winter conditions, and increases in Sweden and the Netherlands seem to support this. However a large population decrease has also been noted in the Baltic. This is a cause for concern and may further reduce winter visitors.

Allan Dawes

Ruddy Duck
Oxyura jamaicensis
Naturalised resident now eradicated

Ruddy Duck are native to North America, and were frequently held in wildfowl collections. The first British breeding record was claimed by the curator at Walcot Lydbury North in 1936.

Following the escape of Ruddy Duck from the collection at Slimbridge in the 1950s, this was one of the earliest counties to be colonised, with a male on Marton Pool Baschurch on 30 April 1962, and two moving between three of the Ellesmere meres on 24 November 1962. Breeding was confirmed, in the wild, at Crose Mere in 1965. Crose Mere continued to be the main site, but in 1975 three pairs with young were also recorded at Oss Mere. By 1979 it was resident on most suitable waters, with winter flocks of up to 77 at Crose Mere.

Breeding habitat is permanent freshwater marshes, ponds and lakes, with emergent vegetation in which to hide the nest. This habitat is mainly to be found in the north.

In winter they were found in flocks at large lakes and reservoirs, with the Mere (Ellesmere) and Crose Mere being favoured, and the highest ever count of 143 was at the latter site in October 1980. The population there remained stable through the 1980s, but around 1990 there was a shift to the Mere. There was no dramatic increase throughout the 1990s, with the main wintering flock at the Mere usually numbering around 30–40. Seventy-four in December 1995 was exceptional.

At the end of the twentieth century, concerns were raised internationally about the Ruddy Duck spreading to Spain and hybridising with the endangered White-headed Duck. Between 1999 and 2005, trials to control Ruddy Duck were carried out at major wintering sites in the West Midlands, Fife and Anglesey when 2,634 were shot. Here, in the first year of the trials (1999), six were shot, 50 in the second year, with nine the following year and two in 2003.

The full eradication programme began in September 2005 at a total

Jim Almond, Venus Pool, 6 May 2007

of 132 sites across England, Scotland and Wales. Here, a further 16 were shot in 2006 and, finally, three in 2007. Nationally, the population was believed to have been reduced to only 60 birds by March 2012, down from 6,000 at its peak in 2001 (Iain Henderson, Animal Health and Veterinary Laboratories Agency (AHVLA), *pers. comm.*), and the cull continued with a view to total eradication by 2015.

The chart shows the WeBS maximum winter period count at the Ellesmere meres between 1965–66 and 2008–09. The number counted builds up from September onwards, so in some periods the maximum was counted in that month or October, and such cases have been counted in the late winter period. The final count was four at White Mere on 21 September 2009.

The breeding distribution change map is a fair reflection of

WeBS Counts at the Ellesmere Meres (1965–66 to 2008–09)

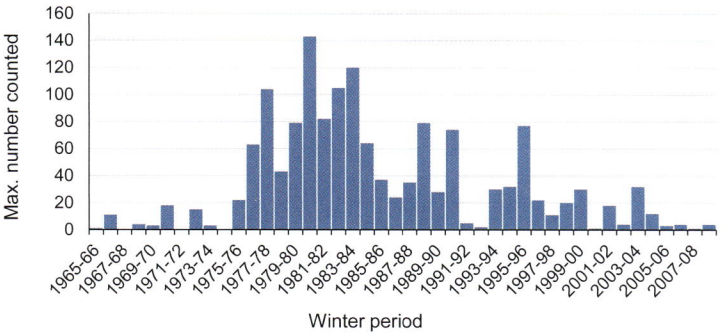

Breeding Distribution Change (1985–90 to 2008–13)

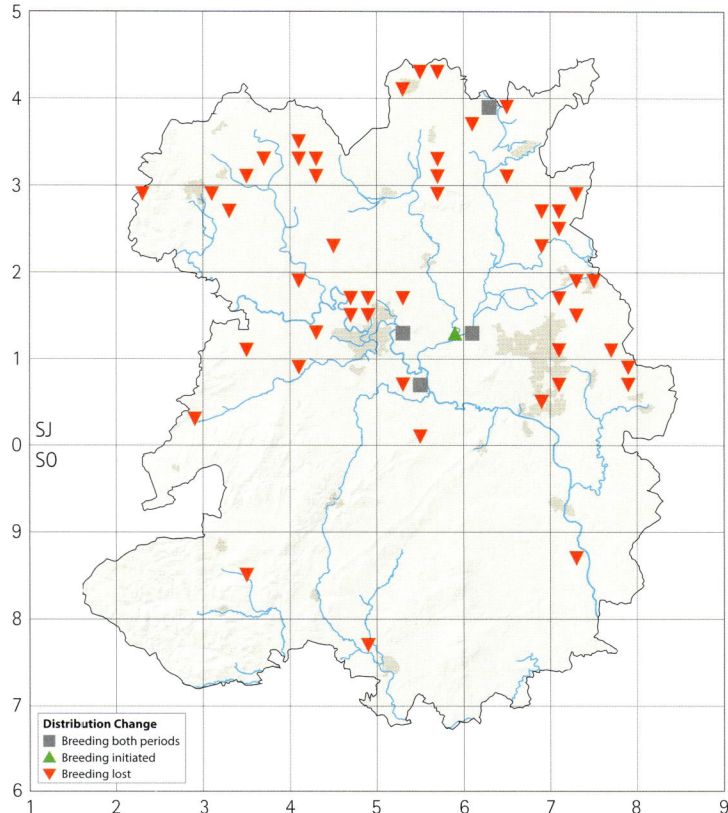

the Ruddy Duck's status close to its peak, around 1990, when the population was estimated 'in the order of 40 pairs', and the effect of the cull. After the second year of the recent Atlas period it was extinct as a breeding species.

However, breeding continued in 2008 at Monkmoor Pool with seven juveniles in July, and probable breeding at Venus Pool and Mirelake. The later site was used for the first time as the ceasing of sugar beet production created suitable conditions for breeding.

In 2009, records were received from six sites. Confirmed breeding was found at Shavington where there were five recently hatched young in July, and this brood was also reported in September. Probable breeding was again reported at Mirelake, Monkmoor Pool and Venus Pool. Non-breeding birds in 2009 were seen at Priorslee Lake (one in January) and White Mere (four in September).

The remaining Atlas years 2010–13 saw a dramatic decline, with no records of attempted breeding, and only single females at Cound Fishery in January 2010, Priorslee Lake in August 2011, the Mere (Ellesmere) in November 2012 and single birds in 2013 at Milford Mill on 26 May and Whixall canal floods on 1 November.

There were no records in 2014, or the following two years, and probably the last ever record was of one at Ash Magna from 3 July 2017, and subsequently culled in late September, so the cull appears to be complete.

Martin Grant

Occupied Tetrads

Atlas period (breeding)	1985–90		2008–13		Change	
	Number	%	Number	%	Number	%
Confirmed	13	1	2	0	−11	−85
Probable	27	3	2	0	−25	−93
Possible	13	1	1	0	−12	−92
TOTAL	53	6	5	1	−48	−91
Tetrads with Winter Records (2007–13): 10 (1%)						

Black Grouse

Lyrurus tetrix

Rare visitor, has bred

Black Grouse thrive in undisturbed country, on wild, open and extensive heaths and moors with patches of woodland and scrub. In modern times they are retreating to high and remote hills elsewhere in England and Wales, and their past distribution is a gauge of the loss of wild Shropshire.

Excavations at Roman Wroxeter produced two bones of Black Grouse dated to the period between the early fifth century and the late seventh century. Of the 19 bird species identified in the bone assemblage, at least 14 can be regarded as game or potential food, Black Grouse among them (Hammon 2011).

The first written account, one of the few records of birds in the sixteenth century, was of Black Grouse on Catherton Common. The

record, in Latin, was translated into Olde English (Smith L.T. 1910) from John Leland's 'Itinerary of England and Wales' of 1535–43, as: 'Ther is anothar [hill] cawllyd Caderton's Cle, and ther be many hethe cokks'. At that time they were probably widespread and even common on the open and part-wooded heaths, wastes, commons and high hills.

Three centuries later Elliott (1911) cited a reference from the Wyre Forest in 1817 when it 'appears to be decreasing in numbers', while, also in the first half of the nineteenth century, Eyton (1838) wrote that it is 'found in most of the extensive heaths'. Eyton lived near Wellington and he was probably referring to the lowland heaths in the north rather than the upland moors in the south.

From the late nineteenth century we are fortunate to have two

The Black Grouse map

The map endeavours to present its past distribution and reflect the rapid contraction of range at the end of the nineteenth century. It does this on the basis of an assessment of the detailed records and other imprecise information in the early accounts, coupled with a working experience and understanding of past habitats and their changes, particularly in the south-west.

In the south, the six upland areas that are still predominantly heathland are mapped in full: the Clun Forest, Stiperstones, Long Mynd, Mortimer Forest (partly in Herefordshire), Titterstone Clee and Brown Clee. Black Grouse were 'still plentiful' at these sites (Beckwith 1893), and the map assumes the whole sites were occupied. The 1903 record of the last one on the Boyne Estate has been notionally mapped in SO68C.

The map further suggests an eastward extension from the Clun Forest onto additional upland areas considered to have been potential habitat at that time. While the precise distribution within these areas is not known, occurrence throughout each of them would have been likely.

In addition, 331 place-names – 'heath', 'common', 'moor' and 'moss' – are mapped to suggest habitat likely to have been utilised during the earlier, even wider, distribution, particularly in the north, as alluded to by Eyton (1838).

Historic Breeding Distribution

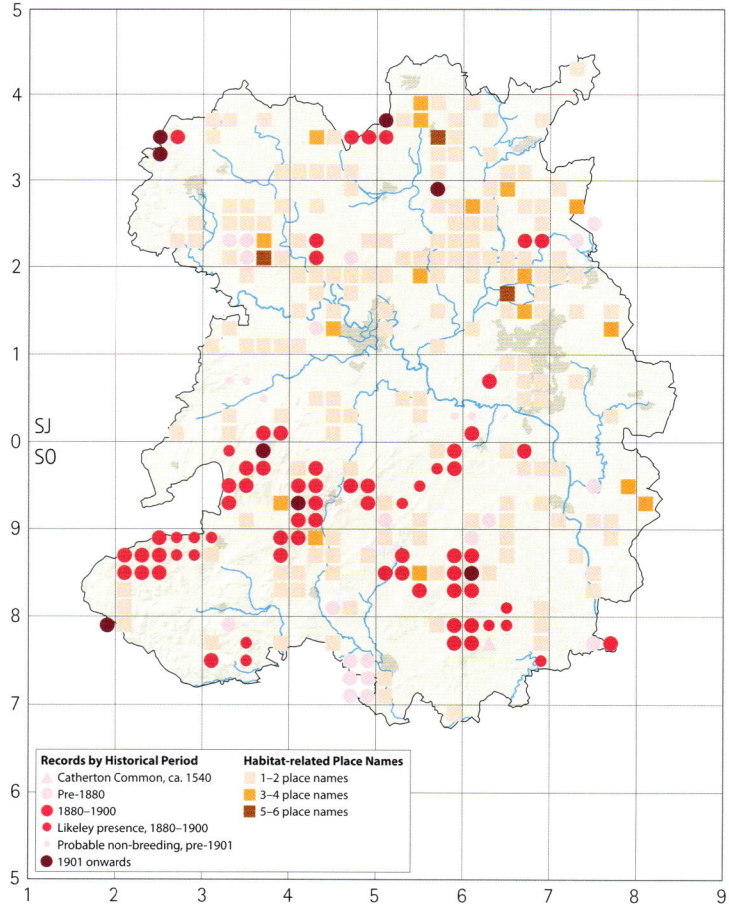

Records by Historical Period
- Catherton Common, ca. 1540
- Pre-1880
- 1880–1900
- Likeley presence, 1880–1900
- Probable non-breeding, pre-1901
- 1901 onwards

Habitat-related Place Names
- 1–2 place names
- 3–4 place names
- 5–6 place names

accounts by Beckwith, the first in 1879, the second in 1893, along with his other casual observations. In the account of 1879 he wrote: 'Frequent on the Longmynds, the Stiperstones, the Clee hills, the Black Hill, and Clun Forest, a few also in Willey Park, and near Ludlow ... in fact it is numerous (in the south), and rare in the north', suggesting a decline in the north since Eyton's account. In an additional article in *The Field* in 1885 he included Corvedale in a similar list of sites, probably on the strength of birds descending to lower levels during winter, though there is a subsequent record of a nest with six eggs among rough pasture at Cold Weston in 1899. The article continued: 'In the south, however, it is common, and in recent years has not only, where preserved, increased in numbers, but become more widely distributed', and he went on to list the usual strongholds and included a few isolated records of what may have been wandering individuals.

In 1886, Rocke cited Black Hill as having Black Grouse 'in considerable numbers' while both Clee hills still 'occasionally boast one or two broods', and Stow Hill, Bucknell and High Vinnalls above Ludlow had them still.

In his 1893 account, Beckwith related records spanning the second half of the century and compared the situations in the north and the south, confirming the decline in the former. It was 'once common ... now nearly extinct (in the north), its last stronghold on Whixall Moss having been ruined partly by the making of the Cambrian Railway in 1862'. Indeed he provided the last record from that site, in May 1882, though 'one or two may exist here still, as, owing to its boggy nature, the moss is very seldom crossed or shot over'. Other records from the north were also listed: 'Formerly ... about Baschurch', near Knockin 'where killed so late as 1872', 'in the summer of 1886 ... reared between Buildwas and Wrekin' and 'a few' seen at the Quinta near Oswestry in the autumn of 1891. In contrast, in the south 'it is still plentiful, and shows a tendency to increase its numbers and to extend its range. Unfortunately, however, very few attempts are made to preserve it'. Beckwith continued 'the Grey hens ... are generally the first to be slain ... and the fatal practice of shooting both old and young on fields of late-standing grain is too common'. He also mentioned several instances of interbreeding with Pheasants.

After centuries of gradual land-use change, especially the progress of agriculture, it was the rise of the shooting estates in the 1890s, documented for at least the estates at Apley and Walcot (Forrest 1908), which dealt the final blows to the species, and the last confirmed breeding records are from the early twentieth century.

In the north, a nest was found by a keeper on the Hawkstone estate in 1903, while in the south the last to be shot on the Boyne Estate, Brown Clee, was in 1903 (Viscount Boyne *in litt.* 1992), Finally, in 1911, Forrest found a brood on 'the top of the Stiperstones' and in the same year one was shot on the neighbouring Gatten Estate, on 23 November, the last recorded there (Hulton-Harrop *in litt.*).

Since then there are a handful of records from our side of the Welsh border, in the north-west at Selattyn Hill in the winter of 1916–17 ('a Greyhen was shot in the young planting at Cefn Coch' on 22 January 1917 on the Brogyntyn Estate west of Oswestry [National Library of Wales]), and at Cefn Coch, in 1962 and 1970, and from the south-west near Beguildy (perhaps not on our side of the border) where seven were recorded on 16 April 1927. Adams (1950) suggested that 'blackcock nested in the brakes' in the Clun Forest until perhaps the

1930s, and a negative report ('none found') was submitted in 1953 from Kerry Forest, where they were previously known (record with SOS, previously unpublished).

There are intriguing tabulated references to Black Grouse being 'seen', 'nesting' and 'common' in the period 1920–24 and 'seen', 'nesting' and 'declining' in 1950–54 in the Bucknell area (SBR 1987), but they lack further substantiation so are best ignored. The *Handlist* stated that 'it was last reported to have bred on Titterhill, Bucknell in 1954' but again, in the absence any supporting records, this must be discounted.

The only accepted post-1950 records are of a female near Earl's Hill in 1981, presumed to derive from an attempted introduction, and, from 2014 onwards, occasional sightings from another site; there is

a strong likelihood these latter grouse originate from an 'unofficial' reintroduction in Wales, which is kept confidential at the request of the landowner. These records are not included on the distribution map.

The map shows the historic distribution of Black Grouse, based on the records referred to above, and postulated on the basis of place-name evidence.

Therefore, Black Grouse were formerly widely distributed, and probably in some numbers, including the lowlands of the north that are now largely arable country and not an area that would be associated with this species. The 'bubbling' of lekking males must have once been a feature of the countryside in the spring, something that is now sorely missed and, unfortunately, unlikely to ever return.

John Tucker

Red Grouse (Willow Ptarmigan)
Lagopus lagopus
Scarce resident

Forrest, writing in 1899, dragged a red herring across the history of Red Grouse, claiming that 'there were few, if any ... before 1840', the year when two pairs from Yorkshire were released on the Long Mynd. According to Forrest, they spread from the Mynd to the Clee hills and Clun Forest. But within a few lines he begins to doubt himself:

> Of course it cannot be stated with certainty that these have all sprung from the Yorkshire birds, as there is no barrier to prevent the influx of others from the Radnorshire and Berwyn Mountains where there have always been Red Grouse. Furthermore it is difficult to imagine why the Longmynd should have been devoid of this species before 1840, seeing how eminently suitable the district is to its habits.

Beckwith (1885) had already been advised of this canard and regarded it as implausible: 'I have heard an opinion expressed that Red Grouse are not natives ... This seems extremely improbable'. And, to clinch the matter, we have the evidence of Eyton who stated in 1838 (i.e. before the Yorkshire birds were released) that Red Grouse were 'common ... on the Stiperstones'.

It is beyond credence that Red Grouse would have failed to reach and move between the Shropshire Hills and their strongholds in Wales. On shoot days in the past, Grouse were observed flying from the Long Mynd to the Stiperstones (Henry Owen *pers. comm.*). More recently, three were seen flying high over Nipstone Rock as if heading from the Long Mynd to Corndon, the nearby Welsh hill (*pers. obs.*), and two were seen flying high from the north-west over Heath Mynd, heading towards the southern end of the Long Mynd (Neil Wainwright *pers. comm.*). So, although, as indicated by the close similarity in distribution between the breeding and winter season maps, Red Grouse are essentially sedentary, they do on occasion move around. These 'roving' movements may explain a scatter of records across the years in widespread and sometimes unlikely locations, including, for example, a covey of eight at Bury Walls on 15 September 1921, a locality described at the time as 'extraordinary'. Even more extraordinary are entries in the Downton Hall game book, held in Shropshire Archives, referring to 61 Red Grouse shot at 'Upper

Letwyche' (presumed to be Upper Ledwyche) in August 1929, and the following day a further 40 nearby at Middleton. Neither location figures in the ornithological literature and it seems inconceivable that they harboured them as recently as this.

Returning to the traditional locations, Rocke (1866) refers to Red Grouse being common on the Long Mynd and Clun Forest, but states that cultivation had driven them from the Black Hill; they must have returned, because the *Church Stretton Advertiser* of 15 August 1912 reported the exclusion of whimberry pickers from the hill while Grouse shooting was in progress. Beckwith (1879) adds both Clee hills to the list of sites, refers to occasional sightings on the High Vinnalls, and mentions a specimen in Lord Hill's collection shot near Prees. This

Jim Almond, Long Mynd, 7 January 2012

Occupied Tetrads

Atlas period (breeding)	1985–90		2008–13		Change	
	Number	%	Number	%	Number	%
Confirmed	8	1	7	1	-1	-13
Probable	5	1	3	0	-2	-40
Possible	3	0	2	0	-1	-33
TOTAL	16	2	12	1	-4	-25
Tetrads with Winter Records (2007–13): 14 (2%)						

Breeding Distribution Change (1985–90 to 2008–13)

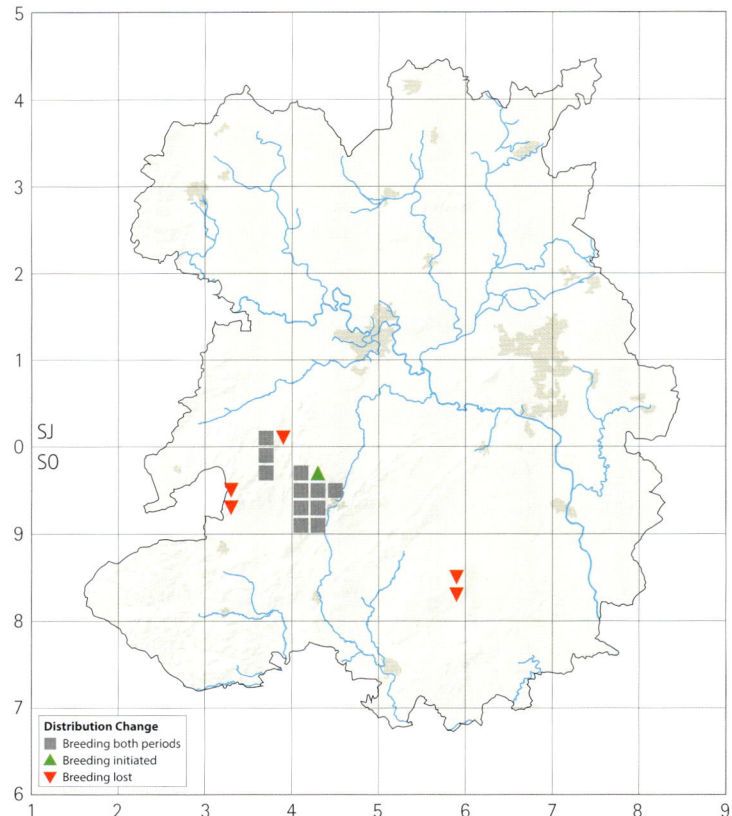

Distribution Change
■ Breeding both periods
▲ Breeding initiated
▼ Breeding lost

is a lowland location, but Red Grouse are tied not to altitude, but to heather, which makes up almost all of their diet. At one time there were extensive lowland heaths in the north, including at Prees, as shown on the map in the Black Grouse species account, and these may perhaps have held Red Grouse.

Some of the earliest records come from the game books of the Brogyntyn Estate, Oswestry; they cover the shooting seasons 1822–23 to 1834–35 (these game books and others for the Estate were consulted courtesy of the National Library of Wales). The books are difficult to interpret, but it seems that Red Grouse were shot in every season bar one, with numbers ranging from 1–18. The next series of game books runs from 1882 to 1933. Red Grouse do not figure again until an isolated record of one shot in 1904. However, from 1908 to 1933 there were only two blank years and the numbers shot ranged from 1–38, including 10 in 1933. Any game books there may be for subsequent years have not been traced. Shooting locations, where named, were Cefn-y-Maes, Cefn Coch (where there is a building named Grouse Lodge), Selattyn Hill and Brogyntyn itself; none of these locations appears in the ornithological literature.

Adams (1950) described Red Grouse as 'numerous' in the Clun Forest in the early 1930s, but absent by the end of the decade (although they were reported in RBF as being present as late as 1953). The *Handlist* (1964) mentions two more sites in the south-west: the Black Mountain, from which there had been no reports 'in recent

Jim Almond, Long Mynd, 7 January 2012

years', and Stow Hill, although local opinion suggests that they may not have lingered here beyond the 1930s (Jonathan Coltman-Rogers *pers. comm.*); neither site is mentioned in SBRs which cover the years from 1955 onwards.

The Hopton Court game books (consulted courtesy of Christopher Woodward) provide a run of records from 1887–92 and discontinuously from 1909–60. The books cover an area which extended from Catherton and Silvington Commons on to Titterstone Clee. Supplementary records for the years 1912 and 1915 for Titterstone Clee come from the Downton Hall game books. The numbers shot were small and clearly this was 'walked up' rather than 'driven' shooting. The nineteenth-century maximum was 14 birds in 1889, and in the first half of the twentieth century the numbers killed ranged from 1–11 per season, the exception being 23, all from Titterstone Clee, in 1915. The Hopton Court books contain records for eight years in the 1950s when an average of 13 and maximum of 23 were shot, but the seven shot in 1960 were the last to be killed, and a covey seen in 1963 (David Martin *pers. comm.*) was the last record apart from two in 1980.

Beckwith (1885) thought there might be 40–50 brace on Brown Clee; he may have underestimated, as 37½ brace were shot in 1901 (10th Viscount Boyne *pers. comm.*). But this was a high point, and by the 1950s the usual bag for the year was less than five birds, though a post-war peak of 16 was recorded in 1974. Thereafter the population declined towards eventual local extinction in the 1990s.

The Gatten Lodge game book (consulted courtesy of Jane Hulton-Harrop) shows a total bag of 115 brace for the Stiperstones in 1911 (the first year for which records survive), and maxima of 55 brace in a year in the 1930s, 32 brace in the 1950s, and 15 brace in the 1980s. The population was estimated to number 160–200 birds in 1986 and 120–150 in 1987 (SBR), but some time thereafter there was a rapid decline. By 1991 there were so few that shooting ceased, and the following year the number of males calling in the spring was thought to be in the range of only five to nine (males engage in territorial calling displays in spring which provide an indication of the population level). From 1986 onwards, the amount of heather management was ramped up by the landowners, the Nature Conservancy Council, as, from 1993, was the control of the principal predators, foxes and Carrion Crows. This may have led to a modest increase in numbers, which reached a peak of 26 calling males in 2007 when there was a late summer count of 88 birds. Some relaxation in predator control thereafter may have been responsible for a decline in numbers, but from 2012 predator control intensified, and perhaps as a consequence, numbers increased to about 36 calling males in 2014, when about 120 birds were counted in late summer (Natural England records). The sporting rights are privately owned and a total of 13 Grouse were shot, or taken by falconry, over the seasons 2005–07, with a further three in 2014.

According to Forrest (1899), bags of 40–50 brace (seemingly per day) were not uncommon on the Long Mynd, with as many as 90 brace having been shot. A decline may have followed, as there were annual releases in the early 1950s of 20–30 Scottish birds, followed by an unknown number in 1962, yet in 1975 as many as 113 brace were shot, the highest tally for the 1970s. For the period 1980–90, the maximum for a year fell considerably, however, to 32 brace (Wall 1992). Indeed, so marked was the decline that by the mid-1990s the number of calling males was estimated to lie in the range of just 16–25. Shooting ceased in 1992, when the sporting rights were taken back in hand by the National Trust, but the burning and cutting of heather in order to stimulate new growth had to be abandoned, as the fresh young shoots, the principal food of the Grouse, were being eaten off by excessive numbers of sheep. Stocking levels, particularly in winter, were cut back significantly following the conclusion in 1999 of an Environmentally Sensitive Area agreement between the Ministry of Agriculture, Fisheries and Food, the Long Mynd Commoners Association and the National Trust, and heather management resumed thereafter. Numbers of Grouse have increased steadily since the late 1990s and systematic surveys in the years 2011–14 suggest that the number of territorial males now lies in the range 52–66, with an estimate of 56–58 in 2014 (Smith 2015).

The *Atlas* (1992) showed confirmed breeding on the Stiperstones and the Long Mynd, but also on Brown Clee. There, by 1994, Red Grouse were 'scarcely maintaining a foothold' (Jordan *et al.* 1995), and, as shown on the breeding distribution change map, that foothold has now been lost. On the Long Mynd, breeding evidence came from one more tetrad in 2008–13 than in 1985–90, but it was wanting in the case of one previously positive tetrad on the Stiperstones and two in the Heath Mynd area (although presence was registered in one of them in winter), and it seems that attempts made there over recent decades to maintain or re-establish a population through releases have failed.

In 2014, the combined total of calling males on the Long Mynd and the Stiperstones was estimated to be in the range 89–96. How does this compare with 100 years ago? Based on the figures given above, 300

Jim Almond, Long Mynd, 7 January 2012

brace would seem to be a reasonable estimate for the numbers shot each year in the first decade of the twentieth century; this includes a notional figure of 35 brace shot in the Clun Forest and adjacent areas, for which no bags are known. Assuming this represents 30% of the population (a typical cull level), some 2,000 birds would have been present at the opening of the shooting season a century ago; nowadays there are probably no more than 350, even in a good year.

The decline has been steep and profound, but it is common to many of our uplands. Nationally, the annual 'bag' was once 2.4 million birds (Leslie 1911), today it is 0.4 million (PACEC 2006). The decline is in part a consequence of the loss and fragmentation of upland heathland through afforestation and conversion to grassland, but the quality of the remaining habitat will have suffered too through inadequate management and excessive grazing; both of these factors have applied locally. In some areas, high parasite burdens and disease can reduce breeding success and cause high rates of mortality, but these are problems of high density populations and have been discounted here where densities are low (Wall 1992). A factor that will, however, have played a part is a relaxation in predator control. Formerly no predatory bird or mammal would have been spared and this was a factor in boosting Red Grouse numbers to artificially high levels. The late Walton Humphrey, whose father William owned the Long Mynd from 1937 to 1963, claimed that 500 foxes were killed there each year (an exaggeration perhaps, but indicative nonetheless) using baits laced with strychnine, while, as late as 1967, long after such practices were made illegal, chicken eggs injected with alpha-chlorolose were being used on the Stiperstones to kill 'bird vermin', with 'good results' (anon. *pers. comm.*).

The Red Grouse is a distinct sub-species of the Willow Ptarmigan; as such it is unique to the British Isles, and today Shropshire is a significant outpost for this quintessential upland bird. To the south, the Black Mountains harbour a small population which spans the England–Wales border, linking with a few sites in South Wales. Otherwise this is the most southerly point in England, apart from Exmoor and Dartmoor, where Red Grouse have been recorded in recent times. The species was introduced to both of these areas (Wesley 1988) and it seems now to have been lost from Exmoor. Current efforts towards the conservation of our native population, through habitat management and the legal control of predators, are to be welcomed.

Tom Wall

Red-legged Partridge

Alectoris rufa

Uncommon naturalised resident; population supplemented by annual releases

Dave Barnes, Long Mynd, 20 June 2010

Two 'Frenchmen' were shot in 1877, one near Wroxeter, the other near Middleton (Beckwith 1879); these are the first records of what is otherwise known as the Red-legged Partridge. The old vernacular name indicates the origin of this game bird which, after a number of failed introductions, finally became established in Suffolk in about 1770 (Lever 2009). Subsequent releases and natural spread followed, until by 1900 it occupied most counties south and east of a line from the Severn Estuary to the Humber (Holloway 1996), and it had edged its way into Shropshire. Its arrival was not welcomed, and Paddock (1890) stated: 'I am sure that sportsmen will not regret that here [the Newport area] it is only a casual wanderer', while Forrest (1908) observed that birds introduced to Willey Park had proved 'so quarrelsome and such indifferent "sport", that as far as possible they were afterwards destroyed'. He added that others had recently bred at Marrington, and near Ludlow, without having been introduced at either location.

Over subsequent decades reports came from widespread areas, but the species' status was summarised only for the Newport area in the mid-1940s, where it was described as 'apparently very uncommon' (Pearson & Watson 1945–46). By 1964 it was reported in the *Handlist* as being as common as the Grey Partridge (then a common bird) in a wide area surrounding Shrewsbury and across to Wellington, and also on the eastern flank, from Newport southwards. It was described

as 'thinly distributed' elsewhere, except for the hill country from which it was 'almost entirely absent', but by the time of the national BTO Atlas (1976) Red-legs were recorded in all but two of the 10km squares wholly or mainly in the County. The national BTO Atlas (1993) showed no overall change, though its more detailed coverage revealed a marked thinning of the distribution towards the west.

Today the core areas, as noted in the *Handlist* (1964), remain largely intact; they correspond, broadly speaking, to the lower, drier and warmer parts, where the land is of better quality for farming and arable acreages are greater. Winter and breeding season distributions are similar, as are the areas of highest relative abundance, and in both seasons the cereal-growing areas north-west and south-east of Telford, shown on the predominant farm types map on p. 27, figure strongly. But while the overall range has not changed materially, it has thinned, as shown on the breeding distribution change map, which illustrates the net loss of breeding evidence from 159 tetrads. Losses have been greatest in the lowland farmland of the east, indeed losses in the upland pastures of the south-west are almost balanced by gains. This thinning of the distribution has occurred despite a major increase in releases. Nationally they have increased four-fold since *Atlas* fieldwork 1985–90, and published figures (Aebischer 2018) indicate that although the world population in the wild is estimated at only 3.25 million breeding pairs (Aebischer & Lucio 1997), in the order of 8.9 million birds are now released in the UK each year, including perhaps 120,000 here, largely in August. Some 48,000 of these will be shot, leaving a balance of 72,000, but the population as estimated from TTV counts is in the range of only 1,000 to 1,500 breeding season territories, indicating that losses over the winter to causes other than shooting must be huge.

In view of the large numbers that are released in late summer, it may seem surprising that fewer tetrads were found to be occupied in winter than in the breeding season: 39% as against 51%. This is essentially a sedentary species, born out by the wing-tagging of 133 captive-bred Red-legs at Attingham in 1969: 32 were recovered, all but one 'locally'. However, the exception travelled a few kilometres to Haughmond, and elsewhere Green (1983) found that between 69% and 79% of wild-bred young females moved more than 0.5km (he was unable to gather information on maximum distances). Such movements may partially explain why, while numbers will be lower in the breeding season, they may be more widely dispersed. Males move less, so the adult male ringed at Cressage in May 1983 and shot five months later 3km away, was somewhat adventurous; this is the only ringing recovery. They are usually inconspicuous in winter, but the territorial call of the male (likened, in its rhythm, to the chuffing of a steam engine) increases detectability at the start of the breeding season.

Why should a species so despised as a quarry 100 years ago, be so popular today? The answer is linked to the fate of the wild Grey Partridge. Sportsmen rated Greys as far more challenging in the field and much tastier on the plate than Red-legs, but numbers fell rapidly from the 1950s onwards, and while Greys can be bred artificially,

Breeding Relative Abundance (2008–13)

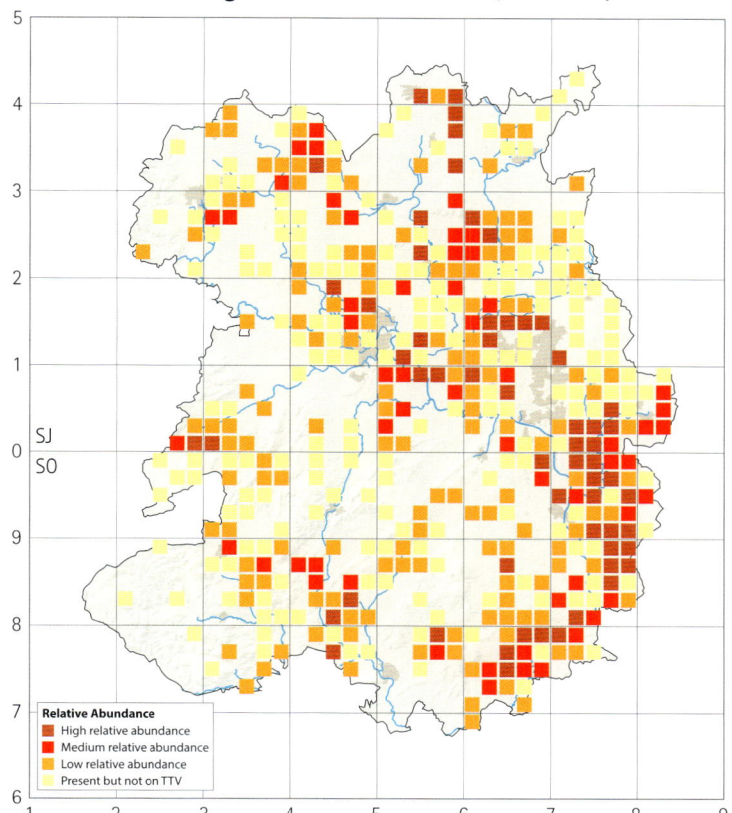

Relative Abundance
- High relative abundance
- Medium relative abundance
- Low relative abundance
- Present but not on TTV

Breeding Distribution Change (1985–90 to 2008–13)

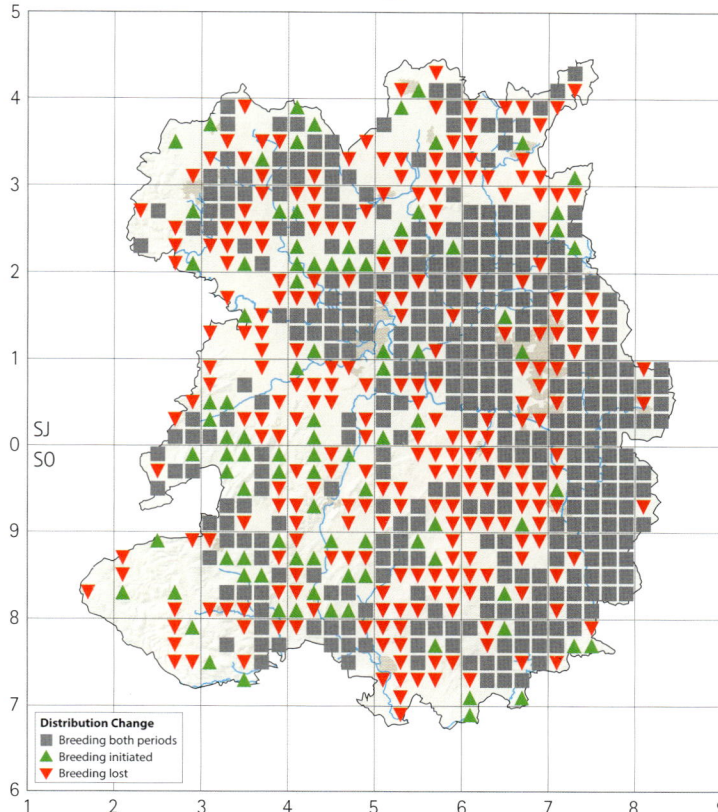

Distribution Change
- Breeding both periods
- Breeding initiated
- Breeding lost

Occupied Tetrads

Atlas period (breeding)	1985–90		2008–13		Change	
	Number	%	Number	%	Number	%
Confirmed	277	32	56	6	-221	-80
Probable	253	29	263	30	10	4
Possible	76	9	128	15	52	68
TOTAL	606	70	447	51	-159	-26
Tetrads with Winter Records (2007–13): 338 (39%)						

Red-legs are more straightforward to rear and release (Tapper 1992). Furthermore, captive birds lay more eggs than Greys, presumably because, in the wild, females often lay two clutches, one of which is hatched by the male (Cramp *et al.* 1980). So Red-legs were the only viable option if partridge shooting was to continue. And it is vital to the economics of shooting, representing more than 10% of the total bag and occupying what would otherwise be the largely fallow months of September and October, before Pheasant shooting starts in earnest in November. Though Red-legged Partridge are a less expensive quarry than Pheasant, nowadays guns on commercial shoots will still be charged in the order of £25 to £40 for each one shot.

But in the late 1960s a new shooting option had emerged. Game-farmers discovered that the closely related Chukar Partridge (*Alectoris chukar*) and Chukar/Red-leg hybrids were nearly twice as prolific in captivity as Red-legs. The first of these new introductions were released in 1970 and they quickly became popular throughout lowland Britain (Tapper 1999). It is assumed that many were put

down here at this time, but the only SBR records are of hybrids released near the Stiperstones in 1986, others seen at Attingham and Bolas Heath in 1989, and of a Chukar reported near Henley in 1992. However, the breeding success of hybrids in the wild is negligible, and hybridisation with wild Red-legs has a damaging effect on the wild stock, so releases were phased out and banned altogether in 1992 (Tapper 1999).

Now that the threat of hybridisation is well in the past, what future is there for Red-legs? The maps show a marked decline in abundance towards the west, and this continues beyond the country boundary, with central Wales being virtually devoid of the species (national BTO Atlas 2013). It would seem that Shropshire is at the edge of the 'natural' range of a bird which prefers a warm, dry climate and well-drained soils, where arable farming predominates (Aebischer & Lucio 1997).

Clearly numbers are greatly augmented by releases; what would happen if they ceased? The population in western England showed a progressive decline from the mid-1970s (Marchant *et al.* 1990) and the results presented here would suggest that this has continued, although this runs counter to an upward trend recorded for the West Midlands as a whole over the years 1995–2014 (Harris *et al.* 2016). Local conditions are probably sub-optimal, and negative factors, notably agricultural intensification (acting principally through the reduction in insects, which are essential chick-food for Red-legs as well as for Greys (Potts 2012)), together with fewer hedgerows and declines in game keeping, may be having a significant impact on the breeding population in the wild. It seems likely that without annual reinforcements there would be few 'Frenchmen' left to shoot.

Tom Wall

Grey Partridge
Perdix perdix
Uncommon resident; population supplemented by annual releases

'What happened in 1972 was to me something that in my wildest dreams seemed impossible. For the first time in the long records of the Apley game book there was not a single entry in the partridge column' (Sharpe 2009). Up until 1962, some 400 Grey Partridge had been shot each year on the Apley Hall Estate near Bridgnorth, but, since then, Norman Sharpe, Head Keeper from 1928 to 1968, had witnessed a steady decline. Today, in spring, there may be as few as 400 individuals in the entire County, with any remnant wild population thought to be significantly outnumbered by releases, and the only records from Apley come from outlying parts of the old estate.

In origin a grassland species, arable farming provided a highly favourable environment for the Grey Partridge, and, when cosseted by gamekeepers, it prospered to such an extent that in 1861, Newton speculated that it might be the commonest bird in England, and Rocke (1866a) stated that 'a dry summer and the vermin kept under seems all that is necessary for the production of this useful bird in almost any quantity'. This 'production' took place naturally, in the wild – these were not released birds – but any creature that might conceivably compromise the annual yield was classed as 'vermin' and rigorously 'kept under'.

Newton and Rocke were writing towards the end of a period of widespread and marked increase which ran, nationally, from circa 1750 to 1880 (Potts 1980). But seemingly, at Walcot (Lydbury North), the increase was belated; here, as revealed by game books held in Shropshire Archives, the shooting bag averaged only 68 birds per year in the 1860s before climbing to 462 in the 1890s and peaking at 486 per year in the first decade of the twentieth century. This was during the heyday of Partridge shooting, when two million were killed annually across the country (Tapper 1999).

Populations and bags remained high into the 1950s, but thereafter, at Apley Hall and elsewhere, a species which had prospered with the spread of farming and game preservation declined, like other farmland birds, in an increasingly intensively managed farmland environment. Herbicides were killing the arable weeds which had previously harboured the insects on which young Grey Partridge depend. As productivity declined, gamekeepers turned to rearing Pheasants or Red-legged Partridge (easier to raise and to drive over the guns than Greys), and the control of the predators of the Grey Partridge became less rigorous, leading to further declines in productivity. At the same time, fields were being enlarged and thereby hedge-bottom nesting-places were lost, and, where hedges were retained, the use of herbicides in field margins reduced nesting cover. The causes of decline were, therefore, threefold: fewer insects in cereal crops, increased levels of predation and a loss of good nesting sites (Sotherton *et al.* 2010).

Although in 1968–72, during fieldwork for the national BTO Atlas (1976), the numbers of Grey Partridge were declining, it was found in all of the County's 33 10km squares, and breeding was confirmed in all but two of them. Despite much more intensive fieldwork in 1985–90 for the *Atlas* (1992), it was not found at all in two of those 10km squares, and breeding remained unconfirmed in five more.

Since then losses have been dramatic: still present in 72% of tetrads in 1985–90, it was found in only 10% of them in 2008–13, a net loss of 540 tetrads, the greatest loss of any species. Breeding was confirmed in 269 tetrads in the earlier survey, but in only eight in the recent one.

In 2005, Brown & Grice reported that nationally some 100,000 were released each year; it seems that similar numbers are likely today (Aebischer 2013), but little is known of the numbers or locations of local releases, apart from the relative abundance maps from the recent Atlas. Given a post-breeding injection of birds, it might be anticipated that the distribution would be wider in winter than during the breeding season, whereas in fact records came from 21 more tetrads in the latter period.

Partridges are normally sedentary. It is no surprise, therefore, that the only ringing recovery, of a male, marked as an adult and found dead in August 1981, was at the very place of ringing eight months earlier; indeed all of the 16 recoveries of 138 juveniles wing-tagged on release at Attingham in 1969 and 1970 were described as 'local'. But it may be that some captive-bred individuals disperse, if only a few kilometres, prior to the breeding season. This would explain why a release site at Wappenshall failed to generate a breeding season

Paul King, Preston on the Weald, 25 April 2010

Breeding Distribution Change (1985–90 to 2008–13)

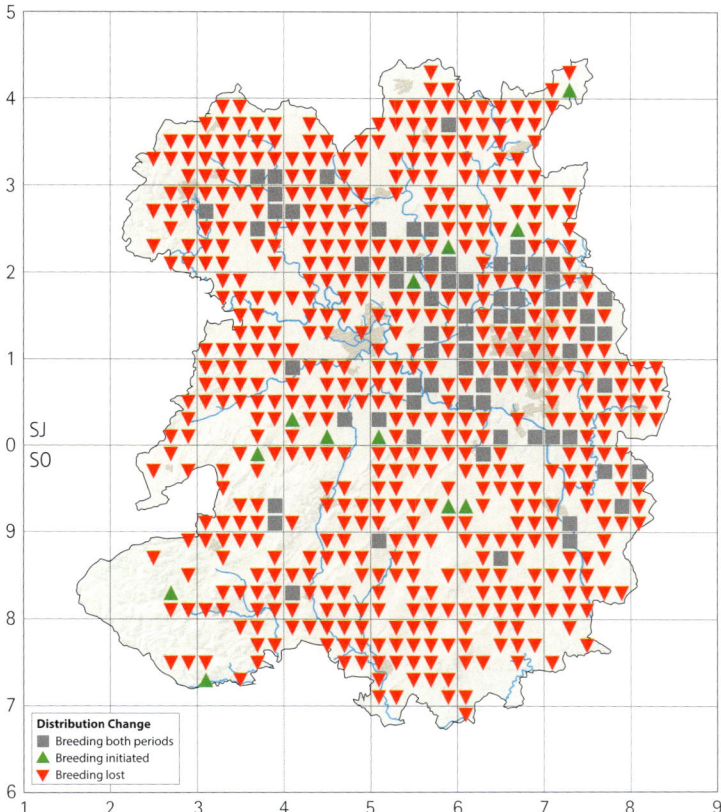

Distribution Change
- ■ Breeding both periods
- ▲ Breeding initiated
- ▼ Breeding lost

Occupied Tetrads

Atlas period (breeding)	1985–90		2008–13		Change	
	Number	%	Number	%	Number	%
Confirmed	269	31	8	1	–261	–97
Probable	261	30	43	5	–218	–84
Possible	93	11	32	4	–61	–66
TOTAL	623	72	83	10	–540	–87
Tetrads with Winter Records (2007–13): 62 (7%)						

shows losses throughout, but the strongest representation is now in an extensive, though markedly discontinuous patch, east of a north–south line from Clive through Shrewsbury to Longnor; elsewhere there are small pockets of records and a scatter of isolated ones. Some of these may be explained by releases; for example, there are three tetrads with winter records in the Stiperstones area, but there were no subsequent breeding season records there. Typically any released birds that dodge the gun are known to suffer heavily from predation, and have a low rate of breeding success (Browne *et al.* 2006). It might be thought that upland ground such as this would in any case be unsuitable, but wild Grey Partridge figured in shooting bags in the 1950s (Gatten Estate Game Book), while the *Handlist* (1964) noted that they could be found up to the 1,500ft contour, and they were still present in hill country during 1985–90 fieldwork for the *Atlas* (1992).

The future of the Grey Partridge looks bleak, particularly as it is dependent on weather, as well as improvements in habitat and a reduction in predation. The summer of 2012 was the worst lowland game bird breeding season of the last 100 years (Buner *et al.* 2014). The following year started badly too, with a cold and late spring, but a brood of 12 at Venus Pool in August showed perhaps that the crops sown there as a winter food-source for seed-eating passerines may serve a dual purpose, and this offers a small ray of hope.

Tom Wall

record, whereas three adjacent tetrads, though blank in the winter, had positive registrations come spring and summer. It may also be that the detection rate of this cryptic, ground-hugging species rises in spring when the males give their 'creaky-gate' calls more often.

Historically, SBRs have alluded to a greater frequency of records in northern and central areas. The breeding distribution change map

Quail (Common Quail)

Coturnix coturnix

Rare summer visitor

'Enigmatic', the customary epithet, is inadequate to describe this most extraordinary and most mysterious of our summer visitors. Having bred in southern Europe in March or early April, many adults (some of which have already migrated from the Sahel region of Africa) then head north, before breeding again. The male's call – 'wet-mi-lips' – was heard here as early as 16 April in 1994 (the next earliest date is 2 May, in 2006), but generally the first records come in the second half of May. Far-carrying and ventriloquial, the call is rapped out repetitively, most often at dusk and usually from cereal crops, though sometimes from mowing grass. Among those that follow on in June there may be recently fledged juveniles: capable of migrating at two months old, they are sexually mature at 12–15 weeks (Marchant 2002).

Records from the distant past are fragmentary, but Moreau (1951), reviewing evidence from across the country, concluded that 'some centuries ago Quail were much more numerous'. They remained

relatively common, however, before a 'marked and progressive decline' in the first half of the nineteenth century led to a low ebb being reached in England by 1865. Local evidence for this is provided by the statements of Eyton in 1838: 'In former times appears to have been met with rather commonly in Shropshire by sportsmen in September, but of late years rarely', and of Rocke, writing in 1866 of Clungunford: 'I do not remember having seen the bird in this parish for nearly 20 years previously to the one recorded in September last', while Beckwith commented in 1885 that 'the Quail, said to have been common early in the century was for many years seldom seen'. Presumably the netting of huge numbers of migrants, notably in North Africa, many of which were sold to France and Britain, was responsible. But although this trade was not curtailed until the late 1930s, as early as 1885 Beckwith was reporting that Quail had become 'a regular summer visitant' to the north, while Forrest (1899)

Breeding Distribution (2008–13)

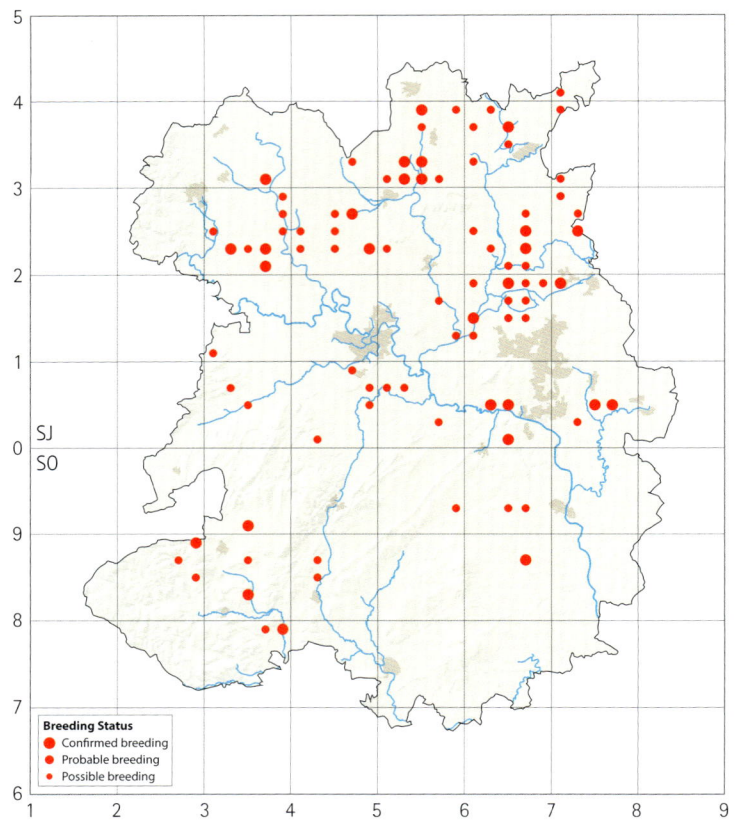

Breeding Status
- Confirmed breeding
- Probable breeding
- Possible breeding

Occupied Tetrads

Atlas period (breeding)	1985–90		2008–13		Change	
	Number	%	Number	%	Number	%
Confirmed	11	1	0	0	-11	-100
Probable	65	7	28	3	-37	-57
Possible	87	10	60	7	-27	-31
TOTAL	163	19	88	10	-75	-46

described it as 'a regular summer visitor, though never numerous'. Writing in 1940, Lloyd concluded that Forrest's assessment held good, indeed it still does today, singing males being heard each year (the last blank year was 1975), but generally only in small numbers. However, sightings are rare, and proof of breeding is exceptional. Over the 25 year period to 2014, such records were received for only seven years and breeding was confirmed in just two of them, most recently in 1994.

While there has been an upward trend in the numbers reported over recent decades, this may reflect better coverage. The nationally notable 'Quail years' (those in which they are unusually abundant) of 1893, 1964 and 1989 generated records here of one, seven and 150 calling males respectively, but the numbers of contributors to annual reports for those years rose markedly too, from five to 35 and finally to 190 (a figure boosted by an appeal through the media for records).

Prolonged anticyclonic conditions with exceptional heat and drought in Spain and France in late spring may trigger these bumper

Breeding Distribution (2008–10 & 2012–13)

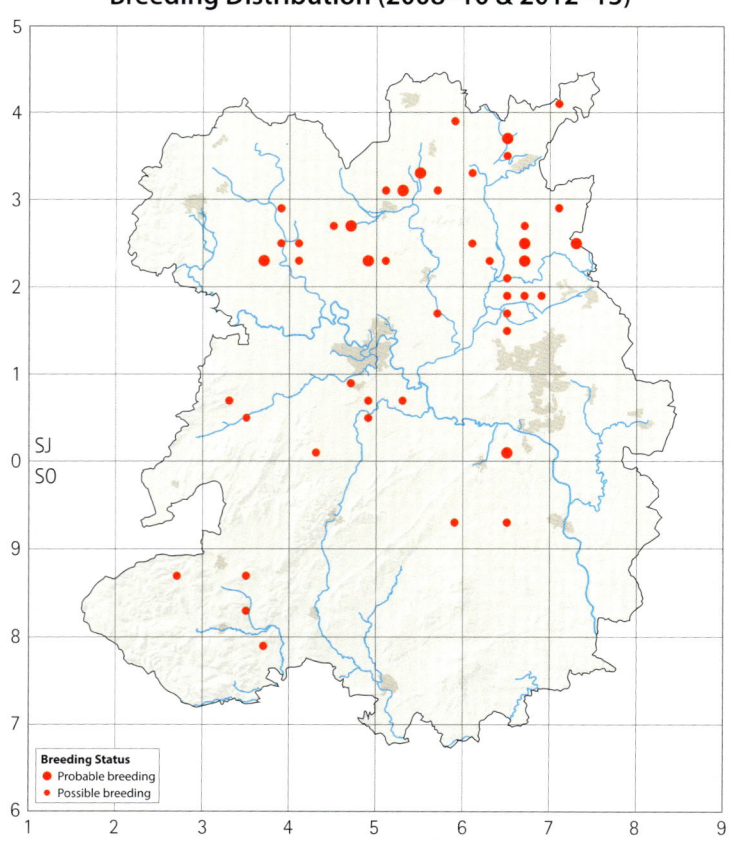

Breeding Status
- Probable breeding
- Possible breeding

Breeding Distribution Change
(1985–90 to 2008–13, excluding 'Quail Years')

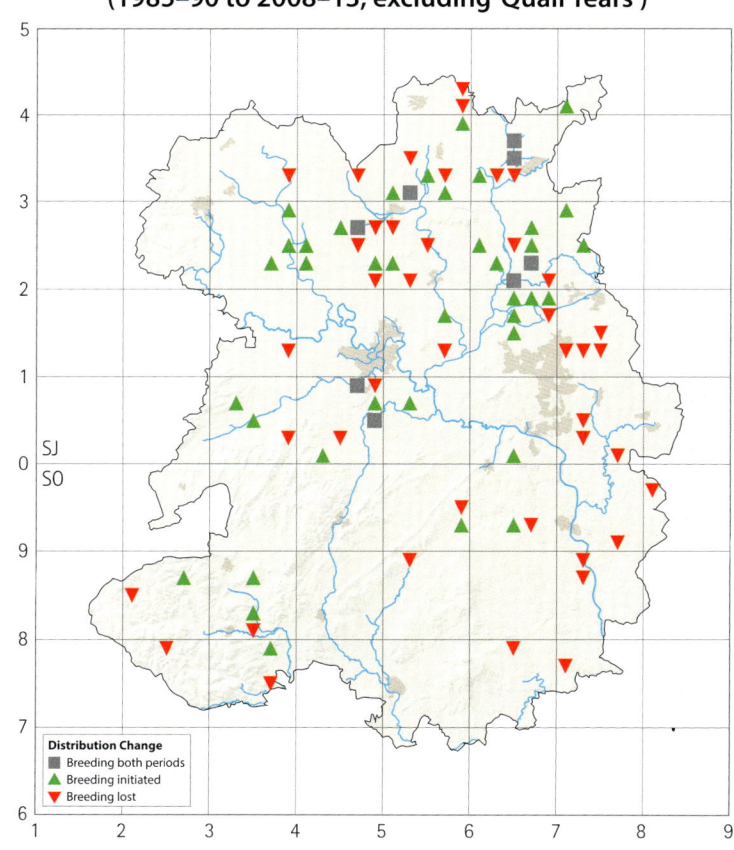

Distribution Change
- Breeding both periods
- Breeding initiated
- Breeding lost

arrivals (Moreau 1951). Such was the case in 2011, when, nationally, even more were reported than in 1989, but there were 'only' 85 calling birds recorded here (Holling *et al.* 2013). While there was no media involvement in 2011, it seems most unlikely that an appeal would have brought the figure near to the 150 of 1989.

'Quail years' are infrequent, however, and over the last 30 years records have generally come from less than a dozen locations each year (indeed there was only one SBR report and three Atlas records in 2013). The first map shows the distribution of records for the period 2008–13, including the 'Quail year' of 2011; the second shows the distribution for the same period, but excluding 2011. However, as most sites are occupied only sporadically, even this map suggests a breadth of presence which is not achieved in any one year. The third map shows the differing distributions in the two survey periods. Records from the 'Quail years' of 1989 and 2011 have been excluded, and once this is done, it is revealing that only eight tetrads generated records in both periods. Only two tetrads had records in more than one year in the recent Atlas period.

It is not possible to estimate with any confidence how many occupied tetrads harbour breeding pairs. To do so simply on the basis of calling males would be wrong. Persistent calling may be by un-mated individuals which may in time move some distance and call persistently elsewhere, while, confusingly, cessation of calling at a location could indicate breeding (Guyomarc'h *et al.* 1998). Furthermore, nests are like needles in haystacks and broods hug cover. A tentative estimate for a typical year would be in the order of five pairs, similar to the estimate of five to eight pairs in the *Atlas* (1992).

It is worrying perhaps that despite the huge increase in observers, two aspects of the earlier records are not replicated today. Firstly, breeding was proved much more frequently in the past. Nests or young were reported in 11 of the 31 years 1875 to 1905, but in only four of the 31 years 1984 to 2014, with none during recent Atlas fieldwork. Historically, the manual nature of agricultural work, and the collection of eggs and 'specimens', will have led to more nests and broods being discovered, but this seems to be an inadequate explanation for all of the discrepancy. Secondly, birds used to stay

longer. Beckwith (1885) states that 'in September and October small parties ... are dotted about'; the latest date known to him was 18 October, although one was found dead on the railway at Ellesmere on 23 December 1909 (Lloyd 1940). Over the 41 years from 1900 to 1940 there were 11 records in September, and Lloyd cites 17 September as 'the average date of its latest appearance'. But despite the proliferation of observers, over the 41 years to 2014 there were only seven September records; the latest was on 22 September, in 1987. The two aspects may be related: chicks which fledge here are more likely perhaps to linger than adults that have failed to breed. And many more may have fledged in the past when arable crops will have been richer in the weed seeds and invertebrates on which the young feed, while mowing grass was harvested later, leading perhaps to fewer losses of eggs and chicks.

Given the species' preference for cereal crops it is no surprise that, as was the case in 1985–1990, most recent records have come from the north, where arable land is more widespread. Some locations in particular are favoured, and these may attract several calling males rather than the usual singles. Such places include Noneley/Sleap (four were heard here in June 2008), Aldersey (six in August 2009), Tilstock (eight in June 2011) and Childs Ercall (10 or more in July 2012). But individuals may turn up, if only briefly, in unlikely places, including, in 2006, one at Wildmoor Pool on the Long Mynd, and two at Abdon Burf on Brown Clee (the highest point), and, in 2014, one at Rhos Fiddle. Why a species known to favour low migratory trajectories should pop up in such elevated places is another of the conundrums surrounding this most mysterious of migrants; others include the scatter of records in seemingly uninviting areas of the south-west, and their virtual absence from apparently more suitable ground in the south-east.

Tom Wall

NB Some of the information published in *British Birds* in the reports of rare breeding birds was queried during the preparation of this account. The RBBP's database has been corrected but not the published record. The corrections appear in the species index on the *Histo* website.

Pheasant (Common Pheasant)

Phasianus colchicus

Very common naturalised resident; population supplemented by annual releases

'The usual situation: a handful of observers (4) submitting records (39) of a bird ignored by all others', was the pithy summary published in SBR 1992. It reflects the disdain felt by birdwatchers for this most domesticated of our game birds. Yet the Pheasant is of real importance: its nurture has a significant impact on other wildlife and their habitats, and on the rural economy. Estimates (based on PACEC 2006) would suggest that habitat and wildlife management for shooting is carried out on more than 5% of the land area of Shropshire, and that sporting shooting directly and indirectly employs here the equivalent of 1,140 full-time workers. Although the data gathering and analysis on which the latter figure is based have been seriously criticised (Cormack & Rotherham 2014), the Pheasant is, unquestionably, the most influential bird of the English countryside.

'The rearing of Pheasants by hand has of late years been carried to such an extent that one begins to look upon them more in the light of poultry than of wild birds'. Though sounding contemporary, this commentary was written 150 years ago (Rocke 1866a). Yet Rocke would be astonished to witness the degree to which Pheasant rearing has now developed. Published figures indicate that numbers released annually in the UK rose from some 4 million in 1961 (the first year for which a figure is available) to 43 million in 2012 (Aebischer 2013; 2018). It is estimated that 859,000 are released here each year, and 20% of these will die before the shooting season begins (GWCT literature). It is thought that nationally over the period 1960–1990, around 50% of those released each year were shot. Subsequently this proportion declined to 35%, but changed little from 2005 onwards (Robertson

Jim Almond, Venus Pool, 3 April 2016

et al. 2017). This would suggest that in the order of 300,000 of the birds released here each year are shot; each one of them is likely to set the sportsman back £30 to £45, depending on the prestige of the shoot. Given the increase in releases but the decline in the proportion shot, the number that are open to predation has increased faster than the number that are being released. Taken together with those that are shot but not retrieved and the many road casualties, this doubtless leads to an increase in the numbers of both mammalian and avian predators and carrion feeders.

Only a third of those that get through the shooting season unscathed will still be alive in July, and in the interim, very few will have bred successfully (Game and Wildlife Conservation Trust (GWCT) literature). Indeed, releases may increase the number of breeding hens without increasing the number of chicks produced (Robertson 1997). Based on TTV counts, the number of hen Pheasants present in the breeding season ('wild' ones, as well as the residue of those released the previous year) is estimated at 44,000; there may be similar numbers of males, but by no means all will be territory-holders; those that are may attract two or three females, sometimes more. The BBS graph shows annual fluctuations. These are difficult to interpret, however

the smoothed trend suggests an increase of approximately 66% over the period, which is similar to the national growth in releases as indicated by GWCT figures (Aebischer 2013).

It is likely that the total biomass of Pheasants in the early autumn now comfortably exceeds the springtime biomass of all the UK's native breeding birds combined (Holling *et al.* 2014b). Locally, based on TTV estimates, Pheasant is the tenth most numerous breeding species, but if ranked by total biomass, it comes a clear first. The effect of this considerable alien presence on the countryside is unknown and there is an urgent need for it to be researched. In the interim, Fuller *et al.* (2005) have provided some pointers. They consider potential negative effects in woodlands to include modification of the field layer, the spread of disease and parasites, and competition for food. There may also be negative impacts on other vertebrates and invertebrates, and on woodland soils too, through enrichment by droppings. More specifically, at a sample of 12 farms in south-east England, Lennon *et al.* (2013) found a higher incidence of infection, albeit sub-clinical, in Woodpigeon and dove species (Turtle, Stock and Collared), by the protozoan parasite *Trichomonas gallinae*, where food was provided for game birds. Such food sources may attract high densities of birds, promoting opportunities for disease transmission and dissemination. Although it was concluded that transmission from game birds to these species was unlikely, a subsequent study (Stockdale *et al.* 2015) suggested that 'parasite spillover' may potentially occur; if it does, it could lead to fatalities.

The huge increase in releases is in part because a substitute quarry was required to take the place of the rapidly declining Grey Partridge. Pheasants made up 15% of the national game bag in 1900, but 80% by 2004, displacing Grey Partridge as the leading lowland game bird (Tapper 1999; PACEC 2006). Yet game books held by Shropshire Archives show that, on some estates at least, Pheasants have long made a major contribution to the game bag. At Walcot (Lydbury North), they already made up 25% of the total bag (including rabbits

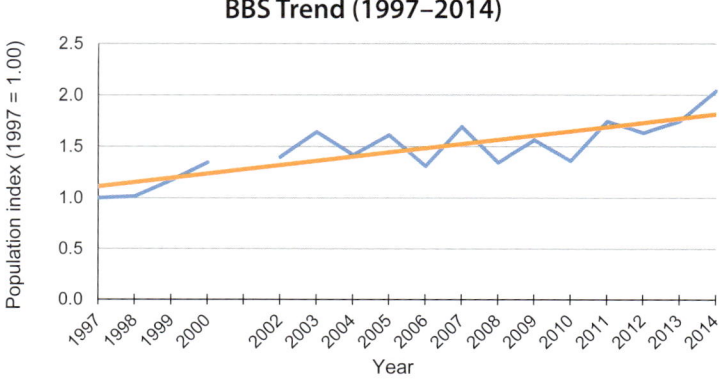

BBS Trend (1997–2014)

Winter Relative Abundance (2007–13)

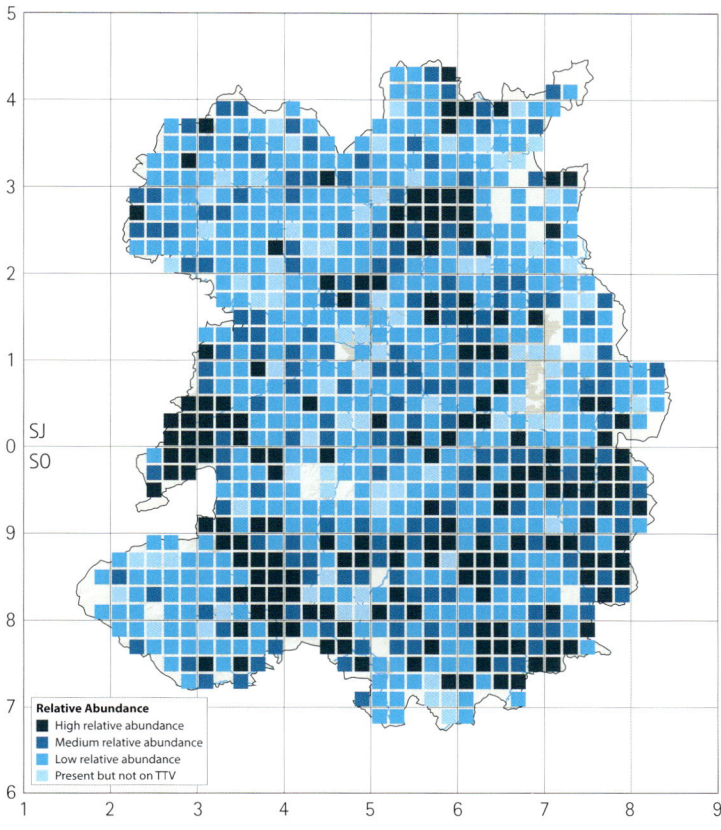

Relative Abundance
- High relative abundance
- Medium relative abundance
- Low relative abundance
- Present but not on TTV

Occupied Tetrads

Atlas period (breeding)	1985–90		2008–13		Change	
	Number	%	Number	%	Number	%
Confirmed	514	59	183	21	-331	-64
Probable	198	23	474	54	276	139
Possible	137	16	197	23	60	44
TOTAL	849	98	854	98	5	1
Tetrads with Winter Records (2007–13): 845 (97%)						

Oswestry Family and Local History Group, show that the Pheasant bag averaged 860 for the years 1910–13, but only 27 for Grey Partridge. At Hawkstone annual bags of over 2,000 were already common in the 1860s and 1870s, but there was to be a substantial and rapid increase such that 'by the 1890s, when 2,000 Pheasants were several times killed in a single day ... there were said to be more Pheasants at Hawkstone than on any other estate in England, not excepting East Anglia' (Gaydon & Price 1973).

Although this is of course an introduced species, fairly well established over much of the English countryside by the end of the sixteenth century (Sharrock 1976), it is capable of sustaining itself in our countryside, and until the 1960s much of the emphasis in game management was on nurturing Pheasants in the wild rather than hand-rearing them. By 1999, however, it was speculated that only 10% of the national bag might be 'wild' birds (Tapper 1999), a proportion which will have continued to decline. This has significance for other species. When the emphasis was on sustaining the populations of 'wild' Pheasants and of Grey Partridges, gamekeepers were under pressure to control all predators, particularly in the breeding season. Nowadays that pressure has eased, and this is one of the main reasons why species beloved of birdwatchers, such as Buzzard and Raven, have increased so dramatically; they also profit from Pheasants as prey or carrion, whether from road-kill or as shooting casualties. Carrion Crows and foxes do too, and thanks in part to this supplement to their diet, and to gamekeepers devoting less time to their control, their numbers are higher than in the past. This adds to the pressure that these common predators exert on populations of ground-nesting birds such as Lapwing and Curlew; it is another aspect of releases that needs researching. By contrast, it is evident that grain put out for Pheasants in the winter months, and the game-crops planted as feed and shelter, sustain many seed-eating species, some of which are in decline; indeed Pheasant shoots are now the place to go to see finches, Tree Sparrows and buntings. So, in a whole range of ways, and despite the general contempt in which it is held by birdwatchers, the Pheasant has become by far and away the most significant of our countryside birds.

Changes in distribution since the *Atlas* (1992) are few and insignificant, and the Pheasant remains virtually ubiquitous. Breeding and winter season distributions are almost identical and the only tetrads without records in both seasons are parts of Telford.

The relative abundance maps are very similar too; both show a strong presence in the more pastoral south where the terrain is more undulating, giving opportunities to provide the challenging 'high birds' much sought after by sportsmen. The clumping of high abundance squares both here, and elsewhere, including around Hawkstone, clearly indicates where shooting estates are releasing particularly large numbers of birds.

This contrasts with the observation in the *Handlist* (1964) that at that time Pheasants were more thinly distributed 'on the hill country away from the areas of corn cultivation'. Indeed, nowadays they may be encountered on the tops of the Stiperstones and Brown Clee, and on the Long Mynd too, where breeding has been confirmed on the highest and wildest terrain. By contrast, Pheasants have become habitual and often unwelcome visitors to some bird-feeding stations in lowland gardens, adding to bird-watchers' disdain.

Tom Wall

and hares) in the 10 shooting seasons 1868–69 to 1877–78, rising to 72% by 1908–09 to 1917–18. At Downton Hall the Pheasant bag averaged just 87 birds over the seasons 1859–60 to 1905–06 (Grey Partridge averaging 120), but jumped to 1,268 in 1906–07 following significant expenditure on rearing facilities and the purchase of 1,000 eggs. From then, up until the First World War, the average seasonal bag was 2,121, with a maximum of 3,045 in 1911–12, while bags of Grey Partridge declined to an average of 51. Lord Harlech's game books for the Brogyntyn Estate, held in the National Library of Wales, show that in the 10 years 1884–1893 more than four times as many Pheasant were being shot on average than Grey Partridge, rising to more than 10 times over the years 1924–1933; the highest annual bag achieved here was 2,787 in 1911. At Acton Burnell, on 3 December 1908, Lord Harlech participated in a 'record shoot' of 1,605 Pheasants and at The Grove, Craven Arms, 1,223 were shot in two days in December 1909 (Downton Hall game book). At Park Hall, Oswestry, particulars for the sale of the estate in 1914 held by the

Red-throated Diver (Red-throated Loon)
Gavia stellata
Very rare winter visitor

This diver has a circumpolar breeding distribution which reaches its most southerly point in the north-west of the British Isles. During the winter it frequents coastal waters, and those found off the north Wales coast are the closest wintering population.

Although some had been obtained previously, the first detailed report was of one picked up dead by a postman near Market Drayton in 1890. A few years later there were two at Shrawardine Pool in April 1897, the only multiple record.

An unusual series of reports came from Bridgnorth: one was shot on the river in the autumn of 1928 and another was shot there the following February. A third was seen on the river in March 1930, but this one probably escaped a similar fate after stones were thrown scaring it away. These are the only local accounts of a river being used.

One was found dead on the frozen Mere (Ellesmere) in January 1939, and the first modern record also came from this site, of one thought to be oiled, seen in March 1955 and later found dead.

Of the 13 modern records, six have been from the Ellesmere meres, four from Chelmarsh, and one each from Priorslee Lake, Priorslee Flash and Cound Fishing Pool. Two arrived in 1976, but in different winter seasons, and no winter season has had two records. Eight appeared during the winter (November–February), three in the spring passage period (March–April) and one in October, but an adult in full breeding plumage at Priorslee Lake in June 1991 was exceptional and must have been a fine sight.

The only recent Atlas records were one at Crose Mere in November 2008 and one at Chelmarsh in November 2011, which remained until January. The latter was unusual, its stay of eight weeks being unprecedented. Previously, one had spent three weeks in the Ellesmere area in February/March 1976, another two weeks at White Mere

Jim Almond, Chelmarsh, 25 November 2011

in March 1958, and one was at Cound Fishing Pool throughout the last week of February in 1996; the rest were reported only on a single day.

Allan Dawes

Black-throated Diver (Black-throated Loon)
Gavia arctica
Very rare winter visitor

Although this diver breeds in north and north-west Scotland, its main breeding range is in boreal and arctic Eurasia. In winter, it is the scarcest of the three divers found regularly around the British coast and there are relatively few inland records.

The historic record is sketchy with just two references: in 1839 Eyton stated that 'a specimen occurred last winter', while 40 years later it was reported that 'several young birds have occurred in winter' (Beckwith 1879).

There are just seven modern records:

Ellesmere	7 December 1950
White Mere	2–5 November 1974
Newton Mere	11–22 December 1985
The Mere (Ellesmere)	28 October–3 November 1990
Chelmarsh	18 December 2000
Chelmarsh	15 October 2002
Chelmarsh	19 October–12 December 2005

The first was found dying about two miles from Ellesmere, on the road to Oswestry, where it was thought to have hit telephone wires. There is no additional information about the individual at White Mere but, from the brief description provided, the one at Newton Mere may have been an adult. The one in 1990 was a first year while the adult at Chelmarsh in 2000 was picked up as a badly oiled corpse. Both the 2002 and 2005 birds were also adults in winter plumage. All seven occurred in the late autumn or early winter, although not obviously directly associated with periods of gales or particularly cold weather.

Graham Walker

Great Northern Diver (Common Loon)

Gavia immer

Very rare winter visitor

Jim Almond, The Mere, 28 December 2013

This is a mainly Nearctic breeding species whose range extends as far east as Iceland. During the autumn, internationally important numbers move south-east to spend the winter off the north-west coasts of Scotland and Ireland.

Several were said to have occurred both on the meres and the River Severn in the mid nineteenth century. One found dead after colliding with telegraph wires at Sansaw near Shrewsbury in 1939 was the first since a female was shot at Ellesmere in 1863, and was the last historic record.

Since 1951 there have been 12 individuals. White Mere and the Mere are the two largest waters in the Ellesmere area, and between them they have accounted for six of these records. The first was at White Mere in April 1958 and the last at the Mere in December 2013. During this period others were at Betton Pool in January 1987, Priorslee Lake in December 1989, Cole Mere in December 1994 and juveniles at Chelmarsh in November 2012 and December 2015, the 2012 individual being the only one during the recent Atlas period.

In October 1999, one crash landed on a wet road at Presthope, Much Wenlock, presumably having mistaken it for a watercourse. After treatment it was released at Cole Mere but it was found dead five days later.

Ten were first noted between November and January but unlike the other divers, this species often stays for a considerable time. One spent eight weeks at the Mere (Ellesmere) in 1965, another nine weeks at White Mere in 1972 and another remained in the Ellesmere area from December 1994 to the following May.

Allan Dawes

Storm Petrel (European Storm Petrel)

Hydrobates pelagicus

Vagrant

This small, relatively common, seabird nests in colonies in the North East Atlantic, a substantial proportion of which are in Britain. It is rarely seen inland, and usually only after severe storms in the autumn and early winter, when it is dispersing from its breeding grounds to winter in the South Atlantic.

The historic record refers to occasional, storm-blown, occurrences, although an adult caught alive at Homer near Much Wenlock on 15 July 1886 was after a period when 'the weather [had] been for some time calm and fine' (Forrest 1907a). Later, two were found, both thought to have been killed by striking wires: the first, at Ironbridge on 28 November 1905, was after 'a fresh, strong or whole gale being experienced in the course of the 26th or 27th on practically every section of our coasts, the wind direction being southerly to westerly to north-westerly', while the second at Donnington on 10 November 1926 also followed a period of high winds and gales.

The latter was the last report until the only modern record:

Longville, Much Wenlock   8 August 1973

Typically, this bird was found dead after 'a day and night of gale-force winds'.

Graham Walker

Leach's Petrel (Leach's Storm Petrel)

Oceanodroma leucorhoa

Vagrant

Also sometimes referred to as the Fork-tailed Petrel, they breed in the far north-west of Scotland and on other north Atlantic coasts, usually on the remotest of islands, and spend the winter months far out in the Atlantic Ocean. Given the combination of their pelagic nature and Shropshire's status as one of the UK's most inland counties, there are a surprising number of sightings in the historical accounts. The earliest was recorded by Eyton in 1839, who stated: 'One specimen only has occurred: it was killed on the Severn near Montford-bridge, and it is in my collection'. Several early records relate to a small number that were either found dead or were deliberately shot after 'fearful' gales in November 1865. Forrest (1899) stated that it had 'been obtained about 10 times ... usually after gales'.

Leaving their breeding grounds, they fly south-west and they are then prone to being blown off-course by deep depressions and strong north-westerly winds, which often push the migrants towards the coast and in to the Irish Sea. Severe gales can result in national 'wrecks' where the petrels can appear almost anywhere inland. One of the biggest occurred in 1952, when there were 11–12 records of 11–12 individual birds, all found dead, during late October and early November (Boyd 1954; Lloyd 1956). There are some discrepancies between the two accounts, for example there were two or possibly three birds found at Attingham, and one found on 15 January 1953, thought to have been dead for a week, may not have arrived as part of the wreck, or it may not have died straight away. However, it is clear that this was a significant event – it alone accounts for over half of the 21 records since 1950.

The remainder are all of occurrences between late September and mid-January.

There was a record in late 1953 when one that had been dead for several days was picked up from Wolf's Head, Nesscliffe on 30 September, then a gap until 1970 where there were again two records, from Cole Mere and the Mere (Ellesmere). Following a single sighting in 1978 (Whitchurch), there were no more records until 2004, again at the Mere (Ellesmere). This record is unusual in that the bird was very much alive; in fact it remained for less than a day before flying off, much to the disappointment of many local birders who missed the opportunity to add this species to their County list.

The most recent records are from 2006, following storms in December. Three were found, at West Felton (10 December), Stirchley (10 December) and Ironbridge Power Station (mid-December), but all were either dead or died whilst being cared for.

Linda Munday

Fulmar (Northern Fulmar)

Fulmarus glacialis

Vagrant

Breeding on Atlantic coasts of north-west Europe, including most of the British coast where suitable nesting cliffs or crags occur, it may visit its nesting sites at any time of the year even though it tends to spend most of its time well out to sea outside the breeding season. It is normally only found inland after severe weather, although this is not always the case as can be seen from two of the recent sightings.

The only historic record is of a bird found dying in a field at Nobold near Shrewsbury in mid-March 1909. It may have succumbed to a period of weather that was 'of a most inclement character, cold rain, snow, sleet or hail being experienced every day over the Country generally' although it was 'far from stormy' (Met Office 1909).

The three modern records are:

Whitchurch	16 September 1976
Hawkstone Park	4 July 1992
Venus Pool	30 June 2001

The Whitchurch individual was typical of a storm-blown, possibly sick, adult that was picked up on a road near the town after a period of severe gales between the 9th and 12th of the month; it was later released at the coast. The two subsequent sightings, however, bucked this trend, both occurring during periods of generally quiet weather in the summer.

Graham Walker

Manx Shearwater

Puffinus puffinus

Vagrant

These shearwaters breed on islands in the north Atlantic, mainly in Britain and Ireland, and they spend most of their lives far out at sea. Unlike other pelagic birds which tend to appear inland after gales, historical records suggest that different conditions played a part for Manx Shearwaters – Beckwith commented in 1886: 'sometimes come inland in autumn, and more frequently in wet foggy weather than after storms'. This view isn't supported by modern sightings, which all appear to have been storm-driven, for example one was found dead in Shrewsbury 'after a night of high wind and rain' (SBR 1980).

Beckwith documented one in October 1877 which 'joined the poultry at the Hay Farm, near Coalport, where it remained for several days, but at last was killed by a cat'. During September 1882, he reported that five were caught – two near Oswestry and one at each of Ludlow, Ditton Priors and Cressage. In the first half of the

twentieth century, there were six records, but only one after 1923, a storm-blown immature found in exhausted condition in All Stretton on 17 September 1946.

Individuals found in the late nineteenth century were specifically targeted for various collections, but those found in more recent times, even in the early 1900s, have fared better, often being taken into care having been found exhausted. Generally, they were released where they were found, but in both 1988 (two) and 1989 they were rehabilitated by the RSPCA then taken to the Welsh coast for release.

Since 1950, there have been 14 records from widely scattered locations, in only 10 years: 1964, 1975, 1980 (2 records), 1987 (two), 1988 (two), 1989, 2004 (two), 2011, 2016 and 2017. All except two were of a single bird found in September, the exceptions being two found grounded near Nesscliffe in April 1989 and one grounded near

Clive in October 2017. All, perhaps excluding the undated record of two in April, were storm-driven.

Of the 15 individuals, three were found dead (1980 in Shrewsbury, 2004 on Black Hill and 2016 at Ruyton-XI-Towns; the Black Hill individual was thought to have been attacked by a Peregrine or Goshawk as the head had been detached). The remainder were all taken into care and all except three were later released. Of these, the first was found in an ornamental pool at Sankey's Works, Hadley, in September 1964, and was kept and fed for 10 days but then died. The second was found in Lydham in September 2011, when the remnants of Hurricane Katia crossed the Atlantic. It was taken into care, but was unable to eat properly and died three days later. The third was the individual at Clive which died shortly after being taken to the vet.

There is a ringing recovery, one ringed as a nestling at Skokholm: (Pembrokeshire) on 27 August 1967, found dead at Selley Hall, only 11 days later, having been driven 170km east-north-east by a storm.

Linda Munday

Little Grebe

Tachybaptus ruficollis

Uncommon resident

Little Grebe is found throughout temperate Europe to North Africa.

'Weedy pools, not uncommon' was the first reference (Leighton 1836). Thirty years later Rocke added more detail: 'Common on almost all our rivers, brooks and ponds, where I fancy they remain the whole year. But for their great powers of diving, and cunning in concealing themselves, they would soon be exterminated in this part of the county, as I fear they have a very bad name amongst our anglers'. Later writers noticed a difference between summer and winter populations: 'less numerous in summer' (Beckwith 1879); 'more numerous in the winter, when residents receive further additions to their numbers' (Paddock 1890), and 'numbers are increased by fresh arrivals in winter' (Forrest 1899).

During frost in February 1912, several appeared on the River Severn at Shrewsbury and two were later found dead. The hard winter of 1962–63 also reduced numbers, according to the *Handlist* (1964), and two were seen diving between ice flows on the river by the West Midlands Show Ground at Shrewsbury in January 1982.

Double figure counts are received occasionally, often in late summer or autumn, when numbers are usually boosted by juveniles bred locally. The highest total of 25 at Allscott Sugar Factory in September 1995 fits into this category. Sixteen at Cleobury North Pool in February 2009 was the largest congregation found during the winter months in the recent Atlas period.

Little Grebe breed mainly on pools and will use quite small ones as long as there is plenty of emergent vegetation. They have not been confirmed breeding on rivers since August 1981, when four young were seen with an adult on the River Tern at Market Drayton, although it is possible that some Atlas tetrad records could include nesting on rivers. A moderate decline from the early 1980s, indicated by the national WBS/WBBS, was thought to be connected with habitat loss on linear waterways, many of which were cleared and straightened at that time. Breeding success is also poorer on rivers and canals, which provide corridors for predators, and nests on rivers also have to contend with fluctuating water levels.

While adults can be elusive when they have nests or small young, large juveniles are more conspicuous. Providing repeat visits were made to suitable waters, few successful attempts should have gone unnoticed, and the tetrad occupancy table shows that 60% of the occupied tetrads had confirmed breeding records, compared with 37% in 1985–90. The *Atlas* (1992) stated 'Little Grebes are easily overlooked,

John Fielding, Atcham, 9 August 2012

and confirmation of breeding can be difficult'. The increase in the proportion of tetrads with confirmed breeding is at odds with the general trend to a lower proportion in fieldwork for the recent Atlas. This could indicate more successful breeding attempts, but there is insufficient evidence to draw firm conclusions.

Hard winters are probably the main controlling factor. Increased over-winter survival during the mild winters prior to 2009–10 may have enabled the increase in range of 16% shown in the tetrad occupancy table. Following the harsh winter of 2010–11 a UK decline of 18% was recorded by BBS.

The breeding distribution change map shows considerable gains and losses, 89 and 70 respectively. Some waters will have deteriorated and some new ones created, but the scale of change appears too large for this to be the entire cause. Small pools may not be occupied every year and those visited in the latter part of the *Atlas* period may not have recovered from the effects of three successive cold winters between 2009 and 2012. Observations at two small pools in 2014 where breeding was confirmed in the years before the hard winters showed them to be absent (*pers. obs.*) even though the UK BBS index for 2014 showed a return to the pre 2010–11 level. Most increases

Breeding Distribution Change (1985–90 to 2008–13)

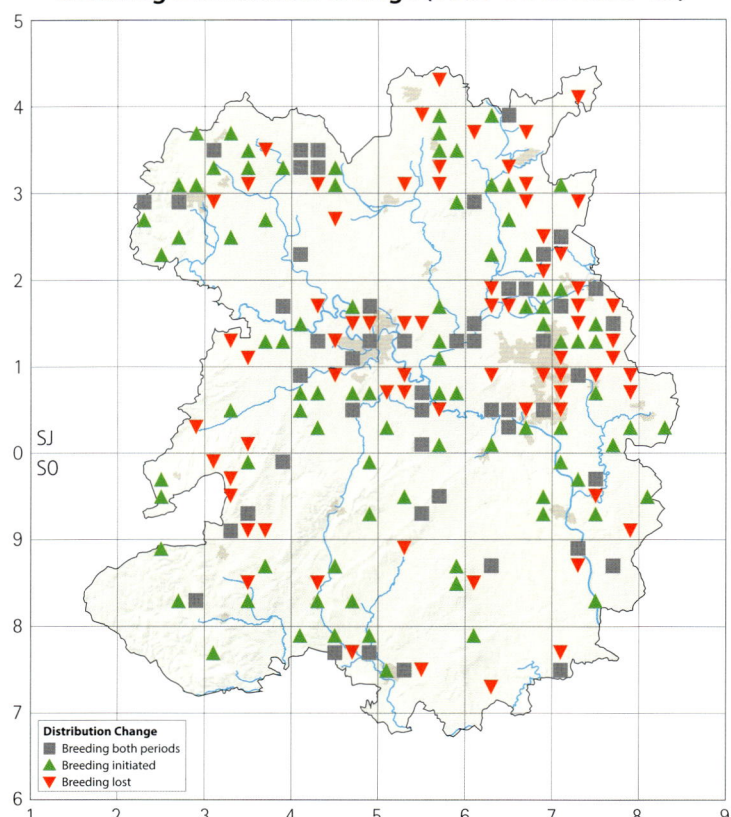

Distribution Change
- ◼ Breeding both periods
- ▲ Breeding initiated
- ▼ Breeding lost

Winter Relative Abundance (2007–13)

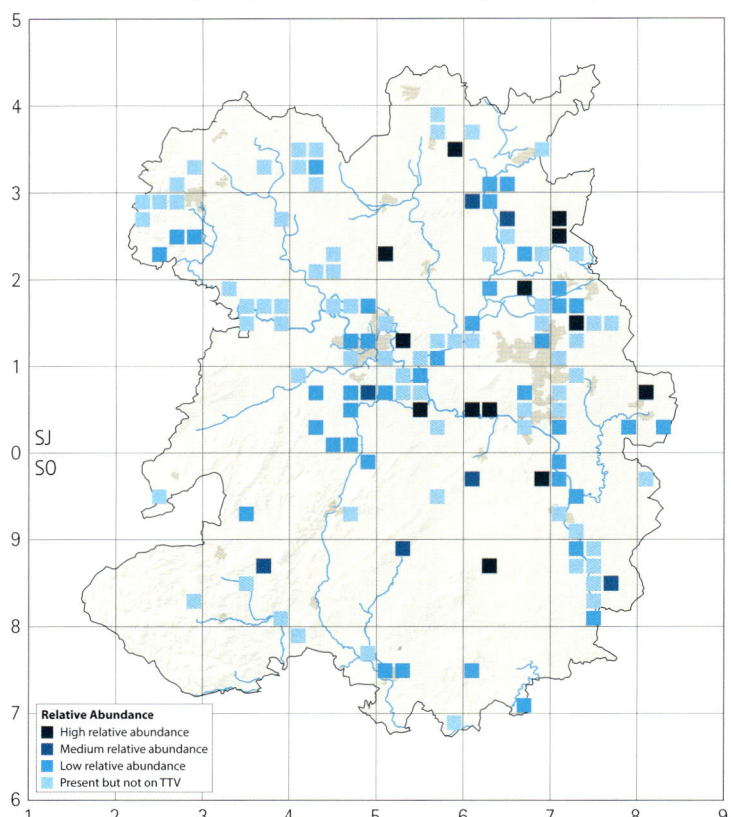

Relative Abundance
- ◼ High relative abundance
- ◼ Medium relative abundance
- ◼ Low relative abundance
- ◼ Present but not on TTV

Occupied Tetrads

Atlas period (breeding)	1985–90		2008–13		Change	
	Number	%	Number	%	Number	%
Confirmed	45	5	84	10	39	87
Probable	38	4	27	3	-11	-29
Possible	38	4	29	3	-9	-24
TOTAL	121	14	140	16	19	16
Tetrads with Winter Records (2007–13): 132 (15%)						

have occurred in the north-west and south-west, but elsewhere gains and losses have cancelled each other out.

Their scarcity on the higher ground reflects the lack of habitat rather than altitude, as shown by successful breeding on Boyne Water at 455m on Brown Clee. In other places, apparently suitable habitat remains unoccupied.

Large waters may have several breeding pairs and, although small pools will only support a single pair, some tetrads will have more than one suitable site. An estimated population of between 210 and 350 pairs is based on 1.5 to 2.5 pairs per occupied tetrad, similar to the estimate of 250–330 pairs in the *Atlas* (1992). There has probably been a small increase since the *Atlas* (1992), as indicated by the increased range and BBS trend, but the estimate of three to four pairs per tetrad with probable or confirmed breeding used in the earlier *Atlas* may have been rather high.

The winter distribution is broadly similar to that in the breeding season. This is not unexpected as many are resident throughout the

year. There were fewer occupied tetrads in the south and west and more occupied tetrads along the Severn Valley. This is particularly evident to the south of Bridgnorth and around Melverley.

Regular winter movements to sheltered coasts and estuaries are known to occur nationally and cold weather movements also take place. Whether those breeding at altitude in the south-west abandon the area each winter or whether they were forced to leave only during the harsh weather is unknown; more frequent observations would be needed to establish this. During the recent Atlas period small pools froze over completely in some winters, as did some of the larger ones, and even some stretches of running water. At a small pool in the north-west, one was sitting on the edge of the ice and, at another, a pair was confined to a few square metres of open water where a small stream fed the pool. Both sites were later abandoned as conditions worsened. Some of the displaced adults could be responsible for increases in the Severn Valley, where conditions are milder, and where they could be joined by juveniles, which are often forced to leave their natal sites, particularly if these are small pools, by their parents who maintain a territory throughout the year.

Because the Little Grebe winters mainly in small numbers at a wide range of locations, only a small proportion of the population is monitored, both locally and nationally, so trends need to be treated with caution. Since it was added to the list of those counted by the WeBS in 1990 the index for England had doubled by 2009–10. A local increase is evidenced by a rise to 30 occupied 10km squares, compared with 23 shown in the national BTO Winter Atlas (1986).

Assuming between 1.5 and 3 per occupied tetrad, the winter population is estimated at between 200 and 400 individuals, rather higher than the estimate based on TTV counts of 100–110. Even so, this is much lower than would be expected if all the breeding birds remained in situ, and the largest counts usually occur during September and October, indicating that some dispersal, possibly to coastal areas, takes place.

There is no local ringing data and little from elsewhere, but a single recovery in Lancashire, of a juvenile from the northern edge of the breeding range in Latvia, supports the belief that these populations move to the west. However, contrary to the situation in the late nineteenth century, when winters were harsher, there is no indication now of increasing numbers during this period. More recent milder winters should provide better conditions enabling resident breeders to stay put, but there may also have been reduced migration from the continent.

A continuation of mild winters should see a recovery of recent losses, but distribution and numbers are restricted by the availability of suitable breeding habitat. This may be close to capacity, although some of the tetrads apparently abandoned since 1992 must still have potential for re-occupancy.

Allan Dawes

Red-necked Grebe

Podiceps grisegena

Very rare winter visitor

A rare winter visitor to Britain from breeding grounds in eastern Europe, it is mainly found along the east coast with only occasional records from inland waters. Of the three grebe winter visitors, it occurs here with the same infrequency as Slavonian, and rather less than Black-necked.

The earliest but undated record is of one on the Severn near Shrewsbury (Ogier Ward 1841). Rocke (1866a) reported that it was occasionally obtained, generally in winter plumage, but gave no details, while the *Wellington Journal* in August 1867 reported one shot at Wrockwardine Wood. Beckwith (1879–81) referred to an undated record of one taken on the Severn near Wroxeter in winter plumage, while Forrest (1899) added another record of one close to Shrewsbury in March 1888. A long gap separated that from the next record, one on the Teme at Ludlow on 14 September 1917.

There were no more sightings for 60 years until 1977, the first of 11 modern records. These records span all winter months October–April, and all except one were of singles. Five were seen on one day only, but two stayed for over three weeks, and another for over three months.

The first modern record was from White Mere on 13 March 1977, and then an intensely cold period of weather across northern Europe in February 1979 brought three to Chelmarsh from 18 February to 3 March 1979.

Since 1979 there have been records in only seven years: a first winter bird at Venus Pool from 19 to 23 December 1989; two in 1991, at Venus Pool from 7 November to 1 December and at Crose Mere on 10 November; another at Crose Mere for most of the 1993–94 winter (20 December to 2 April, not listed in SBRs), and then two more in 1996, an adult in winter plumage at Cole Mere from 30 January to 4 February, and one at Chelmarsh on 31 March. A well-watched individual was at Priorslee Flash from 2 to 26 January 1998, and another at Cole Mere on 30 January 2000. It was then 12 years before the most recent record at the Mere (Ellesmere) on 12 February 2012, the only Red-necked Grebe seen during recent Atlas work.

Colin Wright

Great Crested Grebe

Podiceps cristatus

Uncommon resident

The Great Crested Grebe is widely distributed throughout the Western Palearctic from Fennoscandia to the Mediterranean.

We still have much to learn about many familiar species, although we have come a long way since observations began! One of the first references to Great Crested Grebe reads: 'Common on the meres in the neighbourhood of Ellesmere, where it breeds; its food is entirely vegetable. I have several times tried to keep this bird alive in confinement, but never with success' (Eyton 1839). The belief that the diet was vegetarian may have stemmed from the weed dance that forms part of the courtship display, but it would have foiled any attempt to keep them in captivity.

The beautiful silvery white plumage of the breast, known as grebe fur, was commonly used in muffs and collars during the nineteenth century. Originally these were imported, but in 1851 traders began to target British birds and by 1860 these grebes were thought to be locally extinct as a breeding species. A few colonists returned in 1865 and three meres were occupied in 1866.

The Wild Birds Protection Act 1872 prohibited killing from 1 April to 31 July, but as many adults returned to breeding sites during March they were shot on arrival. There was some refuge on private estates and in 1877, 14 or 15 were seen in full summer plumage at the Ellesmere meres and odd pairs also bred on pools at Hawkstone Park, Alkmund and Betton. The law was amended in 1880, and protection then commenced on 1 March, after which numbers began to grow; further help came in 1887 when it became illegal to collect their eggs.

Fresh arrivals from the north boosted numbers during the winter when they could be found on waters where they were not known to

Breeding Distribution Change (1985–90 to 2008–13)

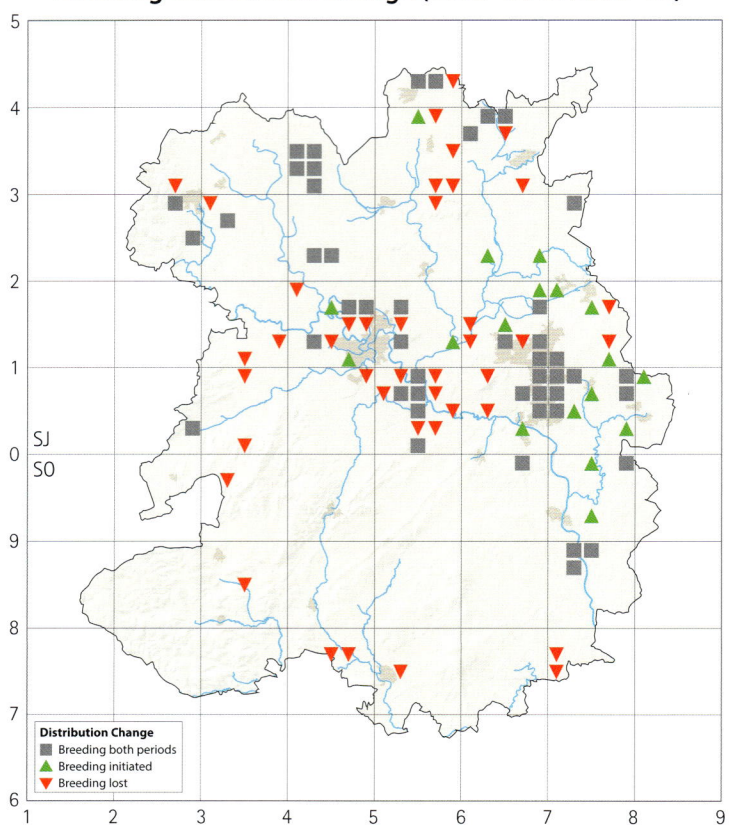

Winter Relative Abundance (2007–13)

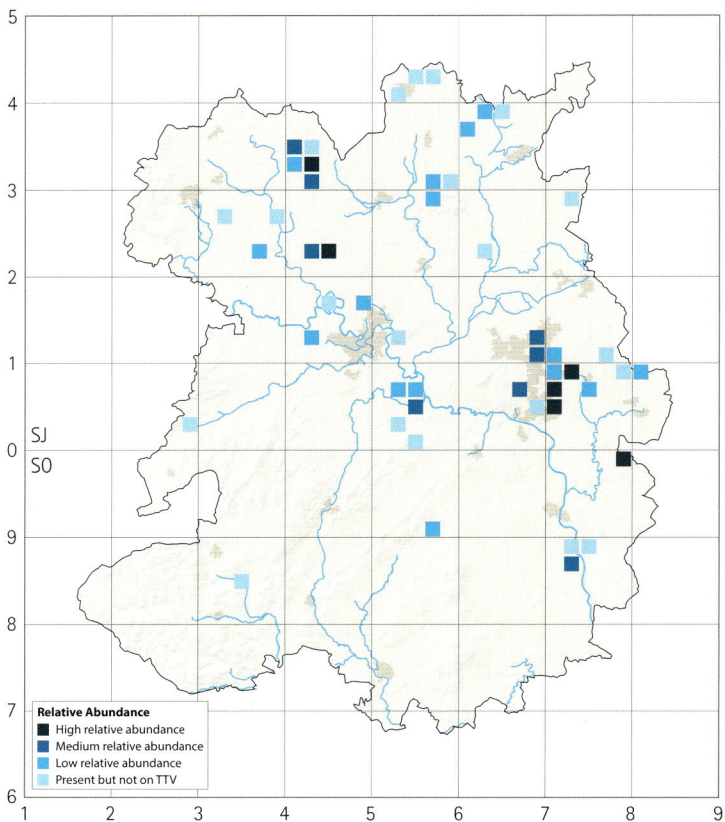

Occupied Tetrads

Atlas period (breeding)	1985–90		2008–13		Change	
	Number	%	Number	%	Number	%
Confirmed	52	6	37	4	-15	-29
Probable	19	2	17	2	-2	-11
Possible	17	2	11	1	-6	-35
TOTAL	88	10	65	7	-23	-26
Tetrads with Winter Records (2007–13): 50 (6%)						

breed, and rivers and streams provided a refuge during cold periods in the late nineteenth century.

A Great Crested Grebe enquiry in 1931, one of the first organised by the BTO, found 48 breeding pairs at 32 locations. This was followed by a sample census from 1946 to 1955, an SOS survey in 1962 and BTO surveys in 1965 and 1975 (Hughes *et al.* 1979). The results are not directly comparable as not all waters were visited each year, but estimates were made to take missed waters into account. The chart shows the total estimated number of individuals at all sites and a comparison of those waters that were visited in each of the main surveys. Non-breeding birds are included so actual breeding pairs would have been lower than these figures suggest. After a period of stability a sharp rise can be seen after 1955.

A number of factors were thought to underlie this change: more effective legislation, increased habitat provided by gravel pits and reservoirs, and a climatic shift across Europe which encouraged a spread to the north-west.

The increase continued and the 1975 census estimated 186 individuals, while the *Atlas* (1992) suggested between 150 and 200 breeding pairs. However, recent Atlas fieldwork shows a 26% decrease in occupied tetrads since 1990, which is rather greater than the population decline of 2% since 1995 recorded by the BBS for England.

Pairs at regular sites are unlikely to have been overlooked, and some sites hold several pairs: between three and 12 individuals were recorded on 16 TTVs, and a further 15 tetrads had casual records of between three and 14. Also, occasional breeding attempts at small pools or along rivers could have been missed. A population of between 110 and 150 breeding pairs now seems likely.

Great Crested Grebes prefer the larger lakes and meres, as small waters lack the space required for the long take off that they need to become airborne. The larger waters are found mainly in the north, and a withdrawal from sub-optimal sites and a contraction to the

Censuses (1931–75)

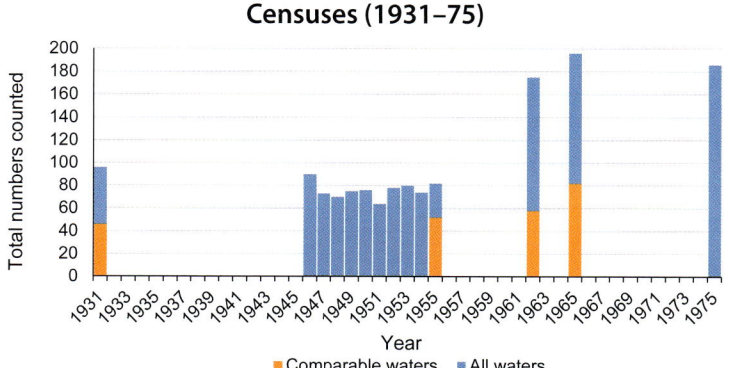

Allan Heath, Wood Lane, 22 April 2011

main waters is indicative of falling numbers. Most losses occurred along the Severn Valley around Shrewsbury, with smaller losses in the north-east. The few sites in the south-west were also abandoned. Prior to 1990, breeding on the River Severn was an annual event, with up to four pairs in some years, mostly between Shrewsbury and Cressage. They were thought to have overflowed on to linear waterways when other sites reached maximum capacity. Since the population has declined, the disappearance of river nesting is not unexpected, and this would explain some of the losses in the Severn Valley. In addition to fluctuating water levels, rivers also form a corridor for predators. These habitat preferences are reflected in the breeding distribution change map.

Seasonal movements are not fully understood, and the single ringing recovery, from Knighton Reservoir in April 1979 of an adult ringed in Lincolnshire in February 1977, does not help to clarify the situation.

The population probably peaks in August, and several high counts from Chelmarsh were made at this time in the late 1990s, 80 in 1995 being the maximum. These are likely to include failed or early breeding birds which have stopped off to moult before dispersing elsewhere, probably to the coast. Most WeBS commence in September so August figures are not widely available, but numbers remain high into September and usually fall to a minimum in January before building again in March, as spring arrivals begin to return. The chart shows the combined monthly total at the Ellesmere meres for the six winters in the recent Atlas period.

Some adults still have dependent young in September, and it is thought that these families remain for the winter. Those that stay may have to contend with icy conditions and if these persist they can be

forced to leave, but those that see out the winter have the advantage of securing the best breeding sites before others return. If conditions are favourable, nesting can begin as early as February; in 2012 a pair were displaying at Venus Pool on the 3rd.

The arrival of continental birds for the winter adds further complications. The strength and direction of movement of these populations is thought to be dependent on weather conditions, and numbers vary greatly between years. Nationally, the underlying trend shows a slow increase up to 2003–04, followed by a shallow decline. A similar trend is evident from the maximum annual WeBS counts at the Ellesmere meres, which may be due to natural fluctuations, or, more recently, 'short-stopping' as a result of climate change. There has been a small increase in range, with three additional 10km squares occupied since fieldwork for the national BTO Winter Atlas (1986).

During the winter, distribution is restricted to larger waters, with rivers being rarely used except during icy spells. The maximum total

Monthly Totals at Ellesmere Meres: Winter Atlas Years (2007–15)

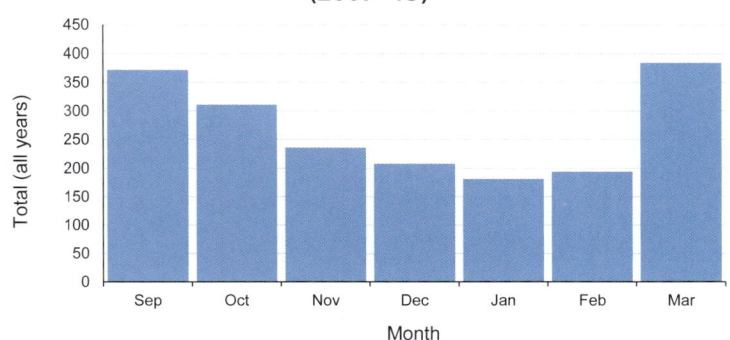

WeBS Counts at the Ellesmere Meres (1978–79 to 2014)

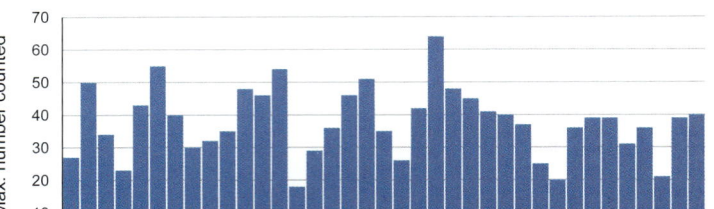

Jim Almond, Venus Pool, 4 March 2008

winter WeBS count was 63 in November 2007, a year which bucked the trend of lower numbers in mid-winter. The minimum count of nine was made in the exceptionally cold December of 2010. Few will have been overlooked by WeBS so the population in an average winter is estimated to be between 30 and 70 individuals.

At present there is no concern over the future of the Great Crested Grebe, although breeding numbers should be monitored so that any further declines can be quickly identified. Annual counts of breeding pairs and the number of fledged young at any site would be helpful.

Allan Dawes

Slavonian Grebe (Horned Grebe)

Podiceps auritus

Very rare winter visitor

Most of the European population breed in countries around the Baltic Sea and further east, with small populations in Iceland and northern Scotland. Many winter around the coasts of Britain, but rarely reach this far inland, and then usually as singles with long gaps between records. Early records were imprecise on dates and locations, but almost all those from the nineteenth century were from different places on the River Severn, as were two out of five from the early twentieth century. However, all the modern records have come from the larger meres and reservoirs, rather than the river.

The earliest but undated record was of one near Shrewsbury (Ward 1841). Rocke, writing in 1866, listed 'Sclavonian Grebe', with immatures not rare in winter, and noted that a specimen in immature plumage in the Clungunford collection had been killed in February 1865. Beckwith (1879) had obtained a specimen at Cressage 'a few years ago' whilst others had been killed in various other places. There were two records from 1890, one shot at Isle Pool in February and one killed by a boy with a stone in the Quarry, Shrewsbury, in December, while another was shot at Montford Bridge in November 1894 (Forrest 1899).

There were five records in the first half of the twentieth century: one on the River Severn at Dowles in December 1909; three in February 1912, all 'taken' during a period of hard weather, at Clun, the River Severn at Shelton, and Ticklerton, the latter 'taken alive'; and lastly one from Ellesmere in December 1936. Amazingly, the observer of this last record wrote in *British Birds* that he had 'very good comparative views of a Slavonian Grebe and a Black-necked Grebe which were on the same water as Great Crested and Little Grebes' (Cohen 1937). This unlikely co-incidence was probably made possible by stormy weather, described by the Met Office report for the month as 'very unsettled at times, particularly during the first three weeks, with some severe gales'.

There was then a long gap until the first of the 13 modern records, probably involving 12 individuals. All were of single birds, in only 12 different years between 1957 and 1998, and have come from only four sites – the Ellesmere meres (eight records from four different meres), Chelmarsh reservoir (three records), and Brown Moss and Marton Pool (Baschurch) (one each). Six were seen on one day only, but the first stayed for 10 days at the Mere from 1 February 1957, the second, in summer plumage, stayed at Brown Moss for 26 days from 7 April 1960, and the third for 12 days at Marton Pool from 22 January 1961. Other long stayers were at Chelmarsh, for 27 days from 5 December 1984 and the longest, at Crose Mere, for 31 days from 19 November 1994. Records have spanned an extended winter period, on dates between 16 October and 9 May, with a gap only between 1–22 January.

Only 1994 has two records, the longest staying individual noted above, and probably the same bird at the Mere (Ellesmere), eight days earlier.

The most recent record was from Crose Mere on 22 February 1998. None were found during the recent Winter Atlas period.

Colin Wright

Black-necked Grebe
Podiceps nigricollis
Rare passage migrant, has bred

Scattered populations of this grebe occur in Western Europe, becoming more abundant to the east. In Britain it is a rare breeding species and scarce winter visitor, mainly to south-east coasts.

It was known as the Eared Grebe in the early literature but, with no sightings in the nineteenth century, and only four in the first half of the twentieth, it did not feature often. A pair at the Mere (Ellesmere) in December 1900 was followed by a single at Halston in April 1932 and another at the Mere in December 1936. A Slavonian Grebe was reported from the Mere on the same day as the latter, by the same observer (Cohen 1937). Finding a rare grebe does not happen often, so finding two scarce species in the same place at the same time is noteworthy.

Since the first modern record in 1956, when one was found at Trench Pool in March, there were a handful of sightings up until 1977, after which they have become more frequent. There have only been three blank years this century. Outside the breeding period, stays have generally been brief: 28 were recorded on one day only, but six singles stayed for two weeks or more, and one stayed for 33 days at Chelmarsh from 23 November 2012. The four that spent several days at Bomere Pool from 30 March 1958 were exceptional, and the remaining sightings were all of singles or pairs. Winter visitors have increased in recent years, with six of the nine records from 1990 onwards.

These local appearances reflect changes in the UK population. After first breeding in 1904, a small but regular breeding population became established. It remained low for much of the twentieth century but later increased, and after peaking at the turn of the century it has stabilised at around 50 pairs, originally centred in southern Scotland but later shifting to the north of England, with the first confirmed breeding at what is now the largest colony, in Cheshire, in 1987.

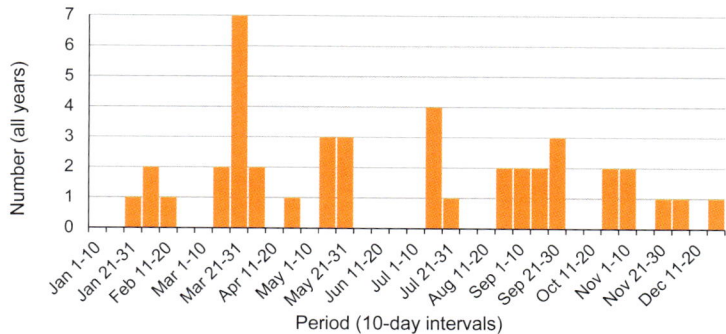

Arrival Date in 10-day Intervals (1950–2014)

The growth of this colony in the adjacent county would explain both the increase in local sightings and their timing, including breeding here at Crose Mere (site not published previously) during the earlier Atlas period, in 1985, 1986 and 1988. Two young were raised in 1985 and four eggs were laid in 1986, but the final outcome of the latter two seasons is unknown. A single in 1982 and a pair in 1983, both in late May, were also at the same location, just preceding the first confirmed breeding record, so possibly earlier attempts were overlooked. After this promising run, it is disappointing that no further attempts have come to light.

The chart, based on records since 1950, suggests a small spring and autumn passage. The former coincides with the return of these grebes to their breeding sites in Cheshire in March and April, where nesting activity normally begins in May. However, spring records of pairs, particularly in May, could indicate prospecting for new breeding sites here. The early peak in late summer may indicate passage of failed breeders, followed by successful pairs or young a few weeks later. In 1995, a pair was seen at Crose Mere in late May, and a single remained at Wood Lane from 16 May to 18 June. The winter records include the long-staying individuals at Chelmarsh, referred to earlier, and Priorslee Lake, referred to later. The breeding pairs in the 1980s are not included because the arrival dates are not known.

A single at Priorslee Lake from 8 December 2009 to 6 January and another at Chelmarsh from 23 November to 26 December 2012 were the only recent winter Atlas sightings. Since then there have been two sightings of singles at Chelmarsh, both for one day only, in September 2013 and February 2014.

Lowland eutrophic waters with extensive emergent vegetation are preferred during the breeding season and there is suitable habitat at several locations, especially in the north. There is also a preference for sites with a Black-headed Gull colony, where the gulls are thought to offer some protection for the grebes, but there are few of these available. Many small pools are not visited regularly, if at all, and breeding Black-necked Grebes are elusive, so hope remains that more breeding pairs will be discovered. However, the national increase appears to have levelled off, so this seems increasingly unlikely.

Jim Almond, Priorslee Lake, 10 December 2009

Allan Dawes

Black Stork
Ciconia nigra
Vagrant

Black Storks breed from central Iberia and eastern France through to the Russian Far East. Most are migratory, wintering in Africa and south and south-east Asia, although the Iberian population appears to be largely sedentary.

Just three records:

Dowles Brook, Wyre Forest	31 May 1956
Upper Teme Valley	30 August–6 September 1990
Whixall Moss	28 April 2014

The Dowles Brook forms the border with Worcestershire for much of its length, and it would be virtually impossible for a Black Stork to take off from the brook without occurring in both counties. As such, and having been accepted for Worcestershire, it constitutes the first Shropshire record (Harrison & Harrison 2005). Interestingly, a Black Stork was reported drifting south east over Haughmond Hill the previous day. However, a review of the 1950–57 British rarities by a sub-committee of BBRC concluded that the latter identification was not considered proven (Wallace *et al.* 2006).

The well-watched adult in the Teme Valley near Dutlas visited both sides of the border with Powys and may have been present since 9 August. It was one of possibly 10 Black Storks in Britain that year (Rogers *et al.* 1991).

The one at Whixall was seen circling over the Moss before drifting south-east towards Shrewsbury.

In addition, there was an unconfirmed report of one at a fish hatchery on the River Ceiriog near Chirk for about two weeks from 20 May 1991, a year when there was an exceptional influx of Black Storks into Britain with perhaps 23 individuals involved; yet another site where the river forms the county boundary.

Graham Walker

White Stork
Ciconia ciconia
Vagrant

The nominate race breeds in Europe, North Africa, and the Middle East. The nearest breeding populations are in the Netherlands and France, two of several European countries that have effected re-introduction programmes in recent years. While the majority of Western European birds winter in tropical Africa, some remain to over-winter in Europe, particularly in Spain and Portugal but including as far north as Denmark. The majority that reach Britain occur in the spring as overshooting migrants and all six modern records are considered to reflect this:

Near Ddôl, St Martins	4 March–3 April 1978
Horsebridge, Minsterley	21–23 April 1997
Great Bolus	c.27–29 May 2008
Pen-yr-estyn/Queen's Head/Rednal	13–20 April 2009
Caynton House near Edgmond	early June 2009
Aston Munslow/Shrewsbury	25 May 2012

The first, near Ddôl, was accepted by BBRC for Clwyd for the dates given, with the comment that it was "thought to have wintered in the area" (Rogers *et al.* 1979); it was subsequently discovered that it crossed the border on a number of occasions to feed (D. Hampson *pers. comm.*). The second was located on a partly ploughed, heavily manured field where it fed avidly on earthworms. Although it appeared to leave once high to the north-east, it returned. On its third day, however, it disappeared off to the south.

The individual at Great Bolus was not reported until some months after it was seen, hence the imprecise dates. Nevertheless, it was photographed while following the gang mowers at a turf farm, making confirmation of the identification very straightforward. The fourth, at Pen-yr-estyn, was also first found in a newly ploughed field and commuted between various sites. The stork at Caynton House was again photographed, but it was not until the following year when the photographs were seen by an experienced birdwatcher that the sighting came to light. The sixth, in 2012, was first seen circling on thermals over Aston Munslow with what was presumably the same individual observed flying over the Shrewsbury bypass later that day.

Graham Walker

Glossy Ibis
Plegadis falcinellus
Very rare visitor

Although this ibis has a worldwide distribution, the nearest breeding colonies are in southern France and northern Italy. There have, however, been several large post-breeding influxes to Britain in recent years, and it is perhaps surprising that the first modern record did not occur until 2012.

Although the historic record includes several accounts, it appears that they all relate to two seen at a pool at Albrighton (near Shrewsbury) in late September/early October 1853. Both were observed coming to the pool at about 9am for several mornings to feed on small 'shell-fish'. They were described as 'by no means wild', although this does not necessarily mean that they were of captive origin.

One, a male 'nearly in adult plumage', was shot on 3 October and assumed to be a black variant of the curlew when described the following day to Mr Franklin, a local taxidermist, in his shop on Mardol, Shrewsbury. He suspected what it may have been and offered a 'liberal sum' for the corpse, which was subsequently brought to him, it not having been eaten by the pigs as the countryman expected! It appears that the second was shot later by the gamekeeper from the Sundorne Estate, but was far too decomposed to be mounted by the time its existence was discovered.

The five modern records are:

Whixall Moss	5 May 2012
Wall Farm, Kynnersley	29 September 2013
Stokesay, Craven Arms	29 September and 5–11 October 2013
Venus Pool	19–20 March 2016
Whixall Canal Floods	2–7 July 2016

Although there is some anecdotal evidence to suggest that the well-watched bird at Stokesay was present from 23 September, it has not been possible to confirm this; where this bird went between 29 September and 5 October remains a mystery.

Graham Walker

Bob Hart, Stokesay, 10 October 2013

Spoonbill (Eurasian Spoonbill)
Platalea leucorodia
Very rare passage migrant

Jim Almond, Venus Pool, 28 September 2013

Spoonbills breed across Eurasia, but in Europe, only the Netherlands, Spain, Austria, Hungary and Greece have sizeable populations.

It became extinct in Britain in the seventeenth century due to hunting and land drainage. There were no references to it in the national *Historical Atlas* (Holloway 1996) or the national BTO breeding Atlas in 1976, although there was a speculative forecast of probable recolonisation in the latter. There were no local accounts by nineteenth-century authors, or in the *Handlist* (1964).

The first record was from near Newport on 26 August 1965. Including that, there have been 12 records from 11 different years, all except one of singletons, and all in the passage periods, usually from late June to November. Four were seen at Claverley on 20 October 1972. Venus Pool has been the most frequented site, with four visits: 25–29 November 1990, 21–22 October 1994, 10 September 1997 and 28 September–3 October 2013. Wall Farm has been graced with two visits, on 18–22 June 2001 and on 24–25 June the following year, while possibly the same individual was at Edgerley a few days later, on 27–29 June 2002. Another visited Wood Lane, Cole Mere and the Mere (Ellesmere) on 2–3 May 2003, roosting in the heronry at the latter site. One was seen at Chelmarsh Reservoir on 22 August 2009.

None were found during Atlas fieldwork in either season. The most recent records were from Venus Pool in 2013, referred to above, and Whixall Canal Floods in 2016, a first summer bird that stayed for nine days from 4 April. Nationally, there has been a major increase in winter range over the last 30 years, sporadic breeding attempts since 1998 and the establishment of a small successful colony in Norfolk from 2010 onwards. The breeding population in Western Europe is growing, so the occasional sightings of Spoonbill may slowly increase in future.

Josie Owen

Bittern (Eurasian Bittern)

Botaurus stellaris

Very rare winter visitor

The sub-species *B.s. stellaris* occurs in Britain. Its range extends right across the temperate parts of Europe and Asia. Being vulnerable to cold weather, some northern populations are migratory.

Occurrence here has reflected its fluctuating status in Britain as a whole. By the early nineteenth century, land drainage had restricted it almost entirely to the Norfolk Broads and fenland, and it had become extinct as a breeding species by 1868. At the start of the twentieth century, legal protection and milder weather helped recolonisation, and it increased until about 1950, largely restricted to East Anglia and spreading to Leighton Moss (Lancashire) in the early 1940s. Pesticides and habitat deterioration then led to a steady decline, with the population falling to a low point of only 11 booming males in 1997. Its specific habitat requirements having been identified (Brown *et al.* 2012), the management of existing reedbeds and creation of new ones encouraged an increase to 104 booming males by 2011, still largely restricted to East Anglia, but with welcome signs of new breeding colonies elsewhere. They are much more widespread in winter, due to dispersal from breeding areas and immigration from the continent, particularly in colder weather.

Occupied Tetrads

Tetrads with Winter Records (2007–13):	6	(1%)

Winter Distribution (2007–13)

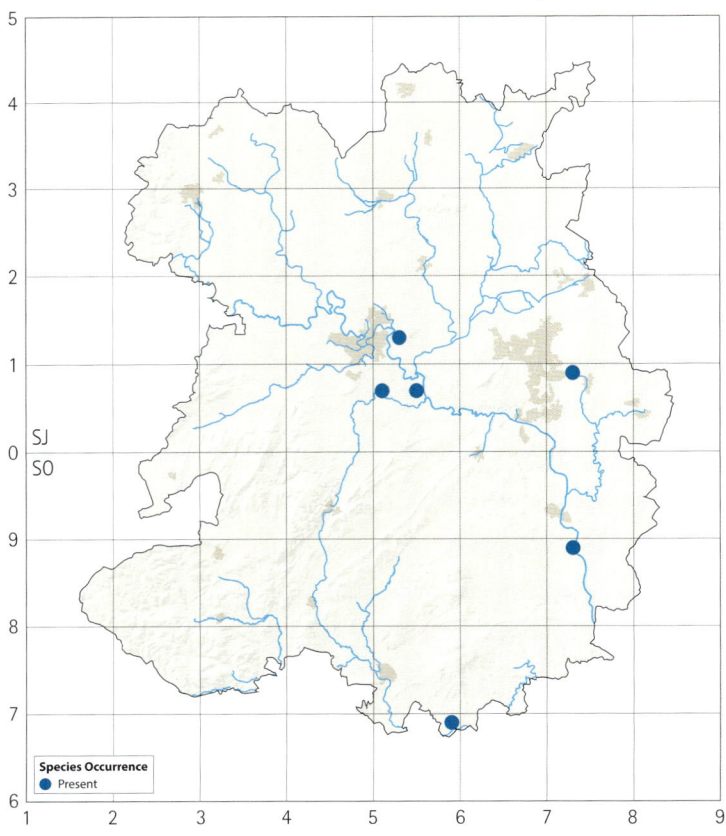

Dave Barnes, Burford, 24 December 2010

It was 'once common in North Shropshire, has not bred here since 1836 when two nests were found near Shifnal' (Beckwith 1885), and 'it is now very rare, but single birds still occur from time to time' (Forrest 1899). In 1917, Mary Webb, in *Gone to Earth* refers to 'voices long gone hence … the love-call of the bittern'.

Between 1900 and 1949, there were reports in 11 years, almost all of singles, except for two together at Ellesmere in 1900, one staying until 28 March, and there were two others in the same year, and two in 1933. Several sightings were at Ellesmere, almost all in January, February or December. One stayed as late as mid-April at Crudgington in 1925 and was heard booming in flight. Another, seen in July and October 1927 at Edgeley, was shot at Brown Moss on 25 October. Another was seen later that year in the same locality. Several of these records refer to very cold weather – 'one was caught alive, numbed by cold' in 1907, while another was seen from time to time between 6 and 25 February 1942: 'during this period the Mere was frozen over'. Several were shot, including most recently in December 1943 at Marton Pool (Chirbury).

The *Handlist* (1964) noted one wintering at Ellesmere in 1950–51, and singles in December, in 1953 and 1962. None were seen for the next 19 years. Then, from 1981 to 1997, there were 12 records in six years, all of singletons in January or December, except for two in October, both from Fenemere. All were present for a single day, except one at Pimley Island from 17 December 1989 to 18 February 1990.

There were three contrasting local sightings in 1992: one caught by a falconer's Goshawk in January at Hortonwood, one at Fenemere in October, and the third flying along the railway at Church Stretton in December; Bitterns were widely recorded in the UK in late 1992 because of cold weather in continental Europe.

After the second October record from Fenemere, in 1996, there were three in a very cold spell in January 1997, from two sites on the Severn and one roosting in a farmyard near Cleobury Mortimer.

A Christmas Day 2001 sighting at Melverley presaged 'a remarkable year' in 2002, with possibly five individuals spread over two winter periods. Between 6 January and 24 March, one was seen on most days at Venus Pool, but two were seen together there on four dates between 10 February and 19 March. At the end of the year, the single that stayed at Wall Farm for 14 days until 22 November was possibly the same one seen at Venus Pool from 29 November up until 22 February and again from 16 to 25 March 2003. A single was also at Priorslee Lake for Christmas week 2002 and into mid-January 2003. A Bittern also turned up at Severn Valley Country Park on 19 and 20 January, and again on 4 and 6 March 2003. The chart estimates three individuals in the 2002–03 winter, but it is possible that one individual accounted for all these records.

One at Venus pool for two days in August was 'unseasonal', but singles at the same place on 16 November and 3 December 2003 were on more typical dates. Frequent sightings of a single from 10 January until 12 April 2004 suggested a bird commuting between Venus Pool, Eaton Mascott and Emstrey, and this individual may also have been the one seen at Severn Valley Country Park on 4 February. Again, the chart estimates three individuals in the 2003–04 winter, but it is possible that one individual accounted for all these records. There were further sightings of singles in February in 2005 and 2006.

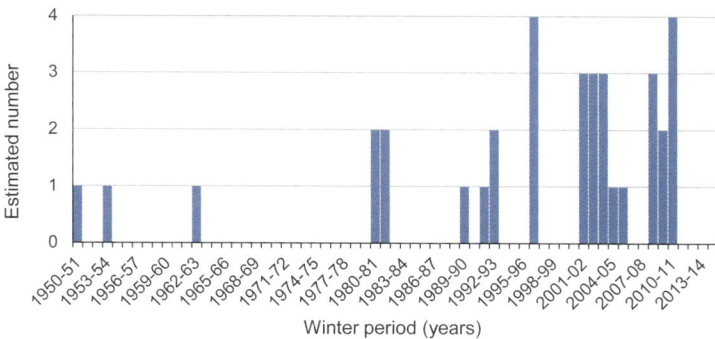

Estimated Winter Period Totals (1950–51 to 2014)

After a gap in 2007, it was found in six tetrads during the recent winter Atlas period. Again, all records were of singletons, with some lengthy stays. Venus Pool only featured in two winter periods, one between 14 October 2008 and 18 March 2009, and again between 9 October and 16 November 2009. Other sightings were at Chelmarsh Reservoir on 8 January 2009, Priorslee Lake on 4 March 2009 and 5–16 January 2010, on the River Teme (near Tenbury Wells) from 7–29 December 2010, Bomere Pool in December 2010, Monkmoor Pool on 1 January 2011 and Cantlop Bridge from the following day until 8 January 2011. This was the most recent record. The chart shows the estimated number of different individuals seen in each winter period since 1950.

There have been no modern breeding season records. With continuing growth and winter dispersal of the British breeding population, the increase in sightings is likely to continue. As the few reedbeds here are small, providing marginal breeding habitat at best, nesting is unlikely until the national population becomes much larger.

Josie Owen

Little Bittern

Ixobrychus minutus

Vagrant

Jim Almond, Chelmarsh Reservoir, 8 July 2018

A summer visitor to southern Europe, the Little Bittern winters in Africa and rarely gets as far north as Britain, although a pair nested in Somerset in 2010 and 2011.

Pennant in his *British Zoology* Volume II of 1768 describes the first British record: 'This species was shot as it perched on one of the trees in the Quarry, or Public Walk, in Shrewsbury, on the banks of the Severn; it is frequent in many other parts of Europe, but this is the only one we have ever heard of in England'.

One was shot at Cole Mere (sometimes mis-attributed to Ellesmere) on 19 May 1880 (Beckwith 1881). In the following July a pair were sent to the Shrewsbury taxidermist Henry Shaw; the male shot at Petton and the female at Marbury 45km further north and just into Cheshire, and Forrest (1899) remarked: 'These were probably a pair and the hen would have laid eggs shortly had she been spared'. The Petton bird is in Ludlow Museum, specimen Z/2006/208 (see p. 501).

One was shot at Ditton Priors on 29 May 1894 (Forrest 1905) and the fifth and final historic record is of a 'fine male' shot at Claverley in September 1897.

There is one modern record:

Chelmarsh Reservoir   6–13 July 2018

On 6 July, a male was heard 'barking' in the reedbed at Chelmarsh Reservoir and, although difficult to see at first, it eventually offered good views to the many observers that came to see and photograph this attractive, but diminutive, heron during its eight-day stay.

This is one of the five species to make their first British appearance in Shropshire (see p. 2).

John Tucker

Night-heron (Black-crowned Night Heron)
Nycticorax nycticorax
Vagrant

This heron has a worldwide distribution with the nearest breeding colonies in the Netherlands, France and Spain.

The historic record refers to three individuals, all sighted within a mile or so of each other. The first, an immature, was shot at Wroxeter, possibly in July 1830, with the second and third coming from Atcham and Attingham and both reported by the same gamekeeper. In May 1912 he obtained an adult bird at Atcham which was subsequently mounted and displayed in the Billiard Room of Belswardyne Hall, Cressage, where it was seen by a field meeting of the Caradoc and Severn Valley Field Club on 22 May 1912. Two years later, on 1 June 1914, he saw and heard another bird at Attingham.

There was then a gap of over 70 years until the only modern record:

Caynham   1 April 1988

This individual, an adult in breeding plumage, was on a small pool near Ludlow and only seen by a limited number of observers.

On 7 April 2017 an adult, also in breeding plumage, was located at Venus Pool and, although often partially obscured by branches as it sat by the small woodland pool in front of the feeding station hide, it was seen by many visitors. Eventually, it moved to a part of the reserve where it was much less visible and was last seen on 18 April. On 22 April the same adult was found in the Dingle, Quarry Gardens, Shrewsbury. Although almost always present in the Dingle, it did occasionally appear on the nearby River Severn and at other local pools. Because it often fished under the boardwalk it afforded excellent views, and it showed characteristics consistent with the American race *N.n. hoactli*; this was subsequently confirmed through DNA analysis. On occasions, it appeared to use bread offered for the pool's ducks to attract fish (J. Almond *pers. comm.*). Eventually, however, it became more approachable having apparently exhausted its food supply, and was taken into care on 27 November by Cuan Wildlife Rescue. After most of the winter, by which time it was rehabilitated, the decision was taken that it couldn't be released back in to the wild and, after a period in quarantine, was taken to a collection elsewhere. At the time of writing, it has been accepted as an individual of the American race by BBRC but the BOU Records Committee are reviewing whether it was likely to be of captive origin or a genuine vagrant because, if accepted, it would be the first occurrence of this race in Britain. Whatever their eventual decision, it did generate a lot of interest locally and, often, quite a debate about its likely origin and welfare.

Graham Walker

Squacco Heron
Ardeola ralloides
No modern records

The single record of this rare vagrant, which breeds in Mediterranean countries and winters in Africa, was first referred to by Rocke in 1866 as 'killed some years ago at Bockleton, under Brown Clee ... I believe it was a male bird, and was in very fine plumage' and Beckwith twice referred to it in similar terms. Only in 1897 was Paddock able to add a date, 24 July 1834. The early fate of the specimen was well documented and it was later recorded in the Ticklerton collection at the home of M.W.S. Buddicom in 1900. The specimen is now in Ludlow Museum, number Z.00106.

John Tucker

Cattle Egret (Western Cattle Egret)
Bubulcus ibis
Very rare visitor

Having undertaken a massive population expansion in the latter half of the last century, the Cattle Egret now has a truly worldwide distribution. It bred in Britain for the first time in 2008, in Somerset, and a continued increase in the number of breeding pairs suggests that it may become established as a breeding species.

There are no historic references and just five modern records:

Jim Almond, Ellesmere, 8 November 2008

Westbury	January 1987
Ellesmere Marina/Wood Lane Nature Reserve	
	6–14 November 2008
Shavington	11 Dec 2011–2 Jan 2012
Northwood	two, 1 April 2017
Venus Pool	26 October 2017

The first was found dead in a field of swedes at Haywood Farm and later mounted. Although the exact date of the find is unknown, it was presumed to be the same bird as seen at Doxey Marshes, Stafford, on 7 January and previously in Derbyshire (Rogers *et al.* 1998). The second coincided with a large influx into the country, mostly in the south-west where the above-mentioned breeding occurred, but reaching as far north as Scotland (Hudson *et al.* 2009). The third was not part of a notable influx, and neither was the individual at Venus Pool, but more likely post-breeding dispersal as part of natural range expansion. The two at Northwood could have been spring arrivals or dispersing over-wintering birds. Whatever their origins, it is very likely that there will be more sightings in the near future if this egret continues to expand its range at the current rate.

Graham Walker

Grey Heron
Ardea cinerea
Uncommon resident

The breeding range covers most of the Old World south of the Arctic Circle. In Britain they are generally non-migratory.

Bones were found in excavations of the Roman settlement at Wroxeter (Hammon 2011). Although there was no direct archaeological evidence that they were exploited for food there, they were eaten at that time. From medieval times to the nineteenth century, Herons were highly valued and protected for food, and as a quarry for falconry. Heronries were sometimes deliberately established and managed on estates. Nestlings were taken to be fattened in pens or stew ponds.

From the early nineteenth century, hawking fell into decline and Herons for the table fell from favour. They went from being protected to being persecuted, as fishing interests increased and egg collecting became fashionable. In 1885 Beckwith observed that:

> the heron, in spite of much persecution, is still common, and besides the heronries at Attingham, Oakley Park, Walcot, and Halston, one or two pairs occasionally breed at Shavington and about Ellesmere, though the heronry on the island in Ellesmere Mere has long ceased to exist. I fear, however, the number of nests in all our heronries is gradually decreasing.

This reduction in heronry size has been widely linked to large-scale land drainage causing loss of feeding grounds.

Forrest (1899) added that 'some of the old heronries are now deserted, but the Heron still breeds at Attingham, Oakley and Walcot Parks; Colemere and Halston', adding Cole Mere to Beckwith's list.

The first known colony, at Halston Hall, was present in the 1820s when it was referred to by the owner, Squire Mytton, as comprising 80 to 100 nests, suggesting that it was, even then, long established. Populations are subject to a variety of challenges. By the 1960s this colony had reduced to about 40 pairs, probably driven by drainage of the marshes around the River Perry and the consequent loss of feeding grounds. After the hard winter of 1962–63 only 14 nests were counted (*Handlist* 1964). Numbers continued to drop, as many trees in the heronry were blown down by winter gales. Disturbance of feeding sites by increasing recreational use of waterways and pesticide poisoning were also taking a toll (Edwards SBR 1971). By the early 1970s, fewer than 10 pairs were breeding, but then the colony began to increase again (Wright SBR 1985) until the hard winter of 1985–86 reduced it to seven pairs (SBR 1986).

The population then grew, both locally and nationally, helped by pesticide bans and mild winters. By 1990 there were 28 pairs at Halston and numbers were fairly steady until 2000. A sharp loss in 2002 was followed by a move, in 2003, from the traditional site on and by the island to larches on the edge of a nearby wood. The move was almost certainly because the lake was opened up to fishing.

The best-known heronry, on Moscow Island, Ellesmere, was reoccupied in 1970. In 1976 there were 12 apparently occupied nests, eight pairs were present in 1985, 25 in 1990. The colony is clearly visible from the lakeside. The RSPB set up 'Focus on Herons', a public observation hide, in 1990, attracting around 18,000 visitors. In 1995 a volunteer group took over, becoming Ellesmere Heronwatch, which continues to offer information and interpretation. Nest cameras operate 24 hours a day throughout the season, giving clear, intimate views on screen in the visitor centre. The detailed information recorded is a very valuable tool for learning about both individual nests and

John Hawkins, Atcham, 1990

colonial life. Watching and learning here is very popular, making the heronry a highly valued local asset and attraction.

Halston and Ellesmere are only four miles apart, and the chart shows their fluctuating populations, and the combined total for the two. It clearly shows the impact of the hard winters in 1962–63, 1981–82 and 1985–86. Discounting foot and mouth year (2001), when no count was possible at Halston, there was a steady increase from 1970 to 2000, and a subsequent decline, in common with the national trend.

The comments of Beckwith and Forrest suggest that heronries come and go, and that has been the case since the longest running national bird survey, the BTO Heronries Census, started in 1928. BTO and SOS records, including the *Handlist* (1964), together list 30 heronries occupied since 1928. Cole Mere and Kinsley Wood (Knighton) appear to have been abandoned before then. Lloyd (1943) noted that censuses of heronries taken 'in recent years have shown a steady decrease'.

The 13 known heronries that ceased to be occupied before the 1985

BTO census and the 1985–90 Atlas are listed in the first table. Two of them were occupied before 1928, Acton Pool for 68 years from 1872 to 1939 (Gandy 1970), with 28 nests counted in one year in the 1930s, and Hinks Wood (Kynnersley) for 32 years between 1914 and 1945.

Two others were first recorded in 1928, perhaps as a result of the first census. Sundorne (near Uffington) was occupied until 1952, and possibly again in 1961, while Oakly Park (Bromfield) was occupied for two years only.

Loss of Eyton was attributed to tree felling, Sundorne was lost to road widening, and the demise of Attingham, which fell from 10 nests in 1954 to only two in 1957, was due to shooting (Perkins SBR 1968).

Four others were apparently occupied for one year each, Sansaw (1939), Trefonen (1951), near Morville (1969), where a pair raised three young in a rookery, and near Buildwas (1980), while two others, Bicton and Diddlebury, were occupied for three and two years respectively. However, the record from near Buildwas might have been the first report of the heronry at nearby Leighton.

Dudmaston (15 nests in 1982, 12 in 1983, and one in 1984), is

Halston and Ellesmere Heronries
Apparently Occupied Nests (1954–2014)

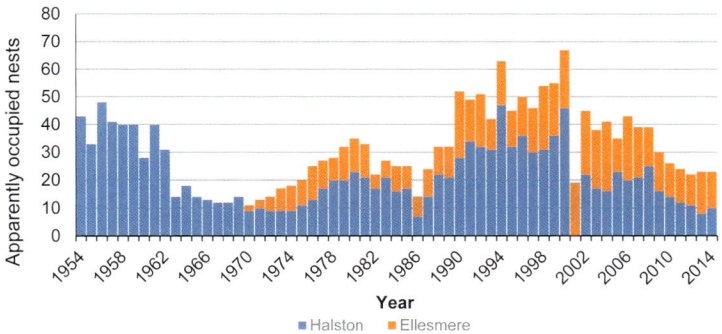

Year

■ Halston   ■ Ellesmere

listed in the first table, but it was probably the temporary location of a heronry displaced from Chelmarsh. The earliest report from the old Chelmarsh site was 1973, following the reservoir opening in 1966. There were up to 18 nests there before felling removed the colony after the 1978 season, but a new heronry was established, where 14 nests were counted in 1984, though this may not have been the first year of occupation.

There was a heronry at Stead Vallets in the 1930s, but felling eliminated the site, and Peplow was apparently occupied only for two years 1959–60, but both were reoccupied prior to the end of the first Atlas period in 1990, and are therefore listed in the second table.

Near Knighton, a heronry moved 10km downstream from a site on the Teme, to a new site on the Welsh side of the river, in the 1940s. This site was abandoned in the late 1960s, as a result of persecution by shooting and throwing stones, and a new heronry was established a kilometre away in a more secluded and private place on our side of the river (local landowners *pers. comm.*) This may have been the site 'near Knighton' where three pairs nested in 1975, but it was reported

as a 'newly discovered heronry in the Teme Valley' in 1987, with four nests, and the same number in 1988 and 1996.

The *Atlas* (1992) mapped confirmed breeding in 22 tetrads. Three of these were adjacent to known heronries, but were presumably not records of different colonies since, in at least one case, the record was of recently fledged young. At two other sites (SO48D Horderley and SO78L Highley), the record was of at least one found nest, but no other information is available. In another instance (SO67T Cleobury Mortimer/Neen Savage), the breeding evidence is not known, so it may or may not have been a found nest. Hopton Heath (SO37Y) was apparently occupied in only one year, 1989. Including this, it appears that in 1985–90 there were 16 heronries, and 2–3 other tetrads with nesting Herons that were probably isolated individuals. The species account noted that 'tree felling has caused the loss of several sites in recent years, and the new ring-road at Shrewsbury has narrowly missed taking out the only established heronry near the town', adding prophetically, 'though the disturbance may still cause its desertion'. The last record of this heronry being occupied was 1990, down to three nests from 14 in 1985.

Between 1990 and 2008, apart from this latter heronry, and Hopton Heath, heronries were also lost at Neenton (SO68N), probably relocated from Aston Eyre, near Bridgnorth (four nests in 1987), and at Hollybush Coppice, Bridgnorth (SO79H), where there were three nests in 1989 and eight in 1991. Considerable efforts were made to locate heronries in the vicinity of each of these lost sites during the recent Atlas period, but no evidence was found to suggest their continued existence.

In 2008–13, confirmed breeding was found in 15 tetrads, shaded red on the map. Two sites, near Beckbury (SJ70Q) and Sambrook (SJ72C), each had a single nest for only one year. There was a heronry near Munslow (SO58I) from 1985 or earlier, but it was last occupied in 2008. The other 11 active heronries were occupied for the whole period 2008–13 (the one near Chelmarsh is shown in two tetrads as it straddles a tetrad boundary).

Apart from heronries, all other 2008–13 observations, whatever the breeding code submitted on the record, have been mapped as pink shaded tetrads to illustrate the widespread foraging range. Any seen in flight, or feeding, are not necessarily indicative of breeding in the vicinity, as they are often encountered at the edge of rivers, streams and lakes, upland pools and garden ponds, or on wet ground. Dry arable land, and upland, tend to be avoided.

There is another heronry, at Chetwynd (SJ72K), near Newport, first occupied in 2010. It is in the County but not in an Atlas tetrad, and is therefore not shown on the map. Heronries at Aqualate Mere in Staffordshire, Combermere in Cheshire and Adley Moor in Herefordshire, all close to the County boundary, will also have contributed records to the foraging range shown.

Recording for the two atlases was not directly comparable. In 1985–90 there was no 'in flight' category, so such observations may have been shown as 'possible breeding', and two together may have been recorded as a breeding pair. Sometimes fledged young were recorded, incorrectly, as having been raised in the square in which they were seen, although they soon feed at considerable distances away from their natal heronry. Such observations may have been shown as evidence of probable or confirmed breeding on the *Atlas* (1992) map, but 'probable breeding' records and those of 'fledged

Heronries lost before 1985

Tetrad	Name (Site Near)	First Year	Last Year
SO38H	Acton Pool	1872	1928
SJ71D	Hinks Wood, Kynnersley	1914	1945
SO47Y	Oakly Park, Bromfield	1928	1929
SJ51H	Sundorne, near Uffington	1928	1941
SJ61M	Eyton upon the Weald Moors	1935	1962
SO37L	Bucknell	1937	1954
SJ52B	Sansaw	1939	1939
SJ51K	Attingham Park	1950	1957
SJ22N	Trefonen	1951	1951
SO28W	Bicton	1954	1956
SO38M	Walcot Park, Lydbury North	1957	1964
SO67X	Morville	1969	1969
SO58C	Diddlebury	1975	1976
SO78P	Dudmaston	1982	1983
SO60M	Buildwas	1980	1980

young' away from known heronries were queried as part of the validation process for the recent Atlas, and most were subsequently downgraded. Immature yearlings also confuse the mapping of breeding distribution, as they do not breed until they are two years old. Because different interpretations were made of similar observations in the two Atlas periods, the figures in the tetrad occupancy table are not directly comparable, and a breeding distribution change map would be misleading.

The 17 known heronries occupied since 1985 are shown in the second table, with the date of the first record, and counts made as part of the BTO national surveys (Wright SBR 1985 and Dawes SBR 2003). Some sites were not counted in each survey, although they were almost certainly active.

Several sites have been counted each year for the BTO Heronries Census, there were 'colony counts' at some during the recent Atlas period, and from 2011 the Shropshire Heronries Project has made a systematic attempt to count every known heronry, although a few have been missed in some years. Apart from a few well-known ones, heronries are not named, to avoid disturbance and because they are on private land.

It can be seen that the number of pairs at each colony fluctuates, and the size of the colony is influenced by the quality of foraging areas within around 25km (Voisin 1991).

It is often very difficult to establish the number of 'apparently occupied nests' counted for the BTO surveys, but comparison of the results of the standard procedure with that of intensive monitoring suggests that the Heronries Census should be finding at least 90% of the active nests at each site (Marchant 2005).

Heronries and March–July Range (2008–13)

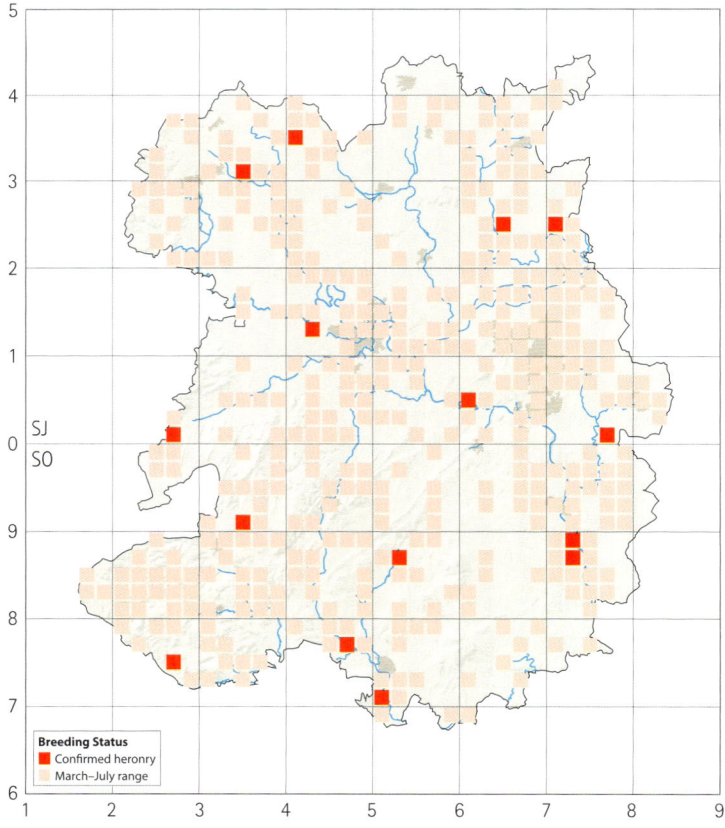

Breeding Status
- ■ Confirmed heronry
- March–July range

Breeding Codes recorded during the *Atlas* (1992) are taken from the Mastercards maintained by the Area Co-ordinators, where they are still available. If they are not, sites with confirmed breeding on the Atlas map are shown with the breeding code 'CB (?)'.

In 1990 the population was estimated to be 100–120 pairs. The 2003 survey counted 114, and adding an estimate for the three active heronries not counted suggests around 130 pairs at that time. The average in 2012–13 for the 11 active heronries was about 109, including Chetwynd. The population estimated from TTV counts is around 120 pairs, but that appears a bit high.

Heronries are usually in isolated places, and can be surprisingly difficult to find, so some may still have been overlooked. Others may have been active for many years before they were 'discovered', and many will have been occupied before the date of the first record in the table.

Since the first BTO census in 1928, the national population reached its highest level in 2001. This was attributed to improvements in water quality, a reduction in persecution and an increase in suitable nesting and feeding sites as gravel pits have been flooded and restored. A 'strong downturn' evident since then is, as yet, unexplained, though recent cold winter weather and spring gales appear to have accelerated the decline. It is almost identical to the local decline from around 130 breeding pairs to 109 in the same period.

Herons nest early. It is not known whether individuals return to the same nest year after year. However, at Ellesmere a bird with a distinguishing feature on its leg returned to the same nest in two consecutive years.

Arrival back at nest sites may be as early as late December, ready to start laying in February. This makes them particularly vulnerable to bad weather in late winter and early spring. For example, in 1998 at least 14 young died at Ellesmere after a fall of snow on 11–12 April, and the following year five nests were blown down at Halston, with young perishing as a result. In March 2013, which was the coldest since 1962, heavy snowfall and freezing cold began on 23 March, only starting to ease on 10 April. At the Teme heronry, prior to the onset of bad weather, five pairs were standing on nests and passing sticks on 8 March; on 7 April, five previously occupied nests were deserted, but heavily splashed white, suggesting breeding had been well underway before the site was abandoned; 10 April, the landowner saw 12 birds return, all together, flying in from the west (the direction of the nearest coast, 90km away); 27 April, five occupied, birds sitting; 20 May, seven occupied nests. Although it was not possible to do accurate counts of chicks, they were apparently present in ones and twos rather than the more usual twos and threes.

In the same year, at Ellesmere, the unique camera monitoring

Occupied Tetrads

Atlas period (breeding)	1985–90		2008–13		Change	
	Number	%	Number	%	Number	%
Confirmed	22	3	15	2	-7	-32
Probable	23	3	10	1	-13	-57
Possible	206	24	140	16	-66	-32
TOTAL	251	29	165	19	-86	-34
Tetrads with Winter Records (2007–13): 471 (54%)						

Counts at Heronries Active 1985–2014

Tetrad	Name (site near)	First record	Last record	Atlas (1992) Breeding Code	Heron-ries	BTO Surveys 1985	2003	Recent Atlas Period 2008	2009	2010	2011	2012	2013	2014	Average 2008–13	Maximum 2008–13
SJ20Q	Chirbury	1975	Still active	CB (?)	1	5			11			7	4	2	7.3	11
SJ33K	Halston Hall	1820s	Still active	CB (?)	1	17	17	25	16	14	12	11	8	10	14.3	25
SJ41G	Bicton Heath	2001	Still active	NY	1		10				3	7	6		5.3	7
SJ43C	The Mere (Ellesmere)	1872	Still active	CB (?)	1	8	21	14	14	12	12	11	15	13	13.0	15
SJ51F	Atcham	1975	1990	NY	1	14	Abandoned in 1991									
SJ60C	Leighton	1990	Still active	FL	1							13	13		13.0	13
SJ62M	Peplow	1959	Still active	CB (?)	1		18					8			8.0	8
SJ72K	Chetwynd	2010	Still active				(new site in 2010)			2	3	5	7	5	4.3	7
SO27S	Knighton (Teme Valley)	1988	Still active	NY	1		4				2	6	7	8	5.0	7
SO37Y	Hopton Heath	1989	1989	NY	1											
SO39K	Lydham	1958	Still active	CB (?)	1		9	8		10	8	14	11	12	10.2	14
SO47T	Bromfield	2003	Still active	CB (?)	1		8				9	20	17	12	15.3	20
SO57A	Ashford Hall	2003	Still active	CB (?)	1		10				2	5	5	5	4.0	5
SO58I	Munslow	1985	2008	NY	1	4	2		Abandoned in 2009							
SO68N	Neenton	1968	1990	FL	1	5										
SO78J	Chelmarsh	1984	Still active	ON	1	14	15			3	2	4	6		3.8	6
SO79H	Bridgnorth	1988	1996	ON	1											
Total Count					16	67	114	47	41	41	53	111	99	67	103.5	138.0

provided a detailed record of the devastating impact of the weather on two nests. Four eggs were laid in each, at 1–2 day intervals, between 13–26 February, hatching before the start of the snow and freezing weather. On 26 March, only two chicks remained in the first nest. On 7 April, the nest was unattended, and then an adult visited, threw out one chick and killed the other. The nest was then reoccupied, but a smaller clutch of three eggs was laid; they hatched but the nest was abandoned the following day (23 May). The second nest was abandoned, and the chicks died on 23–24 March. The nest was reoccupied, with again a smaller clutch of three laid in mid-April, but they were predated on 30 April.

Mobbing of Herons by other species is common, and SBRs include records of one forced down into the water by Black-headed Gulls at Chelmarsh (1977), and being mobbed by Buzzard (2000) and corvids (2001). Herons choose nest sites that are well protected from land and water based predators, but they are vulnerable to predation by other birds. On Good Friday 2004 Ravens predated a nest near the Teme: 'One was on the nest, beak up in defence. The other was flying close by. The Ravens distracted it, moving it away. One of the Ravens then returned and bombed the sitting bird, driving it off. Both Ravens then raided the nest' (landowner *pers. comm.*). At the same site in a later year, Ravens were driven off by a Red Kite circling the Heron tree and its own, immediately neighbouring, nest tree. The following year a pair of Red Kites nested at the centre of the heronry – in the

very middle of the single Corsican pine hosting the whole colony. Carrion Crows were seen predating eggs there in 2014 (landowner *pers. comm.*).

They often feed at night, especially when feeding chicks. Their diet is very varied. There are feeding records of a dead Heron at Cressage with a partially digested rat (perhaps poisoned, 1971) and catching rats (1981). In 1990 a mole was taken at Venus Pool. In hard winter weather in 1979 one visited a farmyard manure heap for three days, while a garden fish pond was raided twice daily for six days. They can tackle fish of considerable size – in 1928, a 24lb salmon in the River Teme was killed by being struck through the eye. At Walcot in 1981 one 'dived from 20 feet' like a Gannet, emerged with a fish and flew to the bank to eat it'. Remains of native Crayfish were found under a nest tree near the Teme in 2014. These, being hard and difficult to swallow, were unlikely to have been fed to youngsters.

Garden ponds offer rewarding hunting. In addition to the example above, a pool at Homer was cleared in 1984, and in 1991 a recently stocked garden pond in Great Bolus yielded seven Koi Carp. Ponds at gardens in Kemberton and Leighton were visited in 1999. This behaviour causes persecution by humans: in April 2012, one was found, shot, at the foot of its nest tree, with a goldfish in its beak. It died of its wounds, despite veterinary care at Cuan Wildlife Hospital.

Small birds are often eaten. In 1932 a young Peewit (Lapwing) was killed and carried off near Shrewsbury. In 1989, 40 Mallard

Jim Almond, Venus Pool, 8 September 2012

ducklings 'were gobbled up just like frogs'. Ducklings were also preyed on at Hanwood (1991). Venus Pool provided 'an almost fully grown juvenile Moorhen' in 2001, and 'a full size juvenile Little Grebe' in 2008. Two Herons were observed predating the Sand Martin colony at Wood Lane in 2000.

Landowners are generally very protective of 'their' heronries and mindful of the threats of public access. Disturbance because of increasing recreational use was almost certainly responsible for the loss of the colony near Munslow. However, living close to humans does not pose a threat if the site is secluded and people and their animals respect the immediate boundary of the nesting area. Sometimes they choose to nest surprisingly close to human habitation. Colonial life continues quite normally within 30m of a working farmyard at one site. The only disturbance here is to the farmers being woken up at breaking dawn by clacking nestlings as parents return with food.

The BTO *Field Guide to Monitoring Nests* refers to the nest site as 'Typically in top of tall trees (up to 25m), especially broadleaf but often conifer; also low tree or bush on lake' (Ferguson-Lees *et al.* 2011). The *Atlas* (1992) stated that it 'usually nests in small colonies in the tops of mature trees close to the feeding areas, although occasional nests have been recorded lower down in bushes'.

Here, the Heronries Project has found an increasing trend to lower nests. In 2012, at six colonies totalling 56 occupied nests, 36 were in broadleaved deciduous trees. Of these 36, two were at approx. 20m, three nests were between approx. 5–10m, while 31 were at approx. 1–5m. The 31 lowest nests were all over water or very marshy ground, in shrubby Willow. Twenty of these nests were in evergreens, but not all were in conifers as one was in Portuguese Laurel, 3m over water. Of the 19 in conifers, nine were at approx. 5–10m, four at 12–15m and six at 20–25m. Six of the highest nests were abandoned by 2014. Where suitable vegetation is available it appears that more nests are being built lower down, at 1–5m. These nests appear to offer better shelter from strong winds. It is possible that nesting high has been of necessity, in the absence of safe sites lower down. Sometimes, wind thrown trees may provide new nesting opportunities. At one site Poplar trees within the heronry were ignored until they were blown over. The shift in the angles of the branches made them suitable for

building a nest on. Two new nests, approx. 4m above the water, were built within weeks of the trees falling.

The Heronries Project has noted the species of nest tree or shrub, the nest height, position, distance from the water, and the surrounding habitat and land use, since 2011, and this monitoring is continuing alongside the BTO census. A colour-ringing scheme is due to start, to learn more about recruitment into colonies, site and pair fidelity, behaviour, movements outside the breeding season and longevity. The research results will be written up for SBR at the appropriate time.

Herons are more widespread in the winter, when they are solitary and more conspicuous as solitary birds, and more likely to be seen in fields. There were winter sightings in almost three times as many tetrads as in the breeding season. Dispersing juveniles supplement the breeding population and visitors come in from the continent, particularly in hard winters such as 2009–10 and 2010–11.

Few are ringed here. There have been 13 recoveries, eight being from this or adjacent counties. The furthest travelled was a nestling ringed near Bishops Castle in May 1979. It was found, long dead, in the East Riding of Yorkshire, 216km north-east, 0y 11m 20d later.

Another nestling, ringed at the same heronry on the same date, was found dead in March 1985, only 3km from the ringing site, 5y 9m 20d later. This is about the typical lifespan for those that reach breeding age. The oldest found here was ringed as a nestling in May 1968 at Gailey Reservoir, Cannock, and found freshly dead in May 1974 at Nobold, 47km distant, 6y 0m 8d later. These are, however, well short of the national longevity record set by one ringed in the Netherlands, found dead in Greater London in 2000, of 23y 9m 2d.

Including the four ringed and recovered here, there have been 30

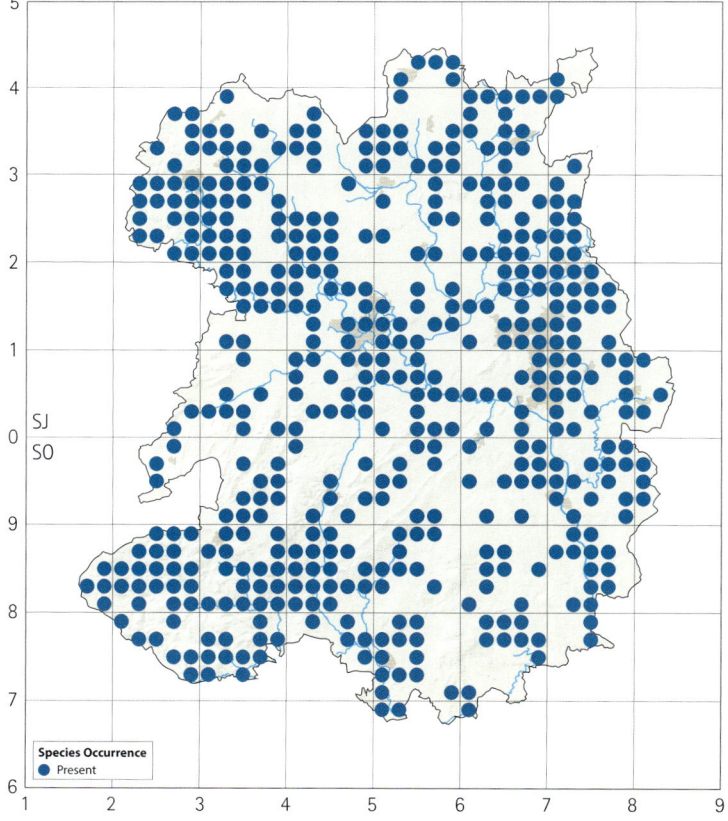

Winter Distribution (2007–13)

recoveries, including 14 ringed in Staffordshire and one each from Cheshire and Herefordshire. Seven have come from further afield within Britain, two each from Cambridgeshire and Lincolnshire, and one each from Buckinghamshire, Essex and Oxfordshire. Three have come from the continent, one each from the Netherlands, Norway and Denmark. All three were ringed as nestlings and found within 7–9 months, in the winters of 1975, 1980 and 1984 respectively.

Periods of very cold weather will always be compromising, so Herons could benefit if the trend to milder winters continues. More frequent strong winds during the breeding season are a challenge, especially to birds nesting high up. It may be that they 'nest typically in the tops of high trees', because there is so little suitable vegetation lower down. A less manicured approach to vegetation management, and possibly some benign neglect, could offer more potential nesting habitat. Hostility from fishing interests will continue, and increased public access for recreation can cause unintentional disturbance, and is potentially harmful. It is important to encourage the landowners who value and protect the privacy of a heronry on their land.

Josie Owen

John Robinson, Cranmere Bog, May 2007

Purple Heron
Ardea purpurea
Vagrant

Jim Almond, Rednal, 10 September 2009

Found from Western Europe through to South-East Asia and Africa, the closest, regular breeding populations are located in the Netherlands and France, although successful breeding occurred in Kent in 2010. It is largely a summer visitor to Europe, with the majority of this population wintering in Africa south of the Sahara. As a result, any reaching Britain are usually either overshooting returning migrants in the spring or dispersing juveniles in the autumn, with most being found south of a line between the Severn Estuary and the Wash. Whilst the five birds seen here are obviously further off-course than usual, they do follow the pattern of being either spring overshoots or autumnal juveniles:

Prees/Lower Heath	30 October–12 November 1995
Venus Pool/Cound Fishery	24 September 2003
Priorslee Lake	27 April 2004
Pen-yr-Estyn, Rednal	9 August–11 September 2009
Whixall	13 August 2015

The long-staying juvenile in 1995, the first for the County, was initially found in the Prees area, where it stayed until 5 November before being rediscovered at Lower Heath. The individual at Venus Pool, also a juvenile, was thought by some observers to be in poor health and, while it provided excellent views for those present, it eventually disappeared into the reedbed in the late evening and was not seen again (Holmes 2003). The following year's adult was initially seen in flight before being disturbed subsequently from the lake's feeder stream: although it did not appear to go far, it could not be relocated. The juvenile in 2009, although first located in early August, was not seen regularly until 5 September. Finally, the individual at Whixall, also a juvenile, was only seen briefly as it visited a small pond in the observer's garden.

Graham Walker

Great White Egret (Great Egret)
Ardea alba
Rare visitor

Jim Almond, The Mere, 30 August 2008

With a worldwide distribution, its spread north and west across Europe since the early 1990s has resulted in a remarkable increase in British records culminating in breeding in Somerset in 2012 (Holling *et al.* 2014a).

Up until the end of 2015 there had been over 20 records, all of singles, since the first in 1995 (Rogers *et al.* 1997). The majority have occurred from 2012 onwards, since when it has been annual; this is perhaps indicative that more sightings can be expected in the future, particularly in and around the numerous wetlands in the north and in the widespread river valleys.

As is often the case, it can be difficult to determine whether one individual has been responsible for more than one sighting. In 2012, for example, while all were treated as separate, one could have visited Chelmarsh, Wood Lane and Venus Pool before returning to Chelmarsh, while the following year another may have been at Priorslee Lake and then Wall Farm. This being the case, as few as 12 birds may have resulted in the 20 or so sightings, or even fewer if the 2012 bird returned for another tour in 2013!

A juvenile present at Ellesmere for 18 days from 27 August 2008 had been ringed as a nestling at Lac de Grand-Lieu, near Nantes, France some 3m 15d previously, a distance of 653km.

Graham Walker

Little Egret
Egretta garzetta
Scarce passage migrant and winter visitor

Gareth Thomas, Temeside, Ludlow, 20 October 2013

Although Little Egret is found in every continent, it was a rare vagrant in Britain even into the 1990s, with records in only one 10km square in the 1981–84 BTO Winter Atlas, and three 10km squares in the 1988–91 national BTO Breeding Atlas, across the country as a whole. It was not mentioned at all in the national BTO Atlas (1976). However, by 2013 its status had changed: 'The colonisation and range expansion of Little Egret represent one of the most phenomenal shifts in abundance and distribution of any bird in Britain and Ireland over the past 20 years'. Coinciding with population increases throughout France and Spain in the 1980s and 1990s, these shifts 'are likely to have been facilitated by subtle climatic changes' (national BTO Atlas 2013). The first successful breeding was found in Dorset as recently as 1996.

There were no mentions of the species at all in the works of Beckwith or Forrest, or in the *Handlist* (1964). The first local record was from Venus Pool on 30 May 1992. An estimated 30 individuals visited between 1992 and 2003, all in the north. Records were largely of single birds seen in the post-breeding dispersal period, although two were seen at Wall Farm in 2002 and two at Venus Pool in 2003 in mid-May and in April respectively. The largest single site count was of three at Wall Farm in September 2003.

This was followed by an 'unprecedented year' in 2004, with records from 14 widespread sites, but only one in the south, and counts of five and six at Venus Pool in September. Sightings increased and it 'occurs so often, so widely ... that summarising a year's records involves making some informed guesses about how many birds might be involved' (SBR 2006). Records fell from 56 in 2006 to 42 in 2007, but there were more sightings in the south.

As recently as 2005 there was only one winter record, but in the recent winter Atlas period it was found in 44 tetrads, mainly in the river valleys, including nine in the south. The concentration of occupied tetrads in the north-west, where small streams are often utilised, represents dispersal from breeding colonies in Cheshire near the Dee and Mersey estuaries. Only five tetrads had Atlas counts of more than one, and only one had a count as large as four, so this is still a scarce winter visitor.

Breeding season fieldwork in the recent Atlas period produced 33 records from 16 tetrads, but there were none between 12 April and 18 May, and only two more before 26 June, with no evidence of breeding. One third came from Venus Pool and Chelmarsh, with several more coming from the immediate vicinity of these two sites. Records have continued to increase; there were 100 in 2013, all from the Severn Valley, but all those in the 'breeding season' were of sightings in the second half of July, indicating the start of post-breeding dispersal, which continued with further sightings in August and September at Bridgnorth, along the River Severn between Buildwas and Leighton, and Severn Valley Country Park, Chelmarsh, Knighton Reservoir, the Teme at Ludlow, Venus Pool and Wood Lane.

Little Egrets are very mobile when dispersing from natal sites, and distribution in winter is much more widespread than the breeding range. The one in the photo was colour-ringed at North Cotes, Lincolnshire, as a nestling, on 10 June 2013. In August it was at Ironbridge, then at Venus Pool and Bridgewater (Somerset), before being photographed at Ludford Bridge (Ludlow) on 20 October. Forty-three days later, it was in Surrey. It visited Wokingham (Berkshire) in March 2014 and on 30 December 2014, it was back at the Surrey site. In March 2016 it was seen at the Surrey site again, then at the London Wetlands Centre. Two other Little Egrets ringed in Lincolnshire have also been found here.

There is no apparent reason why the rapid expansion of winter range over the past 20 years should not continue, so sightings should

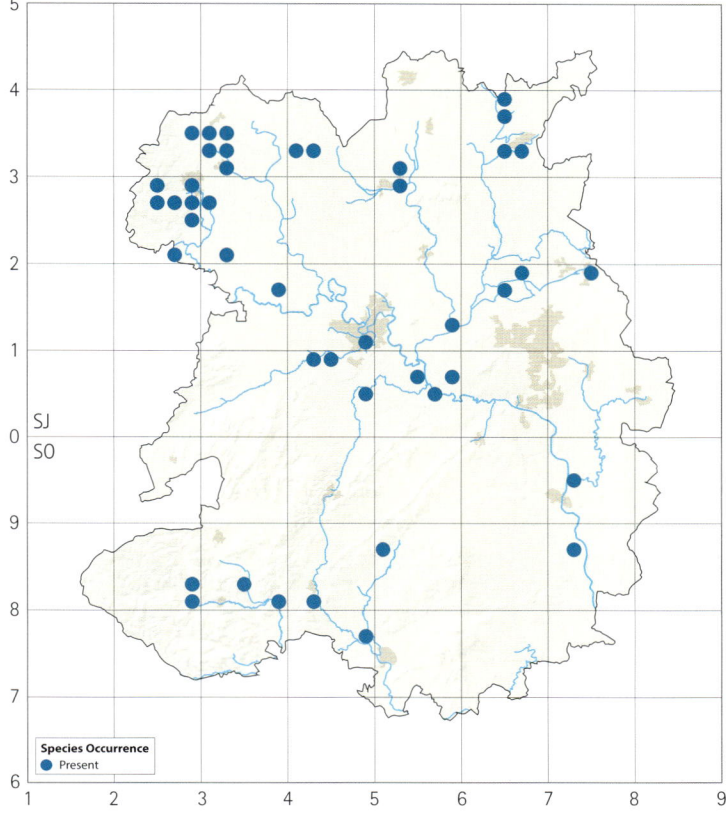

Winter Distribution (2007–13)

Occupied Tetrads

Tetrads with Winter Records (2007–13): 44 (5%)

become more frequent in future. The expansion of the breeding range is progressing more slowly now, and there has been no evidence of any breeding attempt here. As most breeding colonies elsewhere are in heronries, the presence of two at a heronry in the south, in 2016 and 2017, may herald the discovery of their first breeding here in the next few years.

Josie Owen

Magnificent Frigatebird
Fregata magnificens
Vagrant

Normally associated with tropical waters off both the Atlantic and Pacific coasts of the Americas, this exotic seabird is an exceptionally rare visitor to the seas of Western Europe, never mind our inland County! The only record is of an adult male found in a distressed state in a field and taken into care (Holmes 2005b):

Prees Heath, Whitchurch   7 November 2005

It died two days later and was donated to the Natural History Museum, Tring.

Its biometrics suggested that it was from the Caribbean population

and it occurred after a category five hurricane, Hurricane Wilma, had rapidly tracked across the Atlantic in a north-easterly direction. On 6 November, unidentified Frigatebirds had been reported off Porthgwarra, Cornwall and Flat Holm in the Bristol Channel. While it is impossible to determine whether either or both these sightings relate to the individual found at Prees Heath, it is a distinct possibility that the Frigatebird seen off Flat Holm continued up the Bristol Channel and then the Severn Valley to Shropshire.

Previously, a bird found on Tiree in the Inner Hebrides on 10 July 1953 was originally thought to be this species, but was re-identified as a juvenile Ascension Frigatebird *F. aquila* by its louse fauna (Walbridge *et al.* 2003). Therefore, after being accepted by the British Ornithologist's Union Records Committee and British Birds Rarities Committee, this became a new addition to the British list (Bradbury *et al.* 2008). An adult female Magnificent Frigatebird had been found on the Isle of Man on 22 December 1998, but birds from this island do not feature on the British List.

This is one of the five species to make their first British appearance in Shropshire (see p. 2).

Graham Walker

Gannet (Northern Gannet)
Morus bassanus
Vagrant

This pelagic bird breeds mainly on islands around the British coast, and elsewhere in the north Atlantic, and is very rare this far inland. Often referred to historically as a 'Solan Goose', as with other rare birds, the records from the late nineteenth and early twentieth centuries are predominantly of birds which were deliberately killed for collections. Two specimens were held by Rocke at Clungunford – an adult which had been killed near Market Drayton, and an immature, obtained near Shrewsbury. Forrest (1899) lists three further records, the last of 'three at Stapleton on April 24th 1896, one of which was killed'.

There were six records from the first half of the twentieth century, the last being of two immatures fighting on 25 October 1925, which fell to the ground at Weeping Cross; they were captured alive, and released two days later.

There are nine modern records, all of a single bird:

Sleap Airfield	14 August 1959, adult
Melverley	early February 1974
Much Wenlock	8 September 1983, immature
Broome	1 November 2002, adult
Sherrifhales	23 September 2004, adult
Wyre Forrest	1 July 2008, immature
Upton Magna	23 September 2008, immature
Horton	26 September 2010, immature
Chelmarsh Reservoir	3–6 June 2012, immature

These records have occurred in seven months of the year, but only September, with four records, has more than one. All except two (The Rea, Upton Magna, 2008 and Horton, 2010) can be directly attributed to stormy conditions, according to the Met Office records for the time. For example, the report for August 1959, which led to the grounding of the Gannet found at Sleap Airfield, documents thunder storms around this time, described as 'exceptionally severe on the 10th in south-west England ... and some "very rare" falls were recorded'.

Three of these individuals, caught and often exhausted after being blown inland following gales, experienced compassionate care. The one found at Sleap had been ringed as an adult on Grassholm on 26 May 1959, and was found 230km distant, only 80 days later. It was kept alive at Preston Montford Field Centre for five days then taken to Bardsey Island on the Welsh coast where it died a few days later before it could be released. The adult found near Manor Farm, Sherriffhales in 2004 was taken into care at Cuan House Wildlife Rescue in Much Wenlock. It was subsequently released at Llandudno on 26 September. In 2008, an immature bird was found in the Wyre Forest near the Worcestershire border. It was taken to the Bishopwood

Yvonne Chadwick, Chelmarsh Reservoir, 4 June 2012

Rescue Centre near Stourport and released one month later from the north Devon coast.

Of the remainder, the one at Melverley in 1974 was found dead, and the other four flew off without help. The outcome was not recorded in the case of the grounded juvenile at Much Wenlock. Apart from those taken into care, only the last one at Chelmarsh was seen on more than one day. This was the only one seen during the recent Atlas breeding season. An immature, it was only the sixth occurrence this century, providing local birders with the opportunity to add this to their county lists. According to Met Office reports, the weather during late May was fairly settled, but foggy. Towards the end of May, storms moved in from the south-west of the country. This was the precursor to what was to become the wettest June in England and Wales since 1860, where unusually low pressure dominated. This may account for it being so far away from its natural habitat.

Linda Munday

Shag (European Shag)

Phalacrocorax aristotelis

Vagrant

Occasionally referred to as a 'Green Cormorant' in the historical records, this seabird breeds mainly on rocky coasts in western and southern Europe, and North Africa. It does not appear inland as often as its more common cousin, the Cormorant. In 1866, Rocke recorded an immature Shag that was captured at Longville in 1860, where it had 'joined some ducks on a small pond in a garden'. It was brought to him alive, although, as is often the case with these pelagic birds that only appear inland after severe gales and are therefore exhausted, it did not survive for long. Forrest (1899) lists only four occurrences, at Ellesmere, Hawkstone, Longville and Polemere, all without dates.

As with other rare birds, those found during the late nineteenth and early twentieth centuries were often specifically targeted and shot for collections. There were five records between 1900 and 1950: 1903 (shot), 1904 (condition not recorded), 1910 (found injured, outcome not recorded), 1912 (shot) and 1942 (killed by a car).

There have been seven modern records, the most recent in December 1993:

English Bridge, Shrewsbury	mid-March–1 May 1969
Chelmarsh Reservoir	27 April 1980
The Mere (Ellesmere)	26–27 October 1982
Chelmarsh Reservoir	18 September 1989
River Teme (Ludlow)	7 December 1990
Priorslee Lake	2 September 1992
Trench Pool	25–27 December 1993

An immature was present near the English Bridge in Shrewsbury from mid-March to 1 May 1969, a record that is unusual for the longevity of the stay and also for the time of year – more typically, they have been recorded during the autumn and early winter months, usually on one day only. All were individuals, except at Chelmarsh in 1989 and Priorslee Lake in 1992, when there were two. The 1989 (Chelmarsh) and 1990 (Ludlow, where the bird was found dead) records are directly attributable to gales, and in the case of the latter, there was also heavy snowfall. All seven were alive apart from the one in 1990 at Ludlow.

None have been ringed here, but there has been a recovery of a nestling ringed on the Calf of Man in June 1942, and killed by a car in Habberley, six months later and 204km distant.

Linda Munday

Cormorant (Great Cormorant)

Phalacrocorax carbo

Uncommon winter visitor, scarce non-breeding resident

Early records of Cormorant note small, temporary colonies, especially in winter. Between 1820 and 1839 they were found on the Severn near Montford Bridge, then, by the 1930s, at Ellesmere and White Mere. Forrest mentions Cormorants 'obtained' at 'Clungunford, Atcham and other places', and one shot at Shrewsbury in 1897. According to the *Handlist* (1964) Cormorant was 'prior to 1920 ... an uncommon visitor'; by the time of publication, it was present in 'small numbers ... throughout the year on the Dee, Severn, Vyrnwy, and the Ellesmere group of meres'. Listed as a 'miscellaneous species' in the *Atlas* (1992), it appeared as 'winter visitor, small numbers in summer, mainly immatures'. Since 1999, Cormorant has been recorded in all months every year, though rarely at the same site in every month. The highest count to date, 132, was obtained at the Mere (Ellesmere) in January 2004.

In the 1981–84 national BTO Winter Atlas, Cormorant was recorded in only 16 of the County's 33 10km squares; during winter fieldwork for the recent Atlas, it was found in all but one. With migrants from northern European populations swelling its numbers, it was recorded on all major rivers and most sizeable water bodies across 217 tetrads (25%), and in flight over more. Density is greatest in the traditional strongholds around Ellesmere and in the Severn Valley, but other hotspots include pools near Oswestry, Baschurch, Shavington, Sambrook, and, in the south-west, Marton Pool (Chirbury), Walcot

Jim Almond, Venus Pool, 24 August 2008

Winter Relative Abundance (2007–13)

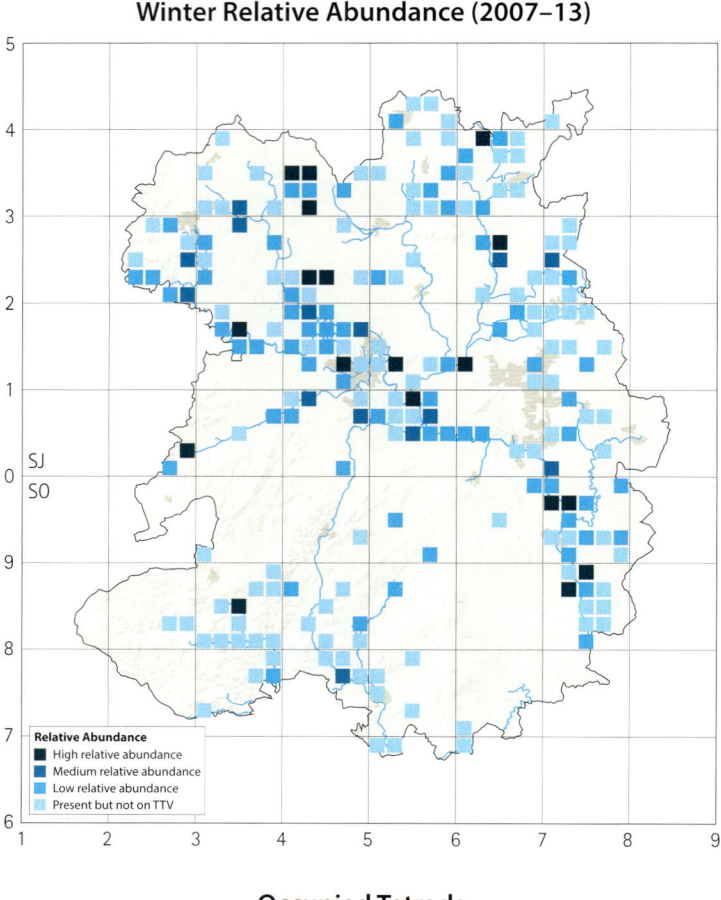

Relative Abundance
■ High relative abundance
■ Medium relative abundance
■ Low relative abundance
■ Present but not on TTV

Occupied Tetrads

Tetrads with Winter Records (2007–13):	217	(25%)

Lake and Stead Vallets. The highest count during the recent winter Atlas period was of over 90 at Ellesmere in January 2012, with over 50 also recorded at Shavington (74 in November 2011), Crosemere (67 in January 2012), and Newton Mere (58 in January 2009). The

highest Atlas count was surpassed in December 2014 by 118 at the same site, though this was still below the 2004 maximum.

Until 1980, breeding was almost completely confined to coastal areas of the country; by 2012 inland sites had been colonised, with breeding recorded at 89 such locations. Numbers also increased, probably as a result of enhanced protection under the Wildlife and Countryside Act 1981. They peaked in 1995, and have since declined slightly, with shooting, both licensed and unlicensed, thought to be a contributory factor. The conflict with fishing interests has led to a number of local applications for licences to control Cormorant in recent years, to unknown effect.

During breeding season Atlas fieldwork, non-breeding birds were found in 100 tetrads, although many April records are likely to be late-departing winter visitors. Around 70 tetrads in which Cormorant was recorded in May, June and July represent the likely range of resident immature birds, or foraging ones travelling from the breeding colony at Aqualate Mere in Staffordshire, near Newport. Breeding has also been confirmed in Worcestershire, very close to the border. Distribution follows a similar pattern to winter, with records heavily weighted towards the north-east, though concentrated in far fewer locations.

Although no unambiguous breeding evidence emerged during fieldwork, it is considered only a matter of time before Cormorant breeds here. Tree-fringed wetlands in lowland areas are favoured for nesting, so any such habitats that regularly attract Cormorant in the breeding season are strong candidates; sites with high winter densities are also possibilities.

No Cormorant has been ringed, but all 15 British-ringed birds recovered here, including five from Anglesey and three from Essex, plus one from Wexford, had fledged at sites on or near the coast, evidence of increasing movement inland. Recoveries of four more ringed as nestlings in Denmark, having travelled between 898 and 994km, confirm the continental origin of part of the winter population. Of seven birds recovered dead, and where the cause was known, four had been shot.

Michelle Frater

Osprey (Western Osprey)
Pandion haliaetus
Rare Passage Migrant

Ospreys are the largest raptors that occur here regularly. They have an almost worldwide distribution, but the British breeding population winters mainly in west Africa. They usually breed for the first time at three to five years old, but young are usually at least two years old before they return to prospect future breeding areas.

Local observations therefore comprise breeding adults, which pass through very quickly on their way north in late March or early April, and immatures, at least two years old, which pass through more leisurely, and occasionally summer here. Return migration of juveniles and breeding adults occurs mainly in late August and September, although migrants from Scandinavia may pass through later.

It was persecuted to extinction in England, with the last nest in Somerset in 1847, and the small remnant Scottish population was

subsequently eliminated, with the last breeding attempt in 1916, resulting in the scarcity of local records up until 1889, and their virtual absence after that. Described as a rare and accidental visitor to lakes and large pools, it usually occurs in the autumn (Beckwith 1887), and Forrest (1899) lists only eight sites where it was 'obtained' between 1833 and 1889. There were only two or three records in the first half of the twentieth century. These exceptional appearances were at Cole Mere, where one stayed from the latter half of May 1909 (it 'was seen there constantly by the keeper up to June 10th, when an otter-hunt, which lasted the greater part of the afternoon, disturbed it' and it disappeared – Forrest 1909b); one on the Wrekin on 1 October 1924; and perhaps another at Pulverbatch on 21 April 1940, which 'appeared to be this species, from (the) description'.

Seasonal Occurrence in 10-day Intervals (1992–2014)

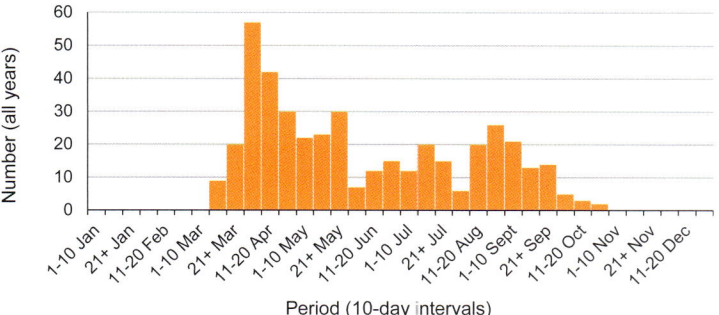

Estimated Number per Year (1950–2014)

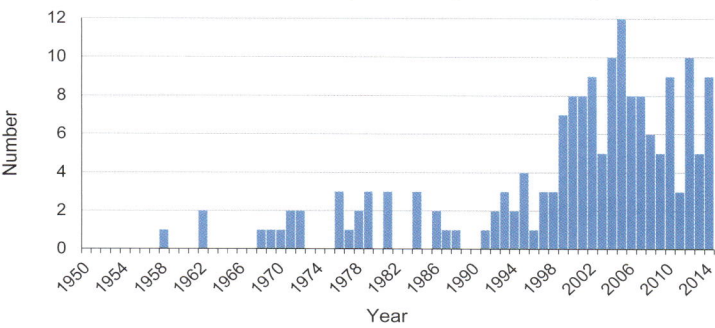

A pair returned to breed in Scotland in 1954, rising to 10 successful breeding pairs by 1973, and local sightings correspondingly increased, with the *Handlist* (1964) describing it as a 'passage migrant, rare', and listing three records – one at Willey Park 4–9 October 1958 and probably the same bird at Acton Burnell on 11 October and Betton Pool on 12 October; another at Marton Pool (Chirbury) on 9 May 1962, and one at Attingham Park on 24 October 1962.

In spite of continued persecution by egg collectors and disturbance by birdwatchers, the Scottish population continued to increase, protected by intense conservation efforts led by the RSPB, and reached over 70 nesting pairs by 1990 and over 200 in 2014. In England, pairs colonised Cumbria in 2001 and Northumberland in 2009, and there were three pairs in each county by 2014. A translocation project started at Rutland Water in 1996, resulting in the first breeding in England for 150 years in 2001, increasing to five nests there in 2014, with a total of 87 fledged young.

A new site was occupied near Welshpool in 2004, just over the border, and one young fledged, the first successful breeding in Wales. A previously unknown nest was found in the adjacent tree, suggesting breeding in 2003. However, only one adult was seen there the following year, and there have been no further breeding attempts (Holt & Williams 2008). Glaslyn was first occupied, unsuccessfully, in 2004, but successfully every year since, the Dyfi Osprey Project site was successfully occupied each year from 2011, and a third pair raised two young at an unnamed site in Powys in 2014. Several of the individuals that have bred in Wales were released or fledged at Rutland Water.

Local sightings increased slowly, reflecting the growing numbers passing through on their way to and from Scotland, but not exceeding three per year until 1993, apart from 1984 when three passed through in spring, and two in the autumn. There was a spectacular increase coinciding with the return of those released at Rutland Water from 1999 onwards, reaching 10 in 2004 and a maximum of 12 in 2005. Numbers decreased again after the end of the first phase of the translocation project, as the surviving wandering birds settled down to breed elsewhere, but fledged young from Rutland Water, and increased numbers passing through to breeding sites further north, provided continuing higher numbers than up to 1998, with a spike in 2010 following a supplementary translocation to Rutland in 2005. The chart shows the estimated number of individual Ospreys passing through each year since 1950.

Between 1992 and 2014, almost all records were of singles, but there were three together at Chelmarsh Reservoir 15–17 July 2000 and four at the same place 17–20 March 2002. Apart from these two records, there was only one more spanning more than one day, an individual at Halfway House, near Ludlow, on 15–16 March 2001. Many of the records note fishing attempts, mostly successful, while others note plumages indicating immature birds.

A large proportion of the spring migrants arrive in the Bristol Channel and proceed up the River Severn, which passes within a mile of both Venus Pool and Chelmarsh. The former accounts for over half the records since 1992 (226 out of 409), and there have been 35 from Chelmarsh. Most of the remainder are from other sites on, or close to, the River Severn, but some are from sites on the more direct northern

Tony Webb, Long Mynd, 24 September 2014

route, above the A49 road south of the Severn, and from the Meres and Wood Lane, north of the Severn.

A pair at Chelmarsh in May 2001, and another in agricultural land south of Venus Pool in April 2002, were seen courtship feeding and carrying sticks, and the 'start of a nest platform (was) evident' at the latter site on 11 April, which was added to on 14 April, but they did not remain at either site. One was seen carrying a stick again at the latter site on 22 April the following year, but it did not take it to the tree occupied in 2002. There have been no records suggesting possible breeding attempts since then.

Up until 1999, all records were in the migration months of March–May and August–October, apart from an individual at Venus Pool on 13 June 1994. The next June record was in 2000, and 29 of the 32 June records occurred between 2000 and 2004, when the released birds from Rutland Water were prospecting.

During the recent six-year Atlas period there were 65 April–July records, over half from Venus Pool, and none in the winter season.

There are usually fewer autumn migrants, as they tend to follow the coast, rather than the direct inland spring route along the Severn Valley. An exception was 2014, with one more September–October record than March–April, and only one seen on 7 July near Sutton Maddock was outside these four months.

The earliest record was on 13 March, in 2007 near Alberbury, and the latest was on 28 October, in 2000 near Market Drayton.

The three ringing recoveries are all from the Scottish population. A female nestling ringed on 9 July near Tain, north of Inverness, was found long dead at Oerley Reservoir (also known as Llanforda Reservoir or High Fawr) only three months later; another female nestling ringed on 6 July 2010 near Feshiebridge in the Cairngorms National Park was found sick near Church Stretton nearly two years later, and one of a brood of three colour-ringed in Loch Ard Forest, near Aberfoyle, Stirlingshire on 27 June 2014, was photographed only three months later, flying over Long Mynd on 24 September, 2014.

The Severn Valley near the Welsh border, and between Shrewsbury and Cressage, which includes several pools including Venus Pool, and Chelmarsh reservoir, all provide suitable breeding habitat, and the meres, with few records to date, also do so. The national population is expanding its range slowly, and if it continues to grow, then breeding here is eventually likely.

Leo Smith

Honey-buzzard (European Honey Buzzard)

Pernis apivorus

Very rare passage migrant, has bred

Britain is at the western edge of the Eurasian breeding range of this summer visitor, with an estimated 100–150 pairs scattered across Britain, mostly near the south coast (Roberts & Law 2014). It winters in north and west Africa. It spends most of its time feeding on the woodland floor, and many seen will be mistaken for Buzzards, so any breeding here may be overlooked.

Beckwith (1887) said it 'occurs at irregular and uncertain intervals, usually in autumn, and during its visits appears to prefer the high extensive woodlands in the south and west', and this is still usually the case today. Beckwith and Forrest give differing accounts of presumably the same event, a pair being trapped at Ferney Hall dingle on 2 June 1865. On dissection of the female, 'one egg was ready for extrusion, and there was little doubt that one or two eggs had been laid previously', and the pair was 'probably breeding in the neighbourhood'. They also listed several other records: one killed near Rushbury in May 1872, one shot in the Edge Wood, and two scratching out wasps' nests in Oakly Park. Although 'decidedly rare' in the north and east, one was 'obtained' at Hatton Grange in 1871. Paddock (1897) listed five other locations where one had been killed. Forrest (1908) summarised the nineteenth century: 'recorded about a dozen times', most recently 'three were seen in the same district (as the breeding pair) in August 1881'.

Almost 30 years later, an immature male was shot near Wem on 28 September 1908, with 'several others obtained in other parts of England about the same time, indicating an immigration ... from the continent'. There was then a long gap until the next, amazingly, 'seen closely by several observers' in November and December 1961, and again in January, March, April, May and October (three dates) in 1962, on and around Linley Hill. Full descriptions were supplied by SOS members, and by a visiting South Manchester Field Group (reproduced on the *Histo* website). The records from November 1961 to May 1962 are listed in 'Early and Late Dates for Summer Migrants' (Hudson 1973), which also includes a further four December and one January twentieth-century records.

Another long gap ended in 1987, and the increase in local records since then reflects an expansion of Honey-buzzards into the upland forests of northern and western Britain in the last 25 years. Even so, there have been casual records in only 10 years since 1987. Individuals seen once only, in June at Wenlock Edge in 1990 and Black Hill in 2009, and in July at Berrington in 1987, Long Mynd in 1989, and Ruyton-XI-Towns in 2002, were all close to suitable breeding habitat. One at Albrighton in September 1990, two at Snailbeach in September 1993, and individuals at Owlbury in late August 1999, Venus Pool in August 2008 and Shawbirch in September 2009 were probably on return migration.

Breeding sites are kept confidential, so the location of a male at Hawkstone from 16 to 23 June 1996 was not published in SBR, as he may have been prospecting for a nest site, but there was no evidence he found a mate.

A pair bred in secret for several years from 1995 in Forestry Commission woods in the south, and raised two young per year until 2000, except in 1998, when one of the two chicks was found dead in the nest, choked on a frog. The first four nests were in Scots pines, and the last two in Grand firs.

The male returned in 2001, but no nest was found. In 2002 a pair moved to a new nest across the valley from the previous site, and two eggs were laid, but they did not hatch. Apparently the nest was predated, and the female was possibly killed by a Goshawk. In 2003 the male built a nest but no female was seen.

In 2004 the regular male was seen from early in the season, but a female was not seen until July, and although a nest was then built, there was no successful breeding. The following year the male again occupied the territory during June and July, and refurbished the nest built in 2004. A new female arrived, and a new summer nest was found just over the border. This was used in 2006, and the pair produced two fledged young. The male returned in May 2007, and built a new nest. He was seen displaying regularly throughout the spring, but 'at no time was a female seen and there was no evidence of breeding'. The nest trees from 2002 onwards were Scots pine (3) and Douglas fir (4), with the failed nest (2002) and the only successful nest (2006) both in the latter.

The same distinctive large pale-phase male was present every year from 1995 to 2007, attracting at least four different females. The Forestry Commission has continued to monitor the site, but he has not been seen since then. While other occasional individuals have been seen, there has been no further evidence of possible breeding.

All fledged young have been ringed in the nest, and colour-ringed since 1998.

A nestling ringed in 1998 was recovered in Manche, France, in August 2003, which, given its age and the date, may perhaps have been part of a breeding population.

A nestling female was colour-ringed 'Green 4' on 24 July 2000. She was later seen paired and holding territory in Glamorgan, south Wales, in 2004, 100km from her natal site. The same pair built a nest there in 2005 and she laid eggs for the first time in 2006. She maintained a territory in each subsequent year, and bred in most years, at one of two nest sites about 1km apart, until her demise during return migration in 2013, when she was found long dead on 14 June at Par Sands (Cornwall). When last seen alive, in August during the 2012 breeding season, she was the oldest known British Honey-buzzard (Roberts & Law, 2014; Steve Roberts *pers. comm.*).

This is an example of the research that has been targeted at this little known and poorly understood raptor for many years, and satellite tracking of three tagged nestlings as they migrate has provided evidence of local passage. One from northern Scotland spent four days near Bridgnorth 29 September–2 October 2001, undetected by local birders, and then moved through Somerset and southern Spain before it made landfall near the Morocco–Algeria border on 25 October. Another from north Scotland left the nesting area on 9 September 2006, and was picked up over Claverley seven days later. The third, a female from north Wales, was last recorded in her natal area in mid August, and was high over Twitchen on 29 August 2008. She continued though Hereford and Worcester, Gloucestershire, Kent (on 1 September), France, Spain and Morocco (19 September) to the Ivory Coast, where she wintered.

Two at Whixall Moss on 12 May 2010, presumably on northbound spring migration, was the most recent record. These, and the one near Black Hill on 13 June 2009, were the only ones recorded during the recent Atlas period.

There are many places, especially in the south, with suitable breeding habitat, so if the national population continues to grow, new breeding sites are likely to be found. The Forestry Commission and the Raptor Group will be looking for them, and birders should carefully check that 'yet another Buzzard' has in fact been correctly identified. Any observation that suggests possible breeding should be reported immediately, in strictest confidence.

Leo Smith

Sparrowhawk (Eurasian Sparrowhawk)
Accipiter nisus
Uncommon resident

'To watch the amazing speed and ruthless savagery of a Sparrowhawk striking a bird in mid-air is indeed to witness nature in the raw' (Sharpe 2009). The emphasis on 'savagery' is perhaps not surprising in a former gamekeeper at Apley Hall, but Sharpe was also a lover and keen observer of birds. He is not alone in his use of colourful language: even Beckwith, an admirer, refers to its 'rapacity'. We can argue about the use of such morally loaded expressions, but they certainly capture the remarkable intensity of the Sparrowhawk's hunting behaviour. Remarkable, but necessary, as it lives almost entirely on small-to-medium-sized birds, and catching them is a very difficult, even dangerous, business. A former County Bird Recorder saw one hawk strike 'within a few feet of my windscreen', while another 'buffeted into my lounge window but escaped unhurt' (Sankey SBR 1979). Others are not so lucky, and County ringing data include Sparrowhawks killed in collision with windows and vehicles. Most potential prey disappear long before the hawk is within striking distance, and most of the rest get away too. At Ryton in 2013 one was seen making five attempts in an hour on a large finch flock, all ending in failure. This is not some super-efficient killer, but one just efficient enough to get by from day to day. As Ian Newton puts it, they 'often go hungry, and … shortage of food is a major factor limiting their numbers and breeding success' (Newton 1986).

A hunting strategy based on concealment and ambush improves its strike rate, and also helps thwart attempts at 'control'. In 1899, after a century in which several other raptor species had been exterminated, Forrest ranked it 'the most numerous of its tribe', in spite of 'unrelenting persecution'. In fact, the fashion for game preservation may have helped it survive the onslaught. Woodland was conserved as cover, and its own predators, especially Pine Marten and Goshawk, were driven to extinction (Holloway 1996).

In the mid-twentieth century, the first SBR still classed it as 'widely distributed despite persecution'; eight years later, the *Handlist* (1964) warned that it was 'now scarce'. The precipitous decline was caused by widespread use of organochlorine pesticides, with the Sparrowhawk especially vulnerable to toxic residues in its prey. Once organochlorine use was restricted, recovery was equally rapid. By 1969, it was already 'well on its way to establishing pre-1959 numbers' (SBR), and fieldwork for the national BTO Atlas 1976 found that it remained absent only from central areas. In 1992, the *Atlas* considered recovery 'virtually complete'. Since then, the local range, like the national one, has contracted; 12% fewer tetrads show evidence of breeding since 1985–90, implying a corresponding drop in population.

Sparrowhawk is especially elusive when nesting. Analysis of SOS records 2009–14 from Venus Pool and Wood Lane, where observer effort is pretty constant through the year, indicates that the number of observations in May and June, when the female would have been incubating, was half the monthly average, and in some years fell to zero. If anything, the chances of detection are likely to be even slimmer in open country, so it is not surprising that 60% of the tetrads where it was found produced no evidence of probable or confirmed breeding. As the local population is sedentary, it is likely that breeding did take place in many, if not most, tetrads with evidence of possible breeding. Similarly, the many ups and downs in the breeding distribution change map are likely to owe a good deal to variation in fieldwork effort, as well as mere chance. However,

clusters of losses, often, but not exclusively, in arable farmland with little woodland, are likely to show real contractions of range worth further investigation.

Sparrowhawks were recorded in 25% more tetrads in winter than in the breeding season. They are more visible then, following flocks of smaller birds out of woodland into the open or into gardens, and more numerous, as they include visitors from the continent, and juveniles, only one-third of which will survive the year.

Recent SOS data make an interesting study: during the years 2009–14, 1,380 records of this species were submitted, an average of 230 a year. Three-fifths came from a small number of well-watched sites, but even there, reported sightings were surprisingly scarce: Venus Pool averaged only 24 a year, Wood Lane 29. At the other end of the scale, more than half the 322 named sites were represented by just a single record in six years. Although these figures exclude some Atlas fieldwork, they suggest that many Sparrowhawks must be staying well below the radar. As a result, it is extremely difficult to arrive at a reliable population estimate. The *Atlas* (1992) proposed a figure of one to three pairs for each tetrad with breeding evidence at any level; if applied to current data this would suggest a range of between 530 and 1,600 pairs. An estimate of around 550 pairs based on TTV counts resembles the lower figure, but it was found on too few TTVs for this to be considered reliable.

Although it is hard to monitor, its diet is relatively easy to study: pluckings from its prey are often conspicuous, particularly those of large or pale-coloured species. In a study of 230 kills in the Much Wenlock area in the 1960s and 1970s the smallest prey recorded was Goldcrest, the largest Woodpigeon. The latter was also the prey

Occupied Tetrads

Atlas period (breeding)	1985–90		2008–13		Change	
	Number	%	Number	%	Number	%
Confirmed	147	17	87	10	-60	-41
Probable	129	15	120	14	-9	-7
Possible	321	37	319	37	-2	-1
TOTAL	597	69	526	60	-71	-12
Tetrads with Winter Records (2007–13): 652 (75%)						

Breeding Distribution (2008–13)

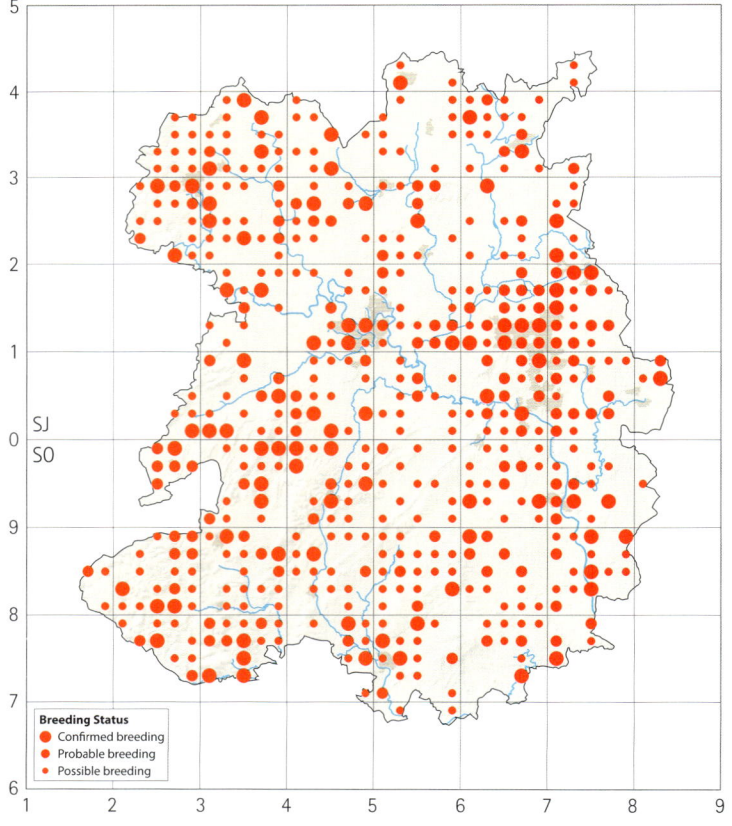

Breeding Distribution Change (1985–90 to 2008–13)

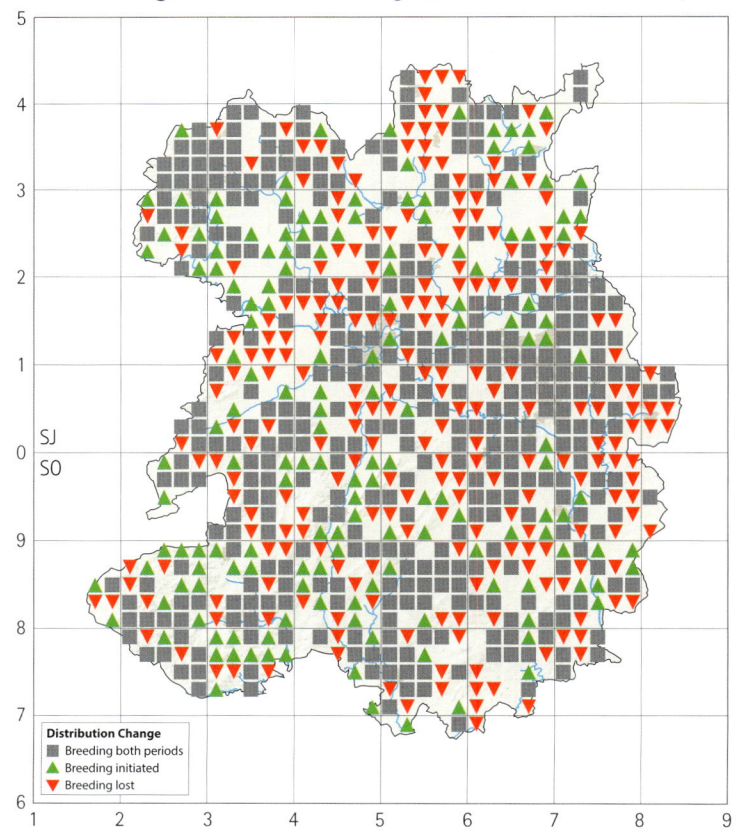

found most frequently, followed by Blackbird and Song Thrush, these three species accounting for over half the kills. The most unusual were Kingfisher, Lesser Spotted Woodpecker, Long-tailed Tit and an 'unfortunate budgie' (Sankey SBR 1979). The author recognised that this was not scientific, smaller species being underrepresented as their remains are harder to find. Nevertheless, it makes the point, confirmed by subsequent research, that common, conspicuous, easily caught species are more likely to be taken than scarce or skulking ones (Newton 1986).

Sparrowhawk, like all predators, follows its prey, quickly learning how to make the most of the man-made landscape. Sankey commented on its 'hunting expeditions' along woodland rides and hedgerows. Recent SOS records show how it takes advantage of farming practices, staking out sheep feeding troughs frequented by small seed-eating species, patrolling stubble fields to raid the flocks that gather there, or ambushing small birds attracted by grain spillages. An early record of predation at a bird table was published in CSVFC in 1935, and no recent SBR is complete without accounts of Woodpigeons pinned down or sparrows snatched from bird feeders. Observers often note the species taken, and one in Ludlow provided a detailed account of the methodical manner in which a hawk plucked its Blackbird prey, tail feathers and larger wing feathers first, before working its way round the rest. Kills are not always viewed so dispassionately, yet they are the natural and inevitable result of attracting birds into the garden.

Over recent decades, life has become much harder for many of the Sparrowhawk's prey species, the lack of winter stubbles being just one example. Versatile feeders like Chaffinch are well enough supported to replace their many losses to predation, while more specialised

John Hawkins, North Shropshire, 4 July 2007

Winter Distribution (2007–13)

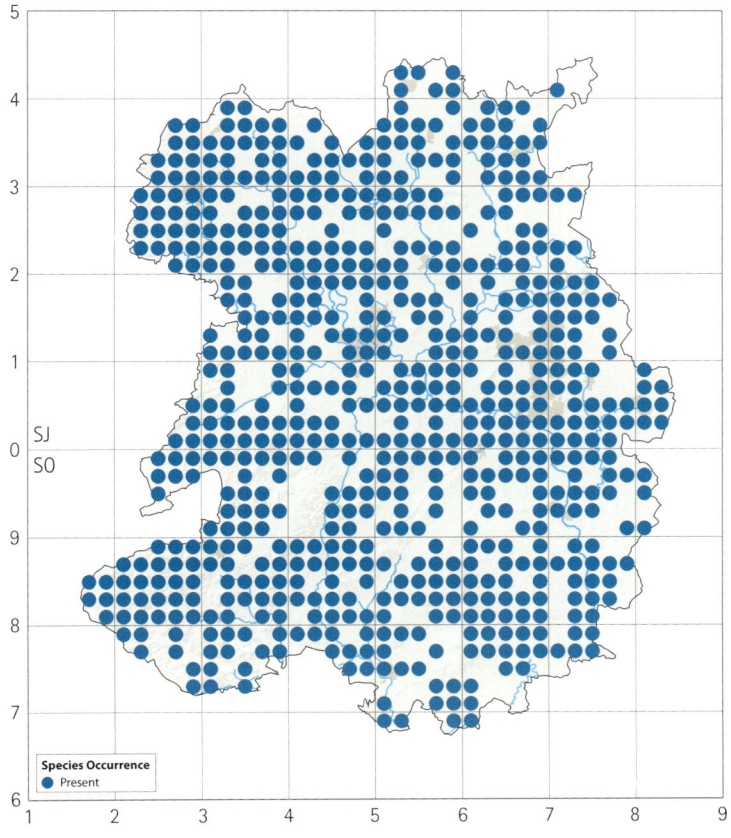

Species Occurrence
● Present

species are increasingly disadvantaged. If some populations that used to be able to shrug off its attentions are becoming more fragile, it is not because there are 'too many' Sparrowhawks: their numbers are controlled by the amount of prey available, not the other way round. It is just one among many warnings that our stewardship of the natural world is disrupting longstanding relationships between species in far-reaching and unpredictable ways.

Around 50 Sparrowhawks have been ringed or recovered here, and around 35 of the recoveries were ringed locally, confirming a strong sedentary tendency. The others all involved nearby English or Welsh counties, and do not include any recoveries of winter visitors to the UK from northern Europe. The three oldest were also recovered close to where they were ringed. The longest lifespan, just over six years, falls well short of the national record of over 17 years, but is half as long again as the typical one. Five, mostly young, birds had moved more than 100km since ringing; the greatest distance was 267km covered by a first-year male retrapped at Chelmarsh five months after ringing on the Suffolk coast. Where the cause of death was known, one had been shot, one struck by a car, and one flew into a window.

Michelle Frater

Goshawk (Northern Goshawk)
Accipiter gentilis
Scarce resident

Gareth Thomas, Shropshire, April 1992

Goshawk was probably once widespread, but it became rare well before the other birds of prey, as a result of habitat loss through woodland clearance and deforestation in the Middle Ages. It was already scarce at the start of the nineteenth century, but then suffered relentless persecution and was exterminated in Britain by the end of that century. The first and only historical record was a specimen collected by H.O. Wilson, but there must be some doubt about the validity of this as a County record (see p. 5).

Falconry became increasingly popular in the post war period, and many Goshawks escaped or were deliberately released so chicks could be taken from the nest. Thus a feral population bred sporadically at widely scattered locations across the country up until the 1960s. One escaped from Willey Park, near Broseley, in 1930, and the first breeding record was 'a pair nested at Hawkstone Park in 1951. They were robbed twice and the female was shot after laying three eggs in her third nest. The female had one jess and was recognised as a bird which escaped from a local falconer in 1949'. 'The only other recent occurrences are a bird over the Camlad Valley [at Chirbury] on 22 February 1959 and one near Oswestry in July 1960' (*Handlist* 1964).

It was not until 1968–72, during fieldwork for the first national BTO Atlas (1976), that regular breeding in Britain was first discovered. The larger Forestry Commission (FC) plantations provided the safest habitat for successful breeding, as there was some protection from relentless illegal persecution from keepers, the stealing of eggs by collectors and chicks by falconers, and other disturbance, and most nests were found in them. The *Atlas* (1992) reported that breeding was first proved here in 1966, and has occurred regularly since, and this comment was repeated in SBR 2004. However, because of the desire to keep nest sites strictly confidential, no FC records were submitted to SOS until the early 1990s. Fieldwork in 1968–72 confirmed breeding in SO77, part of which is in south-east Shropshire, but the dot was removed from the published BTO Atlas map to keep the location confidential. It has been reinstated on the corrected version now published on the Mapstore section of the BTO website, but this nest was in Worcestershire (Ian Newton *pers. comm.*).

Belatedly, in 1988, after Goshawk became protected legally in 1981, FC started regular monitoring of the local population, and a 'Raptor Panel' was set up to oversee this work. Discussion at its meetings was the source of the hearsay report that first breeding occurred here in 1966 (Tom Wall *pers. comm.*). However, recent extensive efforts to confirm this belief have been unsuccessful. If any records were kept, they no longer exist: the local FC office burnt down in the 1980s, and records kept by national FC staff were destroyed when the office was closed (Steve Petty *pers. comm.*). All members of the Raptor Panel, and anyone that any of them suggested, were contacted individually, but none had any information to substantiate the belief. The Panel monitored nests in three counties, and had no need to separate them into county groups, recollections were of discussions more than 20 years earlier, and few if any Panel members were involved in the 1960s, so it is perhaps not surprising that no confirmation could be obtained, but no mention was made of any nest at that time other than the one in SO77, which was in Worcestershire.

It is regrettable that FC reluctance to submit records in confidence to the appropriate body, SOS, or to any other ornithological organisation, means that the hearsay account of early breeding cannot be validated, and is almost certainly incorrect. The *Atlas* (1992) account of breeding being proved here in 1966 is no longer considered acceptable.

After the records in the *Handlist*, there were none in SBR until a falconer's escape was seen on Wenlock Edge in April 1976, but subsequently there were reports almost annually, mainly from the south and, by 1985, for all months of the year except January. The first SBR record of confirmed breeding occurred in the south-east in SO68 in 1979, when a pair with a single juvenile were seen. This is also the first record held by the RBBP, which first started collecting them from across the country in 1973 (Mark Holling *pers. comm.*).

However, research for this account has found previously unpublished records of nests in SO78 visited for ringing, probably in 1978, and another in SO77 in 1979 (Dave Fulton, Jon Lingard, *pers. comms.*). These, and the juvenile in 1979, were the first confirmed breeding records since 1951.

The first known nest on FC land was in 1982, monitored by the Hawk and Owl Trust (Sallie Pittam *pers. comm.*). The first record from FC land received by RBBP was also in 1982, although it was from a different site. Further FC sites were occupied in 1983 and 1984.

SBRs reported successful breeding again in 1982 in the south-east, there was an unconfirmed breeding report in 1983, and there were four confirmed breeding records in 1987 and in 1988 (SBRs). The *Atlas* (1992) stated they 'are now established residents in small numbers at several locations, although they are often overlooked, being elusive and secretive', but cited considerable evidence of continued illegal persecution, and concluded: 'Therefore no distribution map is shown nor is any indication given of the numbers of tetrads with evidence of breeding'.

Atlas records for 1985–90 suggest two confirmed breeding and seven probable breeding sites, but because of the desire to keep locations confidential, no FC data was submitted to the Atlas. However, FC monitoring from 1988 found eight nest sites by 1990, all with successful breeding in at least one year 1988–90, and in tetrads additional to the Atlas records, so the known population in 1990 was around 10–17 breeding pairs.

The *Atlas* correctly predicted that 'several areas have suitable habitat, and favoured prey such as pigeons and corvids abound, so further colonisation can be expected where persecution and disturbance are absent', and the population has grown steadily since 1990. It was supplemented by the female in the photo, which was rescued from near a nest in Herefordshire, and reared and fully rehabilitated at Bourton near Much Wenlock. The photo was taken in April 1992, just prior to her release back into the wild (Gareth Thomas *pers. comm.*).

FC Marches District recognised the need to avoid disturbance at breeding sites, and adopted a management plan in 1987, which included an exclusion zone around nests, preventing work within 400m, and the nest monitoring from 1988. By 2001, nests had been found at 21 different sites (with successful breeding at least once at all but one of them). The number of occupied sites fluctuated from year to year, with an underlying upward trend, reaching 11 nests (all successful) in 2000 and 14 (12 successful) in 2001. In the period 1988–2001 there were 98 successful nests which produced 236 fledged young, an average productivity of 2.56 young per successful nest. The increase in numbers and successful breeding was attributed to legal protection, and the exclusion zone (Smout SBR 2001).

Over the same period, counts were made of almost 3,000 prey items at nest sites. Woodpigeon occurred the most (20.0%), followed by rabbit (17.8%), Feral Pigeon (14.3%) and grey squirrel (10.6%). No other prey exceeded 10%, but Carrion Crow and Magpie were almost 7% each, with all corvids totalling 23.5%. Almost all the remainder were small numbers of 16 other bird species. Of these, Pheasant made up less than 3% of prey items.

Monitoring has continued in FC Marches woods since 2001, and the results are shown in the table. All suitable sites are monitored each year. The variation in numbers reflects the growth of suitable stands of mature timber, and subsequent harvesting (SBRs and FC Marches records). The lower numbers since 2012 also reflect adverse weather conditions, which affected both the Goshawks and their human monitors, and monitoring was also reduced in these years as a result of FC budget cuts, re-organisation and staff uncertainty, so pairs that failed early would probably have been overlooked.

Thirteen different sites were occupied during the recent Atlas period, and the row '% of 2008–13 Sites Occupied' shows the number of occupied sites in each year as a percentage of this 13.

These figures are for County nests in the FC Marches District only and do not include either the FC Wyre District, with a further pair from 1999 and another from 2005, or Forestry Commission Wales, with one pair.

Most nests are in Douglas fir, but this might reflect the dominance of this crop, which grows well in local conditions. A few nests are in larches, and in some years Norway spruce is used.

A further survey of prey remains revealed approximately 30% corvids, 25% pigeons and 20% mammals, with the rest being made up of a variety of bird species including two Tawny Owls (SBR 2005), the slightly different proportions compared to the long-term study reflecting differences between woods as the population expanded. Only rabbit and domestic pigeon have been noted as prey outside FC woods.

The same SBR noted that 'Goshawk numbers appear to have levelled out [in the FC Estate], but it seems certain that there are more breeding territories out there to be found in private woodland'.

Since then, no new FC Marches sites have been found, although some have been re-occupied after lengthy gaps. However, Atlas fieldwork from 2008, and monitoring by the Raptor Study Group from 2010, have confirmed the prediction, and found many more sites in private woodland.

Several examples of persecution have been noted: 'An adult male shot in south Shropshire was taken to the Hawk Trust at Loton Park, but although it recovered it cannot fly' (SBR 1979); 'the male of a pair was shot' (SBR 1987); 'four birds, probably two pairs, were poisoned in the SW in the past two years' (SBR 1991); and two recently fledged young disappeared, believed shot, from an FC site (Alan Reid, *pers.*

Nest Monitoring Results in FC Marches Woods 2002–14

Year	2002	2003	2004	2005	2006	2007	2008	2009	2010	2011	2012	2013	2014	Total
Sites checked	16	16	16	20	19	19	19	18	18	18	18	18	18	**233**
Occupied	14	10	12	13	10	9	10	9	10	10	5	5	6	**123**
Successful nests	8	11	9	8	5	7	10	6	7	7	4	0	3	85
Fledged young	15	21	25	22	12	14	19	12	16	16	8	0	6	**186**
% of all sites occupied	87.5	62.5	75.0	65.0	52.6	47.4	52.6	50.0	55.6	55.6	27.8	27.8	33.3	**52.8**
% of 2008–13 sites occupied							76.9	69.2	76.9	76.9	38.5	38.5		**62.8**
Young/occupied site	1.07	2.10	2.08	1.69	1.20	1.56	1.90	1.33	1.60	1.60	1.60	0.00	1.00	**1.51**
Young/successful nest	1.88	1.91	2.78	2.75	2.40	2.00	1.90	2.00	2.29	2.29	2.00	n/a	2.00	**2.19**

Breeding Distribution (10km) (2008–13)

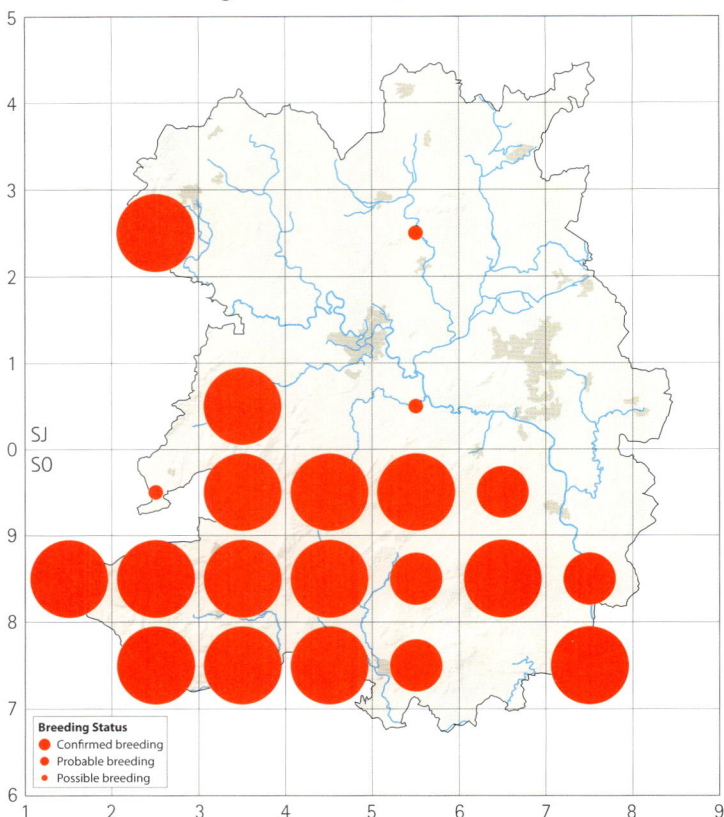

Breeding Status
- ● Confirmed breeding
- ● Probable breeding
- ● Possible breeding

Occupied Tetrads

Atlas period (breeding)	1985–90		2008–13		Change	
	Number	%	Number	%	Number	%
Confirmed	–	–	36	4	–	x
Probable	–	–	17	2	–	x
Possible	–	–	27	3	–	x
TOTAL	–	–	80	9	–	x
Tetrads with Winter Records (2007–13):　63　(7%)						

– 47 occupied tetrads altogether. Records from some of the remaining tetrads will relate to these territories, but a few, particularly the possible breeding records, may relate to previously unknown pairs. There is also a possible breeding record adjacent to where breeding was confirmed in 2007, although no Goshawks were recorded in that square during the Atlas period. Some pairs will have been missed altogether, and a cluster of records on the winter distribution map suggest a further territory. It is therefore estimated that there are at least 50 regularly occupied territories.

Some territories do not appear to be occupied every year, and others sometimes hold only single birds, but it is known that some pairs use different nest sites in the same territory, often several kilometres apart (Hardey at al. 2009), and some of these may have been overlooked. During the Atlas period active nests were found at 15 different FC Marches sites, but two of these were the result of a pair moving during the period (i.e. there were 13 territories). The average occupancy was at least 62.8%, and it was much higher in the first four years of the Atlas, before it was reduced by harvesting and bad weather. Assuming that this average occupancy rate applies to all 50 regularly occupied territories, then there are a minimum of 31 breeding attempts every year, plus perhaps other pairs that breed in other parts of their territory, or which do not come into breeding condition. The population is therefore estimated at 31–50 breeding pairs, about three times the number known in 1990.

The most recent national population estimate, of up to 430 pairs, appears to have been compiled from assessments of each of the home countries, some now over 20 years old (Hardey at al. 2009). It is therefore well out of date and much too low; 622 territories were reported to the Rare Breeding Birds Panel in 2014, and many more are not reported.

In some local areas, density is very high, with at least 13 territories in one 10km square. There were eight instances of nearest neighbour distances between occupied nests of less than 4km, with the shortest being 2.0km. The relative abundance map in the national BTO Atlas 2007–11 shows that some local tetrads are in areas in the top 10% nationally.

Very few Goshawks are known to have been ringed – chicks in one nest were ringed regularly between the late 1970s, perhaps as early as 1978, and the early 1980s, and in another around the same time (Dave Fulton *pers. comm.*); six in two broods in 1988 (SBR), broods from four nests in 1986 and 1987 (RBBP), 7 nestlings in only three of the eight years between 2006 and 2013, and one brood of three in 2014. None have been recovered, and the Raptor Group started a colour-ringing project in 2014, so individuals can

comm.). RSPB has records of three nests being robbed since 1990, one of which resulted in the successful prosecution of a gamekeeper (Guy Sharrock *pers. comm.*), while two nests were both robbed in four successive years from 1991 (Smout SBR 2001). Several other nests were deserted after laying, for reasons unknown.

This continued persecution makes it unwise to publish a tetrad map, but the distribution found in the recent Atlas period is shown at the 10km square level.

All except two of the tetrads with confirmed breeding are in the south-west and south-east quadrants, with the remaining two in the north-west. There were no probable or confirmed breeding records from the north-east, but breeding was confirmed there in 2014 when an unfledged chick found in SJ60 was taken to Cuan House Wildlife Rescue.

Goshawk was found in far fewer tetrads in winter Atlas work, confirming its secretive nature – it is most frequently seen displaying over breeding sites in spring, and while hunting to feed young in the nest. Most records came from near breeding sites, showing its sedentary nature. However, four of the tetrads where it was seen were in the north-east quadrant, and two other sightings were well away from known territories, suggesting either wandering immatures, or perhaps further territories to be discovered.

The occupied FC Marches sites account for 15 of the tetrads with confirmed breeding. The other two FC districts account for a further three. Nests have been found in other woods in another 12 tetrads by the Raptor Group, and a combination of territories located by the Raptor Group, together with other known territories of long standing, account for a further 17 tetrads with probable or confirmed breeding be individually identified, mainly by using cameras at nest sites.

Three ringed in Hereford and Worcester have been found: a nestling female in June 1977 was found 'long dead' in 1995 in the Wyre Forest only 6km distant but 17y 9m 17d later, a nestling male in 1979 was found dead (injured) in the Wyre Forest 8km distant 2y 6m 27d later and a nestling male in 1982 was found dead in November 1984, near Neen Savage, 22km distant 2y 5m 11d later. All three

recoveries illustrate the Goshawk's sedentary nature, and the former was only 11 months short of the oldest ever recorded.

Given the tolerance of close neighbours in good habitat, and continued availability of a plentiful food supply, there is every likelihood that the population of this iconic raptor will continue to increase and spread, provided illegal persecution can be controlled.

Leo Smith

Marsh Harrier (Western Marsh Harrier)

Circus aeruginosus

Rare passage migrant, has bred

Jim Almond, Venus Pool, 11 April 2010

Marsh Harrier breeds widely in suitable wetland habitat across Europe and Asia, but most winter in Spain and north-west Africa. It is mainly a summer visitor to Britain although some now over-winter.

The incidence of local records closely follows its fluctuating national fortunes. Before the nineteenth century it probably bred in many English and Welsh counties, but land drainage severely reduced its numbers, and by the start of that century it remained in only a few of them. Beckwith noted unsubstantiated accounts that it had bred here, which he did not accept: 'This may have been the case in former days, as the extensive heaths and mosses were likely places for such a bird to inhabit ... but in the absence of any record of the discovery of nest or young ... it would scarcely be right to assume that it was ever more than an accidental visitor'. Further drainage, and persecution by gamekeepers and collectors, reduced breeding sites to only Northumberland and Norfolk by the 1870s, and the last known pair in Britain was trapped in 1899.

Not surprisingly, Marsh Harrier locally was 'never numerous, and is now very rare. Specimens were obtained many years ago, at Berwick

and on the Long Mynd, and one was seen near Oswestry in January 1886' (Forrest 1899), the last nineteenth-century record (Forrest 1908).

During the twentieth century, recolonisation of Britain began in Norfolk in 1911, sporadically to start with and annually from 1927; by 1958 there were 15 nests, and breeding had occurred in five English and two Welsh counties, though regularly only in Norfolk and Suffolk. The population collapsed again, as a result of organochlorine pesticides, to one breeding pair in 1971, at Minsmere. Banning of these pesticides, reduced disturbance and persecution, legal protection, and restoration and creation of wetland habitats, especially by the RSPB, allowed the population to recover to over 75 nests in 1988, mainly in East Anglia.

In the first half of the century there were three local records, all of singles: 10 April 1920 at Marshbrook, a female at Middletown on and around 24 March 1921, and near Millichope throughout November 1933.

There was then a long gap until the first modern record, an immature female at Worfield Bog on 12 May 1960, then another 20 years passed before the next, a sub-adult male frequenting standing barley fields near Much Wenlock on 9 August 1980, and, the final modern record before the start of 1985–90 Atlas fieldwork, one at Allscott Sugar Factory in August 1984. During that Atlas period, a juvenile was seen at Cound in September 1985 and an immature was seen at Chelmarsh for three days 16–18 May 1990.

The only known instance of confirmed breeding occurred at Preston upon the Weald Moors in 1988, when a pair built a nest in a field of Italian ryegrass and laid at least two eggs. The grass was cut for silage but the farmer agreed to leave an area uncut around the nest. Apparently the eggs did not hatch and certainly no young were raised. The pair did not return in subsequent years.

Marsh Harrier was still a rare visitor, but, after a further gap of seven years, the twentieth century closed with a female or immature in Corvedale in August 1997 and two spring records from Venus Pool in 1998, an adult female on 23 April and a male on 10 May.

The national population continued to grow rapidly, and reached an estimated 363–429 breeding pairs by 2005. The vast majority were in East Anglia, Lincolnshire and the south-east, but they first nested at Leighton Moss, the RSPB reserve on Morecambe Bay, in 1987, and, after a gap, colonisation of the north-west continued, with confirmed breeding on the Dee Estuary and Merseyside during 2008–11, and at Martin Mere, the WWT reserve in Lancashire, in 2012. There has been a spectacular increase in sightings here, almost certainly due

Estimated Number per Year (1950–2014)

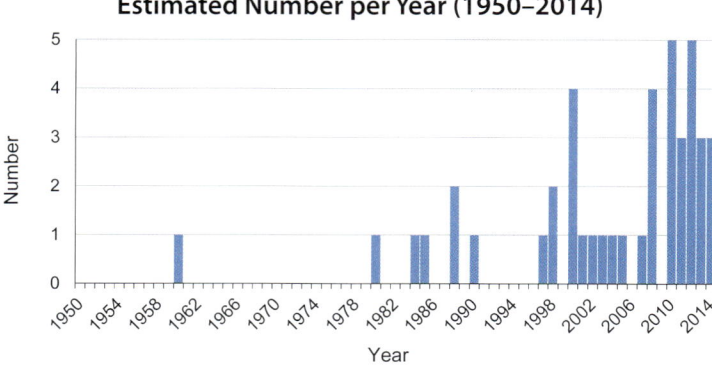

Seasonal Occurrence in 10-day Intervals (1950–2014)

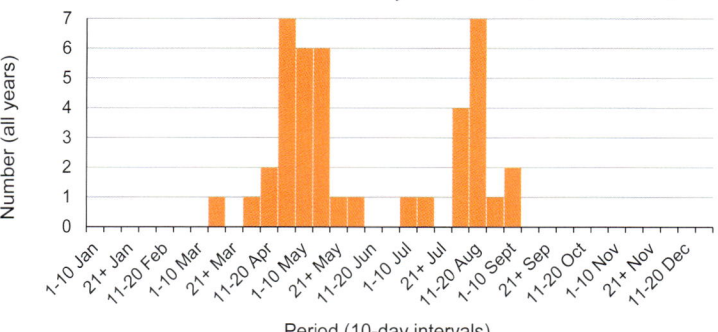

to Marsh Harriers migrating to and from these newly occupied sites in north-west England.

Since 2000 there have been 31 records, with only two years with none, and at least three in each of the last five years. More than three-quarters have come from only four sites, Whixall Moss (12), Venus Pool (6), Chelmarsh Scrape (3) and Wall Farm (3), the remainder being single records from Allscott Sugar Factory, Crudgington Moor, Hollins Farm (Merrington), Mirelake, Overley, Tibberton and Wild Moor.

Apart from the breeding pair in 1988, there have only been two records of two together, both in 2008, at Whixall Moss in April and Wall Farm in the autumn.

Most pass through in the last 10 days of April and first three weeks of May, with return passage mainly in early and mid-August. However, the late May, June and July records are all recent, and all have come from the same site in the north. Breeding is not suspected, but suitable wetland habitat exists there, and a small percentage of

British pairs nest in cereal fields, rather than the more typical extensive reed beds, so breeding in future is quite possible.

During fieldwork for the recent Atlas, individual passage migrants were seen between 5 April and 16 May at Whixall Moss (twice, in 2008 and 2010), Bobbington, in 2010, Venus Pool (three times, in 2011, 2012 and 2013), Chelmarsh (twice, in 2012 and 2013), and Wall Farm in 2013. Return passage was observed in July 2011 at Overley (SJ61A). There were no records during winter fieldwork.

A one year old seen on Wild Moor in May 2014 was wing-tagged. The tags could not be read, but the colour scheme was identified. It did not match the two English schemes (Sheppey and Norfolk), and it probably originated in Northern France or Belgium.

If the national population continues to increase and expand, breeding here is a possibility, in the north, and perhaps at other sites too if nesting in damp cereal fields becomes more frequent.

Leo Smith

Hen Harrier

Circus cyaneus

Rare winter visitor and passage migrant

Jim Almond, Venus Pool, 28 February 2015

Prior to the nineteenth century it was widespread, breeding here and across the UK in a wide range of habitats. Records in the nineteenth and twentieth centuries were scarce, with the first in 1836 at Berwick. A pair was shot near Ticklerton about 1840, four more individuals were shot between then and 1887, a male was seen near Oswestry in 1879 and a female was seen for several days about Betton Pool in 1894. A nest and four eggs were taken on Shawbury Heath in 1890.

By 1900, the British breeding range had largely contracted to Orkney and the Outer Hebrides due to loss of habitat and persecution. However, these harriers were still of almost annual occurrence in the 1920s and early 1930s, and 'apparently two pairs nested in the south-west in 1923' (*Handlist* 1964). Nesting was also reported in the Clun Forest with two pairs breeding each year around 1930, and one still there until just before the Second World War (Adams 1950). Apart from this pair, there were only two records after 1933 in the historic era, of one shot on Haughmond Hill in 1938, and another shot at the same place in 1942.

From 1939 onwards, the main breeding range expanded southwards, as a result of reduced persecution during the war, and taking advantage of nest sites and food in young conifer plantations. It reached northern

England and Wales by the 1960s. Gwynedd and Clwyd have since become the centres of the Welsh breeding population, which has increased from 11 occupied 10km squares in 1968–72, to 28 pairs in 1998 and up to 57 pairs in almost 30 10km squares in 2010.

Local sightings reflect this increase over the same period, with the first modern record in 1962, followed by records in most years up until 1973 and annually since then. Hen Harrier is an irruptive species, and the wintering population in the UK contains residents, and migrants from Scandinavia. Numbers can greatly fluctuate, but migrants tend to stay in south-east Britain, so the increased frequency of wintering birds here suggests that Wales, not the continent, is likely to be the main source. Evidence for this comes from the single ringing recovery, a male found dead near Oswestry on 9 November 1978, which was ringed 29km away in Wales as a nestling on 25 June 1977.

An assessment has been made of the modern records, to estimate the number present in each month of each year. The first chart shows the annual summary of this assessment, an underlying but steadily increasing trend which mirrors the growing Welsh population, with occasional influxes perhaps reflecting good breeding success, abundance of winter food such as voles, and perhaps continental irruptions occasionally reaching this far west.

Most records are of one bird, and records from the same site over a short period are likely to be repeat observations of the same individual. However, individuals seen in two different months have been counted in each month, so the chart exaggerates the actual number of Hen Harriers seen. The recent very high peaks are also partly a reflection of increased observer effort during the Atlas, and of local raptor monitoring, which increased the likelihood of an individual being seen in more than one month. For no obvious reason,

Winter Distribution (2007–13)

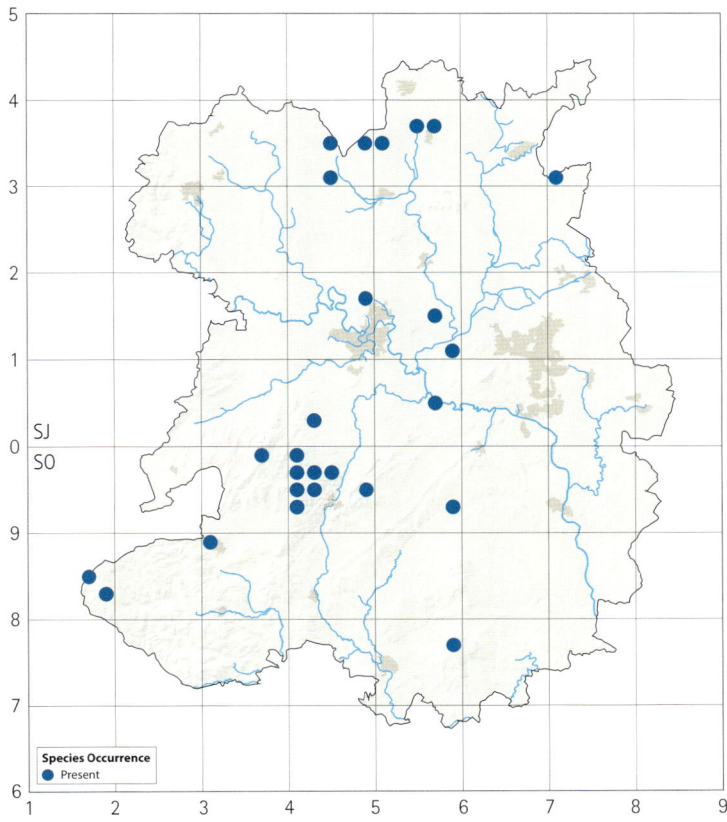

Occupied Tetrads

Tetrads with Winter Records (2007–13): 26 (3%)

Bird Months per Year (1950–2014)

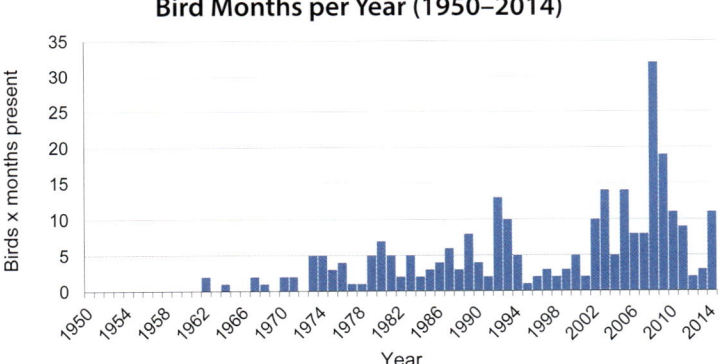

Number Present Each Month (approx.) (1950–2014)

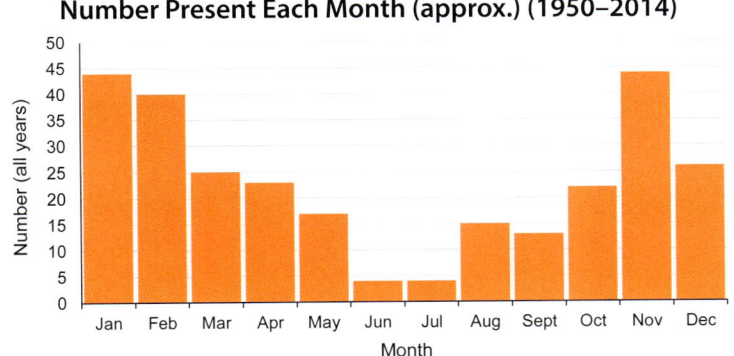

2008 was an exceptionally good year with 92 records, 79 from Long Mynd. Of these, 65 records from Long Mynd in January–April 2008 involved at least five wintering birds, two adult males and three female/juvenile 'ringtails', but there may have been more. The same level of monitoring has occurred there every year since, but similar numbers have not been found (Dave Pearce *pers. comm.*).

The second chart shows the result of the same assessment to give a profile of the months when Hen Harriers have been present. Again, it exaggerates the number of individuals seen. The vast majority were winter visitors, between November and February, with far fewer passage birds and only a handful in June and July.

Although the assessment is imprecise, the two charts accurately reflect the underlying trend of increasing numbers of visitors, the annual fluctuations and the pattern of monthly occurrence.

The distribution of records from the recent Atlas period also shows many more in the winter season (November–February) than the breeding season (April–July). There has been a marked increase from no records in 1981–84 in the BTO winter Atlas to 65 records in 26 tetrads in the winter Atlas 2007–2013. The winter map shows a cluster in the south-west in the Long Mynd – Stiperstones area and a smaller cluster in the north around Whixall Moss, with a scattering of other records from lowland and upland farmland. There were only five March records in 2008, from the same areas, and none in the other Atlas years, so adding March records would make no difference to the map. Three of the Long Mynd tetrads, and one on the Stiperstones, also had records for April or early May, which were spring passage birds. There were also spring passage records from two other tetrads which had no winter records, near Bedstone (SO37S) and Lydbury North (SO38N).

Habitat preference appears to favour heather, taller grasses and rushes which support sufficient levels of prey species. The attraction of Long Mynd may reflect the reduced grazing and improving condition of vegetation, which, together with milder winters, has led to an increase in a variety of species remaining on the plateau in winter. However, while some records were of hunting during the day, most were of harriers coming in to roost at secure sites at dusk, having hunted largely undetected on surrounding farmland.

In 2014 there were a male and a ringtail at Whixall Moss up to 11 January, one on the Stiperstones in March, a male at Berriewood in May, ringtails at Church Stretton in July and on Long Mynd in August, and in October and November, and a ringtail at Whixall Moss in September, presumably different from the ringtail with a male on several dates in November and December.

Known possible breeding attempts in the last 75 years have been few and far between. In 1988 a pair held territory on the Welsh border at an undisclosed site, and appeared as a probable breeding record in the *Atlas* (1992). In 1994 four singles were seen in May at three other undisclosed sites, including a male and a female seen separately at one of them. However, these were all likely to be late passage birds.

The national survey in 2010 reported a mixed picture across the UK. The English breeding population remains perilously low and was restricted almost entirely to the Forest of Bowland, but the Welsh population has continued to grow. This, coupled with a reduction in grazing levels in the uplands due to conservation schemes and consequent improvement in the structure of vegetation, and hence prey availability, may attract breeding pairs in the future. The principal constraint on the UK population is illegal persecution on grouse moors.

Yvonne Chadwick, Titterstone Clee, 28 August 2010

If this ceased and the population was allowed to expand naturally, there is sufficient habitat in England to support around 300 pairs (Fielding *et al.* 2011); this potentially includes around 10 breeding pairs on and around the Long Mynd, Stiperstones, Clun Forest and Clee hills, with perhaps an additional pair at Whixall Moss. However, in England in 2013 no pairs bred successfully, and in 2014 only four did so. Although it is growing, the Welsh population needs to expand further before it might spread this far.

Peter Carty

Montagu's Harrier
Circus pygargus
Very rare passage migrant

This summer visitor to Britain is a rare breeding bird with an average population of less than 15 pairs, most of which nest in the south, south-west and east of England.

Historically, identification of this species, particularly females, must have been difficult without obtaining the specimen, but there are five, perhaps six, records from this period. In 1836, a ringtail '*Falco pygargus*' was seen on 'Westfelton Moors', but there is no other reference to this bird. In July 1859, a female 'Ashcoloured Harrier' was caught at Crose Mere and showed evidence that it had recently been sitting on eggs, although no nest was found or a male seen (Rocke 1865). Later, an immature male was shot at Grafton near Baschurch about 6 June 1902 and was retrieved from 'the gamekeeper's gibbet just in time to save it for the taxidermist'. In 1928 a 'grey male' was seen near Broseley on 4, 19 and 25 May and possibly the same bird was killed at Bridgnorth that summer. Finally, what was originally thought to be a Buzzard hanging on a gibbet on Haughmond Hill in 1938, turned out to be a young male Montagu's Harrier when taken to a taxidermist. The Rev. H.O. Wilson's case of falcons 'collected on or near the Longmynd, between 1848 and 1857' also included a Montagu's Harrier, although the validity of this record is open to question (see p. 5).

There have been two modern records:

| Whixall Moss | 17 June–1 July 2006 |
| Long Mynd | 27 May 2017 |

The second calendar year female in 2006 may have present from 11 June because a 'ringtail' Harrier, assumed to be a Hen Harrier, was reported from the same site on that date. The second, an adult male, was photographed as it flew over the observer.

Graham Walker

Red Kite
Milvus milvus
Scarce resident

Jim Almond, north of Clun, 4 November 2007

There are breeding populations in Europe and north-west Africa, with the UK, Germany, France, Spain and Sweden holding the vast majority. Northern populations migrate south for the winter.

The twenty-first century recolonisation by native Welsh Red Kites is undoubtedly the most exciting of recent ornithological events.

Historically and nationally, Red Kites were frequently encountered in towns, their remains having been found in many archaeological excavations from neolithic times onwards, and they were referred to in Shakespeare's plays. As a scavenger and carrion eater, and one of nature's refuse collectors, they were protected until improving sanitation and public health reduced their usefulness, and they were already in decline in the early eighteenth century, and persecuted vigorously in the nineteenth century.

> The days of this fine denizen of our woods and forests are numbered; common as they used to be ... but I believe that at the present moment there are not more than one or two pairs left. I remember the time when the large woods in this neighbourhood were so well stocked with them that it was no uncommon thing to see a row of them, interspersed with the common buzzard, nailed up against a barn. The great ease with which they were trapped ... has accounted for most of them. (Rocke 1865)

However Rocke went on to highlight another reason for their decline in the wild 'I am happy to say I possess a very fine pair in an aviary, taken some few years ago in Stokes Wood ... here they

bred for many years'. Forrest too noted that 'old writers speak of the Kite as of quite common occurrence, but it must now be regarded as very rare', and gave several examples of them being shot or 'taken'. Beckwith (1879) recorded Kites building nests near Ludlow in the late 1870s, probably again in Stoke Wood, and this was the last known breeding record in England.

Persecution and egg collecting continued to take a toll, and they became restricted to Wales, with only around 10 pairs in 1903, when the first conservation efforts started. By the 1930s, there were still only about 10 pairs, and only one chick fledged in some years, with the result that all Kites were descended from only one female. That is how close Kites came to extinction in Britain.

The Welsh population grew very slowly during the rest of the twentieth century, held back by bad weather, loss of rabbit food following the outbreak of myxomatosis in 1954, robbing of nests by egg collectors and illegal poisoning, but it eventually reached 100 pairs in 1993 (Cross & Davis 2005).

Kites were absent here for most of this period, with the *Handlist* (1964) reporting: 'The only reliable record in the last 60 years is of a bird being mobbed by a carrion crow, 3 miles from Bridgnorth on 9 August 1943'; and there were only three more records up until 1982, singles seen in August 1965 on Long Mynd, in August 1970 near Clun and in June 1974 over the Shrewsbury bypass. There were four records in 1982, including one of two together in September. Prophetically, SBR noted that 'with continued breeding success of the Welsh population, this species could well become an annual visitor', although there were none the following year, and only one in 1984. During intensive Atlas fieldwork between 1985 and 1990, there were only 11 breeding season records over six years, all of individuals, although by 1988 it was 'now almost an annual visitor', with six records that year.

From the late 1990s, the number of records, and locations they were seen in increased rapidly, as the Welsh population continued to grow and spread after reaching the 100 pairs milestone in 1993, as shown in the chart.

Many Kites were wing-tagged as nestlings, and the first one to be identified was found trapped in wire netting near Onibury on 31 July 1990. A female, it had been tagged near Llandovery, 83km

Increase in Records and Locations (1991–2006)

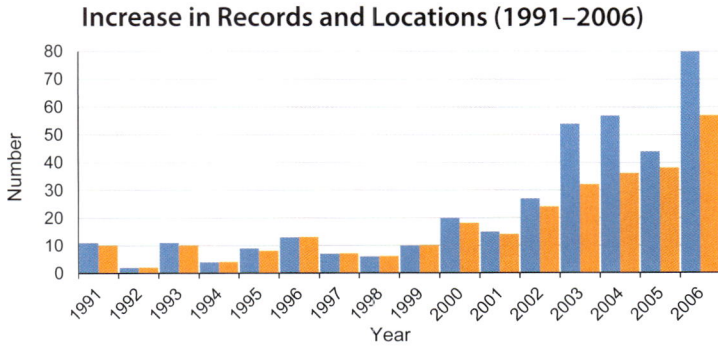

Nests Found, Successful Nests and Fledged Young (2005–14)

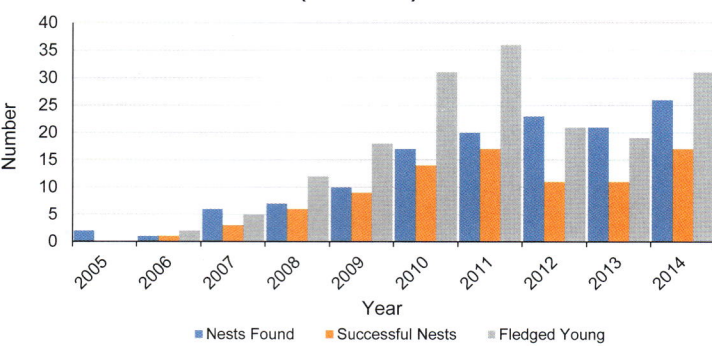

distant, on 12 June 1990, only 49 days previously. This was, and is typical of the rapid long-distance dispersal of recently fledged young. It had returned nearer to home, near Tregaron, by 31 October (Peter Davis *pers. comm.*). Similar movements by other tagged females, here in August–November, then returning to Wales, were noted in 1991 and 1993.

Another early visitor, which was seen several times north-east of Knighton in August 1991, was less fortunate – it was eventually found shot.

Initially most Kites were seen in spring, when they moved into potential breeding areas, and then again in the early autumn when dispersing from breeding sites, but winter records were infrequent. However, by 2000 records were 'fairly evenly spread throughout the year'.

Further evidence of rapid juvenile dispersal was the first ringing recovery, one found dead near Lydbury North on 23 August 2003, only 47 days after ringing near Lampeter, over 80km away.

The first nests were found near Knighton in 2005. Both failed, one very early and the second at about the time the eggs were expected to hatch, but this latter pair returned to the same site the following year and produced two fledged young, the first known to fly from a Shropshire nest for 130 years (Smith SBR 2006). The photo shows these two young, returned to the nest after tagging. Note the remains of Carrion Crow and Magpie, which are frequently fed to nestlings.

Since then, the Welsh Kite Trust and, from 2010, the Raptor Study Group, have tried to find the nests of all breeding pairs. The population has spread rapidly from the south-west, with the first nest found in the north (the SJ grid squares) in 2011, and the first east of the Shrewsbury–Ludlow (A49) road in 2012: in 2014, 34 breeding pairs were found or reported, and 26 nests were found: 17 were successful, and 31 young fledged from them. Twenty of the fledged young were ringed and tagged.

The number of nests found, successful nests and fledged young for each year 2005–14 is shown in the chart. Between 2005 and 2014, 133 nests were found, and 89 (67%) were successful, with 175 fledged young, of which 141 were tagged. Almost two-thirds (64%) of the nests found have been in oak trees, and almost one-fifth (17%) in larches – 81% in these two tree species. Seven other tree species have been used occasionally.

Productivity was much higher in 2014 than in the previous two years, which were badly affected by severe weather. The average productivity 2005–14 was 1.32 fledged young per nest, and 1.97 per successful nest, considerably higher than the comparable Welsh figures, attributed to a less hostile climate and better food supply.

There has been a very high turnover of nest sites, with little loyalty to particular locations. Nests at 63 different sites have been found. Of these, 37 (59%) were used for one year only, and 9 (14%) for two years only. Only seven sites have been used for at least four years, and in every year since they were first found.

Four sites were only used in years before 2008, and 14 nests were used for the first time in 2014, leaving 45 nest sites in 39 tetrads used during the Atlas period. Only three sites were used in all six Atlas years, and 35 (78%) were used for three years or less. Only 13 of the 45 (29%) were still occupied in 2014.

In addition to the 39 tetrads where nests were found, there were two other tetrads with confirmed breeding: one with two recently fledged young where the nest was not found, and one reported by a

Occupied Tetrads

Atlas period (breeding)	1985–90		2008–13		Change	
	Number	%	Number	%	Number	%
Confirmed	0	0	41	5	41	x
Probable	0	0	37	4	37	x
Possible	0	0	137	16	137	x
TOTAL	0	0	215	25	215	x
Tetrads with Winter Records (2007–13): 199 (23%)						

Confirmed & Probable Breeding (2008–13) with Range (2007–13)

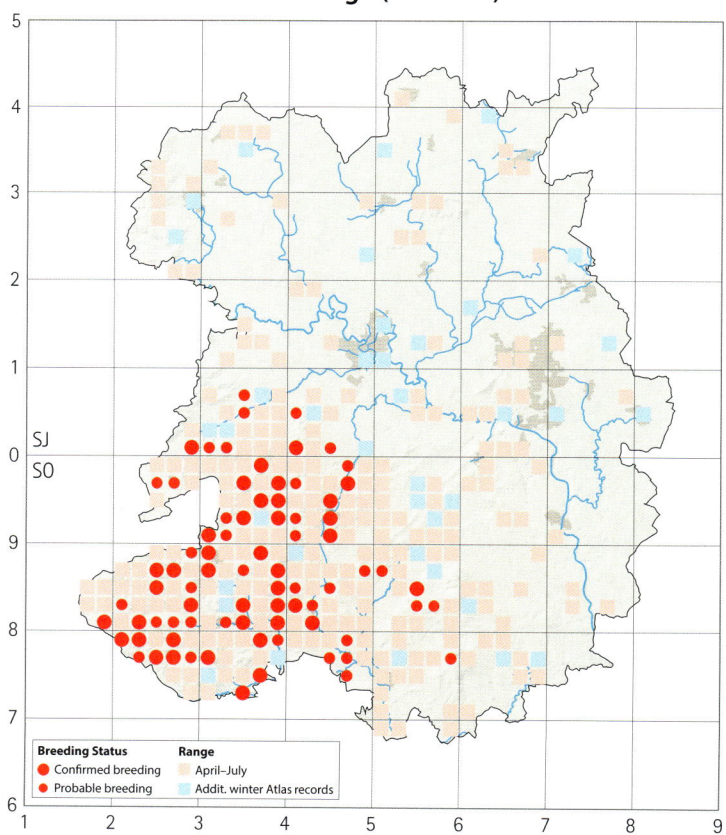

Tony Cross, south-west Shropshire, 5 July 2006

farmer after the end of the season (outcome unknown). All the nests found were in the south-west hills.

Some of the probable breeding records were of pairs displaying in tetrads adjacent to known nests, or near nests that were not found. However, breeding Kites are not territorial, and non-breeding birds, up to three years old, forage over long distances, often in twos, and land rarely. There was no consistency in how such sightings were recorded, and a few wanderers may account for many of the sightings, so a conventional map of breeding evidence might give a false impression. Foraging occurs throughout the year, so all Atlas records apart from nests have been mapped to show the extent of the range found in the whole of the recent Atlas period November 2007–July 2013, with winter season records accounting for several, but by no means all, of those north and east of the River Severn. Although there is little evidence of breeding away from the south-west, the wide distribution of foraging birds suggests that breeding will become more widespread in the near future.

The distribution map shows 41 tetrads with confirmed breeding, but the maximum number of nests found in any one year in the recent Atlas period was 23 in 2012. The map also reflects the increasing number of one and two year olds that have not reached maturity. Only about 40% of fledged young reach breeding age. The map therefore considerably over-represents the number of tetrads with nests, and the breeding range, in any one year.

Some of these nests were very close together: two were only a kilometre apart, and five sets were less than 3km apart.

The steady move eastwards continued in 2014, with two nests very close together on Wenlock Edge, and another almost as far east as Brown Clee. Individual youngsters were also seen on Titterstone Clee, but the pair that probably nested there in 2013 did not return.

As the population increases and spreads, it becomes harder to find all the nests, so the breeding population is undoubtedly well over the 34 pairs located in 2014 – perhaps 50 breeding pairs.

Twenty-four different tagged individuals have been seen at nest sites, and 21 of the tags have been read.

The first seen, in 2007, came from just over the border in Herefordshire, but the next three were all long-distance pioneers from Wales. Another two came from the same site in Herefordshire, arriving in 2009 and 2012, and there were two short-distance travellers from Wales (near Llanbister), arriving in 2010 and 2014. The remaining 12 came from local nest sites, including one of the two that fledged from the first successful nest in 2006.

The first year each of these tagged birds was observed, one, a male, was only one year old, a very rare event. Nine were two years old, and nine were three. Two were four when first seen, but one was known to have bred at a Welsh site as a two year old, and another was seen in Wales, and then near Shelve, in the previous autumn, and probably attempted to breed in Wales as a two or three year old. Excluding these two, the average age of first breeding was 2.4 years.

Excluding the eight still alive at nest sites in 2014, the average age of the other 12 when they were last seen was 4.75. However, some of them may perhaps have lived longer, and changed nest sites, but they were not relocated.

The tagged Kite Yellow/black 37 was photographed near Clun in November 2007 (see p. 173), and found at a nest nearby in the following spring. He was one of the original Welsh pioneers: he nested north of Llangollen, 74km from his natal site near Devil's Bridge, in 2006. He was not observed in 2007, but his 2008 breeding site near Clun was 63km from his 2006 nest. He was at the same site until 2010, and lost one tag during this period, but he was not present at that nest site in 2011 or 2012 (the male there had plumage differences). He was found dead (a road casualty) near Kempton in January 2013, aged 8y 7m 7d, the oldest to date. By then he had lost both tags, so he may have moved to another breeding site without being recognised. His final age is well short of the oldest known Kite, 23y 10 m 18 d.

Another, tagged on 2 July 2006 near Llanbister, was found at different nest sites in 2010 and 2012, the former 12km from his natal site. He was not seen again, but was found dead, with a wing sheared off by a wind turbine, just south of Beguildy on 7 August 2014, 8y 1m 5d later.

In common with other species, such as Raven, female Kites move further from their natal sites than males, presumably a result of natural selection to prevent inbreeding. Eleven of the 21 with read tags were male, and eight female (two unsexed). The average distance from natal to breeding site for all 21 was 17.5km, but the average for the 11 males was 18km and for eight females was 20km (the average for the two unsexed was only 8km). For the 14 tagged and found locally, the average move was 8km, but was 5km for the six males and 11km for the seven females.

Seven are known to have moved nest site. The four males moved only short distances (all less than 4km, average 2.4km) but two of the three females moved further: one nested at three sites, the final one just over the border in Wales, 15km from her natal site; the other moved 10km to her first breeding site in 2011, but returned to within 2km of her natal site in 2014.

There is no evidence from wing tagging that any from the various reintroduction programmes elsewhere in the UK have nested here, although one that fledged in 2005 near Thrapston, in Northamptonshire, was seen at a roost in February 2010 and a month later near Bucknell. However, there are some SBR records of tagged Kites from the Chilterns, and there are many untagged nesting birds, so this evidence is not conclusive, and recent DNA studies (Skujina 2013) provide evidence that an adult breeding south of Clun in 2012 came from one of the reintroduced populations, possibly Dumfries and Galloway, but originally from Spain. The study needs to be extended before its origin can be identified with certainty (Mike Hayward *pers. comm.*).

Although usually the victim of mobbing, by Buzzard, Curlew, Carrion Crow and Raven, recently arrived Kites have displaced Ravens from several long-standing nest sites, and one was seen mobbing and attempting to rob a Peregrine of its kill west of Craven Arms in January 2012.

Recently, SOS members have identified the following tagged Kites:

- Cothercott Hill (20 June 2010) Cerise/black 'H7' from near Abergavenny (72km) in 2009
- Dorrington (7 March 2012) White/black 'W' from near Pen-y-bont (65km) in 2011
- Bridges (21 March 2012) Blue/black 20 from near Aberaeron (95km) in 2010
- Bury Ditches (16 May 2014) Blue/black 'A8' from near Knighton (10km) in 2010

Several Kites often roost together overnight in winter. Around 60 at two roost sites in early 2010 were almost all one year olds, but over half of the tagged birds at the larger roost had come from Wales. The furthest travelled was from Bridgend (104km) and four more were from over 50km away. There were over 30 at a roost in 2012 and in 2014, but few were tagged, suggesting that there were several hundred individuals at that time.

These observations, together with those from the nest and roost sites, and other tags reported to the Welsh Kite Trust, suggest that the large number of youngsters now here comprise a broadly equal mix of local and Welsh Kites, some of which have moved considerable distances.

Some Kites fledged from local nests have also moved considerable distances. One tagged near Clun in 2008 was seen in Pembrokeshire in February 2009, 118km distant (Paddy Jenks *pers. comm.*), while another tagged near Church Stretton in July 2013 was seen in Cornwall in June 2014, 322km distant.

Jim Almond, Stokesay, 8 March 2009

Kites still suffer direct and indirect persecution, and poisoning. The *Shropshire Star* carried a story on 9 February 2013 of one 'peppered with shotgun pellets' near Wem, which eventually died in spite of veterinary treatment. In 2014, the RSPB recovered a dead Kite near the Stiperstones, poisoned by the banned pesticide carbofuran. Also in 2014, at a different location, an adult was found dead on the ground directly under a nest, and two well grown chicks were later found dead in it. There were no visible signs of injury or persecution to the adult, but the landowner had used rat poison in a nearby barn, and post-mortem analysis confirmed this was the cause of death. Rat poison is a common cause of death, and it is found in potentially lethal amounts in most dead Kites sent for post-mortem.

Given the rapid spread from the south-west in only nine years since 2005, the large numbers that are currently too young to breed, and the wide availability of unoccupied habitat, it is likely that Kites will continue to spread, as Buzzards and Ravens have done since the previous Atlas period.

Leo Smith

White-tailed Eagle
Haliaeetus albicilla
Vagrant

This sea-eagle has a breeding range that runs across northern Eurasia, with small numbers in Greenland and Iceland, and a re-introduced population in Scotland. It once bred throughout Britain, but heavy persecution resulted in its extinction by 1918. Indeed, in 1514 the churchwardens' accounts for Worfield show a payment 'to William Hichecox for an yron' suggesting that there was already a bounty on its head 500 years ago.

While a record of one of a pair being shot in 1792 may actually relate to Cannock Chase in Staffordshire, accounts for the nineteenth century provide no evidence of breeding and suggest this species was no more than a sporadic visitor, with perhaps seven records that century, most of which were of immatures, and most were shot. They include four in the winter of 1865–66, at Badger, Chyknell, Halston and in Corvedale, then one at Hawkstone in March 1883, one at Bucknell on 9 March 1892 and one at Dinchope near Craven

Arms on 7 November 1896 'having been swept into Shropshire by the gale'. In the twentieth century, an immature female was shot at Moston, near Hawkstone, in January 1906, an immature was seen at Wellington on 2 April 1911 and one was observed 'by many people, flying over Munslow' on 5 February 1930.

There has only been one modern record:

Venus Pool   26 January 2005

This was also an immature, but it is not known whether it was from the Scottish population or from continental Europe. A sequence of sightings suggest it arrived in Norfolk in early January, was seen in Cambridgeshire, Northamptonshire, and possibly Oxfordshire, before it flew over Venus Pool and continued its tour of central England, finally being tracked to Lincolnshire via Derbyshire (Nickless 2005).

Graham Walker

Rough-legged Buzzard
Buteo lagopus
Very rare winter visitor

This species has a circumpolar distribution, breeding in high latitudes of North America and Eurasia. It is a scarce winter visitor to Britain, mostly along the east coast, particularly East Anglia, where it usually occupies open, lowland habitats; it becomes increasingly scarce towards western Britain. The extreme variability in the plumage of Buzzards (*B. buteo*), including some very pale individuals, can result in identification errors, and the Rarities Review of 'records' of Rough-legged Buzzard proved particularly taxing in this respect, given that many failed to mention most if not all of the key identification features. This, along with other factors such as date of observation and location, resulted in 11 modern records being no longer considered proven and, therefore, they have not been included in this account.

These identification difficulties must also have been the case historically, although specimens were often obtained. At least 10 and perhaps 12 individuals were recorded prior to 1950, with a specimen in Mr Eyton's collection obtained near Ludlow in 1836 being the first. This was followed by mention of a singleton on the Long Mynd in 1841, although it was not reported subsequently, and this may cast some doubt on the record, given that specimens obtained from the Stiperstones and from Vessons Coppice (near Pontesbury), along with the Ludlow individual, are mentioned in subsequent publications (Rocke 1865 and Forrest 1899). In the first week of December 1851 one was killed in Withyford Wood and saved, presumably for preservation, from being hung up to frighten the crows, while in 1877 a pair were taken between Ellesmere and Wrexham, with both being considered birds of the year. One was at Moston, near Hawkstone, in 1889 and a 'fine specimen' was killed near Church Stretton in December 1895. A bird seen near Oswestry on 29 December 1901 was thought to be this species. On 20 February 1904 a wounded male was found and killed at Onslow, having been seen several times in the area over the previous three weeks. Finally, on 9 October 1937, one was seen on the Long Mynd in the company of Buzzards.

There have been just five modern records with sufficient accompanying evidence to support the identification:

Lydham	2 October 1952
Burwarton	3 March 1974
The Stiperstones	3 November 1988
Pen-yr-Estyn, Queen's Head	21 March 2005
Whixall Moss	25 November 2016

The one at Lydham was perched on a tree stump feeding on what appeared to be a rabbit, while the individual at Burwarton coincided with a large influx into Britain (Scott 1978). The last, in 2016, was photographed as it flew over the head of the observer.

In 1967, an individual was reported from the Broseley area for three to four weeks in March/April, a fact that made 'it v. probable that this was a R.L.B.' (SOS record card), and one was picked up dead on the Clun Road out of Knighton on 23 May. These sightings also coincided with a national influx and one Shropshire record, probably the one at Broseley, is listed as one of 10 that were considered to be in addition to the minimum number (57) that 'actually wintered' (Scott 1968). The subsequent review of records decided that both records had elements that cast some doubt about their validity and as a consequence they were removed from the list.

Graham Walker

NB Scott's article talks about birds which wintered remaining until 'late March or even early April' which implies that both Shropshire sightings would have been at the limit in this respect.

Buzzard (Common Buzzard)
Buteo buteo
Fairly Common Resident

The Buzzard inhabits most of Europe and western Asia, but northern populations migrate to Africa for the winter.

In the Middle Ages, Buzzards were common and widespread, but subsequently they were relentlessly persecuted and legally classed as vermin. By 1838 Thomas Eyton wrote that 'now and then I have observed a solitary one, or a pair, in some of the large woodlands'. The same story was reported throughout the 1800s by Beckwith and other observers, and numerous examples were given of birds being 'taken' or sent to taxidermists. Forrest (1899) wrote that it was 'formerly common', that it 'rarely escapes the gun of the gamekeeper' and concluded that 'it is a pity that the Buzzard is being so rapidly exterminated'. Paddock in 1897 even went so far as to class it as 'a resident extinct'.

However, this is one of the few English counties from which the Buzzard was never entirely eliminated, with a few breeding pairs remaining in the Clun Forest at the end of the nineteenth century. The absence of gamekeepers during the First World War enabled it to increase, and the spread to other areas began after the war ended. Nesting was recorded at Ludlow in 1925 and probably as far east as Broseley in 1936. By the 1940s it was not uncommon in the hills in the south-west, Clee hills and Oswestry uplands, and in 1943 Lloyd (using data to 1939) listed the Buzzard among 'species which have showed a marked and general increase'.

This upward trend continued, culminating in a peak in 1954, by which time several pairs were established around Ellesmere. A survey carried out for the BTO located at least 74 nesting pairs, resulting in a population estimate of around 100 nesting pairs.

Unfortunately myxomatosis arrived in 1954, virtually eliminating the rabbit population and leading to a disastrous breeding season for the Buzzard in the following year. The position improved a little in 1956, particularly in the Clun area, owing to a vole plague. A further survey that year located only 20 nests and only 16 young were reared: the population had been reduced to around a quarter of its previous strength.

Forsaking the hills to look for food, many Buzzards came down to lower ground and often perished by the gun, particularly during the organised pigeon shoots. After the crash, numbers slowly increased and the population was estimated at probably 50 nesting pairs in 1963, with most of the breeding records coming from the south and west, but small numbers also bred regularly in the Ellesmere area and at Willey Park, Broseley, where they were protected.

Recovery continued, but was slow due to the adverse effect of organochlorine residues as the use of chemicals in agriculture increased. A further BTO/SOS survey of soaring birds in 1983 concluded that around 80 pairs bred in that year.

Tony Webb, Long Mynd, 3 October 2014

Buzzards are conspicuous due to their large size, effortless soaring and loud mewing call. During Atlas fieldwork, it was often easy to locate territories to establish probable breeding, and breeding was usually confirmed either by seeing adults carrying prey to a nest, or by locating a nest through hearing the distinctive hunger call of the nestlings or recently fledged young. Even so, some pairs can be remarkably elusive for such large birds, and many of the probable records would have related to breeding pairs, while some of the possible records related to non-breeding immatures and prospecting adults.

The *Atlas* (1992) reported that Buzzards had gained further ground since 1983, especially in the east, but also due to infilling in more favoured areas. Breeding was confirmed in five 10km squares where there had been no records for the national BTO Atlas (1976), and in a further nine where only probable or possible breeding had been established previously, reflecting a national increase over this period of around 50%. Though a few tetrads contained two pairs in some years, breeding did not occur every year in every occupied tetrad, which suggested an upper limit of 300 pairs in 1990.

At that time, Buzzards still suffered from irresponsible and illegal persecution. Shropshire was a black spot for the poisoning of raptors, and the Buzzard was the main victim (RSPB 1991). Illegal shooting was still prevalent, and while several injured by shooting were taken in for treatment, those killed outright probably remained undiscovered.

However, fears that the expansion of numbers and range may be limited by a slow recovery of the rabbit population, poor breeding success or continued high persecution levels, have proved unfounded. The number of rabbits rose considerably from 1985 onwards, while stiffer penalties alongside a MAFF-led campaign against illegal poisoning in 1990 reduced the reported incidence of poisoning.

Fieldwork for the recent Atlas found a major expansion of range from their stronghold in the south-west, and Buzzards now breed almost everywhere. The number of tetrads with probable or confirmed breeding has increased from 302 (35%) to 765 (87%).

BBS shows a steady rise in the number of survey squares where Buzzards have been recorded, reflecting the increase in range and a population growth of 85% between 1997 and 2014. They are now more numerous and widespread than at any time in the last 200 years.

Research in south Shropshire and north Herefordshire in the mid-1990s (Sim *et al.* 2000) found that high breeding densities were

Breeding Distribution Change (1985–90 to 2008–13)

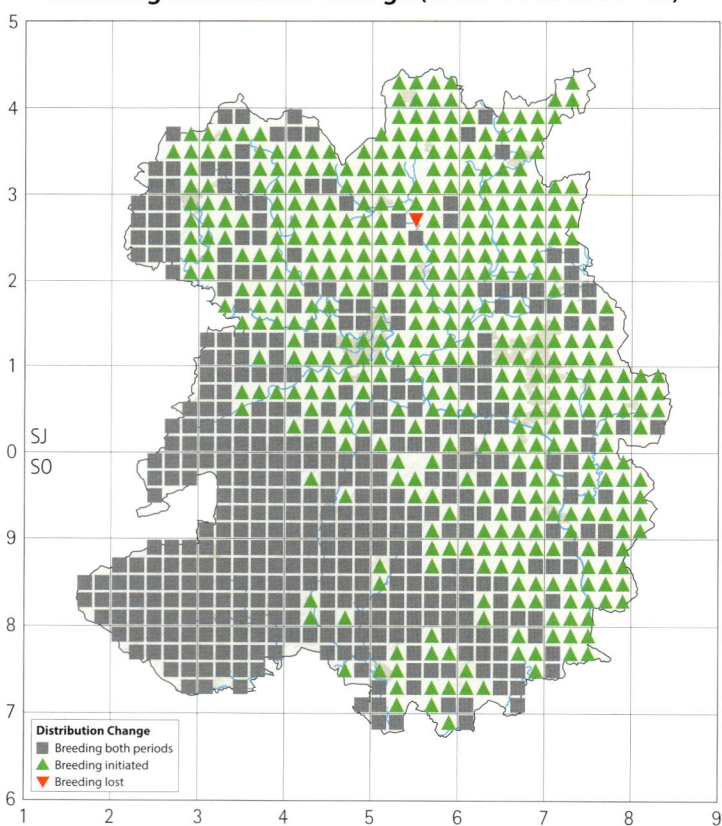

Distribution Change
- ⬛ Breeding both periods
- ▲ Breeding initiated
- ▼ Breeding lost

Occupied Tetrads

Atlas period (breeding)	1985–90		2008–13		Change	
	Number	%	Number	%	Number	%
Confirmed	174	20	448	51	274	157
Probable	128	15	317	36	189	148
Possible	123	14	102	12	-21	-17
TOTAL	425	49	867	100	442	104
Tetrads with Winter Records (2007–13): 870 (100%)						

BBS Trend (1997–2014)

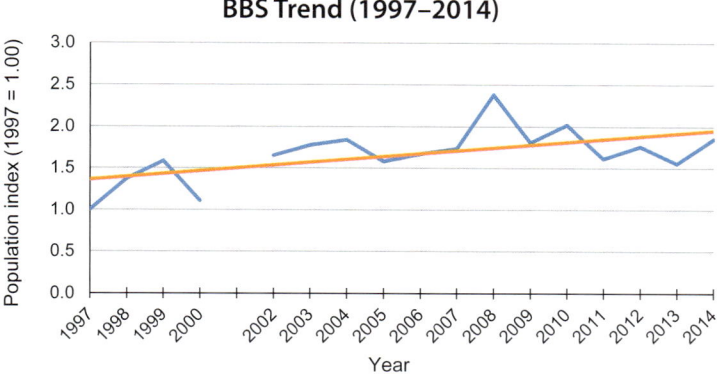

associated with unimproved pasture and mature woodland, with high rabbit abundance close to nests. Rabbits were the main prey item found at nests (56.1%), with corvids (mainly fledgling Crows and Magpies) the second most common (18.2%). In spite of a widespread belief to the contrary, small birds and Pheasants were not common prey items (9.9% and 5.1% respectively), and the latter were probably scavenged road kill, rather than taken as live prey.

The relative abundance map confirms this habitat preference, with much higher densities found in the hills in the south-west, the Oswestry uplands, Clee hills and Wenlock Edge, and much lower densities in the extensive arable areas on the north Shropshire Plain and Severn Valley.

Further research, again in south Shropshire and north Herefordshire in the mid-1990s (Sim *et al.* 2001), repeated the survey of soaring birds by the same method as used in 1983 and found more than twice as many (an increase of 118%) in the core area. All the nests in the survey area were also located, and the research found a good correlation between the number of soaring birds with the number of nests (77 and 88 respectively).

The maximum number of Buzzards seen in a one-hour TTV count will probably be approximately equal to the number that would have been seen on a soaring bird survey. In 2008–11, the sum of the maximum counts was 2,194, which suggests around 2,500 nests.

The same research found 81 territorial pairs and 71 breeding pairs in SO37 in 1994 and 1995, with the BBS results suggesting that this density is now replicated in all areas. The average TTV count per

tetrad per hour in SO37 of 1.53 was the same as that found in the 824 tetrads where Buzzards were located, suggesting a population of 2,670 territorial pairs and 2,340 breeding pairs.

Based on TTV counts, the population estimate for the recent Atlas period is 1,836–2,525 pairs.

As these three different ways of estimating the population produce similar results, this suggests that the number of Buzzards lies within a range of 2,000–2,500 breeding pairs.

Young Buzzards do not usually breed until they are three years old. Nationally, only 63% reach breeding age, and the typical lifespan of those that do is around 12 years. Thus the population each winter includes around 12% of young birds, who forage outside the breeding territories of the adults, and Buzzards were seen in every tetrad during recent Winter Atlas fieldwork. The total number of Buzzards counted during winter TTVs was 12% greater than the number counted in the breeding season, matching the number of birds of the year. The relative abundance map is similar to the breeding season map, but there was a marked lack of Buzzards on the Long Mynd, and fairly high densities in some tetrads containing mainly arable land where few were found in the breeding season. This perhaps reflects the tendency of Buzzards to feed on worms on recently ploughed land, as noted in several SBRs.

Comparison of the distribution and relative abundance maps from the recent Winter Atlas with the relevant part of the national BTO Winter Atlas (1986) map also shows a big increase in range, and in the number of Buzzards in those 10km squares that were occupied in 1981–83.

It is fitting that the species at the centre of the SOS logo is one

Tony Webb, Venus Pool, 2 January 2013

Breeding Relative Abundance (2008–13)

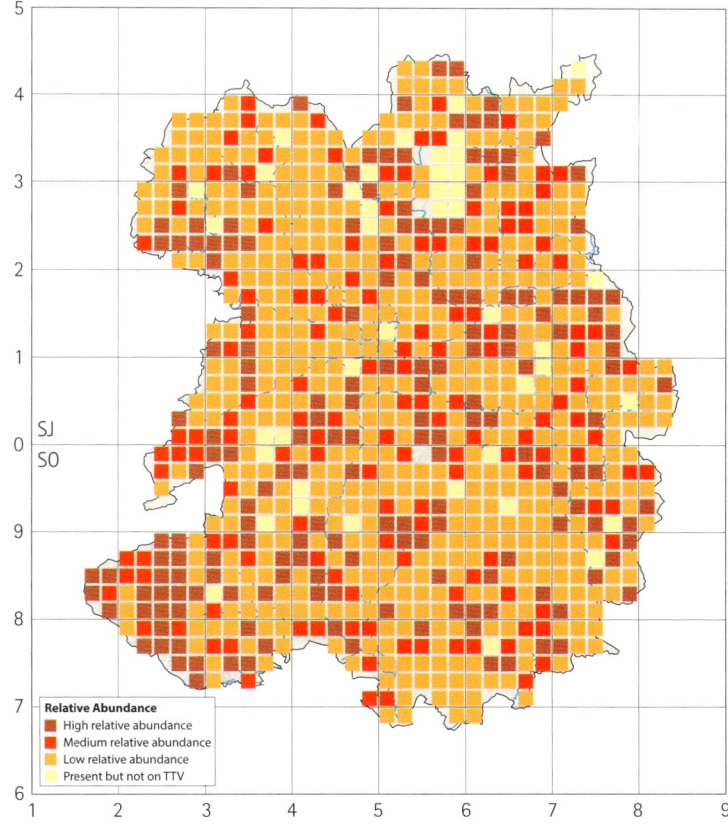

Relative Abundance
High relative abundance
Medium relative abundance
Low relative abundance
Present but not on TTV

of the most frequently reported, in terms of both total records and also the number of different locations, in all months of the year. Up until the 1990s, most records came from the hill country in the south and west, but since then there has been a steady increase in records from the east and north, and by 1998 it was 'almost as common in the northern and central (areas) as it is in its former stronghold in the south'. Most records now come from the north-east quadrant, and reflect the distribution of observers, rather than the Buzzards, but the increasing number of such records confirms the spread.

Buzzards now often flock to soar on thermals, usually in spring, and they form large groups in autumn on recently ploughed fields where worms are easily accessible. The first SBR record of more than 10 together was in 1986, and the number of such records has steadily grown: 12 were received in 2005, 15 in 2006 and 14 in 2009. There have been 12 records of more than 25 together, and largest soaring flock noted was more than 70 over Upper Dinchope with two Kites on 6 April 2010. Notable feeding groups include 'an astounding 63 in a field of recently sown winter wheat at Lower Broughton Farm near Bishops Castle on 27 October 2003, with at least another 10 on a nearby field'.

SOS records also confirm that rabbits are a common prey item, and frequently report mobbing between Buzzards and corvids and other raptors. Plumage is very variable, and some very pale individuals, occasionally almost white, are often seen.

A total of 601 Buzzards have been ringed and, of these, 23 have been recovered, which show the species is very sedentary. Nine were ringed and recovered here, and all except three of the rest have been recovered in adjacent counties; the exceptions went as far as Anglesey, Dyfed and Gloucestershire. Only one Buzzard travelled more than 100km, to Anglesey (121km), in only 3 m 28 d from being ringed as a nestling near Snailbeach in 1970. In addition, four Buzzards ringed in other counties have been found here, of which only one, from Dyfed, did not come from an adjacent county.

Nationally, the longevity record for this species is 25y 6m 26d, set in 2009. The oldest here was a nestling, ringed in 1995 near Clungunford, which was found freshly dead (hit by a car) in 2009 near Leintwardine. The distance travelled was only 2km south, but its age was 13y 10m 7d. The other instance of local longevity was a nestling ringed in 1994 near the Hope, and found dead (sick – natural causes) 24km distant at Wyre Forest, Bewdley (Hereford & Worcester), 12y 1m 18d later.

Buzzards are still persecuted here, and elsewhere some Pheasant shoots are applying for licences to 'control' them, at least two successfully. Near Kempton, the records collected for a criminal prosecution showed that 102 Buzzards were killed by a keeper during part of 2007, only a small proportion of the total number killed on that estate over a longer period. In 2014, another gamekeeper was filmed by RSPB operating an illegal trap designed to catch Buzzards near a Pheasant pen, but was not convicted because the court refused to accept the film as evidence.

The large size of groups of Buzzard feeding in the immediate vicinity of the estate near Kempton, before and after this illegal activity (41 at Kempton on 1 February 2006 and 47 at Lydbury North on 17 October 2010), no doubt partly sustained over the winter by Pheasant carrion and road kill, shows that such persecution has little impact on the number of Buzzards in an area, as others immediately move in to fill any gap if food supplies are good. Attempting 'control' will be equally futile, but this legal and illegal persecution must be resisted if the future of the Buzzard is to be secured.

Leo Smith

John Hawkins, Ellesmere, February 2006

Little Bustard
Tetrax tetrax
No modern records

This rare vagrant breeds from the Iberian Peninsula through southern and central France to central and western Asia. Those in southern Europe are mainly resident, but other populations migrate further south in winter.

The single record is of a female, often referred to as shot 'by a labourer', on Edgmond Marsh in the spring of 1883 (Beckwith 1887). The specimen was passed to Shrewsbury Museum, where it was documented in 1905, and it is now in Shropshire Museums in Ludlow Museum Resource Centre (specimen SHYMS: Z/2006/159).

John Tucker

Water Rail
Rallus aquaticus
Rare resident, scarce winter visitor

Water Rail, or 'water-weasel' as it was known around Ellesmere, is rarely recorded in the breeding season unless pig-like grunts and squeals betray its presence in waterside vegetation. It is so exceptionally difficult to monitor, and its status so poorly understood, that the RBBP has added it to the list of species for which site data should accompany all records submitted.

Forrest, while alive to the difficulties, believed it to be 'not really uncommon', but documented only two instances of confirmed breeding in the last two decades of the nineteenth century. *The Handlist* (1964) noted breeding at two sites, near Oswestry and Bagley, while the

Paul King, Chelmarsh Scrape, 20 September 2010

Confirmed & Probable Breeding Distribution Change (1985–90 to 2008–13)

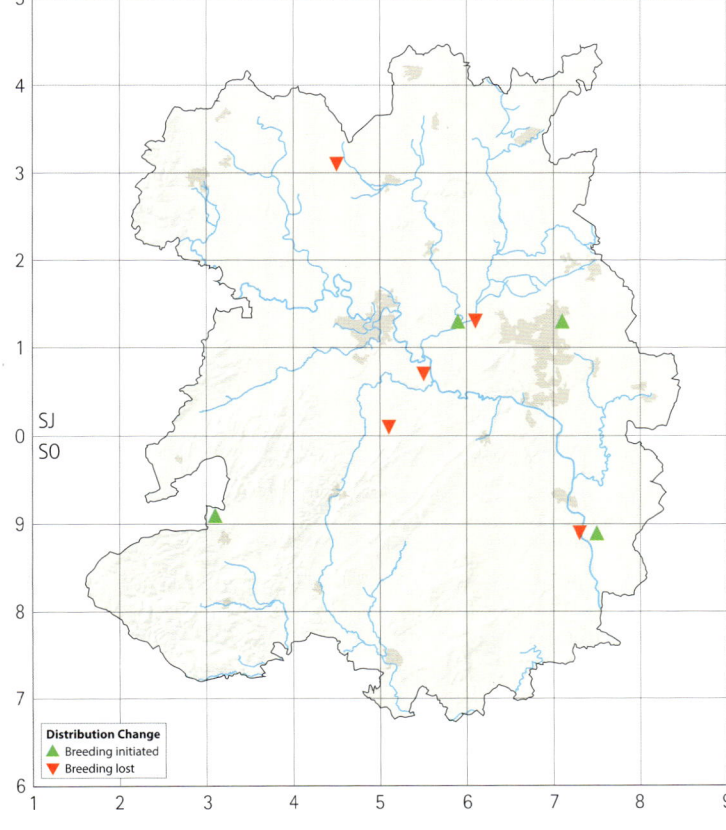

Distribution Change
▲ Breeding initiated
▼ Breeding lost

Atlas (1992) confirmed breeding only at Chelmarsh; at four further sites there was evidence of probable breeding, which may bear extra weight in a species of such low detectability.

During recent Atlas fieldwork, breeding was confirmed at Mirelake, and considered probable near Owlbury Hall (Bishops Castle), Dudmaston and Priorslee.

At Chelmarsh, where breeding was confirmed in four of the six years 1985–90, two Water Rails, possibly a pair, were recorded in late April 2010, and a juvenile seen in July of that year may have hatched there, or nearby. In late May and early June 2014, calls were

Occupied Tetrads

Atlas period (breeding)	1985–90		2008–13		Change	
	Number	%	Number	%	Number	%
Confirmed	1	0	1	0	0	0
Probable	4	0	3	0	-1	-25
TOTAL	5	1	4	0	-1	-20
Tetrads with Winter Records (2007–13): 51 (6%)						

Winter Distribution (2007–13)

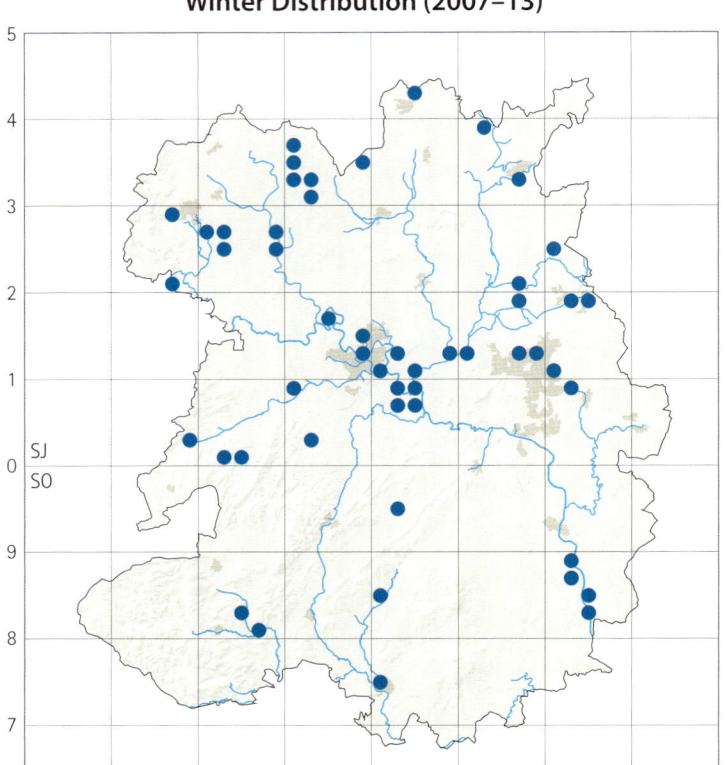

Sites where Water Rails were recorded in April and July have been excluded from the breeding map: the former are likely to have been late-departing migrants, the latter early returners; with such sites yielding only a single observation in early May, and none in June, it seems unlikely that breeding took place. This is a more rigorous interpretation than that applied in the *Atlas* (1992); as the data are not directly comparable, records of 'possible' breeding (including Chelmarsh 2010) have not been included on the breeding distribution change map nor in calculating changes in breeding status in the tetrad occupancy table. Nevertheless, the population estimate of only two to three pairs, based on TTV counts, may be on the conservative side: RBBP's current view is that Water Rail is 'probably more numerous than previously thought'.

In winter, migrants from the continent swell the population, and these rails are more visible when the vegetation has died back, and they are driven into the open by hard weather. Water Rail was recorded in 10 times as many tetrads in winter (51, or 6% of the total) as in the breeding season. Clusters of records came from around Venus Pool, Chelmarsh, Allscott Sugar Factory, Ellesmere, Chetwynd Park, Kempton, Tibberton, reed-fringed lakes in Telford, and the river and pools in Shrewsbury, with additional records from other scattered sites. The highest count during winter Atlas fieldwork was four at Chelmarsh Reservoir on 22 December 2007, with four calling at the same site on 18 March 2009.

As would be expected of such a skulking species, very few are ringed: in recent years, the 2007 total of six was exceptional; one or two, if that, is usual. Recoveries are scant: eight of the nine ringed here were recovered locally, the exception being a first-year female ringed at Shrewsbury Sewage Farm on 31 January 1988 and found dead 20 days later in Norfolk; that was a mild winter, when Water Rails can begin their return migration as early as February.

Michelle Frater

recorded at the same site in two consecutive weeks, followed by a sighting of an unfledged juvenile in July. This confirmed breeding record is outside the recent Atlas period, but reinforces the possibility that breeding did occur in 2010.

Corncrake (Corn Crake)

Crex crex

Very rare passage migrant; has bred

Corncrakes breed mainly in northern Britain and central and eastern Europe, and winter in east Africa.

They are likely to have begun their close association with man in the earliest days of animal husbandry and cereal cultivation, perhaps 10 or 12 thousand years ago. Hay fields, and to a lesser extent, arable crops, provided ideal breeding grounds, and this situation pertained until early in the twentieth century, when Corncrakes were common, especially in the fertile valleys. Apparently numbers varied from year to year, but their unique and unmistakable calls were regarded as a harbinger of spring.

Early authors referred to their ubiquity in meadows, arriving from sub-Saharan African grasslands typically in April and May. Lloyd (1938), summarising all extant data, produced the earliest arrival date of 29 March with an average of 4 May, and the latest recorded on 5 November. This excluded some very late records, which interested Beckwith. He surmised, with some contemporaries, that a few birds over-wintered, perhaps as a result of some infirmity. These records

included four in November, four in December and three in January, but, as Forrest (1899) pointed out, there were then, and still are, no February records to suggest genuine over-wintering here. Beckwith (1879) described how 'When near hatching [they] are singularly tame, and often continue sitting after their nest has been mown over, and while haymakers are working close by, without shewing [*sic*] any signs of alarm'. That was in the days of haymaking by hand and it was the mechanisation of that process, and the general intensification of land use, even in the early twentieth century, which was to cause their demise.

Corncrake was a widespread nesting species up to 1910, but the first mention of a decline in numbers came in 1911, and from then on similar remarks appear, for example in 1928, 'Never heard now Bridgnorth district' and 'A few still occur in Oswestry district', while, rather later, 'None recorded hereabout for many years past' at Posenhall, in 1940.

The *Handlist* (1964) stated that 'records are much more scanty' after 1925, but noted breeding in four areas in the late 1950s: Bishops Castle in 1957, 1959 and 1960, Broseley in 1958, and probably at West Felton in 1959 and Westbury in 1957, 1958 and 1959. Calling was heard in most years from one or two widely scattered areas in May, often for a few days, but it soon ceased and it was 'assumed that the birds pass on northwards'. A farmer at Northwood commented that Corncrakes ceased breeding in a five-acre meadow there in the 1960s when he switched from hay to silage making (Joan Daniels, Natural England, *pers. comm.*). SOS records from 1956 onwards have declined decade by decade: 24 between 1954 and 1959, 15 in the 1960s and 14 in the 1970s, eight in the 1980s and none in the 1990s. There were seven records during the 1985–90 Atlas period (10 were published, but three are now considered not acceptable), all believed to be of passage migrants. The last confirmed breeding was near Wem in 1976 (Atlas 1992). In the twenty-first century there have been only three occurrences, one calling near Wem from 23 May to 6 June 2001, a migrant brought in by a cat at St Georges, Telford, on 12 September 2010, which was taken into care at Cuan House (see photo) but later died, and one found dead near Edgmond in August 2016. None were recorded during recent Atlas fieldwork.

Intense conservation efforts in the Western Isles, where a small natural population has survived, and a successful reintroduction

Jim Almond, Cuan House, 14 September 2010

programme on the Nene Washes in Cambridgeshire, have not reversed the much greater losses in the rest of Britain, so the occasional passage Corncrake appears to be the best we can now hope for.

John Tucker

Little Crake

Porzana parva

No modern records

This diminutive and retiring species of permanent swamps is a summer migrant from Africa, normally restricted to Eastern Europe but sometimes straying west to Britain as a vagrant. The single record, first published in the *Shrewsbury Chronicle*, was of 'a specimen ... shot in November 1898, at Petton Park, seven miles north of Shrewsbury' (Forrest 1900).

John Tucker

Spotted Crake

Porzana porzana

Very rare passage migrant

This inhabitant of fen meadow and sedge swamp was much more common in the nineteenth century prior to the onset of extensive land drainage for agriculture which removed much of its breeding habitat. A similar picture probably applies across temperate Western Europe, where it now only has a thinly scattered distribution with the bulk of the breeding population found in Belarus, Russia and Romania. It is largely migratory, with most Western European birds over-wintering in east Africa, although a few individuals are thought to remain in countries bordering the North Sea. On migration, it is easily overlooked, and the 50–70 birds usually reported in Britain each year, including an average of 25 singing males, is probably below the actual number.

The Spotted Crake's increasing scarcity is reflected in the historic record: while in 1866 it was recorded as being 'often obtained on the banks of the Severn and in the neighbourhood of Shrewsbury', 12 years later it was noted as 'an occasional and rather rare visitor'.

The following year it was recorded as 'a spring and autumn migrant, perhaps sometimes breeding', although in 1899 Forrest stated that it had 'never nested here'.

By 1886 it had become rare 'owing to the reclamation of so many bogs and marshes', although it was still thought to visit the Severn Valley above Shrewsbury, and the areas around West Felton and Kinnersley. Interestingly, in the same year, it was also reported as being occasionally obtained by shooting parties after Snipe in December.

At that time none were thought to over-winter, but in 1908 Forrest mentioned a report he had received of a pair that had 'stayed through a winter at Ludlow, taking up their quarters beneath a wood-stack; from time to time they were seen to run out and capture a worm, retreating again to devour it'. Unfortunately, no date was provided and no other reference to this event has been found, and there must be some doubt about the record, particularly given the unusual behaviour.

By the end of the nineteenth and start of the twentieth centuries, it was largely considered to be an autumn migrant, with several reports of birds being killed. This included one on 15 September 1904, when two out of three birds seen at the Mere Pool, Shrewsbury, were shot by the keeper. Three were also found dead under telegraph wires, one at Rednal on 9 April 1892, one at Montford Bridge on 30 September 1905 and one along the Berwick Road, Shrewsbury on 8 October 1910.

There were then no further reports until 1986, the first of six modern records:

Chelmarsh	29 June–9 August 1986
Chelmarsh	5 August–28 September 1992
Venus Pool	28–30 August 1996

Chelmarsh	15–18 August 1999
Chelmarsh	19 August–30 September 2001
Whixall Moss	25 April–2 May 2014

Apart from the one at Chelmarsh in 1986, which was often heard calling during its extended stay, but seen on only two occasions, and the adult calling at Whixall Moss, the remainder were juveniles. The second, at Chelmarsh in 1992, was regularly seen, while the third at Venus Pool was located when engineering works were being undertaken on the central island. The juvenile at Chelmarsh in 1999 was trapped, ringed and photographed, while the fifth individual was only seen occasionally, even though it was present for six weeks.

Graham Walker

Moorhen (Common Moorhen)

Gallinula chloropus

Common resident

The Moorhen has always been familiar and well liked. In the earliest known accounts, it was described as being 'common on every river, pond and brook' (Rocke 1866b). and 'abundant by all our rivers, brooks and pools, often frequenting the latter when close to houses, and leading a semi-domestic kind of life' (Beckwith 1879). He goes on to describe 'some very pretty Pied Moorhens in Attingham Park, having the whole of the back and wings mottled with white, and the bill and legs bright straw colour'.

The Moorhen in the nineteenth century was also variously known as 'moor-hen', 'moor hen', 'waterhen' and 'water-hen'. Confusingly, 'the name "coot" is very generally applied to the moor-hen, the coot proper being always distinguished by the prefix "bald-headed"' (Forrest 1908), although Forrest himself used the modern 'Moorhen' in his *Fauna* (1899).

It was described as 'resident, common' in the *Handlist* (1964), but 'not so common on the high ground' and 'numbers are recovering from the decrease after the 1962–63 winter'. SBRs contain few noteworthy reports, but in 1978, 120 confirmed territories were recorded along the River Severn from Melverley to Shrewsbury, a pair was breeding high up on a small pool on Brown Clee in 1979, and there were 55 on Brown Moss on 17 September 1980.

SBR (1981) noted that, although common and widespread, it was rarely mentioned by observers; for example, it was reported as present on the meres, which seems to be the earliest record specifying that particular area, although it had presumably been present there for many years. There were also records from the uplands, at Wildmoor

Pool and near the Stiperstones, and some small sites held good numbers, such as 30+ at Weeping Cross Pool on 17 February, and up to 25 at Venus Pool.

The *Atlas* (1992) described it as one of the most characteristic birds of the northern plain, but more scarce south and east of the Severn Valley, due to the absence of small pools and meres, but 'absence of Moorhens from the uplands of the Long Mynd, Stiperstones and Clun Forest is due to their dislike of fast flowing streams rather than

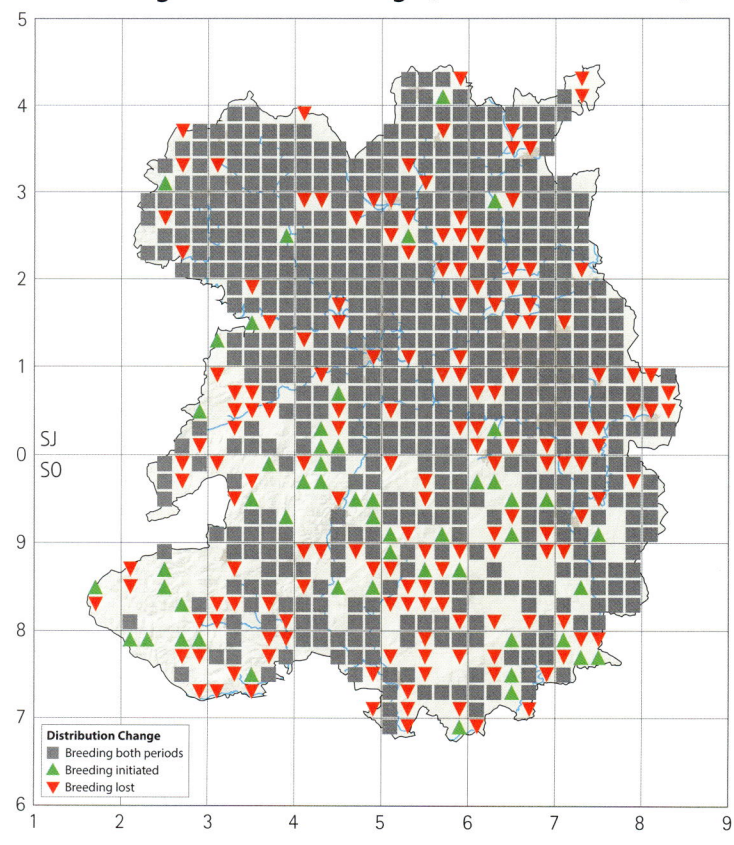

Breeding Distribution Change (1985–90 to 2008–13)

Distribution Change
- ⬛ Breeding both periods
- 🔺 Breeding initiated
- 🔻 Breeding lost

Occupied Tetrads

Atlas period (breeding)	1985–90		2008–13		Change	
	Number	%	Number	%	Number	%
Confirmed	618	71	391	45	-227	-37
Probable	37	4	102	12	65	176
Possible	63	7	114	13	51	81
TOTAL	718	83	607	70	-111	-15
Tetrads with Winter Records (2007–13): 531 (61%)						

Jim Almond, Venus Pool, 5 May 2007

altitude *per se*, as testified by confirmed breeding on Boyne Water at 455m on Brown Clee'.

Breeding at altitude was noted again when three pairs, one with young, were recorded on the five pools near the summit of Brown Clee, at heights of up to 500m (SBR 1994), while a noteworthy 94 were counted in February along a stretch of the River Severn near Shrewsbury (WeBS 1999).

The tetrad occupancy table and the breeding distribution change map indicate a marked decline since 1990. The total number of tetrads with breeding evidence declined by 15%, while the breeding distribution change map shows some gains but more losses, with 163 tetrads (21%) apparently no longer occupied. While the latter tetrads are widely distributed, many of them are grouped, with several clusters of five or six. However, 43 tetrads show gains of confirmed or probable breeding, with over three-quarters of these in the south, at varying altitudes, including at Wildmoor Pool at 420m.

Although on some waters, such as the Mere (Ellesmere), Moorhens will tolerate human presence, they are generally shy and readily seek cover, where they may stay hidden and completely still for long periods. This behaviour seems likely to result in them being under-recorded, particularly on larger areas of water with plentiful surrounding vegetation, and on small farm ponds, streams and even water-filled ditches, especially those away from buildings and public access, where they are likely to have taken cover before being seen.

Nationally, ponds are the most used breeding habitat, and rivers and lakes are also commonly used, with other types of suitable habitat used less frequently. The local distribution of this habitat is reflected on the breeding season relative abundance map.

Wide annual fluctuations in population can result from the Moorhen's potential for high reproduction rates to compensate for its susceptibility to severe winter weather. Recently, the winters of 2009–10, 2010–11, early 2013 and 2013–14 were all periods of harder than average weather, when numbers are likely to have decreased, perhaps contributing to the recent decline.

Although Moorhen is not found in enough survey squares to produce a local BBS trend, there has been a 19% decline in the West Midlands between 1995 and 2014. The net loss in occupied tetrads between the two Atlas periods of 15% suggests a similar reduction in the breeding population here. BTO reports a 'probable shallow decline' in the national long term trend, with an increased failure rate of nests and a shallow decline in the number of fledglings per breeding attempt over recent years. It seems reasonable that these trends apply locally. Several possible causes can be postulated.

Habitat loss noted in the *Atlas* (1992), deterioration or loss of farm ponds and clearance of emergent vegetation, has continued, and drainage of suitable marshy areas may be another cause, but these losses have been partly offset by an increase in newly created areas of water associated with mineral extraction, irrigation and ditches. Dredging of rivers and streams, especially when all bankside vegetation is removed, make them generally unsuitable for nesting, and the growing number of pleasure boats on canals, along with the increased installation of metal sheathing to protect the banks from erosion, make these waterways also much less attractive. Indeed with boating now occurring throughout the year, the 32 reported from 'the Shropshire Union Canal near Colemere' in late November 1982 are certainly not seen now – none have been reported recently on this section of canal.

Large numbers of Moorhens – adults, chicks and eggs – are taken by predators, which include corvids, raptors, foxes and herons, but that has probably always been the case. However, feral American mink may have had a recent impact on the population. Predation by mink was first reported in 1985 along the River Severn in the Atcham area, and a WBS site on the River Tern, 'recently colonised by this mammal, has suffered a 75% reduction in Moorhen numbers' (*Atlas* 1992). Mink was first confirmed breeding in the wild in Britain in 1956 following escapes and deliberate releases from fur farms, and by December 1967 it was present in over half the counties of England and Wales. Numbers have increased rapidly in the last 30–40 years, and it is now common and widespread. Although there is little direct evidence on the extent and growth of the local mink population, it is still regularly seen, for example, at Venus Pool and along the Rea Brook in Shrewsbury (Graham Walker *pers. comm.*). Predation by mink is thought to be one of the main reasons for the decline of the

Water Vole. Vincent Wildlife Trust carried out a national survey in 1989–90 and a repeat in 1996–98 showed a population decline of 88% in only seven years. A local Water Vole survey conducted by SWT in 1992 confirmed this widespread disappearance, with the remaining stronghold in the north, especially around Whitchurch (Shropshire BAP). Moorhens use the same habitats in the same stronghold, so the conclusion in the *Atlas* 1992, that 'the growing population of mink, may however be the most serious threat to the Moorhen' appears to have been prophetic. Based on TTV counts, the breeding population is estimated at 4,000–4,200 pairs, suggesting that it was at the higher end of the range of 3,500–7,000 given in the *Atlas* (1992).

There are more than twice as many occupied tetrads at over 250m in the breeding season than in the winter, reflecting the seasonal movement to lower areas, and congregations on larger areas of water, late in the year. The winter relative abundance map shows these concentrations on larger waters, which lie mainly in the north, with the suitability of the many medium sized lakes across Telford showing up clearly. Based on TTV counts, the winter population is estimated at between 4,250 and 4,500 individuals. However, around 30% more Moorhens were counted per one-hour TTV in winter. Even allowing for easier counting on larger waters at that time, as breeding territories will normally hold two each, and winter counts include birds of the year, this suggests that the winter estimate is much too low. National ringing studies show Moorhens are very sedentary, and there is no evidence for an exodus, or an influx of winter visitors.

Sites where WeBS has recorded in excess of 20 during winter months since 2000 include Allscott Sugar Factory, Whixall Canal Floods, Middle Pool, Polemere, Shrewsbury Sewage Works, Trench

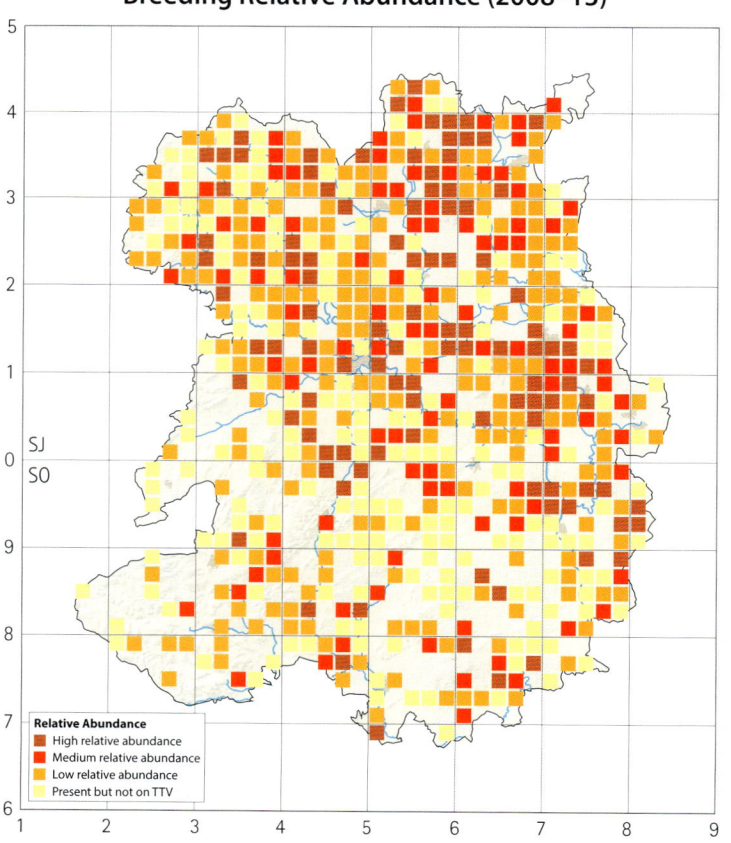

Breeding Relative Abundance (2008–13)

Relative Abundance
- High relative abundance
- Medium relative abundance
- Low relative abundance
- Present but not on TTV

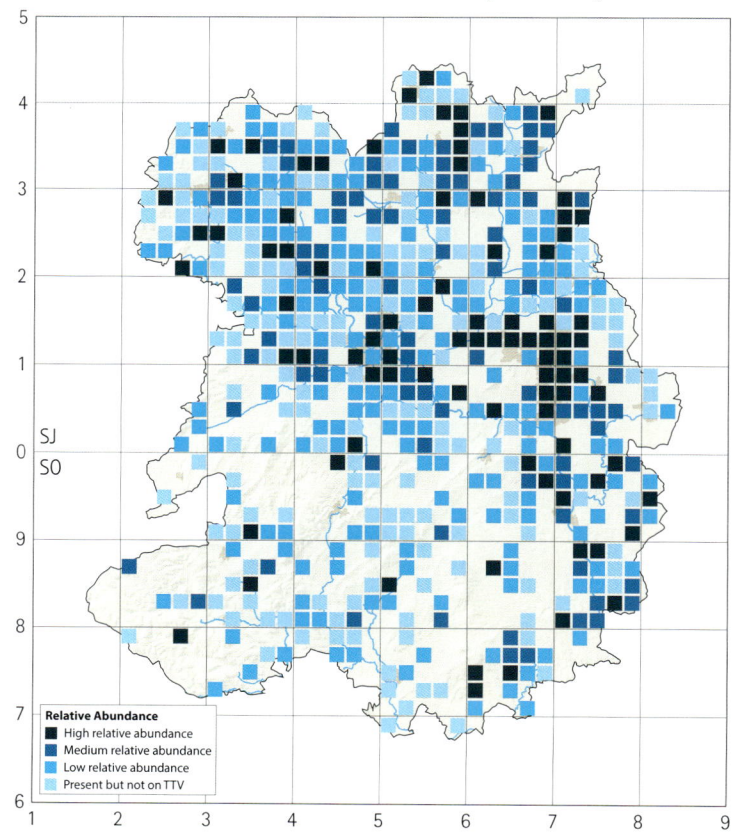

Winter Relative Abundance (2007–13)

Relative Abundance
- High relative abundance
- Medium relative abundance
- Low relative abundance
- Present but not on TTV

Dawn Micklewright, Wood Lane, 22 February 2013

Pool, Walcot Hall Lake and Venus Pool. These same sites produced large counts for the recent winter Atlas too, the largest being 30 at Venus Pool on New Year's Day 2008.

The *Atlas* (1992) stated that 'the many hundreds of small meres and pools (of the north Shropshire plain) each support one or more pairs'. Statements along these lines are found in the earliest written records. Might this idea of 'a pair in every pond' have been repeated down the years without the hard evidence to back it up?

In an unpublished personal survey, 'Moorhens and ponds – a study of habitat requirements', carried out in 2006 between February and July, two or three visits, each lasting about 30 minutes, were made to 54 farm ponds all within 5km of Ellesmere. Although not selected randomly, but for ease of access, the ponds appeared to be representative, varying both in size, and in the amount of emergent and overhanging vegetation. At 25 of the 54 (46%), Moorhen were observed on one or more visits, but none were found to hold more than one pair. Of these, 16 (30% of all ponds) provided confirmed or probable breeding evidence. A total of 40 adults were found, so in some cases both birds in the pair were seen, but at around a third of ponds, only one was found. This study, in one small but probably typical area in one season, in what is considered to be the most frequently used habitat, found that just under half of the pools appeared to hold Moorhen. This result partially reflects the recent decline, but it also confirms how unobtrusive they are, and the likelihood that some are overlooked. It also at least challenges the assumption of 'a pair on every farm pond'.

Few are ringed, but the recoveries confirm the Moorhen's very sedentary nature. Six of the nine were ringed and recovered here, with the other three being found in Lancashire, Montgomeryshire and Staffordshire. None ringed elsewhere have been recovered here.

David Farncombe

Coot

Fulica atra

Fairly common winter visitor, uncommon resident

Found throughout the Palearctic, those from western areas move south and west in winter.

'Few people are aware of the great beauty of the young of this bird when first hatched. Having opportunities of seeing them very often when just leaving the shell, I can fully bear testimony to the correctness of Mr Wolf's beautiful drawing in the "Birds of Great Britain"'. These early comments by Rocke (1866a) seem unusual for a time when most birds were viewed along a gun barrel, and even with modern optics these sentiments are seldom heard today.

Coot were common in the mid-nineteenth century, frequenting larger pools than Moorhen and resorting to rivers during freezing conditions. Most suitable habitat is in the north and its scarcity elsewhere was illustrated by Paddock, who had never met with one near Church Stretton, and wrote 'Mr W.S. Buddicom, some years ago, when walking home from church, noticed one of these birds skulk into the hedge bank on the road side, and secured it, after showing it to some admiring rustics, one of whom remarked "what a big moorhen", he liberated it'.

The *Handlist* (1964) commented 'it breeds on the larger meres and pools throughout ... In winter considerable flocks congregate on larger waters, particularly the Ellesmere group of meres which may hold up to 500'. In addition, Shavington Pool, Trench Pool, Venus Pool and Oerley Reservoir occasionally held over 100, and it suggested that lakes not regularly visited were also likely to have large numbers.

National Wildfowl Counts started at the Ellesmere meres in 1960 and although initially Coot were only counted in January and March, the totals at the Ellesmere meres proved to be higher than previously thought and regularly exceeded 500. From 1982 they were counted monthly and it was soon revealed that numbers generally peaked at the beginning of the winter period when residents are joined by migrants from north-west Europe. Prior to the introduction of monthly counts these influxes may have gone unrecorded. Regular monitoring may

Dawn Micklewright, Wood Lane, 16 February 2014

partly explain the increase shown at the Ellesmere meres compared with the *Handlist* account, but it is a species whose numbers fluctuate greatly between years. Despite annual variations, there has been a marked reduction in those wintering this century; similar decreases in Ireland and Wales pre-dated the local situation, while that for England began later. Climate change, allowing migrants to remain nearer to their breeding grounds, is thought to be responsible, with western wintering populations being the first to show a change. The timing of the local decrease fits this pattern.

The Ellesmere meres are the most important site for Coot and the chart shows the maximum count during each winter period since monthly WeBS monitoring was introduced. The four largest totals, 1,062, 967, 760 and 716, occurred in November 1994, October 1999, September 2004 and October 2007 respectively, and in each of these years the September total was already exceptionally high, ruling out any cold weather movements being responsible. These peaks did not coincide with unusually large numbers elsewhere, although the highest national total was recorded during the 1999–2000 winter.

Many migrant Coot are already present when the WeBS starts in September and maximum numbers are generally found in the early winter period. The fall towards the end of the year suggests onward movement or dispersal, prior to the main departure in February. A colour ringing study noted in the BTO winter Atlas (1986) suggested that Coot make frequent small-scale movements throughout the winter, and local movements to large waters during severe weather are also known to take place. Throughout the 2010–11 winter, numbers at the Ellesmere meres remained low, but the highest count of 191 was made during December, the coldest on record, with a temperature of –12°C recorded during the WeBS.

The chart shows the combined monthly WeBS totals from the

Winter Relative Abundance (2007–13)

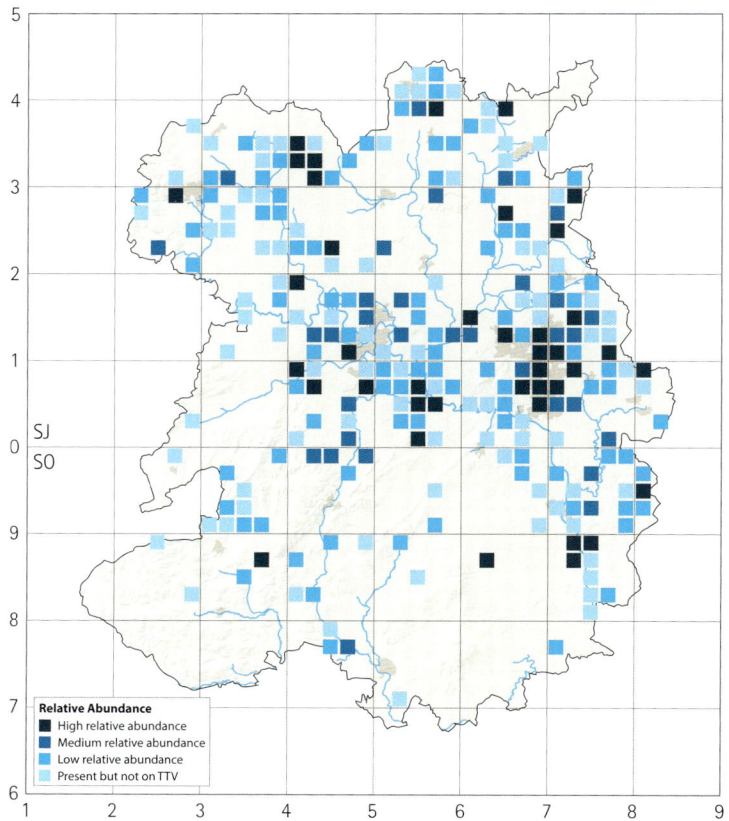

Relative Abundance
- High relative abundance
- Medium relative abundance
- Low relative abundance
- Present but not on TTV

Ellesmere meres during the recent winter Atlas period, and reflects the annual pattern of initial build-up and then fall.

No other locations hold consistently large flocks and the maximum gatherings during the recent Atlas period were 242 at Trench Pool in November 2010, 213 at Priorslee Lake in January 2010, 120 at Venus Pool in November 2007, 108 at Marton Pool (Baschurch) in December 2008 and 101 at Shavington Park in September 2008.

During the winter, Coot are found on most areas of open water, including temporary floods, and a few are reported from the lower reaches of the River Severn. The winter distribution reflects these areas, and the relative abundance map highlights the large number of large pools in the Telford area.

Within this period, maximum WeBS totals from all sites varied between 1,087 in November 2007 and 563 in November 2011. Although all the main waters are included in the WeBS, Coot are widespread and those inhabiting the numerous smaller pools will not have been included, so the actual total will be much higher. TTVs located 1,612 in 118 tetrads during the early visits, a total boosted by a count of 500 at the Mere (Ellesmere) in November 2007, and 1,084 in 138 tetrads during the late visits. However, Coot were not found during TTVs in around half of the 251 tetrads shown on the winter distribution map. Large water bodies are unlikely to have been missed during the TTVs, and tetrads lacking these features averaged 4.4. This would increase the winter population estimate to between 1,600 and 2,200 during the recent Atlas period, although more recent trends suggest that the lower figure may now be more appropriate. This is in keeping with the population estimate based on TTV counts of between 1,300 and 1,600 individuals.

WeBS Counts at the Ellesmere Meres (1982–83 to 2014)

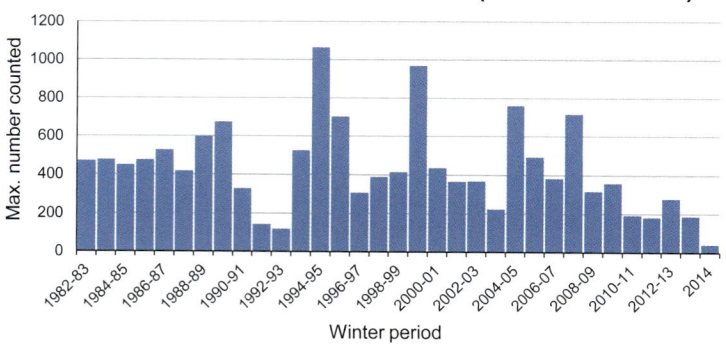

WeBS Counts at Ellesmere Meres: Winter Atlas Years (2007–13)

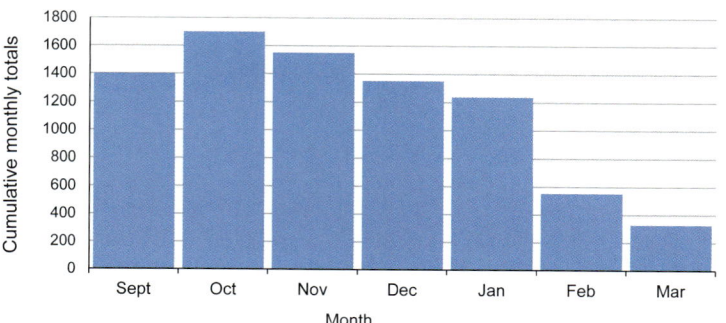

Breeding Distribution Change (1985–90 to 2008–13)

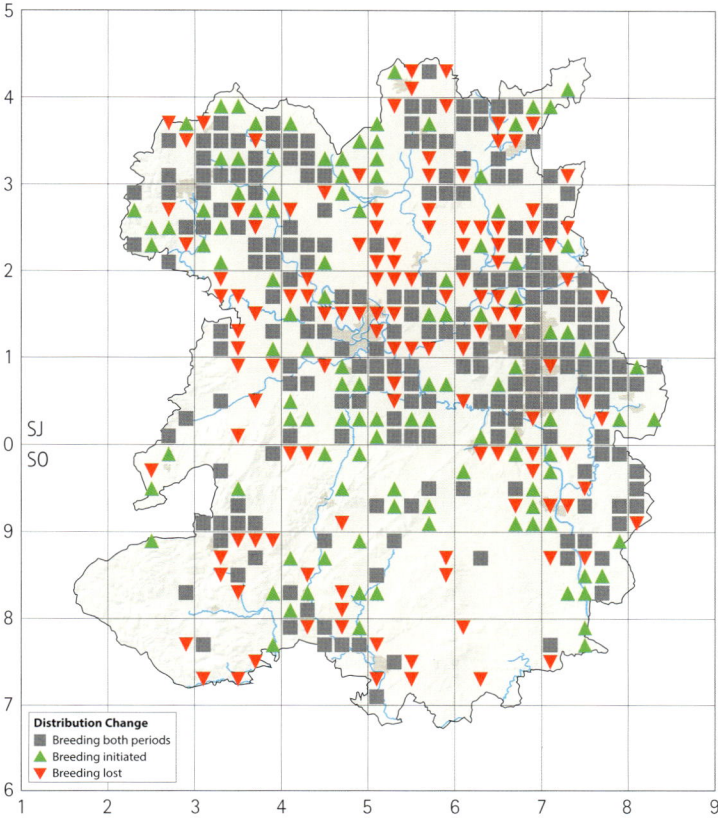

Distribution Change
- ■ Breeding both periods
- ▲ Breeding initiated
- ▼ Breeding lost

Occupied Tetrads

Atlas period (breeding)	1985–90		2008–13		Change	
	Number	%	Number	%	Number	%
Confirmed	276	32	247	28	-29	-11
Probable	37	4	43	5	6	16
Possible	26	3	37	4	11	42
TOTAL	339	39	327	38	-12	-4
Tetrads with Winter Records (2007–13):	251	(29%)				

Jonathan Cartwright, Venus Pool, 16 April 2010

In the breeding season Coot are more widespread, and also occur on some of the smaller marginal waters which may be deserted during the winter, particularly in the south-west. They are very aggressive at this time, so small pools will only accommodate a single pair, although several can be found at larger lakes and meres.

This must be one of the easiest species to confirm breeding and 76% of the breeding records were in this category. Nests are usually quite conspicuous, incubating birds can often be seen from a distance, and once the young have hatched they can be found on open water being fed by their parents.

Although there has been a 4% decrease in occupied tetrads since 1985–90, the turnover is surprisingly large. A third of the tetrads previously occupied have been abandoned while almost as many new ones have been colonised. Some new pools will have been created and others lost but it seems highly unlikely that the distribution of larger pools will have altered significantly during the intervening period. Some of the losses in the central belt and north-east quarter are consistent with those exhibited by other wetland species.

The *Atlas* (1992) suggested three to six pairs per occupied tetrad and a population of at least 1,000 but no more than 2,000 pairs. Breeding season TTVs for the recent Atlas found 725 in 206 tetrads and 849 in 169 tetrads in the early and late periods respectively, suggesting two to four pairs per tetrad. These figures would suggest a population of between 650 and 1,300 pairs. However, a breeding population of 1,000 pairs, along with juveniles and some winter visitors, would be expected to produce much larger counts in the early winter than those found in practice, suggesting the lower figure may be more appropriate. The estimate based on TTV counts, 300–340 pairs, is considered to be too low.

For such a common species, much remains to be learnt about its movements, and little is known about post-breeding dispersal of adults and juveniles. The Migration Atlas shows that some British Coot are resident but some breeding adults and young winter in south-west France and Spain, while at the same time winter visitors arrive from north-west Europe. The numbers involved in each category are not known, and reconciling the figures to estimate breeding and wintering populations is therefore difficult.

Few Coot are ringed and only 20 movements have been recorded. One ringed at Hawkstone in November 1956 was shot in Monaghan, in the north of Ireland, the following February. The remaining movements were all within mainland Britain, and more than half of the recoveries are from this or adjacent counties, consistent with the small-scale movements noted above. The lack of movements from abroad is almost certainly due to the small numbers ringed, here and overseas.

Nationally, the breeding population increased steadily from the early 1970s until the mid-2000s, after which it fell slightly. There have been marked differences across the UK, with relative abundance falling in the north and west, including parts of Shropshire, so there is probably little change since the *Atlas* (1992).

The large turnover in tetrads between the two Atlases suggests that there are many unoccupied waters during the breeding season and that Coot are not limited by availability of habitat. The fall in abundance suggests that other unknown factors may be limiting the population at this time.

Allan Dawes

Crane (Common Crane)
Grus grus
Rare visitor

Yvonne Chadwick, North Shropshire, 6 March 2010

Cranes breed mainly in Scandinavia and eastwards right across Russia, with a few in other parts of Europe. The European population winters mainly in North Africa, with some in south-west Europe. They were widely distributed but uncommon in Britain until they died out in the Middle Ages, probably due to human persecution and drainage of the extensive marshes or reedbeds needed for breeding. The last breeding record was in Norfolk, in 1542.

Remains were found in the archaeological excavation of the Roman city of Wroxeter, dating from around the start of the fifth century AD. Two of the three bones studied showed signs of butchery, but it is not known whether they were hunted for food, or captive bred. Two bones found in an earlier period were 'probably Crane'. Altogether Crane remains were found at 46 out of 172 archaeological sites in the British Isles, ranging from Mesolithic through to post-Medieval dates, but only one of them was after the fourteenth to sixteenth-century period. 'Crane might be counted among the food species, and make an interesting contrast with only 16 records of Heron, which must always have been more numerous, but less palatable' (Yalden 2002).

The Old English 'cranuc' or 'cronuc' appears in two local place names, Cranmoor Gorse, near Ruyton-XI-towns, and Cranmere Bog, near Worfield, suggesting that Cranes were present in Anglo Saxon times between the fifth and eleventh centuries, although this is not certain as 'crane' was also a common name for Grey Heron. Other old field names include Crane Acre (Clun), Cronkhill (Attingham, More and Shawbury) and Cron Leasow (Acton Round) (Foxall 1980).

There were no nineteenth or early twentieth-century records. The first recorded sighting was of one on the border with Cheshire from 6 June to 1 July 1962. 'It had a preference for potato fields. Whilst it was very wary, the possibility of an "escape" cannot be ruled out' (*Handlist* 1964).

There were no further sightings until after one of the most exciting ornithological events in Britain in the late twentieth century; the colonisation of the Norfolk Broads. It began with a pair and an individual arriving in 1979 and staying through 1980. The first nesting attempt came in 1981, followed by the first chick fledging in 1982 (Buxton & Durdin 2011). This population has grown, with further breeding pairs becoming established in Yorkshire (2002 onwards), and in the Fens, at RSPB Lakenheath (2007) and Nene Washes (2010). At least some of the new breeding pairs are believed to have originated in northern Europe, rather than the Broads. The total English resident population in 2014 was 19–26 pairs (Holling *et al.* 2014). As this population has increased, sightings here have become more frequent, starting with singles on passage at Allscott Sugar Factory on 13–14 August 1986, Baggy Moor on 5 May 1995 and Whixall Moss on 3 May 2002, and one wintering in the Higginswood area, near Market Drayton, from 17 November 2000 to 14 January 2001. Other winter records were of four passing over Telford and Broseley on 29 January 2006, and seven over Pen-yr-Estyn, Wrekin and Bridgnorth on 19 January 2013.

More excitingly, two were seen in the breeding season at two sites in the north-east, over a five-week period in April and May 2003. Then, after a five-year gap, there was a 'totally unprecedented series of records' in 2008. Two were together on 17 April, three together on 5 May and then 'a presumed pair' were seen from 28 May until 19 August. They were 'first found feeding on a recently cut silage field. … during their long stay they generally favoured flooded potato fields'. There was no evidence of a breeding attempt but 'some behaviour suggesting display was reported'.

In 2009 most records were of two together, starting on 27 March. There were several sightings, in late April, early July and early August, and 'it seems likely that … the records relate to a single pair which moved around the area'. It 'seems a strong possibility that this pair was the same birds that summered in 2008, but they favoured flooded potato fields' and 'the fact that no similar habitat existed in 2009 may have resulted in them ranging more widely. No sign of breeding activity was reported'. They were joined by two more at Wood Lane on 13–14 August.

In 2010, there was only one at four sites between 12 February and 19 July, and one at one site between 17 March and 2 May in 2011.

All these sites were in the north-east. Cranes forage over large areas, but they are vulnerable to disturbance while breeding, so RBBP recommends that no indication is given of the location of potential breeding sites, and the national BTO Atlas 2013 mapped all these 2008–11 records as probable breeding, at 50km resolution in the south-east quadrant of 100km square SJ.

More definite evidence of probable breeding came from a site in their favourite area in 2012. A single was seen on 15th, two on 23rd and three on 24th of March, then two were heard calling in late April, with 'duetting' on 12 May. Two were seen over three subsequent days, then a single over seven days from 21st to 27th, and one or two on five days between 1–12 June.

Unfortunately, there was only one record from this area in 2013, of a single on 26 April, although two were seen earlier at Berrington Pool on 23 March.

The only sighting in 2014 was of two colour ringed birds at Wall Farm on 4–5 May, both released by the 'Great Crane Project', which has been reintroducing this iconic bird into the Somerset Levels since 2010, with the target of establishing 20 breeding pairs by 2030. None were reported in 2015, but three were seen near Wem in May 2016.

Large expanses of marsh or reedbed are needed for nesting, with arable farmland nearby for feeding. Here, such habitat exists in the north-east. The populations in Norfolk, Yorkshire and the Fens, presumably the origin of most of the Cranes seen here, are growing. So too are the 'Great Crane Project' numbers. This promises the likelihood of sightings increasing in future years. The continuous run of breeding season records over a seven-year period, with 'probable breeding' evidence in three of those years, suggests the possibility of breeding here in the not-too-distant future.

Josie Owen

Stone-curlew (Eurasian Stone-curlew)
Burhinus oedicnemus
Vagrant

The migratory British population, with its stronghold in East Anglia, is at the north-western edge of the breeding range which stretches from Spain and North Africa through to North-west India.

Referred to as the Great Plover, the only historic references are from 1855, when it was listed as being present in the Borough of Oswestry, and 1865, when a single bird was reported as being killed on Ponsart (Pontesford) Hill 'a few years ago'.

There is one modern record:

Wistanstow   1 November 2012

This bird had been ringed as a nestling in Wiltshire on 2 May 2012 before it journeyed to Wistanstow, where it was found freshly dead, a distance of 152km.

Graham Walker

Oystercatcher (Eurasian Oystercatcher)
Haematopus ostralegus
Scarce summer visitor

John Fielding, Venus Pool, 1 April 2014

A familiar noisy species of all coastal habitats in the UK, it bred inland on grass meadows and arable crops only in Scotland in the 1840s (Holloway 1996). Since then, despite extensive land management changes, inland breeding has spread slowly but steadily southwards, assisted also by incursion up some of the major estuaries, and, more importantly, a behavioural change to take advantage of arable habitats. This spread eventually resulted in the first confirmed breeding here in 1981.

Historically, Oystercatchers were recorded in the nineteenth century as occasional visitors, especially along the River Severn and in the Ellesmere area. Beckwith (1879) mentions one in a collection obtained near Wem in 1865, a specimen shot near Atcham on an unspecified date and one shot at Cruckton in 1878. A pair were recorded at Ellesmere on 29 March 1899 (Forrest 1899). The *Handlist* (1964) described it as a scarce passage migrant, but mentioned 10 records between 1955 and 1963 in the Severn Valley, Ellesmere and Oswestry areas, reflecting its national spread inland and southwards during the twentieth century.

After a gap until 1970, there were 19 records in eight of the 11 years 1970–1980, followed by 12 records from nine different sites in 1981, including the first confirmed breeding at Lyneal Wood, a nest with four eggs on 18 June, although the nest was unsuccessful.

A small breeding population had become established by 1985–90, and the Atlas (1992) showed confirmed or probable breeding in nine tetrads and possible breeding in one.

The rapid increase in records (not birds) from 1992 to 2009, reflecting the increasing population, with indications of a stabilisation since then, is shown in the chart.

In the recent Atlas period, confirmed or probable breeding was reported from 36 tetrads with possible breeding from a further seven.

From the first confirmed breeding in 1981, the population increased to at least eight pairs by 1990 and probably to as many as 35–40 by 2013.

Most breeding sites are old or active sand quarries, permanent wetlands or wet meadows, some of which are now bird reserves, and Oystercatchers arrive there early in the year. The first generally appear in late January, with an increase in numbers from mid-February and most have arrived by mid-March. Laying of first clutches is typically in April and successful pairs generally have one brood, but repeat clutches are usual if eggs are lost. Failures are regular, presumably mainly through predation. Some pairs at Venus Pool are thought to have laid three or possibly four clutches in the course of a breeding

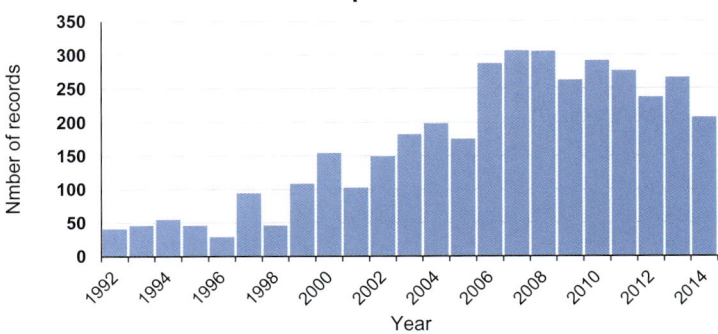

Estimated Number per Year (1992–2014)

season. In spite of the increase in arable farmland aiding the spread southwards from Scotland, breeding has not been confirmed here on such sites, although it may have occurred at Eyton Moor (2005 and 2008) and Ercall Heath (2007).

Oystercatchers are long-lived and relatively slow to mature. Females start to breed at three to six years of age and males on average two years later (Ens *et al.* 1996), so records and the distribution maps are likely to include some non-breeding birds, and the steady colonisation may be set to continue for a few more years yet.

Oystercatchers are well adapted to feeding on soil invertebrates in summer months, especially earthworms, cranefly and beetle larvae (Heppleston 1972), but they desert their inland breeding locations in late summer and only stragglers remain by the end of August. Breeding pairs and their young are thought to revert to their traditional coastal habitats, perhaps to take advantage of the peak numbers of young

Occupied Tetrads

Atlas period (breeding)	1985–90		2008–13		Change	
	Number	%	Number	%	Number	%
Confirmed	7	1	20	2	13	186
Probable	2	0	16	2	14	700
Possible	1	0	7	1	6	600
TOTAL	10	1	43	5	33	330
Tetrads with Winter Records (2007–13): 22 (3%)						

Breeding Distribution Change (1985–90 to 2008–13)

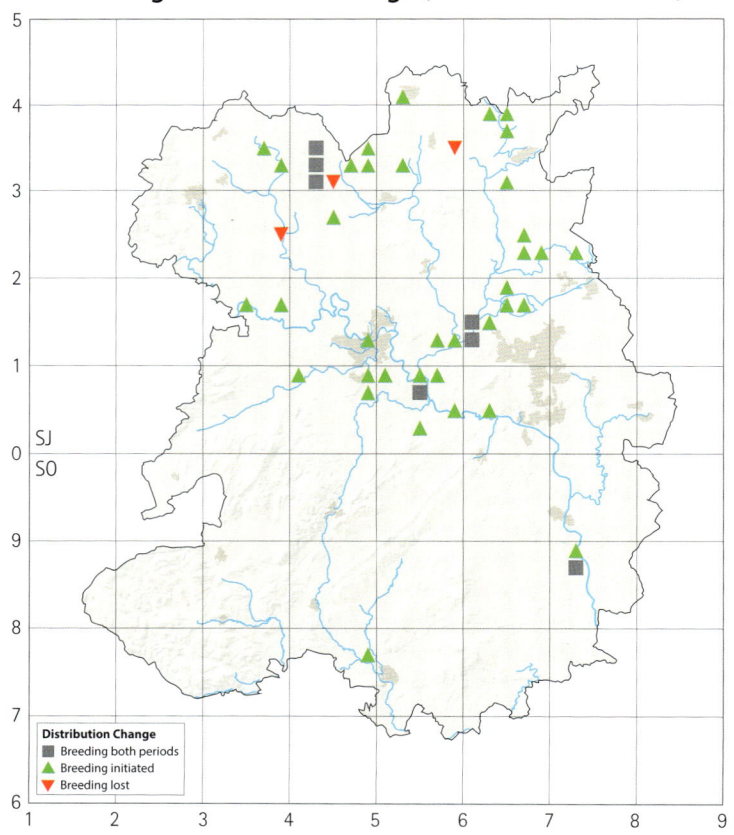

Distribution Change
■ Breeding both periods
▲ Breeding initiated
▼ Breeding lost

Winter Distribution (2007–13)

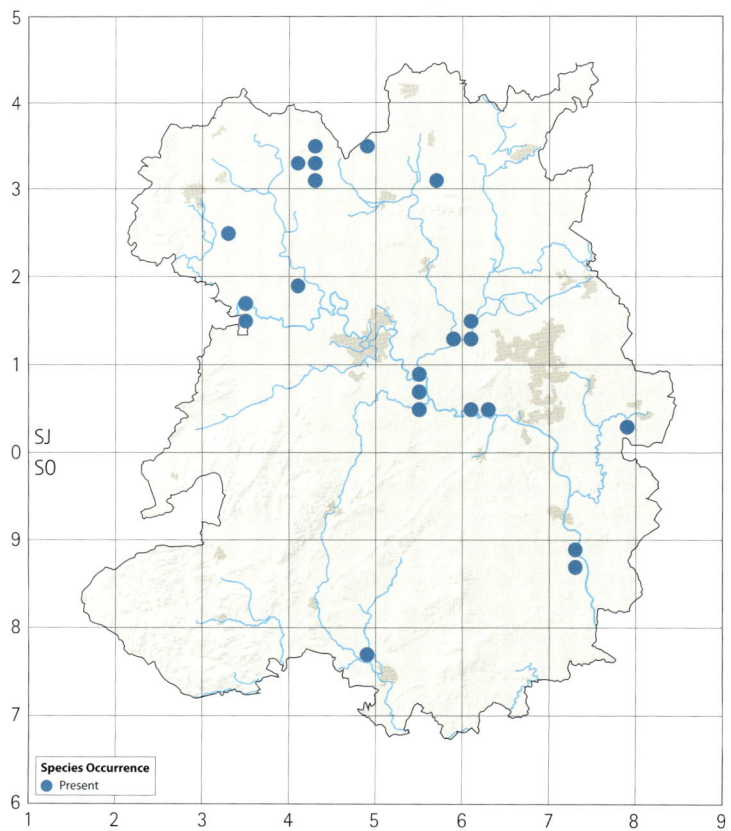

Species Occurrence
● Present

Seasonal Occurrence in 10-day Intervals (1992–2014)

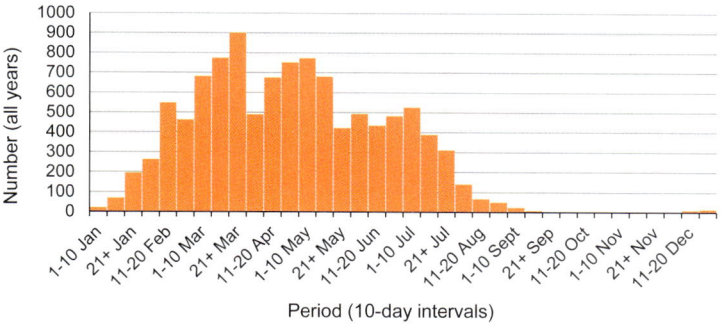

Number (all years) — Period (10-day intervals)

This seasonal occurrence is shown in approximate numbers in 10-day intervals.

The recent Winter Atlas map shows widespread distribution, but with records from only half as many tetrads as in the breeding season. There were only two November and one December Atlas records, and none between 5 December and 20 January. Records after this date reflect early return to breeding grounds. Of the 22 tetrads with winter records, only eight had no breeding season records, and most of these had breeding records in adjacent tetrads.

Only two ringing recoveries have been reported. An adult ringed at Mirelake in July 2002 was found freshly dead at Venus Pool the following August. A nestling ringed at Venus Pool in May 2002 was retrapped and released at Brownsea Island, Dorset in September of that year, lending support to the theory that the annual late summer exodus is to more typical coastal habitats.

shellfish such as common cockles and mussels at this time (MarLIN, The Marine Life Information Network, www.marlin.ac.uk).

The few Oystercatchers which do occur between September and December are thought to be true passage migrants.

Gerry Thomas

Black-winged Stilt

Himantopus himantopus

Vagrant

Breeding from Western Europe through to south-east Asia and southern Africa, the European population of this elegant wader appears to be stable. Northern populations migrate south in the winter and the majority of those that reach Britain (less than 10 in most years) are usually overshooting migrants in the spring. It has bred in Britain on a handful of occasions and more frequently in recent years, including fledged young in 2017.

Typically, the only record is of two birds in the spring:

Venus Pool   two, 20 May 1965.

They were seen to arrive in the evening, but only stayed for about an hour (Harber *et al.* 1966).

Graham Walker

Avocet (Pied Avocet)

Recurvirostra avosetta

Very rare passage migrant

This striking wader, the emblem of the RSPB, has an extensive range across Europe, Asia and Africa. It became extinct in Britain in the nineteenth century, but re-established itself in Suffolk in 1947. The breeding population only grew slowly at first, but has increased considerably in the last 20 years and spread out from its East Anglian base. As well as breeding on the coast, it now nests at a number of inland sites, including in the adjacent county of Worcestershire, where successful breeding was first recorded in 2003 (Harrison & Harrison 2005).

In 1770 Pennant noted this species as occurring 'fontimes on the lakes of Shropfhire', but there were no subsequent historic records, and the first modern sighting was not until 2006. Since then there have been a further seven occurrences which probably reflect the national population increase and range expansion:

Wood Lane	four, 26 May 2006
Venus Pool	26 April 2008
Eyton Moor	two, 28–29 May 2008
Chelmarsh	28 February 2010

Paul King, Venus Pool, 26 April 2008

Allscott Sugar Factory   one (with two on 30 May), 2 May–
                         13 June 2010
Allscott Sugar Factory   24 March–3 April 2011
Wood Lane   two, 18 May 2014
Wall Farm, Kynnersley   31 May 2014

Most records have been in May, with the birds only being present

for one or, at most, two days. They are probably moving from their wintering grounds on the south and east coasts to their breeding sites on the Dee and Mersey estuaries, where breeding commenced in 2001, and in Lancashire. Given the national trend, future breeding is a distinct possibility at sites such as Venus Pool and Wood Lane.

Graham Walker

Lapwing (Northern Lapwing)
Vanellus vanellus
Common winter visitor, uncommon breeding species

Lapwing breeds across much of Eurasia, but vacates central and northern Europe in winter when the ground freezes, and most move to western and south-western Europe, including Britain. Its name comes from its slowly-flapping flight. It is also known as Peewit, reflecting its call, or Green Plover.

Several authors in the second half of the nineteenth century described it variously as very common in most parts, staying the whole year; numbers greatly increase as spring approaches, generally congregating in large flocks until nesting commences; 'very abundant'; and 'very plentiful, much increased in recent years. In autumn, immense flocks in large open fields in the Severn Valley, the home-bred birds joined by migrants, until end October when the greater number leave'; an increase in the Newport area around 1890; and found in large flocks in winter, 'occasionally numbering a thousand or more'.

Nationally, it appears that the Lapwing population has been in decline since the middle of the nineteenth century. Lapwing eggs were a great delicacy, and collected in their thousands for the table, particularly in southern England, and this contributed to the decline. However, Beckwith (1887) attributed 'greatly increased numbers' here to their eggs being 'little sought after', although only 12 years later the position had apparently changed, and 'Plovers eggs are in great demand as a delicacy for the table' (Forrest 1899). Lapwings were perceived as the farmers' friend, because of all the 'pests' they ate, and the Wenlock and District Farmers Association proposed in November 1909 that action should be taken to protect 'plovers eggs', otherwise there would be no root crops left. The trade was eventually made illegal by the Protection of Lapwings Act 1928, and nationally the population recovered as a result, but only to a lower level limited by continuing habitat loss through 'agricultural improvement'. However, it appears that the Act was poorly enforced, and there is a local account from around 1930: 'daily on our way home from school, we could run up and down these ridges [between Bridges and Stitt], never missing an egg, sometimes filling a wicker shopping basket lined with hay'. The eggs were packed into special cases holding 36 each and several each year were posted to Fortnum and Mason's in London for sale the next day. In total, 'hundreds of eggs were collected' (Tuer 2004).

There is little evidence of local pre-war trends, although Lloyd (1943) said Lapwing had 'shown a steady decrease, not yet marked enough to attract wide attention. Where flocks numbering thousands were formerly seen in autumn and winter, now only hundreds are observed'.

The large scale conversion of rough pasture to arable during the

Second World War (see p. 39) would have reduced habitat further, and the *Handlist* (1964) noted that 'some observers report a decrease since the 1930s, but reports are often conflicting': it was described as 'resident, common, breeds throughout … most leave the high ground after the breeding season. In winter, there were "flocks of considerable size"'.

Although 'they suffer severely in hard frosts' (Forrest 1899), and, nationally, the severe winters in 1961–62 and 1962–63 reduced numbers drastically, locally there was only 'some evidence of a decrease following the winter of 1962–63' (*Handlist* 1964), and recovery was complete by 1970, but only to an even lower level limited by continuing habitat loss, which by then included a switch from root crops to cereals.

The 1968–72 national BTO Atlas confirmed the widespread distribution, with breeding confirmed in every local 10km square except one. By 1985–90, Lapwing was still breeding in every 10km square, but several squares, particularly in the south and south-west, had several tetrads that had no evidence of breeding.

National monitoring suggests that the population was fairly stable from 1970 to 1985, after which a rapid decline set in. BTO organised a sample survey in 1987, and repeated it in 1998, which included 34 randomly selected tetrads here (one tetrad in each 10km square, 4% of the County total). In 1987, 119 nesting pairs were found, an average of 3.5. Applying this to the 660 tetrads with probable or confirmed breeding gave the estimated total population of 2,300 pairs published in the Atlas (1992), although it should have been applied to all 870 tetrads, giving an estimate of around 3,000 breeding pairs. When this survey was repeated in 1998, only 24 breeding pairs were found in the 31 tetrads surveyed, suggesting a population around 670 pairs. SOS repeated the survey in 2003 and again in 2008. In 2003, 30 pairs were found, suggesting a population around 768 pairs, and in 2008, 35 and 980 respectively. These results suggested that the population in 1998 was only a quarter (25%) of what it was just 11 years earlier, it increased slightly by 2003, and increased further by 2008. Locally it is not found in enough squares for BBS to calculate a reliable trend, but it initially showed the same pattern, rising to a maximum in 2002, but then it declined by about half between 2002 and 2014. The sample sizes for both surveys are too small to draw reliable conclusions.

The national results of the same surveys found a 49% decline, and Lapwing was added to the Amber list of *Birds of Conservation Concern 2002–07* as a result. Nationally BBS found a decline of 43% in the UK, 26% in England, and 18% in the West Midlands between

Dawn Micklewright, Venus Pool, 15 March 2014

1995 and 2014, the lower percentages indicating that most of the decline occurred in the period before BBS started.

This century, monitoring has been carried out in several local areas, by Community Wildlife Groups, and by the Ruralscapes project, one of the Countryside Agency's land management initiatives covering the Severn–Vyrnwy confluence, the forerunner of the RSPB Lapwing Meadows project on Baggy Moor and the Weald Moors.

In the Upper Onny area (31 tetrads), 19 pairs were found in 2004 in 125sq.km. By 2006 that had declined to 13 pairs, so conservation work with Natural England and some local farmers, aided by the extremely wet summer in 2007 which greatly increased productivity, led to an increase to 26 pairs by 2009, but change in ownership and habitat loss on the most important farm, with half the pairs, led to a steady decline throughout the area, and the population in 2014 (18–19 pairs) was the same as in 2004. In the Upper Clun (110sq.km) there were six pairs in 2004, but they declined by a pair a year and there were none at all in 2011, 2013 or 2014. In Kemp Valley (21 tetrads), there were nine pairs in 2009 and 2010, five to six pairs in the three years 2011–13, and eight in 2014.

The Ruralscapes project found 58 pairs in Baggy Moor, Maesbury Marsh and Melverley in 2003, declining to 41 in 2004 and 38 in 2005. The RSPB Lapwing Meadows project surveyed 22 farms in 2009 and found 47 pairs, 33 farms in 2010 with 73 pairs, in 2011 a partial survey of 15 farms found 36, and eight were found in 2012 and 17 in 2013. The areas overlap, but are not identical. However, the Baggy Moor area is common to both. Numbers there fluctuated from 22 pairs in 2005 down to two in 2006, up to 28 in 2010, and down to two in 2012, but there was no change in the population trend over the 13 years. Interestingly, the Upper Onny and Baggy Moor counts both show the same pattern – they had low points in 2006, a slight increase in 2007, a much larger increase in 2009 as a result of high productivity in the very wet year of 2007, and reached a maximum in 2010 before falling rapidly again. Ruralscapes had been working with farms to help them into agri-environment schemes, and most of the farms surveyed by RSPB in 2009–11 were in such schemes, and the Lapwings, here and in the Upper Onny area, were almost all found in fields which were managed to benefit them through Higher Level Scheme agri-environment agreements (see p. 41). This, taken together with the annual fluctuations in the population, suggests that the agri-environment schemes do deliver benefits for Lapwing, but

Early Breeding Season Relative Abundance (2008–13)

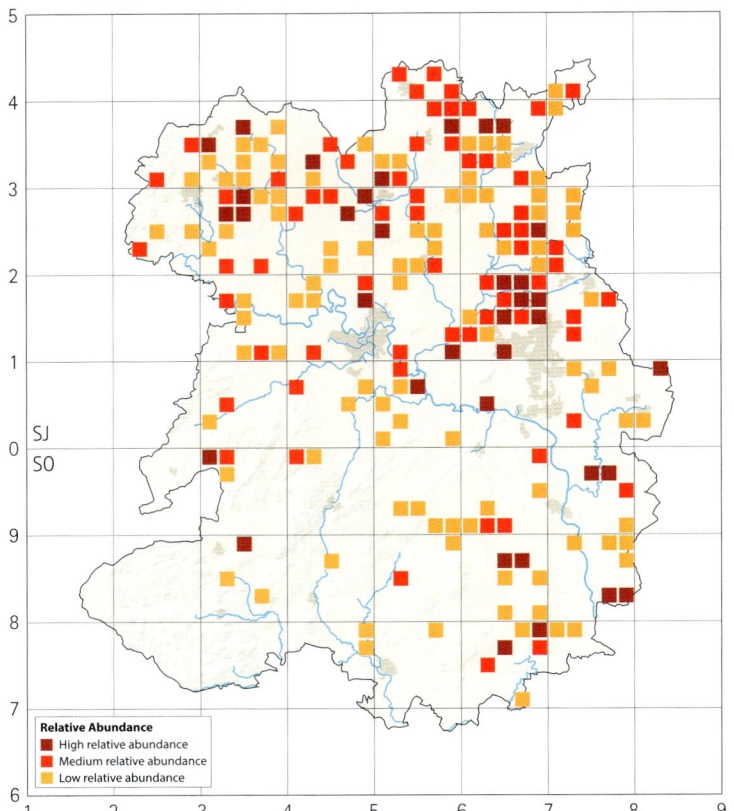

Breeding Distribution Change (1985–90 to 2008–13)

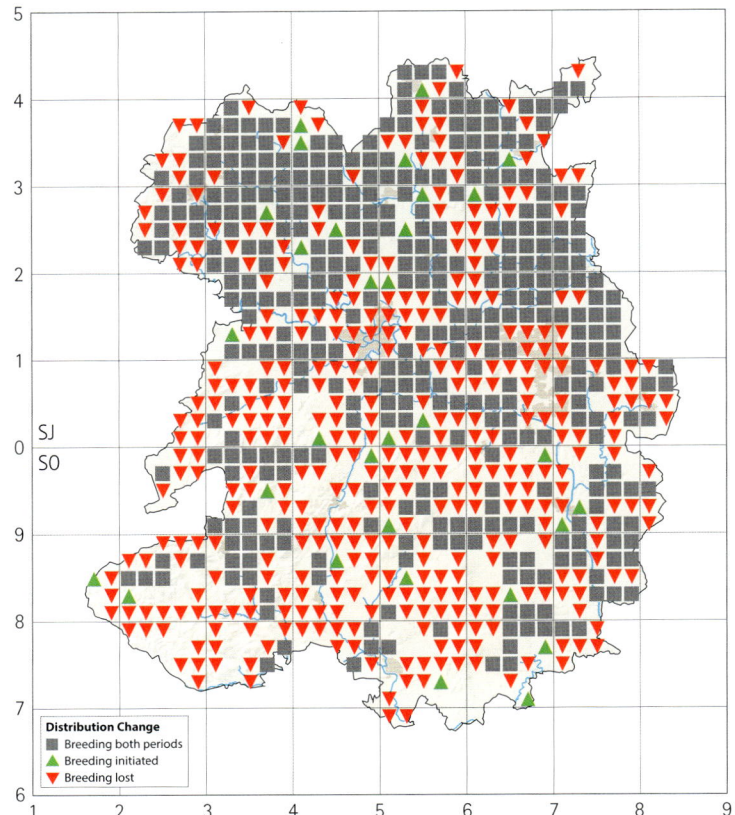

not sufficient to allow a recovery in the population, partly because the ground is still too dry in most years. Provision of wet scrapes helps, but there are insufficient of them at a landscape scale. Predation of nests also limits the population, and it is probably higher in dry years, but the large increase in the population following the extremely wet summer of 2007 suggests that, if there is a will, the population decline can be reversed.

Lapwings are conspicuous at the start of the breeding season, when they display over breeding sites with short vegetation, or bare ground. However, pairs with chicks lead them to better feeding areas, often in nearby grassland, so they are much harder to see then. Failed pairs have often formed large post-breeding flocks well away from breeding sites from mid-June onwards. The early season TTVs therefore provide a much better indication of relative abundance, and the map shows this.

Although small numbers nest on tussocky wet cattle pasture, the vast majority now are on arable land planted with spring crops, particularly on the damp land of the Weald Moors and the flood plains of the major rivers. Sheep pasture is avoided, as the egg pattern provides no camouflage on the uniform close-cropped green carpet.

A total of 829 Lapwings were found on early season TTVs in 207 tetrads, an average of four (at least two pairs) per occupied tetrad. Although not all breeding sites will have been found, and some sitting females will have been overlooked, conversely some one-year olds don't breed, pairs using breeding sites on arable farmland have to follow the rotation of favoured crops, so not every tetrad with records will have been occupied every year, and many of the tetrads with possible breeding and some with probable breeding will have

Occupied Tetrads

Atlas period (breeding)	1985–90		2008–13		Change	
	Number	%	Number	%	Number	%
Confirmed	411	47	153	18	–258	–63
Probable	249	29	170	20	–79	–32
Possible	83	10	77	9	–6	–7
TOTAL	743	85	400	46	–343	–46
Tetrads with Winter Records (2007–13): 425 (49%)						

lower densities than the average, suggesting a population of around 800 breeding pairs. This is consistent with the estimate based on TTV counts of 790–860 pairs.

This estimate is higher than that from the sample surveys in 1998 and 2003, but lower than 2008. During recent Atlas fieldwork, 48 individuals were counted on the first period TTVs in the 34 tetrads included in the sample surveys, probably roughly equivalent to the 30 pairs found in 2003. Although conclusions drawn from this analysis must be treated with caution, as the sample size is small, it appears that the population has been fairly stable, or has perhaps even recovered slightly, since 1998, but the breeding distribution change map shows a withdrawal from 46% of tetrads occupied in 1985–90, which must have largely occurred in the early part of the period. The population has also thinned in the tetrads where it was still found, and the estimated population has declined from around 3,000 pairs in 1990 to 800 in 2013, a massive loss of around 70%.

Described as 'rather thinly distributed' in the Atlas (1992), breeding

Winter Relative Abundance (2007–13)

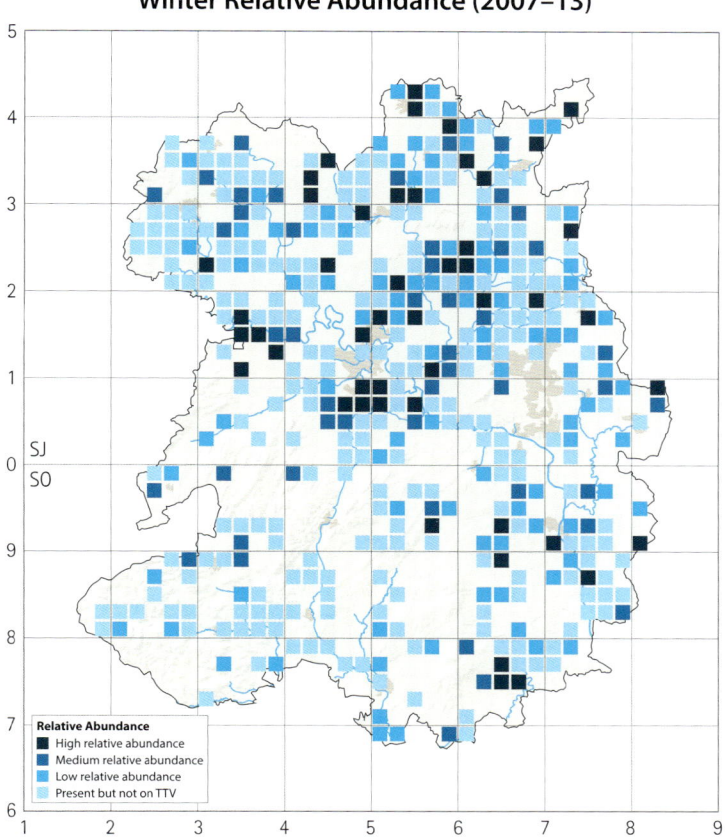

Relative Abundance
- High relative abundance
- Medium relative abundance
- Low relative abundance
- Present but not on TTV

suffered from modern farming practices. The more extensive declines have occurred in the south and south-west, where the grassland habitat has been drained and 'improved', reducing the invertebrate food supply, but greater predation levels may have contributed too, as availability of sheep and Pheasant carrion throughout the year keeps predator numbers high. Local research in the north-east has shown that Lapwing nests are much less likely to survive if they are within 50m of a field boundary or predator perch, such as a tree (Sheldon 2002).

In winter, during the recent Atlas period, Lapwing was found in slightly more tetrads than in the breeding season, but the average TTV count in winter was 59.7/hour, and in total, 10 times more Lapwings were counted in winter than in summer.

The largest flocks were estimated at 3,000 near Bayston Hill in February 2010, 2,225 near Alberbury in February 2013 and 1,250 in the adjacent tetrad in February 2008. The only count of 1,000 from the south was near Morville Heath in December 2007. Seven tetrads had counts of 500 or more, but only one, of 700 near Stanmore (SO69K) was in the south-east. There were 14 counts of over 200 on TTVs, maximum 1,050, average 302, but only one came from the south-east, and none from the south-west. On the winter relative abundance map, tetrads with a high abundance had 90 Lapwing or more on a standardised four-hour TTV, and medium abundance had 28–89.

The average count per one-hour TTV with records was 80 in November, 62 in December, 67 in January and only 46 in February, suggesting that migrants were still passing through in November, and wintering birds were already leaving in February.

Prior to the recent Atlas period, but since 1992, there were larger counts at Cae Howell, almost 3,000 in January 1995, and at Prees Heath, 2,750 in December 2002. Including these, there have been eight counts of 2,000 or more, and a further 51 of 1,000 or more, from 25 different sites altogether, but only one of these sites, Strefford, is in

colonies appear to have got even smaller, with only nine tetrads having counts of more than 20 birds before mid-June, after which larger counts would have been of post-breeding flocks, possibly containing some juveniles. Only one high TTV count was in the south, 25 at Bromfield (SO47Y) on 6 June 2010.

Breeding sites in SOS records from 1992 onwards are fairly widespread, holding small groups and often in fields with standing water or damp rushy areas. Most held only one to three pairs, a few held larger single figures, and only Norbroom Marsh regularly held double figures, with 16 pairs in 2003 and 20 in 2000. The only other record of more than 10 nests was at Hilton in 2002.

Analysis of the results of the 2003 survey, and SOS records in the same year (Dawes & Smith, SBR 2003), show a disproportionately higher decline in the south and south-west, a result reinforced by the subsequent community wildlife group surveys and the breeding distribution change and relative abundance maps. Losses appear concentrated in upland pasture and most of the river valleys.

In the 1987 survey, arable crops supported 87 (73%) pairs, and grassland (including rough wet grassland) 32 (22%). Of those in arable, almost half (40) were on bare earth, 18 in spring cereals and 12 in other crops. Of the 32 in grassland, 20 were on ungrazed land and at risk if silage was cut. Food supply for the remainder was at risk if the livestock were wormed using avermectins. Too few were found in the subsequent surveys to provide a comparison.

All the different aspects of agricultural intensification described in the *Habitats* chapter have contributed to the decline, together with the loss of mixed farming on upland farms, which used to grow their own winter fodder. Lapwing, perhaps more than any other species, has

Jonathan Cartwright, Venus Pool, 16 April 2010

John Hawkins, Wood Lane, 3 May 2012

hard weather in December 1981 and again in February 1987. The distribution and abundance pattern shown in the 1981–84 national winter Atlas is similar to that on the 2007–13 map, with low densities in most southern squares.

The decline in the size of winter flocks is likely to reflect both the decline in the European population as a whole (around 40% between 1980 and 2006), and the trend to milder winters, enabling more to stop over on the continent.

Very few have been ringed. Six of the 12 recovered here were ringed here too, one each came from Cheshire and Cumbria, and four were ringed abroad, in Norway (one), Germany (one) and the Netherlands (two). Ten others ringed here were found in Ireland (three), Cornwall (one), Cumbria (one), Lincolnshire (one) and Northamptonshire (one), and on the continent, one each in France, Portugal, Morocco and Greece.

Almost all fit a very clear pattern. Those ringed here and found in Ireland and Cornwall, and abroad in France, Portugal and Morocco, were all ringed as nestlings, then moved south or to coastal Ireland, and found in the winter months or early March. The four from Europe were also ringed as nestlings, and found here in winter. All had moved more than 500km.

The exception was ringed as a nestling in May 1981 at Cound, and found, appropriately, at Drama in Greece, freshly dead (shot) in September 1983, the furthest travelled and in an atypical direction, 2,394km ESE, 2y 4m 6d later.

Many people find it hard to believe that Lapwing has become an uncommon breeding species, because they see large flocks of continental winter visitors, but the Atlas results, the comparison between breeding season and winter numbers, and the pattern of ringing recoveries, all show that such flocks create a false impression. Agri-environment schemes appear to have allowed the breeding population to stabilise at a much lower level than it was only 30 years ago, but the future is bleak if the lessons are not carried forward into the next generation of schemes, with appropriate levels of funding.

Leo Smith

the south. The largest annual flock size has fluctuated since 1992, but there is no apparent trend.

However, the largest reported flock sizes, 3,000 or more, all occurred before 1992, with up to 4,000 at Hinstock in November 1974 and up to 3,000 at Atcham and at Baschurch in December in the same year; 3,000 at Long Lane a month later; 3,500 near Worfield in January 1976 and up to 3,000 at Allscott in January and February in the same year; 4,000 near Chirbury in January 1977; and a 'vast congregation' of 4,500–5,000 at Sleap airfield in December 1984, the largest reported flock.

The number of counts of 1,000 or more in years before 1992 was also higher, although they were noted as leaving with the onset of

White-tailed Plover (White-tailed Lapwing)

Vanellus leucurus

Vagrant

While the Middle Eastern population of White-tailed Plover is largely sedentary, the Central Asian population is fully migratory, wintering in north-east Africa through to northern India, and the very few reaching Britain probably originate from this migratory population.

There is one record of this vagrant, which is an extreme national rarity:

Allscott   24 May–3 June 1984

It was located on the evening of 24 May on a meadow near Allscott Sugar Factory. The meadow had been flooded during the winter and still held a large area of shallow water, although this was retreating, leaving an open, sparsely vegetated margin. The plover was seen on 26, 27 and 28 May, and 3 June, but not thereafter.

Three days before the initial observation, an adult White-tailed Plover had been found on farmland at Cleadon, Tyne and Wear. The third for Britain, it was watched for about an hour before disappearing to the south-west and, given the movements of another White-tailed Plover in 2010, it would seem reasonable to speculate that this was the same bird that turned up at Allscott. Moreover, it is interesting to note that, while White-tailed Plover is also a very rare vagrant to Western Europe, there were two sightings in the Netherlands in June and July 1984. Whether this was just the one individual observed at four sites or two or more birds cannot be determined with certainty, but one wandering individual would seem to be the most likely explanation (Rogers *et al.* 1985).

Graham Walker

Golden Plover (European Golden Plover)
Pluvialis apricaria
Uncommon winter visitor, has bred

A classic bird of the uplands, its breeding range stretches from Iceland and Britain in the west to central Siberia in the east, placing the English and Welsh populations at the south-westernmost edge. Those breeding in Britain are of the nominate race *apricaria* while winter populations may sometimes include individuals of the more northerly race *altifrons*. Ringing studies indicate that those wintering in Britain are a mix of British and Scandinavian breeders, which in the west will be supplemented by Icelandic breeders of the northern race.

Early records are paltry with little detail and low counts. Eyton (1839) made reference to occasional winter presence while Rocke (1866) noted them on the Long Mynd and 'other high ground'. Beckwith (1885), at least, reported some numbers with winter parties of up to 3–4, either alone or with Lapwings, and spring flocks of 15–20 (occasionally more) along the Severn Valley and other areas with marshy meadows. Paddock (1890) simply stated 'heard on the moors of the neighbourhood in autumn'. Forrest (1899) described it as 'coming in winter on its autumn and spring migrations'. In the spring of 1879 and the winter of 1895–96 he recorded them as being particularly numerous.

Caradoc papers from the late 1800s make reference to individuals or small flocks throughout the winter in most years; for example a group stayed at Roden Hall from September 1898 to March 1899, while in 1889 a flight of about 150 was seen at Weald Moors (Forrest) and the next 'large flock' (no numbers) was at Sharpstones in December 1899. Exceptionally, a solitary bird stayed all summer near Betton Pool in 1899.

Counts of single figures were again consistently reported through the first half of the nineteenth century from isolated locations, often with Lapwings. Occasionally large flocks of Golden Plover alone were seen: 200–300 about Betton and Chilton (1905), 150 at Old Heath (1920), 300 at Longdon on Tern (1931) and Harlescott (1939), and the first flock of about 1,000, near Hadley (1948).

In the late 1950s and early 1960s, a change in status seems to have occurred, with large over-wintering flocks being seen, and in increasing numbers. The *Handlist* (1964) noted the old aerodromes of Sleap, Shawbury, High Ercall, Tern Hill and Condover, and the open fields around Haughmond Hill, Crudgington, the Weald Moors and Oswestry, as favoured haunts. In 1960 and 1962, flock counts

Jim Almond, Long Mynd, 20 May 2013

reached around 1,000 at Sundorne Fields, with similar numbers at High Ercall in 1961 and Kemberton in 1970. In the 1960s and 1970s, records were received from up to 50 locations a year, with counts ranging from individuals to many hundreds. In January 1976, when much of Europe, including Britain, was experiencing gale force winds, the largest single flock – then or since – was recorded with 3,000 at Allscott Sugar Factory. Subsequently, large flocks of over 100 were noted at several places on the plain below Shrewsbury and around the Tern, but only two counts exceeded 500, at Longdon on Tern (about 1,000 in February 1976) and Allscott Sugar Factory (600 in February 1977).

However, the underlying trend in the reported maximum flock size each year has shown a steady decline, and towards the end of the 1990s there were only six records of flocks reaching three figures, all in 1997, the maximum being 300 at the Weald Moors in December, while in 1998 reports were only received from nine sites (Norbroom Marsh, Market Drayton, Tibberton, Venus Pool, Calverhall, Prees Heath, Allscott Sugar Factory, Wall Farm and Donnington). The chart shows the maximum reported flock size every year since 1895.

From the 2000s, SBRs made reference to low numbers returning in the autumn and falling maximum flock sizes, and also to an apparent reduction in site fidelity. Norbroom Marsh, which had established itself over a number of years as a spring migration staging post, returned nil counts and then only small numbers, while the former stronghold around Cosford and Albrighton likewise had small numbers. In contrast to these lowland sites, flocks spending at least part of the winter on the tops of the Long Mynd (about 300 in 2004, 350 in 2005 and 300 in 2006) were also noted.

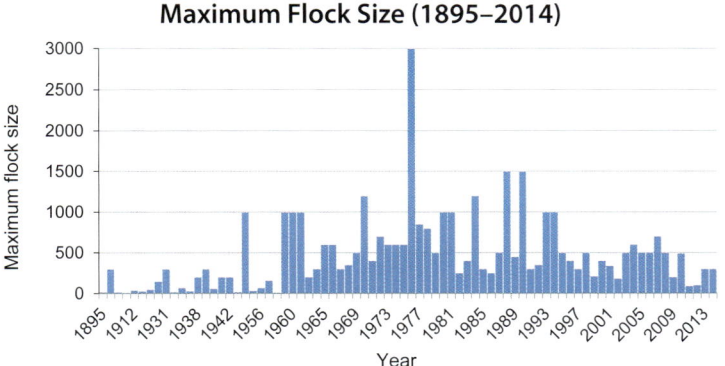

Maximum Flock Size (1895–2014)

Records per Month (1895–2014)

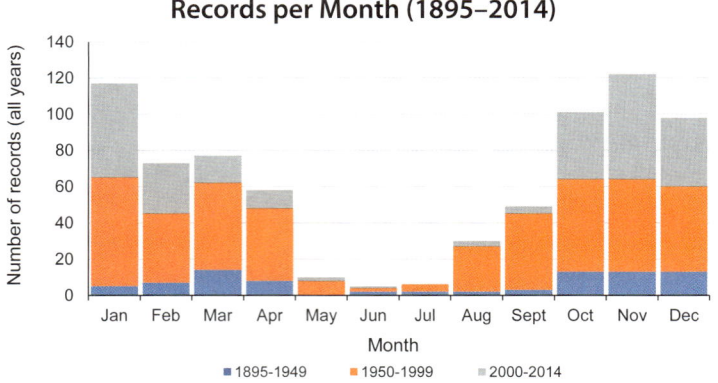

■ 1895-1949   ■ 1950-1999   ■ 2000-2014

Golden Plovers return from mid-July onwards, often associating with Lapwings on a range of habitats including ploughed land, stubbles and mature meadows. Winter numbers peak between October and January but, given their mobile nature, the possibility of double counting or individuals being missed in large flocks of Lapwings, and their susceptibility to the weather – harsh conditions on the continent increasing numbers, while cold winters here will push them further south – it is difficult to determine either the numbers present, or underlying trends. The chart shows the monthly distribution of all CSVFC and SBR records 1895–2014.

Winter numbers nationally are increasing, but loss of suitable winter foraging habitat on fallow ground/pasture and mature, earthworm-rich, meadows is affecting winter flock sizes.

In the recent Atlas period the number of records received each year has increased, although over half of these records have been of counts of less than 100 individuals – big flocks may be something of the past.

The national winter BTO Atlas 1986 maps distribution predominantly in the north-east with almost none further south than the flat fields in the Severn Valley south of Shrewsbury. The recent winter Atlas map highlights the same lowland mainly arable areas to the north and north-east, but also shows extensive use of the uplands of the Long Mynd (mainly on the flat tops above Batch Valley and Jonathan's Hollow) and the wet pastures of the Clun Forest.

The recent Atlas period returned seven TTV counts in excess of 100 individuals, including 300 at Black Mountain in April 2007 (the only such spring record), 370 near Wem in November 2008 and 500 around Condover in late January 2008. Roving records produced a further six counts of flocks greater than 100: 400 at Cwm Ffrydd (November 2007), 500 at Condover during the winter 2007–08, with 500 around Bridgnorth in January 2008. In November and December 2009, up to 1,000 were at Venus Pool, there were 400 at Mirelake in November 2010, and 300 at Soulton in January 2013. In addition SOS records include 250 at Walcot (January 2009) with a similar number at Eyton Moor in October of the same year, and 200 at Plush Hill, 250 at High Park and 300 at Doley in the winter late in 2014. However, the mobility of the flocks, the possibility of double counting, and the number of large flocks not seen on TTVs, limits the value of any relative abundance map.

Plumage variation within the *apricaria* race across its range makes separation in the field of the northern sub-species *altifrons* difficult. Nevertheless, some individuals of the northern race have been recorded. Prior to four at Shawbury and 12 at Haughmond

Hill in March 1962, the *Handlist* stated that there are older records from Bettys-y-Crwyn but gave no details. In March 1976, approximately 100 were at Bletchley, with 60 at Condover in April 1978. An unknown number were again at Condover in April 1980, with five on Long Mountain in May 1985 and 25 at Norbroom Marsh in March 2000.

The wintering flocks on some of the upland hill pastures in the south-west and mid-Wales are potentially vulnerable to proposed windfarm developments. To study this potential impact the regular flock in the Clun uplands has been part of a wider colour-ringing project started in 2011, funded by Ecology Matters Trust (Tony Cross and Paul Leafe, *pers. comm.*).

Although the size of this flock is fairly stable, at around 150, 576 individuals were caught and ringed in fields mostly on our side of the border, between 2011 and the end of December 2014, proving a regular turnover within the flock. There have been 24 retraps or resightings in the same area, including several in subsequent winters, showing a certain degree of site faithfulness, but there have also been a number of retraps and resightings from elsewhere suggesting flocks are nomadic, with movements to and from adjacent Welsh counties, to the Long Mynd in January 2012, and to Belvide Reservoir (Staffordshire) on 2 October 2014.

While one colour-ringed in winter 2011–12 was subsequently identified on breeding grounds in North Wales the following summer,

Winter Distribution (2007–13)

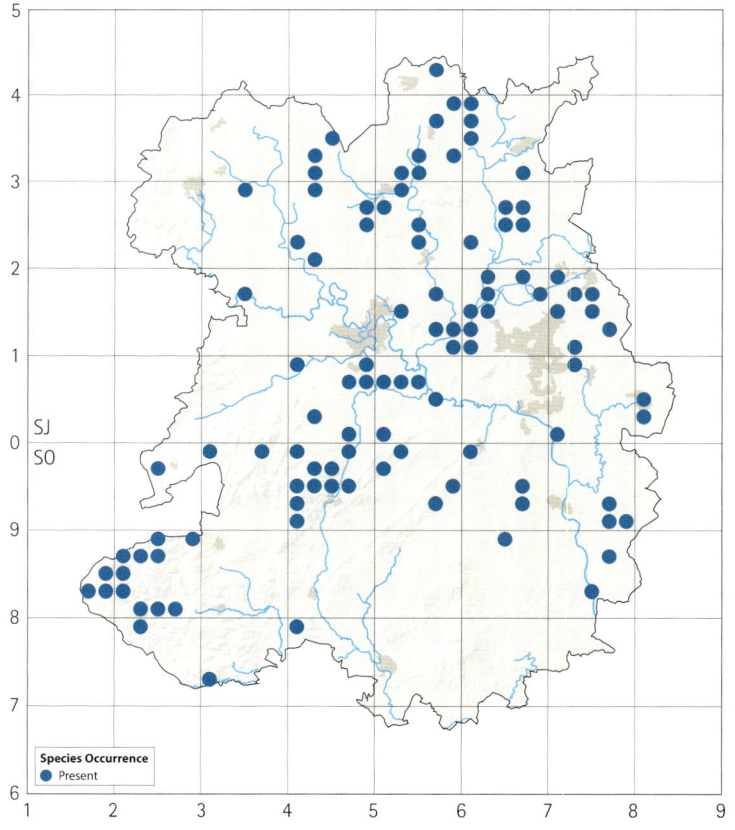

Species Occurrence
● Present

Occupied Tetrads

Tetrads with Winter Records (2007–13):   107   (12%)

Jim Almond, Condover, 9 December 2007

it is thought most of these plovers are of continental origin. A natural assumption would be that most originate in Northern Britain or Iceland, and one ringed in 2014 and seen on Tiree on 25 April 2015 lends some support to this theory. However, of all the ones caught, two caught in nearby counties of Wales were foreign-ringed, one from the Netherlands and one from Belgium, suggesting a migration route through the Low Countries, perhaps indicating origins in eastern Europe or Russia. A sighting of a Welsh-ringed bird in Norway in May 2014 confirms movements to Scandinavia, while two from the Clun Forest seen in north-east Scotland in early autumn suggest this may not be unusual (Tony Cross *pers. comm.*).

In lengthy periods of freezing weather they vacate the upland wintering grounds and move southwards or to the coast. A first year individual ringed on 21 November 2012 was shot 230km distant near Camelford, Cornwall, only 10 days later, while another ringed in April 2012 was found freshly dead in Wiltshire in February 2013.

Two satellite-tagged birds had very different responses to cold snaps. The first, tagged on 11 November 2011 was still present on 1 December 2011, but then quickly relocated to an area just south of Madrid by 11 December 2012 following cold conditions and heavy snowfall. The second, tagged on 5 January 2012 subsequently moved to coastal Carmarthenshire by 16 January 2012. The current colour-ringing study is continuing (Tony Cross *pers. comm.*).

Before the recent study, one ringed at Allscott in November 1986 was forced southwards by bad weather and shot 1,608km distant in Portugal 46 days later.

Records of breeding are rare. At the end of the nineteenth century Forrest stated 'it is not known to have bred' and the *Handlist* commented, with one exception, that up until this time 'no evidence of nesting has been obtained'. In the first half of the twentieth century there was one record of confirmed breeding on Oswestry Racecourse in 1914, with the possibility of a second unproven nest at about the same time.

There are two modern confirmed breeding records. The first was of a pair on Long Mynd with a nest and three eggs found on 3 June 1974 (and not 1976 as previously reported in the *Atlas* 1992) and the second, a previously unpublished record, was of an adult with a flightless youngster near the Gliding Club in the late summer of 1984.

Although wintering flocks and individuals often stay into late April or early May, there have been three June records: Long Mynd, in 1978 and 1996, and Ragleth Hill in 2001. The latter site is not considered to provide suitable breeding habitat; parts of the Long Mynd are suitable, but there is no evidence of breeding attempts in these years.

In fieldwork for the recent Atlas there were 20 breeding season records from 16 tetrads, including four May and one July record, but all except one of them were considered to be migrants. The exception was an individual in suitable breeding habitat, wet rough pasture, at the western end of the Clun Forest beyond Mason's Bank in SO28D on 10 May 2011. A subsequent visit failed to relocate it, so if it did attempt to breed it was unsuccessful. However, a female in full breeding plumage was seen in the same place in the following year, on 16 May 2012, but again it was not relocated on subsequent visits. These two sightings in SO28D were the only possible breeding records in the recent Atlas.

Bob Harris

Grey Plover
Pluvialis squatarola
Very rare passage migrant

Jim Almond, Wall Farm, 15 April 2014

Breeding in the high arctic, mainly in western Siberia, most pass through Britain on the way to wintering grounds in western Africa. It is almost always found on the coast, preferring large, muddy estuaries, but occasional sightings occur inland where it often associates with flocks of Golden Plover and/or Lapwing feeding on farmland or wet meadows.

Arrivals start in late July with non-breeders or failed adults, followed by successful adults in August and their young in September. There may be a further passage in October and November as some move further south, followed by more arriving from the continent. Return migration begins in February and March, but mostly occurs in April, with all but a few first year birds having departed by mid-May. Local records reflect this pattern.

This plover was first recorded in 1913, when one was found feeding with a flock of Lapwing at Ludlow on 14 January. The next was one flushed from a ploughed field at Bridgnorth 'apparently of this species' in 1921. However, lack of further details places this latter observation in doubt.

There have been 23 modern records. Only two have been of more than one bird, and only five have been of birds that stayed more than one day. All of these exceptional occurrences occurred in August or September.

The first was at Allscott Sugar Factory between 26 July and 3 August 1967, which corresponds with the arrival date of adults from their breeding grounds. Eight years were to pass before another was recorded, a lone individual at Longdon on Tern in March 1975. Three years later a singleton was seen at Lythwood Hall Bayston Hill in December 1978.

The 1980s and 1990s witnessed 11 records, with sightings in nine of the 20 years. All but three occurred between June and October, and were of individuals on single dates, except for three at Wood Lane on 22 September 1989, and individuals staying at Allscott Sugar Factory for several days on three occasions, all in the month of September, in 1985 (nine days), 1994 (three days) and 1996 (eight days). The only locations featuring more than once were Allscott Sugar Factory, four times, and Venus Pool, twice. The remaining three records were all of singles; at Venus Pool (March 1985), Sutton Maddock (April 1988) and Alberbury (January 1995).

Between 2000 and 2014, there have been nine additional records. Interestingly, compared to the previous 20 years, two-thirds were of spring migrants: all were individuals, at Venus Pool (April 2000 and March 2003), Whixall Moss (May 2007), Upper Affcot (April 2009), over-flying Cantlop (April 2013) and Wall Farm (Kynnersley) in May 2014. Two of these records occurred during the recent Atlas breeding season.

The first autumn passage record in this period was of one almost in full summer-plumage, which stayed at Allscott Sugar Factory for three days in August–September 2001. The others were both from September, with a day visitor at Venus Pool in September 2011 and three at a confidential site in SJ50 in the Severn Valley south-east of Shrewsbury, in September 2013.

The chart summarises the seasonal occurrence.

Bob Harris

Seasonal Occurrence in 10-day Intervals (1992–2014)

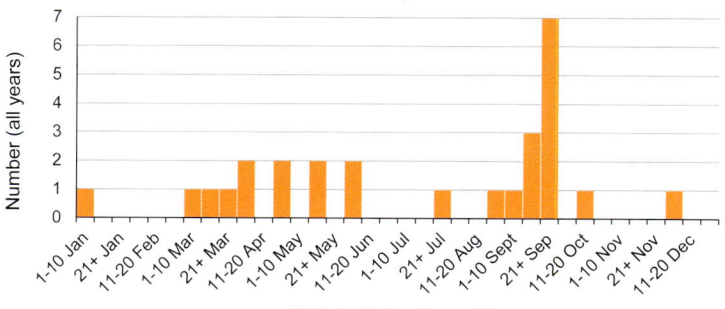

Period (10-day intervals)

Ringed Plover (Common Ringed Plover)
Charadrius hiaticula
Scarce Passage migrant

Jim Almond, Venus Pool, 10 April 2012

The nominate race *hiaticula* breeds from Brittany through Britain and Ireland, southern Fennoscandia, Iceland, Greenland and Canada. Many of those moving through western parts of the UK go to breeding grounds in Iceland, Greenland and as far west as Ellesmere Island in arctic Canada, and return to wintering areas in Spain and west Africa. Breeding inland in Britain occurs on the shingle banks of larger rivers and lochs, and in sand and gravel pits, as far south in central and eastern England as the Thames Valley, but in very small numbers compared to those on the coast.

Beckwith (1876) recorded them occasionally in spring and autumn at Cressage, on the Severn. However it was only with the development of, and regular watching at, Shrewsbury Sewage Farm and the sugar beet ponds at Allscott that the species started to be recorded annually.

The *Handlist* (1964) described Ringed Plover as an annual passage migrant in small numbers, though more frequent in autumn than in spring, and this remains its status today. The charts show the approximate number seen passing through each year, and the pattern of seasonal occurrence in 10-day intervals.

The earliest ever record was of a single at Berriewood on 21 February, in 2003, and there have been a handful of March sightings. Spring passage really gets underway from the first week of April and continues until early June, though with a peak from 10–30 May. Most records have been from the usual wader hotspots of Allscott, Venus Pool and Wood Lane, but the most sizeable spring numbers have been reported from flooded arable fields at Long Lane (up to 21 on both 14 April and 13 May 2013), and Eyton Moor, where a group of 17–21 were recorded from 28–31 May 2008.

Half of the spring passage records are of single individuals with only about 4% of groups of 10 or more.

Very small numbers appear occasionally from mid-June to the end of July, but most autumn passage occurs from the first week in August through to early October. Peak passage dates are usually between 10 August and 30 September with very few recorded after mid-October. The latest record is of two at Fenemere on 15 December, in 1968.

Autumn passage is usually more evident than in spring, because the birds lack the urgency to get to their breeding grounds, and therefore tend to stop longer, particularly if there are good feeding conditions. The population is also swelled by juveniles. Mirelake was particularly important when the water levels were low enough to provide a good expanse of mud for feeding and resting. A flock of 30 there on 29 August 1961 is the largest on record.

There is no documented evidence of breeding attempts here. Although they were described as being 'present in the breeding season' at Venus Pool and Chelmarsh in 1990, the chart shows that this is true of spring passage migrants in most years. Indeed, during the recent Atlas period, there were 25 'breeding season' records from nine tetrads, mostly from sites named above. All except one were April or May records, and all were deemed to be passage migrants. There were no winter Atlas records.

Two ringing recoveries are the only evidence we have of the provenance of passage birds. One ringed at Mirelake on 19 September 1985 was retrapped in Iceland on 7 June 1987. Another ringed in Norway on 25 August 1991 was retrapped at Allscott just 11 days later on 5 September, having travelled 1,200km in that time.

Nationally, ringing has shown that British breeding birds generally move small distances south and west for the winter, and so far there has been no evidence that any of them are involved in passage; in fact the majority are already on breeding grounds by the end of February before spring passage starts here.

Gerry Thomas

Estimated Number per Year (1992–2014)

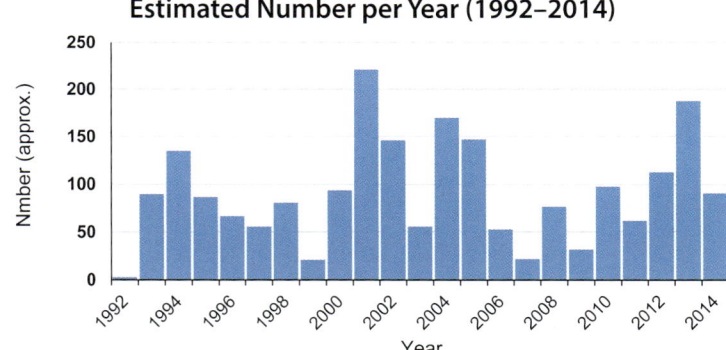

Seasonal Occurrence in 10-day Intervals (1992–2014)

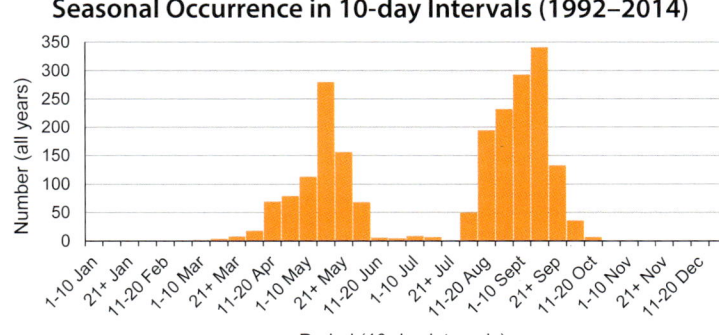

Little Ringed Plover
Charadrius dubius
Scarce summer visitor

Little Ringed Plover breeds across all but the most northerly and north-westerly parts of Europe, and winters in Africa. Britain is at the north-western edge of its range, with first breeding in 1938. Almost 20 years passed before one was first seen here, on 31 July 1957, and a juvenile two weeks later, both at Shrewsbury Sewage Farm.

Two were seen at the same site on 15 May 1958, and autumn passage migrants were then recorded each August at Allscott Sugar Factory in 1960 (one), 1962 (three) and 1963 (one). Between 1964 and 1973 there were three to eight at the same site, almost all in either July or August, every year except 1967. There were April records only in 1968 (one) and 1970 (a party of six).

Not until 1966 was a third site visited, Venus Pool, on 30 July. After 1958, there were no further records from Shrewsbury Sewage Farm until August 1971. The only other record in this period was a passage migrant at Weston Rhyn in May 1973.

After a gap in 1974, Venus Pool produced the first probable breeding record, a pair seen displaying and copulating on 16 May 1975. Breeding was eventually confirmed in 1976, when a pair produced two young that were ringed at Allscott Sugar Factory.

Passage records were occasionally received from other sites, but colonisation after 1976 continued slowly and breeding was confirmed at Venus Pool (two young) and Allscott Sugar Factory (four ringed nestlings) in 1978, and in most years since. In 1981 there were four breeding pairs, including first breeding at a third site, Wood Lane. The 1984 BTO Little Ringed Plover survey confirmed breeding at three sites: Venus Pool, a working quarry at Haughmond Hill (first confirmed breeding there in 1983), and a new site on a shingle bank on the River Severn at Buildwas. Unconfirmed breeding reports from two other unnamed sites in 1984 came in too late to be included in

John Hawkins, Wood Lane, April 2006

the survey results, and there was a pair with two fledged young at Priorslee Lake in July, but it is not known if the young were raised at that site. Unusually, there were no breeding records from Allscott that year.

The plovers continued to spread, and during the 1985–90 Atlas period, Venus Pool, Allscott and the shingle banks on the River Severn appeared to be occupied every year, breeding was confirmed at an open-cast site at Little Wenlock in 1987 and the following years, and at Longdon on Tern in 1989, and they were seen at new sites at a working gravel pit at Bromfield in 1988, and at Knighton Reservoir in 1989, showing that the successful occupation of new sites was a gradual process. There were four pairs at Venus Pool in 1988, and by 1990 approximately 10 pairs occupied six sites. Although it was not referred to in the *Atlas* (1992), breeding was confirmed at Bromfield in 1990 (Jordan *et al.* 1995).

None of the sites with breeding evidence, except Wood Lane in 1981, were named in SBRs until 1989, and the *Atlas* (1992) named only Venus Pool and Allscott Sugar Factory, so much of the detail in this account is previously unpublished.

The repeat BTO survey in 2007 estimated the population at eight pairs, an increase in the local and national UK population from the 1984 survey (Conway *et al.* 2008), and the apparent local decline from the *Atlas* estimate in 1990 may have been due to environmental factors at breeding sites, including poor weather during the spring of 2007, or pairs being overlooked, or poor coverage of outlying sites.

The 2008–13 Atlas survey found 16 occupied tetrads and possible breeding in two more. Breeding was confirmed on shingle banks at four sites within three tetrads on the River Severn between Cressage and Buildwas, with a maximum of three pairs present each year, together with another pair on a wetland at Devil's Dingle, which usually also nested in the Buildwas tetrad, but in the tetrad further east in at least one year. There were usually four pairs in this cluster of four tetrads, but in some years three of the pairs were in the Buildwas tetrad. Several other sites supported a pair each year: Venus Pool (two tetrads), Allscott (including Mirelake: two tetrads), Wood Lane and Bromfield quarry. Other breeding sites were Prees Heath, Priorslee, Nedge Hill, Stokesay, Chelmarsh and a sand quarry near Bridgnorth. Two adults and a juvenile that were seen at Stokesay (SO48F) on the muddy margins of a farmland pool on 29 June 2011 have been counted as a confirmed breeding record, but they may perhaps have bred elsewhere. Allscott and Mirelake (two tetrads) held up to four pairs, Bromfield two pairs and Venus Pool two pairs, adding up to 20 breeding pairs in total.

Territorial activity includes loud, strident piping calls and the furiously fast display flights can be conspicuous, but when they are still and concealed against pebbles, or at isolated sites, they can be difficult to observe, so breeding is difficult to confirm for some pairs. Others may go undetected, particularly at working quarries. Not all sites hold breeding pairs every year, but allowing for some possible breeding records reflecting nest sites and some pairs at private quarries being overlooked gives an estimate of 20–25 pairs.

Breeding Distribution Change (1985–90 to 2008–13)

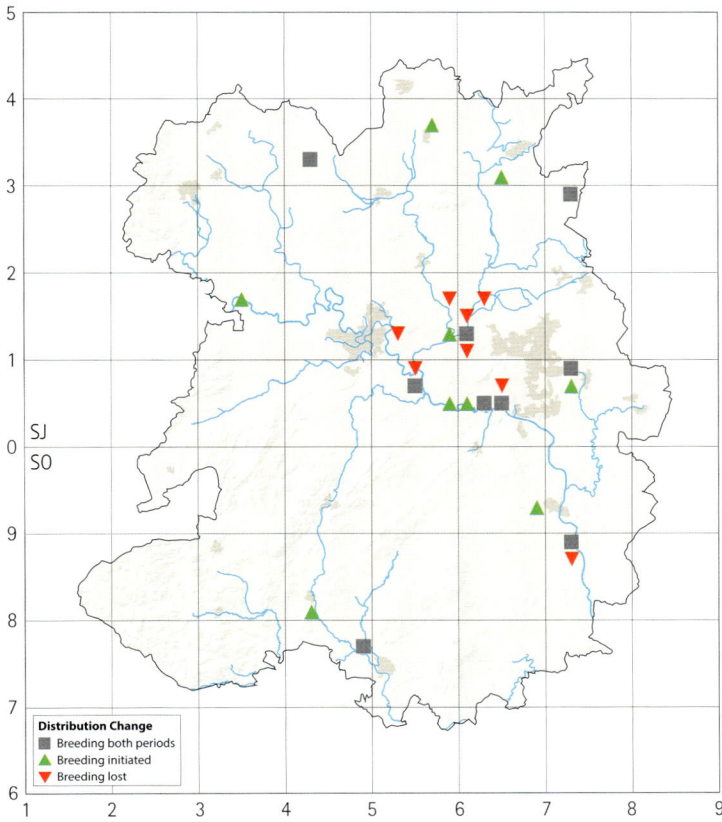

Distribution Change
- ■ Breeding both periods
- ▲ Breeding initiated
- ▼ Breeding lost

Occupied Tetrads

Atlas period (breeding)	1985–90		2008–13		Change	
	Number	%	Number	%	Number	%
Confirmed	7	1	12	1	5	71
Probable	2	0	4	0	2	100
Possible	8	1	2	0	-6	-75
TOTAL	**17**	**2**	**18**	**2**	**1**	**6**

The breeding distribution change map shows gains in five tetrads and losses in five since 1985–90, but the former were confirmed breeding while the latter were possible breeding. This reflects both the increase in breeding pairs, and the transient nature of some sites, but successful breeding on the debris of demolished factories at Allscott and Priorslee emphasises adaptability and an ability to rapidly occupy temporarily available sites.

Most breeding sites are man-made, some of short duration, created through extraction, engineering and reclamation at gravel pits, sand quarries, reservoirs and other industrial wetlands. Farmland and other shallow pools with fluctuating water levels may also be used, but these tend to be more transient. Natural breeding sites include shingle bars on the Rivers Severn and Tern. Within these landscapes they ideally look for dry, level, undisturbed nest sites close to shallow water with accessible shelving banks and a rich food supply. They will usually feed in the mud along the margins of static water bodies and rivers but also among pebbles, shingle and bare earth adjacent to the water.

The earliest date of arrival between 1977 and 2014 was one at

Knighton Reservoir on 15 March, in 2014. The mean arrival date in 2007 based on the previous 10 years was 23 March, and between 2008 and 2014 it remained the same. First clutches are laid in late April or early May. Data on clutch size is sparse, but there were four eggs at Chelmarsh Scrape in July 2003, four eggs at Devil's Dingle, on 24 May 2009 and three at the same site on 5 May 2014. Brood sizes range from one to four young and repeat clutches appear to be common, especially when the first clutch is predated or washed away. The pair at Chelmarsh Scrape were incubating their first clutch in mid-May 2003 and the clutch on 20 July was their third.

Dispersal from breeding areas begins in late June to early July, with autumn passage noticeable in the second half of July. Few are ringed, and there have only been three recoveries. Confirmation of passage movement was a nestling ringed at Cressage on 20 July 1980 and controlled at Walcot on 11 September 1980, and a nestling ringed at Walcot on 20 July 1978 and found dead at Abingdon, Oxfordshire, on 6 May 1979. A female colour-ringed in North Yorkshire on 3 June 1999 was seen at Venus Pool on 2 April 2000, and back in North Yorkshire only seven days later, where she paired with the same male, emphasising the site and breeding fidelity of the birds, when conditions allow. She was seen breeding again in North Yorkshire in 2001, at a different site nearby.

The latest date recorded is 23 September, in 2004 at Allscott Sugar Factory. The mean latest date for the period 2004–14 is 30 August.

They will return year after year to sites that remain suitable, and fortunately several sites, including Venus Pool, Wood Lane and Devil's Dingle, are nature reserves or are managed sympathetically for wildlife. However, the plovers are also opportunists and will nest at new sites where bare ground occurs adjacent to wetlands. Changing water levels and growth of vegetation can create instability in habitat and there is the risk that such sites will be deserted. Natural events, particularly flooding of shingle bars on the River Severn, regularly destroy nests, and fishing and canoeing can also disturb breeding pairs and could possibly result in clutch failure or desertion. They are also susceptible to predation of eggs and young by crows, Sparrowhawk and fox, but their ability to lay replacement clutches enables them to counteract these impacts. Protection cages have been utilised with mixed success at Venus Pool and Wood Lane, where they were found to be effective in protecting the eggs and recently hatched young from marauding crows, but they could also alert crows to the presence of nests and incubating birds were prevented from returning to them. They are no longer used.

Several other species have suffered adverse effects from the long-distance migration from, or over-wintering in, Africa, but there is no evidence of a similar impact on Little Ringed Plover, and the population estimate of 20–25 pairs is double the estimate of 10 pairs in the *Atlas* (1992). Large areas of potential breeding habitat remain unoccupied, so its availability is probably not a limiting factor. Given this, their opportunistic breeding strategy, and the steady increase and expansion of range nationally, it is likely that new sites will be progressively occupied and the Little Ringed Plover population is well placed for continued expansion. This could be assisted by the creation of new static wetlands, especially at redundant gravel and sand extraction sites, and the provision of shingle beds at existing wetlands, which has proved successful in some places elsewhere.

Glenn Bishton

Dotterel (Eurasian Dotterel)

Charadrius morinellus

Rare passage migrant

John Fielding, Long Mynd, 3 May 2010

Dotterel breed mainly in Fennoscandia and east across Arctic Russia, with several hundred males remaining on the mountain tops of Scotland, and the occasional few in northern England. The western European population winters in North Africa. On migration, especially in spring, Dotterel tend to stop off on hilltops in Britain, resulting in records of a group or a single bird (known as a 'trip') on the Long Mynd, and occasionally on Titterstone Clee or the Stiperstones. On the Long Mynd, they are typically found on short grass near the gliding club or on sheep fields, where they stay sometimes for a few days to rest and feed before resuming their journey. A review of records was published (Tucker SBR 1998), and up until 2014 there have been a total of around 63 records representing what are now regarded as the 39 separate trips shown in the table.

There were three nineteenth-century records, the first of one shot on the Long Mynd in about 1840, and the second of one collected at Lutwyche on Wenlock Edge in November 1871. Three were shot from a trip of 13 near Wellington on 12 May 1886 when, not surprisingly, the others 'mounted high into the air and flew away at a great rate' (Forrest 1899). The other two pre-1950 records were of two at Wroxeter 'very tame as usual' on 3 May 1927 and an immature female on the Long Mynd on 16 September 1935, thankfully the last known to have been shot.

The first modern record was of a trip of three males and two females on the Long Mynd on 30 April 1967. In the 1970s there were five records, all of northbound trips in spring, with three from the Long Mynd: nine on 18 May 1973, and singles on 3 and 18 May 1976, all serving to establish this as the place to find Dotterel, and prompting more intensive searching. The other two records in that decade were of perhaps four flying over Shrewsbury at night on 5 May 1976, and the sole Stiperstones record, where on 6 May 1979 there was 'a pair ... with others calling nearby between rocky outcrops on summit ridge'.

By the 1980s, the Long Mynd was being searched regularly. Though some years passed without any sightings, the number of trips increased to 11 in both the 1980s and in the 1990s.

In autumn 1981, there were three remarkable records from the northern plains, where a single (probably the same bird) was found consorting with a flock of Lapwings first near Lilleshall on 2 September, then at Edgebolton (six miles away) on 22nd and at Bratton near Longdon on Tern (seven miles further again) on 23rd.

The other three 1980s records away from the Long Mynd, all of singles in autumn, were from Titterstone Clee on 28 August 1982, the sole Venus Pool record, on 29 September 1988, and on the embankments of a refuse disposal tip near Prees on 29 September 1989. The five Long Mynd records of the 1980s included the largest trip ever recorded, 22 on the early date of 22 April, in 1989, three others in spring (eight in April 1984, 12 in April 1985, and eight in May 1989), and one in autumn (two in September 1985).

In the 1990s there were some 65 birds in 11 trips, 10 of which occurred on the Long Mynd. The one heard over Pole Cottage (Long Mynd) on 13 September 1998 was the sole autumn record. Away from the Long Mynd, there were three on Titterstone Clee on 6 May 1995.

Since 1998 there have been only six trips recorded, all except one on the Long Mynd. The first of the five was seen on 1–2 May 2007, and then there were two on 1–3 May 2010, up to nine on 1–7 May 2011, a single on 10–11 May 2012 and 12 on 17–18 April 2014. The sole record away from the Long Mynd was a single near Broseley on 18 May 2013.

The Long Mynd boasts 26 of the total of 29 trips from upland sites, the others being from Titterstone Clee (two) and the Stiperstones (one). The remaining 10 occurred on lowland sites, mainly in the north.

The chart shows the number of birds seen each year since 1950.

Spring trips since 1967 outnumber those in autumn by 26 (76%) to eight. The earliest in spring was on 17 April, in 2014, the latest on 19 May, in 1996, average 4 May, with a dozen occurring in the first 10 days of May. Those in spring have numbered up to 22 birds and averaged 6.5. Those in autumn have been much smaller and only one was of more than one individual. The earliest was on 28 August, in 1982, the latest on 29 September, in 1989, average 14 September. While spring migrants tend to stop over on high hilltops, more autumn migrants appear to fly to wintering grounds in one go, or to take a coastal route.

Dotterel Trips

Decade	19c.	1920s	1930s	1940s	1950s	1960s	1970s	1980s	1990s	2000s	2010–14
Max. trip	13	2	1			5	9	22	20	5	12
No. of trips	3	1	1	0	0	1	5	11	11	1	5

The chart shows the seasonal occurrence of the total number of birds counted, but only on the first date the trip was present.

Typically, when a trip is found, news spreads and further observers arrive, so the duration of stay tends to be well recorded. Around a third of those in spring, 18 of 26 (64%), appeared to stay for just one day, but some lingered, and separating the records of apparent trips is a matter of judgement. Five were recorded over two days and others have been recorded as apparently staying three, four, seven, eight and, once, 17 days, the latter a maximum of seven present between 26 April and 12 May 1991, during which time there may have been separate trips among those currently assessed as one. Fifteen trips were seen on the Long Mynd in the 18 years from 1981 to 1998 and five in the 16 years since then to 2014, and the birder up there in recent years in early May is more likely to encounter Dotterel-searchers than Dotterel.

Over the past 50 years the number and the maximum size of trips has increased, then decreased, as shown in the charts. This matches changes in the British breeding population, which increased up until the late 1980s, and then halved in the last 25 years. Overgrazing, disturbance and the deposition of atmospheric nitrogen on the breeding grounds, redistribution of the Scottish population to Fennoscandia, climate change and problems on the wintering grounds have all been suggested as possible causes.

John Tucker

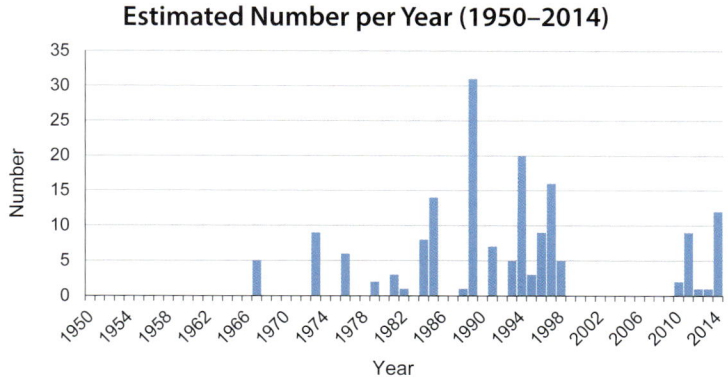

Estimated Number per Year (1950–2014)

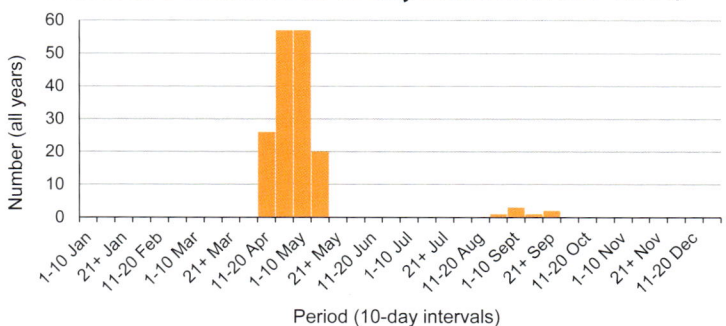

Seasonal Occurrence in 10-day Intervals (1950–2014)

Whimbrel

Numenius phaeopus

Scarce Passage Migrant

Jim Almond, Venus Pool, 8 May 2014

Whimbrel is a long-distance migrant, and the race *phaeopus* which occurs in Britain has a breeding range extending from north-east Greenland east to central Siberia, including Shetland, with smaller populations in Orkney and mainland northern Scotland. They winter mainly along the western coasts of Africa.

Forrest regarded Whimbrel as 'a rare visitor to the Shropshire Moors on its spring and autumn migrations' but 'it has never bred here'. Between 1912 and 1957 there were records in only 1937 and 1939, and it was described as 'never a common visitor, now decidedly rarer than formerly' by Lloyd (1943). The *Handlist* (1964) described it as a 'passage migrant, uncommon', with 'most records of odd birds, or small parties heard flying over'.

The chart shows the estimated number of individuals each year since 1950. After 1957 it was not recorded in eight of the 17 years up until 1974, but it has been recorded every year since. The evocative bubbling call alerts birdwatchers to its presence and fly-over calling birds still account for a high proportion of records.

Notable counts include 20 over Oswestry on 23 August 1973, 35 passing over Muckleton on 5 September 1976, 20+ at Baggy Moor on 17 August 1986, 17 at Venus Pool on 3 May 2008 and 18 the following day, and 23 at Wood Lane on 28 April 2014.

Since 1992 there have only been seven years with more than 10 records, and only 2004 and 2006 produced over 20, the result of extended passage at the well-watched sites of Venus Pool and

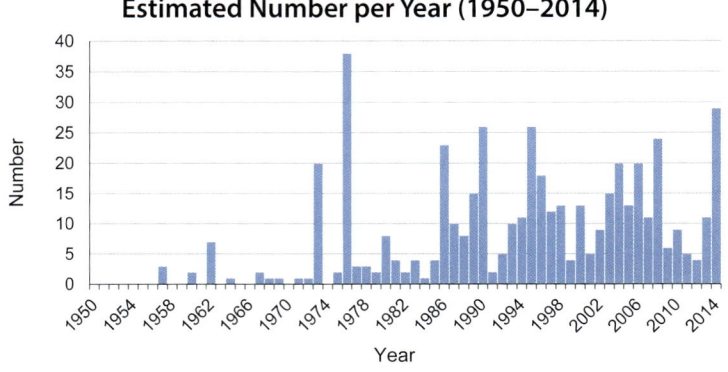

Estimated Number per Year (1950–2014)

Seasonal Occurrence in 10-day Intervals (1950–2014)

Wood Lane, which have accounted for 58% of all records in this period. Allscott Sugar Factory, Chelmarsh and Whixall Moss have also contributed more than 10 each, but the latter would probably produce more if better-watched. As with many waders, it is likely migrating Whimbrel follow the river systems, and the distribution of records reflect this.

Over 70% of the records between 1992 and 2014 were of a single Whimbrel, and only nine records in this period were of more than five. The dates of occurrence chart shows that the peak passage period is at the end of April and beginning of May, with over two-thirds of Whimbrel occurring in those 20 days. There were just 39 records

totalling only 50 birds in autumn (July–September), reflecting the overall pattern of a weaker passage in autumn than spring. Whimbrels passing through Britain mainly breed in Iceland, Fennoscandia and north-west Russia, and tend to take a route along western Britain in the spring, but along the North Sea coasts in the autumn, accounting for the pattern here.

During the recent Atlas period there were spring passage records in four of the six years, nine altogether, between 14 April and 12 May, from Polemere, Whixall, Venus Pool (five), Mirelake and Wall Farm, and one return passage record on 27 July 2008 at Wood Lane.

Dawn Balmer

Curlew (Eurasian Curlew)

Numenius arquata

Uncommon summer visitor, scarce winter visitor

Curlews breed across temperate Eurasia, with some of the northern populations being migratory. The UK holds an estimated 28% of the European breeding population, more when the winter visitors are added, and in response to recent declines here and in Europe, the species has recently been listed by the International Union for the Conservation of Nature as globally near-threatened, one of the few British species on this list. The wintering population in the UK includes a significant proportion of resident breeding birds, but also includes visitors originating largely from Scandinavia.

> Up to about the 1860s the Curlew's breeding range encompassed the moors and uplands of Devonshire, Cornwall, Wales, the Shropshire uplands and, from the lower Pennines, northwards … From about that point the spread to lower areas began … Perhaps the high levels of predator control exercised by gamekeepers during the second half of the 19th century allowed some of the increases in Curlew numbers and distribution. (Holloway 1996)

Locally this distribution was confirmed, with the source of the Morda, above Oswestry, providing the earliest record (Leighton 1836); Beckwith (1886a) noted that:

> during the breeding season, (it) is almost entirely confined to South Shropshire, very few now breeding on Whixall Moss, where once it was common. In the south, however, numbers still breed on Clun Forest, a large tract of moorland … now

much reduced by cultivation, as well as on the Longmynds and other hills about Church Stretton. Of late years, too, there have been several nests in the valley to the north of Wenlock Edge, the birds laying in rough grass fields

In 1901, CSVFC reported: 'Curlews were found breeding on low ground in Corvedale, though previously they were only known to breed on the moorlands of Longmynd, Clun Forest and the adjoining hills'.

Curlew increased in the twentieth century as a breeding species, passage migrant, and as a winter visitor. Lloyd (1943) noted that it had 'shown a general increase' and that it is 'no longer a purely moorland bird; seen frequently in lowland arable and grass, and frequently nests in such situations'. The *Handlist* (1964) provided some detail of this expansion of range and colonisation of new habitats. Prior to 1913 it was considered as nesting regularly only in the hill country, but in that year it bred at Shrawardine and in 1915 near Wem, in the 1920s it was regular at Wem and Whixall, and in 1924 was considered 'numerous' in Corvedale. It reached Newport in 1937 and Bridgnorth and the Wyre Forest about 1941. It was described as 'common', and 'found in suitable habitats throughout'.

The increasingly widespread distribution was confirmed by the national BTO Atlas (1976), which mapped breeding evidence in all the County's 33 10km squares, except one in the south-east, with confirmed breeding in all except six of the remainder. It was found in all these 10km squares in the 1985–90 Atlas, and in three-quarters (76%) of all tetrads. Gaps coincided with the built-up areas, and areas of cereal

Breeding Distribution Change (1985–90 to 2008–13)

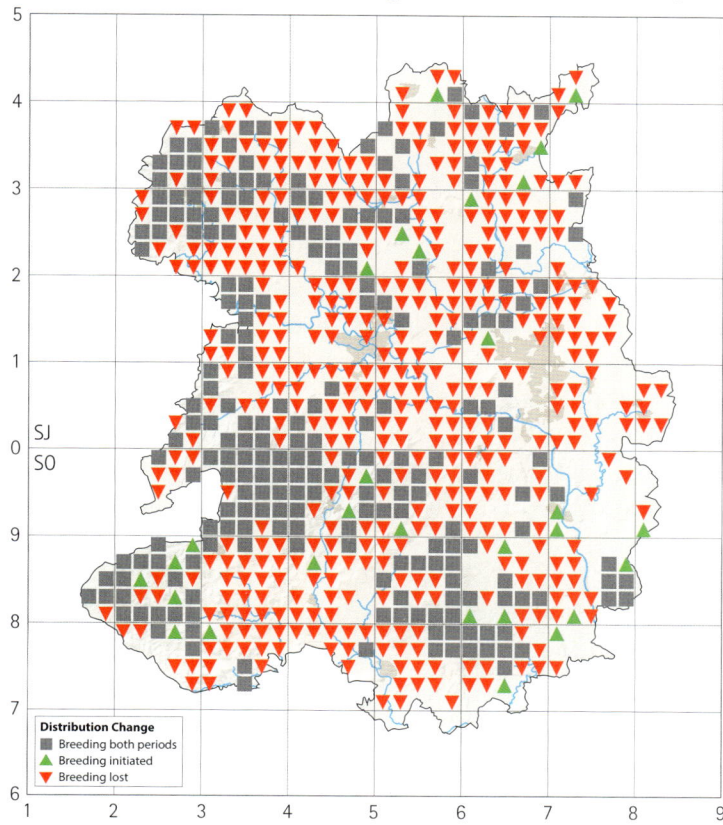

Distribution Change
- ■ Breeding both periods
- ▲ Breeding initiated
- ▼ Breeding lost

Occupied Tetrads

Atlas period (breeding)	1985–90		2008–13		Change	
	Number	%	Number	%	Number	%
Confirmed	184	21	41	5	–143	–78
Probable	322	37	120	14	–202	–63
Possible	155	18	91	10	–64	–41
TOTAL	661	76	252	29	–409	–62
Tetrads with Winter Records (2007–13): 62 (7%)						

crops and low rainfall, almost all in the east. The damp meadows of the Severn–Vyrnwy confluence, the Weald Moors and Tern valley were particular lowland strongholds. The population was estimated at about 700 pairs, one pair per tetrad with breeding evidence.

Monitoring has found some large local declines since the 1980s. The national *Birds of Wet Meadows Survey* 2002 (Wilson *et al.* 2005.) included 11 local sites, mainly in the Severn Valley and the Weald Moors, and found a reduction from 25 pairs to 11 (56% decline) between 1982 and 2002. In the Shropshire hills, breeding bird surveys on the Stiperstones and the Long Mynd have shown a catastrophic decline of Curlew in the last 15 years. The Long Mynd Breeding Bird Project found 11–13 pairs in 1995, declining to seven to eight pairs in 1998, three pairs in 2002 and 2003, and only two pairs in 2004. Very few chicks or young birds were seen, indicating very poor breeding success (Smith 2004a, 2004b). Since 2004 the annual population has usually been two pairs, including 2012 and 2014, but in 2009 it was three, and in 2010, 2011 and 2014 it was only one. On the Stiperstones, five

breeding pairs were recorded in 1995–96, but breeding apparently ceased prior to 2000, and certainly none were found during surveys in 2002 and 2004–05 (Smith 2007).

The recent Atlas confirmed a massive decline, a reduction of 62% in tetrads with breeding evidence, and Curlews have largely disappeared from arable areas.

Specific monitoring programmes are now underway in most of the areas that still have a breeding population. The first, by the Upper Onny Wildlife Group, has been counting an area of 120sq.km (30 tetrads) between the Long Mynd and the Welsh border, and has found an estimated decline from 38 pairs down to 28, a loss of 26% in the 11 years from 2004–14.

The Upper Clun Community Wildlife Group has found at least a 50% decline in 21 tetrads in the eight years from 2007 to 2014, from 20–22 down to eight to ten. Monitoring in other areas started more recently, but around Titterstone Clee there was a decline from 12–14 in 2013 to 10–12 in 2014, and an estimated 13 pairs were found in 2014 in the first year of surveys in the Rea and Camlad valleys.

Other breeding areas with recent survey coverage include the Severn–Vyrnwy confluence, where four pairs were found in a partial survey of four tetrads in 2010; Fenns – Whixall Moss SSSI, where four pairs were found by an RSPB survey in the English part of the site in 2010 and 2011; and Baggy Moor and the Weald Moors, where two and three, and two and none, were found in 2009 and 2010 respectively by an RSPB project. The RSPB project was repeated in 2012 and 2013, but no Curlews were found in either year. Other clusters of occupied tetrads, which have no recent counts, include the Oswestry uplands, and the hills and valleys between Apedale and Corvedale. However, there are an estimated eight to ten pairs in the north-west, including the pairs at Baggy Moor, and another 10 pairs

Jim Almond, Wood Lane, 15 September 2014

Jim Almond, Venus Pool, 17 March 2013

in the Severn–Vyrnwy confluence, including the pairs at Melverley (Allan Dawes *pers. comm.*).

Territories are large, particularly in areas of poor habitat, so many of the possible breeding records will relate to pairs in tetrads with probable and confirmed breeding. The specific species surveys itemised above have found 93 pairs in 125 tetrads with some level of breeding evidence. There are 252 of the latter altogether, which pro-rata suggests a total of 188 breeding pairs. However, this may be over-optimistic: the counts have taken place in the areas of highest density, and only 16% of the tetrads with Curlews in these areas had possible breeding evidence, while 55% of the tetrads with Curlews in other areas had only possible breeding evidence, much of which is likely to have arisen from pairs in adjacent squares. For example, there are six tetrads with probable (two) and possible (four) breeding east of the Severn Valley Country Park at Alveley in SO78 which appear to relate to only one pair, resident near the border of SO78W and X, but foraging into adjacent tetrads, particularly on arrival and before settling down at their breeding site (Keith Bates, Jon Lingard, *pers. comms.*). Therefore the pro-rata estimate is likely to be a bit high, and 160 is a more reasonable estimate.

Although Curlews were not found on sufficient TTVs for it to be reliable, the estimate based on TTV counts is 140–160 breeding pairs. An estimate of 160 in 2010 represents a catastrophic 77% decline in only 20 years. Monitoring in the Upper Onny, Upper Clun and Baggy Moor areas since then has found a continuing decline, and presumably the same decline has continued in other areas, so the population is rather less now.

Agricultural intensification is likely to have started the decline, through 'improvements' to grassland (land drainage; increased use of fertilisers, which accelerates the transfer of ground water into the growing grass, thereby reinforcing the effect of drainage; control of rushes that provide cover for nests; and rolling and chain harrowing, and production of silage, rather than hay, which is cut earlier and more often). These changes have all reduced the available habitat and increased the destruction of eggs and chicks, while higher stocking levels will have increased the number of trampled nests in pasture.

The decline has almost certainly been accelerated by high and increasing levels of nest predation, as the fewer remaining pairs are more vulnerable, particularly to foxes, which are increasing, probably because of the availability of large numbers of released Pheasants as food. The possible impact of another potential predator, Carrion Crow, has been reduced by the substantial reduction in sheep stocking levels (around one-quarter) since 2002, with the crows declining by a similar percentage over the same period (see p. 328). Badgers and birds of prey, which have also increased in recent years, are often cited as culprits, but there is no hard evidence to suggest they have a significant impact.

As a result of the Upper Onny monitoring, the Stiperstones-Corndon Landscape Partnership Scheme, financed largely by the Heritage Lottery Fund up until March 2018, has initiated a nest monitoring project to establish the actual current reasons for poor breeding success, the precursor to an effective conservation strategy. Initial results in 2015 showed that the 12 nests that were found were all predated, nine at the

Decline in the Upper Onny Breeding Population (2004–14)

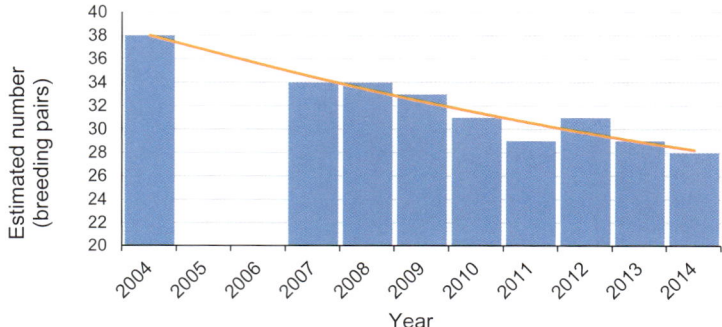

egg stage and three after broods of chicks hatched. Foxes accounted for more than half. The fields containing three further nests were identified, but they were cut before the nest could be found.

The number of passage migrants also increased in the nineteenth and twentieth centuries, then declined. Beckwith commented that 'in spring and autumn, on their way to and from their breeding grounds, Curlews are sometimes seen in other places, near Cressage, Charlton Hill, about the Clee hills, and in the low meadows round Ellesmere, Baschurch and Melverley'. The *Handlist* (1964) account showed an increase in numbers, and more widespread regular occurrence: passage, in the Severn Valley at Shrewsbury Sewage Farm and Ironbridge, and at Allscott, where 'a gathering of some 50 birds takes place in July', and Ellesmere.

Many more passed through in the second half of the twentieth century: a few on their way to and from local breeding sites, but many more to sites further north. The charts show, firstly, the seasonal pattern of movement through the three main passage sites that are regularly monitored, Allscott Sugar Factory (including Mirelake and Isombridge), Venus Pool and Wood Lane; the counts are the cumulative total of the monthly maxima for each site; and, secondly, the sum of the monthly maxima for each year since 1992 for the same three sites. The seasonal chart shows a steady build-up until March, but very few present during the breeding season, then the start of return passage in June, up to a peak in July, then a steady reduction until the year end when very few are present.

The annual chart shows a build-up until 2002, followed by a rapid, fairly even, decline until 2014. The former may reflect an increasing population following the ban on shooting in 1982, as well as a continuation of the increase in numbers passing through, while the latter reflects fewer Curlews passing through as the population has subsequently declined. The same pattern is reflected in maximum

Winter Distribution (2007–13)

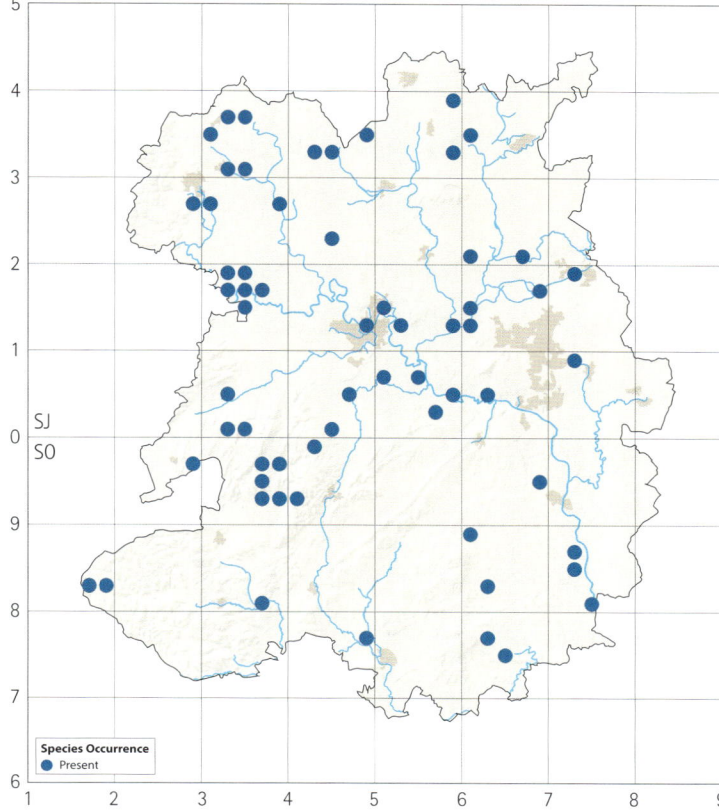

counts at these sites since 1992: all except one of the 34 largest counts at Allscott, and five of the 20 largest counts at Venus Pool, were in 2002 or earlier. While 74 of the top 100 counts at Wood Lane were after 2002, there were only nine in 2008, one in 2009, and none after that.

As a winter visitor it was not noted until 1936, when small flocks were seen in the valley of the Vyrnwy (*Handlist* 1964). The passage starts, and finishes, in the Atlas winter season months November–February, and the three passage sites featured in the charts, plus the less important passage sites at Monkmoor (including Shrewsbury Sewage Farm), Chelmarsh Reservoir, and Whixall Moss and Canal Floods, are all shown on the recent Atlas winter distribution map. Curlews start returning to their breeding sites towards the end of February, accounting for the dots in the Upper Onny, Rea Valley, Clee Hill and Clun Forest areas. The distribution shown in the 1981–84 BTO Winter Atlas is less widespread, but part of the difference may be due to different fieldwork methods missing the early-returning breeders.

However, there is a wintering population too, centred on the floods at the Severn-Vyrnwy confluence near Melverley. The chart again shows the cumulative total of monthly maxima at that site, and that the population does not gather until November, disperses by March, and numbers are fairly constant throughout the winter and do not increase as migration at other sites gets under way. During recent winter Atlas fieldwork there were 20 counts of more than 25, including three of 100 or more and another six of more than 50, maximum 150 at Alberbury on 29 January 2008. All except one of these counts were in the six tetrads in SJ31 with winter Curlew records. The exception was 36 on Whixall Canal Floods on 27 January 2011. The two highest TTV counts, of 40 and 20 respectively, were

Seasonal Passage at Three Sites (1992–2014)

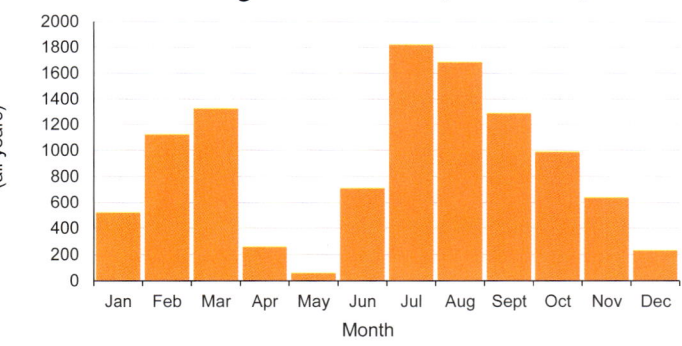

Annual Passage at Three Main Sites (1992–2014)

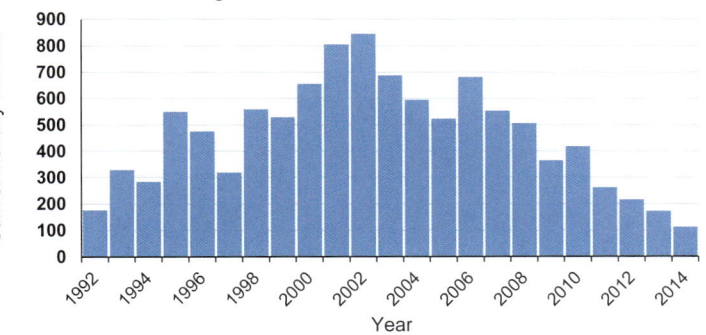

Seasonal Distribution at Severn–Vyrnwy Confluence (1992–2014)

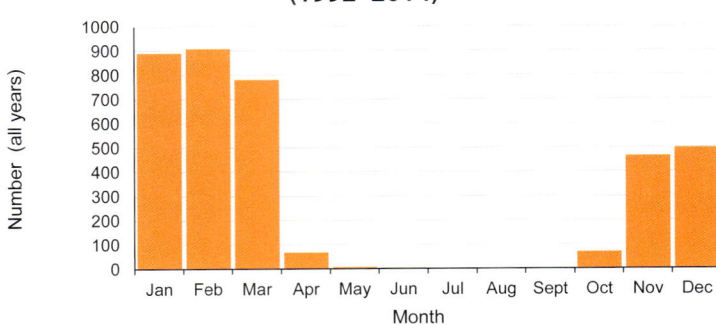

also at the confluence. High counts have persisted here through the first decade of the twenty-first century, but may have been declining since. Nationally, the wintering population has declined by 20% in the last 15 years.

Curlews are long-lived and site faithful. An adult at least one-year old was ringed on passage at Allscott Sugar Factory on 30 September 1965 and caught near Grantham in August 1984 and again at the same site on 7 July 1995, 135km distant, 29y 9m 7d later. This held the national longevity record for a Curlew ringed in Britain until one was found dead at Neston, Cheshire in January 2011, ringed in Lancashire in June 1978, creating a new record of 32y 7m 0d. Three other adults and one first year ringed on autumn passage at Allscott have lived to at least 15–21 years.

All except six of the 70 recovered here were ringed here, the exceptions being two from Ireland, two from Caernarfonshire and one from each of Denmark and Sweden. Another 34 ringed here have been recovered elsewhere, 22 in the British Isles, and 12 abroad, in Channel Islands (one), Denmark (four), Finland (one), France (four), Germany (one) and Sweden (one). The furthest travelled went to Finland, 1,884km distant, and the two involving Sweden had each travelled around 1,300km. Almost all these Curlews were among 752 ringed or caught in the passage period at Allscott Sugar Factory (Julian Langford *pers. comm.*), so many of them are probably migrants from Fennoscandia moving though in the autumn to southern Britain or south-west Europe for the winter, then returning north to breed. The dates for all the continental recoveries are consistent with this pattern. Most of the recoveries probably do not involve local breeding birds, which are believed to winter on British estuaries, mainly in the south, and there is some evidence for this belief: 16 of the 22 recoveries in Britain were found in coastal counties.

The local rapid recent decline shows that there is a real danger that the bubbling evocative call of the Curlew will be lost forever. Nationally, BBS showed that Curlew has declined by 48% in the UK in the period 1995–2014, while monitoring over a longer period showed a decline of 62% since 1969, leading to it being added to the Red list (Eaton *et al.* 2015). In addition to holding an estimated 28% of the European breeding population, the UK holds an estimated 19–27% of the world population (Brown *et al.* 2015), and Curlew is arguably the most pressing bird conservation priority, here and across the UK as a whole.

Leo Smith

Bar-tailed Godwit

Limosa lapponica

Very rare passage migrant

This godwit breeds in Fennoscandia and Russia and is predominantly a winter visitor to British shores. It prefers large sandy estuaries, coastal beaches or sheltered coves, with the nearest large flocks being on the Lancashire and Wirral coasts.

All early records were of birds shot and held in collections (Forrest 1899). One in the Hawkstone collection was shot at Hine Heath in December 1849, with others shot at Barrow in 1870, near Cressage (two birds) in turnip fields in September 1878, and at Buildwas in 1885. Rocke (1866a) reported it simply as 'much oftener met than Black-tailed Godwit'. Nowadays the converse is more likely, due to the population increase in Black-tailed Godwits since the 1920s. As a short-staying passage migrant rather than a breeding season species, records were scant until formal recording began in the 1950s, with apparently only one record in the first half of the twentieth century, three shot out of a flock flying high over Black Mountain in September 1919.

Since 1950, it has been seen less than every other year (32 out of 65 years), there has been more than one record in only seven of those years, no year has more than three records, and there are almost twice as many spring passage records as autumn passage.

This is due to characteristic movements at different times of the year. Autumnal movements occur irregularly involving small numbers. Some adults, still in summer plumage, fly directly to our shores, while other adults and the bulk of juveniles stop off at the Wadden Sea (the Netherlands) to moult and feed before moving on to Britain, arriving later, typically in September and October. Spring

Jim Almond, Wood Lane, 4 May 2010

Seasonal Occurrence in 10-day Intervals (1950–2014)

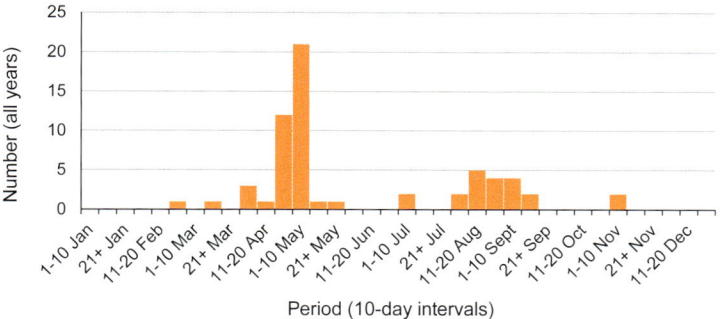

one in 1994 stayed for nine days. Venus Pool hosted one in 1990 and two for two days in 2000. The only other sites with autumn passage records are Knighton Reservoir (1994) and Monkmoor Pool (1999), both individuals on single dates.

While most birds wintering in Britain depart in February or March, there is only one February record (Cressage 1980, a single for a day) and one March record (Chelmarsh 1987, a single over four days). The first records of the second type of characteristic spring movements were from Venus Pool, with individuals seen in April in 1960 and 1962. In May 1978 a flock of 11 were at St George's Flash, Telford (now known as Priorslee Flash), for one day – the highest number ever recorded.

Since then and up until 2014, there have been 20 sightings of late-spring passage birds, predominantly from the well-watched locations of Venus Pool (13 records, 1988–2014) and Wood Lane (three records, 2004–10). Single records have come from Dudlestone Moor and Baggy Moor (1990), Chelmarsh (2007) and Wall Farm (2008).

No records were received in 2012 and 2013, while there was an individual at Venus Pool on 5 May 2014.

The number of inland records has been increasing, but they are typically still of single birds seen on one day only, although some stay longer – 10 days being the maximum here (Venus Pool 1988). There are only three recent records of more than a single bird during this passage period, all from Venus Pool – two in 1994 and 2000, and five in 2011.

The chart shows the pattern of occurrence in 10-day intervals, which reflects the pattern of characteristic movements outlined above.

Bob Harris

migration is more concerted, with wintering birds returning to the Wadden Sea in large numbers to fatten up before departing north to their Siberian breeding areas. This is closely followed by a second movement of those which over-wintered in Africa moving through to Fennoscandia. Consequently more pass through in spring. Modern records reflect this pattern.

The first of these was of two at Shrewsbury Sewage Farm in August 1957 (which stayed for 19 days), with individuals there again on single August dates in 1958 and 1961. Allscott Sugar Factory produced records of both types of autumn movement in the early 1960s, with two present in September 1963 and three in August 1964, all on single dates, with individuals on four other August or September dates up until 1976, one staying four days and another six.

Since 1976 there have been few autumn passage records. Allscott Sugar Factory had single birds on single days in 1986 and 1991, while

Black-tailed Godwit

Limosa limosa

Scarce passage migrant, very rare winter visitor

Although there is a small English breeding population, almost all records are of passage birds moving between breeding areas in Iceland and central and Eastern Europe, and wintering grounds mainly in Africa south of the Sahara, with smaller numbers wintering in France and Iberia. The small number of winter visitors in Britain is increasing.

In 1866 Rocke stated that it was 'not often met so far inland – rarely been obtained'. It was described as a 'very rare spring and autumn visitor', and a group of four seen in spring 1877 near Preston upon the Weald Moors was the only listed record (Forrest 1899). One found and killed on the Stiperstones in September 1901 appears to be the only record from the first half of the twentieth century.

Following a sustained expansion of the *islandica* breeding population in Iceland since the 1920s, thought to be due to a mix of climate and habitat change as well as reduced hunting pressure (Gill *et al.* 2007), the number over-wintering in Britain has increased substantially, including a distinct cluster which has become established in nearby North Wales. Thus, while it was encountered less than Bar-tailed Godwit in the nineteenth century, the number of records here has overtaken those of the latter.

This godwit is a bird of muddy estuaries, wet grassland and fine sediments, rather than the sands favoured by Bar-tailed Godwit. It

can find acceptable feeding areas in flooded farmland and gravel pits, which explains why most records have been from Venus Pool, Allscott Sugar Factory, Wood Lane and Whixall.

Adults will return from their Icelandic breeding grounds as early as June if their breeding attempts have failed. The successful ones and their young appear in a larger movement in the second half of July, peaking in August and September. The first modern record is of such a movement, an individual seen at Shrewsbury Sewage Farm on the 20–21 August 1957. A smaller spring passage occurs in March and early April, and the next records reflected this with 10 individuals, some in summer plumage, seen in 1961 and 1962 at Venus Pool.

From 1963 to 1987 there were only five years without records. Most were of autumn movements through Venus Pool or Allscott Sugar Factory and involved one or two individuals with occasional groups of up to six. Spring movements were fewer but appeared to be increasing. Notable records during this period were of 10 in summer plumage at Cole Mere in May 1966, 10 at Allscott Sugar Factory in August 1972, and 18 at Venus Pool in late July 1979. As with Bar-tailed Godwits, the majority of records are from single dates.

From 1988 to the present day there have been records every year, mostly autumn movements but, increasingly over time, spring

Estimated Number per Year (1950–2014)

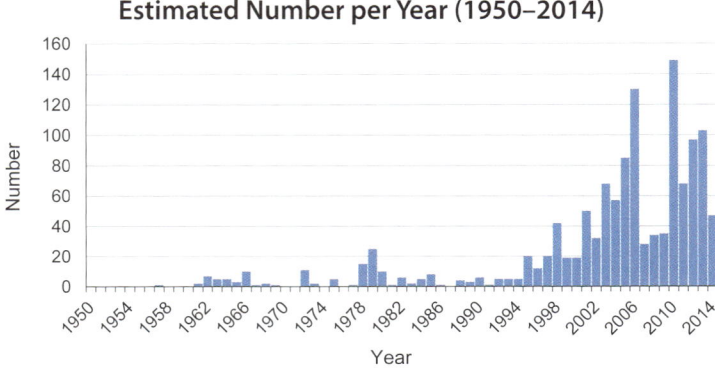

Seasonal Occurrence in 10-day Intervals (1992–2014)

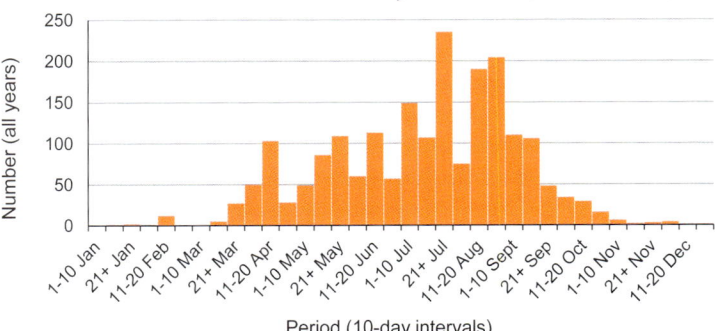

movements, which probably reflect increasing populations as well as increased watching of sites. Venus Pool and Allscott Sugar Factory feature heavily in most years, with other frequent sightings from both Chelmarsh and, since the mid-1990s, Wood Lane. Rarely, records have been received from 13 other sites, including Eyton Moor, Polemere, Crose Mere, Cockshutt and Crudgington.

The chart shows the best estimate of the total number of birds seen each year since 1950. Generally, as a trend, numbers and flock sizes are increasing, with the large flock of 68 at Venus Pool on 11 June 2006 being exceptional. The abnormally low counts in the years 2007–09 may reflect a temporary shift in populations, as numbers at Morecambe Bay and the Severn Estuary increased while those at the nearer Dee, Mersey and Ribble estuaries decreased (WeBS).

During the recent Atlas period, records were received of both spring and autumn movements, mainly from three sites – Wood Lane, Venus Pool and Allscott Sugar Factory – with fewer from Whixall.

As a measure, over the full six-year period, this godwit was at Wood Lane for 100 days (maximum flock count 17), Venus Pool 73 days (maximum count 40), Allscott Sugar Factory 56 days (maximum 37) and Whixall 16 days (maximum 24).

Usually more are seen in autumn than in spring, as native birds migrate south, and large flocks of incoming migrants from Iceland arrive in Britain.

These birds remain, usually at estuarine sites, for several weeks while moulting before moving south to France and Portugal. The large number of birds involved increases the likelihood of some being seen here.

In spring, migration back to Iceland is via the Netherlands, with a peak in mid-April. Winter records are probably of over-wintering birds that move from estuaries to grassland to feed as the mudflats become depleted.

Bob Harris

Jim Almond, Wood Lane, 8 July 2013

Turnstone (Ruddy Turnstone)

Arenaria interpres

Rare passage migrant

Jim Almond, Venus Pool, 4 August 2011

Although a relatively common and widespread winter visitor to coastal Britain from its breeding grounds in the high Arctic, it is relatively scarce inland and usually seen only during the spring and autumn migration periods.

Accounts from the latter half of the nineteenth century state that it was very rare, only having occurred two or three times. There is a specific report of a young bird being shot at Rednal in about 1851, while in 1879 a bird in Mr Bodenham's collection was noted as having been obtained at Atcham, although the year it was shot is not stated.

There were no further reports until one was at Shrewsbury Sewage Farm on 7 August 1957. Subsequently, it has been recorded in 25 of the 57 years, but was annual from 1988 to 1998. Since 2000, however, it has only been recorded in seven of the 15 years. Most reports are of one bird, although two were at Venus Pool on 23 May 1990 and

at Priorslee Lake on 31 July 1991. More noteworthy were three on Whixall Moss on 15 May 1994, and at Allscott Sugar Factory on 1 September in the same year. Not surprisingly, therefore, 1994 had the highest annual count with seven, whereas the average is much closer to two in years when it has occurred, and marginally less than one per year since the first modern record in 1957.

The chart shows that the modern records are split almost equally between the two migration periods, 24 sightings in the spring and 25 in the autumn, although autumn passage is generally spread over a longer period. The earliest spring record was on 29 April, in 1998, a single at Venus Pool. The remaining spring records are from May, peaking in mid-month, except for another single at Allscott Sugar Factory on 8 June 1986. Returning Turnstones appear from the last few days of July onwards, with the majority in August and a few in early September. An exceptionally early return migrant was at Venus Pool on 4 July, in 2000, and a very late one was at Allscott Sugar Factory on 11 November, in 1993. The majority are only present for one day (39 individuals, 33 records), while some are present the next day only (four) or up to nine days (five). Additionally, one ringed at Allscott Sugar Factory was present for a total of 34 days from 20 August 1988. Most that stayed for more than one day were on return passage.

Graham Walker

Seasonal Occurrence in 10-day Intervals (1950–2014)

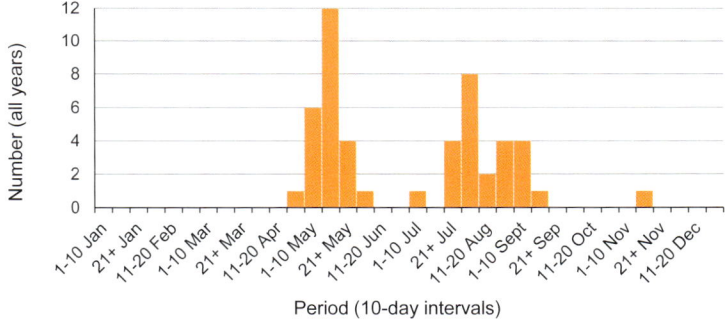

Number (all years) / Period (10-day intervals)

Knot (Red Knot)

Calidris canutus

Rare passage migrant

Knots are one of the group of waders breeding in the tundra zone of the high Arctic. Ringing and population studies over the last 40 years have demonstrated that most of the thousands which spend the winter on Western European estuaries, especially the Wadden Sea, the Wash and Morecambe Bay, are the subspecies *Calidris canutus islandica*, which breed at very high latitudes in northern Greenland and the northern islands of Canada (Piersma & Davidson 1992). The strong association with estuary habitats during passage and winter periods

is linked to their preferred food, small bivalve molluscs, especially *Macoma balthica* (Prater 1972). Therefore Knots occur rarely inland, and here they are sporadic visitors in extremely small numbers.

Rocke (1865) described Knot as 'not infrequently met with' but gave no further details, while Beckwith mentioned birds obtained near the Severn at Cressage, Uffington and Buildwas, and Forrest (1899) added Eyton to the list, but neither gave any dates. In the absence of detail, the *Handlist* (1964) did not include Knot in the list of species

Jim Almond, Allscott, 19 August 2007

recorded before 1950, and stated that 'the only record appears to be of two at Shrewsbury Sewage Farm on 21–22 September 1957'. A single was subsequently recorded at Allscott Sugar Factory on 20 March 1967 (record not published in SBR), and another in the same place on 27 August 1969.

There was then a gap until 1978, since when Knots have occurred in 19 out of 37 years. Thirty-four out of the total of 38 records are of a single bird. They tend to move on quickly but there have been a few extended stays: 1–3 were at Allscott from 8–12 September 1983, one at Allscott stayed from 12–16 September 1987, and one to two lingered there for a week from 1–8 September 1988, while one in its first winter was at Wood Lane between 21 January and 4 February 2007. Only in 1983 and 1988 have Knot bird-days reached double figures.

There were only two records during the recent Winter Atlas period, both in November, one at Chelmarsh in 2007, and the other at Allscott in 2008. There were breeding season Atlas records of passage birds

in April or May at Wood Lane and Whixall Canal Floods in 2008, Mirelake and Wood Lane in 2009, and, the most recent record, from Long Lane on 15 May 2013.

There are records from all months of the year except June, July and December. In autumn, adult Knot arrive from breeding grounds earlier than juveniles and the records in August will mainly be adults migrating to moulting sites (particularly the Wash and Wadden Sea), with peak numbers in September coinciding with the arrival of juveniles. They move between estuaries as winter progresses, predominantly from eastern sites westwards, and records between November and early March probably reflect these movements. In April and May *islandica* Knots move en masse to the Wadden Sea to fatten up before migrating north, and this movement is likely to account for the spring occurrences.

Breeding in the high Arctic, the Knot will be of special interest in monitoring the impacts of global warming in the future. Predictions are that its tundra habitat could reduce in extent by up to 65% due to a combination of rising sea levels and more extensive vegetation cover, while the size of its favoured estuary habitat in passage and winter periods could also be reduced in size by sea level rise (Lindstrom & Agrell 1999). Changes far away are therefore likely to affect the prospects of seeing this wader here in future.

Gerry Thomas

Seasonal Occurrence in 10-day Intervals (1950–2014)

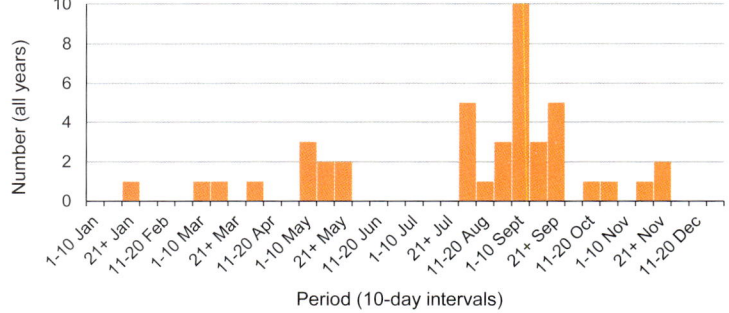

Ruff

Calidris pugnax

Scarce passage migrant, very rare winter visitor

Breeding occurs across northern Europe with important populations in northern Sweden, Finland and Russia, and most Ruff winter in sub-Saharan Africa. Britain is at the north-west extremity of its range, and it was a widespread but uncommon breeding species at the end of the eighteenth century, but became restricted to eastern England before the middle of the nineteenth century, and only sporadic breeding occurred after that, lastly in 1922 (Holloway 1996). There was a short-lived colonisation of the Ouse Washes with breeding first proved in 1963, peaking in the 1970s, and occasional evidence of breeding elsewhere, but the last confirmed breeding, in north-west England, was in 2006. There are no known local confirmed breeding records.

Over several centuries Ruff have suffered from loss of habitat caused by widespread drainage of agricultural land and changes in grassland management. From mediaeval times they were hunted,

and served in large numbers as a great delicacy at banquets, and the resulting high prices 'fuelled a constant harvest across their entire range from Northumberland to south-west England' (Cocker & Mabey 2005). While there are no accounts of such hunting activities locally, it may be assumed that this apparently lucrative trade did occur here, and widespread hunting might also partially explain the extreme scarcity of local records in the nineteenth century.

There were only two such records, both of a Ruff killed along the River Severn during hard frost, one in 1861 near Melverley, the other near Buildwas in 1867 (Forrest 1899). A reeve was shot on 11 May 1881 at Aber-Tanat near Oswestry, being 'especially noteworthy as occurring in summer' (Forrest 1908), but this was presumably on the Welsh side of the border, as no mention was made of it in Forrest (1899), and an immature shot at the old river bed on the north side of

Estimated Number per Year (1950–2014)

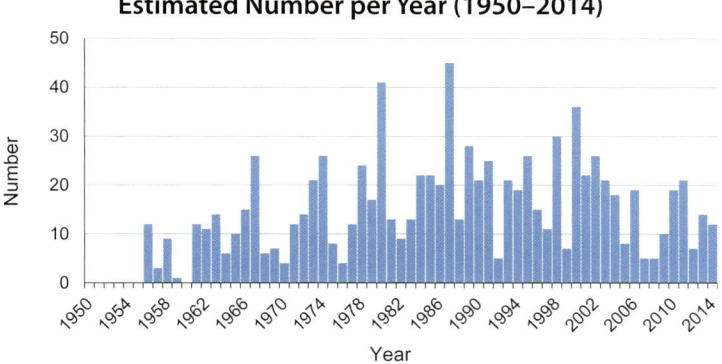

Seasonal Occurrence in 10-day Intervals (1992–2014)

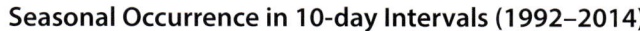
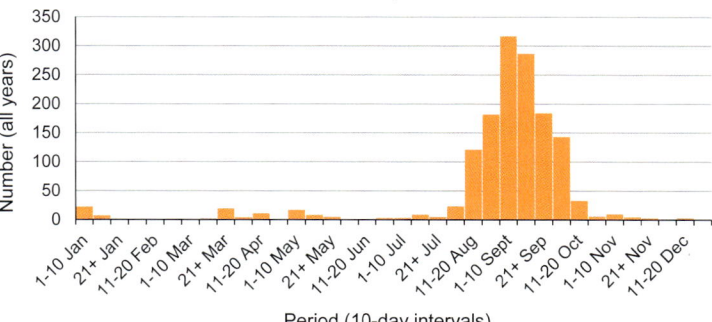

Shrewsbury on 26 August 1922 was noted as the 'third record'. There were no further historic records, and it was 'virtually unknown before 1956 when observations started at Shrewsbury Sewage Farm', after which it was noted as a regular passage migrant numbering five to ten annually from July to October at Venus Pool and Allscott Sugar Factory, and formerly at Shrewsbury Sewage Farm (*Handlist* 1964).

Since then a few records, usually of singles or small groups, have been received every year from freshwater marshes, shallow pools and wet meadows, mainly from Allscott Sugar Factory. Although the maximum annual count fluctuated, the underlying trend steadily increased, with the first double figure count in 1972. Since then there have been 19 such counts, all except one from Allscott Sugar Factory, the last in 1994, and all except one in the months August–October. Maximum counts were 20 on 4 September, in 1974 and again in 1978, and 19 on 15 September 1993. There were also 13 at Cosford Airfield on 26 August 1986.

In the early passage period, there was an 'amazing spring' at Allscott Sugar Factory in 1987, with at least 33 individuals present between 14 April and 20 May, including 22 together on 15 April. A male in full summer plumage at Venus Pool in early June 1980 was exceptional.

The chart shows the estimated number of different individuals observed each year since 1950.

Since 1992, 39% of records have come from Allscott Sugar Factory (including Mirelake), 25% from Venus Pool and 21% from Wood Lane, with the remainder from Chelmarsh, Wall Farm and Whixall Canal Floods, and several other scattered lowland sites with occasional sightings. Few have occurred in the spring migration period, and only 34 (6%) were from late March–early May. All were of one or two individuals, except for eight at Venus Pool in March 2002. The 'amazing spring' of 1987 has not been repeated.

In the autumn, migrants tend to linger for a few days. SBRs estimated a minimum of 29 individuals in 1998 and 36 in 2000 ('bumper years'), and 19 in 2001, although the latter 'may be some way short of the actual total'. Numbers have fallen steadily since then, with less than 15 in most subsequent years. In the five years since 2010, there have been observations of small parties of three or four lingering at the main sites, and only one of a larger party, five at Allscott Sugar Factory in September 2011.

Migration patterns are complex and have been changing in recent years. Autumn migration was generally concentrated along the coastline of western Europe, with the spring migration occurring much further east. Deterioration of staging areas in the Netherlands has led to numbers on the autumn route declining substantially over the last two decades, and the eastern route through Belarus has become more important. As a consequence the breeding population has shifted from the European Arctic to Western Siberia. This change has affected the number occurring here each year, which, apart from abnormal years in 1980 and 1987, reached a peak in 2000, and subsequently declined.

Since 1992, 55% of records have been of singletons, and another 30% were of two or three. The chart showing the pattern of occurrence throughout each year 1992–2014 clearly reflects how the different autumn and spring migration routes affect numbers turning up here, with the much more pronounced autumn migration peaking in early September.

Winter records are rare, with only 32 since 1992, comprising probably 31 different individuals in 11 different years at nine separate sites, with two sites, Allscott Sugar Factory (including Mirelake) and Whixall Canal Floods, providing two-thirds. These two sites also provided all four of the recent winter Atlas records, and the two records of a migrant seen during the recent Atlas period breeding seasons.

The one ringing recovery reflects the different autumn and spring migration routes that were common in the past: a first-year male caught at Allscott Sugar Factory in September 1980, was found shot in Pavia, Italy, in February 1984, 1,160km to the south-east, 3y 5m 19d later. The more recent shift in migration routes suggests that Ruff will become increasingly scarce.

David Farncombe & Leo Smith

Jim Almond, Chelmarsh Scrape, 3 September 2011

Curlew Sandpiper
Calidris ferruginea
Very rare passage migrant

Yvonne Chadwick, Chelmarsh Scrape, 19 September 2010

Curlew Sandpiper is one of the great travellers of the bird world. Breeding in the high tundra of northern Russia, they travel thousands of miles each year to and from wintering grounds in the southern hemisphere, especially South Africa and Australia. From their relatively restricted breeding range, they migrate south on a broad front, with relatively small numbers taking a south-westerly course, passing through the Low Countries and East Anglia.

Appearance here is largely dependent on a successful breeding season, when there are high numbers of juveniles in the autumn population, and on anti-cyclonic conditions over the eastern North Sea and Baltic regions to encourage more to take the south-westerly route. These conditions are rare, and many years go by with no records at all. When they do occur, there are usually only small numbers. Even then, presence also depends on suitable foraging habitat being available, to encourage stop-over. Curlew Sandpipers truly wade, unlike many waders of similar size, so they need extensive very shallow pools.

The only record from the nineteenth century is of one shot at Shrewsbury Old Racecourse in 1836 (Rocke 1865). There were no further records until two were observed at Shrewsbury Sewage Farm on 23 October 1957. Since then, the majority of records have been from Mirelake (over 90% in terms of the length of stay of each individual, expressed as 'bird-days'), at times when water levels have been low enough.

It has been recorded in only 36 years out of 65 (between 1950 and 2014), there have been only 14 years when the number of individuals has reached five or more, and in only four of those years has it reached double figures. Thirty-two at Mirelake on 8 September 1988

is the largest single count, and there were 11 at the same place four days earlier: 1988 stands out as an exceptional season with over 400 bird-days. Other good years in terms of the number of individuals seen were 1983 and 2005, and, in terms of bird-days, 1983, 1985 and 2001.

The second chart confirms that almost all Curlew Sandpipers occur on autumn passage, primarily between 11 August and 20 October, with peak numbers usually in early to mid-September. The latest ever was a single which dallied at Mirelake until 14 November in 1999.

The only spring records were of one at Whixall Moss on 7 May 2006, and another at Mirelake on 5–6 June 2009, and the latter was the only record in the recent Atlas period. Apart from the sites mentioned above, they have occurred at Chelmarsh, Baggy Moor, Wood Lane and Whixall Canal Floods.

Fascinating research (Barshep *et al*. 2011; Summers *et al*. 1998) has shown a significant link between lemming populations in the tundra regions and the breeding success of the Curlew Sandpiper (and indeed of other species in the region). The lemming cycles usually produce high abundance every third to fifth season. In these years the Arctic Fox, the key predator of the tundra zone, has more than enough lemmings to feed on. Ground nesting birds, including Curlew Sandpipers, are left in peace. The high Arctic however remains a precarious place to rear young, and though vast numbers of small insects are present for a short period in midsummer, the weather conditions do change rapidly and inclement weather in that period can still cause significant mortality among chicks (Schekkerman *et al*.

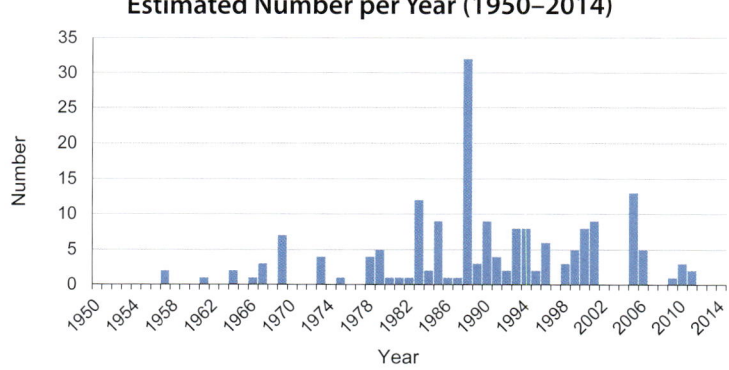

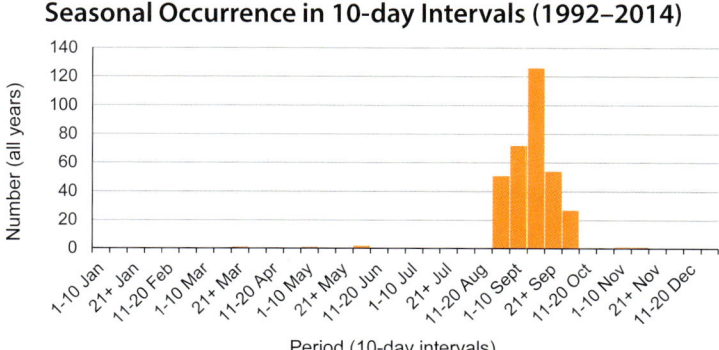

1998). Curlew Sandpipers breed successfully only in years of high lemming abundance and benign weather, and they usually appear here only in such years.

Recent research in parts of the Arctic has shown that the lemming cycles are breaking down, perhaps as a result of global warming, suggesting that the future of many tundra breeding waders looks uncertain. The recent loss of the rich feeding site at Mirelake, part of the Allscott Sugar Factory, is also likely to mean that Curlew Sandpipers will occur less frequently than in the past.

Gerry Thomas

Temminck's Stint
Calidris temminckii
Very rare passage migrant

Jim Almond, Venus Pool, 8 May 2008

The UK is on the western edge of the range of Temminck's Stint. It breeds from Scotland and Norway eastwards across large tracts of Russia and migrates south on a broad front, wintering from Mali and Nigeria right across Africa and Asia.

Unlike many calidrid waders, Temminck's Stints are regularly found inland on passage, but they remain surprisingly scarce here, having occurred in only six of the 65 years 1950–2014, 10 records involving a total of 15 birds.

Those returning to their breeding grounds occasionally overshoot westwards to Ireland, mainly in anticyclonic conditions with easterly winds, and spring records have invariably been in years when they have been recorded in good numbers at this time elsewhere in the UK. They occur more frequently in spring than autumn.

The first record was of one at Allscott on 12–13 May 1986. Two adults at Wood Lane on 18 May 1995 were in summer plumage, as were two at Venus Pool on 22–24 May 2001. Of three birds at Wood Lane on 2–4 May 2003, one had fully moulted into breeding plumage and two were in active moult. Other spring records have been from Venus Pool, with one on 8–9 May 2008 and two on 4 June 2011. These last two records were the only ones received during the recent Atlas period. Subsequently, there was one at Venus Pool 10 May 2015 which was again in the peak appearance period.

Autumn migration is usually on a south-easterly bearing from the breeding grounds, so there are fewer present at this time than in spring, but they tend to stay longer. The first autumn records were in 1995, with a juvenile at Venus Pool on 2 September. This was followed by an un-aged individual at the same site on 22 October–11 November and what was thought to be the same individual on 18 November, a total presence of 28 days.

There is no obvious explanation for these two records in the same year, as there were no unusual influxes elsewhere in the country, and passage at Ottenby in south Sweden, where it is a common passage migrant, was relatively low in the mid-1990s (Hedenstrom 2004).

In 2001 a single, thought to be an adult, was at Wood Lane on 19 August. The most recent autumn record was of an elusive bird at Wall Farm from 16–18 August 2003.

Gerry Thomas

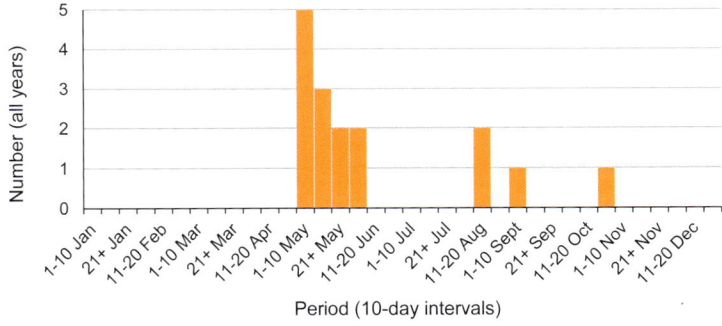

Seasonal Occurrence in 10-day Intervals (1950–2017)

Y-axis: Number (all years)
X-axis: Period (10-day intervals), labelled 1-10 Jan, 21+ Jan, 11-20 Feb, 1-10 Mar, 21+ Mar, 11-20 Apr, 1-10 May, 21+ May, 11-20 Jun, 1-10 Jul, 21+ Jul, 11-20 Aug, 1-10 Sept, 21+ Sep, 11-20 Oct, 1-10 Nov, 21+ Nov, 11-20 Dec

Sanderling
Calidris alba
Rare passage migrant

Sanderlings have a circumpolar, but discontinuous, breeding distribution in the High Arctic. They winter sparsely but widely on beaches and estuaries from northern temperate regions, including the UK, down to some of the most southerly ice-free shores in the world.

One at the old Shrewsbury Sewage Farm from 13 August–2 September 1957 was the first record. Observations were few and far between in the 1960s, 1970s and until the mid-1980s, since when regular spring sightings started to be made at the usual wader sites, especially Allscott and Chelmarsh.

Numbers are generally in just ones and twos with a maximum

of four being seen together on 25 May 1991 at Chelmarsh; and at Allscott on 15 May 1988 and again there on 24 May 2010. However, an unprecedented influx in 2013 included maxima of eight at Chelmarsh on 15 May and seven at Long Lane on the same date. An estimated 21 individuals were present at local sites around this time, which coincided with a general influx into the UK. Interestingly the flock at Long Lane associated with the largest ever flock of Dunlin.

During the recent Atlas period, there were just nine occurrences, all in April or May, although some stayed for several days, at four different sites: Venus Pool, Mirelake, Long Lane and Chelmarsh.

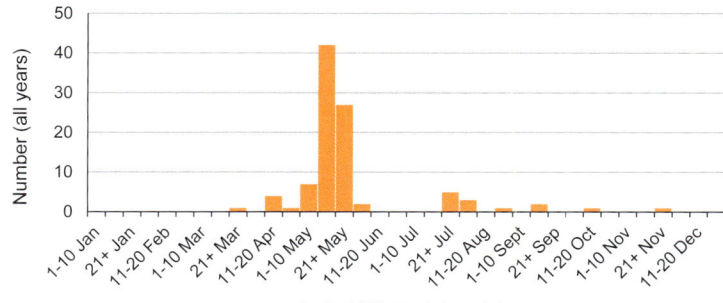

Seasonal Occurrence in 10-day Intervals (1992–2014)

Number (all years) / Period (10-day intervals)

John Hawkins, Wood Lane, May 2010

Most records are of spring passage. One at Wood Lane on 28 March 2011 was unusually early, and the main passage period is in May. Occasionally, they are seen in early June with the latest assumed spring migrant seen at Venus Pool on 8 June, in 2003. They rarely linger for more than a day or two and are soon on their way again. Those seen in May have often been reported as in some stage of pre-breeding moult, with some showing almost complete breeding plumage.

These sightings coincide with a large passage through many estuaries in western Britain in May, with a rapid turnover at all sites (Ferns 1980; Prater 1981; Clark *et al.* 1982), through Iceland, and they rapidly leave on a northerly trajectory (Gudmundsson & Lindstrom 1992), mainly between 25 May and 4 June, apparently to breeding grounds along the north-east Greenland coastal tundra zone, and possibly to the Canadian Arctic beyond.

Autumn migration is much lighter, though more protracted, with most passing along the east coast, rather than the west. Perhaps just 19 individuals have been recorded here over the 65-year period from 1950 to 2014, with sightings between 21 July and 14 October. There have however been no autumn records since 2003. Just two November sightings, at Chelmarsh on 21 November 1991 and at Allscott on 28 November 2004, may have been of birds wintering in the UK.

Gerry Thomas

Dunlin

Calidris alpina

Uncommon passage migrant, rare winter visitor

The Dunlin is the most abundant of the migrant 'calidrid' waders, occurring in both spring and autumn passage periods. There are currently three races found in the UK. *C.a. schinzii* breeds in northern Britain, southern Norway, and Iceland to south-east Greenland, and winters from the Mediterranean basin southwards. It migrates northwards in April and May, favouring the west coast estuaries of the Severn, Dee, Mersey, Morecambe Bay and Solway (Wilson 1973; Eades 1974; Hardy & Minton 1980). *C.a. arctica* is a scarcer race breeding in north-east Greenland, and wintering almost exclusively in West Africa. On spring migration it heads northwards through the same west coast estuaries, with a distinct peak in the last two weeks of May (Eades 1974).

After breeding, adult *schinzii* move south from late June (females) but especially from mid-July to the end of August. Males guard the young and so tend to move south a little later than females. Juveniles

follow from August with peak numbers in September into October. *C.a. arctica* pass through in late July and August.

C.a. alpina is an abundant winter visitor to UK coastal estuaries from breeding grounds in northern Fennoscandia across Russia to western Siberia, arriving in November from moulting grounds in the Wadden Sea off Germany and the Netherlands. Over 500,000 *alpina* winter in Britain most years (Prater 1981; Wernham *et al.* 2002), more than a third of the global population of this race. Inland records are frequent, but in very small numbers compared to the numbers on coasts. They return eastwards from the end of February with many departing in March.

Here, only *C.a. schinzii* has been positively identified so far, as differences between races are subtle (Pienkowski & Dick 1975; Prater *et al.* 1977) and recent DNA studies have cast doubt on the validity of current subspecies' designations (Marthinsen *et al.* 2007).

Approximate Number per Year (1992–2014)

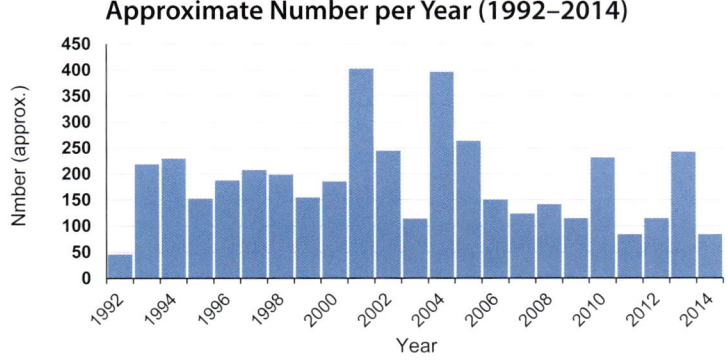

Seasonal Occurrence in 10-day Intervals (1992–2014)

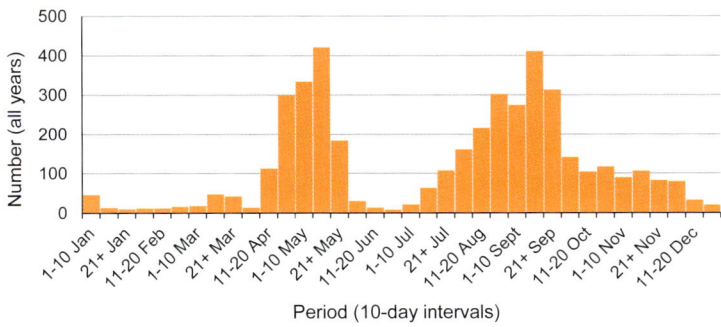

Winter Distribution (2007–13)

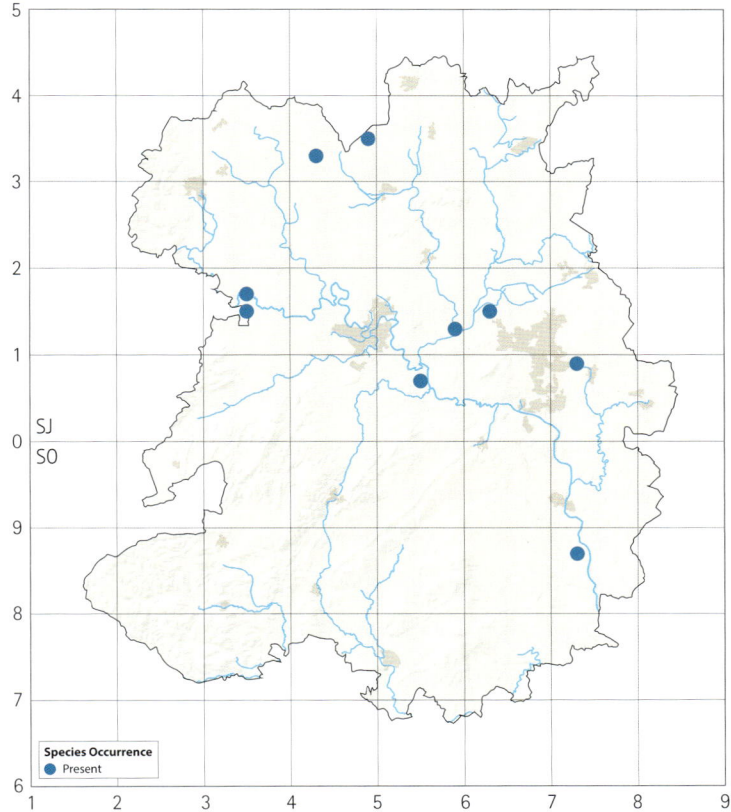

Species Occurrence
● Present

Occupied Tetrads

Tetrads with Winter Records (2007–13): 9 (1%)

There are few historical records. Beckwith (1879) mentioned one or two on the Teme near Ludlow, and Forrest (1899) stated that it 'rarely visits'. He listed four shot, at Ticklerton about 1840, Hawkstone about 1848, Westhope in February 1870, and Madeley in December 1890, and one in 'partial summer dress' flying up the Severn near Shrewsbury on the last day of April in 1888.

Since 1956 Dunlins have passed through in variable numbers every year, except in 1962, the only year they were not recorded. Numbers are generally relatively low and may reflect weather patterns as much as variations in population, but with occasional stand-out years. Up to 31 at the old Shrewsbury Sewage Farm on 24–31 July 1957 was the highest number recorded in the *Handlist* (1964). The largest flock recorded was one of 63 at Long Lane on 15 May 2013, a flock which reduced to 47 next day and just four by 17 May, illustrating the rapid movement while travelling north at that time of year. This flock coincided with the largest ever local count of Sanderling. Other notable counts in spring have been 45 at Wood Lane on 27 April 1986 and 35 at Whixall Canal Floods on 15 May 1994. Sizeable autumn passage flocks have occurred at Allscott Sugar Factory over the years, including 32 on 4–7 July 1986; 35 on 7 August 1980; 43 on August 27 1989 and 32 on 8 September 1988. The largest group recorded in winter months was at the same place, 32 between 4–7 December 1986.

Inclement weather may cause them to pause or interrupt their migration, illustrated by 29 which flew into Allscott Sugar Factory on 29 April 2010 in heavy rain, departing to the east shortly afterwards. The chart shows the estimated number seen each year since 1992.

The second chart shows the pattern of occurrence through the year, which aligns very closely to the peak migration periods of the races *schinzii* and *arctica*, both in spring and in autumn. Peak numbers occur during April–May and again from mid-July through

Jim Almond, Venus Pool, 15 May 2013

to October, as the local sites are on the shortest flight path between the Severn and the Dee estuaries.

The slight increase in occurrence in mid-November and perhaps also in March may be of *alpina* arriving and departing the UK, and winter records most probably are of this subspecies.

The pattern of occurrence in 1992–2014 is very similar to that recorded between 1956 and 1991; though there is an indication that the numbers in spring have increased in recent decades.

Over the years Dunlin records have been primarily from the usual good wader sites, especially Allscott, Shrewsbury Sewage Farm, Venus Pool, Wood Lane and Chelmarsh, but they are more likely than most migrant waders to occur on flooded arable fields and may therefore be seen almost anywhere in passage periods.

During the recent Atlas period, there were 60 records from 14 tetrads during breeding season fieldwork, mainly from the sites named above, but also including Marton Pool (Chirbury), Cross Houses, Cound, Crudgington, Eyton Moor, near Aqualate, Coley Brook Marsh and Bromfield quarry. All except one were April or May records, the exception being on 1 June.

During the winter, it was found in nine tetrads: near Melverley (two), Wood Lane, Whixall: canal floods, Venus Pool, Mirelake, Long Lane, Priorslee Lake and Chelmarsh.

There is only one ringing recovery, one caught at Allscott Sugar Factory on 7 September 1989 which had been ringed three weeks previously at Spurn Point in Yorkshire.

Gerry Thomas

Purple Sandpiper
Calidris maritima
Very rare passage migrant

With a circumpolar distribution, breeding on the Arctic coasts and upland areas of northern Canada and Greenland through to Scandinavia and Arctic Russia, this sandpiper winters on the Atlantic coasts of North America and Western Europe, and is rarely found inland.

In 1865, it was noted by Rocke as being 'occasional, but very rare', although no specific records were listed. By the end of the century no recent occurrences were known. It is unusual that no specific pre-1950 records exist because for most species a 'specimen' had usually been obtained. It will always have been rare inland and it is interesting to speculate whether there had actually been any sightings prior to the bird in 1974.

Given the lack of historic reports, it is perhaps somewhat surprising that there have been four modern records:

Shrewsbury Sewage Farm	29 September 1974
Long Mynd	22 May 1994
Allscott Sugar Factory	18 September 2001
Chelmarsh	27 October 2002

Typically, all records are of individuals present for just one day or part of a day. The one on Long Mynd, the only spring sighting, was observed by a small pool near the gliding station when it was presumably on its way north to its breeding grounds.

Graham Walker

Little Stint
Calidris minuta
Very rare passage migrant

John Hawkins, Wood Lane, 16 September 2006

Little Stints breed in the high Arctic, from northern Sweden right across to the New Siberian Islands in Russia at 140°E, usually between the high latitudes of 72–75°N. Most winter in Africa, predominantly south of the Sahara, especially in the East African Rift Valley (Pearson 1987), with a few remaining in the UK, mainly along southern coasts.

The vast majority occur on autumn passage between August and October, in line with peak numbers in the UK as a whole. 97% of records have come from the four sites with suitable habitat for wading birds, Allscott Sugar Factory, Venus Pool, Wood Lane and Chelmarsh, with records from Shrewsbury Sewage Farm in the late 1950s also reflecting the attractiveness of the old-style sewage farms for waders. Up to four there from 27 July to 29 September 1957 was the first definite record, and a further 12 were seen there the same autumn.

The first chart summarises the occurrence of Little Stint between 1950 and 2014. They have been seen in most years during this period, usually in single figures but with occasional good numbers in the autumn. They breed at very high latitudes and are adapted to rear their single brood very rapidly, chicks reaching their adult weight or slightly above at 15 days old (Schekkerman *et al.* 1998; Tjorve *et al.* 2007). With a very short window of reasonable temperatures at such high latitudes, considerable mortality can occur in spells of colder weather, and breeding success varies considerably from year to year. As with other migratory species breeding a long way to the east, higher numbers here in the autumn reflect a good breeding season coupled with an easterly airstream over the Baltic and North Seas pushing them westwards across the North Sea. There have been nine such exceptional years, with more than 10 individuals seen.

Chicks fledge in July and early August. Adults migrate earlier than birds of the year, reaching the UK from the second week in August, and their African wintering grounds from late August. Juveniles migrate later and form the bulk of those seen here, with peak numbers in the second or third weeks of September. Numbers fall off rapidly in early October though laggards have been recorded in November, with the very latest record one at Mirelake on 20 November, in 1983.

Spring records have been very few, with single birds at Chelmarsh on 28 April 2004, Wood Lane on 31 May 2006, Venus Pool on 1 June 1997, Wood Lane on 3–4 June 1997, and Chelmarsh on 4 June 1984 and again on 5 June 1986. One at Mirelake on 25 April 2008 was the only record during the recent Atlas period. The chart shows the pattern of seasonal occurrence in 10-day intervals since 1992.

Gerry Thomas

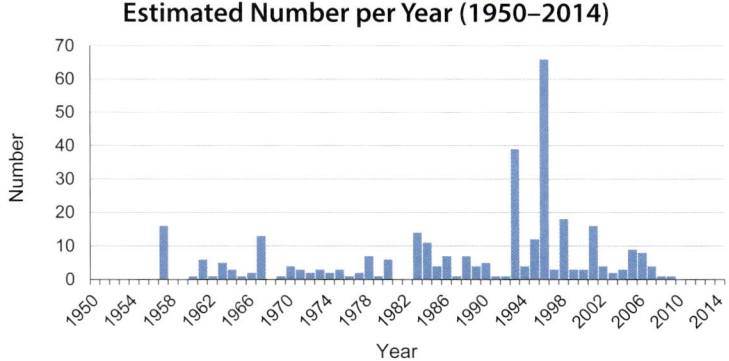

Estimated Number per Year (1950–2014)

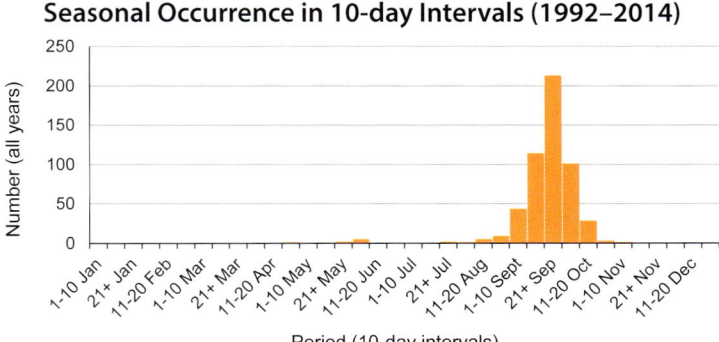

Seasonal Occurrence in 10-day Intervals (1992–2014)

White-rumped Sandpiper

Calidris fuscicollis

Vagrant

The White-rumped Sandpiper breeds on the high tundra in Arctic Canada, winters on the coasts of southern South America as far south as Tierra del Fuego, and is a scarce trans-Atlantic migrant to Britain.

There have been only two records. One, the first recorded in Britain, was shot at Stoke Heath in 1832, and was first reported by Eyton (1839), who referred to it as Schinz's Sandpiper. It was subsequently known as Bonaparte's Sandpiper, and the specimen was in the collection at Hawkstone (Forrest 1899).

This is one of the five species to make their first British appearance in Shropshire (see p. 2).

There is only one modern record, of one photographed at an un-named site kept confidential at the request of the landowner, in 10km square SJ50 (south-east of Shrewsbury).

Confidential site in SJ50   23–24 November 2012

The late date of the record suggests that it was a juvenile (Fraser 2013a), but viewing conditions were not good enough to confirm that.

John Tucker

Buff-breasted Sandpiper

Calidris subruficollis

Vagrant

Breeding on the open Arctic tundra of North America, this attractive sandpiper winters in eastern South America. Hunting in the 1920s massively reduced its population which continues to decline. As a result, it is now placed in the near-threatened category on the IUCN Red list. It is a scarce migrant in Britain, with between 10 and 20 individuals in most years since 1970, but with over 50 in some. It is perhaps surprising,

therefore, that there is only one record, an adult in the autumn.

Allscott Sugar Factory   10–19 October 1986

This well-watched individual was often observed looking skywards during its 10-day stay.

Graham Walker

Pectoral Sandpiper
Calidris melanotos
Very rare passage migrant

Jim Almond, Venus Pool, 21 September 2008

Pectoral Sandpipers have been annual visitors to the UK from their breeding grounds in the Nearctic and extreme eastern Palearctic tundra since the 1960s. They are now classed as scarce annual migrants to Britain and Ireland, rather than vagrants. The majority are thought to have been on a south-easterly migration route through Canada, en route to wintering grounds in Argentina and Paraguay, and caught up in the jet stream and blown across the Atlantic.

The first record was of a single at Allscott on 27 September 1979 in a period when there was a significant influx into the UK. This sandpiper has occurred infrequently, in 16 of the 38 years 1979–2017, most recently in August 2017.

All except one of the 20 records have been in the autumn, with the arrival date in the period 8 August–27 September, with a peak in mid- and late September, an arrival pattern in line with that in the UK as a whole. The exception was a single at Wood Lane on 8–11 June 2004, which was perhaps on reverse migration having crossed the Atlantic in a previous autumn.

All records were of singles. Eight records were of birds staying for just one day, and three others were seen on two days only, but four more stayed for over two weeks and one was present for a month at Allscott in 1983. This was last seen on 11 October, and was one of only two that remained beyond the end of September, the other staying for two weeks in 1994, and being last seen on 6 October.

In total, 12 of the 20 were at Allscott, four at Venus Pool, three at Wood Lane and one at Chelmarsh. None were seen in the recent Atlas period.

Gerry Thomas

Seasonal Occurrence in 10-day Intervals (1992–2014)

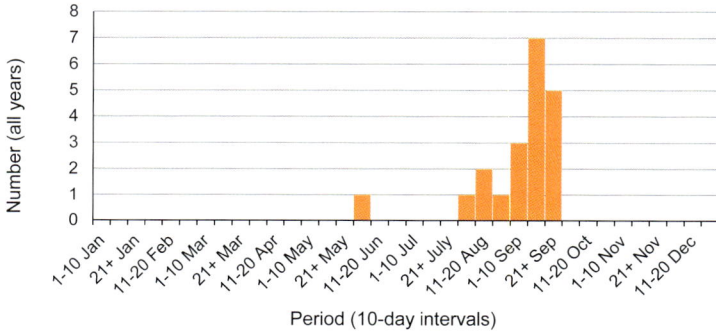

Woodcock (Eurasian Woodcock)
Scolopax rusticola
Common winter visitor, scarce resident

Woodcock breed extensively across the northern Palearctic and those from the western part of the range winter in the southern and western extremities of this biogeographic realm.

Is this the most discreet of British birds? Cryptic and crepuscular, solitary and secretive, in spring it is only the males that reveal themselves, and then just briefly, emitting little grunts and squeaks in their twilight 'rodes' over the tree tops. Numbers are much greater in winter when there is an influx of migrants from Russia and Fennoscandia, but, even then, observations depend largely on individuals being flushed by chance from their daytime roosting sites among woodland undergrowth.

The early history of the Woodcock as a British breeding species is unclear, but Alexander (1945) was convinced that from 1827 onwards, breeding numbers began to increase rapidly. Nevertheless, the implication of Beckwith's account of 1886 is that it was not until about that time that Woodcock had become a regular breeding species

here: he states that there had been 'several instances of breeding … at irregular intervals. And now, in two localities at least, one in the northern the other in the southern portion … there are nests every year'. Forrest (1908) reported an increase, with a good number of nests having been found, mostly in the south, and Alexander (1945) concluded from evidence gathered in 1934–35 that it 'breeds regularly in considerable numbers'. Alexander & Lack (1944) thought that the 'protection of coverts in [the] interest of Pheasants', coupled with the cessation of shooting in the breeding season (probably from the late eighteenth century onwards [Andrew Hoodless *pers. comm.*]), were the main reasons for the nationwide increase. Perhaps it had been by continuing his shooting into the roding season that the Rev. John Purcell, 'the sporting Rector of Sidbury', had chalked up a reputed 2,000 victims in the last 40 years of his life; he died in 1818 (Gaydon & Price 1973).

Ringing returns show that the majority of the country's breeding

Occupied Tetrads

Atlas period (breeding)	1985–90		2008–13		Change	
	Number	%	Number	%	Number	%
Confirmed	36	4	0	0	-36	-100
Probable	72	8	4	0	-68	-94
Possible	51	6	13	1	-38	-75
TOTAL	**159**	**18**	**17**	**2**	**-142**	**-89**
Tetrads with Winter Records (2007–13): 266 (31%)						

Breeding Distribution Change (1985–90 to 2008–13)

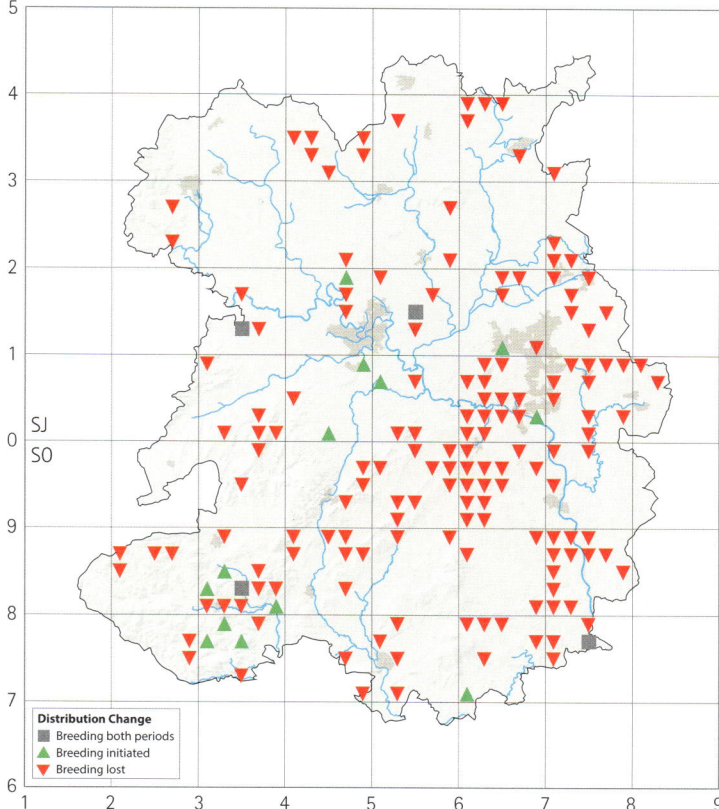

Distribution Change
- ■ Breeding both periods
- ▲ Breeding initiated
- ▼ Breeding lost

Woodcock move little, 68% of recoveries being made within 30km of ringing locations (Hoodless & Coulson 1994). Breeding birds generally favour larger deciduous or mixed woods, those providing both continuous ground cover for feeding and sparse ground cover for nesting, along with worm-rich permanent pastures nearby, although at this time of year there is a progressive switch from nighttime feeding in pastures to daytime feeding within woodland (Hoodless & Hirons 2007).

The breeding distribution seems to have changed little between the national BTO Atlas (1976) and the *Atlas* (1992), but losses since then make for one of the bleakest maps in this book. A steep decline is evident, and the only surviving group of breeding season records comes from the south-west, where, interestingly, while the woods are large, they are predominantly coniferous. In the south-east, the mixed woodland of the Dowles area of the Wyre Forest, a long-documented location, is still represented, but the once oft-reported Haughmond Hill and Wenlock Edge no longer figure.

The only way to monitor numbers is through counts of roding males, and the relationship between the number of roding flight passes and the number of males that are involved has been determined thanks to the individuality of roding calls (Hoodless *et al.* 2008). Working from this relationship, a survey in 2003 of selected 1km squares led to a national estimate of 78,350 males (Hoodless *et al.* 2009). A repeat in 2013 suggested a decline of 29% to an estimated 55,241 males, while a 20% decline in occupancy was registered across the survey's 'South Midlands' region in which Shropshire lies (Heward *et al.* 2015). The sample sizes for these surveys are too small to

John Hawkins, Cockshutt, May 2009

provide definitive conclusions as to local trends, but the results are indicative of the decline. Twenty-four squares were surveyed in 2003, of which seven were found to be occupied, and 20 in 2013, of which four were occupied. Application of the relationship between roding passes and numbers of males gives an estimated 10–14 males in the squares surveyed in 2003, but only four to five in those surveyed in 2013 (Greg Conway *in litt.*).

A comparison between the results for the two tetrad-based Atlas surveys is particularly telling: 159 tetrads were occupied in 1985–90 and breeding was confirmed in 36 of them, but in 2008–13 there were no records of confirmed breeding, and only 17 tetrads yielded records of possible and probable breeding, just 10% of the distribution 25 years earlier. 'A very rough estimate of 150 to 300 "pairs"' was made for the first Atlas period, the inverted commas reflecting recognition that males mate with more than one female, indeed a few dominant males fertilise the eggs of most females within a given area (Hirons 1980; Hoodless *et al.* 2009). The estimate was based on an assumption of one or two nests in each of the 159 occupied tetrads in which Woodcock were recorded. It is now customary to express populations in terms of the numbers of males. Given a presence in 17 tetrads, even at four males per tetrad (an optimistic assumption, though counterbalanced perhaps by tetrads where breeding activity was overlooked), the total now would be no more than 68 males, and is probably substantially fewer.

A number of factors have been advanced as potential causes of the national decline, and all may apply locally to some degree: increases in deer numbers leading to a reduction in the extent of the understorey, intensification of recreational disturbance (notably dog-walking),

Winter Distribution (2007–13)

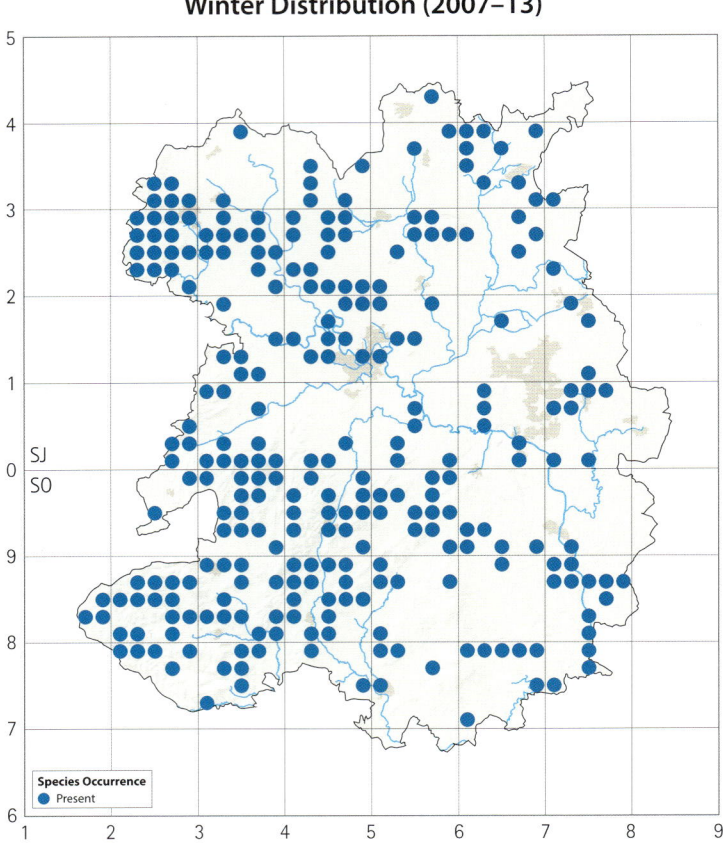

maturation of conifer plantations, reduction in woodland management and the drying-out of soils (Heward *et al.* 2015).

The contrast with the winter distribution map is striking, and comparison of this map with that for the national BTO Winter Atlas (1986) suggests that the distribution at this season is at least as widespread as it was 30 years ago, while the prevalence of local Woodcock-related place-names suggests that the species has long been well-represented in the winter months. Cockshut (or -shutt, or -shot) figures in 10 place-names on Ordnance Survey maps. Forrest (1899) speculates that these were so-named because 'sportsmen waited in these places to shoot the "cocks" as they went by', but Gelling & Foxall (1990), the definitive source, state that it refers to 'a woodland glade where nets were stretched to catch Woodcock'. Indeed Foxall (1980) in his survey of field-names, many of which refer to cleared forest land, states that there is 'hardly a parish' without such a name. Netting was presumably a winter activity, with birds intercepted in their flights at dusk from solitary daytime woodland roosts to nearby fields where they fed overnight, principally on earthworms. Such flights still occur, evidenced by 41 ghosting out of woodland at Hawkstone Park on 21 December 2013. Incidentally, this is a haunt of long standing: Beckwith (1886b) reports that 'the largest bag of [Wood]cocks recently made was at Hawkstone, where Lord Hill's party killed 41 in a day'.

The Hawkstone count of 2013, and reports of significant numbers flushed by beaters on Pheasant shoots, notably some 40 at Crudgington Moor on 14 January 2009, and in excess of 50 at Shavington Park on 21 January 2012, indicate just how numerous Woodcock can be in winter. Their widespread distribution, while related to the prevalence

of woodland, shows a westerly bias, reflecting perhaps the marked increase in abundance from central England into Wales as revealed by the national BTO Atlas (2013). Trends are difficult to determine, but since the 1980s there appears to have been a consolidation in both the wintering range and numbers. As many as 1.4 million may be present in Britain during the winter months (Musgrove *et al.* 2011), and, based on TTV counts, perhaps 11,100–12,000 in Shropshire. It is estimated that 181,000 were shot in the UK in 2004 (PACEC 2006; Musgrove *et al.* 2011), and several hundred of these will have been accounted for here.

As spring arrives, wintering individuals depart, although some may stay as late as April, their presence overlapping with the breeding activity of residents, whose peak egg-laying is in late March or early April (Hoodless & Coulson 1998). The breeding grounds of winter visitors to the British Isles lie principally in Fennoscandia and Russia (Hoodless 2002), as illustrated by ringing recoveries involving three migrants travelling to or from Russia, and singletons to or from Norway, Sweden, Finland and Latvia. Recent satellite-tracking work has shown that some Woodcock wintering in Britain may breed as much as 6,000km away in Siberia (Hoodless 2013), but locally, the longest known movements are of one ringed at Shavington Park in December 2013, which, in April 2014, was shot 2,622km to the east, and another ringed in the Clun Forest in November 2012, which was shot the following May, 2,252km to the east-north-east; both locations are in Russia. The first, sexed at ringing, was a male, and it seems likely that it was shot when roding – the shooting of roding birds is legal in Russia; the second may have been too. Two ringed in the Netherlands, one in October and the other in November, and subsequently shot here, were presumably en route between breeding and wintering areas when caught for ringing.

While currently the future of the Woodcock that winter here seems fairly secure, that of our resident breeders is highly fragile, and unless present trends are reversed, extinction seems likely before, perhaps 10 years hence, fieldwork for a future Atlas begins.

Tom Wall

John Hawkins, Cockshutt, May 2009

Jack Snipe

Lymnocryptes minimus

Scarce winter visitor

Jack Snipe come to Britain for the winter, or pass through on their way to south-west Europe and West Africa, from breeding grounds in north-east Europe and Siberia. They are secretive, unobtrusive and sit tight until nearly underfoot. As a consequence they remain under-recorded, and it is difficult if not impossible to assess their true status.

Beckwith stated they were frequent in winter but never so plentiful as common Snipe. Forrest reiterated the latter comment, but described them as 'an irregular winter visitor. Occasionally seen also in early summer, but has never nested with us'. There were only a handful of records in the first half of the twentieth century, and Lloyd (1943) described them as 'formerly a not uncommon winter visitor; now apparently almost a rarity'.

This status remained unchanged, and the *Handlist* (1964) considered them 'formerly a regular winter visitor in small numbers, nowadays much scarcer or overlooked', and listed only nine records between the first modern record in 1952, and 1962. There are records totalling 579 individuals between 1961 and 2014, summarised by annual total in the first chart and month of the year in the second.

The annual totals show some fluctuation, perhaps reflecting breeding success or winter temperature here and on the continent. Records are usually of single birds or very small numbers. Exceptions include 16 at Cole Mere in December 2002 and 14 at Whitchurch in December 1995. There have been instances of birders deliberately trampling the marsh at Cole Mere to flush this species, which would almost certainly be necessary to record 16. This behaviour is to be deplored.

Jim Almond, Chelmarsh Scrape, 22 December 2007

The monthly totals show a build-up from early arrivals in September to a peak in December, then a steady decline as some move further south in hard winters, up until April when most of the remainder depart. There are very few records outside this period, mainly a few late stayers into May and early arrivals in late July and August.

Records have come from 66 different scattered locations, but birders looking for this species will have visited seven well-watched sites, which have contributed records for 369, 64% of the total: Chelmarsh (114), Cole Mere (89), Allscott Sugar Factory (85), Venus Pool (36), Wood Lane (15), Whixall (15) and Long Mynd (15).

The recent Winter Atlas distribution map shows many more occupied 10km squares than the relevant part of the 1981–84 national Winter Atlas, probably due to higher levels of coverage, rather than an increase in population or change in range.

The well-watched sites listed above all feature, together with a cluster of other sites in the Severn Valley around Chelmarsh.

Other tetrads include two in the Severn–Vyrnwy confluence near Melverley, two on the River Perry, three around Wall Farm and Cherrington Moor, two near Market Drayton and one at Buildwas. These sites are typical habitat, damp lowland wetlands with shallow water and short vegetation. However, tetrads in the south-west, Hope Valley, Picklescott, Comley and Hopesay are higher, and a tetrad on the Long Mynd, and two in the Clun Forest (including Rhos Fiddle) are around 400m. Over the years, most records from the Long Mynd have been early winter arrivals from October onwards, and been seen in flushes at the heads of the valleys, around the moorland pools

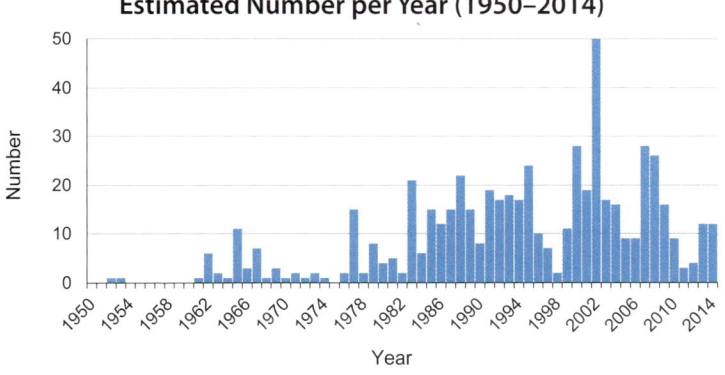

Estimated Number per Year (1950–2014)

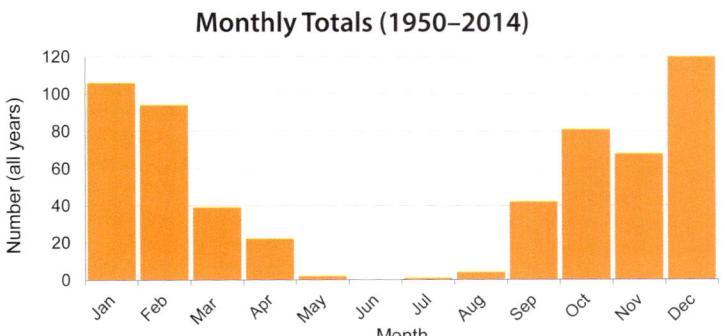

Monthly Totals (1950–2014)

Occupied Tetrads

Tetrads with Winter Records (2007–13):	27	(3%)

and on Wild Moor. Freezing weather clears the hill completely of all snipe, and these may be passage birds on their way south, rather than winter visitors.

Trying to extrapolate the Jack Snipe population and distribution from a comparison with Snipe is inconclusive. The former was found in 28 tetrads, compared with 284 for the latter. Between 2006 and 2013, 30 Jack Snipe were ringed, almost half as many as the 62 Snipe, while nationally for the years 2010–13, the number of Jack Snipe ringed was 46% of the number of Snipe. National shooting records quoted in the BTO Winter Atlas 1986 indicate one Jack Snipe for every eight Snipe shot, though this may exaggerate the number of Jack Snipe, which are easier to shoot once flushed. Similarly, Jack Snipe are more likely to end up in ringers nets, as they do not 'tower' like Snipe when taking off. On Wild Moor, about one in 100 snipe flushed during survey work were Jack Snipe, although this is not ideal Jack Snipe habitat, so numbers may be lower than in lowland habitats. Given this conflicting evidence, the extent to which Jack Snipe is under-recorded is very uncertain.

National ringing records show high level of site faithfulness between years, but there are no local ringing recoveries, apart from three ringed in Clun Forest, one in December 2010, and two in October 2014, all re-caught at the same place within 25 days of ringing.

Peter Carty

Winter Distribution (2007–13)

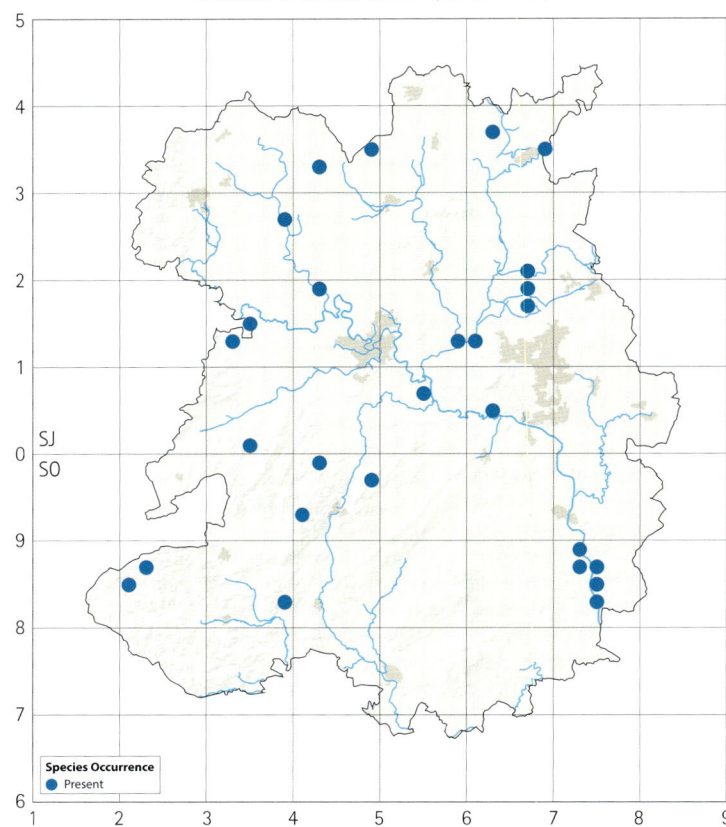

Species Occurrence
● Present

Great Snipe

Gallinago media

No modern records

Great Snipe breed in Scandinavia, north-eastern Europe and western Siberia, and winter in central Africa. There are six reports of this rare vagrant, all from before 1906, with five relating to birds shot perhaps for the pot and unwittingly brought down as 'snipe', rather than for collections. There are other scattered references to the species being 'taken occasionally', deriving perhaps from other instances of them ending up in the game bag but not documented.

Rocke (1866a) reported the first, 'killed at Eaton' but with no date,

while Paddock (1890) shot a pair near Newport in 1879. Two records in 1899 were, first, of one shot near Wenlock on 3 April, and another of two seen at Caer Caradoc (near Clun) on 24 November. One was shot near Minsterley on 23 September 1901, while the last was shot at Brades Farm (near Whitchurch) on 2 September 1905. The latter was passed to Whitchurch Museum and it is now in Ludlow Museum, specimen Z00156.

John Tucker

Snipe (Common Snipe)

Gallinago gallinago

Fairly common winter visitor, rare breeding species

Snipe breed across northern Europe, particularly in Fennoscandia, but leave frozen ground for better feeding conditions in western and southern Europe, including Britain, in winter.

They were well known to the nineteenth-century authors: found in 'wet bogs and springs' around Shrewsbury (Leighton 1836) and 'common' (Eyton 1839), while Rocke (1866a) feared that 'drainage and a better state of cultivation has driven this bird

from its former haunts; except by the river-side, its appearance is now very rare. It continues to breed on the Clun Forest'. Beckwith (1879; 1886a) added Longmynds, Whixall and Wem mosses, 'many small bogs' and a field near Berrington, and then Stiperstones and Clee hills, to the list of breeding sites, while Paddock (1890) had often found nests around Newport, and Forrest (1899) added 'a few nest with us in marshy situations'. The Historical Atlas mapped Snipe as relatively 'uncommon' here by the end of the nineteenth century.

There is little information on trends in the first half of the twentieth century, but Lloyd (1943) commented 'formerly common in suitable localities, now decidedly local and less abundant'.

Certainly by the second part of the twentieth century it was not found in many of the places listed by the early authors. The *Handlist* 1964 said it was a resident, non-breeding visitor and passage migrant, found fairly commonly in marshy areas, moors and sewage farms. 'Some observers are of the opinion that it has decreased in recent years probably as a result of intensive drainage of suitable habitats'. 'At Allscott Sugar Factory, 50–100 are recorded on passage, of which a considerable number have been ringed in recent years. Lack of retrapped birds suggests passage of quite high numbers over some considerable period'.

The widespread breeding distribution was verified by the 1968–72 BTO national Atlas, which mapped it as present in all except four of the County's 33 10km squares, and breeding was confirmed in six of those in the north, and seven in the south. By 1985–90, the more intensive fieldwork for the local Atlas found three and six respectively, evidence that the decline had continued. A few specific breeding sites were listed in SBRs prior to 1985, but not all of these were relocated during fieldwork for that Atlas, when wet hollows and flushes in the uplands and damp meadows in the lowlands, especially the Weald Moors, were occupied. However, Snipe are secretive and elusive, and breeding was difficult to confirm, so the map under-represented the distribution. This was even truer in the recent Atlas period, when densities were much lower.

Between the two Atlas periods, breeding Snipe were reported from several sites. Based on fieldwork between 1994 and 1998, the Long Mynd BBP estimated the population at 20–25 pairs (Smith SBR 2003). One was found on the Stiperstones NNR in 1995, and two on adjacent farmland, including a nest (Cross & Moscrop 1996). One was drumming and displaying at Whixall Moss in mid-June 1995, and there were reports from several sites around Newport, some of which are shared with Staffordshire, including three drumming east of the town in April and May 1996, two drumming and a juvenile on the ground in early July 1996 at Norbroom Marsh and one on a nest in the same place in May 2005, and one chipping at the Sewage Works in May 2004. Other locations with breeding evidence were Beckmoor (one territory in 1999, a nest with four eggs found), a nest with eggs and an adult feeding young at Bow Farm (Lydham) in the same year, one drumming at Rhos Fiddle in mid-May 2005 and one calling repeatedly at Pen-yr-estyn in early May 2006.

BTO and RSPB estimated that the national population had declined by 74% between 1970 and 1998, so Snipe was included in the local Biodiversity Action Plan (BAP) in 2002. The national 'Birds of Wet Meadows survey 2002' (Wilson *et al.* 2005) failed to re-find any of the four pairs found locally in 1982, a 100% loss.

A survey was carried out in 2004 to establish the population, distribution and breeding requirements in the south-west. It included three sites where Snipe were known to have bred previously – Long Mynd, Stiperstones NNR and Rhos Fiddle SWT reserve. In addition, a large number of other sites throughout the south-west hills (total area approximately 580sq.km), which appeared to have potential for breeding Snipe, were assessed, and the 24 judged to have the greatest potential were also surveyed.

On the Long Mynd, an estimated seven to eight pairs were found, only around one-third of the population found in 1994–98. The range had also contracted considerably – Snipe were only found on and around Wild Moor. On the Stiperstones NNR, none were found, although Snipe are known to have bred there in 1995–96, when the population was estimated at up to six pairs. On Rhos Fiddle, an estimated three to four pairs were found (including a possible pair at the adjacent site on Bicton Hill). Of the other 24 sites, plus an additional site adjacent to the Stiperstones, only three were found to hold drumming Snipe. Each of these three sites apparently only held one territory, so the total estimated population of the 25 sites was only three breeding pairs. Most of the sites were very wet in March, but had dried out and become very hard by the end of May, and were therefore not suitable breeding habitat.

The total population of the south-west hills was therefore estimated at 13–15 breeding pairs, with perhaps 20–25 pairs in the whole County. Recommendations were made to safeguard and improve the few remaining breeding sites, and improve the potential of other sites. The likely impact of an increase in potential predators as a result of the increased availability of sheep and Pheasant carrion for food was also discussed (Smith SBR 2004).

The vast majority of Snipe found during the survey were either drumming or chipping in deep twilight, at dusk, well after sunset. Some were flushed, but this required approaching within 5m. If densities are low, there is much less need for drumming to establish and defend territories. If the standard survey methodology (O'Brien & Smith 1992) had been used, far fewer Snipe would have recorded than the number actually found.

The BAP review in 2005 made a commitment to repeat the survey every five years. As a result, in 2009, and again in 2014, the sites where Snipe were found in 2004 were re-surveyed, plus the Stiperstones where drumming and chipping were reported in 2007. Five areas in the north, with breeding season records received by SOS since 2000, were also included: Pen-yr-estyn, several sites around Newport, Whixall Moss, Wood Lane and Wall Farm. All visits were made between sunset and dark. The results are shown in the table. It

Jim Almond, Chelmarsh Scrape, 26 December 2010

Snipe Territories Found on Dusk Surveys

Tetrad	Site	Estimated Population		
		2004	2009	2014
SO49C, D, H & I	Long Mynd	7–9	2–3	2
SO39T	Stiperstones	0	1	1
SO39U	The Hollies Farm	1	1	1
SO39E & J	Stapeley Common	1	2	1
SO28C	Rhos Fiddle	3–4	4	0
SO18W	Black Mountain	1	0	0
SJ32P	Pen-yr-estyn	n/a	1–2	0
Total		13–16	11–13	5

Breeding Territories (2008–13)

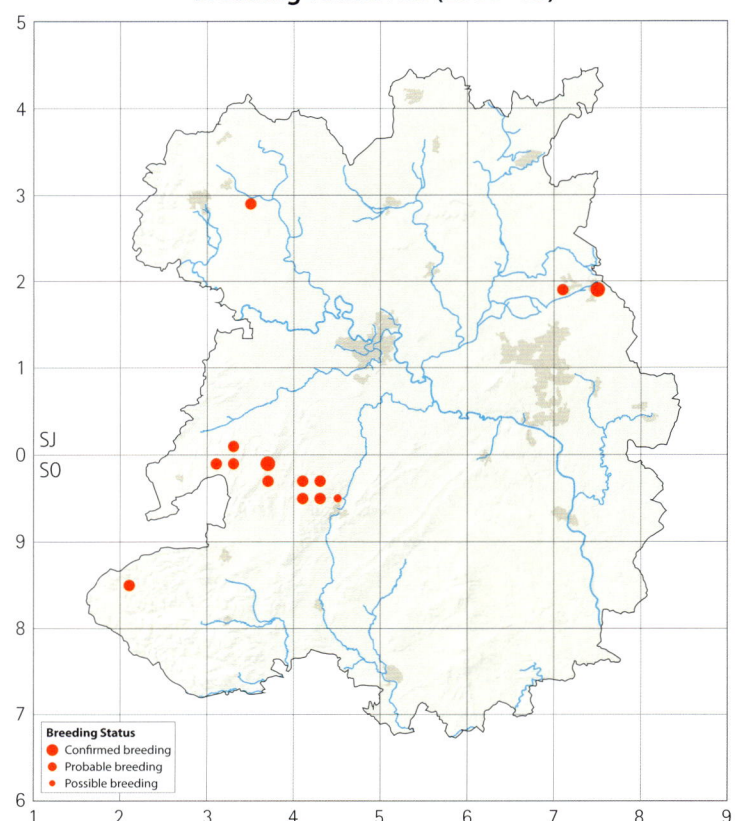

Breeding Status
● Confirmed breeding
● Probable breeding
• Possible breeding

will be seen that the observed population has continued to decline. Snipe were found at only one site in the north in 2009, when one or two were heard chipping on the first two site visits (29 April and 18 May), but no territorial drumming was heard. No Snipe were recorded on the third visit, or on follow-up dusk visits in 2010 to clarify the population, or in 2014.

No Snipe were recorded on any of the visits in the three survey periods to sites around Newport.

Comparison between the results of the two Atlases is difficult. Large numbers of migrants pass through, up until the middle of May, and return passage starts at the end of June. Some observations of passage birds were given breeding codes in both Atlases, particularly the earlier one, and some were classed as a pair (a probable breeding record) if two were seen together. In the recent Atlas, validation procedures treated most of these records as non-breeding passage migrants. Thus the conventional breeding distribution change map would give a misleading impression. A better impression is given by comparing the number of tetrads with confirmed breeding, which shows a dramatic decline, from 11 to two.

During the 2008–13 period, a Natural England Farm Adviser flushed a Snipe from a nest with eggs on a farm north of Newport, on 14 May 2009. Dusk visits were made to the site in 2010, but none were heard. The other confirmed breeding record was of a nest with eggs found at the edge of the Stiperstones on the Hollies farm on 1 May 2013.

RSPB conducted a breeding wader survey on the Weald Moors, and Baggy Moor, starting in 2009. A chipping Snipe was recorded on the Birch Moors on 3 July 2011. Several other Snipe were found, but they are all believed to be passage migrants.

The probable and possible breeding records in the table were not rejected in the validation process, but they were almost all of one or two individuals seen up until 5 May, and they too are believed to be passage migrants. This includes a record of a pair in SJ32S, and another in SJ42D, but there was no other evidence of breeding behaviour, and they were in the passage period. Dusk surveys the following year at both these sites found no evidence of breeding Snipe.

The map shows the two sites where breeding was confirmed, and the sites where drumming or chipping Snipe were heard; these observations have been counted as probable breeding (11). The cluster of records in SO49 come only from Wild Moor or the top of Bilbatch,

Occupied Tetrads

Atlas period (breeding)	1985–90		2008–13		Change	
	Number	%	Number	%	Number	%
Confirmed	11	1	2	0	-9	-82
Probable	47	5	13	1	-34	-72
Possible	70	8	17	2	-53	-76
TOTAL	128	15	32	4	-96	-75
Tetrads with Winter Records (2007–13): 295 (34%)						

a considerable contraction of range since 1990. Only one possible breeding record is shown, from Batch Valley on 29 May 2012, after the passage period, in suitable breeding habitat. The population is therefore estimated at only 15–20 pairs. It was not found on enough TTVs to estimate a population by that method. Although drainage is undoubtedly the main cause of the long-term decline, other aspects of agricultural intensification have contributed, and shooting and increased predation have probably also had an impact. The estimate in the Atlas (1992) assumed that most tetrads with records held breeding Snipe, but the survey work since suggests that most were migrants, and the 1990 estimate was much too high.

The National Trust is cutting soft rush from around the flushes on Wild Moor each year now, to improve habitat, and it appears that this has been beneficial – the population increased to four in 2010, then fell to zero in 2013, but has increased since.

In autumn, large numbers of Snipe from northern Europe pass through, going south for the winter, but others remain and over-winter

Annual Monthly Maxima at Three Sites (1992–2014)

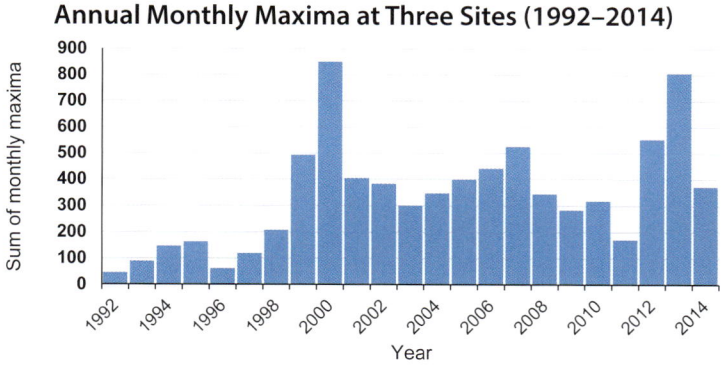

Monthly Maxima at Three Sites (1992–2014)

John Robinson, Chelmarsh Scrape, 5 September 2006

here. In the 1981–84 national BTO winter Atlas, it was found in all 10km squares in the north, but was not recorded in five in the south.

Adding together the monthly maxima in SOS records for each year between 1992 and 2014 for three sites that were well-watched throughout the period, Allscott Sugar Factory (including Mirelake), Venus Pool and Wood Lane, shows that autumn passage starts in July, and numbers increase rapidly to a peak in October. The over-wintering population is fairly constant in November, December and January, then numbers fall as return passage gets underway in February and continues into May.

Numbers each year fluctuate, with a steady growth up to 1999, perhaps reflecting an increase in observers rather than Snipe, a fall until 2003, an increase to 2007, another fall to 2011, followed by another increase, with abnormally high numbers in 2000 and 2013. The winter periods in these years had higher rainfall than usual, but no freezing weather, creating good conditions for Snipe to stay here, rather than move on further south.

Since 1992 there have been 12 counts of 100 or more, the highest being 200 at Venus Pool in December 2012 and the second being 175 flushed from two pools at Whixall Moss by a female Peregrine in February 1995. The others were at Colemere, Chelmarsh Scrape, Isombridge, Beckmoor and Spoonley, together with three further such counts at Venus Pool. Records of small parties predominate – less than 20% are of more than 10, and more than half are of three or less.

The recent Atlas showed that winter distribution is widespread, with Snipe found in more than one-third of tetrads. On TTVs there were only 12 counts of more than 10, with more than half of those from the north-east. The relative abundance map also shows high densities around the meres, Severn–Vyrnwy confluence, Stapeley Hill and Common, Rhos Fiddle and saturated ground in the Clun Forest and upper Corvedale, but the scattering of records show that

Winter Relative Abundance (2007–13)

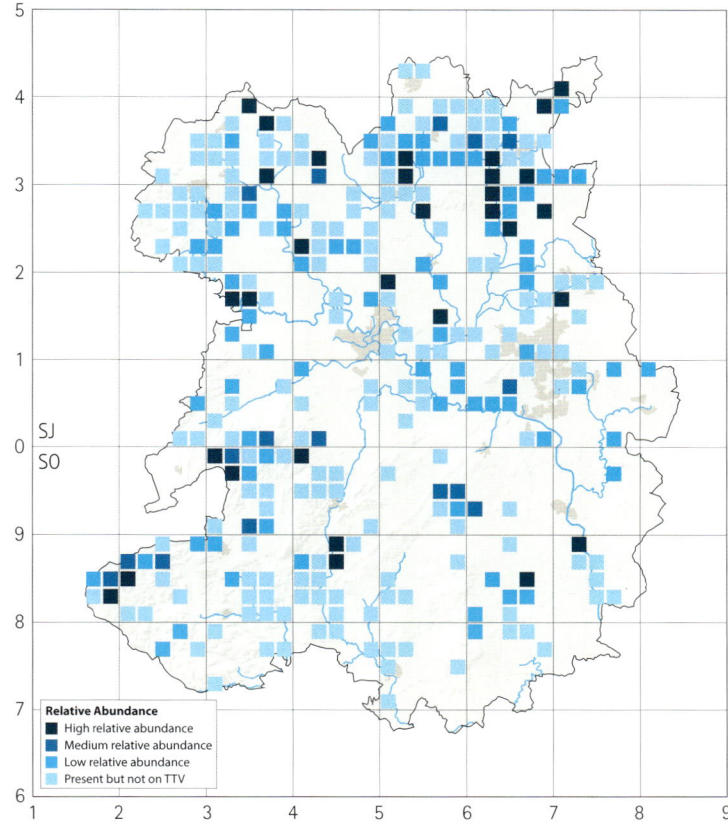

Relative Abundance
High relative abundance
Medium relative abundance
Low relative abundance
Present but not on TTV

widespread wet and flooded fields, ditches, and margins of streams and pools all provide good feeding opportunities.

A total of 928 were ringed at Allscott Sugar Factory between 1961 and 2003, with double figures in all except three years up until 1986, and a maximum of 110 in 1977, almost all during the autumn passage period when water levels were suitably low (Julian Langford *pers. comm.*). One of these holds the national longevity record, a full-grown female ringed on 17 August 1977 that was found freshly dead (shot) on 5 September 1993 in Russia, 2,508km to the east, 16y 0m 19d later. Another adult, ringed there on 4 August 1983 was found freshly dead (shot) on 31 January 1995 near Enniskillen (Fermanagh), 382km WNW, 11y 5m 27d later. One of the earliest Snipe ever ringed, as a nestling on 11 June 1913 near Shrewsbury, was found shot on 20 October 1919 at Lydham Manor, 28km SW, 6y 4m 9d later.

Nine were ringed and recovered here. The three recoveries ringed elsewhere in Britain came from Derbyshire, Leicestershire and Staffordshire, while three ringed abroad, one each from France, Germany and the Netherlands, were all shot here. The 21 ringed here and found elsewhere in the British Isles went to Ireland (eight) Cornwall (three), Somerset (one) and Pembrokeshire (one), presumably all seeking mild winters, while the other eight were found in widely scattered counties. Twelve were found abroad, in France (five), Italy (one), Morocco (three), Portugal (one), Russia (one) and Spain (one). They were almost all shot in the winter months, though one in France and the one in Italy were both found in March and may have been on their return journey. Two of the three in Morocco were more than 2,000km from the ringing site, the furthest 2,130km distant, while the oldest one in Russia was 2,508km distant. This individual was shot on 5 September, so she may have been on the way back from her breeding grounds, perhaps even further away.

More recently, an adult caught in the Clun uplands during the Golden Plover project on 18 October was shot in Finistere, France, 435km distant, only 24 days later.

The ringing records give an indication of the impact of shooting on this increasingly scarce wader, and consideration should be given to removing it from the list of game species that can legally be shot. Its prospects as a breeding species look bleak, but it might possibly be a beneficiary of climate change – increasingly violent storms and flooding may lead to radical action to re-wet the upland river catchments, to hold back the flood waters, reversing past drainage and restoring habitat.

Leo Smith

Red-necked Phalarope

Phalaropus lobatus

Very rare passage migrant

Breeding in the Arctic regions of North America and Eurasia, this phalarope winters at sea off the west coast of Central and South America. Although a very small breeding population occurs on Scotland's Western and Northern isles, it usually appears in Britain as a passage migrant in small numbers, with most occurrences on the coast.

Although Paddock stated that a pair had been shot near Newport in the autumn of 1890, he did not mention them in his book and, as a consequence, Forrest thought the record doubtful. The first record, therefore, may have been a winter plumage adult shot on a small pond in foggy conditions at Boreton on 1 November 1904.

Subsequently, there have been five modern records:

Shrewsbury Sewage Farm	21 September 1957
The Mere (Ellesmere)	24 August 1968
Allscott Sugar Factory	1 June 1987
Whixall Moss	1 and 3 July 1995
Knighton Reservoir	7–8 May 1996

The first, in 1957, coincided with a major influx of both Red-necked and Grey Phalaropes to Britain (Sage & King 1959). Unfortunately, the gender of this individual and the one at Ellesmere were not recorded, but those at both Allscott and Knighton were adult females. The adult male at Whixall spent most of its stay in Wales, although it certainly strayed across the border on at least one occasion.

Graham Walker

Grey Phalarope (Red Phalarope)

Phalaropus fulicarius

Very rare passage migrant

Breeding in the Arctic regions of Eurasia and North America, this is predominantly a coastal passage migrant to Britain when they move to their wintering grounds off the coasts of western and south-western Africa. Most are thought to pass well out to sea west of Britain and Ireland but, as their migration often coincides with late autumn cyclonic weather, they can be blown close inshore, although most inland encounters only occur after severe westerly storms.

The first reference was from 1839 with occurrences noted at Clungunford and Montford Bridge but, while called Grey Phalaropes, the scientific name was given as *Phalaropus lobatus*, the name of the Red-necked Phalarope. This confusion continued through the nineteenth century with sporadic, often undated, reports from Condover, Bayston Hill, Hadnall and Eyton-on-Severn but, since it has always occurred more frequently than Red-necked Phalarope, the common name has been judged to be correct in these early accounts.

The pair at Hadnall 'had been observed for several days running about the puddles, seeking for worms and slugs'. The observer did 'not think they were ever seen swimming, although there was a duck-pond

Jim Almond, Chelmarsh Scrape, 13 September 2011

close by'. More typically, the two birds on the flood meadow below Eyton-on-Severn in October 1881 were observed by Beckwith (1886a) 'swimming about with the greatest ease, reminding one rather of miniature Teal, and were remarkably tame, picking up small objects off the water within a few yards of where I stood'.

In 1886, it was noted that 'although rare, autumn very seldom passes without one or two grey phalaropes being seen', they were 'usually in mixed plumage worn between the two seasons [summer and winter]' and that 'they [along with other marine sandpipers] are of much more frequent occurrence after high westerly gales, especially in September and October', while in 1887 they were considered to 'appear nearly every year'.

In 1896 there were two reports, the first to use the correct scientific name; one found dead under telegraph wires at Walcot on 5 October, and one picked up dead on Ludlow Racecourse on 7 October. These occurrences appear to relate to the 'only disturbance of any real importance' in that month 'which passed north-eastwards along the Irish and Scotch coasts on the 7th–9th'. Similarly, in a month when 'gales were rather numerous', one was found at Great Woolaston on 6 September 1899.

Through the first 30 years of the twentieth century there were a further eight reports, mostly during October. The exceptions to this were two present for over a week at Petton Park in April 1908, with another two on marshy ground at Leaton on 14 April 1921. Surprisingly, individuals at Harmer Hill (17 October 1902), Burlton (9 October 1903), Petton Park, Merrington (3 October 1911), Leaton and Ellesmere (October 1923) were all found within two miles of the A528 Shrewsbury to Ellesmere road. Otherwise, one was shot near Ludlow on 11 November 1916. Most historic and modern records are

of one or, at most, two birds so the report of several on the River Teme near Knighton in 1930 was exceptional.

Thereafter, there were no further reports until the first modern record in 1990. While there is a general paucity of records for all species from the early 1920s to the mid-1950s, it is not clear why there was such a large gap, particularly since it was historically thought of as a regular visitor. The worldwide population is reported as being in decline, but is still extremely large and classified as being of least concern, suggesting that this is not the reason.

There have been a further nine sightings, all of singles, since that first modern record, with only one year, 2011, having two:

The Mere (Ellesmere)	31 October–12 November 1990
Chelmarsh	10–12 November 1996
Battlefield	18 October 2000
Venus Pool	24 November 2001
Priorslee Lake	16 October 2002
The Mere (Ellesmere)	11 October 2003
The Mere (Ellesmere)	13–14 November 2005
Venus Pool	23 October 2008
Chelmarsh	12–14 September 2011
Adeney	18–26 September 2011

All were in the autumn, all except the two in 2011 were between 11 October and 24 November, and most can be linked with a period of strong westerly winds. This is very similar to the trend observed through nineteenth and early part of the twentieth centuries. It can be speculated that the birds at Chelmarsh and Adeney in 2011 were one and the same.

Graham Walker

Common Sandpiper
Actitis hypoleucos
Uncommon passage migrant, rare summer visitor, very rare winter visitor

Common Sandpiper breeds across much of temperate and subtropical Eurasia, and migrates southwards for the winter. Here it is on the eastern edge of the Welsh breeding population, and it occurs most frequently in the Severn Valley on migration to and from the Welsh uplands. The wintering area is uncertain, but is probably in West Africa south of the Sahel.

Nineteenth-century authors considered it widespread, both as a passage migrant and a less common breeding species. Rocke (1865) wrote that it was 'very common on the Teme, [and] appears to be pretty generally distributed on the various rivers and streams across the county', while Beckwith (1878; 1885) described it as 'a summer migrant, breeding by many of our small streams, especially those close to the Welsh borders. On the Severn it is abundant in spring and autumn, but, except on the higher parts of the river, rarely stays to breed'.

> [It] visits our pools and rivers in April, when for a short time it is very numerous. Then the greater number leave for the breeding season, and return again with their young in July; but in South Shropshire, many breed by the large brooks, so common in the hilly parts of that district, and recently one or two pairs have bred by the Severn near Ironbridge. This sandpiper usually migrates south in August and September.

Forrest (1899) summarised its status: it 'may be heard in summer on most of our rivers and brooks, especially near the welsh border. It nests regularly near Shrewsbury, always close to the water'. He reported a sighting of an unfledged young being carried on the back of a female on the Severn near Nesscliffe in June 1892.

However, it was mapped as 'uncommon' in the south-west hills and Oswestry uplands, and 'rare' elsewhere, as a breeding species at the end of the century in the national *Historical Atlas* (1996).

Although considered a summer visitor and passage migrant, there were four winter records between 1879–83.

The Cound brook, between Condover and Leebotwood, was visited in May in seven of the 10 years between 1902 and 1911, when a total of 16 nests and two broods of chicks were found, in spite of 'the winding of the small stream & the thick foliage, which make birds impossible to watch for & nests hard to find'. The location of five nests were noted, four in 1903 ('one in thick weeds, one in reeds, and two in nettles, all on sand reaches') and one in 1911 ('on shingle – a very exposed nest') (Meares & Meares 2018).

It appears from these early accounts that breeding Common Sandpiper was more numerous and widespread than in the second half of the twentieth century.

A lyrical account from Walcot Lakes strongly suggested they were breeding there in the 1930s, and display and at least probable breeding was noted there again, in 1947 or perhaps later (Stevens 1953).

The *Handlist* (1964) described it as a 'breeding visitor, April to September, and passage migrant. Breeds in small numbers in all divisions, but chiefly on the swifter streams of the hill country and the Rivers Severn and Vyrnwy'. However, only one SBR contained

Jim Almond, Venus Pool, 2 May 2007

records to substantiate this: at least two pairs on breeding territory, one on the West Onny near Plowden and the other on the Vyrnwy at Llanymynech (1962). After that, there were breeding records from the south-west, two nests each with four eggs (1970), and 'the BTO Atlas enquiry showed that it breeds in suitable habitats in the west and south' (1972). In fact, the 1968–72 national BTO breeding Atlas mapped confirmed breeding in only one 10km square in the north (SJ41), and in the four most southerly squares which the Teme passes through. Probable breeding was shown in one square in each of the north and south, and possible breeding was shown in four northern squares that the Severn passes through, suggesting that some of these records might be of migrants, rather than breeding birds. There was one possible breeding in the south-west, probably at Walcot.

After 1972, there was confirmed breeding at Ashford Carbonell (1973 and 1974), on the Teme at Ludlow and the Onny at Plowden (1976), on the Teme near Dutlas and a pair on the Onny near Onibury (1978), 'may have bred on the Teme near Quabbs (Clun Forest) and on the Onny at Bromfield' (1979), probable breeding near Bromfield and three pairs on the Onny near Craven Arms (1980), and 'probably breeding pairs between Buildwas and Cressage' (1984).

In 1985–90, breeding was confirmed in a cluster of tetrads on the Onny and Teme near Bromfield, but not at any of the other sites listed in SBRs, although there were probable breeding records from Walcot and the Onny near Plowden.

Most, if not all, of the possible and probable records from the Severn Valley and the north shown on the earlier Atlas map will in fact have related to the singles and pairs which pass through as late as mid-May and return again from late June onwards. There were many other breeding season SBR records which clearly related to migrants and which were not included on the map.

Some pairs attempted to breed in the Severn Valley using the shingle bars and stony shores that are exposed in the summer months but, in addition to nest loss from flash floods caused by heavy rain, they were subjected to considerable disturbance by fishermen and other river users, and to trampling by cattle. For example, young were successfully raised at the Severn–Tern confluence in 1987, but three eggs disappeared from a nest in the same place in 1988.

In the more remote south-west, there is much better breeding habitat on the rocky shores of the faster-flowing streams, and the rivers Teme and Clun, and their tributaries, as well as the Onny, provided several confirmed breeding records, with five pairs in one season on about 20km of the Teme between Bucknell and Felindre (SBR 1987). Adults are fairly conspicuous, but chicks are well camouflaged, so some of the probable breeding records in the south-west were also likely to have related to breeding pairs.

Between the two Atlas periods, breeding records have been few and far between, and far fewer than before 1985. A Waterbird Survey was carried out on a 5.9km stretch of the River Severn, passing through three tetrads near Alberbury, for the nine years 1991–99. There was at least one Common Sandpiper territory each year, with three in 1994 and two in 1997 and 1998, but it is likely that these pairs did not stay to breed, but moved upstream on the Severn or Vyrnwy where more suitable nesting habitat existed (Michael Wallace *pers. comm.*). Though probable breeding was recorded in these tetrads in 1985–90, only migrants were recorded in the recent Atlas period.

A nest with four eggs was found near Bucknell on 22 June 1992, and in 1993 three June records (eight at Bromfield on the 8th, 14 at Bucknell on the 13th, and a single at Ludlow on 28th) were 'worthy of note'. It is likely that the first two are records of four and seven pairs respectively on stretches of river, but the original records are no longer available.

A small flock of 11, including juveniles, was seen on the River Severn near Buildwas on 13 July 2002, but the breeding almost certainly occurred elsewhere. The shingle banks have been checked every year since 1998 for Little Ringed Plover, but no evidence of the sandpipers breeding there has been found. These natural shingle banks are at constant risk from inundation and the chances of successful breeding there is small (Glenn Bishton *pers. comm.*).

The breeding season distribution map shows only one site with confirmed breeding during the recent Atlas period, Bromfield Quarry in SO47Y. There was a nest with four eggs in 2011, and definitely two nesting pairs, probably three, in 2013.

In 2014, there were two territories on the River Onny between Stokesay and Onibury, one in each of SO47P and SO48K, and a local

Occupied Tetrads

Atlas period (breeding)	1985–90		2008–13		Change	
	Number	%	Number	%	Number	%
Confirmed	11	1	1	0	-10	-91
Probable	21	2	3	0	-18	-86
Possible	37	4	8	1	-29	-78
TOTAL	69	8	12	1	-57	-83

Breeding Distribution (2008–13)

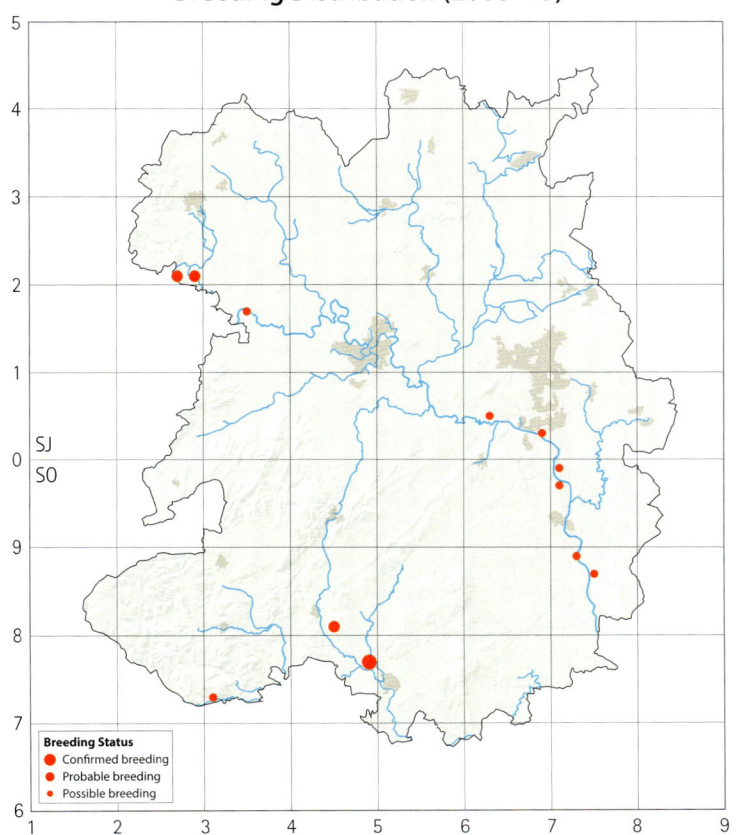

Breeding Distribution Change (Probable & Confirmed Breeding Only) (1985–90 to 2008–13)

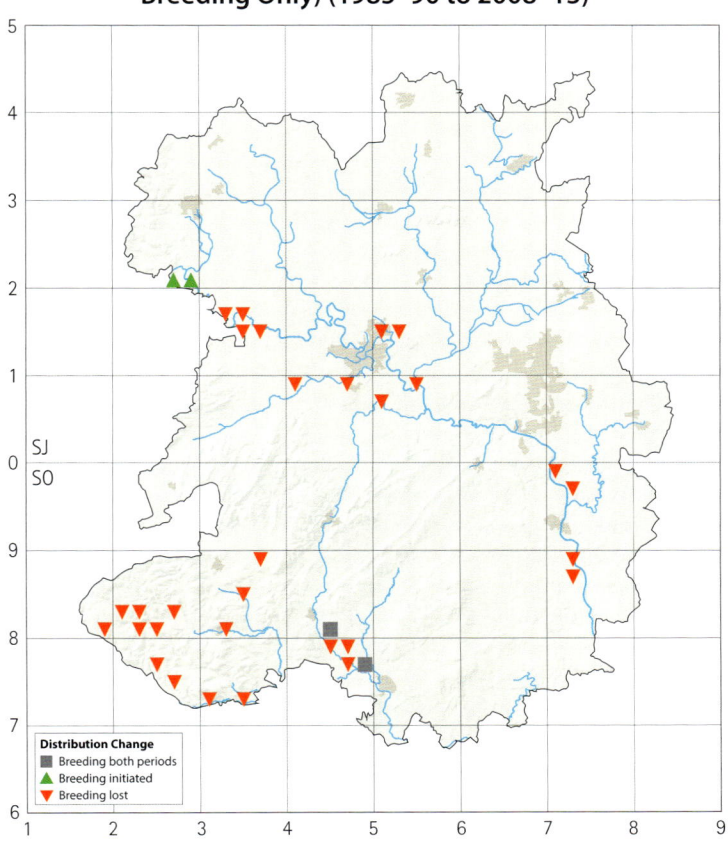

John Robinson, Chelmarsh Scrape, 2007

fisherman confirmed that they occurred 'every year' in the latter, so probable breeding has been mapped in this tetrad. It is likely that the territory in SO47P was also occupied during the Atlas period, but there is no record to justify its inclusion on the map (Tom Wall *pers. comm.*).

The other probable breeding records are from shingle and sand banks in the River Vyrnwy. In SJ22V a pair making alarm calls were heard near Llanymynech on 19 June and 1 July 2011, and a pair was seen near Llwyntidmon on 16 June 2008. A mating pair was seen in an adjacent tetrad in Wales, SJ22K, in 2013, and fledged young were upstream, further into Wales, in SJ22B in 2008.

Breeding was expected on the River Teme, as it was confirmed there in 1985–90, and territories were found on the Herefordshire reach of the river between Lingen Bridge and Downton in the recent Atlas period. However, the shingle banks upstream from Knighton were closely observed, but none were seen. They may perhaps have been overlooked, but the river is more prone to flash flooding now, mink are common, and the national BTO Atlas (2013) shows a loss from many squares in this area, so it is likely they have disappeared.

In Britain, nests are usually well hidden in vegetation, often 10–30m from water and occasionally up to 100m, and in woodland, but some are more exposed on river shingle (Ferguson-Lees *et al.* 2011). Here, most evidence of breeding has come from shingle banks in rivers,

but it is possible that chicks have been taken there from nests nearby. The nest at Bromfield quarry in 2011 was well hidden under a small willow tree about 6ft tall, with the lower branches spread out covering the ground with the nest (Jon Lingard *pers. comm.*).

The breeding distribution change map excludes possible breeding records from both Atlases, as migrants were often recorded as 'H' in the earlier one, but not recently. Even so, it overstates the loss, as two migrants travelling together were often mapped as a pair, although the decline from 11 to one tetrad with confirmed breeding is indisputable evidence of a considerable decline. Breeding pairs are well camouflaged on shingle, and easy to overlook unless they sing or call, and there is no access to many stretches of river, so there may be a few more pairs lurking along the Onny, Teme, Severn and Vyrnwy, but it is unlikely that there are more than five to ten breeding pairs. The site at Bromfield has apparently been lost, as that part of the quarry has been worked out and the pools have been modified – the Sandpipers now arrive, but do not stay to breed (Dave Pearce *pers. comm.*).

In addition, they were recorded as migrants in 58 different tetrads during the recent Atlas period, mainly along the Severn, so they are still seen much more frequently on passage than on breeding grounds.

The chart shows that, between 1992 and 2014, spring passage was more concentrated than autumn, with the vast majority heading north or west in the four weeks between 11 April and 10 May. Autumn passage gets underway in late June and July, peaks in mid-August, and is almost complete by the end of September. The majority of records (61%) were of one individual, and another 20% were of two. Only 34 records (1%) have been of 10 or more, four between 16–20 April, maximum 12 on the last of those dates, and the other 29 have fallen between 9 July and 22 September, the largest flock being 19 on 12 July 1997 at Crose Mere, with an average of 12 for the 29 records. In total, there were almost twice as many sandpipers in the autumn than the spring.

The two large counts in June 1993, itemised above, are not included in the chart.

The analysis in the *Arrival Dates* chapter suggests that, on average, they are now arriving around 10 days earlier than they did in 1900.

The UK is at the northern extremity of the wintering range, and only very small numbers stay over, but the numbers are increasing slowly, presumably in response to milder winters. The 1981–84 BTO winter Atlas showed only one record, from the Severn–Vyrnwy confluence, while there were just two in the recent Atlas period, both singles, on 22 December 2007 from Chelmarsh, and 12 November 2011 from Wood Lane. Excluding these, since 1992 there have been 18 records in the four winter months, all except one of singles.

Nationally, the BTO Atlas (2013) shows a 14% loss in occupied 10km squares since 1968–72, and BBS shows a decline in the UK of 15% between 1995 and 2014, and the losses here since 1985–90 are consistent with that. The reasons for the decline are unclear, but habitat changes in Britain, including increased disturbance from humans and cattle, and increased predation by mink, which have affected breeding productivity, and climatic factors on the wintering grounds, are all likely to be contributing. Although migrants are likely to continue passing through, the small and diminishing breeding population is threatened.

Leo Smith

Seasonal Occurrence in 10-day Intervals (1992–2014)

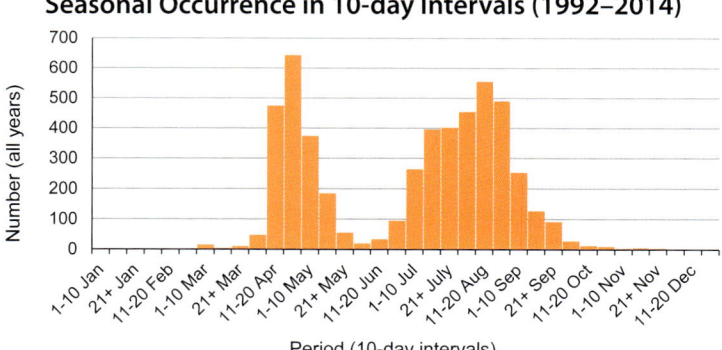

Period (10-day intervals)

Green Sandpiper

Tringa ochropus

Uncommon passage migrant, scarce winter visitor

Green Sandpiper breeds mainly in Fennoscandia eastwards, with a very small number in the Highlands of Scotland. Britain is at the northern edge of the wintering range, which is centred on the Mediterranean basin. It can be found in a wide range of freshwater habitats such as streams, reservoirs, ponds and farm ditches, but also coastal marshes and estuaries.

It has never bred here (the symbol 'B', meaning that the species has bred here, and a record of breeding at Clungunford in 1888, were deleted in an errata replacing page 8 in Forrest 1899, reproduced on the *Histo* website). Forrest describes it as occurring 'pretty often ... in autumn and winter', but one shot in July and another in August, listed as 'summer' records, were actually within the autumn passage period. There were records in most years in the first half of the twentieth century, but the *Handlist* (1964) described it as a 'non-breeding visitor and passage migrant. Recorded more frequently in recent years, though there are a number of old winter records from various localities. Regularly recorded at Shrewsbury Sewage Farm, Venus Pool and Allscott Sugar Factory'. These accounts suggested that it was a scarce species and was seen more frequently as the twentieth century progressed. Over-wintering was first suggested in the early 1960s.

It is not possible to estimate the number of individuals that have occurred each year, but the number of records increased six-fold between 1992 and the maximum in 2004. Allscott and Mirelake contributed one-third of the total records in 2000, and reached over 100, half the total, in 2001. There were four subsequent years with over 80, the fluctuations reflecting whether management of the water levels had created appropriate feeding conditions. The increase reflects a moderate increase of 20% in the European breeding population between 1980 and 2013 (Pan-European Common Bird Monitoring Scheme 2015), and an increase in the over-wintering population through milder winters.

Most are seen during the light spring and protracted autumn passage, in common with the national pattern. This is one of the earliest species to return, from late June onwards, with the peak at the end of August, as breeding adults and young from the Fennoscandia breeding population pass through. Since 1992, 43% of records have been of single birds, but the average of the remaining small groups was 4.6, with only three of more than 20, all from Allscott Sugar

Jim Almond, Chelmarsh Scrape, 12 August 2007

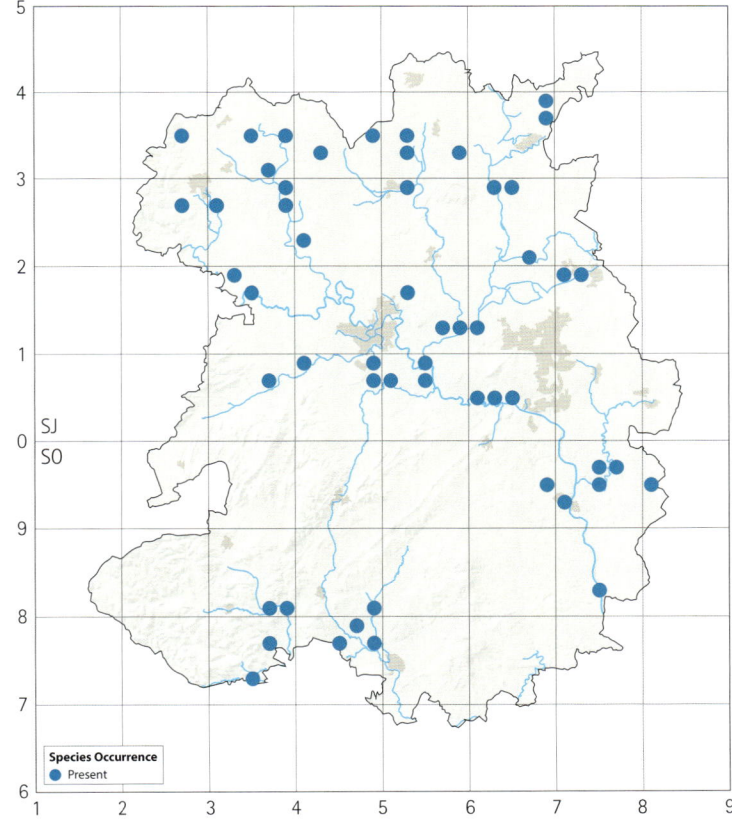

Winter Distribution (2007–13)

Occupied Tetrads

Tetrads with Winter Records (2007–13):   53   (6%)

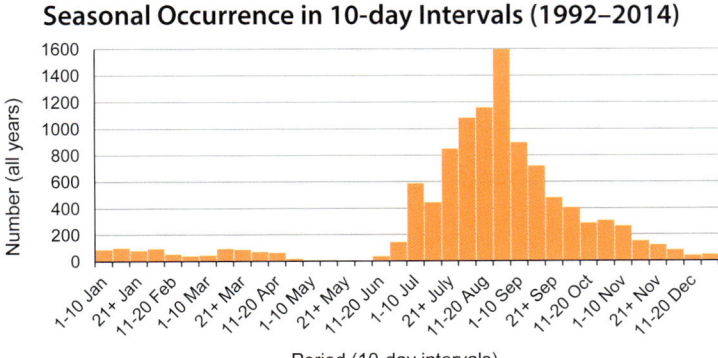

Seasonal Occurrence in 10-day Intervals (1992–2014)

Number (all years)

Period (10-day intervals)

Factory and all in August, 22 on 12th in 1995, 24 on 15th in 2004, and the largest, 25 on 28th in 1993.

In the recent Atlas period, there were 90 breeding season records from 22 tetrads, 19 up until 2 May, and well over two-thirds (71) from 14 June–31 July, but these too were all passage birds, and only Allscott, Mirelake, Venus Pool and Chelmarsh provided more than five of these records.

In winter, Green Sandpiper was still scarce at the time of the 1981–84 national BTO Atlas, when it was mapped in only eight of the County's 33 10km squares, seven in the north in the Severn Valley, and

one in the Teme valley. The distribution had increased considerably by the recent Atlas period, when it was found in all except two of the 19 10km squares in the north, and in six of the 14 squares in the south, at widespread sites in 53 tetrads, almost all in the river valleys. At most sites it occurred infrequently in small numbers, but at the usual key wader sites, Allscott, Chelmarsh, Venus Pool, Whixall Moss and Wood Lane, it was seen more frequently and in larger numbers.

Sightings should become more frequent, if the population growth, and trend to milder winters, both continue.

Dawn Balmer

Lesser Yellowlegs

Tringa flavipes

Vagrant

Although a rare visitor to Britain, it is annual with between five and 10 found in most years. It breeds from Alaska through to Hudson Bay and migrates south to winter in the southern United States and most of Central and South America. In Britain, most are found in the autumn, presumably having been blown off-course on their

southward migration. The only record, a juvenile, was typical of this (Rogers *et al* 1996):

Knighton Reservoir   11–13 November 1995

Graham Walker

Redshank (Common Redshank)

Tringa totanus

Scarce passage migrant, rare winter visitor; has bred

Jim Almond, Polemere, 21 January 2013

Redshank breeds from Iceland across much of Europe and Asia, including coastal marshes, lowland wet grassland, and rough pasture in the uplands of Britain, and most of the European population winters on or near Atlantic coasts south to West Africa. There are several races, and those that breed in Britain are mainly from the nominate

race *T.t. totanus*, while many of the winter visitors are of the *robusta* race from Iceland.

Beckwith (1885) noted that 'solitary examples are sometimes found by pools or streams in autumn' but that it had never occurred in spring. Early references usually related to shot birds. A specimen in Beckwith's collection was shot on the River Severn at Wroxeter (Beckwith 1878) but by the early 1900s records more frequently related to birds observed rather than hunted. At the start of the twentieth century, it was considered 'uncommon' and Forrest (1899) wrote 'it is not known to have bred'. However, the British population had been expanding during the late nineteenth and early twentieth centuries, and a pair at Crudgington on 25 May 1910 was considered by the *Handlist* (1964) to be the first record of breeding, but an 'old pair with three young' at Betton Pool on 29 July 1908 is suggestive of breeding two years earlier. 'Eight or ten' were noted on the marsh at Crudgington in 1914 and breeding was again confirmed there in 1915, where it subsequently became well established.

Breeding occurred at a new site – Halston, on the River Perry floodplain – in 1918, and was followed by breeding at Cressage in 1922 and Atcham in 1924, where 'several pairs' were well established by 1926. A large flock of at least 150 was recorded at Crudgington on 20 April 1930 and 'about a dozen' were resident in the Tern valley between Walcot and Rushmoor in the same year. Consolidation of the core sites at Crudgington and Atcham continued, the Atcham site extending to Sutton and 'within a mile of Bayston Hill' by 1936. At least 14 pairs occupied 10 sites by 1939, mainly along the floodplains of the rivers Severn, Tern, Roden and Perry.

A decline in the breeding population appeared to set in around 1940, probably 'due to excessive drainage of some of the better habitats'. Between 1957 and 1966 it was restricted to four main breeding sites – Venus Pool, various locations on the Weald Moors, Shrewsbury Sewage Farm and Allscott Sugar Factory – and by 1967 it was 'absent from some areas where normally present'. Breeding was sporadically noted at Crudgington, Atcham and some outlying sites throughout the 1980s and 1990s, and the *Atlas* (1992) estimated the breeding population at 30–50 pairs at the most. A pair at Allscott Sugar Factory in June 1996 was possibly the last breeding to occur there, and breeding has not occurred at Venus Pool since a pair was noted incubating on 14 May 1997. Breeding was last confirmed in 1998 when a clutch was believed trampled by stock on 15 June at Beckmoor.

During 2008–13 Atlas fieldwork, records of probable breeding were submitted for four tetrads, all within the traditional area of the floodplains of the Rivers Roden and Tern, on the western edge of the Weald Moors. Of these, pairs in two tetrads, and individuals seen in the same place more than once in another tetrad, were not seen after 13 April, and could not be relocated These were all at well-watched sites, Mirelake, Allscott Sugar Factory and Long Lane. A pair was also noted at Baggy Moor in the north-west on 24 April 2010, but again it was not relocated, and the record was therefore submitted as possible breeding. Two were seen well outside the main passage period at Eyton Moor on 31 May 2008 (submitted as a pair, probable breeding), but there were up to five there in flooded fields from 28–31 May. Four individuals were also seen on single dates in late May at Wood Lane during the Atlas period, in 2008 (two), 2012 and 2013. One was also seen near Chelmarsh in the south-east in mid-June. Those at Eyton Moor could not be relocated and there was no observation of any breeding behaviour. The other two sites are well-watched and, again, there was no evidence of breeding behaviour. Therefore all these records almost certainly relate to late breeders moving to their northern breeding grounds, or early non-breeders moving to their wintering grounds.

In 1985–90 there were confirmed breeding records from near Melverley, Venus Pool, Allscott Sugar Factory, north of Isombridge, Long Lane, Norbroom Marsh, and Hortonwood in northern Telford. All these sites still provide suitable breeding habitat, except Hortonwood, which has been developed, but Redshank was not found at any of them in the recent Atlas period. There were probable breeding records in 1985–90 at several tetrads on the River Severn near Atcham, Buildwas and Chelmarsh, and possible breeding in several tetrads in the north and north-east, but again no Redshank were found there in 2008–13.

The tetrad occupancy table shows the 2008–13 records as observers submitted them, because they meet the technical definition of probable or possible breeding records. However, as detailed above, efforts were made to relocate all these birds, but no further evidence of breeding, or even presence, was found, and the records should not be interpreted to suggest otherwise. With no records of confirmed breeding since 1998, it is believed that Redshank is now extinct as a breeding species, and the breeding distribution change map reflects this.

The British wintering population is more than twice the breeding population, with visitors from Iceland, Scandinavia and continental Europe, so here most Redshank are now spring and autumn passage migrants, with a few non-breeding and winter visitors. Spring passage is more concentrated than autumn passage and peaks between 11 March and 20 April as the British and Icelandic populations both move through to their breeding grounds. Autumn passage shows three distinct smaller peaks, in mid-July, late August and early November, which are likely, respectively, to relate to the return of failed breeders and breeders from northern England and Scotland, the arrival of visitors from Iceland and then more migrants from northern Britain, though there is some overlap in the movements of the different populations. Many, particularly young Redshank, move south to France, Iberia or Morocco.

The pattern of occurrence is summarised in the chart. It is based on all records since 2000 only, so the underlying migration peaks and troughs are not obscured by records of Redshank that bred here. However, Redshanks still occur throughout the year, the summer and winter records reflecting the mobility of non-breeding and failed breeding birds, and movements between coastal wintering areas, particularly in cold weather, respectively.

Allscott Sugar Factory, Mirelake, Venus Pool and Wood Lane

Breeding Distribution Change (1985–90 to 2008–13)

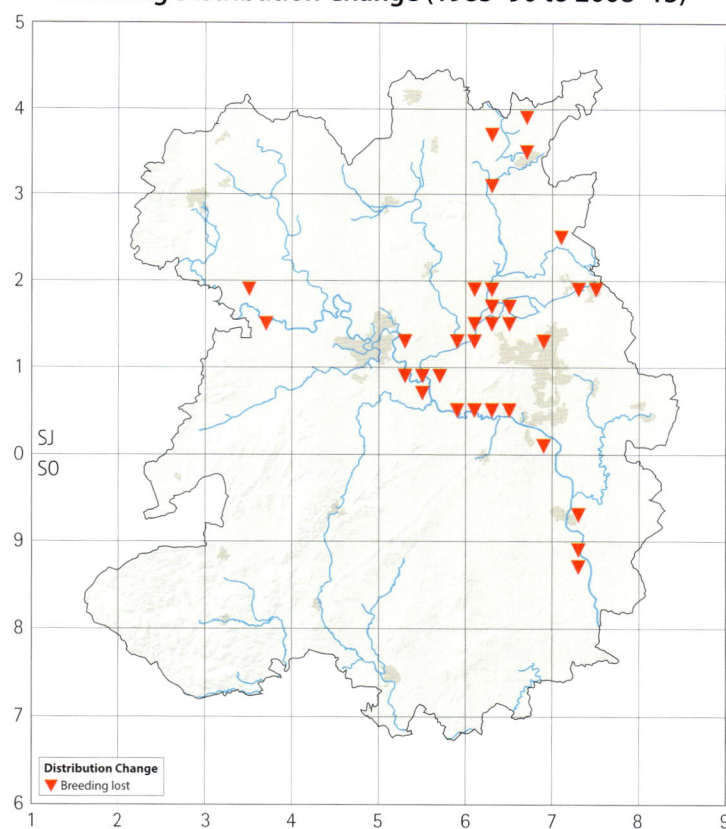

Distribution Change
▼ Breeding lost

Occupied Tetrads

Atlas period (breeding)	1985–90		2008–13		Change	
	Number	%	Number	%	Number	%
Confirmed	8	1	0	0	-8	-100
Probable	10	1	4	0	-6	-60
Possible	14	2	2	0	-12	-86
TOTAL	32	4	6	1	-26	-81
Tetrads with Winter Records (2007–13): 12 (1%)						

Seasonal Occurrence in 10-day Intervals (2000–14)

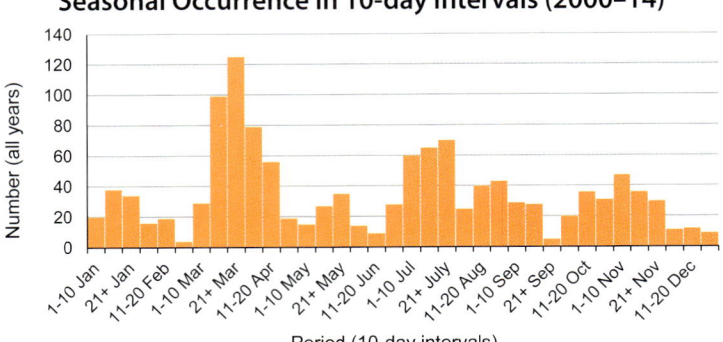

Winter Distribution (2007–13)

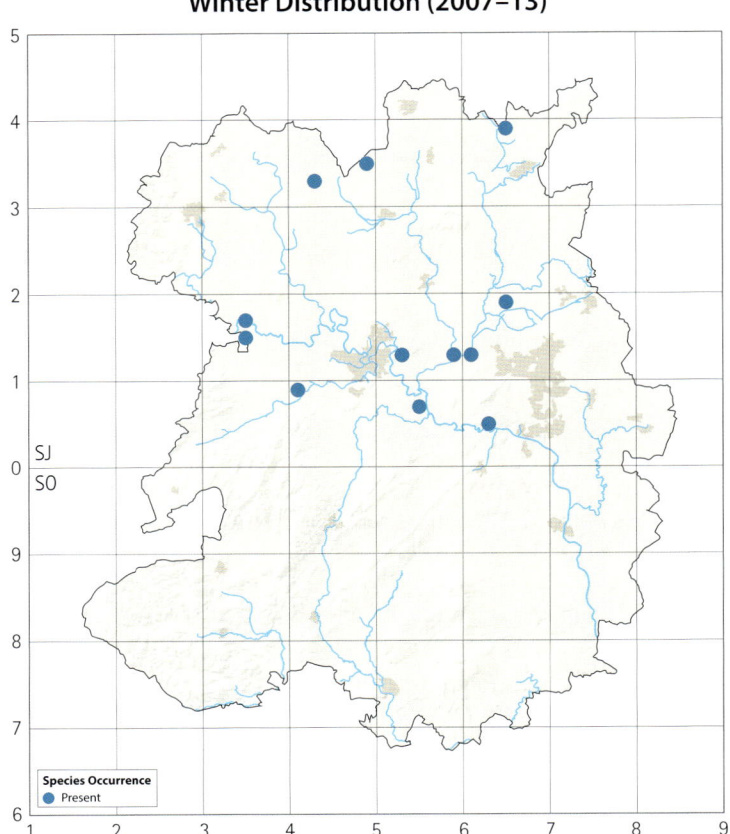

provided 78% of all records between 1992–2014. Other sites regularly frequented included Chelmarsh Reservoir, Polemere, Priorslee Lake, the River Severn at Buildwas and Leighton, Whixall Moss and Whixall Canal Floods. Almost two-thirds of records were of a single bird and there are only three of more than 10 together.

Allscott Sugar Factory and Mirelake was the favoured wintering site with an average of 6.32 birds per visit in the winter periods of 1992 to 2014. Since its closure in 2007, the pools have been drained and the site is now degraded, which will probably adversely affect Redshank numbers in future years. Other wintering sites recorded an average of 1.17 birds per visit during the same period.

During the winter Atlas 2007–13 it was found in 12 tetrads, all in the north, where it was strongly associated with major rivers and wetlands. Apart from the wintering sites named above, the other tetrads include the Severn–Vyrnwy confluence, Crudgington and Tittenley Pool. It is unlikely that any stay for the whole winter and most winter records probably relate to late passage individuals, with almost half the winter records in November, but cold weather movements may account for some dots on this map.

There are few ringing recoveries. An adult ringed at Walcot on 21 July 1964 had the greatest long-distance movement of 559km, the individual recovered dead at Vannes, France, on 4 September 1964. One ringed at Shrewsbury on 5 July 1978 and found dead on the Wirral on 28 February 1979 shows that some move to the coast for the winter. Another ringed as a nestling at Cressage on 31 May 1981 was controlled on passage at Walcot on 13 July 1981. An adult already full grown when ringed near Wellington in January 1965 was caught close by and still alive almost eight years later.

Redshank require wet, invertebrate-rich grassland foraging habitat and grass or rush tussocks for nest sites. Their decline is attributable to drainage, re-seeding and the intensive application of inorganic fertilisers and possible pesticides on lowland farmland. Intensive grazing, mowing, increased human disturbance and predation are all likely to have had an effect (Ausden *et al.* 2003). Even light grazing can cause nest failure (Sharps *et al.* 2015). Their successful reinstatement as a breeding species is probably only feasible if conservation is undertaken on a landscape scale. The RSPB's Wetlands Futurescape project, focused on the restoration and preservation of wetlands on the Weald Moors and Baggy Moor (RSPB website 2014), and the take-up of wet grassland management options within the agri-environment schemes (Eglington *et al.* 2008) offer the best prospects.

Glenn Bishton

Wood Sandpiper

Tringa glareola

Rare passage migrant

While this is a very rare breeding wader in Britain, with 27 pairs reported in 2010 (national BTO Atlas 2013), it is by no means rare globally, breeding throughout northern Eurasia through to the Bering Sea. The bulk of the European population winters in Africa. In Britain, it breeds in boggy habitats in northern Scotland, but is more commonly encountered away from its breeding grounds during the two passage periods, when it stops to feed along the open margins of still water bodies, both large and small.

Although the breeding population at the southern limits of its range in north-west Europe has declined through the nineteenth and twentieth centuries due to habitat loss and, possibly, climate change (Cramp *et al.* 1983), there does appear to have been an expansion of the population within the northern forests zone through the 1950s and 1960s (Fergusson-Lees 1971), including the north of Scotland which it colonised in 1959 (Sharrock *et al.* 1975). Given the relative close proximity of this as a breeding species, and that it is a regular

Jim Almond, Chelmarsh Scrape, 31 July 2009

in years when there have been multiple sightings. Nevertheless, in good 'Wood Sandpiper years' such as 1957, 1961, 1980 and 2004, at least eight and perhaps 10–12 individuals were seen, although the average is nearer to three per year since the first in 1952. The good years are likely to coincide with high populations following successful breeding seasons, and anti-cyclonic conditions over the eastern North Sea and Baltic regions, which encourage more to take a south-westerly route.

The chart shows the seasonal pattern of migration. Since 1992, less than 20% have occurred on spring passage and there have only been two instances of two or more at that time: two at Venus Pool on 19 May 1990, with three at the same site on 18 May 1996. There have been two early records on 2 April, in 1976 and 2012, but the bulk of spring passage occurs in mid-May. It is then difficult to tell whether the records from June through to early July are spring migrants still on their way north or failed breeders returning.

Return passage peaks in mid-August, but is spread from July, and perhaps late June, through to mid-November. The latest November records include singles at Allscott Sugar Factory from the 1st to 12th, in 2007, and 15th to 18th, in 2008. In contrast to the spring, there are more occasions in the autumn when more than one has been present at a site. They also have a tendency to linger as there is no longer the urge to get to their breeding grounds and, with the population swelled by juveniles, the autumn passage is visibly heavier than in the spring.

Graham Walker

passage migrant through Britain, it is perhaps surprising, therefore, that the first record was not until 1952 when one was at Oss Mere on 21 May. The next was at Shrewsbury Sewage Farm where one was present for 12 days from 10 September 1956. Thereafter, it has been virtually annual, only missing two years, 1960 and 1971. Most of the early sightings were from the Sewage Farm and then Venus Pool, but the draw of Allscott Sugar Factory is clearly apparent with over half the records from that site since the first in 1962. They have continued to appear regularly at Venus Pool and, from 1983 onwards, Wood Lane. More recently, the canal-side floods at Whixall have attracted a number of individuals. Unfortunately, records from Shrewsbury Sewage Farm ceased in 1981, partly as a result of reduced access, although the change to filter beds from settling ponds, resulting in the latter becoming overgrown, is more likely to be the major factor; and the draw of the Sugar Factory has also ceased.

Most reports are of one or two birds, but there were six, possibly seven, at Shrewsbury Sewage Farm on 10 August 1961, five at Allscott Sugar Factory on 10 August 1980 and four on Whixall Canal Floods on 21 August 2007. While only one has been recorded in some years, it is much more difficult to determine the total number of individuals

Seasonal Occurrence in 10-day Intervals (1992–2014)

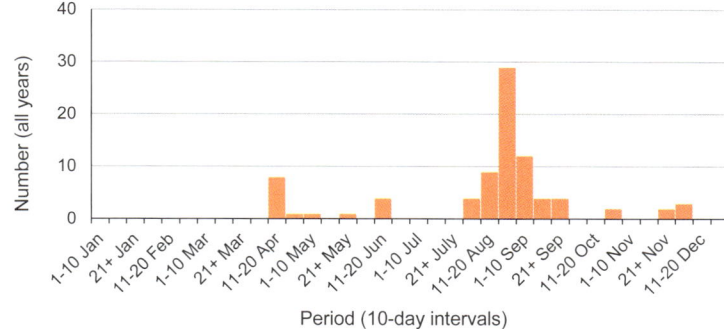

Spotted Redshank

Tringa erythropus

Very rare passage migrant

Breeding in Arctic and sub-Arctic regions from Scandinavia to Siberia, this passage migrant moves through Britain to and from its wintering grounds in equatorial Africa. Small numbers, about 100 each year, remain to winter around the British coast and occasionally occur inland (national BTO Atlas 2013).

Although first recorded in September 1923, when one was collected at Ellesmere, there were no further records until August 1960, when two were seen at Allscott Sugar Factory and one at Venus Pool. Subsequently, it was reported every year for the next 17 years and was more or less annual until 1991 with only two blank years, 1977 and 1984, during that period. From 1992, however, there are only records

for 13 of the 23 years, with seven blank years from 2004 to 2014.

This pattern is also reflected in the average number of birds seen each year. Since 1960, approaching 200 have been recorded, and the overall average has been between three and four, but in the 20 years up to 1980 the average was between four and five, while over the 24 years since 1980 it has been under two. There is no obvious reason for this, but it does correspond with observations in the West Midlands (Harrison & Harrison 2005).

Most have been seen in the autumn. Since 1992, just over 10% have been seen in the spring, and peak spring passage is during mid-April.

While early returners in June are likely to be failed breeders, the

Jim Almond, Whixall Floods, 16 April 2013

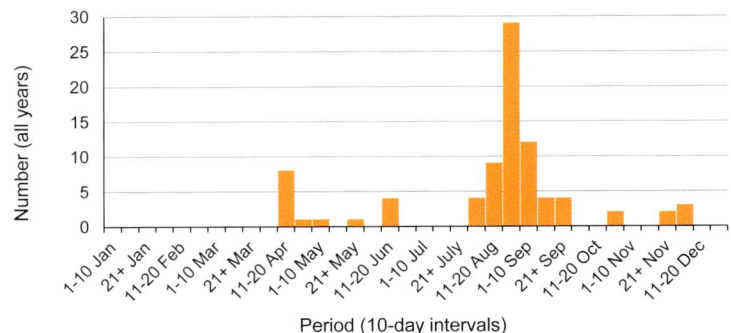

Seasonal Occurrence in 10-day Intervals (1992–2014)

main autumn passage starts during the last week of July, with a peak in the last 10 days in August (over a third of all individuals recorded since 1992), and is largely over by the first week of October. Late records have included two at Allscott Sugar Factory on 27 October, in 2001, two at Venus Pool on 29 November, in 2002, and, exceptionally, one in the Edgerley/Cae Howell area 5–7 December 2012.

Most reports are of one or two birds which are only usually present for a few days. In September 1973, however, a maximum of seven were at Venus Pool, and a flock of eight were observed at Allscott Sugar Factory on 1 September 1978. In 1989, two were at Venus Pool for over a month from the middle of July, during which time they moulted from juvenile into winter plumage.

Graham Walker

Greenshank (Common Greenshank)

Tringa nebularia

Scarce passage migrant

Greenshank is an autumn passage migrant, with occasional records in spring. It breeds in the Scottish Highlands and Scandinavia and winters mainly on the estuaries of southern England, the Bay of Biscay and West Africa.

Rocke (1865) noted that it was 'always' recorded 'in the autumn' and Forrest (1908) described it as a rare autumn visitor with 'recent occurrences at Pontesbury 5 September 1885, near Cressage 5 September 1898 and an adult at Betton Pool 12 October 1898'.

Jim Almond, Venus Pool, 14 September 2014

There was then a dearth of references until the *Handlist* (1964) summarised its status as a regular passage migrant from July to October, mainly at Shrewsbury Sewage Farm, Allscott Sugar Factory and Venus Pool. Up to nine were present at Shrewsbury Sewage Farm between 20 August and 6 September 1956 with a peak of 12 there on 9 August the following year. There were records from only one site in 1960, Allscott Sugar Factory on 5 August and 20 August, attributed to abnormally high rainfall attracting Greenshank to flooded sites elsewhere, which were not under observation.

Additional autumn passage was found at Cole Mere in 1961, Bettisfield Pool and Acton Burnell in 1964, and Fenemere and Oswestry in 1965. Bromfield gravel pit was added in August 1977, and Wood Lane and Chelmarsh in August 1981.

Spring passage was recorded for the first time in 1962 with one at Venus Pool on 10 April, and an unusually early migrant was at the same site on 24 March, in 1978. Spring passage peaks in early May, but it involves smaller numbers than autumn passage and is less prolonged, the Greenshanks possibly adopting a more urgent and direct return to their breeding grounds. Autumn passage on a southerly route to their wintering grounds commences in June with the arrival of adults, mainly females, followed by high numbers of juveniles and the last adults, peaking in late August. The passage periods are shown on the seasonal occurrence chart.

Wintering has not been recorded, but one at Venus Pool on 27 November 1980 was noted as 'unusually late'. Other November and December records include singles at Allscott Sugar Factory on 16 November 1994, Mirelake on 1 November 2007, and one at Venus Pool on 1 November 2010. The absence of January and February records suggests these were late migrants rather than over-wintering. The last two of these records were the only ones received during the recent winter Atlas period.

Almost two-thirds of all records since 1992 are of single birds seen on one day only. Records during this period came from 16 sites with

90% from just four – Allscott Sugar Factory, Venus Pool, Chelmarsh Reservoir and Wood Lane. Apart from these pools and reservoirs, other wetland habitats infrequently utilised include the shoreline of the River Severn, Chelmarsh scrape and Whixall canal floods. Records of 10 or more have been restricted to the autumn and comprise 11 at Allscott Sugar Factory on 13 September 1994, an 'amazing' 27, pursued by a Peregrine over Venus Pool on 20 September 1999, 18 at Allscott Sugar Factory on 20 August 2000 and between 10 and 23 there through to 12 September, 10 at Chelmarsh on 19 August 2001, and 10 at Venus Pool on 12 September 2008. The year 2000 was noted as 'exceptional' with at least 40 individuals passing through Allscott Sugar Factory in August, probably due to optimal water levels, while 2001 was a 'poor year' with high water levels reducing feeding opportunities and detrimentally affecting autumn passage stopover. The 27 over Venus Pool on 20 September 1999 remains the highest number on a single record.

The sole ringing recovery – one ringed at Allscott Sugar Factory on 16 August 1981 and found dead in the Bay of Biscay at Saint Romain, France, on 21 September 1984, 816km distant – supports the contention that they pass along the French Atlantic coast in autumn on route to their wintering grounds further south.

Glenn Bishton

Seasonal Occurrence in 10-day Intervals (1992–2014)

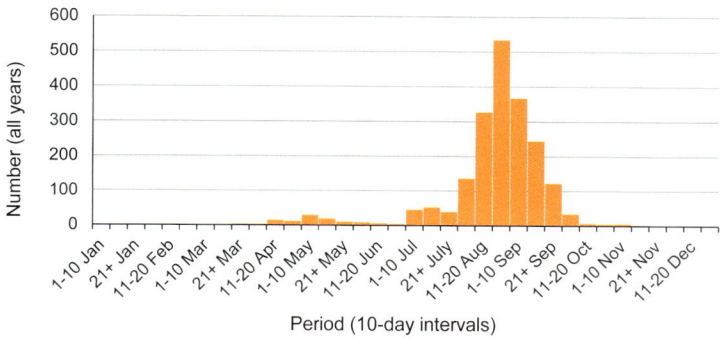

Period (10-day intervals)

Kittiwake (Black-legged Kittiwake)
Rissa tridactyla
Rare passage migrant

Jim Almond, Horsehay Pool, 3 January 2014

The British breeding population of Kittiwakes is widely distributed around the entire coastline, and winters in the north Atlantic.

Rocke mentioned that 'stragglers' were often recorded following heavy gales, often in a thoroughly exhausted state, while Forrest described it as an almost annual winter visitor often arriving after high winds, usually singly, although he only details four individuals that occurred during a single week in Shrewsbury in February 1899. Forrest also noted an unusual specimen in the collection of Mr Harry Shaw in which all the quills had black tips forming a black border to the wings.

There were records in several years between 1900–17, another in each of 1928 and 1929, the latter of a party of seven that stayed on the Severn in Shrewsbury for a week during a hard frost in February, then only one record in each of the next two decades, one in Shrewsbury in September 1935 and finally four at Peaton on 3 February 1945.

The *Handlist* (1964) described it as a vagrant with only four records between 1950–64. It has occurred almost annually since 1977 with 1981, 2005, 2009 and 2010 being the only blank years. In most years only one is seen, but in five years, 1984, 1995–97 and 2014, there were more than five records, with 10 in 1996 and 1997. The majority occur during spring passage, mainly in March, with smaller numbers during autumn and winter.

Four favoured sites have contributed 84% of the 99 records: Chelmarsh (18), the Ellesmere meres (24), Priorslee Lake (19) and Venus Pool (22). Most records relate to single birds with notable exceptions being 35 at Whitemere on 16 March 1996 and 14 at Chelmarsh on 7 April 1984. One at Priorslee Lake on 21 August 2004 was a rare form – it literally stood out due to it showing off bright red legs.

There were only two spring passage records during the recent Atlas period, both in 2013, singles at Priorslee Lake on 1 April and at Wood Lane on 5 May. Two Kittiwakes have been recovered here, a nestling ringed on 2 July 1961 on Puffin Island (Anglesey) was found dead near Shrewsbury nearly four years later on 28 March 1965, 108km distant, and a first-year ringed on 13 July 1999 in Finistere (France) was found near Newport only three months later, 547km distant.

Richard Moores

Sabine's Gull

Xema sabini

Vagrant

This is a long-distance migrant, breeding in the northernmost regions of North America and Eurasia, and wintering in the cold waters of the Humboldt and Benguela currents off Ecuador and Peru, and Southern Africa, respectively. Although considered a scarce migrant to Britain, it is nevertheless seen regularly in the autumn (usually August–November) in coastal counties, particularly at noted 'sea-watching' locations. As may be expected of such a maritime species, it is rare inland and there have only been four records.

A first year was picked up in an emaciated condition on 26 October 1874 at Nobold after a very heavy gale. Originally, this bird was identified as a Little Gull, but this was later corrected. Nineteen years later, on 8 September 1893, a male in full adult plumage was shot at Sandford near Oswestry.

The two modern records are:

Wood Lane	17 July 1983
Venus Pool	14 September 2011

The first summer at Wood Lane was notably early, while the individual at Venus Pool, a juvenile, was on a more typical date.

Graham Walker

Black-headed Gull

Chroicocephalus ridibundus

Common winter visitor, uncommon breeding species

Black-headed Gull is widespread in the northern hemisphere. It is difficult to imagine now that there was a significant risk of it becoming extinct as a breeding species in England in the late nineteenth century. However, large numbers were taken, either as eggs or as adults, for human consumption throughout the 1800s, with ready markets in the emerging industrial towns, and by end of the nineteenth century it was decidedly scarce throughout Britain (Hollom 1940). Land drainage had eliminated many breeding sites, and the colder climate in the second half of the nineteenth century may also have played a part. Here, Rocke (1865) and Beckwith (1879) mention it as no more than a regular visitor, presumably mainly in the passage and winter periods. The British population grew and expanded in the early twentieth century to an estimated 35–40,000 pairs in a national census in 1938, including a single colony at Brampton Bryan just over the border in Herefordshire, but no recorded breeding here (*ibid.*). Significant colonies were present at this time in Scotland and Wales, including one of over 5,000 birds in adjacent Montgomeryshire in 1927, which suggests that absence here may well have been due to human predation.

The first breeding record was of 'a small gullery on pool on Longmynd about three miles on the Shrewsbury side of Church Stretton' (presumably Wildmoor Pool) in 1942. A report from earlier in the year was followed up, and 'although they had left by August (the date of the visit) the unmistakable nests were there. This is the first record of this species breeding' here (CSVFC RBF 1943–44, *British Birds* 37:139). For unknown reasons, the *Handlist* (1964) referred to this as an 'unconfirmed report'.

Following the two World Wars, when again gull eggs made a welcome addition to the restricted diet, there was a rapid increase to over 100,000 pairs nationally by 1973 (Gribble 1962; 1976), which was mirrored here with short-lived colonies: 125 pairs at Stockton Wood Farm (near Chirbury), from 1950–52; five to ten pairs at Acton Pool (near Clun) from 1954–55, where the colony was longer-lived, as some time during the period 1967–70 youngsters recall taking

Dawn Micklewright, Venus Pool, 15 March 2014

eggs from a gullery of ten to twelve nests at the same place; and the same youngsters also visited another previously unrecorded small colony at More Pool (near Bishop's Castle), where fewer than 10 pairs attempted, possibly successfully, to nest in the late 1960s (David Hatfield *pers. comm.*).

The early short-lived colonies were followed by the establishment and growth of permanent colonies. By 1958 there were 20–25 pairs at Cranmere Bog (Worfield), 36 pairs at Allscott Sugar Factory and four pairs at Venus Pool. The *Handlist* (1964) gave maximum figures for these three sites of 40, about 50, and 36 pairs respectively. Since then these locations, together with Polemere and then more recently Wood Lane and Whixall Moss, have been the sites where the majority of pairs have bred over the years. Very small numbers have attempted to breed from time to time at another dozen or so locations but most of these have been isolated occurrences with no repeat attempts in subsequent years. Numbers at the main colonies have fluctuated, often dramatically, with disturbance, predation and low water levels problematic from time to time.

Venus Pool has been occupied continuously from 1957, and this is the only colony that has been systematically counted. Predation especially by fox has been regular since first observed in 1989, resulting in almost total breeding failure in many years. Peak breeding numbers were in the 1980s, with 400 individuals present at the beginning of the 1981 season, and over 100 pairs in many years later in that decade. Ten to 70 pairs were counted most years in the 1990s, with 178 nests counted in 2000, but no chicks were thought to have fledged that year. Two to 40 pairs have attempted to breed annually since, with a low success rate. In the recent Atlas period there were two pairs in 2008, rising to 10 in 2009, 25 in 2010 and 30 in 2011, but the whole colony was wiped out by predation over one night at the end of May in that year, and numbers dropped again to three pairs in 2012 and 5 in 2013. Most of the colony relocated to a newly created pool near Acton Burnell in the adjacent tetrad in 2012.

The colony at Allscott Sugar Factory was also founded in 1957, with an increase in the 1970s, when 200 pairs attempted to breed in 1977 and again in 1979. A few pairs also tried to breed in 1982, but 'the colony ... is apparently now deserted with no nesting since 1988' (*Atlas* 1992). However, 40–50 individuals were present at Mirelake in 2008 and 2009, when sitting females were seen, and two 'juveniles' were recorded in 2008, but no evidence of breeding has been found there since.

Nesting was first recorded at Cranmere Bog in 1958. By 1969, 150 pairs attempted to breed with a peak of 200 nests in 1977, when 150 young fledged. There were 800 individuals early in the 1982 breeding season and 80 young are thought to have fledged, but these numbers were not seen in subsequent years and the colony has since declined. The last breeding record was of five nests in 2005.

A colony at Polemere held 20–30 pairs annually from 1975–91, but now seems extinct.

The UK population continued to grow, to about 150,000 pairs by the late 1980s, contributing to new colonies here. The first recorded breeding on the Shropshire side of Whixall Moss was in 1999, but the absence of earlier records may be due to the lack of regular watching at this site prior to its designation as a National Nature Reserve in 1990. There were up to 40–60 pairs in most years since, up until the late 2000s, but numbers had declined to about 12–15 pairs by the end of the recent Atlas period, as the newly established colony at Wood Lane expanded rapidly, from an estimated three to four breeding pairs in 2008 to 10 times that in 2010, and 50–65, 90–100 and 130–150 pairs in 2011–13 respectively (John Hawkins *pers. comm.*). Nest building was noted there in 2008 and 2009, but the first confirmed breeding at Wood Lane was in 2010 when 11 chicks were recorded. Approximately 20 young were noted in 2011 and 2012 prior to a large increase in numbers in 2013, with over 100 chicks on the large island.

Breeding Colonies & Range (2008–13)

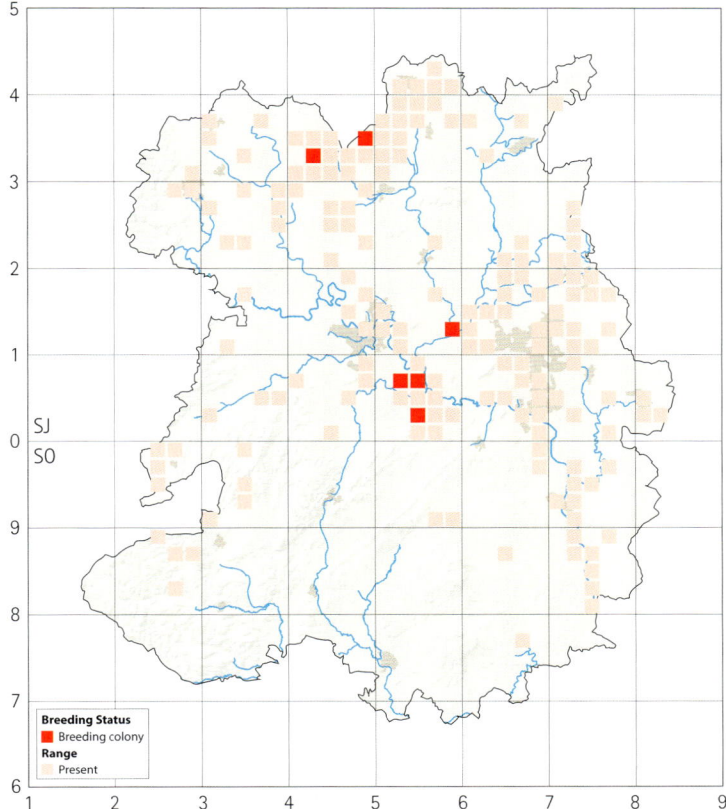

Occupied Tetrads

Atlas period (breeding)	1985–90		2008–13		Change	
	Number	%	Number	%	Number	%
Confirmed	8	1	6	1	-2	-25
Prob+Poss	49	6	18	2	-31	-63
TOTAL	57	7	24	3	-33	-58
Tetrads with Winter Records (2007–13): 418 (48%)						

A range of other sites have been occupied almost incidentally over the years with small colonies at Rorrington, Wall Farm (Kynnersley), Marton Pool (Chirbury), Berrington and Monkmoor Pools present for a few years or apparently intermittently. Breeding success seems to be relatively high when colonies are first established, but it declines over the years, presumably as local predators learn to exploit them.

The 1985–90 Atlas mapped foraging individuals, pairs and fledged young in each category, but such records, particularly of birds in flight, were treated differently in the recent Atlas. The tetrad occupancy table therefore combines probable and possible breeding records, but the standard breeding distribution change map would give a misleading impression of large losses. Black-headed Gulls forage over large distances from breeding colonies, particularly juveniles from late June onwards, so the map shows the breeding colonies (confirmed breeding records of nests only), with all other records, including birds in flight and some juveniles, mapped to show the foraging range during the recent Atlas period. The range is also occupied by gulls from nearby counties, such as the large colony near Welshpool, which produced over 600 fledglings in 2012, and by one-year-old non-breeding birds. Since 1985–90, colonies at Rorrington, Polemere, Allscott and Cranmere Bog have been lost, and Wood Lane, Whixall Moss and near Acton Burnell gained. Only Venus Pool was regularly occupied by colonies in both Atlas periods.

Only six colonies were occupied during the recent Atlas period, and the number of breeding pairs varied considerably from year to year, estimates ranging between 65–75 in 2008 to 170–190 in 2013, similar to the 100–200 estimated in the *Atlas* (1992).

The national BTO Atlas (2013) has demonstrated a shift in breeding populations in Britain with fewer in northern parts and more in the south and east. Declines in northern colonies between 1992–2003 have been attributed at least in part to predation by American Mink, especially in colonies in Scotland (Craik 1997; Defra 2015).

In winter, there were huge numbers of Black-headed Gulls in the British Isles in the 1970s and 1980s, increasing to an estimate in the order of three million in 1986. Ringing has shown the majority of these are from Scandinavia, the Baltic States and European Russia, perhaps involving over 75% of the entire breeding populations at that time. The winter population has steadily declined since then, to around 2.2 million in the mid-2000s. A range of factors may be involved in this decline including the loss of winterfeeding sites (closed and closing landfills, cleaned-up sewage outfalls and enclosed sewage farms), short-stopping on the continent as a result of milder winter conditions, and lower overall European population levels.

Here it is a common winter visitor with numbers steadily building up from August, with a particular increase usually in November and maximum numbers in January and February, except in freezing

Jonathan Cartwright, Venus Pool, 30 May 2010

conditions. Dispersal back towards breeding grounds occurs from late February and early March. Farmland is used widely for daytime feeding. Landfill and urban sites are also extensively used. The gulls are adaptable in their feeding habits and quick to exploit opportunities around farm buildings, towns and villages.

The majority use the larger lakes for roosting, with the Mere (Ellesmere), Cole Mere, White Mere, Chelmarsh and Priorslee Lake being the main sites. The relative abundance map shows most feed in the day in close proximity to these roost sites. The south-west has virtually no suitable roost sites.

Winter counts have been sporadic at these roosts over the years, and the same birds will have moved between the Mere (Ellesmere) and the other meres in the Ellesmere area, making it difficult to estimate detailed local trends. The most comprehensive set of counts since the 1950s has been from the Mere (Ellesmere) and winter numbers here not unexpectedly follow a similar pattern to those across the UK. By 1965 as many as 12,000 roosted at the Mere (Ellesmere) with similar numbers in many years right through until the 1980s. The largest ever winter count was 20,000 at White Mere in February 1976. After a maximum count of about 7,000 at the Mere (Ellesmere) in 1984, numbers declined in the late 1980s and 1990s with counts of 5,000 or more in 1989, 1998, 1999 and 2005, but generally rather fewer. Similar patterns were found at Cole Mere, with maxima of 10,000 in 1982, 9,000 in 1983 and 5,000 in 1997, but no more than 2,000 since.

The pattern of these local counts is reflected in the results of the 10-yearly census organised by BTO, which, starting in 1963 and finishing in 2003, found 8,000, 19,000, 11,500, 5,500 and 3,540 respectively at the Mere (Ellesmere), and 9,000 in 1983, 2,860 in 1993 but only 100 in 2003, at Cole Mere.

Winter maximum counts at Chelmarsh too were up to 8–10,000 in the early 1980s, but declined in the next decade and have since

Winter Relative Abundance (2007–13)

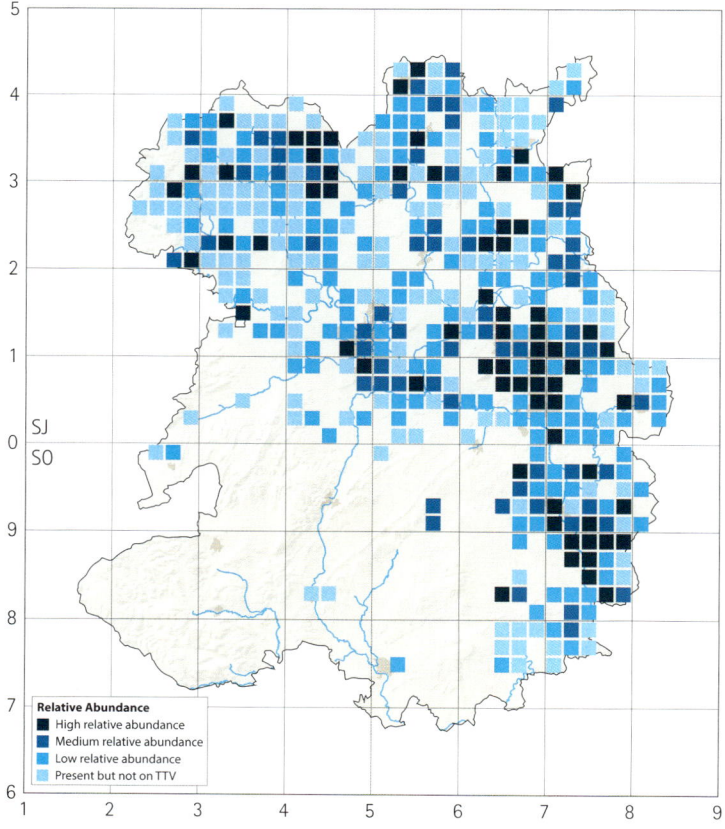

Relative Abundance
- High relative abundance
- Medium relative abundance
- Low relative abundance
- Present but not on TTV

only once (in 2004) exceeded a peak of 3,000. The BTO census found 2,300 in 1983 and 1,915 in 1993.

The local counts and the BTO censuses both suggest a considerable local decline, and the estimated total population based on TTV counts during the recent Winter Atlas was just over 12,000. The BTO census found almost twice that (23,000) at only four sites in 1983.

Black-headed Gulls ringed here as nestlings have predominantly moved south in the UK for the winter months but also to Ireland where 15 have been recovered, with at least four of those found in winter. Including these, the 127 recoveries in the British Isles that were ringed here and found elsewhere, or vice versa, involved 51 different counties, demonstrating their mobility.

Ringing returns have shown a continental origin for 44 recoveries found here between November and February. Nestlings from the Baltic States, Scandinavia and Poland account for 30, while the Netherlands, France and Germany account for 14. Eight of the former had travelled over 1,500km. At least some appear to return winter after winter. A nestling ringed in Lithuania in 1987 was recorded at the Ellesmere roost in 1988 and again in 1992, while one ringed as an adult in Norway (and colour-ringed) was reported from Ellesmere in three successive winters from 2006.

The Black-headed Gull has managed to thrive, and despite suffering heavy predation, including in the past by humans, it has been able to succeed over the last half-century in a landscape hugely changed by man. The recent decline in the wintering population may herald challenging times ahead in the face of further changes, to the climate, more efficient waste disposal and agricultural intensification.

Gerry Thomas

Little Gull
Hydrocoloeus minutus
Rare passage migrant

Jim Almond, Mire Lake, 16 March 2008

The core breeding range of Little Gull is within northern Russia, Finland and the Baltic States. It winters mainly offshore as far south as the Mediterranean and North Africa.

Rocke listed a single record of one in winter plumage obtained at Coalbrookdale (no year given) while Forrest (1899) listed three at Atcham in 1874, and (in the Addenda) a pair in full summer plumage on the Mere (Ellesmere) on 5 April 1899.

There were only two records in the first half of the twentieth century, both in 1939. One was found exhausted at Ellesmere on 6 January, taken indoors and fed for a few days, then released. The other was found dead, apparently oiled, on 30 September in the Wyre Forest.

The *Handlist* (1964) described it as a scarce passage migrant with about seven records between 1958–63 from Cressage, Shrewsbury Sewage Farm, Allscott Sugar Factory and Ellesmere.

Since 1975 Little Gull has been recorded with much greater frequency, annually between 1983–99 and again from 2002–12. Autumn passage is usually heavier than spring although both are highly variable. Spring passage goes unrecorded in some years, but a total of 10 individuals were noted in spring 1994, while a strong passage at Venus Pool between 11 April and 16 May 1990 peaked at four on 1 May and three on 3 May. A flock of seven at the Mere (Ellesmere) on 8 April 1996 represents the record day count. A party of six adults at Allscott Sugar Factory on 17 April 2009 and one at Grindley Brook on 8 May 2009 were the only records during the recent breeding Atlas period.

Mid-summer records are very rare but have included one on the Severn near Chelmarsh on 21 June 1975 and one at Venus Pool on the same date in 2002.

Autumn migrants can arrive anytime from mid-July until late November with the majority normally noted in August and September. A flock of five at Allscott Sugar Factory on 10 November 2002 constitutes the highest autumn day total. A juvenile lingered at Priorslee Lake from 9–23 September 1991.

Mid-winter records are infrequent with most identified in traditional gull roosts such as Chelmarsh and Priorslee Lake. Five second-winter birds at Priorslee Lake on 3 January 2002 constitutes the highest winter count.

There were four records during the recent winter Atlas period; three were from Chelmarsh, singles on 6 January and 10 December 2009, and two on 12 November 2011, plus a single adult at Priorslee Lake on 10 December 2009.

Richard Moores

Mediterranean Gull
Ichthyaetus melanocephalus
Rare passage migrant, very rare breeding species

John Hawkins, Wood Lane, 31 May 2013

Throughout the twentieth century, the Mediterranean Gull has been spreading west from its core breeding area around the Black Sea into Western Europe. It first bred in Britain, in Hampshire, in 1968 and has been slowly establishing breeding pairs and small colonies throughout England since then.

The first record here was of a first-winter bird in the gull roost at Chelmarsh on 21 December 1982. The following year at least six different individuals were found including three together in the gull roost at the Mere (Ellesmere) on 23 January. Apart from blank years in 1986, 1987 and 1993, it has occurred annually since then, with the majority being identified during late February and March, with the Mere (Ellesmere) being a particularly favoured site, especially during the 1990s when three were seen on single dates in March in 1994, 1995 and 1996. While small numbers can be found in gull roosts in early and mid-winter, there are very few records in late summer and autumn. Apart from the Ellesmere meres and Wood Lane, there are only three records of two together, at Whixall Moss on 21 April 2010 and 1 May 2013, and at Venus Pool on 18 April 2014, when both were second years.

The recent Winter Atlas map shows sightings in only five tetrads, the Mere (Ellesmere), Wood Lane, two in eastern Telford (Priorslee Lake and Priorslee Flash) and Venus Pool.

The first indication of this gull trying to establish a breeding foothold came in 2003 when an adult male in breeding plumage appeared to pair up with a female Black-headed Gull at Whixall Moss NNR from 10 May until 1 June. The outcome of any breeding attempt is not known, but no potential hybrids were reported in the area subsequently. No further breeding attempts were recorded until 2013, when Mediterranean Gull finally became an addition to the breeding avifauna when a pair nested successfully at Wood Lane,

raising three young. Although two returned in early March 2014, they were last seen on 7 April and no breeding behaviour was witnessed.

There were several other sightings in 2014, reflecting the increasing occurrence of this species: one at Priorslee Lake on 12 February, two or three at the Mere (Ellesmere) on 8 and 20 February and 1 March, one at Venus Pool on nine days between 4 and 18 April (two on the last date only), and one at Chelmarsh on 18 August, but there was no evidence of any breeding attempts. There was a first-winter individual at the Mere (Ellesmere) and White Mere on three dates from 20 December.

The breeding population in England is still increasing, and it has spread to reservoirs in Staffordshire, so it may well become a regular breeder in the near future. However, inland breeding occurs in Black-headed Gull colonies, so the potential number of sites where it may breed here are very limited.

Adults range widely, evidenced by an adult female, colour-ringed in Poland on 22 May 2005, which was seen at Gailey Reservoir in Staffordshire 18 months later, then again at Priorslee Lake on 4 February 2012, 1,362km from the ringing site, 6y 8m 13d later.

Richard Moores

Occupied Tetrads

Atlas period (breeding)	1985–90		2008–13		Change	
	Number	%	Number	%	Number	%
Confirmed	0	0	1	0	1	x
Probable	0	0	0	0	0	x
Possible	0	0	0	0	0	x
TOTAL	0	0	1	0	1	x
Tetrads with Winter Records (2007–13): 5 (1%)						

Common Gull (Mew Gull)

Larus canus

Uncommon winter visitor; has bred

The Common Gull breeds in north-west Europe and across Russia to Alaska and north-west Canada. Scotland and the west of Ireland are the strongholds of the British Isles breeding population, with a few breeding in England and Wales. There is a large influx from the continent in winter.

Rocke described it as being occasionally encountered but not as commonly as Kittiwake. It was still uncommon in Forrest's time and occurred less frequently than Herring Gull, Lesser Black-backed Gull 'and other species'. Immatures in autumn were encountered most often, and the Severn near Cressage was apparently the most frequently graced site. There were then one or two records of one or two individuals in five of the first 13 years of the twentieth century, when it was described as 'far from common', after which there were records of two killed in frost at Bridgnorth at the end of February 1929 and 'a lot flying up and down the Severn at Atcham' between 1 and 14 April 1932.

There was a gap of over 20 years, after which it became a less rare. The *Handlist* (1964) described it as a non-breeding visitor and passage migrant with 'numbers increasing over the past 10 years', and with most present between September and April. Up to 1,000 roosted on the Mere (Ellesmere) in December 1963. The first of the autumn were often noted on Long Mountain in late June and July.

The north-west is the stronghold and numbers generally build up from July, reaching a peak in December. Dedicated roost counts at the Mere (Ellesmere) in 1964–65 showed that numbers built up from a singleton in August to 3,075 in December, when they comprised 20%

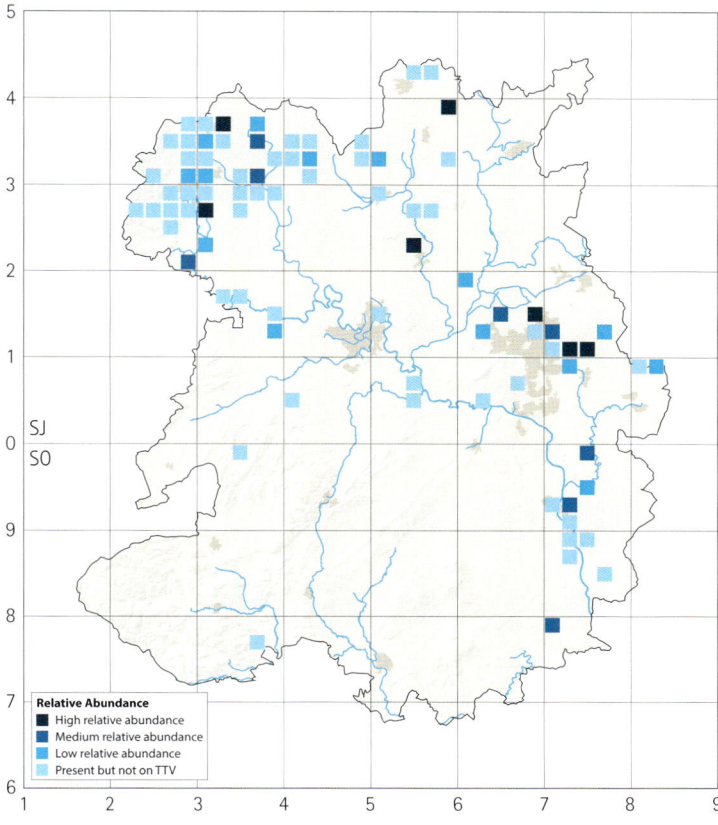

Winter Relative Abundance (2007–13)

Relative Abundance
- High relative abundance
- Medium relative abundance
- Low relative abundance
- Present but not on TTV

Occupied Tetrads

Tetrads with Winter Records (2007–13): 83 (10%)

of the total gull count, before declining rapidly through January–February and only forming 1% of the total roost in March. They have been recorded in every month although few are usually seen from April–June. Peak counts in the Mere (Ellesmere) roost vary widely each year, from a few hundred to a peak of 5,000 on 2 March 1993. However, January counts for the national BTO Gull roost survey undertaken approximately every 10 years between 1963 and 2004 found 2,000, 1,000, 600, 1,200 and 1,800 respectively.

It is much less common further south, and the total in the gull roost at Chelmarsh rarely reaches double figures, with a maximum of 48 on 4 April 1992.

Comparison of the recent Winter Atlas map with the relevant part of the national 1981–84 Winter Atlas suggests little change in range over the last 30 years – this gull was present in 17 of the County's 10km squares in the earlier period and 18 in recent Atlas work. However, it was not found recently in three squares in the north were it had been found earlier, and there have been gains around Telford and Chelmarsh.

Recent Winter Atlas fieldwork found most in the north-west with

Jim Almond, Venus Pool, 22 October 2011

fewer along the Severn Valley and in the Telford area. The south-west is largely unknown territory for this species with just a single record in 2007–13. The highest relative abundance (six counts of over 20 in a one-hour TTV) was found in short grass or recently ploughed fields near the main roost sites. Eight roost counts in the same period at the Mere (Ellesmere) exceeded 50, the highest being 500 on 29 December 2007, 300 six days earlier, and 180 on 22 December 2011. Apart from several at the Ellesmere meres and Wood Lane, no other roost count exceeded 20.

Nationally, WeBS has found a decline of almost 50% between 1999–2000 and 2013–14 in the wintering population, reflecting a reduction in the number of these gulls coming to Britain. The smaller maximum roost counts here since the 1990s are consistent with this decline, exacerbated by withdrawal from the edge of the winter range.

The only ringing recovery provides evidence of the influx of winter visitors. One ringed in Norway in July 1994 was seen alive at the Mere (Ellesmere) more than five years later, 882km distant.

Common Gull has nested only once, in 1976. A pair laid eggs at Cranmere Bog in among the Black-headed Gull colony. The nest was deserted on 9 June and the eggs were eaten by a Moorhen. One adult remained until 18 June.

Richard Moores

Ring-billed Gull

Larus delawarensis

Vagrant

Breeding in the northern United States and Canada and wintering in the southern States, Central America and the West Indies, this gull is a scarce migrant to Britain.

There have been just seven records, all in the winter/early spring period:

Chelmarsh	28 November 1992
The Mere (Ellesmere)	25–26 March 1995
Chelmarsh	12 December 2003
Candles Landfill Site, Horsehay, Telford	23 February 2004
The Mere (Ellesmere)	28 December 2007
Priorslee Lake	29 December 2009
Chelmarsh	29 November 2016

Typically, it is found either on an open-water roost or at a landfill site. The first two at Chelmarsh were identified as first-winter birds, while the third was a second winter. The remainder were considered to be adults, although the second one at Ellesmere may have been in its third winter. The adult at Ellesmere in 1995 was particularly obliging in that it came to bread!

Graham Walker

Great Black-backed Gull

Larus marinus

Scarce winter visitor

In the British Isles, the Great Black-backed Gull nests mainly on northern and western coasts, and is the most maritime and least common of the large gulls. The winter population is supplemented by visitors from Scandinavia and perhaps Iceland.

Rocke speculated that the majority of individuals found here occurred following storms on the Welsh coast, with immature birds predominating. Forrest (1899) described it as a 'rare wanderer' and lists just three records; one obtained on the Severn near Cound in November 1861 and two records near Knockin, in 1891 and subsequently. There were records in nine years in the first quarter of the twentieth century, the last in 1924 when two were seen near Bridgnorth on 13 September.

The *Handlist* (1964) described it as being a non-breeding visitor in small numbers with most records from Ellesmere and Shavington between November and March, with other records from Venus Pool, Shrewsbury Sewage Farm and Trench Pool.

It was still scarce in the 1970s and early 1980s, with 1972 and 1973 producing no records at all. Counts at several waters for the national BTO Gull-roost survey conducted in January every 10 years or so between 1963 and 2004 found no more than three, except for 14 at Chelmarsh in 1983. The maximum annual counts from the gull

Jim Almond, Priorslee Flash, 5 December 2011

Maximum Annual Count at Chelmarsh Roost (1976–2014)

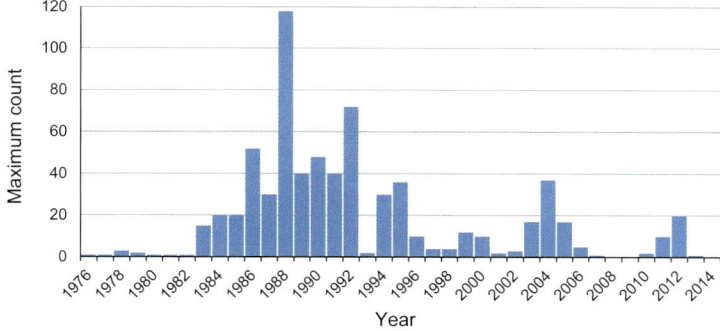

Winter Distribution (2007–13)

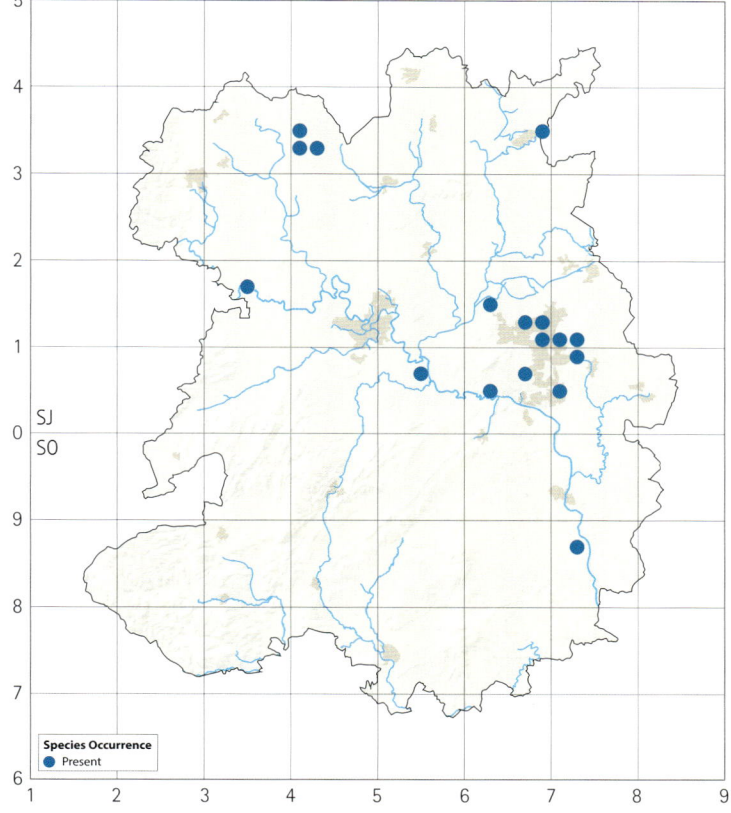

Occupied Tetrads

Tetrads with Winter Records (2007–13):   17   (2%)

roost at Chelmarsh, shown in the chart, highlight the upturn, the highest being 118 in January 1988.

Numbers quickly started to reduce from that peak, perhaps reflecting the decline in the national breeding population (38% between 1986–2012) and by the mid-2000s the roost at Chelmarsh usually comprised single figures, with an occasional higher count, for example, 37 on 17 January 2004 and 20 on 7 December 2012. The only other sites to ever host a count of 100 or more were Cosford Airfield where 105 were roosting on the runway on 2 November 2002, Candles Landfill site, about 100 on 3 January 2013, and Wood Lane, 100 on 23 February 2013.

The Winter Atlas distribution map shows records from 17 tetrads and is typical of recent years with most records from well-watched water bodies, particularly at roosts. There was a concentration of records from the Telford area, particularly Priorslee Lake, but also several from Candles Landfill, Priorslee Flash, Trench Pool and Granville Tip, and occasional records from other less frequented sites, including Leighton, Horsehay and Long Lane. Other records were received from Chelmarsh, Cae Howell, Venus Pool, Almington, the Ellesmere meres and Wood Lane. There were 19 double figure counts. The largest are listed above, and all the others were at Telford sites, except 15 at Venus Pool on 1 January 2009.

In the recent Atlas breeding season there were only four records, including two from Wood Lane and one from Priorslee Lake, but there was no evidence of breeding.

Great Black-backed Gull has always been more common in winter but it may be recorded throughout the year, although usually in very low numbers, and it is occasionally absent in spring and early summer. In 2014 it was reported in seven months of the year, January–March, August, and October–December, almost always from Priorslee Lake. There were only two double figure counts, 11 on 24 December and 27+ on 26 December.

Richard Moores

Glaucous Gull

Larus hyperboreus

Rare winter visitor

Glaucous Gull breeds from the high Arctic to subarctic coasts including northern Iceland, Novaya Zemyla, Spitsbergen and northern Russia. The increasing Iceland population is predominantly resident while other European populations winter as far north as weather conditions allow. The majority of those that reach the British Isles each winter stay in the north of Scotland with comparatively few reaching southern England.

Rocke (1866a) and Forrest (1899) detailed two immature specimens, one obtained at Pradoe in November 1856 that was in the act of feeding upon a dead sheep, and another at Bomere Pool in December 1863.

There were no records in the *Handlist* (1964), and the first modern record was from Cole Mere in November 1969. In common with many large gull species, it has become more frequently recorded and near-annual since 1982, with over half of the 40 records coming from 1999 onwards. Of these more recent records, all have occurred in the period between mid-December and March apart from three November records and a third-summer bird at the Mere (Ellesmere) in April 1996. The chart shows the number recorded in each winter period since 1950–51.

All records have comprised singletons. Initially most were from either gull roosts, with the Mere (Ellesmere) and Chelmarsh being particularly favoured, or from landfill sites and refuse tips, but Priorslee Lake and other Telford sites became the main locations from December 2003 onwards, contributing all except three of the 14 records since then. The gulls feed mainly on the landfill sites at Granville and Candles, and loaf, wash, form pre-roost gatherings and sometimes roost in varying numbers on all of the pools, but they wander widely, here and in adjacent counties.

During recent Winter Atlas fieldwork there were records of 12 individuals in 11 tetrads, at least one in each winter season. Four were

Winter Distribution (2007–13)

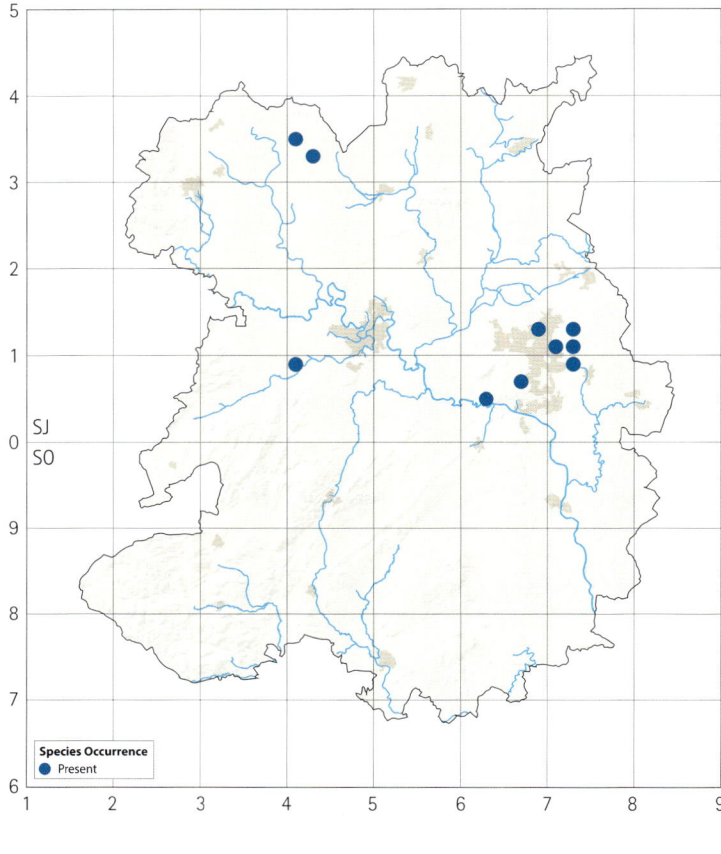

Occupied Tetrads

Tetrads with Winter Records (2007–13):   10   (1%)

Jim Almond, Priorslee Lake, 8 February 2012

seen on one day only, but all except one of the remainder stayed for at least two weeks. Over four winters, nine individuals wandered widely in the Telford area, and some that could be identified individually from age and plumage differences were seen in up to three tetrads, for example the longest stayer, a juvenile, commuted between Priorslee Lake and Flash, and Granville Landfill, between 7 January–14 March 2009, while another juvenile was found at Candles Landfill site on 3 January 2013 and remained in the area until 22 February, and was also seen at Priorslee Lake, on the River Severn floods near Leighton and distantly from Venus Pool (in flight, and therefore not on the Atlas map). Altogether, six Telford tetrads were visited (the others being Horsehay and Trench Pool). The map shows nine tetrads altogether, the three others being Ellesmere, Wood Lane and Polemere.

One at Candles Landfill and Priorslee Lake on 28 March 2013 was the last record before the data cut-off.

A Glaucous x Herring Gull hybrid, known as 'Viking Gull', commuted between Barnsley Tip and Chelmarsh roost in late February 1984, and the five seen in 2013 are thought to have included another of these hybrids, and a Glaucous x Great Black-backed hybrid, both of which were seen at several sites in the Telford area.

Richard Moores

Iceland Gull

Larus glaucoides

Rare winter visitor

The nominate race of Iceland Gull *L.g. glaucoides*, which visits the British Isles in winter, breeds in Greenland. A distinct population in the eastern half of this island moves south-east in the winter, predominantly to Iceland but with varying numbers reaching Britain, mainly in the west. Compared with Glaucous Gull, this species used to be found less frequently inland, but it is now more numerous, with almost twice as many records here in total (73:40). Both these large

gulls do not reach full adult plumage until they are four years old, so they can usually be aged in the field, and some have a distinctive appearance that allows them to be identified individually.

Iceland Gull was first recorded on 29 January 1982, at Ellesmere. The winters of 1982–83 and to a lesser extent 1983–84 saw exceptional influxes into Britain, and, remarkably, 1983 saw four different individuals recorded, an adult at Chelmarsh on 6 January, second-winter

Winter Distribution (2007–13)

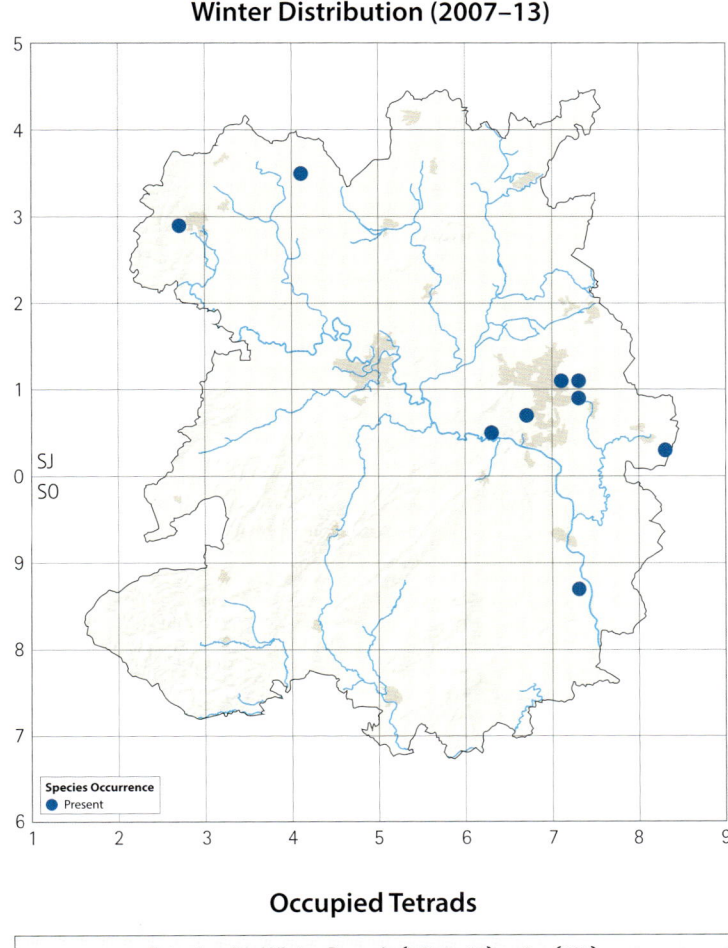

Occupied Tetrads

Tetrads with Winter Records (2007–13): 9 (1%)

Estimated Winter Period Totals (1950–51 to 2014)

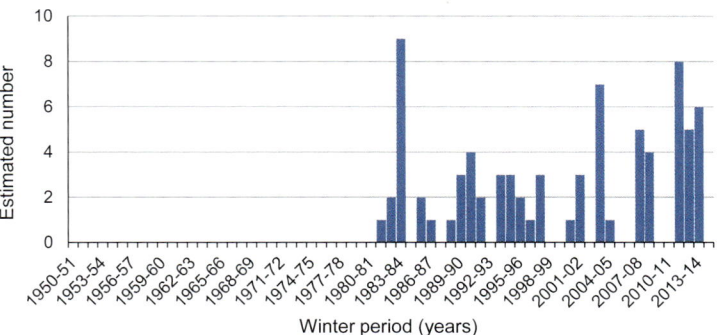

There were several records of some of the more mobile birds, and the chart shows the estimated number of different individuals in each winter period.

Altogether, more than half the records have been observations on a single date only, but 22 stayed for two weeks or more, including eight that stayed more than four weeks, and the longest was seen regularly at the roost over a nine-week period at the Mere (Ellesmere) between 4 January–10 March 2002. The only record of two together was at Chelmarsh on 28–29 December 1989, although there were up to three in the Telford area at the same time in 2009 and again in 2014. All records were in the four winter months, apart from six in March, and two in April, at Chelmarsh on 8 April 1991 and White Mere on 2 April 1995, although two long-staying individuals in 2001 and 2014 at the Mere (Ellesmere) and Wood Lane were last seen on 1 and 5 April respectively.

'Kumlien's Gull' *L.g. kumlieni*, which breeds predominantly on Baffin Island, Canada, is currently treated as a sub-species of Iceland Gull though its taxonomic position remains uncertain. The first records were accepted in 2009, when an adult roosted at Chelmarsh on 8 January and a third-winter bird was at Priorslee Lake on 20 February. Both are included in the species records listed above.

Richard Moores

birds at the Mere (Ellesmere) in both late January and December, and a third-winter bird at Shrewsbury tip also in December.

In the following year, 1984, there were 12 records of an estimated seven different individuals, all in the period January–March. Following a blank year in 1985 three were recorded in 1986, and from 1989–98 it appeared annually in small numbers. After a short gap, there were 13 records from three well-watched roost sites (Priorslee Lake, the Mere [Ellesmere] and Wood Lane, and Chelmarsh) in the four years 2001–04. During the daytime they could often be found at tips such as Candles and Granville landfill sites and, formerly, sites such as Barnsley, Prees and Shrewsbury tips.

There were no records after 2004 until the recent Winter Atlas period, when it was found in nine tetrads, comprising five in the Telford area, and the Mere (Ellesmere), Wood Lane, Oerley Reservoir and Chelmarsh. Almost all records were of single birds, although there were up to three present at the two Priorslee sites between 8 January and 4 March 2009. Some individuals were seen to commute between several of the Telford sites, Granville Landfill, Priorslee Lake, Priorslee Flash, Horsehay pool, Candles Landfill site and Buildwas.

In 2014, there were five records, all in the early part of the year between 8 February and 5 April, at the Mere (Ellesmere) and at several sites in the Telford area. They were all either juvenile or second-winter birds, and were very mobile, spending time in other counties. Some were known to roost at Belvide in Staffordshire.

Jim Almond, Priorslee Lake, 10 February 2012

Herring Gull (European Herring Gull)
Larus argentatus
Uncommon winter visitor

Jim Almond, Horsehay Pool, 16 January 2013

Abundant in northern and north-western Europe, the Herring Gull is a common breeder around Britain's coastline. In winter, most remain within the British Isles but many disperse inland, and it is widely distributed throughout the country, forming concentrations at suitable feeding and roosting sites such as landfill sites and reservoirs. The indigenous population of the sub-species *L.a. argenteus* is supplemented at this time by the Scandinavian sub-species *L.a. argentatus*, the latter making up some 30% of the total in the north and east of the country, but perhaps only 5% elsewhere. Many Scandinavian Herring Gulls are distinguishable from *argenteus* in all plumages but few records detail sub-species, and they are almost certainly under-recorded.

Towards the end of the nineteenth century, Beckwith, Paddock and Forrest all considered it by far the most frequent of the large gulls, often seen in winter during stormy weather, or on passage, flying generally south-east in autumn, and north-west in spring. The *Handlist* (1964) described it as a non-breeding visitor and passage migrant, migrating along the Severn and through Ellesmere in spring and autumn. The wintering population in the Ellesmere area increased rapidly during the middle of the twentieth century, from 50–200 in 1952–56 to a peak of 2,000 in December 1963.

In contrast to the previous species, the fortunes of the Herring Gull seem to have changed little over the last 50 years, and it is only due to the increase in the number of Lesser Black-backed Gulls that it can no longer be considered the more common. Wintering numbers have remained in the region of 2,000, but the centre of the distribution has shifted south-east and is now concentrated around Telford. This is reflected in the counts for the national BTO Gull-roost survey, carried out in January every 10 years or so between 1963 and 2004. At Ellesmere, the counts have been 2,000, 500, 1,300, 40 and 160 respectively.

The recent Winter Atlas recorded it in 90 widespread tetrads, with a majority of those with medium or high relative abundance (13 out of 25) around Telford.

The landfill sites at Candles and Granville, and their associated loafing/bathing pools at Horsehay and Priorslee, are among the hotspots for large gulls. Peak counts in recent years have included 1,000 at Granville landfill in February 2008, 800 at Candles landfill in December 2012 and 1,000 in December 2013, and 500 at Priorslee Lake and 600 at Horsehay Pool, both in January 2013. Detailed or co-ordinated counts have not been made at the landfill sites however, and Atlas TTVs excluded dawn or dusk counts at roost sites, so the actual totals may be much higher. For example, the Chelmarsh tetrad shows only a low relative abundance, and yet may include the largest regular roost, with sample peaks of 310 in December 2010 and 250 in December 2012. The large numbers feeding in the Telford area predominantly roost at Belvide Reservoir in neighbouring Staffordshire.

The Ellesmere area remains important, with regular counts of over 250 roosting on the Mere during October to February. Indeed the recent highest total was recorded there in January 2003 when an exceptional 1,500 were counted in the roost. Four hundred roosted at nearby Cole Mere in January 1997 and 400 were at Wood Lane

Winter Relative Abundance (2007–13)

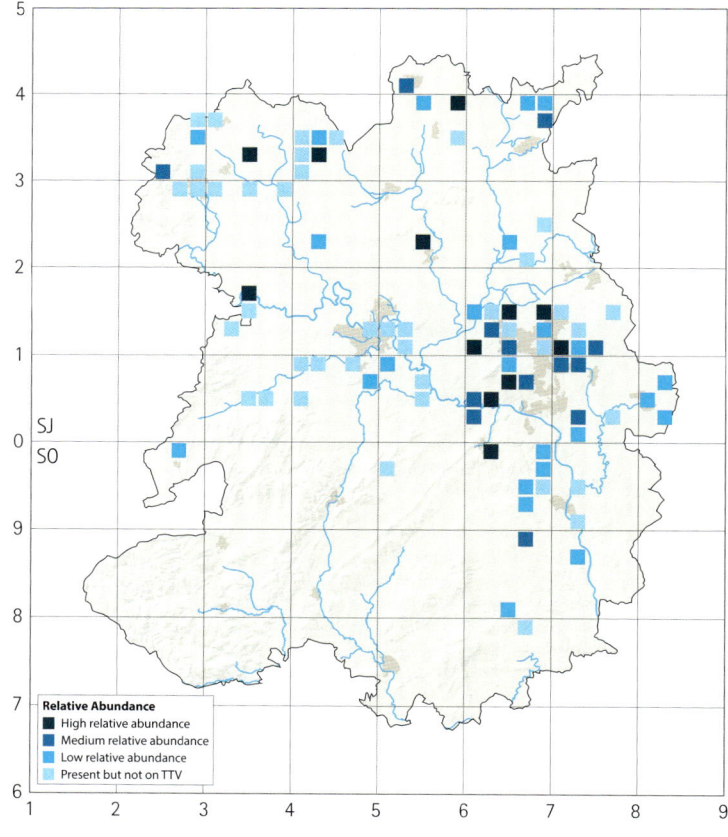

Relative Abundance
- High relative abundance
- Medium relative abundance
- Low relative abundance
- Present but not on TTV

Occupied Tetrads

Tetrads with Winter Records (2007–13):   90   (10%)

in February 2010. The Shrewsbury area and the south-west are little used, reflecting the lack of feeding and roosting opportunities.

Thirty-six have been recovered or re-sighted here, 34 ringed in Britain and Ireland, and two in Norway. Of the former, the vast majority, 22, were ringed in Gloucestershire, five in Anglesey, three in Buckinghamshire and singles in Avon, Worcestershire, North Yorkshire and County Down. One of the Anglesey-ringed birds represents the oldest known here, found dead in Shrewsbury in January 1975, 9y 6m 12d after ringing, but this falls well short of the oldest known Herring Gull, at nearly 33 years. One from Norway originated 893km to the north-east (re-sighted at Priorslee Lake in December 2009, nearly

seven years after ringing), and the other from 1,986km away (ring found at Ellesmere in March 1997 only seven months after ringing).

No evidence of breeding has been recorded, although very small numbers (usually single figures) are present throughout the summer months. Counts of over 20 in Telford in May 2013 at a feeding site at Redhill and a loafing site at Priorslee Lake are noteworthy. Herring Gull may one day follow in the footsteps of Lesser Black-backed Gull and join the breeding avifauna, as rooftop-nesters have been recorded in adjacent Worcestershire annually since 1999 and in Birmingham since 2005.

Tom Lowe

Caspian Gull
Larus cachinnans
Rare winter visitor

Caspian Gull breeds around the coasts of the Black and Caspian Seas, east into Kazakhstan, and increasingly north-west from there into Poland and eastern Germany. As its population and range has expanded, post-breeding dispersal has resulted in some wintering around the southern North Sea and the United Kingdom, and the first British record concerned a first-winter at Radipole Lake in Dorset in December 1992 (Hudson *et al.* 2014). Since then there has been a rapid increase in records both nationally and locally, as a result of an increase in the European population along with improved observer awareness. In 2007 Caspian Gull was granted full species status; it had previously been considered a sub-species of Herring Gull.

The first local record involved an adult photographed at Priorslee Lake on 28–29 December 2006. Another, a fourth-winter, was seen at Granville landfill and Priorslee Lake in early 2007. After a blank year in 2008, three were recorded in 2009, all adult/near-adults, and all at Priorslee Lake: one in January and two in December. The following year saw the first record of one in an immature plumage, when a second-winter was photographed at Priorslee Lake on 4–5 January 2010. An adult initially seen at Priorslee Lake on 5–8 January 2010 was then seen in the roost at Chelmarsh Reservoir on 9 and 14 January, the first record away from Telford. The third record of 2010 was also from Chelmarsh, an adult in the roost on 4 November. Since then, numbers have increased: up to eight in 2011, 10 in 2012, 22 in 2013, and at least 16 in 2014.

During the recent winter Atlas period, it was found in the three gull hotspots of Granville landfill/Priorslee Lake, Candles landfill/Horsehay Pool/Buildwas, and Chelmarsh Reservoir, together with one at the Mere (Ellesmere) in February 2012. Subsequently, and away from the hotspots, records came from Burlton in October 2012, Dorrington, one with 40 Lesser Black-backed Gulls in November 2013, and the Mere (Ellesmere) in November 2014.

The majority of records relate to one, occasionally two individuals, but trios have been recorded on two occasions, at Candles landfill site in February 2013 and February 2014. The first record from this site was of a long-staying adult that remained from 30 November 2012 until 8 February 2013, which was followed by a remarkable run of records that continued until the demise and eventual closure of the site in early 2015. Gulls fed on the landfill during the day between

Jim Almond, Trench Pool, 3 February 2015

bathing and loafing sessions on nearby Horsehay Pool, the River Severn floods at Buildwas, or Priorslee Lake, and then roosted nightly at Chelmarsh Reservoir or at Belvide Reservoir in Staffordshire, where several recognisable individuals were seen. The replacement of landfill sites by refuse incinerators across the country has already resulted in an alteration in the distribution of all gulls, and this is reflected in the return of Caspian Gull annual totals to single figures in 2016–17.

Before 2013, all records came from the period November–March, with a notable concentration between late December–February. In common with much of the Midlands, winter is the traditional 'gulling' season, but increased efforts outside this period have revealed the species to be an autumn passage migrant as well. On 22 August 2013, one in full juvenile plumage was recorded at Candles landfill, and was joined, remarkably, by another a week later. These represent two of only a handful of records of this undoubtedly under-recorded plumage in the Midlands, and were followed by another juvenile at Candles in August 2014.

The central European gull colonies are the focus of large-scale colour-ringing projects, and a fourth-winter seen at Candles landfill on 1 February 2013 bore a yellow darvic ring, and had been ringed as a nestling in southern Poland in May 2009, a distance of 1,551km to the east-north-east. Subsequently, it was recorded in Leicestershire, in December 2013, and again in August and November 2014. A ringed adult at Candles landfill in December 2013 was too distant for the ring to be read, but the green colour indicated an origin on the German/Polish border.

Tom Lowe

Yellow-legged Gull
Larus michahellis
Scarce winter visitor

Yellow-legged Gull was added to the British List as a separate species from Herring Gull (*L. argentatus*) in 2005. The breeding population of Yellow-legged Gull is centred on the Mediterranean basin with smaller numbers on the Atlantic coast of Iberia, and in Switzerland, Austria and through France. Although the majority of the population is generally sedentary, increasing numbers, particularly of immatures, have been seen in northern France and southern Britain since the early 1980s.

The first records here were not until 1989, when singles were at Barnsley Tip on 1 February, and Ellesmere and Wood Lane on 29–30

Jim Almond, Priorslee Flash, 7 December 2011

December. Apart from a blank year in 1992, it has since been noted annually and in increasing numbers. The majority of records come from well-watched gull roosts such as Chelmarsh, Priorslee Lake and the Mere (Ellesmere) during the winter months. They depart by the end of March and are typically absent during spring before numbers begin to build again from July through the autumn and into winter. Counts are usually in single figures although 14 different individuals, including nine adults, were identified at Priorslee Lake on 28–29 December 2009, and 10 were at the same site on 6 January 2012. Other large parties have included 13 at Candles landfill site on 23 August 2013, where several double-figure counts have been made since 2011, seven at Granville Tip in January 2007, and eight at Chelmarsh Reservoir on 21 January 2004.

The Winter Atlas shows records in 13 tetrads: Oerley Reservoir, a cluster around Ellesmere (the Mere, White Mere and Wood Lane), Whixall Moss, six in the Telford area (Candles landfill site, Long Lane, Middle Pool Trench, Trench Pool, Priorslee Lake and Priorslee Flash), Stead Vallets and Chelmarsh. The records from the outlying sites suggest that this species might be overlooked on other waters.

In 2014 there were records in every month except April, May and June, featuring most of the same sites, but only three double figure counts, all in the Telford area.

Richard Moores

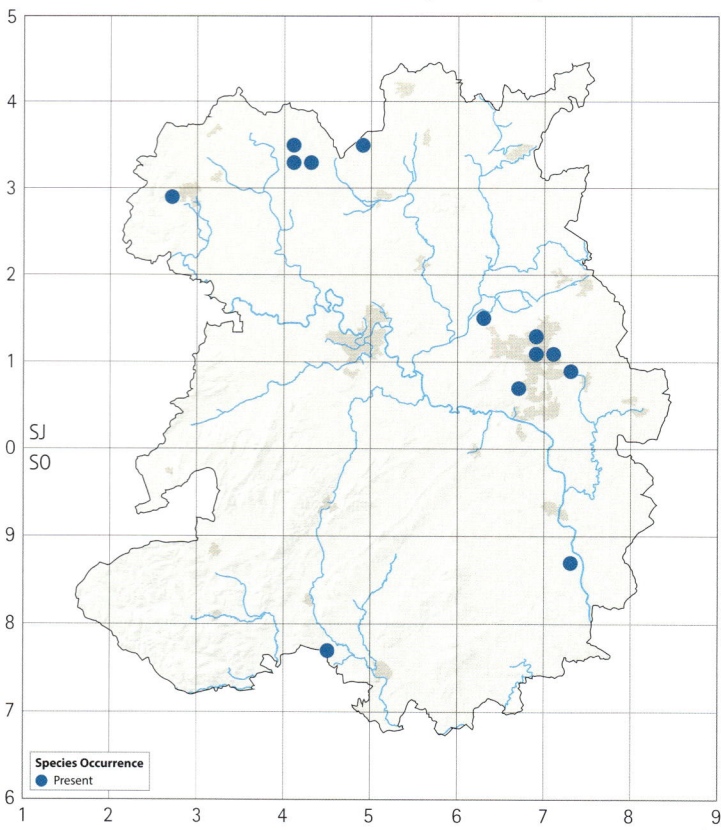

Winter Distribution (2007–13)

Species Occurrence
● Present

Occupied Tetrads

Tetrads with Winter Records (2007–13): 13 (1%)

Lesser Black-backed Gull

Larus fuscus

Fairly common winter visitor, rare breeding species

Lesser Black-backed Gull breeds only in northern and western Europe, from Iceland to the White Sea and south to the Iberian Peninsula. Northern populations are migratory. In the British Isles, it breeds at both coastal and inland sites, and its fortunes have fluctuated considerably in the last 150 years or so.

It was an uncommon visitor according to Rocke, with immatures most frequently recorded following storms. Forrest noted a flock of about 40 'seen by several people about Shrewsbury and Church Stretton' on 12 September 1898, a young one shot on the same date at Marton, and 'several others have been recorded from time to time in various parts', again suggesting that it did not occur frequently, but it was subsequently recorded in over 20 of the first 50 years of the twentieth century.

The *Handlist* (1964) described it as a non-breeding visitor and passage migrant with a fairly regular south to north movement along the Severn of generally small flocks of up to 40 between mid-February and early May. Autumn passage along the Severn was less distinct and occasionally unrecorded, although between 50 and 200 were regularly seen in the Market Drayton area between mid-July and late October. It was rare in winter with odd ones seen at Ellesmere and, 'formerly, small numbers at Trench Pool'.

The status of Lesser Black-backed Gull has changed dramatically over the last 50 years. It is now present throughout the year with the largest numbers typically in late autumn and winter, reflecting a large increase in the population in England and Wales, as only 165 individuals wintered in these countries in 1953. From 1980, numbers in the roost at Chelmarsh regularly topped four figures with a peak of 5,500 there in February 1990. Other regular high counts since 1992 have come from the Mere (Ellesmere) – with a maximum 4,000 on 17 August 1995 and again on 20 August 1998 – various tips and, more recently, Priorslee Lake (maximum 5,000 on 28 December 2005 and

again on 2 February 2012). The highest count was around 6,300, a flock flying over Newport to roost on 24 January 2006.

Since the 1990 peak, numbers at Chelmarsh appear to have remained fairly stable with, for example, 4,200 in the roost on 7 December 2012, and 13 counts of over 3,000 this century. In recent years, a large number of Lesser Black-backed Gulls spend the day in the Telford area, particularly at Candles landfill site (maximum 3,500 on 6 December 2012): they pre-roost at Priorslee Lake before departing to Belvide Reservoir (Staffordshire) to spend the night.

The Winter Atlas relative abundance map displays a lowland distribution throughout the east and north with few records from the hilly south-west, reflecting the difference in feeding and roosting

Winter Relative Abundance (2007–13)

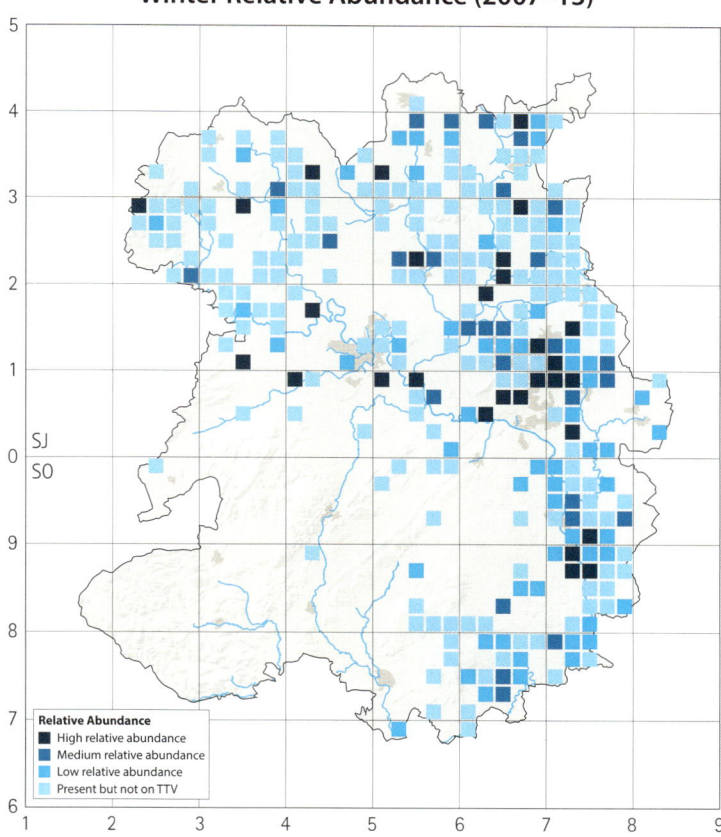

Relative Abundance
- High relative abundance
- Medium relative abundance
- Low relative abundance
- Present but not on TTV

Jim Almond, Shrewsbury, 21 May 2016

Occupied Tetrads

Atlas period (breeding)	1985–90		2008–13		Change	
	Number	%	Number	%	Number	%
Confirmed	0	0	2	0	2	x
Probable	0	0	0	0	0	x
Possible	0	0	0	0	0	x
TOTAL	0	0	2	0	2	x
Tetrads with Winter Records (2007–13): 290 (33%)						

Jim Almond, Venus Pool, 25 July 2010

opportunities between the two areas. TTVs avoided counts near dawn and dusk, so although some of the areas of high relative abundance may include roost sites, most of the gulls were counted in small dispersed flocks at feeding sites, which may be some distance from the roosts. There were only two one-hour TTV counts of more than 500, both at Chelmarsh, 800 in November 2007 and February 2008.

During the Winter Atlas period, there were 25 other counts from seven sites of over 2000, the maximum for each site being: Edgerley (4,000+ on 16 November 2008), Wood Lane (5,000 on 20 November 2009), Priorslee Lake (4,000 on 22 November 2011), Granville Tip (2,000 on 2 February 2008), Chelmarsh Reservoir (2,400 on 4 November 2010); Candles Landfill (2,000 on three dates in late 2012), and Horsehay (2,400 on 11 January 2013).

In the breeding season, there were counts of at least 50 on three one-hour TTVs, all in the Roden Valley near Shawbury. There were several other large counts in April and July, when winter flocks were dispersing and then building up again, but there were five counts of over 250 from three sites in May or June, the largest being 2,000 at Wood Lane on 25 June 2011, 500 at Candles landfill on 11 June 2013 and 275 at Chelmarsh on 23 June 2009.

Lesser Black-backed Gull recently became an addition to the breeding fauna with a pair successfully raising two young at Ludlow in 2012 (and probably one in 2010 as well), and subsequently several pairs nested on the Stadco factory in Battlefield, Shrewsbury in 2013, where four downy chicks from three broods, and a larger juvenile, were seen on 8 July.

In 2014, there were no records with breeding evidence, but it appears that the Shrewsbury site was not checked, and several pairs bred successfully there in 2015. Throughout 2014 there were 36 records of at least 200 from eight sites, but there were more than 1,000 at only two sites: Priorslee Lake (maximum 2,000+ on 16 December) and Shavington (1,250 on 8 November).

None have been ringed here but 14 ringed abroad have been found, from Channel Islands (five), Iceland (one, net distance travelled 1,698km), Norway (five, all over 800km) and the Netherlands (three, all almost 500km). The nestling from Iceland, colour-ringed in July 2000, was seen at Gloucester Landfill site in March 2003 before being seen at Priorslee Lake on 29 December 2009.

From within the UK, 443 have been found, mainly from Gloucestershire (337), Avon (12), Cumbria (49) and Lancashire (29), with a few from Scotland (four), Wales (seven), with five from two other English counties.

Colour-ringing produces additional evidence of the itinerant lifestyle. For example, a nestling ringed in Norway in June 2000 was caught near Gloucester: on 3 February 2007, seen here at Priorslee Lake on 28 December 2009 and four years later at Horsehay Pool, on 26 January 2013, 883km from the natal site, aged 12y 6m 28d, and subsequently, still alive, at Belvide Reservoir (Staffordshire) on 19 October 2013.

Three nestlings ringed in England and seen here have also wandered widely. One from Bristol in July 1996 was seen at Gloucester landfill site in January 2002, December 2009 and March 2011, with a sighting in-between from Belvide Reservoir in November 2003. It was still alive when seen at Horsehay Pool on 24 January 2014, 136km from the natal site, 16y 6m 19d later. Another, ringed in July 1997 on Flat Holm, in the Bristol Channel, was caught in Gloucestershire in 2005, seen alive seven years later at Priorslee Lake on 10 February 2012 age 14y 7m 7d, at Throckmorton Landfill Site (Hereford and Worcester) two days later, and then on Skomer Island, presumably breeding, three months later on 2 May 2012. The third, ringed on the Ribble Estuary in 1998 was seen near

Cannock in December 2007, and at Priorslee Lake on 13 December 2012 aged 14y 5m 16d.

The five oldest were all more than 14, but the oldest, just over 16y 6m, is less than half the age of the oldest known nationally, 34y 10m 27d.

The age of first breeding is usually four, and the summering flocks include many immatures, but given their numbers and range it is likely that the breeding population here will increase and spread.

Richard Moores

Gull-billed Tern

Gelochelidon nilotica
Vagrant

Although this tern has a global distribution, in Europe its breeding range is largely restricted to the warmer southern countries, mainly Spain, with only small isolated populations in northern Germany and Denmark. It is a summer visitor to Europe, migrating south to winter off coastal West Africa, and is rare in Britain, with no more than a handful of accepted records in any year. It is perhaps surprising, therefore, that two have graced local pools:

Venus Pool	19 May 1997
Cranmere Bog	11 June 2003

Unfortunately, both flew off within a few minutes of being found, never to return, but staying long enough for the two observers to note sufficient detail to secure acceptance of their records (Rogers *et al.* 1998; Rogers *et al.* 2004). Interestingly, both were found in association with active Black-headed Gull colonies.

Graham Walker

Sandwich Tern

Thalasseus sandvicensis
Very rare passage migrant

Jim Almond, Venus Pool, 6 May 2007

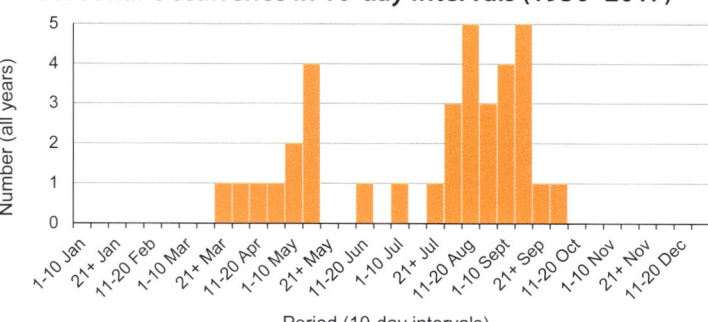

Seasonal Occurrence in 10-day Intervals (1950–2017)

Number (all years) / Period (10-day intervals)

In Britain and Ireland the Sandwich Tern has a scattered distribution, nesting mainly in densely packed colonies in coastal nature reserves. It is rare inland, and most winter on the coasts of West Africa.

Not mentioned prior to Forrest, a male found dead near Shrewsbury in August 1897 appears to be the first record. The next was not until 1963, when there were two, singles at Venus Pool on 15 April and

Ellesmere on 2 October, the former being the first modern record, and the latter being the latest ever. After a further gap, there were two at Chelmarsh on 10 August 1978. Since then, it has become more frequent, occurring in 13 of the 23 years from 1994, most recently in 2017. No year has produced more than two records. Autumn passage records are more numerous than spring records, and more individuals pass through then. There was only one spring passage record during the recent Atlas period, a single at Mirelake on 28 April 2010.

The majority of records involve single migrants at well-watched water bodies, with Chelmarsh, Priorslee Lake and Venus Pool being particularly favoured and accounting for just over half the records. Cound Fishery, Eyton Moor, Knighton Reservoir, Mirelake (three), the Mere (Ellesmere), Wall Farm and Wood Lane have provided the remainder.

Small groups comprising two to four birds are not irregular, with the largest flocks being 10 at Chelmarsh on 20 September 1979, and seven adults through Venus Pool on 12 September 2000. Stays of more than a day are rare, so a group of four, comprising three adults

and a juvenile, that roosted overnight at Priorslee Lake on 18–19 August 2000, and one that stayed at Wood Lane on 13 and 14 May 2004, are notable.

After one at Cound Fishery on 29 September 2010, there was a seven-year gap until one on 29 July 2017 on Priorslee Lake

Richard Moores

Little Tern

Sternula albifrons

Very rare passage migrant

The Little Tern nests colonially, predominantly along coastal shores, with the majority of the UK population found from Lincolnshire south to Hampshire. Most of the European population winters off the coasts of Cameroon and Guinea.

Rocke described it as a very rare visitor, while Forrest suggested it occurred less frequently than either Arctic or Common Tern, and listed a record of one obtained near Alberbury on 3 May 1898.

There were then sightings in nine of the years 1900–40, mostly from the River Severn, usually of a single bird on one day only, and only one year had more than one record. There was a 'party' at Betton Pool on 15 June 1904, in 1926 there was one on the Severn at Dowles on 1 June and another stayed for about a week at Acton Burnell in August, eight were on the Severn below the Mount in Shrewsbury on 25 August 1937, and one stayed for a week between the English and Welsh Bridges in Shrewsbury at the end of August 1940.

There was then a gap, and the *Handlist* (1964) described it as a rare passage migrant, with only two records since 1950, two at Shrewsbury Sewage Farm on 24 April 1958, and one at Ellesmere on 26 August 1962. Occurrences in 10 of the subsequent 54 years have

Jim Almond, Venus Pool, 11 May 2016

produced a further 14 records, three of two birds and the remainder all singles, mainly in spring: a total of 20 individuals in the modern era. The majority have been in the spring between 13 April and 14 June. In addition to the single at Ellesmere in 1962, the only other autumn record was of an immature at Chelmarsh on 22 August 1987. Only four records date from the twenty-first century, singles at Venus Pool on 26 April 2002, 11 May 2007 and 11 May 2016, while the one at Mirelake on 29 May 2010 was the only record in the recent Atlas fieldwork period. Of the other 11 spring records, five were from Venus Pool and three from Priorslee Lake, with the remainder from three separate sites, Monkmoor, Shrewsbury Sewage Farm and Wood Lane. The later spring records may be of immatures in no hurry to return to breeding areas.

Richard Moores

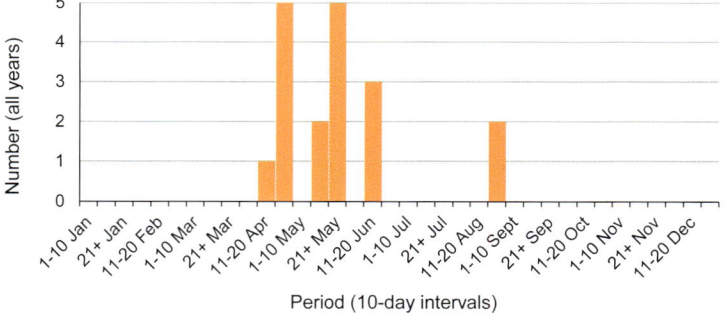

Seasonal Occurrence in 10-day Intervals (1950–2014)

Y-axis: Number (all years), 0 to 5

X-axis (Period, 10-day intervals): 1-10 Jan, 21+ Jan, 11-20 Feb, 1-10 Mar, 21+ Mar, 11-20 Apr, 1-10 May, 21+ May, 11-20 Jun, 1-10 Jul, 21+ Jul, 11-20 Aug, 1-10 Sept, 21+ Sep, 11-20 Oct, 1-10 Nov, 21+ Nov, 11-20 Dec

Roseate Tern

Sterna dougallii

No modern records

There are only two records of this most maritime of British terns, a summer visitor to western European coasts from west Africa. The first specimen was recorded by Eyton (1839) in his collection, killed 'three or four years ago' at Longden Mill on the River Tern. This was reported by subsequent writers as dated about 1830. The second report was

of an immature bird found dead (or obtained, usually meaning shot; the Caradoc *Record* and *Transactions* offer both) near Llanymynech on 21 September 1914.

John Tucker

Common Tern

Sterna hirundo

Scarce passage migrant, very rare breeding species

Common Tern is a summer visitor to central and northern Europe, and winters in western and southern Africa. In Britain it breeds at both coastal and inland sites.

The 'Sea Swallow' occurred regularly in the accounts of several nineteenth-century authors; for example, Forrest (1899) commented that 'this Tern is often seen on or near the Severn in autumn and spring. In the summer of 1898 one of these pretty birds stayed for more than a week ... below the English Bridge, Shrewsbury'.

There were records in most years in the first half of the twentieth century, and the *Handlist* (1964) described it as 'a passage migrant in small numbers, July to October. Most records are of one or two birds at Ellesmere, Crose Mere, Shrewsbury Sewage Farm, Allscott Sugar Factory and Ironbridge'.

The number of records each year has steadily increased, reflecting the growth and expansion of the inland population in England, as these terns take advantage of man-made water bodies, and provision of nesting platforms at some of them.

The *Atlas* (1992) listed it as an 'Annual Passage Migrant, usually in small numbers', but since then breeding has occurred for the first time on several West Midlands waters. This expansion of range has changed its status, as breeding adults are now seen on spring passage as well as autumn, while non-breeding immatures (the age of first breeding is usually three years old) might occasionally be seen foraging at any time during the season on local waters.

Since 1992, there have been no records between the beginning of November and the end of March. Spring migration builds up to a peak in the middle of May, and return passage peaks in the first 10 days of July and the first 10 days of August.

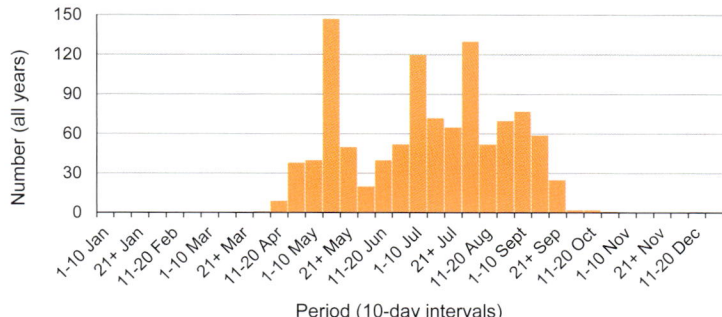

Seasonal Occurrence in 10-day Intervals (1950–2014)

Number (all years) / *Period (10-day intervals)*

There have been more than 10 records every year, with 33 or 34 in 1992, 1996 and 2001, and 29 in 2009. Over half of the 471 records (54%) have been of a single tern, and another quarter have been of two together. There have been only 10 records of 10 or more, the three largest being 64 at Mirelake on 5 August 1993, 16 at the same place on 31 August 2008, and 15 at Venus Pool on 14 May 2001.

During recent Atlas fieldwork it was seen on more than 10 occasions only at Chelmarsh Reservoir, Mirelake, Priorslee Lake, Trench Pool and Venus Pool, but it was seen at 12 sites altogether, the others being Allscott Sugar Factory, Chelmarsh Scrape, Cole Mere, Eyton Moor, the Mere (Ellesmere), White Mere and Wood Lane.

Confirmed breeding was first recorded in 2009, when a pair was present at Trench Pool from 22 June, and one was seen taking fish to his mate on the nest on 4 July. The pair mobbed and drove off a Lesser Black-backed Gull later the same day and incubation of two

Jim Almond, Chelmarsh Reservoir, 22 August 2014

eggs was seen on 7 and 8 July, but there was no sign of the nest the following day. No observations have suggested breeding there in subsequent years.

On 21 July 2011, two adults and two juveniles were seen at Chelmarsh Reservoir. The adults were feeding the juveniles, which were free flying and actively fishing at times. However, there was no evidence that the pair bred there, although five were seen on 14 June, 'two of which lingered'. Juveniles remain dependent upon their parents for two to three months after fledging.

Similar activity was observed in the late summer of both 2012 and 2013, with newly-arrived adults feeding at least one juvenile. They almost certainly bred relatively locally, although not at the Reservoir and probably in Staffordshire.

On 1 July 2014 a nest with two eggs was found at Chelmarsh Reservoir, on the pumping jetty in front of the sailing club, and there were three eggs on 14 July. All three hatched, but one chick was predated just prior to fledging; two fledged, but one was predated almost immediately. The remaining fledged juvenile was seen taking short flights on 11 August for the first time, mostly around the dam, and it was last seen with one adult more than two weeks later, on 29 August, the first successful breeding.

Breeding inland in the West Midlands has increased considerably in recent years, suggesting that breeding may become a regular occurrence here, although there are not many suitable nest sites unless rafts are provided.

Leo Smith

Arctic Tern

Sterna paradisaea

Rare passage migrant

The Arctic Tern has a circumpolar breeding distribution that extends into the boreal zone. In Britain it is most common in the Hebrides, Orkney and Shetland, although colonies are present along the British east coast down to Northumberland and patchily down the west coast to Anglesey, with a small number along the south coast. Most winter in the Antarctic.

Rocke described it as a very rare visitor but Forrest (1899) suggested it to be a frequent visitor with a notable event in May 1842 when 'immense numbers' appeared on the Severn in an exhausted condition. One shot at Kinnerley in April 1898 was the only specific record noted, and there were only four records in the first half of the twentieth century, including one shot on Brown Clee in August 1905, and one found dead at Ticklerton, and 'a party' crossing Corvedale, both in May 1917.

After a long gap, the *Handlist* (1964) described it as a rare passage migrant with only one record between 1950–64, a single at Venus Pool on 30 August 1960.

Arctic Tern was recorded very infrequently during the 1960s and 1970s, with the number of records slowly increasing during the 1980s. It has been seen every year since 1992, with 10 or more records in 1992, 1993, 2004 and 2013, the latter having the highest number with 18 records, all except one in April or May, and all except three of singles, the exceptions being two at Venus Pool on 18 April, three

Jim Almond, Venus Pool, 16 April 2013

at the Mere (Ellesmere) on the same day, and six at the same site the following day.

Spring passage commences in early April, peaking between mid-April and mid-May. The highest count comprises 21 seen at White Mere on 12 May 1993, with other high counts involving 11 at Ellesmere on 29 April 1993, 14 through Priorslee Lake on 28 April 2004 and 12 at Chelmarsh Reservoir on 13 May 2002.

During the recent Atlas period there were 21 spring passage records from six sites (Allscott Sugar Factory, Chelmarsh Reservoir, Cound Fishery, the Mere (Ellesmere), Venus Pool and White Mere), all between mid-April and mid-May.

Autumn passage is generally more prolonged and they can appear anytime from late June, although in some years none are recorded. Numbers are generally fewer in autumn and there have been no double figure counts during this season. The latest record is of a single at Priorslee Lake on 2 November, in 2005.

In 2014 there was one on spring passage, surprisingly near the summit of the Wrekin on 11 May, and a juvenile at Priorslee Lake on 15 October.

Richard Moores

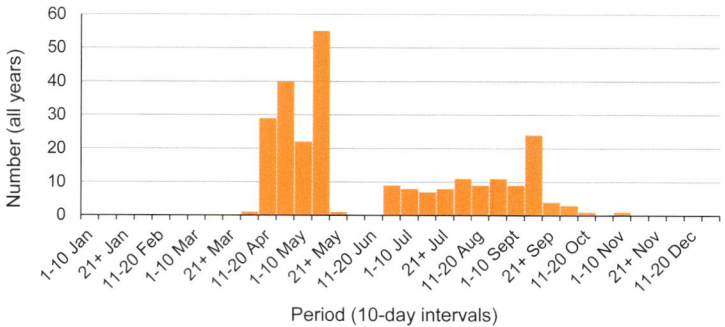

Seasonal Occurrence in 10-day Intervals (1950–2017)

Number (all years) / Period (10-day intervals)

Whiskered Tern

Chlidonias hybrida

Vagrant

The breeding range of this marsh tern extends from Western Europe, through East Asia to Australia, and in Southern Africa. In Europe, it breeds from Iberia through to Poland, usually in small, scattered colonies, and this population winters from the Nile delta to East Africa and in tropical West and Central Africa. The only record was of a juvenile:

Venus Pool	29 August 2010

This bird was considered to be the same as one seen subsequently in Leicestershire, Rutland and Nottinghamshire (Hudson *et al.* 2011).

Graham Walker

Jim Almond, Venus Pool, 29 August 2010

White-winged Black Tern (White-winged Tern)

Chlidonias leucopterus

Vagrant

Breeding from Eastern Europe through to China, this is a migratory marsh tern with the European population wintering in Africa. It is a scarce migrant to Britain, with between 10 and 30 sightings in most years, usually in the southern and eastern counties (Fraser 2013a); it is much rarer in the Midlands and the west. The majority are juveniles in the autumn, but our five records do not reflect this with three spring adults, all in summer plumage, and just two autumn juveniles:

Willey Park Lakes	10 May 1967
Chelmarsh	1 June 1982
The Mere (Ellesmere)	6 November–1 December 1994
Allscott Sugar Factory	6 June 2010
The Mere (Ellesmere)	12–15 October 2015

While the three in spring were, typically, only present for one day, the first juvenile at Ellesmere was unusual in that it stayed for over three weeks into December, becoming Britain's latest ever record (Rogers *et al.* 1984; 1995; 1998). More typically, the second at Ellesmere was present for just four days.

Graham Walker

Black Tern

Chlidonias niger

Rare passage migrant

Jim Almond, Venus Pool, 3 October 2010

The main breeding areas of Black Tern are in eastern Europe and Russia, and the wintering grounds are in southern west Africa. Historically, they used to breed in large colonies in the Fens and Norfolk Broads before numbers fell dramatically in the early nineteenth century, with the last nesting in Cambridgeshire in 1858. Since then only very occasional nesting attempts have taken place within the British Isles. They occur here, often in large numbers, if easterly winds blow continental migrants westwards.

Rocke described it as a very rare visitor but suggested it was more frequently encountered than other species of tern. Forrest listed records at Oxon Pool in May 1871, Rednal in 1873, Gobowen in July 1883 and Walcot Park in the spring of 1894. In the first half of the twentieth century it was reported more frequently, in five years in the first decade, then a gap until five of the years 1925–36, usually of one on one day, occasionally on two days, and sometimes two birds, but unusually six stayed for about a week at Ellesmere in September 1927.

Seasonal Occurrence in 10-day Intervals (1950–2014)

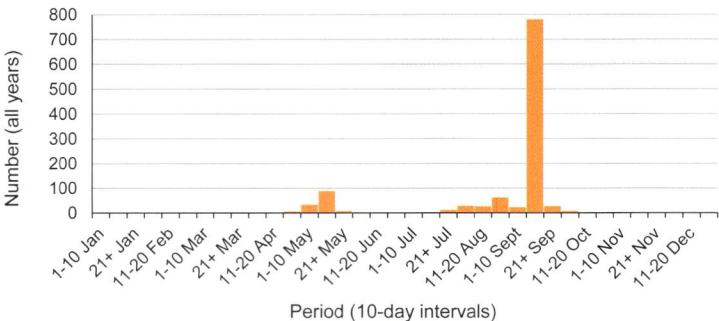

Spring passage in 1948 was exceptional, with seven over Fenemere on May 16, then 15 over Marton Pool (Chirbury) on 17 and 18 May.

The *Handlist* (1964) described it as a passage migrant, regular in small numbers in May, and from August to October, with flock sizes generally fewer than 10. Since then, it has occurred annually in fluctuating numbers, although counts at any one site are usually in single figures. Spring passage is mainly concentrated in early and mid-May, with the biggest spring counts occurring in 1990: 33 at Baggy Moor on 1 and 2 May, 19 at Shrewsbury Sewage Farm on 1 May, and a remarkable 220 passed through Chelmarsh on 2 May, with 50 at Venus Pool the same day. During recent Atlas fieldwork, spring passage birds were seen at Venus Pool, Chelmarsh, Priorslee, Ellesmere, Wood Lane and Mirelake, all in early May.

June records are unusual. Autumn passage is more prolonged, starting in late July. The largest autumn counts occurred on 11 September 1992 when record totals of 300 went through Chelmarsh, with 140 at Allscott Sugar Factory, 80 at Trench Pool and at least 50 at Priorslee Lake all reported on the same day. The latest were two at Ellesmere on 18 November, in 1963.

The chart shows the seasonal occurrence since 1992.

There were only two records in 2014, singles at Venus Pool on 27 August and Trench Pool on 6 September.

Richard Moores

Great Skua

Stercorarius skua

Vagrant

This migratory seabird breeds in Scotland, Iceland, Faroe Islands and Norway and winters off the Atlantic coasts of France and the Iberian Peninsula, reaching as far south as the Cape Verde Islands and Brazil. It is regularly seen around Britain's coasts during the migratory periods, but is rare inland.

The first record, when it was known as the Common Skua or Great Brown Skua, is of an immature female found exhausted in a turnip field at Condover in the spring of 1879. It was kept alive for some time in an aviary where it 'apparently [did] not in the least object to company or confinement', even though it later died! One was shot on the River Severn above Bridgnorth later in the same year.

There have been just two modern records:

Callow Hollow, Long Mynd	9 October 1986
The Manse, near Uffington	1 March 1990

The first was seen flying up the Hollow after a period of gales, while the second was found grounded after severe gales on the 27 and 28 February. It was subsequently ringed and released on the Mere (Ellesmere), but was found dead four days later.

Graham Walker

Pomarine Skua (Pomarine Jaeger)

Stercorarius pomarinus

Vagrant

This trans-equatorial migrant breeds in the far north of Eurasia and North America and winters at sea, largely between the tropics of Cancer and Capricorn, but also along the coasts of Australia and Argentina. In Britain, it is a passage migrant, rarely seen inland, except when storm blown.

It was first reported at Sutton in 1841. In February 1864, a dark phase adult was killed at Marton Pool (Baschurch), and, around this time, one was reported to have been picked up in Shrewsbury after apparently hitting St Alkmund's Church. A young female was captured unable to fly on the Herefordshire border near Downton Castle on 30 March 1866. This account is from Beckwith, but in 1879, he referred to four birds, the St Alkmund's individual and three in Lord Hill's collection at Hawkstone that were from Shifnal, Shrewsbury and Baschurch; the latter was almost certainly the one from Marton Pool. As he did not refer to the one on the Herefordshire border, it is not clear whether this is a Shropshire record or not.

In 1881, another immature was caught in an exhausted state and too weak to fly on the Great Western Railway near Baschurch on 17 October after a gale. The only other historic record is of an immature shot near Shrewsbury on 28 October 1903, following a deep depression which 'lay over the Bristol Channel on the evening of the 27th, travelled northward to Cumbria, then north-westward past the north of Ireland'.

The only modern record followed a period of gales between 24–27 September 1963:

The Mere (Ellesmere)	29 September 1963

This pale phase bird stayed for a few hours during which time it killed a Black-headed Gull.

Graham Walker

Arctic Skua (Parasitic Jaeger)
Stercorarius parasiticus
Vagrant

This marine species has a circumpolar breeding distribution in the northern hemisphere, including northern mainland Scotland and the Western and Northern Isles, and winters in coastal waters south of the equator. Although predominantly coastal, it will migrate over land, and most English records occur during the two migration periods.

The first historic reference was a report in 1841 of an 'Arctic Jager' at Wem. Subsequently, it was noted that by 1866 several immature 'Richardson's Skuas' had been obtained, particularly after gales, but that adults were much scarcer. In November 1878, one was found dead on the Wrekin, while three years later Beckwith reported that:

> the sudden and extraordinary high gale of Oct 14, 1881, blowing as it did directly from the west, and accompanied as it was by torrents of rain, drove a great number of marine birds inland; and besides three or four of this species [later accounts refer to three being picked up], that were found, 'black seagulls' were reported to have been seen in several parts of the county.

In the early twentieth century, an adult was seen flying along the River Severn in Shrewsbury on 13 September 1912, a dark-phase immature was picked up dead at Broadoak near Shrewsbury, on 6 October 1913 and one was found dead near Shrewsbury in 1923.

The three modern records reflect that most inland sightings occur during migration periods:

Jim Almond, Chelmarsh, 26 August 2011

Rhydycroesau	two, 5 May 1976
Minton Batch, Long Mynd	30 April 1989
Chelmarsh	24–30 August 2011

The two at Rhydycroesau were both flying west while the dark-phase bird on the Long Mynd was being mobbed by a Raven in drizzly conditions. The juvenile at Chelmarsh was often observed pursuing the Lesser Black-backed and Black-headed Gulls that were trying to roost on the reservoir, forcing them to disgorge food.

Graham Walker

Long-tailed Skua (Long-tailed Jaeger)
Stercorarius longicaudus
Vagrant

Breeding on Arctic, sub-Arctic and montane tundra in both North America and Eurasia and wintering at sea, usually south of the Equator, this is a scarce passage migrant to Britain and inland records are rare.

There were three reports of Long-tailed or 'Buffon's' Skua prior to 1950. In 1879, an immature in Lord Hill's collection was reported as having been shot at Astley many years ago. A second was shot near Baschurch in October 1881, while a third was found dead at Garmston on 14 October 1891.

Just one modern record:

Bourton Westwood	25 September–4 October 1999

An immature, it was found in a recently harvested potato field to which it remained loyal throughout its stay, mainly feeding on earthworms. The field had not been ploughed because it was saturated, and the bird stayed until ploughing took place, although this coincided with a fine and sunny day which may have encouraged its departure. Low pressure determined the weather prior to its arrival and, although gales were not reported, it was still windy and there were numerous thunderstorms with heavy downpours.

Graham Walker

Little Auk
Alle alle
Vagrant

This High Arctic seabird with a circumpolar distribution was first recorded in 1841, when three separate individuals were found on the Severn in the neighbourhood of Shrewsbury. These were part of a national 'wreck' reported widely in Britain and described by Yarrell (1843) as resulting from 'a violent storm of wind from the N.N.E. which lasted several days' in October.

Singles have occurred on the Severn in Shrewsbury on three occasions: one caught under the Welsh Bridge before 1866 (probably one of those found in 1841), one at the English Bridge in 1910 (Ludlow Museum Z/2006/084), and one in 1912 that was seen swimming and diving in the Quarry on 2 February. The latter was part of the national 'invasion' which 'accompanied the arctic weather of January–February'. Four others were also being seen in February that year, one on the River Severn at Kynnerley on the 1st, another on 2nd at Wem, and two on 5th, at Cressage and Bridgnorth.

Otherwise, in the nineteenth century, six were 'obtained at Shifnal, Acton Scott, Haughmond, Hodnet, Ludlow, and Ellesmere – the last in 1882' (Forrest 1899), while in the first half of the twentieth century, in addition to the five in 1912, one was found 'dead but warm' near Marshbrook on 26 December 1911, and another was found alive at Stokesay on 7 December 1921, but it died the next day.

There have been seven modern records, all except one of singles:

Shawbury	12 February 1950
Ashford Carbonell	2 December 1955
Near Llanymynech	two, early winter 1957
Harley	1 December 1960
Aston Rogers	12 November 1971
Ditton Priors	14 December 1978
Madeley	8 December 1987

Of the eight, five were found alive, although the Madeley individual subsequently died. One of the two found near Llanymynech was released at Porthmadog; it is assumed that the other was either found dead or succumbed shortly thereafter.

All occurred in the middle of winter, and were invariably associated with storms or arctic weather, such as the very severe south-westerly gales on 12–13 December 1978, although it is not always possible to tie sightings to specific storms. Altogether, the dated records were in November (six), December (five), January (one) and February (six).

John Tucker & Graham Walker

Common Guillemot (Common Murre)
Uria aalge
Vagrant

Gareth Thomas, Worm's Head, Gower, (bird released) March 1986

It breeds on Atlantic and Baltic coasts, is common around the British coast, and winters just off-shore, but it is nevertheless rare inland, usually being found after storms.

In 1866, it was acknowledged to be very rare inland, but an occasional specimen had been obtained, while later they were said to occur during the two migration periods and after gales. The first specific record did not occur until 3 February 1885 'when a flock of fourteen visited Ellesmere; and some of them were killed on the canal'. In October 1905, an immature was found on Titterstone Clee, and in January 1909, when there were gales in the middle of the month, another individual was located at Ellesmere. In 1919, one was caught exhausted on the Lily Pond at Acton Burnell; although this find was initially reported as June, a later account said it was in December which is more likely since 'gales and strong winds were frequent' that month.

Neither of the two modern records, however, were of obviously storm blown birds:

Coreley	28 February 1986
Eyton upon the Weald Moors	6 March 1993

The one at Coreley, thought to be a young bird in winter plumage, was found alive and taken to a local veterinary practice. It fed voraciously on the fish it was offered and took great pleasure from a daily swim in the bath! After a week's recuperation, it was released at Worm's Head on the Gower Peninsula (Thomas 1986: see photo); February 1986 was not noted for its gales, although winds of up to 64 knots were reported in south-west England on 26th. The one at Eyton was unfortunately found dead on a ploughed field during a period when high pressure dominated and winds were again not notably strong, although it is not clear how long it had been dead.

Graham Walker

Razorbill
Alca torda
No modern records

There is only one record of this north Atlantic seabird. Despite claims by Rocke (1866a) and Beckwith (1878) that it occasionally occurred, it was only in the winter of 1878–79 that one was documented, 'caught at Bromfield' (Beckwith 1879). It was doubtless storm-driven.

John Tucker

Puffin (Atlantic Puffin)
Fratercula arctica
No modern records

There have been eight reports of this north Atlantic seabird, all storm-driven. The first was from Pontesbury on 14 September 1894, followed by singles at Tasley, in October 1887, Bayston Hill on 29 July 1902, Cross Houses 5 July 1909, Hengoed, 'a young bird', on 28 December 1909, and Montford Bridge on 31 July 1922.

There are no modern records. Recently, however, a verbal account was received of one found dead some time not long before 1987, in what became known by some as 'Puffin Corner', the elevated north-west part of the SWT Nature Reserve at Cramer Gutter. Unfortunately, the account was not documented at the time, details remain sketchy and, consequently, the record is considered unproven.

John Tucker

Pallas's Sandgrouse
Syrrhaptes paradoxus
No modern records

This rather spectacular rarity has irrupted into Britain on a number of occasions from its native range, which extends across the Steppes from Kazakhstan eastwards into China and Mongolia. Large-scale irruptions reached Britain during the nineteenth century in 1859, 1863, 1872, 1876 and 1888, and some reached here in the second and last of these years. On a few occasions, some remained to breed elsewhere in Britain, and more would probably have done so had they not been indiscriminately 'collected'. Irruptions have become less frequent, and more limited, in the twentieth century.

There are seven records, the first two during 1863. Buffon and Wilson, taxidermists in the Stand, London, wrote: 'We have had sent to us, for preservation, a brace of sand grouse, shot from a flock of nearly a hundred birds, near Oswestry' (*The Field*, 13 June).

The source of the second 1863 record, often referenced and quoted as 'about twenty near Ludlow', has been traced to *The Field* (30 May), where it appeared as:

> Partridges. A curious circumstance took place a few days ago. My man put up a pack, I suppose, of partridges, eighteen in number, in a field close to this house. He assured me, in answer to my questionings, that they were certainly partridges, and they went down again in another part of the same field,

flying but a little way. I don't know whether such a case is less remarkable than I think it, but I never heard of such a thing at this time of year. G.V.H. (Hopton Cangeford, Ludlow)

The fact that partridges would not be expected to flock together in May was, first, sufficient to prompt the correspondent to report it to the *Field* and, subsequently, for the local ornithologists at the time to conclude that the birds were in fact a flock of sandgrouse, part of the irruption in that year.

There is a mounted specimen from Edgmond dated 1880 in Shropshire Museums in Ludlow Museum Resource Centre (specimen SHYMS: Z/2006/141), which is surprising as there was no major invasion of Britain that year. The specimen was initially in William Beckwith's collection and passed to his brother, who donated it to the then Shrewsbury Museum in 1925 (Tucker & Tucker 2018); the precise origin of the specimen is not recorded in the literature of its time.

In 1888 there were four records. A male was killed near Neen Savage on 7 June (Horton 1888) and three were seen near Market Drayton by a party of fishermen on 22 June (Mainwearing 1888). One was killed against telegraph wires near Craven Arms (Forrest 1899) and 'others' were seen on Brown Clee (Horton 1888), the latter undated.

John Tucker

Feral Pigeon (Rock Dove)
Columba livia
Fairly common resident

Pure Rock Doves are now restricted to the remote coastlines of north and north-western Scotland, but they used to be much more widespread, and were the ancestor of the domesticated pigeon. Feral Pigeons are their naturalised descendants, and are found in both urban and rural habitats, especially where grain and seeds are readily available, such as around farmyard buildings and in dovecotes, although there are a few records of them visiting garden feeders. Feral

Pigeons have been ignored by most birders, and records have usually been provided by very few individuals. Possibly many observers do not regard them, with their wide variety of plumages and proximity to human habitation, as being truly wild birds, especially where it is clear that some individuals in a flock are non-returning racing pigeons; flocks in flight are easily confused with racing pigeons.

In the late nineteenth century there were several references to

Rock Doves, or Rock Pigeons. Beckwith expressed his doubts about the Rock Dove ever having occurred, 'though it is stated to have done so', as 'the birds alluded to were tame pigeons' (Beckwith 1879–81). Forrest (1908) stated: 'There is some doubt whether the rock-doves occasionally shot are truly wild, or tame pigeons reverted to a wild state', and noted that, in the Ludlow district, there were 'numbers of rock-doves, descended from birds which had escaped from pigeon-shooting matches', and that they nested in trees.

Feral Pigeons are not mentioned in the *Handlist* (1964). The population appears to have increased between 1968–72 and 1985–90, as 21 of the 10km squares with evidence of breeding in the latter period (*Atlas* 1992) had no record in the earlier national BTO Atlas (1976).

Urban populations breed throughout the year. Although it is not known whether this also applies to local populations, some pairs may not breed within the peak periods for Atlas fieldwork within the April–July season, which may contribute to the low proportion of tetrads with confirmed breeding.

They generally nest in farm buildings, and in disused buildings in urban areas, but take advantage of other man-made structures such as bridges, for example the Waterworks Bridge over the River Severn at Hampton Loade. The widespread use of such structures means that changes in their availability may have a significant impact on the breeding population and distribution. In many areas in recent years, older farm buildings have been restored for use as dwellings, or replaced by modern buildings, making them unavailable for nesting, partially accounting for the considerable decrease in the number of tetrads with breeding evidence. Agricultural intensification, reducing the availability of grain over the winter, and more limited access to

Jim Almond, Shrewsbury, 30 May 2010

farm buildings for food, will also have had an impact. As the number of tetrads with winter records is almost identical to the tetrads with evidence of breeding, it seems probable that there has been a similar decrease in winter numbers.

The breeding distribution change map presents a rather startling picture. The 121 tetrads in which breeding evidence was recorded in both Atlas periods tend to be concentrated in or close to urban areas, notably Telford, Shrewsbury and Oswestry. However it is the numbers of losses and gains that are difficult to account for: the 281 tetrads that have lost evidence of breeding are widely distributed, with some 10km squares to the south of Shrewsbury losing the highest numbers. Many tetrad losses are adjacent to the urban areas mentioned above. Some of the 123 tetrads where new breeding evidence has been found are in 10km squares with the highest number of losses. While it is hard to explain such patterns, loss of nest sites, changes in farming practice, variation in fieldwork effort and doubts as to how well Feral Pigeons have been monitored, have probably all played a part.

In most years very few records are received, so Atlas surveys give a much more comprehensive picture. The maximum count was 105, from the riverside walk in Shrewsbury, on 29 June 2013, but the next largest count, 55 on 19 April 2011, came from a rural square at Melverley. During breeding season TTVs, there were six

Breeding Distribution Change (1985–90 to 2008–13)

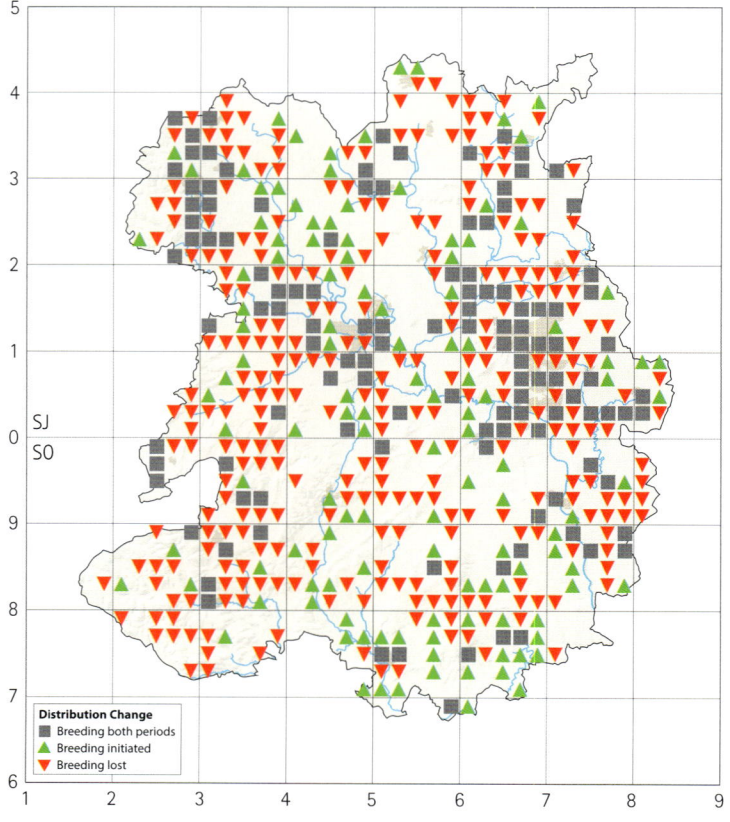

Distribution Change
- ■ Breeding both periods
- ▲ Breeding initiated
- ▼ Breeding lost

Occupied Tetrads

Atlas period (breeding)	1985–90		2008–13		Change	
	Number	%	Number	%	Number	%
Confirmed	88	10	42	5	−46	−52
Probable	152	17	88	10	−64	−42
Possible	162	19	114	13	−48	−30
TOTAL	402	46	244	28	−158	−39
Tetrads with Winter Records (2007–13): 252 (29%)						

one-hour counts over 30, maximum 55, all from different tetrads, and 20 counts of 30–50 in winter from 19 different tetrads. Other counts of over 100 from the last 20 years include 120 at Snailbeach in April 1994; 160 in Ludlow in December 2004; 220 in Shrewsbury in the same month; 152 at Ellesmere in November 2005; and 140 at Oswestry in November 2006. The flock based in Ellesmere, which usually exceeds 50, is regularly seen feeding among the ducks and geese on the Mere-side promenade, taking advantage of the free handouts provided by visitors; they nest in the many older buildings in the town (*pers. obs.*).

The breeding season relative abundance map shows concentrations around Telford, Shrewsbury, Oswestry, Wem and Bridgnorth, and the smaller settlements in the Tern valley, but there are several rural tetrads with equally high numbers, where farms provide good food supplies and nest sites, for example near Stottesdon. The winter map is very similar, confirming the Feral Pigeon's sedentary nature.

Estimating the population is very uncertain. The estimate based on TTV counts is 1,500–1,600, but the national estimate on which it is based is probably not reliable. Based on five to 15 pairs per tetrad the population was estimated at 2,000–6,000 pairs (*Atlas* 1992). The density has probably fallen since then, and it was found in 158 (39%) fewer tetrads, in line with a 47% decline in the English West Midlands between 1995 and 2014. The population is now possibly 1,000–3,000 pairs. Breeding season TTVs counted 912 in 129 tetrads, just over half the tetrads in which it was found, suggesting that this estimate is reasonable.

David Farncombe

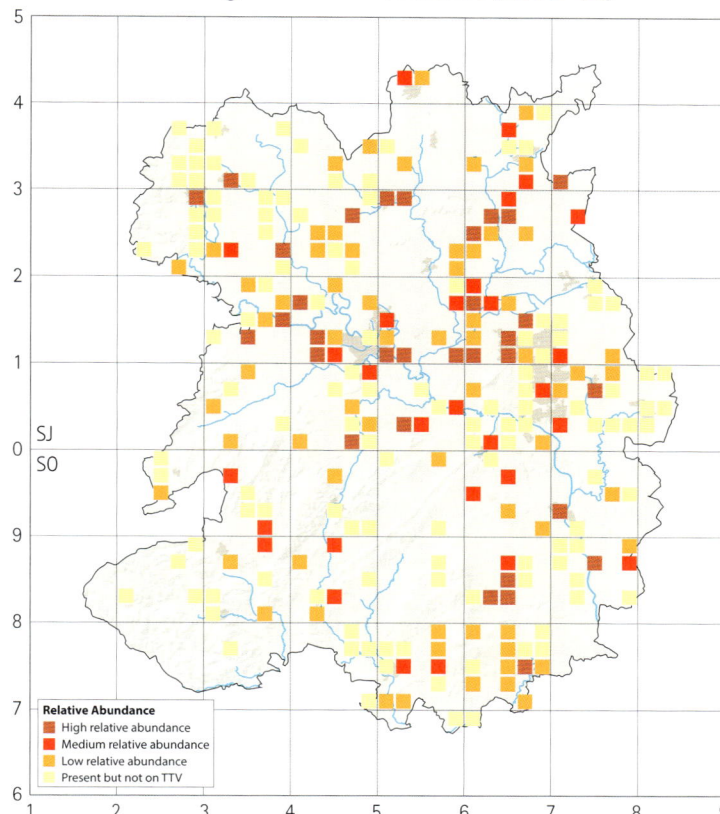

Breeding Relative Abundance (2008–13)

Relative Abundance
- High relative abundance
- Medium relative abundance
- Low relative abundance
- Present but not on TTV

Stock Dove
Columba oenas
Common resident

Widespread and common, Stock Doves are generally more shy, less conspicuous and occur in much lower numbers than their larger relative, the Woodpigeon. Stock Doves were described in the earliest record as 'common' (Eyton 1838), and as 'very common' (Rocke 1866a). At the end of the nineteenth century, nests were often built close to houses in outbuildings, in holes in trees, cliffs and banks of streams, and in ivy against a tree trunk. In the treeless hills of the south-west, they sometimes nested on the ground under a gorse bush, and, further east, in great numbers in cracks and fissures in the limestone of Wenlock Edge and sandstone around Bridgnorth (Beckwith 1893); the old nests of squirrels, owls or other birds were also used (Forrest 1899). The provincial name 'Blue Rock' is attributed in the *Catalogue of Shropshire Birds* (Paddock 1897).

A preference for 'weed' seeds over grain or pulses meant that Stock Doves were regarded as less harmful to agriculture than Woodpigeons (Beckwith 1893), but advances in crop breeding and management from the end of the nineteenth century onwards reduced diversity and the availability of such seeds, and increased the dove's dependence on grain. A range of seeds from farm crops now form a major part of the Stock Dove's diet, and nationally numbers fell in the 1950s and early 1960s following the widespread use of organochlorine pesticide seed dressings. When this substance was banned, they made

Jim Almond, The Rea, 1 June 2007

Breeding Distribution Change (1985–90 to 2008–13)

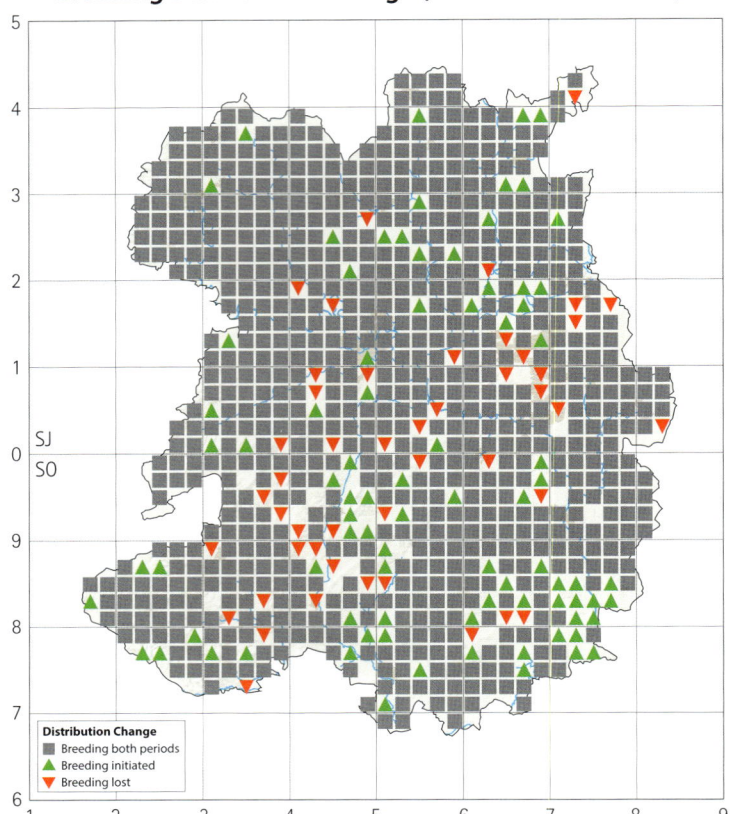

Distribution Change
- ■ Breeding both periods
- ▲ Breeding initiated
- ▼ Breeding lost

Breeding Relative Abundance (2008–13)

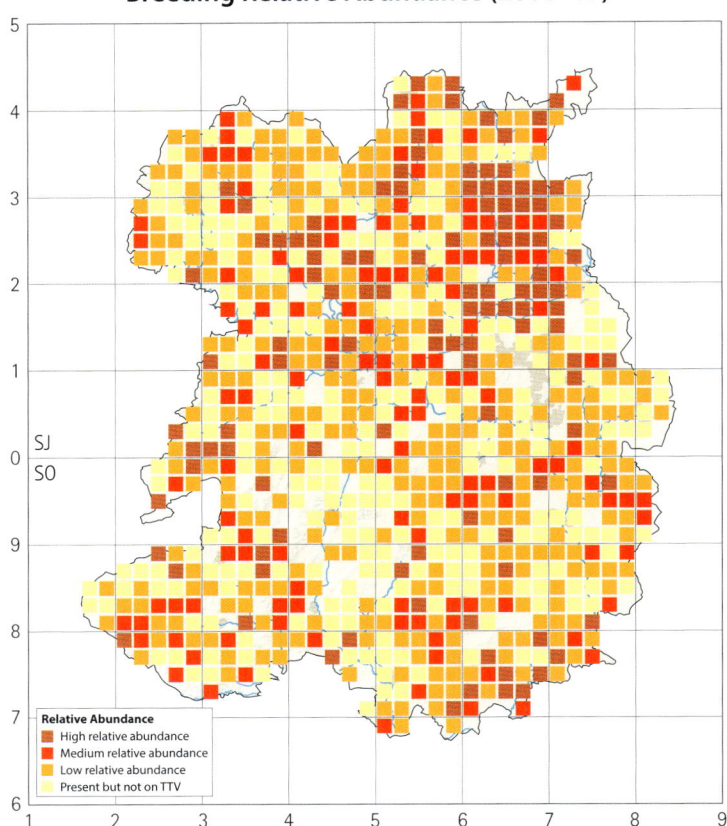

Relative Abundance
- ■ High relative abundance
- ■ Medium relative abundance
- ■ Low relative abundance
- ▢ Present but not on TTV

Occupied Tetrads

Atlas period (breeding)	1985–90		2008–13		Change	
	Number	%	Number	%	Number	%
Confirmed	271	31	143	16	-128	-47
Probable	390	45	530	61	140	36
Possible	106	12	132	15	26	25
TOTAL	767	88	805	93	38	5
Tetrads with Winter Records (2007–13): 709 (81%)						

John Fielding, Shrewsbury, March 2013

a rapid recovery. It seems reasonable to assume that similar trends applied locally; the *Handlist* (1964) described them as widespread and resident in small numbers, rarely reported in flocks exceeding 30, but no longer meriting Forrest's description of 'very common'. Since then numbers have increased steadily, by an estimated 116% in the UK between 1970 and 2014.

They are birds of open parkland, woodland edges and wooded farmland, and such habitats are widespread. They may be confused with Feral Pigeons or, in flight, with racing pigeons, or be overlooked, and may therefore be under-recorded. However, medium sized flocks of up to 100 in autumn and winter are reported from all parts; 300 in a stubble field at Quina Brook was exceptional (SBR 2002). They often associate with Woodpigeons, especially in flight. The majority of records are from open country, with a few from private gardens (SBR 2005).

BBS Trend (1997–2014)

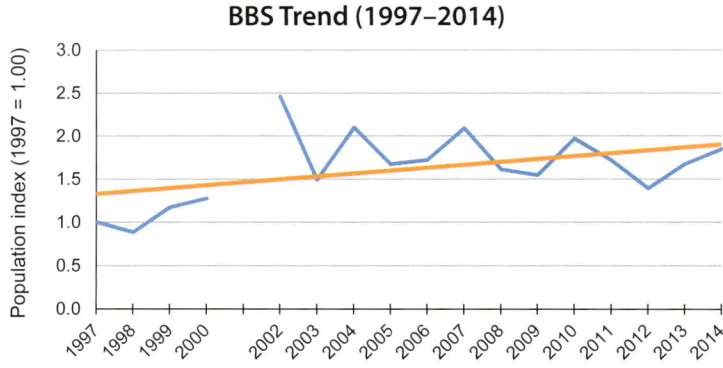

Most nests are built in holes in trees, and they were breeding in quarries, 'such as along Wenlock Edge' in 1985–90 (*Atlas* 1992), but there is little recent reported evidence of nesting in cliffs and quarries. They will use outbuildings and nest boxes, particularly owl boxes, where a pair was found nesting at Wall Farm in 2014. Breeding is widespread, but most of the areas having several tetrads with no evidence of breeding include land over 305m (1,000ft). However breeding occurs in a few tetrads in the Clun Forest with no land *under* 305m, and in a larger number of tetrads with no land under 183m (600ft). Other than parts of Telford, breeding is also recorded in, or close to, all the urban areas. Limiting factors may well be the availability of suitable nest sites, and arable fields providing food.

Changes in breeding distribution since 1990 show losses from 47 (6%) tetrads, mainly in the south, but gains in 85 (11%), mainly in the Stretton hills and the extreme south-east around Cleobury Mortimer, perhaps reflecting an expansion of range as the population grows, or an improvement in the fieldwork in these two areas. Overall, the table shows a 5% increase in tetrads with breeding evidence, reflecting results from BBS, which have shown a steady increase of about 30% since 1997, rather less than the 63% increase between 1995 and 2014 in the West Midlands as a whole.

Based on TTV counts, the population is estimated at around 7,200 pairs. This is at the lower end of the range of 7,000–10,000 pairs suggested in the *Atlas* (1992), implying that the latter was too high at that time.

The breeding season relative abundance map shows a strong bias to the north-east, with areas of arable farmland having the highest densities. The winter map shows the same broad pattern, but Stock Dove was found in 103 fewer tetrads, with most of the gaps lying to the south-west of Shrewsbury, reflecting a move from high ground to more favourable areas in winter.

Of the 115 Stock Doves of all ages ringed since 2006, only one has been recovered. Ringed in March 2011 at Sutton, Shrewsbury, it was recovered, apparently killed by a wild animal over a month later, 3km away in Shrewsbury. Since ringing records began, eight have been ringed and recovered here, with the only additional recovery being of one ringed in adjacent Hereford and Worcester. The overall picture is of a very sedentary species.

David Farncombe

Woodpigeon (Common Wood Pigeon)
Columba palumbus
Very common resident

Rarely on anyone's list of favourites, unwelcome at garden feeders, and a major pest to farmers, the Woodpigeon is highly successful and found in every type of habitat. Alternative names for Woodpigeon include Ring Dove, which has only gradually fallen out of use and, in the past, Quice, Quist or Queest (Forrest 1908).

From the earliest record in 1838, and throughout the nineteenth century, Woodpigeons were described as exceedingly common and widespread, and occurring in large flocks which were very destructive of farm crops. A compensatory benefit was the amount of agricultural weed seeds consumed, often of plant species difficult to eradicate (Beckwith 1879). The preservation of game, leading to 'winged enemies' being 'almost exterminated', and the provision of coverts for Pheasants, providing extra nesting opportunities, were both suggested as possible causes of a very rapid increase in population towards the end of the nineteenth century (Beckwith 1893). Migratory flocks of 500–600 were recorded as appearing in early winter from the north, remaining in years when oak and beech mast were plentiful, but moving on or absent in lean years. These birds were described as smaller and darker, suggesting that they were either first-winter birds (Beckwith 1893), or of an alien race (Paddock 1897; Forest 1908).

By the end of 1964, following the severe winter of 1962–63 when starvation and shooting had a considerable impact, numbers were rising again, with flocks of up to 500 being occasionally reported. Considerable numbers visited the higher hills in July and August to feast on the bilberries, and numbers were augmented in winter (*Handlist* 1964).

The population increased rapidly, and SBRs from 1972 regularly reported winter flocks of up to 5,000, a ten-fold increase in eight years. However, breeding has always been poorly recorded, for example

Dawn Micklewright, Shrewsbury, 30 September 2013

from 2000–06 no more than four confirmed breeding records were received annually.

Nationally, 97% of all farmers shoot Woodpigeons to protect crops, and this level of killing probably goes on locally. There is little evidence from the records that such action has any real effect on the total numbers.

There has been little change in distribution since the previous Atlas. It was found in every tetrad, and in all but two tetrads in winter,

Breeding Relative Abundance (2008–13)

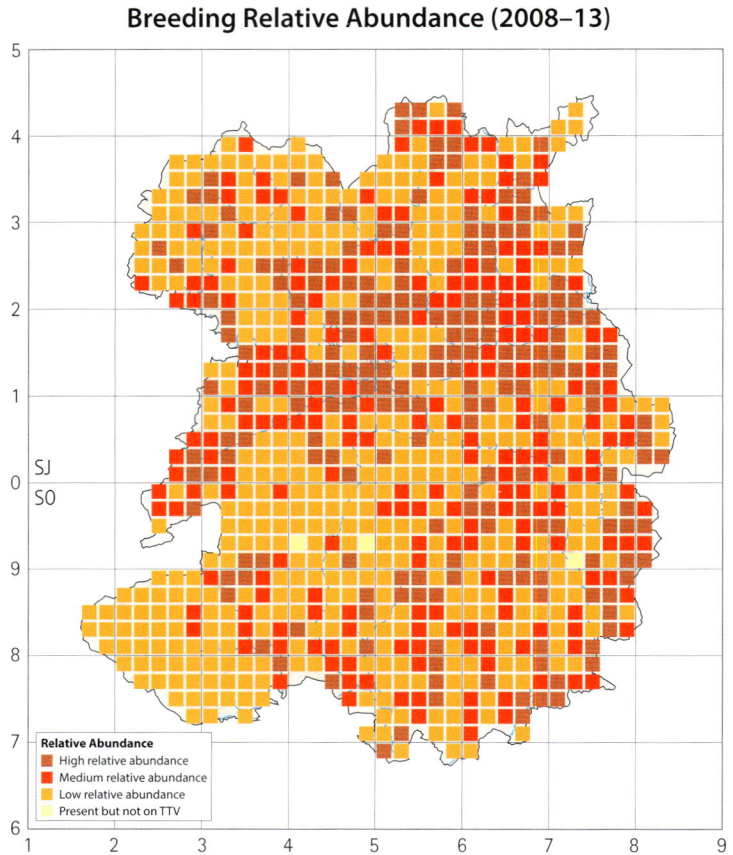

Relative Abundance
- High relative abundance
- Medium relative abundance
- Low relative abundance
- Present but not on TTV

Winter Relative Abundance (2007–13)

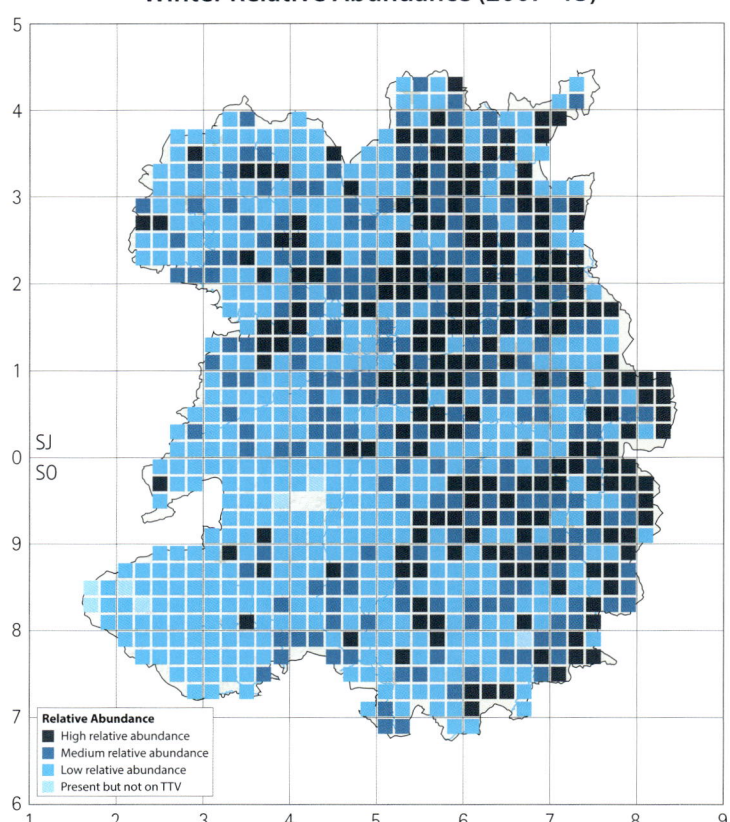

Relative Abundance
- High relative abundance
- Medium relative abundance
- Low relative abundance
- Present but not on TTV

Occupied Tetrads

Atlas period (breeding)	1985–90		2008–13		Change	
	Number	%	Number	%	Number	%
Confirmed	742	85	547	63	-195	-26
Probable	119	14	296	34	177	149
Possible	5	1	27	3	22	440
TOTAL	866	100	870	100	4	0
Tetrads with Winter Records (2007–13): 868 (100%)						

reflecting its great adaptability to a wide range of habitats. In 1985–90 it was overlooked in three tetrads, but may have been absent from one tetrad (SO49H), as it was not noted there during fieldwork on the Long Mynd in 1994–98 either, but it was in 2006–08, so colonisation of this one tetrad since 1990 may perhaps have occurred, a result of the rapidly increasing population (Leo Smith *pers. comm.*).

The relative abundance maps reflect the concentration of arable farming in the Severn Valley, and in northern and eastern areas, while areas of low density occur in the uplands, and in small blocks of tetrads in lowland areas, particularly where pasture predominates. The differences within the arable areas probably reflect which specific crops are planted. In winter, there is a distinct shift in distribution from the uplands to arable areas, particularly in the north-east, where they often congregate in large flocks. During recent winter Atlas fieldwork, flocks of over 1,000 were counted in 19 different tetrads, with 2,000 in five of them, 3,000 in two more and a maximum of 5,000 in another, SJ51R at Rea Farm, Upton Magna.

Based on TTV counts, the current population is estimated at 87,000–89,000 pairs, making it the fifth most common species. A population exceeding 35,000 pairs, based on figures from the national BTO Atlas 1976, was estimated in 1990. However, the long-term CBC/BBS population trend in the UK shows 124% increase from 1970 to 2014, and locally BBS shows an underlying trend of a small but steady increase of around 15% since 1997, so the two estimates are not inconsistent.

Population fluctuations reflected changes in crops being planted, and, where it occurred, the replacement of clover leys by autumn cereals reduced over-winter survival. However, oilseed rape crops provide nutritious winter food, and have more than compensated. They are mostly autumn-sown, and have usually grown enough to extend above snow cover that would bury winter cereals, and nationally the population growth has been correlated with the spread of this one crop since the mid-1970s. A flock of 450 was seen feeding on rape near Lilleshall in February 1992, just before the start of the massive increase in the land devoted to this crop. By November 1998, 1,000 were seen on the same crop near Somerwood. Such a rapid increase in the flock sizes feeding on rape, and the scale of the increase, might suggest that the local population gain recorded by BBS is too low, but national ringing results show many Woodpigeons move relatively large distances to create concentrations on this favoured food.

However, they feed on a range of crops, both grain and brassica leaves, and they have been recorded eating acorns, ash shoots, other brassicas, berries of elder, hawthorn, holly and ivy, seeds in stubble and winter wheat, beet tops and remains in a harvested sugar beet field, fodder parsnips, maize and sunflower hearts.

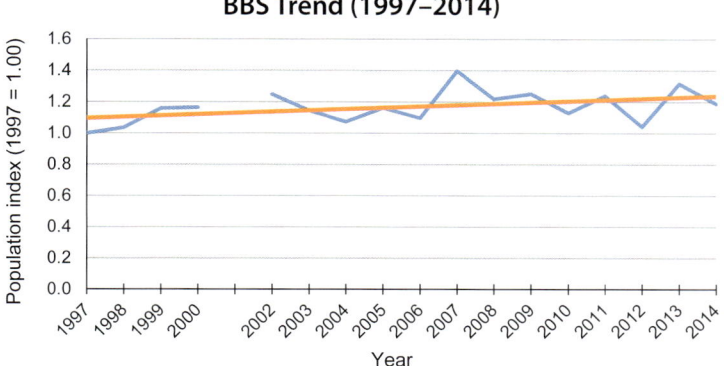

BBS Trend (1997–2014)

The British population is largely sedentary, although many move short distances to congregate at good winter feeding sites, mainly winter wheat and clover leys up until the 1980s, but now more usually oilseed rape. Juveniles disperse in their first winter before returning to their natal area to breed as one-year-olds. Ringing records largely confirm this behaviour. About 70% of those ringed here were recovered locally, but of the eight recovered elsewhere, four went to nearby counties, one to Lothian and three to France. Ten ringed in nearby counties have been found here, with four from Merseyside being the only county involved in more than one recovery, and one from the Isle of Man travelled furthest. Two of the three that went to France

were ringed as nestlings (the third was also a first year), and all were shot within the same calendar year, in Brittany, distance travelled around 500km, with the oldest being only 5m 11d.

An adult ringed near the Lawley in July 1987 is the oldest known. It was shot 84km distant in January 1993 near Burton (Derbyshire), 5y 6m 4d later, but it was well short of the national longevity record of 17y 8m 19d.

Overflying flocks in large numbers are regularly reported in early winter, mainly flying in a southerly direction, but not before the end of October. They may perhaps be from northern Europe, where the breeding population is mostly migratory, but there are no corresponding large return flocks in spring, and local and national ringing results provide no evidence of continental Woodpigeons passing through or wintering in Britain, so it is perhaps more likely that they are mobile residents forming winter feeding flocks and moving to exploit the best food sources.

There are no recent records of smaller and darker Woodpigeons in migratory flocks, as noted by Beckwith, Paddock and Forrest, nor of the considerable numbers visiting the higher hills in July and August to feast on the bilberries, noted in the *Handlist*, presumably because there are a lot fewer bilberries. That, and the more recent adaptation to winter feeding on oilseed rape, reflect the rapid response of this successful species to take advantage of changes in food supplies.

David Farncombe

Turtle Dove (European Turtle Dove)
Streptopelia turtur
Rare summer visitor

This dainty, attractive dove breeds across central and southern Europe, and winters in the Sahel. It is our smallest and only migratory member of the *Columbiformes*, and favours arable farmland, reed-beds, open woodland and villages, with tall, thick hedgerows and scrub for nesting and open ground for feeding on the seeds of arable weeds and cereals. It is said to be particularly fond of the common herb fumitory, although the evidence suggests that this plant does not constitute more than half of its diet. It has also been known at garden bird feeders (e.g. Kemberton, 1996; Somerwood, 2005). Its range underwent an expansion towards the north and the west in the UK during the nineteenth century, from centres in south-eastern and eastern England, although the Welsh Borderlands have always been on its western periphery as it finds uplands unattractive. Beckwith, in 1879, commented that it was formerly scarce but had become common by the period 1854–79 and was becoming more numerous each year. In 1866, Rocke characterised it as 'plentiful' and, by 1893, Beckwith described it as a 'common species and ... very evenly distributed'. So abundant was it around what is now the Telford area that the provincial name of 'Wrekin Dove' was used: Paddock (1897) reported that the name 'Ring Dove' was also applied locally to this species.

Evidence of the Turtle Dove's declining fortunes was apparent by the middle of the twentieth century. It was reported as breeding and common in the period 1920–24, breeding and declining in the period 1950–54 and entirely absent from 1980 onwards, in the Bucknell area

Jonathan Cartwright, Kinlet, 29 May 2010

Occupied Tetrads

Atlas period (breeding)	1985–90		2008–13		Change	
	Number	%	Number	%	Number	%
Confirmed	25	3	0	0	-25	-100
Probable	100	11	5	1	-95	-95
Possible	106	12	10	1	-96	-91
TOTAL	231	27	15	2	–216	–94

Breeding Distribution Change (1985–90 to 2008–13)

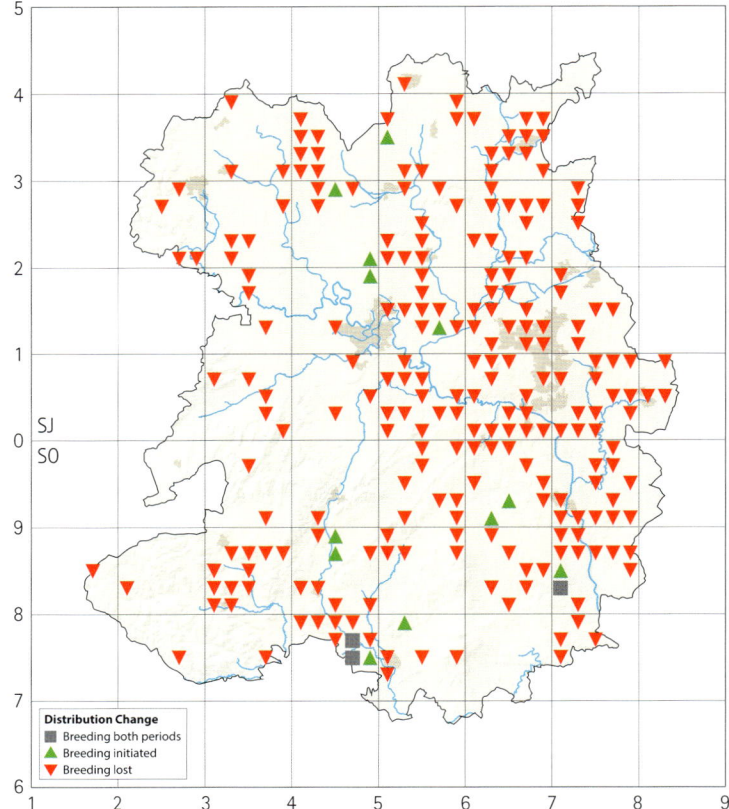

Distribution Change
- Breeding both periods
- Breeding initiated
- Breeding lost

(SBR 1987). As early as 1942, the numbers of Turtle Doves which fed with domestic pigeons in a Bicton Heath garden were diminishing yearly. Low numbers were also reported in the late 1950s. The *Handlist* (1964) stated '[t]hough rather local, it is found throughout the county' but that '[t]here is some evidence of a decrease … in recent years, although numbers are prone to fluctuation from year to year'. In 1978 it was breeding in some numbers in the hawthorn scrub along Chelmarsh Reservoir (Dave Fulton *pers. comm.*).

While observed fluctuations may have been local or ephemeral, evidence of a severe decline became apparent in the UK by the late 1970s and early 1980s, and this is reflected in local records. The *Atlas* (1992) species account stated that a 'comparison with the BTO Atlas (1976) suggests a recent withdrawal from the north-west', consistent with this national decline.

There were confirmed breeding records in SBRs in all but one year from 1981 to 1986, usually from more than one site, and confirmed breeding records from 25 tetrads in the *Atlas* (1992), but no indication whether or not they were equally spread across the fieldwork period.

There were no confirmed breeding records after 1990, apart from the final ones in 1995 (two adults and two juveniles observed together at Granville Country Park on 18 August, and three individuals at Isombridge on 18 June, where one adult and one juvenile were seen on 9 September). By the 2000s, SBR typically listed fewer than 20 records (some of which were passage birds), with no confirmed breeding, and there was a rapid decline in records and sites in the 10-year period from 2005 (SBR 2014), when there was only one record from one site.

The breeding distribution change map depicts the alarming contraction in range. In the 2008–13 period, there were only 48 records of Turtle Dove, and no instances of confirmed breeding. Probable breeding was confined to five locations, with the Rea, Upton Magna, showing most activity. Two probable breeding records (together with a possible breeding record) come from adjacent tetrads near Bromfield, with another about 6km to the north-east (near Downton Hall). The Bromfield area was part of the local stronghold in 1985–90, so these doves may be hanging on there. The fifth probable breeding record came from Lower Netchwood. All other records are widely scattered with no coherent pattern, and will include some passage migrants.

There have been no Turtle Dove observations on BBS routes since 2007 and the population estimate based on TTV counts suggests 20–22 territories, although this estimate is likely to be unreliable for such an infrequently encountered species. Recent SBR accounts tend to be less optimistic and two pairs were reported in only one tetrad during 2008–13 breeding seasons. Therefore, even if all the 15 tetrads with some level of breeding held a breeding pair, which is unlikely as many of the records will relate to passage birds, the population appears to be at a maximum of only 16 pairs.

Turtle Doves are at the edge of their range here, and this contraction is part of a wider disappearance from much of western, central and northern England, and continental Europe.

Possible reasons for the precipitous decline are several, involving factors in the summer and winter range, and in migration between the two:

- Mortality in the wintering grounds due to habitat loss and deterioration resulting from periodic drought and overgrazing in the Sahel.
- Use of more efficient herbicides and fertilisers in the breeding grounds which reduce plant diversity in grass leys and permanent pasture, reducing the availability of weed seeds.
- Changes in the timing of agricultural operations, including the switch from haymaking to silage production.
- Urbanisation in and around the new town of Telford, removing breeding habitat in one of the strongholds shown in the 1985–90 *Atlas*.
- Excessive hunting during migration.

The second and third factors reduce the food supply, resulting in fewer breeding attempts over a shorter breeding season.

The disease trichomonosis may also be a factor. Recent research has highlighted a high prevalence of infection in Turtle Doves by the parasite that causes this disease (Lennon *et al*. 2013), with a suggestion from work in south-east England that this is correlated with food provision for game birds. It is proposed by Stockdale *et al*. (2015) that 'parasite spillover' may occur where game bird densities are high. While clinical signs were present in some infected birds, it is not yet apparent whether population figures are affected to any great extent (through mortality or poor breeding success).

Turtle Doves are late arrivals in spring, when they can be detected relatively easily by their distinctive 'purring' song. Lloyd, in a survey of migrant arrival dates in 1939, gave average dates of 4 May for Shrewsbury (15 years of data), 10 May for Dowles (28 years) and 11 May for Broseley (20 years), although dates varied by roughly two weeks on each side of the average. The *Arrival Dates* chapter suggests an average for the whole County of 5 May in 2014, with little evidence for any advance in arrival date over the prior 114 years, although figures for the later years may be unreliable because of the declining number of observations.

Of 27 ringed locally, 25 have been recovered here, but two were shot on migration. A nestling ringed at the Wrekin in July 1975 was found in May, two years later, in Ceuta, a Spanish city on the North African coast, 1,878km distant. An adult male ringed at the Rea in May 2008 turned up in Albacete (south-east Spain) in September 2014, 6y 3m 29d later. None ringed elsewhere have been recovered here. Together, the recoveries fit the established pattern of fidelity to breeding sites, the long distance and direction of migration, and the hazards encountered on the way.

The current status of Turtle Dove is dire and without positive action to address its plight, it seems inevitable that it will be lost as a breeding species within the next few years, if it is not already too late.

John Arnfield

Collared Dove (Eurasian Collared Dove)
Streptopelia decaocto
Common resident

Originally resident in Asia eastwards from Turkey, Collared Dove colonised Europe at a remarkable rate, spreading rapidly from the near-East from about 1930 onwards, eventually breeding in Britain for the first time in 1955, in Norfolk. Northward and westward expansion continued rapidly, and, following a pair at Atcham in 1961 and two pairs there the following year, the first breeding was confirmed at Ludlow in 1963. In 1964 it was thought to have bred or attempted to breed at several widely scattered localities, from Ludlow in the south, Atcham, Shrewsbury and Wellington, to Llanymynech in the far north-west. Expansion continued throughout the 1960s, and it was widespread but not common (SBR 1965); widespread and increasing in numbers (SBR 1966), and in all areas and recorded by all observers (SBR 1969). By 1971 it merited no more than a cursory listing among species that are 'regularly found', and breeding was confirmed in all except five of the County's 33 10km squares, with probable breeding in one, possible breeding in three, and absent from only SO67 (Clee Hill) during fieldwork for the first BTO Atlas (1976).

By the mid-1970s, just over 10 years since first breeding, it was poorly recorded and overlooked, apart from the regular winter breeding for which it is famous. One fledged on 15 January 1973 in Shrewsbury in a cypress tree, and thereafter Collared Doves have regularly been recorded laying eggs in November or December, and indeed breeding right through the year. This long breeding season, which enables pairs to raise two or three, or even up to five, broods a year is a major factor in the increase in population, which nationally

was still underway when the first local Atlas started in 1985, though it had apparently slowed before then.

There appears to have been a decline here between the two Atlases. In 1985–90, Collared Doves were reported as confirmed or probable breeders in 711 (81%) tetrads and possibly breeding in a further 67 (8%) tetrads. In the recent Atlas period the total number of tetrads in which they were found was virtually unchanged, but the respective figures were 636 (73%) and 143 (16%). Although the ratio of confirmed

Occupied Tetrads

Atlas period (breeding)	1985–90		2008–13		Change	
	Number	%	Number	%	Number	%
Confirmed	430	49	183	21	-247	-57
Probable	281	32	453	52	172	61
Possible	67	8	143	16	76	113
TOTAL	**778**	**89**	**779**	**90**	**1**	**0**
Tetrads with Winter Records (2007–13): 760 (87%)						

Breeding Relative Abundance (2008–13)

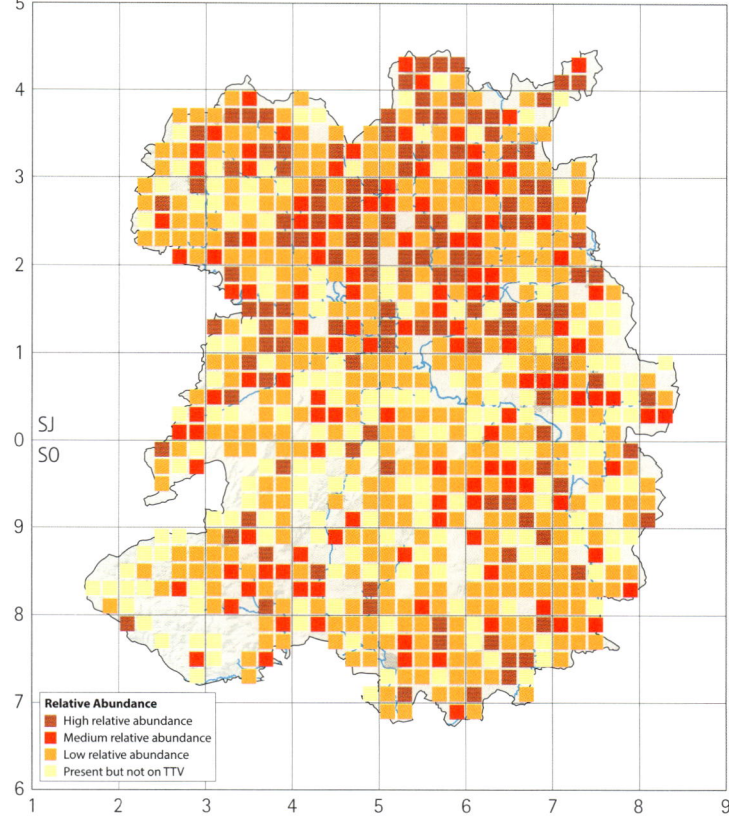

Relative Abundance
- High relative abundance
- Medium relative abundance
- Low relative abundance
- Present but not on TTV

to probable breeding has shifted between the two Atlases for many species, the sum of the two has usually remained similar (see p. 488), so the decline in the two together, and the increase in tetrads with possible breeding, suggests a decline in population density.

Support for this hypothesis comes from BBS data. The population in 2014 was only 6% less than that in 1997, but, although there have been annual fluctuations, the underlying trend is a decline of about 40%. GBW also shows a decline in garden use in the West Midlands, the average annual reporting rate falling from 79% in 1996 to 44.7% in 2014, while in Shropshire the annual average reporting rate in 2006 was 61.9%, but in 2014 it was 44.7%, a 28% decline over nine years (John Arnfield *pers. comm.*).

Based on TTV data, the population is estimated at 14,000–14,500 pairs. This is well above the 2,000–3,000 pairs estimated in 1992. The latter was probably a conservative estimate, but conversely, the current estimate, equivalent to about 16 pairs per tetrad, seems excessive.

The summer and winter relative abundance maps are very similar, and show the population is highest in lowland areas in the north and east, where there are arable and cattle farms providing grain to eat, with fewer in the uplands in the south and south-west. They are absent altogether from the Long Mynd, parts of the Clun Forest and Brown Clee, and large areas of forestry. The number of tetrads

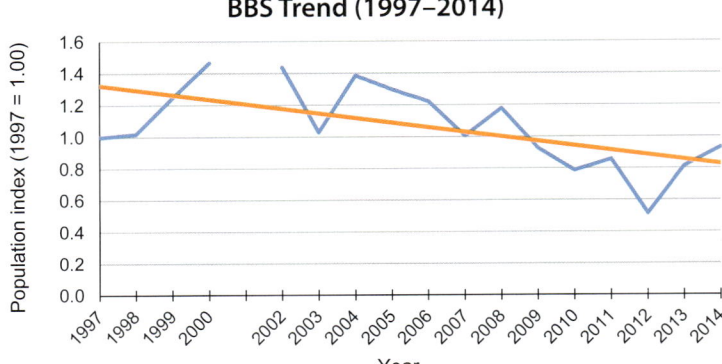

BBS Trend (1997–2014)

Jim Almond, The Rea, 1 June 2007

where Collared Doves were found in the two Atlases was virtually identical, and most gains on the breeding distribution change map are matched by losses in immediately adjacent tetrads, suggesting normal variation in fieldwork results in areas of low density. However, losses in the south-west and in the eastern half of the Severn Valley, which show as gaps on the relative abundance map, are likely to reflect a loss of mixed farming (farms now buying in animal feed, rather than growing their own), a reduction in cattle farming, and increased efficiency of grain storage.

Out of the 20-odd that have been ringed here and recovered, all except four were recovered locally, indicating that many are sedentary. However, of particular interest was one killed by a cat at Shrewsbury in December 1970 which had been ringed as a full-grown bird in Belgium in September 1967, 584km distant, perhaps suggesting that the growth of the local population in the 1960s was being supplemented by continuing immigration from the continent. The only other recovery involving movement overseas was in the opposite direction in 1990; one ringed at Newport in March was shot in Nord, France in October in the same year.

The extraordinary population expansion has been explained by the industrialisation of arable farming and the doves' exploitation of spilt grain from harvesting and stored crops, but the rapidity, and examples of long-distance movements by a largely sedentary species, suggest more than just an ability to disperse and occupy new areas simply because of population growth. After reaching the UK, the range expansion continued to the Iberian Peninsula, North Africa and Ireland. Following escapes from captivity, colonisation of the American continent is in full swing.

Gerry Thomas

Cuckoo (Common Cuckoo)

Cuculus canorus

Scarce summer visitor

Cuckoos breed across the whole of Europe and much of Asia, and most winter in southern and central Africa south of the Sahel.

Few birds are as well known to the general public, both for their bizarre breeding practices and for the distinctive two note song of the male, with its promise of the summer to come. Far more often heard than seen, it is not an easy species to study. As a result, misconceptions about its habits abounded in earlier days. Beckwith (1892) stated

that 'the practice of the Cuckoo ... to lay her egg upon the ground, and then to convey it in her bill to some suitable nest, seems to have been proven beyond doubt', and offers as justification that her size and weight would preclude being able to perch on the fragile nests of host species without damaging them, as well as known cases of parasitism of nests in narrow crevices. It was the pioneering work of Edgar Chance (1922) and slow-motion cinematography of Oliver Pike

John Fielding, Venus Pool, 18 July 2009

her to sit. The egg removed is normally eaten, which may also have given rise to the misconception that Cuckoos feasted on the eggs of other birds.

An analysis of SBRs and all Cuckoo-related articles available through the *Histo* website, covering 1879–2014, reveals that 33 different species have hosted Cuckoo eggs over that period.

The *Atlas* 1992 identified Dunnock, Meadow Pipit and Reed Warbler as the three main hosts, and this correlates well with the post-1950 data in the table below. The earlier results in the listing suggest that a greater diversity of hosts may have been used in the past. In some cases, the explanation for the apparent change may be the decrease in the host itself (e.g. Corn Bunting, Hawfinch) but others are harder to explain, such as the decline in Pied Wagtail as a host. The apparent increase in Cuckoo parasitism of Reed Warbler in the post-1950 period is notable. However, this may be partially an artefact of the efforts of the ringing group at Allscott Sugar Factory which, since the early 1960s, has targeted Reed Warbler nests, increasing the probability that Cuckoo parasitism would be detected. Of the 17 cases of Reed Warbler hosting in the later period, 10 are at that site. Nevertheless, even discounting these cases, Reed Warbler parasitism would appear to have increased in line with the national findings of Brooke and Davies (1987), who also describe the decline in the use of Robin, Pied Wagtail, Dunnock and the 'minor hosts' shown in the table of host species. In addition, the Reed Warbler colony at Crose Mere has been monitored every year since 1962 (apart from 2005). Four Reed Warbler nests were parasitised in 1992 and 1993, and two were parasitised in 1994, and two young Cuckoos fledged in each of

(much of which was carried out at Pound Green Common in the Wyre Forest, just over the border with Worcestershire) that demonstrated conclusively that this was not the case. Hen birds watch potential host nests for long periods of time until laying begins. Then they move in quickly, pick up one of the existing eggs in their bills, turn round and lay their own egg, usually in less than 10 seconds. The female Cuckoo also possesses an extrudable cloaca, permitting it to 'squirt' the rather small egg (relative to the size of the adult) into the nest from on or outside the nest rim when space is not available for

Most Commonly-Reported Hosts for Cuckoo: 1879–2014

Species	Number of Occurrences (1879–2014)	Number of Occurrences (1879–1949)	Number of Occurrences (1950–2014)
Dunnock	49	20	29
Pied Wagtail	35	33	2
Meadow Pipit	19	9	10
Reed Warbler	18	1	17
Robin	8	8	–
Redstart	5	5	–
Willow Warbler	5	5	–
Grey Wagtail	4	4	–
Blackbird	3	3	–
Blackcap	3	3	–
Chiffchaff	3	3	–
House Sparrow	3	2	1
Whinchat	3	3	–
Wood Warbler	3	3	–
Wren	3	–	3
Whitethroat	2	2	–
Corn Bunting	2	2	–

Species	Number of Occurrences (1879–2014)	Number of Occurrences (1879–1949)	Number of Occurrences (1950–2014)
Hawfinch	2	2	–
Reed Bunting	2	1	1
Song Thrush	2	2	–
Spotted Flycatcher	2	2	–
Tree Pipit	2	1	1
Yellowhammer	2	1	1
Bullfinch	1	1	–
Chaffinch	1	1	–
Garden Warbler	1	1	–
Greenfinch	1	1	–
Lesser Whitethroat	1	1	–
Linnet	1	1	–
Sedge Warbler	1	1	–
Shrike sp.	1	1	–
Skylark	1	1	–
Stonechat	1	1	–
All species	190	125	65

This table excludes the previously unpublished 10 occurrences of hosting by Reed Warbler at Crose Mere in the 1990s referred to above.

the three years, but these are the only instances of Cuckoo parasitism found there (previously unpublished data, John Hawkins *pers. comm.*).

As a consequence of the Cuckoo's 'obligate brood parasitism', many early authors were able to describe some unusual cases of hosting. Paddock (1890) reports a nest with four Pied Wagtail and two Cuckoo eggs, of which only one of the latter hatched, the other being ejected from the nest with those of the host. Beckwith described a similar case two years later in a Meadow Pipit nest. Descriptions of three Cuckoo eggs deposited in Meadow Pipit and Pied Wagtail nests come from 1930 and 1932. Cases of Cuckoos removing the previously laid eggs of their own species were described in 1924 and 1930, both involving Pied Wagtail hosts. Finally, that young Cuckoos may be fed by birds other than their foster parents was confirmed by Beckwith, who describes a case from Eaton Constantine in 1881 in which two Dunnocks, two Pied Wagtails and one Spotted Flycatcher were observed responding to the demands of a single chick.

The 1892 study by Beckwith also addressed the issue of (in modern animal behavioural parlance) *gentes*, or host-specific lineages. The female Cuckoo concentrates her parasitism on one host species,

normally the one which was her own foster parent, and lays eggs that usually mimic those of the host. The hypothesis that egg design varies with the host to prevent detection had been proposed about 40 years before Beckwith, and he undertook an experiment to assess its validity. He placed 50 eggs of various origins into the nests of different species, often one with egg size or design that differed far more than do those of the Cuckoo and its host. Of these, 86% ignored the deception and took no measures to eject or destroy the intruding egg. He concluded that 'the Cuckoo's egg has little need to resemble ... those among which she places it'. This position is at variance with modern parasitism studies that suggest that mimetic eggs confer significant evolutionary advantages to the Cuckoo in the arms race with its hosts (Stoddart & Stevens 1986; Brooke & Davies 1988; Davies & Brooke 1989; Davies 2015).

Average first arrival dates are now around 9 April. The data shown in the *Arrival Dates* chapter indicate an advance of six days since 1900. Departure dates are harder to determine as Cuckoos rarely sing after the end of June, requiring visual observation to confirm presence. Most evidence points to a July departure date for adults but with juveniles remaining into August or even September.

There is only one ringing recovery: a juvenile ringed at Dungeness (Kent) in June 1981 was found near Shrewsbury five years later.

The Cuckoo is found in a large range of habitats, including woodland, farmland, heathland, scrub and wetlands, as long as there is an adequate supply of the host species and the insect food (especially hairy caterpillars) that constitute its diet. The breeding distribution map shows clusters on a background of low concentration, and the best evidence for breeding comes from the cores of these clusters in

Occupied Tetrads

Atlas period (breeding)	1985–90		2008–13		Change	
	Number	%	Number	%	Number	%
Confirmed	83	10	12	1	−71	−86
Probable	379	44	76	9	−303	−80
Possible	321	37	255	29	−66	−21
TOTAL	783	90	343	39	−440	−56

Breeding Distribution (2008–13)

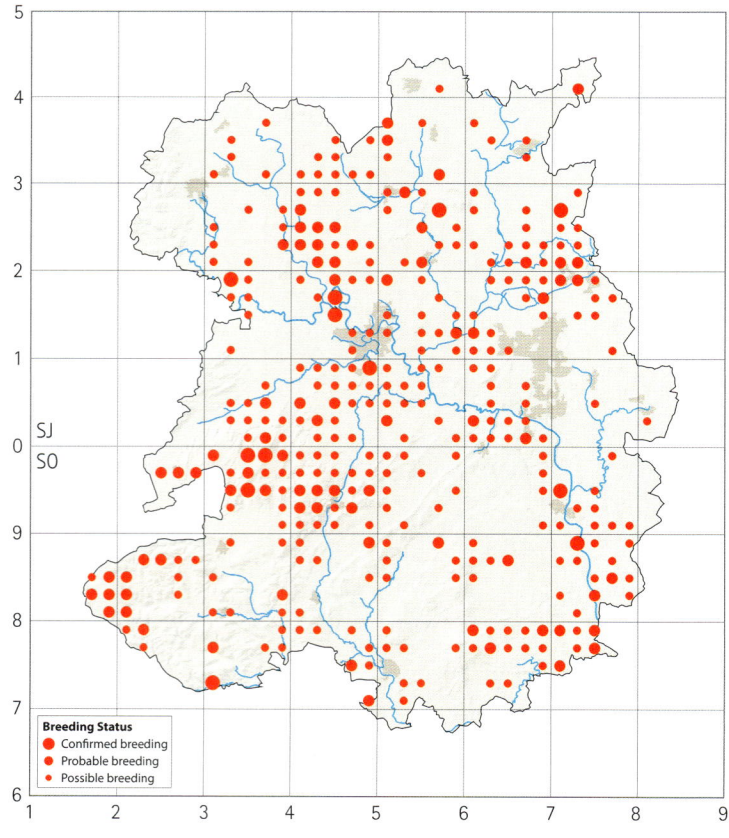

Breeding Status
- Confirmed breeding
- Probable breeding
- Possible breeding

Breeding Distribution Change (1985–90 to 2008–13)

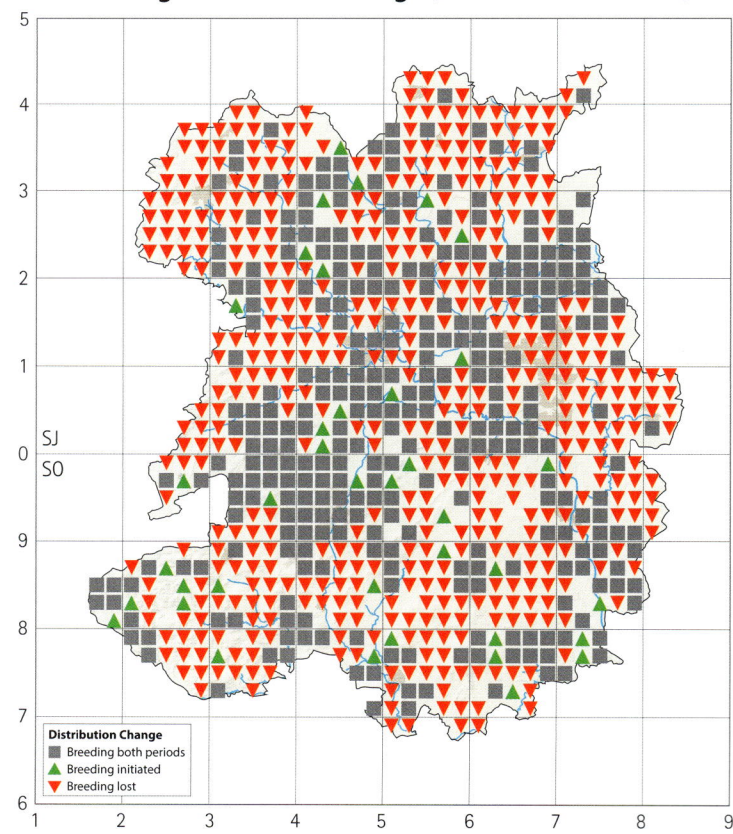

Distribution Change
- Breeding both periods
- Breeding initiated
- Breeding lost

Dave Chapman, Bettisfield Moss, 8 May 2016

drew attention to evidence that it was even then decreasing in lowland areas but maintaining its numbers in the uplands. In the UK overall, Cuckoo populations have shown marked decline since the mid-1980s, and this trend is reflected in local BBS results, which identify a drop in the number of occupied plots over the period from 1994–2014 from 47% in the 1990s to 16% in the period 2010–14.

The breeding distribution change map underlines the population decline. Tetrads showing confirmed or probable breeding decreased from 462 to 88, an 81% reduction, with most of the stability in the clusters of higher concentration mentioned above. Based on TTV counts, the population estimate is 90–95 pairs, in contrast to the estimated 175–350 laying females in 1985–90, an implied decrease of over 50%. The tetrad occupancy table reveals a drop in the percentage of tetrads with any evidence of breeding from 90% to 39%.

The reasons for this decline are unclear. Hypotheses include the following:

- Breeding success has been compromised by declines in host species, such as Meadow Pipit and Skylark.
- Changing arrival dates (perhaps due to climate change) have led to a mismatch with host species' breeding timetables. This is supported by the fact that Reed Warbler (a late-nester) was almost unknown as a host in the analysis mentioned above until the 1970s. Even though some of the apparent increase is likely to be an artefact of an increased rate of discovery of parasitised nests due to Reed Warbler ringing studies, this does not seem to be adequate to explain all of the increase in hosting by Reed Warblers.
- Changing arrival dates have introduced asynchrony with the breeding cycle of prey (food) species.
- Intensification of agriculture, such as increased use of insecticides and herbicides, and reduced field margin areas (partially offset by recent agri-environment schemes), has decreased prey availability, especially moth caterpillars.
- Unknown factors operating in the wintering grounds and/or migration routes, such as decreased food supplies. (Cuckoo is among a number of declining species that winter in the humid habitats of Africa.)

Unfortunately, it is not currently known which of the above hypotheses (or several of them) lie behind the Cuckoo's population decline and range contraction, but concerns about its future are well-founded.

John Arnfield

the Clun Forest, around the Long Mynd and Stiperstones to Chirbury, north-west of Shrewsbury, along the River Perry, Whixall Moss, north-west of Newport and along the Severn Valley from Ironbridge to the Wyre Forest. Several 10km squares have little or no evidence at all of Cuckoo occupancy.

Reports from the late nineteenth and early twentieth century suggest that Cuckoo was both widely distributed and 'very abundant' (Forrest 1899). The *Handlist* in 1964 described it as widespread but

Barn Owl (Western Barn Owl)

Tyto alba

Uncommon resident

As early as 1825 several writers noted that Barn Owl numbers were falling, while Rocke, writing in 1865, stated that it was fast disappearing due to the destructive practice of killing birds for their feathers. Wings were used for fire screens, and feathers in the millinery trade. A decade later Beckwith noted that it was much rarer than it used to be, and added that 'many are shot for the bird-stuffer', while egg collectors and gamekeepers were also taking their toll (Brown & Grice 2005). Game-rearing had become well established in the second half

of the nineteenth century, and although local contemporary evidence of the specific targeting of this owl by gamekeepers is lacking, pole traps employed against other birds of prey would have killed some. In 1890 Paddock emphasised its value to farmers, describing it as a 'useful bird in controlling rats and mice', but observed that continued persecution meant that, though it was once common throughout the district of Newport, it was now 'fast disappearing'. The loss of nest sites, later to become a major factor in its decline, was noted by

John Hawkins, North Shropshire, 19 June 2015

Beckwith around the same time, as 'dilapidated barns, hollow trees and ivy-clad church towers' disappeared.

The high frequency of severe winters during the period 1860–1900 also probably reduced the population through increased winter mortality, but by 1906 Forrest noted that 'persecution was diminishing', probably thanks to increased legislative protection first enacted in 1880. By 1908 he added that it was 'slightly increasing' after the 'senseless persecution in former years' and was 'the most numerous of the owls except in wooded country'. A reduction in the number of gamekeepers during the First World War would have reduced persecution further, and the following years were probably halcyon days for the Barn Owl. Unfortunately, agricultural intensification, which had begun in the mid-1800s and accelerated during the 1940s, resulted in the loss of prey-rich foraging habitat (Cayford 1992).

In 1932, the first national census identified 287 breeding pairs here (Blaker 1933). The *Handlist* (1964) noted that it was resident and well distributed but that it 'does not now merit Forrest's description in 1899 as being "the most plentiful of the owls"'. It added that there was some evidence of a decrease in the north in the previous five years. Regional differences in the breeding distribution became evident in 1962 when it was noted as 'not uncommon in the area between

Shrewsbury and Ellesmere', and in 1971 it was 'thinly distributed'. Records came from 'mostly north and east' in 1977, and in 1981 there were 'few records from the south'. It was mapped as absent from four of the County's 33 10km squares in the BTO national Atlas (1976).

A national census by the Hawk and Owl Trust between 1982 and 1985, which estimated that there were 97 pairs (Shawyer 1987), may have understated the breeding population, as the *Atlas* (1992) estimated over 140 pairs. The species account also noted that it was 'still widely, but thinly, distributed, although it is absent from a large part of the north-east', adding that it was unlikely to regain former high numbers as long as current agricultural practices remained. It was found in three of the four 10km squares where it was apparently absent in 1968–72.

The Shropshire Barn Owl Group (SBOG) was established in 2002 to counteract the persistent loss of nest sites and feeding habitat, the two main drivers of the historical decline, by installing nest boxes and promoting conservation of their habitat with farmers and other landowners. It estimated the population at 121–140 pairs in 2002 (Bishton & Lightfoot 2002), and the content of this account, particularly the number of confirmed breeding records, has been informed by their work.

Breeding Distribution Change (1985–90 to 2008–13)

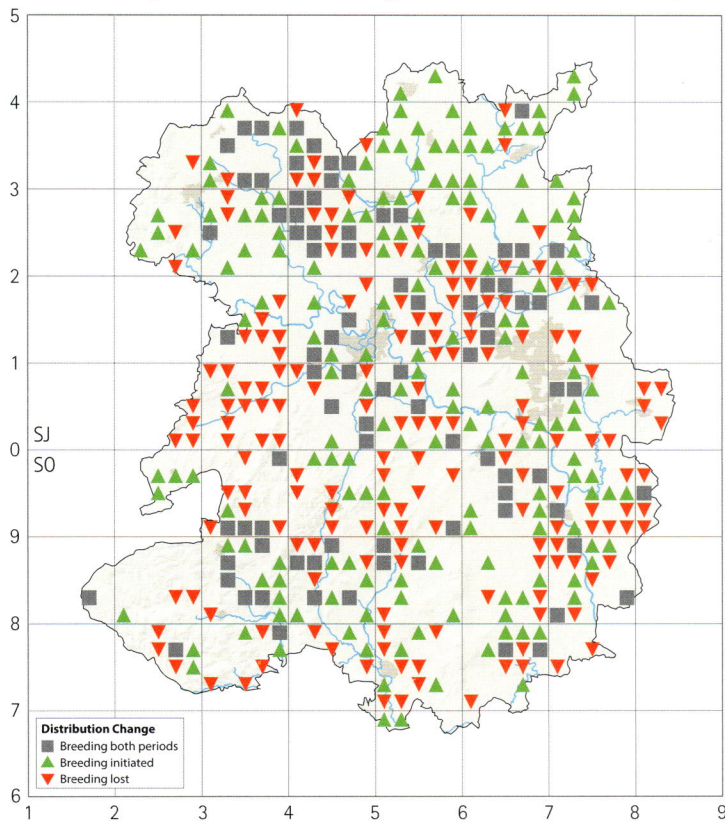

Distribution Change
- ◼ Breeding both periods
- ▲ Breeding initiated
- ▼ Breeding lost

Occupied Tetrads

Atlas period (breeding)	1985–90		2008–13		Change	
	Number	%	Number	%	Number	%
Confirmed	85	10	124	14	39	46
Probable	58	7	31	4	-27	-47
Possible	120	14	114	13	-6	-5
TOTAL	263	30	269	31	6	2
Tetrads with Winter Records (2007–13): 297 (34%)						

However, Barn Owls are difficult to locate and survey because they are nocturnal, thinly distributed and not particularly vocal or territorial. SBOG proactively disseminates information in order to obtain reports of sightings, and it collates records from SOS and other conservation bodies. It determines presence at a specific site from site surveys, local knowledge, systematic searching of suitable habitat and annual monitoring of nest boxes. Nests are notoriously difficult to find and attempting to locate individuals by cold searching is very unproductive, so some breeding pairs go undetected. Fortunately, adults are highly site faithful and, providing the habitat structure remains constant, sites are likely to continue to support a breeding pair indefinitely. SBOG has maintained a database of occupied sites and breeding pairs since 2002, but only those records obtained within the previous five years are used to estimate the annual population. In 2014 it held 195 possible, probable, or confirmed breeding pairs, including those identified during the recent Atlas period, with an average of 208 pairs over the previous five years. The recent Atlas map is much more complete as a result of this database.

Given the increase of 8.4% in tetrads with probable and confirmed breeding between the two Atlas periods, and allowing for a minimum of one breeding pair for each of the 155 such tetrads occupied during 2008–13, with the likelihood that some pairs have gone undetected and some possible breeding records represent breeding pairs, the breeding population is estimated to be in the region of 200–220 pairs. The population will vary from year to year in accordance with the year-on-year cyclical variations in small mammal abundance.

In winter, the Barn Owl is widely distributed but avoids the southern and north-western uplands and also urban areas. It was recorded in a higher proportion of tetrads in the north and was generally more widely and evenly distributed than during the breeding season. Some tetrads in the west and south where it was found were notably unoccupied during the breeding season. The more widespread distribution in winter is due to the dispersal of juveniles from their natal area and to adults extending their home range. Periods of prolonged snow cover, high winds or heavy rain can seriously reduce foraging efficiency and over-winter survival rates, and three recovered dead in the harsh weather of January and February 1979 at Broseley and Sheinton were presumed to have starved.

The majority of reported deaths are road casualties. The first noted in SBRs was killed at Crosshouses on 28 January 1974, followed by two more deaths at the same location on 28 February and 10 September. RSPCA reported '18 injured birds brought in from the Oswestry by-pass'. After adding dead birds found on that stretch, 'it is estimated that 40–50 Barn Owls were killed or injured in less than three years' (SBR 1989). SBOG has recorded 144 road casualties

During the 2008–13 breeding seasons it was recorded in all 33 of the County's 10km squares, an increase in range from 1968–72 and 1985–90, and in 269 tetrads. Most breeding pairs were north of the River Severn. Here, the land is low-lying and dominated by arable and pastoral farmland, and within this landscape there was a marked association with the valleys of the rivers Severn, Perry, Roden and Tern. Sites occupied in the south were similarly located along river valleys and catchment areas around Bishops Castle, east of Clun, Corvedale, Bridgnorth and Cleobury Mortimer. Upland areas above 122m were largely avoided and they were absent from large urban areas.

Breeding was confirmed or probable in 155 tetrads in 2008–13, compared to 143 in 1985–90. Notably, new breeding pairs have been established in the north-east around the rivers Roden and Tern, where none were recorded in 1985–90, but SBOG had been actively installing nest boxes. Highest densities were evident in the north-east, the south-east extremity, and in the Weald Moors area north of Telford, where there was an increase from one to nine pairs between 2002–05 following the installation of 29 nest boxes. Losses were widespread, but particularly in central areas around Westbury and in scattered areas across the south, at least partially due to heavy grazing reducing rough grassland feeding habitat, and loss of nest sites. Losses in the north were relatively lower, due to the nest box programme. Without it, the decline, which accelerated after the Second World War and was still evident at the end of the twentieth century, would undoubtedly have continued. Community Wildlife Groups in the Upper Onny, Upper Clun, Kemp Valley and Clee Hill, all in the south, have promoted nest box schemes in support of SBOG, and boxes in the first three of these areas were occupied in the recent Atlas period.

Road Casualties (2002–14)

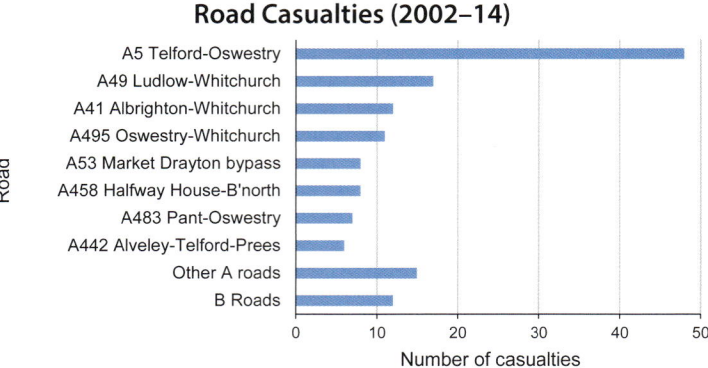

since 2002, an average of 11.0 per year, and 92% have been on trunk roads. The increase in road deaths is attributable to higher volumes of traffic, and more importantly, higher speeds and larger lorries. Adult Barn Owls are site faithful and highly sedentary, but juveniles make short-distance dispersal movements from their natal areas, usually by December, although some make longer movements: 64% of road victims occurred in the winter period October–March, and probably arise from this juvenile dispersal.

Poisoning as a result of eating rodents contaminated with second generation anticoagulant rodenticides (SGARs) increased from 5% to 53% in the UK between 1983–84 and 2002–03. Here, of 31 analysed for SGARs between 1985–2012, four (13%) had detectable residues in their liver, but none displayed signs that rodenticide poisoning was the cause of death (L. Walker, Centre for Ecology and Hydrology, Lancaster *pers. comm.* 2014).

A Barn Owl ringed as a nestling at Ellesmere on 10 June 2007 was recovered dead on a road in Trefechan, Dyfed, 62 days later on 11 August 2007, having travelled 99km, while a chick ringed at Stoke-on-Tern on 15 June 2007 was recovered dead 6km away at Cold Hatton on 31 August 2012. The greatest distance moved involved a female ringed on 18 June 2007 at Achray Forest, Scotland, recovered freshly dead at Shrawardine on 21 February 2009, 1y 8m 4d later, having moved 398km. Other long-distance recoveries included owls ringed in Hampshire (179km), Devon (225km) and Essex (271km). The average dispersal of four ringing recoveries reported to SBOG between 2010 and 2012 was only 6km and the average dispersal

movement in the UK is 12km. These distances are more typical for this sedentary species, and it is possible that the longer distances were all 'lorry-assisted'.

By 2014, SBOG had installed 387 nest boxes. The boxes also provide valuable roosting sites throughout the year. The owls are well established on their breeding sites in February and most eggs are laid in late April and May. The date of laying of the first egg between 2008 and 2014 varied from 15 March to 20 April. Young fledge mainly in July or August, but in favourable years such as the peak years of 2007 and 2014, some pairs will have two broods, resulting in autumn fledging, the latest estimated fledging date being 18 November, in 2007. Late breeding is not a recent phenomenon but was noted in 1880 when a 'young bird' was found at Hatton Grange on 30 November.

Barn Owls are not territorial but occupy a breeding range in which they require a minimum of 4ha of permanent, rough, tussocky grassland with a deep litter-layer averaging not less than 7cm, which supports their primary prey species, the field vole (Barn Owl Trust 2012). Suitable habitat on farmland is now often confined to field margins – arable headlands, hedgerows, fence lines, banks, ditches and riverbanks – and the conservation of these habitats is of paramount importance. Grassy margins sowed around arable fields under the agri-environment schemes are an increasing and valuable hunting habitat. Elsewhere, the grassy banks of rivers and other waterways and wetlands, young plantations, old airfields, heathland and, if they do occur in urban areas, roadside verges, railway embankments, brownfield sites and other unmanaged grassland, provide suitable foraging habitat. Intensively grazed pasture, rough grazing and paddocks are poor habitat.

The total number of chicks produced in monitored nest sites between 2002–14 is 1,243. The table shows that 1,012 were in nest boxes and 231 in natural sites. Although data on the location and number of natural nest sites is limited, it appears that nest boxes now provide most nest sites. Boxes on trees are marginally less productive than those in buildings, and natural tree cavities are the least productive, probably because increased exposure to reduced temperatures, heavy rainfall and predation reduces the survival of young. However, productivity between the different types of nest site is not very different. The mean number of chicks produced per successful brood for the 13 years 2002–14 is 3.0. A long-term average productivity of about 3.2 young per pair is required to maintain viable populations (Taylor 1994), so the breeding data suggests that a viable population now exists here.

Paul King, Kynnersley, 5 October 2009

Number of Chicks Produced According to Type of Nest Site: 2002–14

	Tree Nest Box	Building Nest Box	Pole Nest Box	Tree Cavity Natural	Building Natural	Other Natural	All Sites
Total Broods	221	92	12	63	12	3	403
Total Chicks	678	296	38	183	38	10	1243
Mean No. Chicks	3.0	3.2	3.1	2.9	3.1	3.3	3.0

John Fielding, Atcham, 8 August 2008

pairs may have ceased to incubate due to the scarcity of optimum prey. These breeding cycles are not a recent phenomenon, as Beckwith noted that Barn Owls were 'unusually plentiful' in the autumn of 1887. They are synchronised with their primary prey – the field vole. Analysis by SBOG of 522 pellets collected from 69 sites identified 1,458 separate prey items and confirmed that field vole is the primary prey, comprising 71% of the prey items. Secondary prey includes wood mouse (12%), common shrew (9%) and bank vole (5%).

Trees provide most natural nest sites and 96% of tree nests were in live trees. Three species have been utilised – ash (68%), oak (25%) and sycamore (7%), the occupancy probably reflecting the propensity for ash and oak to develop large cavities. Elms were historically important before Dutch elm disease removed most of them. Some agricultural buildings, both modern and old, and bale stacks, have been utilised, industrial buildings and historical ruins infrequently, with one nest site in a dovecote and one in a fissure in the rock face of a disused quarry. The apparent relatively low usage of agricultural buildings may be the result of under-recording rather than a lower occupancy per se as farm buildings are on private, often inaccessible, land. However, 312 site surveys have been conducted by SBOG and while not all of the sites had suitable outbuildings, where they did they were inspected for occupation and generally found not to hold breeding pairs. Historically, church towers appear to have been important breeding sites, as in Caynham on 23 May 1963 when a young bird was found, but today church towers are not readily accessible because most have been enclosed.

In the future, habitat loss and degradation, decay of old tree nest sites, the renovation and dilapidation of farm buildings, adverse winter conditions, climate change, rodenticides and road casualties, will all continue to have an impact. Some of these factors can be mitigated – the provision of a designated and self-contained loft space within a barn conversion, for example – and, ironically, urbanisation in the larger towns has occasionally had a temporary benefit if grassland on green field development sites is taken out of agriculture and it subsequently remains undisturbed and free to develop naturally over several years. However, urbanisation continues to encroach on hunting habitat and tree nest sites, and the disturbance and development of buildings known to support breeding Barn Owls continues.

In spite of all this, the breeding population now appears to be at a higher level than it was 20 years ago. While SBOG survey work and data collection has undoubtedly contributed to better knowledge, the sustained effort to replenish the loss of natural cavities by the siting of nest boxes in areas of good feeding habitat has promoted a real increase in the number of pairs. This trend can be maintained but there is a caveat. Aside from the inherent problem of maintaining sufficient funding to continually install new nest boxes and replenish ageing ones, existing habitat must be maintained, and continued funding, targeted through agri-environment schemes following the UK's exit from the EU, is essential for grassy headlands and margins to secure the population. In the longer term, oak and ash trees on farmland continue to be lost due to decay and there is an urgent need for a hedgerow tree planting programme to ensure a sustainable supply of natural tree and roosting sites in the future.

The chart shows that a two or three-year cycle in breeding productivity is increasingly evident, as is high productivity following exceptionally poor breeding seasons in 2006 and 2013. In 2006, below average temperatures in March and wet weather in early spring probably inhibited the growth of fresh grass and further depleted the field vole population. Four road casualties in April were also recorded for the first time, suggesting that hunger may have driven some Barn Owls further afield to marginal habitats in search of food. In 2013, there was a relatively large number of nest failures at the incubation stage, and monitoring revealed caches of predominately secondary prey items in the boxes – shrews and wood mice – suggesting that

Glenn Bishton
John & Wendy Lightfoot
Shropshire Barn Owl Group

Number of Chicks Found in Nest Boxes and Natural Sites (2002–14)

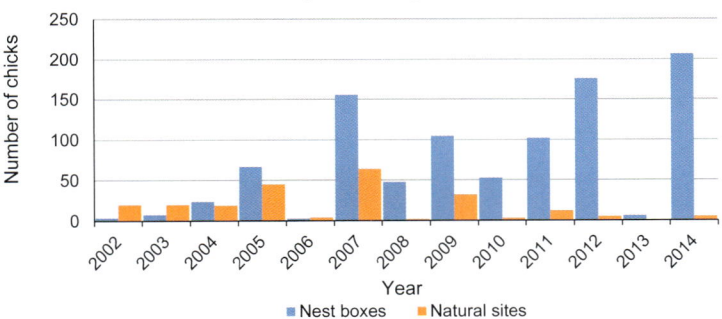

Tawny Owl
Strix aluco
Uncommon resident

Celia Todd, Oswestry, April 2013

Often a harbinger of bad luck, and more commonly heard than seen, it was this owl that caused the downfall of the infamous 'Glutton Club' at Cambridge University, presided over by one Charles Darwin, born and raised in Shrewsbury and renowned for his theory of natural selection. Dining on this owl, its meat was described as 'disgusting and stringy' (Buttery 2011), and the club never convened again. Charles, however, went on to more important matters.

As early as the 1860s, decreases in numbers were being noted nationally (Holloway 1996). This was confirmed locally by Beckwith (1888), who reported it as 'much rarer than in years gone by, and fast disappearing from all its old haunts', but with a few still around the Wrekin, and by Paddock (1897), who believed it to be 'less numerous than formerly'. Beckwith considered that the scarcity of hollow trees and, particularly around the Wrekin, a change in the 'character of the woodland habitat' was partly responsible for the decline, although it is not clear what change he was referring to. Its 'taste for an occasional rabbit' had not gone unnoticed, suggesting that persecution too may have contributed.

Others considered that the activities of gamekeepers and sportsmen, serving their own needs as well as the demands of the taxidermy trade, had pushed numbers lower. Following the Great War, from which many gamekeepers and sportsmen did not return, and a change

in game preservation activity following the Protection of Birds Act 1925, there was a slow recovery (Witherby *et al.* 1938), and it became 'common everywhere' (SBR 1956). The increased use of pesticides and the hard winter of 1962–63 did little to affect its numbers (unlike those of the Little Owl), and the *Handlist* (1964) described it as 'resident, common', 'breeding in built-up areas as well as the country' and 'now the commonest owl'. This was helped by its longer life and broader diet as well as its dominance over other owls, particularly the Long-eared Owl, which it was helping to drive out.

Nationally numbers reached a peak in 1972 (Marchant *et al.* 1990), and the BTO Atlas (1976) showed breeding evidence in every local 10km square.

Fieldwork for the 1985–90 Atlas demonstrated a widespread distribution, but it was apparently missing from the north-west corner due to a lack of suitable habitat, or, more likely, small numbers resulted in less interaction and vocal competition, so it was overlooked. There were also patchy absences and lower densities in upland areas, major conurbations, and open and less wooded agricultural areas. Many apparent absences were likely to be due to insufficient nocturnal survey work. Using figures in the BTO Atlas (1976) as a guide, the population was estimated at 900–1,800 pairs.

Local results from two BTO surveys of Tawny Owl in 1989 and 2005 showed a small decrease in the number of pairs per occupied

Occupied Tetrads

Atlas period (breeding)	1985–90		2008–13		Change	
	Number	%	Number	%	Number	%
Confirmed	237	27	117	13	-120	-51
Probable	197	23	130	15	-67	-34
Possible	219	25	146	17	-73	-33
TOTAL	653	75	393	45	-260	-40
Tetrads with Winter Records (2007–13): 386 (44%)						

Breeding Distribution Change (1985–90 to 2008–13)

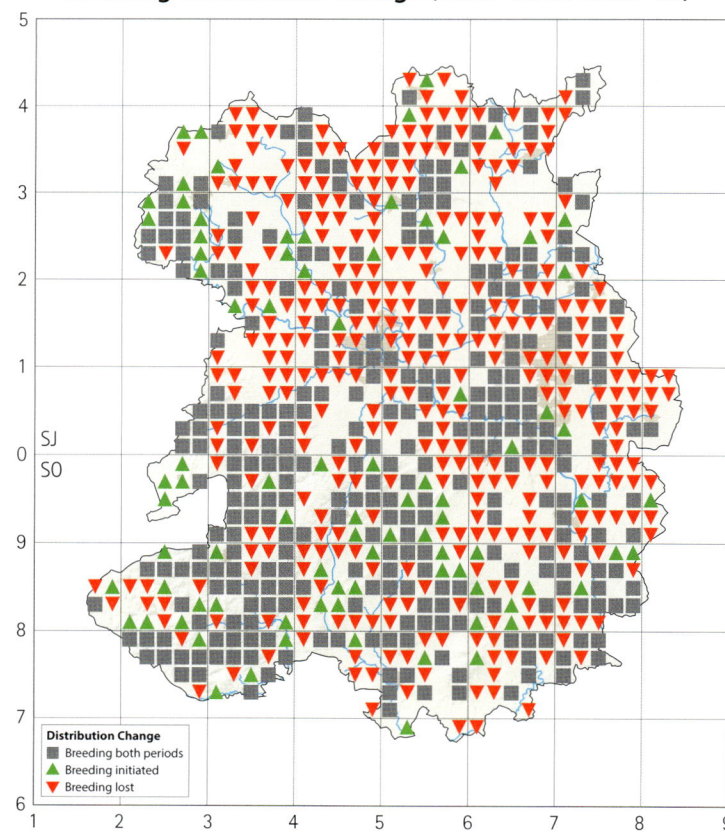

Distribution Change
- Breeding both periods
- Breeding initiated
- Breeding lost

tetrad from 1.42 to 1.36 between surveys, as well as an 8% decrease in occupied tetrads (Dawes SBR 2005). Recalculation of the population, based on these figures, indicated approximately 1,200 pairs. These findings supported the national trend of a shallow decline in numbers (Baillie *et al.* 2014).

The breeding distribution change map reflects these continuing declines with a 40% decrease in the number of tetrads with breeding evidence, including loss of confirmed breeding in 120 tetrads and probable breeding in an additional 67.

Most losses appear to be predominantly above a line running north-west/south-east through Shrewsbury, with numerous pockets of absence elsewhere. These areas are mainly arable farmland, so local land use changes – increased agricultural intensification leading to loss of hedgerows, hedgerow trees, field margins and coppices, and loss or reduction of large woodlands – are all likely to contribute to loss of nest sites and prey. Losses in the urban areas of Shrewsbury and Telford are also striking, probably indicative of changing habitat with the same effects.

Other factors might also be at play as, in parallel with increasing global temperatures, the increased incidence of avian malaria in Tawny Owls has increased nationally from 1–2% in 1996 to 60% in 2010 (Musgrove *et al.* 2013).

Based on TTV counts, the population estimate is 900–1,000 pairs. As this is almost 2% of the estimated national population, it is likely to be much too high. Using the 2005 figure of 1.36 pairs/tetrad for all tetrads with evidence of breeding at any level, gives a population estimate of only 530 pairs, more consistent with the estimate in the *Atlas* (1992) and the local decline found since then.

First-year birds are driven from their natal areas by their parents, and are thus found occupying marginal habitats, but adults are sedentary and remain in their territories for their whole life. Consequently, the breeding and winter distribution maps would be expected to be very similar, with winter maps showing a more extensive range due to a population increased by juveniles, and more vocal adults increasing the likelihood of detection. In fact, the number of tetrads with records is virtually the same for both seasons, suggesting a lower level of nocturnal fieldwork in winter.

The first national BTO Winter Atlas (1986) showed Tawny Owls present in all 10km squares except SO27, 38 and 67, and SJ33. Recent Atlas fieldwork detected them in all 10km squares but, at tetrad level, they were not found in over half (56%) of tetrads, predominantly above the same line running north-west to south-east referred to above.

Few are ringed, but there are three recoveries of note:

- the oldest individual, which lived for at least 18y 6m 20d, was ringed as a first-year male at Kinlet in 1982 and found dead at almost the same place in 2001, supporting evidence for the sedentary nature of this owl. The national longevity record is 21y 5m 13d;
- a nestling ringed near Habberley in May 1976 was found freshly dead 10y 5m 10d later, again only 2km from the ringing site;
- the most travelled, which was an adult ringed in 1983 at Bayston Hill and found dead 3m 18d later, 121km away near Newport in South Wales. Given that road deaths for this owl are often reported locally, and that many long-distance records from elsewhere have been proved to be 'lorry-assisted', this record should be treated with caution.

A total of 28 recovered Tawny Owls have been ringed or recovered here, and 22 of these were both ringed and recovered locally, again confirming just how sedentary this owl is. Five of the other six came from, or went to, adjacent counties, and only the one found in Gwent crossed more than one county boundary.

The future for this owl is unclear in the face of conflicting evidence. BTO BirdTrends reports a long-term shallow decline of 17% over the period 1967–2010 (Baillie *et al.* 2014), but Robinson *et al.* (2014) found no supporting evidence for this decline from analysis of the various stages of its life cycle. The reasons for the decline are unknown; clearly there is a need for more research and targeted surveys of this owl to clarify this uncertainty.

Bob Harris

Little Owl

Athene noctua

Uncommon naturalised resident

The Little Owl was kept as a household pet as early as the 1700s to control cockroaches, but the occasional continental visitor and several deliberate introductions in the nineteenth century established it as a breeding species in south-east England by the late 1880s, after which it spread rapidly.

The first two records were both from 1899, when one was heard in February near Willey and a second was trapped at Wenlock Edge in August. Forrest (1899) made no mention of this owl so presumably it was only just colonising at this time. The next records were two years later when six individuals were found at Wenlock Edge, but there were no further records until 1908. From then until 1916 there were eight records – nearly all of birds found dead or shot – either from central locations: Shrewsbury, Cressage, Petton, Haughmond Hill and Bomere; or from the south: Church Stretton, one trapped at

John Hawkins, Shropshire, May 1989

Occupied Tetrads

Atlas period (breeding)	1985–90		2008–13		Change	
	Number	%	Number	%	Number	%
Confirmed	214	25	40	5	-174	-81
Probable	139	16	39	4	-100	-72
Possible	232	27	89	10	-143	-62
TOTAL	585	67	168	19	-417	-71
Tetrads with Winter Records (2007–13): 107 (12%)						

Breeding Distribution Change (1985–90 to 2008–13)

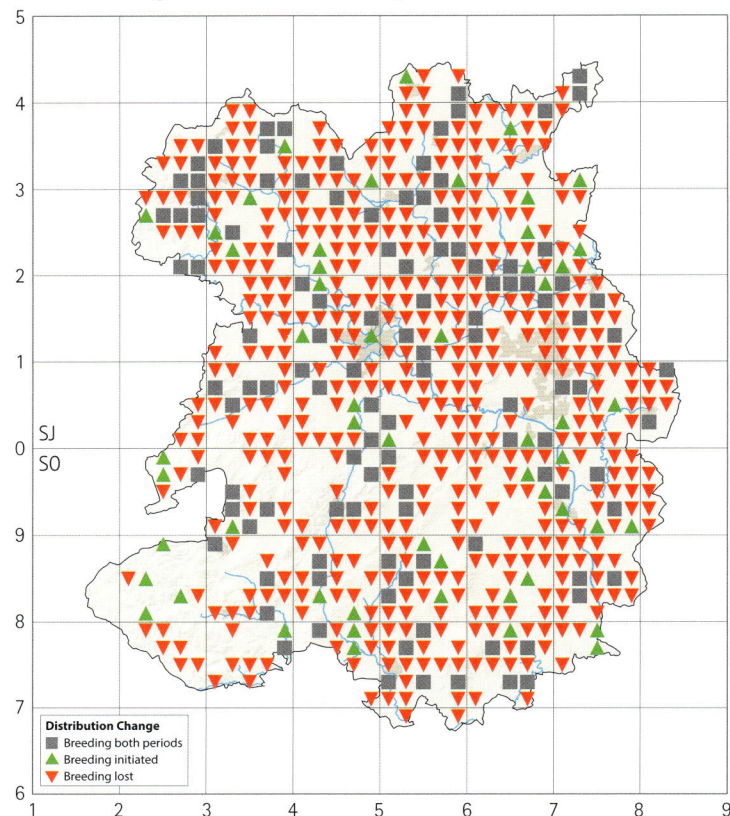

Distribution Change
■ Breeding both periods
▲ Breeding initiated
▼ Breeding lost

Millichope Park, and Craven Arms. The *Handlist* (1964) noted that spring records in 1914 and 1916 'suggest a possibility of breeding' but provided no further detail. The first confirmed breeding was at Stanton Lacy in May 1919, with three eggs found in a nest, and then at Acton Burnell the following year.

The *Handlist* stated that it was well established in the south by 1920 and in the north a year later. Most of England and Wales was colonised by 1925. Records at this time were of birds still being shot, possibly for taxidermy, particularly around Market Drayton, Shifnal, Broseley, Clungunford and Ludlow. Around Weston (near Shifnal), it was being shot as it was 'becoming a pest'. By 1940 it was considered to be possibly the commonest owl, but opinion, even then, was of decreasing numbers over the preceding 10 years.

The severe winters of 1940 and 1960 would have affected survival and almost certainly gave rise to local declines, and few records were received, suggesting the possibility of a decrease (SBR 1960). Research by Prestt (1965) on smaller birds of prey in Britain suggested that habitat loss and pesticide use were also contributing to declines, and there is no reason to believe that different factors were involved here.

The first BTO Atlas (1976) showed some level of breeding evidence in almost all local 10km squares, with absence only from the Clun Forest, around the Stiperstones, and the urban conurbation of Shrewsbury. SBRs from 1978–84 continued to report this owl as widespread, with most records from the months of May–July.

At the start of fieldwork for the 1985–90 Atlas, even taking into account its cyclic pattern of abundance and density, it had reached its highest-ever index of abundance nationally (Marchant *et al.* 1990), so the scene was set for an accurate reflection of the status of this owl

locally. Breeding was confirmed in 25% of tetrads with a further 16% added as probable and 27% as possible breeding (breeding evidence in 67% of all tetrads). The national BTO Atlas (1993) estimated five to 10 pairs per occupied 10km square across the country, but described this as 'conservative', and the map showed Shropshire as having a high relative abundance. The *Atlas* (1992) worked on one to three pairs per tetrad with breeding evidence, and estimated the population at 600–1800 pairs.

Fieldwork for the recent Atlas documented a considerable decline. Breeding evidence from 585 tetrads had dropped to only 168, a loss from 71% of tetrads. This certainly indicates a contraction of range with, almost inevitably, a corresponding population decline.

The breeding distribution change map shows this decline vividly. The few gains have been swamped by widespread losses.

Applying the previous assessment of one to three pairs per tetrad with breeding evidence gives an estimated population of only 170–500 pairs. Given the reduction in population since 1990, evidenced by a 40% decline nationally, particularly in western areas (Balmer *et al.* 2013), and a large reduction in the number of tetrads where it was recorded in recent Atlas fieldwork, it is realistic to take the lower end of this range. Caveats to this are under-recording of a scarce nocturnal species but, conversely, some pairs may have been 'double' recorded in adjacent squares. This suggests a population estimate of, say, 150–200 breeding pairs. Based on TTV counts, the population is estimated at 50–55 breeding pairs, but this appears to be much too low.

The Little Owl is not inconspicuous in winter, being at its most vocal in February when proclaiming territory, and often sunning

Yvonne Chadwick, Cranmere, 6 August 2012

Breeding Distribution (2008–13)

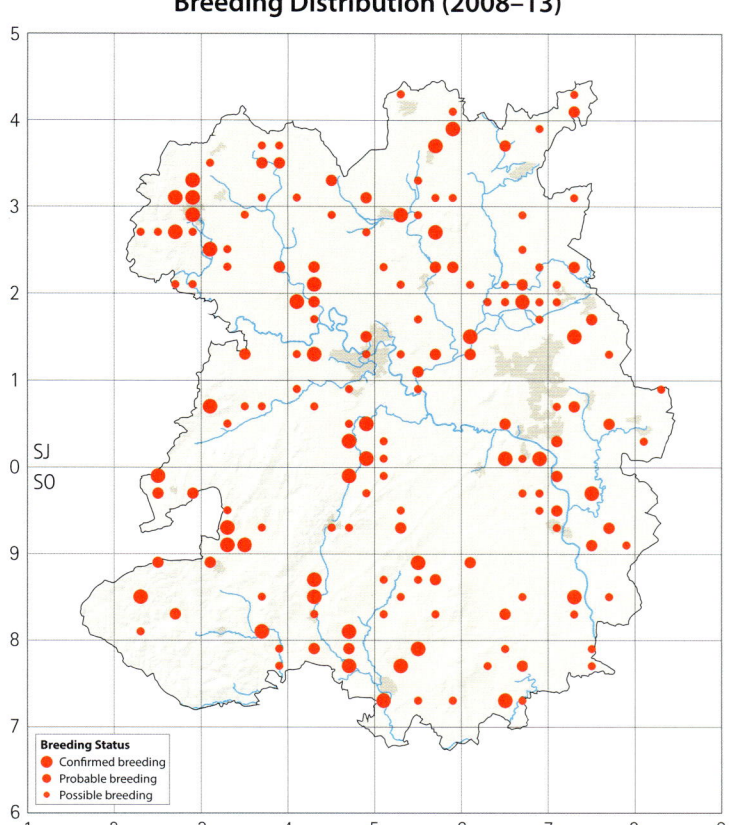

Breeding Status
- Confirmed breeding
- Probable breeding
- Possible breeding

Winter Distribution (2007–13)

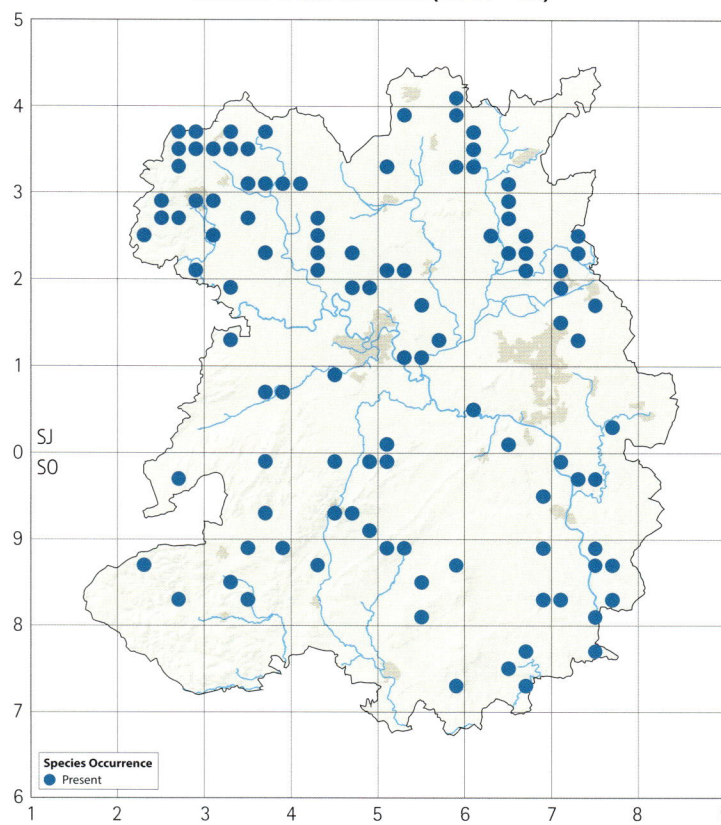

Species Occurrence
- Present

itself, but only a sparse distribution was found in recent fieldwork. It was recorded in only 12% of tetrads, only two-thirds of the breeding season count, predominantly from the lowland northern plain, but in many tetrads with no breeding records.

National ringing studies of a small number of individuals have indicated a possible median adult dispersal of 2km, with first year birds moving up to 7km (Wernham *et al.* 2002). Generally though, the majority of adults are considered to be sedentary, moving little. Data from the first BTO Atlases (1976 and 1986), at 10km resolution, also suggest a sedentary nature with little difference between breeding and winter distribution. As density has decreased, this year-round occurrence may have been lost, as no winter presence was found at many breeding sites; a feature observed by others in an adjacent county (Norman 2008). However, it is more likely that a thinly distributed and declining, largely nocturnal, species has been under-recorded,

and it breeds in tetrads with breeding or winter records.

Eight ringed here have been recovered, four here, three in adjacent counties, and only one further afield, in Leicestershire & Rutland. The oldest, ringed at Cound in 1978, was recovered after being hit by a car 6y 3m 26d later, but in the same place.

Work undertaken nationally has failed to show any demographic parameter responsible for declines, while studies in Europe suggest the main reason is falling rates of juvenile survival, perhaps due to loss of habitat and changes in farming practices, including loss of established hedgerows with mature trees, farm buildings and waste scrubby land. Unfortunately, unless specialist interest groups and targeted nest box projects can be initiated similar to those witnessed for Barn Owls, it appears that we shall continue to see a decline in this owl.

Bob Harris

Tengmalm's Owl (Boreal Owl)
Aegolius funereus
No modern records

This largely sedentary owl mainly inhabits the dense forests of northern and central Europe, so this historic record is particularly unusual.

On the evening of 23 March 1872, 'a labouring man' was out shooting pigeons at Boreatton Park near Ruyton-XI-Towns. He noticed a bird fly 'around the tree under which he was standing, and not knowing what it was he shot it'. It was sent to Henry Shaw for

preservation. Examination found that 'it was a male, and [I] found no traces of captivity about it. The primaries and tail-feathers were clean and perfect [and] ... The plumage and markings ... exhibited a dusky, or rather "smokey" appearance, as if the bird had been living in holes of trees' (Rocke 1872).

John Tucker

Long-eared Owl
Asio otus
Rare resident

John Hawkins, North Shropshire, 8 February 2006

This owl breeds across Europe in all but the most northerly areas. It is strictly nocturnal, elusive, and tight-sitting during incubation. Adults are only vocal in May–July when proclaiming territorial presence, and the continuous night-time begging 'creaking' or 'squeaking gate' calls of young are the only other indicator of presence, so it tends to be naturally under-recorded. However pairs are extremely site faithful so, once found, they can often be found for several years running in almost the same place.

In spite of both persecution and taxidermy in the latter half of the nineteenth century, numbers were still increasing nationally (Holloway 1996). This was considered to be due to falling Tawny Owl numbers, reducing inter-species competition, and the increasing maturity of previously seeded conifer plantations (Holloway 1996; Bull 1888). Beckwith described it as resident and frequently found in woods with evergreen firs, with a presence on the Wrekin where it outnumbered Tawny Owls. Several years later Forrest (1899) described it as 'rather common, especially around the Wrekin'. At various times up until 1924 it was also present in woods at Oswestry, Ellesmere and Worthen, with breeding at Newport, Shawbury Heath, Wenlock Edge, Prees, Hodnet, Haughmond Hill, Higher Ercall, Clun and Ludlow.

The *Handlist* (1964) still documented it as resident, but apparently scarce in the face of insufficient evidence to report otherwise. A pair had been breeding on the Long Mynd since 1955, in either Callow Hollow or Minton Batch, and it was increasing in the Bucknell area in the south, but there were only two other reports of presence – individuals at Ironbridge in April 1956 and White Mere in March 1958.

Two were reported in Callow Hollow (Long Mynd) again in June 1965, with one there in March 1966, while a dead chick was found in an old Carrion Crow's nest on the Long Mynd in 1968 (valley unspecified). These were all breeding season records, and the old crow's nest was probably the owls' nest that year. Another was seen on Long Mynd in December 1971.

In the national BTO Atlas (1976) there were only two confirmed breeding records (SO48 Craven Arms and SO59 Wenlock Edge) and two possible records (SO69 Morville and SO79 Bridgnorth) in the County's 33 10km squares.

From the end of fieldwork for that BTO Atlas until 1985, the beginning of fieldwork for the 1992 Atlas, there were 13 records of presence, typically of single birds at sites in the south and east, but including a pair at Marshbrook in the summer of 1976 and one at Allscott Sugar Factory from the last day in August to early October in the same year.

The Atlas (1992) showed some level of breeding evidence in 20 tetrads, mainly again in the south, and all from small woods of 4–8ha (except one at Knighton). In 1987 a pair bred raising two young at an undisclosed site in the north-east, the first confirmed breeding since 1968. However, the recent review of rarity records judged that some of the 1985–90 records are no longer acceptable, and the tetrad occupancy table and the breeding distribution change map have been revised accordingly. There was some level of breeding evidence in 16 tetrads, and breeding was confirmed in five. The population estimate at the time was a minimum of 20 pairs, but that is now revised to 16.

The BTO Atlas (1993) reported a national loss and thinning of range that was thought to be due to competition with, and dominance by, Tawny Owls. This was reflected locally in confirmed breeding in only two 10km squares (SJ51 Haughmond and SJ71 Newport) and possible breeding in a further three (SO49 Church Stretton, SJ40 Dorrington and SJ41 Montford Bridge). The 1985–90 records from 16 tetrads are within 14 10km squares, the lower figure in the national Atlas reflecting a different, and shorter, fieldwork period 1988–91. However, the two confirmed records and the possible breeding record in SO49 were additional to the 1985–90 Atlas map, and were presumably from 1991.

From 1991 to the start of the recent Atlas period there have been no records of breeding. Indeed, outside the winter months there have only been 10 records (from eight sites). Apart from two seen at Betton Moss in June 1995, three present for three days at Pole Cottage in May 2002, and one at Venus Pool for 14 days in the second half of August 2004, all the records have been of individuals seen on single dates at widely scattered sites.

Recent Atlas fieldwork found confirmed breeding in two tetrads

Breeding Distribution Change (1985–90 to 2008–13)

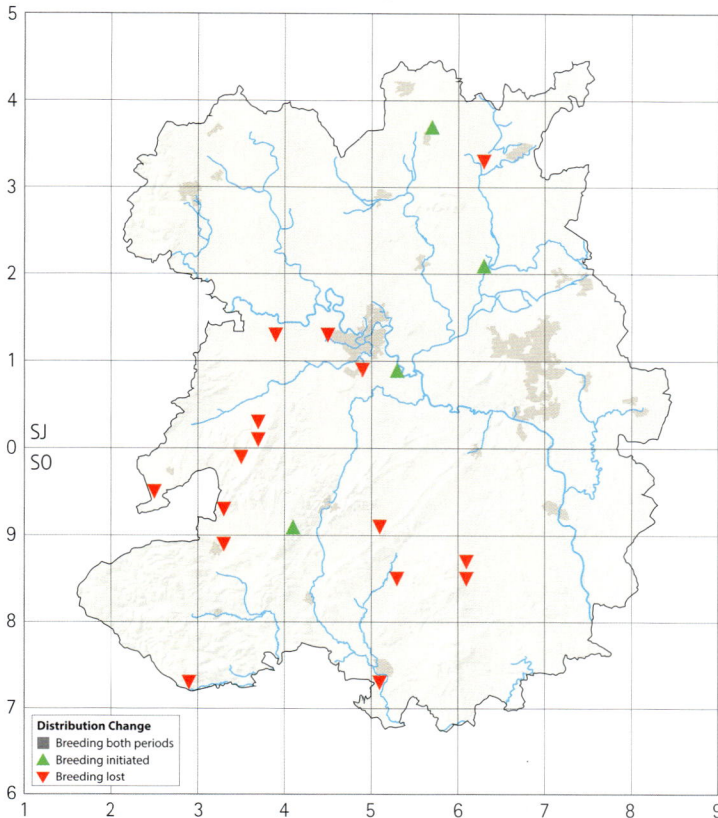

Distribution Change
- ■ Breeding both periods
- ▲ Breeding initiated
- ▼ Breeding lost

Occupied Tetrads

Atlas period (breeding)	1985–90		2008–13		Change	
	Number	%	Number	%	Number	%
Confirmed	5	1	2	0	–3	–60
Probable	2	0	1	0	–1	–50
Possible	9	1	1	0	–8	–89
TOTAL	16	2	4	0	–12	–75
Tetrads with Winter Records (2007–13): 3 (0%)						

winter sightings are infrequent. None were found during fieldwork for the BTO Winter Atlas (1986), although there are three records from the 1981–84 fieldwork period (Hookagate on 14 December 1982, Shrewsbury on 8 December 1982 and Cound Moor on 29 January 1983). Over the last 30 years there have only been 10 records in the winter months November–February, mostly of individual birds on a single date: Rhos Fiddle, Dudmaston, Ruyton-XI-Towns, Shrewsbury (twice), Upton Cressett, and Ridgewardine, with three near Ellesmere for six days in 1996, and two at Spoonley near Market Drayton (twice) in the winter of 2002–03.

These 10 records include recent winter Atlas fieldwork, when it was found in just three tetrads: SO69R (Meadowley) and SJ63U (Adderley), both in 2007, and SO28C (Rhos Fiddle) in 2008, with an additional one being seen just outside the winter period in March in SJ62F (near Waters Upton).

Severe winter weather encourages movements south or west. Two ringing recoveries support this: one ringed in 1968 in Bradfield, West Yorkshire was found dead (shot) near the A458 at Halfway House almost two years later (122km distance travelled) and an adult female ringed in October 2007 in north-west Poland was found freshly dead at Calverhall in February the following year (1,263km distance travelled). Few have been ringed here, most recently one in 2010, and none have been subsequently recovered.

Since being described as 'rather common' at the end of the nineteenth century, Long-eared Owl has suffered a steady decline, and there have been no records since July 2012. Sadly it is probably close to extinction as a breeding species, while nationally it has recently been added to the Rare Breeding Bird Panel monitoring list.

Bob Harris

(SO49A, near the Gliding Club on Long Mynd, and SJ53T, Twemlows Big Wood), probable breeding in SJ62F (near Waters Upton) and possible breeding in SO50J (Lower Betton). There were fledged young on Long Mynd in 2009 and 2010, and in Twemlows Big Wood in 2012. Additional sightings during this period were from Horderley in May 2009, and Lee Brockhurst and Sutton in June 2012, but there was no evidence of breeding.

All the sites occupied in 1985–90 have now been lost.

In winter adults are silent, so overlooked, and do not tend to move far from their breeding territory. Even with an influx of birds from the continent, which remain predominantly on the east coast,

Short-eared Owl

Asio flammeus

Rare winter visitor and passage migrant

The least nocturnal of all the owls, it breeds in the uplands of northern Britain, and numbers increase considerably in winter with visitors from Fennoscandia. Typically it is seen hunting during daylight hours in October to March in the wilder, more remote and least disturbed areas of the countryside. Sightings are fleeting, usually of individuals seen on single days.

In the late nineteenth century, Beckwith (1879) and Paddock (1897) reported it as a frequent winter visitor, not only preferring moorland and rushy places but also turnip fields. The earliest records are from Rocke (1865) who, when hunting in 1864 'saw so many disturbed by

hounds that the singular spectacular presented itself of from fifteen to twenty owls on the wing at once'. Additionally in autumn 1874 it was impossible to drive partridges into a large field of seed clover at Longdon on Tern, in which 'we subsequently found seven Short-eared Owls' (Forrest 1899). Clearly this owl was far more common then than it is now.

Other records were of one at Ellesmere in 1883, another at High Ercall about the same time, and two shot at Aston on Clun in October 1895.

Up until 1918 only seven more were reported, all individuals in

Jim Almond, Whixall Moss, 12 March 2010

There is no evidence that breeding has ever occurred. Rocke (1865) thought they had nested on Black Hill in 1865, although he never found any evidence of a nest or eggs. The comment 'a pair possibly nested on the Longmynd in 1936' in the *Handlist* (1964) was based on the observation, from unknown month(s): 'Flushed one repeatedly from the heather on Longmynd: possibly nests there, but only one seen'. The BTO Atlas (1976) had a 'possible breeding' record in SO39 (which includes the Stiperstones, but which might not have been a Shropshire record). In July 1983 a 'possible family party' was seen at Crudgington Moor but, given that post-breeding dispersal can occur as early as June (Brown & Grice 2005), the provenance of this party (if it was a family) is far from certain. In recent years there have been six records from the months May–July, but wandering or passage birds stay well into the 'breeding season'. In 2008 an individual was in SO49 (Pole Cottage/Boiling Well) for the first two weeks of May, and in 2012 one or possibly two birds were present on Whixall Moss from 12–21 May and two, thought to be young, were seen there on 20 July. In the same year one was at Ellerdine (SJ62) on 11 June, and individuals were present on the Long Mynd (Wild Moor) in April and May 2012, and again on 30 May 2013. Sites where this owl is seen regularly are all fairly well-watched, so it would be surprising if a breeding attempt was successful which produced no records.

Bob Harris

the winter months and all from the end of a gun, and there were only another six records up until 1956, again all of individuals in the winter months.

The *Handlist* (1964) noted five records up until 1963, including 'a pair' in April 1957 at Felindre and one at Anchor in August the following year; and one on several dates in July and August 1957 at Haughmond Hill.

Since then to the present day they have been seen in widely scattered locations. In the earlier years there was a record every couple of years or so with recurrent sightings from Baggy Moor, Haughmond Hill, Long Mynd, Sleap, Chelmarsh, Venus Pool and Allscott Sugar Factory, mostly in winter and usually, but not always, of single birds. Exceptionally, there was a group of five at Dothill from January–March 1989.

During the recent winter Atlas period there were about 15 records per year with the majority being of single birds, in a total of 19 tetrads, at Black Mountain, Rhos Fiddle, Mason's Bank, Linley Hill, two on the Stiperstones (Devil's Chair and Crowsnest Dingle), three on Long Mynd (Round Hill, Pole Bank and Boiling Well), Hungerford, Stanway Manor, Abdon Burf, Whatsill, Edgerley, Nobold, Meole Brace, Whixall Moss, Whixall and Wappenshall Moor. Almost all these sightings are from heathland or rough grassland where the favoured prey, voles and other rodents, can be caught. There were repeated observations from some sites, including Long Mynd, Stiperstones, Brown Clee and, by far the most frequented, Whixall Moss. Two together were seen on seven occasions (Whixall Moss in 2009, 2010, 2012 and 2013, Catherton Common in 2010, and Black Mountain and Mason's Bank in 2012).

These 19 tetrads are in 14 10km squares, compared to records from only two 10km squares in the BTO Winter Atlas (1986). The difference will generally be a reflection of variations in fieldwork intensity and methodology, rather than an increase in numbers. During Atlas fieldwork some individuals will have been recorded in more than one tetrad, and it is likely that a maximum of eight per year were present.

Winter Distribution (2007–13)

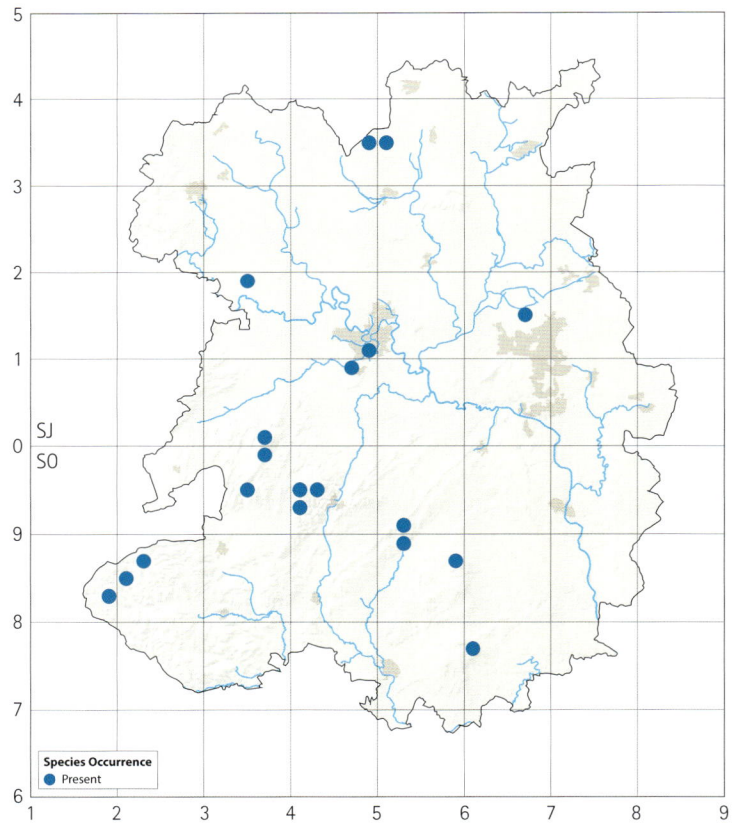

Species Occurrence
● Present

Occupied Tetrads

Tetrads with Winter Records (2007–13): 19 (2%)

Nightjar (European Nightjar)

Caprimulgus europaeus

Rare summer visitor

Britain is at the north-western edge of the Nightjar's range. It winters in tropical and southern Africa.

'Almost as big as a Cuckoo ... speckled like a Woodcock ... sharp little bill ... eyes as big as an owl's ... long hairs on each side of the beak ... white feathers on the wing'. Nightjar is an unlikely candidate for the third-oldest record in the County literature, but this minutely-observed description of an unknown bird captured near Ludford in May 1668 earned it that distinction.

Nearly three centuries on, the first SBR hit the nail on the head when it described the status of this cryptic, crepuscular species as 'imperfectly known'. Earlier observers, when it was widespread, had been similarly reluctant to commit themselves: in the nineteenth century, Eyton, Beckwith and Forrest classed it respectively as 'not very common', 'not unfrequent', and 'not uncommon ... in upland woods', though the last two considered it 'numerous' around the Wrekin and Market Drayton. Paddock was more emphatic: by 1890 he saw Nightjar 'much less commonly than formerly' in the Newport area, a decline he attributed to the cultivation of its former heathland habitat.

SBR (1957) quoted a BTO inquiry into breeding and distribution that had found Nightjars in 'fair numbers' on the northern mosses, Haughmond Hill, and Willey Park, Broseley, and 'numerous' in the young conifer plantations in the Teme Valley, but absent from the Oswestry and Bishops Castle areas 'where they had formerly bred'. BTO's own report concluded that Nightjar was 'uncommon. Status unchanged. Breeds not uncommonly in suitable localities but generally infrequently ... [most usually] among planted conifers ... up to about 9ft'. A summary map of the 1957 survey, reproduced in the national BTO Atlas (1976), p. 460, shows its status here as 'local, no decrease'.

The *Handlist* (1964) felt there were 'small numbers ... present in all parts', drawing attention to the disturbance of Nightjar's 'old haunts' where deciduous woods were felled and replaced with conifers, while acknowledging that in the short term this created opportunities for colonisation.

Up to 1970, Nightjar was found in all years but one; in the national BTO Atlas (1976), breeding evidence was mapped in seven of the County's 33 10km squares. Between 1970 and 2005 its status deteriorated: in about half these years it was not found at all; in the others it was recorded at a total of 19 sites, of which 15 featured only once, and none more than five times, suggesting a struggling, unsettled population. The *Atlas* (1992) showed probable breeding at two sites, and possible at nine, though in different years, remarking

that Nightjar had 'almost disappeared' as part of a national contraction of range away from the north and west.

Periodic BTO surveys of sites selected for habitat suitability located 11 males in 1981, including three at the Stiperstones and two in the Marchamley/Hodnet area, but only one in 1992, and none in 2004. The 1992 survey produced the only local record, in SJ43 (Morris 1994), probably at Whixall Moss, or nearby; in 2004, a churring male was heard at the former Gatten Plantation (Stiperstones), although none was found during the survey.

These results, and SOS data, suggest that Nightjar's local status remained precarious after its national recovery was already underway: a 'catastrophic' 50% contraction of range between the national BTO Atlas fieldwork periods 1968–72 and 1988–91 was followed by a modest net expansion from around 1992, including colonisation of upland clearfell that was becoming available on an unprecedented scale. By 2015, Nightjar's status was considered secure enough to justify moving it from the Red to the Amber list of UK birds of conservation concern.

The local recovery started considerably later, taking hold only around 2006. Since then, Nightjar has been found every year. During recent Atlas fieldwork it was present in the forestry at the southern end of the Long Mynd (altitude over 400m) in all years; in 2010, an adult male and two fledglings with immature plumage were caught

Breeding Distribution Change (1985–90 to 2008–13)

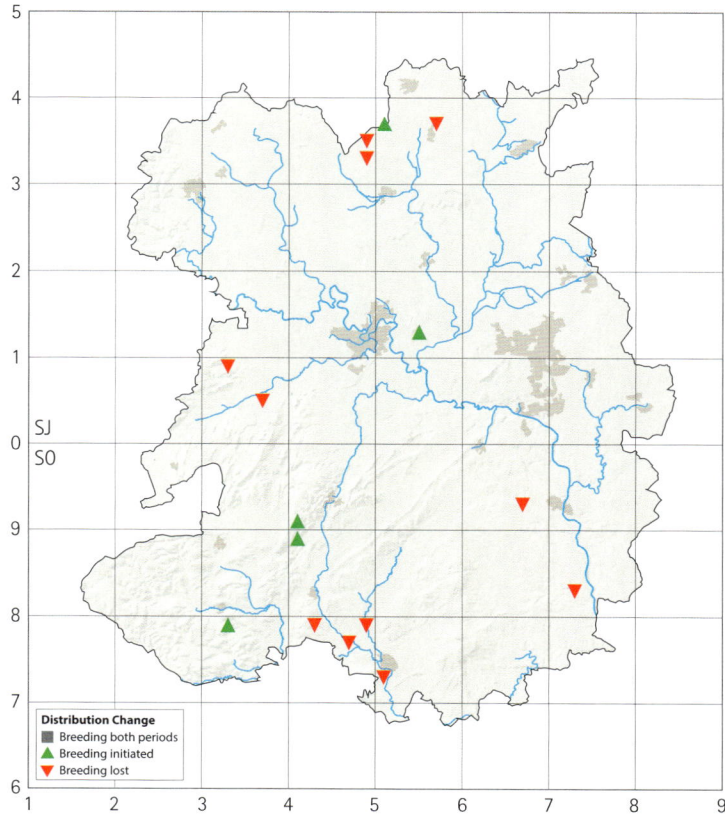

Occupied Tetrads

Atlas period (breeding)	1985–90		2008–13		Change	
	Number	%	Number	%	Number	%
Confirmed	0	0	1	0	1	x
Probable	2	0	3	0	1	50
Possible	9	1	1	0	-8	-89
TOTAL	11	1	5	1	-6	-55

and ringed there, the first confirmed breeding since two chicks were seen on the Wrekin in 1983. Breeding probably took place at Black Hill (*c.*440m) and Haughmond Hill (*c.*140m), and possibly at Fenn's Moss (*c.*90m). All but one of the breeding records were associated with Forestry Commission conifer plantations, both upland and lowland, where the former heathland was regenerating after clearfelling.

Although the breeding distribution change map shows fewer active sites than in 1985–90, four of the five were occupied in more than one year, with confirmed or probable breeding at each, suggesting a degree of consolidation; on the other hand, there was no site at which Nightjar was found in every year. Recorded activity peaked in 2009 and 2010, with records from four and three sites respectively, diminishing to single sites in 2011 and 2012, and two sites in 2013. In 2014, the only record was a male at Black Hill in July.

The breeding population is estimated at one to three pairs based on current knowledge, but with suitable clearfell widely available, especially in the remote south-west, it is quite possible that further pairs have gone undetected. A recent study of Nightjar's use of clearfell and young plantation in south Wales suggests that age and structure are particularly important, the open canopy encouraging regeneration of a variety of native shrubs, herbs and grasses which support good numbers of the moths and beetles that form the bulk of its diet (Jenks *et al.* 2014). There may be grounds for cautious optimism: rotational forestry work will be continuing well into the future, and as long as clearfelling and restocking is managed to produce mixed-age woodland, there should be a steady supply of new habitat for Nightjar to colonise as older stands grow too dense.

Very few Nightjars are ringed; the three ringed in 2010 are the only ones in recent years. There have been no recoveries.

Michelle Frater

Swift (Common Swift)

Apus apus
Fairly common summer visitor

Jim Almond, Venus Pool, 29 April 2006

Swifts breed virtually throughout Europe and across most of Asia, and winter in equatorial and sub-equatorial Africa.

'British' for just a quarter of their lives, Swifts start arriving in late April and begin to depart in late July, with most leaving by mid-August. The average earliest recorded arrival date for the years 1985–2014 was 24 April. Data presented in the *Arrival Dates* chapter suggest that the average earliest arrival dates advanced by eight days between 1900–2014.

The earliest reported date is 15 April, on which single Swifts were seen in both 1949 and 2005. Such early observations may be of birds of passage, but local breeders are normally in evidence by early or mid-May, soon occupying their small, widely dispersed colonies. Arrival at a regularly used nest site in Shrewsbury is usually within a couple of days either side of 3 May, and they appear over the rest of the town at the same time (Graham Walker *pers. comm.*).

Often described as an 'urban bird', the Swift may more accurately be described as a 'bird of buildings', although in 1893 Beckwith reported seeing Swifts 'going in and out of the fissures in the perpendicular face of the High Rock near Bridgnorth', and Paddock (1890) included a report suggesting breeding on an 'eminence' near the Devil's Mouth on the Long Mynd. Nowadays however, all our Swifts resort to buildings, but these do not have to be in towns or even villages, as they can be found nesting in isolated and remote rural settings such as Wilderhope Manor, the Garn, Llanhowell Farm and the Bog Visitor Centre. Nevertheless, the bulk of the population is urban. Beckwith (1878), referring to the Shrewsbury area, stated 'builds in most of the church towers'. 'Most' no longer applies, but where access to suitable towers and to nooks and crannies in masonry, or under the eaves or tiles, has not been blocked, churches remain favoured sites, and among those hosting small colonies in the recent Atlas period were Prees, Wrockwardine, Ightfield, Loppington, Cwm Head, Lydbury North, Bucknell and Ashford Carbonell. Swifts are not known to nest nowadays at St Mary's, Longnor, but Thomas Pennant, in his *British Zoology* of 1776, reported that a pair was found there in a torpid state in February 1766. Another remarkable church record was the discovery in 1836 of the remains of 57 Swifts and Starlings, principally the former, in a cavity in the tower of St Oswald's Church in Oswestry. It was thought very likely that further corpses were in similar cavities which were not examined. The birds appeared to be adults and it was speculated that 'remaining, from some cause or other, beyond their usual period of migration [they] had become torpid and perished' (Salway 1836).

Contrary to widely held assumptions, Swifts do not always nest at height. As Beckwith remarked in 1893, they 'will sometimes content themselves with holes under a low cottage roof'. In 1988 a pair was nesting under eaves only 3m above the pavement of Much Wenlock High Street. In 2014, nests observed on the Old Wood Yard in Ludlow were just 3.07m above the tarmac of a busy public car park across which the Swifts arrived and departed at windscreen height, and a site in Mill Street was at 3.30m, with parked cars, pedestrians, street furniture, a climbing rose and occasional festive bunting as impediments to easy access. In Bishop's Castle, birds emerge from nests only 3.65m above the ground, and close to a road, sometimes clipping parked cars; they have, on occasion, been found grounded, and a juvenile found dead in 2014 may have been a road casualty (Robin Pote, Michael Dawes, Clive Millard *pers. comms.*).

One of the many difficulties in surveying Swifts lies in the interpretation of records of individuals seen in flight at a distance from known nest sites. Are they nesting locally, or foraging far from their breeding location? Are they adults or immatures? Breeding may not occur before the age of four (Perrins 1971) and a good proportion of those recorded will be non-breeders. However, surveyors may, understandably, have recorded a significant number of such flight records as 'possible breeding', and consequently the mapping of all of them as 'possibles' could give a misleading impression. In view of this, while the breeding distribution map identifies tetrads with probable and confirmed breeding records in the usual way, all other tetrads where Swifts were recorded, either as 'possibly breeding' or just 'flying over', are indicated by pink shading, thereby showing their overall range. Poor representation in the upland areas may well be down to a lack of buildings, but elsewhere there are extensive areas

without breeding records and indeed without sightings. Some of these are largely arable farmland, where the abundance of aerial insects may be reduced and nest sites limited, but further surveying would be required in order to confirm that these reflect genuine absences.

BBS suggests a 44% decline in the West Midlands over the period 1995–2014. There is currently no detailed local population information, but a census in south-east Cheshire in 2001 showed a 38% drop in population since 1995–96 (Lythgoe 2001). Interestingly, while the breeding distribution change map shows widespread losses, they appear to have been particularly marked in adjacent parts of north-east Shropshire where, unlike elsewhere, there are virtually no counterbalancing gains. Although overall confirmed and probable breeding evidence came from 138 fewer tetrads than in 1985–90, the map shows a considerable number of tetrads (117) where such evidence was forthcoming for the first time. Given that nest sites are traditional, this is surprising and doubtless reflects the difficulties of surveying this species. Nevertheless, the net gain of 11 confirmed and probable tetrads in the south-west, bucking markedly the trend

Occupied Tetrads

Atlas period (breeding)	1985–90		2008–13		Change	
	Number	%	Number	%	Number	%
Confirmed	290	33	139	16	-151	-52
Probable	150	17	163	19	13	9
Possible	265	30	309	36	44	17
TOTAL	705	81	611	70	-94	-13

Confirmed & Probable Breeding with Range (2008–13)

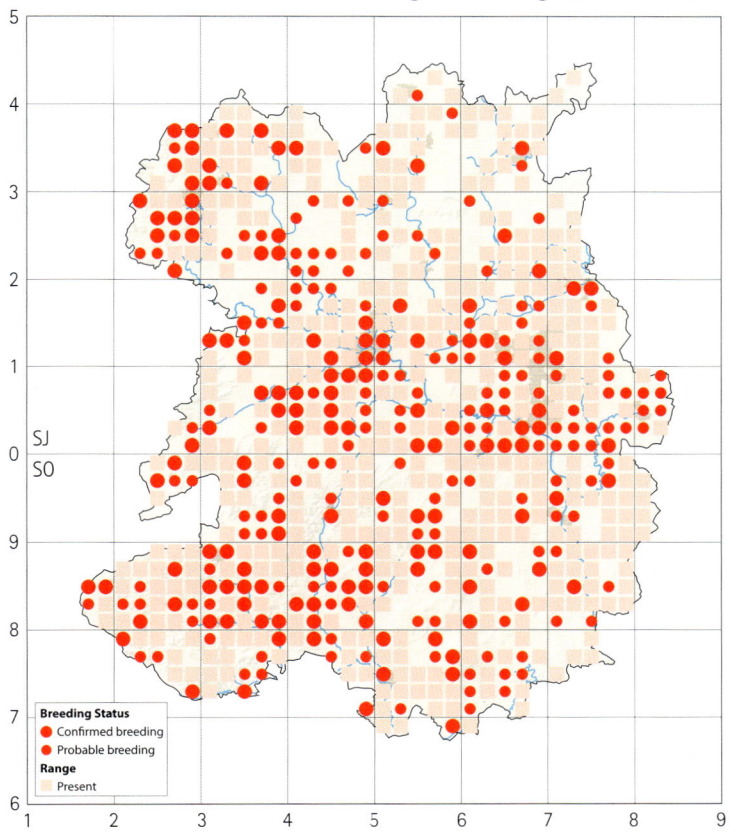

Breeding Distribution Change (Probable & Confirmed Breeding Only) (1985–90 to 2008–13)

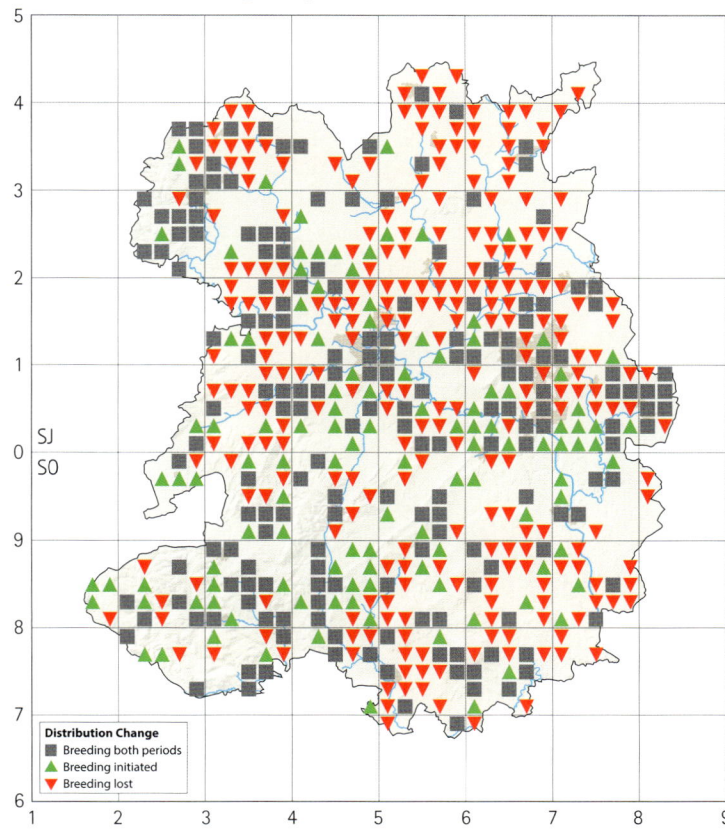

elsewhere, suggests that Swifts may be holding their own in these hillier parts, as indeed they may be in the Oswestry uplands. The table shows the number of tetrads with 'possible breeding' records, but during breeding season fieldwork (2008–13) Swifts were recorded in flight, summering or migrating in a further 85 tetrads; thus 394 tetrads in all are shown with pink shading on the distribution map.

Tetrads in which high relative abundance was recorded are scattered and widespread and it is only in the area running from the eastern half of Shrewsbury across to Longdon on Tern and Wellington that there is any notable concentration of such tetrads, reflecting perhaps the feeding area of birds nesting in the County town and in the older settlements on the western edge of Telford.

Even so, considerable numbers of nest sites have been lost in Wellington: a building in Haygate Road, in which more than 20 pairs nested, was demolished in the early 1980s, as was a loco shed at the station which housed another colony, while the eaves of a Victorian school building in Wrekin Road were wired up, denying access to more than 35 pairs. Such losses are an oft-cited factor in explaining population declines and further examples include Market Drayton, where in the 1990s in excess of 30 pairs were denied access to their nest sites when the eaves of pre-war houses were boarded up, and Craven Arms where a colony nesting at what is now the Land of Lost Content Museum was lost following re-roofing (Julian Langford, Gerry Thomas, Peta Sams *pers. comms.*). While such losses are doubtless more prevalent than in the past, they are not new: Beckwith (1893) reported that a house at Eaton Constantine 'having been repaired during their [the Swifts'] absence, they tried day after day to find their accustomed entrance'. What may be new, however, is a decline in the aerial plankton on which Swifts feed. This has been demonstrated at a study site in Stirling (Perrins 2002) and may be affecting our Swifts too.

While declining prey is a difficult problem to address, commendable efforts are being made, notably through the Ludlow and Shropshire Swift Groups, Community Wildlife Groups, National Trust and Caring for God's Acre, to identify nest sites, preserve them when building works are undertaken, and to provide new ones. At Wilderhope Manor, 38 nest sites were located in 2012 (Star Ecology *in litt.*), the largest known colony. Many are accessed via gaps under the stone tiles and these have been retained by the National Trust during recent re-roofing work. It was essential to retain the existing gaps because, in general, Swifts are faithful to specific nest sites. In Ludlow, a particular stronghold, well over 100 nest sites have been identified around the town centre (Ludlow Swift Group 2013); these are being championed and nest boxes erected. In Bishop's Castle, 37 nest sites have been located on 11 buildings and new potential sites have been purposely created during renovation work on the Town Hall. In Church Stretton, 19 nest sites were identified in 2014, and in Oswestry, boxes have been erected on both Christ Church and Holy Trinity Church. Meanwhile, in Shrewsbury, Historic England, which is undertaking extensive restoration work on the Ditherington Flax Mill, has agreed to preserve nest sites, at least 11 of which were in use in 1999; currently the Swifts access some of them through a network of scaffolding (Clive Millard, Strettons Area Community Wildlife Group, Sarah Gibson, Brian Martin, Peta Sams *pers. comms.*).

Departure is often sudden. Thus, for example, Ludlow was treated to the usual screaming parties on the morning of 26 July 2014, but a day later the town was quiet (Ludlow Swift Group 2014). This was however a year when conditions were favourable and young may have fledged at five weeks old. When the weather is inclement, the adults have difficulty in feeding their young and the fledging period may be as long as eight weeks (Lack 1956), leading to delayed departures. Such may have been the case in 2004 when a pair was seen visiting a presumed nest site on Prees Church on 10 September, being seen regularly thereafter until 25th. This is the latest date for the period 1985–2014, for which the average latest sighting was 6 September, compared with 27 August for 1905–34. The latest date known to Lloyd (1938) was 9 November 1912, but such a laggard could well have been a continental migrant rather than a departing British or Irish bird (Perrins 2002).

On departure Swifts head for their wintering grounds. Some indication of where they spend the majority of their lives has been gained nationally from ringing recoveries, and, more recently, through the use of geolocators (Appleton 2012). Inaccessibility means that very few Swifts are ringed in the nest, but while the high-flying adults might appear even more inaccessible, good numbers can sometimes be caught when they swoop low, as they do during adverse weather such as that of late June and July 1980, when 334 were ringed at Allscott Sugar Factory, including 135 in one afternoon. But it was two caught in June 1983, one at Shrewsbury Sewage Farm, the other at Habberley, that have helped reveal most about where our Swifts over-winter. The first was found 7,094km away in Tanzania in December 1984 (only the second bird from Britain and Ireland recovered in that country), and the second at a distance of 6,670km in the Democratic Republic of Congo in October 1986 (one of 14 from Britain and Ireland now found there). These recoveries fit what seems to be a general pattern of Swifts heading south of the equator to winter in the Congo Basin, Tanzania, and from Malawi to the Cape (Perrins 2002). The two other foreign recoveries (both in May, in France and Algeria) are difficult to interpret: these may well have been of birds on return passage, but both individuals could conceivably have originated from these countries and been ringed here on the very extensive feeding flights said to be taken on occasion by non-breeders (Perrins 2002).

The adverse weather of 1980, which continued into August, appears to have had an effect on nesting success. Two underweight and almost-fledged nestlings were found grounded and moribund under nest sites in Shrewsbury, seemingly after premature attempts to leave the nest. Such weather can impact on adults too: Beckwith (1893) reported that 'numbers' of Swifts were picked up in an exhausted condition during the cold, wet weather of May 1886. In general, however, this is a long-lived bird, and typically an adult will have a lifespan of nine years. One caught at Allscott Sugar Factory in 1986 had been ringed at the same place as an adult 16 years earlier, not far short of the national longevity record of one netted almost 18 years after its first capture, also as an adult.

Loose feeding groups of several hundred Swifts are quite often recorded, particularly over water during inclement weather, but 3,000–4,000 'hawking' over the top of the Long Mynd on 16 June 1948 was exceptional. Non-breeders may travel many hundreds of miles in their search for food (Perrins 2002) and this was perhaps a gathering of such birds. Sizeable numbers, probably on passage, may also be observed early in the season: for example, on 16 May 2013 in excess of 2,000 were at Whixall Moss, and on 19 and 20 May 1962 some 500 were present at Crose Mere. Passage may be witnessed at the end of the season too: on 30 July 1962, over a period of at least 15 minutes, 50 Swifts per minute were watched flying south-west into the wind over Shrewsbury, and on 3 August 1936 'thousands' were seen passing down the Severn Valley at Dowles.

Such parties on passage may exceed in numbers the entire breeding population, as the estimate derived from TTV counts is in the order of only 1,750 breeding pairs. This estimate suggests an average of six pairs for each tetrad where breeding has been recorded as 'probable' or 'confirmed'. This appears plausible, particularly when it is remembered that many of the birds we see will be immatures. Efforts to firm up on

population estimates are being assisted by the Ludlow Swift Group which undertook trial transect walks in 2013 and 2014 as part of a study initiated by the RSPB in three towns in the country in an effort to establish the relationship between the numbers in screaming parties and the number of occupied nest sites. There is little doubt, however, that the population is in decline, and one wonders for how much longer the annual arrival of the Swifts will provide us with the welcome reassurance that, in the words of Ted Hughes (1985), 'the globe's still working ... our summer's still all to come'.

Tom Wall

Kingfisher (Common Kingfisher)
Alcedo atthis
Uncommon resident

Beckwith described Kingfisher as frequent on our streams and pools, while Forrest described it as 'the most brilliantly coloured of British birds, and therefore shot upon every opportunity! In spite of this it is fairly numerous on the Severn and its tributary brooks, and on some pools ... and ... within the town of Shrewsbury'.

Breeding populations fluctuate according to the harshness of preceding winters. The *Handlist* (1964) described Kingfisher as 'resident, well distributed, though absent from the high ground with its fast-flowing streams. The severe winter of 1962–63 brought about a sharp reduction in numbers. It has since been entirely absent from many usual haunts'. However, mild winters in the following years ensured several successful breeding seasons, which resulted in recolonisation of affected waters.

The 1985–90 Atlas found that slower flowing rivers, including the Severn, Tern, Teme and Clun, provided the breeding strongholds, although Kingfishers also bred on many tributary streams and brooks. Sometimes they were seen by lakes, reservoirs and canals during the breeding season, but exposed vertical banks are essential for nest tunnels, so breeding occurred infrequently in these locations. Compared with the national BTO Atlas (1976), Kingfishers had expanded their range slightly by 1990, the most significant advance being along the River Corve.

Recent Atlas fieldwork found the same breeding strongholds, but it was not found in over one-third of tetrads where it was found in 1985–90. An increase was found along the River Perry, which has been recolonised after extensive dredging between 1984 and 1988, when the river was deepened by 1.5m and vertical banks created, which reduced numbers. This has been more than offset by disappearance from large sections of the Rivers Camlad, Roden, Worfe, and Corve, and losses on the Tern, Onny and Rea Brook.

Kingfishers are wary, and some waters are not easily accessible, which might account for some of the large percentage of possible breeding records that were not upgraded in either Atlas period, and other pairs may have been missed.

The losses are probably due to increased pollution and turbidity in the rivers, mainly from agricultural run-off, which reduces fish numbers and makes it very difficult for Kingfishers to see their remaining prey. Recent ecological assessments by the Environment Agency, to assess the condition of all rivers to meet the requirements

John Fielding, Shrewsbury, 3 September 2013

of the European Union's Water Framework Directive (WFD; see p. 33), has confirmed that large stretches of these lowland rivers need considerable improvement.

The changing condition of the River Corve illustrates the issues. It was once noted for its game fishing (Powell 1947), and Kingfishers apparently increased their range in the river between 1972 and 1985, but they have largely disappeared again now. Increased diffuse pollution, exacerbated by a slurry leak in February 2006 which killed fish and left a trail of dead invertebrates, led to the river being classified as good, moderate or poor in approximately equal thirds in the first WFD assessments (Defra and Environment Agency 2009,

Occupied Tetrads

Atlas period (breeding)	1985–90		2008–13		Change	
	Number	%	Number	%	Number	%
Confirmed	81	9	44	5	-37	-46
Probable	56	6	24	3	-32	-57
Possible	117	13	95	11	-22	-19
TOTAL	254	29	163	19	-91	-36
Tetrads with Winter Records (2007–13): 172 (20%)						

see p. 34 in the *Habitats* chapter), and the valley was targeted by the Environment Agency/Natural England Catchment Sensitive Farming initiative in 2010–11, in an attempt to reverse the deterioration.

The impact of ecological deterioration is sometimes reinforced by disturbance by anglers from the start of the coarse fishing season, resulting in the desertion of some nests (SBR 1987), and this problem may have increased since then as angling has become more popular.

Waterbird surveys in prime habitat on the River Severn near Shrewsbury in the 1980s found nests to be around 2–3km apart. Meanders and tributaries mean some tetrads have several kilometres of waterway and supported 2–3 pairs, while other territories include part of two tetrads. The *Atlas* (1992) estimated an average of one to

Breeding Distribution Change (1985–90 to 2008–13)

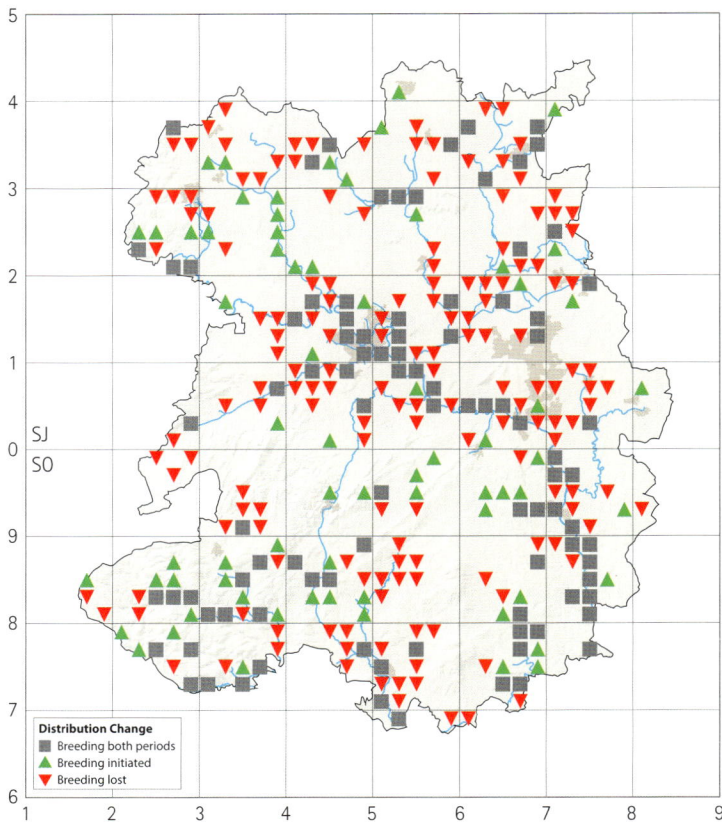

Distribution Change
- ■ Breeding both periods
- ▲ Breeding initiated
- ▼ Breeding lost

2.5 pairs per tetrad with confirmed or probable breeding, giving a population estimate of 140–350 breeding pairs. A similar calculation now gives a population estimate of 68–170, close to the population estimate of 80–140 breeding pairs based on TTV counts, suggesting a population decline of around 50% since 1990.

There is some movement away from breeding territories along rivers and streams after the breeding season, especially by juveniles, and the majority of SOS records are received from late July onwards from pools and meres. Further dispersal occurs after heavy rainfall when rivers are in spate. Records from still waters begin to decline in February, suggesting a return to breeding sites. Despite dispersal, the winter and 'breeding season' distributions do not differ greatly, and there is virtually no difference in the number of tetrads where it was recorded in each of the two seasons. However, many of the winter sightings were in tetrads with only possible breeding records, suggesting some of the latter were due to sightings in July of dispersed juveniles. Other winter records came from tetrads holding the upper reaches of many rivers and streams, with no suitable breeding habitat, such as several of the Long Mynd valleys.

Few Kingfishers have been ringed, and most were ringed and recovered here. Only two, both from neighbouring Staffordshire, were ringed elsewhere and recovered here. Three of the seven ringed here and found elsewhere were recovered in adjacent counties, and only three went further than 100km, the furthest being to Conwy and to Nottinghamshire, both a distance of 106km. All the three longer distance travellers were first-year birds that survived for less than 10 months.

John Fielding, Venus Pool, 18 July 2009

Peter Carty

Bee-eater (European Bee-eater)
Merops apiaster
Vagrant

This colourful summer visitor breeds from southern and eastern Europe, and North Africa, through to Kashmir and Kazakhstan; those breeding in Europe winter in tropical Africa. It is a scarce migrant to Britain with between 20 and 40 individuals in most years, and has occasionally bred. The majority are seen in south-east England and East Anglia with relatively few sightings further north and west (Fraser 2013a). It is not surprising, therefore, that there have only been two records:

Bromfield	three, 9 May 1992
Berriewood, near Condover	29–30 April 2013

Typically, both were spring sightings, suggestive of birds overshooting their breeding grounds further south. True to its name, the one at Berriewood was often seen catching bees.

Graham Walker

Paul King, Condover, 30 April 2013

Hoopoe (Eurasian Hoopoe)
Upupa epops
Rare passage migrant

Typically recorded as an overshoot on spring migration from Africa to southern Europe, a few records are from the reverse migration period between the summer and early winter. Hoopoe has always been a rare visitor, on average occurring less than annually. It has not been known to breed, although a pair nested in adjacent Montgomeryshire in 1996 (Ogilvie 1999).

Beckwith (1892) noted 10 occurrences, the majority meeting their demise via the gun in the hands of their finders. The first record was of one obtained near Grinshill in 1834, followed by one killed at Cold Hatton Heath near Wellington in 1841. Subsequent records included a male and female killed at Acton Reynald in 1855, one seen at Leebotwood in 1880 which 'several times alighted on the road in front of the carriage', and one 'shot as it rose from some carrots' in 1889 near Market Drayton.

Hoopoe continued to be sporadically recorded into the twentieth century, though in more recent decades it has been recorded more regularly. The first modern record was in 1956, after which there were records in 33 of the 59 years up until 2014, with a total of 50 in this period. There have been multiple records in nine years, with at least five in 1957 and four in 1971, but all have been of a single

Ian Butler (ianbutlerphotography.co.uk), Much Wenlock, 10 May 2005

Seasonal Occurrence in 10-day Intervals (1950–2014)

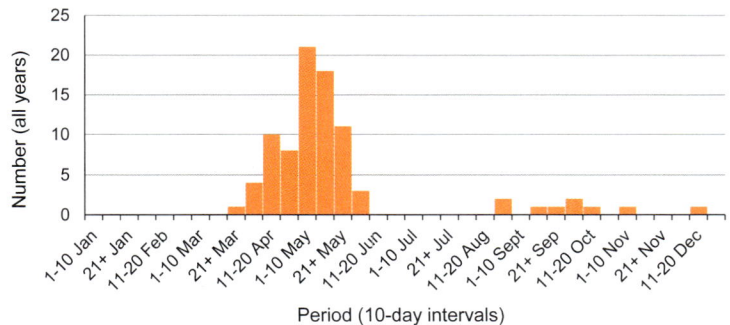

Period (10-day intervals)

bird, usually seen on one day only. Before 2008, 11 were reported on more than one day, but only two on more than three, at Langley, Acton Burnell (four days from 18 May 1978) and, the longest staying, near Much Wenlock (eight days from 3 May 2005).

There were eight records received during the recent Atlas period: three in 2008, at Pentre Hodre on 24–30 April, Adderley on 7–12 May and Badger on 21 May, and annually in the remaining years at Titterstone Clee on 18 April 2009, St Martin's on 19–21 May 2010, Kinlet on 29 April–4 May 2011, Attingham Park on 13 September 2012 and Vron, Felindre on 20 May 2013. Three of these eight individuals

stayed more than three days, compared with only two between 1950 and 2007. There were none in 2014.

Hoopoe is most likely to be encountered in the spring, particularly in April and May and a notable peak in the first 10 days of May. Some stayed for more than one day: the chart shows the first date when each was encountered.

Occurrences are widespread, usually away from typical birding hotspots. As such, a fortunate observer could bump into a Hoopoe anywhere.

Mike Shurmer

Wryneck (Eurasian Wryneck)
Jynx torquilla
Very rare passage migrant

Wryneck mainly breeds in temperate regions of Europe and Asia, and most European populations winter in tropical Africa. In the late nineteenth century, it was welcomed as the forerunner of the Cuckoo, and, as in many other counties, it was referred to as the 'Cuckoo's mate'.

It was noted as a very rare and extremely local summer visitor by Beckwith (1892): 'the only locality where it breeds with any degree of frequency is the valley of the Severn between Buildwas and Bridgnorth'. Forrest (1908) agreed with the limited range, writing that it was 'almost confined to the vicinity of the Severn between Bewdley and Buildwas'; he had found a nest at Broseley in June 1902.

Breeding continued in the south into the early years of the twentieth century, though with decreasing regularity. Nesting was last recorded at Broseley in 1920, Dowles (Wyre Forest) in 1924, and Bucknell in 1952 and 1953. The latter is the most recent confirmed breeding, although it was seen at all these sites in one or more years subsequently. The only other relatively recent record was of a nest at Edgerley in 1941 (*Handlist* 1964).

Observations at Bucknell in four of the five years 1950–54, including the above, were the first modern records, but there were no more for 14 years. Since then it has occurred infrequently on a less than annual basis, with 22 records from widely scattered locations in only 17 of the years between 1968 and 2014, but 1986 was the last year with two, and sightings became less frequent up until 2014, when three were found. All records are of a single bird, and most were seen on one day only. It is now a very rare passage migrant, primarily on autumn migration and mostly in September.

In the Atlas years there were two observations: the first was well-watched, at Titterstone Clee from 22–29 August 2010, the other a brief sighting at Priorslee Flash on 30 September 2011.

There were three in 2014, the only year in the modern era with more than two, all in September: one in a private garden at Yockleton from 1st–4th, one picked up near Hawkstone Park on 7th and released later the same day, and one in a Ludlow garden on 19th.

Mike Shurmer

Seasonal Occurrence in 10-day Intervals (1950–2014)

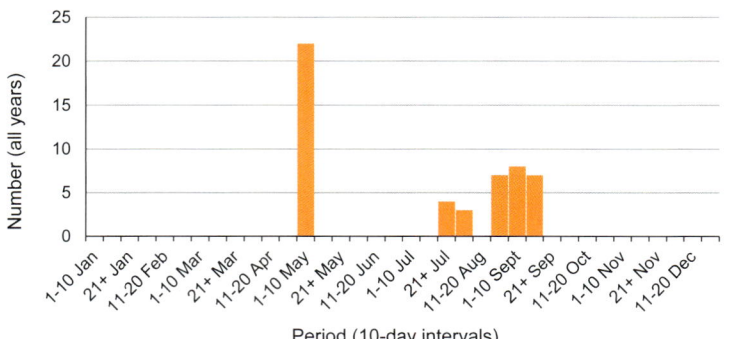

Jim Almond, Titterstone Clee, 28 August 2010

Lesser Spotted Woodpecker
Dryobates minor
Scarce resident

John Hawkins, North Shropshire, 21 February 2012

It is by no means easy to ascertain the population status of this species (occasionally referred to as 'Barred Woodpecker' in early literature) for the period before modern survey procedures. Beckwith described it 'rather rare' in 1879 but by 1892 considered it as 'by no means uncommon'. Rocke, in 1866, described it as very abundant in that year, with three or four pairs nesting within 3km of his home. Forrest (1899) judged it to be as common as the Great Spotted Woodpecker but overlooked more often by virtue of its small size and habit of feeding in the topmost part of the tree canopy. The issue of detectability was also raised in 1964, by the *Handlist*, which considered it to be 'local but thinly distributed ... in lowland areas'. This view is consistent with that of the national BTO Atlas (1976) which described it as 'tolerably common' towards the end of the nineteenth century, and mapped it with evidence of breeding in 17 of the County's 33 10km squares, with five in the north having confirmed breeding.

Despite this uncertainty regarding its absolute abundance, Holloway (1996) maintained that national population figures for the species remained stable throughout the nineteenth and the first two-thirds of the twentieth centuries, apart from local losses due to the removal of dead, dying and unwanted trees, and the cold winters of 1946–47

and 1962–63. A national population increase, starting around 1970 (frequently attributed to the role of Dutch elm disease, leading to a greater supply of dead and dying wood, insect food and nest sites), led to a local range expansion, evident by comparing the national Atlas (1976) and the *Atlas* (1992). The latter showed probable or confirmed breeding in every 10km square, and estimated a population of 250–500 pairs. Nevertheless, it was still regarded as the least abundant of our three woodpeckers.

However, the expansion was not maintained and all available evidence points to a decline in both range and numbers from the late 1980s until the present day. Changes in breeding status since the *Atlas* (1992) are distressingly consistent with other measures of decline. Breeding evidence was found in only 4.4% of tetrads, a decline of 177 (82%).

Based on TTV counts, the population is now estimated at 30–60 pairs, only about 10% of that just over 20 years previously. The drivers of this decline are not clear and, for the UK in general, hypotheses have included competition with other species (especially Great Spotted Woodpecker), predation, declining woodland quality, poor breeding success and the felling of nesting trees killed by Dutch elm disease. The national BTO Atlas (2013) reports that breeding success is low in many areas because of starvation of chicks, suggesting that food availability may be a factor in its decline.

Occupied Tetrads

Atlas period (breeding)	1985–90		2008–13		Change	
	Number	%	Number	%	Number	%
Confirmed	51	6	8	1	-43	-84
Probable	57	7	4	0	-53	-93
Possible	107	12	26	3	-81	-76
TOTAL	**215**	**25**	**38**	**4**	**-177**	**-82**
Tetrads with Winter Records (2007–13): 47 (5%)						

Breeding Distribution Change (1985–90 to 2008–13)

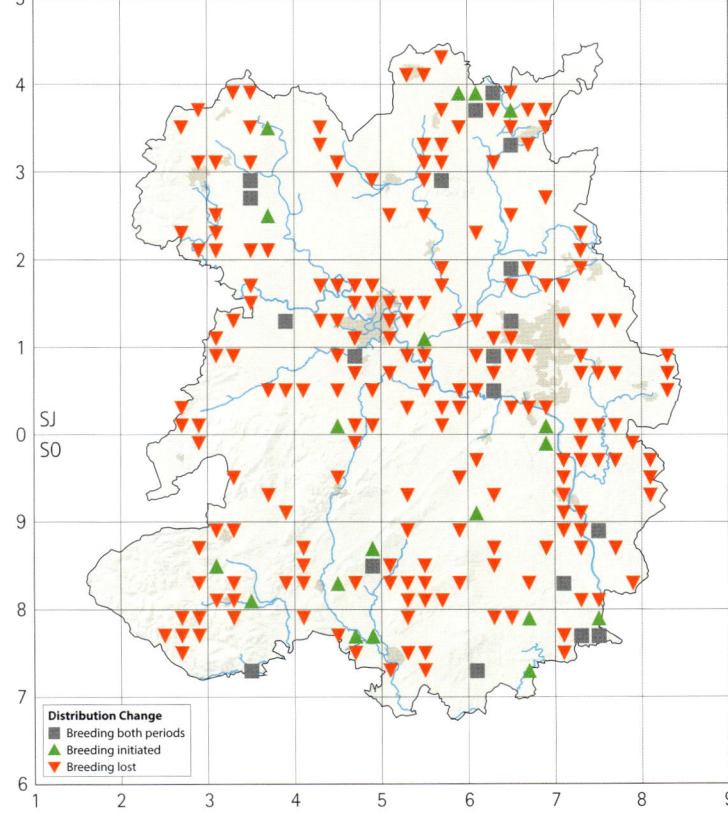

Distribution Change
- ■ Breeding both periods
- ▲ Breeding initiated
- ▼ Breeding lost

The distribution of Lesser Spotted Woodpecker today is scattered and localised. Probable and confirmed breeding is confined to a section of the dip slope of Wenlock Edge and near Corfton, the Wyre Forest, around Neen Sollars, the upper Perry Valley, the area between Market Drayton and Whitchurch, Hawkstone Park, the Leegomery area of Telford and Ironbridge Gorge. While this species favours large deciduous woods, it also exploits copses, overgrown hedges with trees, and lines of willows and alders along stream courses. It rarely strays too far from water so damp treed areas are ideal.

The winter distribution is similar but more dispersed: nevertheless, the species was still found in only 5% of tetrads. The similarity of the distribution in the two seasons suggests its sedentary nature, which is underlined by the fact that all three ringing records refer to birds retrapped in the same location as they were ringed originally, in one case five years later (*SBR* 1964; 1989; 2002).

Unlike the Great Spotted Woodpecker, this species is uncommon in towns, where it is confined to parkland with old trees. Nevertheless, use of peanut feeders in gardens shows signs of increasing. In contrast to its larger cousin, however, over 60% of garden visits are reported in the period May–September, rather than in winter.

Lesser Spotted Woodpecker is hard to detect by virtue of its small size, preferred habitat and short drumming period and, as a result, is likely to be under-reported. Nevertheless, it is without doubt our scarcest woodpecker by a large margin, and is undoubtedly in a critical state, reflecting its position in the country as a whole.

John Arnfield

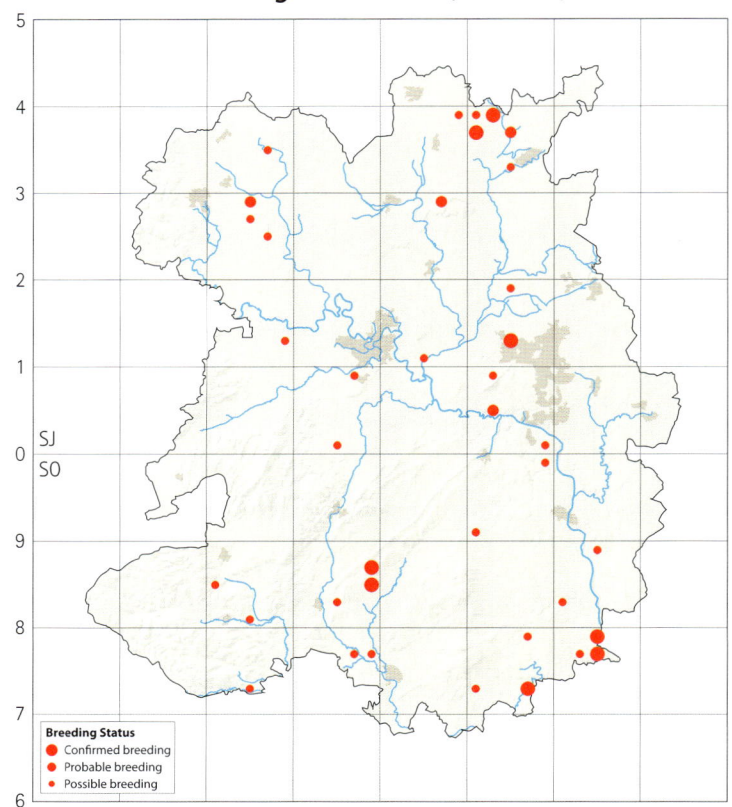

Breeding Distribution (2008–13)

Breeding Status
- Confirmed breeding
- Probable breeding
- Possible breeding

Great Spotted Woodpecker

Dendrocopos major

Common resident

This attractive, boldly plumaged and noisy bird is the commonest of our resident woodpeckers today. This has not always been the case. Rocke, in 1866, described it as 'much more rare' than the Green Woodpecker, a comment echoed by Beckwith 12 years later. However, by 1890, Paddock contended that it had displaced the Green Woodpecker as the commonest member of the *Picidae*, which, he stated, had been the case 'a dozen or more years ago'. Forrest, in 1908, described it as becoming more common. Even allowing for geographical variations in different woodpecker populations, it seems certain that Great Spotted Woodpecker numbers overtook those of Green Woodpecker in the latter half of the nineteenth century, although no reasons for this shift were given by contemporary authors.

Throughout the twentieth century, the population increase persisted, and the *Handlist* (1964) describes it as well distributed, and both commoner and less adversely affected by the severe 1962–63 winter than the Green Woodpecker. During the 1970s, Dutch elm disease increased the availability of dead and dying timber for food sources and nest sites. This explanation was proposed in the *Atlas* (1992) to account for the range expansion over the two decades since the national BTO Atlas (1976), when evidence of confirmed or probable breeding went from 27 to all 33 of the 10km squares wholly or mainly within the County. However, this hypothesis is inadequate

John Fielding, Atcham, 12 December 2012

to account for further population expansion throughout the past 40 years, although decreased competition with Starling for tree cavity nest sites and increased use of garden feeders to offset winter food shortages have both been cited as important. Increased availability of dead wood due to the decline in woodland management may also have been a factor.

BBS Trend (1997–2014)

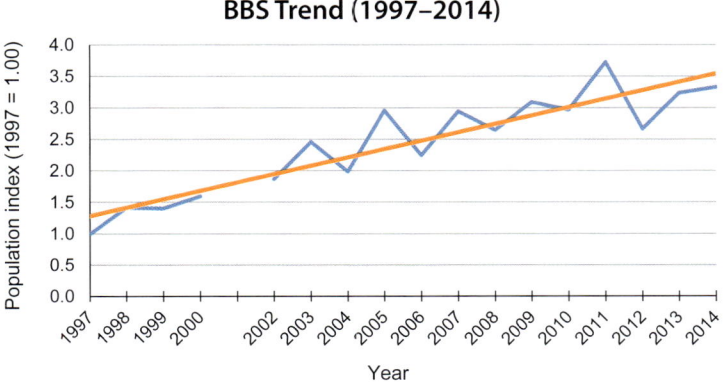

Nesting activity is typically preceded by the familiar 'drumming' territorial advertisement, made on a resonant tree limb or telegraph or electricity pole. This non-vocal 'song', which is practised by both sexes, may be heard mostly between December and May (although most commonly in the period January–April). Interestingly, as late as 1908, Forrest reported that it was still not generally accepted whether the source of this sound (often termed 'jarring' at the time) was percussive or vocal. Nests are mostly in tree cavities, especially in unsound silver birches, usually chiselled out by the mated pair (although cases are known of the commandeering of existing nest holes). More exposed nests in the centres of decayed or pollarded trees or even fissures in trees have been observed. Breeding can be confirmed when young are fed at the nest during May–June, and juvenile birds are often seen in family groups until August.

A UK population increase between 1967 and 2010 of 408% is estimated in the national BTO Atlas (2013), and local figures seem to reflect such a trend. Evidence of breeding was found in 83% of tetrads in the *Atlas* (1992): the equivalent figure in the recent Atlas period is 97%. The BBS chart shows a rapid population increase between 1997 and 2014, with an underlying growth of around 170%.

Based on TTV counts, the population is now estimated at 3,600–3,700 pairs, which is broadly consistent with the lower end of the estimate from the *Atlas* (1992) scaled up by the BBS increase to 2014. This averages 4.1 to 4.2 pairs per tetrad.

Historically, the preferred habitat of this woodpecker is mature mixed woodland with dead and dying trees, especially birch. However, perhaps as a result of the pressure of its increasing population, its distribution in the breeding season suggests that it is quite tolerant of a range of habitats and, while trees are required, these may take the form of smaller woods, coppices and spinneys, as long as these are linked to other such areas by corridors of trees or mature hedgerows, or even open stretches of countryside, if not too extensive. It is now common in old orchards, treed golf courses, conifer plantations and urban parks (Nine pairs were found in the 170ha Telford Town Park in June: Bishton SBR 2003).

The breeding distribution change map suggests that it is exactly these types of sub-optimal habitats that have been exploited over the past quarter century. As a result, Great Spotted Woodpecker is widespread, occupying almost exactly the same numbers and locations of tetrads in both the breeding and winter seasons, with no sizeable gaps in its distribution apart from the Long Mynd, and with similar patterns of relative abundance in both seasons, characteristics which also underline its sedentary nature. While some areas of high relative abundance correspond to extensive woodland (e.g. the Wyre Forest, the area south of Bishop's Castle), others do not (e.g. east of Whitchurch, south of Melverley and along the Staffordshire border south and east of Bridgnorth).

The sedentary habits, alluded to above, are also confirmed by ringing records. Of the 18 birds recovered here, 17 were ringed here and one was from Clwyd. Of the 19 locally ringed recoveries, only two were found elsewhere, both in nearby counties (Cheshire and West Midlands).

It has also adapted to gardens, a trend observed as long ago as the 1890s. It is far commoner in County gardens than in the UK as a whole (here it occurs in 38% of those in the BTO Garden BirdWatch scheme), where it favours peanut and suet feeders. Some 85% of reported garden visits occur between October and March. This garden reporting rate seems likely to increase in the future, as many observers have described adult birds with youngsters during June–July, seemingly 'learning the ropes' of feeding station exploitation.

Breeding Distribution Change (1985–90 to 2008–13)

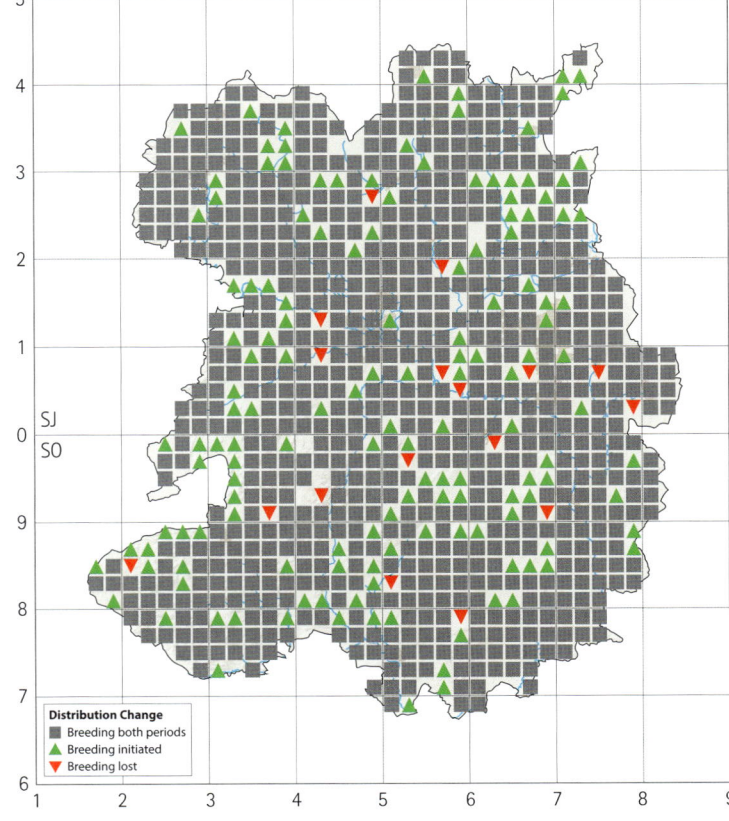

Occupied Tetrads

Atlas period (breeding)	1985–90 Number	1985–90 %	2008–13 Number	2008–13 %	Change Number	Change %
Confirmed	300	34	464	53	164	55
Probable	207	24	228	26	21	10
Possible	211	24	154	18	-57	-27
TOTAL	718	83	846	97	128	18
Tetrads with Winter Records (2007–13): 835 (96%)						

Breeding Relative Abundance (2008–13)

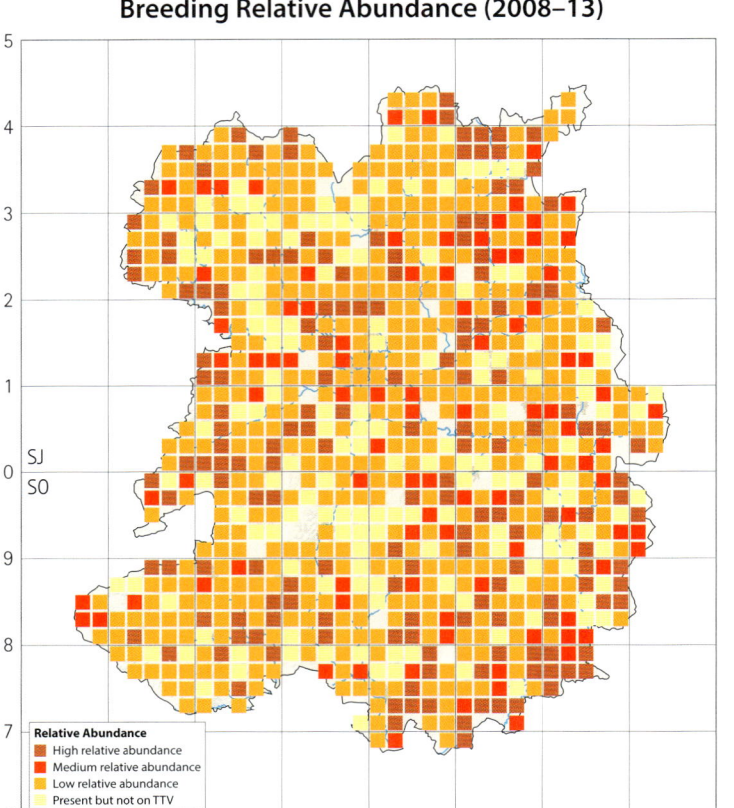

Relative Abundance
- High relative abundance
- Medium relative abundance
- Low relative abundance
- Present but not on TTV

In the wider countryside, this woodpecker feeds on invertebrates taken from rotting timber and from behind the bark of trees. Its tastes are more catholic in winter when these food items are less plentiful, when seeds, nuts and fruit may be consumed. Local observations confirm that both hazelnuts and black sunflower seeds may be taken to a nearby tree trunk, wedged in and hammered open to gain access to the tasty interior. They have also been observed on raspberries (Minsterley), the flowers of mullein (Oswestry) and red-hot poker (Condover) and, most unusually, on a dead adult Blue Tit (High Ercall), the carcass of which the woodpecker carried away. It also predates songbird nests, taking both eggs and young from the nest to feed its own chicks, and may excavate holes in nest boxes to gain access to such food sources. SBRs reports such behaviour in July and August but also outside the breeding season in November, when nest boxes may simply be a source of invertebrate prey. Great Spotted Woodpeckers engaged in 'sap-sucking' behaviour on lime trees in the Ludlow area have been noted, although it is not clear whether this form of drilling is intended to access the sap itself or to attract insects which are consumed later (Gareth Thomas *pers. comm.*).

Surprisingly, the sharp *kick* call has been reported locally in all months except between August–November, with a peak in April.

The relative abundance maps in the national BTO Atlas (2013) show that Shropshire is a stronghold for Great Spotted Woodpecker, where it is common enough to be seen with moderate regularity but scarce and attractive enough always to provoke interest.

John Arnfield

Green Woodpecker (European Green Woodpecker)

Picus viridis

Uncommon resident

Green Woodpecker figures large in British folklore. According to BTO (2012), the seventeenth-century English antiquarian, John Aubrey, reported that the species was used in druidic practice for augury and that 'the country people do divine of raine by their cry', a belief apparently maintained until the nineteenth century (Cocker & Mabey 2005) and probably beyond. For this reason, the name 'Rain Fowl' is used in local late nineteenth- and early twentieth-century accounts. Most other names allude to its loud laughing call, usually uttered in flight, including 'Yawkle' (1899), 'Yaffle' and 'Yarkle' (both 1909), and, perhaps, 'Heighoe' (1891). Finally, some Victorian naturalists (including Forrest 1899) used 'Common Woodpecker', a name consistent with reports from the early nineteenth century, and for about a century afterwards, that this species was both a common resident and was the most abundant of our three woodpeckers.

At the end of the nineteenth century it was common in well-timbered areas, 'especially the neighborhood of Shrewsbury, Ironbridge and along the Church Stretton Valley' (Forrest 1899). Short-term declines in numbers are known to occur as a result of cold winters, when frozen ground limits access to ants. Beckwith (1891) reported population decreases following the severe winters of 1860–61, 1874–75, 1879–80 and 1881–82. Similar losses occurred in the winters 1962–63 and 1981–82. In addition, farmland declines were noted following

the outbreak of myxomatosis, which led to declines in the rabbit population and the reversion of rabbit-grazed turf to a rougher state which is not optimal for ant colonies.

Recovery was largely complete by 1985–90, when breeding was confirmed in five of the six 10km squares with only possible or no evidence of breeding mapped in the national BTO Atlas (1976).

Ants are a favoured food so Green Woodpecker is more often observed feeding on the ground than other woodpeckers. Garden feeding stations are not used, although secluded lawns may be visited rarely if ants are available. Such behaviour has been noted in Albrighton, Blist's Hill, Bridgnorth, Bucknell, Harmer Hill, Leebotwood, Leegomery, Maesbury, Much Wenlock and Shifnal. Individuals have also been seen probing a bunker at Bridgnorth Golf Club, and a stone wall. They also hunt for the larvae and adults of beetles and other insects in decaying timber. SBRs have also reported cases of feeding on building roofs (Hawkstone Hotel, Hoar Edge and Much Wenlock), and probing windfall apples in an orchard (Trefonen). Most unusually, a 2007 record from Venus Pool describes a juvenile Green Woodpecker attempting to steal a newly caught fish from a Kingfisher!

Unlike the *Dendrocopos* woodpeckers, territory is mostly claimed by calls rather than drumming, with about 50% of reported vocal activity in March and April. However, drumming behaviour was

Occupied Tetrads

Atlas period (breeding)	1985–90		2008–13		Change	
	Number	%	Number	%	Number	%
Confirmed	132	15	60	7	-72	-55
Probable	154	18	109	13	-45	-29
Possible	204	23	184	21	-20	-10
TOTAL	490	56	353	41	-137	-28
Tetrads with Winter Records (2007–13): 291 (33%)						

observed at Much Wenlock in March, at Highley in May (on silver birch), and from the Gatten Plantation in August (on a dead branch). Whether this activity represents true drumming or represents nest building or foraging activity is open to question.

Breeding territories are large, which makes locating nests difficult, and the call is far carrying, so several pairs will have been recorded in more than one tetrad, and almost all the possible breeding records were adjacent to clusters of tetrads with more reliable breeding evidence. These clusters occur along the Severn Valley between Buildwas and Highley, in the hills west of Oswestry, south of the River Clun, between Ludlow and the Clee hills and between Church Stretton and Shrewsbury. While much of this territory is well wooded, this species is capable of thriving in a great range of more open habitats (parkland, heaths, meadows, grazed commons, farmland, golf courses, airfields and wetlands), as long as there are scattered trees nearby for nest sites. Pastures grazed by sheep and rabbits are particularly favoured

Breeding Distribution (2008–13)

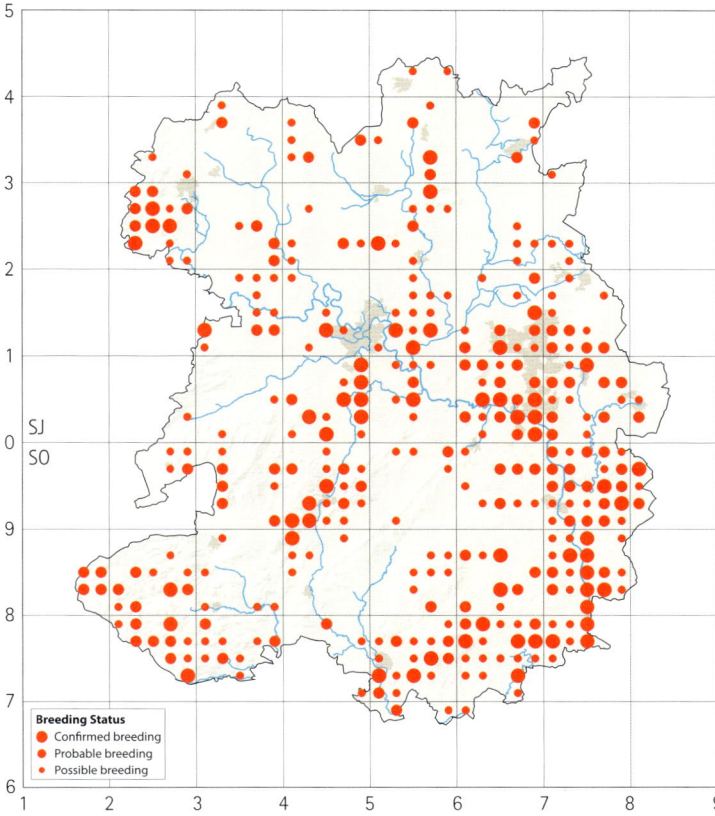

Breeding Status
- Confirmed breeding
- Probable breeding
- Possible breeding

John Robinson, Wyre Forest, 19 December 2008

Breeding Distribution Change (1985–90 to 2008–13)

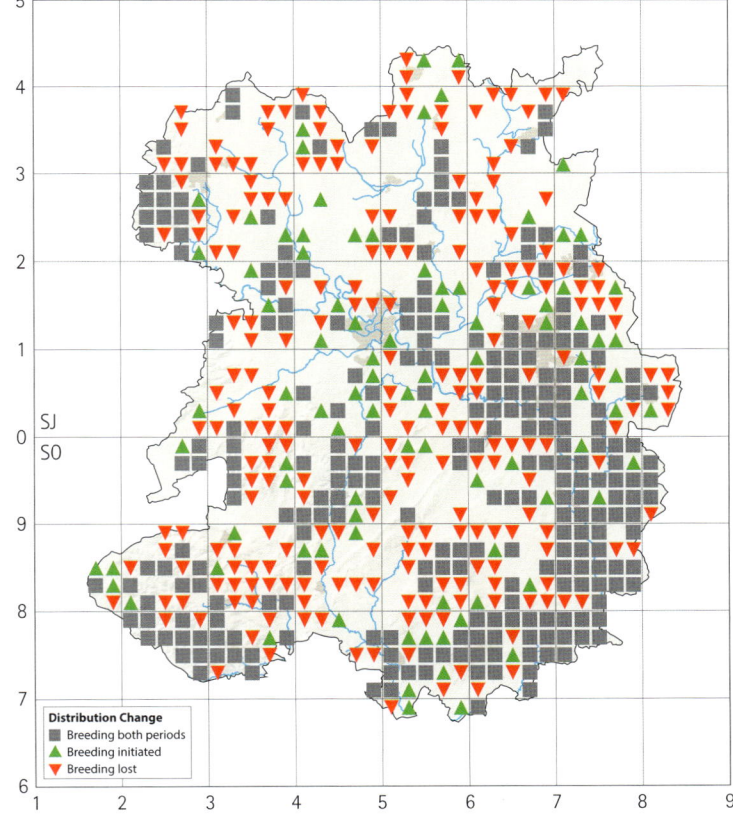

Distribution Change
- Breeding both periods
- Breeding initiated
- Breeding lost

as these possess the short swards and warm soils which support ant colonies. Many records come from open upland areas (e.g. the Long Mynd, Caer Caradoc, the Stiperstones and Brown Clee, the latter at 540m elevation). The clusters of confirmed breeding correspond to the areas of high relative abundance, especially the Severn Valley south of Buildwas.

The winter distribution is essentially the same as that for the breeding season, with clusters of tetrads with high relative abundance, and areas of low concentration, in similar locations, an unremarkable result for such a sedentary species. However, only 33% of tetrads were found to be occupied in winter, in contrast to 41% in the breeding season. This difference may be an artefact of under-recording in winter, when it becomes solitary and does not call as much as in spring. (Only about 15% of SOS records of vocal activity since 1992 occurred in the period November–February.)

Only four have been recovered, all at the site where they were ringed, up to three years later, confirming their sedentary nature.

The changing fortunes of the Green Woodpecker are not easy to assess. Over recent decades, the national BTO Atlas (2013) found range changes in Britain since the national BTO Atlas (1993), with expansion eastwards and contraction in the west, especially Wales, accompanied by a doubling of the population between 1970–2010. This gain does not seem to be reflected at the local level. The *Atlas* (1992) found Green Woodpecker in 56% of tetrads (with confirmed or probable breeding in 33%) and suggested a population of 500–1,000 breeding pairs. Equivalent numbers from the current survey are 41%, 19% and, based on TTV counts, 390–400 pairs. These statistics suggest that this species' fortunes have deteriorated in the quarter century between the two Atlases. This is consistent with the breeding distribution change map, which depicts minimal change in the cores of the areas identified as clusters of breeding activity above. Outside these concentrations, the picture is mixed but, of the tetrads that show breeding change, 70% have lost breeding. BBS coverage is insufficient to provide a local trend, but there has been a 33% increase in the West Midlands, contrasting with an identical decline in adjacent Wales, between 1995 and 2014. The population and distribution of common ants, the main food source, is reduced by farm grassland management, particularly the ploughing and reseeding of permanent pasture, which may be the underlying cause of the decline here.

John Arnfield

Kestrel (Common Kestrel)

Falco tinnunculus

Uncommon resident

Kestrel's old country name, Windhover, captures the almost totemic quality of a creature perfectly in its element. Its near-stationary hunting flight, use of regular 'beats' and perches, and loyalty, where possible, to traditional nest sites, forge a close connection between bird and landscape. When that connection is diminished or broken altogether – and the Kestrel is now absent from about a third of the County – the loss is keenly felt. An observer at Venus Pool, recording only her second of the year in March 2011, remarked that 'they used to be a daily occurrence here, but seem much less common everywhere of late'.

This is not the first time the local population has been under pressure. In 1865, Rocke of Clungunford described it as 'common everywhere, though becoming scarcer every day'. By the end of that century, ornithologists were united in deploring the destruction of Kestrels to protect game birds. Beckwith pointed out that the damage it did in that respect was 'very slight compared with the service it renders farmers ... by killing small mammalia'. Paddock contributed a study of the contents of 13 pellets, finding mostly mammal and insect remains, and bones of a small bird in only one. Forrest went further: finding it 'fairly numerous where it is not molested', he called for both bird and eggs to be protected by law.

Persecution seems to have continued to depress the population. In 1941, two years into the war, CSVFC remarked that the Kestrel already showed an increase 'due to lack of keepers'. By 1956 it was again considered 'common everywhere', recovering well after the era of intensive gamekeeping. Like other raptors, it suffered the effects of organochlorine pesticides in the 1950s and 60s, but much less so here than in intensive arable areas. The *Handlist* (1964) felt that Kestrel, unlike Sparrowhawk, was 'maintaining its numbers',

John Fielding, Long Mynd, 25 September 2010

and this was confirmed over the following decades. In 1985 it was classed as 'very widespread and numerous', with an 'unusually high count' of 31 on Caer Caradoc that August, apparently attracted by flying insects. The *Atlas* (1992) suggested an east-west split, the east seemingly supporting more pairs per tetrad than the west, and yielding far more cases of confirmed or probable breeding. It noted, however, that higher numbers of observers in the east, as well as more arable farming, roadside verges, and urban and suburban areas, might account for these effects.

Breeding Distribution (2008–13)

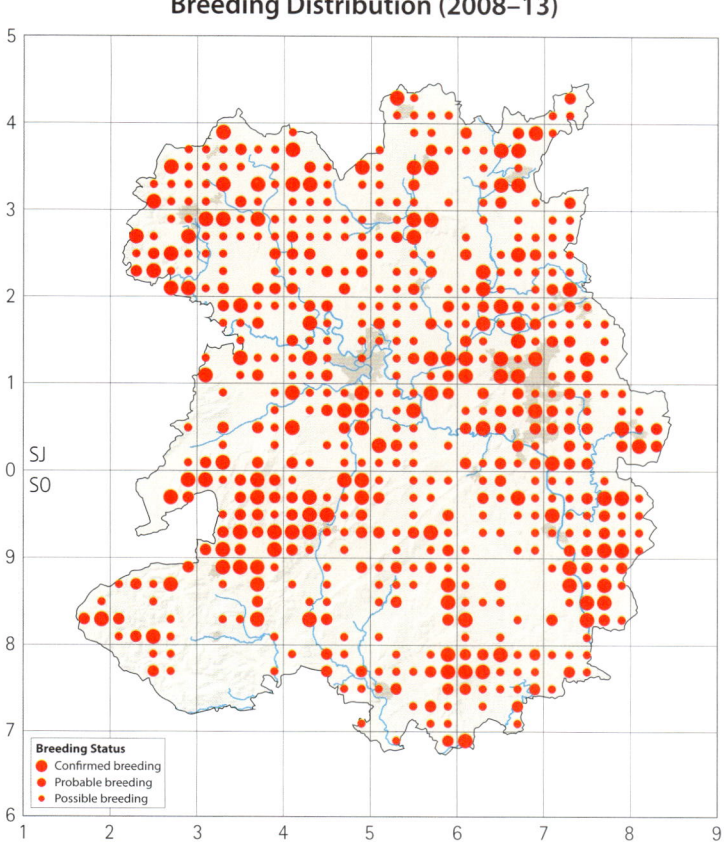

Breeding Status
- Confirmed breeding
- Probable breeding
- Possible breeding

Breeding Distribution Change (1985–90 to 2008–13)

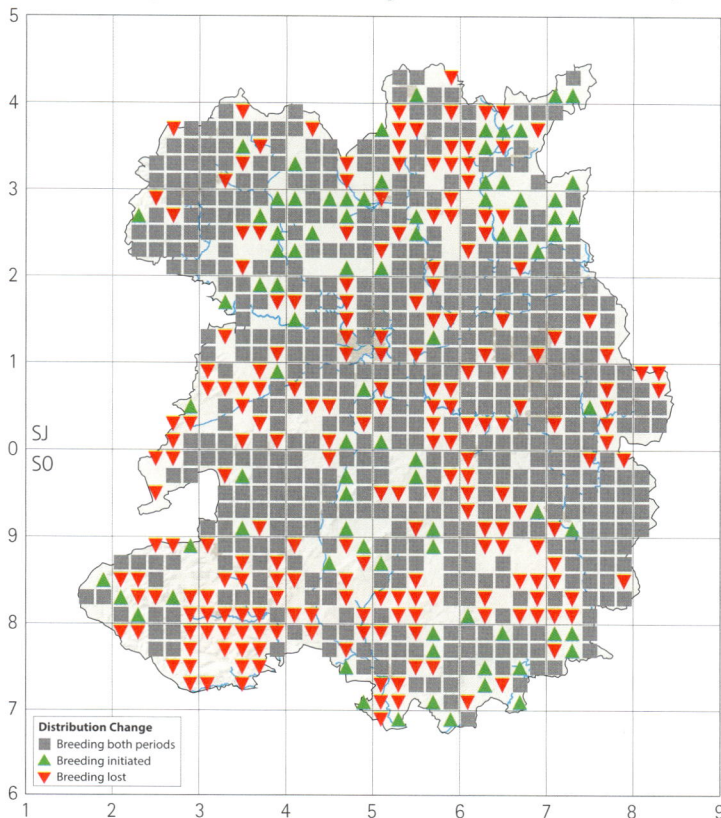

Distribution Change
- Breeding both periods
- Breeding initiated
- Breeding lost

Fieldwork for the recent Atlas revealed a fall of 18% in the number of tetrads with evidence of breeding, very much in line with the national estimate of range contraction. The proportion of tetrads where breeding was probable or confirmed is low, well under half of those where it was present. Breeding season ranges on farmland can average 5–10sq.km, larger in poor habitat or in years in which food is scarce (Village 1990), so some 'possible' records will have been produced by breeding birds whose foraging range extended into adjacent tetrads. All but 30 of the tetrads where breeding was considered possible are contiguous with tetrads where it was probable or confirmed, raising the possibility that the breeding population is even smaller and more thinly spread than already appears.

There are concentrations of tetrads with probable or confirmed breeding in open upland landscapes, such as the hills of the south-west, the Oswestry uplands and the Clee hills, but also in some parts of the valleys of the Severn and its tributaries. The Kestrel's adaptability may account for its relative success, or at any rate survival, in such contrasting habitats. In hill country, where steep or wet ground is more resistant to cultivation, it can profit from a ready supply of voles. On mixed farmland, its diet tends to include a higher proportion of birds and invertebrates (Village 1990), and regularly flooded land in the river valleys is likely to support them in greater diversity and higher numbers. Human control over the landscape, elsewhere almost complete, tends to be less in such areas, allowing greater habitat and species diversity.

In the breeding season the range has contracted by a net 129 tetrads since 1985–90. There have been losses everywhere, the most severe in SO37 in the far south-west, where it was found in only one

Occupied Tetrads

Atlas period (breeding)	1985–90		2008–13		Change	
	Number	%	Number	%	Number	%
Confirmed	174	20	114	13	−60	−34
Probable	174	20	143	16	−31	−18
Possible	370	43	332	38	−38	−10
TOTAL	718	83	589	68	−129	−18
Tetrads with Winter Records (2007–13): 719 (83%)						

tetrad. In four more 10km squares across the south, from Clun Forest to east of the Clee hills, and another on the Welsh border around Minsterley, it was found in fewer than half. There appears to be greater parity between east and west than in 1985–90, but in the form of a levelling-down: there is no 10km square, east or west, where Kestrel was recorded in every tetrad, and the difference in the proportion of tetrads showing higher-level breeding evidence is now much smaller. The pattern of range contraction (or, in fewer cases, expansion) affords little insight into its causes: for each tetrad that has lost or gained the species, another of apparently similar topography and land use can be found where the opposite applies. Variation in fieldwork effort may account for some, though not all, of the changes.

Winter distribution was wider: Kestrel was found in 83% of tetrads, the same percentage as in the breeding season in 1985–90. As well as dispersing young birds, many of which will not survive the year, the winter population includes migrants from the continent, and some seasonal movement from elsewhere in the country, with those

Winter Distribution (2007–13)

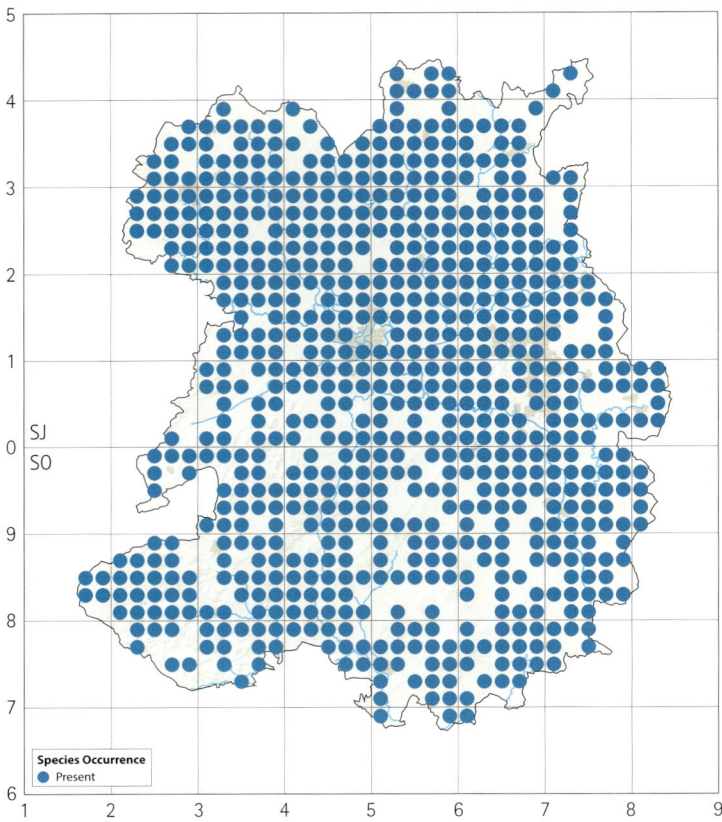

Species Occurrence
● Present

to 'ongoing research' suggesting that 'in England agricultural intensification has played a major role'. Certainly Kestrel's population graph over recent decades, resembling those of other struggling farmland species, points in that direction.

A nest box scheme in SJ63, around Market Drayton, puts this in a local context. There is very little unimproved grassland left in the area. Breeding pairs are usually associated with less intensively managed grassland, itself an increasingly rare commodity. As in many other places, roadside verges have offered alternative foraging opportunities, but in recent years they have been mown throughout the summer, to the probable detriment of the vole population. The overall habitat in the 10km square is now considered marginal at best, and supports only three regular breeding pairs (Gerry Thomas *pers. comm.*).

Barn Owl has been the subject of various conservation measures, such as retention of field margins and conservation headlands through agri-environment schemes, which might be expected also to benefit Kestrel. Atlas data show little evidence of such an effect: a tetrad-by-tetrad analysis of five well-spaced 10km squares where the change in status of either or both species was very marked found few cases where the outcomes coincided, and very many where they diverged. On the other hand, 78% of tetrads which retained Barn Owl also retained or recruited Kestrel, losing it in 22%, only slightly higher than the overall rate of attrition. Since Kestrel is diurnal and Barn Owl (mostly) nocturnal it seems unlikely that there would be much direct competition for food, though at times and places where resources are limited shared habitat may be insufficiently productive to support both species. Lack of suitable nest sites is a known limiting factor for both populations, and the divergence in their success in areas where both appear to be struggling may indicate a degree of competition in that respect, especially when many potential nest sites are taken by grey squirrel, Jackdaw or Stock Dove. There is a case for a Kestrel nest box scheme to complement that of the Shropshire Barn Owl Group.

The breeding distribution change map shows how quickly a 'widespread and numerous' species, as it was described in the *Atlas* 1992, can thin out. The fragmentation of the current population can only accelerate the downward trend. Kestrel's local history has shown its resilience, which makes its struggles in today's landscape

in the north tending to move south and east (Village 1990). Forrest's contention that there were 'fewer … in winter than in summer' (1899) appears no longer to be the case: the average number recorded on TTVs was slightly higher in winter than in summer.

Density appears to have declined even more steeply than range: BBS traced a population decline of 28% in the West Midlands between 1995 and 2014. Local BBS results, like those for Wales, are not published, as Kestrel is recorded in too few squares. Breeding bird surveys on the Stiperstones (2004–07), Long Mynd (2006–08) and Clee Hill (2013–14) found an average of one pair for each tetrad with probable or confirmed breeding (Leo Smith *pers. comm.*), which would suggest a total population of about 257 pairs. Atlas fieldwork discovered Kestrel in a third of tetrads where it had not been recorded during TTVs, hinting that its detectability may have decreased as the population has thinned. With that in mind, and including the 30 tetrads where 'possible' breeding is unlikely to be explained by foraging birds from adjacent tetrads, the upper limit of the population estimate may exceed 300, but is unlikely to reach 350, half the lower estimate of 700–1,400 in the *Atlas* (1992).

The national population decline appears to be driven by poor survival, especially of adults. Causes may include the effects of anticoagulant rodenticides, to which its diet makes it especially vulnerable, predation, sometimes by other raptors, and exposure to traffic when hunting on roadside verges (BTO BirdTrends). However, habitat impoverishment is likely to be an important underlying factor in mortality from other causes, as birds that are malnourished or struggling to find food are more liable to succumb to disease, exposure, accident or predation. *The State of the UK's Birds* (Hayhow *et al.* 2015) refers

John Fielding, Long Mynd, 10 September 2010

all the more troubling. It is not a fussy eater: when voles or other small mammals are scarce, it is perfectly capable of turning to birds. Local targets have included Goldfinch, House Sparrow, Starling, and Lapwing chicks, as well as worms and insects. Equally pragmatic in nest site selection, it has used cliffs, quarries, castle ruins, a Scots pine in central Telford and a 'listed house', as well as the more usual hollow trees and crows' nests.

The suppression of the population in the nineteenth century was intentional, and the means obvious: guns, traps and poison did the job. The causes of the current decline are less stark but more insidious: incremental changes in land use and farming methods over many decades have raised agricultural productivity to an all-time high while simultaneously reducing that of the natural world to a desperately low level. If current practices are making the landscape inhospitable even to a versatile species like the Kestrel, it suggests that loss of biodiversity is fast becoming critical. We would do well to heed the warning.

Forty-seven locally-ringed Kestrels have been recovered, 20 of them here; the remainder were found in 17 different counties, including Moray and Nairn, Londonderry and East Sussex, while three were recovered in France and one at Burgos, in Spain, 1,134km distant. Another 33 have been recovered here: 32 were ringed in 21 different counties scattered across the UK, and one 911km away in Germany.

The furthest-travelled within the British Isles was ringed near Baschurch, and found 'long dead' in Highland region 526km away. Six of the other far-flung recoveries were juveniles dispersing from Scotland and recovered here, or moving south from here to France and Spain. One ringed in Germany was recovered near Newport in late April having moved west during its first year. Two adults ringed here and recovered in south-west France and Spain in their second years may indicate some degree of migration within the population.

The longest-lived was six years three months old, against a typical lifespan of only four years; the national record is almost 16 years.

Michelle Frater

Red-footed Falcon

Falco vespertinus

Vagrant

This migratory falcon breeds in Eastern Europe and west, central and north-central Asia, wintering in southern Africa. It is a scarce migrant in Britain, with 10–15 birds most years, usually during May–July.

There are three, possibly four, historic occurrences of this species. An immature was obtained near Shrewsbury around 1865, and an adult was shot near Ellesmere in 1873. An immature female near Shrewsbury on 18 May 1901, possibly on the Sundorne Estate, was also shot. Finally, the Rev. H.O. Wilson's collection from 'on or near the Longmynd, between 1848 and 1857' included an individual of this species, although the validity of this record is open to question (see p. 5).

The three modern records were on typical dates:

Culmington	24–29 May 1974
Haughmond Hill	13–28 June 1982
Whixall Moss	26 June 1983

The one at Culmington was a male (Smith *et al.* 1975), while the first summer female on Haughmond Hill proved to be very confiding during its stay of over two weeks (Rogers *et al.* 1983). The female on Whixall Moss the following year, however, was only seen on one day (Rogers *et al.* 1984).

Graham Walker

Merlin

Falco columbarius

Rare resident, scarce winter visitor

Initially described as rare, some nineteenth-century authors wrote that the Merlin occasionally bred on the hills and moorlands in the south-west, but Beckwith referred to it as 'rather frequent about the high open ground along the welsh borders, breeding regularly on the Long Mynd and Stiperstones'. A nest was found on the Long Mynd in 1896, but Merlins were encountered more frequently in the autumn and winter, and there are accounts of them being caught in lark nets, or shot.

There was another nest recorded on the Long Mynd in June 1916, without any further information, but records were very infrequent in the first half of the twentieth century.

The *Handlist* (1964) stated that at least two pairs bred annually (on the Long Mynd) and had done so for many years, but no records have been found to substantiate this claim. None were published

between 1956–60, but an unpublished record in 1960 noted a nest in heather with four eggs, and the same site had been occupied in the previous two years. This was the only record of a nest on the ground. In 1961 one pair fledged two young in the same place, but an old Crow's nest was used.

In the next 20 years, at least one Merlin was seen during the breeding season in all years except four, and a pair was seen in 1965, 1969 and 1974. Breeding was confirmed only in 1968–72 (national BTO Atlas 1976), 1973, 1975 and 1976 (SOS records, not published in SBR). A breeding pair was probably present every year, but the population nationally declined considerably during the 1950s and early 1960s, due to organo-chlorine poisoning and habitat loss, and declines continued in Wales and northern England (Ewing *et al.* 2011), so there may have only been a single bird, or they may have

Tony Webb, Long Mynd, 8 October 2015

been absent altogether, in some years. One found in a Shrewsbury garden on 27 March 1983, died next day, 'apparently from alpha-Chloralose poisoning'.

The national BTO Atlas (1976) showed how isolated this population is – the nearest squares with confirmed breeding are about 60km distant in north Wales, and 100km distant in the Peak District.

Two pairs were recorded for the first time in 1980. The national population had recovered almost to its current level by 1983–84, and for four years 1983–86, the RSPB wardened the nests, in the hope of improving the breeding success and hence increasing the local population. Four young fledged in 1983 and three in 1984, but in that year the second nest was deserted when chicks were about a week old. In 1985 the first nest of the only pair was abandoned, with intrusion by large number of bird-watchers being a possible cause of the former, and unnecessarily close attention from a party of four bird-watchers was blamed for the latter. Fortunately the pair in 1985 re-laid, and raised three young. Disturbance by bird-watchers was also noted in 1986, when the RSPB wardens believed that one male was paired with two different females with nests in adjacent valleys, and one and three young were raised (RSPB 1983–86).

During the 1985–90 Atlas period, only 11 young fledged from four nests over the six years, in 1987 there was no confirmed breeding for first time since 1982, and there were no young in 1989 or 1990. There were two pairs in five of the eight years 1983–90 (J. Sankey *pers. comm., Atlas* 1992).

Merlins breed at one year old, and have a typical lifespan of only three years, so a sequence of poor breeding success would have an immediate impact, and, not surprisingly, the population quickly fell

back to only one pair in 1991 and 1992. There were two fledged young in 1992, and two pairs in 1993, and possibly in 1996, when a food pass was seen a considerable distance from the one known nest, but an intensive search did not locate any further evidence of a nest. During the years 1993–98, intensive fieldwork monitoring upland birds was carried out, so if there was more than one pair on a regular basis, some evidence would probably have been found.

Merlin's main prey is Meadow Pipit. The decline to one regular pair since 1993 was also possibly related to a reduction in Meadow Pipit density, which is a likely consequence of overgrazing. Grazing initially opens up the heathland habitat, allowing Meadow Pipit to colonise, but overgrazing removes the long grass that Meadow Pipits favour, and they are replaced by Skylarks. The RSPB wardens' reports highlighted the importance of pipits for breeding Merlins. Comparing their results with other studies, the 1985 report states 'interestingly, the Long Mynd plucking-post samples proved to have the highest proportion of Meadow Pipits at 80%', while the 1986 report showed Meadow Pipits made up 52 (88%) of 59 analysed prey remains. Surprisingly, as Skylarks bred at relatively high density, there were no Skylarks in the 1986 analysis, so increasing the Skylark population may be no help to the Merlins.

Overgrazing was reduced by an ESA agreement (see p. 41) between English Nature and the Long Mynd commoners in 1999, facilitated by the National Trust, which reduced the number of sheep using the moorland, particularly in winter when most damage was done to the vegetation. The basic terms of the ESA agreement were incorporated into an HLS agreement from 2009.

By 2004 the vegetation was recovering, although monitoring of Meadow Pipit numbers has been insufficient to show a corresponding increase, but the Merlin population increased to two pairs, with the same number in 2005, and an unprecedented three nests were found in 2006. Two of these nests were successful, producing seven fledged young, but the third failed. No male was seen at the failed nest, so possibly one male was paired with two females again.

There were two pairs again in 2007 and 2008, but both nests failed in both years, so again the population fell back to only one pair, the number found each year until 2013. Fourteen young fledged in the years 2010–13, including an exceptional five young in 2013.

In 2014 the only nest found was in a new tetrad, and four young were ringed and fledged, but territorial aggression was witnessed in early July against a potential predator 5km from the known nest, suggesting the possibility of a second pair.

There are large numbers of old Crows' nests in the upper reaches of all the main valleys, and the Merlins show no loyalty to the previous year's site, or any particularly valley. Almost all valleys are known to have been used at some time. Illustrating this, breeding was confirmed in three different tetrads in SO49 during the recent Atlas period.

It must be recognised that locating a territory is difficult, as pairs are inconspicuous, hunt close to the ground, and indulge in aerial display infrequently. The male brings food to the female incubating or brooding young only at lengthy intervals, so the late nestling or early fledgling period is the most likely time to confirm breeding. Both male and female defend the nest vigorously against potential predators, particularly when recently hatched young are at their most vulnerable. Crows, Ravens, other raptors, and even Herons will be vigorously attacked and driven away. Observing this behaviour, or seeing or hearing a food pass, is the most likely way of finding a nest. They are almost always in old Crows' nests in hawthorn trees on steep hillsides just below the open heather moorland, although in 1994 an old Magpie's nest was used.

Away from the Long Mynd, the *Handlist* (1964) noted Merlins in another area 'but nesting has not been proved', arising from an unconfirmed report of a pair nesting at Black Hill, near Clun in 1961 (unpublished SOS records), and in spite of a CSVFC report that 'a pair of Merlins bred [in the Clun Forest] in 1952 and 1953'.

A pair was found close to the border near Bettws-y-Crwyn in the 1973 breeding season, but there were no subsequent breeding season reports until a female was seen at Black Mountain on 8 May 1989, although autumn and winter records have been received from this area several times.

An agitated female on Brown Clee on 6 July 1982, a male hunting at Nordy Bank Fort on 16 July 1989, and a pair mobbing a Peregrine on 7 August 1990, together suggest the possibility of another breeding pair on Brown Clee during the first Atlas period, although there has been no evidence of breeding there since.

Other Merlins seen during the first Atlas period were a male near Angel Bank on 9 June 1987, and a female carrying prey near Kinlet on 5 June 1988.

The *Atlas* (1992) also described 'the discovery of perhaps two pairs flourishing in a hitherto unknown area, made up of largely atypical habitats'. The records appear to have come from an area east of Chelmarsh. Given the habitat in that area, and with the advantage of hindsight, it is now believed that these were Hobbies, a species that was still rare at that time, and which casts doubt on the 1988 Kinlet record.

The 1988–91 national BTO Atlas has also mapped 'breeding' (no distinction was made between probable and confirmed breeding) in 10km squares SO57 and SO67. Titterstone Clee spans these two squares, which also include Angel Bank.

During the recent Atlas period, a pair were seen on Titterstone Clee Hill in 2010. Breeding was confirmed in 2011, when two, possibly three young fledged from a nest on the ground, surrounded by heather and gorse. In 2012 the pair were unsuccessful, possibly as a result of

disturbance or trampling of the nest by orienteers, and only one adult was seen in 2013 and in 2014 (Chris Neal *pers. comm.*).

Otherwise, breeding season fieldwork for the 2008–13 Atlas found possible breeding records in three more tetrads in SO49, and one possible breeding record in SO28. There were no records from other suitable breeding habitat, on the Stiperstones or Brown Clee, although Merlins were seen there in the breeding season in years immediately prior to Atlas fieldwork. Given the isolated nature of the population, the continued possibility of disturbance, and the small number of tetrads with records, no breeding season map is published.

The population fluctuates between one to three breeding pairs. The estimate in the *Atlas* (1992), based on six tetrads with confirmed breeding, 'not likely to exceed six pairs', is now considered to have been over-optimistic.

In spite of Beckwith's comment, there have been no modern records suggesting breeding on the Stiperstones.

In winter, Merlins usually vacate the higher ground in northern

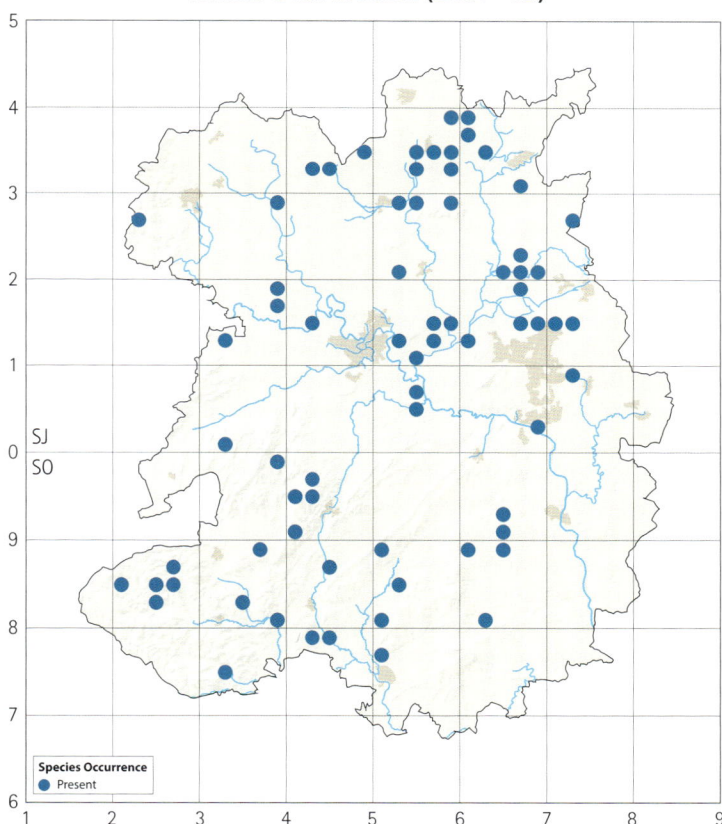

Winter Distribution (2007–13)

Tony Cross, Long Mynd, 15 July 2010

Occupied Tetrads

Atlas period (breeding)	1985–90		2008–13		Change	
	Number	%	Number	%	Number	%
Confirmed	6	1	3	0	-3	-50
Probable	1	0	1	0	0	0
Possible	0	0	4	0	4	x
TOTAL	7	1	8	1	1	14
Tetrads with Winter Records (2007–13): 70 (8%)						

Leo Smith, Long Mynd, 2 July 2006

attributed to the possibility that they might be remaining on the breeding grounds all year, as prey species are also remaining in increasing numbers, presumably as a result of milder winters. Hearsay comments from National Trust staff to this effect in 2004 were subsequently reflected in winter records, two in 2005, and five in 2006 and 2007. Atlas fieldwork increased the number of such records, and three of the Long Mynd tetrads are shown on the winter distribution map.

Several broods of Merlin chicks have been ringed: one nestling in 1986, while nests were being wardened by RSPB, and, more recently, 44 chicks in 13 broods in 12 different years, starting in 1996. One adult has also been caught.

- One of the six young ringed in 1996 was recovered dead in a barn on a farm near Bridgnorth the following April, 284 days after ringing and 21km from the nest site.
- An adult male was caught on 28 July 2010. He was found to be already ringed, and fledged from a nearby nest four years earlier.
- A sibling female, also ringed on 2 July 2006, was found freshly dead on 15 September 2012 at Burnham-on-Crouch (Essex), 267km distant, 6y 2m 13d later.

Four nestlings ringed elsewhere, much further north, have been found here:

- From near Sedbergh (Cumbria), dead having hit wires, found on 10 December 1938 at Norton, 196km distant, aged 5m 10d.
- From Rosedale (North Yorkshire), found dead on 26 April 1985 near Ludlow, 251km distant, 9m 29d.
- From Whickhope, Kielder Water (Northumberland), found long dead on 5 January 1994 at High Ercall: 269km distant, 3y 6m 8d later.
- A male from Tarras Moor, Langholm (Dumfries and Galloway) found sick (injured) on 26 March 2000 near Bridgnorth 293km distant 9m 0d.

These recoveries provide evidence that our wintering population is supplemented by visitors from the north. They also show high first-year mortality, although the oldest here are well short of the national longevity record, 12y 8m 17d (set in 1989).

For the future, ever increasing recreational pressure on the only regular breeding site, and the National Trust's policy of open access, combine to make the Merlin's tenuous talon-hold here rather precarious.

Leo Smith

and western Britain. Most only travel short distances to lowland or coastal areas, but some move further, and the British population is joined by Icelandic birds, so numbers are higher, and they might be seen almost anywhere where small birds can be caught.

Few places appear more than once or twice on the annual list of sightings, apart from the places which birdwatchers visit frequently, although open lowland habitats are favoured, and the distribution shown on the recent Atlas map is typical of any winter period. Merlins have been seen chasing Meadow Pipits, Skylarks, Wagtails and a Redpoll flock, and catching a Swallow, a Fieldfare in flight, and a Chaffinch or Yellowhammer from a stubble field.

A scarcity of winter records was noted in the years 2003–05,

Hobby (Eurasian Hobby)
Falco subbuteo
Scarce summer visitor

Britain is on the north-west corner of the Hobby's breeding range in Eurasia. It winters mainly on the African savannahs, and, as well as breeding here, many pass through on their way to and from breeding areas further north.

Forrest's summary, 'formerly often obtained near Shrewsbury, but now rather rare' echoed Rocke's observation that 'this pretty little falcon is, I fear, becoming rare'. 'It has bred', though no details were provided.

However, there was a six-year sequence of nest records from near Ludlow, starting with one in 1899:

utilizing a Crow's nest several years old in the top of a large oak tree ... Early in May the male was found dead near the tree, but the hen went away and quickly returned with a new mate. The first clutch of three eggs was taken ... but it is believed that the birds bred again, as they remained in the neighbourhood, and were seen repeatedly throughout the summer. (Forest 1908)

The pair returned the following year 'and laid again in the identical spot. The eggs were taken ... but, as last year, they have laid again, and are being allowed to rear the brood. The Hobby is known to return year after year to the same nest'. 'Another pair of Hobbies are haunting the vicinity, and a male was shot five miles away ... These are probably the young reared there last year. I trust they may be spared' (Forrest 1900).

In 1905 they had bred there for six years in succession, and 'four pairs ... bred near Ludlow'.

After that they were noted, on average, every other year up until 1939, including probable nesting near Middletown in 1921, but then only once, in 1947, before the *Handlist* (1964) described Hobby as a rare summer visitor, with the last known nest close to the Herefordshire border in 1950–54; they had also been seen in two 'old nesting haunts' in the south-west, but there was 'no definite proof of nesting'.

At that time Hobbies were largely restricted to south-east England, south of a line from the Severn to the Wash, but their range has rapidly expanded since then, perhaps due to the increasing number of sand and gravel quarries providing more widespread dragonfly prey, and warmer weather resulting from climate change making large day-flying insects available more frequently.

The first SOS records, in 1959, were of one seen on several dates in the south-west, and another in August in the east. Four years elapsed until the next sighting in September 1963, with two more in September 1965, but the late dates and the absence of breeding period records suggests that these were all passage migrants. The first

SOS breeding season record was 6 July 1966, of one chasing Swifts at dusk at Newport.

After another four year gap, a single was seen on autumn passage on the Stiperstones on 30 August 1970, but there were then only five more records until 1975, including one taking Swallows at a roost in Madeley in September 1971, and the first sighting on the Long Mynd, in June 1975. The number of records increased steadily from 1976 onwards, with 12 up until 1979, but it was not until 1978 that two were seen together, on Long Mynd on 25 May and again on 11 June, and in the following year, in the Shrewsbury area from 19 June to 5 July 1979. One was seen to take a Swift on 17 July, the first eye-witness account of the falcon's speed and aerobatic skills. Of the 18 records in the 1970s, most were of single birds during the passage periods.

Hobbies were seen in each of the following years, up until the first confirmed breeding record in 1983, a nest with three fledged young near Harley. They were seen at seven other locations that year. The following year, three were seen together near the nest site, and an uplift in sightings from 1984 suggests an increase in the population from that date.

During the 1985–90 Atlas period 'three pairs bred successfully in widely separated tetrads, and pairs were seen in four other suitable breeding areas. Singles were seen in over 50 tetrads, mostly in a large central area, and in the north, north-east and south-east. Although they

John Hawkins, North Shropshire, 14 July 2011

Confirmed & Probable Breeding with Range (2008–13)

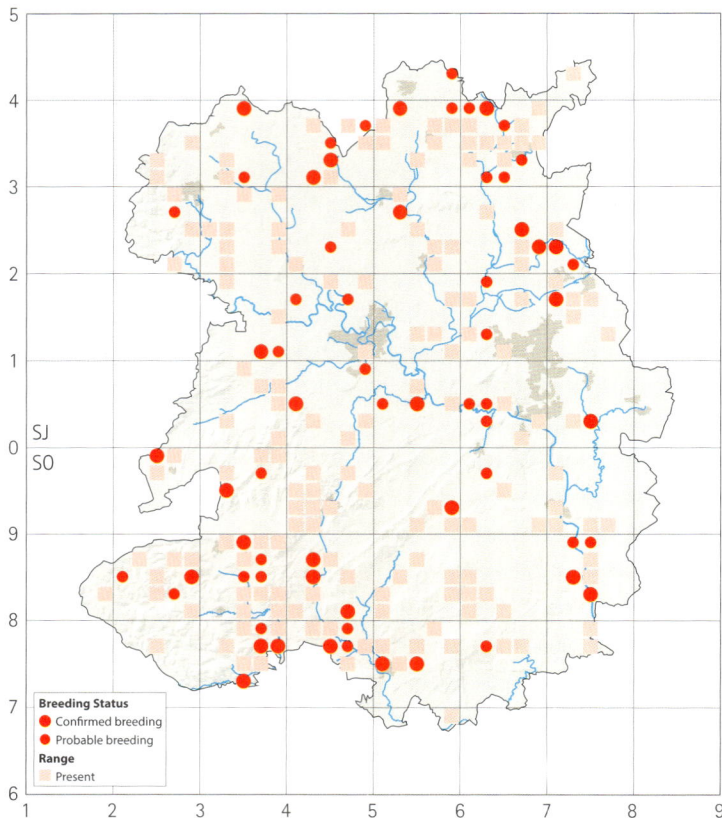

Breeding Status
● Confirmed breeding
● Probable breeding
Range
■ Present

Occupied Tetrads

Atlas period (breeding)	1985–90		2008–13		Change	
	Number	%	Number	%	Number	%
Confirmed	4	0	30	3	26	650
Probable	8	1	35	4	27	338
Range	>50	6	169	19	<119	<238
TOTAL	>62	>7	234	27	<172	<277

usually become noisy until August, when most Atlas fieldwork has finished.

Parents often forage 3–6.5km from the nest, and the distribution map is also complicated by young birds. Most Hobbies do not breed until they are two, so around a quarter of those seen will be non-breeding birds. However, as they may help breeding pairs raise young, seeing them may also be indicative of a nest nearby (Hardy *et al.* 2009).

There is little threat of disturbance now, and RBBP recommended the publication of a tetrad map, not least because 'more information might just encourage birders to look for and record nesting areas and broods' for this very elusive and under-recorded species. The map shows tetrads with confirmed and probable breeding in the usual way, but all tetrads with any other record – birds in breeding habitat, or summering or in flight – have been aggregated to show the observed range. The *Atlas* (1992) noted that single Hobbies had been seen in over 50 tetrads, compared with 169 in recent Atlas fieldwork, a threefold increase.

Some pairs may be found year after year in the same territory, but the actual nest may move between tetrads, and some of the probable breeding records will be of pairs or displaying birds in tetrads adjacent to a nest site. However, two instances were found of two pairs nesting within 1.5km of each other in 2011, one on the

were probably hunting, breeding in these areas is a possibility'. One of these pairs nested for the first time in 1985, and was present every year during that Atlas period. 'It is still uncommon and vulnerable to egg collectors, so locations of nest sites are not published. The population could be up to eight breeding pairs'.

In 1993 confirmed breeding was observed at three sites, all successful, with five fledged young. By 1998, the Hobby was 'undoubtedly becoming more frequent' with four and three noted on the Long Mynd on 24 and 25 April respectively, five on the Stiperstones on 3 May, and breeding at two sites. In 1999, successful breeding was noted at three sites, with confirmed breeding at two others. The threshold of 100 records was crossed in 2000 (103, from 57 sites) but, although Hobbies were well reported up until 2007, there were few confirmed breeding records.

However, SBRs prior to the recent Atlas noted that Hobbies are undoubtedly the most elusive of our breeding raptors, given their tendency to nest very late in scattered trees on arable farmland. As this is a habitat infrequently visited by bird-watchers at the appropriate time, they are almost certainly under-recorded. The hope was expressed that fieldwork which required every tetrad to be visited would provide a clearer picture and, as expected, a large increase in the population of this graceful falcon was discovered.

Proving breeding is difficult. Displaying pairs might indicate a territory in May or early June, but confirmation is most likely to come from noisy adults reacting to an intruder near the nest when the young are at late nestling or early fledgling stage. However, breeding is late, timed to coincide with recently fledged hirundines being available as prey to feed nestlings, so the adults do not

Jim Almond, Chelmarsh, 3 September 2011

border with Montgomery (distance between nests 1.5km), and one on the border with Herefordshire (distance 1.3km), so adjacent tetrads with confirmed or probable breeding may each hold a breeding pair.

An assessment by the Raptor Study Group of distribution in the south-west estimated at least 11 breeding pairs, based on nests found and pairs observed over the Atlas period, and the map shows 11 tetrads with confirmed or probable breeding there. Such specialist fieldwork has not been undertaken elsewhere, but there were 66 tetrads with probable or confirmed breeding records, and several clusters of tetrads where Hobbies have been seen but there is no more definite evidence of breeding. They are also very inconspicuous, so many will have been overlooked. It is therefore likely that the population now exceeds 70 pairs in some good years. However, it fluctuates, with fewer in cold springs, as insect prey will be scarce, and in years following poor breeding seasons, such as the very wet summer in 2011.

As noted by Forrest, some pairs show considerable fidelity to nesting areas. A pair was present at one site throughout the 1985–90 Atlas period, using four nest trees, all within a 100m radius, and was then found each year in the same area right through until the mid-1990s, the last time the site was visited. Although Hobbies were not found in the same area in 2008–13, an adult carrying prey in the adjacent tetrad shows they haven't moved far.

The earliest arrival was 4 April, in 2005, but annual first dates are spread throughout April and early May, with the average around 23 April (SBR 2005). Favoured sites on arrival are Long Mynd, where they can be seen hawking bumble bees, dragonflies and moths throughout May, and Venus Pool. One or other of these two sites have produced the first record in most years. It was not until 2000 that the northward extension of range led to the comment that 'dragonflies at Whixall Moss are a major attraction' with six seen in August, six again the following July, a spectacular 18 on 11 May 2002 (which were mainly passage birds, as only two were present later the same day), seven on 6 May 2006, and 11 on 1 May and 14 on 15 May 2010. Such high numbers have not been found since, perhaps not surprisingly in view of the poor spring weather.

Courtship feeding, provisioning the incubating female, and feeding young in the nest lead to a switch in diet, and Hobbies have been seen to take Swifts, Swallows, House Martins and Sand Martins. Chasing hirundines is usually an indication of a nest close by, and most nests are within a kilometre or two of a colony.

Nestlings usually fledge in early–mid-August, but pairs that have to re-lay may still have young in the nest at the beginning of September. Individuals seen from August onwards may therefore be breeding locally, or be return passage migrants. The average last date is around 1 October, but dates up until 17 or 18 October in 2002 and 2000 respectively are not unusual, with the latest ever on 2 November, in 1985.

Few Hobbies have been ringed, but the Raptor Group has facilitated it, and 17 nestlings have been ringed since 2011. One of these nestlings, ringed on 1 August 2011 near Bucknell, was found freshly dead near Cannock (Staffordshire) 71km distant on 17 May 2013, 1y 9m 16d later. An adult, caught and colour-ringed a few days earlier at the same site, chased a Swallow into the living room of a nearby farmhouse a few weeks later.

A colour-ringed nestling ringed on 28 July 2009 near Eyam (Derbyshire) was seen on 24 July, 2010, at Whixall Moss, 11m 26d later.

As climate change is believed to be a factor in its range expanding northwards, it is likely that the breeding population will continue to increase.

Leo Smith

Gyr Falcon (Gyrfalcon)

Falco rusticolus

No modern records

Gyr Falcon lives in northern latitudes with a circumpolar distribution. The single occurrence was 'just prior to 8 April 1853', when two appeared in the Longnor and Leebotwood area, both eventually shot by a keeper. They were 'both in the same *brown* plumage of the first year, and probably belonged to the same nest ... I should say they belong to the 'Iceland race', but I speak advisedly' (Rocke 1865).

John Tucker

Peregrine (Peregrine Falcon)

Falco peregrinus

Scarce resident, scarce winter visitor

Peregrine is the most widely distributed species in the northern hemisphere.

At the end of the nineteenth century, 'every year a few appear, generally immature birds in the autumn, and more females than males. No recent instance is known of it nesting here'. Two were shot for local collections (Forrest 1899). Its status was unchanged over 60 years later, when it was described as a 'non-breeding visitor, rare' (*Handlist* 1964).

Between these accounts, the national Peregrine saga was one of survival, decline and recovery. In the years prior to the Second World War, the UK population was estimated to be at least 820 pairs. During the war, to protect message-carrying pigeons, it became legal to kill them, and, as a result, numbers fell to about 87% of the 1939 levels. The population increased again after 1946, but plummeted from 1956 onwards as a direct consequence of the effects of organochloride pesticides, then being used extensively on a variety of crops, to a new low in 1963 of about 360 pairs. These toxic agricultural chemicals were banned, and numbers increased again (Ratcliffe 1993). Recovery was helped subsequently by the special protection afforded to vulnerable and threatened species by the Wildlife and Countryside Act 1981.

As part of this recovery, Wales, one of the traditional strongholds, saw a quite dramatic increase during the 1960s and 1970s (Lovegrove *et al.* 1994), and this was the origin of the local population. Individuals started venturing eastwards, so from 1983 there was a sudden and

Jim Almond, Shropshire, 25 June 2008

sustained increase in sightings, mainly of singles, in all months of the year, including many during the breeding season. The majority of reports originated from the western uplands, but a number came from other widespread locations, particularly during the winter months.

The first documented account of breeding was an article in the *Shropshire Magazine* in December 1970 entitled 'The Return of the Peregrine', by Eric Hardy, a well-known writer and ornithologist. It describes two nests, including a pair which bred successfully in the south and fledged two young, one of which was ringed by a 'local ornithologist'. The other, in the north, still had two young in the vicinity of the nest the following November. At least one of the pairs fledged young the following year.

These isolated occurrences were not repeated, and the first nest reported to SOS was discovered in 1987. Three years later the number of pairs had increased to two (*Atlas* 1992). The rock faces used by these pairs, in the south-west and north-west respectively, have been used regularly since, and following a spread eastwards, 10 years after the first breeding known to SOS, there were 10 breeding pairs by 1997, all in working or disused quarries (Tucker 1998). These quarries were created for the post-war construction industry, so they were not available to pre-war Peregrines, and an increase in the number and range of the favoured prey, Feral Pigeons, also facilitated the initial colonisation and rapid population increase.

In 1998, the Shropshire Peregrine Group (SPG) was established in response to repeated incidents of a nest being robbed, and attacks on a different breeding pair. The formation of the Group resulted in a targeted approach to monitoring and protection. The species prefers open country over which to hunt, enough food in the form of other birds, and steep rock faces for nesting. Pairs resorting to traditional cliffs make it one of the easiest species to monitor and census, and 20 young fledged from 13 nests in 1998.

Natural cliffs are scarce and the few that exist were quickly occupied during a further spread eastwards, with working or disused

quarries again being the most favoured new locations. Between 1998 and 2014 breeding was confirmed at a total of 29 locations (including all the 10 used prior to 1998). The number varied annually, increasing to 19 in 2014, but since 2008 successful breeding attempts have been made at only three new sites, and each was occupied for only one or two years. Unsuccessful attempts were made at two more marginal sites.

The number of nests and fledged young for each year since 1987 are shown in the chart, incorporating SPG records since 1998. There were 325 breeding attempts (defined as at least one egg laid), and 240 (73.8%) were successful. At least 848 eggs were laid, 625 hatched, and 535 young fledged. The local increase reflects an increase in the UK population, now estimated at approximately 1,500 pairs (BTO survey 2014).

The distribution in the recent Atlas period is mapped at the 10km square level. No tetrad map is published, as nest locations are confidential.

SPG members, all of whom are licensed by BTO to monitor breeding Peregrines, have also provided physical protection and covert camera surveillance at certain more vulnerable sites, and the Group's records provide a comprehensive account of locations, breeding success, and persecution.

Interaction with Ravens is well documented, and no less than five pairs either use abandoned Raven nests or nest in close proximity to them. The two species exist in a permanent state of rivalry, but appear generally to tolerate each other's presence during the breeding season.

Six busy working quarries have been used for a number of years, and a willingness to tolerate a high level of disturbance from working machinery and the presence of humans has been consistently displayed. For example, a pair which had nested successfully for several years, suddenly vacated the quarry when it ceased operations, and re-appeared the following year in a nearby timber yard where again there was a high level of disturbance.

A particularly unusual and vulnerable nest is located on a rock ledge surrounded by an active moto-cross race track. First discovered in 2004, it has been occupied regularly and has successfully fledged a total of 13 young. Until 2009, the breeding pair tolerated the extreme noise level and general disturbance caused by the racing, only occasionally being forced to abandon. Following representations by SPG to the Amateur Motor Cycle Association, the motocross governing body, the local club agreed to suspend racing during April–June each year, thereby allowing the undisturbed breeding – a minor victory in the cause of raptor conservation!

Since 1998, over 500 young have fledged successfully. This significant and sustained population increase might have been expected

Nesting Attempts and Fledged Young (1987–2014)

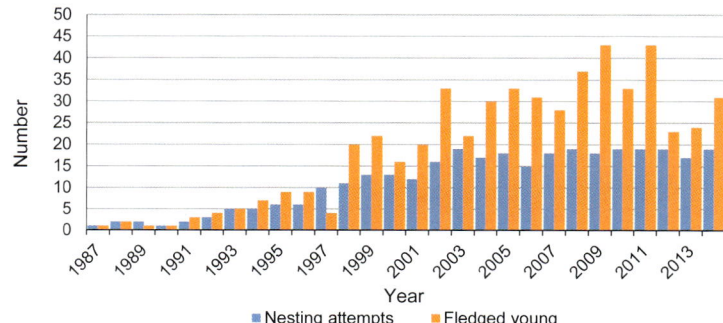

to lead to an even larger increase in the breeding population, but it has been limited by the shortage of suitable rock faces. The growing population has therefore turned to new types of nest site – small cliffs, tall buildings in urban conurbations, and old corvid nests in trees.

In 2010, regular sightings were reported from several town centres, frequently observed feeding on Feral Pigeons. The following year adults were observed for several days carrying prey to their young on a church tower in Shrewsbury town centre, confirming for the first time that successful breeding had occurred. A nest box was installed on the tower in 2012, but although there were regular sightings throughout the breeding period, the box remained unoccupied.

Even more unusually, in the spring of 2002 a local farmer reported that an abandoned Carrion Crow's nest on his land was occupied by what appeared to be young Peregrines. The identification was confirmed, with their parents in regular attendance, and three young fledged successfully a few weeks later. One of the fledged young was found shot shortly afterwards, but was recovered, nursed back to health by a local falconer, and later released back into the wild. Between 2003 and 2011, the 'tree-nesting' pair bred at the same farm every year, occupying either abandoned crows' nests or on one occasion a wire basket, from which three young fledged successfully. During this period a total of nine young fledged. Over the following three years the resident pair were regularly observed engaged in bonding but no further breeding attempt was made.

In the spring of 2012 another pair was discovered occupying an abandoned Magpies' nest on farmland in the north, over 45km from the first tree site. It was abandoned after five weeks, probably due to inclement weather, and the birds have not returned since.

There are only three other authenticated accounts of tree-nesting in the UK, two in Scotland in old Ravens' nests, and the third in mid-Wales; the latter produced two fledged young in 1984 (Ratcliffe

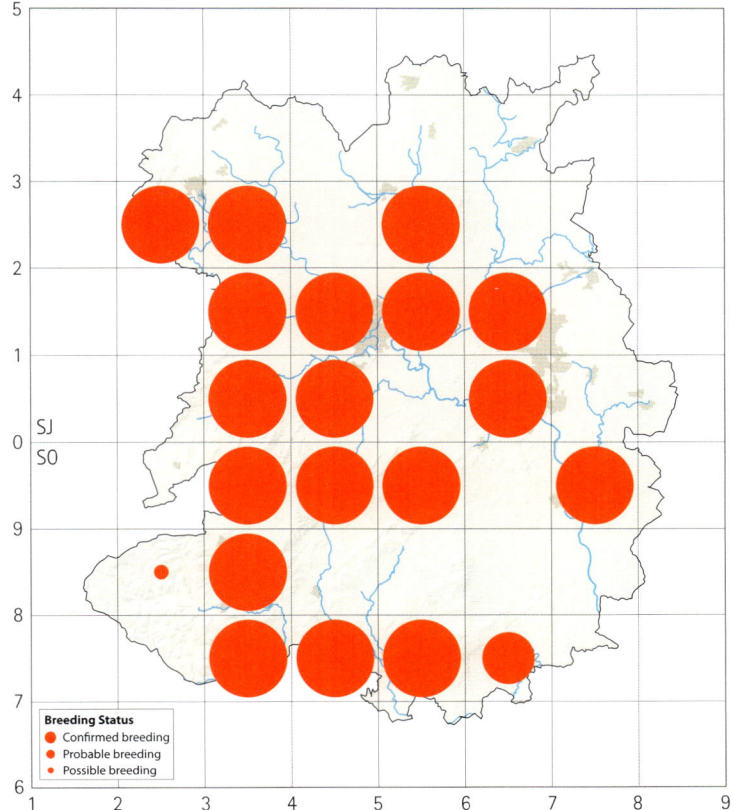

Breeding Distribution (10km) (2008–13)

Breeding Status
- Confirmed breeding
- Probable breeding
- Possible breeding

Occupied Tetrads

Atlas period (breeding)	1985–90		2008–13		Change	
	Number	%	Number	%	Number	%
Confirmed	2	0	31	4	29	1450
Probable	0	0	6	1	6	x
Possible	–	–	21	2	x	x
TOTAL	–	–	58	7	x	x
Tetrads with Winter Records (2007–13): 224 (26%)						

Jim Almond, Shropshire, 25 June 2008

1993). The Shropshire pair are therefore apparently the only Peregrines in England known to have nested successfully in trees, and over a period of 10 years.

Welsh ringed Peregrines have been found dead locally on three occasions, including an 11 year old from Dyfed, hit by a car in Shrewsbury, 78km distant, in August 2001, but the outstanding recovery was of a nestling ringed in Sweden in 1947, and found dead 1,980km distant at Onslow in November 1951, the only one ringed abroad and recovered here, and indicating that occasionally Peregrines from northern latitudes do move here for the winter. The oldest was found dead at Sweeney Mountain, near Oswestry, in 2004, having been ringed as a chick near Pant, 38km to the south-west, over 12 years earlier.

Since 2004, 48 chicks have been colour ringed here. These have helped to increase the relocating rate of ringed birds, and provide tentative insights into the movement of locally bred Peregrines. A male ringed in 2007 attempted to breed two years later in the West

Winter Distribution (2007–13)

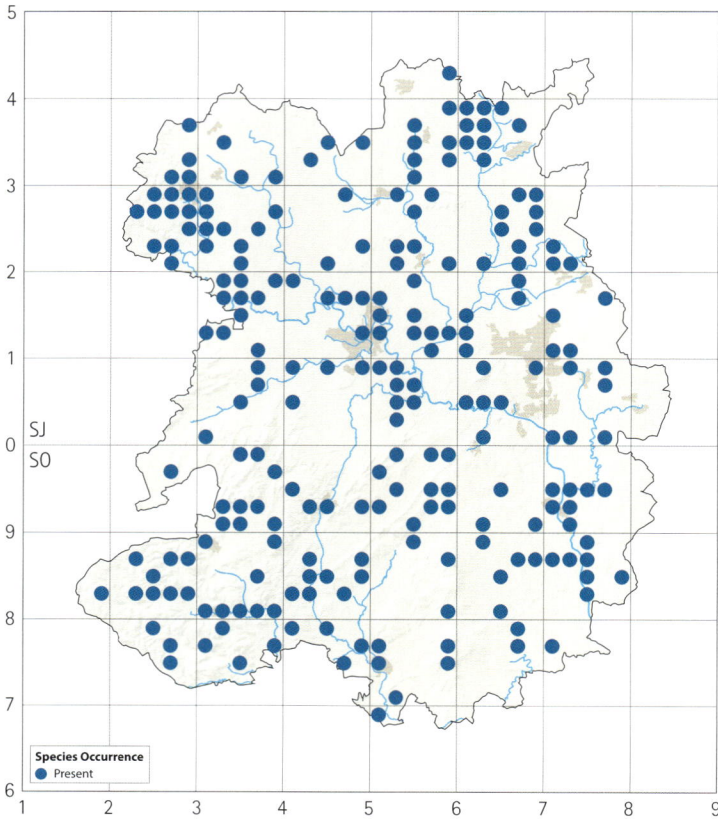

Species Occurrence
● Present

Europe – there is certainly no shortage of abandoned corvids' nests. If these opportunities are not taken up, it is likely that the population will remain fairly stable.

However, it remains under threat from individuals associated with shooting estates, pigeon racing, falconry and egg collecting. Between 1998 and 2006, 13 nests are known to have been robbed of young or eggs. The two most serious incidents of persecution occurred in 1999, when a falcon was shot while brooding three young, and the three chicks also died before they could be rescued, and in 2010, when pigeons baited with poison were used to kill a pair of breeding adults. SPG has subsequently organised a daily patrol of volunteers during the breeding season to protect birds at this latter site.

In 2008, two young Peregrine chicks from a nest on nearby Cannock Chase were rehomed successfully in local eyries containing young of a similar age, after their parents had been illegally killed.

As these incidents show, the ongoing threat to their continued survival remains. As numbers have recovered, some pigeon fanciers and game interests have again called for the removal of legal protection. Maintenance of this protection is essential for the conservation of the Peregrine, because it remains comparatively rare, is extremely susceptible to human activities (including continued illegal persecution), and, once the population has declined, it takes a long time to recover. In the words of Derek Ratcliffe, the UK's foremost authority on this falcon: 'Those to whom the Peregrine is a source of inspiration and wonder have a special duty of vigilance. It will be their responsibility to ensure that it survives in its own right as one of the most spectacular inhabitants of our planet'. The SPG assumed this responsibility in 1998, and will continue with other local conservation groups to safeguard the future of this magnificent bird of prey in the years ahead.

John Turner
Shropshire Peregrine Group

Midlands. A female ringed in 2006 bred in south Breconshire four years in succession from 2009–2012, successfully rearing chicks in each season. Other colour-ringed birds have been reported in winter months, from Cheshire (found dead) and Northamptonshire (road casualty), and one colour-ringed bird was photographed at a falconry display in Cambridgeshire where it had been attracted by the performing falcons. Conversely, a female, ringed as a chick at Worcester Cathedral in 2009, bred successfully in the south in 2013, but did not return in 2014.

In general, the majority of adults are resident at or near their breeding areas throughout the year, although there is some movement, relatively local and at times weather inspired. Young birds become more independent of their parents from October onwards and tend to forage some distance away from their breeding localities, particularly where suitable prey is concentrated, near towns and pools, and along the river valleys, as reflected in the winter distribution map.

In recent years a noticeable change has been observed in the relationship between juveniles and their parents, with many of the former showing an increased tendency to remain with their parents throughout the winter months and even into and during the following spring. In two places juveniles from the previous year's broods have been observed bringing prey to their parents, and even assisting with incubation.

The Peregrine has become an established resident with a breeding population of over 20 pairs. However, the majority of natural rock and quarry faces are now occupied. The move into urban areas may increase, and it is also possible that the number of pairs using trees will grow slightly, in common with the population in parts of northern

Jim Almond, Shropshire, 19 June 2008

Ring-necked Parakeet (Rose-ringed Parakeet)

Psittacula krameri

Rare naturalised visitor or escapee

Jim Almond, Stoke St, Milborough, 26 March 2009

England supports the largest naturalised European population of this parakeet, which originated from escaped or released birds. It is native to sub-Saharan Africa north of the equator, and the Indian subcontinent. It began nesting regularly in the wild in England in 1969 and was admitted to Category C of the British List in 1983, by which time the feral population was estimated at 500–1,000 individuals, the majority in Kent and the Thames Valley. The population and range have increased substantially in the last 20 years, and several other conurbations in central and northern England have been colonised recently.

It has enjoyed a mixed reception, particularly with SBR editors, and it has been moved between the systematic list and the list of escapes on several occasions. While some records undoubtedly relate to escapees of this relatively popular cage bird, many are likely to originate from increasing feral populations.

One feeding on apples at Yockings Gate (Whitchurch) for several days from 3 November 1975 constitutes the first record. Since then it has become a more frequent, though still rare, visitor. Some records are of birds in flight, and all are of singles except two over Shrewsbury on 30 May 2009, and another two near Bridgnorth on 17 December 2011. Records are from widely scattered locations and cover every month of the year, but most were outside the breeding season. There was a gap of nine years until the second record in 1984, with further records in 1991 (two) and 1994. It has been seen in 11 of the 15 years this century, including six records up to 2007, 15 in the six Atlas years, and five in 2014.

In the recent Atlas years, in addition to the two records of two together noted above, singles were seen at Tibberton in January and February 2008; at three further sites in 2009 (Muxton on 6 January, at Stoke St Milborough from 23 March to 6 April and Clunton on 4 May); in gardens in Meole Brace on 2 December 2010 and Aston on Clun on 17 January 2011; near Bridgnorth on three dates (1 April, 18 June and 10 September) in 2012, and again on 26 May 2013; at Walcot on 28 March; Shrewsbury on 9 April and Chelmarsh from 13 August until 21 December, all in 2013.

There were five sightings in 2014, one near Chelmarsh on two occasions, another at nearby Hampton Loade, one on a bird feeder at Cleeton St Mary in April and one at Shrewsbury School in September.

More than half the records up until the end of 2014 have come since 2011, mainly from the Bridgnorth–Chelmarsh area and other sites in the south-east, and probably represent (semi) resident escapes or post-breeding dispersal from the recently colonised West Midlands conurbation, where one or two pairs were confirmed breeding in 2007 and 2008 (Holling *et al.* 2011), and the BTO Bird Atlas 2007–11 shows confirmed breeding in two 10km squares covering Birmingham, Sandwell and Walsall.

The national population is almost certain to continue to increase and spread, so this species is likely to become an increasingly frequent visitor to local gardens.

Richard Moores

Red-backed Shrike

Lanius collurio

Very rare passage migrant, has bred

This striking variety of 'butcher bird' breeds across most of Europe and western Asia. It used to be a regular sight as it arrived each spring from its wintering grounds in tropical Africa. Although it was never common, a small number of nests were recorded each year throughout the nineteenth century. They tended to return to the same nesting area each year, predominantly in the south. Rocke (1865) feared that this predictability may have been contributing to their decline, as it made it easy for eggs to be taken. Beckwith (1888) felt that other factors also contributed to the decline, 'its bright plumage soon attracts attention, so that it is too often killed'.

Their preferred habitat of open scrubland with tall hawthorn trees and tangled hedges allows them both a good viewpoint from which to hunt as well as the opportunity for 'larders' – the act of impaling their prey on thorns for safekeeping is common among

Shrikes. Paddock (1897) reported larders containing 'the remains of small birds, field mice, bumble bees, and a grasshopper'. They are only slightly larger than a House Sparrow, but will also take birds larger than themselves: Paddock (1904) reported with amusement how he had observed a male Shrike attack a Blackbird which was unfortunate enough to land near to the former's nest. The Blackbird survived the encounter, and was able to fly off once the Shrike returned to its lookout post.

In 1901 and 1906 Red-backed Shrike nests were found which also contained Cuckoo eggs, although the final outcome was not documented.

Numbers continued to decline during the twentieth century and the *Handlist* (1964) summarised the work of J.H. Owen, who studied this species on Llynclys Hill. His records date back to 1881, and in one year, not specified, he observed 15 pairs. 'in 1938 they were quite plentiful. During 1940 five pairs were located, rising to nine pairs in

1944. Only three pairs were located [nesting] in 1945–46 after which only two single males were seen in 1952'.

The *Handlist* also summarised the main breeding areas as: Dowles in the Wyre Forrest (1900–33), near Ludlow (1906–25), Benthall and Broseley area (1916–23), the Bucknell area (a number of years up to 1954), and All Stretton and Horderley (occasional records up to 1952). The last definite record of breeding was a pair near Bucknell in 1954.

Since then there have only been four records, all of single passage birds, at Leebotwood in July 1981, Lower Wood in the unusually late month of November in 1987, Carding Mill Valley in June 1988, and most recently in July 1995, when a male was seen at Stoney Hill in Telford. This decline has been matched nationally, and each year there are only a handful of breeding pairs, or none at all. A corresponding decline on the near continent suggests that re-establishment of a breeding population is unlikely in the foreseeable future.

Linda Munday

Great Grey Shrike
Lanius excubitor
Rare winter visitor

Jim Almond, Allscott, 3 April 2011

The race that occurs here breeds in northern Europe and over-winters across central and most of southern Europe. It sometimes occurs on passage, but most records are of winter visitors, probably from Fennoscandia.

Along with other Shrikes, the *Lanius* part of their name, meaning 'butcher', refers to their habit of storing food in a larder, sometimes

impaled on thorns. The *excubitor* means 'sentinel', as they will perch high on trees or telegraph wires to look for prey. In winter their diet consists mainly of small mammals and birds.

Their historical status is unclear, as Rocke (1865) noted that they are 'frequently met with … generally in the winter', whereas Beckwith (1879) described them as 'a rare and uncertain winter visitor'. Forrest (1899) listed the 11 locations where this 'rare winter visitor' had been recorded, namely Shrewsbury, Whitchurch, Harton, Acton Reynald, Hawkstone, Ludlow, Weston (Shifnal), Ellesmere and Westbury. They were specifically targeted and often shot for various collections. In 1897 one was caught near Shrewsbury on limed twigs, something which is now, thankfully, illegal in this country.

There were a handful of records between 1900–03, and only two more in the remaining first half of the twentieth century, one shot near Westbury on 2 December 1922, and one at Merrington Green on 15 January 1939.

The *Handlist* (1964) summarised the modern records from the middle of the twentieth century: Shrewsbury on 7 December 1952; Hadley on 8 and 27 March, and Shifnal on 28 March 1953; Clun on 7 November 1956; Longmynd on 14 April 1963 and Venus Pool from 15 December 1963 for at least a month. All records since 1950 had been of single birds.

There were then 17 records from widely scattered locations between 1966–93, but none in the 1981–84 BTO winter Atlas. In October 1990 one collided with a car while chasing a Dunnock at Stoke Heath. It was kept overnight for observation. The Dunnock did not survive the collision and was fed to the Shrike, which was successfully released the following day.

More recently, they have occurred every year since 1998. One over-wintering at the Abdon Burf summit on Brown Clee in each of the three years 2002–04 was thought to be the same bird returning each year.

Apart from the earlier Brown Clee records, there are other examples

Winter Distribution (2007–13)

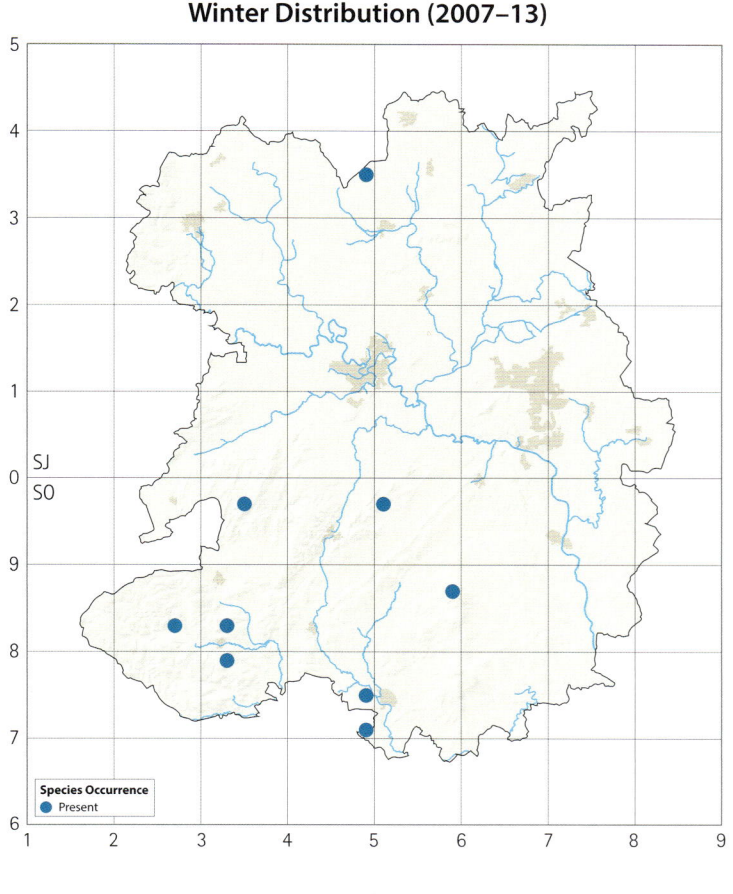

Occupied Tetrads

Tetrads with Winter Records (2007–13):	9	(1%)

from the recent Atlas period suggesting that individuals maintain a winter territory, and return to the same place in subsequent years, in common with those elsewhere in Britain. On Brown Clee again, there was one at the end of March 2008, on 8–9 January 2009, and on 29 January 2011. There was one at Whixall Moss in the winter of 2008–09 for seven days in December and 10 days in March, again in 2009–10, on various dates between 29 October and 7 April, and one just outside the Atlas winter period on 25 October 2010. There was also one at Black Hill for four days in November 2007, and again in 2011–12 on several dates between 15 October and 19 March.

Other locations shown on the map are Whitcott Keysett, Bury Ditches, the southern end of the Stiperstones at the Rock, Lower Whitcliffe, north of Richards Castle, and Yell Bank. The one at Lower Whitcliffe also stayed for 10 days, in January–February 2012.

Although some over-wintering birds stay on until early spring, Great Grey Shrikes can also occur during March and April on passage. During April in both 2009 and 2011, one at Allscott Sugar Factory for one and nine days respectively had not been seen there during the winter months. Similarly, one was found in Hopesay during March of both 2010 and 2011, on four days in each year.

Since the completion of the recent Atlas, they have continued to be found on a regular, if rare, basis. During 2013, one was seen at Whitcliffe in March and April, while during 2014 there was another at Black Hill in November. It was found at these locations during the recent Atlas period as well, again suggesting that individuals return to the same areas each year, over the winter and during spring passage. The birders follow them there, perhaps explaining the steady increase in records during the modern period, although there does not appear to have been a large increase in the national wintering population.

Linda Munday

Steppe Grey Shrike

Lanius meridionalis

Vagrant

Jim Almond, Wall Farm, 3 November 2011

Known as Southern Grey Shrike until the revision of the British List in January 2018, and until recently considered conspecific with Great Grey Shrike,* it occurs in southern Europe, principally the Iberian Peninsula, and in North Africa and Central Asia. The only record was of a first winter of the Central Asian race *L.m. pallidirostris* (Hudson *et al.* 2012):

Wall Farm, Kynnersley   28 October–9 November 2011

This race is migratory, breeding in dry, desert-like conditions from the lower Volga eastwards, and winters from Sudan through the Arabian Peninsula to west Iran. Some authorities consider that it may be a separate species, Steppe Grey Shrike *L. pallidirostris*. Often at a distance, this individual favoured a specific stretch of hedgerow during its two week stay (Latham 2011).

* At the time of going to press, it is once again considered conspecific with Great Grey Shrike.

Graham Walker

Woodchat Shrike

Lanius senator

Vagrant

Breeding from North Africa and Southern Europe to as far north as southern Germany and through to Iran, this shrike winters in central Africa south of the Sahara. It is a scarce migrant to Britain with 20 or so individuals seen in most years, usually in the southern counties (Fraser 2013b). There have been just three records:

Eardiston, near West Felton	23–25 July 1977
Mainstone, near Bishops Castle	17 August 1997
Catherton Common	28–29 July 2007

The ones at Eardiston (Rogers *et al.* 1978) and Catherton were adults, the latter being a female, while the slightly later one at Mainstone was an immature. The individual at Catherton was seen catching bees and other large insects, two of which it impaled on hawthorn in typical shrike fashion.

Graham Walker

Jim Almond, Catherton Common, 29 July 2007

Golden Oriole (Eurasian Golden Oriole)

Oriolus oriolus

Very rare passage migrant, has bred

Golden Orioles breed on mainland Europe and winter in Africa. Within the British Isles, a small population has been largely limited to south-east England. It has been restricted to Suffolk since 2004, and more recently to only one site, RSPB Lakenheath Fen, and it may subsequently have ceased breeding there too. It is therefore not surprising that this species has been such a rare visitor.

The earliest record is from Beckwith (1881), where he documented the sighting of two in 1866 at Harnage, while Forrest (1899) noted one at Neen Savage in May 1886 and 'several other doubtful records'. In 1908, Forrest added that they were 'reported to have nested near Bridgnorth in 1900' but there are no other references to this, and he appeared to have been sceptical himself.

In the early twentieth century, there were apparently five records of individuals, but usually only for a day or two at a time and usually in the south and west. These include Kinnerley in April 1902, Cleobury Mortimer in April 1908, and Ludlow in May 1922. The final historic record was of a male in a garden in Trefonen in May 1923, 'possibly the same bird ... as seen Prees on 2 May [1923]'. Given the dates, all these five were presumably on spring passage.

The *Handlist* (1964) only mentions Golden Oriole in the Appendix relating to species recorded prior to 1950, but there have been eight modern records since then, of 11 individuals, from the following sites (all were of seen on a single day unless otherwise indicated):

Horderley	18–19 July 1964 (a male and a 'possibly immature')
Rhydycroesau	April 1986 (male)
Dolgoch Quarry	7 and 14 June 1988 (male and female seen separately)
Rednall Mill	June 1989 (calling, not seen)
Haughmond Hill	May 1992 (a male and female together, record not published until SBR 2011)
Stanton Lacy	May 1994 (repeated calling from top of large poplar tree)
Poles Coppice	26 May 2000 (male)
Leebotwood	19 June 2000 (male), the most recent record.

The 1964 record from Horderley was interpreted as confirmed breeding, one of only two such records nationally between 1958 and 1967, the other being from Lancashire (Sharrock 1969).

Most of the others were males and seen on one day only, although pairs at Dolgoch Quarry and Haughmond Hill raise the possibility of breeding.

Linda Munday

Jay (Eurasian Jay)
Garrulus glandarius
Fairly common resident

John Fielding, Shrewsbury, 10 December 2014

Jay always seems to have been widespread and well distributed, and although it is the most secretive of the corvids, its loud call, size and unmistakable plumage has made it a notable bird through the ages. Its loud jarring call and showy appearance has meant that the name Jay has been used as a derogatory term for a loose woman.

Historically, it was common in woodlands 'in spite of the relentless war waged by gamekeepers' (Paddock 1897), and Forrest (1899) described it similarly. After the First World War a steady increase was seen, partly due to a reduction in persecution, and partly as a result of the expansion of commercial forestry. Lloyd (1943) listed it among species that have shown a 'less marked or local increase'.

The *Handlist* (1964) description was 'Resident, fairly common in woodlands', and there appears to have been no changes in status since. The recent Atlas map shows little change in distribution when compared to the 1985–90 map. However their secretive nature means that although they are often seen, breeding is only confirmed infrequently, and the possible and probable breeding records are also more than likely to reflect breeding pairs. There have been some losses and gains, probably reflecting chance variation in fieldwork encounters,

but generally there would seem to have been a small decline (3%) in the number of tetrads with some evidence of breeding. This fairly stable picture compares with the national picture where there is evidence that Jay populations are stable in woodlands but increasing in farmland habitats.

The relative abundance map shows a widespread distribution, with Wyre Forest, Wenlock Edge and around the Severn Gorge having clusters of high relative abundance, and gaps in upland areas and where agriculture is more intensive, due to the lack of suitable woodlands in those areas. Jays are generally associated with deciduous woodlands, particularly oak, where their habit of burying acorns helps with the regeneration of this habitat. However they will nest wherever there is thick cover and little disturbance, such as in small copses and plantations.

There are insufficient records to establish a BBS trend here, but in Wales the population has increased by 46% between 1995 and 2014 and in England by 10%. In the English West Midlands it has apparently decreased by 7% over the same period, but the estimate is less reliable.

During recent winter Atlas fieldwork, Jays were found in 90% of tetrads, rather more than in the breeding season. Generally only individuals or pairs are encountered, but in winter they are sometimes recorded in small flocks, usually aggregating around oak trees and

Occupied Tetrads

Atlas period (breeding)	1985–90		2008–13		Change	
	Number	%	Number	%	Number	%
Confirmed	197	23	91	10	-106	-54
Probable	287	33	278	32	-9	-3
Possible	216	25	311	36	95	44
TOTAL	700	80	680	78	-20	-3
Tetrads with Winter Records (2007–13): 785 (90%)						

Breeding Relative Abundance (2008–13)

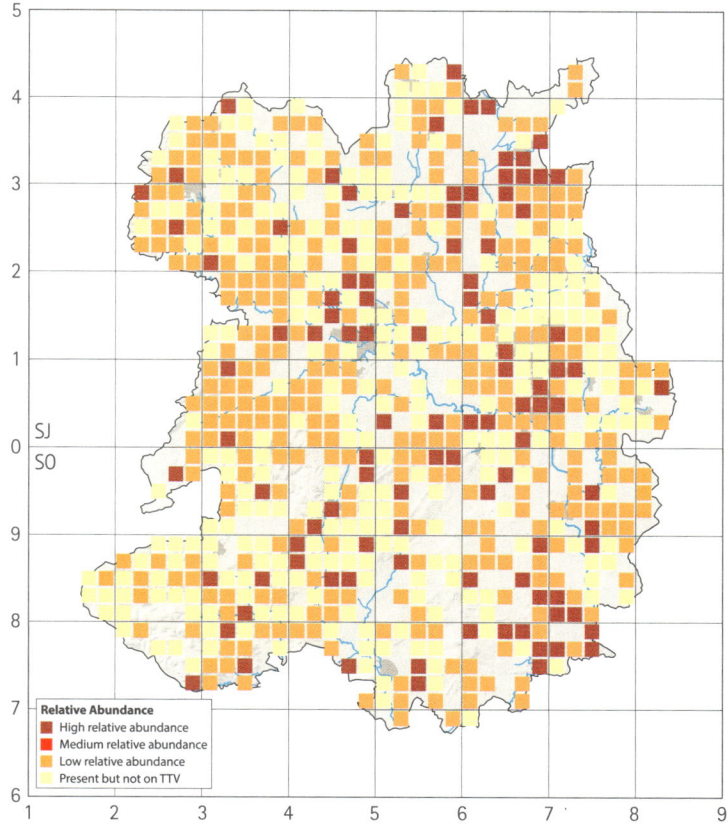

Relative Abundance
- High relative abundance
- Medium relative abundance
- Low relative abundance
- Present but not on TTV

feeding on acorns (one Jay can hoard as many as 2,000 acorns during the autumn). They also forage more widely at this time, young birds disperse, and they are less secretive than when nesting, so they are more easily seen. Also, the BTOs long-running Garden BirdWatch survey shows that they will often visit gardens in the winter.

Influxes of Jays from the continent also occur in some years, with 1972 and 1983 being notable, and another influx in 2012 added to the recent winter Atlas maps.

Breeding records seem to regularly pick out Snailbeach and Minsterley, although this may be due to the close proximity of houses to oak woodland in this area, which does not feature on the relative abundance map.

Based on TTV counts, the population is estimated at around 2,800–2,900 pairs, suggesting that the estimate in the *Atlas* (1992) of 1,600 pairs was perhaps too low.

Few Jays have been ringed, and there have been only 16 recoveries (none from abroad). Ten were ringed and recovered here, and make up the vast majority of the recoveries. Most of the other movements are to or from adjacent Staffordshire (three) and Worcestershire (one), while Gwent and Gloucestershire feature once each.

One ringed at Walcot on 28 August 2003 was recovered on 26 January 2013, 9y 4m 29d later, in Shrewsbury, only 10km away.

These recoveries confirm the sedentary lifestyle of resident Jays, also indicated by the similarity of the winter season relative abundance map to that for the breeding season.

Simon Cooter

Magpie (Eurasian Magpie)
Pica pica
Common resident

Allan Heath, 23 November 2010

Big, bold, handsome and noisy, Magpie is so conspicuous that nothing it does escapes attention. The habit of nest-raiding makes it a natural enemy of poultry-farmers and gamekeepers. Bird-lovers are not always much more sympathetic, the plundering of garden nests leading some to conclude wrongly that Magpies are a major cause of the fall in songbird populations. Ironically, Magpie itself demonstrates that even heavy brood losses do not lead inevitably to species decline, since it suffers a high rate of nest predation.

The Magpie was a prime candidate for persecution by gamekeepers in the eighteenth and nineteenth centuries, although its numbers declined much less here than in the heavily 'keepered eastern counties. Late in the nineteenth century, Beckwith still found that in hard weather it was 'not unusual to take five or six [Magpies] from the same trap in the same day', and Forrest considered the species 'common'.

Nationally, Magpie numbers grew through the twentieth century, the rate accelerating after the Second World War: The *Handlist* (1964) reported that the population had 'greatly increased since 1940'. In 1986, Magpie was rated 'a very successful species', found in 'flocks of up to 20 outside the breeding season', and with a roost of 108 at Brown Moss. This coincided with a peak in the national population, which first stabilised, then declined slightly after 2000. Local BBS data reflect this: the net decline between 1997–2014 is just under 7%, and although the underlying trend shows a steeper fall of almost 20% over the same period, there has been an upturn in the last two years. The current population, estimated on the basis of TTV counts, is around 11,200 pairs.

Magpies were found in all but a handful of tetrads during fieldwork for each of the two breeding season Atlases, and for the recent winter Atlas; as a sedentary species, it probably breeds wherever it occurs. The highest concentrations are around the larger settlements, in the north and the Severn Valley, especially Shrewsbury and Telford; winter distribution follows a similar pattern. Although country dwellers show a marked preference for pasture, Magpies are less numerous in the predominantly pastoral Shropshire Hills, where most tetrads support only low or medium densities.

The reasons for the recent population decrease, and local variations, are not fully understood, though it is suspected that Magpie numbers may have reached carrying capacity, the highest density the habitat can support, in 'intensively-farmed and ... suburban landscapes' (BTO

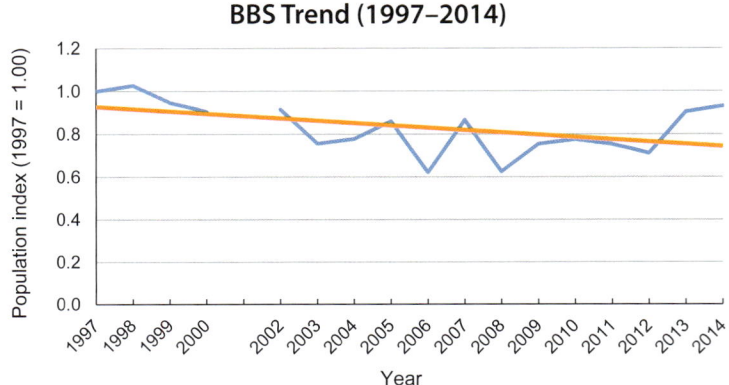

BBS Trend (1997–2014)

Occupied Tetrads

Atlas period (breeding)	1985–90		2008–13		Change	
	Number	%	Number	%	Number	%
Confirmed	767	88	491	56	-276	-36
Probable	79	9	251	29	172	218
Possible	20	2	119	14	99	495
TOTAL	866	100	861	99	–5	–1
Tetrads with Winter Records (2007–13): 865 (99%)						

Breeding Relative Abundance (2008–13)

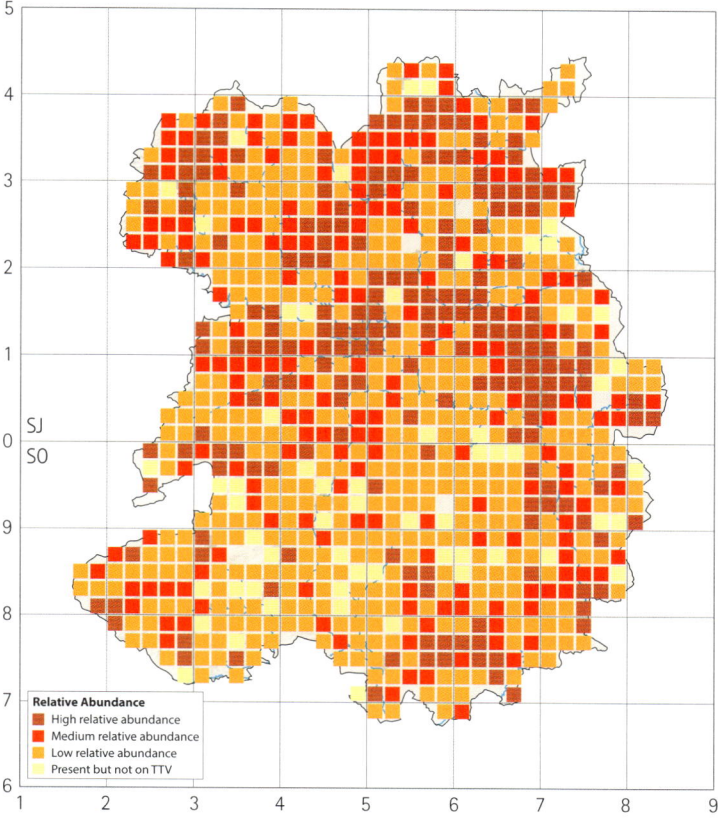

Relative Abundance
- High relative abundance
- Medium relative abundance
- Low relative abundance
- Present but not on TTV

BirdTrends). Around Pheasant shoots, Magpie still suffers vigorous 'control' by gamekeepers; some of the rare gaps in distribution appear to be in such areas. Populations on farmland seem to be faring less well than those in towns and suburbs, possibly because they, like other species, are suffering the effects of agricultural intensification. Removal of hedgerows and trees has reduced the availability of nest sites, while loss of permanent pasture, drainage, intensive grazing and use of pesticides compromise the foraging environment for a species that, despite its reputation, feeds primarily on invertebrates, fruit and seeds.

Ringing data confirm Magpie's sedentary habit: 19 of the 21 recoveries had been ringed here, the remaining two in adjacent counties. The oldest individuals, both around six years old, had moved only four and seven kilometres from where they were ringed as nestlings.

Michelle Frater

Nutcracker (Spotted Nutcracker)

Nucifraga caryocatactes

Very rare irruptive winter visitor

This irruptive species breeds from Scandinavia and the mountains of central and southern Europe, through to Japan. In Britain, it is by no means annual and the only local historic reference is from 1855, when it was reported as an occasional visitor to the area around Oswestry.

In 1968, however, an unprecedented invasion involving some 315 individuals to Britain (Hollyer 1970; Smith *et al.* 1969) resulted in three separate sightings:

Acton Reynald	two, 21 October 1968
Diddlebury	four, 1 November 1968
All Stretton	two, 3 November 1968

Whether or not some of these birds contributed to more than one sighting cannot be determined, but they were nevertheless spectacular finds. While there is insufficient evidence to confirm the race of these particular birds, this invasion largely consisted of the Siberian slender-billed race *N.c. macrorhynchos*, which is known to be irruptive when a good year for their favoured food, pine nuts, is followed by a bad one. The elevated population, as much a result of reduced mortality in the winter as successful breeding, is then forced to move in search of food elsewhere.

Graham Walker

Chough (Red-billed Chough)

Pyrrhocorax pyrrhocorax

No modern records

Resident mainly in the mountains of southern Europe and rocky cliffs around the British coast, not surprisingly there are only three acceptable records. Beckwith wrote of one killed by a keeper near Gobowen in 1862, and then in quick succession came one caught by a pole trap at Ragleth on 24 September 1901 and one watched, 'from 30 yards distance for some time', near Kynnerley on 15 February 1902.

Early sixteenth-century documentary evidence of 'choughs' in fact refer to Jackdaws (Lovegrove 2007) as described in that species' account.

There are no modern records, as three reports from the 1980s are now considered unproven.

John Tucker

Jackdaw (Western Jackdaw)
Coloeus monedula
Very common resident

Jim Almond, Venus Pool, 12 September 2007

Beckwith saw Jackdaws raid dovecotes, take eggs and fledglings, and even suspected them of causing a decline in owls by 'invading' their nest holes, but still had a soft spot for their 'merry' and 'light-hearted' ways. Apparently he was not alone: although Jackdaws are at least as adept at nest-raiding as Carrion Crow and Magpie, even gamekeepers seem to take them less seriously, and the number they kill has declined sharply since the 1960s (Lovegrove 2007, quoting Tapper 1992).

The Jackdaw appears not to have been much persecuted over the centuries for agricultural reasons or for game preservation (*ibid.*), but its nesting habits have sometimes got it into trouble. Beckwith watched the antics of colonies using sandstone cliffs around Bridgnorth and rabbit burrows at Chetwynd Park with interest and amusement. When public buildings were colonised, however, with the attendant mess and disturbance, the authorities took action: clearing Jackdaws from churches, as happened in Ludlow in 1569, was a widespread practice for at least three centuries (*ibid.*). On the whole, however, Jackdaw's

treatment was relatively lenient, and by the end of the nineteenth century Beckwith considered it 'the most abundant of the [crow] family, except the rook'.

Like Carrion Crow and Magpie, Jackdaw flourished throughout the twentieth century, the *Handlist* (1964) describing it as a common resident. However, while the populations of the former have fallen back a little from peaks around the year 2000, Jackdaw has continued to increase: BBS results suggest an astonishing 109% rise between 1997 and 2014. The current population is estimated from TTV counts at around 29,000 pairs, almost three times the 1992 figure of 10,000 pairs. Assuming that the rate of increase was similar in the years between fieldwork for the *Atlas* 1992 and the period covered by the

Breeding Relative Abundance (2008–13)

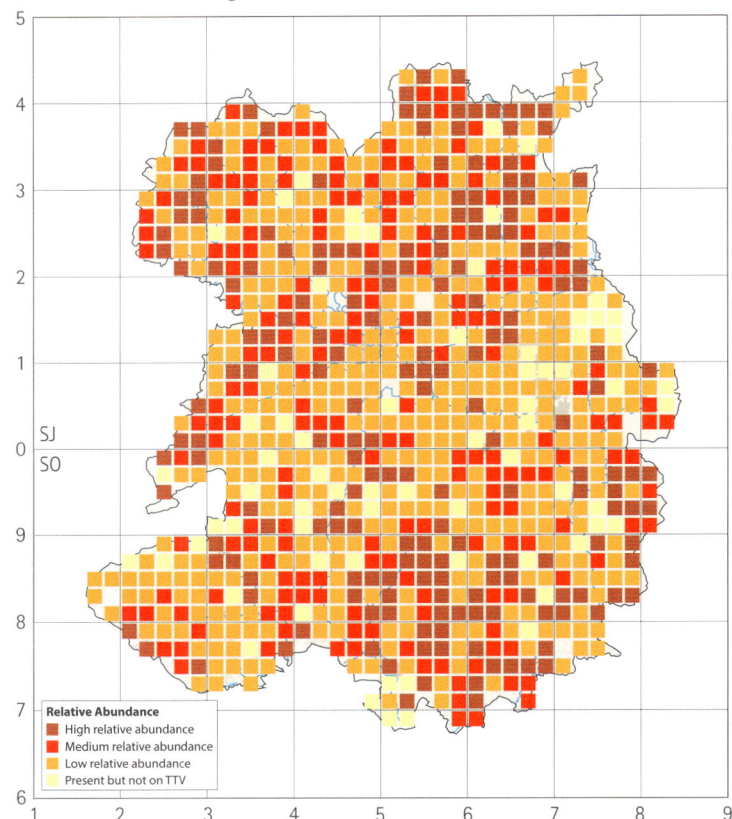

Relative Abundance
- High relative abundance
- Medium relative abundance
- Low relative abundance
- Present but not on TTV

BBS Trend (1997–2014)

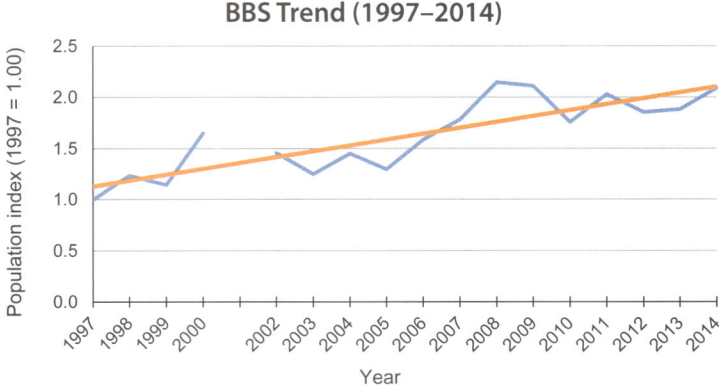

Occupied Tetrads

Atlas period (breeding)	1985–90		2008–13		Change	
	Number	%	Number	%	Number	%
Confirmed	649	75	600	69	-49	-8
Probable	113	13	209	24	96	85
Possible	68	8	54	6	-14	-21
TOTAL	830	95	863	99	33	4
Tetrads with Winter Records (2007–13): 860 (99%)						

graph, the three estimates are broadly consistent. Jackdaw is now the most common corvid, an estimated 24% more numerous than Carrion Crow, and it has surpassed the Rook by 37%.

Jackdaw was found in 4% more tetrads in the recent breeding Atlas than the previous one; it is present in almost every tetrad, summer and winter, and relative abundance is similar in both seasons. Large flocks can gather from autumn onwards near roosts or feeding sites: the highest count during winter Atlas fieldwork was a pre-roost gathering of 1,100 at Puleston Common in January 2009, with a further eight flocks of over 500. In the breeding season, the largest flock, at Romsley in July 2008, was 700 strong, followed by 500 near the canal at Whixall in April 2010. At Priorslee in 2014 winter counts regularly approached 1,000, with a maximum of 1,100 on 29 October.

Distribution is uneven, with the population less dense on average in the south-west quadrant, and across the central belt, than in the north and south-east. High relative abundance in the Oswestry uplands and parts of the Clee hills is not matched in the Stretton hills and Clun Forest, possibly because the average altitude of the latter two areas is higher: Jackdaw is at low density, or absent, in almost every tetrad that includes land over 427m. The explanation may lie in the shortage of sheltered nest sites, such as buildings and mature trees, in these areas, rather than altitude itself, but Jackdaw's tendency to avoid extremes of temperature may play a part.

Relative abundance shows little clear association with land use or settlement: Bishops Castle, Church Stretton, Oswestry and Market Drayton support high densities of Jackdaws, while Shrewsbury and Telford do not, possibly because the proportion of modern, bird-proof buildings is higher, or open feeding areas are less accessible than in smaller towns. Jackdaw's busy, generalist foraging behaviour and readiness to travel in response to fleeting opportunities enable it to make a living almost anywhere. This is likely to be a major factor in its success (BTO BirdTrends), but suggests a weaker link between habitat and density than is the case with many other species.

Nine Jackdaws ringed here have been recovered, eight locally and one in Worcestershire; two Staffordshire-ringed birds were found here. The oldest was ringed at Newport in 1989, and recovered sick in 2002, only 1km away; the national longevity record is just over 17 years, so its age of almost 14, possibly more since it was an adult when ringed, was a very respectable one.

Michelle Frater

Rook

Corvus frugilegus

Very common resident

Rook is resident across the British Isles and much of central and northern Europe. It is familiar to many people, largely because rookeries tend to be associated with human settlements, a habit first commented on by Beckwith: 'From being a wary and unapproachable bird ... as soon as pairing time commences, [it] becomes not only tame and familiar but seems to delight in the society of man'. Lloyd (1939a), in his comprehensive survey of rookeries in the Shrewsbury area in 1939, quantified the habit, demonstrating that 'nearly three-quarters of both nests and groups of nests (70.8%, 71.9%) are situated close to houses'.

The first rookery on record was at Whittington Castle in 1836, and it was still occupied in 2008, although perhaps not continuously.

Normally, Rook eggs have a blueish or greenish background colour, but the clutch of four erythristic (reddish background colour) Rook eggs, shown in the photo, was collected by Tony Waddell (a pupil) at Ellesmere College on 29 March 1958, and is now in the Waddell Oological Collection at the National Museums of Scotland, Edinburgh; at that time the only two other known similar clutches were from Dorset and Germany (Congreve 1959).

Shrewsbury rookeries have received much attention. In the nineteenth century there was one large colony occupying the riverside trees in the Quarry, and in the 1860s and 1870s 'practically the whole of the Rooks in Shrewsbury nested in the Quarry [occupying] the whole of the length of the river walk ... so numerous that it was not safe to walk in the bottom avenue [and they] numbered well over a hundred nests' (Forrest, CSVFC *Transactions* 1912). By the end of the century, the colony was in decline. A move to Kingsland and the grounds of Shrewsbury School had begun in the 1890s, and by 1911 'only three nests were inhabited (at the quarry)'. By the 1920s there were rookeries at St Mary's, near the Free Library, Abbey Foregate and elsewhere in the town 'some thither from the Quarry'. Forest attested 'I believe that nearly, if not quite, all the rookeries in Shrewsbury are branches of the old parent's rookery in the Quarry'.

Colonial breeders tend to be subject to population censuses and Rooks are a good and instructive example. The rookeries of Shrewsbury were initially surveyed for a Ministry of Agriculture national project in 1939, in a rectangle '3.5 miles N.-S. and 4 miles E.-W. centred upon the town'. Seventeen rookeries were identified, often difficult to define when close together, containing 349 nests (Lloyd 1939). He repeated the work in 1944 for a BTO survey and found 248 nests, a decline of 29%, and 'one of the very few areas where the Rook population had

Erythristic Rook eggs (see text). Photo: T. Waddell

decreased during the five years', perhaps because the study area was urban rather than rural. Another look at the same area during the BTO national survey of 1975–76 (Sage & Vernon 1978) produced only four sites, with 133 nests, while later work in 1985–90 for the first Atlas logged four sites (one common to 1975) with 97 nests. Lloyd's area was examined in detail in 2012 (J Tucker *in prep.*) and there were four rookeries (the same as those occupied in 2008) with 119 nests, 34% of Lloyd's 1939 count. The last rookery within the town, by the river at the suspension bridge on Smithfield Road, was abandoned in the late 1980s.

The 1975–76 BTO national survey covered all local 10km squares, and results were augmented by additional counts from other recorders. A total of 460 rookeries was located and 12,092 nests counted (average 26.3). These rookeries were found in 292 tetrads, an average of 41.4 pairs per tetrad. Fifteen rookeries were found with 100+ nests, the largest being at Downton Hall near Ludlow with 175+, and 50 more each contained 51–100 nests. Eight rookeries were above 1,000 feet (*c.*300m).

A re-survey of 27 tetrads in 1996, for comparison by BTO with the corresponding 1975–76 data, showed a 78% increase in the number of rookeries (14 to 25) and 195% in nests (240 to 708) (SBR 1996). However 'change estimates were partly dependent on assumptions about the thoroughness of coverage in 1975–77, and some estimates of increase may be too high' (Marchant & Gregory 1999); by implication the increase may have been exaggerated by incomplete cover during the earlier work. Indeed the 1975–77 survey probably did under-record sites (Colin Wright, local organizer *pers. comm.*), and comparisons between different surveys are not straightforward.

The Atlas (1992) suggested a breeding population of 25,000 pairs, assuming an average of 45 pairs in each of the 545 tetrads with confirmed breeding.

The breeding season relative abundance map produced by recent Atlas work demonstrates that, despite an apparent ubiquity, Rooks are absent from considerable tracts of land, notably the urban areas of Telford south to Broseley, Shrewsbury, and around several other towns, together with upland areas, with gaps in the Clee hills, Clun Forest, Long Mynd and Stiperstones, and some of the more intensive (and treeless) arable areas. The areas of low relative abundance are generally on the edges of these gaps. There seems to be little correlation between the locations of the largest 20 or so rookeries and the highest densities.

The cumulative results of the surveys in 1975, during 1985–90 for the previous Atlas, and 2008–13 for the recent Atlas (including the 2008 survey that aimed to cover all rookeries) leave 298 tetrads (34.3%) with no reports of rookeries.

There is broad concurrence between the breeding and winter

Occupied Tetrads

Atlas period (breeding)	1985–90		2008–13		Change	
	Number	%	Number	%	Number	%
Confirmed	545	63	453	52	-92	-17
Probable	48	6	36	4	-12	-25
Possible	142	16	136	16	-6	-4
TOTAL	**735**	**84**	**625**	**72**	**-110**	**-15**
Tetrads with Winter Records (2007–13): 783 (90%)						

Breeding Relative Abundance (2008–13)

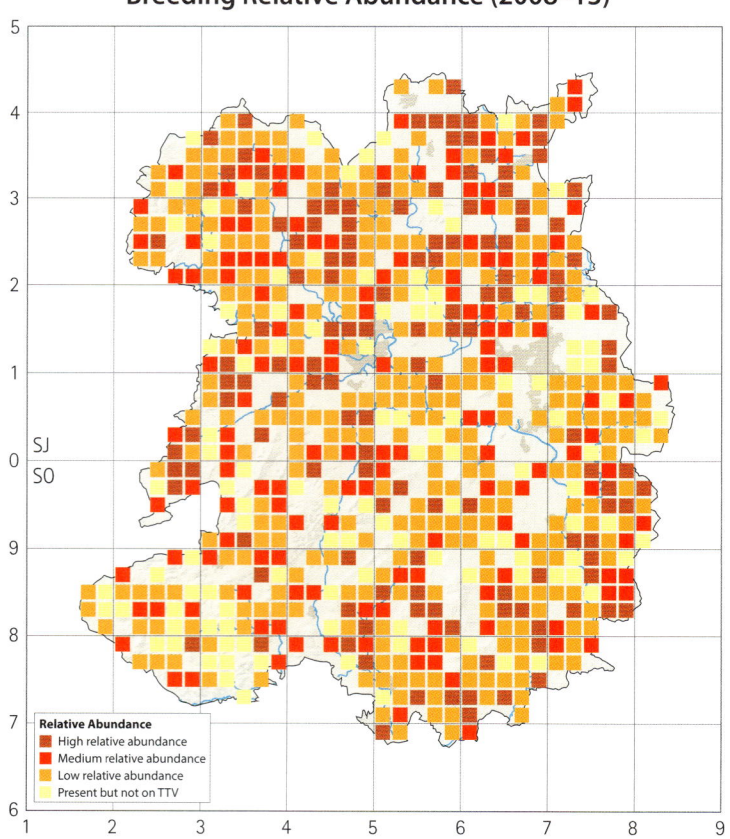

Breeding Distribution Change (1985–90 to 2008–13)

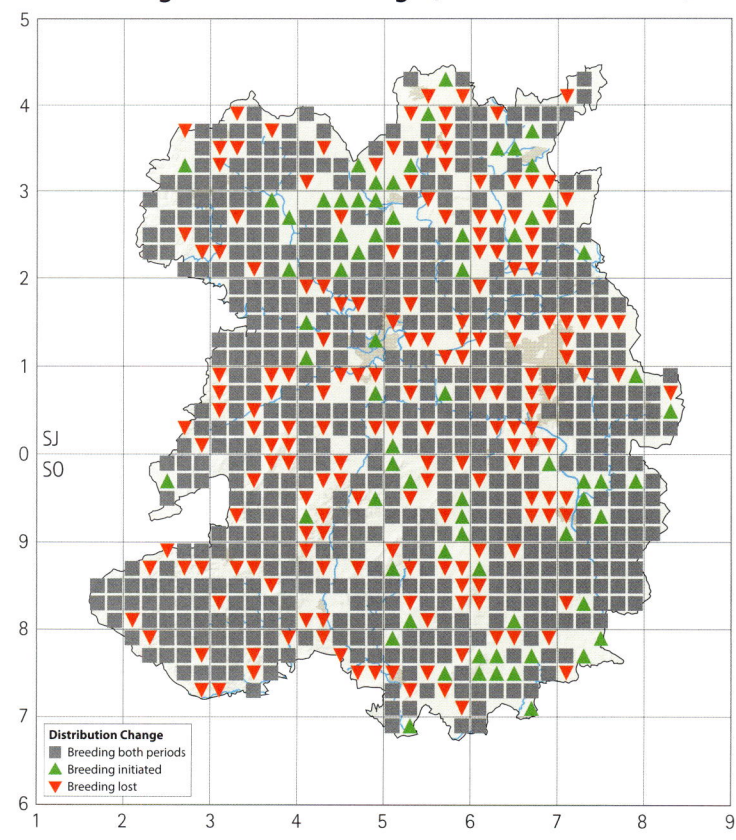

Jim Almond, Venus Pool, 6 June 2011

relative abundance maps, but dispersal of Rooks into suitable winter-feeding areas led to almost 25% more tetrads being occupied in winter, and only 10% were unoccupied.

The breeding distribution change map shows a 15% reduction in the number of tetrads with evidence of breeding, and rookeries have generally disappeared on the edge of the range (i.e. the gaps between them have grown).

Recent Atlas fieldwork, including the 2008 survey, confirmed breeding in 453 tetrads at a density of 1.60 rookeries per occupied tetrad with an average of 28.9 nests in each (average 46.2 nests per tetrad with confirmed breeding). This suggests an adult breeding population of around 20,950 pairs, considerably higher (by 28%) than the estimate of 16,100–16,750 based on TTV counts. However, as noted above, the largest rookeries are not apparent on the relative abundance map, and the average count on TTVs in 199 tetrads with colonies was 25.4, considerably less than the 46.2 found on the 2008 survey, indicating that many rookeries were not visited and counted during TTVs.

The *Atlas* (1992) suggested a breeding population of 25,000 pairs, larger than the current estimate of 20,935, a 16% decrease. This is consistent with the decline of 21%, with an underlying trend of about a 15% decline found by BBS between 1997 and 2014, and mirrors declines of 12% in England and 35% in Wales between 1995 and 2014 found by BBS nationally. The number of pairs in each tetrad

BBS Trend (1997–2014)

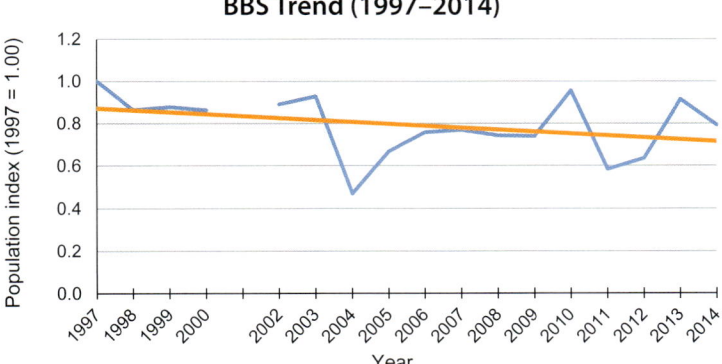

with rookeries appears to have been more or less constant since 1975, indeed it appears to have increased slightly, so the decline has been driven by a decrease in the number of occupied squares, including a decline of 17% in tetrads with confirmed breeding. Diet consists of grain and invertebrates, so agricultural intensification, which has reduced the availability of both, has probably been the main cause of this contraction in range (Marchant *et al.* 1990).

Rooks do not breed until they are two years old, so, assuming that 10–25% of the adult population is comprised of non-breeding birds (B.W. Tucker 1935) and applying a mean of 17.5% (as did Lloyd in 1939), then the total population might be around 48,800 adults, with considerably more immediately after fledging.

The 2008 study (J. Tucker, unpublished) produced site-specific data for many rookeries, including colony size, tree species and altitude (derived from grid references):

- *Colony size.* In 2008 eight sites had 100 or more nests (average 121.6), the largest at Kynnersley with 195 nests. In the 1975–76 survey, 15 sites were reported with 100 or more nests (average 123.7), the largest being that at Downton Hall (5km north of Ludlow) with 175 nests. The current slightly lower number of large colonies should not mask the fact that more than one in 10 of all nests (12.1%) was in a colony of 100 or more; large rookeries appear to be significant feature in the species' population ecology. Grouping rookeries into ten nest-size classes and comparing 1975 results with 2008 shows an apparent upward change to the peak colony size-class. The peak size-class in 1975 was 1–9 nests (27.1% of colonies) whereas in 2008 the peak size-class was 11–20 nests (33.3% of colonies).

- *Tree species used.* Prior to the impact of disease, elm was the favoured tree for rookeries, but now a variety of species is used. The following proportions were in use at the 350 sites (62%) out of 561 where the tree species were reported: oak (28.2%), ash (20.7%), Scots/Corsican pine (11.3%) and sycamore/maple (10.8%), followed by alder, beech, conifers and others all with less than 5% each. One colony was in a monkey puzzle, *Araucaria araucana* (in Bishop's Castle, subsequently felled), and the only power pylon site, at Ironbridge power station in 1975, was not reported in 2008.

- *Altitude.* It has been possible to compare rookery altitudes with the mean of all 3,750 1km squares (Tucker *in prep.*). Placing rookeries into 50m vertical zones, the highest altitudes are shunned, probably because the few trees there are highly exposed; the three 351–500m highest zones have no rookeries and there were only four rookeries between 301–350m. The lowest altitude zones were preferred disproportionately; 1.0% of 1km squares are in the 1–50m zone but 2.8% of rookeries occurred there, while 41.2% of 1km squares are in the 51–100m zone and 47.6% of rookeries were recorded within it.

It appears that Rooks are very sedentary. Few are ringed, and only three have been recovered. Two were found at the ringing site, one only 21 days old in 1915, but the other over four years and two months after being caught and ringed when already an adult in February 1913. The only modern recovery was of an adult male ringed in Newport in April 1990, and found dead at Chetwynd Park in May 1997, over seven years later and all of 3km distant.

John Tucker

Carrion Crow

Corvus corone

Very common resident

These days, Carrion Crow might almost be said to hide in plain sight, so widespread and common that it merges into the background. That wasn't always the case: during the nineteenth century, the crow came under severe pressure from farmers and gamekeepers. In 1836, Leighton described it as 'everywhere'; by 1891, Beckwith noted a more uneven distribution: 'in the enclosed districts of north Shropshire the Crow … has become rather scarce; but in the wild and more open localities in the south it is still common'. A link between its density and the level of human intervention in the landscape was already apparent.

Intelligent, adaptable and resilient, the crow needed only the relaxation of 'control' after the First World War to recover lost ground. According to the *Handlist* (1964), numbers had 'increased considerably in recent years', with 'some 300 crows … shot on the Duke of Westminster's estate at Ellesmere in 1956'. Local attitudes to Carrion Crow during this period of rapid growth were, at best, ambivalent: a Mr Adams of Colebatch, writing in 1950, was moved to describe it as the 'black terror' of the Clun Forest, and 'probably the greatest enemy of bird life throughout the country'. An undercurrent of disapproval runs through many SBRs: examples of the crow's 'predatory and opportunist nature' include 'stealing Black-headed Gull's eggs at Allscott Sugar Factory', and killing an injured Black-headed Gull at Venus Pool. One householder finds the crow 'very bold and persistent … constant nuisance vandalising the bird tables'.

Dawn Micklewright, Venus Pool, 8 February 2014

There is the occasional spike of interest in breeding or behaviour. In 1985, an observer near the Long Mountain counted all the local nests. On 7 May, they totalled 240; a week later, adults were feeding 'large young' at 138 nests; by June, 600 crows were feeding on a meadow nearby at Westbury. Forty nests were found in the upper reaches of valleys on the Long Mynd in 2006. An observer at Priorslee Lake in 1997 captured the crow's versatility: one bird was seen 'paddling in the lake picking food from the surface … and one taking a fish from the middle of the lake using a very slow wingbeat to almost stand on the surface'.

The *Atlas* (1992) produced a population estimate of 17,500–22,000 pairs, postulating a doubling of local densities between 1976–90 in line with national BTO population trends. Between 1997–2014, numbers decreased by nearly 9%: after peaking in 2006, they fell by 20% over the following eight years. Based on TTV counts, the current population is estimated at 23,100–23,700 breeding pairs, close to the 1990 upper figure, and consistent with the rise and fall charted by BBS in the intervening period.

Carrion Crow is present in every tetrad, winter and summer, and probably breeds in all of them. The 38 tetrads where only possible breeding was recorded are widely scattered, and include habitats in which it is almost inconceivable that breeding did not take place.

Breeding Relative Abundance (2008–13)

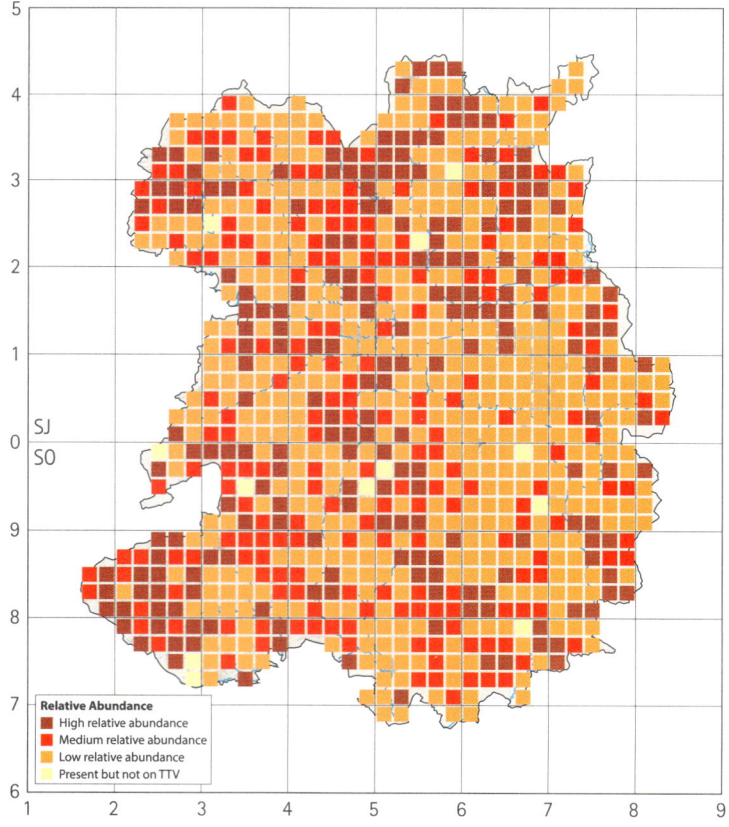

Relative Abundance
- High relative abundance
- Medium relative abundance
- Low relative abundance
- Present but not on TTV

Occupied Tetrads

Atlas period (breeding)	1985–90		2008–13		Change	
	Number	%	Number	%	Number	%
Confirmed	767	88	672	77	-95	-12
Probable	73	8	160	18	87	119
Possible	26	3	38	4	12	46
TOTAL	866	100	870	100	4	0
Tetrads with Winter Records (2007–13):			870	(100%)		

BBS Trend (1997–2014)

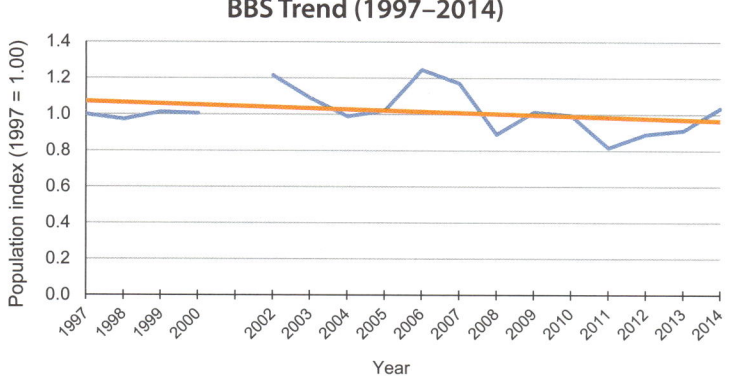

Relative abundance is difficult to assess: the caveat in the introduction to *Species Accounts* as to the reliability of TTV returns applies particularly to Carrion Crow. Its restlessness can make it hard to count accurately; local, temporary effects such as ploughing or silage-cutting may produce exceptionally high counts, with flocks of young, non-breeding birds an added complication. On the whole, abundance tends to be lower in arable areas of the north and east, with the sheep pastures of the Shropshire Hills and Oswestry uplands supporting higher numbers. Winter abundance is similar, small variations between the seasons accommodated by a necessarily elastic interpretation of TTV counts.

Overall, abundance is high by national standards; annual maximum counts, such as 235 near Bishops Castle in 2007, routinely reach treble figures. A TTV near Kynnersley (SJ61T) in winter 2008 logged 800 crows; though no other total came close to this, 40 winter counts, and 20 in the breeding season, exceeded 100. This is not universally welcome: crows can still be found gibbeted on farmland, and are controlled around shoots. The conservation of ground-nesting birds such as Lapwing and Curlew creates an ethical dilemma: while they suffer the effects of habitat loss and degradation, their predators are making the most of opportunities as diverse as Pheasant releases and food waste, livestock farming and roadkill. Native species such as Carrion Crow, whose raids they have withstood for millennia, can now contribute to local extinctions, with conservationists left struggling to reconcile the competing claims of species whose status has been profoundly affected by human activity.

Few Carrion Crows are ringed. Altogether, there have been 10 recoveries, of which nine were ringed here, and one in adjacent Cheshire; although a very small sample, it is consistent with the sedentary habit of the species.

Michelle Frater

Hooded Crow

Corvus cornix

Vagrant

Once considered to be a subspecies of Carrion Crow, this crow's range includes Ireland, the Isle of Man, and north and west Scotland. It is also found from northern and eastern Europe through to Turkey and Iran. Although it readily hybridises with Carrion Crow where their respective ranges overlap, their offspring do not usually roam far and any occurrences here are likely to be genuine Hooded Crows (Parkin *et al.* 2003).

It was first recorded in 1836 at Weston Lullingfields, while in 1838 Eyton reported that he had caught a Hooded or 'Royston' Crow some years earlier by setting a trap near a dead sheep. Subsequently, it was noted as being 'occasionally met with, but not very common', and had 'been killed here in one instance', perhaps Eyton's bird. Until about 1915, it was reported as being seen in most winters with up to four in some years. This included one that was apparently killed by a labourer with a stone at Maesbrook on 25 October 1899, and two caught in a rabbit warren at Shirlett in the same year. After 1915 there were only sporadic reports and there are only seven modern records:

Onslow	19 February 1962
Long Mynd	29 November 1975
Shrewsbury Tip, Betton Abbots	8 December 1976
Bagley Marsh	2 April 1979
Newport	27–28 February 1985
Long Mynd	28 May 2009
Earls Hill near Pontesbury	24 April 2010

Typically, it is a winter visitor to those parts of Britain outside its normal range, although three of the most recent sightings were in the spring, which may have been individuals returning to their breeding sites after wintering elsewhere.

Graham Walker

Raven (Northern Raven)

Corvus corax

Uncommon resident

Raven is a highly adaptable species found through most of the Northern hemisphere from the high Arctic to Death Valley in California, south to Central America and across Asia northwards from the Himalayas.

It was widespread in historic times, and produced a number of archaeological records, including several from excavations of Roman ruins at Wroxeter, where remains were found from the late third to late seventh centuries. More Raven bones were found than for any other non-domesticated bird except Woodcock, although the latter was probably hunted for food (Yalden 2002; Hammon 2011).

There are several former coaching inns, long-established hotels and public houses, and field and road names, incorporating the name 'Raven', suggesting that it was widespread in the middle ages and later, and one of the earliest written local accounts of any species, from Hawkstone, describes 'a high point on the Grotto Rock, called the Raven's Shelf, because time immemorial the Ravens have annually made a nest there' (Rodenhurst 1783).

At the start of the nineteenth century it bred in almost every county, and local authors chronicled its initial widespread distribution and its subsequent rapid disappearance. It was first noted at Wrekin and Aston (Leighton 1836), and Eyton (1838) reported that 'a few pair breed, and have been known to build in the same trees since time immemorial, in spite of the nest being robbed every year'. By the 1860s it still existed 'in considerable numbers in Clun Forest, the Stiperstones and other high localities ... but it has been driven from most of its former breeding places', and less than 20 years later it had become 'very rare, although a few are found about the Longmynds, and along the Welsh borders'. A pair or two frequented the Wrekin, but one had not been seen 'for years in that neighbourhood' (Beckwith 1879). Ravens were relentlessly persecuted, and 'the persistent use of poisoned flesh by gamekeepers and shepherds has, however, almost exterminated it, and if a pair attempt to breed, their eggs or young are taken' (Beckwith 1891).

By the end of the century, Raven was 'now very rare. It used to be seen on the Longmynd, Wrekin, and on the Welsh border, and nested regularly in Hawkstone Park, on the Longmynd, and at Linley near Bishop's Castle. The last instance was in 1884, when some young ones were taken from a nest in a quarry, near Church Stretton' (Forrest 1899).

It was rarely seen for the next quarter century, but a pair haunted the old breeding place near Church Stretton throughout the spring in 1910, and in 1918 a nest with four well-fledged young was found in an old quarry in Ashes Hollow on the Long Mynd, 'the identical spot where the last Shropshire nest was recorded 42 years ago' (Forrest 1918). It may not be coincidence that the recovery started in the last year of the First World War, as so many gamekeepers had left their jobs to join the war effort.

Another brood was raised near Church Stretton in 1920, there was 'a welcome (recent) increase in numbers' in the west (Lloyd 1943) and 'a pair nested every year in a secluded dingle' in the Clun Forest (Adams 1950).

The *Handlist* (1964) recounted that breeding was sporadic between 1918 and 1939, but that thereafter numbers increased steadily. By 1964 'about 20 pairs' were breeding regularly in the Clun Forest upland and the other south-western hills; the Wrekin and Clee hills were thought to be occupied too, while Ravens were 'regularly seen' in the Oswestry upland where nesting had occurred in the 1940s. The national BTO Atlas (1976) mapped confirmed breeding in the two 10km squares encompassing the Oswestry uplands and one including Middletown Hill, and probable breeding in two adjacent squares in the north, and confirmed or probable breeding in all except three of the 14 squares mainly in south Shropshire. These squares include the Clee hills, but not the Wrekin.

By 1985–90, the distribution showed only a few peripheral changes. A group of records suggested a new site to the east of Clungunford, and breeding was confirmed at Loton Park, a new site, and on the Wrekin. The Oswestry uplands, Clee hills and Millichope areas were all represented on the Atlas map, which showed that the core area remained in the south-western hills. At that time Ravens principally inhabited the uplands where intensive sheep farming is widespread. For nesting they favoured conifers, especially Scots pines, and no cliff site was then used.

The *Atlas* (1992) expressed the fear that future population 'trends may be downward, as improved animal husbandry is progressively reducing the sheep carrion upon which Ravens depend'. As a result, the Shropshire Raven Study Group (SRSG) was established in 1993 to:

- research history and present status
- assess productivity
- publish findings
- promote conservation.

Ringing and colour-ringing nestlings was added to the objectives in 1995.

Up until 1999, the Group aimed to find and monitor every nest. The results are shown in the table.

The study coincided with a rapid increase in population and expansion of range. The *Atlas* (1992) estimated the breeding population at 30–35 pairs, but with the advantage of hindsight this was probably an underestimate, with 50 being nearer the mark. By 1999 it was estimated at 175, and over 100 nest sites and 160 territories were actually found. The range expanded north-eastwards, and while the 1985–90 map showed breeding evidence north of the River Severn only from the Wrekin and the Oswestry uplands, by 1999 there were nine more nests found north of the Severn. This included re-occupation of the Grotto Rock at Hawkstone Park in 1997.

This was the only natural rock nest site. Two others were on quarry faces, but all the other nests were in trees: in 1999, 85% were in conifers, mainly Scots pine (28%) and Douglas fir (27%). Nests were mainly in clumps or small copses, or on the edge of a wood. Isolated trees were rarely used, and nests within a wood were also infrequent, and only occurred on the edge of a wide ride – in effect, a woodland edge. The tree was usually the most substantial available, and the average nest height was 18m, with the highest being 37m. All were 10m high or over, except for one in a birch tree, only 6m high and occupied only in 1999. Although most sites were quiet and remote, one in a quarry was within 100m of a face being worked, thousands of people each year visited a viewing platform at Grotto Rock only 10m above the nest ledge, and many hundreds of visitors

Nest Monitoring Results 1994–99

| Year | Totals For Each Year | | | | | Cumulative Total of all Territories |
	Pairs Found	Active Nests	Nests Found	Broods Ringed	Nestlings Ringed	
1994	32	27	23	7	27	32
1995	47	44	39	21	67	56
1996	70	57	55	36	130	75
1997	95	80	78	50	170	102
1998	104	89	85	47	175	123
1999	132	106	101	62	214	160
Total	–	403	381	223	783	160

John Hawkins, Pentrecoed, 22 February 2008

walked within 10m of a nest in a cedar near Attingham Hall where young fledged for three years in succession.

The average altitude of the 177 different nests found was 244m. The vast majority were between 175m and 325m, with only five below 100m, and seven above 400m. There was a high degree of nest site fidelity, with 52% of the previous year's nests being reused, and 85% reusing the same site (i.e. within 100m, in the same landscape feature). Ravens are highly territorial, and nests each year were at regular nearest neighbour distances. In a core area of 220sq.km, the average, and the most commonly occurring, distance at the start of the study was 2.5km, but as the population increased new pairs squeezed in between established pairs, leading to an increase in nests with a nearest neighbour distances of around 1.2km, and the average nearest neighbour distance reduced each year, to 1.86km in 1999.

Ravens breed early, to take advantage of the peak food supply – the high level of sheep carrion and afterbirth from outdoor lambing – and the average first egg date was estimated from the age of ringed nestlings at 2 March. However, this included some second clutches from re-lays after failed first nests, and, after eliminating known second clutches, the average was 27 February. The range over all six years was 25 February–2 March. The success rate of the nests with a known outcome was high – 285 (81.9%), including re-lays, out of 348.

The population increase was attributed to milder winters and a decline in persecution, particularly poisoning (see p. 57), and also a large increase in sheep carrion. The latter derived from a rapid increase in sheep numbers and the increasing tendency for ewes to have two (or more) lambs, clearly evident in the graph (see p. 41). Research at the end of the study showed that SRSG's original fear, that the trend to indoor lambing would reduce the food supply, particularly of afterbirth, was unfounded: there had been virtually no change in the level of indoor lambing in the preceding 20 years. However, as a result of legislation to combat the spread of BSE (mad cow disease),

which eliminated the market in sheep byproducts, there was a big increase in the number sheep carcasses left out on the open hillsides. Farmers were unable or unwilling to meet the resulting cost of collection, and the number of carcasses taken to four local knackers yards in February–April, the Raven breeding season, dropped from several hundred per week in the years prior to 1990 to close to zero from 1993 onwards, and a hunt which collected as many as 200 sheep a week at lambing time ended up collecting as few as 30 per month. The rapidly increasing number of Pheasants released by local shoots (see p. 131) has added to the supply of carrion. A simple population model, taking into account the increase in brood size and the rate of population increase, suggested an increase in annual survival rate of about 8%.

Average brood size at ringing was 3.51, substantially higher than the national average of 3.10 in BTO nest record cards, and the national figure at the time of the local study was lower than this, as nationally the average brood size was in decline.

Interestingly, breeding success was better at lower altitudes, suggesting that this is not primarily an upland species, but was driven there by persecution.

The detailed results of the study have been written up in an unpublished report (Smith 2002), and summarised in an article in *British Wildlife* (Wall 2003).

From 2000 onwards, Phase II of the study aimed to visit nests to find and read colour-rings, and monitor the increase in population. Another 46 new tetrads were occupied by 2005, including a further 14 north or east of the River Severn. By then, breeding density was higher than that in mid-Wales in the 1980s, the highest in Europe at that time.

Colour-rings were fitted to 783 chicks by climbing to the nest. The ring colour changed each year, so the age of the bird could be identified, and each ring had a unique two-letter combination that could usually be read in the field.

The table shows the number of colour-ringed birds by year of birth, and the number of individuals re-sighted. In a few cases a ring was seen, but it could not be read. It has been assumed that, if an un-read ring was the same colour at the same site on a bird of the same sex as a ring that was read in an adjacent year, then it was the same ring.

The number of otherwise un-read rings may appear high, but most were seen in one year only, and the 18 un-read rings were seen on a

Number of Colour-Ringed Ravens, and Re-Sightings

| Year | Number Colour-Ringed | Number Re-sighted | | | | | | |
| | | Ring Read | | | | Ring Un-read | Total | % |
		Male	Female	??	Total			
1994	27	3	1		4	2	6	22.2
1995	67	6	2	1	9	4	13	19.4
1996	130	10	4		14	6	20	15.4
1997	170	5	9		14	3	17	10.0
1998	175	10	6		16	0	16	9.1
1999	214	8	5	2	15	3	18	8.4
Total	783	42	27	3	72	18	90	11.5

Age of Colour-Ringed Ravens When Last Seen
(Average = 8.6 Years)

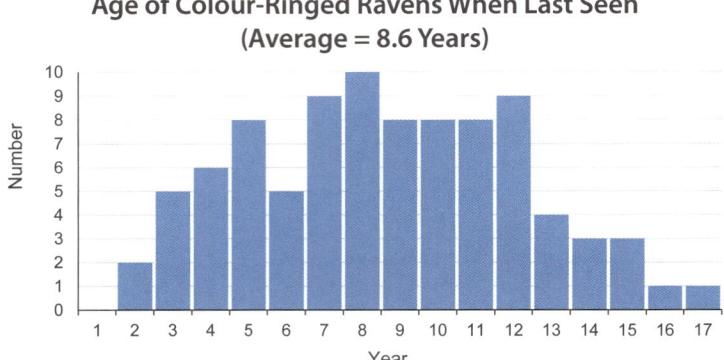

total of only 22 occasions. In comparison, the 72 read rings were read annually for an average of 4.9 years each. The table includes two read rings reported from the Peak District and one from Herefordshire, and three and one un-read rings respectively from the same two counties. It does not include any Ravens found dead, including the ringing recoveries referred to below.

The table also shows that the proportion of ringed birds re-sighted declined year on year. This may reflect lower survival rates of youngsters as the population, and hence competition, increased, but it is more likely to reflect increased dispersal into adjacent areas to avoid competition. For example, Ravens were not found breeding in the Peak District until 1992, but there were 14 breeding pairs by 1999, and at least four colour-ringed in the local study (two read, two or three un-read) had bred there for at least one year between 1998 and 2003 (Mick Lacey, South Peak Raptor Group *pers. comm.*).

Fieldwork was concentrated in the south-west hills, and around Brown Clee, where the ringing took place, so if there were colour-ringed Ravens at nest sites further afield they were unlikely to have been discovered. More effort was put into finding them from 2003 onwards, and over 40 were found in each of the three years 2003–05, then 34 in 2006 and 2007, and a steady decline to single figures after 2012. No new colour-ringed birds were found after 2007, and from 2010 onwards, only sites that held such birds in the previous year were revisited.

Three were found breeding at two years old, and 15 at three, probably the usual age of first breeding: 61% were found breeding at five years or less. Some of those that were several years old when found for the first time may have been breeding previously at undiscovered sites.

The average age of Ravens with rings that were read, when seen for the last time, was 9.4 years, but including the ones with un-read rings, which were usually in the breeding population at an early age

for one year only, the average was 8.6. The age profile of the former is shown in the chart. Five were still alive in 2013, but only two in 2014, a pair at the same site on the western slopes of the Long Mynd. They were seen for the last time in 2015, the male 17 years old and the female 16. Forty-two of the 72 reached 10 years or older.

A summary of the key findings of the colour-ring study (Smith and Cross *in prep.*) include:

- Females moved much further than males from their natal site to their breeding site. This applied to the average distance for all colour-ringed birds, and to every single case where siblings were found at adult breeding sites, and at nests where both birds were colour-ringed. The table also shows that a lower proportion of females was re-sighted, suggesting that more moved beyond the area.

- A few pairs produced a high proportion of the breeding population, and 48 of the 69 colour-ringed birds (69.6%) came from only 15 nest sites.

- It appears that both birds in the pair are needed to defend the territory. If one dies, it is unusual for the survivor to hold on to it with a new mate, although there are examples of this happening. However, there was one example of rapid re-mating, when a new partner was recruited at an active nest to help raise a brood in 2003. The likelihood of any similar examples from later years being observed was small, as visits became more concentrated earlier in the breeding cycle.

- A high proportion of the breeding population in the south-west Shropshire hills was locally bred. For example, in 2005, at 94 nest

Occupied Tetrads

Atlas period (breeding)	1985–90		2008–13		Change	
	Number	%	Number	%	Number	%
Confirmed	27	3	245	28	218	807
Probable	72	8	135	16	63	88
Possible	57	7	161	19	104	182
TOTAL	156	18	541	62	385	247
Tetrads with Winter Records (2007–13): 729 (84%)						

Breeding Distribution Change (1985–90 to 2008–13)

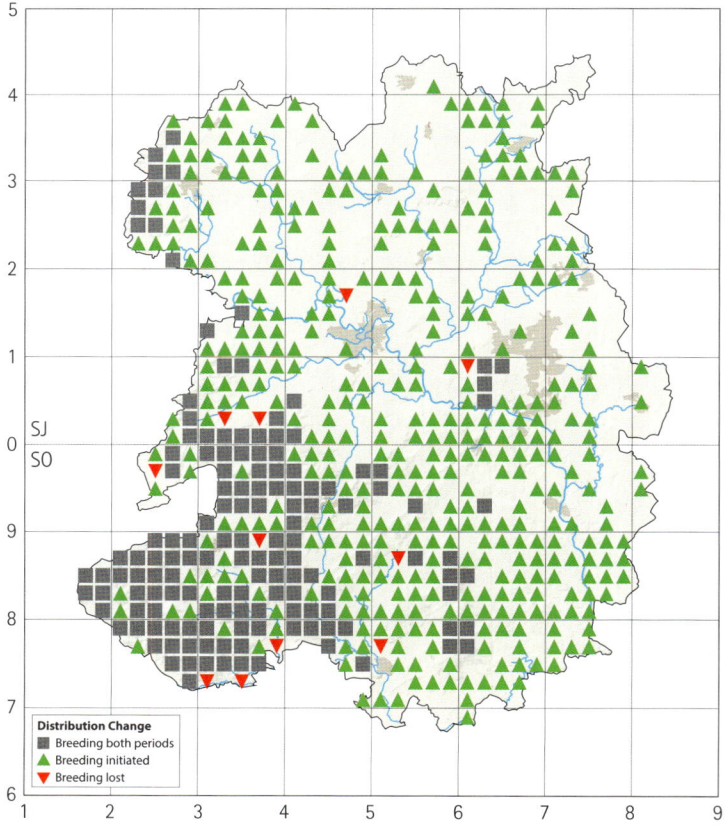

Distribution Change
- Breeding both periods
- Breeding initiated
- Breeding lost

sites, 43 (25%) out of 171 observed had colour-rings. (It is not known whether the other 17 birds at these sites had rings or not.)

- Males are bigger than females, and assessing sex from measurements of a nestling, and comparing it with the sex of the same individual determined by observation of the adult at the nest, gave a very high level of agreement.
- Although Ravens mate for life, infidelity was witnessed, with a red-ringed female mating on her own nest with a green-ringed male from the adjacent nest, rather than with her own red-ringed mate. Such examples of 'sperm competition', which increase the likelihood of the female's own genes making it into the breeding population, are well known in other species.

Apart from the ones seen in the Peak District, there have been three recoveries from elsewhere of nestlings ringed in this study and found more than 50km from their nest site, all dead. Two were ringed in 1996: one was found near Newcastle (Staffs.) in June 1997, 71km from its birthplace, and the other, the longest movement, was ringed on Titterstone Clee Hill and found in May 1998 in the Hope Valley (Derbyshire), 123km NNE, 2y 0m 12d later. The third was a 1999 nestling from near Black Rhadley Hill, found freshly dead 100km SSW, near Newtown (Powys), 2y 3m 27d later.

The expansion of range found by the study continued, and by the end of the recent Atlas period there had been a massive increase in the tetrads where breeding evidence was found, and Ravens had also spread across most of England.

Ravens forage over a large area, and many were seen carrying food, or with fledged young, in tetrads adjacent to the nest site. Although this may have led to over-recording of tetrads with confirmed breeding,

Gareth Thomas, Stokesay Court, 29 April 2005

an occupied nest was found in 172 (70.2%) of the 245 such tetrads. There would also have been nests in some of the other squares with confirmed breeding, and in most of the squares with probable breeding, and some of those with possible breeding. Two nests were found in at least 30 of the tetrads in the south and south-west. Assuming that the number of tetrads with two nests is approximately equal to the number with confirmed or probable breeding records which did not contain a nest, and allowing for a further, say, 20 nests in tetrads with possible breeding records, suggests a population of around 400 breeding pairs.

An effort was made during the recent Atlas period to find all the nests in the south-west, and 121 were found in 110 tetrads. There was strong evidence for a few more pairs in the same area, although the nests were not found. The relative abundance map shows the highest densities still occur in the south-west, and analysis of the abundance bands for all tetrads suggested that the south-west had a third of the population, giving a pro-rata total of almost 400.

Both these methods of calculating the estimated population give the same result, around 400 breeding pairs in 2013. The estimate based on TTV counts of 230 is much too low.

Ravens do not usually breed until they are three or older, but most do not survive that long, with BTO data showing the average longevity of fledged young as 14 months (Ratcliffe 1997). At the start of each breeding season there are many more non-breeding birds than breeding pairs, which tend to stay in their breeding territory all year, and these youngsters form non-breeding flocks. The size of these flocks has increased too, with 11 on the Stiperstones in 1988 being the largest recorded by 1990, followed by 21 in October 1994 at the same place, and 53 in July 2000 on the Long Mynd. Subsequent counts of over 50 were 55 at Cefn Coch on 10 April and 84 at Stow Hill on 4 October (both in 2004), and, the largest flock ever recorded, 93 around the summit of the Lawley on 4 May 2008. However, the size of flocks, and the frequency of their occurrence, appears to have decreased since then, reflecting the decrease in the sheep population. The total population at the start of the breeding season in 2013 would have been somewhere around 1,800 individuals, including 800 breeding adults.

Breeding Relative Abundance (2008–13)

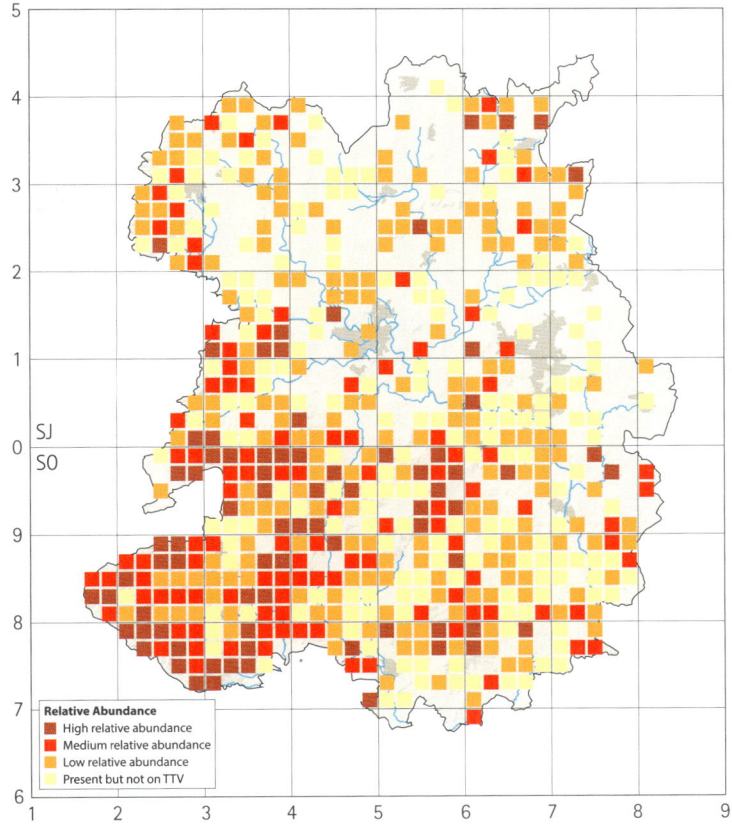

Relative Abundance
- High relative abundance
- Medium relative abundance
- Low relative abundance
- Present but not on TTV

John Hawkins, Ellesmere, 22 December 2006

Ringing studies summarised by Tony Cross in the BTO Migration Atlas show that adult Ravens are largely sedentary, but birds of the year forage away from their natal areas, and this is reflected in recent Atlas results. Distribution in winter (733 tetrads, including four with flight records only) is more widespread than in the breeding season (626, including 85 tetrads with records without any breeding code, all except two in the north and south-east). Although upland sheep farming areas are still the stronghold, the expansion of range since the 1990s, and colonisation of adjacent counties, shows that many pairs are no longer dependent on sheep. The increase in more widely available Pheasant carrion has probably provided the necessary alternative source of food.

Ringing recoveries confirm their sedentary nature. Only 14 ringed here have been recovered elsewhere, all in adjacent counties except those in Derbyshire referred to above. Fourteen ringed elsewhere and recovered here were all from Wales, except two from Herefordshire. In addition, a red colour-ringed female which bred near Asterton for six years up until 2007 had fledged 64km away near Blaenau Ffestiniog in 1996.

The population will be determined by the available food supply, and the Raven study has shown that the size of each breeding territory gets smaller as the food supply improves and the population increases. The recent reduction in sheep stocking rates is likely to lead to a reduction in the number of Ravens in the areas of highest density, and, while this has already affected the size of the non-breeding flocks, it is not clear whether it will also lead to a reduction in the breeding population, or only reduce the number that never breed. However, apart from the possibility of a few new sites being occupied in the north and east, it is likely that the period of rapid population growth is at an end, unless the future economics of sheep farming reverses the reduction in stocking rates.

Leo Smith

The Shropshire Raven Study Group comprised Tom Wall (Convenor), Tony Cross, Allan Heath, Simon Holloway, Caroline Moscrop, Leo Smith, Ken Stott and Gareth Thomas. The ringing and colour-ringing was carried out by Tony Cross.

The Group wishes to acknowledge two BTO Research Grants, the first in 1997–98 for nest finding and the second in 2003–05 for looking for colour-ringed birds at nest sites, another from Forest Enterprise, sponsorship from Swarovski Optic, who provided a telescope for reading colour-rings, and permission from landowners to visit nest sites.

Waxwing (Bohemian Waxwing)
Bombycilla garrulus
Rare irruptive winter visitor

Waxwings breed in loose colonies throughout the coniferous forests of northern Europe, Asia and western North America. They are partial migrants and withdraw from a large part of their breeding range in winter, to lower elevations, roadsides, gardens and the edges of deciduous forest, to seek out food. Their winter range is dictated by the abundance or shortage of their main source of food, rowan berries. In some years a high population coincides with a poor rowan crop, and then a large number move towards the UK, normally at the south-western edge of their range. Sometimes the irruption is so huge that large flocks spread across most of the country. Occurring mostly between October and April, their preferred foods here are the berries of rowan, pyracantha, cotoneaster and hawthorn, but apples and crab-apples are also eaten. Their tameness, distinctive plumage and habit of feeding in urban car parks have made the Waxwing one of our most eagerly awaited winter visitors.

In the nineteenth century they were classified as a very rare and uncertain visitor. Four out of a flock of seven or eight killed near Clungunford in February 1829 were noted as being the only record of more than one being seen at a time. Even when Waxwings were plentiful in some years in eastern and northern counties, they seldom reached here. A male obtained at Hawkstone was listed in the Peplow catalogue, which noted that small parties of Waxwings had occurred, especially during severe weather in mid-winter (Forrest 1907), and there were perhaps three irruptions before 1950.

The sparse historical records suggest that Waxwings have probably become more common in recent years. The *Handlist* (1964) classified it as 'a non-breeding visitor, irregular. In recent years there have been two or three recorded each winter, mainly in February at widely scattered localities'.

The winter of 1965–66 was the first irruption in the modern era. Approximately 42 were reported, including flocks of 20 at Pontesbury on 4 January 1966 and 14 at Monkmoor on 18 January 1966, although the latter report could relate to the same birds. The only record in the three years following that irruption was of a single in Shrewsbury in November 1967. During the winter of 1970–71 a total of 18 were recorded from seven scattered locations between Richards Castle and Ludlow in the south-west, and Shrewsbury and Telford in the central areas, with five at Shray Hill, Telford, first recorded in January and staying until 4 April 1971, staying an unusually long time. They were found in only three of the next nine years up to the 1981–84 national BTO Winter Atlas, with the only sizeable flock being 15 at Haughmond fields, Shrewsbury, in February 1973.

That Winter Atlas showed only two to four found in the 10km square that includes Telford, and one in two other adjacent 10km squares, but none of these were irruption years.

There was then a five season gap until up to 20 Waxwings arrived in December 1988 and stayed into March 1989.

Since then numbers have shown an upward trend, although they still vary greatly from year to year. They were recorded in five of the 10 years 1990–2000, but only the 25 recorded from Wem, Market Drayton and Ludlow during the winter 1995–96 were notable. Influxes totalling 35 were reported during January 2001. Winter 2004–05, although starting slowly with five in Madeley during December, quickly emerged as the best to date when up to 350 were present, including flocks over 200 in the Shrewsbury area over the Christmas period and numerous small flocks reported elsewhere.

The first winter of the recent Atlas period, 2007–08, seemed to have returned to earlier times with no records at all. Subsequent years have varied widely, with 50 seen during a minor irruption in 2008–09, followed by a single at Harmer Hill on 24 January 2010, the only record during that winter. First indications that the winter 2010–11 was going to be a 'Waxwing year' began in October with large influxes appearing in Shetland and mainland Scotland. Our first record was of a single bird at Walford Heath on 7 November. From then until the end of the year, 167 records were received from over 50 observers. Although numbers were difficult to assess accurately because flocks ranged widely, 200+ were seen in both the Ludlow and the Oswestry areas within a two-day period. At the same time numerous smaller flocks were also recorded, so there must have been over 500 birds at 40 different widely scattered sites in early December 2010, the highest total count ever recorded. Sizeable flocks continued to be reported during January 2011 with the last ones finishing the rapidly diminishing food supply by early February, before moving further south.

The only 2011–12 records involved six birds in the Highley area on 5 February 2012. The most recent irruption occurred in 2012–13, with 170+ in late November and early December 2012, which included a wide-ranging flock in the Shrewsbury area, peaking at 132 on 9

Jim Almond, Shrewsbury, 18 January 2011

Estimated Winter Period Totals (1950–51 to 2014)

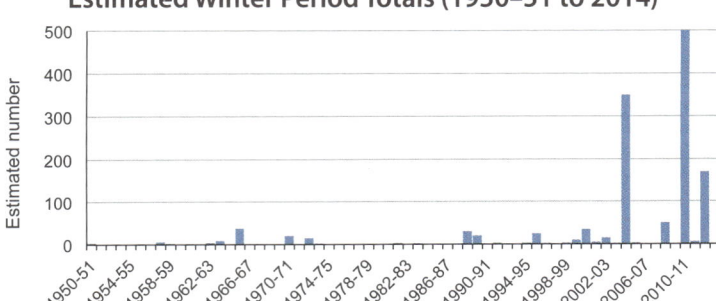

Winter Distribution (2007–13)

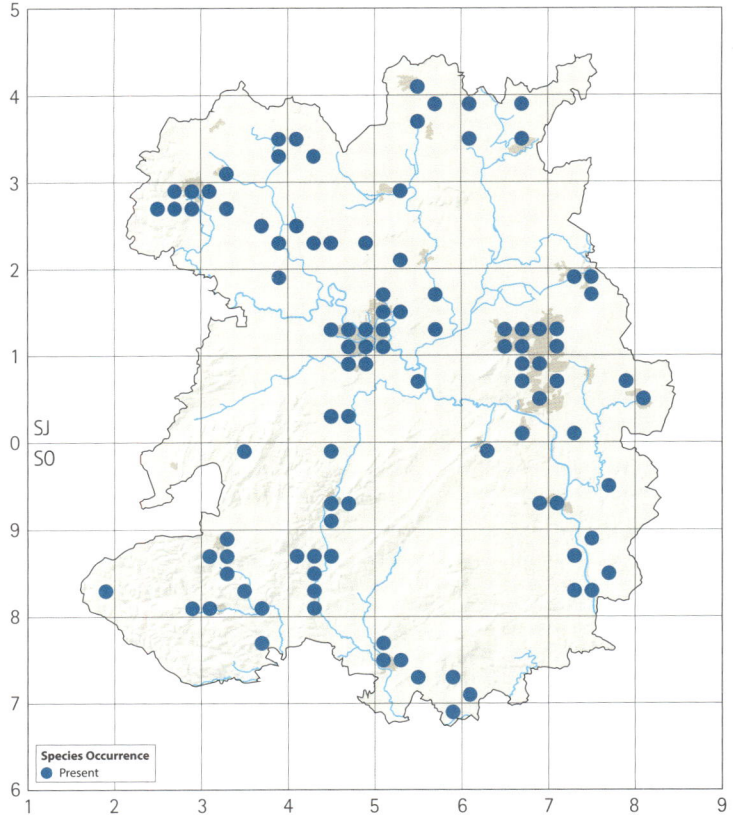

December in Sundorne, and 40 at Newport on 2 December, 30 at Horsehay on 30 November, 12 at Craven Arms on 28 November and 40 at Ludlow on the same date.

Following the memorable winters of 2010–11 and 2012–13 none were reported during the winter 2013–14, or in late 2014.

The chart shows the approximate number of birds recorded during each winter period since 1950, and clearly illustrates both the irruptive behaviour, with totals varying between zero and 500, and the increase in recent years.

The advent of planting berry-laden trees and shrubs around car parks and alongside roads within our towns and cities has provided food for large flocks. Sites around Shrewsbury, Ludlow, Oswestry and Telford are used fairly regularly at these times, with subsequent public and media attention.

The winter Atlas distribution map shows their presence in most other towns too, including Market Drayton, Newport, Telford and Bridgnorth. Even towns in the upland south-west, Church Stretton, Craven Arms, Bishop's Castle and Clun, can be picked out on the map. Whitchurch is the exception, although flocks have occurred there in earlier years. The lowland areas between Oswestry in the north-west, Shrewsbury, and Ludlow in the south, also have records, but upland areas are usually avoided, as Redwings and Fieldfares, which arrive earlier, will have already stripped most of the rowans there.

Breeding season records during the recent Atlas period were 15 at Tilstock on 4 April 2009, 10 at Bromfield on 11 April 2009 and 11 at Telford railway station on 15 April 2011, but these were of birds preparing to move back to their breeding areas. In suitable weather in April they can sometimes be observed fly-catching, their major source of food during the summer.

None ringed in Shropshire have been recovered. However, a young male which was one of three colour-ringed in an Aboyne, Aberdeen garden on 7 November 2010, and seen in again in Horwich, Greater Manchester on 27 November 2010, was observed among a flock of 31 at Belvedere, Shrewsbury on 9 January 2011.

Nationally, during the 'invasion' years of 2010–11 and 2012–13 more than 1,600 were ringed of which 77% were juveniles, while in 2011–12, a poor year for Waxwing records, 160 were ringed of which only one was a juvenile. This could be an indication that, in years of high breeding success and limited berry resources close to their breeding grounds, the juveniles are forced to forage further afield while the adults claim the local food. The high percentage of juveniles does indicate that ringing recoveries in the UK will not give an accurate representation of the longevity of the species.

Pete Nickless

Occupied Tetrads

Tetrads with Winter Records (2007–13):   101   (12%)

Jim Almond, Newport, 7 December 2010

Coal Tit
Periparus ater
Common resident

Coal Tit has a large range across Eurasia and north-west Africa. Here, it is a fairly common resident, usually associated with coniferous and mixed woodland, but also found in mature gardens and churchyards with large conifers, and is a regular visitor to garden feeders in winter.

In the first available historical reference (Leighton 1836), it was called 'coal titmouse' and described as 'not uncommon'. The name was generally written as 'cole tit' until 1890 when it was pointed out (Beckwith 1887–93) that 'Merrett, in 1667, called this species the Coalmouse, giving it the Latin name *carbonarius*'. Early historical references describe it as common, frequent and widespread, particularly in areas of conifer woodland, and refer to its liking for larch and yew. Beckwith (1879) described it as 'very common about the Wrekin, where I often see small flocks of Tits, comprising all five members of the family'; this, of course, pre-dates the separation of Willow and Marsh Tit into two species.

Early SBRs recorded it as common and frequent in areas of suitable habitat, and usually contained more records for winter than for the breeding season. These included a flock of approximately 200 Tits and Goldcrests passing through Clarepool Moss on 16 August 1960, 60 of which were Coal Tits.

The *Handlist* (1964) described it as 'resident, fairly widespread, especially in areas where there are mature conifers', and noted that it is 'nowhere as common' as Blue Tit or Great Tit.

Celia Todd, Pant Glas, 2 February 2013

The *Atlas* (1992) estimated the population to be 'somewhat lower than 8,000 pairs', based on national CBC figures, but went on to comment that recent increased planting of conifers had 'improved its breeding status and densities', and gave a further estimate, based on an assumed conifer cover of 4%, suggesting that the population 'may be as high as 20,000 pairs'.

Later SBRs continued to record it as common and widespread, both in winter and in the breeding season. Good yields of beech mast attracted large feeding flocks in 1995, 20 on Kerry Ridgeway on 14 October, and 30 at Colstey Wood on 15 November. An interesting breeding record from 1995 was of a pair nesting in a dry stone wall amid arable farmland at Berriewood, with only a few hawthorn bushes nearby. Three breeding records were received for 1998, of nests in 'a box, a conifer and a stone wall'.

During the recent Atlas period, breeding records included six young fledging from one nest in Craig Sychtyn in 2009, and eight from another nest at the same site in 2011. Larger counts included 20+ on Black Hill in March 2012, 20 at Shavington Big Pool in November 2011, 20+ in Wyre Forest in June 2010, and 31 at Breakneck Bank on 1 March 2011; and, after the conclusion of the Atlas, around 20 at Postenplain on 1 December 2013.

Breeding Relative Abundance (2008–13)

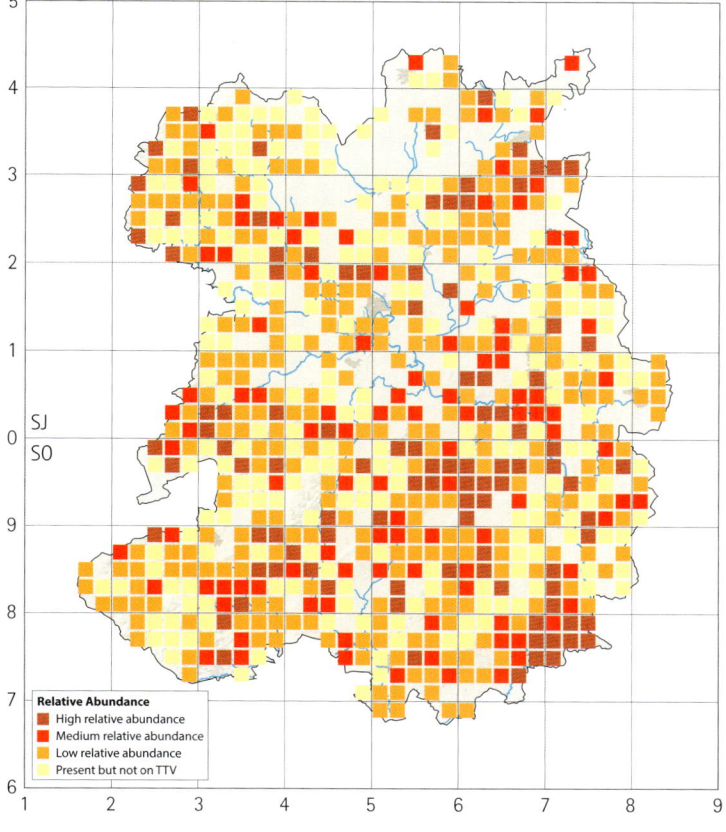

Relative Abundance
- High relative abundance
- Medium relative abundance
- Low relative abundance
- Present but not on TTV

Occupied Tetrads

Atlas period (breeding)	1985–90		2008–13		Change	
	Number	%	Number	%	Number	%
Confirmed	344	40	315	36	-29	-8
Probable	147	17	182	21	35	24
Possible	148	17	196	23	48	32
TOTAL	639	73	693	80	54	8
Tetrads with Winter Records (2007–13):			773	(89%)		

Coal Tits were found in 693 tetrads (80%) in the breeding season, compared with 639 tetrads (73%) in 1985–90. Based on TTV counts, the estimated breeding population is 8,100 to 8,300 pairs, suggesting that an increase to the higher estimate due to increased areas of conifer plantation, postulated in the *Atlas* (1992), has not occurred. This estimate also confirms that Coal Tit is far less numerous than Great Tit and Blue Tit.

Coniferous woodland still has the highest densities, and comparing the abundance map with that showing forestry and woodland (p. 27) shows a close correlation. There is a lower density or gaps in distribution on arable land in the north and east, and on higher ground in the south-west, particularly the heather moorland of the Long Mynd.

In winter, numbers increase in some years due to an influx from further north. They were generally more evenly distributed and were present in 773 tetrads (89%), suggesting that the resident population needs to forage away from breeding areas during the winter months. They are regularly seen at bird tables, and are ranked twelfth in the BTO Garden BirdWatch, with a reporting rate since 2006 of 52%.

The long-term trend in UK is a moderate increase of 20% between 1970 and 2014, and a 3% increase between 1995 and 2014. Population change has been fairly uniform across the UK, although some effect of cold winters is evident in the figures from the 1960s and 1980s, and it is likely that this situation is reflected here too. BBS shows a 60% increase in the West Midlands, but an 18% decrease in Wales, from 1995 to 2014, the latter probably reflecting the same run of recent bad winters and poor breeding seasons that has affected some other, more common, species here.

Its sedentary nature is confirmed by ringing records, as the vast majority have been both ringed and recovered here. Only 11 involve other counties, and only one found in the West Midlands was not in an adjacent county. The oldest was 6y 3m 22d, ringed as a first year on 27 October 2003 at Radlith, Pontesbury and recovered, freshly dead, in Pontesbury on 18 February 2010, a distance of only 2km.

Helen J. Griffiths

Marsh Tit
Poecile palustris
Fairly common resident

Marsh Tit is distributed locally in western Europe and eastern Asia, and is a fairly common resident here, although becoming increasingly less so. Its name is something of a misnomer because it is not restricted to marshy areas, but found in mixed and deciduous woodland, hedgerows and areas of scrub.

It was first recorded 'near Shrewsbury' (Leighton 1836), and Beckwith (1879) described it as the least common of the five species of tit, but 'by no means uncommon'; this comment, of course, pre-dates the recognition, in 1897, that Marsh and Willow Tit were different species. Both Paddock (1897) and Forrest (1899) recorded it as common but not numerous, and noted that it was not restricted to marshy places. They seemed 'reduced in numbers' by the severe winter of 1939–40. Noted as early breeders, there were several records of them in song in January and February. It seems likely that some of these early records were actually of Willow Tit, particularly given that reference is made to behaviour or habitat more typical of that species (for comparisons between Marsh and Willow Tit, see the next account).

The only local name mentioned is 'black cap', but this was also used for Reed Bunting (as well as the species Blackcap).

It was described as a 'regular and widely distributed breeding species ... generally more numerous than Willow Tit' (SBR 1956). A flock of approximately 200 Tits and Goldcrests passed through Clarepool Moss on 16 August 1960, about 40 of which were Marsh Tits. Although numbers were down following the severe winter of 1962–63, it appeared that the effects of that winter were generally less serious than had been feared (SBR 1964). The *Handlist* (1964) described it as 'fairly widespread', and probably commoner than Coal Tit.

SBRs for the years up to 1992 contain records from many areas of suitable habitat, both in winter and in the breeding season, but almost always as singles or in pairs, and often in mixed tit flocks; flocks of 12 at Chelmarsh in November 1986 and 15 at Bury Ditches

Gareth Thomas, Whitcliffe, 3 May 2009

in January 1987 were exceptional counts. The first reports of them visiting garden bird tables were in 1976.

The *Atlas* (1992) mapped it as thinly, but widely distributed, being found in every 10km square, and reaching its highest densities in deciduous woodland. Based on a national average of 50 to 100 pairs per 10km square (national BTO Atlas 1972), and assuming no significant change, the population was estimated to be between 1,750 and 3,500 pairs, considerably less than Coal Tit.

Breeding Distribution Change (1985–90 to 2008–13)

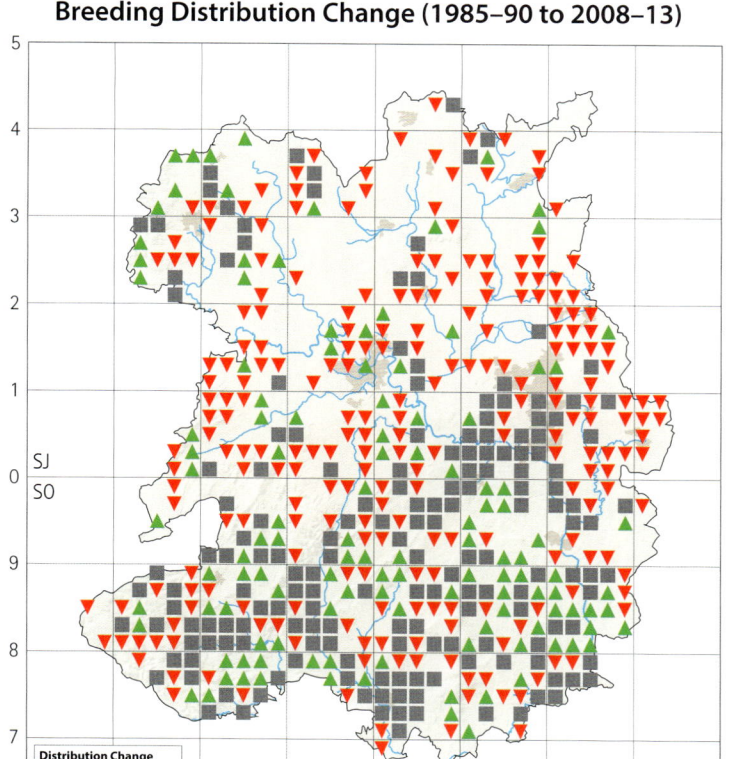

Winter Distribution (2007–13)

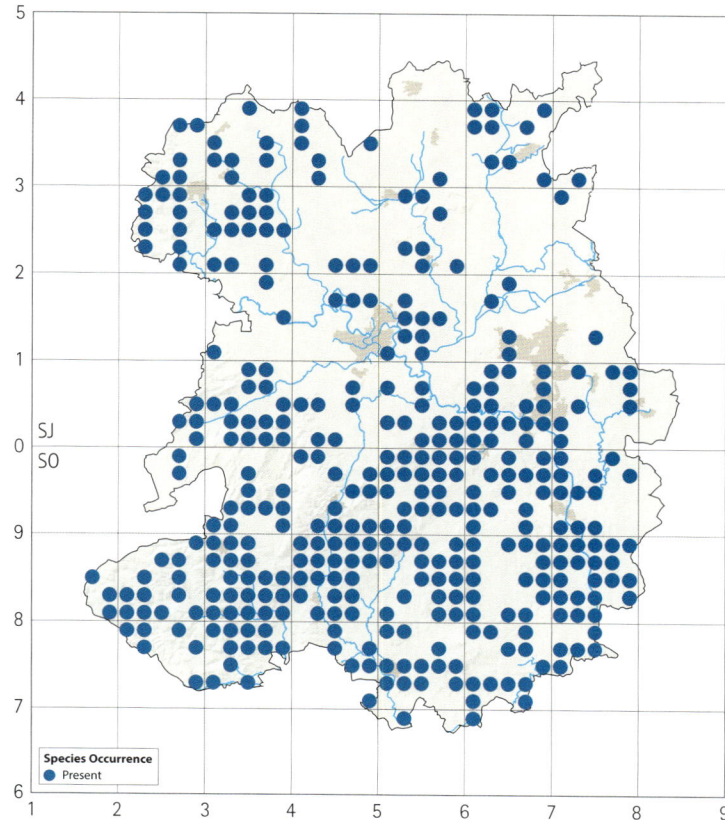

Later SBRs continued to report them as frequent, usually in ones and twos, although during some years in the 1990s no breeding records were received. SBR (1996) noted that more records of Willow than of Marsh Tit were received, and wondered whether this was due to a change in fortune of either species; however, the usual pattern was restored in the following year. Small numbers continued to be recorded from widespread locations, and a few breeding records were received each year. Visits were made to many feeding stations, including regular sightings at Venus Pool. In 2003, 11 pairs were located in a survey of Telford and Wrekin woods. There was a flock of 10 at Shavington in February 2007, 10 were seen at Dudmaston in February 2008 and 14 at Benthall in June 2008.

In the recent Atlas period, they were found in 294 tetrads (34%) in the breeding season, compared with 409 tetrads (47%) in the previous Atlas, an apparent loss from 28% of the tetrads where it was found in the earlier period. In winter, they were more widely distributed, being present in 365 tetrads (42%). Based on TTV counts, the breeding population is estimated to be 1,600–1,700 pairs.

Both maps show clearly that the stronghold for this species is still in the south, with the breeding distribution change map showing an almost equal number of losses and gains there, suggesting it was under-recorded in both Atlases. There were many more losses than gains in central and eastern areas, which are predominately arable farmland.

The winter distribution is more widespread, as the population is swelled by juveniles, and they forage away from their woodland breeding sites, often in mixed flocks and visiting bird tables. Areas of predominantly arable farmland are still apparently avoided.

Occupied Tetrads

Atlas period (breeding)	1985–90		2008–13		Change	
	Number	%	Number	%	Number	%
Confirmed	159	18	116	13	-43	-27
Probable	131	15	83	10	-48	-37
Possible	119	14	95	11	-24	-20
TOTAL	409	47	294	34	-115	-28
Tetrads with Winter Records (2007–13): 371 (43%)						

In the UK, the long-term trend is a rapid decline, by 72% between 1970 and 2014, and 32% since 1995. BBS data for England shows a 35% decline between 1995 and 2014, comparable to the 28% decline here in occupied tetrads between the two Atlas periods.

Recent research elsewhere (Broughton & Hinsley 2015) indicates that the decline of different woodland species does not have a single underlying cause, and in this case they discounted poor breeding success, predation, climate change and changes in habitat quality as drivers of the decline, although deterioration of habitat had been suggested as a likely cause by earlier research summarised in the national BTO Atlas (2013). Attention was focused on reduction in survival rates, and loss of connectivity between woodlands through hedgerow removal and tree loss, which affects this species more than most because of its sedentary nature, unwillingness to cross open ground, and the distances that juvenile females in particular need to disperse to find mates in new territories. The breeding distribution change map shows that it has disappeared from large areas of arable farmland, where removal of hedges and hedgerow trees are most likely to have occurred, suggesting that similar mechanisms may be at work here.

The slight increase in the Coal Tit population since 1990, and the large decline in Marsh Tit and the catastrophic decline of Willow Tit over the same period, leave no room to doubt the relative abundance of the three less common tits now.

There are few ringing recoveries, but an example of its sedentary nature was a first-year male ringed at Shavington Park, Calverhall on 27 March 2004, and caught again at the same location in April 2004, March 2007 and March 2010, 5y 11m 22d later.

Helen J. Griffiths & Leo Smith

Willow Tit

Poecile montanus

Scarce resident

Willow Tit is resident throughout temperate and subarctic Europe and northern Asia. It is now a scarce resident, and its long-term decline seems to be continuing at an alarming rate. It is the only tit to excavate its own nest hole, and it is therefore more likely to be found in damp woodland with rotting tree stumps, and softwood conifers.

There were no early references because it was not identified as a separate species from Marsh Tit until 1897. However, it seems likely that some of the early records of Marsh Tit were actually Willow Tits, given that reference was made to behaviour or habitat more typical of Willow. Writing of Marsh Tit nests, Beckwith (1890) referred to a correspondent from Astley Abbots observing that 'when it could not find a hole suitable for its purposes, it excavated one in some rotten stump', and there was a similar reference to nesting in a rotten stump at Bomere Pool in 1900. In The *Victoria County History of Shropshire*, Forrest (1908) still referred to Marsh Tit 'sometimes excavating a hole for itself', and there was no species account for Willow Tit.

The first mention of it here came in 1937. In an attempt to gain a better understanding of its distribution and status in Britain, all records were collated, and the resulting paper stated that 'information from *Shropshire* is extremely inadequate, but the Willow Tit has occurred in the north and probably also in Wyre Forest' (Witherby & Nicholson 1937). However, the *Handlist* (1964) stated that it appears not to have been recorded until 1945, when it was 'occurring in most woods between November and March in the Whitchurch district, about as commonly as Marsh Tit', although the CSVFC *Transactions* for 1945–46 seem to question this, noting that 'of the Willow Tit we have no satisfactory definite observation'.

Early SBRs described it as 'locally distributed' in widespread localities, including along the River Severn between Leighton and Bridgnorth, but Marsh Tit was 'generally more numerous' (1956), in 'most woods around Ellesmere' (1957), and near the Lawley (1957), Crose Mere and Ellesmere (1959), Whixall Moss and Clarepool Moss (1960), and the Hollies and Ashford Carbonell (1962). In 1959, a pair was feeding young in a nest in a fence post at Peplow, while in 1960 a flock of approximately 200 Tits and Goldcrests passed through Clarepool Moss on 16 August, when four to six were heard.

The *Handlist* summarised its status as local, but fairly widespread, and described it as 'perhaps most common in the Severn Valley' and 'on the mosses in the Ellesmere/Wem area', but added 'possibly as numerous as the Marsh Tit and they often occur together', apparently contradicting early SBRs and suggesting that the relative abundance of these two tits was unclear.

The status was clarified by the national BTO Atlas (1976), which

John Hawkins, North Shropshire, 16 January 2008

mapped it in 27 of the County's 33 10km squares, with confirmed breeding in half (17), whereas Marsh Tit was mapped in every square except one, with confirmed breeding in 22, reflecting the CBC evidence that generally Willow Tit was rather less common at that time.

It was first noted visiting garden bird tables (in Much Wenlock and Ashford Carbonell, in winter) in 1976–77. Between 1982 and 1990, the Marsh Tit records submitted annually always outnumbered those for Willow Tit; and the latter were usually noted as one or two birds, or occasional family parties, and most frequently from the north and east, although this reflected the distribution of observers rather than the species. There were seven breeding records in 1988, and it was noted as 'numerous on Haughmond Hill in December' in that year.

The *Atlas* (1992) noted its preference for damp woodland and carr, where there are rotten tree stumps in which to excavate nest holes. River valleys are perhaps more likely to provide such habitat, particularly when alder, birch and willow are present, all trees which decay readily in a damp environment; however, there were few records from the areas surrounding the meres and mosses. Marsh Tit was usually found to be more numerous and widespread, but the reverse was found in the south-west, where coniferous plantations were occupied; in SO39 (the area west of the Long Mynd) it was present in 17 tetrads, but Marsh Tit in only nine, perhaps because of the prevalence of conifers, and wetter conditions resulting in more rotten tree stumps.

Distinguishing between Willow and Marsh Tits is very difficult, even for experienced observers, unless the distinctive calls are heard.

In addition, some features traditionally used for identification, and quoted in field guides, have proved to be unreliable (Broughton 2009). As a result there are likely to be some errors on each map in both Atlases.

After 1990, it was reported regularly, and from over 20 sites in 2001, 2004 and 2005, mostly from widespread localities in winter. Very few breeding records were received, but attempted breeding at Betton (near Market Drayton) in 1994 was 'interrupted by a grey squirrel'. Nine pairs were found during a survey of woodland in Telford and Wrekin in 2003, and breeding was confirmed at Whixall Moss (a pair with six fledged young on 28 May 2003), and at Priorslee Lake, Whixall Moss and Betton Moss in 2005.

In the recent Atlas period it was found in 68 tetrads (8%) in the breeding season, compared with 337 (39%) in the 1985–90 Atlas, an apparent disappearance from 80% of the tetrads where it was found in the earlier period.

It was found in every 10km square in 1985–90, and the breeding

distribution change map clearly demonstrates the rapid and dramatic decline in only 20 years. Few areas have escaped, but it has disappeared altogether from five of the County's 10km squares in the north, and one in the south-east, but none in the south-west, where coniferous woods and wetter conditions still provide more suitable habitat.

The local decline up until 2014 reflects the long-term UK trend of a 93% decline since 1970, and 77% since 1995. In England the decline has been 78% since 1995. It no longer occurs in sufficient survey squares for BBS trends to be calculated for the West Midlands, or Wales, but the 80% reduction in tetrads where it was found indicates a comparable decline here.

However, Willow Tits are inconspicuous and difficult to find, so the map almost certainly under-represents their distribution. They were not found in 1985–90 in half the tetrads where they were found in 2008–13, in spite of the large population decline, suggesting they were very under-recorded in both periods.

Based on TTV counts, the breeding population is estimated at 70–80 pairs, only 5% or less of the estimated population in 1990. It was found on too few TTV counts for this to be a reliable estimate, and the 1992 estimate was calculated by applying a national average figure taken from the national BTO Atlas (1976), which would have been too high by then, but an average of two to three pairs per tetrad with records (which includes an allowance for some under-recording), would give an estimate of 140–200 breeding pairs.

The national decline has been accompanied by a striking shift in distribution away from south-east England, attributed to the combined effect of climate change and decline in habitat quality due to agricultural intensification, which leaves no room to adapt to

Occupied Tetrads

Atlas period (breeding)	1985–90		2008–13		Change	
	Number	%	Number	%	Number	%
Confirmed	104	12	14	2	-90	-87
Probable	119	14	19	2	-100	-84
Possible	114	13	35	4	-79	-69
TOTAL	337	39	68	8	-269	-80
Tetrads with Winter Records (2007–13): 119 (14%)						

Breeding Distribution Change (1985–90 to 2008–13)

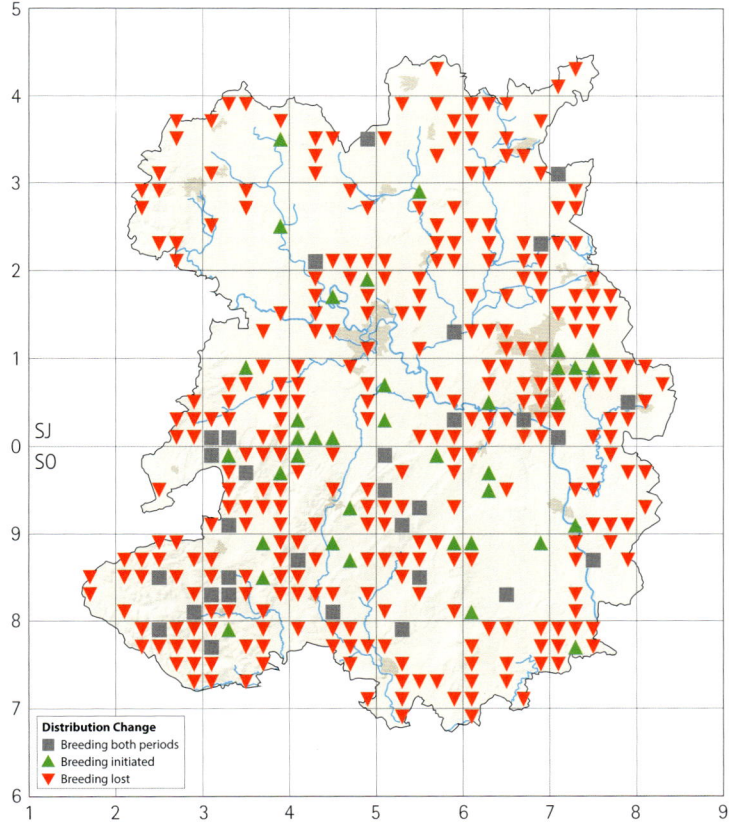

Distribution Change
- Breeding both periods
- Breeding initiated
- Breeding lost

Winter Distribution (2007–13)

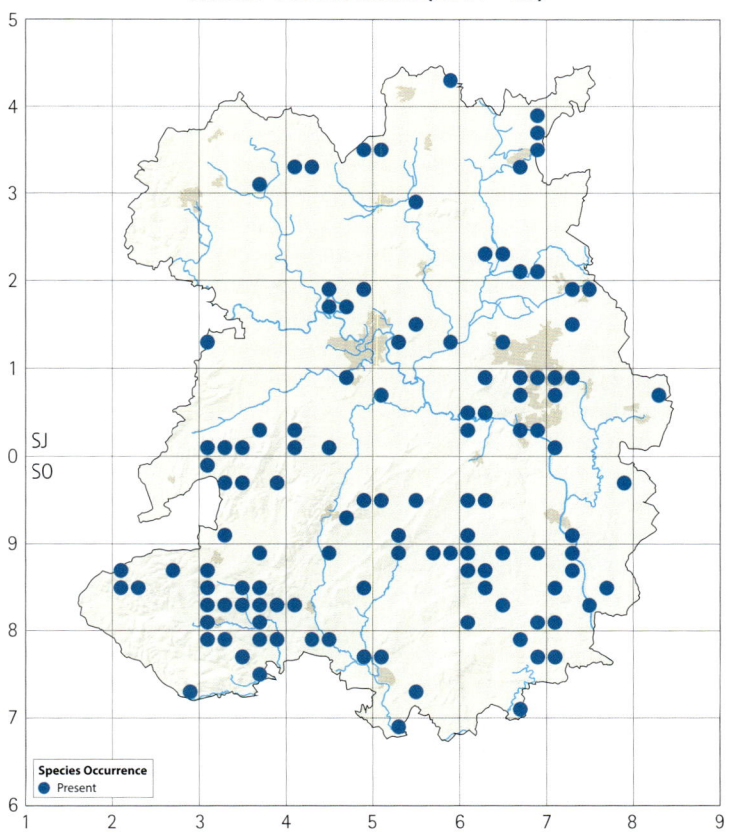

Species Occurrence
- Present

rising temperatures (Oliver *et al.* 2017). Drying out of wet woodland, and loss of the shrub layer in many woods due to shading and deer grazing, are also contributing. The rate of decline appears to have increased recently, probably at least partly attributable to a succession of cold winters (2009–11) which affected numbers of many species.

In winter, they were more widely distributed and were found in 119 tetrads (14%). This may reflect a higher population in early winter, including birds of the year undergoing post-breeding dispersal, and easier observation due to less leaves and undergrowth in woods, visits to bird tables, and because they are most vocal in late winter and early spring.

Few are ringed, but the three recoveries are all examples of its sedentary nature: two were ringed and recovered at the same place,

while the third moved 4km. An adult ringed in Pontesbury on 3 March 1968 was recovered freshly dead there 27 days later; one first-year ringed at Shelve Pool on 23 August 1981 was recovered freshly dead near Minsterley, on 21 September 1981 (killed by a cat); and one first-year ringed at Hinstock on 6 November 1982 was found freshly dead there on 18 September 1983 (also killed by a cat).

A Willow Tit survey started in 2016 (Jonathan Groom *pers. comm.*). Species such as this, which are localised, elusive and difficult to identify, are likely to be under-recorded in Atlas surveys, and the overall decline would lead to a lower detection rate in many tetrads. The Atlas has therefore provided a baseline, but this more intensive specific survey will provided a valuable addition to the Atlas results.

Helen J. Griffiths & Leo Smith

Blue Tit (Eurasian Blue Tit)
Cyanistes caeruleus
Very common resident

Blue Tits seem always to have been common and familiar, and the 'Tom Tit', as it was known, was the 'commonest of all the tits, and a general favourite' (Forrest 1899). They long ago learned to pierce the then silver-foil tops of milk bottles to gain access to the cream. The first records of this behaviour were in the winter of 1942–43, and the most recent was in Wellington on 10 November 1981: the lack of subsequent records chronicle the almost complete disappearance of the home milk delivery and the introduction of homogenised, semi-skimmed and skimmed milk, rather than a change in bird behaviour.

Usually described throughout the life of SOS as abundant and widespread, they breed in every tetrad. There has been no change in distribution since the previous Atlas. Nationally, relative abundance reaches some of its highest densities here.

They feed mainly on insects in the breeding season, feed chicks predominantly on caterpillars, and nest in holes, so they occur most frequently in deciduous woodland, and are also common in pasture farmland wherever hedgerows survive, and in villages and gardens in towns, provided there are mature trees or nest boxes. The breeding season relative abundance map reflects these habitat preferences, and shows much lower densities in arable farmland on the Shropshire plain and the Severn Valley, and in the uplands in the south-west and the Clee hills. The relative abundance in winter is similar.

Blue Tits first breed at one year old. Nationally, in an average year, only 38% of the fledged young survive until breeding age, and just over half the adults survive from one year until the next. However,

brood sizes are large, so, in common with many other species, there are many more Blue Tits in early winter than there are at the start of the following breeding season. In addition, leaves on trees and hidden incubating females reduce the numbers seen in the breeding season, but they become more conspicuous when they leave woodland to search for food, so not surprisingly 48% more were counted on winter TTVs. However flock sizes are rarely large. Although no one-hour

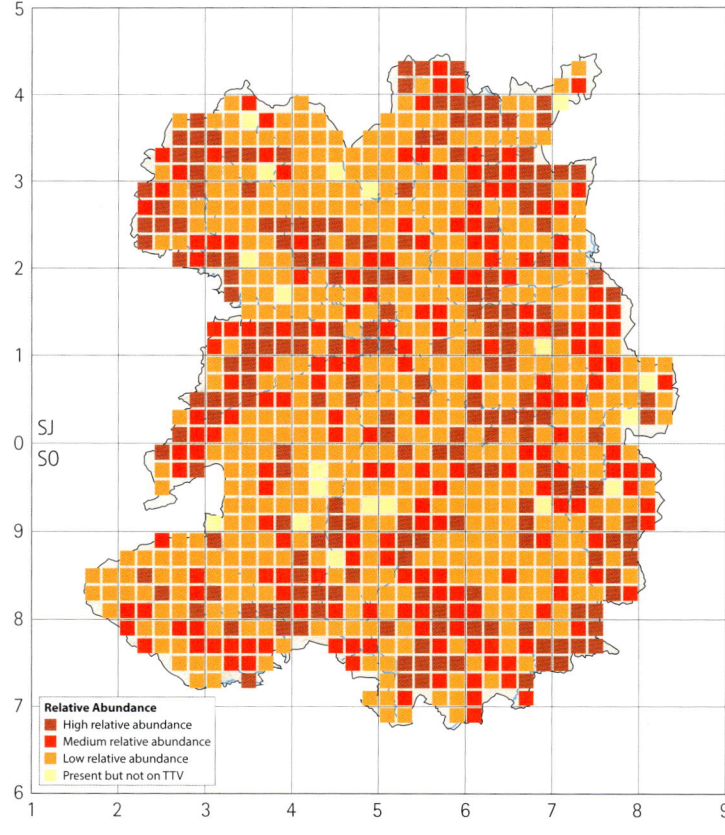

Breeding Relative Abundance (2008–13)

Relative Abundance
- High relative abundance
- Medium relative abundance
- Low relative abundance
- Present but not on TTV

Occupied Tetrads

Atlas period (breeding)	1985–90		2008–13		Change	
	Number	%	Number	%	Number	%
Confirmed	843	97	814	94	-29	-3
Probable	19	2	45	5	26	137
Possible	7	1	11	1	4	57
TOTAL	869	100	870	100	1	0
Tetrads with Winter Records (2007–13): 869 (100%)						

BBS Trend (1997–2014)

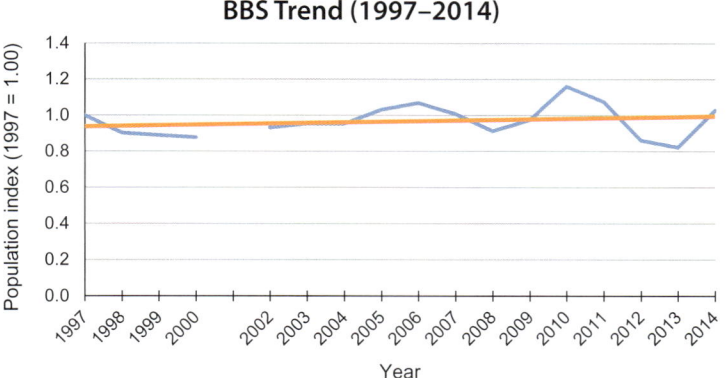

TTV count in the breeding season reached 50, only two exceeded 50 in the winter (maximum 58).

The population fluctuates each year, reflecting productivity the previous year and subsequent winter weather conditions. There are numerous SBR reports of very variable clutch sizes and numbers of fledged young at the same sites in successive years, attributed to changing weather conditions and food supply in the breeding season. In most years post-breeding and winter flocks are reported. There are usually several double figure counts, occasionally reaching up to 60. Exceptional counts of 100 or more have come from a mixed tit flock in 1980, 160 at Cressage on 23 August 1988, and 'an incredible 150 amongst a large mixed tit flock' at Madeley on 18 November 2002.

Clutches are usually 7–12 eggs, so a brood of 13 fledged from a nest box at Hinstock in 1971, and an average brood size of 10.1 from 19 pairs at Purlogue in 1983, were also exceptional.

High counts are made on BBS in most years, and it was recorded in more than 90% of survey squares since 1997, while in three years it reached 100%. BBS results reflect the annual fluctuations, but show virtually no underlying population trend since 1997.

Based on TTV counts, the breeding population is estimated at 75,300–76,500 territories.

This is the seventh most common species, but it has the second highest GBW reporting rate, an average of 90%. There is no apparent annual trend in use of gardens, but use in winter is slightly higher than during the breeding season.

Blue Tits take readily to nest boxes. Several hundred have been retrapped or recovered here, and almost all of them were ringed here too. Almost all the 40 exceptions came from adjacent counties, with only one from Scotland and none from overseas. Almost all the 53 ringed elsewhere and found here came from adjacent counties, with none from Scotland or overseas. This shows that the species is

Celia Todd, Pant Glas, 10 April 2013

extremely sedentary, but there are a few examples of Blue Tits moving more than 100km, the furthest being a first-year ringed in December 2009 in south Ayrshire which was caught at Shavington in February 2013, 316km distant and 3y 1m 22d later. Another, a nestling male ringed in 2011 at Farlington Marsh, Portsmouth (Hampshire) was caught 8m 16d later at Sutton, Shrewsbury, 237km NNW.

Although the typical lifespan of adult breeding Blue Tits is around three years, the oldest known is 7y 8m old (compared with the oldest known nationally of just over 9y 9 m). There are five local examples of living to more than six years old. All were found at the same place as they were ringed, confirming the extremely sedentary nature of the species.

Leo Smith

Great Tit

Parus major

Very common resident

Great Tit has a very large range, covering Europe, Asia and north-west Africa. Here, it favours deciduous woodland and parkland, but it is also often found in coniferous woodland, gardens, scrub and farm hedgerows. As with other species of tit, it is a regular visitor to garden feeding stations.

The first historical record described 'greater titmouse' as very common around Shrewsbury (Leighton 1836), and in 1837 Hulbert said that 'the whole of the tribe of Tomtits are constantly fitting about among the trees in the garden'. Its presence as 'common' and 'frequent' is recorded several times in early accounts, often being

Celia Todd, Pant Glas, 16 May 2005

seen in the company of other tit species. A detailed account from 1890 (Beckwith 1887–93) described their liking for yew and holly berries, and beech nuts, and their useful role in removing garden pests (unspecified insects and caterpillars). The account goes on to reveal their propensity for disembowelling honey bees, although it seemed that this usually occurred early in the year when there was a lack of other insect food (Beckwith 1887–93). The local name was 'Ox-eye' (Forrest 1899), apparently derived from the bird's call.

They can often take a novel approach when choosing nest sites, and there are a number of records of nests in upturned flower pots, letter boxes, metal gate posts, and in more recent years, in plastic tree guards (for example, in Twemloes Big Wood, where a pair reared five young at ground level in 2002). There is a record from 1885 of a nest in a disused beehive at Leaton Knolls (Beckwith 1887–93), and in Ludlow in 1900 a nest was found inside an active beehive, where a pair laid 13 eggs; both bees and birds used the same exit and entrance hole, showing a high degree of tolerance towards each other. Unfortunately, as was so often the case at the time, the eggs were taken – this time by the Secretary of the Shropshire Beekeepers' Association (Forrest 1900). Another record from Beckwith's species account noted a Great Tit on a nest in Astley Abbots in 1882 on seven eggs of her own and 13 of either a Coal or Blue Tit (Beckwith 1887–93). The severe winter of 1939–40 apparently reduced numbers, but they did not seem to have suffered as greatly as other small birds.

Early SBRs usually recorded it only as present or 'known to breed', but there was a record from January 1957 of about 100, 'searching beech mast, with Chaffinches' in Ellesmere. The *Handlist* (1964) described it as 'resident, common, even in the highest hill country,

where there are suitable habitats'. SBRs for the years up to 1992 continued to describe it as common and a regular breeder, particularly in areas of deciduous woodland, for example a flock of 30 in Wrekin Woods in February 1980, and, in the same year, 19 pairs in nest boxes in Lilleshall which laid 179 eggs and fledged 179 chicks. SBR (1987) noted that it was less widespread and far less numerous than Blue Tit, and ringing records from Leebotwood (SBR 1990) showed 'a ratio of about a third of the numbers of Blue Tit, comparing favourably with reports generally'.

In the *Atlas* (1992), it was described as 'a well known and common resident', recorded in 98% of tetrads, but it was unevenly distributed and much scarcer, or even absent, in areas of unsuitable habitat. Based on an estimated UK population of two million pairs, averaging eight pairs per sq.km, the population was estimated to be 30,000 pairs.

Later SBRs include a record from 1996 of two mixed broods of Great and Blue Tits, one being fed by Great Tits and one by Blue Tits. Breeding records continued to be reported from many locations, including various nest box schemes. Larger flocks of up to 30 were recorded during the 1990s, with 32 at Hawkstone Park in March 2001, 41 at Venus Pool in August 2002, and 100 among a huge tit flock at Madeley in November of that year.

In the recent Atlas period they were found in all 870 tetrads in the breeding season, compared with 853 tetrads (98%) in 1985–90, suggesting that the population had grown in the intervening period. Wintering birds were equally widely distributed and were present in 869 tetrads and not found in one tetrad only, on the top of the Long Mynd. Based on TTV counts, the estimated breeding population is 51,600–52,400 pairs.

Occupied Tetrads

Atlas period (breeding)	1985–90		2008–13		Change	
	Number	%	Number	%	Number	%
Confirmed	759	87	762	88	3	0
Probable	60	7	85	10	25	42
Possible	34	4	23	3	-11	-32
TOTAL	853	98	870	100	17	2
Tetrads with Winter Records (2007–13): 869 (100%)						

Breeding Relative Abundance (2008–13)

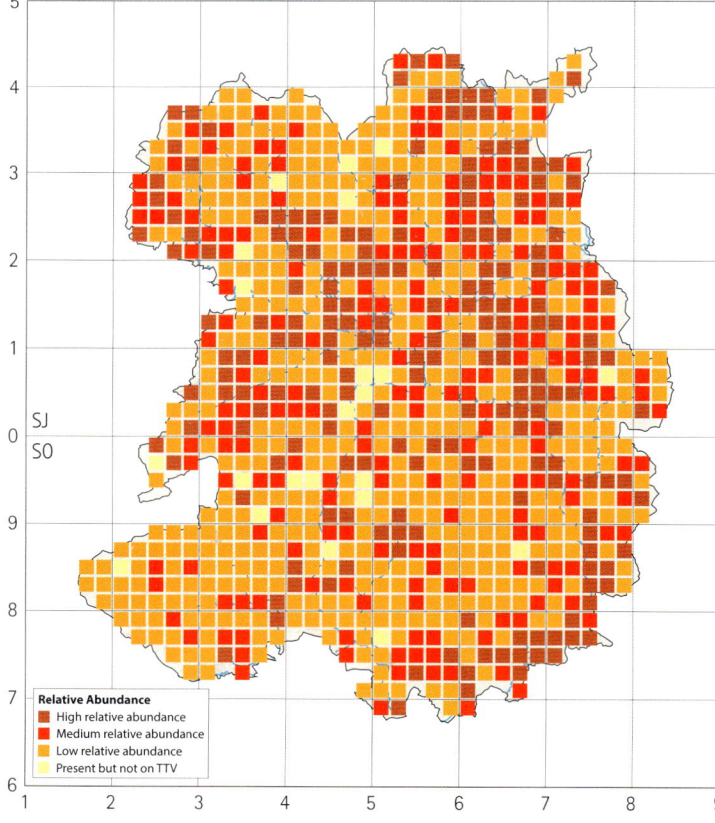

Relative Abundance
- High relative abundance
- Medium relative abundance
- Low relative abundance
- Present but not on TTV

BBS Trend (1997–2014)

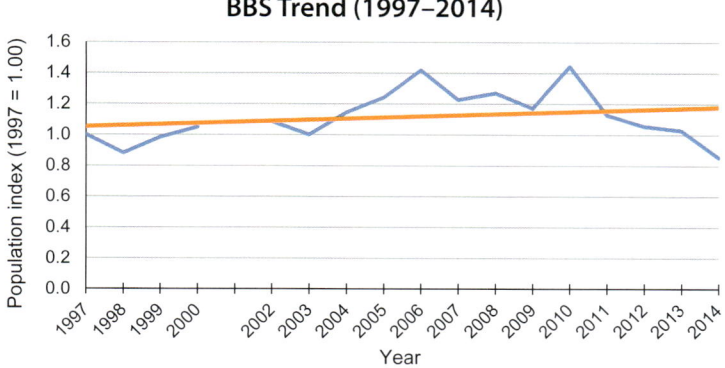

Comparing the total counts of Great Tit and Blue Tit on all the breeding season TTVs carried out during the recent Atlas period, the ratio was 0.6:1, i.e. there were 60% more Blue Tits. The total population estimates based on TTV counts indicate that there are 45% more Blue Tits, suggesting that Great Tits are more conspicuous.

The relative abundance map gives a more detailed picture, showing that distribution is far from uniform, with fewer found in the higher, hilly country of the south and south-west, and in tracts of arable land, with higher densities in wooded areas, such as Wyre Forest and Wenlock Edge, and the main river valleys (including the more populated areas).

The long-term trend in UK is a fairly steady increase since the 1960s, with an increase of 40% found by BBS between 1995 and 2014, with a 41% increase in Wales, and a 30% increase in the English West Midlands, over the same period.

The local BBS trend shows a steady increase up to 2010, but a steep decline since, reflecting a series of mild winters, then hard winters in 2009–10 and 2010–11 followed by poor breeding seasons affected by wet and cold springs.

Great Tit is a regular visitor to bird tables, and is ranked fifth in the Garden BirdWatch reporting rate (77%) since 2006.

Ringing records show that this is largely a sedentary species, with only about 15% of recoveries involving a different county, and none involving an overseas journey. A juvenile ringed in Newport on 10 July 2003 was caught on 14 November 2011, also in Newport; a first year female was caught at Cross Lane Head, Bridgnorth, on 9 October 2005 and the same bird was caught there by the same ringer in 2006, 2009, 2010, 2012 and 2013; a first-year female was caught in Sutton, Shrewsbury on 23 November 1997 and caught again there on 27 December 2004; and a nestling was ringed in Hawkstone Park on 4 June 2005 and the ring found on 7 January 2012, between Hodnet and Weston, only 2km away.

The birds moving the greatest distance were: a first-year female ringed at Keith (Moray) on 27 December 2009 and caught in Severn Valley Country Park on 31 January 2010, a separation of 567km; an adult female caught at Calverhall, near Market Drayton, on 10 March 1984 and caught at New Scone (Perth) on 12 April 1986, 391km distant; and an adult female ringed in Pontesbury on 3 February 1968 and found dead near Sidmouth (Devon) on 28 May 1969, 217km.

The oldest was 9y 4m, a first-year male ringed in Wellington on 7 January 1985 and recovered after being hit by a car on the M6 in the Coventry area, on 7 May 1994, a distance of 73km.

Helen J. Griffiths

Bearded Tit (Bearded Reedling)

Panurus biarmicus

Very rare irruptive visitor

Although its range stretches from Europe though to China, in Britain the bulk of the breeding population is found along the south and east coast from Sussex through to the Humber. Its preferred habitat is extensive stands of Common Reed and it only occasionally irrupts from its strongholds, hence its rarity here with no historic reports and just eight modern records:

Marton Pool (Chirbury)	two, 9–16 January 1960
Boyne Water	four, 22 October 1971
Allscott Sugar Factory	two, 31 October 1973
Allscott Sugar Factory	two, 14–20 October 1979
Roden	19 December 1983
Allscott Sugar Factory	21–22 October 2012
Chelmarsh, two	22 November 2015
Wood Lane	two, 31 October 2016, 21 February and 5 April 2017

Although only two were definitely seen at Marton Pool following the large dispersal from East Anglia that year, one observer thought that there may have been up to 12 on 14 January. The party at Boyne Water included at least one male and coincided with a period when Dutch and East Anglian birds had started to irrupt annually from their breeding areas. The pair at Allscott in 1973, and the pair at the same site six years later (both the latter were caught and ringed), also occurred during this period of irruptive behaviour (O'Sullivan 1976). Interestingly, the male at Roden which was found dead, having been taken by a bird of prey, had been ringed as an adult at Ousefleet on the Humber two months and 20 days earlier and 164km away. The male and female at Chelmarsh were trapped and ringed while the male and female seen at Wood Lane on three occasions over a period of some five months were in an inaccessible part of the reserve.

Graham Walker

Woodlark

Lullula arborea

Very rare winter visitor, has bred

Woodlark is resident in south and south-west Europe, including England, and breeding occurs across central Europe.

Beckwith summed up the nineteenth century situation well in 1890:

> This is a rare and extremely local bird … and when Rocke [in 1865] stated that it was '*not nearly so common as the Sky Lark, though a good many are at times to be found*' … it is greatly to be regretted that he did not mention where and under what conditions they occurred.

During that time Woodlark featured on lists and in general statements but, as Beckwith lamented, locations were rarely given.

Nevertheless 14 nineteenth-century records do give localities, indicating that Woodlark was resident, although the dates tend to relate to the publication rather than the specimens or sightings: Oswestry Old Racecourse (nest, about 1808); West Felton (before 1836); near Walford ('not uncommon', 1838); two specimens in the Clungunford collection, possibly of local origin (1866); Eaton district (1879); Aldenham Park, Baschurch ('several places in the neighbourhood'), Moelydd Hill, and Oakley (now Oakly) Park (all about 1890); near Craven Arms (a nest, about 1891); Trench (one among other caged larks, summer 1893); Caynton (November 1896); and Whitcliffe Wood (a nest with five eggs) and Shirlett, both before 1897.

At Moelydd they were 'rather numerous on some rough broken ground where low oaks and underwood grow', Beckwith himself adding 'the nature of the ground at Aldenham and Baschurch … is very much the same'.

Reporting increased in the first half of the twentieth century when there were 24 records, including five from the Knighton area (precise locations not specified, and perhaps not County records). Locations in the north were Boreatton Park, Hawkstone, the Shrewsbury area ('by no means rare in the vicinity') and Haughmond Hill, while, in the south, Aldenham, the Clun district ('doubtless nesting there'), near Knighton (several nests), Ludlow, Gatten on the Stiperstones and the Wyre Forest (records in four different years, including two nests). The last of these records was from 1934.

In 1924 Elliott, commenting from Dowles in the Wyre Forest, wrote of a 'marked increase, 8 or 10 pairs in district', while Lloyd, reviewing the changing status of several species in 1943, wrote 'has shown some increase in certain districts' though regrettably, rather in the manner of Rocke a century earlier, he gave no localities.

In contrast, the second half of the twentieth century brought only 16 records despite the increase in recording generally. There were six records in the 1950s, from Llynclys, Shirlett and Snailbeach in 1952, Willey Park (1955), Horderley (1956), the Ercall (1957) and two from undisclosed sites (SBR 1958; 1959). In addition, a table of species present in the Bucknell area (SBR 1987) listed Woodlark as 'breeding, increasing' in 1950–54 but offered no further details.

The *Handlist* (1964) described Woodlark as 'resident, local in small numbers in the south and west' and 'probably often overlooked' and there were 10 records in the 1960s, all from the south: Stapeley Hill (1961), Linley Hill (twice in 1962), the Long Mynd and Wolverton (1962), Wenlock Edge and Wynett's Coppice (1963), the Bucknell area (1965) and Black Rhadley Hill (1967). The last three decades of the twentieth century produced only one record, a single near Shifnal, a road casualty found dying on 12 February 1985.

This century has produced a mere four records, starting when farmland transect fieldwork near Sheinton found one on a stubble field on 4 December 2002 and another, perhaps the same bird, at Woodcote 20km north-east on similar ground the following day. One graced Venus Pool from 22 January to 13 February 2005, and two were seen at Coopers Mill Cottage near Dowles Brook in the Wyre Forest on 10 February 2008, during woodland clearance being carried out to help conserve the butterfly population there.

This last record was the only one received during the recent Atlas period, and there were no breeding season records. Neither was Woodlark recorded during the 1985–90 *Atlas*, and the last confirmed breeding was a 'pair with young' on the Ercall on 29 July 1957.

The fact that all of the last five records, spanning almost half a century, have occurred in winter, and only one has been of more than a single bird, suggests that there is no longer a breeding population here, and those seen were wanderers from breeding areas elsewhere in this country or, perhaps, continental migrants.

Nationally, the Woodlark population enjoyed something of a renaissance between the 1920s and the 1960s, followed by a low in the mid-1970s to mid-1980s, perhaps as a result of hard winters in the 1960s (Sitters *et al.* 1996); here, six of the 10 records from the 1960s pre-date the hard winter of 1962–63.

The BTO Atlas (2013) documents an increase in the size and range of the national breeding population since the 1968–72 Atlas some four decades earlier, including expansion into the West Midlands, so perhaps the Cannock Chase population will increase and spread to our local heathlands.

John Tucker

Skylark (Eurasian Skylark)
Alauda arvensis
Common resident

Skylark breeds across Europe, but those in central, eastern and northern Europe move south for the winter. Here, it was abundant and widespread throughout the 1800s. In 1838 Eyton described the 'lark as common' and Beckwith noted in 1890 that it was 'common everywhere'. He remarked that in October and November they congregated in large flocks augmented later by migrants, and added that the flocks fed 'on blades of wheat where they caused considerable damage but also fed on the troublesome weed-knotgrass'. Forrest (1899) stated that it was 'very plentiful' and 'is killed for the table, and is very good eating' although he argued it was 'a thousand pities that it should be destroyed for this purpose'. By 1906 he wrote that it was now rarely killed for the table and was 'very numerous especially in the open country to the north'.

Its status appeared to remain unchanged throughout the first half of the twentieth century, but following agricultural intensification after the Second World War, the population began to decline. The *Handlist* (1964) asserted that the Skylark was still 'resident, common ... also a winter visitor and passage migrant. Winter flocks of up to 100 birds frequent the low ground in the north and east'. However, between 1970–2013 there was a decline of 60% in the UK population, with a particularly rapid decline between the mid-1970s until the mid-1980s, after which the rate of decline slowed.

This decline was evident here by the time the Atlas (1992) was published, forced largely by agricultural intensification, particularly a shift to autumn-sown cereal and consequent loss of winter stubbles, and increased application of pesticides and herbicides. The breeding population was estimated at around 14,000 pairs and the *Handlist*'s earlier reference to the Skylark as common 'no longer applies'. BBS data indicates that the breeding population declined by a further 4.5% in the period 1997–2014.

Annual breeding populations fluctuate and can be influenced by seasonal mortality rates, which may be highest in severe winters such as November–December 2010. Population declines between 2007–11 may also have been partially driven by the effect of severe rainfall during the breeding season in the summers of 2007 and 2008.

During 2008–13 Atlas fieldwork, evidence of breeding was found in 723 (83%) tetrads, compared to 764 (88%) in 1985–90, a decrease of 5%. Relative abundance is highest in the north-east but substantial

John Hawkins, North Shropshire, 26 January 2009

losses have been incurred in a broad band across the northern border and north-west, and losses far outweigh gains. This is likely to be due to a decrease in cereal crops, with pasture for heavy sheep grazing, or silage, now dominating. There were some gains in the south, but even there, losses have exceeded them. The detrimental impact of urbanisation is clearly exemplified in Telford, where they have been lost from eight tetrads, and the development of brownfield sites and open farmland for business parks and housing has undoubtedly destroyed once-suitable habitat.

Breeding activity begins with sporadic bursts of song, mainly from aerial song flights, but also from the ground or a post, as early as January. Increasingly from March they are recorded as singles, or in twos and threes, as they begin to settle on their breeding grounds and winter migrants depart. The main breeding habitat is lowland farmland and within this landscape Skylarks predominantly occupy fields of autumn and spring-sown cereal, oilseed rape, beans and similar crops, but also rough grazing. Elsewhere, unmanaged rough grassland, and grassland on moorland, commons and heathland, is occupied, often on high ground. A study on farmland at Dudmaston suggested that breeding densities were higher on ungrazed short rank permanent grass than among cereal crops (Tucker 2011). The Long Mynd Breeding Bird Project estimated a fluctuating breeding population between 1994 and 1998 of 140–230 pairs on the moorland plateau (Smith 2004). In 1997 an average density of 6.6 to 6.8 pairs per sq.km was found, substantially higher than the national average of 2.69 per sq.km in upland landscapes. An estimated 54 breeding

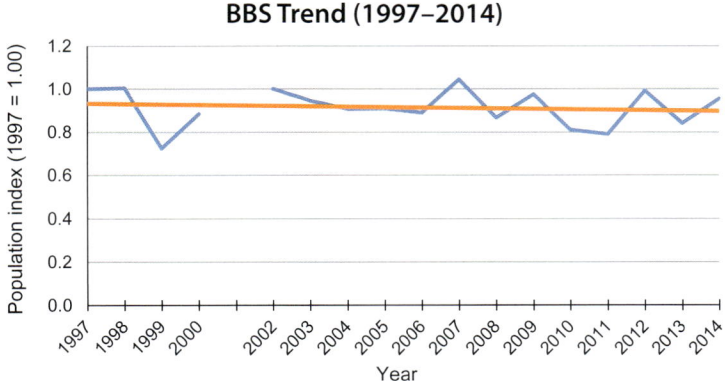

BBS Trend (1997–2014)

Breeding Distribution Change (1985–90 to 2008–13)

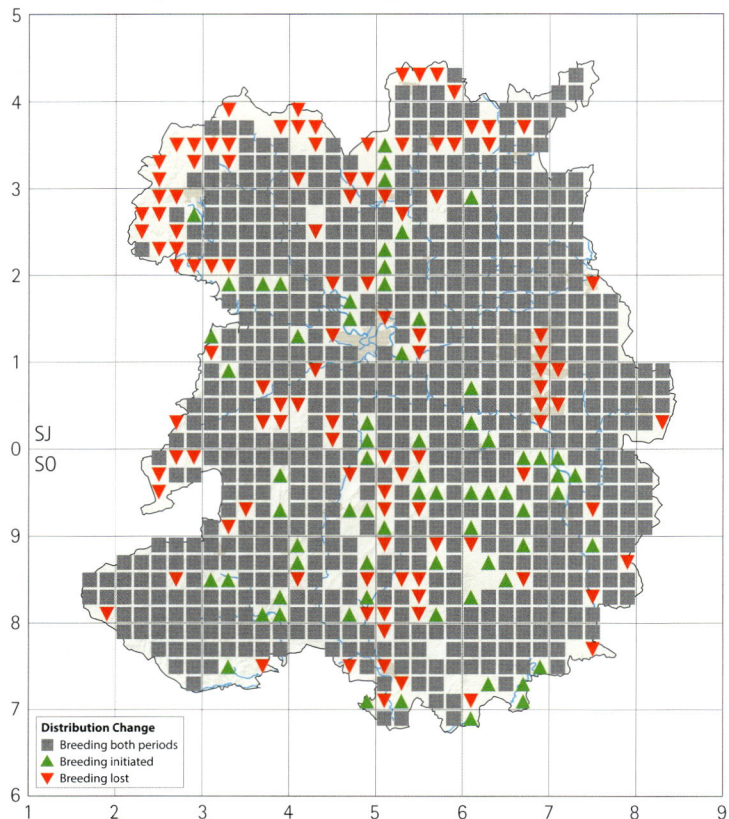

Breeding Relative Abundance (2008–13)

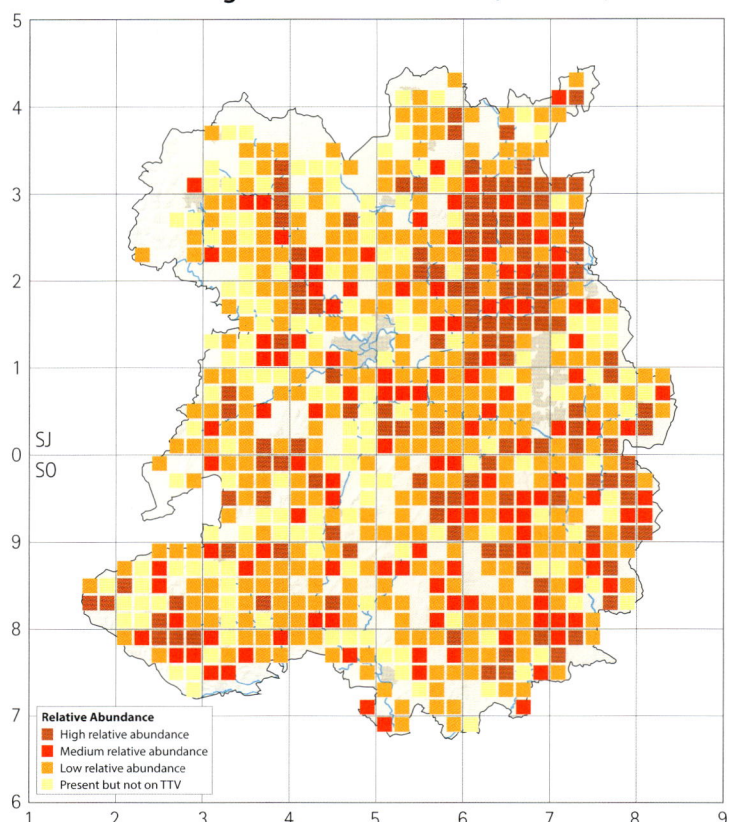

Occupied Tetrads

Atlas period (breeding)	1985–90		2008–13		Change	
	Number	%	Number	%	Number	%
Confirmed	224	26	153	18	-71	-32
Probable	377	43	406	47	29	8
Possible	163	19	164	19	1	1
TOTAL	764	88	723	83	-41	-5
Tetrads with Winter Records (2007–13): 533 (61%)						

pairs on the Stiperstones in 2002 (11 pairs per sq.km) was similarly high by national standards (Smith 2004).

Based on TTV counts the breeding population is estimated at 13,700–14,000 pairs. Given the population estimate of around 14,000 pairs in the Atlas (1992), loss from 5% of tetrads since then in the recent Atlas, and the BBS evidence also pointing to an underlying decline with annual fluctuations, the population is likely to be around the lower end of the TTV estimate, or perhaps a bit lower, in most years. It is certainly less than it was 20–25 years ago.

In winter, Skylarks exhibit a southerly migration from Scotland and an influx from northern and western Europe from mid-September to early November, and increasingly congregate in flocks from September. An adult ringed on 23 October 1981 at Westduinen, the Netherlands, was recovered dead less than three months later at Meole Brace on 10 January 1982, a distance of 479km, after being killed by a cat. This is one of only three ringing recoveries, the other two being recovered at the place where they were ringed, one and three years later.

Passage movement is noticeable from October, continuing throughout November into December. In a diurnal migration watch over the Brown Clee in autumn 1994, numbers were highest in the first half of October and 63.3% were flying in a southerly direction (Blunt 1995). 57% of SOS records in October 2009 specifically related to Skylarks passing over sites and, where recorded, in a westerly direction. There were 'few seen' after the cold weather of December 1981, probably indicative of cold weather movements and also increased mortality.

Skylarks were widely distributed during the winters of the 2007–13 Atlas, but found in far fewer tetrads (60%) than in the breeding season. The highest abundance levels were again in the north-east and east, but upland areas were largely vacated, with notable absence from the high ground of the Long Mynd, Clee hills and Oswestry uplands. The northern fringes and large urban areas, where they were absent in the breeding season, were again vacant.

Over 200 fed in stubble at Redhill, St Georges on 20 February 1961, and where it still exists, cereal stubble is an important feeding habitat during the winter, but autumn-sown crops, winter fodder, and weedy fields also provide food-grain, seeds and fresh green plant material. Counts of wintering flocks mainly below 50 in 1999 illustrated the continued problems. Flocks in excess of 100 are now unusual, and 400 at Wall Farm in November 2002, 300 at Upton Magna, and 280 feeding in stubble and an autumn-sown field at Arlescott on 14 Nov 2010 and then 560 in the same field on 28 Nov 2010, were exceptional. Most flocks recorded in recent years are now relatively small with an average flock size of 24 in 2010. TTV counts

Winter Relative Abundance (2007–13)

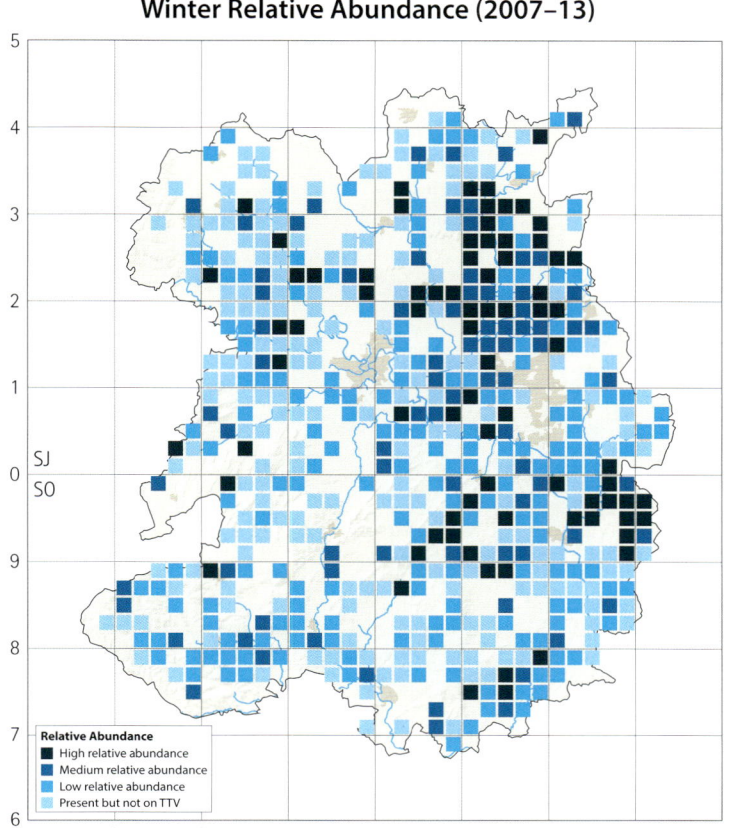

Relative Abundance
- High relative abundance
- Medium relative abundance
- Low relative abundance
- Present but not on TTV

suggest Shropshire supports an estimated 2.0 % of the British winter population, but this is unlikely.

The population decline is largely attributable to agricultural intensification, particularly an increase in autumn-sown cereals reducing the availability of winter food in stubbles and then nest sites, use of pesticides, and an increase in unfavourable habitats such as oilseed rape and intensively managed or grazed grass. Set-aside provided a welcome refuge, but this has been lost. The optimum sward height for breeding is 60cm (Donald 2004), and autumn sowing results in crops that are too tall and dense to allow late-season nesting attempts, so the number of broods per year has fallen. Crop development also forces Skylark to nest closer to tramlines, with a consequent increase in nest predation.

Agri-environment schemes can be effective. The mean monthly maxima wintering at Harper Adams University College Farm, Edgmond, increased from 0.6 in 2004–05 to 34.1 in 2011–12 following the implementation of an HLS scheme which included stubble retention, beetle banks, grassy and unsprayed headlands, and the planting of wild bird-seed mixes (Bishton 2012).

Pairs nesting in fields containing Skylark plots – small unsown patches within crop fields – raised more chicks per breeding attempt than those in conventional fields (RSPB 2012). Available as an ELS agri-environment option, in 2014 there were 21 local agreements delivering 393 Skylark plots (Denise Howe, NE *pers. comm.*). The rate of take-up must be improved and Skylark plots adopted on a landscape scale if the decline is to be reversed.

Glenn Bishton

Shore Lark (Horned Lark)

Eremophila alpestris

Vagrant

Jim Almond, Titterstone Clee, 8 April 2017

Although this lark has a global distribution, it is the Eurasian race *E.a. flava* that is a scarce winter visitor to Britain. It is largely restricted to open habitats along the east coast of England, with very few records inland. There has been just one record:

Titterstone Clee   13 March–1 May 2017

This long-staying individual favoured an area of sloping ground with tracks and other bare patches on the edge of the former quarry. It could disappear for lengthy periods of time, but where it went when it was absent was never determined.

As this was a first for the County, it is one of only a handful of records for 2017 in this publication.

Graham Walker

Short-toed Lark (Greater Short-toed Lark)

Calandrella brachydactyla

No modern records

This species of open dry areas, after wintering in Africa, rarely ventures further north than the Mediterranean countries, although it is a scarce annual migrant in Britain during spring, and especially in autumn.

A man with a lark net at the Isle, 5km north-west of Shrewsbury, captured a bird on 25 October 1841 and, presumably noticing that it was not the usual Skylark, took it to Henry Shaw in Shrewsbury. Shaw recognised the bird and subsequently passed it to Yarrell, who used it for the figure in his *A History of British Birds* (1843). The specimen passed to Lord Hill's collection at Hawkstone, and was later sold to Peplow Museum where it was catalogued (Forrest 1907). Its subsequent fate is not known.

This is one of the five species to make their first British appearance in Shropshire (see p. 2).

John Tucker

Sand Martin

Riparia riparia

Fairly common summer visitor

Sand Martins have a global distribution. They breed extensively in Europe and most winter in sub-Saharan Africa.

Eyton's statement of 1838 still holds good: 'arrives the first of the Swallow tribe'. The average earliest date over the years 1986–2014 was 16 March. The earliest date known to Lloyd (1938) was 21 March 1902; the earliest of recent years was 2 March 2003. More observers and more open waters enhance the chances of early arrivals being spotted, but it seems clear that Sand Martins are arriving earlier. Indeed, according to the data presented in the *Arrival Dates* chapter, the average first arrival date was around 19 days earlier in 2014 than in 1900.

The story at the other end of the breeding season is interesting too; it starts with a further summary by Lloyd (1938). His latest date for the period 1892–1937 was 21 October, in 1905; the latest over the 30 years to 2014 was 11 October, in 1985. Twelve records made at Dowles over the years 1904–30 indicated an average 'latest appearance date' of 29 September whereas the average latest SBR date over the 30 years to 2014 was 21 September. The difference of eight days is comparable to national data which suggest that departure dates in the 2000s were some seven days earlier than in the 1960s (Newson *et al.* 2016).

And to where are they bound? It is the western part of the Sahel region, notably Senegal, which has figured in 16 recoveries. Their route can be plotted through a further 130 recoveries involving in particular France and Spain, but also the Netherlands, Belgium, the Channel Islands, Portugal, Algeria, Morocco and Western Sahara.

Beckwith (1893) considered the Sand Martin to be 'far more numerous ... than either the Swallow or the House Martin'; that is not the case today. He noted that 'in early spring it frequents in considerable numbers, the meres and large pools' in the north, visiting them again (as it still does) after the breeding season. Perhaps Beckwith's judgement was coloured by seeing a flock similar in size to that of 'at least 2,000' feeding at the Mere (Ellesmere), on 5 April 2008; 1,000 were still present next day.

They may be even more numerous at pre-migration roosts, such as the 'immense assembly' in osiers along the Severn opposite Melverley on 5 August 1911, and the 3,000–4,000 in common reed at Chelmarsh on 15 July 2000. Itinerants, sometimes from distant locations, will visit these roosts and they visit breeding colonies too. Many have been caught by ringers: 293 from other British counties have been netted or found here; vice versa the tally is 452. These recoveries demonstrate a widespread range of origins and destinations, a range which stretches from the Highland region in the north to Kent in the south, and Anglesey in the west to Norfolk in the east. But Hampshire data suggest that only one in 200 adults and one in 50 juveniles will breed more than 99km from the colony where they were caught as breeding adults, or as recently fledged young (Mead 1979), so the youngster ringed on 9 July 1977 at Coundarbour Quarry, from where it had probably fledged, and caught a year later as a breeding adult 620km to the east, in a colony in Germany, was exceptional.

The Coundarbour Quarry, which included what are now the Cound Fishery and the SOS reserve at Venus Pool, was a significant site for the Shropshire Ringing Group, and, spurred on by the BTO's Sand Martin Enquiry, the Group, which had already ringed a total of 9,500 at various sites over the years 1965–67, achieved a staggering 9,274 in 1968. The quarry then held a colony of 400–500 pairs, while the one at Wood Lane Quarry was estimated to number 1,250 pairs, many more than any other colony counted before or since. Thereafter numbers plummeted everywhere, and over the years 1968–75 the Wood Lane colony suffered an 80% loss (Cowley 1979), as did the one at Coundarbour Quarry. A prolonged drought in the Sahel region was the major cause of the decline, from which there has never been a complete recovery. However, a series of cool, late springs in Britain, which depressed breeding success, may also have been a contributory factor, because rainfall here may have had a negative influence on over-winter survival (Cowley & Siriwardena 2005). This is in any case a short-lived species: the typical life span of an individual which reaches breeding age is only two years, so the one caught near Cound five years and 11 months after being ringed as a juvenile in Chichester, was a veteran, though it was still well short of the national record of seven years nine months.

Following the drought of 1968–69, Sahel rainfall remained below the long-term average and was particularly low in 1983–84, impacting on numbers returning to the UK that spring; it left a population that

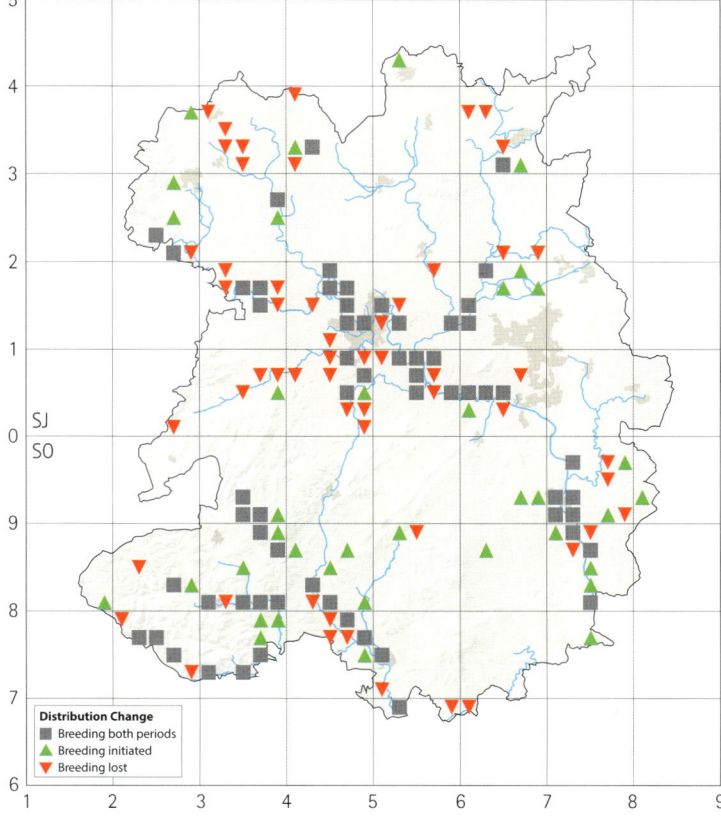

John Fielding, Shrewsbury, 14 January 2004

was perhaps less than 10% of its mid-1960s level (Mead 1984). Local evidence for this second decline comes from two sources. Firstly, the annual ringing total: this had recovered to 2,114 in 1981, but it fell to 465 in 1984 and to an 'all time low' of 184 in 1985. And secondly, counts at a sample of colonies: estimates totalling 400 occupied burrows were made in 1981, rising to 600 in 1982, but they fell to 200 in 1983 and 140 in 1984 (Mead 1984).

Given that Sand Martin numbers had reached this low ebb, it is unsurprising that during *Atlas* fieldwork 1985–90, observers reported confirmed breeding from only 16 of the County's 26 10km squares in which breeding had been confirmed during fieldwork 1968–72 for the national BTO Atlas (1976). Since 1985–90 there has been a further contraction: the breeding distribution change map shows a net loss of 22 tetrads in respect of confirmed and probable breeding. In the absence of indications that the same numbers are concentrating in fewer sites, it seems that this contraction reflects a reduction in population; furthermore, it appears that numbers have fallen at surviving colonies.

Occupied Tetrads

Atlas period (breeding)	1985–90		2008–13		Change	
	Number	%	Number	%	Number	%
Confirmed	64	7	42	5	-22	-34
Probable	18	2	18	2	0	0
Possible	35	4	39	4	4	11
TOTAL	**117**	**13**	**99**	**11**	**-18**	**-15**

Breeding Distribution Change (1985–90 to 2008–13)

Distribution Change
- ■ Breeding both periods
- ▲ Breeding initiated
- ▼ Breeding lost

A good starting point for an assessment of the population is an estimate of average colony size. Those counted since 2005 have ranged from four pairs (River Severn at Lower Wood, Town Wall in Ludlow, and River Clun near Aston on Clun) to 450 pairs (Wood Lane). The last is atypical, indeed among quarry sites that have been counted during the recent Atlas period, only those at Wood Lane, Ternhill (200 pairs) and Bromfield (170 pairs) have exceeded 100. If these three large colonies are excluded from the tally, the average of the 31 colonies at which counts have been made since 2005 is 25 pairs. Assuming that each of the 60 tetrads in which breeding was confirmed, or regarded as probable, holds one colony of this size, and an estimate of an additional 1,000 pairs is made to cover Wood Lane, Ternhill, Bromfield and any large uncounted quarry and river sites, the population would be in the order of 2,500 pairs. This is well below the estimate of 4,000 pairs made following *Atlas* fieldwork 1985–90. At that time, 22 more tetrads were occupied at the confirmed and probable level, with an estimated average of 50 pairs per colony; the post-2005 average for all colonies, including the biggest ones, is 45.

Ringers have concentrated their efforts at breeding sites and in this have benefited from the predilection of this gregarious species for establishing big colonies in sand and gravel quarries. Sand Martins have long resorted to such places, and the expansion of quarrying since 1945 has provided additional and larger sites, possibly drawing them away from some traditional haunts. However, once the faces being used are worked out, they are no longer suitable. Since 1990, the colonies at Hilton (205 nest holes in 1989), Condover (163 in 1987) and Bromfield (170 pairs in 2011) have been lost in this way.

Beckwith (1893) refers to nests in 'the limestone cliffs along Wenlock Edge, the soft sandstone rocks about Bridgnorth and Shifnal, together with innumerable railway cuttings and gravel pits'. Other cliff sites were Shelton Rough (Johnson 1910) and the sandstone cliffs at Tyrley exposed in the construction of the Shropshire Union Canal. The latter were abandoned in the 1970s and it seems that no cliff sites are occupied nowadays. In suitable substrates, Sand Martins prove adept at excavating burrows, but they are not averse to adopting ready-made sites, including chinks and holes in walls at Moreton Corbet (1900), Ironbridge (1934), Priorslee and Snedshill (both 1951), and pipes too, of which four were in occupation in 2014 in the Town Wall above the Linney in Ludlow (Gareth Thomas *pers. comm.*).

Riverbanks provide the 'classic' sites, however, and the association with rivers is very evident on the distribution map, most notably along extensive reaches of the River Severn. In the south-west, the rivers Onny, Clun and Teme are well represented too, and in the north-west the Ceiriog, Tanat and Vyrnwy all figure. At some of these sites there may be just a handful of nests, particularly on smaller watercourses such as the East and West Onny, but there are bigger groupings on the Severn, including the 95 nests between Atcham and Wroxeter in 1988, and the 96 fresh holes counted on the Uffington bend in 2011. But colonies at some of the quarry sites are much larger. Nothing compares to the Wood Lane colony at its height, but, in addition to the large counts listed above, there were 300 at Blodwel Quarry in 2000. The 30 nest holes counted in a spoil heap at Callow Quarry in 2011 was a modest tally, but demonstrated that Sand Martins will sometimes nest well away from any significant areas of water.

Riverbank sites may be lost or significantly modified by winter floods, and in response to this Sand Martins have become adaptable and opportunistic, characteristics which also serve them well at quarries, where major change is equally likely. And these characteristics lead to the use not just of walls and pipes but other artificial sites too. Examples include a large heap of black sand at Oswestry sewage works (1908), an ash tip at Buildwas Power Station (1959), the banks of a settlement pool at Allscott Sugar Factory (1962) and a freshly excavated bank on a building site at Wykey (2011). Beckwith (1893) reports on other opportunists: 'the Sparrow takes possession of the Sand Martin's nest … and in more remote places the Tree Sparrow makes use of its industry'. Both are included in the list of 16 species using Sand Martin burrows drawn up by Mead & Pepler (1975), but the House Martins seemingly nesting in a burrow at Wood Lane in 2009 suggest the possibility of a remarkable addition.

Writing in 1866, Rocke offered a warning: 'I should advise any one in search of the eggs of the Sand Martin to beware of fleas' (1866a). Two subspecies of Sand Martin flea are found in the UK: *Ceratophyllus styx styx* in southern and eastern counties and *C.s. jordani* which has a more northerly and westerly distribution. Riddoch *et al.* (1984) reported the two as overlapping in a band running between the Wash and the Severn Estuary, but sampling in 2012, including at quarry sites at Ternhill and Wood Lane, showed that, presumably as a consequence of climate change, the zone of overlap is now further north, embracing Shropshire (Hannah Wickenden, Oxford Brookes University PhD student *pers. comm.*).

Inclement weather, such as the very wet spring of 2012 and the remarkably late and cold spring of 2013, can adversely affect feeding conditions, and river levels may over-top nesting sites. On 21 May 2013, along the Leighton to Buildwas reach of the Severn, some 60 Sand Martins were seen 'recovering their nest tunnels' which had been inundated by a flash flood the previous week, when a number of nests would have held complete clutches. But such calamitous events will have tested the species on many previous occasions, not least in the spring of 1886, which was also 'remarkably cold and wet'. On 29 April Beckwith 'found several large flocks … sitting on the pebbles so benumbed with cold, that after flying a few yards they again settled around me' (Beckwith 1893), and, in the second week of May, Sand Martins 'lay dead under the banks of their nesting haunts' (Paddock 1897). Let us hope that they continue to withstand such setbacks.

Tom Wall

Jim Almond, Venus Pool, 27 April 2006

Swallow (Barn Swallow)

Hirundo rustica

Common summer visitor

John Hawkins, Dudleston, 23 July 2010

This globally distributed hirundine breeds across most of Europe and winters in southern Africa.

'Quo abis a Salopia?' were the words engraved by J.F.M. Dovaston on the piece of copper he hung round the neck of a Swallow before its autumn departure from a nest site at West Felton (Dovaston 1839). Clearly he hoped that a fellow enquirer equally conversant with Latin might catch his envoy in its winter quarters and engrave the answer before releasing the messenger for a safe return: some hope! Dovaston, an early pioneer of 'ringing', had, in his own words, been 'emboldened' by the 'kindness and constancy' of the Swallows which nested at his home. Over the preceding years he had provided them with plain wire necklaces prior to their annual migration. Some returned still bearing their neckwear, demonstrating the 'constancy' to nest sites which Dovaston had suspected. But he never saw the wearer of the copper pendant again.

It was more than 100 years before Dovaston's question, 'Where hast thou gone to from Shropshire?', was answered, thanks to a ring put on the leg of a nestling in June 1943 by Captain and Mrs Hirst, 10km away in Oswestry. This time the inscription would have

been 'Inform British Museum Nat Hist London'. The response came from the Orange Free State, South Africa, where, in March 1945, the wearer of the ring had been found dead after a storm. This was not the first British recovery in South Africa, but it appears to have been the first for Shropshire.

The Swallows ringed by Dovaston and the Hirsts probably headed off in September, but some are dilatory, like the 12 on 17 October 1997 flying high to the south with Redwings over Leebotwood, indeed the average latest recorded date over the 30 years 1985–2014 is 21 October. Interestingly, this is marginally earlier than the average of 25 October for the years 1905–1934, running counter to evidence from elsewhere in the country, where Swallows are departing markedly later than they used to (Turner 2009). Furthermore, the latest departures were recorded long ago. Among November sightings recorded over the years, the latest was on 26 November, in 1932, and the two December records are old ones too: in 1941 one found its way into Oswestry Parish Church on Christmas Eve, and in 1900 two were at Ironbridge on New Year's Eve. Such late dates are suggestive of attempted over-wintering, a phenomenon of which there are several reports from elsewhere in the country (Witherby *et al.* 1938; Hudson 1973), but, perhaps through incredulity, an extraordinary example at Plaish Hall appears to have been overlooked by the ornithological establishment in its national listings: about 30 Swallows from a colony occupying the attics of the Hall reportedly remained during the winter of 1929–30, as did 10 or 12 in 1930–31, although these were not seen after 10 January. They are said to have fed on insects in the roof space, flying out in fine weather.

Prior to departure, Swallows may gather in sizeable roosts. Paddock (1890) reported thousands roosting in an osier bed by Polly's Lock, and it was another osier bed, by Moreton Mill, that was used by 'very large numbers' in 1934, with smaller numbers the following year. There are no longer any significant osier beds, but stands of common reed and bulrush provide alternative roosting sites, and, since the introduction of mist-nets in 1956, ringers have caught many Swallows among these wetland plants. Such was the case at Dawley in 1967 when, during August and September, 11,382 were netted out of a roost estimated at its peak to have exceeded 40,000 (Julian Langford *pers. comm.*); 70 of those caught already carried rings, of which four had been fitted in South Africa. Subsequently, recoveries from this Dawley roost came from France, and 5,291km away in Nigeria, the latter less than six weeks after ringing. In 1967 the catches were so great that the kitty for ring purchases ran out, and a halt had to be called on 30 September, but an estimated 11,000 were still using the roost on 4 October. A more affordable number (fewer than 3,500), were caught there the following year, including individuals ringed as nestlings that year at Hinstock, Tibberton, Shawbury, Allscott, Claverley and Shifnal, as well as juveniles caught earlier in the year at Carnforth, Lancashire, and the Calf of Man, confirming that passage visitors were present too. Remarkably, nine Swallows from the roost went on to be reported from South Africa: Dovaston's question had received a comprehensive answer. Indeed, more than half (39 out of 62) of the recoveries relating to overseas countries have now involved South Africa.

Sizeable roosts recorded elsewhere include those at Shrewsbury Sewage Farm (1,500 on 27 August 1959), near Wellington (3,000 on 8 September 1976), Newport Canal (10,000 in August 1990) and Venus Pool (15,000 on 10 September 1971 and 1,500 on 27 August 1990). A roost site at Allscott Sugar Factory which had been occupied throughout the 1960s was barely used after 1971 before being re-occupied in 1987, when numbers peaked at 25,000 on 31 August. However, since 1990 no roost of more than 1,000 has been reported from any location. Is this suggestive of a decline in numbers, or are Swallows resorting perhaps to new, out-of-the-way sites, such as stands of maize, a crop from which roosts have been reported elsewhere (Rolls *et al.* 1977; Ford & Elphick 1993)?

Swallows continue to occupy roosts as they migrate through

Africa, and at their winter destinations, as well as on their return journeys, but these are now undertaken rather earlier. Over the years 1986–2014 the average earliest arrival date was 25 March. The earliest date known to Lloyd (1938) was 13 March, in 1913, which has since been exceeded only by 12 March, in 1993. The data presented in the *Arrival Dates* chapter suggests that the average first arrival date for Swallows advanced by around 13 days between 1900–2014.

Males generally arrive first, and the earliest get a head-start in the breeding cycle (Turner 2006), but early arrival carries risks: Swallows reaching Ellesmere on 6 April 1911 were welcomed by six degrees of frost! In 1886, later arrivals were affected too: the second week of May was particularly cold and wet and many Swallows died (Paddock 1897). These are, in any case, short-lived creatures: two years would be a typical life span for a Swallow which reaches breeding age; so the one caught at Allscott six years after it had been ringed had attained a good age, although well short of the national record of 11 years.

Once a site is chosen, the breeding cycle is soon underway. Incubation is entirely by the female and it may be that nesting within buildings serves to reduce the risk of the eggs becoming chilled while she is away foraging (Turner 2006; 2009). There are exceptions however: Lloyd (1939c) reported a nest under the eaves of his Shrewsbury home (it was a mere foot from that of House Martins); three nests in external sites were found in the Prees area in 1988; and at Upton Magna one has been occupied every year since 2009, albeit in a well-sheltered location. In southern Europe, nearly a quarter of pairs nest on the outside of buildings and more may do so here as the climate warms (Turner 2009).

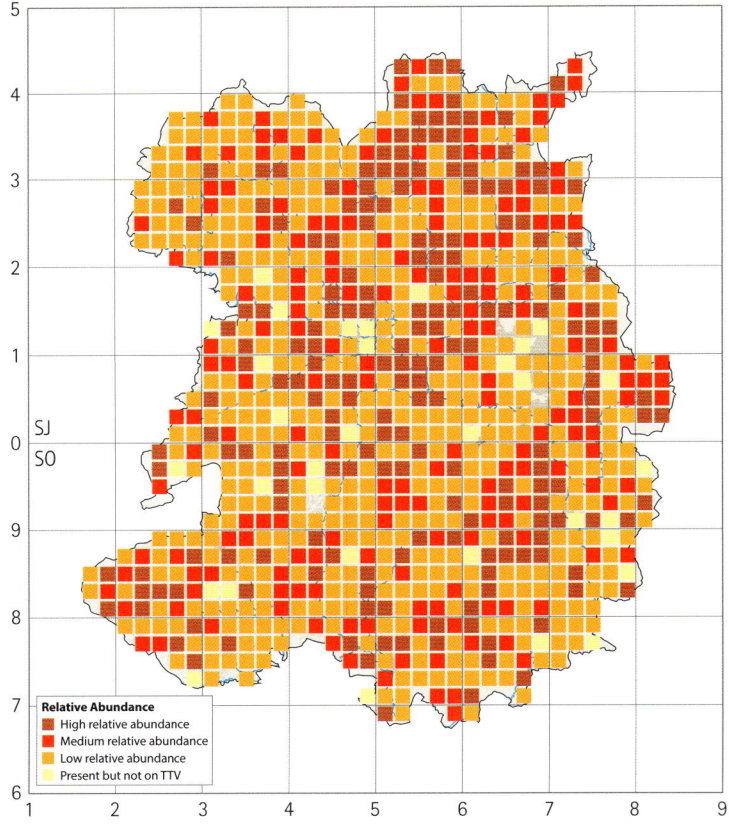

Breeding Relative Abundance (2008–13)

Relative Abundance
- High relative abundance
- Medium relative abundance
- Low relative abundance
- Present but not on TTV

Occupied Tetrads

Atlas period (breeding)	1985–90		2008–13		Change	
	Number	%	Number	%	Number	%
Confirmed	795	91	690	79	-105	-13
Probable	46	5	138	16	92	200
Possible	19	2	37	4	18	95
TOTAL	860	99	865	99	5	1

BBS Trend (1997–2014)

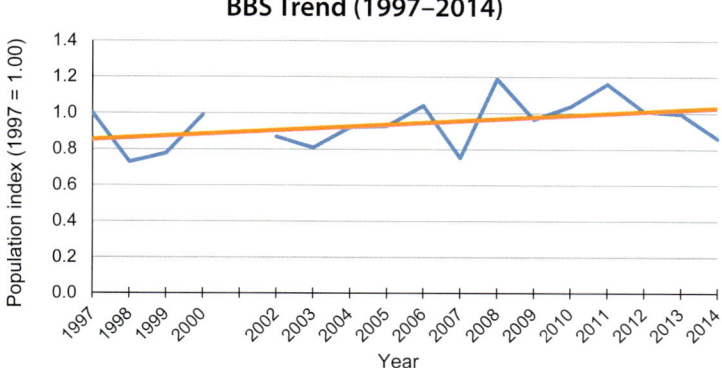

Nevertheless, inclement weather can reduce productivity and may lead to nest failures, as during cold and wet weather in July 1909, when, at Bayston Hill, a brood of four died of starvation due to the lack of insects. By contrast, a pair reared 12 from three broods in the same village in 1997; the successful rearing of three broods is uncommon, but not exceptional. Having started earlier than the House Martin, and having less onerous nest-building commitments, it is unsurprising that in general the Swallow finishes its breeding cycle sooner. The latest record of fledging was on 20 September, in 1900, whereas on occasion young Martins are still in the nest in October.

Many nests are in agricultural buildings, particularly buildings on livestock farms, and typically the birds hunt within a few hundred metres of the nest. Studies elsewhere have shown that the best habitat is grazed grassland with cattle; sheep-grazed pasture is less favoured, and while mixed farms, particularly with hedges and trees, provide excellent habitat, arable farms do not (Henderson *et al.* 2007; Turner 2009). At first sight the relative abundance map does not reflect these preferences consistently. The relative abundance map shows that there are many tetrads with high abundance in the north, and to the south of Telford, where, as shown on the Predominant Farm Types map (p. 27) there are significant areas of dairy farming. But there are many tetrads with high relative abundance between Shrewsbury

and Telford where arable farming predominates. The Farm Types map does however show some dairy farming here too, and where winter flooding in the river valleys occurs there are areas of pasture where stock-rearing may contribute to higher densities.

As to the few blank tetrads, Swallows do not penetrate deep into urban areas and it appears that Telford is large enough to exclude them. By contrast the absence of breeding records from the Long Mynd may be explained by the scarcity of buildings as well as less than ideal habitat; it is not a question of altitude because a nest site has been used at 430m on the Stiperstones.

Little is known about population levels in England before the intensification of farming, but there are no quantitative data to suggest that they differed significantly from the numbers present today (Brown & Grice 2005), and oft-expressed anxieties about significant population declines over recent decades are not supported by evidence from the population monitoring that has been carried out nationally since the mid-1960s (Robinson *et al.* 2003).

An estimate of 3,000–6,000 pairs was suggested in the *Atlas* (1992). This was based on an extrapolation from the national BTO Atlas (1976) and assumed a fairly even density across England. However, the abundance map in the national BTO Atlas (1993) showed that there was a high relative abundance in this part of the country, suggesting that the population in the early 1990s would have been above the upper end of the estimated range. Since then the population has grown: the combined CBC/BBS graph for England shows a population increase of around 60% between 1985–2009 (BTO website), and the 20% increase between 1997–2014 suggested by the trend line on the local BBS chart is consistent with this. In the light of these factors the population estimate of 14,000 breeding pairs for the recent Atlas period, based on TTV counts, does not seem out of order.

Climate change is likely to be the biggest influence on future population levels, but its impact on Swallows here and in Africa is hard to predict. Difficult questions did not end when the one posed by Dovaston was answered.

Tom Wall

House Martin (Common House Martin)

Delichon urbicum

Common summer visitor

House Martins breed across most of Europe and the northern part of Asia, and they winter south of the Sahara but in areas that have yet to be determined.

This is the last of the Swallow family to arrive, but, as with the others, average earliest arrival dates are now markedly earlier. The average earliest recorded date for the years 1986–2014 was 1 April, 11 days earlier than the estimated average date for 1900 (see the *Arrival Dates* chapter). However, the earliest ever date, 17 March, was recorded in 1886 (Lloyd 1938); the next earliest was 21 March, in 2009. These are the exceptions, and it is the second half of April before most take up residence'.

There is no marked pattern in relative abundance, and just eight tetrads where breeding evidence was wanting in both Atlas periods. Buildings are absent in one of these, on the Long Mynd; they are

very sparse in a further four, and these may not offer suitable nest sites. The other blanks will indicate either a genuine lack of House Martins or observer oversight.

A key factor determining the choice of nest sites is the ready availability of mud for nest construction. It has been suggested that this needs to be sourced within 200m of the nest site; any further and the mud will dry out in transit (Piersma 2016). The margins of running and standing waters are an important source; good use is also made of muddy puddles, which are now rare in urban areas but will doubtless have been widespread in the past.

An estimate of 7,000 pairs was given in the *Atlas* (1992) but this now looks to have been too low. It was based on an assumption of a national average of two nests per sq.km. The population estimate based on TTV counts in 2007–11 is 12,700–13,300 pairs, indicating

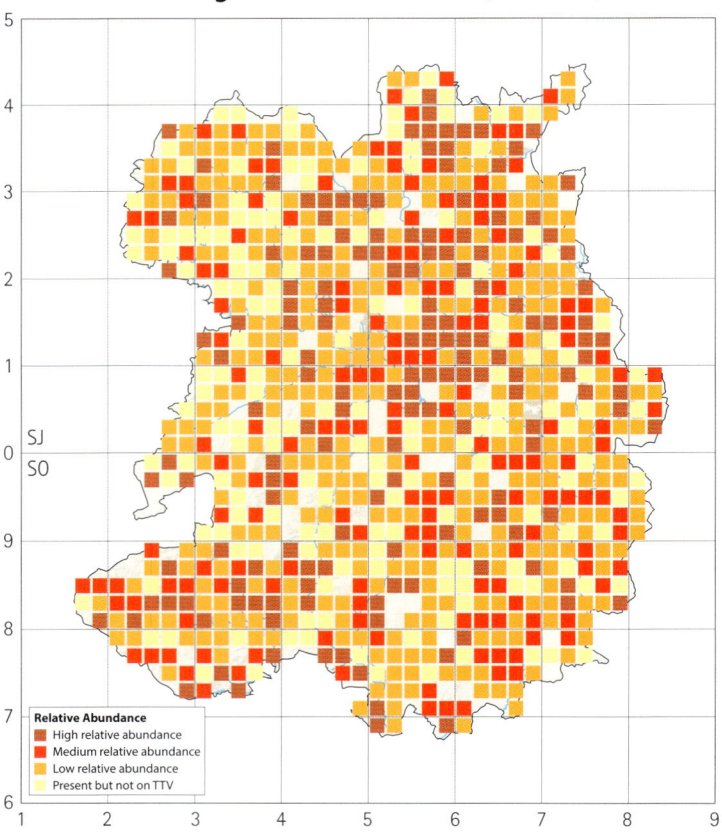

Jim Almond, South Shropshire, 22 August 2013

Breeding Relative Abundance (2008–13)

Relative Abundance
- High relative abundance
- Medium relative abundance
- Low relative abundance
- Present but not on TTV

an average of four pairs per sq.km. Most evidence points to a decline: in 2008–13 breeding evidence was recorded in 19 fewer tetrads than for the *Atlas* fieldwork of 1985–90, and the reduction in tetrads with confirmed or probable breeding was greater. National data, though sparse, suggest that numbers may have fallen by as much as 65% since the early 1980s (Baillie *et al.* 2014); and BBS points to a 36% decline in the West Midlands over the period 1995–2014. Local BBS data, albeit based on a small sample which might be unreliable, suggest a population that, while fluctuating widely, is effectively flatlining.

Time will tell, but declines have been feared in the past, when the finger of blame has sometimes, as elsewhere in the country, been pointed at the House Sparrow, including by Forrest (1908): 'Of late years [the House Martin] has diminished considerably, due in part to the House Sparrow taking over its nests'. It may now seem implausible that this behaviour should have an impact at the population level, but House Sparrow numbers increased dramatically during the eighteenth and nineteenth centuries (Holloway 1996), and they were doubtless far

Occupied Tetrads

Atlas period (breeding)	1985–90		2008–13		Change	
	Number	%	Number	%	Number	%
Confirmed	758	87	615	71	-143	-19
Probable	27	3	118	14	91	337
Possible	47	5	80	9	33	70
TOTAL	832	96	813	93	-19	-2

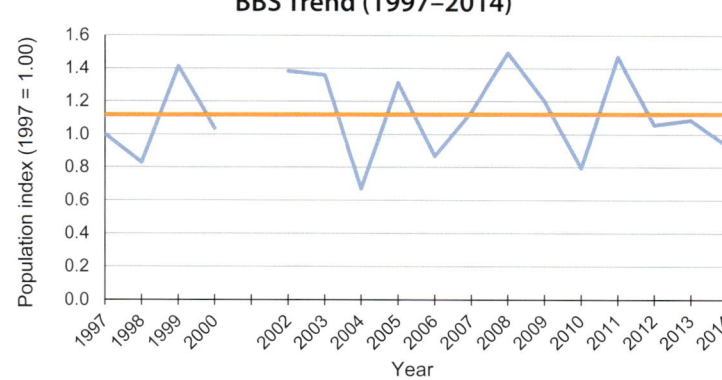

more numerous than today. Interestingly, the *Handlist* (1964) noted that nests at the important Atcham Bridge colony were not usurped by House Sparrows, speculating that this was because the nests were over water, but Wrens and Pied Wagtails were not deterred, as both were noted occupying nests there. Reports have come from elsewhere of Tree Sparrows and Spotted Flycatchers occupying Martins' nests, but it is likely that these other species took over unoccupied nests rather than evicting the original occupants.

Short-term declines may be a consequence of extreme weather conditions. 'Wrecks' have been found in southern Africa, probably involving House Martins from central Europe (Hill 2002), and Paddock (1897) reported that 'the severe weather during the second week of May 1886, will long be remembered by ornithologists for the great destruction of life among the summer migrants. I saw 16 dead and dying House Martins taken one morning out of the draught hole … at Old Caynton Mill where they had crept for shelter'. But even in more normal circumstances the typical lifespan of a House Martin that reaches breeding age would be only two years.

While small groups of nests are the norm, large colonies do occur. Their fortunes are said to be a very unreliable guide to population trends over a wider area (Marchant *et al.* 1990), but the demise of the largest colonies is disquieting. In 2014 there were no nests on the Belvidere Railway Bridge, where there was a 'large colony' in 1935; none on the old Atcham Bridge, where there had been 335 in 1963; none at Haughmond Abbey, where 53 pairs were reported in 1973; and only eight intact, and possibly occupied nests, on the Chirk Railway Viaduct (shared between Shropshire and Wales), where Forrest (1907) reported 'numbers of nests', and 114 were counted in 1966.

The history of the colony occupying the old Atcham Bridge has been researched by Tucker & Wright (2014). In 1908 Forrest reported 'an almost continuous row of nests from end to end' of the bridge; this suggests far more nests than the estimate of 100 made in 1938 (SSOS 1939; they are presumed to have been on the old rather than the new bridge, which was opened in 1929). The colony on the old bridge numbered 168 nests in 1957 growing to a maximum of 335 in 1963, when it may have been the largest in the country. A decade later only 50 pairs were present, and although numbers rose to 159 nests in 1985, they declined to just 10 in 1995. The site was deserted thereafter, but its history overlaps that of an individual nest on the nearby stable block at Attingham Hall which was reported in 2011 as having been occupied for 22 of the previous 23 years.

House Martins show much flexibility in site selection. A nest at Kempton in 2014 was a mere 2.1m (6ft 11in) above the ground (*pers. obs.*); in marked contrast, Beckwith (1893) reported nesting 'under the square block of stone that supports Lord Hill's statue in Shrewsbury 116 feet [35m] from the ground'. However, there were usurpers there too, and Forrest observed in 1908 that 'to attain such a height required strenuous efforts on the part of the sparrow'. Such dispossession seems particularly iniquitous when it is realised that some 1,000 pellets of mud are needed to build a nest (Cramp 1988). On occasion, however, these most consummate of architects take a short-cut: at Whixall in 1966 a pair built no nest but used a gap in the eaves instead, a ploy also adopted at Prees Parish Church in 1988; and in 2009, at Wood Lane, House Martins appeared to be nesting in a Sand Martin burrow.

Many householders welcome these entertaining visitors, indeed some encourage them, and for many years so-called House Martin 'nest boxes' – they are in fact ready-made nests – have been used in a number of places, including at Hinstock, where, in 1977, a colony consisted of five pairs in surrogate nests and six in natural ones. Other householders are less welcoming, and at Adderley in 2003 several nests were destroyed, an illegal act if the nests were occupied at the time. And the season can be a long one: there have been many reports of young still being fed in the nest into October, the latest being at Isombridge in 1966, where two adults were still feeding young on the 19th.

Large numbers are sometimes recorded at the end of the breeding season, but unlike Sand Martins and Swallows, House Martins do not habitually gather in big late-summer roosts. In fact, little is known of the roosting habits of House Martins, and there is uncertainty as to whether, in general, they roost terrestrially or aerially, in the breeding season (other than when using their nests) or in their wintering areas

Paul King, Acton Burnell, 17 June 2010

(Hill 2002), so the record in the late summer of 1933 of a mixed House Martin and Swallow roost in an osier bed by Moreton Mill dam is of particular interest. The following year 'very large numbers' returned, mostly House Martins, roosting there nightly in August and September. Gatherings of this size are not often reported, exceptions being more than 500 at Shrewsbury Sewage Farm on 4 September 1985, 1,000 feeding in the Cardingmill Valley on 22 September 2000, in excess of 2,000 on wires at Wellington on 15 September 1983, and 2,000–3,000 hawking over maize at Walford College on 3 September 1991. These large September counts were presumably of flocks pausing while on passage, and it is also in the autumn that the only records of sizeable numbers actually on the move have been made: 300 flying south at Cosford on 2 September 1988, 600 south-west at Chelmarsh on 11 October 1987 and 1,250 travelling south along the Severn near Hampton Loade in one hour on 2 October 2006.

Over the 30 years to 2014, the average latest date recorded was 17 October; interestingly it was 18 October for the years 1905–1934. The latest date cited by Lloyd (1938) was 10 November, in 1911 (he overlooked a record on 21 November 1906), but in 1941 one was seen on 10 December. Since then the latest dates recorded have been 13 November, in 1986, and 21 November, in 1979.

Twenty-eight of the 40 recoveries of locally ringed House Martins showed little in the way of movement, suggesting a measure of site fidelity. But despite having been our close and confiding companions from spring through to autumn, we then lose track of them through the winter. Gilbert White reflected in 1789 that 'after all our pains and inquiries, we are yet not quite certain to what regions they [House Martins] do migrate' (White 1789). Two hundred and fifty years later, despite many more enquires, uncertainty persists. Records of ringed individuals at coastal sites in Dorset, Sussex and Kent suggest that, initially at least, our House Martins migrate on a fairly broad front. Overseas, there has been one recovery in central France, and one ringed on the Wrekin in 1970 and trapped the following year in Algeria was the first recovery in Africa of a House Martin from Britain and Ireland. It is known that from North Africa they head on to winter south of the Sahara, but although over 390,000 House Martins have been ringed in the British Isles, there is only one recovery south of the desert: ringed in Hertfordshire in 1983 it was trapped in Nigeria the following winter. So, although in many ways very familiar, the House Martin remains something of a mystery.

Tom Wall

Red-rumped Swallow

Cecropis daurica

Vagrant

The breeding range extends from the Iberian Peninsula in the west to Japan in the east. While there are some sedentary populations, most are migratory, with the southern European and North African population wintering south of the Sahara. Numbers reaching Britain appear to be increasing, with the majority occurring in spring along the south and east coasts (Fraser 2013b). Inland sightings are much rarer, but the dates for the two records are typical:

River Severn, Monkmoor, Shrewsbury   30 April–5 May 1978
Priorslee Lake, Telford   27–28 April 2003

Both were located in flocks with other hirundines and, while the first was the only one reported that year (Rogers *et al.* 1979), the second occurred at a time when there was a small influx into the country (Rogers *et al.* 2004).

Graham Walker

Cetti's Warbler

Cettia cetti

Rare, non-breeding, resident

Although Cetti's Warbler first colonised Britain in the late 1960s and early 1970s (Sharrock *et al.* 1975), the bulk of the breeding population remains to the south and east of a line between the Severn and Humber estuaries. There are small populations further north, such as on Anglesey and in north Cheshire but, perhaps surprisingly, breeding has yet to be proven here, particularly since there are numerous wetlands that would appear to provide suitable habitat.

There are just 14 records, although birds at Priorslee Lake (2015–16) and Chelmarsh (2016–17) may have been continuously present, but went undetected (or unreported) for certain periods. The first, at Fenemere (14–27 December 1975), occurred not long after the initial colonisation (Dymond *et al.* 1976), but there was then a gap of 32 years until a long-staying individual at Shrewsbury Sewage Farm (29 October 2007–3 July 2008) put in an appearance. The

wait for the third at Chelmarsh (30 October–5 December 2009) was considerably shorter (15 months), while it was another two years before the fourth was heard, but rarely seen, at Venus Pool (14–26 September 2011).

From 2014 onwards, however, it has been recorded annually with singles at two sites that year, one site in 2015, four sites in 2016 and three sites in 2017. These sites were in the east and included Chelmarsh, Norbroom Marsh, Priorslee Lake, Shifnal and Wall Farm. Most were present during the autumn and winter, but the individuals at Priorslee Lake (2016) and Norbroom Marsh (2016–17) were present well into the breeding season. If it has not already done so, this is a species we can expect to breed in the near future, providing there is not a return to hard winters.

Graham Walker

Long-tailed Tit

Aegithalos caudatus

Common resident

Long-tailed Tit has a very large range across Eurasia, but excluding the Indian subcontinent. Here, it is a common resident which may be present in any type of woodland, but is most frequent in thick, thorny hedges and scrub, disused gravel pits, quarries and commons with thorny bushes, and open canopy deciduous woodland.

Leighton (1836), in the first historical record, noted 'long-tailed titmouse' in 'hedges and bushes' and in 1837 Hulbert said that 'the whole of the tribe of Tomtits are constantly flitting about among the trees in the garden'. Beckwith (1879) described it as common, with family parties often seen, and although they were believed to lay up to 24 eggs, he had 'examined great numbers of their nests and never found more than thirteen', although a nest containing 16 eggs was reported in 1890. Nests were found in various prickly bushes, particularly gorse, rose and hawthorn. Several local names were mentioned – 'Canbottle' (Cathrall 1855; Beckwith 1890), 'Huggen muffin' (Eyton 1838), 'Canbottlin' (Forrest 1899c), and 'Feather poke' (Forrest 1909c); these are likely to have been derived from the nest shape (like a soda-water bottle of the time) and material (many feathers).

The vulnerability of these tits to cold winters was recorded in 1917, when they were 'exterminated in many districts by frost', and in 1918, when the first 'since the severe frost of early 1917' were seen at Dowles Brook in February. After a recovery during the 1920s,

Terry Arch, Venus Pool, 16 November 2012

another severe winter in 1939–40 had a serious effect on numbers, followed by another recovery recorded in 1943.

Early SBRs listed it as common and widespread, and recorded the regular sight of family parties. SBR (1960) noted about 60 at Clarepool Moss on 16 August in a flock of about 200 other tits and Goldcrests. The *Handlist* (1964) described it as resident and common throughout, and often a dominant member of autumn and winter tit flocks. It noted that they are particularly susceptible to severe winters, such as the long, hard winter of 1962–63, but SBR (1964) commented that 'many reports were received showing recovery in numbers during the year'.

Subsequent SBRs contained regular accounts of larger flocks of up to 20 in many locations, but a count of 200+ near Shifnal in December 1973 was exceptional. SBR (1986) stated that is was 'not the most numerous tit, but always the most reported', while SBR (1987) noted new behaviour, taking peanuts from a bird table at Bayston Hill, from a 'nut basket' in Albrighton (near Shifnal); and, in 1988, two were seen on a garden feeder in Snailbeach, three relatively widespread locations. By 1992 the 'trend ... appears to be escalating'. Larger counts outside the breeding season included 51+ between Pimley Manor and Uffington, and 35 at Cosford, in 1989, 30+ at Cole Mere in 1992, and 40 at Monkmoor Pool in 1997. Larger flocks were still

Breeding Distribution Change (1985–90 to 2008–13)

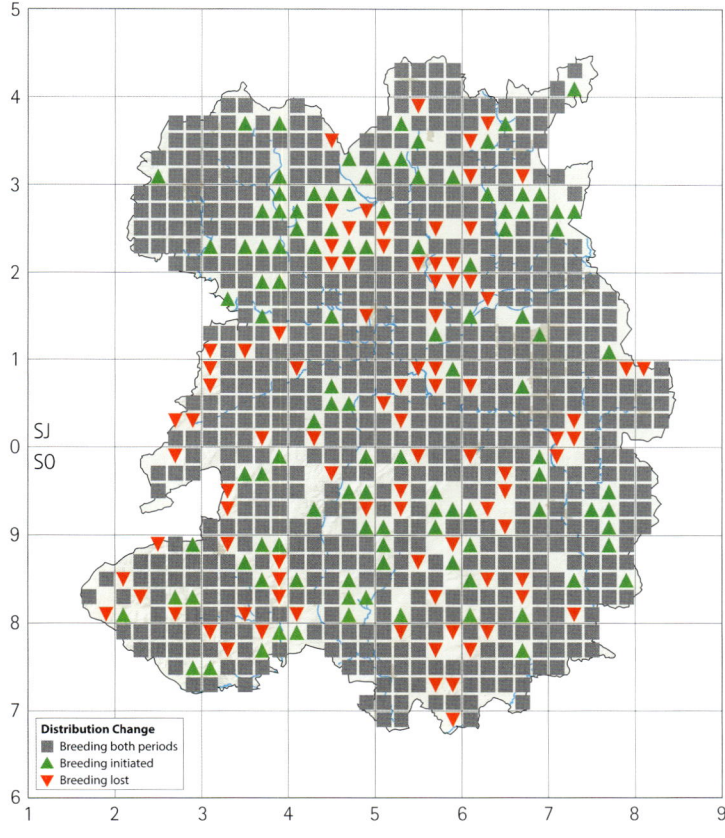

Distribution Change
- ▪ Breeding both periods
- ▲ Breeding initiated
- ▼ Breeding lost

Occupied Tetrads

Atlas period (breeding)	1985–90		2008–13		Change	
	Number	%	Number	%	Number	%
Confirmed	478	55	475	55	-3	-1
Probable	147	17	162	19	15	10
Possible	102	12	111	13	9	9
TOTAL	727	84	748	86	21	3
Tetrads with Winter Records (2007–13): 833 (96%)						

Breeding Relative Abundance (2008–13)

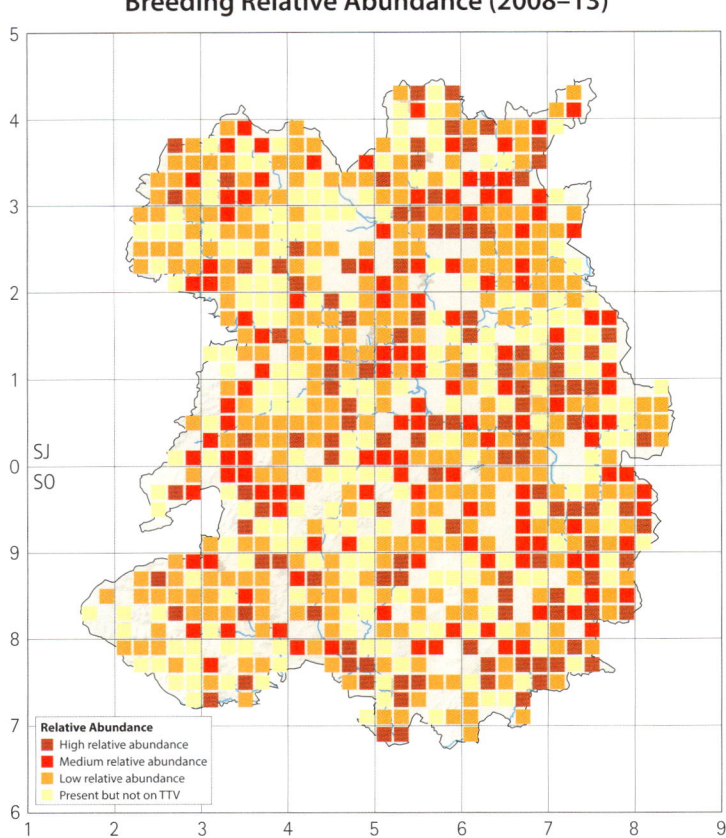

Relative Abundance
- High relative abundance
- Medium relative abundance
- Low relative abundance
- Present but not on TTV

BBS Trend (1997–2014)

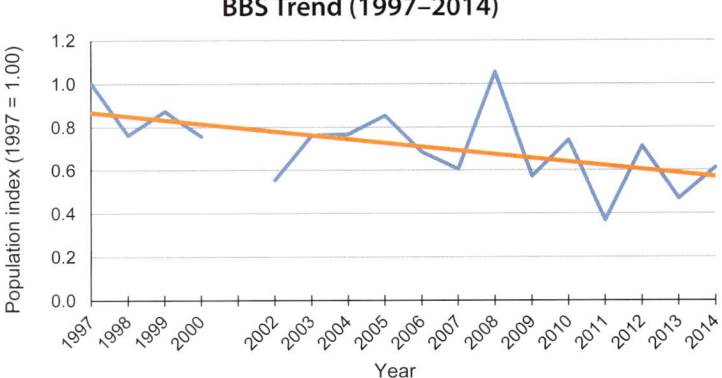

winters of 2009–10 and 2010–11. Here, the winters in 1985–86 and 1986–87 also had a severe impact.

Relative abundance and population trends vary considerably across the country, with this County having a fairly high relative abundance, and change in relative abundance since 1988–91. However, BBS shows a decline in the West Midlands of 15% between 1995 and 2014, rather less than the local decline of nearly 40% since 1997.

Estimation of the breeding population is complex. In addition to the marked fluctuations caused by harsh winters, each territory holds an adult pair plus 'other adult helpers', thought to be a result of high nest predation, and so estimates are made of territories, rather than pairs.

In 1990, based on an estimate of 200,000 territories in Britain, an average of about 100 per 10sq.km, it was estimated that there were 2,500 to 3,700 territories. Based on TTV counts, the estimated breeding population in 2010 was 6,700–7,000 territories, suggesting that the earlier estimate was too low, perhaps because the national estimate on which it was based was too low, or the density here was higher than the national average.

In winter they were more widely distributed, being present in 833 tetrads (96%), reflecting the large number of juveniles in the early winter population, as well as foraging outside the breeding areas. They just make the BTO Garden BirdWatch top 20, with a reporting rate since 2006 of 23%.

Ringing records show that the vast majority of those ringed here are also recovered here. Only 13 movements involve different counties, mostly adjacent ones, and none involve overseas movements, confirming their sedentary nature.

This sedentary nature is illustrated even more clearly by the oldest individuals. All were ringed as full-grown birds and caught again at the original ringing site, one in Market Drayton in December 2014, 7y 2m 28d later, one at Hawkstone Park in February 2009 and again in February 2010, the latter occasion 6y 1m 19d after ringing, one in Sutton (Shrewsbury) 5y 9m 29d later and another in Hawkstone Park, caught in November 2003, April 2005, January 2006 and May 2008, the last occasion 5y 6m 1d after ringing.

However, a particularly adventurous individual travelled all of 6km and was killed by a cat 5y 11m 5d after ringing, while just two, both full-grown when ringed, travelled more than 100km, one in Shifnal on 12 January 1997 was caught near Swansea on 15 January 2000, 3y 3d later, a distance of 159km SSW, and the other ringed in Northumberland on 14 October 2012 and caught on Whixall Moss on 2 November 2013, 1y 0m 19d later and a distance of 278km SSW.

Helen J. Griffiths

recorded regularly, usually comprising eight to 20, but larger flocks of 30 or more include 46 trapped and ringed at Whixall Moss on 16 October 2010, 37 at Priorslee Lake in July 2010, 31 on the Wrekin in January 2011, 40 at Venus Pool in October 2013, and 30 at Attingham Park in October 2013.

In fieldwork for the recent Atlas, they were found in 748 tetrads (86%) in the breeding season, compared with 727 tetrads (83.6%) in the 1985–90 Atlas. However, the harsh winters in the first years of the 1985–90 fieldwork and the middle years of the 2008–13 fieldwork were likely to have had an impact on numbers, affecting the likelihood of this less common tit being found in both periods, and it may have been overlooked in some other tetrads. Local changes in suitable breeding habitat may also have occurred. These factors are reflected in the breeding distribution change map, with losses and gains almost equal in number.

The Atlas (1992) commented that the gaps in distribution 'correspond to towns, higher ground and arable farmland', and this is still the case. The relative abundance map shows that distribution is patchy, and higher densities are associated with mature deciduous woodland, scrub and mature tall thick hedges. The pattern of tetrads with low density immediately adjacent to ones with high density may indicate local differences in habitat, with suitable breeding habitat adjacent to unsuitable farmland, upland or conifer woodland.

It undergoes wide fluctuations in numbers between breeding seasons in the UK, suffering heavy mortality when winters are severe, but usually recovering quickly due to its high breeding potential. In addition to the early 1960s, numbers were low after a series of relatively cold winters in the late 1970s, and after the recent severe

Willow Warbler

Phylloscopus trochilus

Scarce summer visitor

Willow Warbler is one of our commonest summer visitors, a long-distance migrant from wintering grounds in sub-Saharan Africa. Although there are usually a few late March records, most arrive in the second week of April, announced by the beautiful fluent song of wistful descending notes, heard from most areas with small trees, tall bushes or other scrubby habitats. Willow Warbler avoids areas with a closed canopy, though it can be found along rides or on the margins of woodland.

Once known as the Peggy Whitethroat and the Wood Wren, it was formerly extremely common. It was even considered by gardeners as a pest species in the nineteenth century, said to damage fruit and young peas, though Beckwith correctly believed they were more likely to be picking insects off the vegetation (Beckwith 1889).

There does not seem to have been any major change in numbers into the twentieth century. The *Handlist* (1964) simply noted that Willow Warbler was very common, with no further detail given, while five years later it was described as 'plentiful in all areas, the commonest warbler' (SBR 1969).

Willow Warbler remains widely distributed and found in the vast majority of tetrads. However, comparison with the 1992 Atlas shows a 16% contraction in range, with particular losses in the north-east. In these areas the loss of scrubby habitats and hedgerows associated with agricultural intensification, along with a reduction in insect food, may be factors explaining this trend. The large increase in the number of tetrads with only possible breeding evidence suggests that they were also harder to find, reflecting a reduced population in the tetrads where they still occur.

There are notable patterns of abundance with highest concentrations occurring in the south-west in the Shropshire Hills, where they frequent scrubby hillsides and copses. The Oswestry uplands, Brown Clee and Titterstone Clee also show up clearly as areas of high relative abundance. These less intensively managed habitats, with a higher availability of insect-rich scrubby habitats, may explain this pattern, together with perhaps a preference for cooler, wetter, climates. Apart from Wyre Forest, there are no large areas of high abundance in the lowlands.

Occupied Tetrads

Atlas period (breeding)	1985–90		2008–13		Change	
	Number	%	Number	%	Number	%
Confirmed	554	64	217	25	-337	-61
Probable	218	25	240	28	22	10
Possible	77	9	255	29	178	231
TOTAL	849	98	712	82	-137	-16

Breeding Distribution Change (1985–90 to 2008–13)

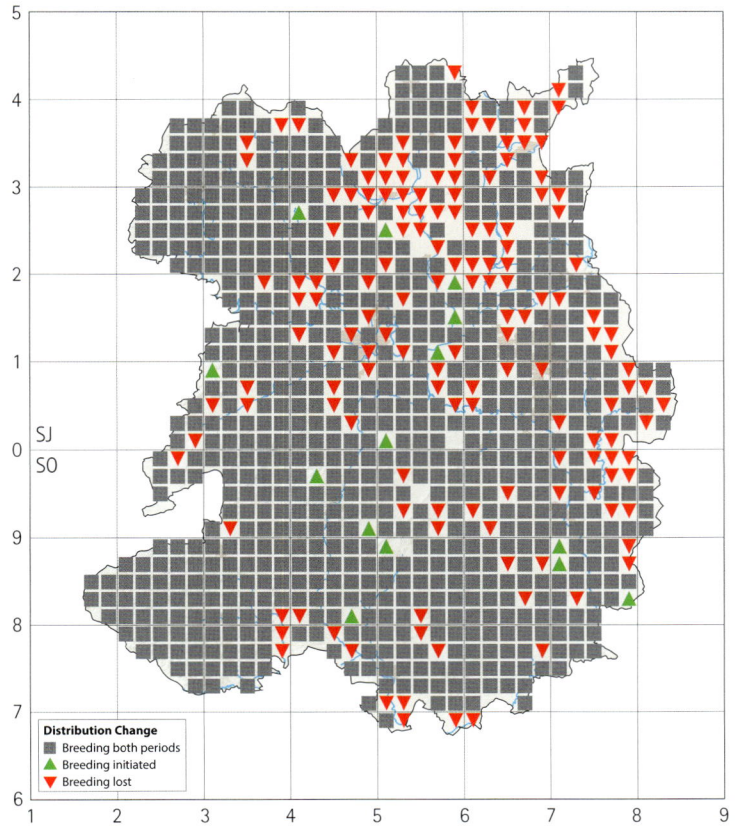

Distribution Change
- Breeding both periods
- Breeding initiated
- Breeding lost

Breeding Relative Abundance (2008–13)

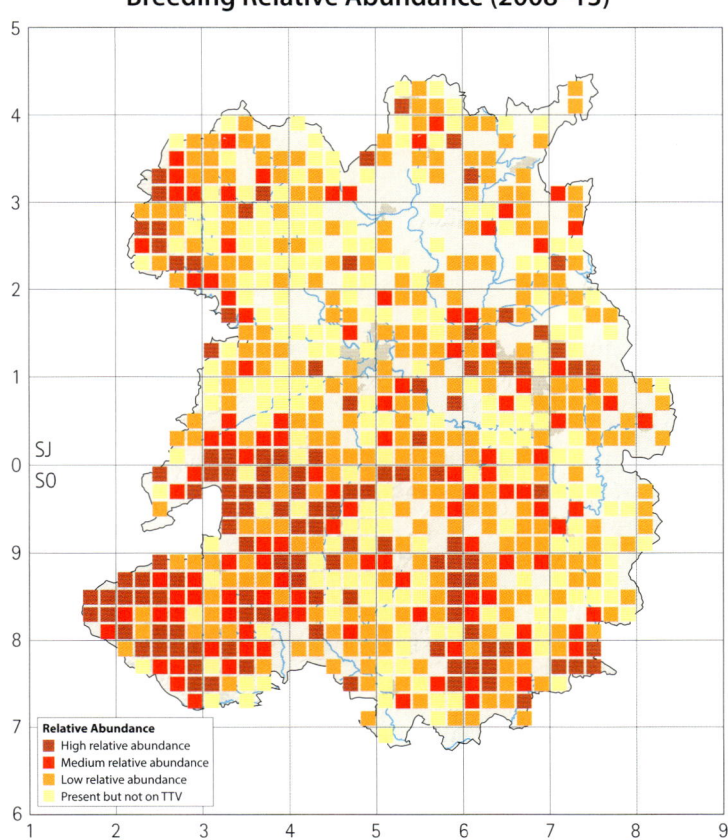

Relative Abundance
- High relative abundance
- Medium relative abundance
- Low relative abundance
- Present but not on TTV

Jim Almond, Stiperstones, 2 May 2009

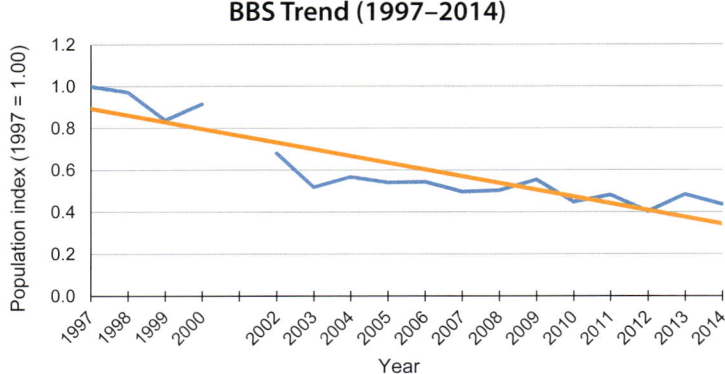

The declines in range are matched by the BBS trends, with Willow Warbler having suffered a decline of 56% since 1997. The population estimate, based on TTV counts, is 13,500–13,800 territories. This is considerably below the 40,000 estimated in the *Atlas* (1992), but, if there was a similar rate of decline between 1990 and 1997, the two population estimates are consistent. Though it used to be the most common warbler, that is no longer the case, with Chiffchaff now being the more common of the two species, as described more fully in that account.

During the Atlas period there was a winter record, from Minsterley on 8 November 2012, accepted by the Rarities Committee, though it is not known if this was a late migrant or a bird attempting to over-winter.

Ringing recoveries suggest many return to the same area year after year. The oldest, ringed as a first-year in July 1974 in Wellington, was found one day short of five years and 11 months later, only 2km from the ringing site. Twenty-four have shown movement between here and 18 other widely-scattered counties, either being ringed here and recovered elsewhere or recovered here after being ringed elsewhere. They include two in Scotland and two on the south coast, suggesting these were mainly passing through. They included a nestling ringed at St Andrews in Fife in June 1986 and then found dead the following summer in Ludlow, 443km distant, and an adult ringed in August 1987 at Walcot and found the following June at Inverlochy in Highland, some 485km away, although in both cases the bird presumably made the 5,000+km round trip south of the Sahara.

Three have been recovered abroad. The most interesting is one ringed at Cressage in July 1980, which was subsequently caught on the German island of Heligoland, the following April, some 718km east-north-east. Two others have been recovered in France, both over 500km to the south.

Additionally, two ringed in the Channel Islands have been subsequently found here. The most interesting was a full-grown female ringed on Guernsey on 26 April 1985, which was hit by a car near Clun just six days later on 2 May, 331km from the ringing site.

None of the recoveries has involved a country to the south of France, although the wintering area is well beyond that.

Mike Shurmer

Chiffchaff (Common Chiffchaff)

Phylloscopus collybita

Very common summer visitor, scarce winter visitor

Though usually regarded as a summer visitor to northern and temperate Europe and Asia, mostly from wintering grounds in the Mediterranean basin and North Africa, an increasing number of Chiffchaffs are now recorded here in the winter period. They are common, widespread and familiar, with the distinctive disyllabic song heard from mid-March onwards, primarily in mature deciduous woodland with undergrowth, overgrown hedgerows and mature gardens.

In the nineteenth century, Chiffchaff was noted as being very common in woods, copses and shrubberies, although not as plentiful as Willow Warbler (Beckwith 1889). It was said to be found where there were trees of a moderate height and thick undergrowth.

The breeding population is widely distributed, and common wherever it is found, apart from in some arable farming districts and upland areas where there is less suitable habitat. It has shown an increase in range of 96 tetrads since 1985–90, and was recently found in all but four tetrads. There appears to be no obvious pattern in relative abundance.

Numbers in the breeding season have shown a steady increase since 1997. BBS data reveal that the population has more than doubled over this period, and it has now overtaken Willow Warbler as our most common leaf warbler. All assessments up to and including the *Atlas* (1992) considered Willow Warbler to be more common, but a

BBS Trend (1997–2014)

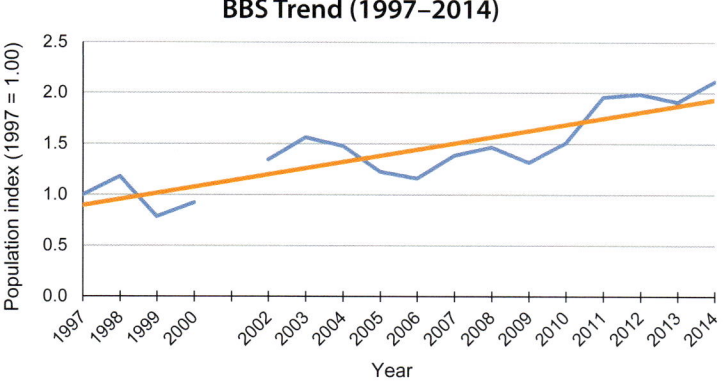

the winter, but, unlike Blackcap, availability of invertebrate food near water is also important, and few visit gardens to feed. In the four winter months during the whole of the recent Atlas period, the GBW reporting rate (see p. 70) for Chiffchaff was only 3% of the reporting rate for the four months of the breeding season, and less than 1% of that for wintering Blackcap.

Wintering Chiffchaffs in Britain appear to include some that have bred in this country, larger numbers of the nominate race from elsewhere in Europe, and a few individuals of one of the eastern races. One showing characteristics of Siberian Chiffchaff (*Phylloscopus collybita tristis*), which breed in the Taiga forest from the Urals eastwards, was present at Venus Pool from 11 February to 9 March 2014, although reports of its calls were not conclusive and the Rarities Committee did not think it safe to accept a non-typical individual. However, the taxonomy of Chiffchaffs, and the criteria for their identification, is still evolving so the record may be revisited in future. There was no doubt about the one caught, ringed and photographed in the hand, and which called on release, at Whixall Moss

corresponding decline of over 50% found by BBS in the same period means this is no longer the case. The Chiffchaff population estimate, based on TTV counts, is 27,900–28,400 territories, making it now twice as numerous as Willow Warbler. During breeding season TTVs, 2.44 times as many Chiffchaffs as Willow Warblers were counted (7,337 and 3,001 respectively).

There is no mention in the historic literature of Chiffchaffs being seen in the winter period, and wintering was not mentioned in the *Handlist* (1964). The first winter record was at Ludlow on 18 February 1973, with further records following in the immediately subsequent years. It is still scarce at this time of year, but was found in 34 widely scattered tetrads in the recent winter Atlas period, all except two in the north or east. Like the more numerous and widespread wintering Blackcap, there is an apparent preference for river valleys, suggesting that it seeks areas which are slightly warmer than average to spend

Occupied Tetrads

Atlas period (breeding)	1985–90		2008–13		Change	
	Number	%	Number	%	Number	%
Confirmed	219	25	339	39	120	55
Probable	361	41	439	50	78	22
Possible	190	22	88	10	-102	-54
TOTAL	770	89	866	100	96	12
Tetrads with Winter Records (2007–13): 39 (4%)						

Breeding Distribution Change (1985–90 to 2008–13)

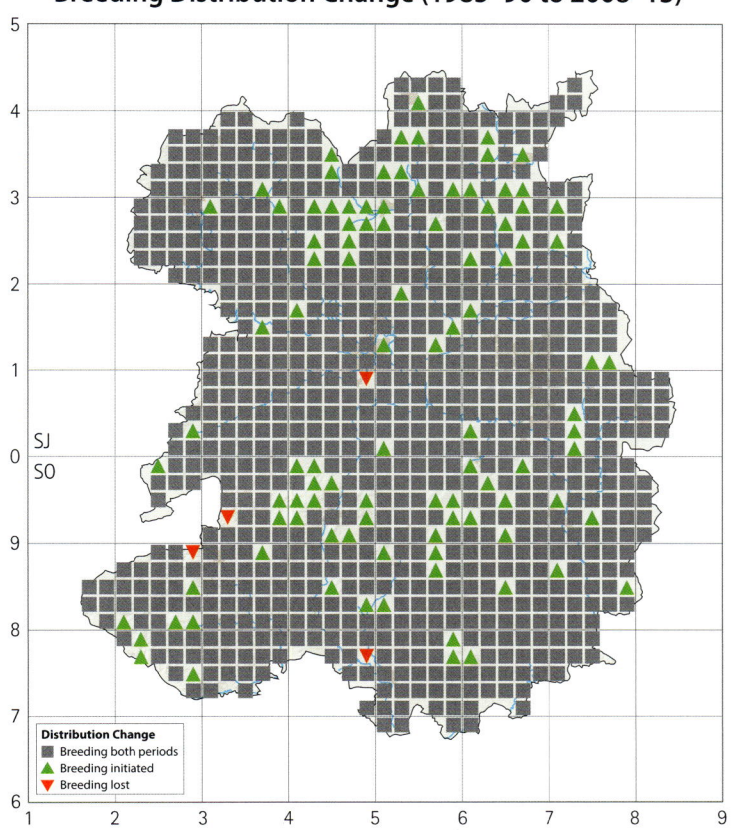

Winter Distribution (2007–13)

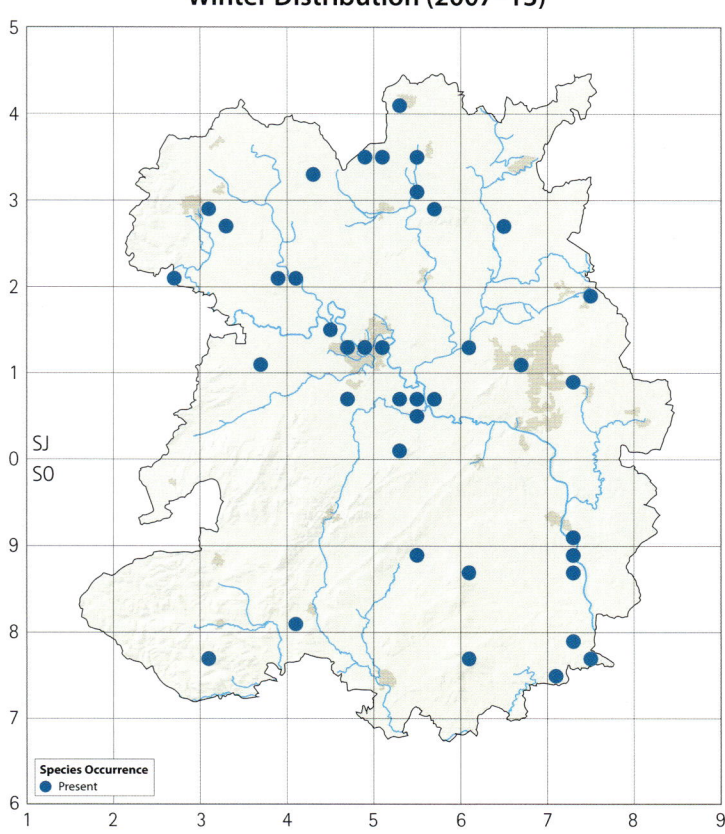

Allan Heath, North Shropshire, 21 April 2011

later in the year on 9 November. One at Priorslee Flash on 14 April 2016 also showed characteristics of this race.

One-quarter of the 24 ringed here and recovered elsewhere were found in Sussex, one only seven days later, with another five in other counties well to the south, at or on route to south and south-east coasts, while three ringed in Sussex were found here. Four, all first-year individuals, were recovered abroad, closer to the wintering grounds, in France, southern Spain, Portugal and Senegal. The furthest movement involves one ringed at Walcot in August 2006 and found in Senegal on 26 January 2006, a movement of 4,217km. The other three all moved more than 1,000km, and two were caught only one month and a few days after ringing. Coming the other way, one ringed as a first year in Portugal in October 2007 was caught in the following August at Chelmarsh Reservoir.

Mike Shurmer

Iberian Chiffchaff

Phylloscopus ibericus

Vagrant

This typical leaf-warbler breeds on the Iberian Peninsula and in North Africa, and winters in tropical Africa (Svensson 2001). Previously considered to be a race of Chiffchaff (*P. collybita*), it was recognised as a distinct species in 1998 (Clement & Helbig 1998; BOU 1999). It is an overshoot migrant to northern Europe and, although numbers in Britain appear to be increasing, identification remains problematic with its distinctive song still the key feature to note (Collinson & Melling 2008). There has been just one record, although it returned for a second year:

Granville Country Park, Telford   5 April–2 May 2016 and
3–7 April 2017

As might be expected, it was its song from a patch of open birch woodland on the edge of the Park that first attracted the finder's attention. It remained faithful to this location throughout its stay and was enjoyed by a large number of observers. As this was a first for the County, it is one of the few records for 2016 included in this publication (Holt *et al.* 2017). In 2017, the bird was again present at the same locality (British Birds *in prep.*).

Graham Walker

Jim Almond, Granville Country Park, 10 April 2016

Wood Warbler

Phylloscopus sibilatrix

Scarce summer visitor

Wood Warbler is one of Europe's long-distance migrants which winter in sub-Saharan Africa. A declining summer visitor, males arrive from late April with their shivering song heard in mature woodland with sparse understorey, mainly in the south-west and the Wyre Forest. The local population is on the eastern edge of the main UK range, with Wood Warblers now being much less frequent in counties to the east.

Once known as the Yellow Wren (Eyton 1838) or Willow Wren

(Beckwith 1889), Wood Warbler was noted as being found in suitable habitats of mature open woodlands and clumps of trees, particularly oak, beech and ash woodlands. It was generally distributed, and plentiful in favoured localities, and 'abundant' at the end of the nineteenth century.

Separation of this species from the closely related Willow Warbler and Chiffchaff was an identification challenge for early ornithologists.

Occupied Tetrads

Atlas period (breeding)	1985–90		2008–13		Change	
	Number	%	Number	%	Number	%
Confirmed	71	8	12	1	-59	-83
Probable	68	8	26	3	-42	-62
Possible	86	10	34	4	-52	-60
TOTAL	225	26	72	8	-153	-68

Breeding Distribution Change (1985–90 to 2008–13)

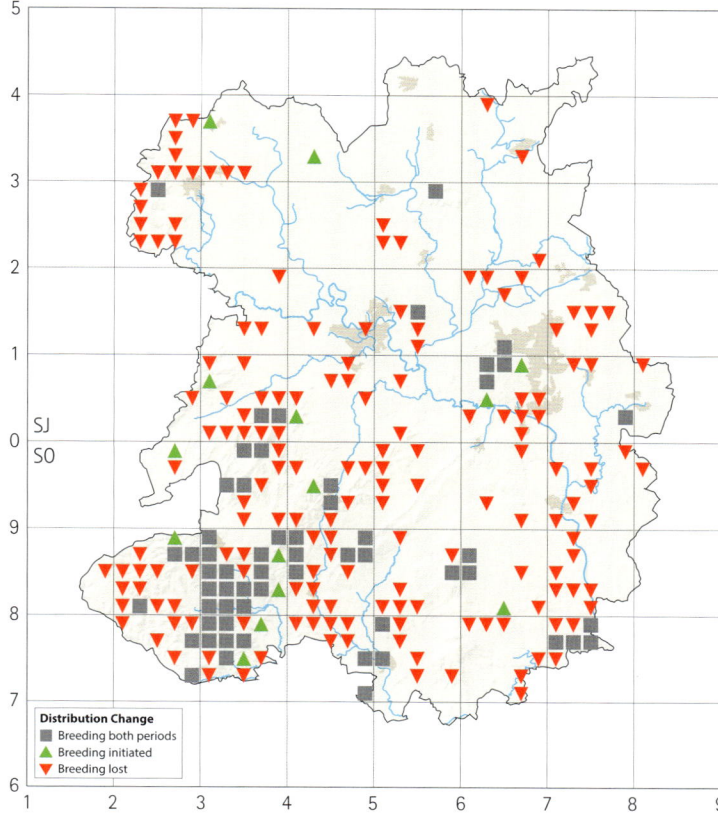

Distribution Change
- ■ Breeding both periods
- ▲ Breeding initiated
- ▼ Breeding lost

Beckwith described how each species utilised different parts of woodlands, with Wood Warblers characteristically singing from the treetops, but identification of non-singing individuals was problematic. On one occasion, to identify a nest he caught the female in a light butterfly net, returning it to the nest once he had identified it.

The *Handlist* (1964) indicated that Wood Warbler was 'much more scarce than either Chiffchaff or Willow Warbler', except in the Wyre Forest, where it was possibly as common as the latter, but felling of oak woodlands and replacement with conifers resulted in a decrease in areas where it was formerly found. It continued to be found in suitable habitat, although under-recorded, and the 1968–72 national Atlas found it was widespread 'where suitable habitat was available' (SBR 1972). Breeding evidence was found in all except four of the County's 33 10km squares, all of them in the north. By the 1980s there were warning signs of decline, for example in SBR 1985 numbers were described as being much reduced from the previous year. The *Atlas* (1992) noted that Wood Warbler was locally distributed due to its specialised habitat requirements, and mainly found on the western hillsides, and on Whitcliffe, Benthall Edge, the Wrekin and Wyre Forest.

Since then, Wood Warblers have shown a notable contraction in breeding range. The breeding distribution change map shows a marked decline in many areas, and in 2008–13 they were found in 72 tetrads, only one-third of the 225 tetrads in 1985–90. They have nearly completely disappeared from the north, with particularly dramatic losses in the north-west. The formerly strong populations along the Severn Valley, such as at Benthall Edge and the Severn Gorge woodlands in Telford, have also been lost.

While the distinctive song makes Wood Warblers relatively easy to locate in suitable habitats, confirmation of breeding can be difficult. They nest close to the ground in domed nests hidden among vegetation, and parents feeding young are difficult to observe. Also they occur in some of the more remote and difficult areas to access. A study of the woodlands in and around Telford found that Wood Warbler favour hillside sites with gradients of 10° to 75°, with the optimum being 40° and flat areas being avoided (Bishton SBR 1983). These factors may partly explain the small number of confirmed breeding records in the tetrads in which it is found.

They are now predominately found in the south-west, in the woodlands of the Shropshire Hills. There is also a stronghold in the Wyre Forest, which continues into Worcestershire, and smaller populations persist around the Wrekin and Brown Clee, and at a few other scattered localities in the west and north. However, breeding densities have also declined. For example, there were five territories at Whitcliffe in the 1990s, but only one in recent years. Similarly, Candy Wood was their stronghold in the north-west, and nine singing males were found during a BTO survey in 1984 (SBR), but only one was present during the recent Atlas period (Allan Dawes *pers. comm.*).

As unmated males sing to defend territories, and polygamy occurs, the population is best expressed as territorial males, rather than pairs. Some of the possible breeding records will relate to passage migrants. An average of two males per tetrad with probable or confirmed

Gareth Thomas, Whitcliffe Common, 2 May 2013

breeding, and one per tetrad with possible breeding, suggests a population of around 100 males, rather higher than the estimate of 55–65 based on a small number of TTV counts. This can be compared to the population estimate of 400 males in the *Atlas* (1992), which underlines the severe declines in the last two decades. The causes of the Wood Warbler's decline in population and range are unknown and may be driven by factors at their breeding grounds, wintering sites, on migration routes or a combination of these.

Few are ringed, and two of the four recoveries were found here, but the other two indicate the southward migration route. A nestling ringed at the Hurst, Clun on 12 June 2012, was caught 305km away at Pett Level in Sussex on 6 August 2012, and a nestling also ringed at the Hurst, on 12 June 1988, was recovered in unknown circumstances at Ras El Ma in northeast Morocco on 15 August 1989, a movement of 2,031km.

Mike Shurmer

Pallas's Warbler (Pallas's Leaf Warbler)

Phylloscopus proregulus

Vagrant

Although this diminutive warbler breeds in southern Siberia, northern Mongolia and north-eastern China, and winters in southern China and north-eastern Indochina, it is nevertheless a scarce, but regular, migrant to Britain with over 50 reports in most of the last 20 years (Fraser 2013b). The majority, however, are located in the coastal counties of eastern and southern Britain, with relatively few found inland. The only record is of one trapped and ringed at a roost in Rhododendron (*Rhododendron* spp.):

Pell Wall, Market Drayton   18 November 1987

The late autumn date was typical for this eastern vagrant and, perhaps surprisingly, coincided with another in the Midlands, an immature at Westwood Park, Worcestershire, the previous day (Rogers *et al.* 1988, Harrison & Harrison 2005).

Graham Walker

Yellow-browed Warbler

Phylloscopus inornatus

Very rare passage migrant

The breeding range extends from just west of the Urals through to north-eastern China, although there has been an expansion of the population westwards in recent years (Snow & Perrins 1998). It winters in tropical south-east Asia, but is a regular and increasing autumn migrant to Britain, with several hundred appearing in most years (Fraser 2013b). The majority arrive from late September through to early November along the eastern side of the country with some evidence of onward movement towards the south-west thereafter. It is much scarcer inland with few records in the English Midlands (Harrison & Harrison 2005). There have been just four records, all since 2008 and with three in the last three years:

Shawbirch, Telford	12 October 2008
Whixall Moss	10 October 2015
Whixall Moss	15 October 2016
Shawbirch, Telford	4–7 April 2017

The first, on a typical date, was located in the morning by call at the rear of the finder's garden, but was then seen fly-catching from a willow (*Salix* spp.). It disappeared about midday and could not be re-found (Latham 2008). The second and third were caught during ringing sessions on the NNR while the fourth individual roamed between a small copse and a nearby residential estate, feeding both low down and in the tree canopy. While it is perhaps understandable that regular ringing at a site has turned up two birds, there are no obvious reasons for the two sightings in the Shawbirch part of Telford, particularly when all the well-watched sites have produced no records.

Graham Walker

Jim Almond, Shawbirch, 4 April 2017

Great Reed Warbler

Acrocephalus arundinaceus

No modern records

There is a single record of this summer visitor to the reed beds of southern and central mainland Europe, which winters in sub-Saharan Africa. It was shot at Ellesmere 'about 1866' (Forrest 1899; Witherby & Ticehurst 1908). The specimen passed into the collection of Mr W.S. Brocklehurst of Kempston in Bedfordshire.

John Tucker

Sedge Warbler

Acrocephalus schoenobaenus

Uncommon summer visitor

Sedge Warbler breeds across Europe and winters in sub-Saharan Africa.

Some early writers confused this species with Reed Warbler but, gradually, the true status of each was revealed. Described as 'very common' (Rocke 1865), 'abundant' (Beckwith 1879), 'numerous near water' (Paddock 1890) and 'common' (Forrest 1899), a picture emerged of a species that was familiar, and Forrest added that 'although it prefers the neighbourhood of water, it occurs, and sometimes nests, far from any pool'.

There were records in most years of the first half of the twentieth century, but Lloyd (1943) listed it among the species that have shown 'a slight decrease generally: not now to be seen on the banks of some rivers and streams which were formerly regular haunts'. The *Handlist* (1964) noted that it was 'fairly common ... in suitable habitats, though less frequent in the hill country', but no sites or numbers were listed. The national BTO Atlas (1976) confirmed this distribution; it was mapped in 20 of the County's 33 10km squares, but only four of them were in the south.

Nationally, the population declined sharply (over 50%) between 1969–73, attributed to drought in the wintering area in the Sahel, but it recovered to about 70% of the 1969 level by 1977, then another decline occurred from 1980 to the lowest national population in the last 50 years in 1985, only about one-third of the peak in 1969, again due to failure of the rains in the Sahel.

Too few records were submitted throughout the 1960s and 1970s for SBRs to identify the local effects of the decline, and in some years it was only listed as present, but 'numbers were reported to be down in the Shifnal and Weston Park area' (SBR 1973). The later lowest point was reflected in a '50% decrease in breeding pairs' at Allscott (SBR 1984), and, while they were 'more numerous' at that site the following year, 'in contrast, reports from riverside sites stated numbers to be lower than 1984' (SBR 1985).

Although breeding was noted at Allscott Sugar Factory in several earlier years, no counts were given until 1976, when an estimated 42 pairs were breeding. There were 16 singing males in 1979, a 'reduction on 1978', but an increase in breeding pairs in 1980, 'the first since 1976 and now back to 1976 levels'. In some of these years 'several' were also breeding at various other named waterside sites, but '10 to 12 along ½ mile of the Severn at Cressage' (1979), seven territories at the same place in 1980 and six in 1984, and about 30 pairs along the Severn between Atcham and Cressage and five+ at Shrewsbury Sewage Farm in 1982 were the only records of more than

John Hawkins, North Shropshire, 6 June 2007

five pairs, apart from Marton Pool (Chirbury), where the cessation of water-skiing led to an increase from about 12 pairs in 1980 to 22 singing males in 1981. Predation by mink was noted at Allscott Sugar Factory in 1979, and a swimming mink was mobbed by Sedge Warblers at Cressage in 1984.

Fieldwork between 1985–90 for the first local Atlas therefore provided the first systematic assessment of status. There were few reports of more than a couple of pairs, but seven pairs bred at Chelmarsh in 1986 and 1988, and at Cranmere Bog there were four in 1988 and five in 1990. A BTO Waterbird survey on 5.5km of the River Severn between Atcham and Wroxeter (three tetrads) counted 20 pairs in 1985, increasing to a peak of 33 in 1987, before falling to 30 in 1990, the fluctuations matching those in the national population. Some nests were destroyed by floods on the river in 1985. The *Atlas* (1992) also revealed a healthy population along the Strine Brook and its ditch tributaries on the Weald Moors. There was little evidence of breeding away from wet or damp areas.

This fieldwork was carried out against a background of a rising national population from the low point between 1984–87, with the 1990 population equal to that in 1976; then, after a further trough, a higher population was reached in 2001. After another fluctuation, numbers nationally in the second half of the recent Atlas period were similar to those in the same part of the previous Atlas, but those in the first half were much higher. These more recent troughs also reflect low rainfall in the Sahel in the previous winter. Numbers here have fluctuated too, with several SBRs since 1956 reporting an increase or a decrease compared with the previous year at a specific site.

Another Waterbird Survey was carried out on a 5.9km stretch of the River Severn, passing through three tetrads near Alberbury, for the nine years 1991–99. Between 1992–96, 20–25 territories were found each year, but in 1997–99, only 13–15 were found (Michael Wallace *pers. comm.*).

Otherwise, records between the two Atlas periods remained confined to well-watched areas by interested observers, but most were of small numbers. More than six territories were found at Chelmarsh (13 in 2004), Wood Lane (seven singing males in 1997), and Lower Brompton (12 in 2000), but only three other records between 1991 and 2007 gave counts which may have been of six or more territories.

The recent Atlas found increases of 11% in the number of tetrads with probable or confirmed breeding, and 18% with evidence of breeding, suggesting a population increase, and also a redistribution. Losses from the damp areas along the Strine Brook and its man-made tributaries are probably due to changes in land management and local housing development. Ongoing efforts to drain that landscape since the 1970s have included regular clearance of ditches and the consequent loss of scrubby vegetation. Losses from the north-east and around Shrewsbury are probably also due to agricultural change and housing development. Conversely, the maturation of the margins of the artificial water bodies in and around Telford has created new breeding habitat, and the appearance of a significant population along the River Perry and Montgomery Canal is particularly noteworthy. The river was being canalised during the first half of the 1985–90

Atlas period, and then when work finished it was shallow and fast-flowing, with little bankside vegetation, but it has now had time to grow in some places, particularly by the pools and backwaters that have developed. Other work nearby to restore habitat for wading birds, carried out by SWT and RSPB, has also helped. On the canal, restoration work carried out by British Waterways and the Canal and River Trust included the creation of a number of reserves for plant species protected under schedule 8 of the Wildlife and Countryside Act 1981, which, happily, also appear to provide good breeding habitat for this warbler (Allan Dawes *pers. comm.*). Some of the new possible breeding records come from areas far from wet areas and may relate to birds on passage. One or two isolated pairs have, however, been confirmed breeding in moist depressions far from obvious water bodies, and others may have bred in oilseed rape crops, a habitat which is utilised elsewhere. Probable breeding was found at 300m around Shelve Pool.

Most counts were small, and during the recent Atlas period there were counts of more than six singing males or territories only at Allscott (26) and Leighton (seven), both SOS records in 2010, and TTV counts exceeding six were made in only eight tetrads (largest count 14 near Cherrington (SJ61U) in 2009).

A BTO WBBS survey during the recent Atlas period along 5km of the River Severn near Alberbury found only 4–5 territories/year, so the decline noted in 1997–99 appears to have continued. There were no apparent changes in habitat, so the lower counts appear to reflect the recent decline in the national population, attributed to lower over-winter survival. However, the different survey methodologies may have also led to fewer of those present being found (WBBS involves only two visits, rather than up to 10 for WBS, and an earlier start means part of the survey takes place before the day warms up sufficiently to encourage these warblers to sing). Even so, there has been 'a clear decline in numbers' (Michael Wallace *pers. comm.*).

The *Atlas* (1992) estimated the population at three to six pairs per occupied tetrad. Males generally defend a territory of 0.1 to 0.2ha, so each kilometre of riverbank may hold up to five to ten pairs (Cramp *et al.*, Vol. VI 1992). Meanders and tributaries provide several kilometres of riverbank in some tetrads, but there is no evidence that such high densities occur here, so there is no reason to change the earlier estimate. Including tetrads with possible breeding gives an estimate of 400–800 breeding pairs. Based on TTV counts, the population estimate is 1,000–1,100 breeding pairs, but this may be unreliable, as Sedge Warblers were found on TTVs in only 63 tetrads. Even allowing for a recovery from a low point in 1985, and an expansion of range, this latter estimate appears to be too high.

Analysis presented in the *Arrival Dates* chapter suggests that the first Sedge Warblers arrive 11 days earlier than they did in 1950.

Breeding Distribution Change (1985–90 to 2008–13)

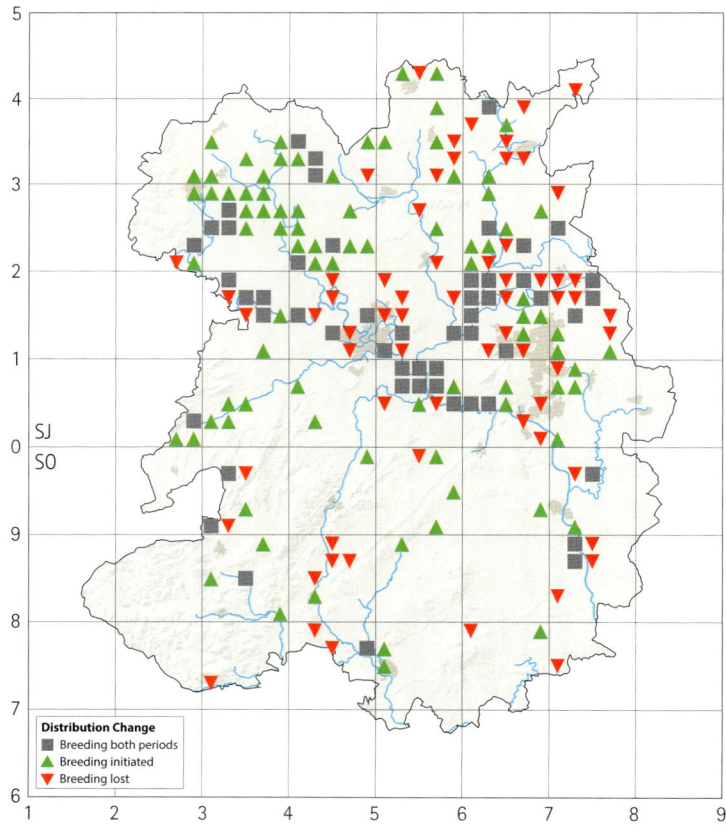

Distribution Change
- ■ Breeding both periods
- ▲ Breeding initiated
- ▼ Breeding lost

Occupied Tetrads

Atlas period (breeding)	1985–90		2008–13		Change	
	Number	%	Number	%	Number	%
Confirmed	40	5	39	4	-1	-3
Probable	36	4	45	5	9	25
Possible	40	5	53	6	13	33
TOTAL	**116**	**13**	**137**	**16**	**21**	**18**

Paul King, Shropshire, 18 May 2008

Regular ringing began in 1961 with the establishment of the Walcot Ringing Station at Allscott Sugar Factory. Sedge Warblers put on substantial fat reserves before making their long migration flights, and a study at Allscott showed that this started in August, before they left. Ones caught in August weighed an average of 14.2g, 20% more than the average of 11.8g in July, and some individuals increased by more than that (Wright, SBR 1967). By the end of 1966, 310 had been ringed there and elsewhere by the Ringing Group, rising to over 1,100 by 1973, the year of the Group's first recovery (in Hertfordshire), and 4,833 were ringed at Allscott between 1961 and the closure of the ringing station in 2012 (Julian Langford *pers. comm.*). The number ringed has varied over the years with population size and ringing effort, and has ranged from less than 10 to 200 per year.

Other ringing sites include Chelmarsh and Shrewsbury Sewage Farm. Most of the 175 ringing recoveries were both ringed and recovered here, but the remainder show the migration route. The 40 recovered elsewhere in Britain or Ireland involve 23 other counties, mostly to the south, and Kent is the only one with more than two. The 21 recoveries involving other countries show movement along the Atlantic coast of Europe, crossing the Mediterranean to Morocco and then across the Sahara into western Africa. The final winter destination has been confirmed by a female ringed in Senegal in March 1992 and found breeding in June 1993 at Walcot, 4,210km distant, and the discovery of one ringed here in August 1982 found dead in Burkina Faso in April of the following year, eight months later and 4,286km distant.

An adult female caught at Allscott on 15 August 1988 was retrapped at the same place on 14 July 1990, 27 June 1992 and 29 July 1995 (Julian Langford *pers. comm.*), when it was more than eight years old, possibly exceeding the national longevity record of 8y 8m 8d. With a round trip of 8,000km per year to western Africa, this individual must have flown well over 60,000km in her life.

Apart from the hazards of migration, and drought and habitat loss in the west African wintering grounds, there are increasing threats to the Sedge Warbler here too, through disturbance or removal of the tangled, dense riverside vegetation that it prefers. Bankside clearance due to agricultural demand to maximise the planted area for crops, flooding and flood mitigation efforts, and fishing, have all had an impact. In some areas large stands of Himalayan balsam and Japanese knotweed (neither of which are used for nesting) have swamped and replaced native vegetation suitable for nest building.

Simon Holloway & Leo Smith

Reed Warbler (Eurasian Reed Warbler)
Acrocephalus scirpaceus
Uncommon summer visitor

Yvonne Chadwick, Chelmarsh, 21 July 2010

Britain is at the western edge of the Reed Warbler's European range. It winters in sub-Saharan Africa.

It was not properly identified as a distinct species until 1804 by a French naturalist, and it was many years before it was described in accessible print and came to the notice of English naturalists. Eyton's comment in 1838 that it was 'common' was directly refuted by Beckwith in 1889 (and indirectly by others) and seems to demonstrate that confusion with Sedge Warbler cleared as the nineteenth century progressed. Considered a common breeder in suitable habitats by all Victorian writers, some of the nesting sites mentioned then will be very familiar to modern birdwatchers (even if the spelling is not). Leighton (1836) mentioned a site on the Severn in Shrewsbury, Rocke (1865) noted Ormond Park pool, Crowsmere, Hawkstone Park and Polemere. Beckwith (1878, 1879 and 1889) added Almond, Bomere, Oakley Park and Walcot pools, Hencott, Berrington, Shrawardine, Fennemere, Marton, the Berth, Osmere and Blackmere, together with the Weald Moors near Eyton and Kinnersley and banks along the River Tern (all original spellings).

An interesting change through the twentieth century was the increase in pairs breeding in common reed stands along the abandoned canals in the north and north-west. This habitat was not mentioned during the previous century as canals were, presumably, kept clear of bankside vegetation that may have impeded traffic. The *Handlist* (1964) specifically noted records from 'parts of the disused canals', but otherwise considered that 'Its breeding haunts are confined to the reed fringed lakes and meres in the north'. Eleven sites with records in the previous 10 years were listed, almost all still in use today. The colony at Marton Pool (Chirbury) was noted as 'probably the largest' and this status is still enshrined in the Marton Pool SSSI citation.

The national BTO Atlas (1976) mapped it in 15 of the County's 33 10km squares, including only two in the south, both with only

possible breeding. The largest colonies listed in SBRs up until the start of the first Atlas period in 1985 were at Allscott Sugar Factory and Crose Mere. The former was first noted in 1963, when two colonies, each of four to five pairs, were established, then there were 24 pairs in 1974, 23 in 1976, only four in 1977 when a Black-headed Gull colony established itself in the phragmites bed, a 'much recovered' colony of 15 in 1978 and 1979, and 'a 40 % decrease in 1984'. The latter, the 'largest breeding colony', had about 40 pairs in 1956, 21 nests and 11 more being built in 1959, but 'much reduced numbers, five nests noted' in 1960, and eight pairs in 1961, 'presumably due to disturbance of reed beds by fishermen'. The colony has been counted every year since 1962, with eight or nine each year up until 1985 (John Hawkins *pers. comm.*). The old river bed at Shrewsbury (10 in 1973) and Fenemere (about 21 singing males in 1981) were also worthy of note, and the cessation of water-skiing at Marton Pool (Chirbury) led to an increase from about 15 pairs in 1980 to 40+ singing males in 1981. Breeding pairs were noted at several other sites in this period, but only in single figures, and it was frequently 'under-reported'.

The *Atlas* (1992) noted the continued importance of the Allscott site, where '80 adults were netted, putting the colony at a minimum of 40 pairs' in 1986, and increased ringing totals at Chelmarsh following work to extend the reed bed, but only one other site was listed in SBRs as having more than 10 pairs or singing males in that period,

Breeding Distribution Change (1985–90 to 2008–13)

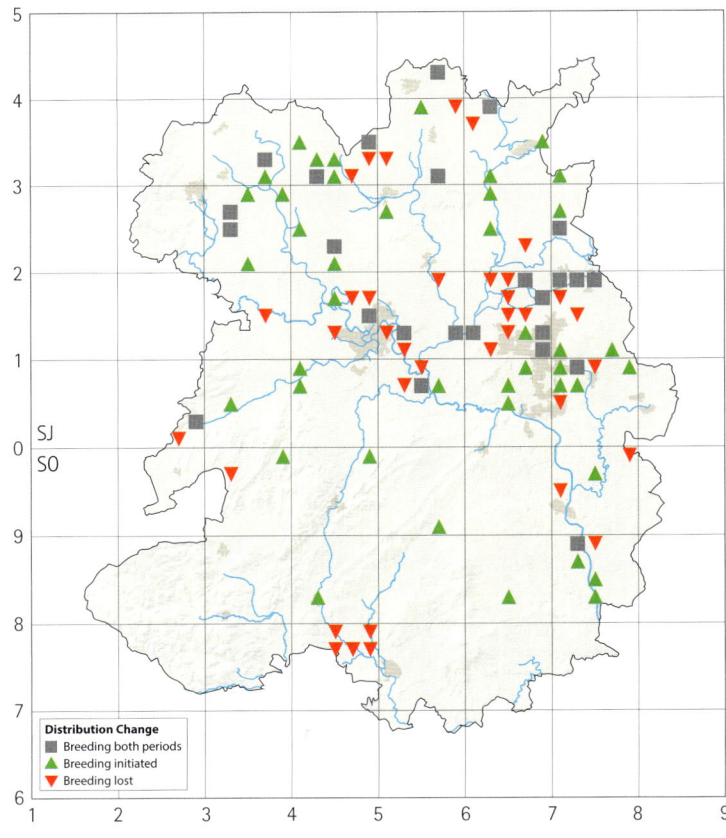

Distribution Change
■ Breeding both periods
▲ Breeding initiated
▼ Breeding lost

Fenemere (22 in 1989). New colonies were noted around Highley along the Severn and Borle Brook, and along the Onny just upstream from its confluence with the Teme at Bromfield. A map showing the distribution of common reed showed that 'there are tetrads with common reed but no records of Reed Warbler, especially in the north-west where the latter is at the western edge of its range in the Midlands'.

The national population has doubled over the last 40 years, and increased by about 50% between the two local Atlas periods, with an expansion of range northwards and westwards, driven in part by improved breeding success. The decline of the Cuckoo will have had some limited effect, while nationally egg laying was nine days earlier in 2010 than it was in 1966, attributed to climate change, which has extended the breeding season and increased the possibility of second broods. This corresponds to first arrival here dates being 19 days earlier in 2014 compared to 1950 (see Table 2 on p. 19). Unlike the Sedge Warbler, it does not suffer large fluctuations due to drought in the Sahel.

This population increase and expansion of range is reflected in the local breeding distribution change map, as these warblers have colonised many of the reed beds that were unoccupied in 1985–90, including several at the Ellesmere meres. In addition, reed beds have become established, and occupied, around the artificial lakes in Telford, and around a fishing lake and a fish farm in the south, but none were found at Shelve Pool. Reeds along the Strine Brook and associated ditches have been cleared, where Sedge Warbler too have been lost. Other losses since 1985–90 appear to coincide with areas of river prone to flooding and the consequent loss of stands of reed, at the Onny–Teme confluence at Bromfield, and around Shrewsbury. Counts of five or more singing males in SOS records during the recent Atlas period came from Allscott Sugar Factory (26), Aston Locks (Queens Head, six), Norbroom Marsh (Newport, 12), Priorslee Lake (maximum 10), Sambrook Mill (six) and Stirchley (five), while TTV counts of five or more came from only 13 tetrads altogether, including these named sites, plus nine at Marton Pool (Chirbury) and eight at Chelmarsh.

There were 16 singing males at Fenemere in 1995, and up to 25 in some years in the late 1980s and early 1990s, and the population appears to have increased since then, with a minimum of 23 in the reedbeds on three of the four sides of the pool during the recent Atlas period, and an unknown number in the inaccessible fourth side. There were also around six pairs at the adjacent Marton Pool (Baschurch), in the same tetrad (Michael Wallace *pers. comm.*).

At Chelmarsh, numbers appear fairly constant. Ringers estimated about 30 pairs in 2000 and 2003, and caught 30 pairs in 2014. 'We obviously do not catch all of them and we only target half of the reedbed, so actual breeding numbers are likely to be a lot higher' (Dave Fulton *pers. comm.*).

In the Ellesmere area, part of the edge of the Mere was fenced

several years ago and the protected strip has since been colonised by reeds and in turn by Reed Warblers, with three to four pairs present in recent years. There were also around 10–12 pairs annually at Wood Lane, and eight to 14 pairs annually at Crose Mere from 1985 onwards, with double-figure counts in seven of the years 1988–2001, maximum 14 in 1993 and 1994 (John Hawkins *pers. comm.*). This latter colony has not recovered to its 1950s levels.

In addition, there were around 16 pairs in two tetrads along the Montgomery Canal between Queens Head and Maesbury (Allan Dawes *pers. comm.*), and the reeds appear to be spreading, so fears that they might be lost following restoration and re-opening of some sections of the canal appear unfounded so far. However, as more of the canal is restored, there could be pressure to ease the restrictions on the number of boats, so the situation needs to be monitored.

The *Atlas* (1992) concluded that 'apart from the main sites at Allscott, Chelmarsh and Fenemere, most sites have few singing males, suggesting an average of 5–10 pairs per occupied tetrad'. Counting only tetrads with probable and confirmed breeding produced an estimate of 250–500 pairs. With only five sites currently known to have more than 10 pairs, it seems reasonable to apply the same range, but including tetrads with possible breeding, the estimate is 335–670 breeding pairs. The population estimate based on TTV counts

John Hawkins, North Shropshire, 11 July 2006

Occupied Tetrads

Atlas period (breeding)	1985–90		2008–13		Change	
	Number	%	Number	%	Number	%
Confirmed	28	3	26	3	-2	-7
Probable	14	2	22	3	8	57
Possible	19	2	19	2	0	0
TOTAL	61	7	67	8	6	10

is 1,200–1,350 pairs, but it is unreliable, as Reed Warblers were found on TTVs in only 33 tetrads, and it appears too high.

The Cuckoo species account shows a total of 20 records of parasitism on Reed Warbler since the first modern record in 1972, 10 of them at Allscott Sugar Factory where ringers have looked for Reed Warbler nests, but none has occurred there since 1998. At Crose Mere, four nests were parasitised in 1992 and 1993, and two in 1994, with two Cuckoos fledging in each of the three years, but there has been no occurrence since (John Hawkins *pers. comm.*). More recently, there have been only three records, all in July, at Chelmarsh Scrape in 2009 and 2010, and Venus Pool in 2014.

Of the 540 that have been ringed or recovered here, 373 of them were both ringed and recovered, while 84 involve five adjacent counties, 67 involve 26 other English or Welsh counties, mainly to the south (but none with Scotland or Ireland, reflecting the restricted range) and 16 overseas. The latter, from the Channel Isles, France, Spain, Cueta (north Africa) and Morocco confirm the first part of the migratory route, but none have yet been recovered on the wintering grounds, probably in west Africa south of the Sahara.

Ringing started at Allscott Sugar Factory in 1961, and 6,255 were ringed up until 2012, including all three of the most distant overseas recoveries, the furthest being in the Atlas Mountains, in Morocco,

2,267km distant. Reed Warblers put on substantial fat reserves before making their long migration flights, but a study at Allscott found no evidence that this started before they left in August, unlike Sedge Warbler (Wright, SBR 1967).

There is a high level of site fidelity, shown by the number of retraps at Allscott. The length of time between the initial ringing and the retraps also gives an indication of longevity. In 2004, 313 were ringed, the highest annual total, but another 30 were retrapped, 10 first ringed the previous year in 2003, and five, seven, four, three and one in each preceding year back to 1998. In 2005, 160 were ringed, and another 54 retrapped, 35 from 2004, and zero, three, four, one, two respectively each year back to 1999, and two more from 1996. A juvenile ringed in July 1987 was caught in six further years, the last time in August 1997, 10y 3d after ringing, and a juvenile female ringed in July 1980 was caught each year from 1985–89, finally at an age just short of nine years (Julian Langford *pers. comm.*). These compare with the national longevity record of 12y 11m 21d.

The *Arrival Dates* chapter indicates that Reed Warbler is arriving here 19 days earlier than it did in 1950.

If the increased productivity and expansion of range continues, there are still unoccupied reed beds to be colonised in the near future.

Simon Holloway & Leo Smith

Marsh Warbler

Acrocephalus palustris

Vagrant

Jim Almond, Priorslee Lake, 7 June 2015

This summer visitor to temperate Europe and western Asia winters mainly in south-east Africa. Britain is at the extreme north-western edge of its range, with less than 10 territories occupied in most years, and it is rare for breeding to occur at the same location in successive years (BTO BirdTrends; Robinson 2017).

The only historic report was of a singing Marsh Warbler seen at

close quarters at Eaton Constantine on 10 April 1930. This, however, is a very early date compared to the usual arrival period, from the end of May into early June, and there must, therefore, be some doubt about the reliability of this record.

There has been one confirmed sighting subsequently:

Priorslee Lake   6–12 June 2015

The distinctive song of this warbler assisted observers in pinpointing its location as it regularly circulated around a territory of tall herb fen, with isolated small bushes, trying to attract a mate.

In addition to the Priorslee Lake individual, there are two other modern reports that are worthy of note. The first is of a specimen in the Birmingham Museum and Art Gallery labelled 'Marsh Warbler, 13 September 1971, Much Wenlock, H Southern'. Visual examination of the skin suggested it was a Marsh Warbler and the biometrics, although not totally conclusive, were largely within the acceptable range (A. Latham *pers. comm.*). Unfortunately, subsequent DNA analysis of a sample did not confirm the visual identification and determined that the mother of the specimen was a Garden Warbler (*Sylvia borin*) instead (Prof. J.M. Collinson, Aberdeen University *pers. comm.*). However, the biometrics rule out the identification of the specimen itself as a Garden Warbler (A Latham *pers. comm.*). It appears that the uncertainty will not be resolved, as the museum is unwilling to provide another sample for analysis. The background to the specimen is also intriguing, in that it is assumed that H. Southern is the eminent Oxford ornithologist H.N. Southern. He is not known, however, to have had any connection with Much Wenlock, nor is there any apparent reason why he would have a specimen of a Marsh Warbler which he donated to Birmingham Museum, although there is also a specimen of a Bullfinch (*Pyrrhula pyrrhula*) in the museum that is attributed to the same person from the same year (Luanne Meehitiya, Natural Sciences Curator, Birmingham Museums Trust *pers. comm.*). Nevertheless, 1971 did coincide with a period when there was a small but widespread breeding aggregation in the Avon Valley, Worcestershire (Harrison & Harrison 2005).

The second notable report was of a bird at Tong Lodge on 1 June 1972. Unfortunately, the information on this individual is limited to that which appears in the SBR, with no account from the observer in the archive. It was only seen and heard on the one occasion, despite subsequent visits, and an element of what was reported casts some doubt over the identification. Nevertheless, the date was appropriate for an individual arriving on its breeding grounds and it again coincided with the period when breeding was regular in Worcestershire.

As there is some doubt over the records from 1930, 1971 and 1972 it was questionable whether Marsh Warbler had actually occurred here prior to 2015. However, there was no doubt about this record, and it is one of the few records from that year to be included in this publication.

Graham Walker

Icterine Warbler

Hippolais icterina

No modern records

The only record of this summer visitor to central and eastern Europe, which winters in sub-Saharan Africa, was in June 1941, when 'two [were] searching for insects on a row of runner beans in a garden in Corvedale'. The observer recognised them by the yellow underparts and characteristic angry song. While on active service in 1917 a pair had a nest, with eggs, in a bush only a few yards from the trench where he was stationed, so he could study them at close quarters.

John Tucker

Grasshopper Warbler (Common Grasshopper Warbler)

Locustella naevia

Scarce summer visitor

Grasshopper Warblers are scarce and declining summer visitors and passage migrants from their wintering grounds in West Africa.

Interestingly, Beckwith's 1889 account noted that it was first described by Pennant as a British species in 1768 from a specimen collected in Shropshire. In the late nineteenth century it was said to be evenly distributed, albeit sparingly. Nests, which were either in a tuft of grass or under low rank herbage, were found regularly at that time in the gorse and heather of the Stretton hills and along the Severn Valley, generally in damp situations. In 1908 it was described as thinly scattered but nowhere numerous. It was recorded in most years in the first half of the twentieth century, but described by Lloyd (1943) as 'never exactly common, now rare, and hardly ever observed in haunts where it used to occur regularly'.

The *Handlist* (1964) described Grasshopper Warbler as a breeding visitor and passage migrant, April–September, 'rather local' but fairly widespread. Its main habitats were described as open scrubland with rough grass and brambles, the mosses of the north-east and north-west, and conifer plantations until they reach 10 years of age; it has been recorded in such plantations up to the 1,400ft contour (425m) on the Long Mynd (including six there on 1 May 1960). There had been a 'considerable decrease in lowland areas, owing to more efficient farming, but ... a corresponding increase in upland areas owing to forestry operations' in the previous 20 years (Ireson SBR 1967).

Although numbers fluctuated from one year to the next it continued to be well reported up to the mid-1970s. Widespread examples include 1965, when eight were located along a three-mile forest track at Ludford; 1967, when it was 'present in all areas in moderate numbers'; 1971, when there were several pairs at Much Wenlock and records from eight other sites; and 1973, when they were first reported at Prees on 24 April followed by several at Much Wenlock on 25 April, with other notable multiple reports including four at Nesscliffe and four at Llynclys in early May, and six on the old river bed (Shrewsbury) on 30 May. There were June records from four sites, and two or three pairs at Haughmond Hill 'during the season'. While many of these reports up until mid-May could have been of migrants passing through, those from late May onwards almost certainly related to breeding birds.

Jim Almond, Mire Lake, 22 April 2010

Breeding Distribution (2008–13)

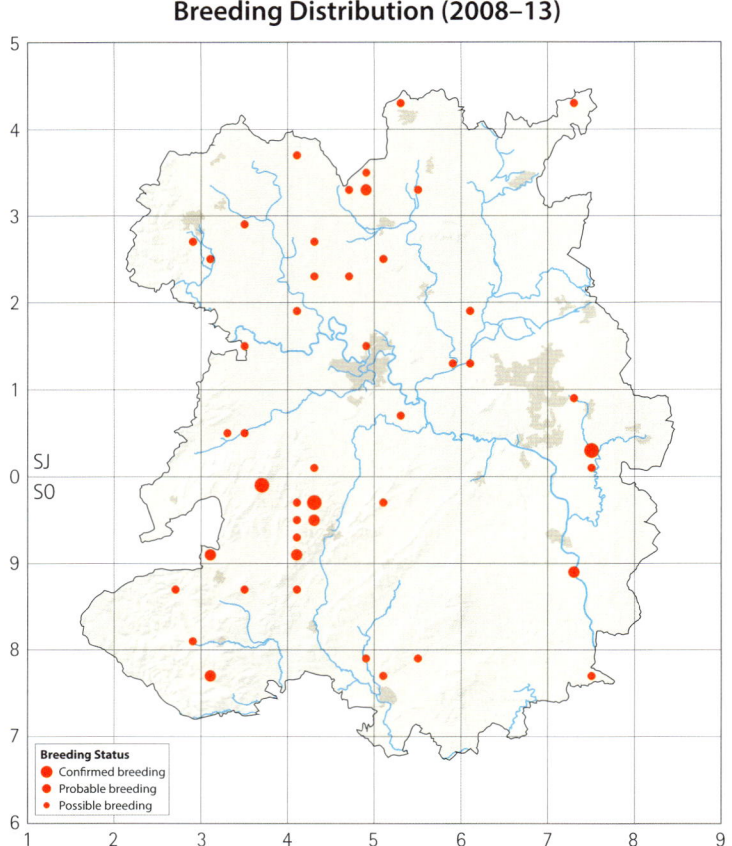

Breeding Distribution Change (1985–90 to 2008–13)

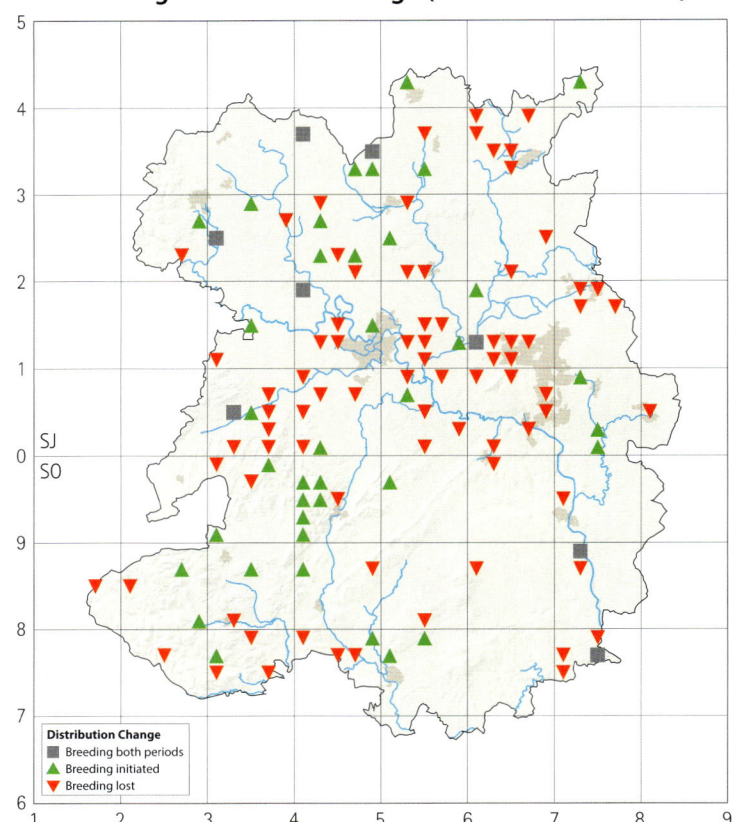

The long-term decline probably began around this time, with fewer records than in recent years being noted in 1976 and again in 1978. This pattern continued up to 1983 when there was a temporary increase which included five or six along the old river bed in Shrewsbury and three in Telford Park, both in April, and 'breeding season records' from seven sites, including up to seven on Haughmond Hill. The most records for 10 years coincided with Atlas work in 1986 and 1988, with Haughmond Hill again being noted as the stronghold. By 1990 reports were back to pre-1983 levels with the downward trend continuing to this day.

Since then, SBRs noted that it 'seems to have undergone a decline, being absent from many of its former haunts' (2000), and 'many sites used in the previous 10 years were no longer suitable' (2005).

During Atlas work Grasshopper Warblers are difficult to find, and evidence of breeding even more so. They skulk low down in thick vegetation, so the characteristic reeling song is often the only means of detection, but they sing rarely or not at all once nesting has commenced, and then often only at dawn and dusk. Some might be detected later in the season when males sing again before the second clutch is laid.

They sing regularly on passage, so many of the possible breeding records will be migrants, and the breeding distribution map from recent Atlas fieldwork probably overestimates the number breeding in any one year, although this perhaps compensates for those missed altogether due to their secretive behaviour.

It shows three confirmed and six probable breeding records. A bird carrying food into, and faecal sack out of, an oilseed rape crop on 13 July 2008 at Grindle had been recorded in May at the same

Occupied Tetrads

Atlas period (breeding)	1985–90		2008–13		Change	
	Number	%	Number	%	Number	%
Confirmed	8	1	3	0	–5	–63
Probable	26	3	6	1	–20	–77
Possible	53	6	37	4	–16	–30
TOTAL	87	10	46	5	–41	–47

site, when it was presumed to be a migrant. Breeding in this habitat is unusual but not unprecedented, as they have been found in this crop in Lincolnshire (BTO Research report 171, 1996) and in 'most years' since 1998 in south Nottinghamshire (Pendleton 2013). The other two confirmed breeding records were from more typical damp heathland habitat. An adult feeding a recently fledged young was seen on Wild Moor on 3 July 2011, and a youngster barely able to fly was seen on the Stiperstones on 2 July 2013, where a male had proclaimed territory all season. A recently fledged young caught by ringers at Chelmarsh on 25 July 2010 was perhaps another (unmapped) confirmed breeding record, although it might have moved some distance. Probable breeding was found at only five other sites: two in tetrads on the Long Mynd, another in wet willow scrub near Bishop's Castle, one at Whixall Moss and one at Chelmarsh. They were reported most years from Whixall Moss, the Long Mynd and Chelmarsh.

The number of territories found on Wild Moor during dusk surveys for Snipe fluctuated in the five years 2010–14, with three, four, four, three and two respectively. Several other flushes have been occupied in these years as well, records coming from late evening surveys in

April and May for Red Grouse. Many of these were of birds heard only once, in early May, and these are probably passing through, but some were heard several times later in the season, giving a population on the Long Mynd of at least six territories in 2011 and seven in 2012. The Stiperstones held at least two territories in 2013. Several suitable breeding areas of damp heathland on both these hills, particularly on the Long Mynd, are rarely visited at the right time, so the population is probably higher.

Fluctuating numbers, many on passage, are also shown by Natural England data for the English part of Fenn's, Whixall and Bettisfield Mosses NNR, which listed them in five of the six years 2008–13. There were 10 records, probably involving 10 individuals, in 2010 with most in late April, but the latest on 17 May could have been breeding. There was one record in 2008, three in 2011, four in 2012 and one in 2013. However, a late June record in 2012 indicated probable breeding.

After the Atlas period, all except one of 12 records in 2014 were of single birds, mostly during the spring passage period, but singing at Horderley on 17 June, Telford Town Park on 11 July and Priorslee Lake on 21 July may indicate additional breeding sites.

The recent decline is starkly illustrated by the breeding distribution change map. Some of the possible breeding records will be of passage birds, so it exaggerates the number of squares where it was found to breed, in both periods, but the number of tetrads with confirmed and probable breeding has declined from 34 to nine, a loss of 74%. There were no records from many of the sites where Grasshopper Warbler occurred in earlier years, listed above, including the former 'stronghold', Haughmond Hill. The decrease in the number of tetrads with possible breeding is indicative of a decline in these warblers passing through, if it does not reflect a decline in the breeding population.

It is likely that Grasshopper Warblers have always bred on the Long Mynd, and the apparent colonisation there is more likely to reflect records from recently introduced dusk surveys for other species.

Comparing Atlas results at the 10km square level provides further evidence of a major decline. The national BTO Atlas (1976) showed evidence of breeding in all but three of the County's 33 10km squares, with confirmed or probable breeding in 26 of those squares, and possible breeding in a further four. Comparable figures for the 1985–90 Atlas were absence from five, 17 and 11 respectively, and for 2008–13 were 10, 6 and 17 respectively. The decrease in squares with confirmed and probable breeding reflects fewer staying to breed.

Although many BBS transects do not contain any suitable habitat, only five have been counted during BBS fieldwork since 1994, another indication of current scarcity.

The population was estimated at 90–180 in 1990. Now, based on TTV counts, it is estimated at 20–25 pairs. Given the number on the Long Mynd, this may be too low in good years, but a population estimate for a species found on so few TTVs is unlikely to be reliable.

Possible causes of the decline are habitat loss due to suitable vegetation being overgrown by scrub or removed by overgrazing, and drainage, here and on the wintering grounds, as well as climate change, particularly in Africa. However, climate change here might be having an effect in the UK too, as the relative abundance change map in the national BTO Atlas 2013 shows a substantial shift westwards and northwards.

The only ringing recovery was a bird ringed as a juvenile on 3 August 1985 at Walcot, Wellington, found long dead on 21 November 1989 in Sussex, by which time it had presumably gone to Africa and back four times.

Pete Nickless

Blackcap (Eurasian Blackcap)
Sylvia atricapilla
Very common summer visitor, uncommon winter visitor

John Hawkins, Dudleston Heath, 30 January 2007

The Blackcap breeds across most of Europe. It can be seen here in all months of the year, with a large population of summer visitors from the western Mediterranean basin and West Africa being replaced by a much smaller wintering population from the Low Countries of Belgium and Germany, and elsewhere in north-west Europe. It is one of our best songsters, its rich melodic fluting song, delivered from a favoured song-post or from treetops or scrub, having earned it the name of 'northern nightingale'.

Beckwith (1889) noted that the Blackcap was found in most woods and copses. Though always plentiful, it had increased in numbers in the preceding years. There is no evidence of any large change since then, and the *Handlist* (1964) described it as 'fairly common' in most areas. Blackcaps are now well distributed in the breeding season, having been found in all but 18 tetrads in the recent Atlas period. The net increase of 56 tetrads with breeding evidence reflects an increasing population moving into sub-optimal habitat at low densities. The relative abundance map shows they are more common in the east, and are less common in the large-scale arable farming areas of the northern plain and the upland areas of the south-west. This is a

Breeding Relative Abundance (2008–13)

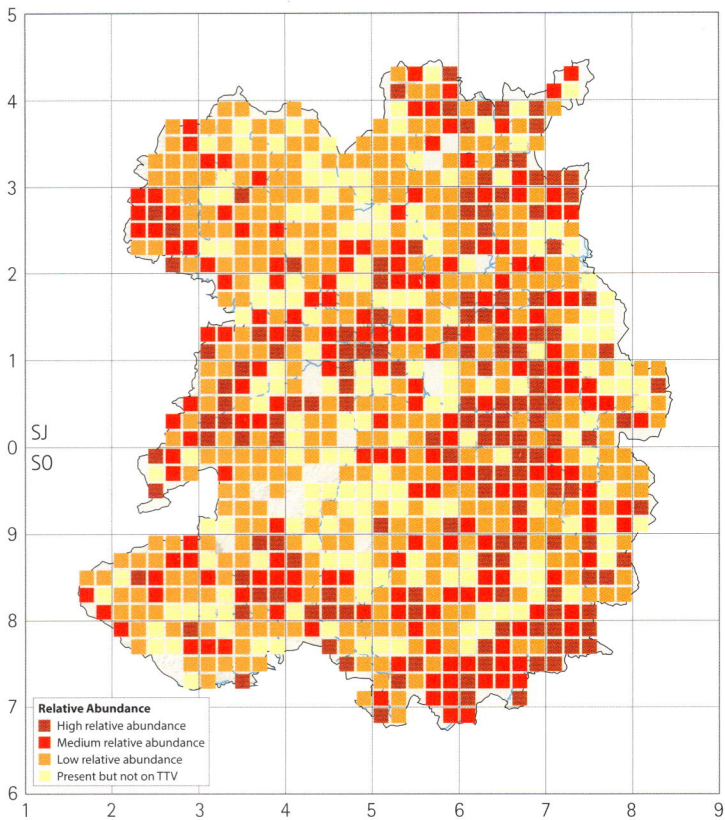

Relative Abundance
- High relative abundance
- Medium relative abundance
- Low relative abundance
- Present but not on TTV

Occupied Tetrads

Atlas period (breeding)	1985–90		2008–13		Change	
	Number	%	Number	%	Number	%
Confirmed	335	39	265	30	–70	–21
Probable	311	36	444	51	133	43
Possible	150	17	143	16	–7	–5
TOTAL	796	91	852	98	56	7
Tetrads with Winter Records (2007–13): 121 (14%)						

128 respectively). These tetrads are part of the '110 tetrads south and west of SO39', and generally have low relative abundance of Blackcap, and high relative abundance of Garden Warbler. Blackcap is a short-distance migrant which arrives back earlier, often before spring arrives in the uplands. The national relative abundance maps show that Blackcap occurs at highest densities in lowland areas in southern Britain, while Garden Warbler has a western bias, with highest densities in Wales, so it is more tolerant of the late spring and higher rainfall found in these areas.

Over-wintering was initially noted by Beckwith, although no details were given, and it was not mentioned in any of the other historic literature. Lloyd (1944) noted a bird of the year on 6 December 1916, a male on 10–14 March 1934 which 'had perhaps wintered here', and two in the winter of 1943–44, a male present for about a month from December, and another in February. The first three were near Shrewsbury, the fourth near Bridgnorth. 'So far as I know ... these are the only records of wintering Blackcaps'.

The first modern record was in 1958, a male in Shrewsbury on two February dates, followed by a female 'probably over-wintering' in March 1961, with one in Shrewsbury and at least four at Ashford Carbonell, in 1962. Records have steadily increased since, rising to 87 in the four winter months in 2006, the last year before the recent Winter Atlas started. These records included five, four and three separate individuals at Copthorne, Belle Vue and Madeley respectively.

The winter Atlas map shows they were found in 121 widespread tetrads, with a preference for milder conditions, typically in gardens around larger towns and in the river valleys, where the climate is slightly warmer.

Recent research by BTO (Plummer *et al.* 2015) has shown that this winter population growth reflects a new migration strategy, with a discrete population breeding in central Europe coming to Britain to

pattern shared with some other warbler species and is likely to relate to the more diverse habitats, particularly with shrubby undergrowth with blocks of scrub, found more frequently away from intensively farmed and upland landscapes.

The population estimate, based on TTV counts is 19,300–19,700 territories, compared to the estimate of 6,500–10,000 pairs in the *Atlas* (1992). The two estimates are consistent, as the BBS trend has shown an increase of 128% since 1997.

Although Blackcap and Garden Warbler are closely related, and occupy similar habitats, the relative abundance maps are very different. Comparison of the respective population and distribution has been made, in the *Handlist* (1964), which stated Blackcap was the more numerous except around Bucknell and Bishop's Castle, and around Newport, where Garden Warbler predominates; and in the *Atlas* (1992), which compared tetrads with probable or confirmed breeding of one of the two species but not the other, and showed that generally Blackcap was more widespread, 'but in the 110 tetrads south and west of SO39 [including Bishop's Castle and Bucknell], the pattern is reversed, with 15 tetrads occupied by Blackcap but not Garden Warbler, and 24 *vice versa*'. Though the *Atlas* (1992) did not make a similar comparison with the area around Newport, inspection of the respective maps shows a cluster of confirmed breeding records for Garden Warbler which were not replicated for Blackcap.

The population of Blackcap has more than doubled since 1990, but that of Garden Warbler has declined considerably, in common with other trans-Saharan migrants. Even so there were almost as many Garden Warblers counted as Blackcap in TTVs in the 40 tetrads south-west of Bishop's Castle, in SO18, SO27 and SO28 (113 and

BBS Trend (1997–2014)

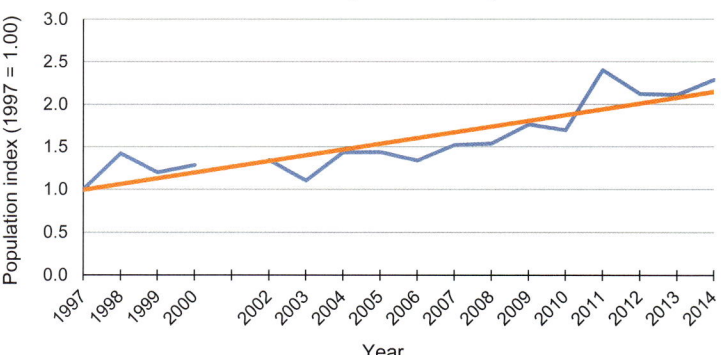

Winter Distribution (2007–13)

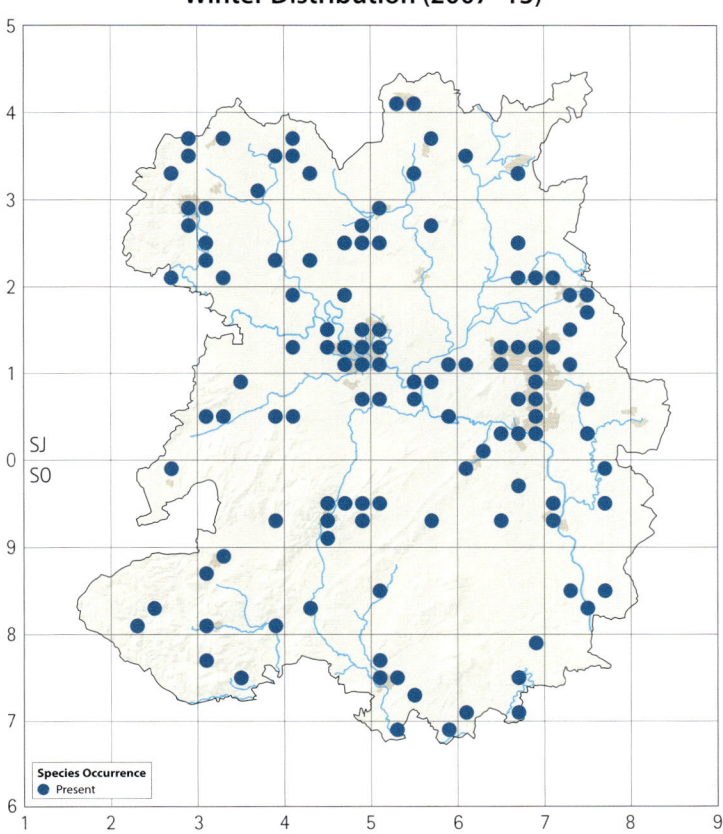

Garden BirdWatch Reporting Rate (1995–2014)

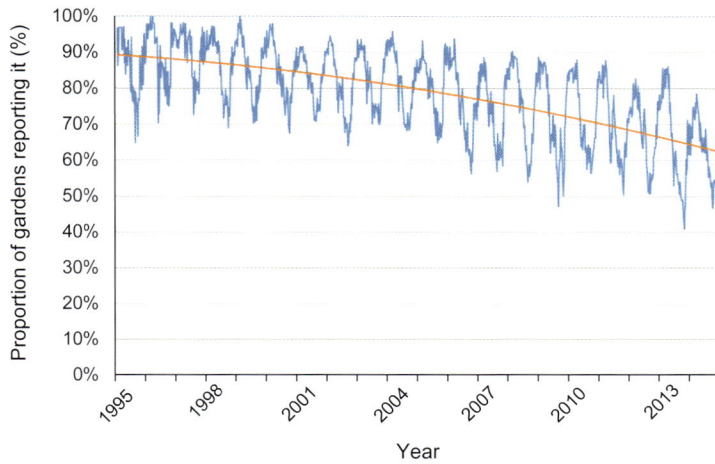

take advantage of the milder climate, but also the increasing amount of bird food, particularly fats and sunflower hearts, put out in gardens in recent years. The occurrence of Blackcaps has become more strongly associated with provision of this bird food over time. The new strategy is successful, and the discrete population is increasing.

The peaks in the Garden BirdWatch graph occur in winter months, and, although the winter population is very much smaller, the reporting rate of Blackcaps in gardens in the four winter months was more than twice that in the four breeding season months during the six years of the recent Atlas period. The very large peaks correspond with cold

weather and heavy snowfall in March 1996 and April 2013.

The ringing recoveries illustrate the migration routes and wintering area of the breeding population. Around one-quarter of the Blackcaps ringed here have also been recovered here. Around two-fifths (22) have been recovered in Britain and Ireland, almost all in counties to the south, and around two-fifths (19) have been recovered abroad. This includes two caught in Gibraltar (over 1,850km), and seven records from North Africa (six from Morocco and one from Algeria, a Blackcap ringed at Walcot in July 1986 and trapped alive in March 1987, 1,849km distant). The other recoveries are from Europe, six from Spain and singles from the Netherlands, France, Belgium and Germany. At least five were recovered on the south coast, or abroad, within two and a half months of ringing. The oldest was trapped at Cadiz in Spain, 6y 5m 11d after it was ringed as a first-year male in September 1979, 1,798km away.

Only one ringing recovery involves a wintering Blackcap. It was found dead at Pant in February 1996, having been ringed as a first-year male at Antwerp in Belgium in October 1993, a movement of 579km WNW from the ringing site, reflecting the origin of those wintering in the UK in the Low Countries.

Mike Shurmer

Garden Warbler

Sylvia borin

Common summer visitor

Garden Warbler is a summer visitor which winters in tropical western and southern Africa. The first few appear typically in mid to late April, but most arrivals are noted in early to mid-May. Its preferred habitat is open broad-leaved woodland with a tangled understorey, shrubby overgrown hedges, and scrub.

The first historical reference, from Leighton (1836), described it as 'common in high leafy sycamores' in the 'neighbourhood' of Shrewsbury. More typically, Rocke (1865) noted that it was shy and easily overlooked, but often heard 'pouring forth its song from the thick foliage of some tree or bush'. In various accounts Beckwith also described it as common, but less so than Blackcap, which was

found in similar habitat. He noted its vocal powers, particularly in the dingles around the Wrekin, often heard at night; retiring during the breeding season, but more commonly seen in autumn, feeding on elderberries and blackberries. In contrast, Paddock (1904) believed that it appeared to be more numerous than Blackcap in the Church Stretton area where it was 'vulgarly known as Nettle-creeper', although this name was also used for Whitethroat (Forrest 1909c). It was 'fairly common and probably most common in Wyre Forest' (Forrest 1908), and it continued to be described as 'common' and 'frequent' through the first half of the twentieth century.

Early SBRs usually listed it only among the species present, or as 'common' and 'numerous', but it was 'exceeding blackcap' in the Bucknell area (SBR 1960). The *Handlist* (1964) described it as a widespread 'breeding visitor and passage migrant'; and generally scarcer than Blackcap, except in parts of the north-east and south-west. However, it also stated that more records were needed to show its true distribution. It was later described as 'well distributed' but 'appears to have declined, particularly over the past five years' (Ireson SBR 1967), while a later report noted that it had been steadily decreasing over the previous 20 years (SBR 1969).

Nationally, the population suffered a steep decline in the early 1970s, but it was not as dramatic as that of other trans-Saharan migrants, and its recovery was quicker. It winters well south of the Sahel, and its difficulties arose while migrating across it, rather than in its wintering grounds. However, no reference was made to this in SBRs, and Garden Warbler was still described as widespread and common in the 1970s, although under-recorded and 'overlooked by many observers'. Few breeding records were received, exceptions being at Marshbrook in 1976, a pair feeding young at Howle Pool in June 1977, and juveniles seen with adults at Chelmarsh in August

of that year. During the 1980s it seems to have been better recorded, and there were several breeding records, and instances of it feeding on elderberries and blackberries in autumn – Kingsland in October 1980, Cosford in October 1985, and Alkmund in August 1988.

The national recovery was largely complete by the time of the 1985–90 Atlas, which showed it to be well distributed, but more likely to be absent from central and eastern areas which have more arable farming, and therefore less scrub to provide suitable breeding habitat. The decline and recovery was also reflected in counts on the CBC plot at Oswestry Old Racecourse, where there was one territory in 1970 and 1971, and gaps in some years until 1985, when there were two. By the late 1980s there were six, and in 1991 there were nine. After 1991, the number fluctuated between three and seven, indicating a fairly stable population (SBR 1999), while 2005 was noted as a particularly good year overall, with 90 reports from 49 sites, including two breeding records, adults carrying food at Chipnall Lees in June and a juvenile at Priorslee Lake in August.

In the recent Atlas period, it was found in 556 tetrads (64%), a contraction in range compared with 644 tetrads (74%) in 1985–90. The increase in the proportion of possible breeding records, and the decrease in probable and confirmed breeding, suggest that it was more difficult to find almost everywhere. Both observations suggest a considerable population decline since 1990.

The breeding distribution change map shows a number of ups and downs which are probably due to it being under-recorded for one or other of the Atlases, but a net disappearance from about 14% of tetrads where it was previously found, attributable to continued scrub and hedgerow removal for intensive farming and the temporary

Occupied Tetrads

Atlas period (breeding)	1985–90		2008–13		Change	
	Number	%	Number	%	Number	%
Confirmed	216	25	84	10	-132	-61
Probable	245	28	206	24	-39	-16
Possible	183	21	266	31	83	45
TOTAL	**644**	**74**	**556**	**64**	**-88**	**-14**

Breeding Distribution Change (1985–90 to 2008–13)

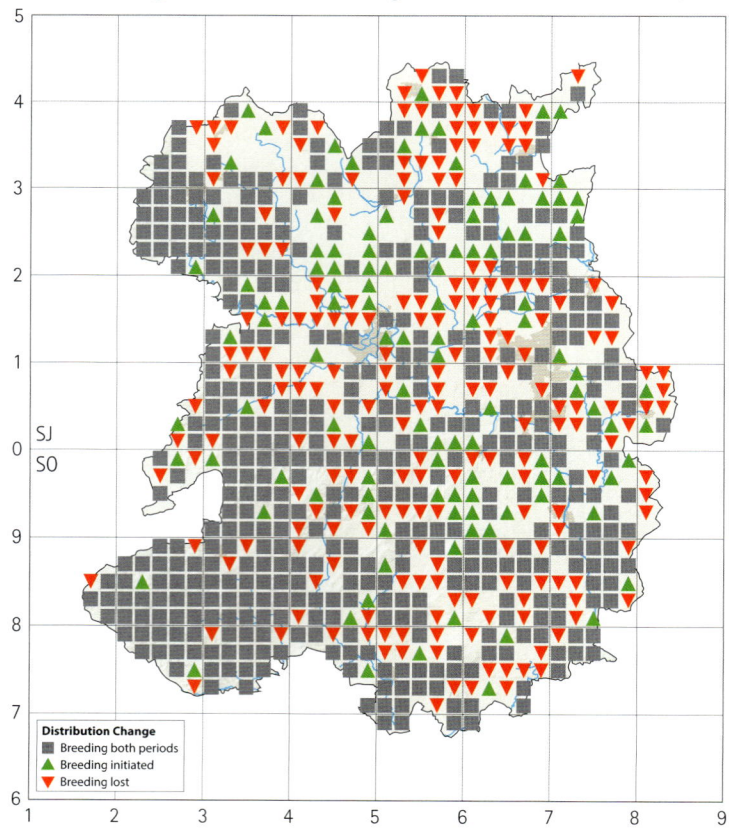

Distribution Change
- Breeding both periods
- Breeding initiated
- Breeding lost

Breeding Relative Abundance (2008–13)

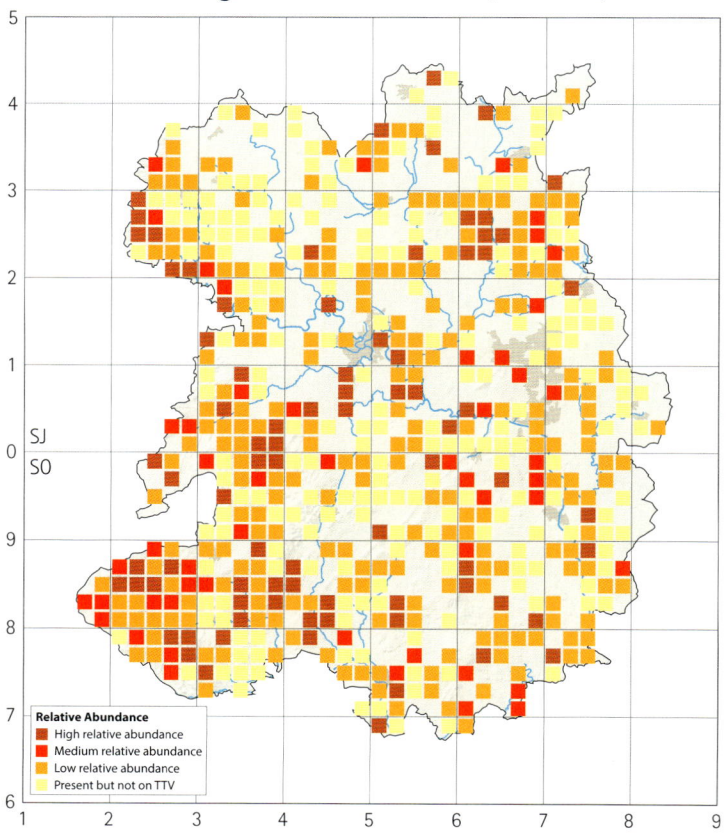

Relative Abundance
- High relative abundance
- Medium relative abundance
- Low relative abundance
- Present but not on TTV

nature and lack of management of other scrubby habitat, together with the underlying population decline.

The relative abundance map shows it to be still widely distributed in upland and rural areas, with the highest densities in the west, south-east, and west of Newport. This pattern reflects the western bias of relative abundance nationally, where the highest densities are in Wales. It avoids the more built-up areas. Its preference for hiding in dense vegetation, and the consequent difficulty of identifying it, together with the challenge of differentiating its song from that of Blackcap, all suggest that it may be under-recorded. A comparison with the habitat and distribution of the closely related but generally more common Blackcap is given in that account.

Garden Warbler occurs in insufficient BBS squares to produce a valid local population trend, but the data suggest a decline of around 50% since 1997. BBS shows a 27% decline in the West Midlands between 1995–2014 and 28% in England as a whole. In common with several other trans-Saharan migrants, this recent decline is again attributed to drought in the Sahel, and to habitat changes in their wintering areas. Based on TTV counts, the estimated breeding population here is around 4,000–4,200 pairs. The lower estimate of 1,600–2,600 pairs published in the *Atlas* (1992) did not include any weighting for the nationally high relative abundance here.

Earliest arrival dates have been noted in many SBRs, and the results shown in the *Arrival Dates* chapter indicate that Garden Warbler is arriving here around 18 days earlier than it did in 1900, with two-thirds of the advance since 1950.

The few recoveries largely reflect the migration route. Five of the six involve counties to the south – Gloucestershire (two), Kent, London and Wiltshire. The only one from abroad, a juvenile, ringed at Dothill, Telford in August 1983, was trapped near Cordoba, Spain in April 1985, 1,679km to the south, presumably on its return migration.

Helen J. Griffiths

Paul King, Nedge Hill, 9 May 2010

Lesser Whitethroat

Sylvia curruca

Uncommon summer visitor

Lesser Whitethroat winters mainly in north-east Africa, and is a summer visitor to Britain, arriving in mid to late April.

The first account was from Eyton (1838), who described it as 'somewhat rare', but usually arriving before Whitethroat. Rocke (1865b) said that, although it was thought to be 'exceedingly rare', careful observation had confirmed that this was not the case, and that, although easily overlooked, it was in fact quite common. Beckwith (1878) added that it 'often nests in gardens' in the Shrewsbury area, and later said it was rarer than Whitethroat, although widespread, and particularly plentiful in Clungunford where it frequently nested in 'gardens and shrubberies', while in 1889 he stated that its numbers had increased in recent years, to become 'a common summer visitor'. The vernacular name was 'Peggy Whitethroat' (Forrest 1899) but, confusingly, this was also used for Whitethroat, Willow Warbler and Chiffchaff.

In the first half of the twentieth century there were a small number of records of presence and of breeding, from widespread localities,

and early SBRs recorded it as 'plentiful near Ironbridge' and 'well distributed in lowland areas' (1956), but 'under-recorded' (1962). The *Handlist* (1964) said it was found in 'all divisions' and 'much less common than the Whitethroat, though probably overlooked'. Ireson (SBR 1967) believed that it was 'very widely distributed' but 'very thin on the ground', and probably most common in the Teme Valley.

SBRs for the 1970s and 1980s included a handful of records, relating to arrival dates, location and breeding. However, they suggested that it was generally more widely distributed in northern and north-western areas, an impression confirmed by the 1985–90 Atlas, and perhaps indicating an expansion of range since Ireson's observations. It was found to be widely, but unevenly, distributed, more common in the mixed farming areas of the north and south, absent from some central and eastern arable areas, and 'almost certainly under-recorded'. None were found in nine of Shropshire's 33 10km squares during fieldwork for the first national BTO Atlas (1976), but all had some level of breeding evidence in 1985–90, 'consistent with the national expansion in range and increase in population density since then'. Extrapolation of national CBC data suggested an average of two to three pairs per tetrad and a population of 700–1,000 pairs. The account concluded that, as they migrated in a south-easterly direction across Europe to winter quarters in north-east Africa, they were not affected by droughts in the Sahel region.

Since 1990, SBRs have continued to report a small but increasing number of records from widespread places, reaching a maximum of 96 from 45 sites in 2003, including confirmed breeding from a dozen scattered locations.

In the recent Atlas period it was found in 321 tetrads (37%), virtually identical to the 320 tetrads (37%) where it was found in 1985–90.

Although widespread, it was absent from higher ground, mature woodland and arable farmland where hedges have been cut back or removed, particularly in the south. It is also largely absent from the urban areas, but its skulking nature continues to result in under-recording, and this may contribute to its apparently sparse distribution.

The breeding distribution change map is one of the most intriguing. The number of tetrads where it was found in both Atlas periods (137) is exceeded by tetrads where it was found only in 1985–90 (183) or only in 2008–13 (184). The stronghold in the recent Atlas period was in the north-west, with breeding evidence found in 57 out of 75 tetrads, and confirmed breeding in 17.

Favoured habitat is good quality hedgerows with occasional trees, particularly the long-established ones with dense and overgrown hawthorn and blackthorn, and brambles and dense scrub. The song is characteristic and far carrying, but the males are skulking when not

Allan Heath, North Shropshire, 7 June 2010

singing, and once the territory is established and nesting starts, they become quiet and are much more difficult to find. However, when a pair has young their alarm calls are very loud and distinctive, so confirming breeding should be relatively easy, but that occurred in only 18% of tetrads where they were found in the recent Atlas period (almost one-third of which were in the north-west). Late April or early May visits for song and July visits for young are more likely to be successful than those in-between, and if visits are not made in these periods they are likely to be overlooked, especially in areas with low density.

Some apparent losses in the Severn Valley and the north-east can be attributed to further loss of hedgerows in arable areas (bigger fields for bigger machinery), with heavier cutting back of those that remain, and loss of food through a reduction in field margins and use of pesticides. However, such losses appear to have been compensated for by an equal number of gains. Some will be due to under-recording for one or other Atlas. However, it is at the edge of its range here, and, although nationally its range is expanding, it is still uncommon. Potentially suitable habitat is still widespread, particularly in the north, so the few returning males have considerable choice and little competition when establishing territory. As they are short-lived, with a typical lifespan for those reaching breeding age of only two years, there has been little opportunity, or necessity, to develop regular re-use of the same territories. Thus, while many may return to their natal area, other tetrads where they were found will not have been occupied every year.

Breeding Distribution Change (1985–90 to 2008–13)

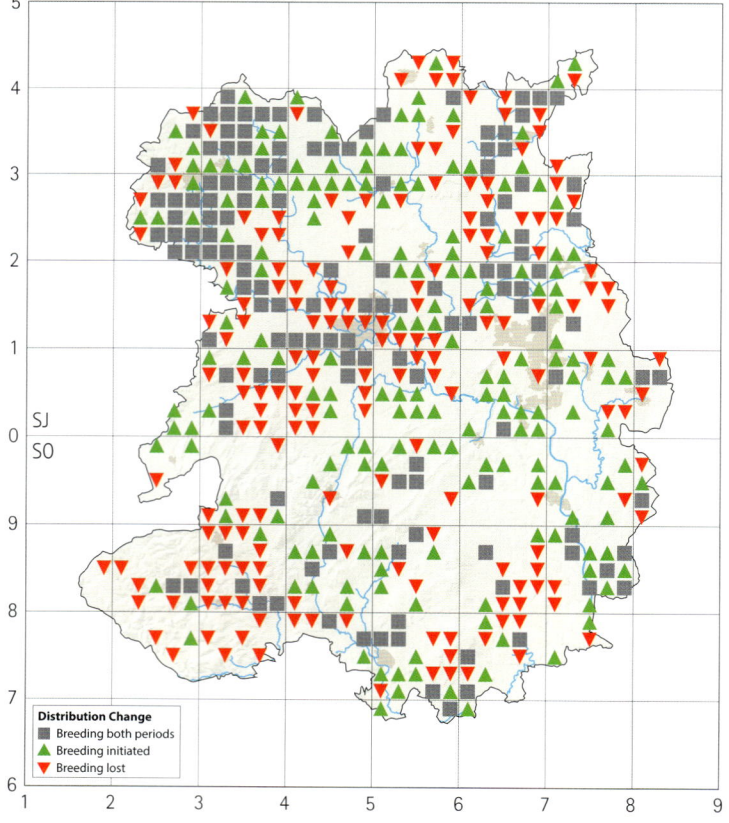

Distribution Change
- ⬛ Breeding both periods
- 🔺 Breeding initiated
- 🔻 Breeding lost

Occupied Tetrads

Atlas period (breeding)	1985–90		2008–13		Change	
	Number	%	Number	%	Number	%
Confirmed	72	8	57	7	-15	-21
Probable	112	13	98	11	-14	-13
Possible	136	16	166	19	30	22
TOTAL	320	37	321	37	1	0

Analysis of the results from the long-running CBC at Oswestry Racecourse, and the raw BBS data since 1997, also indicates that the species is easily overlooked, it has little tendency to use the same sites year after year, and the population has increased somewhat over the last 10 years or so.

The change in relative abundance shown in the national BTO Atlas (2013) suggests losses in the uplands in the south-west and gains in the Ellesmere/Wem area as part of a northerly (climatic?) shift.

BTO data indicate a fluctuating population with no long-term trend. It occurs in insufficient BBS squares to produce a valid local or regional trend, but the data for England suggest virtually no change (an overall 2% increase) between 1995–2014. Based on TTV counts the population is estimated at 1,100–1,200 territories, slightly more than the 1992 estimate of 700–1,000 (an average of two to three pairs per tetrad where it was found), but this figure is often exceeded in the north-west stronghold (Allan Dawes *pers. comm.*).

The *Arrival Dates* chapter suggests that Lesser Whitethroat is arriving here around six days earlier than in 1900.

There was one winter record, of one in a Shrewsbury garden, first recorded on 21 December 2008 and seen intermittently in this and adjoining gardens until 7 April 2009. It is rare in the UK in winter and most are believed to be of eastern races; the record was considered by the SOS Rarities Committee who thought it likely that this bird was of an eastern race (SBR 2008).

Very few are ringed, and there have only been four recoveries. One was ringed and recovered here, and one was found in Leicestershire, but the other recoveries reflect the migration route. A first-year ringed in late August 1988 at Dover was killed by a cat in Snailbeach during the breeding season the following June; another first-year caught in September 1977 at Beachy Head was re-caught in Cressage at the end of August 1980. The only recovery of one ringed abroad reflects the migration route across Europe to north-east Africa. An adult ringed in Bergamo, northern Italy, in early September 1997, was killed by a 'domestic animal' the following June, near Diddlebury, a movement of 1,172km.

Helen J. Griffiths

Whitethroat (Common Whitethroat)

Sylvia communis

Common summer visitor

Whitethroat breeds across most of Europe, and winters in tropical western and southern Africa. Males typically arrive here in mid to late April and females a fortnight or so later. It is found most commonly in hedgerows bordering lanes and farmland, areas of scrub and gorse, and regenerating woodland.

Leighton (1836) described it as 'common in woods' and Eyton (1838) also noted it was common, usually arriving during the first fortnight of April. Rocke (1865) observed that it was 'probably the most numerous of all our spring birds', and it was 'abundant' in the Shrewsbury area (Beckwith 1878), often nesting 'not withstanding many dangers, in ditches by the sides of public roads' (Beckwith 1889). Paddock (1890; 1897) also noted that it was abundant everywhere, and that it visited his garden to feed on raspberry, cherry and currant, while a different Paddock (1904) reported it as an 'abundant summer visitor' in the Church Stretton area.

Vernacular names included 'Nettle Creeper' due to its 'fondness for beds of nettles', 'Jack-Straw' 'in allusion to its nest of loosely built grass stalks' (Forrest 1899), and 'Flaxey' (Paddock 1890).

It was still described as common, abundant and widespread throughout the first half of the twentieth century, but a 'marked decrease' was noted in 1930, with a 'marked increase' by 1935, although no explanation was suggested. It was 'the most common of the nesting warblers' in the Clun Forest in 1950 (Adams), and it was still 'common' (*Handlist* 1964), 'numerous in all areas' (SBR 1966), and the second most common warbler, after Willow Warbler, although a decrease in numbers breeding alongside roadside verges was noted 'owing to the increase in traffic and roadside spraying' (Ireson SBR 1967).

Although it was mapped with probable or confirmed breeding in every 10km square in the BTO Atlas (1976), a dramatic decline occurred from 1969 onwards. The national CBC farmland index fell by about

Terry Arch, Broseley, 28 April 2014

70% in one year, and only nine records were received (SBR 1969), due to a severe drought in its wintering grounds, the Sahel region. SBR 1971 noted only two territories in the Oswestry Racecourse CBC plot, compared with an average of seven before 1969. Nationally, by 1974 the CBC farmland index had declined to only one-sixth of the 1968 level, but locally numbers appeared to show a slight recovery. The 'improvement in numbers' was 'apparently maintained', with 10 pairs breeding at Allscott Sugar Factory (SBR 1974), a 25% increase on 1973, and by 1975 it was 'well distributed in small numbers', while by 1979 it was 'widespread after mid-May', with an increase over 1978, with 21 singing at Allscott Sugar Factory on 12 May, and a good breeding season there. Many other fledged young were reported that year.

Breeding Distribution Change (1985–90 to 2008–13)

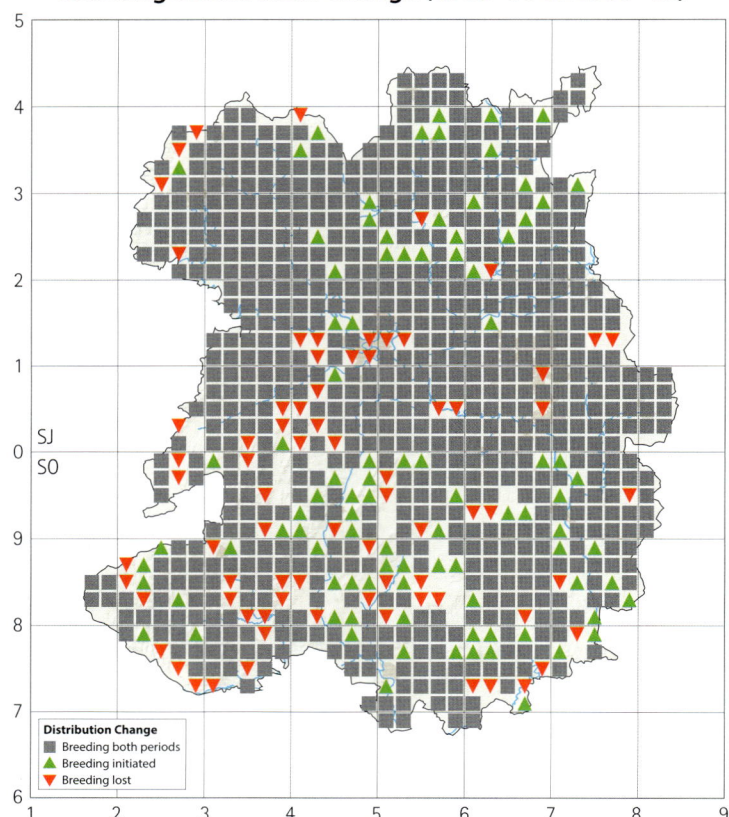

Distribution Change
- ■ Breeding both periods
- ▲ Breeding initiated
- ▼ Breeding lost

Breeding Relative Abundance (2008–13)

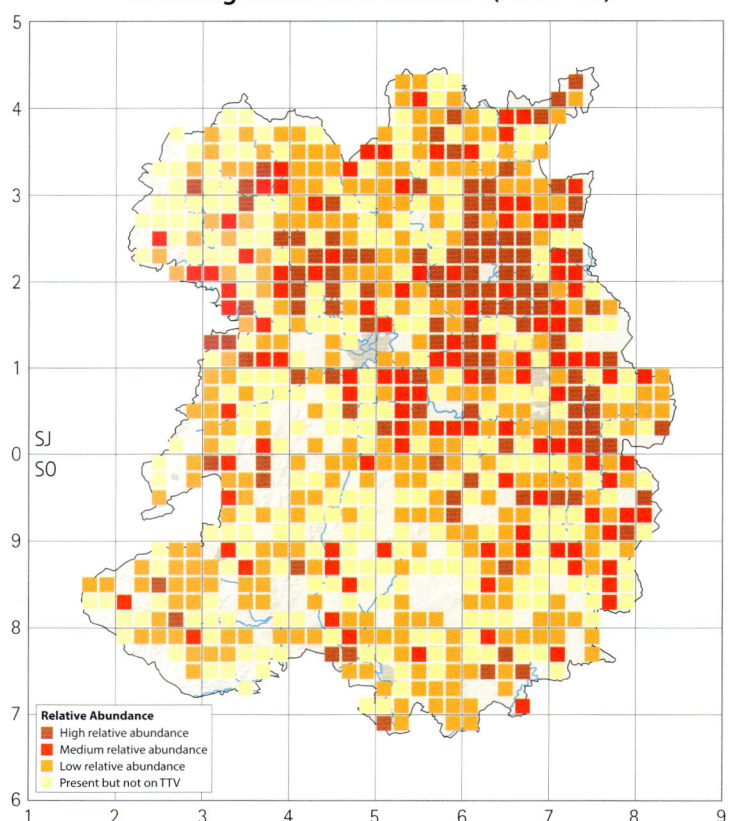

Relative Abundance
- ■ High relative abundance
- ■ Medium relative abundance
- ■ Low relative abundance
- □ Present but not on TTV

Occupied Tetrads

Atlas period (breeding)	1985–90		2008–13		Change	
	Number	%	Number	%	Number	%
Confirmed	374	43	306	35	-68	-18
Probable	238	27	311	36	73	31
Possible	139	16	154	18	15	11
TOTAL	751	86	771	89	20	3

Numbers fluctuated during the 1980s, with 1985 being identified as a poor breeding season, but the *Atlas* (1992) confirmed they were widespread, particularly in the north, but largely absent from areas of mature woodland and the uplands. The *Atlas* account concluded that 'conditions in winter quarters determine the population levels more than breeding success and habitats here'.

In the recent Atlas period, it was found in 771 tetrads (89%), compared with 751 tetrads (86%) in 1985–90. This very modest increase is likely to reflect its continued slow recovery. The breeding distribution change map shows losses in the growing built-up areas of Shrewsbury and Telford, but also from parts of the Teme Valley and from an area to the south-west of Shrewsbury. However, there is an increase in several areas, particularly in the south and north-east, reflecting a growing population moving into new habitat. Further evidence of this comes from monitoring on the Long Mynd 1994–98, where the first upland territory was not found until the third year, but by 1998 there were four territories in the upper reaches of three valleys, all associated with gorse bushes on heathland above 400m. In 2006–08, a repeat survey found 25–30 pairs widely distributed

in the upper valleys and open heath. Also, a survey of the whole Stiperstones ridge in 2004–07 found a total population estimated at 59–65 pairs, including 12–14 pairs in the southern half of the NNR, where comparison with the results of a similar survey in 1994–95 suggested a fourfold increase (Leo Smith *pers. comm.*).

The relative abundance map shows it to be widespread, but a line drawn from north-west to south-east shows very few unoccupied tetrads and much higher densities north of the line, where more hedges and areas of scrub provide suitable habitat, while south of the line the more open upland habitats, and mature woodland, is less suitable, with corresponding lower densities.

The BBS graph shows a population increase of 28% between 1997 and 2014, with a stronger underlying trend.

National CBC and BBS data show this consistent shallow recovery has been sustained since the mid-1980s. In 2014 the UK population

BBS Trend (1997–2014)

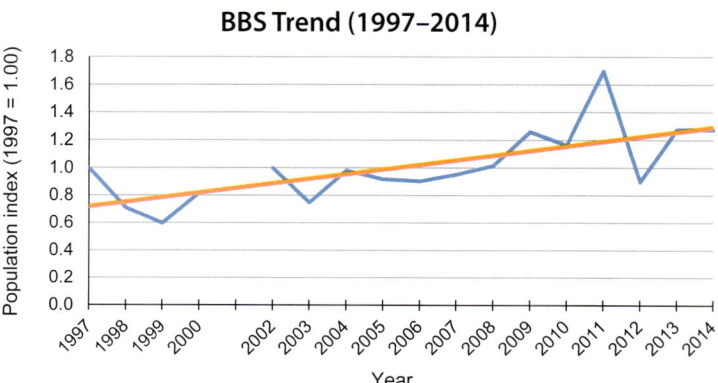

was 9% higher than its 1970 level, but only about one-third of its 1968 level.

Based on TTV counts, the estimated breeding population here is 12,000–12,350 pairs. Allowing for the BBS trend, this is consistent with the estimate in the *Atlas* (1992), a population fluctuating between 5,000–10,000 pairs, and towards the upper end of that range in 1990.

Earliest arrival dates have been noted in many SBRs, and the *Arrival Dates* chapter indicates that Whitethroat is arriving here seven days earlier than it did in 1900.

One ringed at Leegomery in August 1965 was found in Portugal 1,254km SSW, 2y 0m 25d later, and another ringed at Beachy Head in August 1975 was caught near Bowbrook in May 1977, both reflecting the migration route.

Helen J. Griffiths

Dartford Warbler

Sylvia undata

Vagrant

It is resident in southern England and Wales, which are at the northern edge of its range in south-west Europe.

In the autumn of 1902 'some small birds in a gorsy patch near Ludlow' were seen. The following year, in May, the observer relocated the birds, confirmed that they were Dartford Warblers, and that there were two pairs. On 17 June a nest was found with five eggs by a local collector who 'secured the eggs'. The second pair was found with a brood of 'four young at the beginning of September' (Forrest 1903; 1908). One pair 'bred' at the same location in 1904 but there is no record of the outcome, and there were no subsequent records.

The breeding locality was evidently in the Teme Valley, close to the border with Herefordshire. The patch of gorse where the breeding occurred was 'burnt down two years ago and the birds have not been observed since' (Page 1908).

Apart from such isolated records, they were found only in southern England at the start of the twentieth century, and habitat loss and occasional severe winters confined them there until, more recently, milder winters have enabled them to spread rapidly northwards in the last 40 years.

After breeding ceased, a century passed before the one modern record:

The Stiperstones (Mytton Dingle) 23 March–8 April 2000

This record of a male, sometimes singing, coupled with confirmed breeding in Staffordshire, raised hopes that they might breed here too during the recent Atlas period. However these hopes were dashed by several severe winters, which have reversed the increase in numbers and range, postponing the hoped-for colonisation.

John Tucker

Firecrest (Common Firecrest)

Regulus ignicapilla

Rare passage migrant and winter visitor, very rare breeding species

On the morning of 16 December 1882, a young male Firecrest flew into a shop on the Wyle Cop, Shrewsbury, an error of navigation compounded by proximity to the premises of John Shaw, a well-known 'bird stuffer', to whom the itinerant was taken. It was an unfortunate end, but the lives of all eight nineteenth-century Firecrests listed by Forrest (1907; 1908), the first being in 1843, were foreshortened by human hand. Where specified, all were winter occurrences, but the next was on 14 September 1905, at Cockshutt; it was killed by a boy with a stone. Thereafter, up until 1971, there were just three records (in 1906, 1913 and 1939), but all, it seems, were of birds safe in the bush.

From 1971 onwards, some 57 further individuals have been recorded, and there have only been 11 blank years. November and March are the months in which most were seen; these are likely to be on passage, leaving central Europe with the onset of cold weather and returning there in the spring (Riddiford & Finley 1981). Although some may linger with us over winter; no new arrivals have been spotted in the months June–August.

The chart covers the years 1971–2014 and shows the month in which individuals were first recorded (it has been assumed that both members of the pair that bred at Kempton in 2012 were present in

March when song was first heard). Each column comprises records for the entire time span, with the orange area representing those made during the recent Atlas period.

Marchant (1997) described the significant north-westward expansion of the breeding range in Europe throughout the twentieth century, and it is presumably individuals from this extended range that visit on passage. Furthermore, the national BTO Atlas (2013) reported that the winter distribution has more than doubled in extent over the last 30

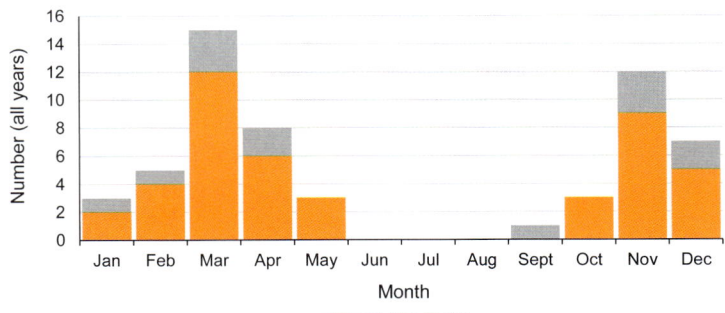

Month When First Recorded, 57 Individuals (1971–2014)

Number (all years) / *Month*

■ Recent Atlas Period

Occupied Tetrads

Atlas period (breeding)	1985–90		2008–13		Change	
	Number	%	Number	%	Number	%
Confirmed	0	0	1	0	1	x
Probable	0	0	1	0	1	x
Possible	2	0	1	0	-1	-50
TOTAL	**2**	**0**	**3**	**0**	**1**	**50**

Distribution & Breeding Status (2007–13)

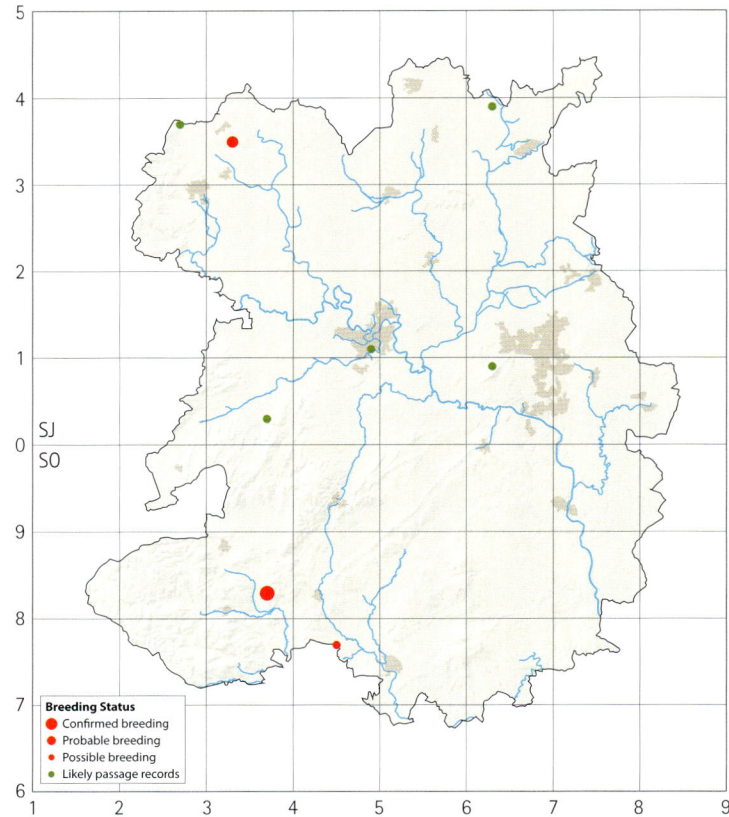

Breeding Status
- Confirmed breeding
- Probable breeding
- Possible breeding
- Likely passage records

years, with a substantial shift both inland and northwards, although detection rates will have been enhanced by a progressive increase in the number of observers, and by the use, from January 1966, of mist-nets in woodland, since when 11 individuals which might otherwise have gone undetected, have been caught by ringers; interestingly only one of the seven for which they determined the sex was a female.

The records are widespread, but there have been four from each of three locations: Hinstock, where, in each case, individuals were mist-netted, Market Drayton (one netted) and Whitcliffe Woods. The habitat at these locations may be particularly favourable, but these are sites well worked by observers and ringers; presumably Firecrests go unrecorded in other, less frequented, places.

England and Wales are at the northern edge of the breeding range, but Firecrests have flourished and spread northwards since breeding was first confirmed in the New Forest in 1962. In excess of 1,000 territories were estimated by 2010 (Holling *et al.* 2012), a few of which were in Wales and central England, where breeding was confirmed in Worcestershire as early as 1975 (Harrison & Harrison 2005) and in Montgomeryshire in 1982 (Lovegrove *et al.* 1994). Males in song in the Wyre Forest in 1982 and again in 1989, during Atlas fieldwork 1985–90, when a singing male was noted too at Oxon Pool (1990), raised hopes that Shropshire would follow suit, but in the end the *Atlas* (1992) yielded just these two 'possible breeding' records, in SJ51L and SO77I (note that the published map showed possible breeding in SO78J, and probable breeding in SO77I, but the former, and one of the two records that made up the latter, have since been considered not acceptable by the recent review of rarities records).

Two at Brown Clee on 25 April 2004, one of them in song, was highly suggestive of breeding, particularly as song was heard again on 8 June and presence suspected on 22nd, but it was not until 2012 that breeding was proved. In that year, a pair was seen at New Marton on 7 and 9 April but breeding was not confirmed, nor was it to the west of Stead Vallets, where a single was present on 20 May (this was in the Herefordshire part of the tetrad, so it is not included in the chart), but it was eventually proved at Kempton, where song

was heard on 29 March and on many subsequent dates, with a pair observed on 5 July and young on 10 July.

The map shows these three 2012 breeding season records and all other tetrads for which there were records during the Atlas years, these others being interpreted as birds on passage. There has only been one subsequent record, a single at Knuck Wood on 13 April 2014.

The Kempton nesting site was an undistinguished hotchpotch and provides few pointers for future searches: a patch of trees, including mature Turkey oak, with younger yew, cypress, beech and wild cherry, bisected by a busy 'B' road and adjacent to a lay-by with phone and letter boxes. Surely there must be very many equally or more suitable sites, but given the size and restlessness of Firecrests, and a song which, like that of the Goldcrest, is too high-pitched for some to hear (Tucker *et al.* 2014), it is likely that in the breeding season and indeed at all times of year, many go undetected.

Tom Wall

Goldcrest

Regulus regulus

Common resident

Early nineteenth-century scientific accounts of this species (often then referred to as 'Gold-crested Wren', 'Golden-crested Wren' or 'Golden-crested Regulus', sometimes unhyphenated) report it as familiar and common, especially in areas with coniferous trees. Additionally, Victorian ornithologists commented on its status as

the smallest British bird, a focus of much early commentary on its population dynamics. Fatalities due to hypothermia and starvation during extended periods of cold weather (when food availability is reduced) are magnified by its lack of adaptability in diet, and winter population 'crashes' were reported after the cold winters of 1916–17,

John Hawkins, Pentrecoed, 13 April 2008

1929–30, 1939–40, 1962–63, 1985–86 and 2009–10, with authors often reporting very few or none in the following breeding seasons.

Large winter losses are offset by a breeding strategy that results in rapid recovery in numbers, including large clutches, multiple broods and serial nesting (starting a second nest while the first has young). Consequently, an 'encouraging number' was reported in October 1920 after the population crash of winter 1916–17, in which the Goldcrest was described as 'almost exterminated' in many areas. SBR (1964) records increasing numbers throughout the year after the winter of 1962–63, during which it 'suffered badly', according to the *Handlist*. Likewise, SBR (1987) reported larger numbers compared to the previous year after the 1985–86 severe winter. As a result of these and similar recoveries, population numbers fluctuate widely from year to year.

SBR accounts describe Goldcrest primarily as an inhabitant of coniferous woodland, a habitat offering little competition from other species, and in which its size and agility are well adapted to seeking food among the smallest shoots and needles of the canopy. Favoured trees are yew, pine, spruce and fir, although many others are used for foraging, including many deciduous species, especially outside the breeding season.

The breeding period extends from April until August. The timing of stages in the nesting cycle at Dowles was reported by Elliott (1914), who found nest building took 18 days, incubation 14–16 days and unfledged young were present in the nest for 16–19 days, periods broadly consistent with modern BTO BirdFacts data.

Twentieth-century afforestation with conifer species was predicted by the *Handlist* to favour a regional population increase, and the six 10km squares with only possible breeding evidence mapped in the 1968–72 national BTO Atlas all had several tetrads with confirmed breeding in 1985–90. This increase was largely complete by then, as a comparison with recent Atlas results shows only a trivial further increase in tetrads occupied for breeding (from 64% to 67%).

The distribution is essentially unchanged between the two local Atlas periods, with the greatest stability in breeding status in the west and, to a lesser extent, the south-west.

The largest area with few breeding records is mainly arable farmland in the north, circumscribed by Ellesmere, Baschurch, Myddle, Wem, Market Drayton, Prees Heath, Whitchurch and the Cheshire border, which lacks extensive tree cover. High relative abundance occurs in wooded areas, especially those with high proportions of conifers (the uplands west of Oswestry, the area between Pontesbury and Knighton, around Ludlow, the Wyre Forest and Severn Valley to the north, a strip from Wellington to Brown Clee, and south-west of Market Drayton). The presence of breeding Goldcrest in areas without large areas of coniferous woodland suggests that it is well able to thrive in mixed and broad-leaved woods, and in more open areas with a moderate number of scattered trees in hedgerows, gardens, coppices, game coverts and the like. The 22 breeding territories found by Bishton (2003), in a survey of the woodlands of Telford and Wrekin, occurred in deciduous, coniferous and mixed woodland, at a density of about 17 nests per sq.km (equivalent to an average territory size of 6ha).

BBS results are insufficient to calculate a local population trend, but the percentage of routes where Goldcrest was detected averaged 25% between 1994 and 2014 but with a short-lived peak of 40% in 2005 (following several mild winters and favourable summers) and a minimum of only 12% in 2009, after a colder than average winter.

Occupied Tetrads

Atlas period (breeding)	1985–90		2008–13		Change	
	Number	%	Number	%	Number	%
Confirmed	177	20	119	14	-58	-33
Probable	204	23	245	28	41	20
Possible	172	20	219	25	47	27
TOTAL	**553**	**64**	**583**	**67**	**30**	**5**
Tetrads with Winter Records (2007–13): 635 (73%)						

Breeding Relative Abundance (2008–13)

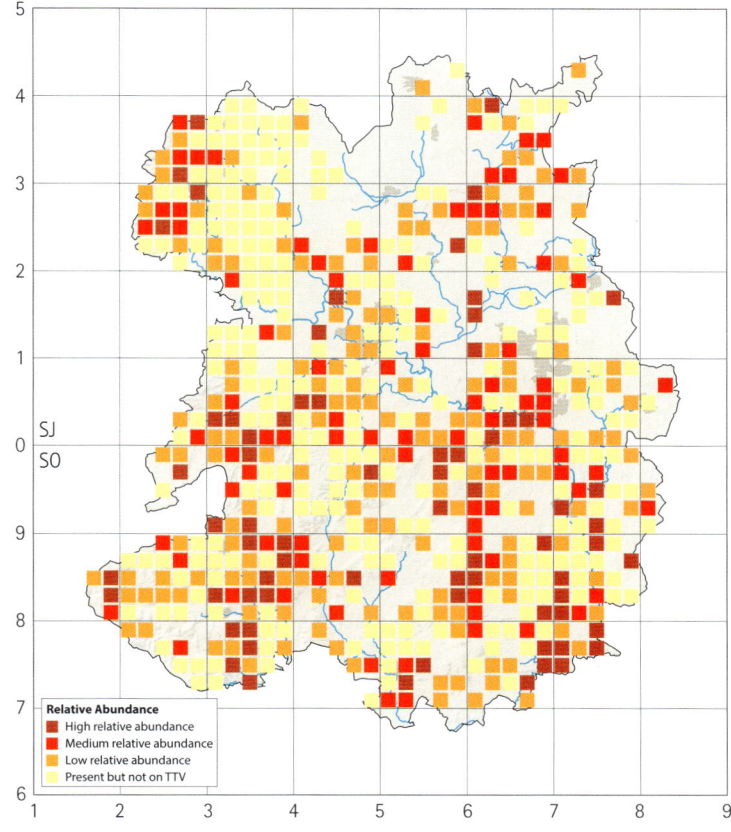

Relative Abundance
- High relative abundance
- Medium relative abundance
- Low relative abundance
- Present but not on TTV

The *Atlas* (1992) suggested a population of 30,000–50,000 pairs, but the calculation erroneously applied the density in woodland to the whole County area, not the 8% with woodland. The population estimate based on TTV counts is 5,600–5,800 territories. Smoothed trends for the UK as a whole (to remove weather-related population fluctuations) show a relatively stable population, while there has been an 83% gain in the West Midlands, and a 33% decline in Wales, between 1995 and 2014. These discrepancies may result from inadequate sampling of such a small, inconspicuous bird with a penchant for deep, dark conifer forests, or regional variation in weather patterns, rather than any real change in abundance.

Goldcrest joins post-breeding and wintertime mixed foraging parties in woodland and other habitats, most commonly with Coal, Blue, Great and Long-tailed Tit, and Treecreeper, with Blackcap, Chiffchaff, Great and Lesser Spotted Woodpecker and Siskin as less frequent flock participants.

In winter, it occupies roughly the same areas as in the breeding season, but evidence of local dispersal and immigration is apparent, with tetrads occupied increasing from 67% to 72%, and much of the area of breeding absence in the north is occupied. Areas of high relative abundance remain in approximately the same locations, but densities are higher by 10–20%, reflecting both the previous summer's breeding successes and winter visitors from the countries around the North and Baltic Seas, with increases in flock counts that start in mid-October, and with much of the influx remaining for the winter season.

Use of gardens was first reported as early as 1908 (in Shrewsbury), and this behaviour is now widespread, especially during severe winters, with most garden observations occurring from October–April. Peanut and suet feeders are the major attractions at garden feeding stations. Nevertheless, only 2.2% of local gardens reported Goldcrest between 2006 and 2014 (although reporting rate rises to 10% in a cold winter) (BTO Garden BirdWatch).

A total of 540 individuals was ringed between 2006–14, of which 17% were caught again within 5km of the ringing location. Nevertheless, the Goldcrest is hardly sedentary. In total, of the 11 recovered here, only three were ringed locally, and, in addition to these three, three ringed here were found elsewhere. Furthermore, of the 10 counties involved in these movements, only one (Staffordshire) is contiguous, but all are in England or Wales. The greatest distance travelled by a bird ringed here was 271km (Pontesbury to Maidstone, Kent) while, for recoveries, it was 300km (Hollesley, Suffolk to Pontesbury).

John Arnfield

Wren (Eurasian Wren)

Troglodytes troglodytes

Very common resident

Ubiquitous, but often inconspicuous by sight given its ground-hugging habits, the presence of a Wren is often first signalled through its explosive song, which is wholly incommensurate with its tiny physical proportions. Once spotted, it is unmistakable given its creeping movements and cocked-up tail, suggesting a state of constant alertness. Perhaps this perky little bird has not forgotten the ritual of 'Wrenning', when it was stoned to death on St Stephen's Day, for obscure reasons from both religious and pagan history and folklore over the centuries. It can be thought of 'as a bird of crevices and crannies, of stems, and twigs and branches, of woodpiles and fallen trees, of hedge-bottoms and banks, walls and boulders, wherever these may occur. Wrens, therefore, cut across, or rather scramble under, the imaginary boundaries which we are accustomed to draw between different types of country' (Cocker & Mabey 2005), a general account which rings true here today.

In the earliest historical accounts Leighton (1836) referred to the Wren as being 'common (in) outbuildings', Beckwith (1878) described it as being 'everywhere' near Shrewsbury, and Forrest (1899) gave the 'provincial name, Jenny Wren'. Along with an ability to adapt to habitat, no doubt the Wren's ability to nest opportunistically is also a factor in its success, it being 'no less varied in its choice of building sites than in habitat' (Paddock 1904). In fact, the male is known to build several nests before a female makes her choice and lines one nest for laying the eggs. From the start of the twentieth century there are records of Swallow and Great Tit nests being taken over, and nests innovatively crafted into old coat pockets, a horse-shoe hung in a tree and even in the body of a crow on a keeper's gibbet. More

John Fielding, Long Mynd, 30 May 2014

BBS Trend (1997–2014)

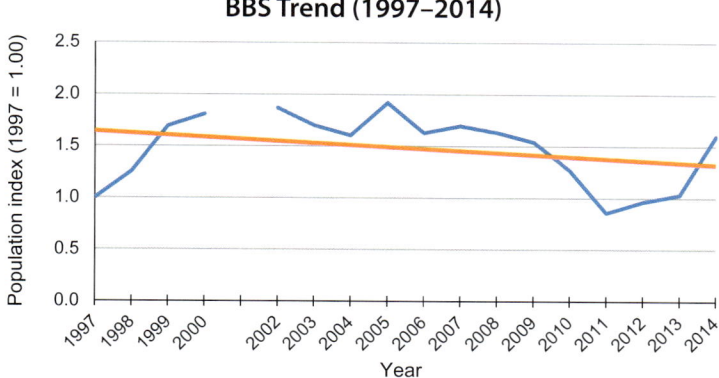

recently, one nested side by side with a Spotted Flycatcher on an old Swallow's nest near Chelmarsh, and another occupied an old House Martin's nest at Atcham Bridge (SBR 1989). Nest records range from January through until November.

Numbers fluctuate enormously according to the severity or otherwise of the winter, an observation made as early as 1892 by Beckwith when he noted: 'Although usually considered a hardy little bird, the Wren suffers terribly in severe winters; and notwithstanding ... places of shelter which it seeks in such seasons, numbers of its family die'. A run of bad winters with frost and snow in the 1860s through to the early 1890s had the result that 'so many Wrens perished that for some time their scarcity was commonly remarked'.

The *Handlist* (1964) described this resident as 'common throughout ... even occurring on the hills up to the 1,500 foot contour', and confirmed its vulnerability to hard weather: 'very few survived the winter of 1962–63. It appeared to be the species worst affected by this winter'. However, numbers can and do bounce back quickly, as the Wren typically has two broods and an average of five to six young on each occasion. 'Numbers also fell, to a much lesser extent, in the hard winters of 1978–79, 1981–82 and 1985–86, with subsequent recovery. With winters towards the end of (the 1985–90 Atlas period) being relatively mild numbers are, once again, high' (Atlas 1992).

Not surprisingly the Wren is found in every tetrad, although it was overlooked in one in the 1985–90 Atlas, with probable or confirmed breeding in almost all of them.

BBS annually records Wren in over 90% of the survey squares, more often than not above 95%, with very high counts in most squares. The population has fluctuated considerably from year to year, with the recent decline reflecting big freezes and deep snow in 2008–09, 2009–10 and 2010–11, but recovery is already underway, and the underlying trend is a small decline of around 15% between 1997 and 2014.

The breeding season relative abundance map shows higher densities in areas of extensive woodland, scrub, hedgerows and gardens. In the larger arable areas, or at higher altitudes, suitable habitat for nesting and feeding may be limited to corridors following hedges or streams, so densities are lower there. The pattern of abundance in the winter is very similar to that in the breeding season, reflecting the sedentary nature of the species.

Based on TTV counts, the population is estimated at 123,500–125,000 pairs, making Wren the most common species.

Perhaps surprisingly, the BTO Garden BirdWatch reporting rate for Wren since 2006 is only an average of around 30% of all gardens, with the highest values of around 40% near the beginning of November and the lowest (23%) in early June, at the height of the breeding season. Although population estimates indicate Wren is the most common species, it is ranked only 17th by GBW reporting rate, which may reflect its skulking habits and infrequent forays onto the bird table itself. Year on year, the decline in reporting rate mirrors the population decline found by BBS since 2006.

It is short-lived, typically just a couple of years. Its sedentary nature is confirmed by the ringing record, with 253 of the 257 recoveries ringed here also being recovered here, but three of the other four, in Kent (315km), Cornwall (260km) and Sussex (268km), all counties on the south coast, hint at some partial migration by first-year birds. Only individuals ringed in Cheshire, Denbighshire, Lancashire and Norfolk, one from each county, have been recovered here.

Stuart Cowper & Leo Smith

Occupied Tetrads

Atlas period (breeding)	1985–90		2008–13		Change	
	Number	%	Number	%	Number	%
Confirmed	740	85	528	61	-212	-29
Probable	102	12	300	34	198	194
Possible	27	3	42	5	15	56
TOTAL	869	100	870	100	1	0
Tetrads with Winter Records (2007–13): 870 (100%)						

Breeding Relative Abundance (2008–13)

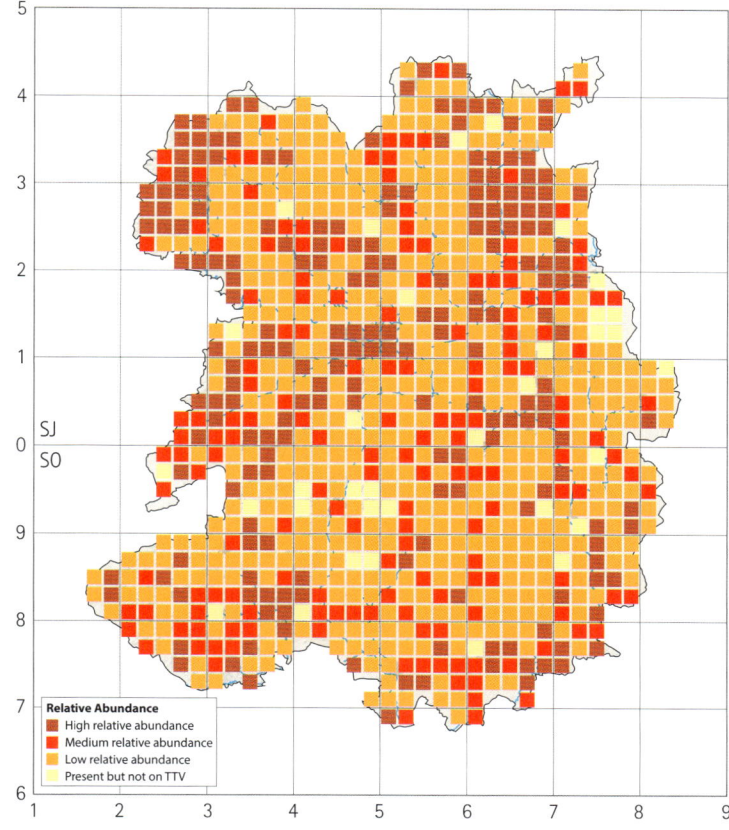

Relative Abundance
High relative abundance
Medium relative abundance
Low relative abundance
Present but not on TTV

Nuthatch (Eurasian Nuthatch)

Sitta europaea

Common resident

Celia Todd, Pant Glas, 5 May 2013

Initially documented in the nineteenth century, when Leighton (1836) recorded it in Shrewsbury Quarry and Rocke (1866a) described it as 'very common in this neighbourhood' from his Clungunford home, the Nuthatch was abundant at the end of the nineteenth century. It must have been a constant member of our avifauna before historical recording, as its preference for mature deciduous woods with oak or beech with hazel understorey means it is ideally suited to the wildwood habitat which was dominant here for thousands of years.

The *Handlist* (1964) described it as 'resident, common ... wherever suitable habitats occur', adding it 'apparently did not suffer badly in the winter of 1962–63'.

Nesting in natural holes or crevices in trees, it is surprising that the first breeding in nest boxes was not recorded until 1981, at Lilleshall, but this behaviour has been recorded in most years since. Most nest box schemes have the occasional or indeed a regular pair of Nuthatch, but their territoriality limits the number of pairs in all but the largest schemes. SBRs have documented a number of occurrences of regular

use of specific nest boxes or trees and one tree at Ashford Carbonell had an occupied nest for 12 years in succession.

Breeding evidence is easy to obtain to at least probable level. It is extremely vocal during the breeding season, and once the single brood has fledged, family parties forage in close proximity, calling constantly.

The BTO Atlas (1976) showed evidence of breeding in all but two of the County's 33 10km squares (confirmed breeding in 22, probable breeding in two and possible breeding in seven), and nationally the population has continued to increase and expand since then.

Fieldwork for the 1985–90 Atlas found Nuthatch in 72% of tetrads.

Gaps in the distribution in the Atlas (1992) were predominantly in mainly arable farmland in the north and east, especially a band of the North Shropshire Plain from Loppington to Market Drayton and south to Wellington. This is a region with large fields, and small or non-existent woodlands with poor wooded corridors between them. Recent Atlas fieldwork found that half of the previously-vacant tetrads have now been colonised, and Nuthatch occupied 83% of tetrads in the breeding season, over 100 more than in 1985–90. This reflects an increasing population, locally and nationally.

While local BBS surveys have not found Nuthatch in sufficient squares to calculate a statistically valid trend, nationally they have produced evidence of a large population increase, an estimated at 93%

Occupied Tetrads

Atlas period (breeding)	1985–90		2008–13		Change	
	Number	%	Number	%	Number	%
Confirmed	283	33	313	36	30	11
Probable	169	19	220	25	51	30
Possible	171	20	193	22	22	13
TOTAL	623	72	726	83	103	17
Tetrads with Winter Records (2007–13): 760 (87%)						

Breeding Distribution Change (1985–90 to 2008–13)

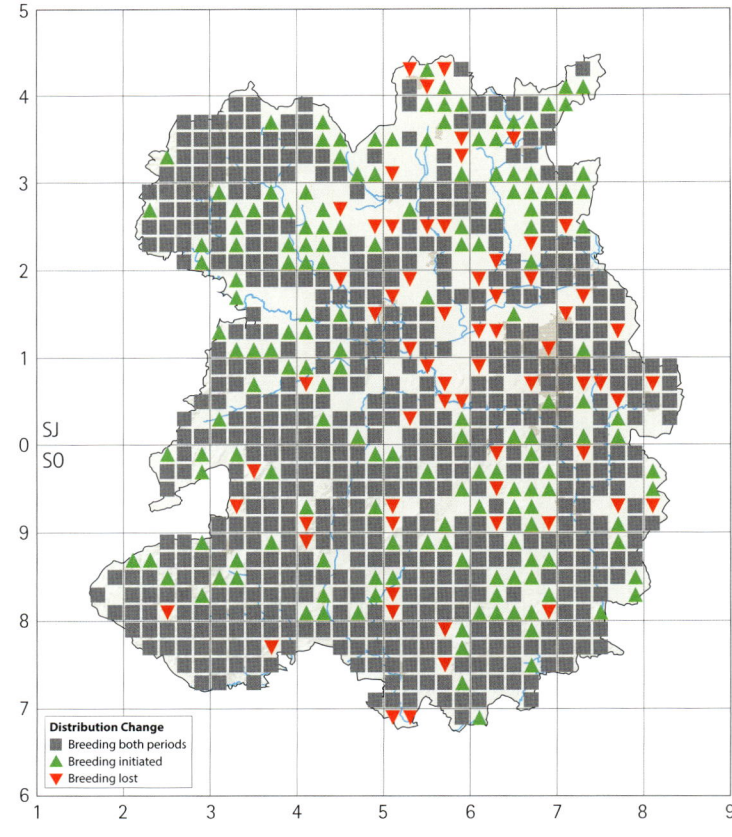

Distribution Change
- Breeding both periods
- ▲ Breeding initiated
- ▼ Breeding lost

Breeding Relative Abundance (2008–13)

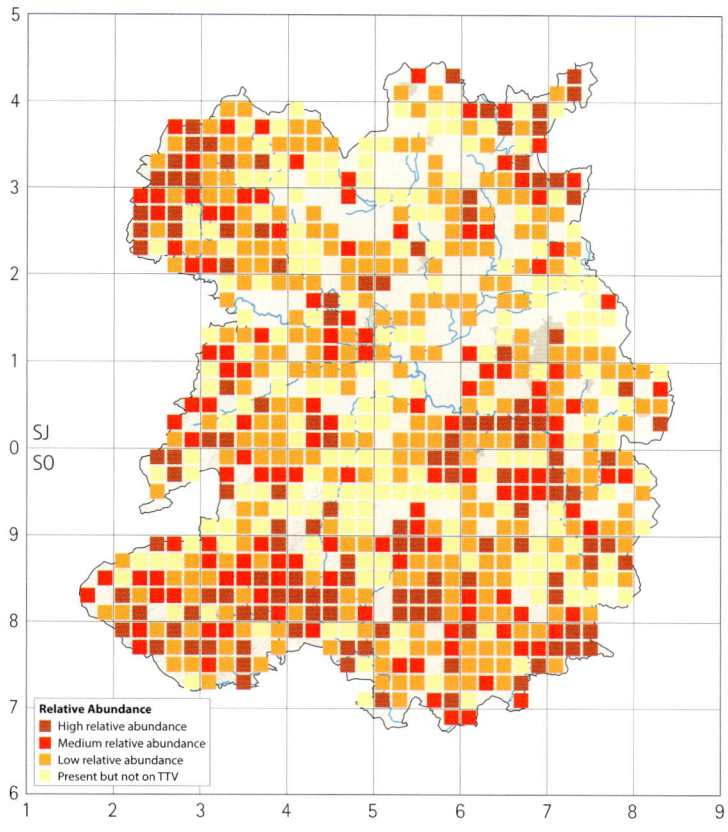

Relative Abundance
- High relative abundance
- Medium relative abundance
- Low relative abundance
- Present but not on TTV

in England, 159% in the English West Midlands and 54% in Wales, between 1995–2014. Based on TTV counts, the population is now estimated at 7,900–8,300 territories; twice that estimated in 1992.

Nuthatch occurs at its highest densities in areas of good woodland cover, and the Oswestry uplands, Severn Gorge, parts of Wenlock Edge, Wyre Forest, and large oakwoods in the south and south-west stand out on the relative abundance map. Surveys in woodland of various ages and types in Telford and Wrekin have found that it is dependent there on mature oak trees (Glenn Bishton *pers. comm.*).

Nuthatches have a strong pair-bond and are strongly territorial all year round. After the single brood fledges, the young birds disperse and find their own territories a few weeks later (Enoksson 1990), perhaps accounting for the 4% higher number of tetrads where they were found in winter.

Various theories have been put forward for the population expansion in recent decades, including better over-winter survival as a result of milder winters, and perhaps more extensive garden feeding. Nuthatches are now commonly seen at feeding stations within villages and in the better wooded parts of towns.

Ringing has demonstrated its sedentary nature. Of 20 or so recoveries of Nuthatches ringed locally, the majority have been found within 1–2km of the ringing location, and only two have crossed the County boundary. All five of those ringed elsewhere and recovered here came from adjacent counties. However, a female ringed near Shifnal in February 2006 was caught by a cat in June of the same year in Glamorgan, an unusually long movement of 118km.

Gerry Thomas

Treecreeper (Eurasian Treecreeper)
Certhia familiaris
Fairly common resident

Treecreeper breeds in woodland throughout temperate Eurasia, and is sedentary. It was first documented in Shrewsbury Quarry (Leighton 1836), even then a landmark of the County town. It was evidently common in the nineteenth century; Beckwith (1879) described the 'Common Creeper' as 'very frequent and often found in the company of various kinds of tits', and it was 'not uncommon' at the end of that century. It was described as 'common in suitable habitats, numbers were reduced after the winter of 1962–63' in the *Handlist* (1964), and mapped as confirmed breeding in all except five of the County's 33 10km squares in the national BTO Atlas (1976), with possible breeding in four, and absence in one.

The Treecreeper is relatively unobtrusive, but it is not difficult to locate in the breeding season and find at least 'probable' breeding evidence, primarily due to its distinctive, high-pitched song, heard mainly from March–May. It is found in all types of woodland, although here it is more abundant in deciduous habitats. In the south especially it frequents woodland strips, and is frequently found in alders along streams and rivers.

One of the more surprising findings of the recent breeding Atlas was a 26% reduction in the number of occupied tetrads compared to 1985–90, although it remains widely distributed. It is not recorded on sufficient surveys to provide a local population trend, but BBS

has found virtually no population change in England, and an 18% gain in Wales, between 1995–2014. In contrast, the national BTO Atlas 2007–11 shows extensive local decreases in relative abundance since 1988–91.

Though its range in Europe does extend to high latitudes, it is known to be vulnerable to prolonged periods of cold, and especially to ice glaze and hoar frosts on branches and tree trunks, which prevent access to the small insects and spiders which make up its food in winter months. A particularly cold winter in 2010–11 occurred halfway through the recent Atlas period, and this may have reduced the overall population sufficiently to contribute to the reduction in tetrad occupancy. Although cold winters also occurred in the 1985–90 Atlas period, numbers were 'not apparently' reduced then. The recent cold winter was followed by three poor weather-affected breeding seasons, which further reduced the opportunity to confirm breeding, as fewer pairs reached the 'feeding young' or 'fledged young' stages. Even so, these weather conditions seem insufficient to explain the scale of the losses found, and it is likely that habitat loss and degradation has also contributed.

Hinsley *et al.* (1995) showed that Treecreepers avoid isolated woodlands and are most abundant where woodland patches are relatively close together. Here, about 8.5% of the land area is still

John Hawkins, Pentrecoed, 16 June 2006

Breeding Distribution Change (1985–90 to 2008–13)

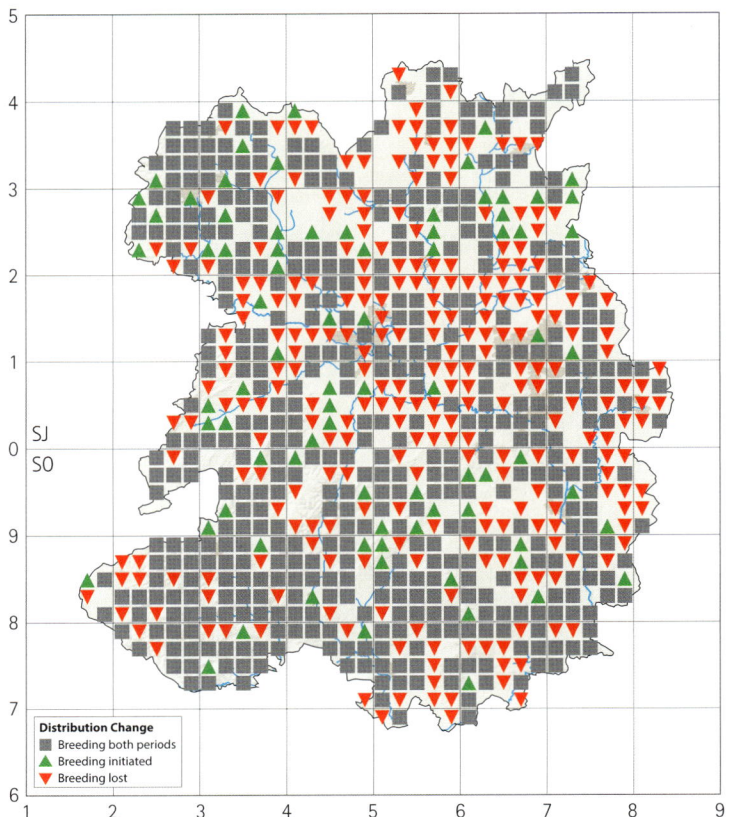

Breeding Relative Abundance (2008–13)

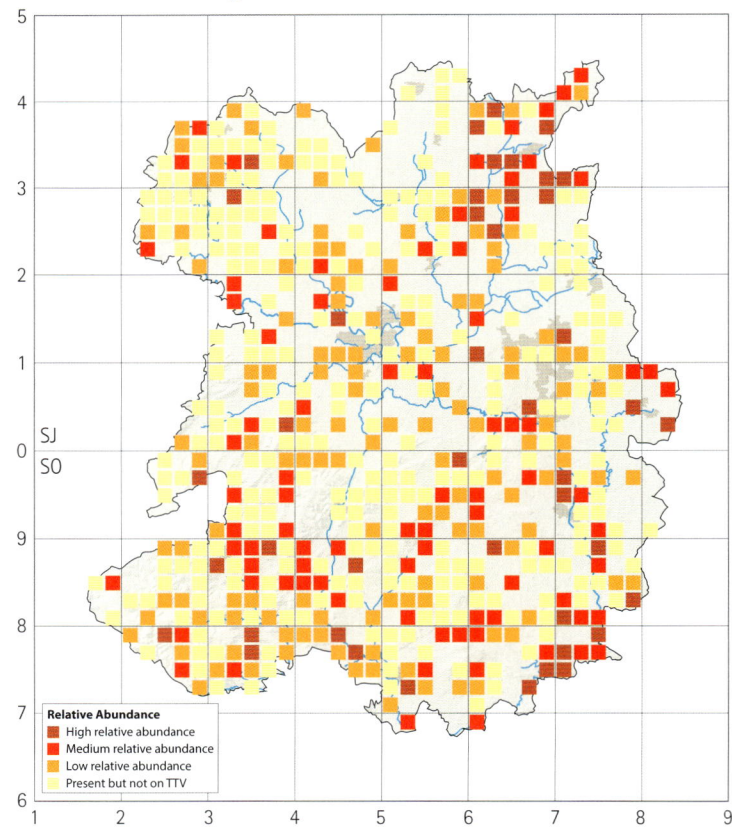

covered by woodland, and the lowlands and river valleys have traditionally been well endowed with hedgerows, many of which have mature trees as key components. Treecreepers are able to use these hedgerows as corridors for movement between woods, and for foraging. There has been no extensive change in woodland cover at a landscape scale in recent decades, though it is likely that suitable habitat has been reduced, especially in those areas where agricultural intensification has resulted in hedgerow removal or degradation through poor management, or loss of small-scale woodland and copses, particularly in northern and eastern areas, and where land has been used for housing and other development, particularly in the Shrewsbury and Telford areas. These areas show the greatest losses on the breeding distribution change map.

Recent research on Marsh Tit (Broughton & Hinsley 2015) has suggested that reductions in connectivity between woodland patches has created increased barriers to dispersal and are a factor in the decline of that species. Similar research for the Treecreeper would be of interest.

Few were found during TTVs, in spite of the widespread distribution, but the breeding season relative abundance map highlights

Occupied Tetrads

Atlas period (breeding)	1985–90		2008–13		Change	
	Number	%	Number	%	Number	%
Confirmed	313	36	171	20	-142	-45
Probable	175	20	132	15	-43	-25
Possible	230	26	231	27	1	0
TOTAL	718	83	534	61	-184	-26
Tetrads with Winter Records (2007–13): 589 (68%)						

woodland in the north-east, Severn Gorge, Wenlock Edge, Wyre Forest and the south-west. Low abundance, or absence, occurs in arable areas, particularly in the north and east, and in the uplands, where mature trees and woodland cover are less frequent.

Treecreepers are resident and sedentary, so the distribution and relative abundance maps for the breeding and winter seasons are very similar. Post-breeding dispersal will partly account for the higher number of tetrads where they were found in winter, though they will also be more detectable at this time of year following leaf fall.

Based on TTV counts, the breeding population is estimated at 3,000–3,100 pairs, compared with an estimate of 5,000–10,000 pairs in 1992 (10–20 pairs per occupied tetrad), suggesting a considerable decline in the population of occupied tetrads, as well as a contraction of range.

Treecreeper is one of the most sedentary of all British species, and all 13 recoveries were both ringed and recovered here. Many recoveries were found within 2km of the ringing location, but one ringed at Calverhall in January 2011 was retrapped at Whixall Moss in November of the same year, a movement of 14km, the furthest recorded. Only one was found more than 10 months after ringing, 18 months later, but still less than 3km from the ringing location.

Nationally, the overall distribution has shown subtle changes, decreasing in southern and eastern areas of the UK and increasing in Ireland and the north, since the 1968–72 BTO Atlas. High relative abundance has shifted markedly northwards and westwards since 1988–91, with large declines here and in much of the West Midlands. Although the changes are partially linked to new woodland in some areas, and loss in arable areas, perhaps climate change is also involved, so the results of the next Atlas will be of particular interest for this species.

Gerry Thomas

Rose-coloured Starling (Rosy Starling)
Pastor roseus
Vagrant

Jim Almond, Porth-y-wain, 18 November 2005

This scarce migrant to Britain breeds from the very south-east of Europe through Kazakhstan to the western edge of China, and winters on the Indian subcontinent. It is nomadic, searching out places to breed with a high population of grasshoppers, and adults regularly reach western Europe as a result. Most British records come from Devon and Cornwall or the Northern Isles, with adults usually occurring in the late spring or early summer and juveniles in the autumn.

Sometimes known as the Rosy or Rose-coloured Pastor, this starling was first recorded in 1841 at Meole when perhaps two birds were involved, including a male which was killed. Another was killed at Brockton near Lydbury North in the autumn of 1857. Thereafter, there was only an unconfirmed report of one in the latter half of the nineteenth century, and none at all in the twentieth century.

The two modern occurrences, an adult male in June and an immature in October–November, were typical records:

| Minsterley | 16–17 June 2002 |
| Porth-y-wain | 28 October–24 November 2005 |

The one at Minsterley was located in a private garden, and coincided with an unprecedented influx to Britain and Ireland with approaching 200 individuals reported (Fraser 2013b). The Porth-y-wain starling was also in a private garden where it fed on fallen apples. It was quite territorial, protecting its food supply from Fieldfares and Blackbirds but, as the weather became much colder, the area it was able to protect diminished until it was just protecting the foot of a tree. On its last day, the day before the first snows of the winter, it spent much of its time in the porch, only making brief forays to feed.

Graham Walker

Starling (Common Starling)

Sturnus vulgaris

Very common winter visitor, common resident

The Starling was not always common, but its range and density began to increase nationally around 1830. Changes in land use during the agricultural revolution, and intense persecution of raptors, may have helped create conditions in which it could flourish (Holloway 1996). As early as 1838, Eyton found it 'common', and by 1900 Forrest considered it 'exceedingly numerous everywhere'. Beckwith described a rampantly successful breeding species, 'rapidly increasing', colonising every available nook and cranny, in buildings, ivy and haystacks, and regularly evicting woodpeckers from their holes.

In the mid-twentieth century, the *Handlist* (1964) recorded post-breeding flocks of up to 20,000 between July–September, distinguishing those from larger numbers of winter migrants arriving from October onwards. According to SBR 1985, Starling was 'very numerous, and bred widely'; only seven years later, the *Atlas* (1992) warned that though still 'abundant', there was an accelerating population decline driven by changes in agriculture. The population estimate was between 27,000–54,000 breeding pairs; estimates based on TTV counts suggest there are currently around 12,000. An already declining Starling population plummeted by 61% between 1997 and 2014 (BBS), and shows no sign of stabilising.

Starling was found breeding in almost all tetrads in 1985–90, but only in 79% in 2008–13. The six-fold increase in tetrads where Starling has only 'possible' breeding status, with no higher-level evidence, also suggests a smaller, thinly spread, and less detectable population. The breeding distribution change map shows that while no area has escaped losses, the most extensive have occurred in the south-west. Relative abundance follows a similar pattern: the highest densities are in the centre and north, with scattered hotspots in the east and south; in the south-west, there is a patchwork of low-abundance squares, a sprinkling of moderate and high abundance, and many vacant squares beginning, ominously, to cluster together. Starling appears to be sliding rapidly towards extinction in some places: in SO49, around Church Stretton, Starling was not found at all in over half the tetrads; in 1985–90, only three were apparently vacant.

Starling's decline as a breeding species is thought to be driven by poor over-winter survival of first-year birds, linked to agricultural intensification, especially on pasture land (BTO). Its dagger-like bill is adapted to probe the sward for invertebrates, and loss of permanent pasture, land drainage, pesticide use, high stocking densities, and compaction by sheep have all reduced the quality of the foraging environment. The disproportionately high losses of breeding Starling from areas dominated by pastoral farming are consistent with BTO's assessment of the likely causes of decline.

In winter, great flocks of migrants from Scandinavia and eastern Europe dominate recording. During Atlas fieldwork, winter timed counts logged 95,219 Starlings, almost 16 times as many as in the breeding seasons. Nine one-hour TTV counts topped 1,000, with another 22 over 500. Nevertheless, winter flock sizes have shrunk over time in step with the breeding population. An estimated 1,400,000 birds roosted at Onslow Park near Shrewsbury in winter 1957–58; in 1966,

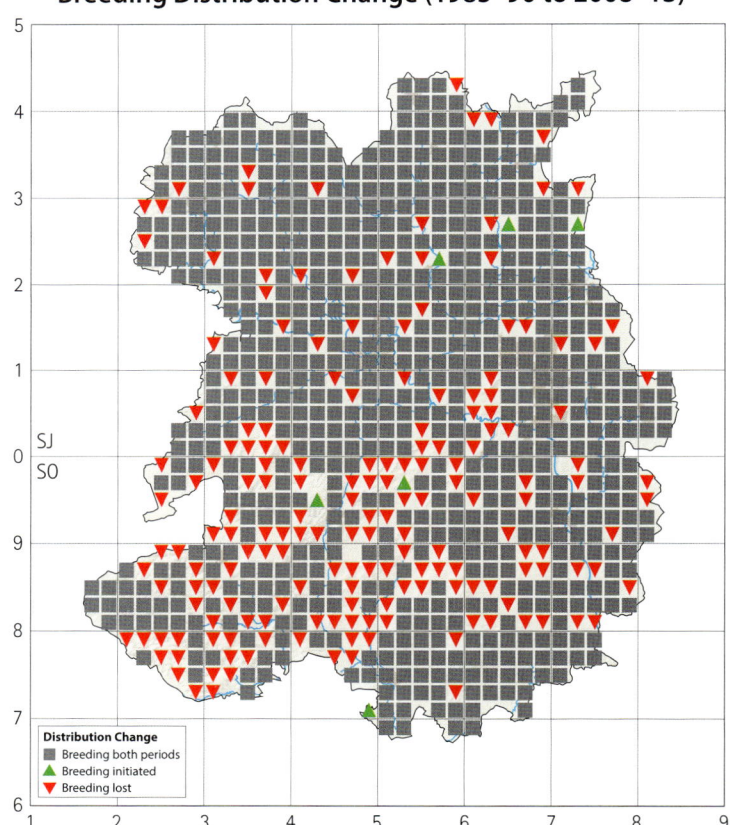

Breeding Distribution Change (1985–90 to 2008–13)

Distribution Change
- Breeding both periods
- Breeding initiated
- Breeding lost

BBS Trend (1997–2014)

Occupied Tetrads

Atlas period (breeding)	1985–90		2008–13		Change	
	Number	%	Number	%	Number	%
Confirmed	818	94	469	54	-349	-43
Probable	20	2	85	10	65	325
Possible	22	3	132	15	110	500
TOTAL	860	99	686	79	-174	-20
Tetrads with Winter Records (2007–13): 822 (94%)						

Breeding Relative Abundance (2008–13)

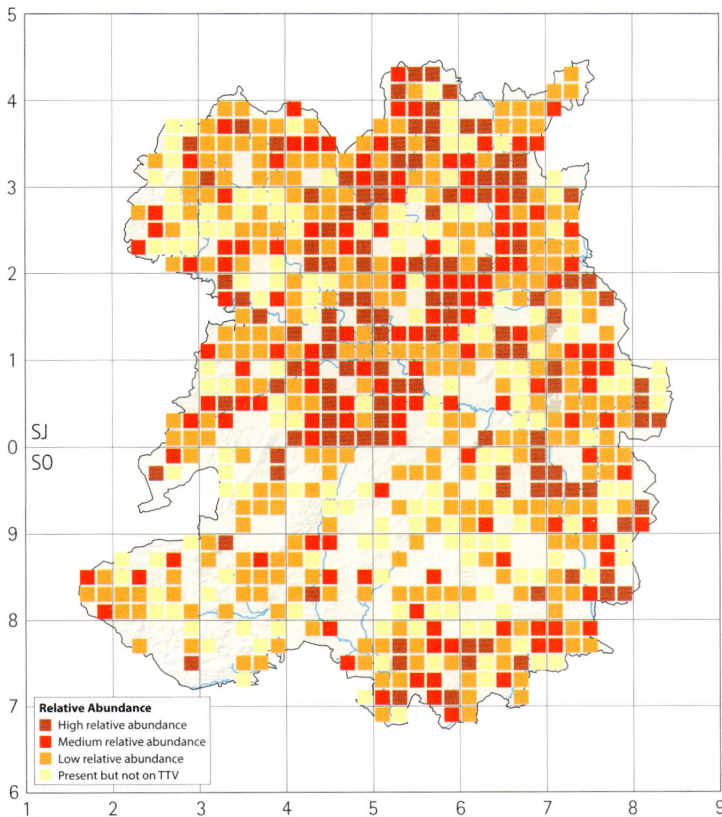

Relative Abundance
- High relative abundance
- Medium relative abundance
- Low relative abundance
- Present but not on TTV

the maximum flock was 1,000,000 strong. Since 2000, flock sizes, while varying widely from year to year, have never reached that scale: the largest count in 2001 was only 1,500, though the maximum generally hovers around 50,000. A murmuration of 'hundreds of thousands' of Starlings at Westbury in March 2014 was exceptional, and there was another of 40,000–50,000 birds at Bridgnorth during the same period.

The number of winter visitors, and thus roost sizes, is influenced by continental weather, so climate change may play a part in this trend, but it also reflects the general decline of Starling populations throughout Europe.

The highest concentrations are in the north in both seasons; in the south, however, the far south-west supports much higher densities in winter, whereas in the south-east the reverse tends to be the case. The greatest contrast is in the Clun Forest: SO27, west of Knighton, for example, contains one of the highest winter concentrations, but one of the lowest breeding densities. Here, high rainfall with slow evaporation rates in winter and early spring create favourable feeding conditions, bringing invertebrates closer to the surface and temporarily reducing soil compaction; the ground, however, is often hard again by May. Winter sheep-feeding stations in pastoral areas are an added inducement, and can attract thousands of Starlings in hard winters.

Almost two-thirds of locally-ringed Starlings were recovered here. Those found in neighbouring counties (including the West Midlands) take the total to just under 80%, suggesting that much of the national population is sedentary. Some, however, are very mobile: the remainder were found in 39 counties throughout the British Isles, including four in Ireland and three in northern Scotland. The 66 Starlings ringed while wintering here and recovered abroad were found in 15 countries extending across northern and eastern Europe. The 41 local recoveries of birds ringed abroad are from a similar spread of countries. Most follow the migration routes from Russia, Belarus and Poland, or Scandinavia and the Baltic states, through Germany and the Low Countries. In many cases the ringing and recovery locations were over 1,500km apart, and two of the four found in Belarus were over 2,000km away. The greatest separation is shown by an adult ringed at Leegomery on 10 January 1963, and found killed by a cat just over four months later in Karelia (Russian Federation), 2,424km east-north-east. A few of these long-distance travellers had lived long enough to make the return journey more than once.

Some of the data are unexpectedly colourful: a Starling on

Jim Almond, Venus Pool, 26 October 2007

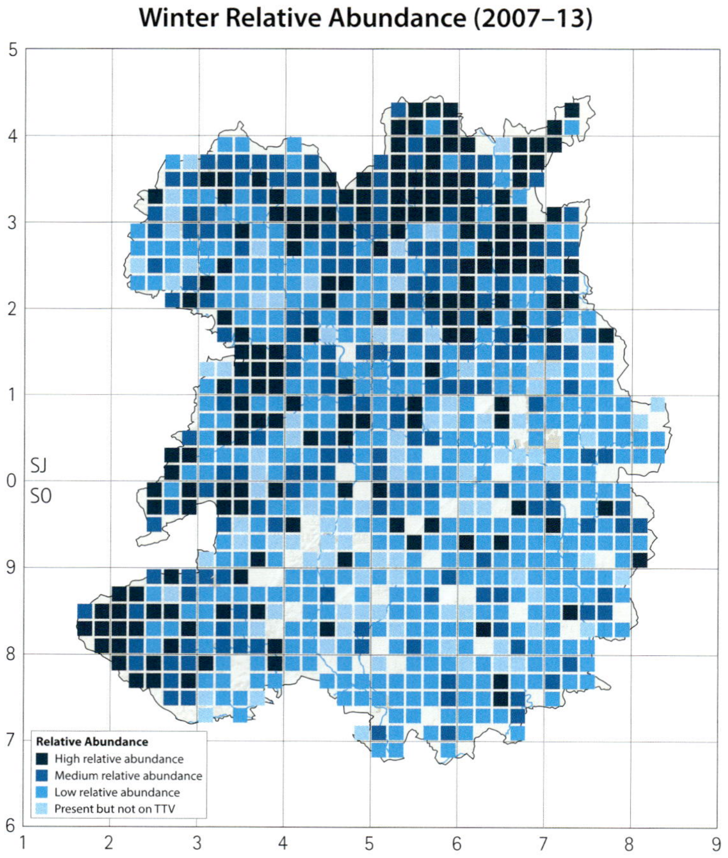

Winter Relative Abundance (2007–13)

Relative Abundance
- High relative abundance
- Medium relative abundance
- Low relative abundance
- Present but not on TTV

Dawn Micklewright, Shrewsbury, 30 December 2013

migration was ringed on a light vessel in the North Sea, and shot near Wellington two years later; another, ringed at Leegomery, fell victim to a ship's cat at sea off the Somerset coast. The oldest Starling, ringed in Boston, Lincolnshire, was just over 13 years old when it too was killed by a cat at Walcot. A young female ringed near Hopton Wafers in January 1979 flew 1,000km in only two days, presumably with a strong following wind; it did her no good, however, as she was shot on arrival in south-west France.

Michelle Frater

White's Thrush

Zoothera aurea

No modern records

The sole record of this rare vagrant, which breeds in Siberia and eastern Asia, and winters in south-east Asia, is of one shot near Morton Corbet on 14 January 1892. The bird, apparently not sexed but 'very fat, with a crop full of worms', was preserved. The specimen passed into William Beckwith's collection and it is now in Shropshire Museums in Ludlow Museum Resource Centre (specimen SHYMS: Z/2006/123).

Paddock in his *Catalogue of Shropshire Birds* of 1897 adds an account of a bird resembling a White's Thrush, but no date or locality was given.

John Tucker

Ring Ouzel

Turdus torquatus

Scarce passage migrant, has bred

The nominate race which occurs here is a summer visitor, and winters mainly in southern Spain and the Atlas mountains of Morocco and Algeria.

Although earlier authors mention this 'Mountain Blackbird', the first account specifically referring to its occurrence here said 'most of the high uncultivated grounds produce this bird in tolerable abundance. I have obtained them and their nests on the Black Hill, the Longmynd, the Steperstones, and Clee hills. Wherever the Mountain ash grows, they will generally be found' (Rocke 1865b).

The first wide-ranging account described Ring Ouzel as a rather rare spring and autumn migrant, 'while to the high hills and moorlands in the south it is a common summer visitor'. Haughmond Hill, the Wrekin, and heaths and mosses were noted as regular passage sites, with a preference for 'high and uncultivated places where gorse and heather grows', but, of these, only Haughmond Hill was a regular breeding site. However, 'in the south it nests on all the high hills, and is very numerous on the Longmynds, the Stiperstones, and the hills around Bishop's Castle and Clun'. While authors in nearby counties described it as resident, and noted occasional instances of families staying all winter, here, in spite of several occurrences in November, and a few in December, there was no evidence that it over-wintered, or had been seen in the first two months in the year (Beckwith 1888b). 'Several pairs' were seen in Townbrook in June 1885 (Paddock 1890), Forrest confirmed that it bred 'regularly on the Longmynd and hills on the Welsh border', and added that a nest had been found at Myddle. The British population appeared to have been stable during the nineteenth century, and it was mapped as 'abundant' in the Oswestry uplands and Shropshire hills. However, it started a 'long and steady decline early in the twentieth century' (Holloway 1996), and these early accounts show that it was far more common and widespread in the nineteenth century than the twentieth. The frequency of records declined through the first half of the twentieth century, but Bucknell and Knighton were added to the list of breeding sites.

The *Handlist* (1964) referred to breeding on the Oswestry uplands, Clun Forest and Stow Hill, 'but not in recent years', and added 'up to 10 pairs breed regularly on the Long Mynd, though numbers vary from year to year'. The 1968–72 national BTO Atlas mapped breeding evidence in only three of the County's 33 10km squares, restricted to the Long Mynd and the Stiperstones. Since 1964 all confirmed breeding records have come from the Long Mynd, except for two from the Stiperstones, firstly in 1968 (two pairs, including a nest with one egg, precise location not specified) and secondly in 1982, when one pair in Mytton Dingle successfully fledged a brood (J. Sankey *pers. comm.*). However, the species is believed to have bred at Rigmore Oak during the 1970s, although there is no documentary evidence to substantiate this (Tom Wall *pers. comm.*).

The 10 or so pairs on Long Mynd estimated by the *Handlist* was little different from that found in 1985–90: 'at least seven pairs of which four were confirmed breeding' (SBR 1987), and 'probably a dozen pairs' (SBR 1989). Otherwise, the 1985–90 Atlas map showed probable breeding records on Titterstone Clee and Stapeley Hill (a pair in April 1985), and possible breeding on the Stiperstones. The account also noted records from Brown Clee: a pair on 20 April and a single on 4 June 1987, and a further two pairs in April 1990 (SBRs), all of which were almost certainly passage migrants.

The Long Mynd BBP estimated the population at 13 pairs in 1998, based on a range of 11–16 pairs found during five years of intensive fieldwork. Two-thirds of the 65 territories monitored were in heath and scattered bracken (the most favoured habitat) or heath, but short grass was also needed for feeding (Smith, SBR 2003).

Ringing during this period had found a high degree of site fidelity (one female at least two years old when first found in 1997 was caught three years running at the top of Callow Hollow at nests separated by

Decline in the Long Mynd Breeding Population (2004–14)

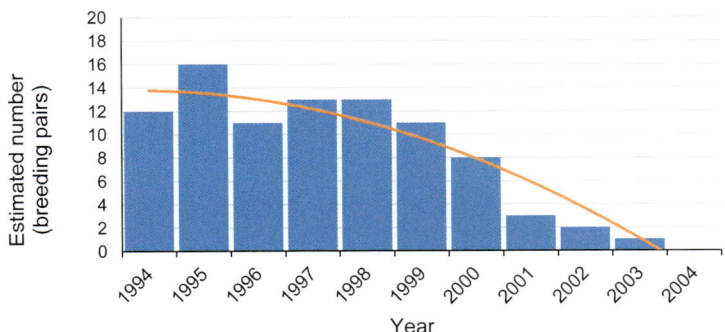

only 250m), so a colour-ringing project to assess this and monitor the population was started in 1999. This project witnessed a catastrophic decline from 13 pairs in 1997 and 1998, down to 11 in 1999, eight in 2000, three in 2001, two in 2002 and one in 2003.

Between 1995–2003, 45 nests were found, 35 in heather, five in gorse, three in bilberry and two in bracken. All were on the ground, but most were on top of a small outcrop or vertical bank. Most were on the steep hillsides of the upper valleys, but 11 were on the flat open moorland plateau, an unusual if not unique habitat. The lowest altitude was 310m, 30 (67%) were above 400m and the highest was 490m. The average clutch size of 30 nests was 4.27. The total number of pairs observed 1995–2003 was 74, and 31 (42%) were observed to have two clutches.

The predation rate was very high – 17 nests altogether, 11 at the egg stage and six broods of chicks, out of 41 found before predation. For example, one pair in 2002 built five nests, eggs were laid in at least three (the other two were empty when found), and all three were predated.

Sixty-two nestlings and six adults were ringed at 18 nests between 1996 and 2003, and six adults were retrapped. Of these, between 1999 and 2003, 30 nestlings and seven adults were colour-ringed, and 10–12 of the nestlings were seen again the following year. Four of the seven colour-ringed adults were retraps of birds fitted with a BTO ring as nestlings between 1996–98. Two were seen in one following year, and one was seen for three following years. Of the seven, and depending on whether they were ringed as nestlings (known age) or adults (unknown age), one was two years old when last seen, three were at least two, one was four and another at least four, and one was nearly seven. She was ringed in a nest in Ashes Hollow on 11 July 1996, caught in Light Spout Hollow (1.2km distant) and colour-ringed in 2000 (RRW – see photo), and last seen on 16 June 2003, aged 6y 11m 5d. She too showed remarkable nest site fidelity, returning to Light Spout Hollow for at least four years running. The

Jonathan Cartwright, Titterstone Clee, 7 April 2013

round trip to wintering grounds in southern Spain or North Africa is 4,000–6,000km, so she probably flew well over 30,000km in her lifetime. She held the national longevity record for six years, but then a male ringed in 2002 was seen close to his natal site in Glen Effock, Angus, in June 2009 and again on 19 June 2010, making him 7y 11m 28d (Mike Nicholl *pers. comm.*).

The last breeding male was unringed when he arrived in 2003, and it is virtually certain that he did not originate on the Long Mynd. It is believed that no unringed birds fledged since 1998, and no other unringed adults had been seen since 2000. Ring Ouzel populations have strong local dialects, but his song was unlike any other heard previously, and his plumage was more striking, the contrast between the white tips and the remainder of the black feathers on the body and wings was unusually strong. Passage birds usually pass through quickly, so the last female would have had little time to recruit him here if she arrived back unpaired, suggesting that pairing took place before arriving back on the breeding grounds, either in the winter quarters or on migration.

The potential causes of the decline were investigated. The rate of

Breeding Distribution Change (1985–90 to 2008–13)

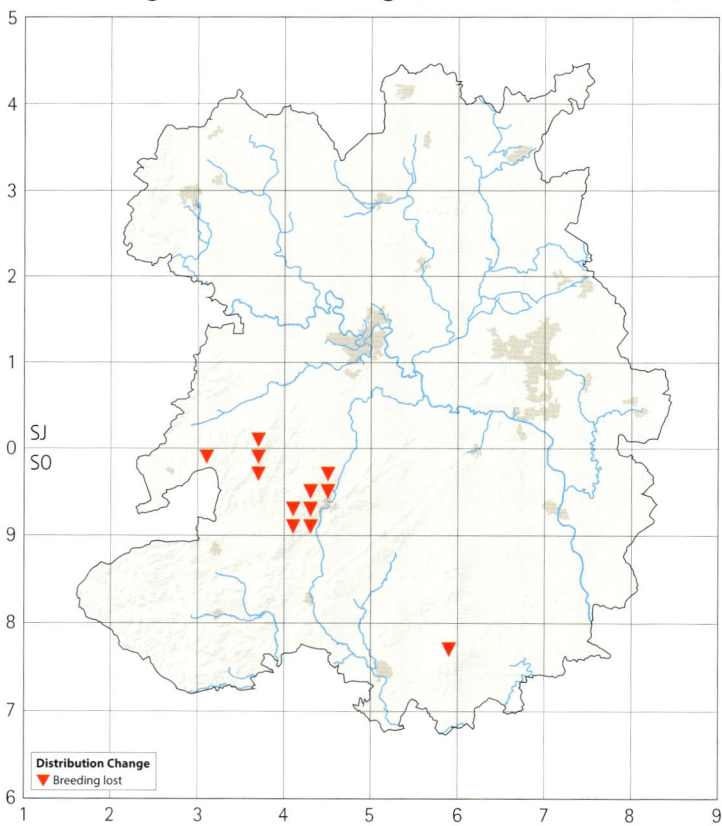

Leo Smith, Long Mynd, 10 July 2000

return of colour-ringed birds was sufficient to sustain the population, eliminating problems on migration, or on the wintering grounds. Sampling of invertebrates in areas where Ring Ouzels had been seen feeding, and an equal number of apparently similar patches of short grass, showed no loss of food or shortage of foraging areas, but the rate of nest predation was much higher than in eight other comparable studies (Burfield 2002), and the number of second broods was lower (no nests were found for some pairs, probably because they were predated early), so the number of fledged broods per pair was not sufficient to sustain the population.

Articles in 1999 and 2000, and the annual species accounts, in SBRs provided more detail, and the Ring Ouzel Study 1994–2003 (Smith 2003) collated and analysed the results.

All places where Ring Ouzels were seen between 1994–2003 were revisited each year 2004–06, but after a single colour-ringed bird was seen in 2004 no others were found, nor were any seen during intensive fieldwork for the Long Mynd BBP over the three years 2006–08. In view of the effort made to find them, the fact that none were seen or heard after the end of April in any of the five years suggests that there were no breeding attempts made, and 'the species is now apparently extinct as a breeding bird on the Long Mynd' (SBR 2008). This systematic project work was not continued after 2008, but

Occupied Tetrads

Atlas period (breeding)	1985–90		2008–13		Change	
	Number	%	Number	%	Number	%
Confirmed	2	0	0	0	-2	-100
Probable	4	0	1	0	-3	-75
Possible	6	1	1	0	-5	-83
TOTAL	12	1	2	0	-10	-83

none were found when all previously known nest sites were checked again for the national Ring Ouzel survey in 2011.

Research on the Stiperstones in 2002, to complement that on the Long Mynd, reviewed all SOS records, but Ring Ouzel had been reported in only eight years since 1956, apart from the two confirmed breeding records, and English Nature staff had recorded them in only nine years since 1986. These additional records were all from the March–early May or September–October passage periods. However, there was one probable breeding record, in 1999, when a pair was seen in early April and again in early May, and one was singing in mid-May, in Mytton Dingle. A survey specifically designed to detect any breeding Ring Ouzels, comprising six full-day visits between 3 June and 7 July 2002, thoroughly searched all the sites with previous breeding season records, at least twice, but none were found.

Forty-two records were submitted during recent Atlas fieldwork, all for the period on or before 7 May, and all were judged to be passage migrants.

The only possible exception was a pair seen gathering nesting material above the former Gatten Plantation on the Stiperstones on 7 April 2009, and again on 14 April when 'the pair appeared to be collecting nest material from nearby woodland and flying to a section of thick heather'. A male, perhaps the same one, was feeding and singing in the nearby Brook Vessons area on 12 April. Subsequent attempts to relocate the pair, by the original observer and others, were unsuccessful. These are technically probable breeding records and may reflect a failed nesting attempt, the first since 2003, but it was very early (the earliest estimated first egg date on Long Mynd 1995–99 was 19 April) and the pair may have been engaged in pair bonding while still on migration. The tetrad occupancy table includes these two 2009 records (Brook Vessons is in a different tetrad), but they are not included on the breeding distribution change map, which reflects the local extinction of Ring Ouzel as a breeding species.

There have been no further indications of breeding, and Titterstone Clee is now the most favoured spring passage site, with other sightings coming regularly from Brown Clee, the Long Mynd valleys, the Stiperstones, Caer Caradoc and Whixall Moss. Most sightings are of small flocks of up to half a dozen, though larger flocks are occasionally seen on return passage in September or October. There were two records, each of a single bird, during the recent winter Atlas period, on Titterstone Clee in very early November, in 2011 and in 2012, judged to be late passage migrants.

Declines elsewhere have been attributed to increased human disturbance on increasingly accessible moorlands, afforestation, and climate change, but the Long Mynd population seemed tolerant of people, with females sitting tight until nests were approached to within a few metres, and the slow process of climate change cannot account for extinction in only four years. Afforestation might explain why no Ring Ouzels were found in Minton Batch during the Long Mynd surveys, but do not explain the disappearance from other valleys, reinforcing the conclusion that predation was the major cause of the local extinction here. These factors have led to a contraction of range of 43% in Britain since 1968–72, local extinctions in Exmoor and much of central Wales, an estimated 72% decline over 25 years in the UK (Eaton *et al.* 2015) and a substantial decline in all except one of the other study areas monitored by the national Ring Ouzel Study Group (Sim *et al.* 2010; www.ringouzel.info), and they would also have had a local underlying effect.

The ongoing national decline suggests that recolonisation is unlikely.

Leo Smith

Blackbird (Common Blackbird)

Turdus merula

Very common resident

Blackbirds breed across all but the most northerly parts of Europe, but the more northern and north-eastern populations migrate southwards for the winter. They have been common here throughout recorded history; indeed, Rocke stated in 1865 that they are 'too common to require any remarks'. This author will endeavour to find more to write about them, despite their continuing abundance.

Blackbirds are prone to variation, and this at least gave some of the historical naturalists a reason to comment on them. A melanistic individual was described in 1897 as 'a male with beak, legs etc, perfectly black', and there is also a reference to a 'Golden-coloured variety' in 1906.

The records contain many sightings of various 'pied' individuals, sometimes including full albinism. The distinctive patterning meant that some were easily recognisable and one individual in Broseley was thought to be at least six years old in 1936. The patterning also allowed some birds to form the basis of informal studies before ringing started. The earliest was in the grounds of Atcham Vicarage, 'a pair of Blackbirds, both of which were partly white, produced one or more pied young ones every year' (Beckwith 1888).

Dawn Micklewright, Shrewsbury, 5 December 2009

Occupied Tetrads

Atlas period (breeding)	1985–90		2008–13		Change	
	Number	%	Number	%	Number	%
Confirmed	832	96	784	90	-48	-6
Probable	34	4	81	9	47	138
Possible	4	0	5	1	1	25
TOTAL	870	100	870	100	0	0
Tetrads with Winter Records (2007–13): 870 (100%)						

Breeding Relative Abundance (2008–13)

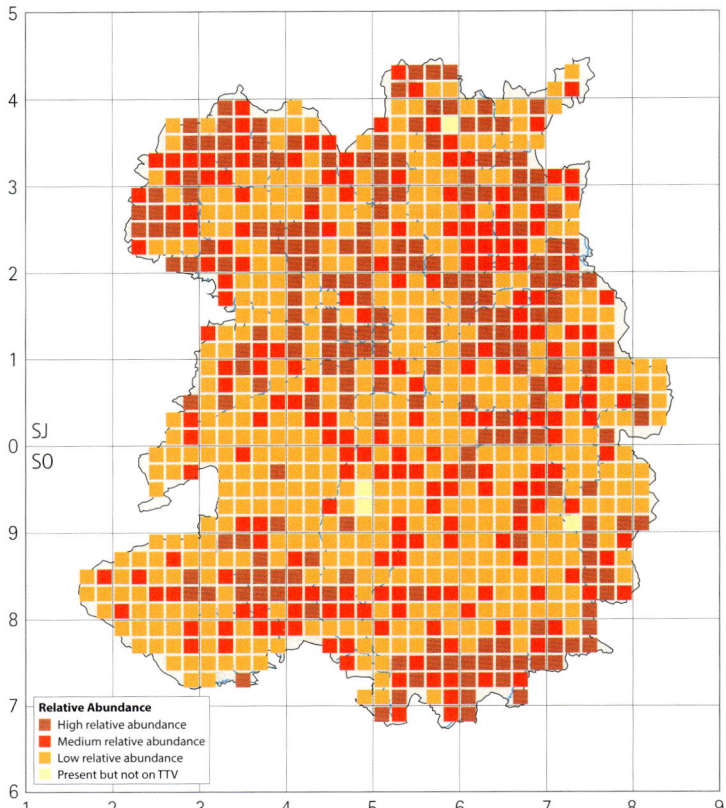

Relative Abundance
- High relative abundance
- Medium relative abundance
- Low relative abundance
- Present but not on TTV

They feed on worms and other invertebrates, switching to berries and fruit in the autumn. During spells of dry weather, when their normal food can be hard to get at, they can be quite adaptable. Forrest (1908) recounted one seen to 'catch a small fish in the Tanat near Oswestry. It carried the fish to a stone against which it banged it several times after the manner of a thrush with a snail, and then swallowed it'.

A pair of Blackbirds helped rear a brood of Robins. They were frequently seen feeding the chicks, in fact the female would often brood the young as well. The young fledged successfully, which was hardly surprising given the full attention of four adults, and both sets of parents continued to feed the young once they left the nest (SBR 2000).

The Blackbird can truly be described as ubiquitous, having been recorded in every tetrad during both the breeding and winter Atlas periods. Breeding was confirmed or regarded as probable in all but five tetrads.

BBS data show an estimated 57% increase in the breeding population since 1997. Based on TTV counts, the current population estimate is 104,000–105,500 breeding pairs, making Blackbird the fourth most numerous breeding species. It is the most frequent user of gardens, with an annual GBW reporting rate of 92%. The reporting rate falls in the autumn, after the breeding season, partly because they become less obvious during the post-breeding moult, and partly because of the shift to the countryside to avail themselves of the bounty of the fruit and seed crops there.

There are higher densities in suburban areas, woodland and scrub, and lower densities on arable farmland and higher ground, where the terrain tends to be more open.

Colonisation of the higher ground was documented in the *Handlist* (1964), with the observation that Blackbirds were 'well distributed ... and apparently nesting at higher altitudes on the more open hill

country than previously'. They have spread away from their typical ancestral breeding sites of woodland and hedges, a behaviour that was recorded in 1937 where one was seen on a nest 'in middle of a grass field, no bush or tree near it'.

The pattern of relative abundance in the winter is very similar to that shown in the breeding season, with higher densities generally occurring in the north. The winter population is supplemented by birds of the year and winter visitors from northern Europe, and so the population density is higher than during the breeding season. On average, winter TTV counts were 27% higher than breeding season counts.

A few hundred are ringed every year, and it is worth noting that all of the interesting recoveries relate to birds either ringed or recovered during winter months. This confirms the seasonal inward migration, and indicates that the breeding population is generally sedentary. One ringed here was found on an oil rig and another on a ship, both in the North Sea, typical of the migration route across to northern Europe. Countries with the highest number of recoveries abroad are all in Scandinavia, with Denmark (11), Norway (10) and Sweden (17) being the only ones to reach double figures, while four found in Finland had moved over 2,000km. The longest distance travelled was an adult male ringed near Bridgnorth in January 2010, which was found dead nine months later, in Pollakka, Finland, where it had hit a window, having travelled 2,155km. Others were found abroad in Germany, the Netherlands and France. Birds ringed abroad and subsequently recovered here came from Belgium, Denmark, Germany, Finland, Norway, Sweden and the Netherlands.

Movements have been to or from 26 counties in England and

BBS Trend (1997–2014)

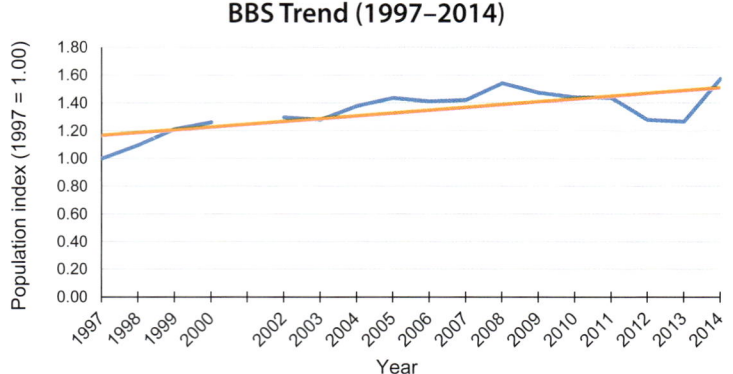

Redwing

Turdus iliacus

Common winter visitor

For many of us, the high 'tseep' call of migrating Redwings is the first indication that autumn is moving into winter, their arrival marking the changing seasons as much as Swallows herald the coming of spring. They commonly migrate at night, from their breeding grounds in Fennoscandia and eastwards across Siberia, and on cold, clear evenings their call can be heard over towns and cities as well as more rural locations. The much smaller Icelandic population winters mainly in Ireland, north-west Scotland and Iberia, and it is unlikely that many reach here.

The only colloquial name that seems to have been used is 'Swinepipe', quoted from a text by Robert A. Slaney M.P., one of 'Two Old Shropshire Naturalists', who described the name as 'taken from a singular note that he has – a sort of inward deep-drawn sigh, like an attempt at ventriloquism' (Forrest 1910).

Redwing were described as 'Common' in the earliest historical record in 1836. The earliest arrival date recorded historically was 8 September 1904 in West Felton, and the latest departure was 24 April 1938 in Bathcote (Lloyd 1938). The earliest recent reported arrival date was 6 September, in 2013, when a single bird was heard flying over Prees. Early to mid-October is more typical, and the data in the *Arrival Dates* chapter suggest that Redwing is arriving here 21 days earlier than it did in 1900.

A 'Pied Redwing' was seen near Bridgnorth in March 1918 and reported in *The Naturalist*. Presumably what we would now refer to as leucistic, the individual 'appeared to have a white tail and primaries' and was described as 'very noticeable on the wing'.

The 1937 Shrewsbury School Ornithological Society Annual Report documents a passage migration consisting of 'vast flocks' on 31 October, estimated at 1,000 per minute between 19:45 and 20:20, with smaller numbers still flying over at 23:00. Redwing commonly feed in open fields and woodland leaf litter, often in mixed, nomadic flocks with other thrushes. They are unable to eat some of the larger berries taken by bigger thrushes (Snow B & D. 1988), but turn to them during harsh weather: in the severe 1939–40 winter at Shrewsbury, a flock stripped a holly hedge of its berries – 'a diet ordinarily disdained' (Lloyd 1939b).

The *Handlist* (1964) stated that 'numbers are lower in the middle of winter, as in the case of Fieldfare, but it is on the whole more common'. As noted in the Fieldfare account, this is no longer the case, as Fieldfares are more numerous, and don't move on. The *Handlist* (1964) also noted a regular roost at Lea Wood, near to Welshampton, where up to 3,000 had been recorded.

They are widespread throughout the winter months and during the recent Atlas they were not observed in only 11 tetrads. There are many records of large flocks, the largest of which was estimated to be around 750, at Edgerley on 30 December 2007. There were 10 other occurrences of flocks of 250 or more, the largest being 400 at Atcham in November 2008, at Harlescott in November 2008 and Venus Pool in November 2013, and about 500 at Venus Pool in November 2012.

Large total numbers were also recorded as part of TTV surveys, the

Allan Heath, Dudleston Heath, 28 November 2008

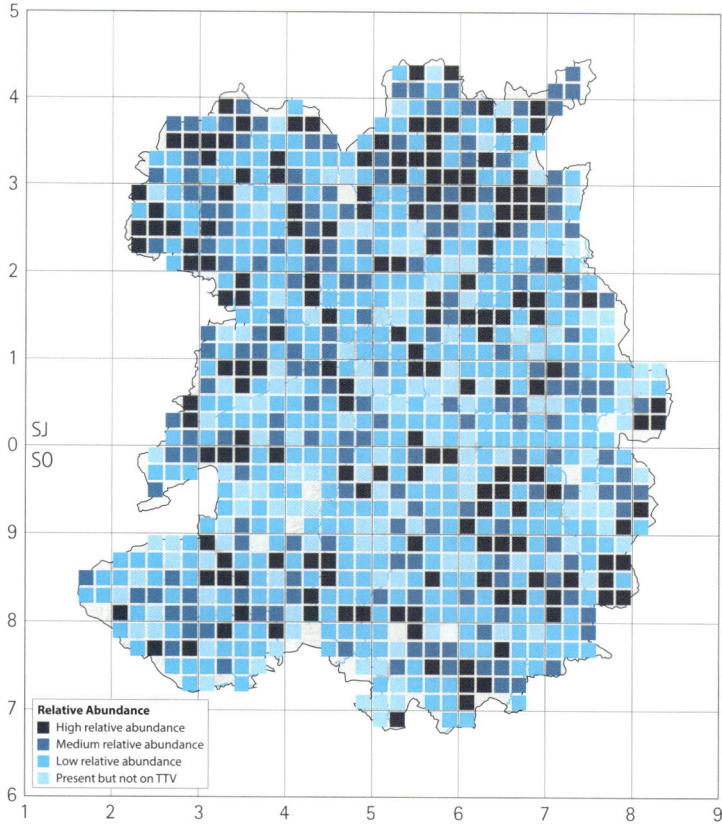

Winter Relative Abundance (2007–13)

Relative Abundance
- High relative abundance
- Medium relative abundance
- Low relative abundance
- Present but not on TTV

Occupied Tetrads

Tetrads with Winter Records (2007–13): 859 (99%)

most notable being 556 between Crudgington and Hodnet (SJ62G) in January 2008 and 315 near Clungunford in November 2008.

Flocks of over 300 recorded in flight were 321 at Hampton Loade in November 2013 and 300+ at Maesbury in February 2014.

The relative abundance map shows that they generally occur in higher densities in the north, perhaps again reflecting the direction of their arrival and departure, rather than the best feeding areas.

Although they are rare garden visitors, the harsh winter of 2010–11 brought them in to feed on unpicked apples. This probably accounts for some of the records in the more urban areas where perhaps they would not usually be present.

Based on TTV counts, the population estimate is around 20,000 individuals, rather higher than the estimate for Fieldfare. However, as shown in the Fieldfare account, there appear to be almost three Fieldfares for every two Redwings. The total count during the earlier of the two-hour TTVs was just over 22,000. There would have been considerable double counting of many of them, moving around within and between squares, but this would be counterbalanced by many others, which would not have been seen at all.

Comparing the total TTV counts in the November–December period with those in January–February, there was an average decline of almost 40% by the second period, suggesting considerable onward migration, in some at least of the four winters 2007–11 when TTVs were carried out. The population and movements of the two winter Thrushes are discussed more fully in the Fieldfare account.

In the breeding season, Redwings were recorded during the recent Atlas period in 34 widely scattered tetrads, in small numbers and only one flock of 50 or more, a flock of 100 on 10 April 2013 near Chorley. They were even heard singing, but there is no evidence to suggest that this was anything other than delayed departure back to their breeding grounds, as there were few records beyond the middle of April.

More recently, a mixed flock with Fieldfares was seen at Venus Pool on 16 November 2013, estimated at 2,000 birds; there were around 500 Redwings together with around 250 Fieldfares three days later.

Few are caught and ringed each year. Despite the low numbers, there are some interesting recoveries. Some, such as those found in Devon, Cornwall and Kildare in the British Isles, and Spain and Portugal overseas, reflect onward passage southwards. Those from Denmark, Finland, Norway, Sweden and Russia were usually found in the breeding season, while those found in France, Belgium and the Netherlands reflect the gateway to the continent, and Italy, Greece, Poland, Germany, Georgia and Syria reveal complex movements between breeding grounds across northern Europe and Asia, and wintering grounds across much of central and southern Europe, sometimes involving the same individual in different widespread locations in different winters. The furthest travelled was shot in Tartus, Syria, in April 1984, a distance of 3,685km from where it was ringed in January 1983 in Minsterley.

Linda Munday

Song Thrush
Turdus philomelos
Very common resident

The earliest documented reference in 1836 (Leighton) is to a 'Throstle', a name which seems to have been used commonly until the end of the nineteenth century, when it was described as 'plentiful everywhere'.

Although it was never a common host species, a Cuckoo egg was found in a Song Thrush nest, near to Ludlow, in 1900 and 1901, and the Cuckoo species account shows these are the only known local examples.

Paddock (1904) said that we should readily forgive any damage done to our garden fruit by Song Thrushes because of the service they do in destroying slugs, but this view may not have been shared by everybody. For example, large numbers were caught in the strawberry nets, sometimes as many as 40 a day, at Marrington Hall (Horton 1902). It is hard to imagine such numbers now, as they are rarely seen in large groups – there were no SBR records of large numbers, and during recent Atlas fieldwork there were only a handful of records of groups into double figures.

As with some other members of the Thrush family, pied (what we now refer to as leucistic) and other coloured varieties appear to have occurred fairly frequently, and, because they tend to stand out, many such birds have made it into the historical records. Forrest (1899) reported one shot near Shrewsbury in 1898, whose 'whole plumage was suffused with bright buff colour'. Faring better than this, a buff-coloured bird was resident at Boreatton Park, near Baschurch, from 1911–14. The records do not document whether it bred each year, but

John Hawkins, Pentrecoed, 17 December 2008

Breeding Relative Abundance (2008–13)

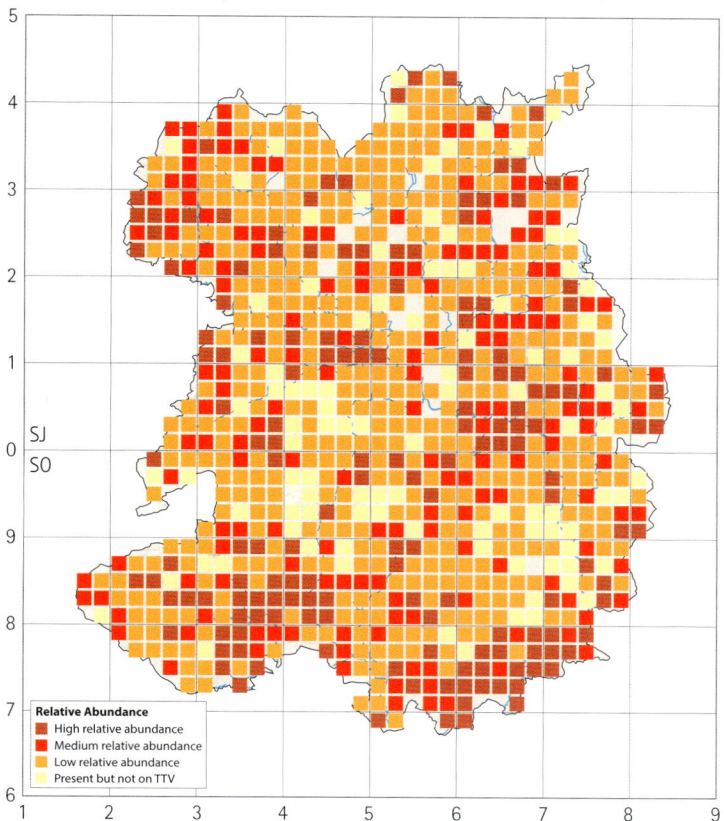

Relative Abundance
- High relative abundance
- Medium relative abundance
- Low relative abundance
- Present but not on TTV

Occupied Tetrads

Atlas period (breeding)	1985–90		2008–13		Change	
	Number	%	Number	%	Number	%
Confirmed	572	66	403	46	-169	-30
Probable	147	17	297	34	150	102
Possible	107	12	150	17	43	40
TOTAL	826	95	850	98	24	3
Tetrads with Winter Records (2007–13): 816 (94%)						

However, numbers picked up in the late 1980s, perhaps due to a run of milder winters, and this upward trend has continued, as the table indicates that they were found in 3% more tetrads compared to the previous Atlas. They were not recorded in only 20.

The BBS data also shows a promising overall increasing trend, but with a decline since 2008, probably weather-related, due to hard winters in 2009–10 and 2010–11, and very wet breeding seasons in 2007, 2008 and 2012. Based on TTV counts, the breeding population is estimated at 20,000–20,350 pairs, compared with a much broader estimated range of 17,500–35,000 pairs in 1990.

There is no obvious pattern to the few overall gains in the breeding distribution, and the changes are likely to reflect local variation in fieldwork effort, coupled with the underlying population increase, rather than any particular environmental factors.

The relative abundance map shows low density or absence in the uplands and some agricultural areas in the north and south-east, and higher densities in woodland and suburbia. Research elsewhere has shown the decline is continuing on arable farmland, following the loss of hedgerows and wet ditches, which removed feeding and nesting sites, while increased land drainage, tillage and use of pesticides have reduced the number of snails and earthworms (RSPB website). Gardens now hold much higher densities than such farmland, although the continued use of slug pellets is still a threat in this habitat.

They were found in 36 fewer tetrads in winter, but they were probably under-recorded, as they are usually far less conspicuous than during spring and summer, when their song proclaims their territory, although they have been heard to sing during mild winter periods.

The relative abundance was similar to the breeding season, but they were not found in some of the higher tetrads, for example around the Long Mynd, and there appears to be a shift to more open ground in the north, although that might just reflect lower detectability in woodland in winter.

Their winter status is complex, as the residents are certainly supplemented by migrants and passage birds from the continent, although there is only one local example, one ringed in Denmark in October 1988 which was re-caught in Wellington in January 1989, 868km distant. It is also likely that some of residents move away in very cold weather. Ringing recoveries reported in the *Handlist* (1964) show that some moved to Ireland during the harsh winter of 1962–63 (one in Wexford and one in Kerry). The only other overseas movement was from the Isle of Man. In more usual conditions the population is largely sedentary, and 104 (87%) out of 119 ringed here were also recovered here, and most of the remaining movements were to or from nearby counties.

it did in 1913, although it was not successful as the nest was predated by a Jay. In 1924, an albino bird was 'obtained' from the same location.

The Song Thrush is well known for its habit of mimicking other bird song and this was first referred to in 1908 (Forrest). He describes 'one district where the green woodpecker was exceedingly numerous, I noticed several thrushes interpolating now and again a palpable imitation of the harsh laughing cry of that bird'.

The *Atlas* (1992) noted that the population was hit badly by a series of particularly cold winters during the middle of the twentieth century, but the decline was poorly recorded, as early SBRs through into the 1980s listed Song Thrush only as an 'Additional Species' or as 'under-reported'. This decline continued to be evident during the early fieldwork for that Atlas, but 'steeper than expected from cold winters alone, and increased use of pesticides to kill slugs and snails may be having an impact'.

BBS Trend (1997–2014)

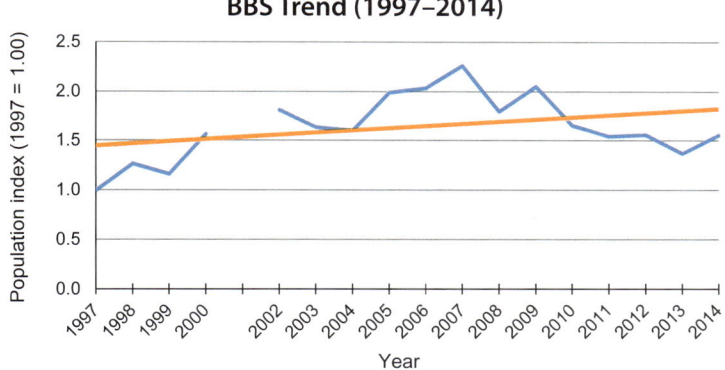

Linda Munday

Mistle Thrush

Turdus viscivorus

Fairly common resident

Often spelt as 'Missel' Thrush during the nineteenth century, it was also commonly referred to as a Storm Cock, 'from its being said to utter its peculiar chattering note before rain' (Eyton 1838). Other observers described it as singing during storms and other inclement weather, again helping to justify this name. Its other common name was Mistletoe Thrush: 'The ancients held the absurd idea that the mistletoe plant could not vegetate without having passed through the stomach of this bird' (Paddock 1890). It does eat mistletoe berries, and will actively defend a good supply of this and other berry species, but they do not form a major part of its diet (Snow B. & D. 1988), especially in this County where mistletoe has a restricted distribution. Beckwith (1888) wrote a good deal about the diet, highlighting a fondness for cherries and wild berries, such as holly, hawthorn and yew, during the autumn. Slugs, snails and other invertebrates are the preferred food through most of the year.

Beckwith (1879) stated that 'in Autumn these thrushes assemble in large flocks'. Paddock (1904) also documented that 'towards the end of June, large flocks consisting of both old and young, may be met with on the Longmynd, roaming over the open fields and hillsides for food'. As with other thrushes, their numbers were severely affected by harsh winters: Paddock (1897) referred to several such occurrences late in the nineteenth century. They were similarly affected by the winter of 1939–40, when they 'suffered considerably, disappearing from many places', and numbers were reduced again in the winter of 1962–63 (*Handlist* 1964).

During the first half of the twentieth century, autumn flocks of around 100 birds were regularly reported, but SBRs from the 1960s onwards record decreasing flock sizes through to the present day. There are some exceptions, and flocks of 130 were recorded in both 1974 (Hoar Edge, 13 October) and 1989 (Ashes Hollow, 16 July), while the comment accompanying the record of 70 at Venus Pool on 28 August 2006 described 'an exceptional count'. During the recent Atlas years, there were only five records of flocks of more than 20, all in July and August (23 at Betton Moss in 2008, 25 at Cound Stank in

Occupied Tetrads

Atlas period (breeding)	1985–90		2008–13		Change	
	Number	%	Number	%	Number	%
Confirmed	524	60	263	30	-261	-50
Probable	159	18	232	27	73	46
Possible	101	12	218	25	117	116
TOTAL	784	90	713	82	-71	-9
Tetrads with Winter Records (2007–13): 780 (90%)						

Breeding Relative Abundance (2008–13)

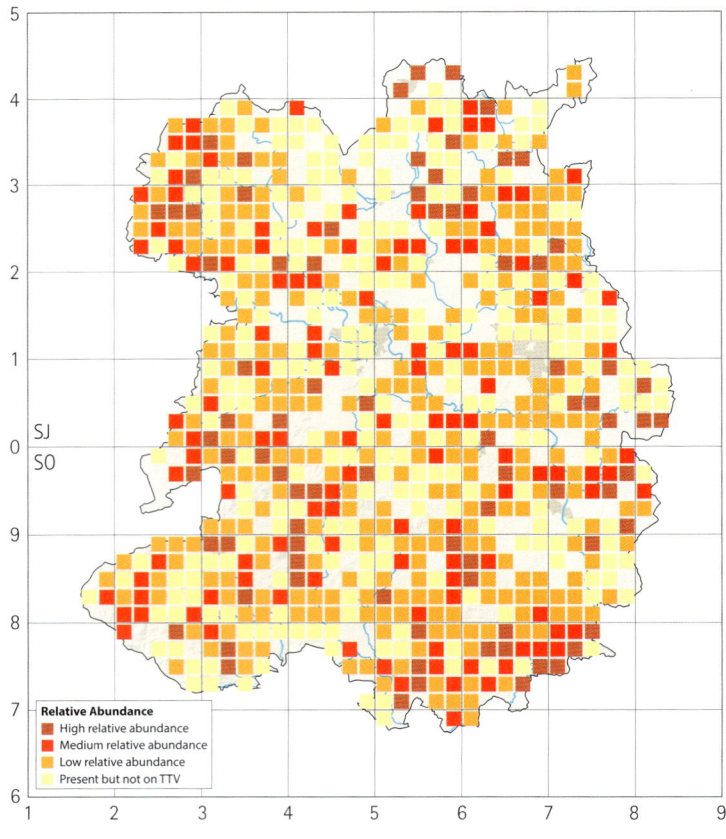

Breeding Distribution Change (1985–90 to 2008–13)

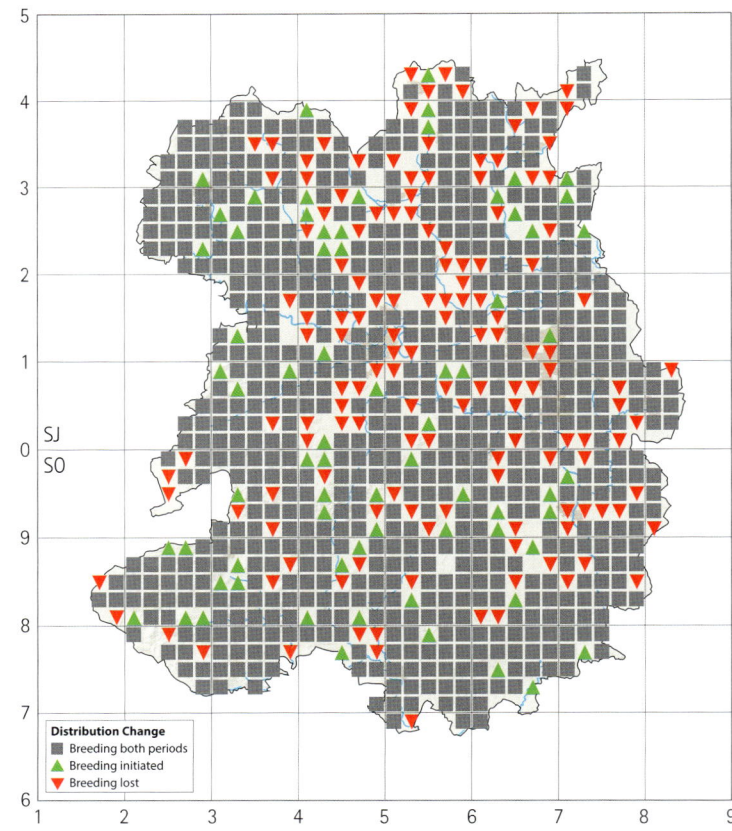

BBS Trend (1997–2014)

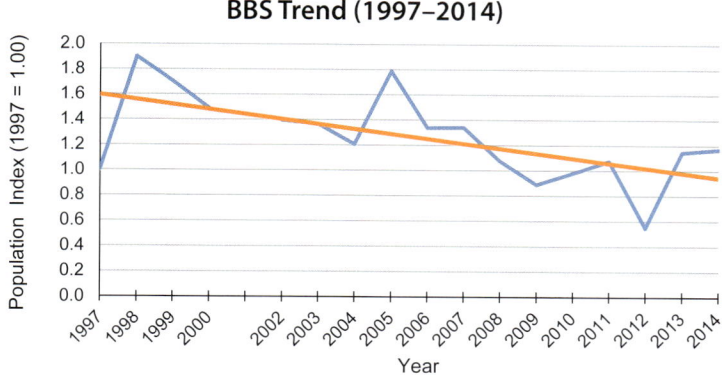

Population Index (1997 = 1.00) vs *Year*

2009, 21 at Lilleshall Hall in 2012, 52 at Adeney in 2012 and 28 at Tibberton Moor in 2013.

The maximum count recorded as part of a TTV was 23, north-east of Onibury (in SO48Q) on 15 November 2008.

During the breeding season, they are widely distributed, and were found in 82% of tetrads in the recent Atlas, a reduction of 71 tetrads compared to 1985–90. They don't occur at high densities anywhere, but they need suitable trees for nesting and open areas with short vegetation for feeding, and are more numerous in woodland and settlements in the south-east and north-west, and absent from some arable areas, uplands and larger settlements.

The breeding distribution change map shows losses in some arable areas, notably the northern river valleys, and in the towns of Shrewsbury, Telford, Whitchurch and Bridgnorth, probably a result of

John Fielding, Atcham, 12 January 2015

a loss of food supply (slugs and snails) through use of molluscicides, in agriculture and in slug pellets in gardens.

The population was estimated at 4,800–5,500 breeding pairs in 1990, whereas the current estimate, based on TTV counts, is around half that, 2,400–2,500. This reduction is confirmed by the BBS, which shows an underlying downward trend of about 40% since 1997.

In winter they are apparently more widespread, having been recorded in 89% of tetrads, but the relative abundance is very similar.

Many species are quieter during the winter months, so could have been under-recorded, but this is not the case for Mistle Thrush as it breeds early in the season, laying eggs in late February or March, so it holds territory during the mid to late winter months. Nest building started on 21 November in 1909. They have been observed singing even during frosty days, so it is likely that they will have been relatively easy to find during winter fieldwork, perhaps more so than in the breeding season.

Very few have been ringed here, so it is not surprising that there have been few recoveries, but all eight that have been recovered were also ringed here, providing further evidence for the sedentary nature of this thrush.

Linda Munday

Winter Relative Abundance (2007–13)

Relative Abundance
- High relative abundance
- Medium relative abundance
- Low relative abundance
- Present but not on TTV

American Robin

Turdus migratorius

No modern record

The American Robin is migratory, breeding throughout North America, and wintering in the southern US and Mexico.

The single record, of one seen in a Shrewsbury garden on 15 September 1927, was initially considered as 'doubtless an escape', perhaps from the failed introduction attempt by Lord Northcliffe near Guildford in the early twentieth century, 'to improve Anglo-American relations'. However, a review of the then known records, including the Shrewsbury bird, concluded that they were 'more likely to have been genuine drift migrants from across the Atlantic' (Fitter 1959). In that month, 'from the 5th to the 10th, unsettled weather associated with low pressure systems accompanied by frequent rain and high winds, reaching gale force locally on the 8th and 9th, was experienced widely'.

Subsequently, the 24 British records between 1958 and 2012 were analysed, and the winter period October–January was identified as the peak period of occurrence, with four to six records per month, with none in September (Bond 2014). While the absence of a recent September record might cast doubt, the gales that preceded the event suggest that the Shrewsbury bird may indeed have been a true transatlantic vagrant.

John Tucker

Spotted Flycatcher

Muscicapa striata

Uncommon summer visitor

Spotted Flycatcher is a summer visitor which winters in sub-Saharan Africa, south of the Sahel. It is one of the last summer migrants to arrive, typically not until the second week of May. A few September records are received every year, although most seem to have departed by the end of August. Here, it frequents deciduous woodland, woodland edges, orchards, mature gardens, parks and churchyards, usually close to water, and can be seen returning to the same perch time and time again while fly-catching.

It was first described by Leighton (1836) as 'very common – about houses'. Rocke (1865) called it 'a neat and sociable little bird' and 'common everywhere', and, commenting on its confiding nature, he noted a nest in a summer house within a couple of inches of people passing in and out. Beckwith (1888) confirmed that it 'nests near habitation', with the young often 'falling victims to the cat', and added that it usually raised two broods in spite of being here for only three months, and was 'most useful in the garden' because it feeds entirely on 'insects and their larvae'. Paddock (1890) found it to be 'well distributed and common' in the Newport area, and reported a pair nesting on the hinge of a garden door and raising a brood, despite the door being opened and closed several times a day. Vernacular names included 'Beam-bird', possibly a reference to preferred nesting locations, and 'Miller', apparently because it looked as though it was dusted with flour.

In the first half of the twentieth century, it continued to be reported annually as common, frequent and widespread, and several interesting and often exposed nest locations were noted. In 1900, the decision to build a nest inside that of a Mistle Thrush did not end well: the eggs were taken and put in Shrewsbury museum, a sign of the times. Paddock (1904) noted that it often repaired an old nest rather than built a new one, while other sites were on top of a large fungus on an oak tree at Merrington, in an old bird cage hanging on a nail in Llanymynech, and two pairs nesting on Atcham Bridge in 1936.

Early SBRs continued to report it as common and widespread, but the first early warning appeared in 1958, the 'lowest numbers this year that I can remember' in the Wellington area. The *Handlist* (1964) confirmed this picture, saying that it was 'fairly common' and widespread, although 'numbers seem to fluctuate from year to year' and 'several observers are of the opinion that a general decrease has been taking place over the past 20 years'.

SBRs for the 1970s continued to describe it as widespread and common, with some fluctuations. There were 'more records than in recent years' and it 'appeared to have had a good year in many parts' (1972). Breeding records from the 1970s include two fledging from an empty tin nailed to a tree in Shrewsbury (1972), a pair taking over a Robin's nest and laying eggs among those of the Robin, successfully raising their own young (1975), and nests in an old bean tin and a drain pipe (1979). It was still described as 'plentiful', 'numerous' and 'widespread' in the 1980s, and breeding was reported from 33 sites in 1986, with some locations having more than one pair, and 40 breeding pairs were confirmed in 1987 during 'an exceptionally successful breeding season'. There was little published evidence of any decline during this time, although it was noted that a few traditional sites were no longer in use; one nest in a porch in Prees was used for 22 consecutive years, the last being 1984.

The *Atlas* (1992) confirmed the widespread distribution, although it was perhaps overlooked in some areas away from human habitation. Gaps appeared to 'coincide with well-drained arable land' where hedgerows and copses 'may not hold sufficient flies for successful breeding', but noted that the use of insecticides will further reduce the availability of prey. The *Atlas* noted a gradual long-term decline, particularly on farmland in southern Britain, and, based on the relatively low population density in western England, estimated the population here to be about 2,000 pairs, or three pairs per occupied tetrad.

In 1990, some observers thought that they were still numerous, but others thought they were 'down in numbers'; breeding was confirmed at 17 sites with five nests in one Munslow garden. Throughout the 1990s, SBRs listed few confirmed breeding records, and commented that they were 'under-recorded'. No major decline was noted until 1999, when only 72 records were received, many from gardens, churchyards and well-visited sites; the SBR for that year suggested that it is 'becoming localised and generally in decline'.

SBRs for the 2000s documented a fluctuating, but decreasing, number of records. Its absence was noted from several more traditional sites, with most breeding records continuing to come from areas near human habitation, unrepresentative of the widespread variety of habitats utilised. The continued decline reflected the national situation, and SBR (2007) quantified it, noting that BTO data showed a decrease greater than 50% since the 1960s, attributable to 'deterioration of woodland habitat and declines in numbers of large flying insects'.

In the recent Atlas period, it was found in 418 tetrads, compared with 748 tetrads in 1985–90, disappearing as a breeding species from almost half (44%) of the tetrads where it was found previously.

The breeding distribution change map reflects this alarming decline. Remaining strongholds seem to be the villages and wooded uplands of the south-west and north-west, Clee hills, Wenlock Edge, the Wrekin and the Severn Valley, but it was largely absent from the two large towns, and arable land in the north and the south-east.

The relative abundance map shows the patchiness of this widespread but scattered distribution, with clusters of high abundance in the remaining strongholds, in isolated patches of woodland, and in rural areas with other preferred habitats, such as gardens, churchyards and wooded streams. However this map gives an incomplete picture, because the low proportion of records from TTVs reflects its late arrival date (so the number of early TTVs carried out before its

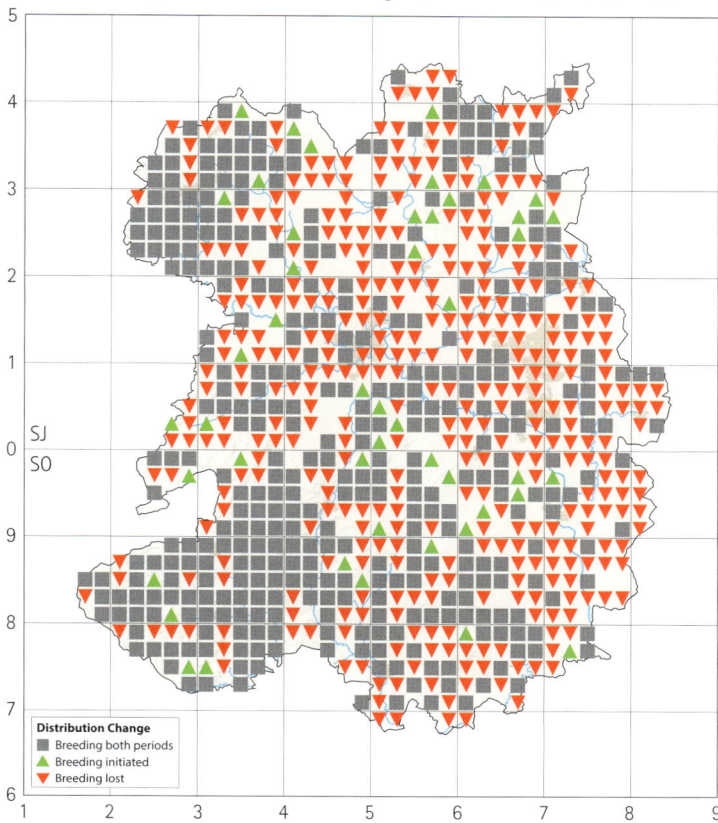

Breeding Distribution Change (1985–90 to 2008–13)

Distribution Change
■ Breeding both periods
▲ Breeding initiated
▼ Breeding lost

John Hawkins, Ellesmere, 2 July 2009

Breeding Relative Abundance (2008–13)

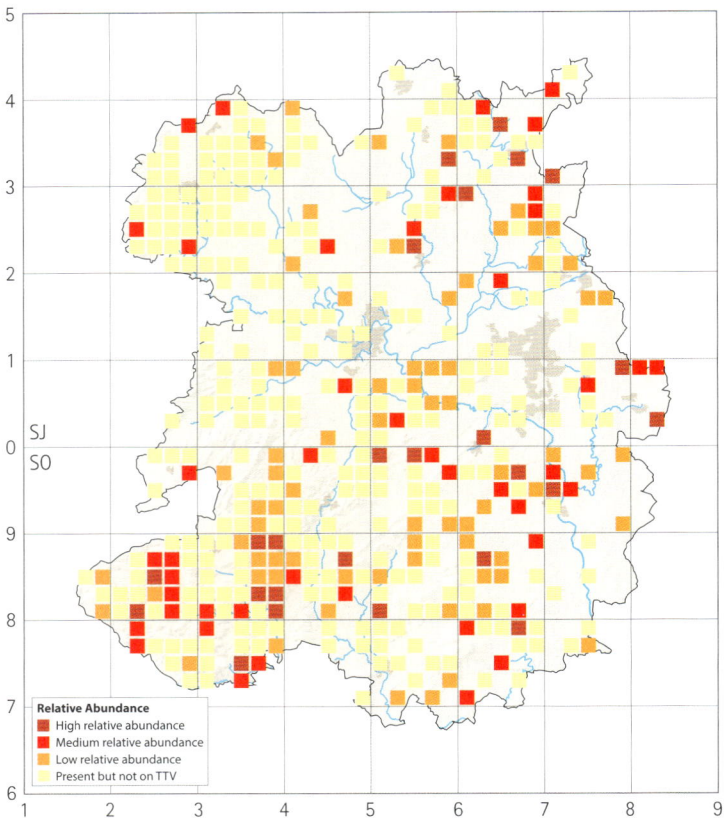

Relative Abundance
- High relative abundance
- Medium relative abundance
- Low relative abundance
- Present but not on TTV

Occupied Tetrads

Atlas period (breeding)	1985–90		2008–13		Change	
	Number	%	Number	%	Number	%
Confirmed	525	60	191	22	-334	-64
Probable	127	15	93	11	-34	-27
Possible	96	11	134	15	38	40
TOTAL	748	86	418	48	-330	-44

of deterioration of woodland habitat and less invertebrate food as a result of pesticide use. Based on TTV counts, the estimated breeding population is approximately 450 pairs. However, it was found on TTVs in only 16% of tetrads, so this estimate may be unreliable. It is equivalent to only one pair per occupied tetrad, but some hold more. The estimate in 1992 of three pairs per occupied tetrad will now be too high, but an average of two pairs per tetrad with probable or confirmed breeding, and one pair per tetrad with possible breeding, gives a more plausible estimate of around 700, just over one-third of the 1992 estimate.

Earliest arrival dates have been noted in many SBRs, but the *Arrival Dates* chapter suggests that it is arriving here no earlier than it did in 1950, on average around 8 May. This result is unreliable, as it is likely to reflect the decreasing likelihood of the first arrivals being encountered and reported, as the population disappears, but if it is the case, this too may be a contributory factor in the decline, in that its breeding cycle may no longer coincide with the earlier date of peak availability of insect food.

Very few have been ringed, and there are only four recoveries: one was also recovered here, but the others, one on the Isle of Wight, and two abroad, reflect the migration route – a nestling ringed in Sutton Maddock in July 1975 was caught in September 1975 in Sark (Channel Islands), 350km south; and another nestling ringed in Wellington in June 1973 was trapped as an adult in Morocco in September 1982, 1,955km south and 9y 2m 21d later. This individual holds the national longevity record.

Helen J. Griffiths

arrival would have been higher than for most summer visitors) and low detectability away from areas near human habitation.

It occurs in insufficient BBS squares to produce a reliable local or regional population trend, but in the UK, the decline up to 2014 has been 86% since 1970, and 44% since 1995. In England, the decline has been faster than that, 61% since 1995. In common with several other trans-Saharan migrants, the decline is attributed to habitat changes on passage and in their wintering areas, as well as problems here

---

Robin (European Robin)

Erithacus rubecula

Very common resident

Much loved and familiar to just about anyone with the most basic interest in and knowledge of British avifauna, the almost ubiquitous Robin is a common sight, not shy of the gardener at work, willing to nest close to people, a regular visitor to the bird table, and frequently depicted on the nation's Christmas cards. The Robin's song is also a common sound, heralding not just the spring and breeding season, but also the period of autumn into winter when both male and female sing to establish and then defend winter territories.

The earliest historical records refer to the 'Robin Redbreast' being 'every where' (Leighton 1836) and 'very common' (Forrest 1899), and many reports in the early twentieth century pointed to its opportunistic choice of nesting site, including old watering-cans, kettles, flower pots and even a bookshelf in the office of the Army Pay Corps in 1913,

where two successful broods were raised despite the coming and going of personnel. A more surprising record, in 1929, reported that during a frosty period, a Robin 'first attacked a weakly Chaffinch, after dead, fed on its flesh', taking its well-known aggression to a new level!

The *Handlist* (1964) described it as 'resident, very common', and in the 18-year period 1950–67, there was a steady increase in numbers in the Wyre Forest as the oak and beech trees matured (Yapp 1969). The Atlas (1992) stated that, although 'numbers are greatly reduced following severe winters such as 1981–82, when they may disappear from marginal sites ... following several mild winters Robins have increased steadily, and in 1989 they reached their highest level on a local CBC plot near Oswestry since recording began in the mid-1960s'. It was found in every tetrad except one during 1985–90 fieldwork.

Celia Todd, Pant Glas, 21 February 2013

The BBS chart shows that the population has increased by around 15% since 1997, but it fluctuates from year to year, and the weather has been unhelpful in the recent past, with hard winters in 2010–11 and 2011–12, record wet summer conditions in 2012 and a record cold and dry spring in 2013, but recovery was again underway in 2014.

In the recent Atlas period, Robin was confirmed as breeding or probable breeding in 97% of tetrads, and overlooked in only two. The population estimate based on TTV counts is around 122,000 pairs,

which, given the rise shown by BBS, is broadly consistent with the estimate of 'probably near 100,000 pairs' made for 1985–90. This makes it the second most common species, exceeded only by Wren.

The breeding season relative abundance map shows that the Robin is particularly common in areas with mature gardens and deciduous woodland, and even on higher ground where there are valleys with suitable habitat cut into the landscape. In more intensely farmed areas, or where the higher ground is more extensive, then densities are lower.

In the recent winter Atlas period, Robin was found in every tetrad, and the relative abundance pattern was very similar to that in the breeding season, pointing to the sedentary nature of the species.

Frequent use of gardens is reflected in the BTO Garden BirdWatch reporting rate: it is ranked third highest, with an annual average of 88%, fluctuating between 76% in early July and 95% in December. There is little apparent change in the annual average since 2006, and the decline to a low point in 2012 and subsequent recovery found by BBS is not reflected in use of gardens, emphasising the value of garden feeding when the wider environment is more hostile.

Early experiments in marking had some virtue despite not exactly fitting the modern day criteria for unusual methods stipulated by BTO! Dovaston (1839) wrote: 'I was working in a wood at distance from my house; and while I was eating my bread and cheese, a Robin lit on the handle of my spade: by a sudden sweep of the arm I caught him, and in a frolic, cut off his tail. I have since observed that bob-tail bird never wanders far from that spot'.

The ringing record confirms that Robins rarely travel far, with only 11 ringed here recovered in different counties in the UK, and only one abroad, shot in Badajoz (Spain), 1,612km from the ringing

BBS Trend (1997–2014)

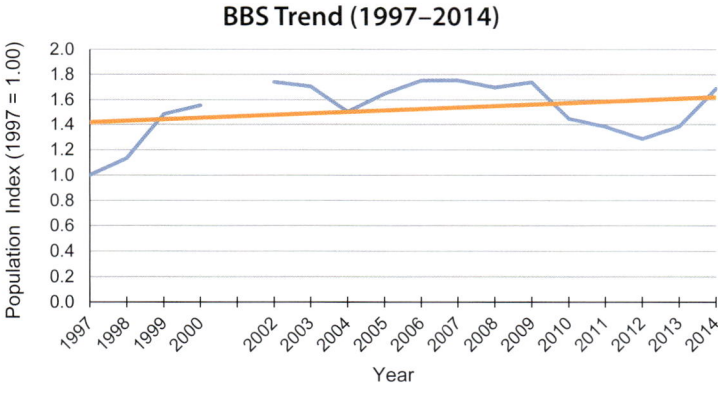

Occupied Tetrads

Atlas period (breeding)	1985–90		2008–13		Change	
	Number	%	Number	%	Number	%
Confirmed	794	91	700	80	-94	-12
Probable	63	7	151	17	88	140
Possible	12	1	17	2	5	42
TOTAL	869	100	868	100	-1	0
Tetrads with Winter Records (2007–13): 870 (100%)						

Breeding Relative Abundance (2008–13)

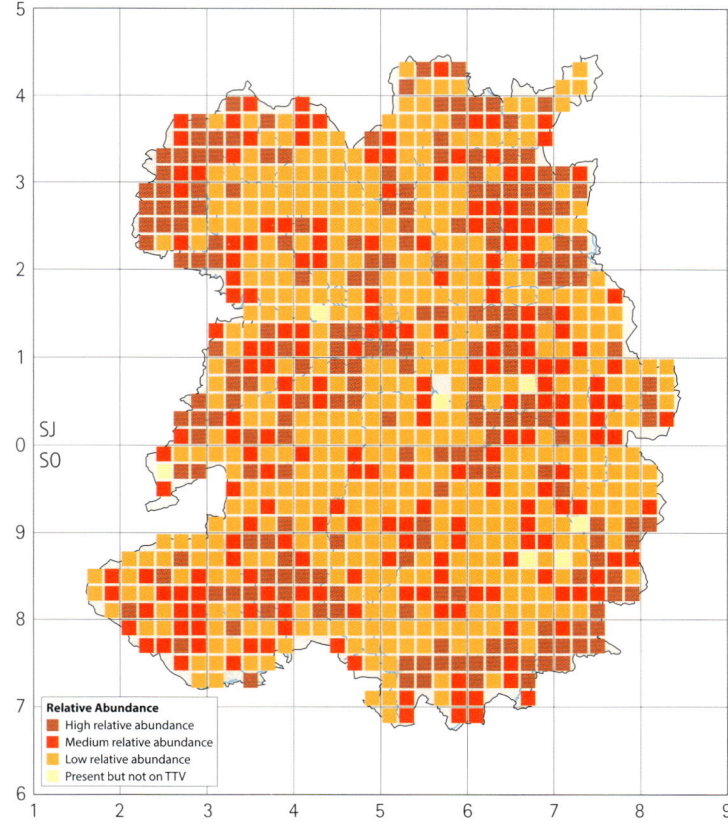

Relative Abundance
■ High relative abundance
■ Medium relative abundance
■ Low relative abundance
□ Present but not on TTV

site at Porth-Y-Waen, 0y 8m 6d after ringing as a nestling on 21 May 1966. Similarly, only 16 ringed elsewhere have been recovered here.

Two first-years ringed at Cross Lane Head, Bridgnorth, were caught several times at the same garden, the first 10 times between August 2006 and January 2013, and the second seven times between July 2004 and September 2010. These two were both more than six years old, as was a first-year ringed at Sutton, Shrewsbury in November 2005, and caught three more times at the same site, in August 2009 and twice in December 2011.

While the race that inhabits Britain and Ireland is largely sedentary, Robins from Fennoscandia are almost totally migratory, many passing along the British east coast, and wintering as far south as Iberia and North Africa. A first-year Robin caught twice within five days in January 2005 in a Bridgnorth garden, but not earlier or subsequently

(suggesting a winter visitor, rather than one holding a local winter territory), and then found dead in Caithness the following January, 678km distant, suggests the intriguing possibility of such a migrant being caught while passing through, returning to northern Europe, and then losing its life on its second southward winter migration.

In total, only five travelled more than 100km, to or from Scotland (three), the Scillies (one) and Cornwall (one), and the dates suggest some of these too were found on migration. Three of these recoveries have been courtesy of a cat (two), or hit by a car.

The longevity record was one from Shrewsbury that died in 1991 at the age of 7y 3m 8d, a year short of the national record, at the same place where it was first caught as a first-year, again having been taken by a cat.

Stuart Cowper & Leo Smith

Bluethroat

Luscinia svecica

Vagrant

The Bluethroat occurs widely across Eurasia and winters in southern Europe and Africa south of the Sahara through to South-east Asia. In Britain, it is one of the more common 'scarce migrants', with between 40 and 100 in most years, sometimes many more, with sightings split between the spring and the autumn migration periods (Fraser 2013b). The majority are found in the Northern Isles, with most of the rest scattered along the eastern and southern coasts. It is rarer inland and there have only been two records:

Gobowen	8 April 1987
Pontesbury	21 May 1996

The first, at Gobowen, was found dead beneath a window and noted to be of the white-spotted race *L.s. cyanecula* of southern and central Europe; well over 200 birds were recorded in Britain that year. Unfortunately, the details of the individual at Pontesbury have been lost and the race, therefore, cannot be determined.

Graham Walker

Nightingale (Common Nightingale)

Luscinia megarhynchos

Very rare summer visitor, has bred

Nightingale breeds in woodland and scrub in Europe and south-west Asia, and winters in sub-Saharan Africa. In Britain, it is on the north-western edge of its range. It has never been common, and had a specific local range. Eyton, in 1838, stated it could not be found breeding north of the Wrekin and its regular haunts were in the south-east, generally along the Severn Valley. Passage birds were also noted in other places, and some males even appeared to take up territories and started singing, but rarely would they stay to attempt to breed.

It is difficult to get a view of the population size during the late nineteenth century, as authors such as Beckwith (1889) expressed doubt about some of the earlier records, suggesting that many observers would have mistaken the song for that of other species such as Song Thrush. Also, many of the records of the time include non-committal statements such as 'more numerous than usual'.

During the early twentieth century, both the breeding population and range increased; there were records from around Shrewsbury, and a small breeding colony was established on Pulley Common, near Bayston Hill, in 1913, while in 1915, at least three pairs nested near Ludlow, 'quite outside their usual range'.

By 1935, the breeding range had started to shrink as, for the first time in recent years, breeding did not take place on Pulley Common. Other than a male singing in 1945, there was no further evidence of possible breeding there again, nor anywhere else around Shrewsbury.

However, some other areas along the Severn supported regular breeding populations, such as Dowles in the Wyre Forest, which appears to have been used until 1941. The most densely populated breeding area was 'The Lloyds', between Ironbridge and Coalport, with six to eight pairs in 1897 (Paddock), increasing to a peak in 1943, 'a particularly good Nightingale year'; when an estimated 18 or 20 were in song at the same time, 'creating pandemonium'. Such a musical performance is hard to imagine occurring in the few remaining strongholds nationally, let alone here, and which of us could read this account without experiencing some feelings of envy?

Nightingales appear to have reached their peak during the period 1938–50. In addition to the Ironbridge area, they were found at a number of other locations, often including the presence of singing males, but evidence sufficient to confirm breeding, such as a nest being found or recently fledged young being present, was reported only from:

- Pontesford (precise date unknown)
- Bishop's Castle (1938)
- Much Wenlock (1939)
- Wrekin and Ercall (Forrest Glenn area) (1939, 1940 and 1941)
- Llanymynech (1942 and 1943)

The total population then was perhaps around 40 breeding pairs.

The Lloyds remained the stronghold, with 15 pairs between Buildwas and Coalport, 'a good year for them' (SBR 1956), but the *Handlist* (1964) reported only five or six pairs there in 1963, where 'numbers appear to have decreased in recent years'. The *Handlist* also recorded 'outside this area, odd pairs are regularly found' at Buildwas, Leighton, Wroxeter, Attingham, Broseley, Bridgnorth, Quatford and Morville, suggesting a total population estimate of around 15 pairs in 1963.

Subsequent SBRs make for sad reading as they chart the demise of the Nightingale. By 1968, only two males were reported singing in the Ironbridge area, and 'the regrettable decline apparently continues'. In 1969, two nests were found near Highley, one with five eggs and the other with eggs 'sucked by a predator', and these were the last confirmed breeding records until 1990. In 1973, there was a glimmer of hope as more appeared than in recent years at Ironbridge, so

a survey was organised, and 'all observers' were asked to report sightings. Twelve singing males were found near Ironbridge, but none anywhere else.

In 1980, the last singing male was reported from the same area, and although there were other sightings, there was no evidence of breeding. There were further occasional sightings of individuals until fieldwork for 1985–90 Atlas, when breeding was again confirmed in 1990 near the River Onny (SO47T), where young were observed. There had been records from the same area in 1988 and 1989. Nightingales were recorded in 10 tetrads, although most of the possible breeding records would have been of passage migrants.

There has been no evidence of breeding since 1990, and the last record was of a single Nightingale seen and heard calling at Wood Lane during the spring passage period on 14 May 2000, but it could not be found again later that same day. There were no records in the recent Atlas period, and it is believed that they are now extinct as a breeding species.

Locally, much of the Nightingale's preferred habitat of dense, thick scrub has now disappeared as part of the continued urbanisation. In 1964, the *Handlist* referred to the loss of Pulley Common, 'much of which is now a housing estate', while the 'last outpost' at Ironbridge 'will be coming under increasing pressure from the New Town' (SBR 1971). The local decline in the breeding population mirrors that which has occurred nationally, as the Nightingale's British and European range contracts. Habitat loss, reduction in quality of woodland habitat due to changes in management and removal of dense understorey by browsing deer, pressures on migration and degradation of the African wintering grounds, partly a result of climate change, have all appeared to play a part. With these factors occurring locally, it is probably too much to hope that they will ever return to breed.

Linda Munday

Occupied Tetrads

Atlas period (breeding)	1985–90		2008–13		Change	
	Number	%	Number	%	Number	%
Confirmed	1	0	0	0	-1	-100
Probable	1	0	0	0	-1	-100
Possible	8	1	0	0	-8	-100
TOTAL	10	1	0	0	-10	-100

Pied Flycatcher (European Pied Flycatcher)
Ficedula hypoleuca
Uncommon summer visitor

With a habitat requirement of oak woodland or alder-lined streams, the Pied Flycatcher is generally restricted to the hill country of the south and west. There are exceptions, including the wooded areas of the Wrekin and Hawkstone Park, and some areas along the Severn Valley. Sightings away from these areas are probably passage migrants and are quite infrequent.

The males arrive from the sub-Saharan African wintering grounds early in April, followed a week or so later by the females. Usually single brooded, although replacement of clutches lost early has been noted, the young leave the nest by the end of June or early July, and they apparently disappear during August as they prepare to migrate south to the wintering grounds.

The analysis in the *Arrival Dates* chapter indicates that the Pied Flycatcher is arriving around 18 days earlier than it did in 1900, with almost all the change occurring since 1950.

Pied Flycatcher only started breeding here following an expansion of range towards the end of the nineteenth century, and Beckwith (1888) wrote that it was a rare summer visitor breeding in limited

Jim Almond, Bridges, 10 June 2013

John Fielding, Long Mynd, 27 May 2011

numbers, with nests at Willey Park in 1880, and more frequent in the west, breeding along the Teme and its tributaries, and along the Ceiriog near Chirk. Forrest (1899) noted that it was a 'rather uncommon summer visitor; partial to hawthorn trees in parks; breeds near Ludlow, Shrewsbury, Wroxeter and Hawkstone'. Recorded annually from 1900–11 in the areas mentioned by Beckwith, records resumed in 1918, with a brood at Church Stretton in 1919 being the first recorded in a nest box. By 1947 there were suggestions that the range was increasing with more records from the south Shropshire hills. Although this might be attributed to an increase in observers, the *Handlist* (1964), after noting that records came from widespread areas but breeding regularly occurred only in the Oswestry hills, the south and the south-west, added that 'it has evidently increased its range considerably' since Forrest's description, and was 'well established in suitable habitats'. A partial survey in 1962 found 30 pairs, but a decrease was noted in the previous five years at Willey Park, Bucknell and Bishops Castle, 'otherwise it appears to be at least holding its

own'. The *Atlas* (1992) recorded a 'significant increase in range and numbers', with breeding at several sites in the 'lowland north', and, in comparison with the national BTO Breeding Atlas (1976), 'including those on the borders, a total of 21 10km squares were occupied then compared with 37 [in 1985–90]'.

Pied Flycatchers readily take to nest boxes in their preferred habitats. Commencing in 1979, nest box schemes were set up in the upper and lower Clun Valley, around the Long Mynd and in several other woods in the south-west. By 1992 it was estimated that some 2,700 nest boxes were available in these areas. The schemes were largely responsible for the expansion of range between 1968–72 and 1985–90, quickly infilling the area between Herefordshire and the Severn Valley, and also increasing the population where natural holes are hard to find.

They were followed by new schemes in the Wrekin woodlands, at Craig Sychtyn, at Hawkstone Park and several smaller sites. They provide nest boxes in excess of requirements, reducing competition with Great Tits, Blue Tits and other hole nesting species.

Cold or wet weather during the fledgling period in June is probably the main reason for nest failure, and whole broods are sometimes found dead in the boxes, with contrasting reports in 1997 when the Hawkstone scheme had its most successful season, while the Craig Sychtyn boxes had their worst season for four years. Craig Sychtyn reported losses again in 1998 due to a week of heavy rain, a problem that has also been noted in the Clun Valley. The arrival in the valley in the mid-1990s of Cypermethrin pesticides, used in sheep dip, led to various pollution incidents which reduced the insect populations along the Clun, and this also contributed to losses among the young in the nest. The use of Cypermethrin was banned in 2006 and a slight improvement in success rates has been noted in more recent years.

The Clun Valley nest box scheme has involved ringing as many as possible each year. The total of new birds ringed annually since 1990 has steadily declined, but the average brood size has remained constant at 6.0 throughout the period, indicating that the number of broods ringed each year has declined. The chart shows a decline of around 71%, from 234 in 1990 to 67 in 2013. The underlying trend line smooths out the annual fluctuations, and shows a reduction of 55% over this period, confirming the large decline. It was not possible to complete some visits in 1993 and 1994, accounting for the lower totals in those years. Broods are a measure of successful nests, but the total population of breeding pairs is not known: failure rates fluctuate for reasons outlined above, and 1999 and 2000 were poor years. There was no access to the area in 2001 because of foot and mouth disease. No monitoring took place in 2014, and the project has now been taken over by the Shropshire Ringing Group.

Nest box schemes in other areas have also found a big decline. Breeding occurred in several tetrads in the Severn Gorge woodlands in 1985–90, and over 100 nest boxes were installed there from 1999 onwards by the Severn Gorge Countryside Trust (SGCT) to encourage conservation, but only two pairs nested in the boxes from 2001–05, in Benthall Edge wood and Loamhole and Lydebrook Dingle, and Pied Flycatchers last bred there, at the former site, in 2005. At Hawkstone Park, 23 broods were ringed in 1997, declining to five in 2010 and only one in 2013, followed by a slight recovery to four in 2014, while at Craig Sychtyn, there were 16 nests in boxes in 1994 and 1995, followed by a gradual decline to a low point in 2010, with only four

Broods per Year (Clun Valley Next Box Scheme) (1990–2013)

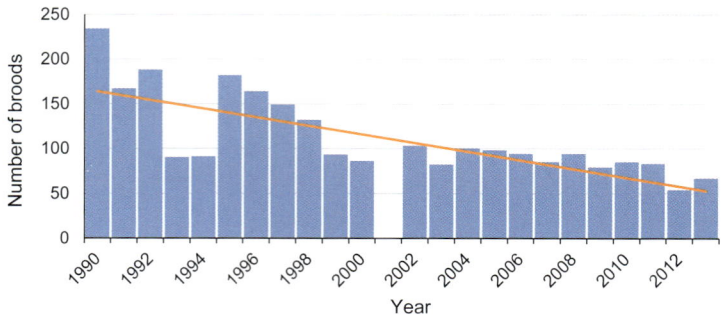

nests, rising to 9–10 in 2013. Another scheme with boxes in several woods near Purlogue declined from 64 pairs in 2002 to 13 broods in 2010, when 82 pulli (nestlings) were ringed, 18 pairs in 2011 and only nine pairs in 2013, after which the ringing was abandoned.

Other schemes have been started more recently by the Upper Onny and Upper Clun Community Wildlife Groups in the south-west, the National Trust and Chelmarsh Ringing Group on Wenlock Edge, and SWT with the help of Ricoh UK Ltd volunteers on the Wrekin and the Ercall, and some newly occupied tetrads in the south are due to these schemes. In 2010, in addition to 85 broods in the Clun Valley boxes, there were another 20 in the Upper Clun, 38 in the Upper Onny, eight in the Wrekin and Ercall, four at Craig Sychtyn, five at Hawkstone Park and 13 at Purlogue, totalling at least 173 pairs in these schemes. However, the population fluctuates from year to year: during the recent Atlas period, the main part of the Upper Onny scheme varied between 30 and 42 pairs; boxes on the Wrekin and the Ercall were first installed in 2009, and there were 13 pairs in 2012, only six in 2013, but 14 in 2014; and boxes on Wenlock Edge were first occupied in 2012, when there were four pairs, but only one in

2013 and in 2014 (Glenn Bishton, Bob Harris, Allan Dawes, Andy Latham, Michelle Frater, Andy Spencer, Linda Munday *pers. comms.*).

The breeding distribution change map shows that similar losses to those suffered in the nest box scheme areas have occurred throughout the range. Some of the losses and gains will reflect variation in fieldwork effort between the two fieldwork periods, but there has been a net decline of 54% in the tetrads with probable and confirmed breeding, and a contraction of range back towards the core areas.

The restricted distribution means that BBS is unable to calculate a local trend, but this flycatcher was found on eight plots in 1998, four in 2006 and only one in each of the three years 2011–14.

Areas of high relative abundance include places where the nest box schemes have provided additional nesting opportunities.

The population in 1990 was estimated at an average of 10 pairs per tetrad with probable or confirmed breeding, about 2,000 pairs. Assuming no change in density per tetrad where it was found, the population in 2008–13 would have been 920. However, in the Clun Valley the density has reduced as well, by around 55%, suggesting an estimate of about 500 breeding pairs. Based on TTV counts, the population is estimated at 550–650 breeding pairs, but it was only found on TTVs in 71 tetrads, less than half of those where it was recorded, so this estimate is not reliable. It is probably too high now, given the continuing losses.

The population decline is attributed mainly to climate change. On the breeding grounds, earlier and warmer springs have led to the main food supply, caterpillars, peaking earlier. Although the flycatchers are also arriving earlier, there is a growing mismatch between their breeding cycle and peak food, increasing nest failure rates at the

Occupied Tetrads

Atlas period (breeding)	1985–90		2008–13		Change	
	Number	%	Number	%	Number	%
Confirmed	154	18	66	8	-88	-57
Probable	44	5	25	3	-19	-43
Possible	61	7	59	7	-2	-3
TOTAL	259	30	150	17	-109	-42

Breeding Distribution Change (1985–90 to 2008–13)

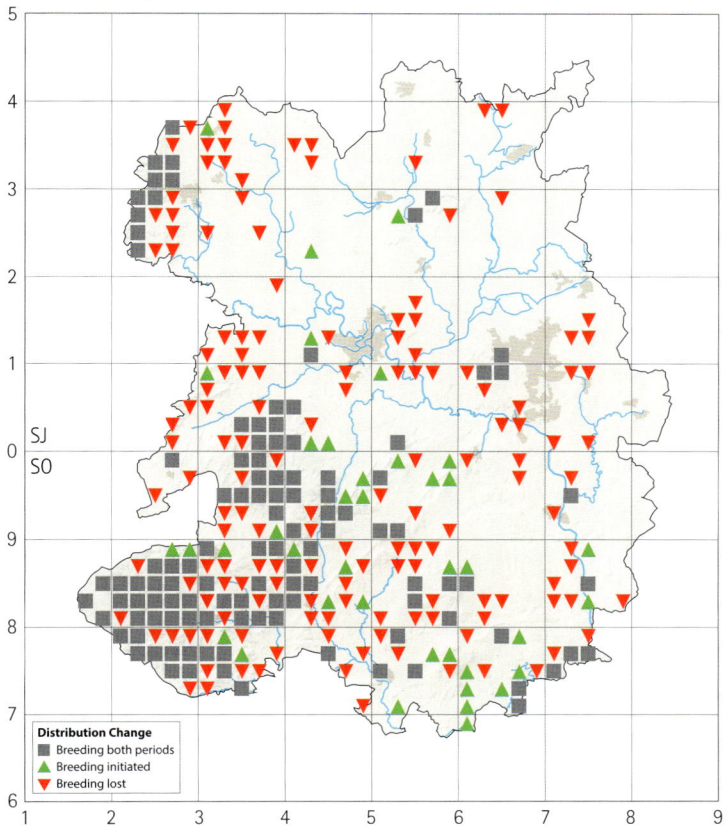

Distribution Change
- ■ Breeding both periods
- ▲ Breeding initiated
- ▼ Breeding lost

Breeding Relative Abundance (2008–13)

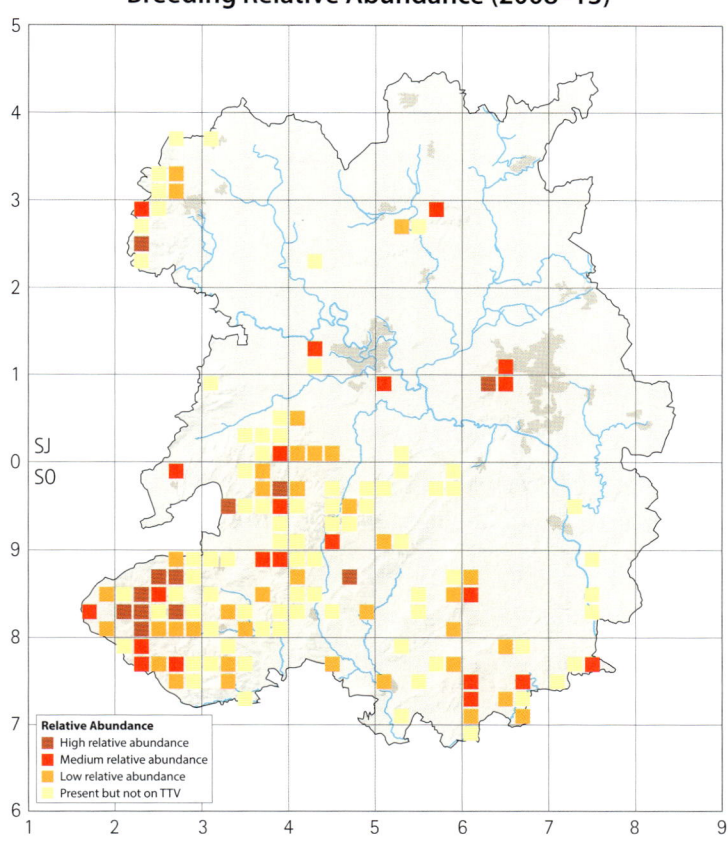

Relative Abundance
- High relative abundance
- Medium relative abundance
- Low relative abundance
- Present but not on TTV

chick stage. More importantly, in Africa, and on migration routes, the changing weather patterns have reduced over-winter survival. The increasing size of the Sahara desert, and overgrazing of the wintering grounds by goats, may also be having an effect.

In the Clun Valley, in addition to the broods, many of the sitting females have been ringed, but the males are harder to catch. Between 1990–2013, a total of 17,536 Pied Flycatchers were ringed, mainly nestlings but including 1,926 females and 153 males caught in the nest boxes.

Small passerines marked with rings generally have a low recovery rate, especially of fledglings which have high mortality in the first few months after leaving the nest, but there are many nest box schemes around the country, so many of these flycatchers have been recovered at other sites as well as back in the Clun Valley. Of those returning to the valley, many of the females return to where they were born or where they had bred successfully in previous years, and 28 returned to the same nest box. Females with known ages were 11 five-year-olds, six six-year-olds, four seven-year-olds and one eight-year-old. Another ringed at Church Stretton returned to the Habberley area and

Origins & Destinations

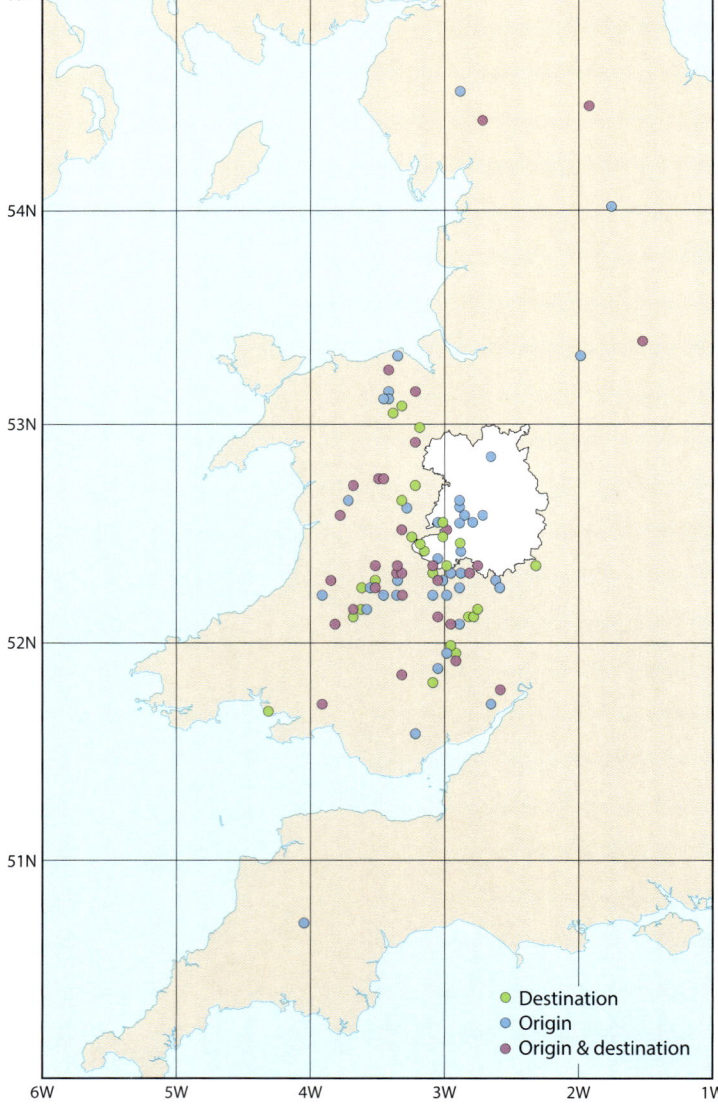

Overseas Ringing Recoveries

Date Ringed	Site	Date Recovered	Place	Distance km
06/06/70	Snailbeach	12/04/71	(Bouches-du-Rhone) France	1203
07/06/72	Pontesbury	18/11/72	Corruna, Spain	1105
01/06/85	Near Clun	01/02/86	North-west coastal plain*, Morocco	2091
16/06/85	The Hurst, Clun	11/08/86	Cap Gris Nez, France	361
22/06/86	Badger Moor	09/09/86	North-west coastal plain*, Morocco	2117
08/06/87	The Hurst, Clun	09/05/88	Westplaat, Oostvoorne, Holland	484
03/06/88	Selley Hall, nr Purlogue	21/04/89	North-west coastal plain*, Morocco	2187
11/06/88	Newcastle, nr Clun	05/05/89	Weelde, Belgium	572
04/06/89	The Hurst, Clun	13/05/91	Imintanoute, Morocco	2447
11/06/89	Newcastle, nr Clun	20/08/89	Angoulins sur Mer, France	720
12/06/89	Selley Hall, nr Purlogue	06/05/90	Groda, Farsund, Norway	883
12/06/89	Garbett Hall, nr Purlogue	30/03/91	Eauite Talsint, North-east Morocco	2204
30/05/90	Selley Hall, nr Purlogue	25/04/91	Messaad, Algeria	2090
02/06/90	Badger Moor	31/08/90	Southern* Morocco	2527
10/06/90	Badger Moor	10/04/92	Central* Morocco	2244
30/05/92	Selley Hall, nr Purlogue	20/04/93	Souk Tleta de Tagmoute, Tata, Morocco	2580
28/05/95	The Gogin, nr Newcastle	20/04/96	Pre En-Pail, France	490
05/06/00	The Hurst, Clun	29/08/00	Vigo, Spain	1208
09/06/00	Garbett Hall, nr Purlogue	13/08/00	Lege Cap-Ferret, nr Arcachon, France	854
04/09/03	Nouakchott, Mauritania	31/05/04	Duffryn, near Newcastle, Clun Valley	4023

Place names were not provided for most of the Moroccan recoveries, marked *, and approximate locations are given based on the map co-ordinates.

was in its eighth year when last seen at a next box. These last two were close to the national longevity record of 9y 7d.

The map shows the destinations in later years of flycatchers leaving the Clun Valley and the origins of ones found there but initially ringed elsewhere. Most of the movements are to or from local sites, or sites nearby in Wales. In some cases different individuals have moved in each direction, and several exchanges have occurred with nearby sites in Powys. There have also been exchanges with a site in Devon to the south, and sites as far north as Cumbria (two) and County Durham (two). There have been no exchanges with any Scottish sites, and the decline in relative abundance there has been less than in England and Wales, suggesting that these are separate populations (Chris Whittles *pers. comm.*).

In addition, two nestlings ringed in the Clun Valley were found on migration. One was handled at Portland Bill Bird Observatory in April 2004, as it returned to the UK almost a year after being ringed, while another was killed about nine weeks after ringing when it hit a window in Poole, Dorset on 10 August 1994, on its southward migration. Another was ringed at Portland Bill on 3 May 2001, and was found a year later breeding in the Clun Valley on 24 May 2002. Also on its southward migration, a nestling ringed in another nest box scheme near Pontesbury on 6 June 1987 was handled at Dungeness on 8 August 1987.

Altogether, including the examples above, about 500 have been ringed and recovered here, and 703 recoveries involve other counties or overseas countries. The three adjacent counties of Clwyd, Powys, and Hereford and Worcester account for three-quarters of these, with approximately equal numbers moving to and from local schemes. No other British county is involved in as many as 20 recoveries, but 19 have been recovered abroad. Those found overseas reflect the general pattern of migration to and from the African wintering areas, with eight of the 19 being found in Morocco. Most of them were on the north-west coastal plain, north of the Atlas Mountains, but three had crossed the Atlas. Two of them were still near the coast, but the third was well inland on the edge of the Sahara Desert. The former two were both over 2,500km from the ringing site. Although none ringed here have been found south of the Sahara, one ringed in Mauritania at the end of the autumn migration in September 2003 was found in the Clun Valley the following May, having travelled over 4,000km. Only one was found to the north, a nestling male found in Norway at the start of the following breeding season. Such a movement is very unusual, but not unique.

The overseas recoveries are listed in the table. All except four were found with 12 months of ringing, and none more than two years after ringing. All except three were ringed as nestlings.

There is no evidence that the factors driving the population decline are weakening, so the future for Pied Flycatcher is probably bleak.

Colin Wright

Black Redstart

Phoenicurus ochruros

Rare passage migrant and winter visitor, has bred

Widespread across central and southern Europe, small numbers breed in the south-east of England. In the UK they winter largely around the coast.

In the spring of 1878, one was brought to the taxidermist John Shaw by a labouring man who had killed it near Wem. This was believed to be the only one taken locally (Beckwith 1879). The second was at Hengoed in March 1922 and only three more were seen prior to 1950.

The first modern record came from Welshampton in April 1958, then in 1963 two pairs were found breeding near Buildwas, fledging six young between them. The only other instance of breeding occurred nearby at Ironbridge in 1978, when four young were successfully reared in a disused building at Castle Green. A pair had also been seen there in April 1975 so other breeding attempts may have been overlooked. Since then only one May and two June records have been received, and there has been no further suggestion of breeding.

Despite successfully rearing broods on at least three occasions, they remained very scarce and only five others were noted up until 1978.

Following a gap up until 1982, sightings have become an almost annual event with between one and three individuals in most years up until 1992, and rather more since, with a minimum of seven in 2003. There was a small increase in both the UK and European breeding populations at that time and they are beginning to extend their range from south-east England. The chart shows the number of individual adults seen each year, with the few over-wintering birds shown in their first year only.

Since 1992, 29 of the 55 arrivals have been first found in the period from mid-March to late April, suggesting spring passage or overshooting migrants. Over half of these (15) have been seen at Titterstone Clee or Brown Clee, where the habitat appears similar to mountainous areas used for breeding in Europe. A minimum of four were at Brown Clee in the last week of March in 2003 but, in common with most other spring arrivals, they did not stay for more than a few days. Multiple sightings are extremely rare: two have been seen on the Clee hills on three other occasions and two were at the Royal Shrewsbury Hospital on 8 April 1988.

Another smaller autumn peak occurs from late October to late November. Two that were first found during this period, at Knockin in 1983 and Shawbirch in 1989, remained until March and February respectively. Others found during January and February at Ellesmere

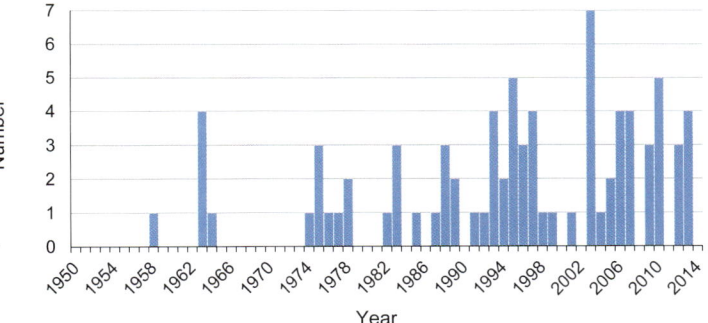

Estimated Number per Year (1950–2014)

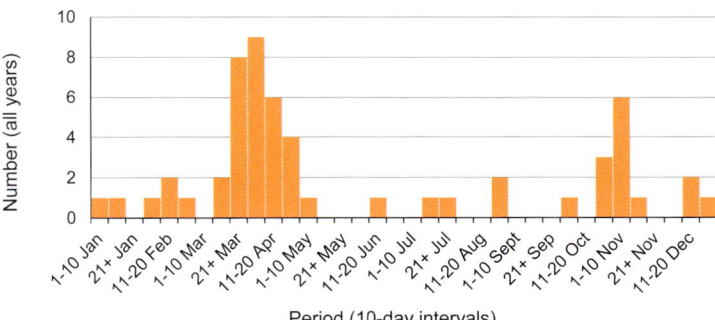

Seasonal Occurrence in 10-day Intervals (1992–2014)

Winter Distribution (2007–13)

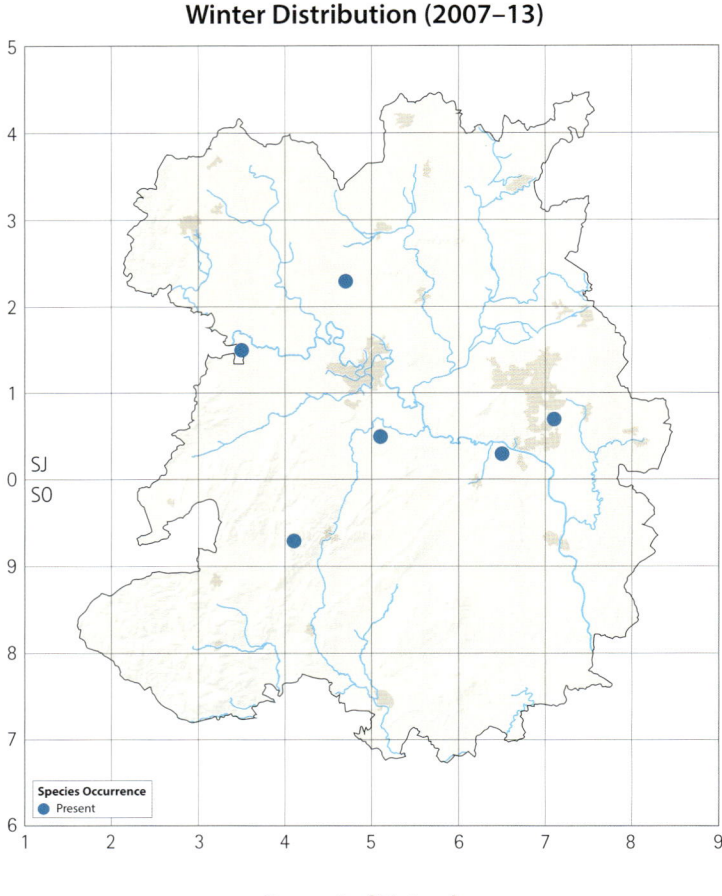

Occupied Tetrads

Tetrads with Winter Records (2007–13):   6   (1%)

Jim Almond, Cantlop, 11 November 2012

December 2007 and February 2010, the remainder were all in the first two weeks of November and may have been on passage. In the winter, Black Redstarts are often associated with buildings and these can range from isolated farms to towns, so they could appear anywhere and the map reflects this.

During the recent Atlas breeding period three April records, from Pole Cottage and Pipe Gate in 2009, and Titterstone Clee in 2010, were noted as migrants. There were single day records in 2013 from Albright Hussey in May and Meeson in June, but no indication of breeding.

One at Farlow in October 2013 was the only record after the recent Atlas period. If the underlying trend of an increase in annual occurrence shown in the chart is being driven by climate change then it seems set to continue, and further breeding records may only be a matter of time. It will be interesting to see whether they use the rocky terrain of the Clee hills or buildings on lower ground.

Allan Dawes

in 2003, Oswestry in 2007 and Alberbury in 2013, were still present in March.

The only long stayer during the recent winter Atlas period was the one at Alberbury mentioned above. None of the others were present for more than three days and, apart from singles at Ironbridge in

Redstart (Common Redstart)

Phoenicurus phoenicurus

Fairly common summer visitor

Redstart breeds throughout most of Europe; in mainland Britain it is chiefly found in the north and west. Autumn migration takes them to the Sahel area of Africa.

In 1888 Beckwith wrote:

Few of our summer visitors are more locally distributed than the Redstart; and it is only about parks and similar places, where the large partly hollow trees in which it delights are still allowed to stand, that it is to be found in any numbers. One or two pairs are often also to be seen, in the vicinity of some retired homestead, especially of an old fashioned one, whose usual surroundings of ruined walls or neglected orchards afford them ample choice of nesting places.

Suitable habitat would have been more extensive in those days, and it was said to be common around Shrewsbury, Haughmond, Berwick and Newport. A successful nest was found in the Quarry in Shrewsbury: the Redstart is not generally associated with towns although Shrewsbury was much smaller in those days. The species was common enough to be known locally as Fiery-brand-tail. According to Forrest in 1908 it was 'generally distributed ... particularly in the wooded south'.

The *Handlist* (1964) described it as 'well distributed ... in woodlands ... often found in quite open habitats around farm buildings. On the Longmynd it occurs in the thorn scrub and bracken up to 1200ft contour'.

The Redstart is found in similar habitats today but also frequents other upland areas, where many neglected hawthorn hedges have matured into rows of old trees which now provide ideal nest sites. Singing males are easy to locate in suitable habitat and although adults can be wary when feeding young at the nest, probable and confirmed breeding records were frequently obtained from the core breeding areas.

Most are now found to the south-west of a line drawn from Oswestry to Bridgnorth, which corresponds well with the altitude map showing land above 210m (see Map 3, p. 25). The distribution map also highlights the higher ground outside this area, such as the Wrekin and Hawkstone Park, which retain isolated breeding populations. They can also be found in low lying areas which separate the hills, with the flood plain at Melverley, which divides the Oswestry uplands from the more extensive southern hills, being a good example.

The breeding distribution change map shows an expansion of range since 1985–90, both by infill and an eastern extension. This has not occurred in the north of the range, as the boundary between the Oswestry uplands and the agricultural plain is a sharp one and suitable habitat there was already occupied before the earlier Atlas period. However, there have been some losses. They have disappeared from all the tetrads near Lilleshall Hill, and the occupied area around the Wrekin and the Severn Gorge has been reduced. The habitat near Ellesmere, and to the north and east of Telford, is mainly arable farmland and the apparent losses here may be a result of erroneous recording in the earlier Atlas of migrants in spring, which can occur anywhere, or of independent juveniles, which wander extensively later in the summer. However, some of the possible breeding records

John Fielding, Long Mynd, 1 April 2013

to the east of the main range may represent breeding at very low densities.

Redstarts reach their highest densities in the Oswestry uplands, Clun Forest and around the Stiperstones. These areas are close to the Welsh heartland where they are at their most abundant and where the population has shown a steep increase in recent years (BBS in Wales shows a 41% increase between 1995 and 2014).

Nationally, after a population crash between 1969–73, thought to be due to drought conditions in the Sahel, there was a steady increase prior to fieldwork for the *Atlas* (1992), which has continued slowly with a sharper rise over the past few years. An increase in the number of fledglings raised per pair is thought to have been responsible. Data presented in the *Arrival Dates* chapter suggests that Redstarts are arriving around 12 days earlier than they did in 1900, with all of the change since 1950. Nationally, eggs are now being laid up to two weeks earlier than previously, and this may also have helped improve productivity, although the exact mechanism operating is

Breeding Distribution Change (1985–90 to 2008–13)

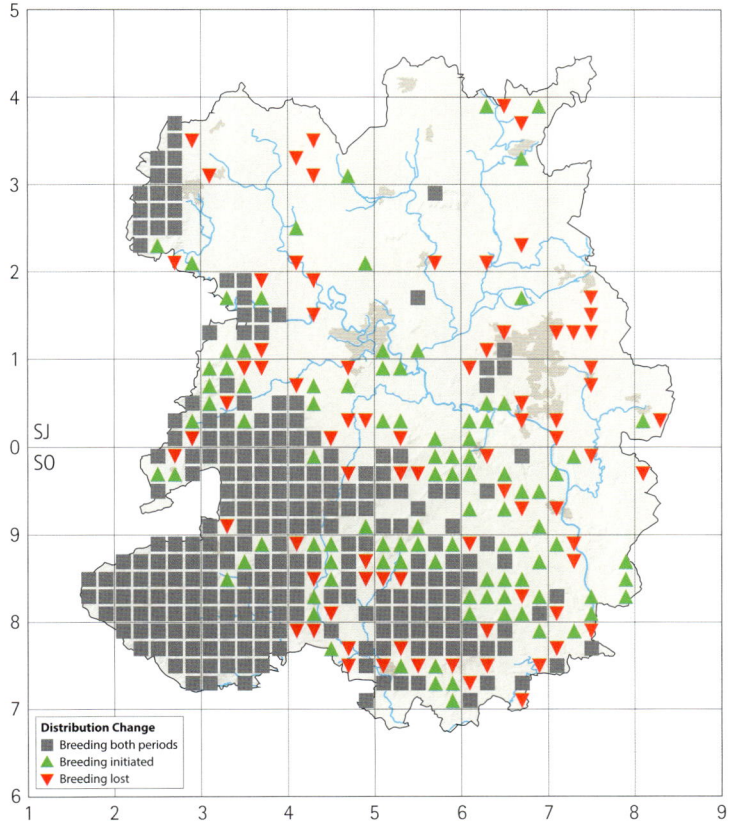

Distribution Change
- ■ Breeding both periods
- ▲ Breeding initiated
- ▼ Breeding lost

Occupied Tetrads

Atlas period (breeding)	1985–90		2008–13		Change	
	Number	%	Number	%	Number	%
Confirmed	196	23	177	20	-19	-10
Probable	70	8	79	9	9	13
Possible	58	7	87	10	29	50
TOTAL	324	37	343	39	19	6

Breeding Relative Abundance (2008–13)

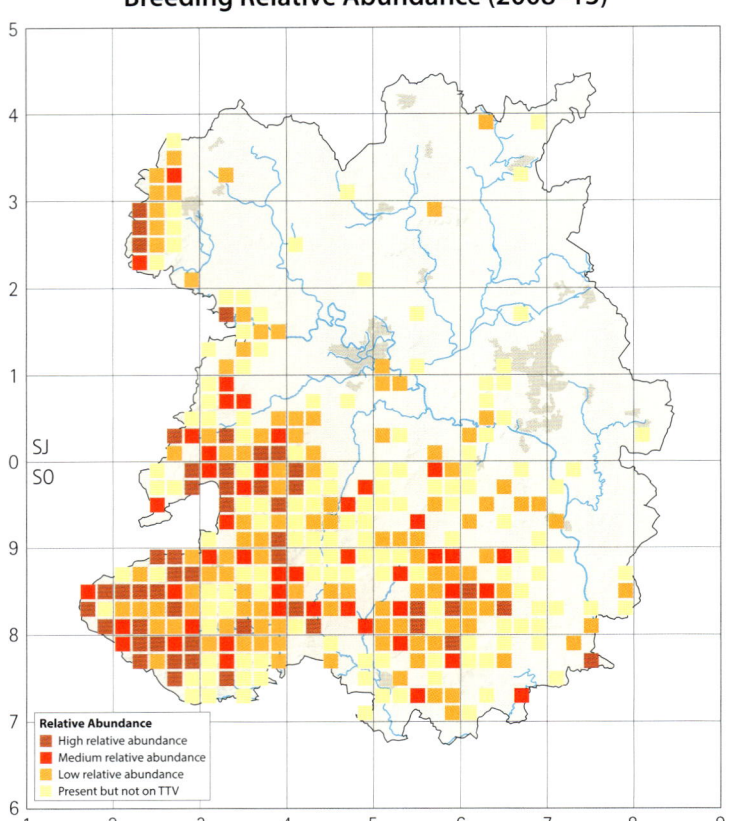

Relative Abundance
■ High relative abundance
■ Medium relative abundance
■ Low relative abundance
■ Present but not on TTV

unclear. Locally, this has probably resulted in both a higher density in previously occupied areas and the expansion in range evident in the breeding distribution change map.

Based on TTV counts, the estimated population is between 3,100 and 3,300 pairs, equal to nine to ten pairs per occupied tetrad. This is just above the midpoint of the large range quoted in the *Atlas* (1992). There were an estimated 75–85 pairs on the National Trust Long Mynd property (excluding pairs in the boundary hawthorns) in 1998, in an area equal to less than 10sq.km, but spanning 12 tetrads. Taking unsuitable habitat in these tetrads into account this equates to six to seven pairs per tetrad (Smith SBR 2003).

A number of Redstart broods are ringed each year in nest boxes. There have only been three overseas recoveries of these nestlings. The first was in Portugal at the end of September 1972, almost four months after fledging. Another ringed near Church Stretton in 1982 was found in Morocco, 2,462km away, and one from near Clun in 2000 was recovered in France. The latter two were almost two years old, and were on return passage in April ready for their second breeding season. The longest movement within the UK was a one-year-old male from a nest near Knighton which was re-trapped at Dungeness in April 1994. Hopefully he completed his first round trip successfully. A few other movements were noted within this or adjacent counties.

If numbers continue to rise they may recolonise some of their former outposts, but lack of suitable habitat will always restrict their range.

Allan Dawes

Whinchat

Saxicola rubetra

Scarce summer visitor

Whinchat is a summer visitor to most of Europe and western Asia, and it winters in sub-Saharan Africa. It was common in the UK until the middle of the twentieth century, but has declined dramatically since then.

It was first described as 'common on gorse bushes' (Leighton 1836), 'common' without any reference to gorse (Eyton 1838; Rocke 1865) and 'common in summer wherever gorse bushes are found, and on meadows on the Severn Valley. Its nest is generally placed under a furze [gorse] bush' (Forrest 1899). However, Beckwith (1888) painted a picture which is hard to believe today:

> The pretty Whinchat, so often to be seen flying along a roadside hedge, or perched upon a thistle or other tall plant in a field … found on hills, heaths and moors, but is also common throughout the cultivated parts … Meadows along the banks of rivers are peculiarly attractive to it … and it is also extremely fond of frequenting the sides of railways, where it finds secure breeding places on the embankments.

Holloway (1996) mapped it as 'common' here at the turn of the nineteenth century, and suggested that little change in population and distribution occurred nationally throughout that century, and through to the 1940s.

Early SBRs included breeding records from widespread locations, and the *Handlist* (1964) described it as 'well distributed throughout … wherever suitable localities occur, though more common on the high ground', but in the national BTO 1968–72 Atlas it was mapped as breeding in all except the three most easterly of the County's 14 10km squares in the south and in 10 of the 19 10km squares in the north, suggesting some losses. That Atlas described Whinchats as 'now basically upland birds in Britain'. The decline was attributed partly to habitat loss, a result of the trend to increased tidiness and mowing of roadside verges. It continued, and several breeding sites used since then (SBRs) were no longer occupied by the time of the 1985–90 Atlas, and the long-standing CBC plot near Oswestry had nine pairs in 1968, six in 1977, and only one when it was last occupied in 1982.

The 1985–90 Atlas showed a loss from two of the 10km squares in the north, and three in the south-east, and 'the range has contracted south-westwards compared with that shown in the BTO Atlas (1976)': 'with few exceptions it is now restricted to the uplands'. Although some of the occupied 10km squares had few occupied tetrads, the species account referred to arrival 'in strength in early May in their western stronghold – Bryn Shop, Rose Grove, Riddings, Mason's Bank, Black Bank and Cefn Hepreas in the Clun Forest; and Stapeley Hill, Heath Mynd, Norbury Hill, Black Rhadley Hill, Stiperstones, Long Mynd,

Earl's Hill and Caer Caradoc in the uplands – and also on Brown Clee, Titterstone Clee and Catherton Common. Thick bracken is the favoured habitat, usually on steep hillsides. Other tall herb plants and young conifer plantations are used by some pairs, and in the north they are found in the damp lowlands of Whixall Moss. There were only two other isolated confirmed breeding records from the lowlands, from near Culmington and near Brompton.

The Long Mynd BBP carried out a full survey between 1994–98, and estimated a population of 110–130 breeding pairs (Smith SBR 2003). A repeat survey in 2006–08 found 59–65, and estimated a population of around 70 breeding pairs, a decline of about 40% (Smith *in prep.*). All pairs on both surveys were found in the upper reaches of the steep sided valleys, with no territories on the flat open moorland, although some pairs did take recently fledged young up onto the moor to feed.

A transect in Callow Hollow is the only BBS to regularly record Whinchat. Counts from the 1990s through to the late 2000s were fairly consistent, suggesting a population of around seven pairs, but this fell to six in 2010–12, five in 2013 and only four in 2014. This decline has occurred in prime habitat, which does not appear to have changed over 21 years.

During the 1994–98 project, the habitat occupied by each Whinchat was noted. A total of 254 records were made over the five years, and bracken was present in every single one. Correlation with the distribution of the main habitats, comprising 16 different vegetation mixes, showed that 38% of territories were in bracken with bracken litter understorey. However, this was the most widespread habitat, and the breeding density was higher where heath, rather than bracken, was the understorey, and the highest densities, twice that in bracken with bracken litter understorey, were reached where the heath understorey is less thick, and is interspersed with grass, or where heath is dominant, but it is interspersed with bracken. All the territories were in the upper reaches of steep-sided valleys, all were close to streams, and many were adjacent to flushes where springs emerge. Although the results have not yet been quantified, all territories in 2006–08 also contained bracken and wet areas.

Bracken is usually perceived as an alien invasive nuisance, and habitat management plans often attempt to eradicate it through expensive spraying. However, Whinchats are clearly wholly dependent on bracken on the Long Mynd, and presumably other uplands as well. Certainly this is the case in the North York Moors and the Eastern Highlands of Scotland (Allen 1995), while a survey in Pembrokeshire National Park in 2012 found 29 breeding pairs, all of which occupied a mosaic of bracken, low bushes of various species and a wet area such as a gully or flush (Paddy Jenks *pers. comm.*).

A survey of the Stiperstones for English Nature in 1995–96 found eight pairs in the southern half of the NNR and two in the northern half (Cross & Moscrop 1996). A similar survey in 2004–05 found five to six and two pairs respectively (Smith SBR 2007). Five territories in the south were in the same place in both surveys, four in bracken and wet flushes east and north-east of Devil's Chair, and one in similar habit 200m east of Cranberry Rock. A territory just south of the former Gatten plantation, and two on the western slopes above Rigmore Oak, had been lost. The 2004 survey specifically searched for Whinchat, and checked all previous sites, so the decline was undoubtedly real. Although two territories were found in the north in both 1996 and

2005, there was no overlap between them. All these territories were also dominated by bracken, with wet flushes nearby.

NNR staff started a BBS survey in 2003, and a single Whinchat was recorded each year until 2006. In 2005 and 2006 there were June sightings, suggesting a breeding bird, although no breeding evidence was looked for. None were seen in 2007. During the recent Atlas period, one was seen in May 2008, but none was found on the BBS in either 2009 or 2010, while in 2011, a male Whinchat was singing in early May and seen nearby on the June survey. There were no sightings in 2012, and, reflecting concern that Whinchat might have disappeared altogether, Natural England commissioned a species survey in 2013. Every site where it had been found in either 1994–95 or 2004–05 was visited in both May and June. Only one male was seen, at the flush east of Cranberry Rock, and no evidence above possible breeding was found. However, that year had the coldest spring nationally for over 50 years, and locally it was wet and windy as well. It is therefore possible that some Whinchats did arrive, but either abandoned any attempt at breeding, or delayed breeding so they were inconspicuous on the survey dates. All these Stiperstones records add up to probable breeding in one tetrad, and possible breeding in two more, as shown on the Atlas map. It is likely that Whinchat was absent in some years.

A Whinchat was also singing near the Rock, at the southern end of the Stiperstones ridge, in late May 2008, in habitat newly created through the Back to Purple project.

After the Atlas period, there was a probable breeding record of an agitated pair in the flush north-east of Shepherd's Rock on 19 June 2014.

Two Whinchats were seen on the northern part of Stapeley Hill in June 2010, and an agitated pair was seen on the southern part in

Jim Almond, Long Mynd, 2 May 2009

June 2011, but none have been seen there in subsequent years (Ron Kinrade *pers. comm.*).

Occurrence in the north is now a rare event. Breeding was confirmed in two tetrads at Whixall and Fen's Moss in 1985–90, and Whinchat certainly bred there until the late 1990s and probably up to 2003 when eight were recorded singing on 12 May. Since then only a handful of records have been received, all in late April, and all were likely to have been of passage birds. In the recent Atlas period, there were records of one in June 2008, and two pairs up until 2010, but these were in border tetrads not included in the Atlas. There was only one record for an Atlas tetrad, a singing male on 19 July 2009, and Natural England has no records at all, in any tetrad, for the final three Atlas years. Habitat has changed on much of the moss since then, and is probably now unsuitable: scrub has been cleared, and the water table has been raised. The breeding distribution change map suggests that breeding still occurs there, but that is not the case. A single on 6 July 2014 is likely to have been an early returning passage migrant.

Otherwise there was only one probable breeding record, a pair

behaving as if there was a nest nearby in SJ32N on 2 June 2008, but there is no evidence that this has ever been a regularly used site.

Of all the sites listed in the *Atlas* (1992) where they 'appear in strength', only the Long Mynd, Stiperstones and Stapeley Hill had probable or confirmed breeding records in 2008–13. The breeding distribution change map suggests that breeding still occurs at Whixall Moss, Stapeley Hill, Caer Caradoc, Clun Forest and Titterstone Clee, but that is not the case. In 1985–90, breeding was confirmed in one tetrad on Brown Clee, and in four on Titterstone Clee, but in the recent Atlas period there were no records for the former, and one record only from each of four tetrads in the latter, none later than 6 May, and these were all almost certainly passage birds. Other possible breeding records on the distribution map are also passage birds, except one in SO49Y, which contains part of Caer Caradoc, seen on 15 June 2010. Otherwise, apart from the sites with probable or confirmed breeding records listed above, there were no records between 13 May and 17 July.

The *Atlas* (1992) noted that 'Clun Forest, The Long Mynd, Titterstone Clee and Stiperstones have a minimum total of 36 pairs (SBR 1989), which, coupled with experience of Atlas fieldwork, suggests an average of 2–5 pairs per occupied tetrad, giving a population estimate of only 110–275 pairs'. The 1994–98 Long Mynd results showed that one site on its own exceeded the lower end of this estimate, and it was revised upwards, to around 300 pairs, in 1998 (Smith SBR 2003). Now breeding Whinchat are restricted to the Long Mynd, with occasional pairs on the Stiperstones, and the estimate in 2008 was around 75, only one-quarter of the earlier figure. The population estimate based on TTV counts, 200–220, is considered to be much

Occupied Tetrads

Atlas period (breeding)	1985–90		2008–13		Change	
	Number	%	Number	%	Number	%
Confirmed	35	4	9	1	-26	-74
Probable	20	2	5	1	-15	-75
Possible	20	2	12	1	-8	-40
TOTAL	75	9	26	3	-49	-65

Breeding Distribution (2008–13)

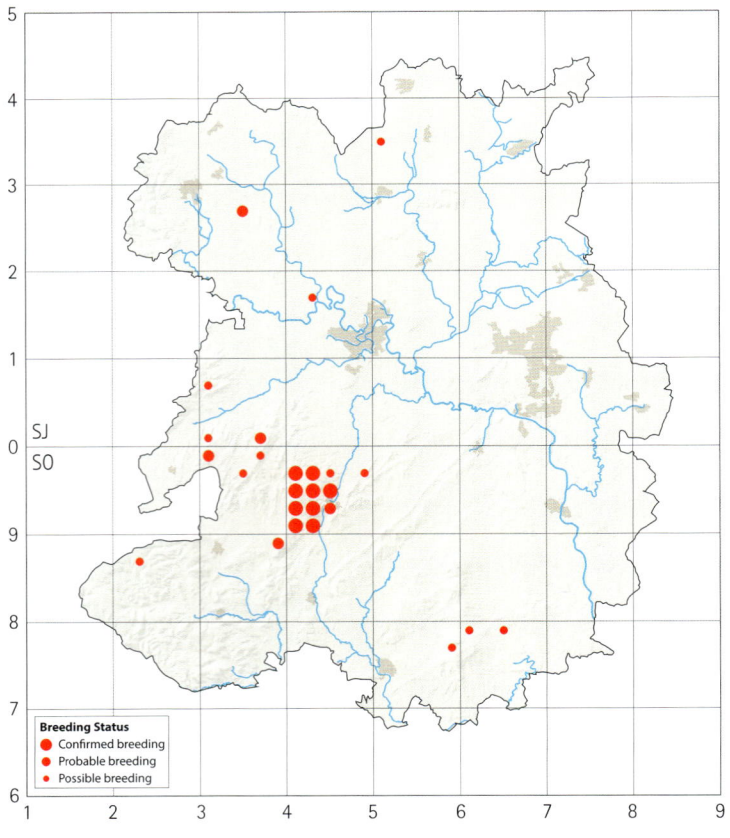

Breeding Distribution Change (1985–90 to 2008–13)

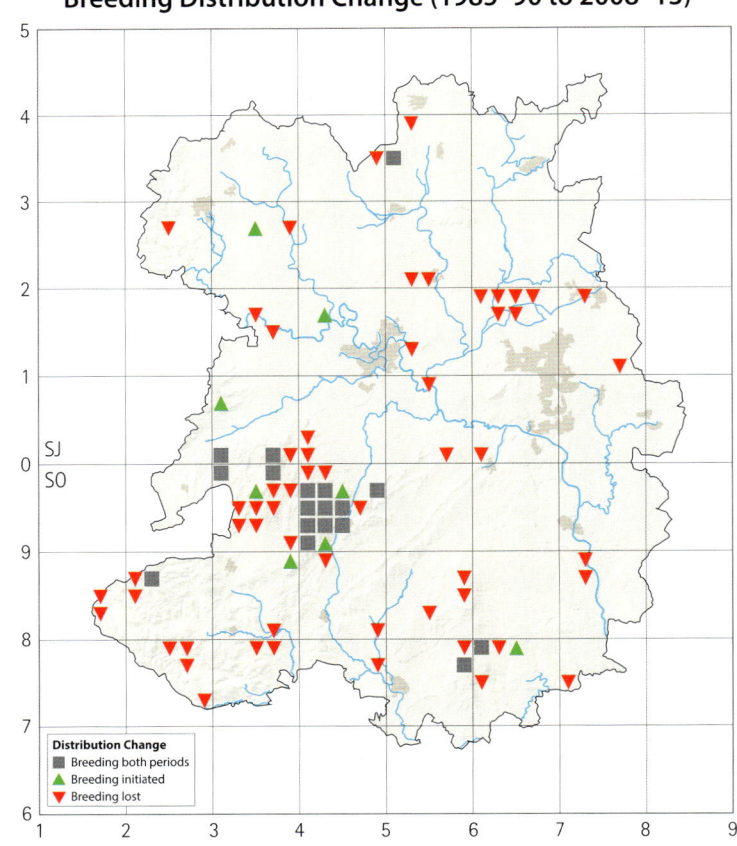

too high. Whinchat was found on TTVs in only 11 tetrads, with only one count of more than three, but this exception was a count of 10 in the Long Mynd stronghold, which presumably distorted the TTV estimate. The Callow Hollow BBS results suggest that the population has continued to decline.

This decline is reflected in the national picture, a contraction of range of 47% in Britain since 1968–72, increasingly confining it to marginal uplands, while BBS shows a 53% decline 1995–2014 in the UK. The Whinchat has declined over much of lowland England since the 1950s or earlier, and the current strong local dependence on upland bracken is probably the result of destruction of other suitable lowland tall herbs and grasses described in *Losing Ground in Shropshire* (SWT 1989). BTO (Marchant *et al.* 1990) extended the list of habitat loss to include roadside verges, railway embankments, derelict land and rough farmland. However, the declines have accelerated since the 1990s, and reduced over-winter survival has been implicated. As Whinchat over-winter in sub-tropical Africa, periodic droughts and climate change in the Sahel appear to be taking a toll too. Predation may also be a factor in the local population decline, in view of a study on Salisbury Plain, which found that most nests failed to fledge any young, mainly due to nocturnal (mammalian) predators, and recruitment of juveniles was too low to sustain the population, which is maintained by immigration (Taylor *et al.* 2015).

The *Arrival Dates* chapter suggests that Whinchats now arrive around two days earlier than they did in 1900. This is a very small advance in dates, and probably reflects the decreasingly likelihood of a declining species being recorded when it first arrives.

There is one recovery, a nestling female ringed on the Long Mynd on 16 June 2007 as part of a colour-ringing project, was caught in East Sussex, 304km south-east, 2m 4d later on her southward migration.

RSPB has initiated a long-term study on the causes of Whinchat decline, which will include the Long Mynd, and the local population monitoring project is being repeated, from 2017 onwards.

Leo Smith

Stonechat (European Stonechat)
Saxicola rubicola
Uncommon resident

Dave Barnes, Clee Hill, 11 May 2013

Stonechat was described as common, breeding, and not so numerous as Whinchat (Rocke 1865a), but Beckwith's was the first account of its habitat and distribution: 'in summer, wherever there are one or two acres of gorse or heather, one or two pairs of Stonechats are almost certain to be found', while over the 'more extensive heaths and mosses' in the north, and 'on the hills and moorlands in the south, it is common and generally distributed', but 'they never breed as Whinchat does in the middle of cultivated ground'. 'In winter only a few remain, the great majority migrating south in autumn' and while Paddock (1890) added that they are 'fairly common' on the higher ground' but not seen on the 'flatter and cultivated ground', he did report a family party on gorse bushes near Oakengates, where a 'few are to be found breeding' on the hilly ground around the town. Forrest (1899) confirmed it was 'common in summer on moorlands, but not so plentiful as the Whinchat', and 'a few stay 'though the winter in sheltered places'.

It was mapped as 'common' in the hills in the north-west and south-west, and 'uncommon' elsewhere, at the end of the nineteenth century (Holloway 1996), after which there was a national decline. The combined effects of enclosure of wasteland, and clearance and burning of gorse, for agriculture, followed by the ploughing up of marginal land during the 1939–45 war (see p. 39), reduced the extent of suitable habitat and resulted in fragmented populations which were then vulnerable to harsh winter weather.

By 1964, the *Handlist* described Stonechat as 'resident in small numbers and passage migrant. As a breeding bird it has decreased considerably' since the turn of the nineteenth century. It was of regular occurrence at Dowles, in the Wyre Forest, until 1925 and at Llanymynech until 1939. 'These areas have changed little in character so the decline is due to causes unknown, though hard winters are known to reduce the population seriously'. It was found breeding sparingly in the area that includes the Stretton hills and the Stiperstones, and occasionally an odd pair was found on Haughmond Hill. In 1961 pairs were noted at Redwith and at Clun, a year in which there were more records than usual following the exceptionally mild winter. 'On passage the Stonechat occurs in all areas particularly in November and March'.

The hard winter of 1962–63 may have caused a further decline, because the 1968–72 national BTO Atlas showed confirmed breeding only in SO37 (perhaps not in the County), probable breeding only in the two 10km squares that include the Long Mynd and Brown Clee, and possible breeding only in SJ53, which includes part of Whixall Moss, and Prees Heath.

That Atlas probably occurred at the Stonechat's lowest point, as by the 1985–90 Atlas, numbers and range were clearly much greater, and breeding was confirmed in nine 10km squares, with probable breeding in two more, though not on Brown Clee. In the 13 intervening years up to 1985, SBRs noted sporadic breeding in several other widely scattered areas not occupied during the 1968–72 Atlas period. The account concluded that perhaps there were only one to two pairs in each regularly used tetrad, suggesting a population of less than 25 pairs in 1990.

The population has continued to grow. The Long Mynd BBP found an estimated 22 pairs in 1995 and 1998, but only 15 in 1996 after a preceding hard winter, with several prolonged cold spells and snowfalls. Regular visits to Ashes and Callow Hollow, and Minton Batch, in the 1970s and 1980s, did not find more than one pair until 1977, rising to three in 1980 and four in 1984, but after the hard winters of 1984–85 and 1985–86, there was only one in 1985–88 and two in 1989 (Jack Sankey *pers. comm.*). The BBP survey found several more pairs in these valleys. The rapid increase was attributed to a succession of mild winters.

Most were in the upper reaches of the valleys, though several pairs were on the moorland plateau. In 1997, of the 20 territories, only one was below 420m, 10 were above 450m, and two were at 480m. The higher altitude habitats appear to be preferred – the lower territories were only occupied when the population was at its highest. The habitat where Stonechats were seen was noted, and heath was the most frequented, one-third of all observations, but heather and bracken mixes also held one-third. Heather was present in their immediate vicinity in 82% of cases, but most territories were also close to acidic flushes, and it appears the Stonechats nest close to the boundaries between habitats, so they can feed in each.

The repeat survey in 2006–08 found that the population had almost doubled, to around 42 pairs.

Surveys on the Stiperstones showed a similar growth. Only possible breeding was found in the tetrad incorporating the southern part of the NNR during fieldwork for the 1985–90 Atlas, but in 1995–96 a site survey found four pairs there, and six to eight in the northern part of the NNR (Cross & Moscrop 1996). A repeat survey in 2004–05 found the population had more than doubled, to 10 and 15–17 respectively (Smith SBR 2007). Many pairs were found associating with gorse.

Gorse is also the favoured habitat in the strongholds on coasts and lowland heaths in southern Britain, so the apparently low association with gorse on the Long Mynd was a surprise. However, gorse there is usually scattered on heather slopes, and may not occur sufficiently to be included in the definitions of the vegetation types. It is also inconspicuous during the breeding season, as it is mainly western gorse *Ulex gallii*, which flowers in the late summer or autumn. The Stiperstones is a dry heath, with mainly common gorse *Ulex europaeus*, which is

Breeding Distribution Change (1985–90 to 2008–13)

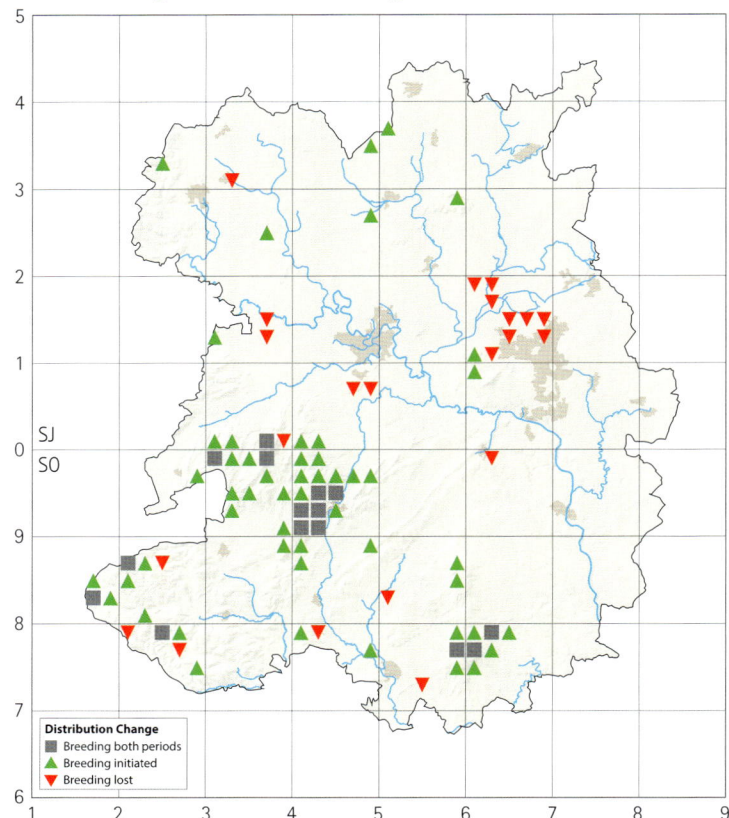

Occupied Tetrads

Atlas period (breeding)	1985–90		2008–13		Change	
	Number	%	Number	%	Number	%
Confirmed	13	1	40	5	27	208
Probable	10	1	16	2	6	60
Possible	14	2	11	1	-3	-21
TOTAL	37	4	67	8	30	81
Tetrads with Winter Records (2007–13): 100 (11%)						

Nests Found: Titterstone Clee Hill (1998–2014)

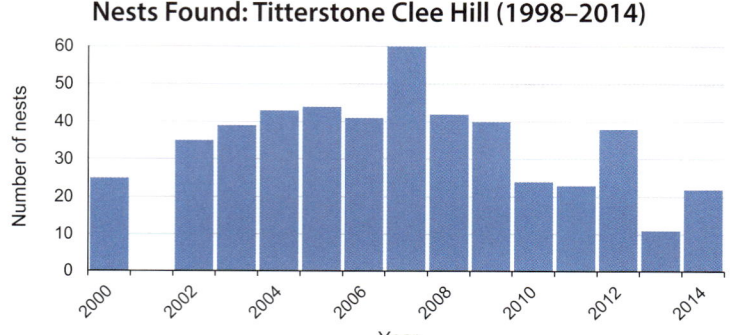

also widespread on coasts and lowland heaths. It flowers in May and June, so insects are abundant on it when Stonechat are breeding.

The Chelmarsh Ringing Group has been colour-ringing Stonechats on Titterstone Clee since 2000, as part of a BTO Retrapping Adults for Survival project. The population there also grew rapidly from 25 pairs in 2000 to a maximum of 60 in 2007, after which it returned to previous levels, then suffered a sharp decline after the hard winter in 2009–10. The Stonechat's vulnerability to hard winters, and its ability to bounce back, was again demonstrated by the much greater decline from 38 pairs in 2012 to only 11 pairs in 2013, then a recovery to 22 pairs in 2014 and 38 again in 2015 (Dave Fulton *pers. comm.*). The annual population is shown in the chart. There was no access to the site at the appropriate time due to foot and mouth disease restrictions in 2001.

Most nests (about 90%) were in late-flowering western gorse, and the remainder were in heather, apart from the odd one or two in grass. Between 2000–14 (excluding 2001), 690 adults were colour-ringed or

Winter Distribution (2007–13)

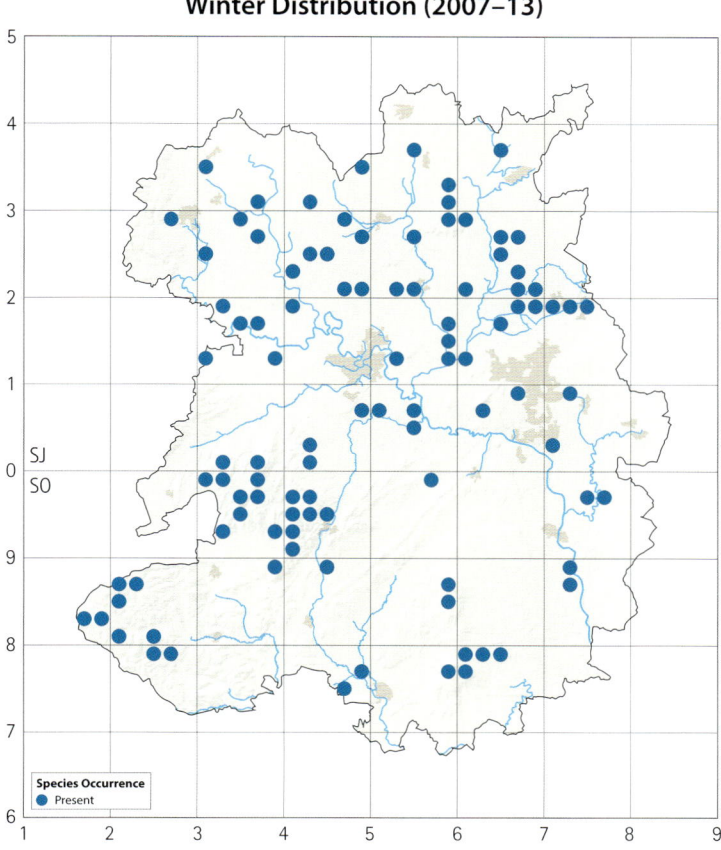

Species Occurrence
● Present

enough TTVs for it to be reliable, the estimate based on TTV counts is the same, 200–210 breeding pairs.

In winter too, the Stonechat population has increased considerably since the national 1981–84 Atlas found it in only eight of the County's 33 10km squares. It was predominantly a lowland species at that time – only one of the eight included any of the uplands (Titterstone Clee) where Stonechat were found breeding in 1985–90. It was not found in only three of these 10km squares in the recent winter Atlas period. Distribution is much more widespread than in the breeding season. While there is an increasing tendency for some adults to remain on their upland breeding grounds all year, as milder winters increase, others leave for lowland areas, particularly in harder weather, and may migrate as far as Iberia and the Mediterranean. The local winter population is probably supplemented by migrants from more northerly parts of Britain.

This mobility provides the secret of the Stonechat's recent success. Individuals that stay on the breeding grounds all winter and survive can lay earlier, and raise three broods each year, while the migrants arrive back later and generally raise two broods. Thus both the initial population and breeding success are usually higher after mild winters, but the population falls after a hard winter, and takes around three years to fully recover, longer if there are a succession of hard winters.

However, little evidence for this mobility has been found by local ringing recoveries. Apart from several hundred ringed and recaught on Clee Hill, only about a dozen Stonechats have been ringed and recovered here, and there have been no reported resightings from elsewhere of colour-ringed birds from Clee Hill. However, a male ringed as a nestling on the Stiperstones in June 1996 was caught in Ashes Hollow (Long Mynd) just over two years later. There will be few older than this – no individual ringed in Britain and Ireland has been recovered more than three years after ringing. None have been ringed elsewhere and recovered here, and there is only one example of one ringed here and recovered elsewhere, a male ringed as a nestling on Clee Hill in April 2005 and caught near Shotley, Suffolk, 6m 5d later and 266km to the east.

If climate change leads to a succession of increasingly mild winters, the prospects for Stonechat are good.

Leo Smith

retrapped/resighted and 1977 pulli (chicks in the nest) were ringed, including 91 adults and 318 pulli in 2007, the highest annual totals. Only nine of those retrapped or resighted were three years old (all the others were only one or two years), and none have been seen at other sites.

The population increase found by the local studies is reflected in the increase in range found by the recent breeding Atlas, and the number of tetrads with confirmed or probable breeding has more than doubled since 1990.

This mainly reflects expansion of the increasing population around the Long Mynd, Stiperstones, Stapeley Hill, Clun Forest and Titterstone Clee, a recolonisation of Brown Clee and colonisation of sites near the Wrekin and Whixall Moss. Only the area north of Telford has shown a decline, the result of urban expansion, removing the remaining pit mounds, and drainage and agricultural 'improvement' on the Weald Moors. The map perhaps exaggerates the distribution after cold winters.

It is likely that the populations around the Long Mynd and Stiperstones have continued to grow since the detailed surveys in the mid to late 2000s. Titterstone Clee has around 40 pairs, and the Clun Forest and the area to the south of that have around 30 pairs (Michelle Frater *pers. comm.*), adding up to around 150 pairs after a run of mild winters, more if the run extends for several years, but less after hard winters. There were another eight to 12 on Cefn Gunthly, Heath Mynd and Black Rhadley Hill in 2006 (Smith SBR 2007) and three on Whixall Moss (SBRs), and allowing for an average of two pairs per tetrad not included in the listed sites, the population must be over 200 pairs in good years. Although Stonechat was not found on

Dave Barnes, Clee Hill, 11 November 2012

Wheatear (Northern Wheatear)
Oenanthe oenanthe
Uncommon passage migrant, scarce summer visitor

Jim Almond, Titterstone Clee, 2 April 2011

The Wheatear breeds across most of Eurasia, Iceland, Greenland, northern Canada and Alaska, and winters in sub-Saharan Africa. It is one of the earliest migrants to return, from early March onwards.

It was first noted at Haughmond Hill and Dovaston (Leighton 1836), while Beckwith described it as 'a common summer visitor to the Clee hills and the adjoining high ground, the Longmynds, the Stiperstones and all the hills and moorlands in the south-west'. Forrest (1899) emphasised a different breeding habitat: 'They prefer low hills, and nest there in holes in the ground, frequently selecting a rabbit burrow', prefaced by 'in numbers on spring and autumn migrations. Only a portion of these spend the summer with us'.

The national *Historical Atlas* (1996) mapped it as 'common' in the Oswestry uplands and the south-western hills, and 'uncommon' elsewhere, but Forrest's description suggests it was much more common in the nineteenth century than in the second half of the twentieth. The ploughing up of marginal land in the Second World War (see p. 39), and the loss of rabbit burrow nest sites following the outbreak of myxomatosis, would have contributed to the reduction.

The *Handlist* (1964) referred to it as a 'breeding visitor and passage migrant from late March to October. Fairly common in the Oswestry uplands, the south-western hills and the Clee hills' and a few pairs bred locally on the pit mounds in the old coalfield around Wellington. In autumn considerable numbers, up to 100, had been recorded on the Clee hills. The decrease due to myxomatosis was described as 'temporary', occurring between 1956–58.

This distribution was reflected in the national BTO Atlas 1968–72, which showed breeding evidence in 17 of the County's 33 10km squares, including confirmed breeding records from five squares in the north that had no breeding evidence at all (two), or only probable (one) or possible (two) breeding, in the 1985–90 Atlas, suggesting the decline continued, particularly outside the main breeding stronghold.

Short turf, created by grazing sheep or rabbits, or poor soil on steep rocky slopes, and rich in insects and larvae, is essential for feeding; and nests are made in rock crevices, drystone walls or rabbit burrows. In 1985–90 the main concentration was on Titterstone Clee, where 26–40 pairs nested in the rocks on the summit and along Hoar Edge. Rock habitat – drystone walls, and old quarries and mine workings – was also favoured by the few pairs on Brown Clee. Close-cropped grass and scree with rabbit burrows was the main habitat elsewhere: Cefn Gunthly had several pairs, but the other hills of the south-west – Mucklewick Hill, Linley Hill, Black Rhadley Hill, Grit Hill and Stapeley Hill – had only a few. Black Mountain and other hills in the Clun Forest, the Stiperstones, Ragleth Hill, Caer Caradoc and the Lawley also had a few pairs. Although none of the Long Mynd valleys had large numbers, each had a few, adding up to this site being the second stronghold.

The *Atlas* (1992) suggested an average of only three to five pairs in the 60 or so tetrads with probable or confirmed breeding in the south and south-west, giving an estimated population of 180–300 pairs in 1990.

Subsequent local studies in the two strongholds quantified their populations, but then found a sharp decline.

On Titterstone Clee Hill, the Chelmarsh Ringing Group has been locating nests and ringing pulli (chicks), and catching adults and colour-ringing them, since 1998. A total of 596 nests have been found, ranging between 41–46 annually up to 2004, except 36 in 2001 and 35 in 2002 (2001 was not greatly affected by foot and mouth restrictions, which were lifted just before the ringing period). The highest number (66) was found in 2005, but a rapid decline then set in, and the population is now less than half what it was then (Fulton 2010; Dave Fulton *pers. comm.*).

On the Long Mynd there were an estimated 50–60 pairs in 1995–98 (Smith SBR 2003), where the slopes favoured by Wheatear were usually steep, interspersed with scree, and so badly eroded that there were areas

Nests Found: Titterstone Clee Hill (1998–2014)

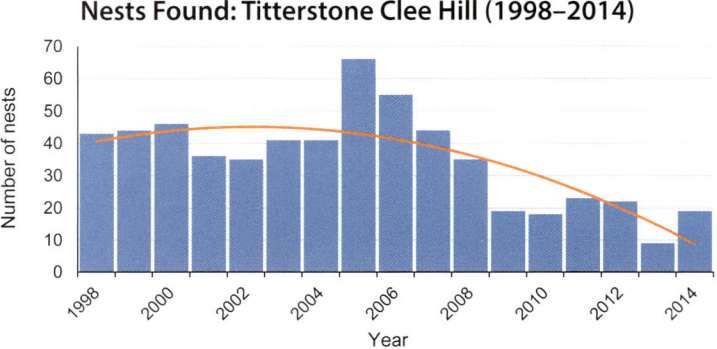

Breeding Distribution Change (1985–90 to 2008–13)

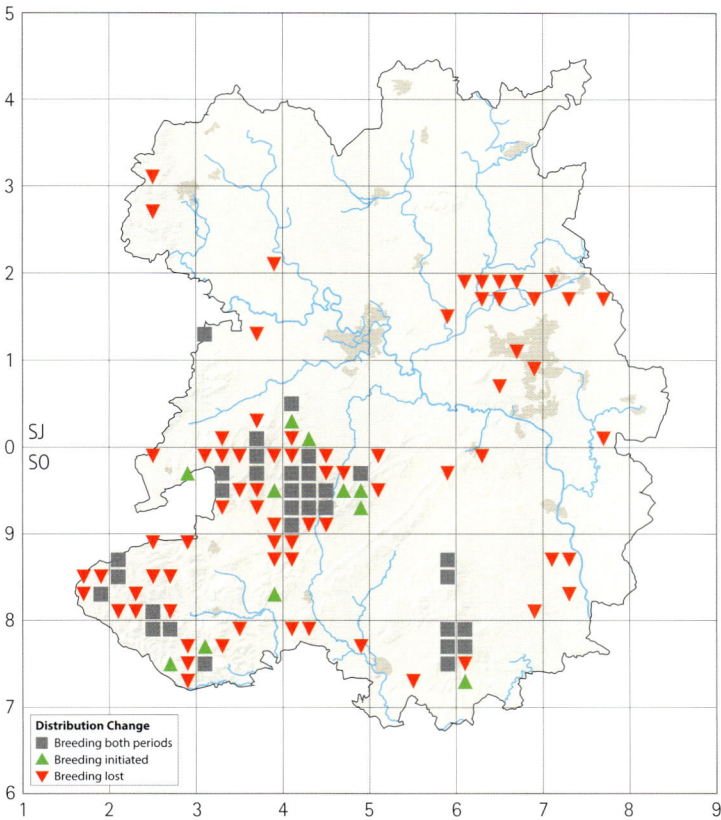

Distribution Change
- ■ Breeding both periods
- ▲ Breeding initiated
- ▼ Breeding lost

Occupied Tetrads

Atlas period (breeding)	1985–90		2008–13		Change	
	Number	%	Number	%	Number	%
Confirmed	43	5	19	2	−24	−56
Probable	33	4	10	1	−23	−70
Possible	27	3	14	2	−13	−48
TOTAL	103	12	43	5	−60	−58

and boulder fields, and there is only limited grazing, controlled by the NNR staff for heathland management, so it is possible that the increase there was due to relocation of pairs from Long Mynd as the latter habitat became less suitable.

During 2008–13 Atlas fieldwork, none were found on several hills listed as having breeding pairs in 1985–90: Linley Hill, Black Rhadley Hill, Grit Hill and Stapeley Hill. Only one pair was found on Cefn Gunthly, where 'several' were found during fieldwork for the earlier Atlas, and other tetrads around Long Mynd and Stiperstones in the south-west also had confirmed or probable breeding in the earlier period, but not the later one. Seven tetrads west of Clun had confirmed breeding, and 12 had probable breeding in 1985–90, compared with only two and two respectively in 2008–13. These changes are reflected in the breeding distribution change map.

There was only one breeding record from the north, fledged young from a nest in a rabbit burrow near the top of Middletown Hill in 2010, where possible breeding was recorded in 1985–90. The occasional pairs that nested in the north-west and on the old pit mounds around Telford have also been lost, even though Wheatears were described as 'fairly common' in the Oswestry uplands as recently as 1964 (*Handlist*). It is likely that ploughing and reseeding to 'improve' grassland pasture since then, which reduces invertebrates in the soil, has contributed to the losses in many of these areas, reinforced by contraction of range associated with the population decline.

Most of the other apparent losses since 1985–90, especially on the Weald Moors in a favoured passage area, will have been migrants mapped then as 'possible breeding'. In the recent Atlas such observations were coded as migrants, a category not used in fieldwork for the earlier Atlas. Passage migrants might be seen anywhere, and in the recent Atlas period Wheatears were recorded as such in 96 different tetrads in the north, and in 33 tetrads in the south, where no evidence of breeding was found.

Nationally, numbers are known to fluctuate, but the reasons are unclear, and monitoring of upland birds has been inadequate, although partially improved by the introduction of BBS in 1994. The UK population peaked in 1996, declined to a low point in 2004, increased again to another peak in 2010, followed by a further steep decline, with a net loss of 11% between 1995 and 2014. BBS in both England and Wales also show a rise from a low point in 2004, with no net change in England but a 16% decline in Wales since 1995. As Wheatear winters in the sub-Saharan Sahel region, these fluctuations may reflect changing drought conditions there, which have also affected several other species.

Fluctuations and trends on Titterstone Clee Hill and Long Mynd do not reflect these national trends: their populations have declined by much more because of the local habitat changes. Clee Hill averaged 21 pairs over the recent Atlas period. Assuming no further change in

of bare rock and earth. The short grass (and much of the erosion), was created by sheep, ably assisted by rabbits. All nests located were in rabbit burrows. By 2006–08, there were an estimated 25 pairs, fewer than half the number in 1995–98. The population in most of the valleys had declined, particularly on the upper slopes of Carding Mill Valley, and they had disappeared altogether from Long Batch, Jonathan's Hollow and the north-east slopes of Grindle and Packetstone Hill.

The specialist habitat required by Wheatears at both the major sites deteriorated due to a reduction in sheep grazing, as a result of the abolition of the headage payment to farmers in 2005, leading to an 18% reduction in sheep between 1999 and 2013 (p. 41). On the Long Mynd, this has been reinforced since 1999 by an ESA and subsequent HLS agri-environment agreement between Natural England, the National Trust and the commoners to reduce sheep density, particularly in winter, to allow regeneration of the heathland vegetation (p. 50). While this has benefited many species, much of the Wheatear habitat has become overgrown.

On Titterstone Clee Hill, the habitat has been even more seriously affected by 'restoration' work as quarrying at each seam has come to an end. The ground has been levelled, and rocks and other rough ground have been removed, and quick-growing coarse grass has been planted, which has been largely ungrazed by the remaining sheep, thereby removing both nest sites and feeding areas.

This population trend has not been found on the Stiperstones. There was one, possibly two pairs on the NNR in 1995–96, but five to eight pairs (probably six) in 2004–05, including four on the summit ridge, where none were found in 1995–96. The Stiperstones is much more rocky than the Long Mynd, with large natural scree slopes

Seasonal Occurrence in 10-day Intervals (1992–2014)

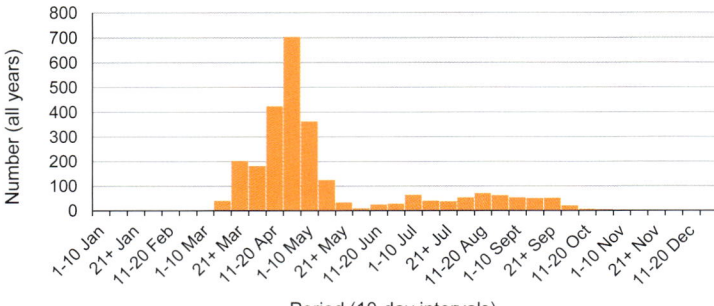

the estimated 22 pairs on Long Mynd in 2006–08, and the six pairs on the Stiperstones in 2004–05, and allowing an average of two pairs for each of the occupied tetrads on Middletown Hill, Cefn Gunthly, Mucklewick Hill, Caer Caradoc, the Lawley, Clun Forest and Brown Clee (40 altogether, probably an over-estimate), the population is estimated at around 90 pairs, a decline of more than 50% in around 20 years.

In 2014, breeding was confirmed at a new site in the Clun Forest, not occupied in either local Atlas period, where felling a coniferous plantation and regeneration of the heather on the recently acquired SWT reserve at Mason's Bank (SO28I) created new habitat.

The Clee Hill study found an average clutch size for 297 nests used in the analysis of 5.42, and an average brood size of 5.06. First egg date ranged between 1–8 May. There were only four examples of double broods in the 505 nests found. Adults returned more closely to their nest sites the following year (average distance: males 250m, females 341m) than first-year birds to their natal site (males 1,041m, females 1,013m). Resighting rates for 156 males were 37.2% the following year, and 28.4% for 204 females, steadily declining until two five-year-olds were seen of each sex, but only one six-year-old male, and none older. No evidence has been found of Clee Hill Wheatears moving to other breeding locations (Fulton 2010).

However, on the Long Mynd, two peaks in the observation of family parties, in mid-June and again in mid-July, suggested that two broods were fairly common during the 1994–98 survey. All Wheatears seen between 2006–08 were checked for colour-rings, but none were seen.

Many more pass through than breed here, and passage birds might be seen anywhere. Although the first arrivals are seen in early March, spring passage does not peak until late April. Since 1992, more than half the records have been of singles, and 90% are of four or fewer. Less than 2% are of flocks of more than 10. Far fewer are seen on return passage, which peaks in late August. The larger and more brightly coloured Greenland race *O.o. leucorohoa* pass through, and return, later, usually in mid-April to early May, and mid-August to mid-September, respectively. Generally there are no favourable winds helping the northward spring migration, so this race tends to favour overland routes rather than risk sea crossings at this time, largely accounting for the difference between the two passage periods, although absence of observers in likely areas in late summer may also be a factor.

The chart shows the numbers recorded passing through in 10-day periods since 1992. It excludes all records from the main breeding site at Titterstone Clee, and occasional records from other breeding sites.

There are only four recoveries, all of Wheatears ringed on Clee Hill, one from each of Brixham (Devon), Hayling Island (Hampshire), Bardsey Island (Gwynedd) and northern Morocco, which illustrate the migration route. The last of these was ringed as a nestling in May 1997 and recovered 1y 3m 18d later, 2,011km to the south.

A male ringed as a nestling on 28 May 1995 was retrapped on 31 May 2003, and held the longevity record of 8y 3d until 2010. Four more over five years old have been seen at the Clee Hill ringing site.

The *Arrival Dates* chapter estimates that Wheatears arrive 21 days earlier than they did in 1900, the largest change for any species listed.

It is likely that the detrimental local factors affecting the habitat in the two strongholds will continue, and the outlook for this increasingly scarce summer visitor is not good.

Leo Smith

Desert Wheatear

Oenanthe deserti

Vagrant

Widespread and common in arid habitats throughout its breeding range, stretching from North Africa through the Middle East to China, most are migratory, wintering in the Sahel/Sahara region of Africa, the Arabian Peninsula and North-West India. In Britain, the peak period for occurrence of this rare vagrant is October and November, and the only record was typical in this respect:

Titterstone Clee   25 November–10 December 2011

This female (Hudson *et al.* 2012) was located in cold windy conditions in and around the quarry at the summit of the hill, and probably roosted in one of derelict buildings (Kernohan 2011).

Graham Walker

Dave Barnes, Titterstone Clee, 9 December 2011

Dipper (White-throated Dipper)
Cinclus cinclus
Uncommon resident

Dipper is resident on the fast-flowing rivers and streams of upland Europe, and parts of Asia.

Several nineteenth-century authors referred to the presence of the 'Water Ouzel', summarised by Beckwith (1879) with the comment: 'Generally distributed along rivers and large brooks ... becoming very numerous towards the Welsh borders. Along the River Teme, and its tributaries, it is plentiful, but along the River Severn ... I rarely see it'. Rocke (1866a) referred to heavy persecution because of the 'mistaken idea that it is a destroyer of the spawn of trout and salmon' while Beckwith mounted the defence that mayfly larvae are the main predator of the spawn, and as Dipper feed on the mayflies, they do good rather than harm.

There is one record of the Black-bellied race found in parts of continental Europe, of one shot at Church Stretton on 15 December 1895.

A pair raised four broods near Ludlow in 1901, 'a very unusual event', but records in the first half of the twentieth century gave no indication of any change in status.

The *Handlist* (1964) described it as 'well distributed on the hill streams and rivers' of the Oswestry uplands, and the southern hills south and west of the Severn, but 'thinly distributed' in the north,

with small numbers occurring on Shell Brook, Rea Brook, Cound Brook and the upper reaches of the Tern.

A similar distribution was mapped in the 1968–72 national BTO Atlas, which showed evidence of breeding in all except two of the 19 10km squares south of Shrewsbury, but in only two of the 14 squares north of that. Breeding was confirmed in all except three of the squares with breeding evidence. Not surprisingly for a resident sedentary species, the distribution mapped in the 1981–84 national BTO winter Atlas was almost identical.

The *Atlas* (1992) stated that 'the stronghold is in the south-western hills, mainly along the rivers Onny, Clun, Teme, Rea and their tributaries, including the upper reaches of the Long Mynd valleys. Smaller numbers are found in the Oswestry uplands, on the upper reaches of the River Tern, along the Rea, Cound, Harley, Farley and Borle Brooks, on the rivers Worfe and Corve, and occasionally elsewhere'.

The Long Mynd Biological Survey, prepared for the National Trust in 1984, records Dipper as present in all the valleys, but the last confirmed breeding was in 1996 (probable breeding in 1997) in Carding Mill valley, and the last breeding in other valleys was in 1994 at the waterfall at the bottom of Callow Hollow.

Dippers have disappeared from other sites too, for example, 'they

John Robinson, Shropshire, 21 October 2011

returned to breed on the River Tern for the first time in more than 15 years' (SBR 2007), but they have not done so since, and only possible breeding was found there in the recent Atlas period.

The Upper Onny Wildlife Group started a nest box project in 2005, and the first two years suggested that a local decline had occurred.

The population in the River Teme catchment had been monitored extensively in the late 1980s and 1990s, up until 2000, by Tony Cross. Concern about the apparent decline led to a reinstatement of this monitoring, together with action to improve breeding success by increasing the nest box project in the Upper Onny, and subsequently extending it through the Upper Clun Community Wildlife Group, from 2007 onwards, and the Kemp Valley Community Wildlife Group (including part of the Lower Clun), from 2010 onwards.

All this activity was consolidated into a Dipper project, which was extended further, to the Teme, the Redlake, the Lower Onny and the remainder of the Lower Clun, and then to the River Rea and its tributaries near Cleobury Mortimer, in 2014, completing coverage of the whole Teme catchment. This project consists of four complementary activities:

- Monitoring the overall population and survival rate by catching birds at night-time roosts during the winter. Around 70 bridges were surveyed 1987–92, and all of these were re-surveyed 2006–14, together with an increasing number of new sites made suitable for roosting by the provision of nest boxes (a total of 149 sites by the end of 2014).
- Ringing nestlings (and adults when they can be caught), and ringing adults and first-year birds at winter roost sites.
- From 2010 onwards, colour-ringing all the adults that are caught at nest sites, and all the adults and first-year birds caught at winter roost sites. Each colour-ring has a unique combination of letters and numbers, so breeding birds at nest sites can be individually identified without the need to catch them. This is one of BTO's Retrapping Adults for Survival (RAS) projects. Over 700 adults and first-years had been colour-ringed by the end of 2014.
- Installing specially designed nest boxes under all bridges and other suitable structures, to improve breeding success, and monitor population levels and productivity. This has increased the population, as nest boxes have been installed and used on bridges where there is no other suitable ledge or cavity for a nest. Analysis has shown that the productivity of pairs nesting in boxes is slightly higher than that of other pairs. By the start of the 2012 breeding season, boxes had been installed at around 150 sites. In 2011 and 2012, 65 and 76 nesting pairs respectively were found, a total of 141, with 85 (60.3%) nesting in boxes, with 62% in boxes in 2014.

Monitoring has also provided information to the ringer about which sites to visit when, and consequently more broods have been ringed from 2010 onwards.

As a result, since 2006, the number of potential breeding sites to monitor has increased from less than 90 to over 190, the number of Dipper chicks ringed in the nest increased from 85 in 21 nests to 202 in 54 nests, and the number of winter roost sites checked increased from 67 to 149.

Comparison of results obtained in 2006–09 with those from the 1980s and 1990s show an initial overall decline in the number of Dippers, with a much greater decline on the lower reaches of the rivers than on the upper reaches, and a deterioration in the condition of the birds (measured by average body weight). This has been attributed to a reduction in food supply as a result of poorer quality rivers, primarily due to pollution from, and silting up by, agricultural activities.

However, more Dippers were found at roost sites in 2009 and 2010 than in any previous year, due to an increase in the number of nest sites in the upper reaches of the rivers, and improved breeding success, as a result of the nest box scheme. Numbers declined in 2011 and 2012, attributed to unusual weather conditions, which lead to very low water levels in 2011, and very high water levels, with particularly fast-flowing turbid water, in 2012. Both types of conditions make it difficult for Dippers to feed, so brood sizes were smaller than those found in most of the previous years, and the survival rates of young were also very low.

Numbers were also very low at the end of the previous Atlas period, attributed to 'two very dry summers in 1989 and 1990 when stream levels were very low. Under these conditions pollution and

Occupied Tetrads

Atlas period (breeding)	1985–90		2008–13		Change	
	Number	%	Number	%	Number	%
Confirmed	132	15	94	11	−38	−29
Probable	34	4	11	1	−23	−68
Possible	36	4	31	4	−5	−14
TOTAL	202	23	136	16	−66	−33
Tetrads with Winter Records (2007–13): 112 (13%)						

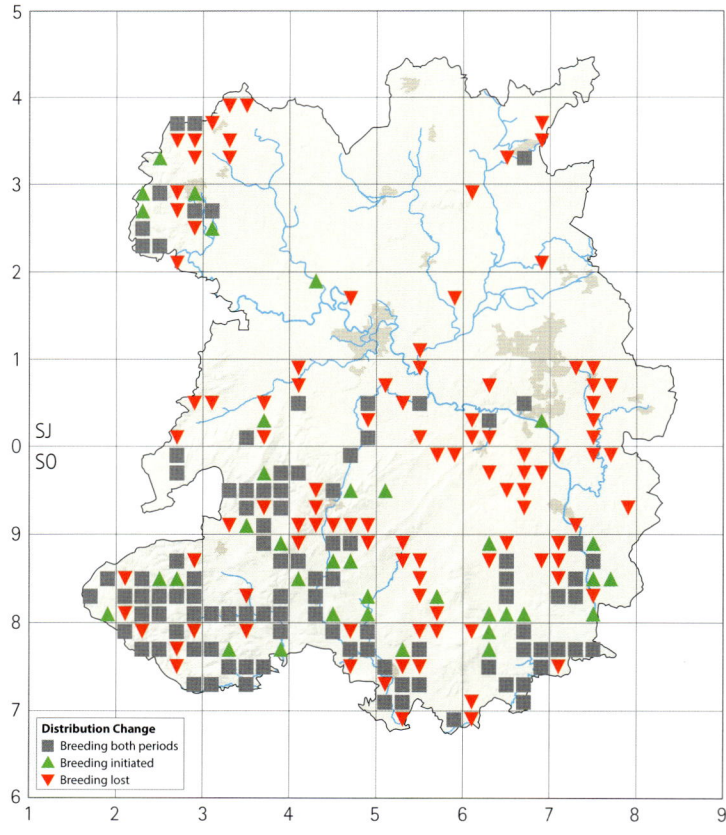

Breeding Distribution Change (1985–90 to 2008–13)

Distribution Change
- Breeding both periods
- Breeding initiated
- Breeding lost

siltation levels increase, causing a drop in the abundance of oxygen-loving invertebrates such as mayflies, stoneflies and caddis flies on which Dippers feed. The low water level may also lead to increased rates of predation'.

Numbers recovered in 2013 and 2014, and a total of 196 individuals was found at roost sites in 2014, much higher than in any previous year.

Detailed annual reports have been produced, including analysis of cumulative data, and, where appropriate, comparison of results from 2006 onwards with those from the late 1980s and 1990s, most recently *Dippers in the Teme Catchment 2011–12*, which can be found on the Severn Rivers Trust and the *Histo* websites. The project is continuing, and a further report on *Dippers in the Teme Catchment* (Cross & Smith *in prep.*), covering the years after 2012, is now being produced.

The Dipper project was therefore developing concurrently with recent Atlas fieldwork, and almost all of the 23 tetrads in the Teme catchment with new evidence of breeding were at sites found by project fieldworkers, often in nest boxes installed by the project. Apart from in the Severn catchment south of Bridgnorth, and around Oswestry, there were only three other tetrads with new evidence of breeding, and two of these were only of possible breeding evidence.

The Oswestry uplands continue to be a stronghold, but the breeding distribution change map suggests loss of several pairs from the more marginal habitat at the edge of the range.

Twenty-five years ago Dippers were widespread, but they have disappeared from many rivers and streams, especially in the Severn Valley, where the species has disappeared entirely from the River Worfe (and the streams north and west of Bridgnorth), and almost entirely from streams around Much Wenlock, the Rea Brook (south-west of Shrewsbury), and the headwaters of the Tern. It has also declined on Cound Brook, where it is now restricted to the upper reaches. Following analysis of a provisional distribution map, these rivers and streams were searched for Dippers before the end of the recent Atlas period, and none were found, so the decline is undoubtedly real. There has been a decline of one-third in the number of tetrads with breeding evidence, similar to the 32% decline in England between 1995 and 2014 found by BBS.

This reflects a major reduction in the water quality of those rivers and streams, confirmed by monitoring for the Water Framework Directive, which shows that they are all in 'failing' environmental condition (see p. 34).

The *Atlas* (1992) stated that 'during the breeding season Dippers are highly territorial, with lengths of 0.5–1.0 km usual for rivers of similar morphology and water quality to those found locally (BWP). Most occupied tetrads would therefore contain only 1–3 pairs, giving a population of 160–480 pairs'. Similar neighbour distances have been found in the upper reaches of rivers in the Teme catchment, where the population is close to two pairs per tetrad with probable or confirmed breeding, suggesting that the 1990 population was closer to the lower end of the range. Allowing for the fact that the river is the County boundary, so perhaps 10 nests on the Teme in the south-west are not in the Atlas area, the population is estimated at around 180 breeding pairs in the Teme catchment. This includes 41 in the south-east: 16 on the Rea, nine on Mill Brook and two on Farlow Brook (both tributaries of the Rea), six on Corn Brook, six on Ledwyche Brook, and two on Benson's Brook. Monitoring in this

John Fielding, Longnor, 19.March 2012

area largely started in 2014, so some of these sites do not appear on the Atlas map (Jon Lingard *pers. comm.*). There are also around 20 in the Severn catchment (almost all in the south-east, including three on Dowles Brook) and 10 in the Oswestry uplands, adding up to a total of around 210 pairs, though the number fluctuates from year to year depending on river levels. Dipper was found on too few TTVs for it to be reliable, but the population estimate based on TTV counts is rather less, 45–140 breeding pairs.

In spite of being restricted to upland streams where water pollution is less damaging than in lowland areas, the use of the sheep dip Cypermethrin caused problems with invertebrates on some water-courses, having a direct and dramatic effect on Dippers' food sources, but this is now banned. Other pesticides, such as avermectins and neonicotinoids, may also be having an adverse effect. It probably will not be long before new inadequately tested agricultural chemicals add to these effects. Climate change is forecast to produce drier summers and wetter winters, probably creating more frequently the conditions in 1989 and 1990, and in 2011 and 2012, that reduced the population. Like many other species in niche habitats, Dippers are very vulnerable to the impact of humans on the environment.

Tony Cross & Leo Smith

House Sparrow

Passer domesticus

Very common resident

Dawn Micklewright, Shrewsbury, 30 September 2013

Well known for living noisily in close proximity to human habitation in town and country, most people who have any environmental concern or awareness no longer take the House Sparrow for granted, given its well-documented national decline which led to it being added to the Red list of *Birds of Conservation Concern* in 2002. This characterful bird, anthropomorphically set in our folklore and culture, has at various times been persecuted locally and nationally as a pest species, and at other times considered a welcome neighbour. In return, the House Sparrow's approach to living cheek-by-jowl to people seems equally ambivalent in the sense of being both brave and trusting, yet shy and wary.

The early historical record suggests that the House Sparrow's great numbers were unwelcome. When not described summarily in a word or two as being '(very) abundant', writers went into great detail about ways of reducing the population to manageable proportions. Around the 1880s, Beckwith blamed the persecution of the hawk for burgeoning House Sparrow numbers and their 'mischief ... beyond calculation' with concomitant depredations on crops in gardens and in the field. The cost of paying for hunting parties to keep numbers in check was also lamented. Forrest (1899) described it as 'far *too* numerous, owing to the ruthless slaughter of its natural enemy, the Sparrowhawk. Sparrows destroy quantities of seeds, both in gardens and farms and drive away many of the birds, smaller or weaker than themselves, which are useful to man in keeping in check the many insect pests', and by 1908 little had changed with Forrest again calling it 'a pest, infinitely too numerous', and lambasting policy towards reducing its natural predators. So deep it seems was the antipathy that Forrest even took particular and detailed exception to a pair of House Sparrows that had gone to great efforts to evict a pair of House Martins at the top of Lord Hill's Column in Shrewsbury. In colourfully emotive language, he went so far as to suggest that this was a wilful and physically demanding effort to displace the summer visitors rather than find an easier and more obvious nesting site. Reports thereafter are more matter-of-fact, with common themes

being great abundance, opportunistic and scruffy nesting, instances of winter egg-laying, catching mayflies over the Severn and a number of albino individuals.

The *Handlist* (1964) referred to it as 'resident, common', without giving any indication that it was *too* common, and there appears to be no firm evidence for a decline in population at that time. However, a big reduction in the availability of food supplies in the first half of the twentieth century, in towns and on farmland, due to replacement of horses by vehicles and mechanical harvesting, with the dual impact of no stable blocks or grain stores for the Sparrows to forage in, and less spillage of seeds while harvesting, must have had a major impact. This was reinforced after the war by continued agricultural intensification, particularly the final phasing out of horses, loss of winter stubbles, tidier farmyards and secure grain stores, and the use of toxic seed treatments. A reduction in the relentless persecution of raptors would also have had some effect, although this would have been offset in the short term by the disastrous impact of the insecticide DDT on raptor populations.

It was not until the CBC was finally able to produce a national index in 1976 that a baseline for monitoring was established. Prior to that, censusing had proved difficult, because the distribution is closely linked to human habitations, and nesting is semi-colonial. After reaching a peak in 1979–80, the population fell by around 40% by 1984 (Marchant *et al.* 1990). The 1985–90 Atlas fieldwork was carried out against a backdrop of a rising national population from this low point, but House Sparrow was not found in 16 tetrads. Three of these were on the Long Mynd, and two more were on the highest part of the Clun Forest. In other parts of the south-west hills, and in extensive arable farmland, few people live and dwellings are very isolated, so Sparrows may indeed be absent from these few tetrads, but it is possible that they are very scarce, and overlooked.

BBS usually records House Sparrow in 69–75% of survey squares, and shows an increase of around 15% between 1997–2014. Annual fluctuations will be weather related, but there is no evidence of wholesale population crashes such as that in the East of England, not least London where there was a 75% decline in just six years up until the millennium.

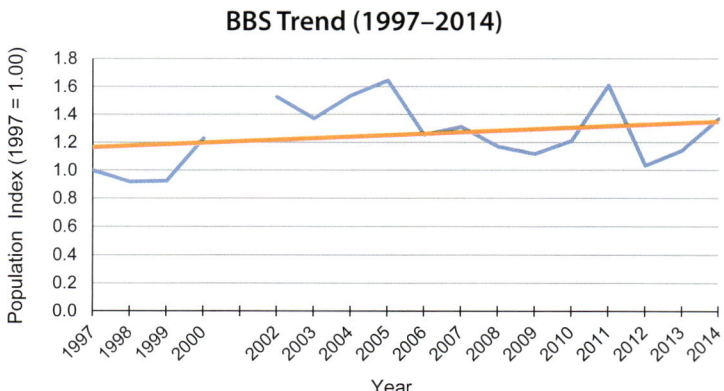

BBS Trend (1997–2014)

Breeding Relative Abundance (2008–13)

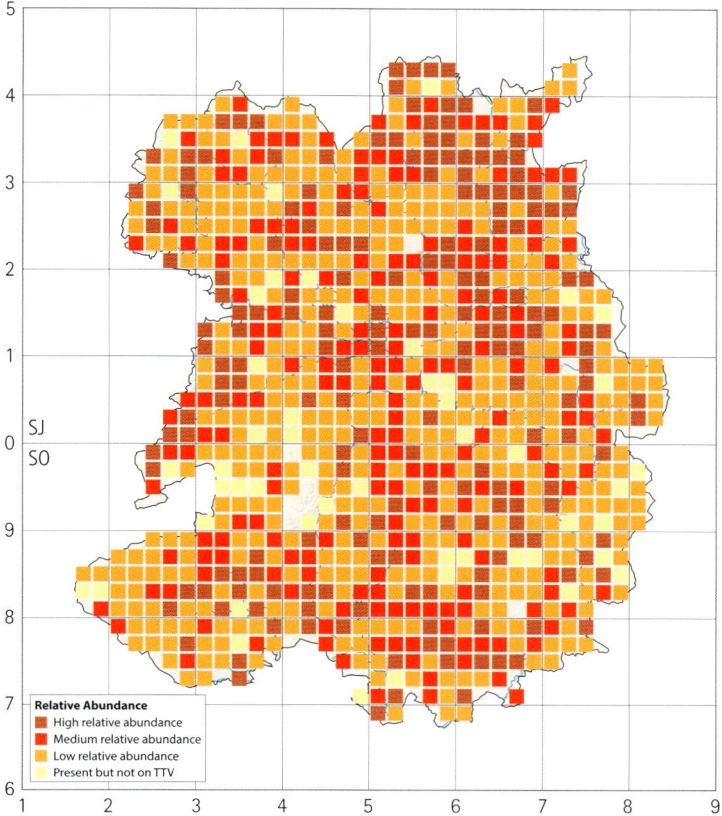

Relative Abundance
- High relative abundance
- Medium relative abundance
- Low relative abundance
- Present but not on TTV

Occupied Tetrads

Atlas period (breeding)	1985–90		2008–13		Change	
	Number	%	Number	%	Number	%
Confirmed	813	93	754	87	-59	-7
Probable	32	4	96	11	64	200
Possible	9	1	10	1	1	11
TOTAL	854	98	860	99	6	1
Tetrads with Winter Records (2007–13): 847 (97%)						

small increase found by BBS suggests that the estimate of 'not more than 70,000 pairs' (*Atlas* 1992), based on estimated densities from limited CBC results given in the national BTO Atlas (1976), was much too low.

The breeding season relative abundance map shows high densities where there is more opportunity for benefiting from human presence both for food and nesting sites, with the most populous areas in the north-east being highlighted. The House Sparrow is less common on higher ground and in extensive tracts of farmland. The winter season relative abundance map shows close parallels, not surprisingly for this sedentary species.

Although It is the third most common species, it is ranked 7th in the BTO Garden BirdWatch reporting rate, with an annual average of 76%. The peak (84%) is in June, about the time when first broods fledge, and the minimum (72%) is in December. Unlike the BBS, the reporting rate annual trend from 2006 shows a small decline.

Whether in town or country, populations can be very localised. In Shrewsbury, for example, in a particular area of SY3, House Sparrows were not seen by the author (SC) in over 10 years while living there, despite apparently good feeding and nesting being available. Short distances away in various directions, the familiar chirp could soon be heard.

Their sedentary nature is confirmed by ringing recoveries: all those ringed here were recovered here, and only one ringed anywhere else, in Flintshire, has been found here. On average a House Sparrow will live for three years, with common causes of mortality of those found being the cat and the car. Four over five years old have been recorded (oldest 5y 8m 8d). Three of these were finally found at the site where they were first ringed, and the fourth had moved 3km. A long distance traveller moved 4km over nearly two years from 2011, from the Rea, Upton Magna to Lees, at Walcot.

Stuart Cowper & Leo Smith

Fieldwork for the recent Atlas did not find breeding evidence in 10 tetrads, again in sparsely populated areas, but there are only two tetrads, both on the Long Mynd (SO49B & H) where House Sparrows were not found during fieldwork for either breeding Atlas. Human habitation is absent only from the latter, and it was not found there during the recent winter Atlas period either. In both Atlas periods, the proportion of tetrads with confirmed breeding was very high. It was not found in 23 tetrads in the recent winter Atlas period, again in the highest uplands and sparsely populated areas.

Based on TTV counts, the population is estimated at between 120,300–123,600 pairs, making it the third most common species. New habitat has been created since 1990 through the increase in housebuilding and urbanisation (leisure, retail, commercial and industrial development), although the design of modern buildings has limited nesting opportunities, but the relatively

Tree Sparrow (Eurasian Tree Sparrow)

Passer montanus

Fairly common resident

The Tree Sparrows that William Beckwith knew in the late nineteenth century spent their time in the open countryside 'about stubbles and weedy fields', readily finding nest sites in tree holes, thatch on barns and crevices in masonry. Forrest, too, described them as 'frequent' in stubble fields. Both men would have been surprised to find them using garden feeders and nest boxes: Beckwith noted their 'retiring

and shy habit', Forrest that they were 'not generally found near houses except in severe weather'. They appear to have become bolder in the intervening century, possibly because stubbles, weeds and nest sites are in short supply in today's farmland.

The first SBR (1956) noted 'many flocks of 20 to 100' at Ellesmere, though in most early editions Tree Sparrow figures only as an

John Hawkins, North Shropshire, 7 April 2006

'additional breeding species'. Where there is an entry, it is almost always a record of flock sizes, such as the impressive 'several hundred' at Brompton in 1961. Some accounts convey unease over the lack of data, referring to 'few reports' (1966), Tree Sparrow being 'overlooked' (1978–89), and 'presumably widespread' (1980–81).

By contrast, recent SBRs refer to the large volume of records, as monitoring of Tree Sparrow, and other declining farmland birds, has acquired an urgency it previously lacked. Flock counts still feature, with 150 feeding on maize stubble at Quina Brook in 2002, but they are fewer and smaller in recent years. Venus Pool, where species-rich hay meadows were established in 2001, attracted flocks of up to

150 until numbers crashed inexplicably in 2006. Since then, annual maximum counts from all sites have varied between 30–59, with totals in the thirties most frequent.

The *Handlist* (1964) described Tree Sparrow as 'common in mixed flocks ... often in excess of 100', but added that more work on distribution was needed 'to work out its true status'. The *Atlas* (1992) delivered clear evidence of a contraction of range: confirmed or probable breeding was detected in fewer than half of all tetrads. National data indicate that the population 'nose-dived spectacularly' (BTO) between the late 1970s and early 1990s, falling by over 90%. The precipitous decline was attributed to poor over-winter survival, caused by shortage of food: stubbles were ploughed up for autumn sowing, and 'weeds' that might have helped replace them suppressed by widespread use of herbicides on intensively cultivated land.

Fieldwork for the recent Atlas found a further contraction of range: breeding evidence at any level was obtained in fewer than one in three tetrads, and probable or confirmed breeding in less than a quarter.

Occupied Tetrads

Atlas period (breeding)	1985–90		2008–13		Change	
	Number	%	Number	%	Number	%
Confirmed	216	25	102	12	–114	–53
Probable	178	20	85	10	–93	–52
Possible	114	13	78	9	–36	–32
TOTAL	508	58	265	30	–243	–48
Tetrads with Winter Records (2007–13): 250 (29%)						

Breeding Distribution Change (1985–90 to 2008–13)

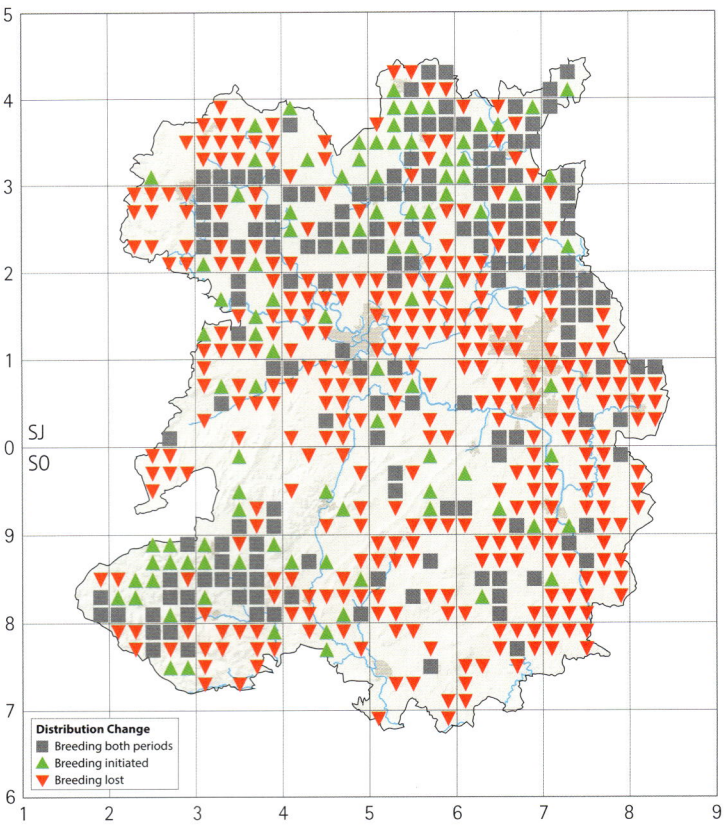

Breeding Relative Abundance (2008–13)

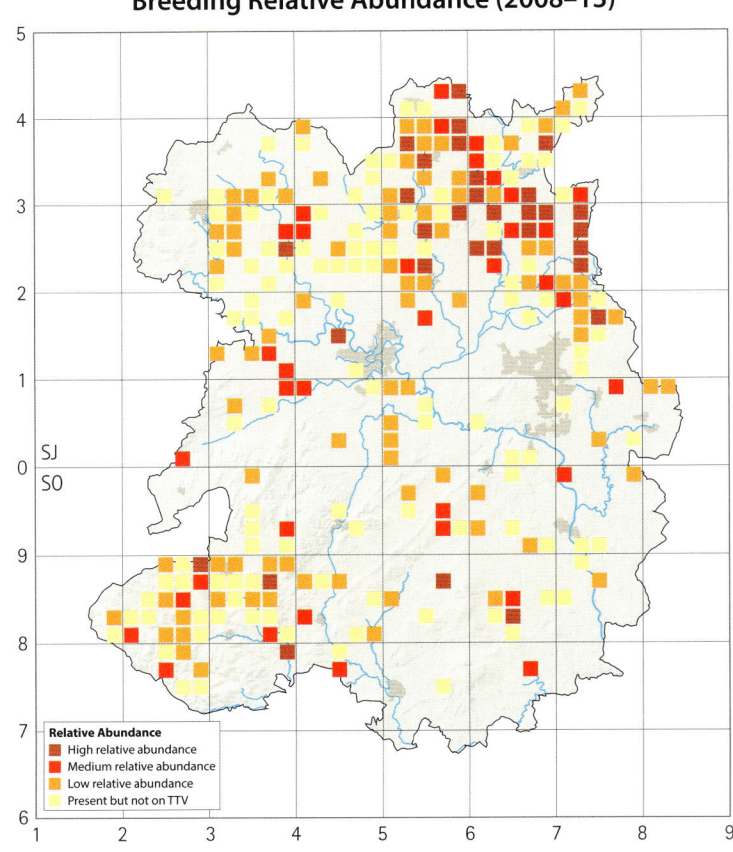

There have been widespread losses of breeding Tree Sparrow, especially adjacent to gaps in the 1985–90 distribution along the Teme valley and Corvedale, up the Severn Valley and around Oswestry, where arable has reduced with the decline in mixed farming. By contrast, about 10% of tetrads appear to have been newly occupied, or reoccupied, since 1990. Many of these cases are likely to be products of variations in fieldwork, though some may reflect Tree Sparrow's known volatility, with sudden fluctuations in both numbers and range.

Based on TTV counts, the population is estimated at 1,850–2,000 territories, substantially less than the estimate of about 5,000 pairs in 1990.

As regards relative abundance, the north-east emerges clearly as the stronghold, with most high-abundance squares in a band from Newport through Market Drayton to Whitchurch. The winter distribution and relative abundance are very similar, as would be expected of such a sedentary species.

Tree Sparrow is currently found in only about 15% of BBS plots, and although not recorded in sufficient numbers to produce population trends for either Wales or the West Midlands, the population is likely to have seen a decline corresponding to the approximately 50% contraction in range in the period between the two Atlases. BBS recorded a population increase in England of 72% between 1995–2014, but from such a low base that it is still less than 10% of its pre-1970s level. The map showing the change in relative abundance since 1988–91 in the national BTO Atlas (2013) also shows clearly that, far from experiencing even this modest recovery, the local population has not stabilised, and continues to suffer high losses.

The population crash between the 1970s and early 1990s also affected other farmland birds, such as Yellowhammer and Linnet.

Agricultural intensification (see pp. 38–41) has made farmland a less hospitable environment, and Tree Sparrow suffered particularly severely. During the breeding season, adults must find a nest site and feed themselves and up to three broods within a small local area: Field & Anderson (2004) found that birds feeding chicks rarely travel more than 300m from the nest. Diversity within the habitat may be especially important: as invertebrate life cycles are short, successions of species are needed to provide continuous food over several months. There is some evidence that Tree Sparrow is attracted to wetland edges, where vegetation and invertebrate life is more diverse than on most cultivated land (*ibid.*). Similarities with local Reed Bunting distribution, both species concentrated in the north-east, and thinnest, or absent, across much of the south-east, tend to support this view.

Nine of the 10 locally ringed Tree Sparrows were recovered here and the tenth was found in adjacent Montgomeryshire, confirming their sedentary habit.

Michelle Frater

Alpine Accentor
Prunella collaris
No modern records

There has been a single occurrence of this very rare vagrant from the mountains of central Europe. The bird was 'caught in a brick trap' at Boreatton Park in 1891 and identified 'but unfortunately not preserved' (Forrest 1908).

John Tucker

Dunnock
Prunella modularis
Very common resident

Dunnock breeds over much of Europe, but vacates eastern areas for the winter.

Leighton (1836) noted that the 'hedge warbler' was 'everywhere' while Rocke (1865) wrote that the 'hedge accentor' was 'one of the prettiest, as well as the most common, of our native warblers'. Beckwith (1878) used the same name, but Forrest (1899) referred to the 'hedge sparrow'. Both described it as 'common', and the latter praised 'this soberly clad bird, with its short sweet song, and nest with lovely blue eggs', while H. Paddock (1890) described it common everywhere in the neighbourhood of Newport, and G. Paddock (1904) wrote the same in relation to Church Stretton, the latter adding 'around dwellings but also to be found in gorse on the hillsides from which it could be heard at any time of year providing the weather was suitable'.

Its status and distribution appears to have remained unchanged throughout the first half of the twentieth century, and the *Handlist* (1964) noted that the 'Hedge-Sparrow' was common except on very open high ground. However, its abundance meant it suffered from under-reporting, with early SBRs either omitting it from the species accounts or simply noting it as breeding without detailing specific records. The 'Dunnock' was mapped with confirmed breeding in every 10km square except one in the BTO Atlas (1976), and the modern name was adopted by SBRs in 1960. In 1981 it was recorded from the lowlands in the north to 1,300 feet in Callow Hollow. Harsh winters could take their toll, the winter of 1940, for example, notably reduced the number in the north-west. Local comments that it appeared to have survived the hard winters of 1962–63 'very well' and 1980–81 'reasonably well' have to be read in the context of the national results of CBC monitoring, which indicate a steady population decline since the mid-1970s (Marchant *et al.* 1990). It was considered to be generally widespread and a regular garden bird throughout the 1980s and 1990s.

The widespread distribution was confirmed in the *Atlas* (1992), when it was mapped in 99% of tetrads. Unobtrusive and skulking, often overlooked and described as dull plumaged, it possibly has the most complex social system of any British bird. As well as breeding in conventional pairs (monogamy) other strategies include two or three males and a female (polyandry), one male and two females (polygyny) and two or three males with two, three or four females (polygynandry) (Davies & Lundberg 1984). This is because both males and females compete to increase their reproductive success by gaining extra mates. The mating system of a population of Dunnocks at Nedge Hill,

John Fielding, Atcham, 10 January 2013

Telford, was less complex than populations studied in other habitat types, with only monogamy and polyandry exhibited, possibly due to the increased defensibility of linear hedgerows by males against other males and a lower population density (Bishton 2001). The mean number of young produced in pairs was 4.00, compared with 3.45 in polyandrous units. This difference was not found to be statistically significant. Clutch size, provisioning rates, nestling weights and nest failure rates also did not differ significantly.

Males begin establishing breeding territories in February, though bursts of song can be given as early as December on sunny, mild days, and by the end of March they are well defined (Bishton 2001). The sweet but high-pitched warbling song is issued frequently from territory boundaries and assists in the locating of breeding pairs. Males engaged in wing-flicking and chasing displays in the presence of a female can also be conspicuous in more open habitat.

Dunnocks were recorded in 100% of the tetrads during the 2008–13 Atlas period. Their preferred habitat structure is one of low-level, dense, thorny vegetation, such as hawthorn and blackthorn hedgerows, and the highest levels of abundance were found in lowland areas, especially in the north, extending from the Weald Moors northwards to Market Drayton and also around Oswestry, Ellesmere and Shrewsbury. Southern areas around Bridgnorth, Cleobury Mortimer, Clun and Ludlow also supported relatively high densities. Overgrown thorn hedgerows that support dense outgrowths of bramble, dog rose and nettle patches, within or in their immediate proximity, are preferred (Bishton 2001). Young and remnant hedgerows lacking outgrowths and with large gaps comprise only a small area of territories. In linear

scrub habitats such as hedgerows they are well spaced, but high densities can occur in other scrub habitats with 15 singing males noted at Priorslee Lake on 27 February 2012 and 25 at Bicton Heath on 24 April 2012.

Gorse on heathland, commons and moorland on the high ground of the Long Mynd and other upland areas in the west held relatively low populations, demonstrating a fairly distinctive preference for lower ground with a general decrease in abundance from east to west. Scrub, woodland edge, young plantations, wetland edge and similar marginal habitats are frequently occupied and, in urban areas, gardens and planted shrubberies in and around business estates and retail parks are too.

In the recent Winter Atlas period they were found in 99% of tetrads. Relative abundance sharply mirrors that in summer, with higher levels of abundance across the eastern lowlands. The Stiperstones, Long Mynd, Clee hills and other high ground exhibited low relative abundance.

Analysis of 202 faecal samples at Nedge Hill, Telford, revealed a transition from a predominately invertebrate diet dominated by beetles (especially weevils, rove beetles, ground beetles and leaf beetles), spiders, false scorpions, snails, earthworms, flies and springtails in the summer to one dominated by small seeds in winter, mainly nettle, grasses and elder (Bishton 1985; 1986). This physiological adaption probably enhances their winter survival and promotes a widespread distribution. They feed mainly on the ground, especially in winter, on exposed soil and leaf litter beneath hedgerows, nettle and bramble patches and similar low-level vegetation, but will pick larvae, leaf beetles and weevils directly from the foliage of trees and

Occupied Tetrads

Atlas period (breeding)	1985–90		2008–13		Change	
	Number	%	Number	%	Number	%
Confirmed	665	76	484	56	-181	-27
Probable	139	16	323	37	184	132
Possible	59	7	59	7	0	0
TOTAL	863	99	866	100	3	0
Tetrads with Winter Records (2007–13): 866 (100%)						

Breeding Relative Abundance (2008–13)

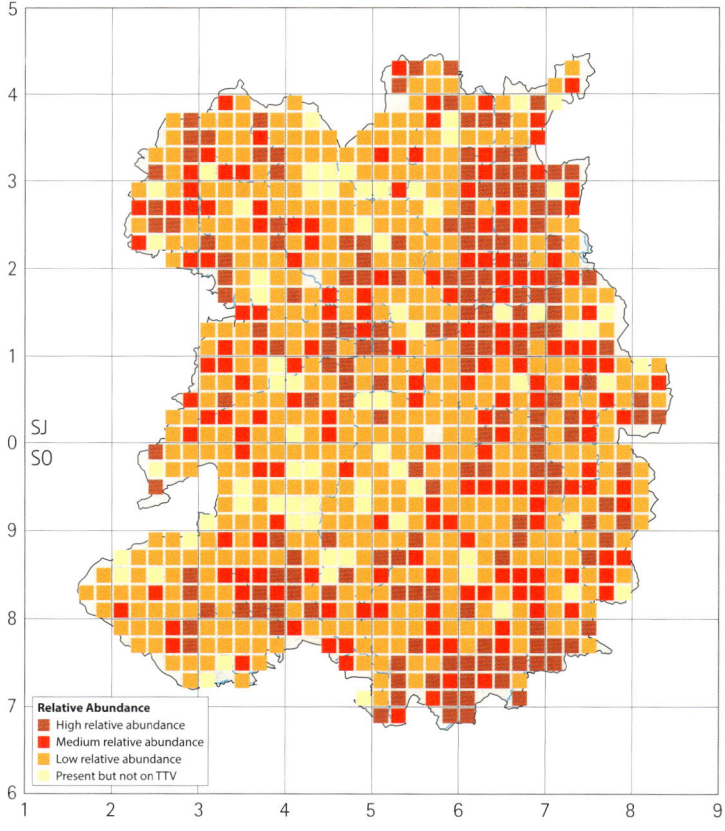

Relative Abundance
High relative abundance
Medium relative abundance
Low relative abundance
Present but not on TTV

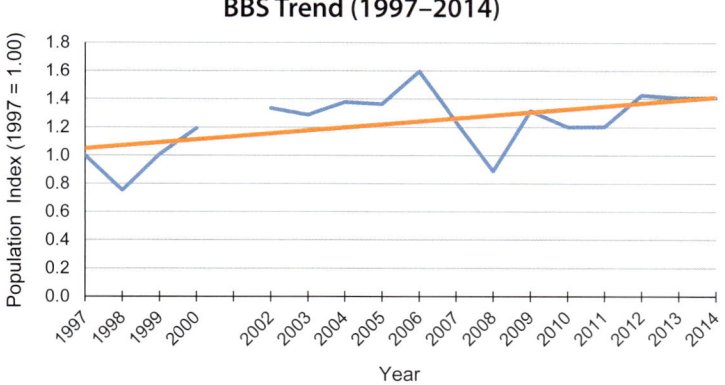

BBS Trend (1997–2014)

Population Index (1997 = 1.00) — Year

other vegetation in summer, and seeds from plants in autumn. Most feed as singles or in twos in the winter period, but larger numbers will congregate where the food source is concentrated. Eight were recorded feeding in close proximity along a lane at Cherrington in February 2010 and seven fed around a manure heap along a 30m stretch of hedgerow at Woodcote Hall in February 2012. TTV counts suggest Shropshire supports an estimated 2.1 % of the British population of Dunnocks in winter.

BBS data indicates an increase of 40% for the period 1997–2014, which is consistent with a UK-wide recovery from long-term decline since the late 1990s. Nationally, the UK population increased by 22% between 1995–2014, but it was still 30% less than it was in 1970.

The recovery may be partially associated with the implementation of Entry Level and Higher Level Environmental Stewardship schemes on farmland since 2005, which has enhanced breeding and foraging habitat. Initiatives likely to benefit Dunnock are the planting and management of hedgerows, unsprayed and wildflower headlands, the

provision of wild bird seed mixes and the retention of winter stubble. However, many farms did not join either scheme, and their benefit for farmland birds at a landscape scale has been small (Baker *et al.* 2012).

Population declines in specific years during the BBS period are likely to be due to increased mortality as a result of harsh winter weather, mitigated to some extent by its adaption to a diet dominated by seed. Inclement weather during the breeding season, as for example, heavy rainfall in the summer of 2007 and 2008, possibly contributed to a fall in breeding productivity and fledgling survival, and reduced the breeding population in subsequent seasons.

Because of the complex structure of the Dunnock's breeding system, the population is expressed as a number of breeding territories, not pairs. Based on TTV counts, the population estimate is 50,100–51,100 territories, a large increase on the *Atlas* (1992) estimate of 22,000–26,000 territories.

They are highly sedentary and post-natal movements very short, in the region of a few hundred metres (Davies 1992). In a study of a colour-ringed population at Nedge Hill, Telford, only one male ringed as an adult was subsequently recorded away from the study site, at a distance of 3km (Bishton 2001). All except two of the 56 ringed have been recovered here, and the other two were found in adjacent counties. There is no evidence of long-distance movements.

A positive approach within the farmed landscape to the retention of mature hedgerows with outgrowths, planting of new hedgerows and the retention of scrub corners and nettle and bramble patches, through Natural England's agri-environment schemes, might assist in the recovery of the Dunnock population to earlier levels. In its woodland habitat the creation of low-level scrub through coppicing at the woodland edge and along woodland rides should be considered.

Glenn Bishton

Yellow Wagtail (Western Yellow Wagtail)

Motacilla flava

Uncommon summer visitor

Races of Yellow Wagtail breed from Fennoscandia, Britain and Iberia eastwards across the Palearctic, and winter mainly in sub-Saharan Africa. The British yellow-headed sub-species *M.f. flavissima* breeds largely in England, but also elsewhere in Britain and coastal mainland Europe, and winters in West Africa.

It was 'rather common on ploughed fields around Shrewsbury' (Leighton 1836), 'common' and 'generally observed to arrive the 29th April' (Eyton 1838), and locally distributed and 'tolerably plentiful' (Rocke 1865). Beckwith writing in 1878 stated that Ray's Wagtail (as it was then known, after the naturalist John Ray 1627–1705) was 'very common in spring and autumn about the Severn Meadows of Shrewsbury, especially where there are cattle' and added in 1890 that it frequented upland fields and meadows rather than the vicinity of water. Paddock in the same year agreed that it was 'the least aquatic of the wagtails and a regular summer visitor around Newport'.

At the turn of the century it was 'usually seen in the pastures with sheep and cattle' around Church Stretton, and nests were 'generally built in cornfields either in the furrows or in a slight depression of the ground' (Paddock 1904). It was fairly common and generally

distributed in 1908, and in 1957 it was noted as widely distributed in the north. The *Handlist* (1964) stated that it was 'fairly common in the river flood plains and marshy pastures', but absent from hill country, suggesting a contraction in range since Beckwith 1890.

By 1968 there were further indications that the population was in decline, when it was noted as local with few reports from the east, but it was mapped with breeding evidence in all the 10km squares in the north, but only five out of 14 in the south, in the BTO Atlas (1976). By 1974 'fewer reports' were received and the *Atlas* (1992) mapped it predominately north and east of the River Severn, and confined to two distinct habitats, damp pasture on the floodplains of the main rivers, but predominantly in fields of root crops and cereals. Its population was estimated at 1,150–2,300 pairs. An SOS survey in 2003 and 2004 of a sample of tetrads where it was shown as breeding in the Atlas (1992) found fewer than 10% of the pairs in wet meadows. Most were recorded on arable land, with cereal and potato accounting for nearly 70% of pairs. The survey also showed a 57% decline in the number of occupied tetrads and estimated the population at a much reduced 400–600 pairs. However, the extent of the main breeding range remained unchanged (Dawes SBR 2004).

The 2008–13 Atlas found that the main breeding populations are still concentrated on the lowlands of the Northern Plain, especially the Weald Moors, the catchment area of the rivers Roden and Tern, along the River Perry and east of Oswestry, and also on the Eastern Sandstone Plain between Newport and Bridgnorth. An isolated population exists to the south of Shrewsbury and west of Much Wenlock, and the only notable pockets in the south are small populations on the relatively low arable ground around Bishops Castle, Clun and Bucknell. Apart from that, the whole of the south, much of the west and north-west, and the large urban areas of Telford and Shrewsbury, are devoid of Yellow Wagtail. The occupied land is predominately lowland arable farmland, and habitat within this landscape comprises mainly fields of potato, cereals, peas, beans, cabbage and oilseed rape.

Some tetrads are not occupied every year as suitable habitat moves around farms through crop rotation, although pairs may move between nesting attempts as the structure of the different crops changes within the season. The distribution map therefore probably exaggerates the position in any one year.

Since the previous Atlas, substantial losses have been incurred in tetrads in the north, between the rivers Perry and Roden, and around Market Drayton and Newport; and, in the south, in Corvedale, and the lower Severn and Teme valleys, but the extent of the breeding range remains essentially the same as shown in the Atlas (1992).

Yellow Wagtail usually breeds at low densities, but high numbers found during recent Atlas fieldwork included eight singing males at Eyton Moor on 23 June 2009, six singing males at Allscott Sugar Factory on 27 April 2010, and six pairs in a field of peas and potatoes at Beckbury on 18 June 2011. The highest relative abundance was found in scattered tetrads in the north-east between Telford and Market Drayton, and on the Eastern Sandstone Plain around Sutton Maddock and Beckbury, south to Bridgnorth.

Based on TTV counts, the population estimate is 200–230 pairs. This is consistent with the decline recorded between the Atlas (1992) and the 2003 and 2004 surveys, and a continuing decline since, but it is only one pair per tetrad with breeding evidence. They appear to form loose colonies, and TTV counts and other fieldwork observations indicate that some tetrads have as many as four to five pairs. An average of two to three pairs in each tetrad with confirmed or

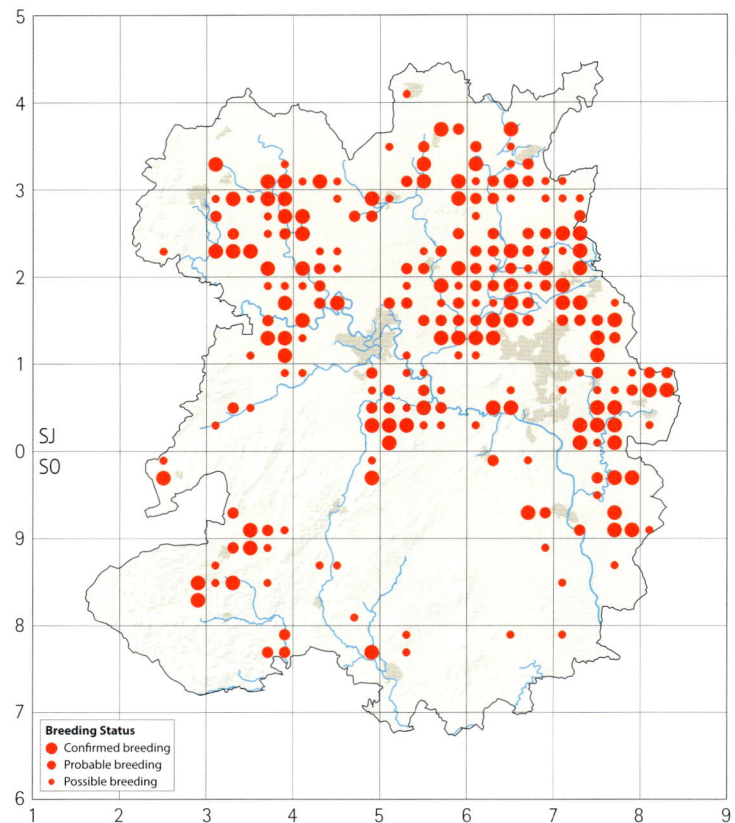

Breeding Distribution (2008–13)

Breeding Status
- ● Confirmed breeding
- ● Probable breeding
- • Possible breeding

Occupied Tetrads

Atlas period (breeding)	1985–90		2008–13		Change	
	Number	%	Number	%	Number	%
Confirmed	146	17	82	9	-64	-44
Probable	88	10	68	8	-20	-23
Possible	82	9	85	10	3	4
TOTAL	**316**	**36**	**235**	**27**	**-81**	**-26**

probable breeding, plus one in half the tetrads with possible breeding, would give a more realistic 350–500 pairs.

There are several contributory factors to the steep decline of this delightful long-distant African summer migrant. Much wet meadow habitat has disappeared, for example a loss in the area of 'grazing marsh' of 29% between 1979 and 1993 (Dargie 1993), with a further subsequent loss which is difficult to quantify because of changing definitions of habitat types (NE *Priority Habitat Inventory* 2011). This has been accompanied by a more general loss of pasture as a result of a decrease in the number of cattle (down 26% from 1985 to 2013, including 11% from 2000–13), and the value of what remains has been reduced by drainage, increased inputs of nitrogen fertiliser and a consequent decrease in sward diversity and structure, and the use of chemicals such as avermectins in farm animal husbandry and consequent reduction in invertebrates in dung. Arable land is also becoming less suitable as it is farmed more intensively. Research elsewhere shows that the rapid growth of autumn-sown cereals is limiting the breeding season and the choice of available nest sites, so the wagtails nest near crop 'tramlines', where access to the ground is

Jim Almond, Venus Pool, 15 April 2010

Breeding Distribution Change (1985–90 to 2008–13)

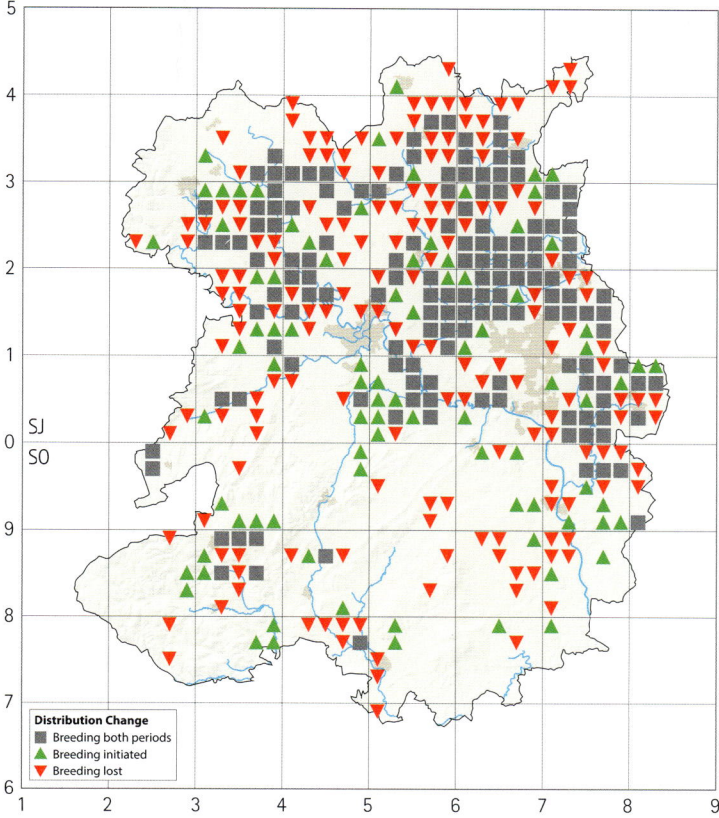

Distribution Change
- ■ Breeding both periods
- ▲ Breeding initiated
- ▼ Breeding lost

change. Initially on arrival many occupy wetland areas such as Venus Pool and Allscott Sugar Factory where 16 and 14 individuals were recorded on 18 April 2011 and 11 April 2010 respectively. Thirty, probably passage migrants, were feeding in a recently tilled field on the Weald Moors on 26 April 2009. The average final record date for the years between 1992–2003 was 26 September. Between 2004–14 it was 19 September, the latest in this period being one recorded unusually late at Chelmarsh Scrape on 19 October, in 2005.

There are only five ringing recoveries and of these only two involved foreign recoveries – an adult male ringed at Dothill, Telford, on 20 June 1965 and recovered dead at Bilbao, Spain, on 15 December 1965, 1,050km distant, and a first-year ringed near the Wrekin on 13 August 1967 and recovered dead at Finistere, France, on 11 September 1967, 558km distant and only 29 days later. One first-year ringed at Leegomery, Telford, on 2 August 1965 was recovered dead at Coalmoor on 15 June 1967. Indicative of a passage movement from further north was one ringed in Greater Manchester on 25 August 1984 and recovered freshly dead in Shrewsbury on 17 October 1984.

Uncommon continental races are seen occasionally. Between 1939–45, several pairs of Blue-headed Wagtail were recorded in the area of the River Perry near Whittington. Probable breeding occurred between 1939–46 at Perry Moor and West Felton, where several pairs were noted on marshy ground with 'no possibility of mistaken identity'. During the recent Atlas period, breeding was confirmed of a Blue-headed Wagtail near Market Drayton, but the race of the female was unproven. Another male was seen carrying food at Tong Hill on 15 June 2013. A Blue-headed x Yellow Wagtail hybrid, known as a Channel Wagtail *Motacilla flava* subsp. *Flava x flavissima*, was present at Venus Pool between 8 April and 1 May 2011, and another was present at the same site between 11 and 29 April 2012. The Atlas data and maps include all records for the species, whether or not a sub-species was identified. After the Atlas period, a Blue-headed or perhaps a hybrid was at Polemere on 29 April 2014.

Our data are consistent with the UK-wide trend that the majority of Yellow Wagtails now probably breed in arable areas. Because of the large landscape scale issues involved, appropriate conservation initiatives require political will, and the continued promotion and intervention of agri-environments schemes at every opportunity, to create water features, fallow plots, spring crops, invertebrate rich headlands, 'skylark plots' and beetle banks in autumn-sown cereals. The current distribution map will prove invaluable in highlighting the breeding hotspots and should assist the targeting of agri-environment schemes to help improve the habitat for this enigmatic wagtail.

Glenn Bishton

better but exposure to predators is higher (Gilroy *et al.* 2010). Later breeding attempts from June onwards show a shift to spring-sown broad-leaved crops such as beans, peas and especially potatoes, as they try to maximise the number of broods each year. Pairs nesting in silage are rarely successful due to the frequency of cutting. Farmland drainage, soil degradation and increased use of pesticides, with the associated loss of invertebrates, together with possible problems on wintering grounds and migration routes, are also probably contributing to the decline. It shows no signs of abating.

A high-pitched 'tseep' from somewhere across a distant potato field or a fluttering flash of vivid yellow over a cereal crop reveals their presence in spring. The average first arrival date between 1906–37 was 26 April, the earliest date being 14 April, in 1937, and the latest 6 May, in 1910. The average arrival date was 9 April between 1992–2003, and 6 April between 2004–14, indicating a marked advance in the last 100 years, which is continuing and probably linked to climate

Grey Wagtail
Motacilla cinerea
Uncommon resident

This is arguably the most charismatic of our three regularly occurring species of wagtail. Traditionally associated with fast-flowing upland streams and rivers during the breeding season, it can be found during the winter in all environments where water is present, including town centres and garden ponds.

Beckwith observed that in northern areas this elegant bird breeds by the small brook running from the Wrekin, and along the banks of the river Worfe and some of its tributaries, as well as by the numerous streams in the neighbourhood of Oswestry and Llanymynech. A pair was occasionally found in other places, but in these areas it was rather rare during the summer months. In the south, the Grey Wagtail was much more evenly distributed and there were few streams by which it did not nest. This wagtail was, however best known as a winter visitor with numbers greatly increased by arrivals from further north.

It was recorded in most years in the first half of the twentieth century, with no evidence of a change in status, apart from being 'considerably reduced' after the hard winter of 1939–40.

'Markedly fewer [were seen] than last year' (SBR 1963, following the hard winter of 1962–63). The *Handlist* (1964) described Grey Wagtail as a resident and probably a passage migrant, frequently met with on both upland and lowland streams, particularly near bridges and weirs, and often reported in urban areas in winter.

The BTO *Atlas* (1976) showed no evidence of breeding in 10 of the County's 33 10km squares, but breeding was confirmed in all of them in the *Atlas* (1992). These squares were mainly along the southern edge of the northern plain, and some of them had several occupied tetrads in 1985–90, indicating an expansion of range since 1968–72, contrary to the national trend at that time.

The *Atlas* (1992) showed a breeding distribution centred on the south-west uplands, particularly the tributaries of the Clun, Teme and Onny, with smaller concentrations in the Oswestry uplands, Clee hills and the streams flowing into the Severn in the Telford area. However most of the major rivers in the north and the Severn below Bridgnorth, and their tributaries, also held some breeding Grey Wagtails. Even in these lowland situations they are closely associated with stretches of 'riffly' water such as weirs, river confluences or outfalls. Canals too may be used, favoured locations being not too distant from locks or feeder channels where some fast-moving water occurs. In 1990, four pairs nested near weirs and sewage works in northern areas.

Recent Atlas fieldwork found a similar basic pattern, but a contraction of range, and the breeding distribution change map shows a net reduction in tetrads with any level of breeding evidence of 27%. Gains have occurred in tetrads at the northern edge of the southern hills, and gains have cancelled out losses in the south-west, but these wagtails have largely disappeared from the headwaters of the Perry, the Roden and the Tern, and some of their tributaries, and especially from the Weald Moors around Telford, and Oswestry.

Some of the losses in the strongholds on the upper reaches of many of the rivers may be temporary. Grey Wagtails are particularly susceptible to bad weather, which can have a dramatic effect on numbers, and the cold and late winters in the early years of the recent Atlas period, along with a mix of some very dry and some very wet summers, probably reduced the population. However, it may bounce back after good breeding seasons. Supporting evidence for these losses has come from BBS and WBBS, which led to it being added to the Red list in 2015 because of a severe decline in the national breeding

Occupied Tetrads

Atlas period (breeding)	1985–90		2008–13		Change	
	Number	%	Number	%	Number	%
Confirmed	150	17	112	13	-38	-25
Probable	74	9	53	6	-21	-28
Possible	106	12	75	9	-31	-29
TOTAL	330	38	240	28	-90	-27
Tetrads with Winter Records (2007–13): 369 (42%)						

Breeding Distribution Change (1985–90 to 2008–13)

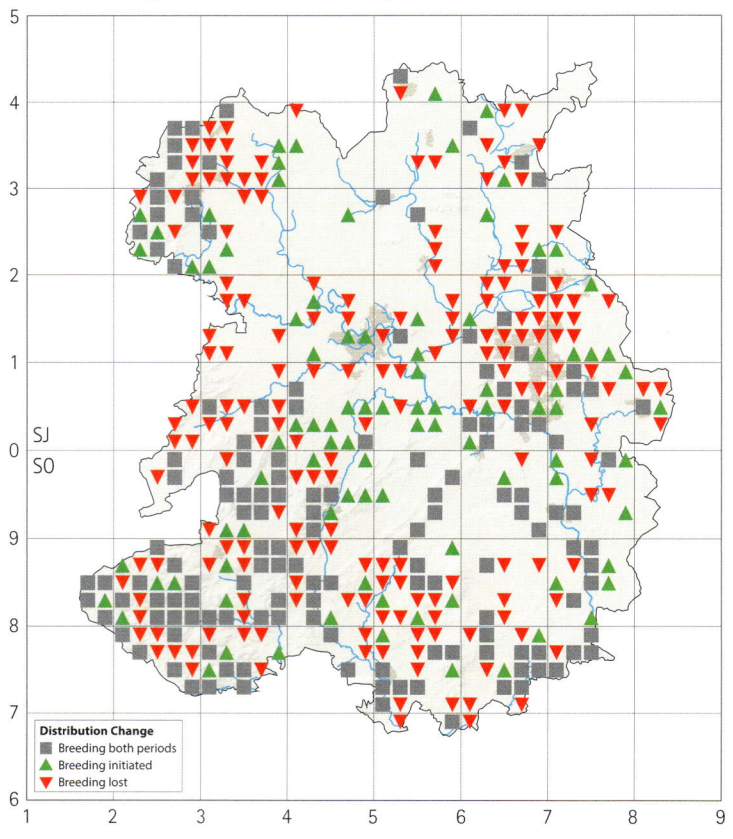

Distribution Change
- ■ Breeding both periods
- ▲ Breeding initiated
- ▼ Breeding lost

Winter Relative Abundance (2007–13)

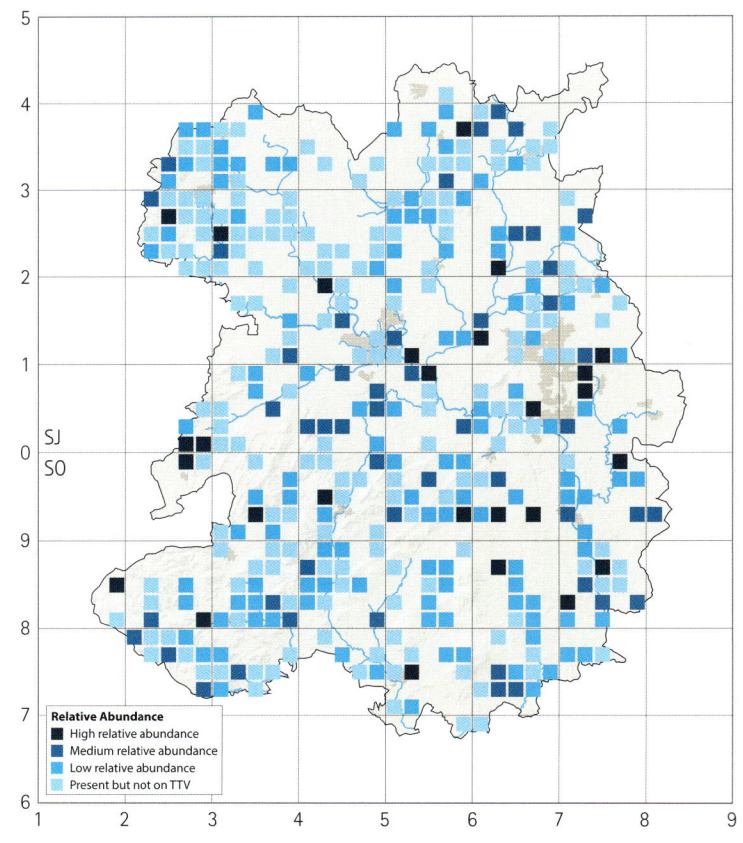

Relative Abundance
- ■ High relative abundance
- ■ Medium relative abundance
- ■ Low relative abundance
- ■ Present but not on TTV

population. Locally, in 2013 on the river Clun, none were seen along a regular breeding stretch until August.

Grey Wagtails share the upland habitat of fast-flowing streams with Dippers, but they feed on non-aquatic insect prey caught on the banks and in trees, as well as in the streams, making them less vulnerable to changes in water levels or quality. Even so, assessments to meet the requirements of the EU Water Framework Directive suggest that many rivers and streams where Grey Wagtails used to be found are now of poor ecological quality, probably also contributing to the reduction in the number of tetrads with records since 1985–90.

Based on TTV counts, the breeding population is estimated at 390–410 pairs. This figure is broadly similar to that in the Atlas (1992), which estimated 250–500 breeding pairs, although the large reduction in the number of tetrads with breeding evidence suggests that the population has declined considerably since 1990.

SBRs do not reflect the breeding status of this wagtail, as the vast majority of records come in the winter months from sites in the north that do not have breeding season records. Usually, less than 10 sites are reported each year, representing a very small proportion of the breeding population. Notable winter records since 1990 include 12 at the Mere (Ellesmere) in October 1993, 18 there again in October 1995 and 20 at Shrewsbury Sewage Farm in October 1999.

The winter relative abundance map clearly shows a more widespread distribution compared to the breeding season, with 50% more occupied tetrads, and the move to lowland areas is also reflected in absence from the Clee hills, Stiperstones and Long Mynd uplands. The highest numbers were found close to the central towns and other lowland areas, which have very few breeding season records. Numbers are augmented in winter when our upland breeding birds are joined at lower elevations by migrants from the northern UK and northern Europe. Farmyards, sewage plants and school playing fields are among the favoured habitats where they can sometimes be found.

Very few of those ringed have been recovered. A nestling ringed

Dave Barnes, Market Drayton, 16 May 2015

on the Long Mynd on 19 May 1984 was caught 3y 1m 4d later at Upper Cochran, Powys, 65km distant. Four ringed elsewhere have been recovered here, with three from the neighbouring counties of Clwyd and Powys. The notable exception involved a first-year ringed at Kindrogan, Tayside on 1 July 1998 which died hitting glass 460km south just over four months later at Haughton, near Shifnal, on 16 November 1998.

Hopefully Grey Wagtails will return to their former breeding sites, and the population will recover, particularly if the requirements of the EU Water Framework Directive, for all water bodies to be in good ecological condition by 2027, are met.

Pete Nickless

Pied Wagtail (White Wagtail)
Motacilla alba
Common resident

The resident sub-species *M.a. yarrelli*, which is largely restricted to the UK and Ireland, is the most common of our wagtails. After the breeding season, migrants from the highlands of northern Scotland and England pass through to winter in the lowlands to the south, as far as Iberia and north-west Africa. The pale grey nominate sub-species 'White Wagtail' *M.a. alba* is common throughout its wide Eurasian and north-west African range, but northern populations are migratory. Here it a scarce passage migrant in spring and autumn.

It was our most common wagtail in historical times too, but the early authors disagreed on its status. Beckwith initially considered it to be a partial migrant with a few remaining in winter but the greater portion moving south in autumn. He revised this assessment in 1890, when he considered it to be common in winter although less numerous than in summer, while in August and September large and small parties were found either flitting along the banks of streams or tripping about around cattle and sheep, catching insects disturbed

by the animals. Paddock (1897) agreed with Beckwith's assessment, but Forrest (1899) described it as 'common in summer, and nests here regularly. The numbers are less in winter owing, doubtless, to emigration'.

The next review of its status came when the *Handlist* (1964) noted Pied Wagtail as a 'resident and passage migrant, fairly common … especially about farms in the river valleys and open hill country … on passage in spring and autumn along the Severn valley, formerly in some numbers at Shrewsbury Sewage Farm'. Subsequently in 1979 it was noted as being sufficiently common in the breeding season to be overlooked by many recorders, and in 1980 and 1983 it was described as being common but less often recorded than the other wagtails.

The 1985–90 Atlas provided the first detailed review of its breeding status. It uses a wide variety of habitats where suitable nest sites in buildings and rocks, and insect food, can be found, including pasture farmland, and villages and urban areas, usually near to water. Flat

Breeding Relative Abundance (2008–13)

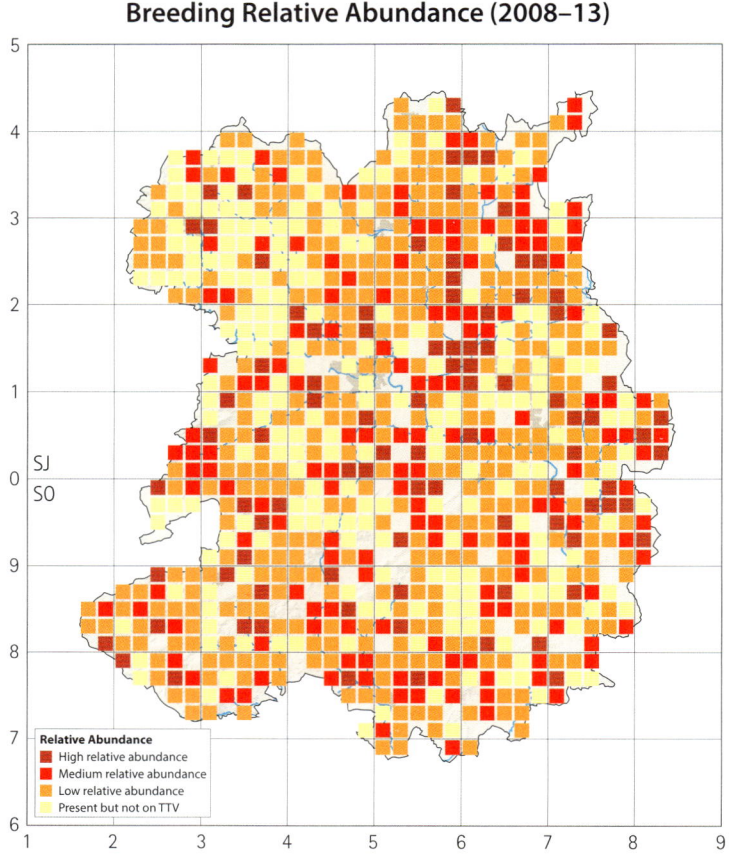

Relative Abundance
- High relative abundance
- Medium relative abundance
- Low relative abundance
- Present but not on TTV

Winter Relative Abundance (2007–13)

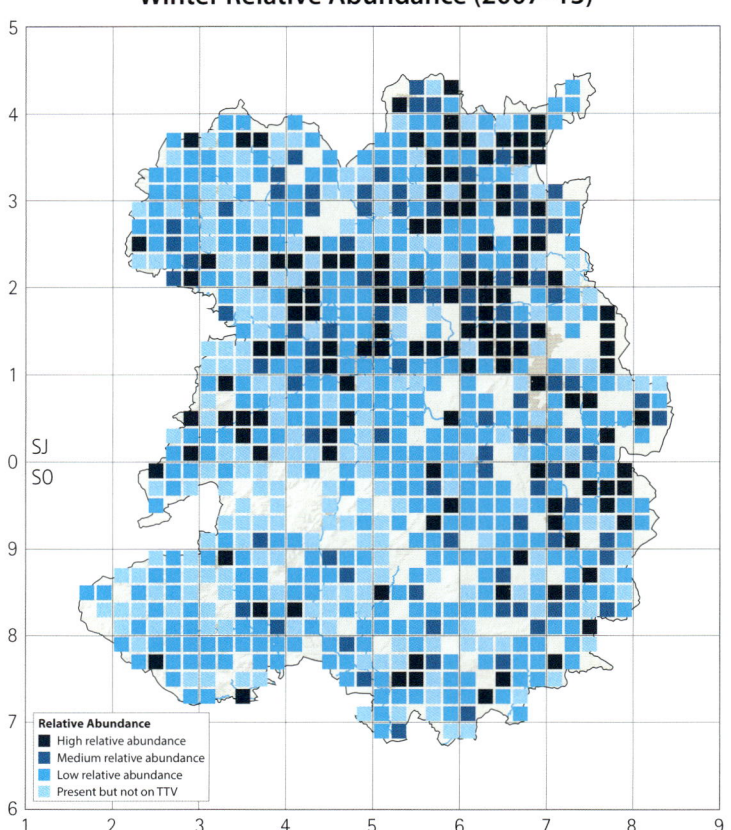

Relative Abundance
- High relative abundance
- Medium relative abundance
- Low relative abundance
- Present but not on TTV

Occupied Tetrads

Atlas period (breeding)	1985–90		2008–13		Change	
	Number	%	Number	%	Number	%
Confirmed	518	60	408	47	-110	-21
Probable	150	17	187	21	37	25
Possible	129	15	214	25	85	66
TOTAL	797	92	809	93	12	2
Tetrads with Winter Records (2007–13): 774 (89%)						

open areas to catch prey are essential. In 1985 a pair bred in an old House Martin's nest on Atcham Bridge.

Recent Atlas fieldwork found little change in distribution, or in the proportion of tetrads with evidence of breeding, although there have been more gains than losses in the north, and the large increase in possible breeding records suggests it has become harder to find as the population has declined.

The gaps, areas where only possible breeding has been established, and clusters of tetrads where it was not found during TTVs, are mostly associated with either intensive agricultural areas, particularly in the north and east, or the large conifer plantations in the south, both of which lack nest sites and flat open areas with insectivorous prey. Conversely, fieldwork found few clusters of tetrads with high relative abundance.

BBS shows an overall 30% decline between 1997–2014, but a large decrease of over 40% since a peak in 2000. This could be associated with a loss of summer food, resulting from a decrease in

cattle (down by 26% from 1985 to 2013 and 11% from 2000–13), and in the quantity of invertebrates in dung, due to the use of chemicals such as avermectins in farm animal husbandry, as well as from the more general increased use of pesticides, and their cumulative effect.

Based on TTV counts, the estimated breeding population is 5,000–5,100 pairs, at the upper end of the estimate in the *Atlas* (1992) of three to six pairs per tetrad, a population of 2,500–5,000 pairs. Given the scale of the decline found by BBS, it is likely that the population in 1990 was higher than the range quoted.

After breeding, and throughout the winter period, Pied Wagtails often gather in large communal roosts, which provide warmth and security from predators. Often several hundred strong, these roosts are commonly in reed beds but are also found in town centres, sewage farms and around other man-made structures.

The *Handlist* (1964) noted two roosts, one on the glass and girder roof of Sentinel Works, Shrewsbury, with up to 100 each year from September–February, and another of about 1,000 in reed mace at Wombridge, from 18 October to mid-November 1958.

From August–October between 1987–93, Cosford Airfield held large flocks, with 1,000+ passage migrants present during September in 1987, 1991 and 1992. Notable winter roosts have included 400 on the GKN Sankey roof at Hadley in 1976, 750 on factory buildings at Stirchley in 1978, 500 on the Creamery at Minsterley in 2000 and over 500 in the Baker Street area of Shrewsbury in 2005.

In recent winter Atlas fieldwork, Pied Wagtails were found in fewer tetrads than in the breeding season, as they vacate many arable areas, and the uplands, and congregate where winter food and suitable roost sites are available. However, the total TTV counts were almost 40%

BBS Trend (1997–2014)

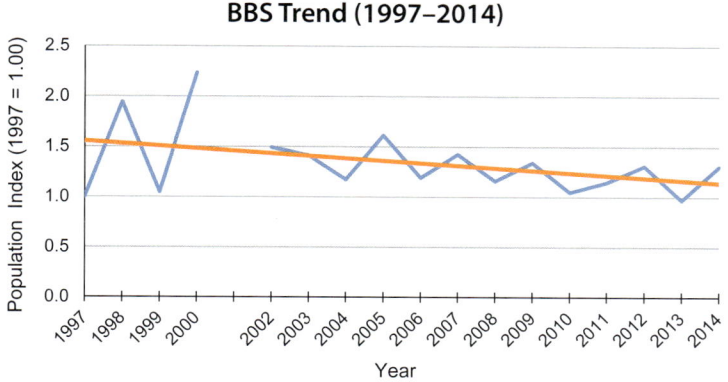

Terry Arch, Broseley, 20 March 2010

higher in winter than in the breeding season, as a result of additional juveniles and winter visitors.

The size of regular recent urban roosts vary from year to year, but those over 100 include Tesco's roof at Ludlow, peaking at 300 on 26 January 2009, 105 on Harry Tuffin's roof at Craven Arms on 24 September 2012, and 250+ in Shrewsbury Town Centre in February and March 2012, where they favoured a small birch tree. The Shrewsbury roost moved to Tesco at Harlescott in 2014, with 200–300 on 4 November.

Away from the more built-up areas, regular flocks up to 90 each winter were found at the Rea, Upton Magna, while smaller ones included those at Priorslee and Whixall floods. The reed bed at Chelmarsh regularly supports roosts before and after the breeding season, with up to 300 during March and April 2010, and 200–300 on 14 July the same year.

About half of the ones ringed here have been recovered locally and several of the remainder in nearby counties, reflecting the sedentary nature of the resident population. However, four were found on the south coast, and eight in winter months on the continent, in France (three), Portugal (two) and Spain (three), while two were found in the breeding season in Scotland. There is a similar pattern in the six ringed elsewhere and found here, suggesting that many pass through on their way between breeding sites in Scotland and wintering areas to the south.

All five found in Spain and Portugal were over 1,000km from their ringing site, the furthest being a first-year ringed in July 1968 at Leegomery, trapped in November the same year at Faro (Portugal), 1,776km distant, 4m 9d later.

The continental race White Wagtail *M.a. alba* has been recorded occasionally on passage in spring, with six sites noted in the *Handlist* (1964): 'Most records are of male birds'. This race was also recorded during the recent Atlas period on migration each spring and autumn, mainly in the Severn Valley. Spring migration produced most records, especially in April, which would be expected as there was no Atlas fieldwork in the autumn. Sporting a pale grey mantle, they are easily distinguishable from dark-backed residents. Later in the year, with paler juveniles to contend with, identification is not as obvious, but nevertheless small numbers were found between August–November each year. Although flocks are never as large as those passing through some of the country's migration hotspots, there were six at Allscott Sugar Factory on 20 April 2008, seven at the same site on 20 April 2010, and seven at Venus Pool on 16 April 2012.

Pete Nickless

Richard's Pipit
Anthus richardi
No modern records

The single record of this large pipit, a rare Siberian vagrant, is of one caught in a lark net at Shrawardine on 24 October 1866. It was taken alive to the taxidermist John Shaw in Shrewsbury and passed into the collection of Mr T. Bodenham (Beckwith 1878).

John Tucker

Meadow Pipit
Anthus pratensis
Fairly common resident

Meadow Pipit is the ultimate common 'LBJ' ('little brown job', with few distinguishing features), and the species that, while fell walking, first got me interested in birds. It is increasingly restricted to upland heath, moorland and bog, and, although widespread in central and northern Europe, its IUCN status is now 'near threatened'.

It was referred to by all the early authors, and Beckwith (1879) described it as 'resident, though the greater portion migrate during the autumn, and few only remain with us for the winter', adding in 1890 that 'in summer, it abounds on high hilly ground, and the mosses and bogs about Whixall and Ellesmere, and it is very plentiful on Whixall Moss, the Clee Hills, the Longmynds and other such wastes, where for five months it is the most numerous bird inhabitant', although Paddock (1890) said it was 'not common around Newport' in the breeding season.

Dave Barnes, Clee Hill, 27 May 2013

Meadow Pipit is undoubtedly the most numerous species on the Long Mynd, and transect surveys were carried out in the three years 1996–98 to estimate numbers. Eight transects, each 1km long, were surveyed three times each year, and the results analysed using CBC methodology: 68, 47 and 48 territories were estimated for the three years respectively, with the average territory size 1.2–1.7ha. The transects effectively represented around 8% of the total plateau area, giving a population estimate of 550–950 territories for the plateau. This is consistent with the estimate obtained by multiplying the estimate for Skylark with the Meadow Pipit:Skylark ratio of about 4:1 found on the transects. Adding in an allowance for those in the heathland and wet flushes in the upper reaches of the valleys gives an estimate of 600–1,050 breeding pairs for the whole of Long Mynd. The transect surveys were repeated in 2002–04, and again found large fluctuations, but the average for each period was broadly the same (Tucker 1998; 2004).

A less thorough count of birds, not territories, on four 1km transects on the Stiperstones in 1995 and again in 2004 found little difference between the two years, and a comparable density to that on Long Mynd (Smith 2007). Suitable breeding habitat is managed heathland (about 200ha) and acid grassland (about 39ha) on the NNR, and 34ha for the former Gatten Plantation, giving an estimate of 170–230 pairs. Otherwise, Meadow Pipit was not surveyed on the NNR, but at least 32 territories were found on Cefn Gunthly, Heath Mynd and Black Rhadley.

The breeding distribution change map shows that the Long Mynd and the Stiperstones ridge are still the strongholds, and they, together with other parts of the south-west hills, the Clun Forest, Brown Clee

Several early authors cite examples of Meadow Pipit being a host to Cuckoo, and it is the third most numerous in records from 1879–2014.

These early descriptions suggest that Meadow Pipit was more common in the nineteenth century than in the second half of the twentieth. Initially it would have been affected by land drainage and the loss of lowland heath, then by the switch from largely unimproved pasture to arable farming in the war years (see p. 39), and then by 'improvement' of upland pasture, where in the Clun Forest 'the laying down of new leys has embarrassed even the tenacious Meadow Pipit, which now nests in small colonies in the cramped areas which have so far escaped the "blitz"' (Adams 1950).

The *Handlist* (1964) described it as a resident and passage migrant, common on the high ground in the breeding season but less common elsewhere. 'Several observers consider it has decreased as a breeding species in recent years'. Formation of flocks on high ground in August and September, and an influx onto low ground and large migratory flocks in the latter month, were noted. Winter records were not mentioned at all.

The 1968–72 BTO Breeding Atlas showed it absent from two 10km squares in the north-west and three in the south-east, but by 1985–90 it was absent from four more in the north, so the contraction of range through loss of habitat had continued. The population is likely to have declined too during this period, as, nationally, CBC showed a decline of 39% between 1970 and 1994, although the estimate is not considered reliable because of poor coverage of upland areas.

Breeding Distribution Change (1985–90 to 2008–13)

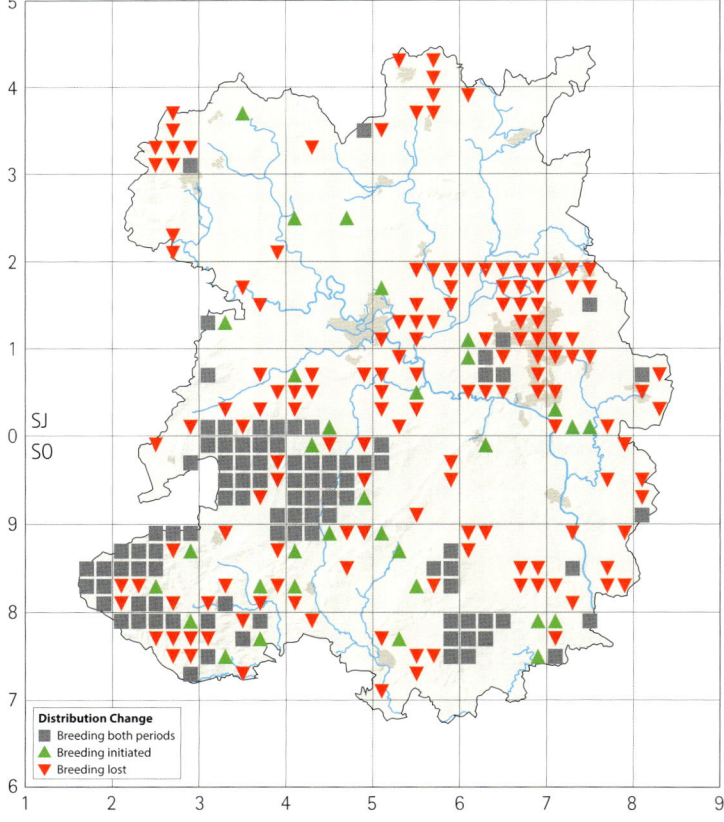

Distribution Change
■ Breeding both periods
▲ Breeding initiated
▼ Breeding lost

and Titterstone Clee, have the highest abundance. However the relative abundance map also shows high densities in several tetrads in the north and east where large migratory flocks were counted in early April, so to include it would give a misleading impression.

In the north (not including tetrads at the northern end of the south-west hills), confirmed breeding was found only at Oswestry Hill Fort, Whixall Moss, Long Mountain and the Wrekin, with probable breeding at Lilleshall. Densities were low, for example four to eight pairs at the Hill Fort. All other records from the north were in early April, and probably of migrants, including some tetrads shown as a gain on the distribution change map. Although some of the apparent losses are due to migrants having been mapped as possibly breeding in 1985–90, other sites such as Oswestry Racecourse, the Weald Moors, Roden Valley and derelict land in and around Telford had confirmed breeding records, and these have been lost. There were three to four pairs at Oswestry Racecourse in the late 1980s and for most of the 1990s, but gradual succession from open grassland to scrub and

woodland has removed the habitat. Other examples of habitat loss include drainage and 'improvement' for agriculture, and, in the south-west, for pasture or fodder crops, as well as building development on derelict land in Telford.

Conversely, numbers on Long Mynd and Stiperstones are likely to have increased since 1985–90 through habitat improvement and re-creation, the reduction in grazing on the former (see p. 50) and the Back to Purple project at the latter (p. 51). Meadow Pipit is the staple diet of Merlin, and the regular high productivity of the raptor on the Long Mynd suggests that the population there is healthy.

Confirmation of breeding is relatively easy, so discounting possible breeding records in both Atlases, most of which were probably migrants, there has been a 44% reduction in range.

The population fluctuates, with hard winters and cool wet springs depressing numbers. Areas with the highest abundance in uniformly suitable habitat may have 100 pairs per tetrad, but the average will be much lower, giving an estimate of 1,500–2,000 breeding pairs, consistent with the estimates for Long Mynd and Stiperstones, but substantially less than the estimate of 2,500–5,000 in 1990. The estimate based on TTV counts is 4,400 breeding pairs, which is likely to be too high because of the inclusion of migratory flocks in April.

Meadow Pipit occurs on too few local BBS squares to calculate a trend, but nationally there has been little change, declines of 9% in the UK and 7% in England, and a rise of 5% in Wales, between 1995–2014. This supports the belief that populations are not likely to have declined considerably in areas where habitat has not been lost.

In winter, the British breeding population moves south, mainly to Iberia but as far as Morocco, but there is no evidence from ringing that continental populations winter here, suggesting that the much more widespread distribution, with almost four times as many occupied tetrads, is due to movements from the uplands onto lowland arable farmland in the east and north-east, which has the highest relative abundance. There has been no apparent change in distribution since the 1981–84 BTO winter Atlas, which mapped this Pipit as present in all 10km squares wholly or partly in the County, mostly at low density.

Flocks form to forage on stubbles. There were counts of 20 or more in 33 tetrads (22 of them in the north) on one-hour TTVs, and similar numbers on casual records from 14 other tetrads. Highest counts were 100 on a TTV, and 63 on a casual record. Some remain in the uplands, and this habit has increased in recent years, certainly on the Long Mynd, attributed to milder winters, but there were counts of 20 or over in only four tetrads in the south-west.

As few remained for the winter in the nineteenth century, the *Handlist* did not give any winter records, and the BTO winter Atlas (1986) mapped it in only half the 10km squares, and then at only the lowest density, it appears that the current widespread distribution is the highest for many years, and reflects the trend to milder winters.

Few are ringed, and the two recoveries shed little light on movements or longevity. A full-grown individual was ringed at Allscott in January 1963, and found dead at the same site two days later, while a nestling ringed at Clee Hill, in July 1998, was killed by a cat in Ludlow, 9km WSW, 21 days later.

Hopefully the remaining breeding habitat in the south and south-west hills is relatively safe from further agricultural 'improvement', and the rate of population decline will slow down.

Leo Smith

Occupied Tetrads

Atlas period (breeding)	1985–90		2008–13		Change	
	Number	%	Number	%	Number	%
Confirmed	88	10	62	7	-26	-30
Probable	68	8	25	3	-43	-63
Possible	87	10	40	5	-47	-54
TOTAL	243	28	127	15	-116	-48

Tetrads with Winter Records (2007–13): 457 (53%)

Winter Relative Abundance (2007–13)

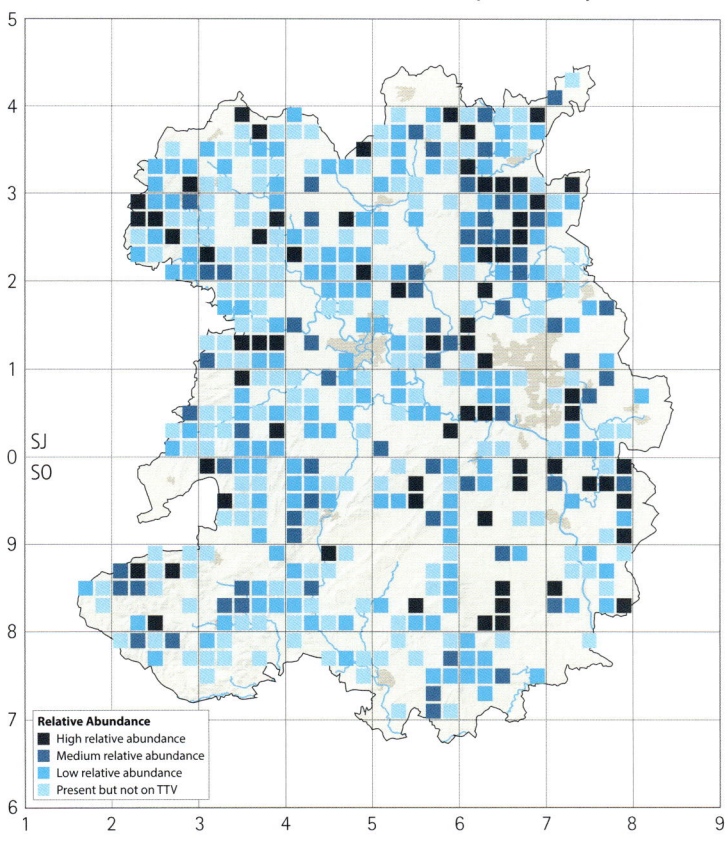

Relative Abundance
- High relative abundance
- Medium relative abundance
- Low relative abundance
- Present but not on TTV

Tree Pipit
Anthus trivialis
Uncommon summer visitor

Jim Almond, Long Mynd, 6 June 2010

Tree Pipit breeds across most of Europe and temperate western and central Asia, and winters in Africa in wooded country south of the Sahel. It returns here in small numbers from mid-March onwards, becoming widespread by the end of April.

In the nineteenth century it was described as 'frequent' in the vicinity of Shrewsbury, and a summer migrant, frequenting all cultivated areas, and valleys, parks, along hedges and in fertile areas where there are numerous scattered trees (Beckwith 1878; 1879; 1890). It was 'plentifully distributed' (Paddock 1890; 1897), and 'common in summer, but less plentiful than the Meadow Pipit' (Forrest 1899). It was mapped as 'common' here at the end of the century (Holloway 1996).

Holloway suggested a decline over much of central and south-east Britain from the late nineteenth century onwards, perhaps reflecting a decline in coppicing. This decline had not set in locally, as Beckwith commented in 1890 that 'perhaps the Tree Pipit is now more numerous than in former days. Eyton, writing of it some 50 years ago, did not consider it common in this district', but after another half-century Lloyd (1943) included it in a list of 'species that have shown a less marked or local decrease', and described it as 'still numerous in some districts, but has shown a definite diminution in others'.

The *Handlist* (1964) provided little detail, and referred to it only as a 'breeding visitor, well distributed in suitable habitats', but seldom above the 1,200ft contour, while the 1968–72 BTO Breeding Atlas mapped it with breeding evidence in all except five of the County's 33 10km squares, four of which were in the north. During this period it was 'less plentiful than previously' (SBR 1970) and there were 'fewer records' (SBR 1971), and the gaps shown in the BTO Atlas suggested

a further decline compared with the descriptions of the nineteenth-century authors.

The decline continued, with the 1985–90 Atlas showing it absent from a further two 10km squares in the north, including for example SJ62, where it had previously been 'common in suitable habitats in the Hodnet area' (SBR 1956) and it was also absent from SJ43, except at Whixall Moss, although it had been 'common in new plantations east of Ellesmere' (SBR 1957). By then, most occupied tetrads were in the south and south-west, where the upland topography with copses, wooded valleys and rough pasture with scattered trees provided ideal habitat. Although Tree Pipit and Meadow Pipit occupy many of the same tetrads, and both nest on the ground and need open areas for feeding, the habitat requirement of the Tree Pipit for prominent song posts is more restrictive.

The populations of Clun Forest, Stiperstones, Long Mynd and Stretton hills merged to form the large concentration in the south-west. Further east they were found on Wenlock Edge, Catherton Common and Brown Clee. To the north, the Wrekin, Haughmond Hill and Oswestry uplands were the main centres, with lower numbers on the smaller hills at Lilleshall, Edgmond, Nesscliffe, Middletown and Loton. In the south-east they occurred in the Wyre Forest and some of the smaller woods nearby. Whixall Moss, represented by a single tetrad in the north, provided a different habitat with birch scrub encroaching onto lowland bog.

Densities can be high in suitable habitat, for example 15 singing males at Stow Hill and at least 20 more at Black Mountain (SBR 1989), and about 30 pairs at Brown Clee and 25 pairs at Long Plantation (Clun Forest) (SBR 1990).

The Long Mynd BBP confirmed the habitat preference as scattered trees (including hawthorns) on the steep valley sides, with ground vegetation of mixed bracken, heath and grass having the highest density. This density was 50% higher than bracken with some heath, and twice that found in thick bracken, or heath with scattered bracken, consistent with the requirement for areas with only sparse vegetation cover for feeding. The population was estimated at 90–100 pairs in 1998 (excluding those using hedgerow trees and bushes along the National Trust boundary), and around 70 in 2006–08, a 20% decline. However, the number counted each year on a BBS transect in Callow Hollow over the same period fluctuated considerably, between one and seven individuals, but the overall trend showed no decline, and neither has there been a decline since.

The long-running CBC plot at Oswestry racecourse first recorded Tree Pipit in 1969. Numbers built up to nine pairs in 1987, but dwindled during the 1990s, with none at all in 1997 and 1999, largely due to habitat loss through gradual scrub encroachment.

There was little change on the Stiperstones NNR between 1995–96 and 2004–05, with around 24 being found in each period, but the total population on the ridge increased by about 15 pairs (60%) due to new habitat being created on the heathland restoration areas by the felling of mature conifers on Gatten Plantation, and around Nipstone Rock and the Rock, as part of the *Back to Purple* project

(p. 51). A further 12–13 were found on Black Rhadley Hill, three on Cefn Gunthly and five on Heath Mynd in 2006, and five on the Hollies and six on Brook Vessons in 2007, giving a total of 67–70 on the whole Stiperstones ridge.

Tree Pipits respond quickly to new nesting opportunities, such as new plantations or clear-felled woodland, provided that suitable song posts are available, as evidenced by the colonisation of the heathland restoration areas. Also, there were three displaying males in a triangle of clearfell in the Long Mynd plantation in 2009, separated by neighbour distances of 150m (Alan Reid *pers. comm.*). Neighbour distances on the most densely populated parts of the Long Mynd and Stiperstones were rather less than this. They will leave again if conditions become unfavourable, as trees become too dense, or open feeding areas are lost. While habitat on Forestry Commission clearfell will eventually be lost, that on the Stiperstones is likely to be maintained through management by light grazing. On the Long Mynd, encroachment by increasingly thick bracken may have contributed to the decline between 1998–2008, and may be a continuing threat in future.

The 2008–13 Atlas found a 50% reduction in range compared with

the 1985–90 Atlas. The strongholds around Long Mynd, Stiperstones, Brown Clee, Titterstone Clee and Catherton Common, the Wrekin, and Wyre Forest, remained, but there were considerable losses around the edges of the Clun Forest, and in the hills north and north-east of the Stiperstones, and the Oswestry uplands, while populations in the Severn Valley and on Wenlock Edge, the Weald Moors and the outcrops on the northern plain have been lost altogether. Only possible breeding was found on Haughmond Hill, from where there have been no other records for many years, although several pairs used to breed there. Similarly there were only records of passage birds from Whixall Moss, so another breeding site has probably been lost, although there was a record of a single pipit on 26 May 2010 from SJ43Y, a tetrad largely in Wales and not included in the Atlas (although this individual was on our side of the border). The general decline has been exacerbated by habitat loss on the NNR because of an increase in thick purple moor-grass, the result of three times the recommended amount of atmospheric ammonia being deposited on the reserve from local sources, such as poultry farms and dairy farms (David Tompkins *pers. comm.*).

Far fewer records have been received in the years in the twenty-first century, compared with previous decades, and none include such high counts as those at the end of the previous Atlas period, leading to the conclusion that 'the limited evidence we have seems to suggest a serious decline locally' (SBR 2002). Only 10 tetrads had counts of more than five in the recent Atlas period (maximum only seven), a far cry from the numbers only 20–25 years ago.

Most of the apparent gains shown on the breeding distribution change map were only possible breeding records, and were therefore

Occupied Tetrads

Atlas period (breeding)	1985–90		2008–13		Change	
	Number	%	Number	%	Number	%
Confirmed	76	9	25	3	-51	-67
Probable	74	9	50	6	-24	-32
Possible	73	8	37	4	-36	-49
TOTAL	223	26	112	13	-111	-50

Breeding Distribution (2008–13)

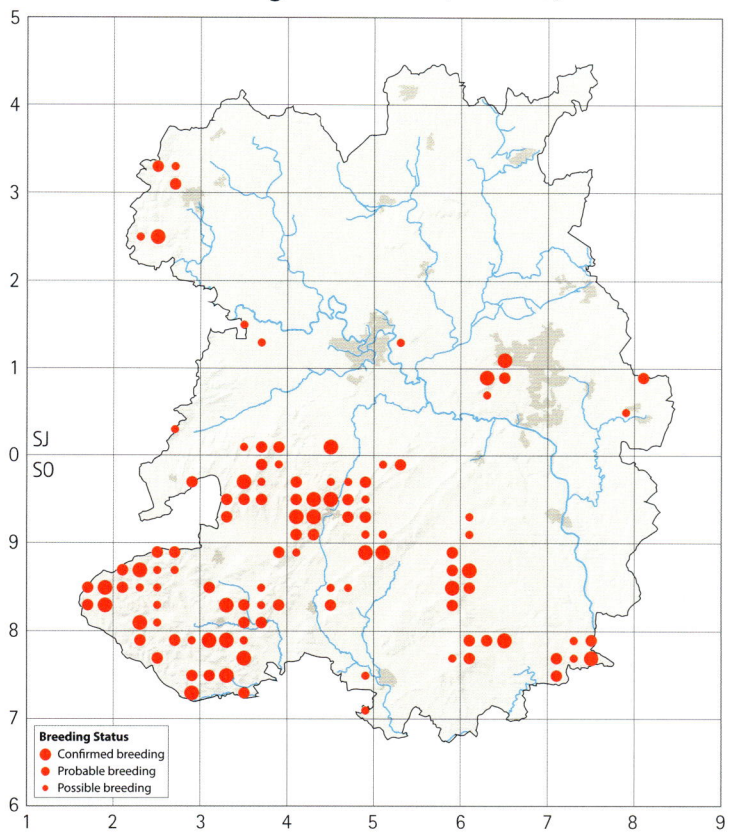

Breeding Status
● Confirmed breeding
● Probable breeding
• Possible breeding

Breeding Distribution Change (1985–90 to 2008–13)

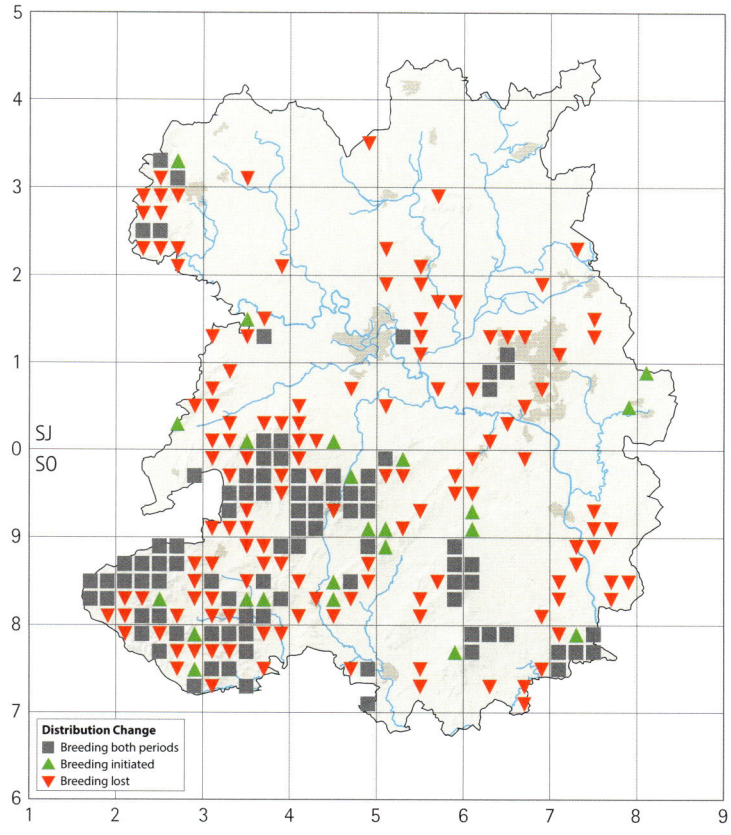

Distribution Change
■ Breeding both periods
▲ Breeding initiated
▼ Breeding lost

likely to be passage, rather than breeding, birds. Due to the transient nature of some breeding sites, the map may slightly exaggerate the distribution in any one year.

Nationally, Tree Pipit has suffered a steep population decline since the 1980s. Population change monitored by BBS shows a considerable variation across the country, but a 44% loss in England and 4% loss in Wales during the period 1995–2014. The population fluctuates for unknown reasons, but it does not appear to be affected by droughts in the Sahel itself. However, it is one of several long-distance migrants that have experienced a large decline in recent years, so factors on migration routes or wintering grounds are likely to be important. The *Arrival Dates* chapter suggests that this species now arrives around four days earlier than it did in 1900. This is a relatively small advance, so climate change may have contributed to the decline, if its arrival is no longer synchronised with peak food supply. Nationally, the areas of highest relative abundance have shifted to northern Scotland, also suggesting that climate change may be a factor. Encroachment of scrub and bracken onto open feedings areas, reducing breeding density, may also be having an impact.

Atlas fieldwork 1985–90 suggested an average of six to 12 pairs for each tetrad with probable or confirmed breeding, giving a population estimate of around 900–1,800 pairs in 1990. The comparable figure in 2013 is half that, 450–900 pairs. The populations on the Long Mynd and the Stiperstones fall into nine tetrads each, a minimum density of around eight pairs per tetrad. Applying this to the 112 tetrads with records suggests a population in the order of 900 pairs, although most tetrads will have less and some of the possible breeding records will be of migrants, so this figure will be at the upper end of the current range. Applying it to the 75 tetrads with probable and confirmed breeding gives a more realistic 600. The estimate based on TTV counts is 700–800 pairs.

The national ringing scheme started in 1909, and one of the earliest recoveries was of a nestling ringed on 11 June 1914 near Shrewsbury

and found dead on 29 September 1916 at Braga in Portugal, 1,302km SSW, 2y 3m 18d later.

The only modern recovery was of an adult ringed on 4 June 1983 near Kinlet, caught on its return migration the following May at Dungeness, 288km south-east and 11m 6d later, by which time it had presumably travelled to central Africa and back.

Until the reasons for the large decline are better understood, it is likely that the decline will continue, nationally and locally.

Leo Smith

Jim Almond, Black Hill, 26 April 2008

Water Pipit
Anthus spinoletta
Very rare passage migrant and winter visitor

Water Pipit breeds mainly in Alpine regions in Europe and southern Asia, and migrates short distances to lower altitudes or wet open lowlands in winter, when small numbers visit British coasts. It is less common inland. It was only recognised as a separate species, rather than a sub-species of Rock Pipit, as recently as 1987. Separation of the records submitted prior to that date is outlined in the Rock Pipit account. All records included in this account have been positively identified as Water Pipit, but it is possible that some seen before 1987 are still listed as Rock Pipit records.

Water Pipit occurs very infrequently, with only 22 records since 1950. All except two have been of singles, the exceptions being the first record, two together at Shrewsbury Sewage Farm on 6 October 1957, and two again on the sixth, from Allscott Sugar Factory on 29 January 1983. Eight of the nine records from before 1987 came from these two sites, four from each, the exception being the last, from

Chelmarsh in 1986. More than half the sightings are from the 1980s, and only four from this century. This pattern, shown in the annual occurrence chart, probably reflects the trend to milder winters, with fewer crossing the channel.

All were seen on one day only until the tenth record in 1987, but five have stayed for three weeks or longer since then: at Chelmarsh from 8 February until 20 April (72 days), and at Allscott Sugar Factory from 28 February to 29 March (30), both in 1987, at Shrewsbury Sewage Farm from 30 December 1988 to 11 March 1989 (72), and two at Allscott Sugar Factory, from 21 October to 10 November 2001 (21) and from 25 October 2003 to 11 January 2004 (79 days, the longest stay). Of the other eight recorded since 1987, five were seen on one date only, and none of the other three stayed for more than five days.

All except three records have come from the three sites named above, the exceptions being singles at Leighton in January 1995,

Estimated Number per Year (1950–2014)

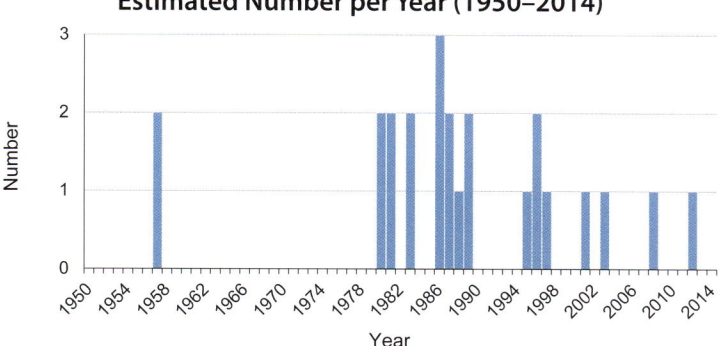

Venus Pool in April 1997, and Titterstone Clee in April 2012.

All arrivals have fallen into three periods, 8 February–14 April (eight on spring passage, including two long-stayers), 6 October–22 November (eight on autumn passage, including two long-stayers, one of which stayed for 79 days and well into the winter period), and 30 December–29 January (eight winter visitors). All the five that stayed for three weeks or more arrived in the years 1987–2003.

Leo Smith

Jim Almond, Whixall Floods, 26 March 2016

Rock Pipit (Eurasian Rock Pipit)

Anthus petrosus

Very rare passage migrant

Rock Pipit inhabits rocky coastlines in Western Europe, and the British resident population is supplemented by winter visitors from the continent. Until their split in 1987, Water Pipit was considered to be a sub-species of the Rock Pipit. The two species can be separated in the field, and most records from before 1987 identified the sub-species, or provided sufficient detail to enable the SOS Rarities Review to determine the species. However, insufficient detail was provided in

Jim Almond, Chelmarsh Reservoir, 7 October 2011

some cases, so some of the nine individuals described as Rock Pipit below, but seen before 1987, may possibly have been Water Pipit.

Forrest (1899) described Rock Pipit as 'properly a shore bird, but an immature specimen was killed at Berwick' in November 1877. There appears to be a long gap up until the first of 22 modern records, from Marton Pool (Chirbury) on 5 February 1961, with other singles at Shrewsbury Sewage Farm in October and Allscott Sugar Factory in November, both in the same year. Only four of the records are of two individuals, with no higher counts: three from Allscott Sugar Factory, in 1963, 1967 and 1988, and the most recent from Chelmarsh in October 1989.

All except three were seen on one day only, and none was seen for more than two days. This contrasts with Water Pipit, a few of which have been long-stayers. Seventeen of the 26 individual Rock Pipits were seen in October, five in November, none in December, one each in January and February, and two in March. All except three of the 12 seen since 1992 were in October.

Only nine of the 26 individuals occurred before 1987, and all except the first two were at Allscott Sugar Factory. From 1989, there has been a slight increase in the frequency of sightings, from Allscott Sugar Factory (eight), Chelmarsh Reservoir (six), Titterstone Clee, White Mere and Wood Lane. The penultimate record, from Titterstone Clee, was in March 2009, otherwise all records this century were from the two main sites, with Chelmarsh providing the most recent, on 7 October 2011.

The British resident population is sedentary, so the prevalence of October records suggests that the few seen here are of Fennoscandian origin moving through to Welsh coasts.

Leo Smith

Chaffinch (Common Chaffinch)
Fringilla coelebs
Very common resident

The breeding range extends across central and southern Europe into North Africa. It is a partial migrant, those from the north joining residents further south for the winter.

In the nineteenth century, the Chaffinch was described as common or abundant in both the breeding and wintering seasons, and the commonest of the finch species. Two writers referred to winter flocks being dominated by males, which was thought to be caused by females leaving the area rather than males arriving from further north, as is now known to be the case.

Today the Chaffinch is still the commonest of the finches. During the breeding season it is found in a wide range of habitats, equally at home in woods, farmland and built up areas; indeed anywhere that has trees or bushes is likely to provide a home. Breeding is easily confirmed by watching adults carrying food but even a quick visit to a square should produce records of singing males on territory.

One of our most widespread species, they were found breeding in every tetrad in the recent Atlas. Although they were missing from two squares in the Atlas (1992), it seems likely that they were overlooked then and there has been no change in distribution.

They are most abundant in the north, west and south-west, and are scarcer in the south-east quadrant roughly between Shrewsbury, Telford and Highley, although there is much variation within these areas. This may be a reflection of the amount of trees and scrub

Celia Todd, Pant Glas, 20 November 2013

within each individual tetrad, as areas with large fields and few or poor quality hedgerows are likely to have lower densities.

Since 1997, the BBS trend has fluctuated but it reached its highest point in 2005 and 2006. A sharp fall in the index in 2008 was triggered by an outbreak of trichomonosis, reported by many garden bird-watchers at that time. The downward trend has continued and in 2014 numbers were only 68% of the level prior to the outbreak.

The Chaffinch is a regular visitor to gardens, particularly during the winter months, when food is often provided. At this time, garden numbers are influenced by both the weather and the availability of natural food supplies. As spring progresses, gardens lacking suitable nesting opportunities may become deserted. The Garden BirdWatch graph shows a similar decline to the BBS, but during both the winter and breeding periods, following the onset of disease.

Breeding Relative Abundance (2008–13)

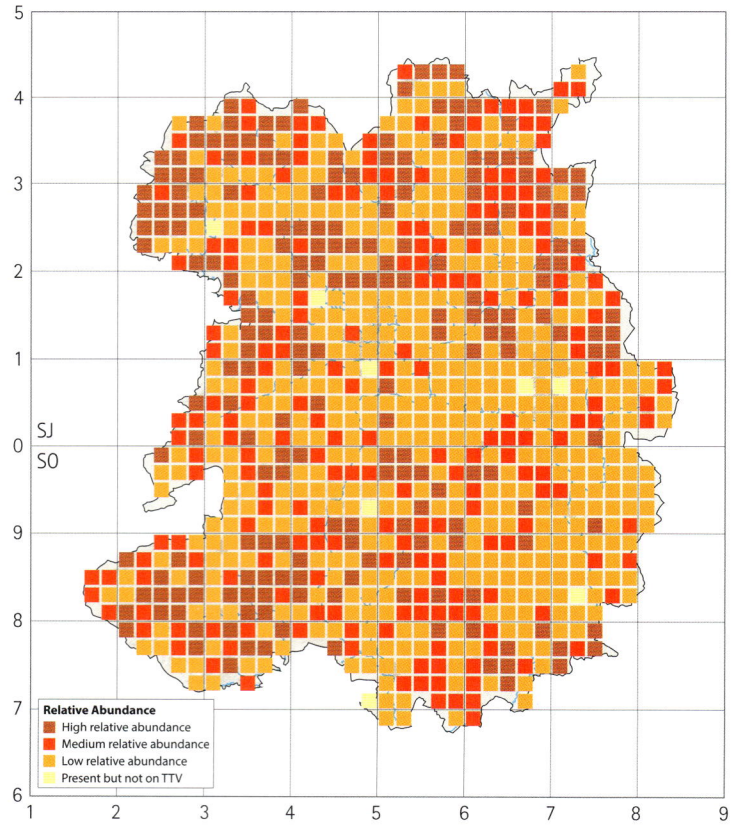

Relative Abundance
- High relative abundance
- Medium relative abundance
- Low relative abundance
- Present but not on TTV

Occupied Tetrads

Atlas period (breeding)	1985–90		2008–13		Change	
	Number	%	Number	%	Number	%
Confirmed	747	86	523	60	-224	-30
Probable	109	13	326	37	217	199
Possible	12	1	21	2	9	75
TOTAL	868	100	870	100	2	0
Tetrads with Winter Records (2007–13): 869 (100%)						

BBS Trend (1997–2014)

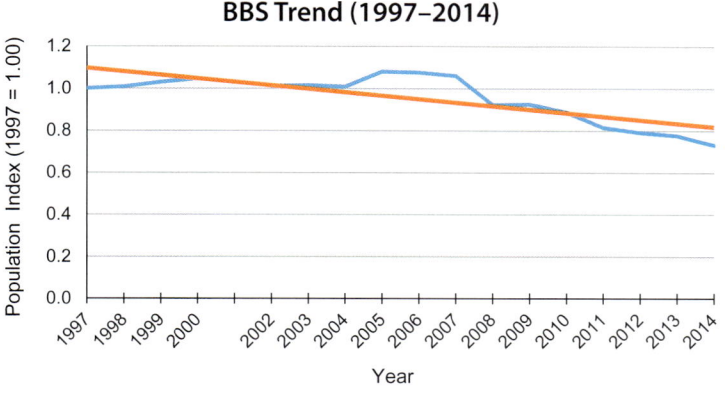

Garden BirdWatch Reporting Rate (1995–2014)

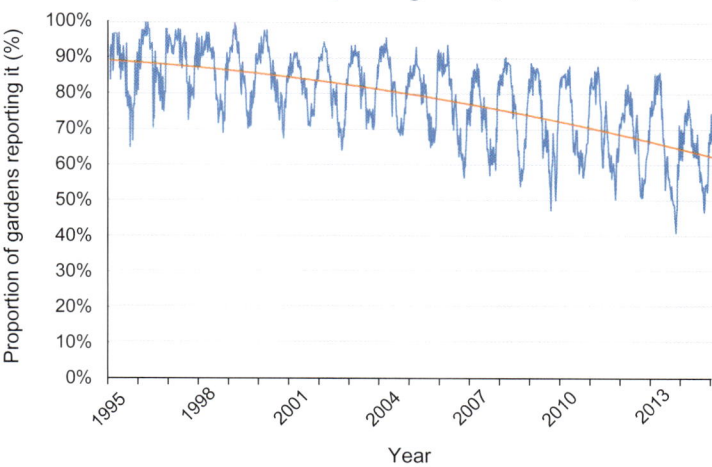

Based on TTV counts the population was estimated to be between 85,000–90,000 pairs in 2009, but due to the continuing effects of disease, as illustrated by the BBS, a figure of between 66,500–71,000 in 2014 is more appropriate. This equates to 19–20 pairs per sq.km, half the estimate used in the *Atlas* (1992), suggesting that the previous estimate was too high, exaggerating the apparent loss.

During the winter, residents are joined by visitors from northern Europe but, despite the increased numbers, they can be more difficult to find at this time. Although easily found in towns and villages, flocks feeding in large fields or under trees can be missed, depending on the route taken. Despite this the only tetrad to draw a blank was on the top of the Long Mynd, and they were only missed by the TTVs in one other square. During the winter the TTV totals were 160% higher than those in the breeding season which would suggest a population

Winter Relative Abundance (2007–13)

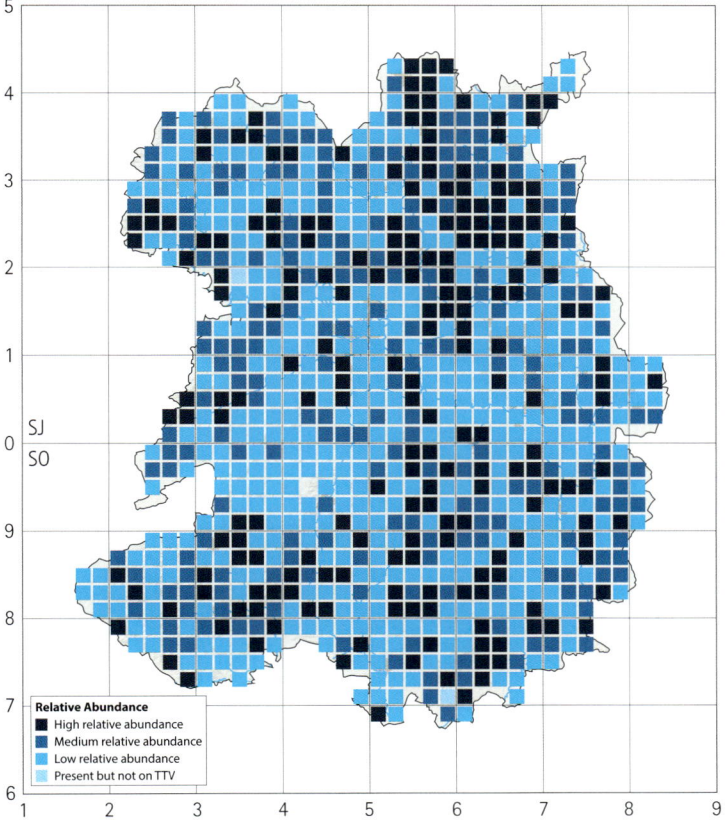

Relative Abundance
- High relative abundance
- Medium relative abundance
- Low relative abundance
- Present but not on TTV

of between 213,000 and 227,000 individuals. This will include both young of the year and migrants.

Winter flocks often congregate on arable farmland to feed, especially in years when the seed crop in woodland is poor or has been depleted as winter progresses. Because of this they were more evenly distributed during the winter and, although they remained abundant in the north-east, relative abundance in the south-east also increased.

Movements to and from Norway, Sweden and Finland indicate that these countries are the likely origin of our winter visitors, and they account for 43 (30%) of the overseas recoveries. Four ringed here have been found over 1,800km away in Finland, the furthest moving 2,079km from Hinstock in 1986. Although the majority, 86 (60%) of the 142 overseas movements recorded, have involved Belgium, the Netherlands and Germany, during their spring and autumn migration the Scandinavian breeding population crosses the North Sea where it is at its narrowest, which funnels them through these countries.

The resident population is mostly sedentary, and the vast majority of recoveries were ringed locally. Of the remainder, 92 were found abroad and 63 were in Britain from Cornwall to the Borders. The paucity of recoveries from further north is surprising, but it adds further evidence that most migrants cross the North Sea to the south-east, where it is narrowest. There may be an exception, a first-year female ringed in Shetland in October 1976 and re-trapped at Shrewsbury in February 1978, but it could have been a Scottish breeding bird, as was the only other recovery from Scotland, a nestling from South Lanarkshire in June 1985, also re-trapped at Shrewsbury in December 1986.

The oldest was ringed at Shifnal in September 1978 and re-trapped there 9y 4m later still going strong, but well short of the national longevity record of just over 12 years. A first-year female ringed in Market Drayton in January 1981 was found still alive 7y 6m 26d later in arctic Norway 1,806km distant, and an adult male ringed in the same place in January 1990 was recovered 812km distant, in Denmark 8y 2m 4d later.

At home in a wide variety of habitats, the Chaffinch should do well, and hopefully the effects of trichomonosis will be short lived.

Allan Dawes

Brambling

Fringilla montifringilla

Fairly common, occasionally irruptive, winter visitor

John Hawkins, Dudleston Heath, 1 December 2008

Brambling takes the place of the Chaffinch in northern latitudes where it breeds from Norway to the Bering Sea.

Formerly known as the Mountain Finch it was described as 'a regular and sometimes plentiful winter visitor; a few arriving in October, and many more arriving in November, when they resort to beech woods to feed upon the mast'. This description still holds true today. They were also known to feed alongside Chaffinch in open fields and resort to stackyards in severe weather, but were considered less likely to be found in the vicinity of houses. The advent of modern bird food seed mixes, and a reduction in winter stubbles and arable weeds, has modified this behaviour.

Most are from the Fennoscandian breeding population, although others breeding further east in Russia have been found in the UK and may occur locally. They begin to move south and south-west in early September, travelling slowly and stopping to feed on route wherever a good food supply is available, rather than migrating in a single long flight. A poor beech mast crop, or deep snow which prevents access to food, will encourage them to move on and first arrivals can sometimes be found in September. However the main arrival period varies; if there is a plentiful supply of food and mild conditions on the continent they will remain as far north as they can, but a sudden period of harsh weather can drive them further south at any time. For this reason numbers vary enormously between years. In 1974, between 2,000 and 3,000 were in fields surrounding Allscott Sugar Factory in January and February, 1,000 were coming to roost at Hinstock in February 1976, and 1,000 were feeding on an unspecified crop at Grindle, near Shifnal, in January 1984. Since then numbers have been much smaller, and the winter of 2000–01 was particularly poor with only six being reported.

Although beech mast is their preferred food, Bramblings will take advantage of other opportunities and may be found wherever a good food source is available. Scanning mixed groups of seed eaters on farmland often reveals a few, and large single species flocks are occasionally found. They often linger well into April and occasionally

May, and were frequently encountered early in the breeding season during recent Atlas fieldwork. One was still to be found at Church Stretton during the first week of May in 2012, where it visited a feeder for several days. In early spring, males begin to acquire their breeding plumage and are a splendid sight.

There has never been evidence of confirmed breeding, although a pair was reported from Church Stretton in July 1898 and a singing male from Ightfield in June 1987. The latter record came at a time when Brambling looked like becoming a regular breeder in the UK, although most breeding attempts were in Scotland and these have since petered out (Spencer *et al.* 1993).

In recent Winter Atlas fieldwork they were found in almost 40% of tetrads. Although they were widely distributed, the map shows several quite large gaps. Some of these are low-lying arable areas, where some feeding opportunities are probably present, but if their preferred woodland habitat is lacking they may not be attracted to the vicinity. Other gaps will be due to fieldwork being carried out only in poor 'Brambling years'. The chart shows how the total number of Atlas records varied across the six winters during the recent Atlas period.

Winter Distribution (2007–13)

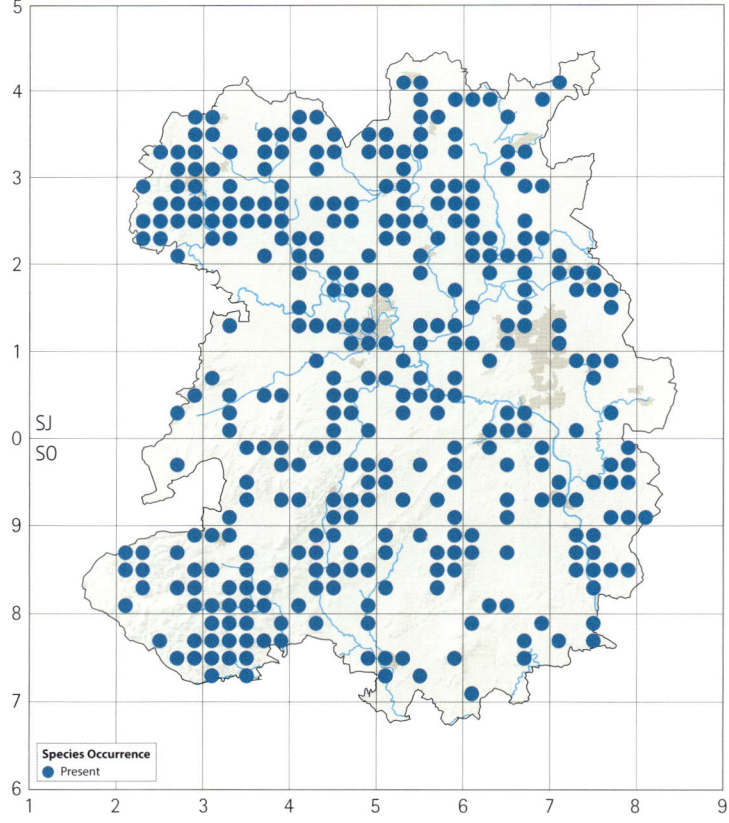

Species Occurrence
● Present

Occupied Tetrads

Tetrads with Winter Records (2007–13):   342   (39%)

Number of Records: Winter Atlas Years (2007–13)

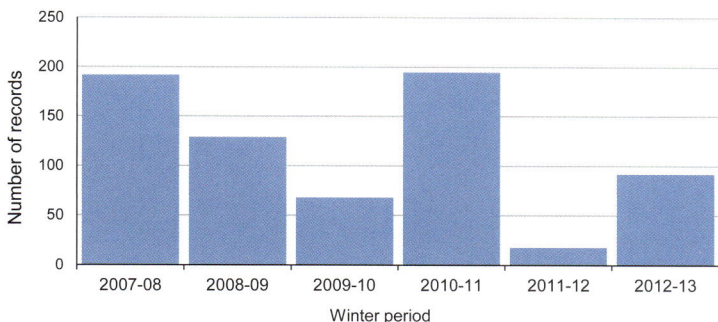

Garden BirdWatch Reporting Rate (1995–2014)

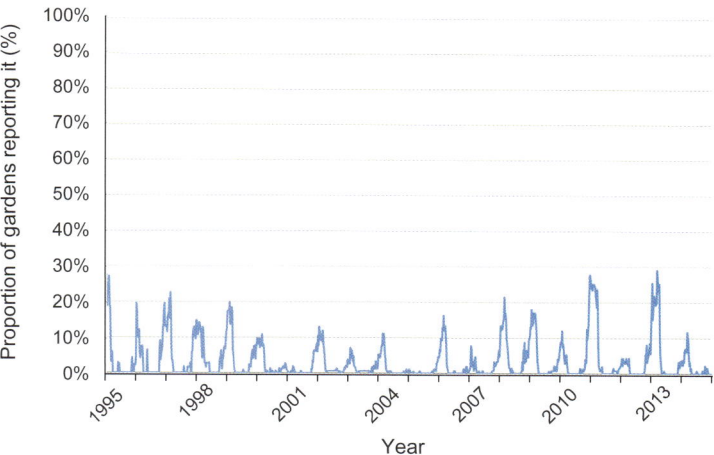

Brambling feed in a variety of widespread different habitats, such as woodland, farmland and gardens although, apart from in harsh winter weather, the main period for garden use falls outside the winter recording period. The Garden BirdWatch chart shows good and bad Brambling years over a longer period than the Atlas chart, and a tendency for garden use to increase throughout the winter as natural food supplies diminish, with numbers peaking in late March and early April. This tendency is shown more clearly in results for the West Midlands region, published on the Garden BirdWatch Results page on the BTO website (www.bto.org).

Only eight sites held flocks of 100 or more in the recent Atlas period. In December 2007, 100 were counted during a TTV at Bucknell Wood; in the following February 150 were at Warren Farm (Stoke Heath), and in February 2009, 250 were at Nib Green in the Wyre forest. The winter of 2010–11 proved to be a good one for Brambling as shown by the number of Atlas records received. The wild bird food crop at Venus Pool attracted good numbers throughout December 2010, with a maximum count of 300, and in the same month 300 were also found feeding in a weedy brassica crop near Llanyblodwel. In

January 2011 there were 100 near Ratlinghope and 150 at Rea Farm, Upton Magna, while in February there were 150 at Clunton Coppice.

The variability of the food supply means that the quality of feeding sites can change between years, and food sources can be consumed quickly, making flocks highly mobile. A reliable and plentiful food supply such as that provided at Venus Pool is an exception. Relative abundance is therefore likely to vary enormously both within and between years, so any map would be misleading. The distribution map overstates the position in any one year.

Ringing recoveries include 50 overseas movements, involving 10 different countries, with Norway (16) leading the way, followed closely by Belgium (10), the Netherlands (eight) and Germany (six). Unlike the Chaffinch, migrants often cross directly over the North Sea rather than funnelling though the Low Countries.

Tree seed is not produced annually and crop failures can be

John Fielding, Venus Pool, 25 January 2014

extensive, so Brambling are often found wintering in widely differing locations. One shot in Italy in October 1967 and another re-trapped in Luxembourg in February 1968 had both been ringed at Market Drayton in the late winter of 1966. Of the others, mainly ringed in different winters at Hinstock, Market Drayton or Allscott Sugar Factory, three have subsequently been found wintering in the south of France, two in the Netherlands and one in Denmark. Reverse movements have also been seen with three ringed in the Netherlands and one in Germany wintering here.

Of nine found overseas during passage periods, seven have been at coastal or island locations from the Netherlands to northern Germany, with the other two from inland Belgium, both in the autumn. This is a small sample, but it suggests that spring passage follows a more direct route, and this is supported by movements within the UK, which have mainly been to the north and east, and include the

longest movement, one ringed at Market Drayton in November 1977 and re-trapped on Shetland in the following April, while one found dead at an unknown site in the North Sea to the west of Norway in April 1989 had been ringed at Walcot two years earlier. The latest spring dates were also from the greatest distances, from Finland on 16 May, in 1985, distance travelled 1,872km and from Sweden on 8 May, in 1991, 1,297km away, and both could have reached their breeding grounds. No recoveries have been made between mid-May and the start of autumn passage.

The phrase 'another poor year for this species' is repeated time and time again in SBRs, and poor years would appear to be the rule rather than the exception. With milder winters being predicted it seems highly likely that we will be seeing less of this delightful visitor in the future.

Allan Dawes

Hawfinch

Coccothraustes coccothraustes

Rare, occasionally irruptive, winter visitor, has bred

Found throughout central Europe, the Hawfinch is absent from northern latitudes and has a patchy distribution in the south.

It is a mysterious species with a chequered past and an uncertain future. In the mid-nineteenth century it was thought to be a winter visitor, and during the long frost of 1878–79 many were seen and taken. Having written in 1879 that he could find no evidence of nesting, Beckwith soon received many such accounts. A natural expansion had occurred following the first proven British breeding in Epping Forest in about 1830, and the timing of Beckwith's writing seems to have coincided with this. It is not known whether the influx of the previous hard winter encouraged some to stay and breed, as often happens with crossbill species, increasing both the numbers and spread. In 1890 Beckwith wrote: 'And this is not an apparent increase, due to the greater numbers of observers and greater consequent attention paid to Natural History, for the damage done by the old and the young birds to green peas forces itself upon the attention of even the most unobservant gardener'. Eighteen were shot in one Broseley garden in 1901, and many others in neighbouring gardens. Forest too commented in 1899 'but of late years it has multiplied very much, nearly a dozen nests having been found in one year close to Shrewsbury' and in 1905 that '30 years ago it was considered rare … in some districts it is now quite numerous'.

The *Handlist* (1964) stated:

Infrequently recorded in recent years, but possibly overlooked … It continued to be reported as a breeding bird up until about 20 years ago in a number of widely scattered localities … Recent records suggesting breeding are from Leighton (near Buildwas) in 1956–57, Shirlett 1957, Apley Castle (Wellington) 1957–58, Bucknell bred until 1954. Other recent records are from Fishmore Hall (Ludlow), Fullway, Attingham and Berrington.

Hawfinches were discovered wintering at Whitcliffe in January 1966 and a flock of 50 in February 1968 provided the largest modern

record. Whitcliffe became the place to see this species and, with the exception of 1974, it was seen annually and the site dominated the number of both reports and birds in SBRs up to 2004. Although this may have been self-sustaining, it enabled their decline and ultimately the end of this sequence to be documented. During this time a small number continued to be found elsewhere and occasional breeding records were received.

There were records from 14 widely scattered tetrads in the *Atlas* (1992), five of which were of confirmed breeding, ranging from Oswestry in the north-west to the Wyre Forest in the south-east, with two close to Whitcliffe. Habitats were parkland (eight), woodland (four) and large overgrown gardens (two).

Shortly after fieldwork for the *Atlas* (1992) was completed, adults with three young were seen at Woolstaston in June 1991, the last instance of confirmed breeding. The subsequent drop in sightings

John Robinson, Button Oak, 6 February 2011

Breeding Distribution Change (1985–90 to 2008–13)

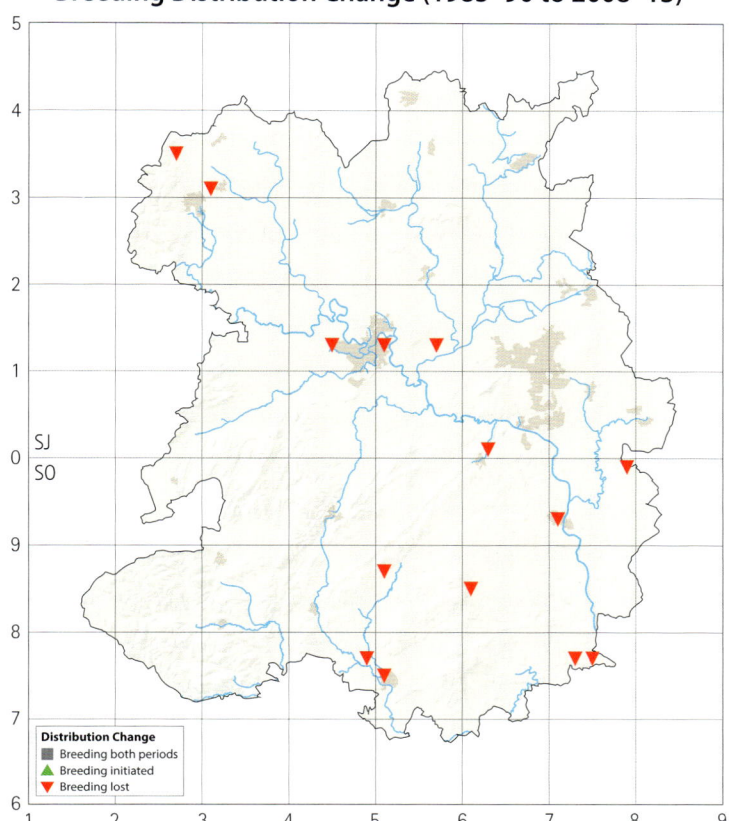

Distribution Change
- ■ Breeding both periods
- ▲ Breeding initiated
- ▼ Breeding lost

Winter Distribution (2007–13)

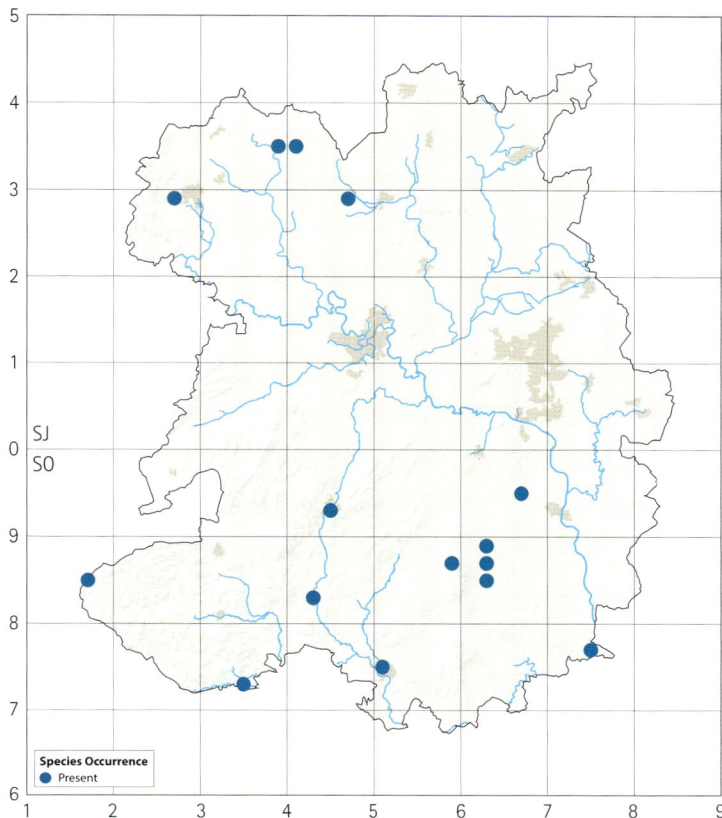

Species Occurrence
- ● Present

Occupied Tetrads

Atlas period (breeding)	1985–90		2008–13		Change	
	Number	%	Number	%	Number	%
Confirmed	5	1	0	0	-5	-100
Probable	6	1	0	0	-6	-100
Possible	3	0	0	0	-3	-100
TOTAL	14	2	0	0	-14	-100
Tetrads with Winter Records (2007–13): 15 (2%)						

resulted in the Hawfinch being added to the list of species considered by the local Rarities Committee in 2006.

There were just four records during the recent Atlas breeding period. The timing and location of one near the summit of Brown Clee in early April 2009 suggests a passage bird. Others were near Craven Arms in June and Ellesmere in July, both in 2012, and Hopesay Common in May 2013. None were known to be present for more than two days and they were not considered to have been breeding. While its elusive nature could easily result in a breeding attempt going unnoticed, the Hawfinch can no longer be considered a resident, as it has only occurred sporadically during the breeding season since it was last known to have bred in 1991. The breeding distribution change map reflects this apparent local extinction.

Recent winter Atlas reports came from 15 tetrads, mostly in the south, but with four in the north-west. The largest flocks were in the Wyre Forest and Llanforda (near Oswestry), where eight were reported, and at the latter site some were present in three consecutive winters from 2010–11. There were five near Aldenham in December 2008.

Two groups of tetrads elsewhere also had wintering Hawfinches in more than one year: singles were at Ellesmere in January 2009 and January 2011, and four tetrads surrounding Brown Clee had records in the three years 2009–11, including three together in SO58Y, and two in SO68H near Cleobury North, early in 2009. Four were also seen on 30 October 2010 at Woolston SO48I, the day before the start of the winter season, and one was near Craven Arms SO48G later in the same winter season. After an absence of six years, Hawfinches made a welcome return to Whitcliffe, with at least one seen in January 2010, three in January 2011, when they were seen feeding on beech mast and hornbeam, and, after the recent Atlas period, three were present there again in October 2014. These clusters may indicate small, but regular, wintering sites. However, numbers are low and Hawfinches are elusive, so this remains speculative.

The timing of all records from recent Atlas fieldwork, along with earlier reports from the well-watched Whitcliffe site, suggest that Hawfinches are primarily winter visitors, although they are known still to breed in some adjacent counties. There are no local ringing recoveries, but a project in Gloucestershire targeting Hawfinch has recaptured one from Sweden and one from Norway, both in March, but as yet the numbers involved are too small to enable any firm conclusions to be drawn.

The national BTO Atlas (2013) shows a large contraction of the breeding range, and the lack of any local breeding records since 1991 fits with the national picture. The reasons behind the decline are unknown. In contrast, the European population is doing well. A long frost and winter influx such as that in 1878–79 is urgently needed.

Allan Dawes

Bullfinch (Eurasian Bullfinch)

Pyrrhula pyrrhula

Common resident

Bullfinch is a sedentary species with a large range across Eurasia. Here, it is a common resident of deciduous woodland, tall and overgrown hedgerows, scrub, parks and mature gardens, and is widely distributed.

Although it has, historically, been regarded as a 'pest species' in commercial orchards because of its liking for the buds of fruit trees and bushes, this has never really been an issue here, although Paddock (1890) noted that, as well as being partial to larch buds in spring, and often feeding in nettle patches in autumn and winter, it regularly visited gardens and orchards, causing damage to plum and damson trees.

Earlier evidence for a lack of serious damage to fruit orchards here comes from a map of England and Wales, identifying parishes where churchwardens' accounts list high levels of bounty payments for killing Bullfinches during the seventeenth or eighteenth century, as it shows only one parish, in east Shropshire (which looks to be Worfield), that was killing Bullfinches at that time. This persecution was likely to have been the result of 'An Acte for the preservation of Grayne' (reviewed in 1572 and 1598, and not repealed until 1863); this act allowed serious and intensive 'vermin' control of many species throughout England and Wales, and sanctioned payment of a bounty of 'one peny' for 'the Heade of every Bulfynche or other Byrd that devoureth the blowth of Fruite'. The analysis of the large numbers killed in parishes elsewhere show that the species then 'occurred in an abundance that is not equalled today' (Lovegrove 2007).

Leighton (1836) noted that it was a common garden bird, Beckwith (1879) described it as 'frequent in thick woods' and particularly frequent around the Wrekin, while it was mapped as 'common' at the end of the nineteenth century in the national *Historical Atlas* (1996). Beckwith (1890) commented on its perceived destructive tendencies, and stated that the extent of damage it causes, especially to fruit trees, is a matter of some debate, and advised careful observation, rather than 'hasty and inaccurate observation'.

Forrest (1908) noted a flock near Shrewsbury 'so numerous as to impart a red tinge to the yew trees on which they settled'. During the remainder of the first half of the twentieth century, it was a widespread and frequent breeder, but most often seen in autumn and winter, which appeared to confirm its secretive behaviour when breeding.

In the second half of the twentieth century it was described as 'numerous in and out of the breeding season' (SBR 1958), and 'widely distributed' (SBR 1963), while the *Handlist* (1964) noted Bullfinch as 'common ... except in the open hill country', and it 'was considered to have increased in numbers in recent years'. It also contained an 'unusual record' of 13 seen in hawthorn scrub on the top of the Long Mynd in February 1962. Although most SBRs during the 1970s only listed it as a frequent or common breeder, SBR (1971) recorded flocks of '20+ feeding on seeds in Abbey Foregate' and of '40 in Granville Quarry'. In SBRs for the 1980s, most records were of pairs or small groups, but some larger flocks were recorded in winter, including 25 from Catherton Common in 1988.

The *Atlas* (1992) confirmed its widespread distribution, and based its population estimate on average CBC densities of 2.4 pairs per sq.km on farmland and 7.1 pairs per sq.km in deciduous woodland, reported

John Hawkins, Pentrecoed, 8 April 2006

in the BTO Atlas (1976), but noted that 'since then, CBC indices have demonstrated a 70% drop in population density on farmland and 40% in woodland'. It concluded that 'although Bullfinch is never numerous, its widespread occurrence suggests a population of 1,500–3,000 pairs based on an assessment of around 2 to 4 pairs per tetrad'.

Although reported regularly in the breeding season in SBRs from 1990 onwards, few confirmed breeding records were received, and five in 2004 (from Belle Vue, Bromfield, Bury Ditches, Castlefields, and Venus Pool) was typical. Forty-three breeding territories were found at six different sites during a survey of Telford and Wrekin Council woods (SBR 2003). Autumn and winter produced more records than the breeding season, usually in numbers of fewer than 10, but larger flocks included 17 on Haughmond Hill (4 February 1998), 12 at Hawkstone Park (21 September 2002), 20 in Severn Valley Country Park (25 October 2003), and 30 at Buttercross in November 2005, which were attracted by set-aside. SBR 2005 reported that local BBS data seems to show that the population was stable at that time, although the trend in England as a whole was a decline. Widespread records in 2013 included several sightings early in the year of feeding in gardens on sunflower hearts. Counts were mainly of ones and twos, but larger flocks included 15 on the Stiperstones NNR (20 November), 12+ at Priorslee Lake (20 January), nine in Granville Country Park (8 December), and six+ at Oswestry Old Racecourse (12 October).

In the recent Atlas period, Bullfinches were found in 692 tetrads (80%) in the breeding season, compared with 709 tetrads (81%) in the previous Atlas. Winter distribution was generally more widespread, with presence in 749 tetrads (86%). Their secretive nature when breeding, contrasted with a higher population including birds of the year in winter, and a tendency to move from woodland in search of food, accounts for the difference.

Based on TTV counts, the estimated breeding population is 4,100–4,250 pairs, suggesting that the 1992 estimate was too low.

The breeding distribution change map shows Bullfinch is still fairly evenly and widely distributed, but apparently still absent from the Long Mynd and some areas of arable farmland. It appears to have been lost from around 100 tetrads, but found in around 80 new tetrads, a net loss of only 2%.

Despite their striking and distinctive appearance, Bullfinches can be surprisingly easy to overlook because of their unobtrusive and secretive behaviour, particularly during the breeding season, so many of the apparent losses and gains will reflect variation in fieldwork effort, and chance, rather than real change in numbers and distribution. However, clusters of tetrads with losses will reflect loss of suitable habitat, such as removal of hedgerows and scrub on arable farmland.

Also, although they are not difficult to locate by their distinctive call, their nests are notoriously hard to find, and food being taken back to a nest for young is very difficult to see in their beaks, and this contributes to the high number of probable (rather than confirmed) breeding records received for both Atlases.

There tends to be a higher density, most marked during the breeding season, in areas of deciduous woodland, large hedgerows and scrub, particularly to the south of Telford, and in western areas, with a lower density or gaps in distribution on arable land in the north and east, and on higher ground, such as the Long Mynd, reflecting the availability of suitable breeding habitat. Densities may have increased slightly in the tetrads where it does occur. Although it is not found in sufficient squares here to produce a statistically valid BBS trend graph, there have been population increases in adjacent areas, the West Midlands as a whole, and Wales, between 1995–2014 (27% and 6% respectively).

These welcome recent trends are not sufficient to start a recovery. Nationally, numbers fell steeply between 1977–82, especially on farmland. The decline eased during the mid-1980s and there has been an upturn since 2000, but the UK population declined by 40% over the whole period 1970–2014. It is believed that deteriorating habitat quality, caused by agricultural intensification and reduced structure and diversity in woodland, has played a part in this decline.

Ringing recoveries confirm the sedentary nature of this species. None ringed elsewhere have been recovered here, only five ringed here have been recovered elsewhere, all in nearby counties, and the vast majority, around 40, have been both ringed and recovered here.

Helen J. Griffiths

Occupied Tetrads

Atlas period (breeding)	1985–90		2008–13		Change	
	Number	%	Number	%	Number	%
Confirmed	217	25	164	19	-53	-24
Probable	359	41	349	40	-10	-3
Possible	133	15	179	21	46	35
TOTAL	**709**	**81**	**692**	**80**	**-17**	**-2**
Tetrads with Winter Records (2007–13): 749 (86%)						

Breeding Distribution Change (1985–90 to 2008–13)

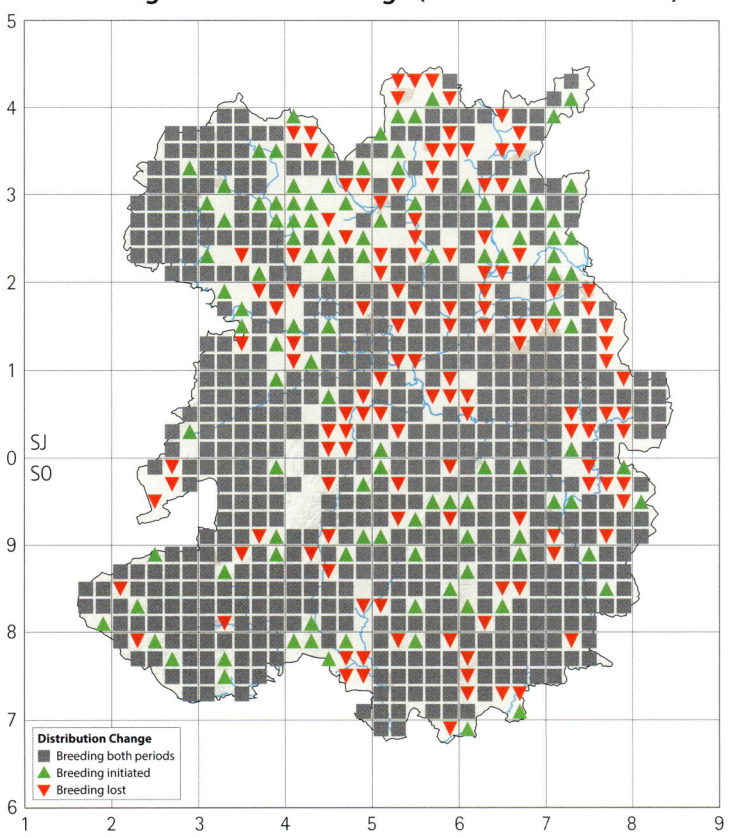

Distribution Change
- Breeding both periods
- Breeding initiated
- Breeding lost

Breeding Relative Abundance (2008–13)

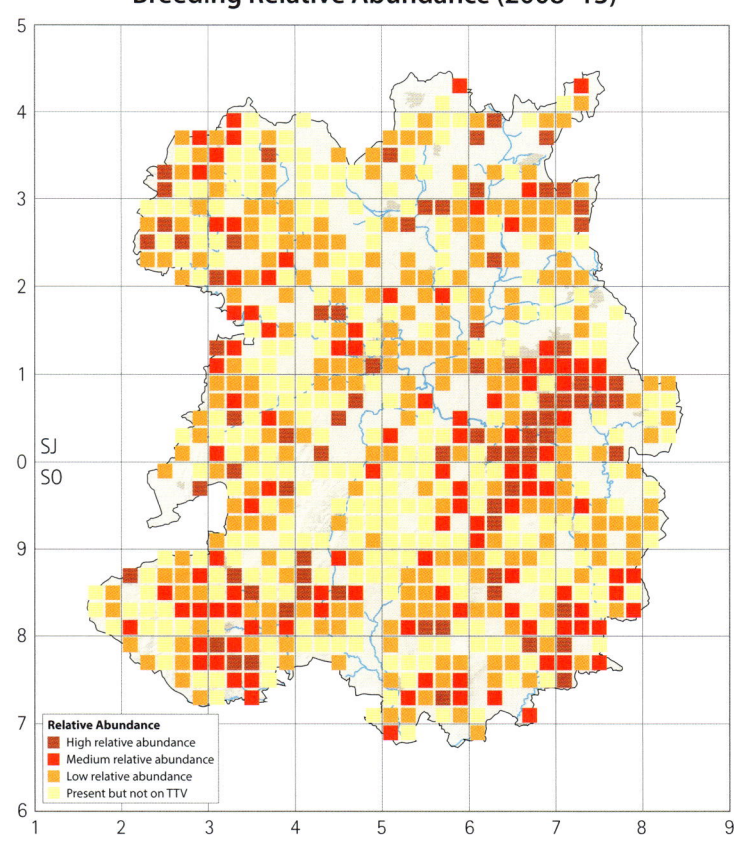

Relative Abundance
- High relative abundance
- Medium relative abundance
- Low relative abundance
- Present but not on TTV

Greenfinch (European Greenfinch)
Chloris chloris
Very common resident

Greenfinch is found throughout temperate Europe, into Asia and North Africa. Formerly known as the Green Grosbeak or Green Linnet, in the late nineteenth century it was considered to be common in all seasons, a lover of richly cultivated districts, and rarely seen on high or waste ground. Some migration was noted in autumn and large flocks of bright coloured males were said to be very noticeable on their return in February and March. In addition to its fondness for wild berries, it was also known to be fond of sunflower seeds, which at that time it obtained from garden plants. By the time of the *Handlist* (1964), flocks of 100 or more were not uncommon during the winter, but it was considered by several observers to have decreased in the previous 20 years. The *Atlas* (1992) suggested the use of organochlorine pesticides, and severe winters in 1961–62 and 1962–63, as possible causes, but indicated that the population had recovered by 1966.

Greenfinches are now a familiar garden visitor, having readily taken to sunflower seeds and other bird seed mixes provided for them by householders. During the early years of the recent Atlas period they were easily found in the vicinity of towns and villages or isolated homes with well-stocked feeders. As the Atlas period progressed they became more difficult to find as the disease trichomonosis devastated the population in some areas.

They are widely distributed during the breeding season, but are absent from higher ground in parts of the Clun Forest and the Long Mynd, and a few isolated tetrads elsewhere. Although they utilise a wide variety of habitats, during fieldwork they were found more easily near human settlements than in woodland or farmland. Singing and territorial males were easy to locate, but confirming breeding usually depended on finding recently fledged young. There has been a mix of gains and losses, but with an overall increase of 7% in occupied squares. These are widely scattered, and on the whole this probably indicates better coverage, although increased numbers will have made it easier to locate them. A small number of gains have been made in upland areas in the south-west; clusters show on the map around the Stiperstones, and Clun Forest. Many conifers have been planted in these areas and as these plantations mature, new habitat is created in previously unsuitable sites.

Breeding abundance is very variable and higher densities probably indicate local hotspots based around villages. The higher ground in the south-west, and large tracts of arable farmland, generally have lower densities.

By 1985–90 the population was increasing steadily, and the local BBS results since 1997 show a continuation of this trend with numbers almost doubling by the mid-2000s. In 2007, numbers fell sharply due to an outbreak of trichomonosis. This disease had previously been known mainly in doves and pigeons, but it started to be reported locally in Chaffinch and Greenfinch, although the Greenfinch decline preceded the Chaffinch by a year. After rallying slightly in 2008, the fall has continued and over 50% of the population has been lost in only six years.

Based on TTV counts, the estimated population was between 28,000–30,000 breeding pairs in 2009. Since then a further decline of 35% is evident from the BBS, almost certainly due to the continued effect of trichomonosis, and an estimate of 18,000–19,500 pairs in 2014 may now be more appropriate. As the population increased considerably between 1990–2005, this is still above the estimate quoted in the *Atlas* (1992). It is possible that by 2014 Greenfinch no longer occupied some tetrads where it was found earlier in the recent Atlas period.

The winter distribution is similar to that in the breeding season but there were a few more isolated tetrads where Greenfinches were not located. The odd gaps coincide with sparsely populated parts of the countryside, and the larger gaps with high ground such as the Long Mynd and Clun Forest. Abundance is greatest in the north-east, which

Breeding Distribution Change (1985–90 to 2008–13)

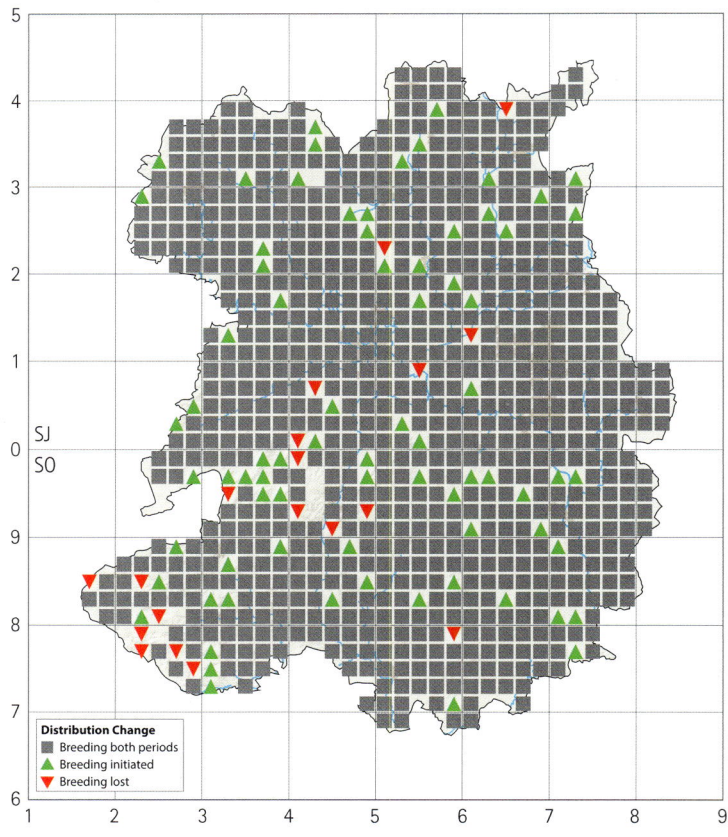

Occupied Tetrads

Atlas period (breeding)	1985–90		2008–13		Change	
	Number	%	Number	%	Number	%
Confirmed	368	42	354	41	-14	-4
Probable	290	33	382	44	92	32
Possible	129	15	107	12	-22	-17
TOTAL	787	90	843	97	56	7
Tetrads with Winter Records (2007–13): 800 (92%)						

BBS Trend (1997–2014)

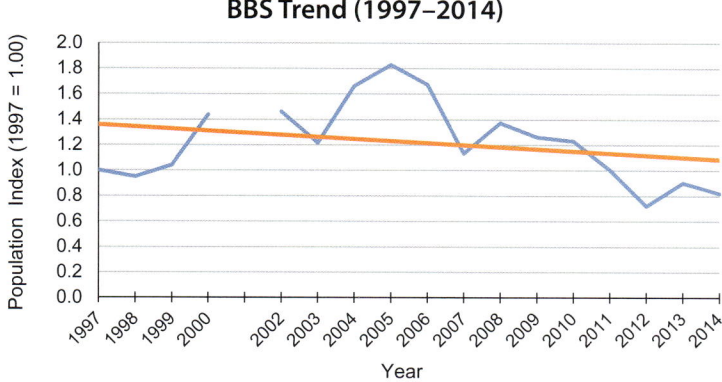

John Hawkins, Pentrecoed, 6 February 2006

fits well with the comments of early authors. However, there are far fewer weed seeds available in heavily cultivated districts these days, and the increased abundance is likely to reflect the higher human population and garden feeding opportunities.

The Garden BirdWatch graph shows that gardens are visited throughout the year, with peaks occurring in early spring when natural foods are in short supply.

In contrast to the BBS trend, the GBW results show a slight decline from the beginning of the survey in 1995, becoming more pronounced after the onset of disease. However, GBW monitors birds throughout the year, not just the breeding season, and the graph shows presence per garden, rather than total numbers. In a Trefonen garden, 40 were regularly seen at feeders every week during the winters prior to 2007, but since then they have been less frequent visitors and usually in ones or twos (*pers. obs.*).

Greenfinches are largely sedentary, and around two-thirds of movements have been within this or to or from immediately neighbouring counties. Of the remainder, most have been spread across England and Wales, with three involving Ireland and one from Scotland at Inverness. Only two involve continental countries. One ringed in Leegomery in January 1964 was found long dead in central Germany four years later and another ringed in Norway in October 1993 was re-trapped in Shrewsbury in December 1996. The oldest, an adult when ringed at Newport in April 1987, was found freshly dead 7y 2m later in the same town.

Human settlements play an important part in the life of the Greenfinch by providing ample breeding habitat and a good winter food supply. Their future should be secure, but much will depend on the long term effects of trichomonosis. Those of us who encourage wildlife into our gardens by providing food need to ensure that we maintain good hygiene and do not harm the birds we are trying to protect.

Allan Dawes

Breeding Relative Abundance (2008–13)

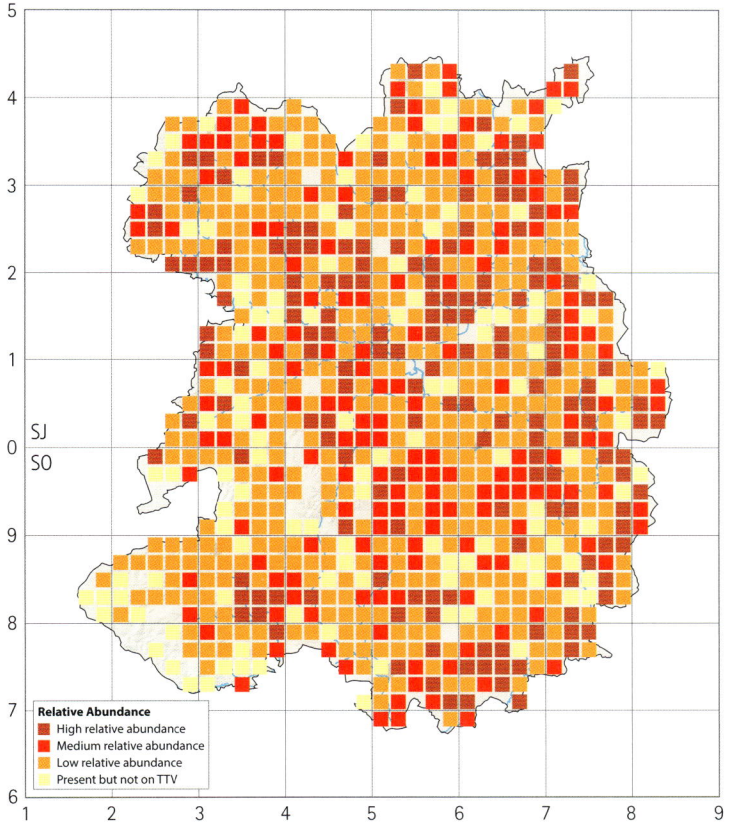

Garden BirdWatch Reporting Rate (1995–2014)

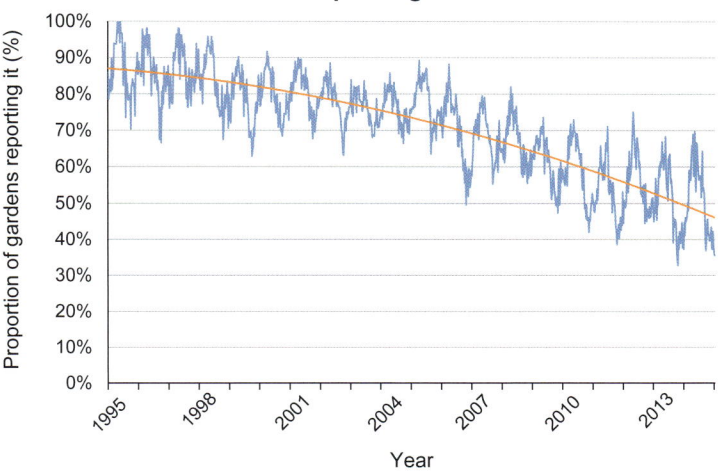

Twite

Linaria flavirostris

Very rare winter visitor

In Europe, Twite breed mainly in Scandinavia, and the stronghold of the British population is in Scotland, with smaller, isolated populations in the mountains of North Wales and the Pennines of northern England. In winter, those breeding on high moorland usually move to lower ground, mainly to the coast, but occasionally elsewhere such as the margins of large water bodies.

In 1836, Eyton reported that the 'Mountain Linnet' was 'occasionally found' while in 1865 it was considered to nest 'occasionally on the Longmynd', although Rocke goes on to suggest that it could be more common, but overlooked, given the difficulty in separating it from Linnet. Later, in 1879, it was thought to be very rare although, again, the possibility of confusion with Linnet meant that it 'probably, occasionally occurs all along the mountainous parts ... bordering Wales'.

In 1897, a 'large flock' was reported near Oswestry on 6 April, but there must be some doubt about the identification, given the date, numbers involved and paucity of records before and since. A report of one near Clive on 18 December 1899 may be more acceptable given that all modern records are in the late autumn or winter.

By 1904, doubt was being expressed about it ever having bred on the Long Mynd, and this was repeated in 1908. On 3 October 1907 an unknown number were 'heard and seen flying over Ludlow', an interesting but, again, somewhat doubtful, record. Later, a pair was seen in suitable breeding habitat on Glyn Common, Middletown on 30 July 1914, and on 16 May 1929 'a party' was reported on the Long Mynd, with one seen subsequently on 30 May.

Interestingly, the historic record contains no reference to obtaining specimens, which is unusual given that the species has apparently always been rare and would have been in demand by those with collections. As a consequence, and given the difficulties in separating it from Linnet, there must be considerable doubt about it ever having bred here. Also, some of the other records must be questioned, particularly those involving large numbers or outside the winter period.

The three recent occurrences are on typical dates and sites:

Shrewsbury Sewage Farm	ten (approximate), 11–12 February 1967
Chelmarsh	21 October 1974
Chelmarsh	four, 23 December 2004

Those in 2004 were associating with a small flock of Linnets. Initially picked out on call, the presence of the Linnets enabled direct comparison and confirmation of the key features.

Graham Walker

Linnet (Common Linnet)

Linaria cannabina

Common resident

Dave Barnes, Catherton Common, 10 June 2012

The Linnet is found throughout Europe, western Asia and North Africa, except in the most northerly latitudes.

According to Beckwith, during the winter the Linnet was to be found in gardens and stubble fields feeding on arable weeds, often in mixed flocks. Come the spring he wrote:

> these assemblies disperse, and enclosures are deserted for hills and moorland, where in company with the Stonechat and Meadow Pipit, the Linnet reared its young, the numbers of the bird being regulated by the abundance or scarcity of the golden gorse amongst which it so delights to place its nest ... The Linnet also builds in hedges and low shrubs, but rarely far away from unenclosed or high land.

The implication that breeding was restricted to these areas is supported by Paddock (1897) 'numerous in the summer, wherever gorse grows' and Forrest (1899) 'resident and numerous, nesting on hillsides in low bushes as well as hedgerows'. While the countryside has changed considerably since these lines were written in the late nineteenth century, the association with gorse and uplands still rings true, but their apparent absence from the wider countryside at that time is a surprise.

The presence of gorse is an indication of less intensively managed land rather than a pre-requisite for breeding. Gorse provides both

Breeding Distribution Change (1985–90 to 2008–13)

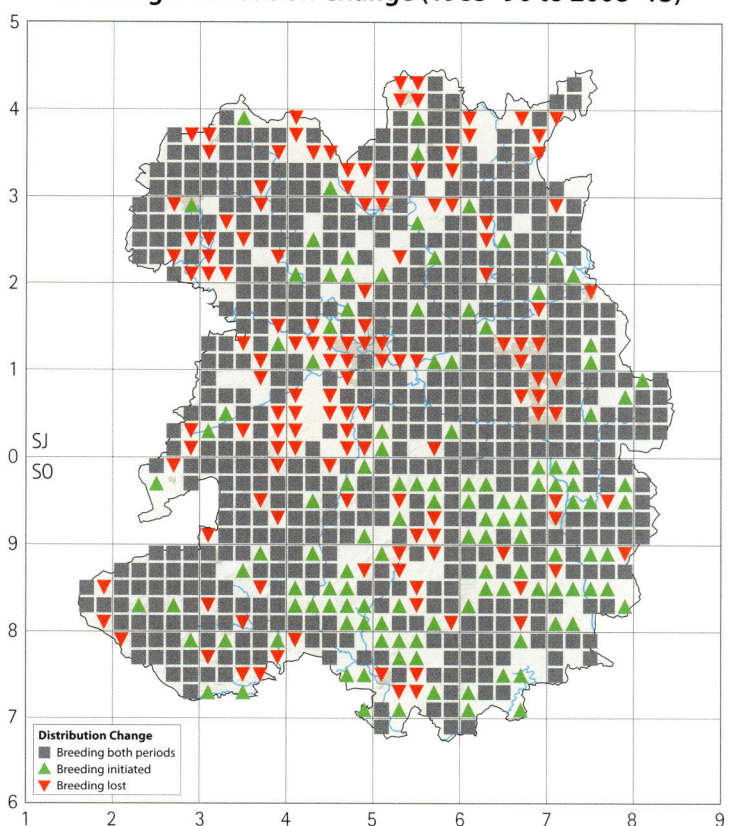

Distribution Change
- ⬛ Breeding both periods
- 🔺 Breeding initiated
- 🔻 Breeding lost

Breeding Relative Abundance (2008–13)

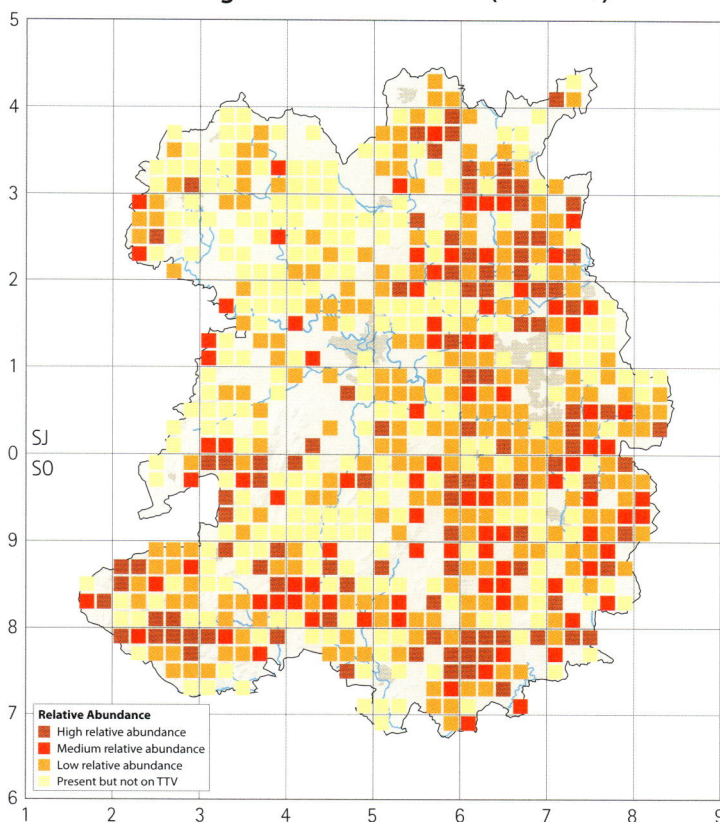

Relative Abundance
- High relative abundance
- Medium relative abundance
- Low relative abundance
- Present but not on TTV

Occupied Tetrads

Atlas period (breeding)	1985–90		2008–13		Change	
	Number	%	Number	%	Number	%
Confirmed	236	27	190	22	-46	-19
Probable	334	38	379	44	45	13
Possible	120	14	109	13	-11	-9
TOTAL	**690**	**79**	**678**	**78**	**-12**	**-2**
Tetrads with Winter Records (2007–13): 367 (42%)						

protection for the nest and a song post, but plays no other part in the breeding cycle; it is the surrounding habitat which is important. This was demonstrated at the Old Hill Fort at Oswestry, where many pairs breed in an isolated oasis between the town and surrounding farmland. The area became overgrown, and to preserve the important historic heritage of the site all the scrub was cleared in 2009–10. The following year the Linnets adapted to the change by nesting in dead bracken, honeysuckle and bramble regrowth early in the season, and in new bracken growth as the season progressed. As the scrub has recolonised, gorse and hawthorn stunted by sheep nibbling have been used regularly as nest sites, although bracken and bramble were also still widely used.

The *Handlist* (1964) said 'considered to have decreased in the last 20 years by many observers', and the national CBC also registered a sharp decrease from its beginning in 1966. The steep decline which affected many farmland species during the 1970s hit the Linnet particularly hard, and it reached its lowest ebb in 1987. After rallying slightly up to 1995, numbers slumped again and although there have

been fluctuations, the West Midlands BBS index fell by 31% between 1995–2014.

An increase in nest failure rates was evident during the population crash and this was considered to be the main cause of the decline. Agricultural intensification resulting in the disappearance of several arable weeds, the seeds of which used to form an important part of the diet of nestlings, and other habitat loss, is thought to be responsible for these nest failures. Winter stubbles are important too, and reduction in them will have also contributed to the decline.

In good habitat, breeding Linnets are semi-colonial and nests can be built quite close together, enabling large numbers to breed in small areas of gorse and rough grassland that have been neglected by modern agriculture. In these circumstances, they are easy to locate, but at lower densities scattered along hedgerows in farmland they are less conspicuous and breeding is harder to confirm.

The Linnet remains widespread but the breeding distribution change map shows an interesting pattern of gains in the south and south-east and losses in central and northern areas. Disappearances around Shrewsbury, Telford and Whitchurch can be explained by increased urbanisation, and other individual tetrads may have suffered from habitat loss or a change in farming practices. Studies into the diet of nestlings have shown that, where available, the unripe seeds of oilseed rape can form a large part of their diet from June onwards, with adults flying up to 2km to reach the crop. The growth rate and condition of young is also higher the closer the nest is to a rape field (Moorcroft *et al.* 2006, Bradbury *et al.* 2003). There has been a 557% increase between 1985–2013 in the area where oilseed rape is grown (see p. 40), which may have been responsible for the gains in the south and south-east, and it would also explain the higher relative abundance in both this area and similar habitat in the north-east. Of all the habitats recorded by the national BBS, only arable farmland shows a positive trend although it is small. The relative abundance map also highlights the traditional upland sites where gorse is still common, including the Oswestry uplands, Clun Forest, the Stiperstones, the Long Mynd and Clee hills.

Based on TTV counts, the estimated population is between 4,000–4,300 pairs, at the upper limit quoted in the *Atlas* (1992), which estimated three to six pairs per occupied tetrad. At favoured sites, the numbers will be much higher so the upper figure is probably nearer the mark.

Wintering Linnets feeding in large arable fields can easily be missed, and even large flocks can go unnoticed unless they are put to flight, explaining some of the apparent gaps in distribution. Other gaps are real, the result of higher ground being abandoned. Relative abundance was very patchy, but arable land surrounding Telford and Shrewsbury tended to have higher densities. They were not recorded in seven 10km squares in the national Winter Atlas (1986), but they were found in every one in recent Atlas fieldwork. While this suggests some recovery since the low ebb in the 1980s, recent fieldwork was more concentrated at the tetrad level and spread over twice as many years, so detailed comparisons cannot be made.

Few Linnets have been ringed in comparison to other finches, as they do not frequent feeding stations, making capture more difficult. They are a partial migrant, with some spending the winter in southern Europe. Recoveries have come from France (nine), Spain (two) and Portugal (one), while one ringed in Jersey and another at Portland Bill have been found here. In addition to the local recoveries, other movements have involved Cheshire and Wirral (three), Greater Manchester (one), Staffordshire (two) and the West Midlands (one). It would be interesting to know whether there is any difference in migration strategy between the upland and lowland populations. Do those vacating the hills just go to lower altitudes or do they move further afield?

The Linnet has yet to discover the delights and easy living offered

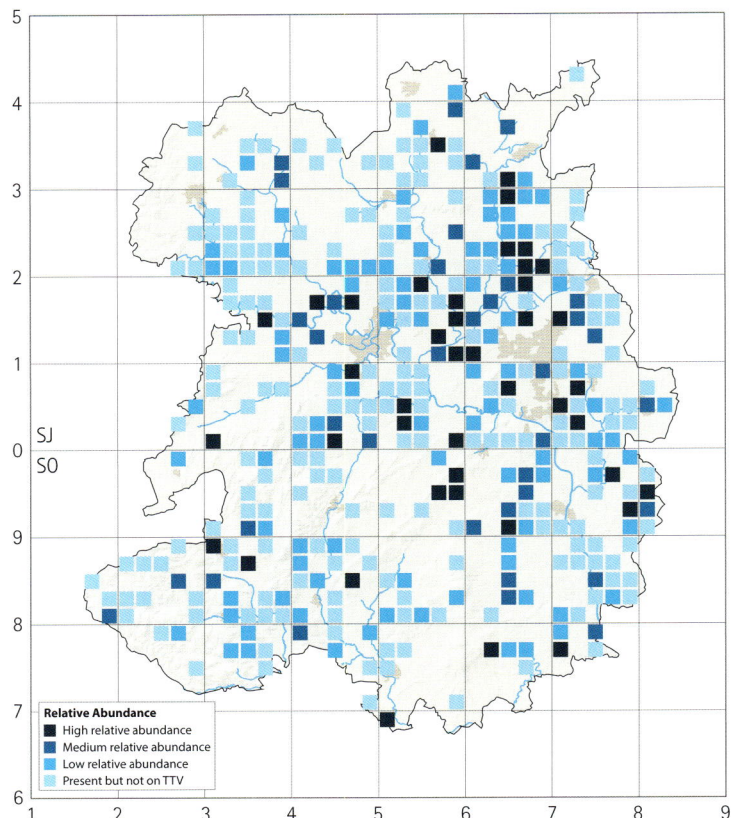

Winter Relative Abundance (2007–13)

Relative Abundance
- ■ High relative abundance
- ■ Medium relative abundance
- ■ Low relative abundance
- ■ Present but not on TTV

by garden feeders and, compared with other finches, it remains essentially a bird of the countryside. The need for a good seed supply, essential for its over-winter survival and raising young, will continue to provide a challenge. Agri-environment schemes may help, but have so far been found wanting. A trend towards a neat and tidy countryside, with regular hedge and verge cutting eliminating many seed sources, does little to help. A future on the Red list seems assured.

Allan Dawes

Common Redpoll
Acanthis flammea
Rare winter visitor

This redpoll species breeds in northern Scandinavia and eastwards throughout northern Europe.

Different races of redpoll have been known for a long time, as too has been the difficulty of identification. It was first mentioned in 1836 when Eyton wrote 'it has occurred'. One shot in a small dingle near Eaton Constantine in 1865 was assigned to this species, known as Mealy Redpoll in the nineteenth century, but others shot in later years, thought at first to be the same race, showed mixed characteristics (Beckwith 1890). Only three other records were listed by Forrest (1908), who described it as 'rare, or generally overlooked', probably an understatement as two near the River Severn above Shrewsbury in 1912 were the last reported for over 80 years.

Some time in the first half of the twentieth century this species

and the Mealy/Lesser Redpoll were merged into one species, and all SBRs, and the *Handlist* (1964), referred to one species, known simply as 'Redpoll', up until 2001, when it was split into two, and the races then known as *Carduelis cabaret*, Lesser Redpoll and *C. flammea*, Common Redpoll were each given species status by the BOU.

Prior to the split the race was seldom noted on records, and the recent increase may be due to the change in species status rather than a change in occurrences.

In common with other species that breed in boreal forests and rely heavily on tree seed for food during the winter, they are subject to periodic irruptions. Such an event occurred in the 1995–96 winter when 150 were found in the Wyre Forest in January and smaller flocks were noted at several other locations from December to early April.

Since then, 23 were found in four winter periods prior to recent Atlas fieldwork, 2000–01 (four), 2001–02 (16), 2003–04 (one) and 2005–06 (two). Those in 2001–02 were at three locations; Hawkstone Park (11), Twemlows Big Wood (three) and a Shrewsbury garden (two), but each of the others were seen at separate locations.

Many records for Common Redpoll and 'Common/Lesser Redpoll' were received during the recent Atlas period, especially in the early years, because observers were not aware of the split, and the resulting correct names for the species in this group. Observers were informed that the one generally encountered was the Lesser Redpoll and that Common Redpoll was considered to be a local rarity, records of which needed to be submitted to the SOS Rarities Committee for acceptance. Observers were asked to review and if necessary edit their records to ensure they showed the correct species, and submit a rarity form to substantiate any sightings of Common Redpoll. Records of the latter were then assessed as any other rarity would be. If no form, or response to queries, was forthcoming, the records of Common were rejected and records of Common/Lesser treated as Lesser. It is appreciated that some genuine records of Common Redpoll may have been lost by this process.

This review resulted in Atlas records from four sites, singles at Bancroft Lodge in February 2011 (SO58I) and Shrewsbury in January 2012 (SJ51B), three at Bobbington in January 2013 (SO89A), and two in November 2012 during ringing at Whixall Moss (SJ43X). In March, two at Madeley in 2009 and four at Broseley in 2011 were just outside the winter Atlas recording period.

Autumn ringing at Fenns and Whixall Mosses caught the two in November 2012 referred to above, and 14 in 2013 (one in late October and 13 in November), but none were caught there in 2014. Species that feed on tree seeds do not follow a regular migration pattern so their absence was not unexpected. We eagerly await the outcome of future ringing at this site, which will help to unravel their movements. Within the redpoll flocks the Common remain together and are often caught at the same time (Gerry Thomas *pers. comm.*). Even with birds in the hand, the identification of some individuals can prove challenging. The only overseas movement recorded was of a first-year female ringed in Norway as *C. flammea* and controlled 835km away, on the Moss, 40 days later. The bird showed mixed characteristics and was initially thought to be Lesser Redpoll *C. cabaret*, illustrating the problems of identification (these scientific names were changed to *Acanthis cabaret* and *A. flammea* in 2018). The problems are magnified when flocks are feeding in tree tops and individuals can be hard to follow. No single feature confirms identification and a range of variable features need to be carefully noted.

Since the Atlas period, in addition to the 14 caught at Fenns and Whixall Mosses in October and November 2013, a single was at Venus Pool in November 2013 and eight were ringed at Market Drayton in February and March 2014.

It is interesting to note that most records are either from ringing or visitors to garden feeders. Close views obtained in this way may be necessary to convince both the observer and the Rarity Committee of the identification, and it seems likely that many others go unrecorded.

Allan Dawes

Lesser Redpoll
Acanthis cabaret
Fairly common winter visitor, uncommon resident

In the mid-nineteenth century the Lesser Redpoll was mainly confined as a breeding species to the north of the UK and the Alps. During the next century the British population expanded its range to include lowland Britain, the Netherlands, Denmark and Germany, and at the same time a similar increase in the Alps saw it move into adjacent countries.

Prior to 2001, the Lesser Redpoll was treated as a sub-species of 'Common Redpoll'. All historic and modern records have been assumed to be of this species if other races were not specified.

In the nineteenth century it was described as common or abundant in winter, often feeding on alder and birch in the company of Siskin. There was initially some doubt over its breeding status, but correspondence between peers in the 1870s indicated that it bred regularly, but was sparingly distributed and possibly overlooked. Nests were reported from widespread locations, from Clungunford in the south to Hawkstone in the north and from Ford in the west to Newport in the east. At this time, the expansion of the breeding range may have been starting to impact locally and this could have been responsible for the earlier uncertainty.

Following tree felling during the Second World War, birch trees dominated the early woodland successional stages and, along with the widespread planting of new coniferous woods, ideal breeding conditions for Lesser Redpoll were established. It was quick to take advantage of these new habitats and numbers flourished. The *Handlist* (1964) described it as resident in small numbers and a non-breeding visitor, but the upsurge was to be short lived. As deciduous woods grew up, the pioneering birch trees were out-competed, and as the conifer plantations matured they also became less suitable. This loss of winter food and breeding sites were both implicated as causes of the decline. After peaking in the mid-1970s, national numbers plummeted and by 1985–90 they were approaching their lowest point.

Rob Stokes, Shrewsbury, 6 April 2013

The population began to stabilise in 1995, but it remained at a low ebb until 2009. A small increase (2%) in the BBS trend for England between 1995 and 2014 is positive news, but levels have yet to reach those during the *Atlas* (1992) fieldwork period.

Atlas records for 'Redpoll' in 1985–90 and for 'Common/Lesser Redpoll' in the recent Atlas are assumed to refer to this species and have been included as such in the maps (see Common Redpoll species account for more detail).

Young conifer plantations and moorland with encroaching scrub are typical breeding habitats. The main areas where breeding was recorded, Clun Forest, Stiperstones, Wyre Forest, Oswestry uplands and the Wrekin all contain such habitat, and their continued presence at these sites is shown in the breeding distribution change map. Losses and gains nearby may reflect local changes in distribution driven by forestry operations and the natural succession of vegetation. Changes elsewhere mostly concern isolated records, some of which will indicate breeding in small pockets of suitable habitat, but some will be due to late passage or non-breeding birds, which sometimes visit garden feeders well into April, and which might have been incorrectly recorded during either the recent or the previous Atlas. These records will have exaggerated the actual breeding range in both Atlases, but less so in the recent Atlas when some were removed during the verification process.

This species is very unobtrusive and easily overlooked when only one or two pairs are present. Even where they are numerous, it is very difficult to prove breeding, and they have the lowest proportion of confirmed records of any finch.

Where conditions are favourable densities can reach 50 pairs per sq.km, but due to the natural succession of woodland these conditions are short lived. These key sites will hold the bulk of the population and it will fluctuate in line with the number of optimum sites, which is in turn dependent on forestry cycles. Other tetrads with little or poor quality habitat may only have one or two pairs and they will have little impact on the overall total. Based on TTV counts, the estimated population is 370–410 pairs, but this estimate may be unreliable for such an unobtrusive and thinly distributed species. While this number could be found in a few key sites, it is probably too high, and between 150–300 pairs would be more appropriate. This is higher than the maximum of 100 pairs estimated in the *Atlas* (1992), but the increase in occupied tetrads in the south-west – where there has been an increase in prime habitat as conifer forests planted after the war have been harvested and replanted – suggests that the population has increased since then.

Most of the flocks recorded during recent winter Atlas fieldwork were in single figures, with only a few numbering over 50. When feeding they can be unobtrusive and small parties can easily be

Occupied Tetrads

Atlas period (breeding)	1985–90		2008–13		Change	
	Number	%	Number	%	Number	%
Confirmed	7	1	5	1	-2	-29
Probable	33	4	29	3	-4	-12
Possible	48	6	26	3	-22	-46
TOTAL	88	10	60	7	-28	-32
Tetrads with Winter Records (2007–13): 305 (35%)						

Breeding Distribution Change (1985–90 to 2008–13)

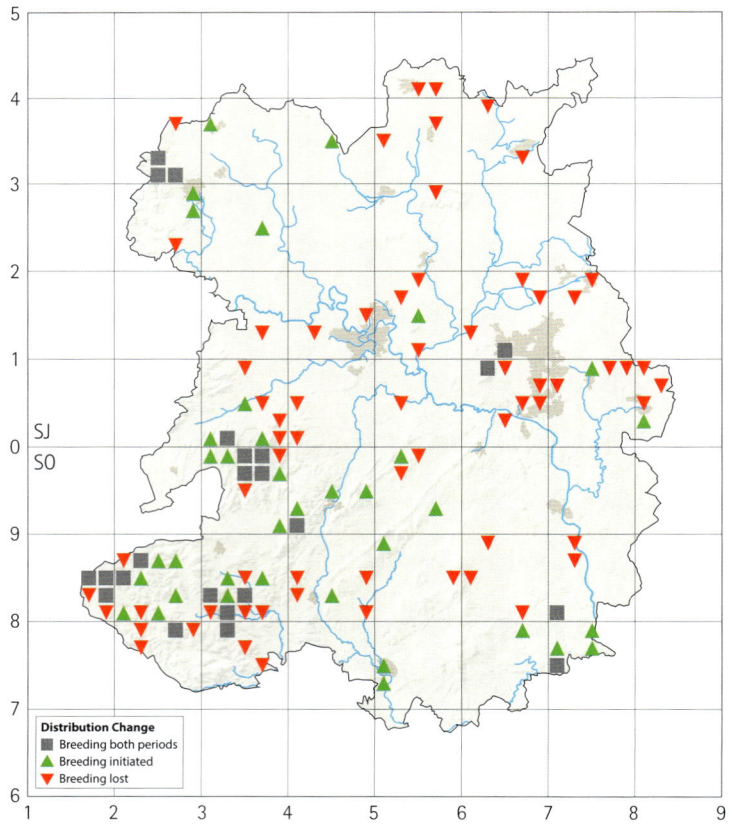

Distribution Change
- ■ Breeding both periods
- ▲ Breeding initiated
- ▼ Breeding lost

Winter Distribution (2007–13)

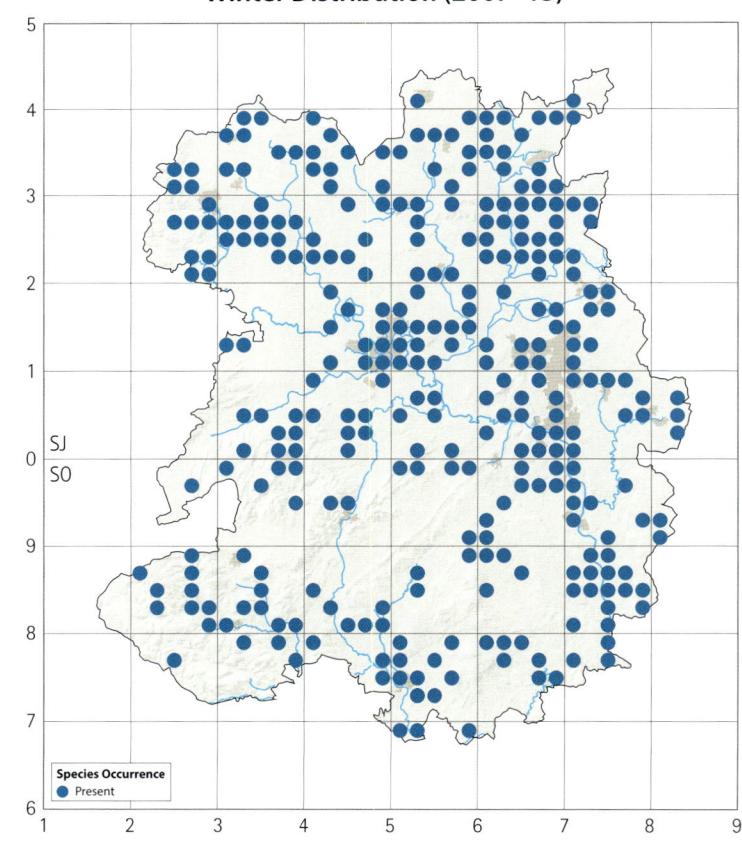

Species Occurrence
- ● Present

missed, even at close range, so they are probably under-recorded in the countryside. However, they have recently started to visit garden bird feeders in late winter and this will have added many dots to the map, and introduced a bias towards tetrads with feeders. Although patchily distributed, they are widespread, the north-east having a more contiguous distribution and highest densities. Elsewhere, there appears to be a correlation with the main rivers, which are often lined by alders, the seeds of which form an important part of the diet of redpoll species. Compared with the national Winter BTO Atlas (1986), more detailed and prolonged fieldwork for the recent Atlas is likely to have produced more records and could account for the slight increase in range, but a small increase was also recorded by the national BTO Atlas (2013) in other parts of the country.

Once the breeding season is over, and moulting is completed, there is a general movement to the south-east to wintering grounds in southern England or the near continent. Evidence for this comes from local ringing, as movements between Ireland (two), Scotland (seven) and the Isle of Man (one) have been recorded, along with 59 to and from other counties, but none have involved those to the south and west of Gloucestershire. One ringed at Chelmarsh in February 2012 was re-trapped in South Lanarkshire in May 2014 and again in May 2015, and another ringed at Highley in April 2013 was re-trapped at Morar, Highland, two months later in June; these movements suggest that they breed in Scotland and over-winter here, or pass through.

A ringing project at Fenns and Whixall Mosses NNR, started in 2010, has discovered a strong autumn passage between late September and early December, the exact timing varying between years depending on the size of the post-breeding population and availability of tree seeds. Only two have been re-trapped during the same autumn, indicating a rapid turnover with birds moving on quite quickly (Gerry Thomas *pers. comm.*). One ringed in October 2013 was controlled in Sussex 34 days later. The only overseas recovery was made in Belgium in March 2011, of one ringed at Whixall the previous October. In the opposite direction, one from the west coast of Norway in August 2013 was re-trapped at Shavington Park the following April.

Continuing forestry operations creating new habitat, albeit temporary, and new garden bird food mixes that help improve over-winter survival, enabling breeding pairs to start the season in good condition, will be beneficial to the small population.

Allan Dawes

Arctic Redpoll
Acanthis hornemanni
Vagrant

Breeding in tundra birch forests of northern North America and Eurasia, this diminutive finch is only partly migratory and, therefore, a scarce migrant to Britain with a handful of records in most years (Fraser 2013b). There has only been one record:

Whixall Moss   9 November 2013

Regular ringing of the autumn and, to a lesser extent, spring passage of Lesser Redpolls through Fenn's, Whixall and Bettisfield Mosses NNR was rewarded when this juvenile Coues's Arctic Redpoll *A.h. exilipes*, was trapped.

Graham Walker

Parrot Crossbill
Loxia pytyopsittacus
Very rare irruptive winter visitor

This crossbill mainly inhabits Scandinavia and northern Russia, although there was a Scottish population of perhaps 131 birds in 2008 (national BTO Atlas 2013).

Pennant (1776) described receiving 'a male and a female out of Shropshire', but no further details were provided.

There are two nineteenth-century records and one in the twenty-first century. Specimens of a male and female 'were obtained near Oswestry ... about the year 1852; but as both birds were purchased by Henry Shaw, and eventually placed in the Clungunford collection [of J. Rocke], their identity cannot be questioned'. Another was killed near Shifnal in February 1862 and added to the collection of T. Bodenham (Beckwith 1891). However, the BOU in their *List of British Birds* (1883) did not include Shropshire in a list of counties where Parrot Crossbill had been recorded, although the above records provided strong evidence that it had occurred here, and which 'justify [it] being retained in our county avifauna' (Beckwith 1891). However, a review of the status of 'Parrot Crossbill in Britain' (Catley & Hursthouse 1985) stated that 'many past references to the status of the Parrot Crossbill in Britain were confounded by the earlier classification of the Scottish Crossbill *L. scotica* as a race of Parrot Crossbill', and none of the above County records were included in a list of 'apparently acceptable records of Parrot Crossbills *Loxia pytyopsittacus* in Britain before 1958'.

The one modern record is of a female:

Postenplain, Wyre Forest   18 February 2014

She was feeding with Crossbills, large numbers of which were present, part of a widespread influx into the UK in the winter of 2013–14; three Two-barred Crossbills were also loosely associated with this Crossbill flock. One or more Parrot Crossbills may have been present over much of that winter as there were several unsubstantiated reports between 2 November 2013–13 March 2014.

John Tucker

Crossbill (Red Crossbill)

Loxia curvirostra

Uncommon resident, scarce irruptive winter visitor

Jim Almond, Bury Ditches, 17 February 2008

The Crossbill is resident in conifer woods throughout Europe, Asia and North America. Numbers here have varied considerably from year to year and are boosted periodically by irruptions from taiga forests in Scandinavia and western Russia. These influxes occur when there is a widespread failure in the conifer seed crop, and numbers are magnified if this occurs after a productive breeding season. Local movements can also be triggered by seed failure and a wood can be suddenly vacated until the next seed crop has matured.

Reflecting this irruptive behaviour, in the nineteenth century the Crossbill was an erratic visitor in very variable numbers and long periods could pass without any sightings. Its fondness for larch seed was often mentioned and most of those reported were associated with conifer plantations. Two birds eating peas in a garden at Kinnerley in July 1888 were shot and on inspection were found to be this species, although the Reverend gentleman that shot them had suspected Hawfinch, being the usual culprit. I would gladly sacrifice my peas to either species.

Beckwith reported that the first breeding record occurred in 1879 following an irruption the previous year. A nest in ivy growing on a fir tree at Llanyblodwel contained eggs when first found and the young were then watched until fledged. Unusually, the nest was started in September, so it may have been an early first brood or a late brood, depending on the type of conifer seed present, with mid-winter being a more typical time for laying. Breeding was suspected at the same spot the following year, as a pair was seen and heard regularly.

A few pairs bred at Clive between 1913–16 following a large irruption in 1909, and, after three irruptions in the late 1920s, possible breeding was reported in the Broseley/Wenlock area in 1928 and 1931. Further breeding was suspected, though not confirmed, at several widespread locations following irruptions in 1956, 1958 and 1962.

The Crossbill is rarely seen far from conifer plantations, many of which were planted after the Second World War. As the trees mature they produce the cones on which crossbills feed, extracting the seeds using their specially adapted bills. Thus entries in SBRs prior to 1976 show only small numbers, and for most of the 1960s no records were received. Since then, they have become established in many maturing conifer plantations and breeding, though hard to confirm, was often suspected. As a result of this increase in suitable feeding and breeding habitat, there were many more records in the recent winter and breeding Atlas periods than in the previous corresponding Atlas periods.

Local population trends are not available, but there has been a big increase in range since fieldwork for the *Atlas* (1992). At that time it was thought that local breeding was sporadic and no established population existed. This is no longer the case, as populations are readily found in the more extensive conifer forests and, although a crop failure or clear felling would cause them to move on, there are now more alternative sites available locally.

Nests are usually in the top or outer branches of conifer trees, but plantations can be hard to access and obtaining good views of the canopy can be challenging. The distinctive call attracts attention, but

Occupied Tetrads

Atlas period (breeding)	1985–90		2008–13		Change	
	Number	%	Number	%	Number	%
Confirmed	6	1	11	1	5	83
Probable	4	0	14	2	10	250
Possible	15	2	30	3	15	100
TOTAL	25	3	55	6	30	120
Tetrads with Winter Records (2007–13): 84 (10%)						

Breeding Season (2008–13) & Winter (2007–13) Distribution

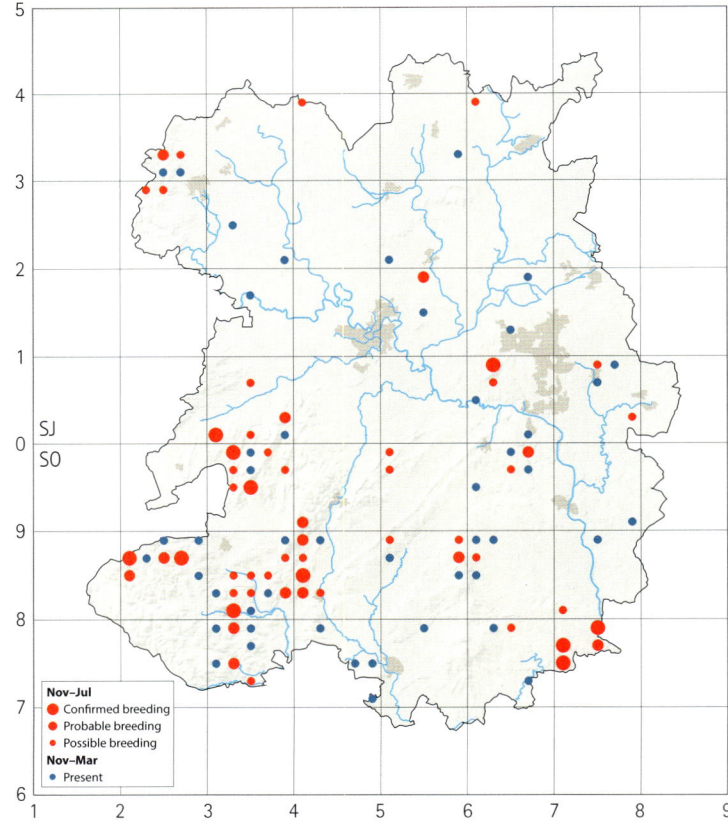

Nov–Jul
- Confirmed breeding
- Probable breeding
- Possible breeding

Nov–Mar
- Present

they often remain unseen or are only glimpsed as they fly overhead. The breeding season starts very early, as eggs can be laid in January or February, and juveniles seen in early spring. By the time most species are settling down to breed, Crossbill flocks, including fledged young, may have already moved considerable distances in search of new feeding areas. In irruption years, flocks from the continental far north often appear in July and have been recorded earlier.

This unusual breeding timetable, therefore, makes presentation of normal breeding and winter season maps difficult, so a special combined map has been produced. All records submitted with breeding codes are included, and such observations made in the winter period, or March, were particularly useful. These records have been mapped using the three levels of breeding evidence, irrespective of the month recorded, in the breeding season colour. As other records from the winter period are potentially possible breeding records, they have been shown on the map as blue dots, indicating presence during November–February, but only if there is not a breeding record for the tetrad. All of the confirmed breeding records were coded FL (recently fledged young). Three dated after the end of May, which is rather late for this species, have been included: in one tetrad earlier probable breeding records had been submitted, and the other two were each close to tetrads with breeding codes. In the latter cases the young could have originated from adjacent tetrads.

This map highlights the main known breeding locations, and also picks out smaller sites where breeding may have gone unnoticed. Many of the isolated red and blue dots probably represent wandering birds in late summer and winter but, if a good cone crop was available, opportunistic breeding attempts cannot be discounted. Irruption years are more difficult to identify now that a resident population exists, but none were known during the recent Atlas period.

The Oswestry uplands, Clun Forest, Wyre Forest, the Stiperstones and other forestry blocks in the south-west all have clusters of records in both seasons. Breeding was confirmed at the first location prior to the recent Atlas fieldwork, but access to the woodland concerned was withdrawn before the Atlas period. Breeding was confirmed at the other four locations, and these are the main areas for Crossbill. Brown Clee, Shirlett and the Wrekin also had records in both seasons and are likely to have breeding populations. The first two of these sites also had May confirmed breeding records in 1985–90. Breeding was also confirmed in some isolated tetrads, and others with records in both seasons are more likely to have breeding pairs than those with either winter or breeding season records only.

The increase in range, indirectly reflecting the increasing breeding population taking advantage of maturing coniferous plantations, is shown by the growth in the number of the County's 33 10km squares that were occupied. In 1968–72 none of them had records but, in 1985–90, 12 were occupied and in 2008–13, 26 were (although this includes some with winter only records).

The seed supply necessary for successful breeding will be more reliable in larger plantations, especially those with a mix of conifer species and different ages. Smaller woods are unlikely to sustain regular breeding because cones are not produced every year. Following an irruption these birds are more widespread and, if conditions are good, they will take advantage of the opportunity to breed so they may be found temporarily in new sites.

The national population is not easily monitored due to the timing of the breeding period, but the UK BBS indicates a 16% increase since 1995. The national BTO Atlas (2013) shows a dramatic change in both the breeding and winter periods. Most of central Wales has been colonised, with high relative abundances found in both seasons.

Between two and five pairs per occupied tetrad would give a population estimate in the range of 100–250 breeding pairs, but this will vary enormously from year to year. Both range and numbers are limited by habitat, but the Crossbill should continue to be a highlight of conifer plantations for the foreseeable future.

Allan Dawes

Two-barred Crossbill
Loxia leucoptera
Very rare irruptive winter visitor

Jim Almond, Wyre Forest, 8 January 2013

A resident of the coniferous forests of North America, north-eastern Europe and Asia, this Crossbill will 'irrupt' south if its food source fails. The two records were both of long-staying, over-wintering birds:

Walcot Forest	8 November 1972–3 February 1973
Wyre Forest	three, 28 November 2013–31 March 2014

The male in Walcot Forest was first located in coniferous forest on Sunnyhill in 1972 then re-found, again in conifers, in Blakeridge Wood, also part of Walcot Forest, in 1973 (Smith *et al.* 1974). It was usually with Crossbills *L. curvirostra*, and was considered to be a 'by-product of a marked (but not massive) invasion of Crossbills in late summer' of 1972 (Smith *et al.* 1973). The two males and a female in 2013 were part of an unprecedented influx of both this species and Parrot Crossbill *L. pytyopsittacus* into Britain. Throughout their stay they were usually loosely associated with Crossbills and Siskins, and often found feeding in Larches after being located by their distinctive call.

Graham Walker

Goldfinch (European Goldfinch)

Carduelis carduelis

Very common resident

Goldfinch is a partial migrant throughout central Europe, absent from northern latitudes above around 60°.

The 'seven-coloured linnet' was a popular name for this species, and one which I grew up with. Proud Tailor, Red-cap, Sheriff's man, Nichol and Nichws were also listed as former names by Forrest (1909c). These local names reflect its beauty, song and call, and indicate how popular and well known it was. The combination of a brightly coloured plumage and sweet song contributed to a downfall in its fortunes, as it was frequently kept as a cage bird and much sought after by bird-trappers. The Wild Birds Protection Act of 1880 provided protection during the breeding season and in 1908 Forrest wrote: 'It is gratifying to state that this pretty bird, which was rapidly disappearing due to the havoc wrought by bird-catchers, is now, thanks to the protection of the law, reinstating itself in our midst in something like its old numbers'.

The recovery continued, and the *Handlist* (1964) described it as 'widely distributed' and 'increased considerably in the last twenty to thirty years'.

More recently, a few were noted beginning to feed from peanut holders (SBR 1989). This new habit, coupled with improved and better quality seed mixes (Goldfinches are particularly fond of sunflower hearts and niger seed), has seen an upward trend in garden feeding. The maximum GBW recording rate was just 16% in 1995 but by 2011 it had reached 55%. The percentage has since fallen slightly and recent results suggest that the increase may now have peaked. The use of gardens builds up from late autumn when natural food supplies begin to run down and reaches a peak at the end of April. This enabled fieldworkers to obtain both late winter and early breeding records from town and village gardens, and also ensures that the birds are in good condition when they begin their breeding cycle.

In the breeding season, Goldfinches are frequently seen in pairs as, during incubation, the male will accompany the female when she leaves the nest to feed. Probable breeding is therefore quite

Paul King, Venus Pool, 18 June 2010

straightforward to prove, but confirming breeding normally depends on seeing recently fledged young.

The *Atlas* (1992) showed a wide distribution, but with many gaps, and they were not found in 10% of tetrads. Recent fieldwork shows an almost complete distribution. They were not found in three tetrads on the Long Mynd, which have very limited suitable habitat, and two other tetrads where few additional visits were made to improve coverage. The increase in the number of occupied tetrads is undoubtedly real and in line with the findings of both the BBS and GBW.

A high relative abundance was found in the north-east, but there are also many other similar pockets, including in the south-west and north-west where upland valleys and hill sides, often deserted during the winter, are less intensively managed and provide ideal habitat for breeding.

During 1985–90, numbers were beginning to recover from a low point in the mid-1980s, attributed to agricultural intensification resulting in fewer weed seeds being available in winter. Since that time there has been a steady increase. The BBS trend shows the population has more than doubled since 1997. This is almost entirely explained by improved annual survival, and a major factor will be the exploitation of garden bird food.

Based on TTV counts, the estimated population is between 22,000–22,500 pairs. This is equal to 25 pairs per tetrad, a fourfold increase on the estimate in the *Atlas* (1992), and broadly consistent with national and local BBS trends.

During the autumn, large flocks can gather where feeding opportunities are good. The wild bird seed crop of chicory at Venus Pool attracted about 300 in August 2009, and this was the largest flock recorded during the recent Atlas period. These flocks tend to break up once the glut of seeds has been reduced.

The largest winter TTV count of 140 came from Church Preen in

Garden BirdWatch Reporting Rate (1995–2014)

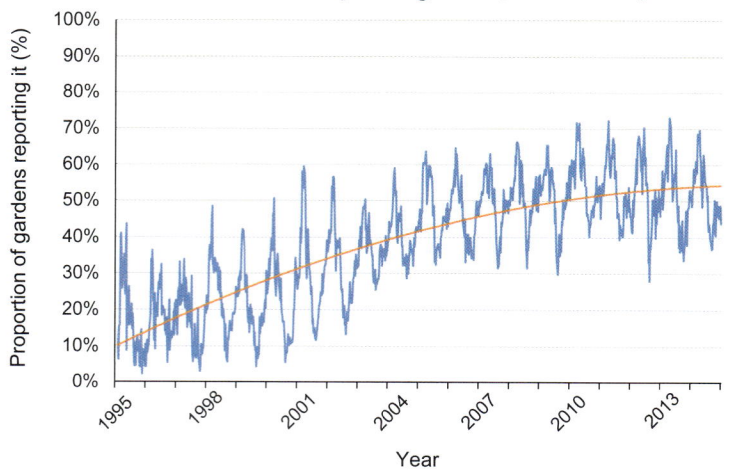

December 2009, but the average number recorded during each hour was only five, and they were not always found in a two-hour visit. Goldfinches have a more restricted distribution at this time as they abandon some of the upland areas such as the Long Mynd, Clun Forest and the Clee hills, but are otherwise widespread. Their absence from the high ground during the winter months is not surprising, but the few gaps in distribution elsewhere on lower ground would probably have been filled with more fieldwork.

Although some tetrads in the south had a high winter relative abundance, many more were found in the north-east, a change in distribution which can be clearly seen when comparing the two relative abundance maps. Lower altitude may be more beneficial during the winter and there is likely to be more natural food available, but there are also many more gardens.

The vast majority of Goldfinch recoveries have been of birds ringed and recovered here. This is a surprisingly high proportion, suggesting a resident population. However, 60 other movements have been recorded between other parts of the British Isles including:

BBS Trend (1997–2014)

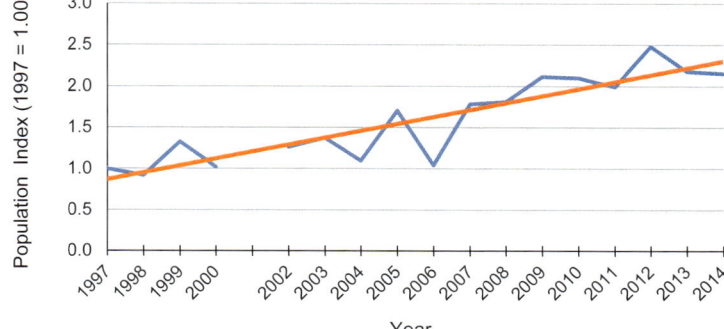

Ireland (two), the Isle of Man (three) and Scotland (nine). The five furthest movements from Scotland for which details are available are indicative of Scottish breeders, ringed in April and May, and wintering locally. Some overseas migration for the winter is confirmed by nine movements, involving Belgium (one), France (six) and Spain (two). Four of these were ringed here in late summer and found abroad later in the same autumn, including the furthest traveller, a nestling ringed at Shrewsbury on 12 July 1967 and re-trapped in Caceres, Spain on 10 November, 1,434km distant. Three others were ringed locally during the winter and found abroad in a subsequent winter, indicating that they may adopt a different strategy from year to year. This may be dependent on local weather conditions as one of these was caught twice in France, in February and March 2010, during a very cold winter.

Occupied Tetrads

Atlas period (breeding)	1985–90		2008–13		Change	
	Number	%	Number	%	Number	%
Confirmed	340	39	407	47	67	20
Probable	344	40	419	48	75	22
Possible	97	11	39	4	-58	-60
TOTAL	781	90	865	99	84	11
Tetrads with Winter Records (2007–13): 823 (95%)						

Breeding Relative Abundance (2008–13)

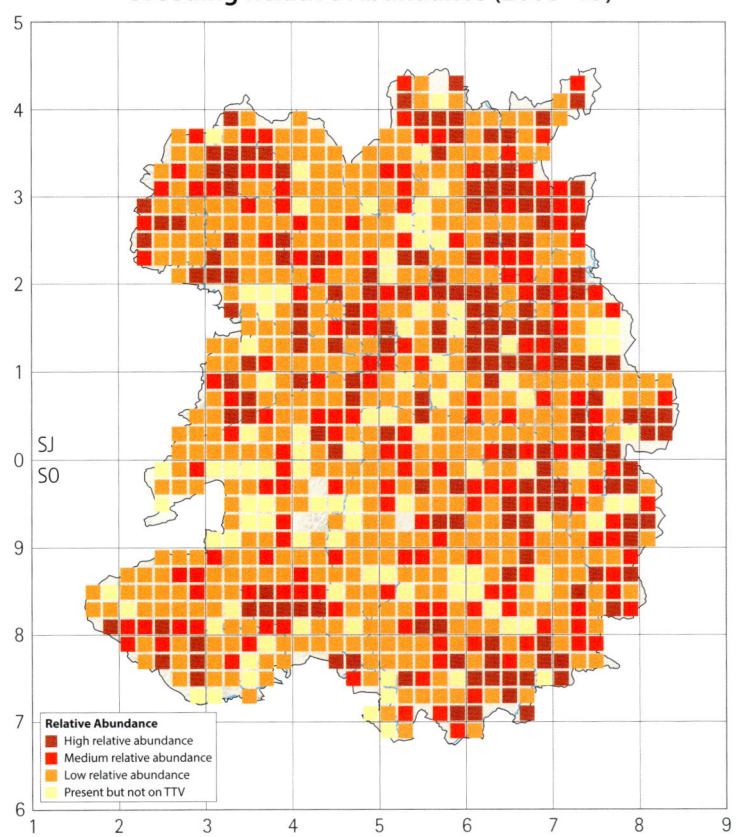

Winter Relative Abundance (2007–13)

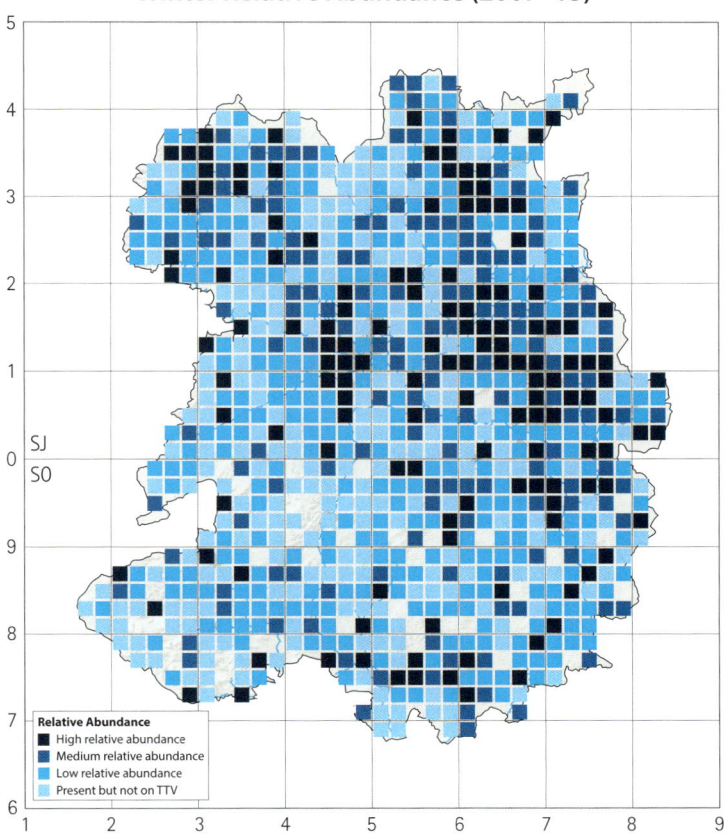

The oldest at 8y 5d was ringed as a first-year male in October 2004 at Cross Lane Head and re-trapped at the same site in November 2012. It was approaching the longevity record of 8y 8m 4d and may have gone on to surpass it. Another ringed at the same site in December 2006 was caught in each subsequent winter and was still alive in April 2013. These, along with others from this site and from the Rea, Upton Magna, demonstrate great fidelity to a regular wintering area. Different strategies are clearly present within the population, and if milder winters continue those remaining may be at an advantage, so it will be interesting to see what the future holds. The effect that supplementary feeding may be playing has

been the subject of an investigation by the BTO. They asked Garden BirdWatchers to collect information on numbers and food taken by Goldfinches and what other supplementary and natural foods were available between November 2015 and February 2016. Some preliminary results from this survey have been broadcast, but the main issue, whether garden feeding is driving population growth, is ongoing.

At present, Goldfinches are doing well although it looks as if the population may have reached its limit. Hopefully, they will remain immune to the disease affecting Chaffinch and Greenfinch.

Allan Dawes

Siskin

Spinus spinus

Uncommon resident and fairly common winter visitor

Rob Stokes, Radbrook, 10 April 2013

In addition to breeding throughout northern Europe, isolated populations are present in many mountainous areas to the south.

In the nineteenth century, Siskin was known exclusively as a winter visitor, occurring in very irregular numbers with flocks of 20 to 30 representing a good year. It was known to be fond of alder and birch seed and could regularly be found feeding in the company of Lesser Redpoll. It was often very tame, as Beckwith noted 'I shot one out of a flock of fourteen or fifteen. To my surprise the others flew around and again alighted in the same tree, and on my firing for a second time, they again came back, after which I left them in peace'. He went on to say 'luckily for it, bird catchers do not care to ensnare it, and men who eagerly pursue Redpolls leave Siskins unmolested'.

An unsuccessful breeding attempt was noted in 1898 at Grinshill. A nest was built in a fir tree but, after a single egg was laid, the nest was predated by a Jay. In 1931, a pair reared a brood in a damson tree in an orchard at Bucknell and two pairs successfully reared broods there the following year. Despite their success, no further breeding attempts were noted and it was not recorded in any of the County's 33 10km squares during the 1968–72 national BTO Atlas. There

were no breeding season records until 1975 when a male was heard singing in Clun Forest in July. Summering individuals were located at other sites in the following two years, and since 1983 they have become an annual occurrence. The first modern confirmed breeding record came from the Wyre Forest in 1986, during fieldwork for the 1985–90 breeding Atlas, when the population was becoming established. The map in the *Atlas* (1992) showed it in 21 of the County's 33 10km squares, although it was still thinly distributed and found in only 56 tetrads.

Nesting occasionally occurs in large gardens or villages where mature conifer trees are present, but conifer plantations are the main breeding habitat. These are often difficult to access and unpopular with birders, and although song can be prolific inside these plantations, confirming breeding can be hard. Most records are, therefore, of probable or possible breeding. The onset of breeding varies between years, influenced by the scale of the previous year's cone crop, but the situation is further complicated because Siskin are still to be found visiting gardens in April and May. Some probable and possible breeding records in each Atlas period are likely to refer to such birds, perhaps more so in 1985–90, so the breeding distribution change map needs to be treated with caution, especially the apparent losses in the north-east.

However, it is clear from the map and tetrad occupancy table that a substantial population increase has taken place, with a 191% increase in tetrads with breeding evidence, mainly in the south and west where the main conifer plantations are located. It was not found in only three 10km squares.

Although the UK BBS trend showed a dip in the early 2000s, there has subsequently been a steady rise (59% between 1995–2014), and Siskin numbers are at an all-time high. The national BTO Atlas 2013 shows an increase in both distribution and relative abundance along the adjoining Welsh border since 1991, which will have provided an additional source of recruitment to the local breeding population. The increase is thought to be a result of maturing conifer plantations providing ideal breeding habitat, and garden feeding aiding over-winter survival and breeding fitness when natural foods are in short supply.

The *Atlas* (1992) considered colonisation to be still at an early stage and estimated the population at less than 100 pairs. Based

Breeding Distribution Change (1985–90 to 2008–13)

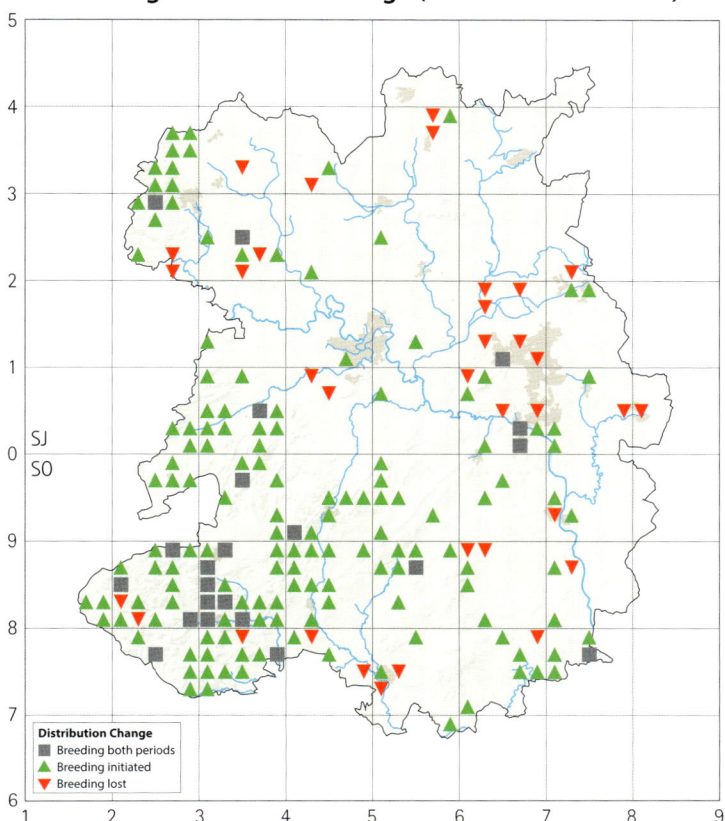

Distribution Change
- ■ Breeding both periods
- ▲ Breeding initiated
- ▼ Breeding lost

Winter Distribution (2007–13)

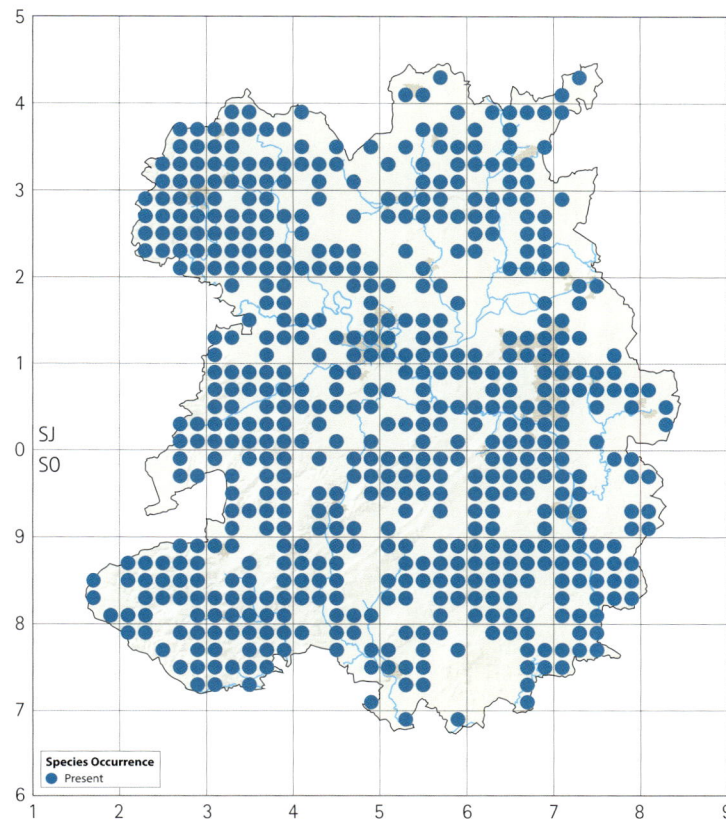

Species Occurrence
- ● Present

on TTV counts, the estimated population now is between 700–750 pairs. While such estimates may be unreliable for species with small populations, this equates to four to five pairs per occupied tetrad which seems appropriate. Siskin can certainly reach high densities in some plantations, but are very sparse in others.

In winter, they were first noted taking food from a bird table in Ludlow during February 1974, and two years later feeding on peanuts was said to be increasing with several records from Shrewsbury, Church Stretton and Stoke Heath. Their use of gardens has continued to increase, but it varies considerably from year to year, depending upon the amount of natural food available, as shown in the GBW

Occupied Tetrads

Atlas period (breeding)	1985–90		2008–13		Change	
	Number	%	Number	%	Number	%
Confirmed	5	1	23	3	18	360
Probable	18	2	56	6	38	211
Possible	33	4	84	10	51	155
TOTAL	56	6	163	19	107	191
Tetrads with Winter Records (2007–13): 567 (65%)						

Garden BirdWatch Reporting Rate (1995–2014)

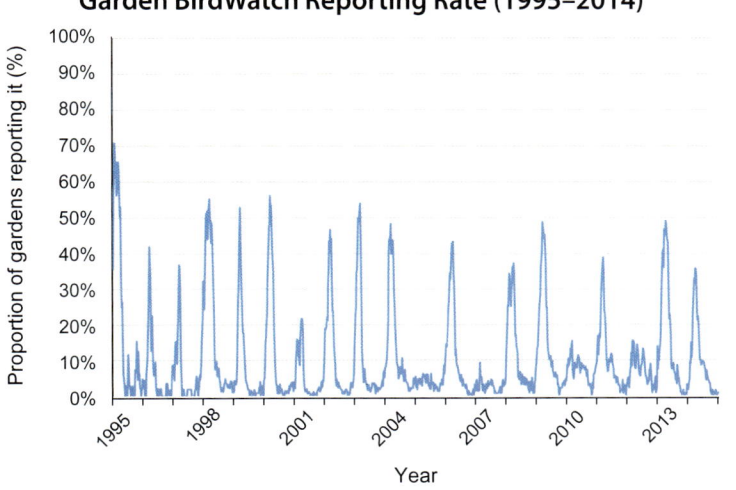

graph. Garden use tends to increase as winter progresses, with few visiting before the end of the year, then a steady rise which reaches a peak in March. Good numbers can often be found throughout April and some continue to visit feeders into May.

Away from gardens, they can be found feeding on alder and birch seeds, particularly along water courses or fringing wetlands, and they can also found in conifer plantations where they utilise cone seeds. The winter distribution map shows them to be widespread, but there are large gaps which mainly coincide with intensively arable areas with fewer feeding opportunities. Siskins were found in every 10km square, whereas the national Winter Atlas (1986) showed them to be absent from two. The number of winter visitors varies each year, mainly influenced by the success of the cone crop elsewhere, but the local breeding population now provides a permanent winter presence.

Siskin are easily trapped at garden feeders and over 4,000 have been ringed here since 2007. Annual numbers caught are very variable, with less than 40 in three of these years and over 1,100 in two. As is the case with some other species, many more pass through gardens than are counted at any one time. No more than 10 were seen in a Wellington garden in 1982 but 69 were ringed and a further 97 the following year. Feeders are also used by some ringers to trap Siskins once they have returned to their breeding woods, and this may have influenced the number of re-traps in some areas.

Only about one quarter of those ringed here are recovered here. Some demonstrated great site fidelity, as four were re-trapped at the original ringing location over five years later, and one of these was also found at the same site on the Isle of Skye in two different springs: Highley in February 2006 and February 2008, Isle of Skye in May 2008 and April 2009, and Highley again in February 2011. Over 40% of the remaining movements were to or from Scotland and 4% from Northumberland, and the timing suggests that many of our winter visitors breed in these northern areas. However, some will have been on passage, as shown by one re-trapped on Fair Isle in October 2012 where they do not breed, and another re-trapped on South Ronaldsay on 25 April 1986, just 21 days after being caught at Telford.

Belgium dominates overseas recoveries, but many of these will be passage birds moving to and from Scandinavian and Baltic breeding sites, which involve 27 of the recoveries. The furthest movement was of a male ringed near Leningrad (St Petersburg) in late May 1985, presumably breeding in the area, which travelled 2,331km to spend the following winter in Wellington. They sometimes winter in widely differing areas; one ringed in Spain in March 2004 was re-trapped at Chelmarsh in December the following year.

Some mature conifer plantations are now being harvested and replanted, but there is still plenty of suitable habitat available. With supplementary feeding during lean times the future for the Siskin looks promising.

Allan Dawes

Corn Bunting
Emberiza calandra
Uncommon resident

Jim Almond, Venus Pool, 27 February 2015

The Corn Bunting breeds mainly across central and southern Europe, and the Asian Steppes, and reaches its most northerly latitude in Scotland.

Historical records all refer to this species as the common bunting, and although vernacular names are often misleading, in this case it seemed to be apt, at least until the 1870s. Prior to this date, 'common' was mentioned by several writers and in 1890 Beckwith wrote that the late Henry Shaw '[told me] this bunting used to be found on all the high ground round Shrewsbury, and in winter, numbers were killed in stackyards, and brought to him with other small birds as food for tame hawks and owls'.

By the end of the 1870s, it was described as rare. Beckwith had only seen it on one occasion and in 1890 wrote: 'At a period when most small birds are increasing in numbers, more especially the granivorous ones, as their natural enemies have decreased, and they do not suffer from starvation in winter, the scarcity of the present bird, which 50 years ago was common, is a curious yet inexplicable fact'. In 1904 Paddock attributed the decline to 'the decrease in the grain growing area'. By 1908 the population had rallied and although local, it could be found in several places. By 1940 it was described as decidedly rare, again the result of a decline in arable farming which reached a low point between the two world wars. Following a BTO survey in 1952, which covered 15 5km squares, it was classed as 'observed but probably not breeding'.

Many of the early SBRs have no mention of this species at all. In 1960 a pair, possibly two, near High Ercall provided the first modern breeding record. Recent breeding by one or two pairs to the north-west of Wellington and at the Weald Moors was noted in the *Handlist* (1964), which suggested that up to 10 pairs may have been present in what had 'apparently always been a strongly favoured area'. A few years later, in 1967, it was well established in the north-east, and the next three SBRs all commented on the expanding population. During the Second World War arable production, particularly barley, increased and this continued until the mid-1970s, enabling the expansion of the Corn Bunting population. In the national BTO Atlas (1976) it was shown to be breeding in 10 of the County's 33 10km squares, but, while the national BTO Atlas (1993) documented a rapid contraction, this was one of the few places to show an increase, with breeding records from a further seven 10km squares, probably because the local population was still expanding when fieldwork for the earlier Atlas was being conducted.

The jangling song, usually delivered from one of the highest perches available, is distinctive and carries well, enabling singing males to be readily located, and the frequent use of the same song post is handy for recording territories. Follow-up visits often need to be made to confirm breeding, when adults carrying food can easily be observed in the open arable fields where they usually nest. They breed late into the season, and young are often being fed in August, so a number of confirmed records may have been missed, as some observers will have stopped looking at the end of the main recording period.

A sample of tetrads with breeding records in the *Atlas* (1992) was re-surveyed by SOS in 2004, with the findings added to SBR records 2002–04. In total, results were received from 90 tetrads. This suggested that Corn Buntings were being lost from the periphery of the three main populations (Dawes SBR 2004).

The loss has continued, and the breeding distribution change map shows a marked contraction in range in both the north and south-east since 1985–90, while the smaller groups in the west and south-east are becoming isolated from the main central population, as previously suggested by the SOS survey. The continued expansion at the western edge of the range, also noted during the winter period, is in stark contrast to the position elsewhere and there is no obvious explanation for it. Corn Bunting have a close association with cereals, particularly barley, and there has been a continuing switch away from barley to wheat since 1985. Many of those breeding in SJ31 and SJ32 were observed breeding in cereal fields (*pers. obs.*), and their distribution may reflect different cropping regimes. Crop details were not recorded, but those in SJ32 were on sandy soils and barley does better than wheat in these conditions.

The fluctuations in the Corn Bunting population in the past have been driven by agriculture. The present plight is believed to be due to the reduced availability of seeds and invertebrates, which affects both productivity and over-winter survival. Corn Buntings were found during TTVs in 70 tetrads, and using the highest count from either the early or late period gives a total of 294 individuals. Singing males will have predominated, so for these tetrads a population of around 300 pairs seems likely. However, breeding records were obtained from a further 59 tetrads, so the population may be nearer 400 pairs, which is consistent with the results of the 2002–04 survey, but at the lower end of the estimate in the *Atlas* (1992). This is to be expected following the reduction in range since that time. The national BBS trend continues to fall, down 34% between 1995–2014 in both the UK and England. Locally, with the exception of the north-west, the contraction of range is a sign of a declining species.

Corn Buntings are not easy to find during the winter. Flocks resting in trees or bushes at field margins provide the best chance of an encounter, but small numbers actively feeding in large fields will certainly have been missed. They were found in just 64 tetrads during the winter, compared with 128 during the breeding season, but 23% of these tetrads had no breeding season records. Although they are mainly sedentary some local movements do occur. Since December 2010, a few have been seen feeding in the wild bird food crop at Venus Pool during the winter, but despite occasional singing in April they have not stayed to breed.

The main locations, based roughly on Weald Moors, Beckbury and Queens Head, all had groups of between 50 and 125 in several

Occupied Tetrads

Atlas period (breeding)	1985–90		2008–13		Change	
	Number	%	Number	%	Number	%
Confirmed	39	4	17	2	-22	-56
Probable	91	10	65	7	-26	-29
Possible	58	7	46	5	-12	-21
TOTAL	188	22	128	15	-60	-32
Tetrads with Winter Records (2007–13): 64 (7%)						

Breeding Distribution Change (1985–90 to 2008–13)

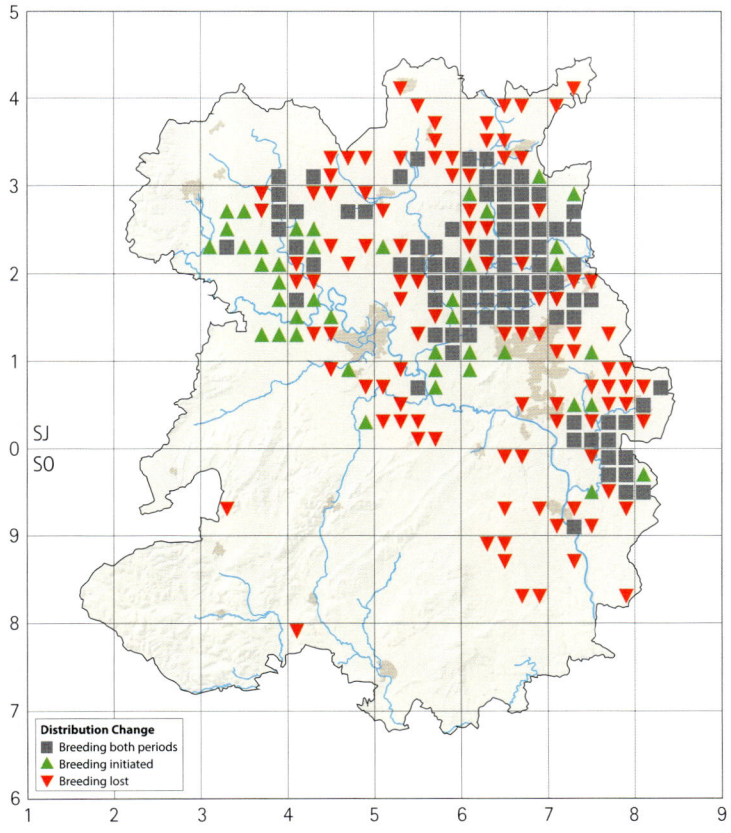

Distribution Change
- Breeding both periods
- Breeding initiated
- Breeding lost

Winter Distribution (2007–13)

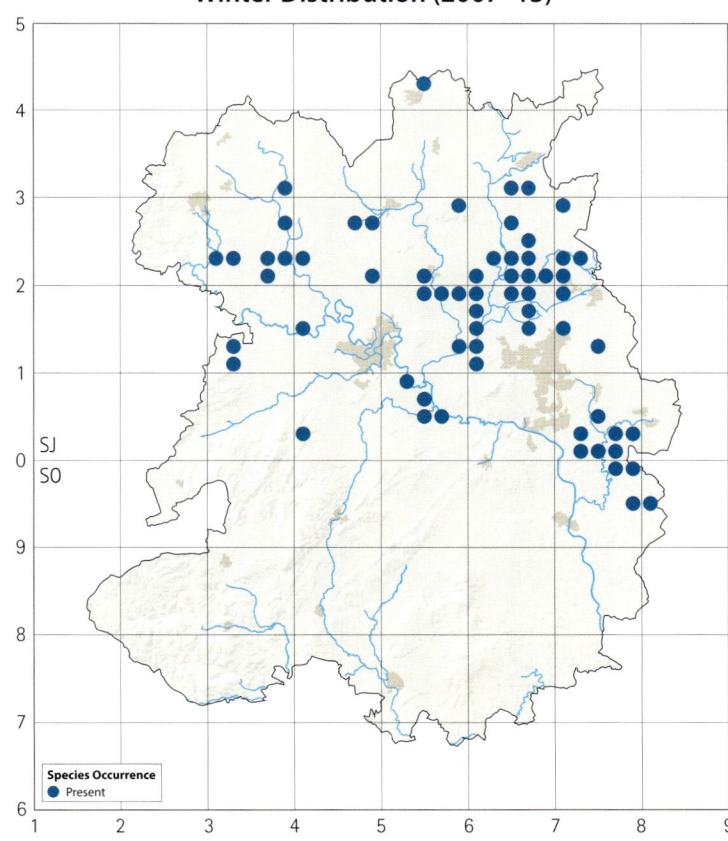

Species Occurrence
- Present

winter periods. It would be interesting to know what proportion of the local population these flocks contain. If it is concentrated in a few flocks they will be easier to find, but the area that they forage over may be limited which, combined with the difficulty of locating them, may explain the lower number of tetrads where it was found in winter compared with the breeding season. The national BTO Atlas (2013) shows an increase in the north-west, where six additional 10km squares have been occupied since the national Winter Atlas 1986. This is consistent with the increases in the breeding season since the national BTO Atlas (1976).

Only small numbers of Corn Bunting have been ringed and there are no local recoveries. National findings, although based on a small sample, indicate that they only move short distances. The future looks bleak. The national BTO Atlas (2013) shows that the local population is already virtually isolated, so if it dies out the chances of re-colonisation look slim. Maintaining it by providing a winter seed supply and better quality breeding habitat is crucial and, for this species more than any other, its future depends almost entirely upon what happens within this County.

Allan Dawes

Yellowhammer
Emberiza citrinella
Common resident

Yellowhammer is one of our most familiar residents, found in open habitats, particularly arable areas and farmland with mature hedgerows, bracken-covered hillsides, commons, and patches of scrub, as long as there are plenty of song posts.

Leighton (1836) called it 'Yellow Bunting' in the first historical record, and described it as 'common in high-roads'. Eyton (1839) described 'Yellow-hammer' as 'equally common everywhere', and noted that specimens from Anglesey, obtained at the same time of year, were brighter than those obtained locally. Rocke (1865) said that 'Yellow Bunting' was 'too common to require any remarks', and Beckwith (1879) wrote that it was 'very abundant, and locally known as "the Goldfinch", the true Goldfinch being called a "Seven-coloured Linnet"'. He added in 1889 that some often nest late into the summer and referred to three nests still active in August. Paddock (1897) said they were common and abundant, and 'gregarious in winter', often feeding with finches in 'stackyards' and manure heaps in severe weather. Local names included 'Writing Master', probably due to scribble-like marks on the eggs, although a clutch of white eggs was found near Ludlow (Forrest 1899; 1908).

In the first half of the twentieth century it was still called common, abundant and frequent, both in the breeding season and in winter. Most early SBRs only reported it, with no comment, on the list of regular breeders, but one was still sitting on three eggs (in Bishop's Castle) on 5 September 1956. The *Handlist* (1964) summarised its status as 'resident, common' with 'flocks of up to 50' in 'early spring, particularly on the high ground'.

There was another very late nest with young at Horton on 9 September 1974, unfortunately later destroyed. Flocks of 50+ at Chelmarsh and about 50 at Cranmere Bog were seen in 1976, and the first reports of visits to bird tables were in 1979, but no locations were mentioned. High counts in the 1980s included an exceptional 400+ on freshly-spread manure near Shifnal on 17 January 1984, and over a dozen flocks of more than 50 in 1986. A nest containing five eggs, plus a Cuckoo's egg, was reported in the same year: the young Cuckoo was later seen being fed by the adults.

The *Atlas* (1992) identified it as 'distributed widely', and absent from only 10 tetrads covering 'some urban areas' and 'high open moorland in the southern hills'. The very secretive behaviour during breeding was mentioned, and this may partly explain the small number of breeding records after 1990.

Notable winter flocks of 100 or more were 150 at Exfords Green in January 1992, 130 feeding on stubble in Little Drayton in December 1995, 100 at Kingsnordley in the same month, and 160 in Albrighton in February 1999. The first mention of declining numbers was of a national decrease of between 25% and 50% over the previous decade found by BTO, and, in the same year, an absence from the CBC plot at Oswestry Old Racecourse for the first time since the survey began in 1964 (SBR 1998). Counts had been in double figures on the CBC plot for almost half of the years up until 1988. Despite the reported declines, it continued to be widespread and well reported in SBRs for the 2000s, perhaps partly due to raised observer awareness of the plight of our farmland birds and of the importance of submitting records; however, the size of all flocks were generally noted as being smaller. Larger flocks were of 120 at Wall Farm Kinnersley in March 2000, 105 at Tedsmere, in January 2002, and 110 at the same place 13 months later, and 130 at Venus Pool in February 2004.

In the recent Atlas period, they were found in 765 tetrads (88%) in the breeding season, compared with 860 tetrads (99%) in the previous Atlas. There were large areas where they were not found at all, mainly along the Teme valley in the south-west, the Oswestry uplands in the

Allan Heath, North Shropshire, 23 March 2005

Breeding Distribution Change (1985–90 to 2008–13)

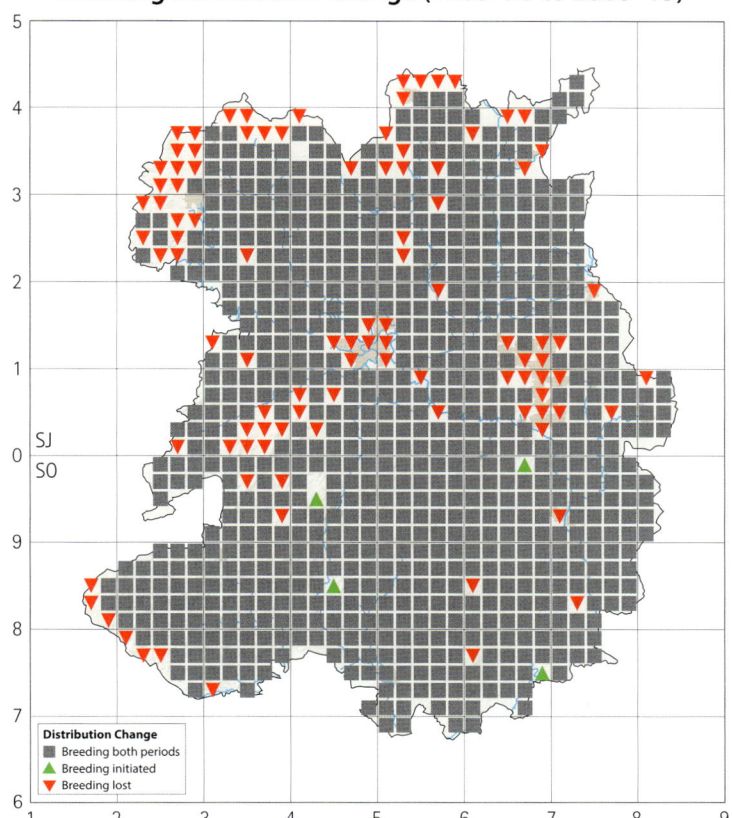

Distribution Change
- ■ Breeding both periods
- ▲ Breeding initiated
- ▼ Breeding lost

Breeding Relative Abundance (2008–13)

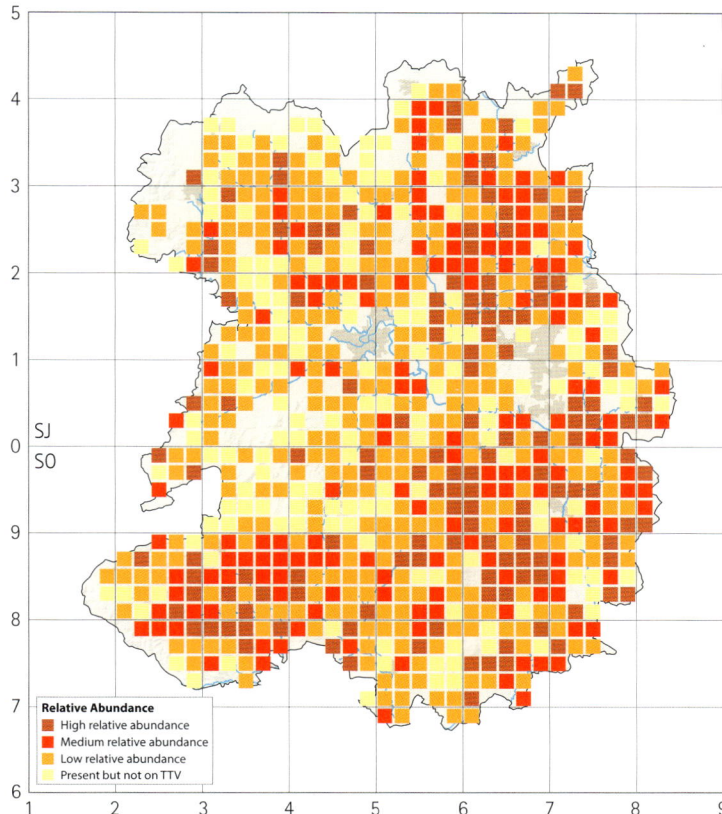

Relative Abundance
- ■ High relative abundance
- ■ Medium relative abundance
- ■ Low relative abundance
- ■ Present but not on TTV

Occupied Tetrads

Atlas period (breeding)	1985–90		2008–13		Change	
	Number	%	Number	%	Number	%
Confirmed	541	62	251	29	-290	-54
Probable	269	31	385	44	116	43
Possible	50	6	129	15	79	158
TOTAL	**860**	**99**	**765**	**88**	**-95**	**-11**
Tetrads with Winter Records (2007–13): 635 (73%)						

north-west, arable areas in the north, the upper Rea valley around Minsterley, the northern edge of the hills south-west of there, and the two main urban areas of Shrewsbury and Telford, and Bridgnorth. The breeding distribution change map shows that it has disappeared from all those areas since 1985–90, from 11% of the tetrads where they were found previously. The four widely scattered gains are likely to be due to increased fieldwork effort in the later period.

The relative abundance map shows areas of highest density in the east, and in river valleys around Clun, associated with availability of grass and cereal seeds, and invertebrates, for summer food, together with hedgerows and trees for nest sites. There are low densities around the Long Mynd and other hills in the south-west, as well as in scattered areas of pasture elsewhere.

BBS shows a population decline of about 25% since 1997, indicating a reduction in densities where it still occurs, as well as a contraction of range.

Based on TTV counts, the estimated breeding population here is 13,900–14,300 pairs, compared with an estimate of around 35,000 in

1990, based on national figures from the BTO Atlas (1976) and a local CBC plot, both of which averaged 10 pairs per sq.km. However, the national figures were actually much lower by 1990, as Yellowhammer declined by 55% between 1970 and 2014, but only 14% since 1995. After making the necessary reduction in the 1990 estimate, by about 40%, then applying the decline found by BBS, it is broadly compatible with the current estimate.

In winter, they were found in fewer tetrads, 635 (73%). The relative abundance map is similar to the breeding season map, as expected for a sedentary species, but shows a shift away from areas of low abundance, particularly the uplands in the south-west, to the arable areas in the north and east, in search of winter food, especially in stubble fields where they still exist, and wild bird and game crops. An impressively large flock was present at Venus Pool during December 2014, when there were up to 400, together with a similar number of

BBS Trend (1997–2014)

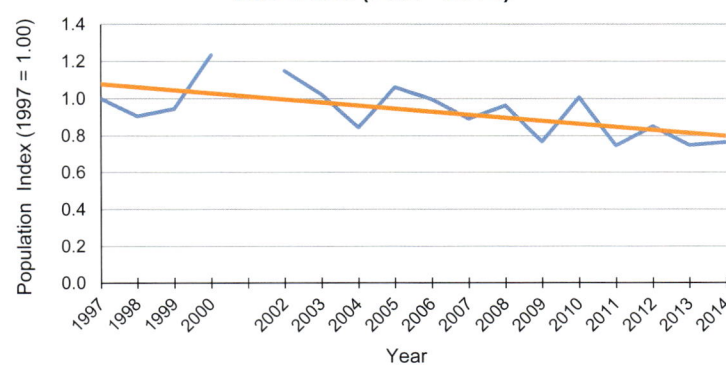

Winter Relative Abundance (2007–13)

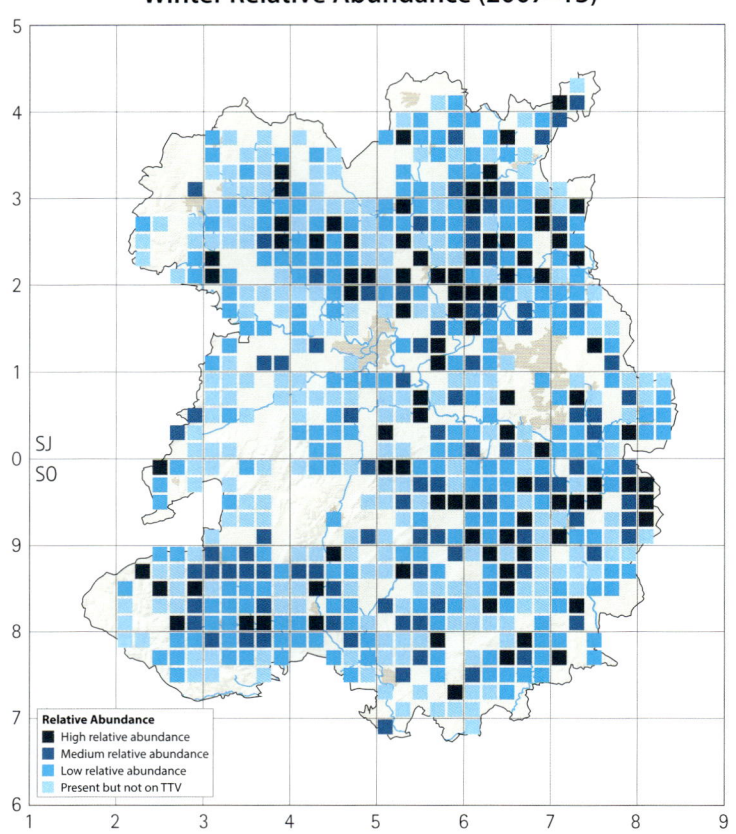

Relative Abundance
- High relative abundance
- Medium relative abundance
- Low relative abundance
- Present but not on TTV

Reed Buntings, and up to 22 Corn Buntings, feeding on bird-seed crops in the arable field on the reserve.

Apart from the large numbers in several recent years at Venus Pool, on the specially created habitat, large flocks are a thing of the past, providing further evidence of the decline. There have been only two records of 100 or more since 2003, 100 at Woolston in February 2009, and 200 at Clunton Coppice in February 2011. During six years of fieldwork for the recent winter Atlas, there were no counts of 100 or more, and only six of more than 50. There were only 19 counts of more than 25 on one-hour TTVs.

The lack of availability of food has contributed to the decline, mainly as a result of the switch from spring to autumn cereals, leaving no seed-rich winter stubble fields, and also pesticide use reducing the availability of invertebrates to feed the chicks in the breeding season. Effective use of the relevant provisions in Natural England's agri-environment schemes could make a big difference, though their future is uncertain. Recent changes to hedgerow regulations for farms, banning cutting in August, will also help, although some very late September nests will continue to be destroyed.

Few are ringed, but the eight recoveries confirm its sedentary behaviour, in that all eight were ringed here too. The oldest, ringed as a nestling in May 2002 at Wollaston, was caught at the same site in January 2008, 5y 7m 29d later.

Yellowhammer is largely associated with cereal farming, so if the pressures for ever-higher yields are not mitigated by habitat restoration and improved winter food supplies, by provision of wild bird crops and weedy stubbles through agri-environment schemes, the future for the species is bleak.

Helen J. Griffiths

Pine Bunting
Emberiza leucocephalos
Vagrant

Closely related to the Yellowhammer (*E. citrinella*), this bunting breeds across much of Siberia from the fringes of Eastern Europe, where 50–120 pairs breed, to the Sea of Okhotsk. It migrates south to winter from Turkestan through to eastern and southern China and is a major rarity in Britain.

There has been just one record:

Venus Pool   1–6 January 2017

The autumn of 2016 saw a major influx of eastern vagrants to Western Europe, including at least ten other Pine Buntings (BBRC 2017 *in prep.*) and it is thought likely that this individual was part of that movement. It was found amongst a mixed flock of birds dominated by Yellowhammer and Reed Bunting that were feeding on a crop of oats and kale grown specifically as a winter food source for birds. Throughout its stay, it only occasionally appeared on top of the crop or in the adjoining hedge, and many visitors either failed to see it or had only the briefest of views.

As this was a first for the County, it is one of only a few records for 2017 in this publication.

Graham Walker

Jim Almond, Venus Pool, 1 January 2017

Cirl Bunting
Emberiza cirlus
No modern records

Britain is at the north-west corner of the Cirl Bunting's range in southern Europe. It was widespread in England and Wales up until the 1930s, but it subsequently suffered a rapid decline and contraction of range, mainly due to agricultural change including loss of winter foraging habitat (stubbles and fallow fields).

Rocke (1865) referred to the possibility of Cirl Bunting being overlooked or mistaken for Yellowhammer (always a potential problem) and Beckwith (1879) reiterated it.

There are 21 records. The first was of a specimen collected by Beckwith near his home in Eaton Constantine 'during the severe frost of January 1879', which is listed in the Peplow Hall Museum catalogue (Forrest 1907). Subsequently Beckwith (1890) noted a report of a singing male in a hedgerow at Whitcliffe Wood on 23 June 1882. The observer wrote that 'I think it had a nest near but I could not find it'.

Paddock (1897) reproduced an earlier report, again from Whitcliffe Common and from some time before 1882: 'I saw the birds, and found the nest with one egg in it. I took the egg, replacing it with a Yellow Hammer's, but the nest was so exposed, that it was taken by someone else in a few hours ... I have the egg in my possession.' The last nineteenth-century record is a specimen from Loton Park, known to be in the Ticklerton Collection in 1900, from 'many years ago' (Forrest 1908).

At the turn of the century there was an extraordinary spell of breeding records. In 1900 a pair was located at Worlds End, Church Stretton, and on 8 May another pair was seen at Kinnerley 10km south-east of Oswestry. In 1901 confirmed breeding was reported from five localities. The first, again close to Oswestry, a pair with young at Sweeney Park on 6 July, were 'probably stragglers from the old-established colony at Glyn-ceiriog', near Wrexham in Wales (Forrest 1908). The three other 1901 records were of two near Ludlow and one from Bridgnorth, and the last was of a nest and eggs at Marrington.

In 1904 there was possible breeding at two sites at Dowles in the Wyre Forest, and 'two or three pairs bred Ludlow district', the last a confirmed breeding record. Occasional further records came from Dowles, a single seen on 11 April 1905, a pair on two dates in February 1906, song heard on 28 March 1908 and 3 April 1910, and finally a pair on 2 May 1917.

This bout of breeding occurred around the same time as the only recorded nesting of Dartford Warbler, similarly near Ludlow. Two pairs of the warbler were present in 1903, and in 1904 when one pair bred successfully and a clutch of five eggs was collected.

The history of pre-1950 records ends with 'several at Eaton Constantine for a few days' at the end of January 1928, and song at Sharpstone ridge on 27 June 193(3?) (SSOS 1937). Five reports received in the period 1950–1961, some of which were noted in the *Handlist* (1964), are not now considered acceptable.

It is now restricted to south-west England, and the last record was around 1933, so recolonisation is not imminent.

John Tucker

Reed Bunting (Common Reed Bunting)
Emberiza schoeniclus
Uncommon resident

Dave Barnes, Catherton Common, 19 April 2012

The Reed Bunting is found across most of central and northern Europe, but is patchily distributed in the south. Here, it has always been closely linked to riparian habitats. Consequently, it was commoner in the north, and nineteenth-century accounts noted that it was particularly numerous in the areas around Ellesmere, Wem and Whitchurch. South of the River Severn, Berrington and Bomere Pools were the only places that it was at all numerous. This quiet unobtrusive species seems to have kept a low profile, and no obvious changes can be found in the historic record. The *Handlist* (1964) description, 'well distributed ... along rivers, canals and the borders of lakes', mirrors that of the early authors, only adding 'it is found in marshy areas at the top of the highest hills during the summer months'.

In the early 1970s, a national increase occurred which was associated with a gradual spread into drier habitats, especially farmland, but it was to be short lived. Agricultural intensification was gathering pace at this time and many drainage schemes were being implemented so, by the early 1980s, numbers had fallen and were well below their previous levels. Since then the population has fluctuated, with a slight peak at the time of fieldwork for the *Atlas* (1992), but it remains low. Detailed analysis suggests that the fall was caused by decreased over-winter survival rates caused by agricultural changes, and that increased nest losses have since prevented a recovery. Continuing habitat loss will have exacerbated the situation.

The monotonous song ensures they are easily found in suitable habitat, usually wet or marshy ground bordered by tall marginal vegetation and scrub. The number of tetrads with breeding evidence has remained very similar in the two Atlas periods, but changes in distribution since the *Atlas* (1992) are extensive and intriguing. Losses and gains are similar in number, but the former are concentrated in the central belt around Shrewsbury and Telford, and in the north-east. Urban sprawl has engulfed some tetrads surrounding both towns, but losses further afield, and in the north-east, where densities are otherwise high, are puzzling. It may be that the slight peak recorded at the end of the 1980s encouraged more in to sub-optimal habitats and these have since been deserted. Crop rotation could also explain some of the changes in the central and eastern arable areas. Reed Buntings have recently taken advantage of oilseed rape fields, and singing and breeding in them was noted several times. This crop is now widely planted and has increased the breeding habitat available.

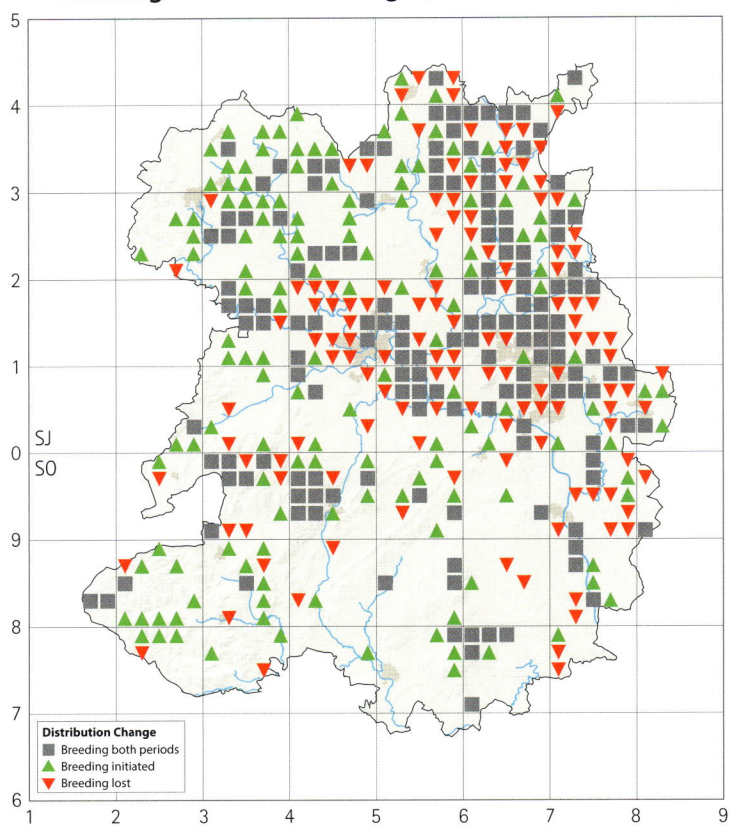

Breeding Distribution Change (1985–90 to 2008–13)

Distribution Change
- ■ Breeding both periods
- ▲ Breeding initiated
- ▼ Breeding lost

Occupied Tetrads

Atlas period (breeding)	1985–90		2008–13		Change	
	Number	%	Number	%	Number	%
Confirmed	91	10	70	8	-21	-23
Probable	99	11	110	13	11	11
Possible	90	10	115	13	25	28
TOTAL	280	32	295	34	15	5
Tetrads with Winter Records (2007–13): 252 (29%)						

John Hawkins, Tetchill, April 2008

Rape fields are often large and species using them could well be overlooked. Understandably, most birders pay little attention to these fields, but observation of them and any new crops should be encouraged in the future.

In the north-west, restoration of the Montgomery Canal has included some additional wetland habitat creation, mainly for rare aquatic plants, and some wetland habitat has been restored along the River Perry at Baggy Moor as part of an RSPB wader project; both these areas are now utilised by Reed Buntings. In the south-west, the upland populations on the Stiperstones, the Long Mynd and Clee hills have all expanded; in addition to the small number of pools present, boggy areas dominated by rushes fulfil the breeding requirements there. Much of the Clun Forest has apparently been colonised since the previous Atlas, and this is thought to be due to ESA agreements which encouraged a reversion to rough grassland and an increase in rushes, and the restoration of some wet areas, although more thorough fieldwork in this area may also have contributed. Combining Atlas

Breeding Relative Abundance (2008–13)

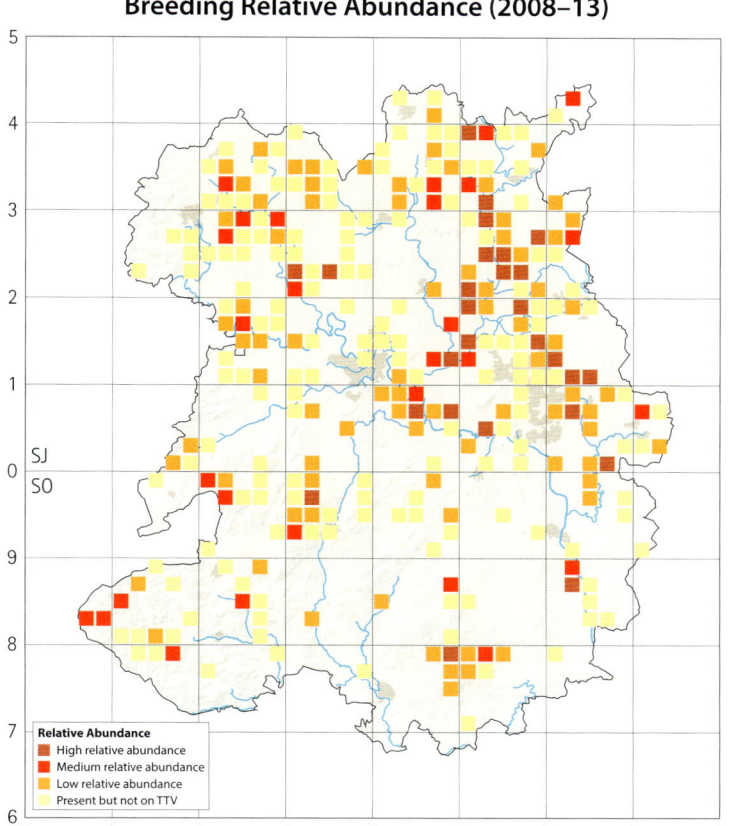

Relative Abundance
- High relative abundance
- Medium relative abundance
- Low relative abundance
- Present but not on TTV

Winter Relative Abundance (2007–13)

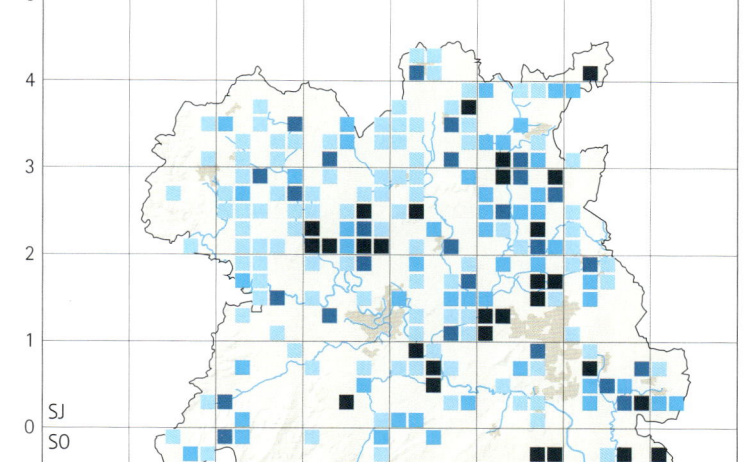

Relative Abundance
- High relative abundance
- Medium relative abundance
- Low relative abundance
- Present but not on TTV

records with those of the Upper Clun Community Wildlife Group shows at least 26–31 pairs in 14 tetrads in SO18, SO27 and SO28 during the period 2008–13 (Michelle Frater *pers. comm.*).

The importance of the uplands for this species was confirmed by a survey on the Long Mynd in 1997 and 1998 which found between 45–51 breeding pairs in seven tetrads, although fewer, 29–31 pairs, were found in the same area in 2006–08 (Leo Smith *pers. comm.*).

Based on TTV counts, the estimated population is between 1,400–1,500 pairs, in the higher part of the range quoted in the *Atlas* (1992). They are found on insufficient BBS plots for any local population trend to be produced, but the similar tetrad occupancy counts from the two Atlas periods do not suggest that any changes have occurred, in spite of the substantial redistribution. This is supported by the BBS trend for England; although it shows a 34% increase between 1995–2014, the latter level is very close to that in 1990 when fieldwork for the previous Atlas was completed. During the intervening period it has fluctuated without any clear trend.

Although superficially similar, there are subtle differences in the recent Atlas winter and breeding season distribution maps. Large areas north of Shrewsbury with no breeding records are occupied during the winter. Here they can often be found in mixed flocks feeding on seeds in arable fields and gardens, especially in hard weather, but these sites generally lack the abundant invertebrates necessary to rear broods successfully and they will have been abandoned in the spring for more productive locations. Changes can also be seen to the south of the River Severn with many upland sites being deserted and more records from the lower valleys.

Mainly occurring in low numbers, they can be overlooked,

especially in arable areas, if observers are not familiar with the call note. The wild bird seed crop at Venus Pool provides excellent feeding opportunities and the highest counts of the recent Atlas period came from this site in January and February 2011, when 100 were present. Estimating numbers feeding in the tall vegetation is difficult, but 83 were trapped and ringed over the New Year period in 2008–09 and many more will have escaped the nets. Catches in the following two winters were 57 and 41, the latter batch including only one ringed in the previous year.

During the winter, reed beds are often used as communal roost sites. None were reported during the recent Atlas period, although it seems likely that those at Venus Pool also roosted there. At Chelmarsh, 148 were ringed at the roost during 1990, and at Allscott Sugar Factory 91 were counted going to roost in February 1999, but there is no recent information from either site, and there must be many other roosts to be found. Counts of those entering or leaving a roost are extremely useful as they give an indication of local numbers.

The sedentary nature of this species is illustrated by ringing; with about two-thirds of all recoveries involving only this or adjacent counties. Small numbers from Scandinavia do winter here, and two movements have been recorded from Norway, and one from Sweden. Single recoveries have also been made in both Belgium and France, both ringed in winter and found in subsequent winters, and they too are likely to have been from the Scandinavian breeding population.

After its historic decline, the Reed Bunting population now appears to be stable, but a return to its previous level seems unlikely, unless major changes take place in the countryside.

Allan Dawes

Lapland Bunting (Lapland Longspur)
Calcarius lapponicus
Very rare passage migrant

The few that winter in Britain are mainly found on the east coast and originate from the Scandinavian breeding population. A small number which arrive on the north and west coasts earlier in the autumn are believed to be from Greenland.

Two good specimens obtained near Shrewsbury were in Lord Hill's collection at Hawkstone. One of them, an immature male, was taken in a lark-net in November 1852.

There is only one modern record:

Brown Clee   22 September 1979

One was seen at close range near the summit with a flock of Meadow Pipits. It took flight from within a few yards, and circled around uttering two different types of call before landing 50m away. Although the observer was familiar with the species, neither the age nor the sex could be determined.

Occasional influxes of Lapland Bunting occur, notably one in late 2010 which produced many inland records, including from neighbouring counties, but not here. In contrast, 1979 was not recognised as an exceptional year.

Allan Dawes

Snow Bunting
Plectrophenax nivalis
Very rare passage migrant and winter visitor

Jim Almond, Earls Hill, 10 April 2011

The Snow Bunting has the most northerly circumpolar breeding distribution of any songbird, with a few in northern Scotland. During the winter they move south, when they frequent higher ground in northern England and Scotland, but more are found at coastal locations in southern England and Wales.

Prior to 1950, eight detailed records, all of singles, came from a wide range of sites: Habberley (possibly Earls Hill) in 1888, Wroxeter, Ruyton-XI-Towns, Ellesmere, and Cressage, all in 1899; Haughmond

Hill in 1905; Wenlock in 1927; and the Long Mynd in 1941; an equal mix of upland and lowland habitats.

Since then, two at Prees Heath in 1968 and two at Titterstone Clee in 1971 were the only records up until 1993, but a change in the frequency of Snow Bunting sightings has occurred since.

A further 19 have been noted subsequently, with 15 of these in the twenty-first century, including six in 2011. Apart from singles at Knighton Reservoir and Allscott Sugar Factory, the southern hills, particularly the Stiperstones and the Long Mynd, have provided the bulk of the records.

Since 1950, 14 occurred in October and November, and four in March and April, but none were known to be present for more than a week, suggesting passage movements. Four others have been found during January; a flock of three on the Long Mynd in 2003 and a long stayer (16 days) at Round Hill in 2006, and one in February, at Brown Clee in 2003, so over-wintering may be a possibility. The uplands can be inhospitable for watchers at this time and it is likely that some are overlooked.

It was found in only five tetrads during recent winter Atlas fieldwork, in 2007 at Wild Moor (two tetrads, SO49D and I) and the Stiperstones (SO39U), and in 2011 at Mason's Bank (Clun Forest – SO28I) and Titterstone Clee (SO57Y), where two were present. In addition, one was observed in flight over Allscott Sugar Factory. All were seen in November, but those at Titterstone remained for six days and were last seen on 2 December. In 2011, one was at Wild Moor for three days in late October, but it left prior to the winter Atlas period. Earlier in the same year, one on passage at Earls Hill on 10 April was the only other record during the recent Atlas period.

The national BTO Atlas (2013) found an increase of 34% in occupied 10km squares during the winter and this was particularly evident in upland areas which accords with local events.

The European population is stable, but with milder winters Snow Bunting may become even scarcer here.

Allan Dawes

Releases, Escapes, Birds from Populations That Are Not Self-Sustaining, and Hybrids

In both the historic and the modern record there are a number of reports of birds that are either deliberate releases, escapes or species that have bred in Shropshire or elsewhere, but are from populations that are not considered to be self-sustaining. They are listed below, and are the equivalent of the BOU's Category E species.

Two, Golden Pheasant and Lady Amherst's Pheasant have had, at one time or another, self-sustaining feral populations in Britain, but all local records are considered to be escapes or, more likely, releases. Lesser White-fronted Goose, Snow Goose, Ruddy Shelduck and Barrow's Goldeneye have been recorded in the wild in Britain, but all our records are considered to relate to individuals of captive origin.

Seventeen of these species were recorded during fieldwork for the 2007–13 Atlas: they are marked with an asterisk. Only Black Swan (13) and Snow Goose (six) were recorded in more than five tetrads, and no other species was recorded in more than two. They may occur more widely, but observers may not have recorded all sightings. Probable breeding records were submitted for Black Swan (one tetrad), Helmeted Guineafowl (one) and Swan Goose (two). There was a probable breeding record for Snow Goose in the 1985–90 Atlas, but not in the recent Atlas period, so the prospect of this goose establishing a feral population has receded.

One particularly interesting report relates to the finding of a dead Eurasian Eagle-owl under power lines near Church Stretton in March 2005. It had been ringed as a pullus at a nest in North Yorkshire the previous year. Both parents were considered to be of captive origin and the female certainly was because it originally wore jesses. Between 1997–2005 this pair raised 23 young, all of which were ringed (Melling *et al.* 2008).

Hybrids, particularly of waterfowl, have occasionally occurred. Greylag x Canada Goose and Canada x Bar-headed Goose hybrids are the most common, although Canada x Barnacle Goose, Canada x Snow Goose and Brent x Barnacle Goose have also been reported along with two *Aythya* hybrids: Tufted Duck x Scaup (on two occasions) and Tufted Duck x New Zealand Scaup *Aythya novaeseelandiae* (one). Gull hybrids have included three Viking Gulls (Glaucous x Herring Gull), and one each of Mediterranean x Black-headed Gull, Glaucous x Great Black-backed Gull and Ring-billed x Lesser Black-backed Gull. There have been three cases of House x Tree Sparrow hybrids, including one during surveys for the recent Atlas.

Finally, domesticated forms of waterfowl – particularly Greylag Goose and Mallard – are also regularly reported, including during surveys undertaken for the Atlas. This includes specimens exhibiting odd plumage patterns as a result of matings between individuals that are obviously domesticated and those exhibiting typical plumage which, in some cases, will include true wild birds.

Graham Walker

Whistling Duck *Dendrocygna* spp. #
Cackling Goose *Branta hutchinsii* *
Bar-headed Goose *Anser indicus* *
Ross's Goose *Anser rossii*
Snow Goose *Anser caerulescens* *
Swan (Chinese) Goose *Anser cygnoides* *
Lesser White-fronted Goose *Anser erythropus*
Black Swan *Cygnus atratus* *
Black-necked Swan *Cygnus melancoryphus* #
Blue-winged Goose *Cyanochen cyanoptera*
Ruddy Shelduck *Tadorna ferruginea* *
South African Shelduck (Cape Shelduck) *Tadorna cana* *
Muscovy Duck *Cairina moschata* *
Wood Duck *Aix sponsa* *
Ringed Teal *Callonetta leucophrys*
Hottentot Teal *Spatula hottentota* #
Silver Teal *Spatula versicolor* #
Chiloe Wigeon *Mareca sibilatrix* *
Philippine Duck *Anas luzonica*
White-cheeked Pintail *Anas bahamensis*
Yellow-billed Teal *Anas flavirostris* *
Chestnut Teal *Anas castenea*
Barrow's Goldeneye *Bucephala islandica* #
White-headed Duck *Oxyura leucocephala*
Helmeted Guineafowl *Numida meleagris* *
Northern Bobwhite *Colinus virginianus* #
Chukar Partridge *Alectoris chukar*
California Quail *Callipepla californica* *
Reeve's Pheasant *Syrmaticus reevesii* *
Golden Pheasant *Chrysolophus pictus* *
Lady Amherst's Pheasant *Chrysolophus amherstiae*
Indian Peafowl *Pavo cristatus* *
Greater Flamingo *Phoenicopterus roseus*
Chilean Flamingo *Phoenicopterus chilensis*
Sacred Ibis *Threskiornis aethiopicus*
American White Ibis *Eudocimus albus*
Harris's Hawk *Parabuteo unicinctus* *
Crowned Crane *Balearica* spp.
Diamond Dove *Geopelia cuneata*
Eurasian Eagle-owl *Bubo bubo*
Cockatiel *Nymphicus hollandicus*
Eastern Rosella *Platycercus eximius*
Budgerigar *Melopsittacus undulatus*
Village Weaver *Ploceus cucullatus*
Zanzibar Red Bishop *Euplectes nigroventris*
African Waxbill *Estrilda* spp. #
Zebra Finch *Taeniopygia guttata*
Domestic Canary *Serinus canaria domestica*
Red-winged Blackbird *Agelaius phoeniceus* #

* Recorded during fieldwork for the 2007–13 Atlas
# Historic species only recorded before 1 January 1950

Changes in the Status of Breeding Species

Leo Smith

Confirmed breeding evidence was found for a total of 127 breeding species in fieldwork for one or other of the 1985–90 or 2008–13 Atlas projects, or both. The recent *Atlas* results build on the baseline provided by the *Atlas* (1992), and have again clearly demonstrated the dependence of most breeding species on specific habitats, particularly by showing just how few species are so adaptable that they can breed in every tetrad.

Widespread and Common Species

The recent Atlas project found breeding evidence for only seven species in every tetrad: Blackbird, Blue Tit, Carrion Crow, Chaffinch, Great Tit, Woodpigeon and Wren.

Dunnock and Robin are also believed to breed in every tetrad: Magpie was listed among the species thought to breed in every tetrad in the *Atlas* (1992), but it was not found at all in 2008–13 in some tetrads that are close to Pheasant release sites, so persecution may now be limiting its range.

Even some of the most widespread and numerous species appear to be missing from a few tetrads, where suitable habitat is lacking. This analysis ignores gaps on the distribution maps that appear to reflect the vagaries of fieldwork, where the species was overlooked, rather than its real absence.

Of the species listed in a similar table in the *Atlas* (1992), the population and distribution of Great Tit has increased, and it has been added to this category. House Martin has declined, and Willow Warbler, Starling and Yellowhammer have suffered substantial population declines and contraction of range. Pheasant has increased, and is now found on high ground without copses, but it is still apparently absent from the largest urban area, Telford. House Sparrow, Song Thrush and Swallow have increased their range slightly. Buzzard and Goldfinch have increased considerably, and have been added to Table 1.

Most species are so dependent on specific habitats that changing land use may have a major effect on their population level, range and distribution. Climate is another factor, including that on the migration routes and in the wintering areas of our summer visitors.

Recent Rapid Change of Range

The most striking result of the recent Atlas project is the rapid change in the distribution (the number of tetrads occupied) of many breeding species since 1985–90 – a period of less than 25 years. Some of the gains come from conservation measures, reduction in persecution, creation of new man-made habitats such as gravel pits, and climate change (warmer springs and/or milder winters). There have been many more losses, usually the indirect result of one or more aspects of 'agricultural intensification', particularly a greater use of herbicides and pesticides, removal of hedgerows and hedgerow trees, and autumn sowing of cereal crops which removes the food supplies in winter stubbles. These changes have been more fully described in the *Habitats* chapter.

Confirmed breeding evidence was found for a total of 127 breeding species in fieldwork for one or other of the two Atlas projects, or both. The table of tetrad occupancy in the respective species accounts show that 39 of these breeding species increased their range by more than five occupied tetrads, but far more, 60, were apparently lost from more than five tetrads. However, in the case of a few species, the losses can be attributed to more strict validation, or introduction of the 'F' code (birds seen in flight only) in 2008–13, when similar observations may have been recorded as 'H' (possible breeding) in 1985–90.

Table 1. Status of Widespread and Numerous Species

Species Believed to Breed in Every Tetrad	Species Which Do Not Appear to Breed in Every Tetrad	
	Species	Absent from Tetrads Consisting Primarily Of
Blackbird	Buzzard	Intense arable with few large trees
Blue Tit	Chiffchaff	High ground and intense arable with few large trees
Carrion Crow	Goldfinch	Upland (Long Mynd)
Chaffinch	House Sparrow	Moorland with no buildings
Dunnock	Jackdaw	High ground and intense arable with few large trees
Great Tit	Pheasant	Large urban areas
Robin	Song Thrush	Moorland with no buildings
Woodpigeon	Swallow	Moorland with no buildings
Wren		

Table 2. Species with a Substantial Distribution Change Since 1985–90

Gained in More Than 50 Tetrads			Lost from More Than 50 Tetrads			
201+	101–200	51–100	51–100	101–200	201–400	401+
Buzzard (442) Raven (385)	Canada Goose Great Spotted Woodpecker Greylag Goose Nuthatch Siskin	Blackcap Chiffchaff Goldfinch Goosander Goshawk Greenfinch Mute Swan Red Kite	Corn Bunting Dipper Garden Warbler Grey Wagtail Kingfisher Mistle Thrush Quail Sparrowhawk Wheatear Yellow Wagtail Yellowhammer	Feral Pigeon Green Woodpecker Kestrel Lesser Spotted Woodpecker Marsh Tit Meadow Pipit Moorhen Pied Flycatcher Red-legged Partridge Rook Starling Swift Tree Pipit Treecreeper Willow Warbler Wood Warbler Woodcock	Lapwing Spotted Flycatcher Tawny Owl Tree Sparrow Turtle Dove Willow Tit	Curlew (409) Little Owl (417) Cuckoo (440) Grey Partridge (540)

Table 2 shows the major changes, gains and losses of more than 50 tetrads. Gains total 15 species, but losses total 38. Most of these latter species have shown a substantial population change corresponding with the change in range. Possible breeding records for three species (Common Sandpiper, Grey Heron and Snipe), which would otherwise have appeared in the table, and for Swift, which still appears in the table but in a different column, have been discounted.

No table of tetrad occupancy was provided in 1985–90 for two species shown as gains in more than 50 tetrads: Goshawk and Red Kite. The Goshawk population was still becoming established in 1990, and there were probably around a dozen pairs then, but Atlas workers didn't know where to look for them, so it is unlikely that there were as many as 30 occupied tetrads recorded. There were only 'a total of 11 breeding season records' for Red Kite in five of the six years 1985–90. The table of tetrad occupancy for two more species, Hobby and Peregrine, contained no possible breeding records. All four of these species forage over large distances, so possible breeding records have been discounted for all of them.

Changes in Status of Breeding Species Since 1950

Red Kite was a regular breeder in the first half of the nineteenth century, with the last breeding record in 1876, but it was locally extinct for almost 130 years. It has re-colonised, and the first modern confirmed breeding record occurred in 2005.

Egyptian Goose, Common Tern, Lesser Black-backed Gull, Firecrest and Mediterranean Gull were all confirmed breeding for the first time during the recent Atlas period, but Nightingale, Hawfinch, Redshank and Ring Ouzel have been lost since 1985–90. Barnacle Goose has also been lost since 1985–90, but this feral population never became established.

Species that bred for the first time since 1950 (including Red Kite), and were breeding in 2008–13, are listed in Table 3. Note that the table includes the five species listed above that bred for the first time during the recent Atlas period.

Some other species (Barnacle Goose, Black Redstart, Black-necked Grebe, Common Gull, Golden Oriole, Golden Plover and Marsh Harrier) have bred very occasionally since 1950, but not since 1990, and Honey-buzzard bred regularly (but not every year) between 1995 and 2007, but there was no evidence that any of these species might have bred in 2008–13.

Garganey bred for the first time in 2014 since 'about 1888', the only previous breeding record.

Species that have bred regularly in previous years, but which were not found breeding in 2008–13, are listed in Table 4. These were all

Table 3. Species That Bred for the First Time Since 1950, and During 2008–13

Species	Date of First Breeding
Goshawk	1951
Shelduck	1963
Collared Dove	1963
Ruddy Duck	1965
Greylag Goose	1969
Peregrine	1970
Little Ringed Plover	1976
Gadwall	1980
Oystercatcher	1981
Hobby	1983
Goosander	1987
Mandarin Duck	1988
Red Kite	2005
Egyptian Goose	2009
Common Tern	2009
Lesser Black-backed Gull	2012
Firecrest	2012
Mediterranean Gull	2013

Table 4. Species That No Longer Breed

Species	Last Confirmed Breeding	
	Lost 1950–84	Lost Since 1985–90
Wryneck	1953	
Black Grouse	1954	
Red-backed Shrike	1954	
Woodlark	1957	
Corncrake	1975	
Nightingale		1990
Hawfinch		1991
Redshank		1998
Ring Ouzel		2003
Ruddy Duck		2009

lost since 1950. In the previous 100 years, between 1850 and 1949, a few species bred sporadically in ones and twos, but no known regular breeding species was lost in that period.

Ruddy Duck was found breeding in the recent Atlas period, but the last confirmed breeding record was in July 2009, and there have been no probable breeding records since that year. It has apparently been exterminated, with the last record, of one subsequently culled, in 2017.

Pochard and Turtle Dove may also have been lost.

The last confirmed breeding of the former was in 1995, and in 2008–13 there was probable breeding at one site only (Allscott Sugar Factory – two tetrads) and only in 2008 and 2009. The breeding site is no longer available.

There were probable breeding records of the latter from five tetrads in 2008–13, but there were records of only one bird at one site in 2014, 2015 and 2016, and no records at all in 2017. The last confirmed breeding record was in 1995.

Changes in Status of Breeding Species Between the Two Atlas Periods

There were confirmed breeding records for 127 species from fieldwork in one or other of the 1985–90 or 2008–13 Atlas projects, and breeding was confirmed in both periods for 112.

There were probable breeding records for Firecrest and Nightjar, and no breeding records of any description, for Common Tern, Red Kite, Lesser Black-backed Gull and Mediterranean Gull in 1985–90, all of which were confirmed breeding in 2008–13.

There were probable breeding records for Pochard, Quail, Redshank, Ring Ouzel, Turtle Dove, Water Rail and Woodcock, and no breeding records of any description for Barnacle Goose, Black-necked Grebe, Hawfinch or Nightingale in 2008–13, all of which were confirmed breeding in 1985–90.

These changes reflect the population changes of the species concerned, as described in the individual species accounts, and summarised in Tables 3 and 4 above.

Conclusions and Further Action

Leo Smith

This Avifauna, and particularly the results of the recent Atlas project, published here for the first time, represents the greatest contribution to our knowledge of the County's birds since the formation of the Shropshire Ornithological Society in 1956.

Conservation Action

The *Atlas* (1992) clearly achieved its original aim of providing a benchmark against which future trends could be assessed, and the 2008–13 Atlas project has found that the population and distribution of many species has undergone remarkable change in a very short timescale, usually for the worse.

This has highlighted the species that need urgent conservation action if they are not to be lost from the County altogether, and SOS intends to take a lead on this, starting with a 'Save our Curlews' campaign.

It is intended to produce a County Red list of *Birds of Conservation Concern* to help guide and prioritise this work, and the species lost from more than 50 tetrads, shown in the *Breeding Status* chapter, Table 2, together with species concentrated in uncommon habitats which have disappeared from slightly fewer tetrads, such as Whinchat, are prime candidates for consideration, as well as the scarce breeding species identified in the heading to each account and summarised in Appendix 6.

The fieldwork has also identified important sites, and these are now being surveyed so they can be incorporated into the list of Local (County) Wildlife Sites.

The SOS Conservation Sub-committee is leading this work, and will organise future fieldwork activities, involve the members and seek to answer some of the many questions highlighted by the species accounts and interpretation of the maps.

Conclusion

The *Atlas* (1992) concluded that:

> One of the major lessons of the Atlas is the need for more detailed records to be submitted by observers for the Shropshire Bird Report (SBR). In trying to assess the current population and establish whether there has been any change in numbers or distribution, the most useful SBR records were those that gave counts, especially of singing males or pairs; a precise location, so the record could be related to one or more specific tetrads; and an indication of any evidence of breeding, particularly whether two birds were a pair, or parties included recently fledged young or juveniles. Hopefully the experience gained from fieldwork will encourage more useful records in future. Records for species in tetrads where they are missing from the *Atlas* maps, or that improve the level of evidence of breeding, will be especially welcome.

In the light of experience in compiling this Avifauna, this comment is reiterated even more strongly.

Since writing the Atlas (1992) we have learnt a considerable amount about Shropshire's breeding birds, and fieldwork for the recent Atlas project has added considerably to that, and to knowledge of our winter visitors. Writing the accounts of the breeding and wintering species, and of the passage migrants, has consolidated our knowledge. The Society intends to build upon the results to continually increase and update our knowledge and understanding of this important topic.

The status of every species that has been recorded here since the early part of the nineteenth century, and changes over that period, are summarised in the *History* chapter (p. 12) and Appendix 6.

Atlas Fieldwork and Results, and Comparison with 1985–90

Allan Dawes & Leo Smith

Survey Areas

To create manageable survey areas, and in common with similar atlases elsewhere, Shropshire was divided into 2km squares, known as 'tetrads', using the National Grid printed on all Ordnance Survey (OS) maps. On the 1:50,000 series OS maps, the grid lines are shown in pale blue and define 10km and 1km squares.

Each 10km square has a unique number, and those in Shropshire are shown on Map A1.1.

Every 10km square is then further divided into 25 tetrads, each with a separate letter taken from the tetrad letter key shown in Figure A1.1.

Each tetrad therefore has a unique number, consisting of the 10km square number, followed by the tetrad letter. For example, Shrewsbury Castle is in tetrad SJ41W.

At the edges of the County, tetrads were included in the Atlas if more than half of their area is in Shropshire. In such cases the whole tetrad was surveyed, not just the part inside the County boundary. If more than half the tetrad is in the neighbouring county, it was excluded. Thus a few records will have come from just outside Shropshire, and small fragments at the edge of the County have not been covered.

Shropshire is England's largest landlocked county, and the survey covered 870 tetrads (3,480sq.km). The same recording areas were used in both 1985–90 and 2007–13. A separate survey was carried out for each tetrad, in the breeding season during both Atlas periods, and in winter for the recent Atlas project only, and a dot on a distribution map means that species was recorded in the appropriate habitat somewhere within the 2km square at that position on the map.

Map A1.1 10km squares in Shropshire

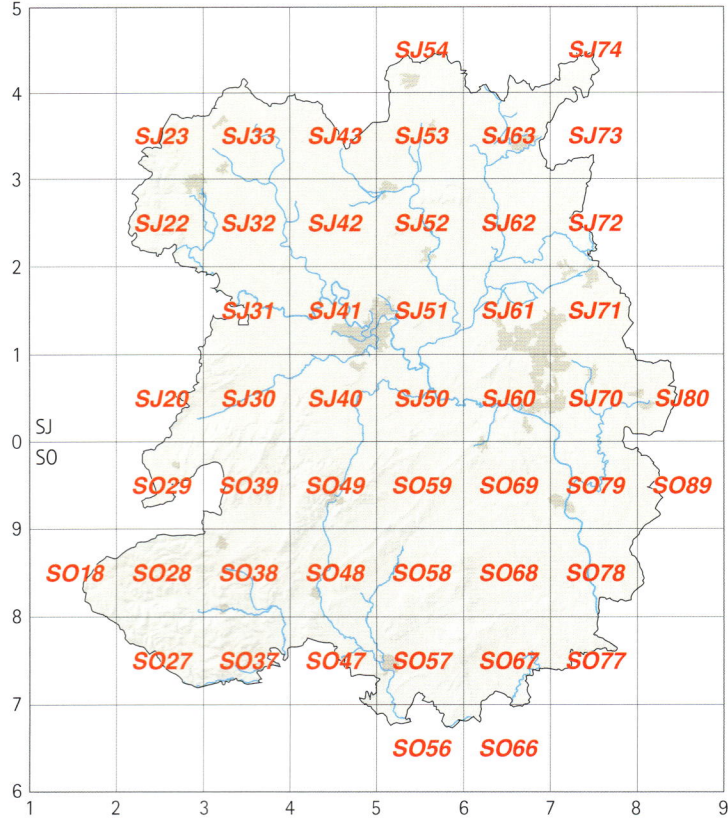

Figure A1.1 Tetrad Letter Key

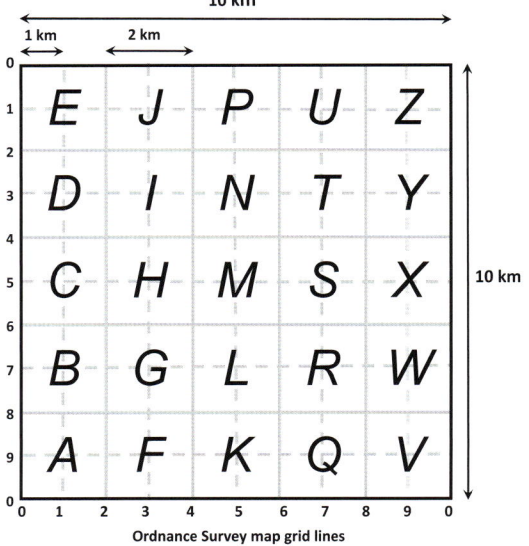

Place Names and Gazetteer

The species accounts frequently refer to places where particular birds were found. Appendix 2 lists every place mentioned in the accounts, with a four-figure grid reference which identifies the 1km square on the map where that place can be located. The first two numbers in the grid reference are those printed at the top and bottom of the map to define lines that run north to south, known as 'eastings'; and the second two numbers are those printed on each side of the map to define lines that run east to west, known as 'northings'. Where the two lines defined in the grid reference intersect is the bottom left-hand corner of the appropriate 1km square. As the 1km grid lines divide each 10km square into 100 1km squares, the number of the 10km square is the first and third number in the four-figure grid reference of the 1km square. For example, Shrewsbury Castle is immediately to the east of 'easting' 49, and to the north of 'northing' 12, so its four-figure grid reference is SJ4912, in 10-km square SJ41.

Fieldwork

The County was divided into nine areas, each with an Area Co-ordinator (ACO) to organise the fieldwork. In 1985–90, there were 10 areas, but Area 4 (West) was split into two, and the northern half (tetrads in SJ20, SJ30 and SJ31) was added to Area 1 (the North West) and the southern half (tetrads in SO29 and SO39) was added to Area 7, the South West.

Detailed information and instruction sheets produced by BTO were sent to all fieldworkers, including maps showing tetrad boundaries, so records could be mapped accurately. Printed record cards listing all the species likely to be found in the County, with three columns for the categories of breeding evidence, were also offered to all fieldworkers, but most chose to use notebooks or other means of recording.

The methodology was precisely the same as that used for the national *BTO Atlas 2007–11*, as set out in chapters 2 and 3 of that publication.

Volunteers, mainly members of the SOS or SWT, were initially asked to survey one or more of their local tetrads, in winter, and in the breeding season, as described below. They were advised to identify all the different habitats using an OS map (preferably 1:25,000 scale) and visit them all. Fieldworkers covering an unfamiliar tetrad were advised to seek out knowledgeable local people and accept records from them if they were certain the species and location had been correctly identified.

Recording the Evidence of Breeding

Fieldworkers were asked to visit each square at least twice, and several times if possible, to record the highest possible category of breeding evidence for each species seen or heard, as defined in Table A1.1. Visits were also requested at dusk and night to find the crepuscular and nocturnal species.

The main breeding season was defined as 1 April to 31 July, though earlier visits were requested for Crossbill, owls and Mistle Thrush, and later ones for pigeons, doves and Corn Bunting. Where appropriate, early records were also accepted for other resident early breeders such as Raven, and late records were accepted for species that raise second and third broods, and might still have young in the nest as late as August or September. All accepted records that were submitted with breeding codes have been included on the maps and tetrad occupancy tables, whatever the date, but records from outside the main breeding season were carefully scrutinised.

Proof of breeding was only required once for each species in each tetrad at any time over the whole Atlas period. No attempt was made during fieldwork either to establish population levels, or check presence every year.

The categories for possible, probable and confirmed breeding were the same as those used for the local *Atlas* (1992), but the addition of the three non-breeding codes was new, and introduced some inconsistency of interpretation: if a breeding species was seen flying over, should it be coded 'F', or, if there was breeding habitat or a potential nest site in the tetrad, should it be coded 'H'? In fieldwork for the *Atlas* (1992) such observations were often coded 'H', and appeared on the Atlas maps as possible breeding records. In the recent Atlas, 'F' codes were not mapped. In the case of a few species that forage a long way from their nest site – Red Kite, Hobby, Grey Heron, Black-headed Gull and Swift – the dilemma has been solved by producing maps of foraging areas that include both 'H' and 'F' records.

Table A1.1. Categories of Breeding Evidence

NON–BREEDER	
F	Flying over
M	Migrant
U	SUmmering
POSSIBLE BREEDER	
H	Observed in suitable Nesting Habitat
S	Singing male
PROBABLE BREEDER	
P	Pair in suitable nesting habitat
T	Permanent Territory (many individuals on one day or one individual over one + weeks.)
D	Courtship and Display
N	Visiting probable Nest site
A	Agitated behaviour
I	Brood patch of Incubating bird
B	Nest Building or excavating
CONFIRMED BREEDER	
DD	Distraction Display
UN	Used Nest or eggshells found from this season
FL	Recently FLedged young or downy young
ON	Adults entering or leaving nest site indicating Occupied Nest
FF	Adults carrying Faecal sac or Food for young
NE	Nest containing Eggs
NY	Nest with Young seen or heard

Timed Tetrad Visits

Fieldworkers were asked to complete four timed tetrad visits (TTVs) in each square, one in each of the periods November–December, January–February, April–May and June–July, and visit each different habitat in the square on each visit. Ideally each visit should last two hours, with the counts from each of the two hours recorded separately. Although two-hour TTVs were strongly encouraged, a small proportion of squares only had one-hour TTVs.

In winter, TTVs were not to be conducted within an hour of sunset, so as not to record birds travelling to, or at, large roosts.

Winter

Fieldworkers were asked to revisit their tetrads between 1 November and the end of February, and record all species seen in addition to those recorded on TTVs, except those in flight and not using the habitat in the square, to compile a complete record of species in the tetrad.

Submission of Records

BTO provided an online data entry facility, and the vast majority of records were submitted that way by the observers. A few submitted paper records, which were entered into the online database either by BTO or the ACO. Each observer had access to their own records, and records submitted by other observers for the same square(s), except for confidential species.

A summary of the best breeding evidence for each tetrad, and the species recorded in winter, was supplied by BTO to the local Atlas organisers, to assist with organising coverage and validation of records.

Validation

BTO asked the local Atlas organisers to validate all records. This usually involved the ACO identifying any that seemed unlikely, and querying them with the observer. Many instances of incorrect data entry, and some instances of uncertain identification, were rectified.

Early breeding season 'H' records for several species that have much higher wintering or passage numbers than breeding populations were also downgraded to 'M' records in this process, if it was believed that the bird was unlikely to stay and breed. There was no formal validation process in 1985–90, although some ACOs queried this type of record and encouraged the observer to withdraw them, but some were shown as possible breeding on maps in the *Atlas* (1992), so in a few cases e.g. Water Rail, the distribution maps are not directly comparable, and including 'H' records from the earlier Atlas on the breeding distribution change map would create a misleading impression.

Some 'H' records, for example for Cormorant, were also downgraded. Many summering birds were immatures so not likely to breed and observers generally withdrew the record when queried. If the age category was unknown or the observer failed to respond to the query the record was downgraded.

Progress

Area Co-ordinators attempted to persuade fieldworkers to cover different tetrads in the second and subsequent years, to ensure total coverage. New fieldworkers were also recruited, and training meetings were held each year. While some workers covered several new tetrads each year over the whole period, others were reluctant to do more than one or two home squares. While the Atlas project could not have been completed without the latter, around half of the records were submitted by 27 observers who contributed each year and who each supplied over 1,000 records in either the winter or breeding season, with 22 of them submitting over 1,000 in both periods. These 27 were responsible for 48% of winter records and 54% of breeding records.

Comparison of the Breeding Results from Both Atlas Periods

Fieldwork Targets

At the beginning of the recent Atlas project, it was agreed that the fieldwork targets set for the previous breeding *Atlas* (1992) would be used again. During the fourth year it became clear that these targets would not be met. One of the main reasons for the new Atlas project was to see how species distribution had changed since 1992, and to make valid comparisons it was necessary to extend the fieldwork by an extra two years to achieve comparable coverage. Despite this it was apparent that the higher target would not be met in many tetrads with limited variety of habitat. However coverage in those tetrads that were poorly recorded for the 1985–90 Atlas was improved upon.

The targets set were:

- Achieved = 50 species with 30 confirmed or probable breeding records.
- Acceptable = 40 species with 30 confirmed or probable breeding records.
- Acceptable minimum = 40 species with 20 confirmed or probable breeding records.
- Below target = less than 40 species or less than 20 Confirmed or Probable breeding records.

Note that each target level includes two criteria, and both had to be met.

Maps A1.2 and A1.3 show those tetrads which fell below the 'achieved' target during each Atlas period. Blank squares indicate that the target was achieved.

Table A1.2 summarises the number of tetrads in each target category (total tetrads = 870).

Although the number of tetrads reaching the 'target achieved' level has fallen by 136, this is believed to be largely due to the

Table A1.2. Number of Tetrads in Each Target Category

Target	Atlas 1985–90	Atlas 2008–13	Change
Achieved	718	582	−136
Acceptable	114	241	+127
Acceptable minimum	20	46	+26
Below target	18	1	−17

Map A1.2. Fieldwork Coverage Below Target 1985–90

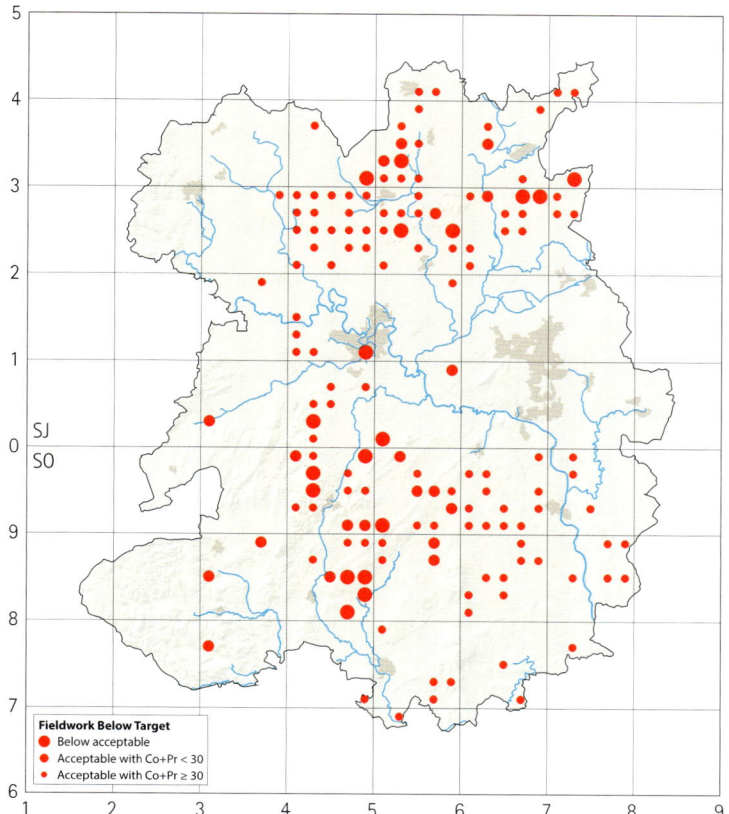

Map A1.3. Fieldwork Coverage Below Target 2008–13

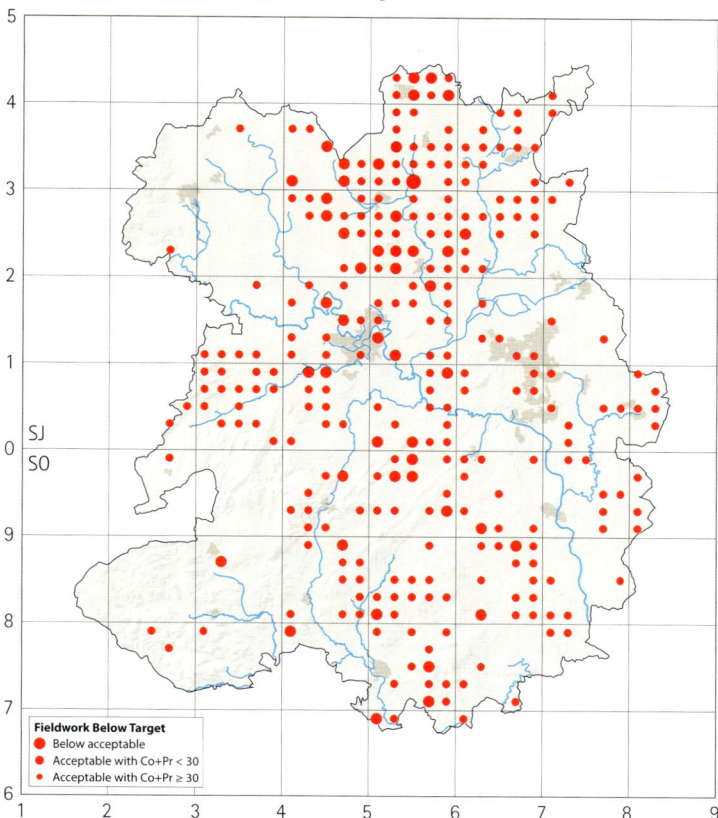

widespread decline of several species. An increase of just three species found in each tetrad would have promoted an additional 141 tetrads into the 'achieved' category. Conversely, a reduction of three species per tetrad during fieldwork for the *Atlas* 1992 would have reduced those in the 'target achieved' level by 110. Thus a reduction of three species per tetrad since 1985–90 accounts for the difference in the targets achieved. This, together with an improvement in the number falling below any target level, suggests that coverage in the recent Atlas fieldwork was at least as good as in 1985–90. The one that fell below the minimum target slipped through the mopping-up process: it was registered with 40 breeding species, but on validation a possible breeding Little Egret record was removed, and this dropped the tetrad into the 'below target' category.

The increase in those in the 'minimum acceptable' level is disappointing. In most cases it was the lack of probable and confirmed breeding records, rather than the total number of species found, which resulted in their inclusion in this category. It indicates a lack of fieldwork effort or experience, as a further visit should have enabled several of the species already found, but with only possible breeding records, to be promoted to a higher level.

Two other indicators suggest that coverage was at least as good in 2008–13. Firstly, seven species were found in every tetrad, Blackbird, Blue Tit, Carrion Crow, Chaffinch, Great Tit, Woodpigeon and Wren, compared to only one (Blackbird) in 1985–90. Secondly, there were 486 observers credited for submitting records in the *Atlas* (1992), compared with 650 credited in Appendix 5 for submitting records for the recent Atlas. Although some of these observers might have only submitted winter records, or undertaken TTVs or entered BirdTrack

records without submitting breeding codes, collectively they have contributed records that would otherwise be missing. For both Atlases the bulk of the breeding records were submitted by a smaller number of committed observers visiting many tetrads over the six-year period, but the additional records were essential to the total coverage level obtained.

Changes in Levels of Breeding Evidence

For the purpose of both analysis and mapping described in the remainder of this chapter, only the highest breeding code submitted for each species in each tetrad has been used, a confirmed record overriding a probable or possible one.

Overall fewer such records were received in 2008–13, down from 50,252 to 45,613, an average of 5.3 less per tetrad, again in line with the decline of several formerly widespread species. A more stringent approach to validation rejected some records. Late wintering wildfowl and waders, and species that occur widely on passage outside of their normal breeding range, were generally not included in the breeding data if 'H' or 'P' records were submitted, unless the observer thought breeding likely when the record was queried. Removing these rejected records from the calculations altered figures by less than 1%.

Comparable observations of some other species may have been submitted as possible breeding records in 1985–90, but as non-breeding 'F' records in 2008–13. Hobby, Grey Heron, Black-headed Gull, Common Sandpiper, Snipe and Swift changes were exaggerated by these methodology changes, and Hobby, Peregrine and Goshawk records were not fully documented in the *Atlas* (1992). The individual species accounts provide more detail.

Figure A1.2. Change in the Number of Tetrads Occupied by Each Species (Gains and Losses)

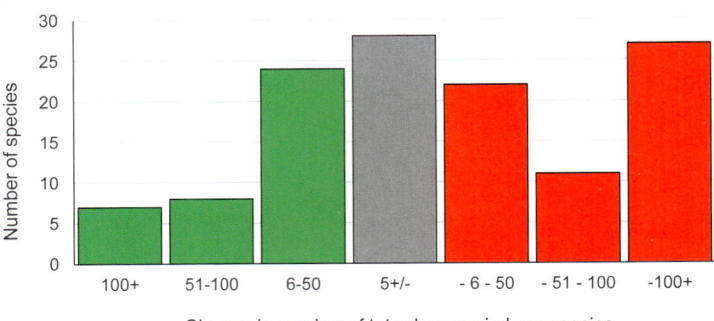

Figure A1.2. Change in the Number of Tetrads Occupied by Each Species (Gains and Losses)

Gains and Losses

Figure A1.2 looks more closely at gains and losses, in terms of the number of occupied tetrads, for each of the 127 species with confirmed breeding evidence found in either or both of the two Atlas periods. Of these, 112 were confirmed breeding in both periods.

A gain occurs in a tetrad if a species has any breeding code where none was shown on the map in the *Atlas* (1992). A loss occurs in a tetrad if a species has no breeding code in the recent Atlas period but a breeding code in 1992. No distinction has been made between changes from one category to another, i.e. between confirmed, probable or possible breeding. Figures are taken from the summary tetrad occupancy table in each species account. For example, nine species were found in over 100 more tetrads, 26 species were lost from over 100 tetrads and 30 remained stable (gains or losses of five or fewer tetrads, which includes all species, from common ones such as Blackbird found in most or all tetrads to rare species found in only one or two).

The species that have been gained or lost from more than 50 tetrads are listed in Table 2 of the *Breeding Status* chapter on p. 481.

It is clear that the losses outweigh the gains and this was widely expected. For most species the changes are in line with BBS trends, which suggests that they are real and not caused by insufficient or poor fieldwork. Taking both gains and losses into consideration, an average of five species has been lost from each tetrad, although in practice some habitats and tetrads will have been affected more than others.

Geographic Distribution of the Gains and Losses

Map A1.4 shows the geographic distribution of the changes, using the same 50:25:25% bands used on the relative abundance maps for each species, with the darker shades of green (gains) and red (losses), reflecting the bands with the greatest changes, and dark grey showing tetrads with little change.

The change shown for some tetrads will be due to a difference in fieldwork effort in one or other of the two periods, but most will reflect real changes in the number of species present. By comparing this map with the predominant farm types (Map 7, p. 27), it will be seen that the areas of greatest loss, the darker red tetrads, are largely arable farmland, especially in the Weald Moors north and north-west of Telford.

Map A1.4. Change in Number of Breeding Species Found Per Tetrad

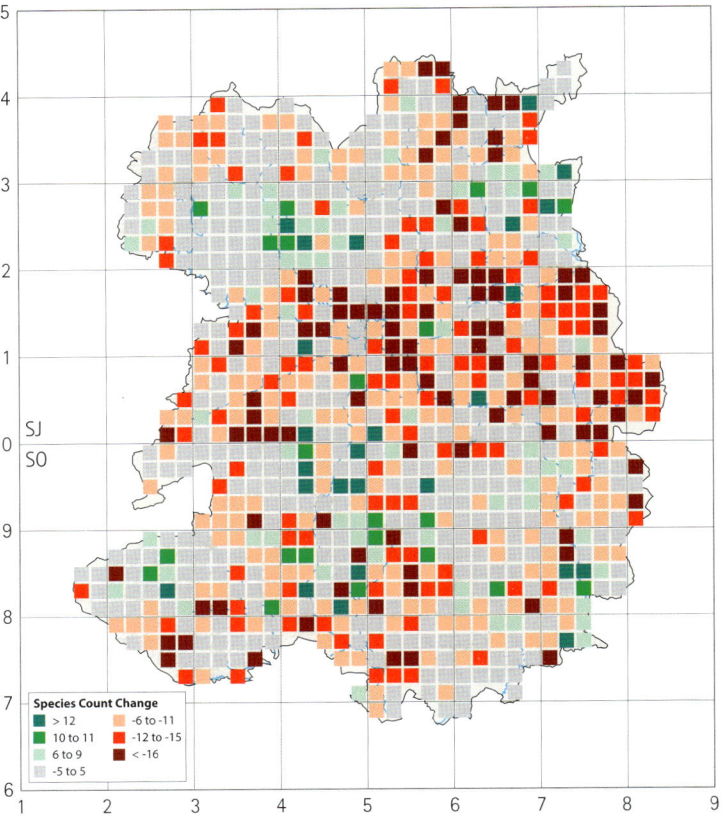

Comparison Between Categories of Best Breeding Evidence

The total number of records with breeding evidence declined, but, while the number of confirmed breeding records also declined, the number of probable breeding records increased. Figure A1.3 shows the totals in each category of breeding evidence. The drop in confirmed records and a corresponding increase in probable records is apparent for many species, as shown in the individual tetrad occupancy tables. This has been a common result in several recent tetrad atlases, such as Kent (BB Vol 109 March 2016).

The number of species with confirmed breeding records per tetrad for each of the two Atlas projects, is shown on Maps A1.5 and A1.6, and the reduction in these records for the recent Atlas can be clearly seen. The tetrads with high numbers of confirmed records are often

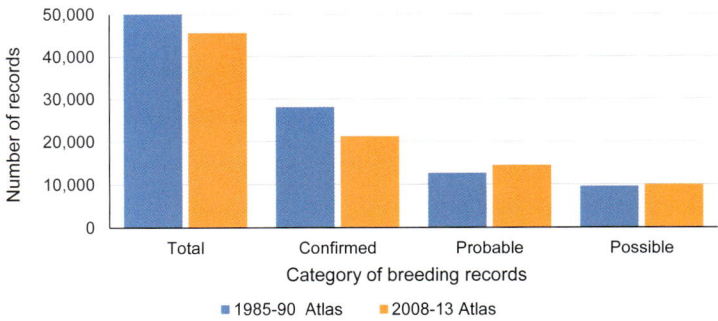

Figure A1.3. Comparison Between Categories of Best Breeding Evidence Recorded (All Species)

Map A1.5. Number of Species with Confirmed Breeding 1985–90

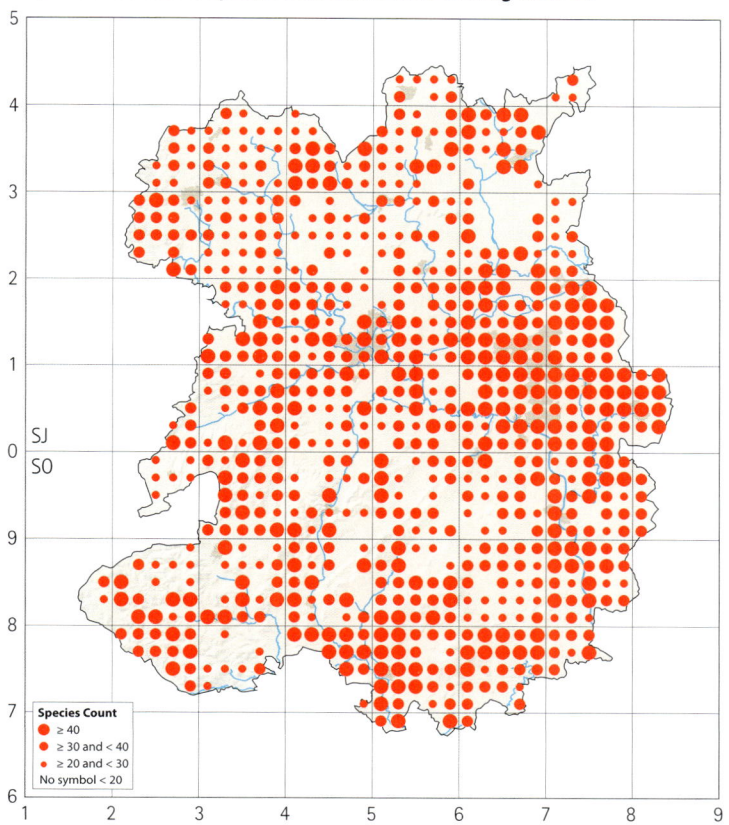

Map A1.6. Number of Species with Confirmed Breeding 2008–13

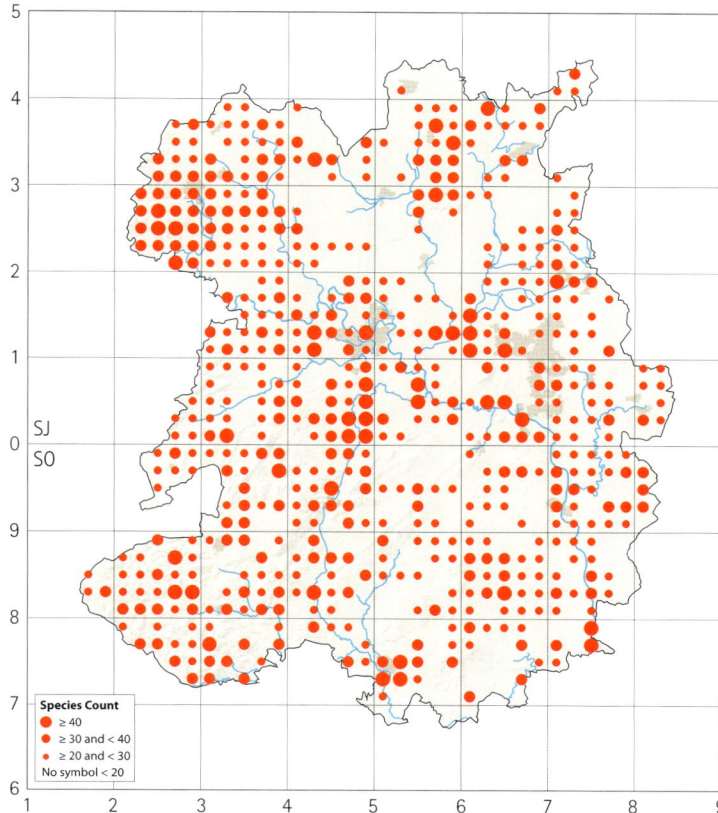

in clusters, suggesting that some observers spent more time and effort to confirm breeding.

There were several different contributory factors to the lower proportion of tetrads with confirmed breeding evidence. Again, the decline of many species did not help. These species would have been encountered less often, reducing the chance of obtaining the necessary evidence. However, results for common and increasing species outlined below suggest that this is not the major factor.

The introduction of TTVs complicated the simple instruction in 1985–90 'to find and record the best breeding evidence'. Volunteers carrying out TTVs were encouraged not to spend time obtaining breeding records during the counts. Many of the submitted TTV record sheets lacked any breeding codes, and a small number of these observers failed to make any additional visits. Devoting part of the fieldwork effort to counts, rather than looking for breeding evidence, was perhaps the major factor.

Many volunteers were new to Atlas fieldwork and some may have lacked the confidence or skill required to confirm breeding. While this applied in both Atlas periods, the recent Atlas made more effort to recruit helpers from outside the bird-watching community, who were perhaps less experienced.

Other possible reasons include the effect of a shift in the emphasis of bird-watching and recording for the experienced observers, such as listing, twitching and *BirdTrack*, perhaps reducing the patience needed to confirm breeding for some common species; and worse weather in some breeding seasons, leading to higher rates of early nest failure, reducing the likelihood of FF and FL observations.

Figure A1.4 shows the percentage change in the number of

confirmed breeding records for all the species in each of the gains and losses bands shown in Figure A1.2.

Only the species that increased by more than 100 tetrads showed an increase in the proportion of confirmed records. These included Canada Goose, Greylag Goose, Buzzard, Great Spotted Woodpecker and Raven, which are relatively easy to confirm. The largest decrease in confirmed records was shown by species in the +/- 5 tetrad group that had not altered range significantly, suggesting that decreasing range was not a prime cause of the decrease in confirmed breeding records.

Local BBS data shows eight species in the +/-5 tetrad range that have increased in numbers since 1997. These are Blackbird, Blue Tit, Dunnock, House Sparrow, Pheasant, Robin, Swallow and Woodpigeon (see BBS charts in respective species accounts). Figure A1.5 shows the combined results for these eight species. The total number of

Figure A1.4. Percentage Change in Confirmed Breeding Record Within Each Band

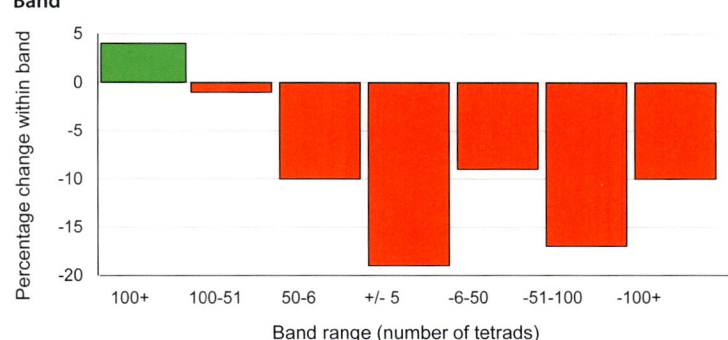

Figure A1.5. Comparison of Level of Breeding Evidence Recorded (8 Increasing Species)

occupied tetrads with a breeding code for these eight species has increased slightly from 6,900 to 6,923 but the proportion of tetrads with confirmed breeding has fallen from 87% to 72%. This indicates that reduced population sizes are not the only factor responsible for the fall in confirmed breeding records.

Of the seven species found in every tetrad, only one (Great Tit) had more confirmed breeding records in 2008–13 than in 1985–90.

Summary: Comparison Between the Two Atlas Periods

The level of coverage, standard of fieldwork and the results achieved for the recent Atlas project are robust enough to enable meaningful comparisons to be made with the distribution maps in the *Atlas* (1992), and for the breeding distribution change maps to be valid for most species. In a few cases where validation has been more stringent, the distribution maps are not directly comparable, and this has been explained in the relevant species accounts. Due to the relatively low level of confirmed breeding records for the recent Atlas project, direct comparisons with the *Atlas* (1992) using only these records should not be made, but comparisons made by combining confirmed and probable or confirmed, probable and possible records will be valid. The breeding distribution change maps in the species accounts are based on all three categories of breeding evidence in each of the two Atlas periods.

Winter Coverage

As this was the first local winter Atlas project, no previous targets existed, so an arbitrary figure of 40 species per tetrad was set as a minimum. This figure was easily reached in tetrads which contained a mix of habitats but proved testing in many which lacked diversity. Only eight tetrads failed to reach the 40 mark, six of these were in the uplands where between 26 and 38 species were counted, one in a mainly arable area had 36 and the other with 38 contained a mix of forestry plantations and pasture. In addition to the 40 target, ACOs and observers were asked to check that common and widespread species had been found in every tetrad, and if not they were asked to revisit them to fill any obvious gaps. This worked well and the resulting maps provide an accurate picture of winter distribution for most species, but for some scarce and erratic visitors the maps reflect chance encounters.

Timed Tetrad Visits (TTVs)

Although the Atlas working group was unsure that it could be achieved, it was decided to try to get two-hour TTVs completed in both the early and late periods for all tetrads in both the winter and breeding seasons, during the period of the BTO national Atlas 2007–11. Unfortunately some observers only completed the minimum one-hour TTV, of which there were 267 during the winter and 265 during the breeding season, but TTVs were eventually completed in every tetrad for early and late periods in both seasons, and two-hour TTVs were completed in 85% of tetrads in each season. In addition to calculating results from a mix of single and two-hour TTVs, many other factors will have affected them: experience of the observer, date of visit, weather (both during a single season and between years), access to all habitats in the square, and, during the winter period, the roaming nature of many flocks. For these reasons the relative abundances in each individual tetrad need to be treated with caution, but, taken together, clusters of tetrads with similar levels of relative abundance are a reliable guide to the concentrations of the more common species.

Bias in the Results

Several factors influence the results obtained through this type of survey, involving many people over several years.

Uneven Fieldwork Coverage

There is a tendency for the maps to reflect the distribution of observers and timing of their fieldwork visits, rather than the distribution of the birds. Almost half the human population lives in Telford or Shrewsbury, and the number in each of the other settlements is listed on p. 29. There are many more people in east Shropshire than south, west or north. Not surprisingly, the distribution of fieldworkers largely mirrors the pattern of human settlement. Although 650 people contributed records to the recent Atlas project, most helpers covered one or two local squares for part of the survey period, often in great depth, but around half of the records were contributed by 27 experienced active birdwatchers, many of whom travelled widely. With 870 tetrads, many have not received thorough coverage.

Tetrads in the more remote parts of the County, especially in the hills to the west and south, and the heavily agricultural areas in the north, may have been visited by only one observer in only one season. Even the most skilled and dedicated observers are not going to find every species in just a few visits. Such visits are unlikely to be made during the evening or night, particularly if a long journey is necessary, and if they happen to be made in a bad season, either in general or for specific species, then the result will be even less complete. The more frequently a tetrad is visited, the more comprehensive the final record will have been.

Visits from different people also improve results. Fieldworkers survey tetrads in different ways and at different times. Holidays at the end of the season may prevent a visit at the best time to confirm breeding of many species, but a late visitor will hear little song and might find access to the only pool prevented by overgrown nettles and brambles. People also have preferences for particular habitats, and different recognition skills, especially for the secretive small active species best identified by song. Some observers may even overlook the extremely common species, which they consider uninteresting.

Though most tetrads have been covered reasonably well, comparison of the final coverage maps (Maps A1.3 and A1.6) with those of each individual species shows that some of the gaps on the maps in the species accounts are due to inadequate coverage. This is particularly true for a few very widespread species that are believed to breed in every tetrad.

Taking the coverage maps and fieldwork method together, it is clear that most species are under-recorded to some extent, but this is less likely to occur in the north-eastern quadrant.

Mapping Error

Though fieldworkers were given clear guidance in identifying tetrad boundaries, a few records will inevitably be wrongly mapped either because the tetrad was incorrectly recorded on the online data entry form, or the fieldworker unknowingly crossed the boundary. Though the ACOs trapped most of these errors, some will have escaped.

The birds too are mobile, and do not remain in their 'home' tetrad. Raptors carry prey great distances to their nestlings, water birds may nest some distance from the pools where they eventually raise their young, and fledglings too may cross tetrad boundaries once family parties leave the nest site. Even pairs with small territories may occupy part of two tetrads, and breed in both in different years during the Atlas period. Fieldworkers were advised to record such observations in the tetrads in which they were seen, rather than guess where the nest may have been.

A few errors will arise from incorrect identification in the field, particularly of Marsh and Willow Tits and, to a lesser extent, Willow Warbler and Chiffchaff, and Blackcap and Garden Warbler if, as often occurs, the singing bird is not seen.

In spite of thorough checks, it is unlikely that data entry errors were totally eliminated.

Many records were submitted to the County Bird Recorder by observers not directly engaged in Atlas fieldwork, and for both Atlas periods a check was made of records and SBRs for the less common species, to ensure that there were no other records missing from the appropriate Atlas map. Many such records did not include an accurate map reference, so they have been included in the Atlas map at the most likely location, but again a few of them may not be in the correct tetrad.

Though the above factors may mean that a few dots should have been mapped in nearby tetrads, and the occasional one perhaps omitted, they are more than offset by the under-recording inherent in the fieldwork. They should not affect the broad interpretation of the maps.

The Effect of Weather

The weather influences results in two ways. Firstly, more fieldworkers will be involved, and each will make more field trips, in good summers.

Secondly, the ease of obtaining records is affected. The vast majority of confirmed breeding records arise from seeing either adults carrying food to their nest, or fledglings with their parents. Weather that reduces the food supply or the number of young raised for any particular species will reduce the chance of observers proving breeding that year. The number of breeding pairs of some resident species is reduced, perhaps considerably, if the preceding winter was hard, especially if there are several bad winters in succession. The numbers of summer visitors may be reduced by weather conditions in wintering quarters or on migration routes.

Also, the number of young birds raised depends on the abundance of the right food supply. Insect numbers will be reduced by cold springs or summers, their breeding cycle may be delayed, and they may spend less time on the wing, making it harder for insect-eating birds to gather enough of them.

Hot summers may dry out the ponds or mud in which various insects breed, reduce the water level of streams, ponds and marshes, and make the ground much harder, affecting the food supply or breeding habitat of other birds. This effect is exaggerated if the preceding winter and spring have been dry.

Table A1.3. Weather Conditions Each Year 2007–13

Year	Winter (Dec*, Jan, Feb) * Previous year	Spring (Mar, Apr, May)	Summer (Jun, Jul, Aug)	Autumn (Sep, Oct, Nov)	Year
2007	Very warm with little frost and very wet	Warm with few frosty periods but with showery periods	Cool and wet	Very dry	Warm and moist
2008	Very warm days and plentiful sunshine	Cloudy with near normal temperatures	Temperatures near normal but with warm nights and cloudy skies	Cool and wet	Near normal temperatures and precipitation but cloudy
2009	Close to normal temperatures but wet	Mild nights with few frosts	Near normal	Mild with few frosts	Near normal
2010	Very cold and frosty but dry and sunny	Cold nights but dry, with plentiful sunshine	Mild and rather dry	Very cold and frosty but sunny	Cold, dry and cloudy
2011	Cold but very cloudy and dry	Warm, especially by day but sunny and with very dry conditions	Very cool but dry	Very warm, with few frosts and very dry	Warm, especially by day, with low precipitation
2012	Warm	Wet	Cool days and mild nights, wet and cloudy	Very cold, frosty and wet (yet with sunny skies by day)	Temperatures near normal but cloudy and wet
2013	Cloudy and very wet	Very cold and frosty, with sunny days	Warm days but cool nights, dry and sunny	Mild and very cloudy	Markedly cooler (with more frost) than normal

Table A1.4. Overall Weather Conditions 2007–13

Daily Maximum Temperature	Daily Minimum Temperature	Daily mean temperature	Frost	Precipitation	Sunshine
+0.32°C	+0.22°C	+0.27°C	+0.06 days	+2.12mm	–3.42 hrs./month

Table A1.5. Prevailing Weather Conditions During the 1985–90 Atlas Period

Breeding Season	Weather	Previous Winter	Period			
			April	May	June	July
1985	Temperature	Cold	Cold	Cold	Cool	Cool
	Precipitation	Lying snow	High	Average	High	Mixed
1986	Temperature	Very cold	Cold	Cold	Mixed	Mixed
	Precipitation	Low	High	High	Mixed	Mixed
1987	Temperature	Average	Cool	Cool	Cool	Cool
	Precipitation	High Lying snow	High	High	High	High
1988	Temperature	Mild	Cold	Warm	Mixed	Cool
	Precipitation	High	Average	High	Mixed	V. High
1989	Temperature	Mild	Cold	Warm	Hot	Hot
	Precipitation	Low	High	Low	Low	V. Low
1990	Temperature	Mild	Cool	Average	Average	V. Hot
	Precipitation	High	Average	Low	Average	Low

Note: 'Mixed' indicates a considerable fluctuation, rather than a steady pattern.

Heavy rain may damage some nests, the consequent floods may wash away nests at the water's edge, and if adults become wet while collecting food for nestlings they will carry moisture into the nest and the bedraggled young may die of cold.

While annual fluctuations due to the weather should not have affected results in tetrads which were visited several times in each year of the survey, they may have a significant effect on finding particular species in tetrads only visited a few times in one year.

The prevailing weather conditions for each year of the survey are summarised in Table A1.3.

Weather Conditions During the Recent Atlas Period

Table A1.3, based on records from the UK Met Office Shawbury station (Met Office, 2015b), summarises weather conditions both seasonally (using standard meteorological seasons) and for the year as a whole. Use of terms like 'mild', 'wet' etc. are based on an analysis of the variability of daily maximum, minimum and mean temperatures, precipitation and sunshine data over the observation period.

Table A1.4 characterises the weather conditions during the observations period in relation to the whole available period of record from Shawbury, shown as observation period value minus long-term average value. As is apparent, data were collected during a slightly warmer, wetter and cloudier period than average. A more detailed summary of the weather conditions each year can be found in the respective SBRs.

Weather Conditions During the 1985–90 Atlas Period

Table A1.5 summarises the weather conditions reported in the respective SBRs, and is reproduced from the *Atlas* (1992). Terms are somewhat subjective, and used relative to the County average for that time of year, to indicate the likely impact of the weather on population levels and breeding success for each species. The first four summers were relatively cold and wet, and only the last two were hot and dry.

Interpreting the Results

To summarise, each map must be interpreted in the light of variable fieldwork coverage and conditions, the possibility of a small number of mapping errors, and the natural fluctuations of bird populations and nest sites over the six years of fieldwork.

Similar caveats apply to the maps from the 1985–90 Atlas fieldwork, but the comparison of fieldwork results from the two Atlas periods outlined above indicates that the breeding distribution change maps summarise fairly the substantial and rapid changes that have taken place between the two Atlas project periods.

Both of the Atlas projects aimed to show the broad distribution of each species breeding in Shropshire, and the factors that influence it, and this has undoubtedly been achieved.

Place	NGR (4 fig.)
Doley	SJ7429
Dolgoch Quarry NR	SJ2724
Donnington	SJ7113
Donnington Wood (Telford)	SJ7012
Dorrington	SJ4702
Dothill	SJ6412
Dovaston	SJ3420
Dowles (Wyre Forest)	SO7776
Dowles Brook	SO7576
Downton Hall	SO5279
Dudleston Heath	SJ3636
Dudlestone Moor	SJ3636
Dudmaston	SJ3535
Dudmaston (Hall & pools)	SO7488
Duffryn	SO2282
Dutlas	SO2177
Dyffryd	SJ2919
Eardiston	SJ3625
Earl's Hill	SJ4004
East Onny, River	SO3996
Eaton	SO4989
Eaton Constantine	SJ5906
Eaton Mascott	SJ5305
Edge Wood	SJ6100
Edgebolton	SJ5721
Edgeley	SJ5540
Edgerley	SJ3518
Edgmond	SJ7219
Edgmond Marsh	SJ7220
Ellerton Mill Pools	SJ7126
Ellesmere	SJ4034
Ellesmere College	SJ4033
Emstrey	SJ5210
English Bridge (Shrewsbury)	SJ4912
English Frankton	SJ4529
Ercall	SJ5817
Ercall Heath	SJ6822
Ercall, The	SJ6409
Exfords Green	SJ4505
Eyton Moor	SJ6515
Eyton on Severn	SJ5706
Eyton upon the Weald Moors	SJ6415
Fallway	n/a
Farley Brook	SJ6302
Felindre	SO1781
Felton Butler Pool	SJ3917
Fenemere	SJ4422
Fenn's Moss	SJ4836
Ferney Hall Dingle	SO4277
Folly Pool – Alderton	SJ3816
Ford	SJ4113
Frodesley	SJ5101
Fullway	SJ5503

Place	NGR (4 fig.)
Furber's Flood (Whixall Canal Floods)	SJ4935
Garbett Hall	SO2676
Garmston	SJ6006
Garn, The	SO2381
Gatten	SO3998
Gatten Lodge	SO3899
Gatten Plantation (Stiperstones)	SO3798
Gatten Wood	SO3798
Gliding Club (Long Mynd)	SO4091
Glyn Common	SJ3011
Gobowen	SJ3033
Gogin, The	SO2384
Grafton	SJ4318
Granville Country Park	SJ7112
Granville landfill	SJ7211
Great Bolus	SJ6421
Great Hay	SJ7002
Great Woolaston	SJ3212
Greystones	SJ2722
Grindle (Long Mynd)	SO4392
Grindle (nr Telford)	SJ7502
Grindley Brook	SJ5143
Grinshill	SJ5223
Grit Hill	SO3398
Grouse Lodge	SJ2333
Grove, The (Craven Arms)	SO4384
Guilden Down	SO3082
Habberley	SJ3903
Haddon Hill	SO4395
Hadley	SJ6812
Hadnall	SJ5220
Halfway House	SJ3411
Halston	SO4107
Halston (Hall)	SJ3431
Hampton Loade	SO7486
Hanmer Mere	SJ4539
Hanwood	SJ4409
Hardwick Pool	SO3690
Harlescott	SJ5115
Harley	SJ5916
Harley Brook	SJ5900
Harmer Hill	SJ4822
Harnage (Cound)	SJ5604
Harper Adams Uni. Coll. Farm (Edgmond)	SJ7120
Harton	SO4888
Hatton Grange	SJ7604
Haughmond Abbey	SJ5415
Haughmond Hill	SJ5414
Haughton (nr Shifnal)	SJ7408
Hawkstone Park	SJ5729
Hayes Farm	SJ3515
Heath Mynd	SO3393

Place	NGR (4 fig.)
Hem Flash	SJ3506
Hem, The (Shifnal)	SJ7305
Hencott Pool	SJ4916
Hengoed	SJ2933
Henley	SO5476
High Ercall	SJ5917
High Park	SO4397
High Rock	SO7293
High Vinnalls	SO4772
Higher Heath	SJ5635
Highley	SO7483
Hilton	SO7795
Hine Heath	SJ5825
Hinks Wood	SJ7016
Hinstock	SJ6926
Hoar Edge	SO6077
Hodnet	SJ6128
Hodnet Heath	SJ6126
Hollies, The (Stiperstones SSSI, SWT reserve)	SJ3901
Hollins Farm (Merrington)	SJ4621
Hollybush Coppice (Bridgnorth)	SO7295
Homer	SJ6101
Hookagate	SJ4609
Hope, The	SO5178
Hope Valley	SJ3300
Hopesay	SO3983
Hopton Cangeford	SO5480
Hopton Court	SO6476
Hopton Heath	SO3876
Hopton Wafers	SO6376
Horderley	SJ4087
Horsebridge	SJ3606
Horsehay (Telford)	SJ6707
Horton	SJ6814
Hortonwood (Telford)	SJ6813
Howle Pool	SJ6923
Hungerford	SO5389
Hurst, The	SO3180
Ightfield	SJ5938
Ironbridge	SJ6703
Ironbridge Gorge	SJ6902
Ironbridge Power Station	SJ6503
Isle, The	SJ4517
Isle Pool	SJ4617
Isombridge	SJ6013
Jonathan's Hollow (Long Mynd)	SO4496
Kemberton	SJ7304
Kemp Valley (and River)	SO3582
Kempton	SO3682
Kerry Ridgeway	SO3083
Kingsland	SJ4811
Kingsnordley	SO7787

Place	NGR (4 fig.)
Kinlet	SO7180
Kinnerley	SJ3320
Kinnersley Moor	SJ6816
Kinsley Wood (Knighton)	SO2973
Kinver	SO8383
Kite's Nest Farm/Kitesnest Farm	SO6094
Knighton	SO2872
Knighton Reservoir	SJ7328
Knockin	SJ3322
Knockin Heath	SO3521
Knuck Wood	SO2786
Kynnerley, see Kinnerley	
Kynnersley	SJ6716
Langley (Acton Burnell)	SJ5302
Lawley, The	SO4997
Lea Wood (Welshampton)	SJ4235
Leaton	SJ4618
Leaton Knolls	SJ4716
Ledwyche Brook	SO5472
Lee	SJ4032
Lee Brockhurst	SJ5427
Leebotwood	SO4798
Leegomery (Telford)	SJ6612
Leighton	SJ6105
Leighton Flats	SJ2305
Lilleshall	SJ7615
Lilleshall Hall	SJ7414
Lilleshall Hill	SJ7315
Linley	SO3492
Linley Big Wood	SO3494
Linley Hill	SO3594
Little Drayton	SJ6633
Little Wenlock	SJ6406
Llanforda	SJ2629
Llanhowell Farm	SO3479
Llanyblodwel	SJ2422
Llanymynech	SJ2620
Lloyds Coppice	SJ6903
Llwyntidmon	SJ2920
Llyn Rhuddwyn	SJ2328
Llynclys	SJ2823
Llynclys Hill	SJ2723
Loamhole & Lydebrook Dingle	SJ6605
Long Batch (Long Mynd)	SO4496
Long Lane (Sleapford)	SJ6315
Long Mountain	SJ2707
Long Mynd, The	SO4194
Long Mynd plantation (FC)	SO4090
Longden Mill	SJ4406
Longdon on Tern	SJ6115
Longnor	SJ4900
Longville	SO5493
Loppington	SJ4629

Place	NGR (4 fig.)
Loton Park	SJ3514
Lower Betton	SJ5208
Lower Brompton Farm	SJ5508
Lower Netchwood	SO6291
Lower Whitcliffe	SO4874
Lower Wood (nr Alberbury)	SJ3614
Lower Wood (nr All Stretton)	SO4697
Ludford	SO5173
Ludford Bridge (Ludlow)	SO5074
Ludlow	SO5175
Lutwyche (Wenlock Edge)	SO5594
Lydbury North	SO3586
Lydham	SO3391
Lydham Heath	SO3490
Lydham Manor	SO3389
Lyneal Wood	SJ4531
Madeley (Telford)	SJ6904
Madeley Court	SJ6905
Maesbrook	SJ3021
Maesbury	SJ3025
Maesbury Marsh	SJ3125
Mainstone	SO2787
Marchamley	SJ5929
Market Drayton	SJ6734
Marrington	SO2797
Marrington Hall	SO2797
Marsh Green	SJ6014
Marshbrook	SO4489
Marton Pool (Baschurch)	SJ4423
Marton Pool (nr Chirbury)	SJ2902
Mason's Bank	SO2287
Meadowley	SO6692
Meeson	SJ6520
Melverley	SJ6119
Melverley	SJ3316
Meole	SJ4810
Meole Brace (Shrewsbury)	SJ4809
Mere Pool, Shrewsbury	SJ5110
Mere, The (Ellesmere)	SJ4034
Merrington	SJ4720
Merrington Green	SJ4620
Middle Pool, Trench	SJ6811
Middleton	SO5377
Middletown	SJ3012
Middletown Hill	SJ3013
Milford Mill	SJ4221
Millichope Park	SO5288
Minsterley	SJ3705
Minton Batch (Long Mynd)	SO4190
Mirelake (Allscott Sugar Factory)	SJ5912
Moelydd Hill	SJ2425
Monkmoor Pool (Shrewsbury)	SJ5213
Montford Bridge	SJ4315

Place	NGR (4 fig.)
Montgomery Canal	SJ2621
Mor Brook	SO6694
Morda	SJ2827
More	SO3491
More Farm	SO3492
Moreton Corbet	SJ5623
Moreton Mill	SJ5722
Morville	SO6794
Morville Heath	SO6893
Moston	SJ5626
Mountford Bridge	SJ3917
Much Wenlock	SO6299
Muckleton	SJ5921
Mucklewick Hill	SO3397
Munslow	SO5287
Myddle	SJ4723
Mytton Dingle (Stiperstones)	SJ3600
Nedge Hill	SJ7107
Neen Savage	SO6777
Neen Sollars	SO6672
Neenton	SO6587
Nesscliffe	SJ3819
New Marton	SJ3334
Newcastle	SO2482
Newport	SJ7418
Newport Sewage works	SJ7319
Newton Mere	SJ4234
Nib Green	SO7579
Nib Heath	SJ4118
Nipstone Rock	SO3597
Nobold	SJ4710
Noneley	SJ4727
Norbroom Marsh (nr Newport)	SJ7519
Norbury Hill	SO3594
Norbury Pool	SO3692
Nordy Bank Fort	SO5784
Northwood	SJ4633
Norton	SJ7200
Oakengates (Telford)	SJ7011
Oakly Park	SO4876
Oerley Reservoir	SJ2729
Old Caynton Mill	SJ6921
Old Heath	SJ5115
Old Racecourse (Oswestry)	SJ2631
Onibury	SO4579
Onny, River	SO3987
Onslow & Hall	SJ4312
Oss Mere	SJ5643
Oswestry	SJ2929
Oswestry Hill Fort	SJ2930
Oswestry Old Racecourse	SJ2631
Overton	SO5072
Owlbury	SO3191

Place	NGR (4 fig.)
Oxon Pool	SJ4513
Packetstone Hill (Long Mynd)	SO4291
Pant	SJ2722
Park Hall & Camp (Oswestry)	SJ3131
Parrs Pool (Bayston Hill)	SJ4708
Pell Wall	SJ6733
Pentre Hodre	SO3276
Pen-yr-Estyn	SJ3527
Peplow	SJ6324
Perry, River	SJ3926
Petton	SJ4426
Petton Park	SJ4326
Picklescott	SO4399
Pikes End Farm	SJ4431
Pim Hill	SJ4821
Pimley Island	SJ5214
Pimley Manor	SJ5214
Pipe Gate	SJ7340
Plaish Hall	SO5396
Plowden	SO3787
Plush Hill	SO4596
Pole Cottage (Long Mynd)	SO4193
Polemere	SJ4109
Poles Coppice	SJ3904
Polly's Lock	SJ7319
Pontesbury	SJ4006
Pontesford	SJ4006
Ponthen	SJ3317
Porth-y-waen	SJ2523
Posenhall (nr Broseley)	SJ6501
Postenplain (Wyre Forest)	SO7479
Pound Green Common	SO7578
Powis Castle	SJ2106
Pradoe	SJ3528
Prees	SJ5533
Prees Heath	SJ5637
Prees Lower Heath	SJ5732
Presthope	SO5897
Preston Montford	SJ4214
Preston upon the Weald Moors	SJ6815
Priorslee	SJ7009
Priorslee Flash	SJ7110
Priorslee Lake	SJ7109
Puleston Common	SJ7323
Pulley Common (no longer exists)	SJ4709
Pulverbatch	SJ4202
Purlogue	SO2877
Purslow	SO3680
Quabbs (Clun Forest)	SO2080
Quarry, The (Shrewsbury)	SJ4812
Quatford	SO7390
Queen's Head	SJ3426
Quina Brook	SJ5233

Place	NGR (4 fig.)
Radbrook Pools	SJ4711
Radlith (nr Pontesbury)	SJ4105
Ragleth	SO4592
Ragleth Hill	SO4592
Ratlinghope	SO4096
Rea Brook	SJ3404
Rea Valley	SJ3404
Rea, River	SO6773
Rea, The (Upton Magna)	SJ5612
Red Wood	SO3183
Redhill (St Georges, Telford)	SJ7310
Redlake, River	SO3275
Rednal	SJ3628
Rednall Mill	SJ3729
Redwith	SJ2821
Rhos Fiddle	SO2085
Rhydycroesau	SJ2430
Richards Castle	SO4969
Riddings	SO1986
Ridgewardine	SJ6838
Rigmore Oak (Stiperstones)	SO3698
River Camlad	SO3092
River Ceiriog	SJ2438
River Clun	SO3381
River Corve	SO5691
River East Onny	SO3996
River Kemp	SO3582
River Onny	SO3987
River Perry	SJ3926
River Rea	SO6773
River Redlake	SO3275
River Roden	SJ5624
River Severn	SJ4014
River Tanat	SJ2421
River Teme	SO2674
River Tern	SJ6229
River Vyrnwy	SJ3216
River West Onny	SO3494
River Worfe	SO7594
Rock, The	SJ6809
Rock, The (Stiperstones)	SO3596
Roden (Hall)	SJ5716
Roden, River	SJ5624
Rodington	SJ5914
Romsley	SO7882
Rorrington	SJ2900
Rose Grove (nr Anchor)	SO1885
Round Hill (Long Mynd)	SO4192
Rushbury	SO5191
Rushmoor	SJ6113
Ruyton Brook	n/a
Ruyton-XI-Towns	SJ3922
Ryton	SJ7602

Place	NGR (4 fig.)
Sambrook	SJ7124
Sambrook Mill	SJ7125
Sandford (nr Oswestry)	SJ3423
Sandford Pool (Whitchurch)	SJ5834
Sansaw	SJ5023
Selattyn Hill	SJ2534
Selley Hall (nr Knighton)	SO2676
Severn Gorge	SJ6703
Severn Valley Country Park	SO7583
Severn, River	SJ4014
Severn–Tern confluence	SJ5509
Severn–Vyrnwy confluence	SJ3215
Sharpstones	SJ4909
Shavington	SJ6337
Shavington Big Pool	SJ6338
Shavington Hall	SJ6338
Shavington Park	SJ6338
Shavington Park, Claverhall	SJ6139
Shavington Tittenley Pool	SJ6438
Shawbirch	SJ6413
Shawbury	SJ5422
Shawbury Heath	SJ5420
Sheinton	SJ6103
Shell Brook	SJ3539
Shelton	SJ4613
Shelton Rough	SJ4613
Shelve	SO3399
Shelve Pool	SO3397
Sherriffhales	SJ7612
Shifnal	SJ7507
Shirlett	SO6597
Showell Mill Pool	SJ7124
Shrawardine	SJ3915
Shrewsbury	SJ4912
Shrewsbury Old Racecourse	SJ5012
Shrewsbury Quarry	SJ4812
Shrewsbury School	SJ4812
Shrewsbury Sewage Farm	SJ5213
Sidbury	SO6885
Silvington Common	SO6279
Sleap	SJ4826
Sleap Airfield	SJ4826
Smethcote Farm	SJ5610
Smethcott	SO4599
Smithfield Roads, Shrewsbury	SJ4812
Snailbeach	SJ3702
Snedshill	SJ7010
Somerwood	SJ5614
Soulton	SJ5430
Sowdley Wood	SO3280
Spoonley (nr Market Drayton)	SJ6635
Springfield Mere (Shrewsbury)	SJ5110
St George's (Telford)	SJ7011

Place	NGR (4 fig.)
St Martin's	SJ3236
Stanmore	SO7492
Stanton Lacy	SO4978
Stanway Manor	SO5291
Stapeley Common	SO3299
Stapeley Hill	SO3199
Stapleton	SJ4604
Stead Vallets	SO4676
Steel Heath	SJ5436
Stiperstones, The	SO3698
Stirchley	SJ7006
Stitt	SO4098
Stockton Wood Farm (Chirbury)	SJ2601
Stoke Heath	SJ6529
Stoke Wood	SO4381
Stoke-on-Tern	SJ6428
Stokesay	SO4381
Stoney Hill (Telford)	SJ6606
Stottesdon	SO6782
Stow Hill	SO3074
Strefford	SO4485
Stretton hills	SO49
Strine Brook	SO6818
Sundorne	SJ5214
Sunnyhill	SO3283
Sutton (Shrewsbury)	SJ5010
Sutton Farm Pool Market Drayton	SJ6731
Sutton Maddock	SJ7201
Swan Farm mine	SJ6406
Sweeney Mountain (nr Oswestry)	SJ2725
Tanat, River	SJ2421
Tasley	SO6994
Tedsmere (Baschurch)	SJ4223
Telford	SJ6909
Telford Town Park	SJ6908
Teme, River	SO2674
Tenbury Wells	SO5967
Tern Hill	SJ6332
Tern, River	SJ6229
Ternhill Quarry	SJ6530
The Berth	SJ4223
The Birch Moors	SO7019
The Bog Visitor Centre (Stiperstones)	SO3597
The Bog, Cranmere	SO7597
The Ercall	SJ6409
The Garn	SO2381
The Gogin	SO2384
The Grove (Craven Arms)	SO4384
The Hem (Shifnal)	SJ7305
The Hollies (Stiperstones, SWT reserve)	SJ3901
The Hope	SO5178
The Hurst, Clun	SO3180

Place	NGR (4 fig.)
The Isle	SJ4517
The Lawley	SO4997
The Long Mynd	SO4194
The Mere (Ellesmere)	SJ4034
The Quarry (Shrewsbury)	SJ4812
The Rea (Upton Magna)	SJ5612
The Rock	SJ6809
The Rock (Stiperstones)	SJ6809
The Stiperstones	SO3698
The Weald Moors	SJ6715
The Wrekin	SJ6208
Tibberton	SJ6820
Ticklerton	SO4890
Tilstock	SJ5437
Tittenley Pool	SJ6438
Tittenley Pool, Shavington	SO6438
Titterhill	SO3577
Titterstone Clee hill	SO5978
Tong Lake	SJ7907
Tong Lodge	SJ7806
Townbrook (Long Mynd)	SO4494
Trefonen	SJ2526
Trench Pool	SJ6812
Twemlows Big Wood	SJ5636
Twitchen	SO3779
Tyrley (and Tyrley Locks)	SJ6932
Uffington	SJ5213
Upper Affcot	SO4486
Upper Clun valley	SO2083
Upper Dinchope	SO4583
Upper Ledwyche	SO5579
Upper Onny valley	SO3893
Upton Cressett	SO6592
Upton Magna	SJ5512
Vennington	SJ3309
Venus Pool	SJ5406
Vessons Coppice, Pontesbury	SJ3902
Vron (nr Felindre)	SO1682
Vyrnwy, River	SJ3216
Walcot (Telford and Wrekin)	SJ5912
Walcot Hall Lakes (Lydbury North)	SO3485
Walford	SJ4320
Wall Farm (Kynnersley)	SJ6817
Wappenshall	SJ6614
Waters Upton	SJ6319
Weald Moors, The	SJ6715
Weeping Cross Pool	SJ5110
Wellington	SJ6411
Welsh Bridge (Shrewsbury)	SJ4912
Welsh Frankton	SJ3632
Welshampton	SJ4435
Wem	SJ5128

Place	NGR (4 fig.)
Wem Moss	SJ4734
Wenlock Edge	SO5089
West Felton	SJ3425
West Onny, River	SO3494
Westbury	SJ3509
Westhope	SO4786
Weston (nr Marchamley)	SJ5628
Weston Lullingfields	SJ4224
Weston Park (nr Shifnal)	SJ8010
Westwood, Oldbury	SO7091
Whatsill	SO6176
Whitchurch	SJ5441
Whitcliffe	SO5074
Whitcott Keysett	SO2782
White Mere	SJ4132
Whittington	SJ3231
Whittington Castle	SJ3231
Whixall	SJ5134
Whixall canal floods	SJ4935
Whixall Moss	SJ4935
Wilcott Marsh	SJ3817
Wild Moor (Long Mynd)	SO4296
Wilderhope Manor	SO5492
Wildmoor Pool (Long Mynd)	SO4296
Willey Park (nr Broseley)	SO6699
Winsley Hall	SJ3507
Wistanstow	SO4385
Wolf's Head Moss	SJ3621
Wollaston Pond	SJ3312
Wolverton	SO4687
Wombridge (Oakengates, Telford)	SJ6811
Wood Lane (Nature Reserve)	SJ4232
Woodcote (nr Bicton Heath)	SJ4511
Woodcote (nr Lilleshall)	SJ7615
Woolaston & Pond	SJ3312
Woolstaston	SO4598
Woolston (nr Wistanstow)	SO4287
Worfe, River	SO7594
Worfield	SO7595
Worfield Bog (see Cranmere Bog)	SO7597
Worthen	SJ3204
Wrekin, The	SJ6208
Wrockwardine	SJ6211
Wrockwardine Wood	SJ7011
Wroxeter	SJ5608
Wykey	SJ3924
Wynett Coppice	SO5485
Wyre Forest	SO7576
Wytheford Wood	SJ5720
Yell Bank	SO5497
Yockings Gate (Whitchurch)	SJ5542
Yockleton	SJ3910

APPENDIX 3

Scientific Names of Plants and Animals

The scientific name of each bird species is included in the heading to each account, and in the Shropshire List table in Appendix 6. Many of the accounts also refer to plants (trees, bushes and crops) and animals. The scientific name of all these other taxa are listed below.

Plants

Common Name	Scientific Name
Alder	Alnus glutinosa
Apple (domestic)	Malus spp.
Ash	Fraxinus excelsior
Barley	Hordeum vulgare
Beech	Fagus sylvatica
Bilberry (Whimberry)	Vaccinium myrtillus
Birch	Betula spp.
Birch, Silver	Betula pendula
Blackthorn	Prunus spinosa
Bramble	Rubus fruticosus
Bulrush	Typha latifolia
Butterfly Bush (Buddleia)	Buddleja spp.
Conifers	Coniferae
Corsican Pine	Pinus nigra
Cotoneaster	Cotoneaster spp.
Crab Apple	Malus sylvestris
Cypress	Cupressaceae
Dog Rose	Rosa canina
Douglas Fir	Pseudotsuga menziesii
Elder	Sambucus nigra
Elm	Ulmus spp.
Fir, Douglas	Pseudotsuga menziesii
Fir, Grand	Abies grandis
Fumitory	Fumaria spp.
Gorse	Ulex spp.
Gorse, common	Ulex europaeus

Common Name	Scientific Name
Gorse, western	Ulex gallii
Grand Fir	Abies grandis
Hawthorn	Cratageus monogyna
Hazel	Corylus avellana
Heather	Calluna vulgaris
Himalayan balsam	Impatiens glandulifera
Hornbeam	Carpinus betulus
Ivy	Hedera helix
Japanese knotweed	Fallopia japonica
Kale	Brassica oleracea
Larch	Larix spp.
Lime	Tilia spp.
Maize	Zea mays
Maple	Acer spp.
Monkey puzzle	Araucaria araucana
Mullein	Verbascum spp.
Nettle	Urtica dioica
Norway Spruce	Picea abies
Oak	Quercus spp.
Oak, Sessile	Quercus petraea
Oak, Turkey	Quercus cerris
Oats	Avena sativa
Oilseed rape	Brassica napus
Osier	Salix viminalis
Pea	Pisum sativum
Pine	Pinus spp.

Common Name	Scientific Name
Pine, Corsican	Pinus nigra
Pine, Scots	Pinus sylvestris
Poplar	Populus spp.
Potato	Solanum tuberosum
Purple Moor-grass	Molinia caerulea
Pyracantha	Pyracantha spp.
Raspberry	Rubus idaeus
Red-hot poker	Kniphofia spp.
Reed, Common	Phragmites australis
Rhododendron	Rhododendron spp.
Rowan	Sorbus aucuparia
Scots Pine	Pinus sylvestris
Sessile Oak	Quercus petraea
Silver Birch	Betula pendula
Spruce	Picea spp.
Spruce, Norway	Picea abies
Sunflower	Helianthus annuus
Sweet pea	Lathyrus odoratus
Sycamore	Acer pseudoplatanus
Turkey Oak	Quercus cerris
Wheat	Triticum spp.
Whimberry (Bilberry)	Vaccinium myrtillus
Wild Cherry	Prunus avium
Willow	Salix spp.
Yew	Taxus baccata

Animals

Common Name	Scientific Name
American Mink	*Neovison vison*
Ant	*Formicidae*
Badger	*Meles meles*
Beetle	*Coleoptera*
Bloodworm (midge larvae)	*Chironomidae*
Butterfly	*Lepidoptera*
Caddis Fly	*Trichoptera*
Caterpillar	*Lepidoptera / Symphyta*
Cockle, common	*Cerastoderma edule*
Cranefly	*Tipulidae*
Deer	*Cervidae*
Dragonfly	*Odonata*

Common Name	Scientific Name
Earthworm (or worm)	*Oligochaeta*
Fly	*Diptera*
Fox, Red	*Vulpes vulpes*
Freshwater Pearl Mussel	*Margaritifera margaritifera*
Grey Squirrel	*Sciurus carolinensis*
Mayfly	*Ephemeroptera*
Mink, American	*Neovison vison*
Moth	*Lepidoptera*
Mussel, common	*Mytilus edulis*
Mussel, Zebra	*Dreissena polymorpha*
Otter	*Lutra lutra*
Pine Marten	*Martes martes*

Common Name	Scientific Name
Rabbit (European)	*Oryctolagus cuniculus*
Rat (Brown)	*Rattus norvegicus*
Slug	*Pulmonata*
Snail	*Pulmonata*
Spider	*Arachnida*
Springtail	*Collembola*
Squirrel, Grey	*Sciurus carolinensis*
White-headed Duck	*Oxyura leucocephala*
Zebra Mussel	*Dreissena polymorpha*

Only the main family groups of invertebrates are listed, although a few accounts name some of the subgroups.

References

Compiled by Leo Smith & Graham Walker

All references referred to in the preceding chapters are in the form author (or lead author *et al.*) (and year of publication). The full reference, including title of the work, name of publication and publisher, are set out below, in alphabetical order of author.

However, as explained on p. 71,

To minimise interruptions to the flow of the narrative, in most accounts the quoting of references has been kept to the minimum necessary. Therefore, material from the *Handlist* (1964), *An Atlas of the Breeding Birds of Shropshire* (1992), the annual Shropshire Bird Reports from 1956 (SBRs), the national BTO Atlases, 'Bird Facts' and Ringing Data on the BTO website, and records on the *Histo* website (including various publications and the works of Beckwith and Forrest, but particularly the annual reports produced by CSVFC from 1892 onwards, the 'Record of Bare Facts' and the 'Transactions') has largely been quoted or reproduced without reference to the source. In the case of SBR and CSVFC records, this only applies if the record appears in the publication for the same year, so if, for example, an SBR article which summarises survey work and records over several years has been cited in a species account, it has been referenced as (SBR year).

The *Handlist* (1964) and all SBRs can also be found on the *Histo* website (see the box on p. 6), which can be accessed from the menu on the home page on the SOS website by clicking on 'Historical Ornithology of Shropshire'.

Other source material, including the BTO online ringing report for 2014 and the agricultural statistics, can be found on the Avifauna part of the SOS website by selecting 'Avifauna Supplement' from the main menu.

Some accounts also cite 'Shropshire Archives', the archives and local studies service for the historic county of Shropshire (including the borough of Telford and Wrekin). It is based in Shrewsbury, and preserves and make accessible documents, books, maps, photographs, plans and drawings relating to Shropshire past and present. The accounts contain the archive reference, so these specific references are not included in the list below (see www.shropshirearchives.org.uk).

Other accounts refer to 'specimens' in 'Shropshire Museums in Ludlow Museum Resource Centre', also with their reference number, which again are not duplicated below (see www.ludlow.org.uk/ludlow-museum.html).

Adams, G.D. (1950) Birds on a forest. *Birds and Country Magazine*, 3:15, 83–85.

Aebischer, N. & Lucio, A. (1997) Red-legged Partridge. In: *The EBCC Atlas of European Breeding Birds* (eds Hagemeijer W.J.M. & Blair, M.J.), T. & A.D. Poyser, London.

Aebischer, N. (2013) National Gamebag Census: released game species. *Game and Wildlife Conservation Trust Review of 2012*, 34–37.

Aebischer, N. (2018) How many birds are shot in the UK? *Game and Wildlife Conservation Trust Review of 2017*, 42–43.

Alexander, W.B. (1945) The Woodcock in the British Isles. Publication of the British Trust for Ornithology, based on a Report on the Inquiry, 1934–35. *Ibis*, 87.4, 512–550.

Alexander, W.B. & Lack, D. (1944) Changes in status among British breeding birds. *British Birds* 38, 62–69.

Allen, D.S. (1995) Habitat selection by Whinchats: A case for bracken in the uplands? In Thompson, D.B.A., Hester, A.J. & Usher, M.B. (eds) *Heaths and Moorland: Cultural Landscapes*, 200–205. HMSO, Edinburgh.

Anon (1900) Excursion to Ticklerton. *Transactions of the Caradoc and Severn Valley Field Club*, 236–237.

Anon (1904) Little Bittern. In: *Caradoc and Severn Valley Field Club: Record of Bare Facts 1904*, 35.

Anon (1987) Over the years. *The Shropshire Bird Report 1987*, 5–9.

Appleton, G. (2012) Swifts start to share their secrets. *BTO News* May–June 2012, 16–17.

Arnfield, A.J. (2015a) Climate Classification. Encyclopaedia Britannica Online. http://www.britannica.com/topic/classification-1703397.

Arnfield, A.J. (2015b) Encyclopaedia Britannica Online. http://www.britannica.com/science/Koppen-climate-classification.

Ausden, M., Rowlands, A., Sutherland, W.J. and James, R. (2003) Diet of breeding Lapwing *Vanellus vanellus* and Redshank *Tringa totanus* on coastal grazing marsh and implications for habitat management. *Bird Study*, 50:3, 285–293.

Baillie, S.R., Marchant, J.H., Leech, D.I., Massimino, D., Eglington, S.M., Johnston, A., Noble, D.G., Barimore, C., Kew, A.J., Downie, I.S., Risely, K. & Robinson, R.A. (2014) *BirdTrends 2013: trends in numbers, breeding success and survival for UK breeding birds*. BTO Research Report 652, British Trust for Ornithology, Thetford.

Baker, D.J., Freeman, S.N., Grice, P.V. & Siriwardena, G.M. (2012) Landscape-scale responses of birds to agri-environment management: a test of the English Environmental Stewardship scheme. *Journal of Applied Ecology*, 49, 871–882.

Balmer, D., Gillings, S., Caffrey, B., Swann, B., Downie, I. & Fuller R. (2013) *Bird Atlas 2007–11 The breeding and wintering birds of Britain and Ireland*. BTO Books, Thetford.

Barn Owl Trust (2012) *Barn Owl Conservation Handbook. A comprehensive guide for ecologists, surveyors, land managers and ornithologists.* Pelagic Publishing, Exeter.

Barshep, Y., Hedenstrom, A. & Underhill, L.G. (2011) Impact of Climate and Predation on Autumn Migration of the Curlew Sandpiper. *Waterbirds* 34:1, 1–9.

BBS (Breeding Bird Survey) *The population trends of the UK's breeding birds*. Published annually since 1994 by BTO for BTO/RSPB/JNCC. Various authors.

Béchet, A. (2009) *European Union Management Plan 2009–11: Golden Plover Pluvialis apricaria*. Technical Report 2009-34. Office for Official Publications of the European Communities, Luxembourg.

Beckwith, W.E. (1878) List of Birds found near Shrewsbury. In: *A guide to the Botany, Ornithology and Geology of Shrewsbury and its vicinity* (ed Phillips, W.), 2–9.

Beckwith, W.E. (1879) Birds of Shropshire. *Transactions of the Shropshire Archaeological and Natural History Society*, Series I, volume II: 365–395.

Beckwith, W.E. (1881) Birds of Shropshire. *Transactions of the Shropshire Natural History & Philosophical Society*, Series I. Volume IV, 326–328.

Beckwith, W.E. (1885) Notes on Shropshire birds (Part 1) *The Field*, 876.

Beckwith, W.E. (1886a) Notes on Shropshire Birds (Part 2) *The Field*, 31.

Beckwith, W.E. (1886b) Sea Birds Inland. *The Field*, 872.

Beckwith, W.E. (1887) Notes on the birds of Shropshire. *Transactions of the Shropshire Natural History & Philosophical Society*, Series I, Volume X, 383–398.

Beckwith, W.E. (1888a) Notes on the birds of Shropshire. *Transactions of the Shropshire Natural History & Philosophical Society*, Series I, Volume XI, 223–238.

Beckwith, W.E. (1888b) Notes on the birds of Shropshire. *Transactions of the Shropshire Natural History & Philosophical Society*, Series I, Volume II, 387–402.

Beckwith, W.E. (1889) Notes on the birds of Shropshire. *Transactions of the Shropshire Natural History & Philosophical Society*, Series II, Volume I, 201–216.

Beckwith, W.E. (1890a) Notes on the birds of Shropshire. *Transactions of the Shropshire Natural History & Philosophical Society*, Series II, Volume II, 1–16.

Beckwith, W.E. (1890b) Notes on the birds of Shropshire. *Transactions of the Shropshire Natural History & Philosophical Society*, Series II, Volume II, 303–318.

Beckwith, W.E. (1891) Notes on the birds of Shropshire. *Transactions of the Shropshire Natural History & Philosophical Society*, Series II, Volume III, 313–328.

Beckwith, W.E. (1892) Notes on the birds of Shropshire. *Transactions of the Shropshire Natural History & Philosophical Society*, Series II, Volume IV, 183–198.

Beckwith, W.E. (1893) Notes on the birds of Shropshire. *Transactions of the Shropshire Natural History & Philosophical Society*, Series II, Volume V, 31–48.

Bibby, C. (2000) More than enough exotics? *British Birds* 93: 2–3.

BirdLife International (2017) IUCN Red List for birds. Downloaded from http://www.birdlife.org on various dates.

Bishton, G. (1983) Distribution etc of Wood Warbler in Telford *The Shropshire Bird Report 1983*, 12–16.

Bishton, G. (1985) The diet of nestling Dunnocks *Prunella modularis*. *Bird Study*, 32:2, 113–115.

Bishton, G. (1986) The diet and foraging behaviour of the Dunnock *Prunella modularis* in a hedgerow habitat. *Ibis*, 128:4, 526–529.

Bishton, G. (2001) Social structure, habitat use and breeding biology of hedgerow Dunnocks *Prunella modularis*. *Bird Study*, 48:2, 188–193.

Bishton, G. (2003) Breeding birds of Telford and Wrekin woodland. March–June 2003. A survey and evaluation. *The Shropshire Bird Report 2003*, 16–30.

Bishton, G. (2012) Birds of Harper Adams University College Farm. April 2011–March 2012. Unpublished report.

Bishton, G. & Lightfoot, J. (2002) An estimate of the breeding population of Barn Owl in Shropshire. *The Shropshire Bird Report 2002*, 3–9.

Blaker, G.B. (1933) The Barn Owl in England – Results of the Census. *Bird Notes and News*, 15, 169–172 & 207–211.

Blunt, A.G. (1995) Diurnal migration over Brown Clee, Autumn 1994. *Bulletin of Shropshire Naturalist' Union*, 22–28.

Bond, T. (2014) The occurrence and arrival routes of North American landbirds in Britain. *British Birds*, 107:2, 66–82.

BOU (British Ornithologists' Union) Records Committee (1999) 25th Report (October 1998) *Ibis*, 141:1, 175–180.

Bowes, A.F., Lack P.C., & Fletcher M.R. (1984) Wintering gulls in Britain, January 1983. *Bird Study*, 31:3, 161–170.

Boyd, H. (1954) The "Wreck" of Leach's Petrels in the Autumn of 1952. *British Birds*, 47:5, 137–163.

Bradbury, R., Eaton, M., Bowden, C. & Jordan, M. (2008) Magnificent Frigatebird in Shropshire: new to Britain. *British Birds*, 101, 317–321.

Bradbury, R.B., Wilson, J.D., Moorcroft, D., Morris, A.J. & Perkins, A.J. (2003) Habitat and weather are weak correlates of nestling condition and growth rates of four UK farmland passerines. *Ibis* 145:2, 295–306.

Briggs, K. (1984) The breeding ecology of coastal and inland Oystercatchers in north Lancashire. *Bird Study*, 31:2, 141–147.

Brook, M. de L. & Davies, N.B. (1987) Recent changes in host usage by cuckoos *Cuculus canorus* in Britain. *Journal of Animal Ecology*, 56, 873–883.

Brook, M. de L. & Davies, N.B. (1988) Egg mimicry by cuckoos *Cuculus canorus* in relation to discrimination by hosts. *Nature*, 335, 630–632.

Broughton, R.K. (2009) Separation of Willow Tit and Marsh Tit in Britain: a review. *British Birds*, 102: 604–661.

Broughton, R.K. & Hinsley, S.A. (2015) The ecology and conservation of the Marsh Tit in Britain. *British Birds*, 108:1, 12–28.

Brown, A. & Grice, P. (2005) *Birds in England*. T. & A.D. Poyser, London.

Brown, A., Gilbert, G. & Wotton, S. (2012) Bitterns and Bittern Conservation in the UK. *British Birds*, 105:2, 58–87.

Brown, D., Wilson, J., Douglas, D., Thompson, P., Foster, S., McCulloch, N., Phillips, J., Stroud, D., Whitehead, S., Crockford, N. & Sheldon, R. (2015) The Eurasian Curlew – the most pressing bird conservation priority in the UK? *British Birds*, 108, 660–668.

Brown, P. (2018) Shropshire Duck Decoys. *Transactions of the Shropshire Archaeological and Historical Society*, 93, 7–20.

Browne, S.J., Buner, F. & Aebischer, N. (2006) A review of Gray Partridge restocking in the UK and its implications for the Biodiversity Action Plan. *Gamebird*, 380–390.

BTO (2012) Laughing bird: Green Woodpecker. *Bird Table*, 71, 18–19.

BTO (2016) http://app.bto.org/bbs-results/results/county_lists/bbscountylist-GBSA.html.

Bull, H.G. (1888) *Notes on the Birds of Herefordshire*. Jackman & Carver, London.

Buner, F., Brockless, M. & Aebischer, N. (2014) The Rotherfield Demonstration Project. *Game and Wildlife Conservation Trust Review of 2013*, 28–29.

Burfield, I. (2002) *The Breeding Ecology and Conservation of the Ring Ouzel Turdus torquatus in Britain.* PhD thesis Queen's College University of Cambridge.

Burton, N.H.K., Watts, P.N., Crick, H.Q.P. & Carter, N. (1996) The effects of pre-harvesting operations on birds nesting in Oilseed Rape (*Brassica napus*) *BTO Research Report 171*, British Trust for Ornithology, Thetford.

Butter, A. (2003) Ferruginous Duck at Cole Mere on 4 Dec 2002, *Shropshire Bird Report 2003*, 176–177.

Buttery (2011) *Charles Darwin and the Owl.* URL: https://britishfoodhistory. wordpress.com/2011/07/26/charles-darwin-and-the-owl/.

Buxton, J. & Durdin C. (2011) *The Norfolk Cranes' Story.* Wren Publishing.

Calbrade, N., Holt, C., Austin, G., Mellan, H., Hearn, R., Stroud, D., Wotton, S. & Musgrove, A. (2010) Waterbirds in the UK 2008/09: *The Wetland Bird Survey.* BTO, RSPB and JNCC, in association with WWT. British Trust for Ornithology, Thetford.

Campbell, B. (1960) The Mute Swan census in England and Wales, 1955–56. *Bird Study*, 7, 208–223.

Cathrall, W. (1855) *The History of Oswestry.* George Lewis, Oswestry.

Catley, G.P. & Hursthouse, D. (1985) Parrot Crossbills in Britain. *British Birds*, 78: 482–505, October 1985].

Cayford, J. (1992) Barn Owl Ecology on East Anglian Farmland. *RSPB Conservation Review*, 6, 45–48.

Chance, E.P. (1922) *The Cuckoo's Secret.* Sidgewick & Jackson, London.

Clark, N.A., Turner B.S. & Young J.F. (1982) Spring passage of Sanderlings *Calidris alba* on the Solway Firth. *Wader Study Group Bulletin*, 36, 10–11.

Clement, P. & Helbig, A.J. (1998) Taxonomy and identification of chiffchaffs in the Western Palearctic. *British Birds*, 91, 361–376.

Cocker, M. & Mabey, R. (2005) *Birds Britannica.* Chatto and Windus, London.

Cohen, E. (1937) *British Birds*, Volume 30.

Collingwood, R.W. (2012) Noyfull Fowles and Vermin' The Statutory Control of Wildlife in Shropshire: 1532–1861. *Transactions of the Shropshire Archaeological and Historical Society*, 87, 47–80.

Collinson, J.M. & Melling, T. (2008) Identification of vagrant Iberian Chiffchaffs – pointers, pitfalls and problem birds. *British Birds*, 101, 174–188.

Community Wildlife Group annual reports available on www.ShropsCWGs. org.uk, and summarised in the annual *Shropshire Bird Reports*.

Congreve, W.M. (1959) *The Oologists' Record.* 33:1,195. Harrison & Sons, Ltd., London, W.C.2.

Conway, G.J., Burton, N.H.K., Handschuh, M. & Austin, G.E. (2008) *UK population estimates from the 2007 Breeding Little Ringed Plover and Ringed plover surveys.* BTO Research Report 510, British Trust for Ornithology, Thetford.

Cooke, G. (1892) White's Thrush near Shrewsbury. *Field*, 133.

Cormack, P. & Rotherham, I.D. (2014) *A review of the PACEC reports (2006 & 2014) estimating net economic benefits from shooting sports in the UK.* Commissioned by the League Against Cruel Sports.

Cotton, P.A. (2003) Avian migration phenology and global climate change. *Proceedings of the National Academy of Sciences of the United States of America*, 100, 12219–12222.

Cowley, E. (1979) Sand Martin population trends in Britain, 1965–1978. *Bird Study*, 26, 113–116.

Cowley, E. & Siriwardena, G.M. (2005) Long-term variation in the survival rates of Sand Martins *Riparia riparia*: dependence on breeding and wintering ground weather, age and sex, and their population consequences. *Bird Study*, 52, 237–251.

Craik, C. (1997) Long-term effects of North American Mink *Mustela vison* on seabirds in western Scotland. *Bird Study*, 44, 303–309.

Cramp, S. *et al* (eds) (1977–1994) *Handbook of Birds of Europe, the Middle East and North Africa: The Birds of the Western Palearctic.* Vols I–IX. Oxford University Press, Oxford.

Cross, A. & Davis, P. (2005) *The Red Kites of Wales.* Subbuteo Books, Shrewsbury.

Cross, A.V. and Moscrop, C. (1996) *The Stiperstones National Nature Reserve: Bird Survey 1995–96.* Report submitted to English Nature.

Curator, The (1938) Waterfowl at Walcot (Breeding Season, 1937) *Avicultural Magazine*, 104–106.

Dargie, T.C. (1993) *The Distribution of Lowland Wet Grassland.* English Nature Research Report No. 49. English Nature, Peterborough.

Davies, A.K. (1988) The distribution and status of the Mandarin Duck *Aix galericulata* in Britain. *Bird Study* 35: 203–208.

Davies, N. (2015) *Cuckoo: Cheating by Nature.* Bloomsbury Publishing, London.

Davies, N.B. (1992) *Dunnock behaviour and Social Evolution.* Oxford University Press, Oxford.

Davies, N.B & Lundberg, A. (1984) Food distribution and a variable mating system in the Dunnock *Prunella modularis. Journal of Animal Ecology*, 53, 895–912.

Davies, N.B. & Brooke, M. de L. (1989) An experimental study of co-evolution between the Cuckoo *Cuculus canorus*, and its hosts. I. Host egg discrimination. *Journal of Animal Ecology* 58: 207–224.

Dawes, A. (2004) Shropshire Corn Bunting Survey 2003–2004. *The Shropshire Bird Report 2004*, 18–20.

Dawes, A. (2003) BTO Heronries Survey. *The Shropshire Bird Report 2003*, 44–45.

Dawes, A. (2004) *Shropshire Yellow Wagtail Survey 2003–2004* 15.

Dawes, A. (2005) BTO Tawny Owl Survey 2005 *The Shropshire Bird Report 2005*, 12–13.

Deans, P., Sankey, J., Smith, L., Tucker, J., Whittles, C. & Wright, C. (eds) (1992) *An Atlas of the Breeding Birds of Shropshire.* Shropshire Ornithological Society.

Defra (2015) http://jncc.defra.gov.uk/page-2882.

Donald, P. (2004) *The Skylark.* T. & A.D. Poyser, London.

Dovaston, J.F.M. (1839) *Mr Dovaston's Lectures on Natural History and National Melody.* John Davies, Shrewsbury.

Dymond, J.N. and the Rarities Committee (1976) Report on rare birds in Great Britain in 1975 (with additions for nine previous years) *British Birds*, 69, 321–368.

Eades, R.A. (1974) Monthly variation in foreign-ringed Dunlins on the Dee Estuary. *Bird Study*, 21:155–157.

Eaton, M.A., Aebischer, N.J., Brown, A.F., Hearn, R.D., Lock, L., Musgrove, A.J., Noble, D.G., Stroud, D.A. & Gregory, R.D. (2015) Birds of Conservation Concern 4: the population status of birds in the United Kingdom, Channel Islands and Isle of Man. *British Birds*, 108, 708–746.

Edwards, W. (1971) The Heronries at Ellesmere and Halston. *The Shropshire Bird Report 1971*, 9–10.

Eglington, S.M., Gill, J.A., Bolton, M., Smart, M.A., Sutherland, W.J. and Watkinson, A.R. (2008) Restoration of wet features for breeding waders on lowland grassland. *Journal of Applied Ecology*, 45, 305–314.

Elliott, J.S. (1911) Former occurrence of Black Grouse in Wyre Forest. *Zoologist*, 4:15, 387–88.

Elliott, J.S. (1914) Nesting of the Golden-crested Wren (Regulus cristatus) *The Zoologist (Series 4)*, 18, 273.

Enoksson, B. (1990) Autumn territories and population regulation in the Nuthatch Sitta europaea: an experimental study. *Journal of Animal Ecology*, 59:3, 1047–1062.

Ens, B.J., Briggs, K.B., Safriel, U.N. & Smit, C.J. (1996) Life history decisions during the breeding season. In: *The Oystercatcher: from individuals to population* (ed. Goss-Custard, J.D.) Oxford University Press, Oxford.

Environment Agency (nd) Data on WFD. http://environment.data.gov.uk/catchment-planning/.

Environment Agency & Natural Resources Wales (2009) *River Basin Management Plan: Severn River Basin District.* Environment Agency, Bristol.

Ewing, S.R., Rebecca, G.W., Heavisides, A., Court, I.R., Lindley, P., Ruddock, M., Cohen, S. & Eaton, M.A. (2011) Breeding status of Merlins *Falco columbarius* in the UK in 2008. *Bird Study,* 58:4, 379–389.

Eyton, T.C. (1838) An attempt to ascertain the Fauna of Shropshire and North Wales. *Annals and Magazine of Natural History,* 2:2, 285–293.

Eyton, T.C. (1839) An attempt to ascertain the Fauna of Shropshire and North Wales. *Annals and Magazine of Natural History,* 3:2, 52–56.

Ferguson-Lees, I.J. (1971) Studies of less familiar birds 164, Wood Sandpiper. *British Birds,* 64, 114–117.

Ferguson-Lees *et al.* (2011) *A Field Guide to Monitoring Nests,* BTO, Thetford.

Ferns, P. (1980) The spring migration of Sanderlings Calidris alba through Britain in (1979) *Wader Study Group Bulletin,* 30, 22–25.

Field, R.H. & Anderson, G.Q.A. (2004) Habitat use by breeding Tree Sparrows *Passer montanus. Ibis,* 146, 60–68.

Fielding, A., Haworth, P., Whitfield, P., McLeod, D. & Riley, H. (2011) *A Conservation Framework for Hen Harriers in the United Kingdom.* JNCC Report 441, Joint Nature Conservation Committee, Peterborough.

Fitter, R.S.R. (1959) *The ark in our midst.* Collins, London.

Ford, A.A. & Elphick, D. (1993) Barn Swallows roosting in maize. *British Birds,* 86, 95–96.

Forestry Commission (2002) *National Inventory of Woodland and Trees.*

Forrest, H.E. (1899) *The Fauna of Shropshire.* Wilding, Shrewsbury.

Forrest, H.E. (1900) Little Crake in Shropshire. *Zoologist,* 280.

Forrest, H.E. (1903) Dartford Warbler in Shropshire. *Zoologist.*

Forrest, H.E. (1905) Museum lecture. *Transactions of Caradoc and Severn Valley Field Club,* 43–47.

Forrest, H.E. (1907a) *Peplow Hall Museum Catalogue.* Privately published. URL http://www.pgt7.uk/sos/general/Resources/1907%20Peplow%20Collection%20Forrest.pdf.

Forrest, H.E. (1907b) *The Vertebrate Fauna of North Wales.* Witherby, London.

Forrest, H.E. (1908) Birds. In: *The Victoria County History of Shropshire (Vol. 1)* (ed. Page, W.) Archibald Constable, London.

Forrest, H.E. (1909a) Additions to the Shropshire avifauna. *British Birds,* 3, 165.

Forrest, H.E. (1909b) Osprey in Shropshire. *British Birds,* 3, 165.

Forrest, H.E. (1909c) Shropshire names of bird and beast. *Transactions of the Caradoc and Severn Valley Field Club 1909,* 75–82.

Fox, A.D. (2009) What makes a good alien? Dealing with the problems of non-native wildfowl. *British Birds* 102: 660–679.

Foxall, H.D.G. (1980) *Shropshire Field-Names.* Shropshire Archaeological Society, Shrewsbury.

Fraser, P.A. (2013a) Report on scarce migrant birds in Britain in 2004–2007, Part 1: Non-passerines. *British Birds,* 106, 368–404.

Fraser, P.A. (2013b) Report on scarce migrant birds in Britain in 2004–2007, Part 2: Passerines. *British Birds,* 106, 448–476.

Fraser, P.A., Rogers, M.J. and the Rarities Committee (2007a) Report on rare birds in Great Britain in 2005, Part 1: Non-passerines. *British Birds,* 100, 16–61.

Fraser, P.A. and the Rarities Committee (2007b) Report on rare birds in Great Britain in 2006. *British Birds,* 100, 694–754.

Fuller, R.J., Noble, D.G., Smith, K.W. & Vanhinsbergh, D. (2005) Recent declines in populations of woodland birds in Britain: a review of possible causes. *British Birds,* 98, 116–43.

Fulton, D. (2010) The breeding population of Northern Wheatears at Clee Hill, Shropshire, 1998–2009. *British Birds,* 103, 223–228.

G.V.H. (1863) Partridges. *Field.*

Gandy, I. (1970) *An idler on the Shropshire Borders.* Wilding and Son, Pentre, Shropshire.

Gaydon, A.T. & Price, D.T.W. (1973) Shooting. In: *The Victoria County History of Shropshire – Volume 2* (ed Pugh, R.B.) Oxford University Press, Oxford.

Gelling, M. & Foxall, H.D.G. (1990) *The Place-names of Shropshire: Part One – The Major Names of Shropshire.* English Place-Name Society.

Gibbons, D.W., Reid, J.B. & Chapman, R.A. (1993) *The New Atlas of Breeding Birds in Britain and Ireland 1988–1991.* T. & A.D. Poyser, London.

Gill, J.A. plus 23 other named individuals (2007) Contrasting trends in two Black-tailed Godwit populations: a review of causes and recommendations. *Wader Study Group Bulletin,* 114, 43–50.

Gillings, S., Balmer, D.E. & Fuller, R.J. (2015) Directionality of recent bird distribution shifts and climate change in Great Britain. Global Change Biology, doi: 10.1111/gcb.12823.

Gilroy, J.J., Anderson, G.Q.A., Grice, P.V., Vickery, J.A., Sutherland, W.J. (2010) Mid-season shifts in the habitat associations of Yellow Wagtail *Motacilla flava* breeding in arable farmland. *Ibis,* 152, 90–104.

Green, R. (1983) Spring dispersal and agonistic behaviour of the Red-legged Partridge (*Alectoris rufa*) *Journal of Zoology,* 201, 541–555.

Gribble, F.C. (1962) Census of Black-headed Gull colonies in England and Wales, 1958. *Bird Study,* 9, 56–71.

Gribble, F.C. (1976) A Census of Black-headed Gull colonies in England and Wales in 1973. *Bird Study,* 23, 135–145.

Gribble, F.C., Harrison, G., Griffiths, H.J., Winsper, J. & Coney, S. (2007) *Where to Watch Birds in the West Midlands: Herefordshire, Shropshire, Staffordshire, Warwickshire, Worcestershire and the Former West Midlands County* Christopher Helm.

Gudmundsson, G.A. & Lindstrom, C. (1992) Spring migration of Sanderlings through SW Iceland: where from and where-to? *Ardea,* 80, 315–325.

Guyomarc'h, J.C., Combreau, O., Puigcerver, M., Fontoura, P., Aebischer, N. & Wallace, D.I.M. (1998) *Coturnix coturnix.* Quail. BWP Update *The Journal of Birds of the Western Palearctic* 2: 27–46.

Hammon, A. (2011) Understanding the Romano-British–Early Medieval Transition: A Zooarchaeological Perspective from Wroxeter (Viroconium Cornoviorum) *Britannia,* 42, 275–305, esp. 295–96.

Harber, D.D. and the Rarities Committee (1966) Report on rare birds in Great Britain in 1965 (with 1958, 1959, 1961, 1962, 1963 and 1964 additions) *British Birds,* 59, 280–305.

Hardey, J., Crick, H., Wernham, C., Riley, H., Etheridge, B. & Thompson, D. (2009) *Raptors – A Field Guide for Surveys and Monitoring (Second Edition)* Scottish Natural Heritage, The Stationery Office, Edinburgh.

Hardy, A.R., Minton C.D.T. (1980) Dunlin migration in Britain and Ireland, *Bird Study* 27:81–92.

Hardy, E. (1970) The Return of the Peregrine. *Shropshire Magazine.*

Harris, S.J., Massimino, D., Newson, S.E., Eaton, M.A., Marchant, J.H., Balmer, D.E., Noble, D.G., Gillings, S., Procter, D. & Pearce-Higgins, J.W. (2016) *The Breeding Bird Survey 2015.* BTO Research Report 687. British Trust for Ornithology, Thetford.

Harrison, G. & Harrison, J. (2005) *The New Birds of the West Midlands.* West Midland Bird Club, Studley, Warwickshire.

Harrop, A.H.J., Collinson, J.M. and Melling, T. (2012) What the eye doesn't see: the prevalence of fraud in ornithology. *British Birds,* 105, 236–257.

Hayhow, D.B., Bond, A.L., Eaton, M.A., Grice, P.V., Hall, C., Hall, J., Harris, S.J., Hearn, R.D., Holt, C.A., Noble, D.G., Stroud, D.A. & Wotton, S. (2015) *The state of the UK's birds 2015*. RSPB, BTO, WWT, JNCC, NE, NIEA, NRW and SNH, Sandy, Bedfordshire.

Heaton, A. (2001) *Duck Decoys*. Shire Publications.

Hedenstrom, A. (2004) Migration and morphometrics of Temminck's Stint *Calidris temminckii* at Ottenby, south Sweden. *Ringing and Migration*, 22, 51–58.

Henderson, I., Holt, C. & Vickery, J. (2007) National and regional patterns of habitat association with foraging Barn Swallows *Hirundo rustica* in the UK. *Bird Study*, 54, 371–377.

Heppleston, P.B. (1972) The comparative breeding ecology of Oystercatchers (*Haematopus ostralegus*) in inland and coastal habitats. *Journal of Animal Ecology*, 41, 23–51.

Heward, C.J., Hoodless, A.N., Conway, G.J., Aebischer, N.J., Gillings, S. & Fuller, R.J. (2015) Current status and recent trend of the Eurasian Woodcock *Scolopax rusticola* as a breeding bird in Britain. *Bird Study*, 62, 535–551.

Hill, L.A. (2002) House Martin. In: *The Migration Atlas: movements of the birds of Britain and Ireland* (eds Wernham, C., Toms, M., Marchant, J., Clark, J., Siriwardena, G. & Baillie, S.) 465–467. T. & A.D. Poyser, London.

Hinsley, S.A., Bellamy, P.E., Newton, I. & Sparks, T.H. (1995) Habitat and landscape factors influencing the presence of individual breeding bird species in woodland fragments. *Journal of Avian Biology*, 26:2, 94–104.

Hirons, G. (1980) The significance of roding by Woodcock *Scolopax rusticola*: An alternative explanation based on observations of marked birds. *Ibis*, 122, 350–354.

Holling, M. & the Rare Breeding Birds Panel (2011) Non-native breeding birds in the United Kingdom in 2006, 2007 and 2008. *British Birds*, 104, 114–138.

Holling, M. & the Rare Breeding Birds Panel (2012) Rare breeding birds in the United Kingdom in 2010. *British Birds*, 105, 352–416.

Holling, M. & the Rare Breeding Birds Panel (2013) Rare breeding birds in the United Kingdom in 2012. *British Birds*, 106, 496–554.

Holling, M. & the Rare Breeding Birds Panel (2014a) Rare breeding birds in the United Kingdom in 2012. *British Birds*, 107, 504–560.

Holling, M. & the Rare Breeding Birds Panel (2014b) Non-native breeding birds in the UK, 2009–11. *British Birds*, 107, 122–141.

Holling, M. & the Rare Breeding Birds Panel (2016) Rare breeding birds in the United Kingdom in 2014. *British Birds*, 109, 491–545.

Hollom, P.A.D. (1940) Report on the 1938 survey of Black-headed Gull colonies. *British Birds*, 33, 202–221.

Holloway, S. (1996) *The Historical Atlas of Breeding Birds in Britain and Ireland 1875–1900*. T. & A.D. Poyser, London.

Hollyer, J.N. (1970) The invasion of Nutcrackers in autumn 1968. *British Birds*, 63, 353–379.

Holmes, G.E. (2003) Purple Heron at Venus Pool 24 September 2003. *The Shropshire Bird Report 2003*, 178–179.

Holmes, G.E. (2005a) Lesser Scaup at Monkmoor Pool from 6–26 June 2005. *The Shropshire Bird Report 2005*, 8–9.

Holmes, G.E. (2005b) Magnificent Frigatebird at Whitchurch. *The Shropshire Bird Report 2005*, 14–15.

Holmes, G. & Walker, G. A Review of the Shropshire Record between 1950 and 2010. *The Shropshire Bird Report 2011*, 19–30.

Holt, B. & Williams, G. (2008) *The Birds of Montgomeryshire*. Published by the authors, Montgomeryshire.

Holt, C. & the Rarities Committee (2017) Report on Rare birds in Great Britain in 2016. *British Birds*, 110, 562–631.

Holt, C., Austin, G., Calbrade, N., Mellan, H., Hearn, R., Stroud, D., Wotton, S. & Musgrove, A. (2012) *Waterbirds in the UK 2010/11: The Wetland Bird Survey*. BTO, RSPB and JNCC, in association with WWT. British Trust for Ornithology, Thetford.

Holt, C.A., Austin, G.E., Calbrade, N.A., Mellan, H.J., Hearn, R.D., Stroud, D.A., Wotton, S.R. & Musgrove, A.J. (2012) *Waterbirds in the UK 2010/11: The Wetland Bird Survey*. BTO/RSPB/JNCC, Thetford.

Hoodless, A.N. (2002) Eurasian Woodcock. In: *The Migration Atlas: movements of the birds of Britain and Ireland* (eds Wernham, C., Toms, M., Marchant, J., Clark, J., Siriwardena, G. & Baillie, S.) T. & A.D. Poyser, London.

Hoodless, A.N. (2013) Unmasking migrations: tracking woodcock. *Game and Wildlife Conservation Trust Review of 2012*, 24–25.

Hoodless, A.N., Lang, D., Aebischer, N.J., Fuller, R.J. & Ewald, J.A. (2009) Densities and population estimates of breeding Eurasian Woodcock *Scolopax rusticola* in Britain in 2003. *Bird Study*, 56, 15–25.

Hoodless, A.N. & Coulson, J.C. (1994) Survival rates and movements of British and continental woodcock *Scolopax rusticola* in the British Isles. *Bird Study*, 41, 48–60.

Hoodless, A.N. & Coulson, J.C. (1998) Breeding biology of the woodcock *Scolopax rusticola* in Britain. *Bird Study*, 45, 195–204.

Hoodless, A.N. & Hirons, G.J.M. (2007) Habitat selection and foraging behaviour of breeding Eurasian Woodcock *Scolopax rusticola*: a comparison between contrasting landscapes. *Ibis*, 149 (Supplement 2), 234–249.

Hoodless, A.N., Inglis, J.G., Doucet, J-P & Aebischer, N.J. (2008) Vocal individuality in the roding calls of Woodcock *Scolopax rusticola* and their use to validate a survey method. *Ibis*, 150, 80–89.

Horton, A.R. (1902) Birds and beasts in 1901, At Marrington Hall. *Transactions of the Caradoc and Severn Valley Field Club*, 37–46.

Horton, J.R. (1888) Pallas's Sandgrouse. *Field*, 854.

Hudson, N. & the Rarities Committee (2008) Report on rare birds in Great Britain in 2007. *British Birds*, 101, 516–577.

Hudson, N. & the Rarities Committee (2009) Report on rare birds in Great Britain in 2008. *British Birds*, 102, 528–601.

Hudson, N. & the Rarities Committee (2011) Report on rare birds in Great Britain in 2010 *British Birds*, 104, 557–629.

Hudson, N. & the Rarities Committee (2012) Report on rare birds in Great Britain in 2011, *British Birds*, 105, 556–625.

Hudson, N. & the Rarities Committee (2014) Report on rare birds in Great Britain in 2013, *British Birds*, 107, 579–653.

Hudson, R. (1973) *Early and Late Dates for Summer Migrants*. BTO Guide 15, British Trust for Ornithology, Thetford.

Hughes, S.W.M., Bacon, P.J. & Flegg, J.J.M. (1979) The 1975 census of the Great Crested Grebe in Britain. *Bird Study*, 26, 201–226.

Hughes, T. (1985) *Season Songs*. Faber & Faber, London.

Hulbert, C. (1837) *The History and Description of the County of Shropshire*, 2 Vols, Charles Hulbert, Hadnall.

Ireland, W.F. (1943) Black-headed Gulls nesting in Shropshire. *British Birds (Letters)*, 37, 139.

Ireson, G.M. (1967) Breeding Warblers in Shropshire over 20 years. *The Shropshire Bird Report 1967*, 6–10.

Jenks, P., Green, M. & Cross, A.V. (2014) Foraging activity and habitat use by European Nightjars in South Wales. *British Birds*, 107, 413–419.

Jepson, P. (1991) *Shrewsbury Countryside Strategy*, Shrewsbury and Atcham Borough Council.

JNCC (2015) Seabirds and Seaduck, latest population trends: Black-headed Gull *Chroicocephalus ridibundus*. URL: http://jncc.defra.gov.uk/page-2882.

Johnson, G. (1910) Birds and Flowers of Shelton Rough. *Transactions of the Caradoc and Severn Valley Field Club*, 165–168.

Jordan, J.D., Smith, M.S & Webb, M.J. (1995) Breeding birds of Plymouth estate gravel pits, Bromfield, 1993. *Bulletin of Shropshire Naturalists' Union* (its single volume), 1–6.

Jordan, J.G., Smith, M.S. & Webb, M.J. (1995) A survey of the breeding birds of Brown Clee, 1994. *Bulletin of Shropshire Naturalists' Union*, 15–21.

Kernohan, J. (2011) Finders Accounts: Desert Wheatear. *The Shropshire Bird Report 2011*, 17–18.

Kington, J. (ed.) (1988) *The Weather Journals of a Rutland Squire. Thomas Barker of Lyndon Hall.* Rutland County Museum, Oakham. .

Kirby, J.S., Waters, R. & Prys-Jones R.P. (1990) *Wildfowl and Wader Counts 1989–90.* British Trust for Ornithology, Thetford.

Lack, P. (ed.) (1986) *An Atlas of Wintering Birds in Britain and Ireland.* T. & A.D. Poyser, London.

Latham, A. (2006) Montagu's Harrier at Whixall and Fenn's Moss. *The Shropshire Bird Report 2006*, 8–9.

Latham, A. (2007) Finder's Account: Black-throated Thrush *Turdus atrogularis*, Walcot, 8 Apr 2007. *The Shropshire Bird Report 2007*, 17–18.

Latham, A. (2008) Finders account: Yellow-browed Warbler. *The Shropshire Bird Report 2008*, 10.

Latham, A. (2011) Finders accounts: Steppe Grey Shrike. *The Shropshire Bird Report 2011*, 16–17.

Lawton, J.H., Brotherton, P.N.M., Brown, V.K., Elphick, C., Fitter, A.H., Forshaw, J., Haddow, R.W., Hilborne, S., Leafe, R.N., Mace, G.M., Southgate, M.P., Sutherland, W.J., Tew, T.E., Varley, J., & Wynne, G.R. (2010) *Making space for nature: a review of England's wildlife sites and ecological network.* Report. Submitted to the Secretary of State, Defra on 16 September 2010.

Leighton, W.A. (1836) *A Guide through the Town of Shrewsbury etc.* John Davies, Shrewsbury.

Lennon, R.J., Dunn, J.C., Stockdale, J.E., Goodman, S.J., Morris, A. & Hamer, K.C. (2013) Trichomonad parasite infection in four species of Columbidae in the UK. *Parasitology*, 140:11, 1368–1376.

Leslie, A.S. (ed.) (1911) *The Grouse in Health and Disease.* Smith, Elder & Co., London.

Lever, C. (2009) *The Naturalized Animals of Britain and Ireland.* New Holland, London.

Lever, C. (2013) *The Mandarin Duck.* Poyser, London.

Lindstrom, A. & Agrell, L. (1999) Global change and possible effects on the migration and reproduction of Arctic-breeding waders. *Ecological Bulletins*, 47, 145–159.

Lloyd, L.C. (1938) Migratory Birds in Shropshire. *Transactions of the Caradoc and Severn Valley Field Club*, 213–236.

Lloyd, L.C. (1939a) A Survey of Rookeries in the Shrewsbury District. *Transactions of the Caradoc and Severn Valley Field Club*, 76–93.

Lloyd, L.C. (1939b) Effects on Bird-life of the Severe Winter of 1939–40. *Transactions of the Caradoc and Severn Valley Field Club*, 207–215.

Lloyd, L.C. (1939c) Nesting association of Swallow and House Martin. *British Birds*, 33, 109.

Lloyd, L.C. (1940) The Quail in Shropshire. *Transactions of the Caradoc and Severn Valley Field Club*, 11:2, 148–152.

Lloyd, L.C. (1943) Changes in the bird population of Shropshire in the present century. *Transactions of the Caradoc and Severn Valley Field Club*, 58.

Lloyd, L.C. (1944) Winter Blackcaps in Shropshire. *The North-western Naturalist*, 19, 58.

Lloyd, L.C. (1951–56) CSVFC *Recorders Report.*

Lockton, A. and Whild, S. (eds.) (2015) *The Flora and Vegetation of Shropshire* Shropshire Botanical Society, Shrewsbury.

Lovegrove, R. (2007) *Silent Fields, the long decline of a nation's wildlife.* Oxford University Press, Oxford.

Lovegrove, R., Williams, G. & Williams, I. (1994) Birds in Wales. T. & A.D. Poyser, London.

Ludlow Swift Group (2014) *October Newsletter.*

Lythgoe, C. (2001) 2001 SECOS Swift Survey. *South East Cheshire Ornithological Society Bird Report 2001*, 30–31.

MacLean, I., Austin, G., Rehfisch, M.M., Blew, J., Crowe, O., Delany, S., Devos, K., Deceuninck, B., Günther, K., Laursen, K., van Roomen, M. & Wahl, J. 2008. Climate change causes rapid changes in the distribution and site abundance of birds in winter. *Global Change Biology*, 14, 2489–2500.

MAFF (1988) *MAFF in the Midlands & Western Region*, MAFF.

Mainwearing, R.K. (1888) Pallas's Sandgrouse. *Field*, 935.

Marchant, J. (1997) Firecrest. In: *The EBCC Atlas of European Breeding Birds* (eds Hagemeijer W.J.M. & Blair, M.J.), T. & A.D. Poyser, London.

Marchant, J. (2005) Charting the Success of UK Herons. *BTO News*, 257, 12–13.

Marchant, J.H. & Gregory, R.D. (1999) Numbers of nesting rooks *Corvus frugilegus* in the United Kingdom in 1996. *Bird Study*, 46:3, 258–273.

Marchant, J.H. (2002) Common Quail. In: *The Migration Atlas: movements of the birds of Britain and Ireland* (eds Wernham, C., Toms, M., Marchant, J., Clark, J., Siriwardena, G. & Baillie, S.) T. & A.D. Poyser, London.

Marchant, J.H., Clark, J.A., Siriwardena, G.M. & Baillie, S.R. (eds) (2002) *The Migration Atlas: movements of the birds of Britain and Ireland.* T. & A.D. Poyser, London.

Marchant, J.H., Hudson, R., Carter, S.P. & Whittington, P.A. (1990) *Population trends in British breeding birds.* British Trust for Ornithology, Tring.

MarLIN. The Marine Life Information Network, The Marine Biological Association of the UK, www.marlin.ac.uk.

Marthinsen, G., Wennerberg, I. & Lifjeld, J.T. 2007. Phylogeography and subspecies taxonomy of Dunlins (*Calidris alpina*) in western Palearctic analysed by DNA microsatellites and amplified fragment length polymorphism markers. *Biol. Journal Linnaean Society* 92: 713–726.

Massimino, D., Johnston, A. & Pearce-Higgins, J.W. (2015) The geographical range of British birds expands during 15 years of warming, *Bird Study* 62, 523–534.

Mathysen, E. & Schmidt, K-H. (1987) Natal dispersal in the Nuthatch. *Ornis Scandinavia*, 18, 313–316.

Mead, C.J. (1979) Colony fidelity and interchange in the Sand Martin. *Bird Study*, 26, 99–106.

Mead, C.J. (1984) Sand Martins Slump. *BTO News*, 133, 1.

Mead, C.J. & Pepler, G.R.M. (1975) Birds and other animals at Sand Martin colonies. *British Birds*, 68, 89–99.

Meares, C.S. & Meares, D.H. (2018) *Bird-nesting and other notes. Vol. 1 & 2. 1892 to 1949* (ed. T.A. Waddell) Privately Printed by Orphans Press at Leominster; 50 copies. Full details on *Histo* website.

Meddens, B. (1987) *Assessment of the Animal Bone Work from Wroxeter Roman City, Shropshire.* English Heritage Ancient Monuments Laboratory Report 171.

Meissner, W. (2006) Timing and phenology of Curlew Sandpiper on southward migration through Puck Bay, Poland. *International Wader Studies*, 19, 121–124.

Melling, T., Dudley, S. and Doherty, P. (2008) The Eagle Owl in Britain. *British Birds*, 101, 478–490.

Met Office (Various dates) *Met Office Monthly Weather Reports*: © Crown Copyright 1884–2017. www.metoffice.gov.uk.

Met Office (2015a) http://www.metoffice.gov.uk/public/weather/climate/gcqh76ug7#?region=midlands.

Met Office (2015b) http://www.metoffice.gov.uk/public/weather/climate-historic/#?tab=climateHistoric.

Met Office (2015c) http://www.metoffice.gov.uk/hadobs/hadcet/.

Met Office. (2017) http://www.metoffice.gov.uk/research/monitoring/climate/surface-temperature.

Montagu, G. & Rennie, J. (1832) *Ornithological dictionary of British Birds*. Hurst, Chance & Co., London.

Moorcroft, D., Wilson, J.D. & Bradbury, R.B. (2006) Diet of nestling Linnets *Carduelis cannabina* on lowland farmland before and after agricultural intensification. *Bird Study*, 53:2, 156–162.

Moreau, R.E. (1951) The British status of the Quail and some problems of its biology. *British Birds*, 44, 257–276.

Morris, A., Burges, D. & Fuller, R.J. (1994) The status and distribution of Nightjars *Caprimulgus europaeus* in Britain in 1992. *Bird Study*, 41:3, 181–191.

Morris, A.J., Holland, J.M., Smith, B. & Jones N.E. (2004) Sustainable Arable farming for Improved Environment (SAFFIE): managing winter wheat sward structure for Skylarks *Alauda arvensis*. *Ibis*, 146:2, 155–162.

Musgrove, A.J., Aebischer, N.J., Eaton, M.A., Hearn, R.D., Newson, S.E., Noble, D.G., Parsons, M., Risely, K. & Stroud, D. (2013) 'Population estimates of birds in Great Britain and the United Kingdom'. *British Birds*, 106, 64–100.

Musgrove, A.J., Austin, G.E., Hearn, R.D., Holt, C.A., Stroud, D.A. & Wotton, S.R. (2011) Overwinter population estimates of British waterbirds. *British Birds*, 104, 364–397.

National Library of Wales, Brogyntyn Estate and Family Records, Game Books, EAC7/1–5.

NE (Natural England) (1993) *Lowland Wet Grassland In England: Distribution of the Resource (Vol 1)* Project Report Contract No F72–08–17.

NE (Natural England) (2008–13) *Birds England Summary for 2008–13* Fenn's, Whixall and Bettisfield Mosses NNR.

NE (Natural England) (2011) (subsequently updated) *Priority Habitat Inventory (England)* Natural England.

NE (Natural England) (2012) *Shropshire: Agri-Environment Schemes: Key information, scheme uptake and expenditure data – November 2012*, Natural England.

Newson, S.E., Moran, N.J., Musgrove, A.J., Pearce-Higgins, J.W., Gillings, S., Atkinson, P.W., Miller, R., Grantham M.J. & Baillie, S.R. (2016) Long-term changes in the migration phenology of UK breeding birds detected by large-scale citizen science recording schemes. *Ibis*, 158, 481–495.

Newton, A. (1861) On the possibility of taking an ornithological census. *Ibis*, 3, 190–196.

Newton, I. (1986) *The Sparrowhawk*. T. & A.D. Poyser, London.

Nickless, P. (2005) White-tailed Eagle at Venus Pool. *The Shropshire Bird Report 2005*, 16–18.

Norman, D. (2008) *Birds in Cheshire and Wirral: A breeding and wintering atlas*. Liverpool University Press, Liverpool.

O'Brien, M. & Smith, K.W. (1992) Changes in the Status of Waders Breeding on Wet Lowland Grassland in England and Wales Between 1982 and 1989. *Bird Study*, 39, 165–176.

O'Sullivan, J.M. (1976) Bearded Tits in Britain and Ireland, 1966–1974. *British Birds*, 69, 473–489.

Ockendon, N., Baker, D.J., Carr, J.A., White, E.C., Almond, R.E.A., Amano, T., Bertram, E., Bradbury, R.B., Bradley, C., Butchart, S.H.M., Doswald, N., Foden, W., Gill, D.J.C., Green, R.E., Sutherland, W.J., Tanner, E.V.J. & Pearce-Higgins, J.W. 2014. Mechanisms underpinning climatic impacts on natural populations: altered species interactions are more important than direct effects. *Global Change Biology*, 20, 1365–2486.

Ogier Ward, T. (1841) The Medical Topography of Shrewsbury and its Neighbourhood. *The Transactions of the Provincial Medical and Surgical Association*, Appendix No.2.

Ogilvie, M. (1996) Rare Breeding Birds in the United Kingdom in 1996. *British Birds*, 92, 120–154.

Ogilvie, M. (1999) Rare Breeding Birds in the United Kingdom in 1996. *British Birds* 92: 120–154.

Ogilvie, M.A. (1981) The Mute Swan in Britain, 1978. *Bird Study*, 28, 87–106.

Oliver, T.H., Gillings, S., Pearce-Higgins, J.W., Brereton, T., Crick, H.Q.P., Duffield, S.J., Morecroft, M.D. & Roy, D.B. (2017) Large extents of intensive land use limit community reorganization during climate warming. *Global Change Biology*.

ONS (2018) Agriculture in the United Kingdom 2017 Office for National Statistics Crown Copyright 2018.

PACEC (2006) *The Economic and Environmental Impact of Sporting Shooting*. Public and Corporate Economic Consultants, Cambridge.

Paddock, G.H. (1890) *Notes on Some of the Birds Found in the Neighbourhood of Newport, Salop*. Hobson & Co., Wellington.

Paddock, G.H. (1897) *Catalogue of Shropshire Birds*. H.R. Lunn, Newport.

Paddock, G.H. (1904) A Catalogue of the Birds met within the District of Church Stretton. In: *Church Stretton: some results of local scientific research* (eds Campbell-Hyslop, C.W. & Cobbold, E.S.), 2, 1–50.

Page, W. (ed) (1908) *The Victoria History of the County of Hereford*. Archibald Constable & Co. Ltd., London.

Parkin, D.T., Collinson, M., Helbig, A.J., Knox, A.G. and Sangster, G. (2003) The taxonomic status of Carrion and Hooded Crows. *British Birds*, 96, 274–290.

Parslow, J.L.F. (1967) Changes in status among breeding birds in Britain and Ireland. *British Birds*, 60, 2–47, 97–123, 177–202, 261–285, 396–404 & 493–508; 61, 49–64 & 241–255.

Payne-Gallwey, R. (1886) *The book of Duck Decoys. Their construction, management, and history*. Kessinger Legacy Reprints.

Pearce-Higgins, J. (2011a) When spring has sprung. *BTO News*, 294, 12–14.

Pearce-Higgins, J. (2011b) Highs & Lows. *BTO News*, 297, 12–14.

Pearce-Higgins, J. (2011c) How bad can it get? *BTO News*, 297, 12–14.

Pearson, D.J. (1987) The status, migrations and seasonality of the Little Stint in Kenya. *Ringing and Migration*, 8, 91–108.

Pearson, A.G. & Watson, E.V. (1945–46) Birds observed on the Shropshire-Staffordshire Border, 1942–1946. *Transaction of the Caradoc and Severn Valley Field Club*, 131–138.

Pendleton, T. & D. (2013) Bird association with Oil-seed Rape. *Eakring Birds* URL: www.eakringbirds.com.

Pennant, T. (1768) *British Zoology* (Second edition, Volume II).

Pennant, T. (1770) *British Zoology* (Third edition)

Pennant, T. (1776) *British Zoology* (Fourth edition, Volume II) 106. Benjamin White, London.

Perkins, H.R. (1968) The Decline of the Heron in Shropshire. *The Shropshire Bird Report 1968*, 7–8.

Perrins, C. (1971) Age of first breeding and adult survival rates in the Swift. *Bird Study*, 18, 61–70.

Perrins, C. (2002) Swift. In: *The Migration Atlas: movements of the birds of Britain and Ireland* (eds Wernham, C. *et al.*), T. & A.D. Poyser, London.

Pienkowski, M.W. & Dick, W.J.A. (1975) The migration and wintering of Dunlin *Calidris alpina* in north-west Africa. *Ornis Scandinavica* 6:151–167.

Piersma, T. & Davidson, N.C. (1992) The migration of Knots. *Wader Study Group Bulletin*, 64 (suppl.), 209.

Piersma, T. (2016) *Guests of Summer: A House Martin love story*. British Trust for Ornithology, Thetford.

Pitt, F. (1918) Pied Redwing. *Naturalist*, 335.

Plummer, K.E., Siriwardena, G.M., Conway, G.J., Risely, K. & Toms, M.P. (2015) Is supplementary feeding in gardens a driver of evolutionary change in a migratory bird species? *Global Change Biology*, 4353–4363.

Plymley, J.A. (1803) *General Review of the Agriculture of Shropshire*. Shropshire Archives.

Potts, G.R. (1980) Grey Partridge Population. In: Cramp, S. *et al. The Birds of the Western Palearctic*, Vol 2, 487.

Potts, G.R. (2012) *Partridges*. Collins, London.

Powell, T.A. (1947) *Here and there a lusty trout*. Faber & Faber, London.

Prater, A.J. (1981) *Estuary Birds of Britain and Ireland*. T. & A.D. Poyser, Calton.

Prater, A.J., Marchant J.H., Vuorinen J. (1977) *Guide to the identification and ageing of Holarctic Waders*. BTO Guide 17, BTO, Tring.

Prater, A.J. (1972) The Ecology of Morecambe Bay. III. The Food and Feeding Habits of Knot (Calidris canutus L.) in Morecambe Bay. *Journal of Applied Ecology*, 9:1, 179–194.

Prestt, I. (1965) An enquiry into the recent breeding status of some of the smaller birds of prey and crows in Britain. *Bird Study*, 12, 196–221.

Rarities Committee (2018) Report on Rare Birds in Great Britain in 2017. *British Birds in prep.*

Ratcliffe, D. (1984) Tree-nesting by Peregrines in Britain & Ireland. *Bird Study*, 31, 232–233.

Ratcliffe, D. (1993) *The Peregrine Falcon (Second Edition)*. T. & A.D. Poyser, London.

Ratcliffe, D. (1997) *The Raven*. T. & A.D. Poyser London.

Riddiford, N. & Findley, P. (1981) *Seasonal Movements of Summer Migrants*. BTO Guide 18, British Trust for Ornithology, Thetford.

Riddoch, B.J., Greenwood, M.T. & Ward, R.D. (1984) Aspects of the population structure of the sand martin flea *Ceratophyllus styx*, in Britain. *Journal of Natural History*, 18, 475–484.

Ringing Report. Published annually by BTO. Various authors.

Roberts, S.J. & Law, C. (2014) Honey-buzzards in Britain. *British Birds*, 107, 668–691.

Robertson, P. (1997) *A Natural History of the Pheasant*. Swan Hill Press, Shrewsbury.

Robertson, P.A., Mill, A.C., Rushton S.P., McKenzie A.J., Sage R.B. & Aebischer N.J. (2017) Pheasant release in Great Britain: long-term and large-scale changes in the survival of a managed bird. *European Journal of Wildlife Research* 63: 100.

Robinson, R.A. (2017) *BirdFacts: profiles of birds occurring in Britain & Ireland*. BTO Research Report 407, British Trust for Ornithology, Thetford, http://www.bto.org/birdfacts.

Robinson, R.A. (2018) *BirdFacts: profiles of birds occurring in Britain & Ireland* (BTO Research Report 407) BTO, Thetford, http://www.bto.org/birdfacts.

Robinson, R.A., Crick, H.Q.P. & Peach, W.J. (2003) Population trends of Swallows *Hirundo rustica* breeding in Britain. *Bird Study*, 50, 1–7.

Robinson, R.A., Leech, D.I. & Clark, J.A. (2015) *The Online Demography Report: Bird ringing and nest recording in Britain & Ireland in 2014*. British Trust for Ornithology, Thetford, www.bto.org/ringing-report.

Robinson, R.A., Morrison, C.A. & Baillie, S.R. (2014) Integrating demographic data: towards a framework for monitoring wildlife populations at large spatial scales. *Methods in Ecology and Evolution*, 5, 1361–1372.

Rocke, J. (1865) Ornithological Notes from Shropshire. *Zoologist*, 23, 9683–9688 & 9775–9781.

Rocke, J. (1866a) Ornithological notes from Shropshire. *Zoologist*, 24, 76–84 & 161–166.

Rocke, J. (1866b) Catalogue of the bird collection at Clungunford House.

Rocke, J. (1872) *Zoologist*. 2.7.3111.

Rodenhurst (1783) *A description of Hawkstone*. Shropshire Archives.

Rogers, M.J. & the Rarities Committee (1978) Report on rare birds in Great Britain in 1977. *British Birds*, 71, 481–532.

Rogers, M.J. & the Rarities Committee (1979) Report on Rare birds in Great Britain in 1978. *British Birds*, 72, 543–549.

Rogers, M.J. & the Rarities Committee (1983) Report on Rare birds in Great Britain in 1982. *British Birds*, 76, 476–529.

Rogers, M.J. & the Rarities Committee (1984) Report on rare birds in Great Britain in 1983. *British Birds*, 77, 506–562.

Rogers, M.J. & the Rarities Committee (1985) Report on rare birds in Great Britain in 1984. *British Birds*, 78, 529–589.

Rogers, M.J. & the Rarities Committee (1990) Report on Rare birds in Great Britain in 1989, *British Birds*, 83, 439–496.

Rogers, M.J. & the Rarities Committee (1991) Report on Rare birds in Great Britain in 1990. *British Birds*, 84, 449–505.

Rogers, M.J. & the Rarities Committee (1995) Report on Rare birds in Great Britain in 1994. *British Birds*, 88, 493–558.

Rogers, M.J. & the Rarities Committee (1996) Report on rare birds in Great Britain in 1995. *British Birds*, 89, 481–531.

Rogers, M.J. & the Rarities Committee (1997) Report on Rare birds in Great Britain in 1996. *British Birds*, 90, 453–522.

Rogers, M.J. & the Rarities Committee (1998) Report on Rare birds in Great Britain in 1997. *British Birds*, 91, 455–517.

Rogers, M.J. & the Rarities Committee (2004) Report on rare birds in Great Britain in 2003. *British Birds*, 97, 558–625.

Rogers, M.J. & the Rarities Committee (2005) Report on rare birds in Great Britain in 2004. *British Birds*, 98, 628–694.

Rolls, J.C., Rolls, M.J. & Youngman, R.E. (1977) Swallows and Sand Martins roosting in maize. *British Birds*, 70, 393.

Rowley, T. (1972) *The Shropshire Landscape*. Hodder & Stoughton.

Royan, A., Hannah, D.M., Reynolds, S.J., Noble, D.G., Sadler, J.P. (2013) Avian community responses to variability in river hydrology. *PLoS ONE* 8(12): e83221. doi:10.1371/journal.pone.0083221.

RSPB (1983, 1984, 1985 and 1986) *Merlin Report (Long Mynd Shropshire)* RSPB Midland Office.

RSPB (1991) *Death by Design (The persecution of Birds of Prey and Owls in the UK 1979-89)* Royal Society for the Protection of Birds, Sandy.

RSPB (2012) *Hope Farm: Farming for food, profit and wildlife*. Royal Society for the Protection of Birds, Sandy.

RSPB (2015) *Legal Eagle*, 77, 4. Royal Society for the Protection of Birds, Sandy.

RSPB *et al.* (2012) *The State of the UK's Birds 2012*, Royal Society for the Protection of Birds, Sandy.

Rutter, E.M., Gribble, F.C. & Pemberton T.W. (1964) *A Handlist of the Birds of Shropshire*. Shropshire Ornithological Society.

Sage, B.L. & King, B. (1959) The Influx of Phalaropes in Autumn 1957. *British Birds*, 52, 33–42.

Sage, B.L. & Vernon, D.R. (1978) The 1975 National Survey of Rookeries. *Bird Study*, 25(2), 54–86.

Salway, T. (1836) A Notice of the discovery of the skeletons of Swifts and Starlings in the Tower of the Church at Oswestry. *Loudon's Magazine of Natural History*, 9, 350–352.

Sangster, G., Collinson, M., Helbig, A.J., Knox, A.G., Parkin, D.T. & Prater, T. (2001) The taxonomic status of Green-winged Teal *Anas carolinensis*. *British Birds*, 94, 218–226.

Schekkerman, H., Nehls, G., Hotker, H., Tomkovich, P.S., Kania, W., Chylarecki, P., Soloviev, M., & van Roomen, M. (1998) Growth of Little Stint *Calidris minuta* chicks on the Taimyr Peninsula. *Bird Study*, 45, 77–84.

Schekkerman, H., van Roomen, M.W.J. & Underhill, L.G. (1998) Growth, behaviour and weather-related variation in breeding productivity of Curlew Sandpipers. *Ardea*, 86, 153–168.

Scott, R.E. (1968) Rough-legged Buzzards in Britain in the winter of 1966/67. *British Birds*, 61, 449–455.

Scott, R.E. (1978) Rough-legged Buzzards in Britain in 1973–74 and 1974–75. *British Birds*, 71, 325–338.

Sharpe, N. (2009) *Apley Hall: The Golden Years of a Sporting Estate.* Merlin Unwin, Ludlow.

Sharps, E., Smart, J., Skov, M.W., Garbutt, A., Hiddink, J.G. (2015) Light grazing of saltmarshes is a direct and indirect cause of nest failure in Common Redshank *Tringa totanus. Ibis*, 157, 239–249.

Sharrock, J.T.R, Ferguson-Lees, I,J. & the Rare Breeding Birds Panel (1975) Rare breeding birds in the United Kingdom in 1973. *British Birds*, 68, 5–23.

Sharrock, J.T.R. (1969) Scarce Migrants in Britain and Ireland during 1958–67. *British Birds*, 62:5, 169–189.

Sharrock, J.T.R. (1971) Scarce migrants in Britain and Ireland during 1958–67. *British Birds*, 64:3, 93–113.

Sharrock, J.T.R. (1976) *The Atlas of Breeding Birds in Britain and Ireland* (BTO/IWC) T. & A.D. Poyser, London.

Shaw, J. (1861) Rare water birds occurring near Shrewsbury. *Zoologist*, 19, 7388.

Shawyer, C.A. (1987) *The Barn Owl in the British Isles; its Past, Present and Future.* The Hawk and Owl Trust.

Sheldon, R.D. (2002) *The breeding success and chick survival of Lapwing Vanellus vanellus in arable landscapes, with reference to The Arable Stewardship Pilot Scheme.* Unpublished PhD Thesis, Harper Adams University College.

Shrewsbury School Natural History Society and Wildfowl Collection (1964) *4th Annual Report.*

Shropshire and North Wales History Society (1835) General Meeting Notes, *Salopian Journal*, 18 November 1835.

Shropshire Barn Owl Group (2003–2014) *Shropshire Barn Owl Group Annual Report.*

Shropshire Council (2002) *Biodiversity Action Plan.* Shropshire Council website www.shropshire.gov.uk.

Shropshire Ornithological Society (1956–2017) *The Shropshire Bird Reports*

Sim, I. *et al.* (2010) The decline of the Ring Ouzel in Britain. *British Birds*, 103. 229–239.

Sim, I.M.W., Campbell, L., Pain, D.J. & Wilson, J.D. (2000) Correlates of the population increase of Common Buzzards *Buteo buteo* in the West Midlands between 1983 and 1996. *Bird Study*, 47:2, 154–164.

Sim, I.M.W., Cross, A.V., Lamacraft, D.L. & Pain, D.J. (2001) Correlates of Common Buzzard *Buteo buteo* density and breeding success in the West Midlands. *Bird Study*, 48:3, 317–329.

Sinker, C.A., Packham, J.R., Trueman, I.C., Oswald, P.H., Perring, F.H. & Prestwood, W.V. (1985) *The Ecological Flora of the Shropshire Region.* Shropshire Trust for Nature Conservation, Shrewsbury.

Sitters, H.P., Fuller, R.J., Holbyn, R.A., Wright, M.T., Cowie, N. & Bowden, C.G.R. (1996) The Woodlark *Lullula arborea* in Britain: population trends, distribution and habitat occupancy. *Bird Study*, 34, 172–187.

Skujina, I. (2013) *Population genetics of an endangered bird of prey: the Red Kite in Wales.* A thesis submitted in partial fulfilment of the degree of Master of Philosophy, Aberystwyth University.

Slaney, R.A. (1832) *An Outline of the Smaller British Birds, intended for the use of Ladies and Young Persons.* London.

Smith, F.R. & the Rarities Committee (1969) Report on rare birds in Great Britain in 1968 (with 1964 and 1967 additions) *British Birds*, 62, 457–492.

Smith, F.R. & the Rarities Committee (1973) Report on rare birds in Great Britain in 1972. *British Birds*, 66, 331–360.

Smith, F.R. & the Rarities Committee (1974) Report on rare birds in Great Britain in 1973. *British Birds*, 67, 310–348.

Smith, F.R. & the Rarities Committee (1975) Report on rare birds in Great Britain in 1974 (with additions for 1961 and 1968–73) *British Birds*, 68, 306–338.

Smith, L. (2002) *The Raven in Shropshire.* Unpublished report, 2002 (available on the *Histo* website)

Smith, L. (2003) Upland Birds of the Long Mynd. *The Shropshire Bird Report 2003*, 4–15.

Smith, L. (2004) Breeding Snipe in the South-west Shropshire Hills. *The Shropshire Bird Report 2004*, 11–14.

Smith, L. (2007) The Stiperstones Breeding Bird Survey 2004–07 *The Shropshire Bird Report 2007* 9–13.

Smith, L. (2015) *Red Grouse on The Long Mynd. Survey and Population Estimate 2014.* Report submitted to National Trust.

Smith, L., Carty, P. & Uff, C. (2007) *Wild Mynd: Birds and Wildlife of the Long Mynd.* Published for the National Trust, Shropshire Hills by Hobby Publications.

Smith, L.T. (1910) *The itinerary of John Leland in or about the years 1535–1543* by Leland, J. (1535–43) Volume 5. Parts IX, X and XI. Edition translated and edited by Toulmin Smith, L. (1910)

Smout, C. (2001) Goshawk and the Forestry Commission in South Shropshire. *The Shropshire Bird Report 2001*, 3–7.

Snow, B. & D. (1988) *Birds and Berries.* T. & A.D. Poyser, London.

Snow, D.W. & Perrins, C.M. (1998) *The Birds of the Western Palearctic* (Concise ed.) Oxford: Oxford University Press.

Sotherton, N.W., Aebischer, N.J. & Ewald, J.A. (2010) The conservation of the Grey Partridge. In: *Silent Summer: The State of Wildlife in Britain and Ireland* (ed. Maclean, N.) Cambridge University Press, Cambridge.

Sparks, T.H., Huber, K, Bland, R.L., Crick, H.Q.P., Croxton, P.J., Flood, J., Loxton, R.J., Mason, C.F., Newnham, J.A. & Tryjanowski, P. (2007) How consistent are trends in arrival (and departure) dates of migrant birds in the UK? *Journal of Ornithology*, 148, 503–511.

Sparks, T.H., Roberts, D.R. & Crick, H.Q.P. (2001) What is the value of first arrival dates of spring migrants in phenology? *Avian Ecology and Behaviour*, 7, 75–85.

Spencer, R. & the Rare Breeding Birds Panel (1993) Rare breeding birds in the United Kingdom in 1990. *British Birds*, 86, 62–90.

SSOS (Shrewsbury School Ornithological Society) (1937) *Annual Report, September 1936 to July 1937.* Shrewsbury School.

SSOS (Shrewsbury School Ornithological Society) (1939) *Annual Report, September 1937 to July 1939.* Shrewsbury School.

Stevens, R. (1953) *Laggard.* Faber & Faber.

Stockdale, J.E., Dunn, J.C., Goodman, S.J., Morris, A.J., Sheehan, D.K., Grice, P.V. & Hamer, K.C. (2015) The protozoan parasite *Trichomonas gallinae* causes adult and nestling mortality in a declining population of European Turtle Doves *Streptopelia turtur. Parasitology*, 142:3, 490–498.

Stoddard, M.C. & Stevens, M. (1986) Pattern mimicry of host eggs by the common cuckoo, as seen through a bird's eye. *Proceedings of the Royal Society B: Biological Sciences* 277: 1387–1393.

Summers, R.W., Underhill, L.G. & Syroechkovski, E.E. (1998) The breeding productivity of dark-bellied brent geese and curlew sandpipers in relation to changes in the numbers of arctic foxes and lemmings on the Taimyr Peninsula, Siberia. *Ecography*, 21, 573–580.

Sustainability West Midlands. 2003. *The Potential Impacts of Climate Change in the West Midlands.* Technical Report by Entec UK Ltd. (Available online at http://www.ukcip.org.uk/wordpress/wp-content/PDFs/WM_tech.pdf.)

Svensson, L. (2001) The correct name of the Iberian Chiffchaff *Phylloscopus ibericus* Ticehurst 1937, its identification and new evidence of its winter grounds. *Bulletin of the British Ornithologists' Club*, 121, 281–296.

SWT (Shropshire Wildlife Trust) (1989) *Losing Ground in Shropshire*, SWT.

Tapper, S. (1992) *Game Heritage: An ecological review from shooting and gamekeeping records*. The Game Conservancy Ltd, Fordingbridge.

Tapper, S. (ed.) (1999) *A Question of Balance: Game animals and their role in the British countryside*. The Game Conservancy Trust, Fordingbridge.

Taylor, J.A., Henderson, I.G. & Hartley, I.R. (2015) Breeding Whinchats (*Saxicola rubetra*) on Salisbury Plain: Evidence that carrying capacity is not currently limited by habitat or food availability. In: Bastian H-V, Feulner J (Eds.): *Living on the Edge of Extinction in Europe*. Proc. 1st European Whinchat Symposium: 211–218. LBV Hof, Helmbrechts.

Taylor, I. (1994) *Barn Owls: Predator-prey relationships and conservation*. Cambridge University Press, Cambridge.

Telford Development Corporation (1981) *Wildlife in Telford*.

Thomas, G.B. (1986) in Soper, A. (1988) *A Passion for Birds*, David & Charles Publishers, Newton Abbot.

Tjorve, K.M.C., Schekkerman, H., Tulp, I., Underhill, L.G., de Leeuw, J.J. & Visser, J.J.H. (2007) Growth and energetics of a small shorebird species in a cold environment: The Little Stint *Calidris minuta* on the Taimyr Peninsula, Siberia. *Journal of Avian Biology*, 38:5, 552–563.

Toghill, P. (2006) *Geology of Shropshire (Second Edition)*. Crowood Press, Wiltshire.

Tomkovich, P.S., & Soloviev, M.Y. (1996) Distribution, migration and biometrics of Knot and Little Stints breeding on the Taimyr Peninsula. *Ardea*, 84, 85–98.

Toulmin Smith, L. (1910) *The Itinerary of John Leland in or about the years 1535–1543*. Vol. 5. G. Bell and Sons Ltd, London.

Trueman, I.C. (2015) Changes in the vascular plant flora of Shropshire in Lockton, A. and Whild, S. (2015) *The Flora and Vegetation of Shropshire* Shropshire Botanical Society, Shrewsbury.

Tucker, B.W. (1935) The rookeries of Somerset. *Proceedings of the Somerset Archaeological and Natural History Society*, 81, 149–240.

Tucker, J. (1997) The Peregrine Falcon in Shropshire. *The Shropshire Bird Report 1997*, 3–20.

Tucker, J. (1998) The Dotterel in Shropshire. *The Shropshire Bird Report* 27–35.

Tucker, J. (1998) The Peregrine Falcon in Shropshire – whatever next? *British Wildlife*, 9:4, 227–231.

Tucker, J. (1998; 2004) *Breeding Meadow Pipits and Skylarks on eight transects on the Long Mynd, Shropshire*. Report for the National Trust.

Tucker, J. (2011) *Bird Population change of the National Trust estate at Dudmaston, Bridgnorth, Shropshire over the course of a Countryside Stewardship Scheme 2002–2011: Summary report: 2002–2011*. Unpublished report.

Tucker, J. (2016) Shropshire's Migrant Arrivals Database, SMAD. http://www.lanius.org.uk/sos/general/Resources/2016b%20Tucker%26R%20SMADss.pdf.

Tucker, J., Musgrove, A. & Reese, A. (2014) The ability to hear Goldcrest song and the implications for bird surveys. *British Birds*, 107, 232–233.

Tucker, J. & Tucker, P.E. (2012) *A Historical Ornithology of Shropshire*. URL: http://www.pgt7.uk/sos/index.php.

Tucker, J. & Tucker, P.E. (2017) *Beckwith's Nineteenth Century Birds of Shropshire* (Published by authors)

Tucker, J. & Wright, C. (2014) The House Martin colony on Atcham Bridges. *The Shropshire Bird Report 2009*, 15–21.

Tuer, B. (2004) *A Prince among Poachers*. Cheshire Country Publishing, Chester.

Turner, A. (2006) *The Barn Swallow*. T. & A.D. Poyser, London.

Turner, A. (2009) Climate change: a Swallow's eye view. *British Birds*, 102, 3–16.

Vickery, J.A., Tallowin, J.R., Feber, R.E., Asteraki, E.J., Atkinson, P.W., Fuller, R.J. & Brown, V.K. (2001) The management of lowland neutral grasslands in Britain: effects of agricultural practices on birds and their food resources. *Journal of Applied Ecology*, 38, 647–664.

Village, A. (1990) *The Kestrel*. T. & A.D. Poyser, London.

Voisin, C. (1991) *The Herons of Europe*. T. & A.D. Poyser London.

Walbridge, G., Small, B. & McGowan, R.Y. (2003) From the Rarities Committee's files: Ascension Frigatebird on Tiree – new to the Western Palearctic. *British Birds*, 96, 58–73.

Walker, R.H., Robinson, R.A., Leech, D.I., Moss, D., Kew, A.J., Barber, L.J., Barimore, C.J., Blackburn, J.R., De Palacio, D.X., Grantham, M.J., Griffin, B.M., Schäfer, S., Clark, J.A. (2014) Bird ringing and nest recording in Britain and Ireland in 2013. *Ringing and Migration* 29, 90–150.

Wall, T. (1992) The Red Grouse in Shropshire. *Shropshire Naturalist*, 1:2, 18–23.

Wall, T. (2003) Shropshire Ravens On A Roll. *British Wildlife*, 14:3, 160.

Wallace, D.I.M., Bradshaw, C. and Rogers, M.J. (2006) A review of the 1950–57 British rarities. *British Birds*, 99, 460–464.

Webb, M.G. (1917) *Gone to Earth*. Constable, London. Reprinted 1979 by Virago Modern Classics, London.

WeBS (Waterbirds in the UK) Published annually since 1993 by BTO for BTO/WWT/RSPB/JNCC, with forerunners back to 1947. Various authors.

Wernham, C.V., Toms, M.P., Marchant, J.H., Clark, J.A., Siriwardena, G.M. & Baillie, S.R. (eds) (2002) *The Migration Atlas: movements of the birds of Britain and Ireland*. T. & A.D. Poyser, London.

Wesley, N.A. (1988) Red Grouse. In: *Tetrad Atlas of the Breeding Birds of Devon* (ed Sitters, H.P.) Devon Bird Watching & Preservation Society, Yelverton.

White, G. (1789) *The Natural History of Selborne*. London.

White, S. & Kehoe, C. (2014) Report on scarce migrant birds in Britain in 2008–10: non-passerines. *British Birds*, 107.

Wilson, A.M., Vickery, J.A., Brown, A., Langston, R.H.W., Smallshire, D., Wotton, S. & Vanhinsbergh, D. (2005) Changes in the numbers of breeding waders on lowland wet grasslands in England and Wales between 1982 and 2002. *Bird Study*, 51, 55–69.

Wilson, J. (1973) *Wader populations of Morecambe Bay*, Lancashire. Bird Study, 20:9–21.

Witherby, H.F. & Nicholson, E.M. (1937) On the distribution and status of the British Willow Tit. *British Birds*, 30, 358.

Witherby, H.F. & Ticehurst, N.F. (1908) On the more important additions to our knowledge of British birds since 1899. *British Birds*, 2:12, 406–421.

Witherby, H.F., Jourdain, F.C.R., Ticehurst, N.F. & Tucker, B.W. (1938) *The Handbook of British Birds* (Volume 2) Witherby, London.

Wood, A.C. (ed.) (1958) *The Continuation of the History of The Willoughby Family*. The University of Nottingham.

Wright, C.E. (1967) A study of the Allscott Reed and Sedge Warblers – 1967 *The Shropshire Bird Report* 1967 11–12.

Wright, C.E. (1985) The Heron in Shropshire 1970 to 1985. *The Shropshire Bird Report 1985*, 10–12.

Yalden, D.W. (2002) Place-name and archaeological evidence on the recent history of birds in Britain (In: Proceedings of the 4th Meeting of the ICAZ Bird Working Group Kraków, Poland, 11–15 September 2001) *Acta zoologica cracoviensia*, 45 (special issue), 415–429.

Yapp, W.B. (1969) The bird population of an oakwood (Wyre Forest) over eighteen years. *Proceedings of the Birmingham Natural History Society*, 21:3, 199–216.

Yarrell, W. (1843) *A History of British Birds*. John Van Voorst, Paternoster Row, London.

Acknowledgements
Leo Smith

The Avifauna Working Party

The Birds of Shropshire has been co-ordinated and edited by Leo Smith for the Shropshire Ornithological Society, but he has been supported in that role by an Avifauna Working Party of six other members, all of whom have made a major contribution to the finished publication. It could not have been completed without all their efforts.

Leo Smith was born and raised in Manchester, and moved to Shropshire in 1986. He rapidly stepped into a vacancy for fieldwork organiser in the East area for the 1985–90 Breeding Bird Atlas, and was Editor of the final book, published by SOS in 1992. Leo has surveyed the upland birds of the Long Mynd over many years, and was a founder member of the Raven Study Group in 1993, undertaking fieldwork until 2016. He founded the Raptor Study Group in 2010, and has been Convenor since then, and a Trustee of the Welsh Kite Trust since 2011. He has been supporting Lapwing and Curlew surveys by Community Wildlife Groups since 2004, and there are now 11 such groups contributing to the Shropshire Wildlife Trust/SOS Save our Curlew Campaign, which he co-ordinates. A member of the SOS Conservation Sub-committee, he has co-ordinated the production of this Avifauna, commissioned and edited the content, liaised with the other authors, contributed 23 species accounts and finished off several more, and written several of the other chapters.

Jim Almond has lived in Shrewsbury all his life and developed an interest in all aspects of nature and photography during childhood. He has specialised in birds for the past 18 years, travelling widely and gaining a reputation for capturing high quality images. These appear regularly in books and bird magazines for which he occasionally writes articles. He is an experienced and popular speaker, travelling throughout the UK giving talks which cover a wide variety of themes including the birds of Shropshire or Venus Pool. A member of the SOS Management Board since 2007, he currently writes the quarterly 'Bird Notes' for the Society. He has been a member of the Shropshire Peregrine Group since 2005 and is currently the group Co-ordinator. Jim assembled almost 600 photos of 225 species from 21 local photographers for this Avifauna, and worked with the book designer on the selection of images for publication.

John Arnfield is an ex-academic, living in Church Stretton. Having spent 40 years living in North America (until 2005), he can distinguish Louisiana from Northern Waterthrush faster than he can separate Meadow and Tree Pipits! John was Shropshire Ambassador for the BTO's Garden BirdWatch survey until recently and is currently Chairman of the Shropshire Ornithological Society. He has also contributed to the activities of the Church Stretton branch of the SOS and the Strettons Area Community Wildlife Group. His role in the creation of this volume has largely been in the statistical manipulation of the bird data and preparation of all the standard and special distribution maps of these statistics. John also contributed six species accounts, the section on climate change in the *Habitats* chapter, and the statistical analysis of the arrival dates of migrants.

Allan Dawes has had a lifelong interest in birds and natural history. After moving to the Oswestry area in 1974 and joining the local RSPB group he became more actively involved in conservation and recording, counting wildfowl at the Ellesmere meres and Oerley reservoir for the WeBS since 1978. Allan was organiser for fieldwork in the north-west area for the 1985–90 Atlas project, and contributed several species accounts to the *Atlas* 1992. He has taken part in the BTO BBS since the pilot project began in 1992. He took on the voluntary role of BTO Regional Rep. for Shropshire in 1995, rapidly built up the local BBS coverage to over 50 squares a year, organised the local contribution to several national BTO surveys, and was Shropshire Atlas Organiser for both the recent national and local Atlas projects. He handed on the role in 2016, after 21 years. Allan contributed 42 species accounts, commented on drafts of all the others, compiled the data for many of the charts, and commented on the other chapters for this Avifauna.

Geoff Holmes is a retired Chartered Surveyor. He was born in Shrewsbury where he still lives. His interest in birding began in the mid-1970s and soon developed into a passion to see new species, with much of the 1980s and 1990s spent travelling the country to see rarities. More recently, his UK birding has been confined to Shropshire and especially Venus Pool, but still with the occasional twitch and annual visits to the Isles of Scilly in October. He has travelled and birded extensively worldwide. He was Shropshire Bird Recorder from 1997 to 2014, and Bird Report Editor from 1997 to 2004 and again from 2011 to 2014. Geoff (together with Graham Walker) led the Rarities Review, which ensured that all records included in this Avifauna have been validated. He commented on many of the species accounts and checked a large number of records referred to in them. A married man with two children and four grandchildren, his other interests are Shrewsbury Town FC, real ale and all things St Agnes.

John Tucker has been interested in birds from boyhood in Portsmouth, and a BTO member since 1964, developing an interest in the history of bird recording. He spent six years in Zambia and became a ringer there. After returning to Britain, and two biological degrees with a particular interest in environmental change and historical ecology, he followed a career in nature conservation, initially as an ecologist on Hartlebury Common, Worcestershire, where he was senior author of *Hartlebury Common: A Social and Natural History* (1986). John moved to Shropshire in 1983 to be Conservation Officer with Shropshire Wildlife Trust until 2003, after which he became an ecologist at the Shropshire Hills AONB Partnership for five years. John is currently on the SOS Conservation Sub-committee and is author of the unique website *The Historical Ornithology of Shropshire*, and *Beckwith's Nineteenth Century Birds of Shropshire* (Tucker & Tucker 2018). The website was of considerable help to the authors of the species accounts, 31 of which John contributed.

Graham Walker was born in Lincolnshire, and joined the Nature Conservancy Council, a forerunner of Natural England, in 1981 after obtaining an honours degree in Applied Biology from the University of Bath. After some six years working on ancient woodland inventories in Scotland, he moved to Shropshire in 1988 to undertake County-based conservation work; for the majority of his career he was Natural England's Conservation Officer for Staffordshire. He became a member of SOS Conservation Sub-committee at its inception in 1992, becoming Chairman a year later, a position he still occupies. This role includes responsibility for drawing up and implementing the management plan for the SOS reserve at Venus Pool. In addition to his work on the Rarities Review, Graham contributed 65 species accounts to this Avifauna, including 60 rarities, and commented on most of the others. He also commented extensively on the other chapters. He is widely travelled, enjoying all aspects of the natural world on his regular trips.

Authors

The species accounts have been written by 27 different authors: John Arnfield, Dawn Balmer, Glenn Bishton, Pete Carty, Simon Cooter, Stuart Cowper, Tony Cross, Allan Dawes, David Farncombe, Michelle Frater, Martin Grant, Helen Griffiths, Bob Harris, Simon Holloway, Tom Lowe, Richard Moores, Linda Munday, Peter Nickless, Josie Owen, Mike Shurmer, Leo Smith, Gerry Thomas, John Tucker, John Turner, Graham Walker, Tom Wall and Colin Wright. The author(s) of each account are credited at the end of it.

Authors of other chapters are named in the chapter heading.

Photographers

The images in the *Species Accounts* have been contributed by 21 different photographers: Jim Almond, Terry Arch, Dave Barnes, Ian Butler, Jon Cartwright, Yvonne Chadwick, Dave Chapman, Tony Cross, John Fielding, Bob Hart, John Hawkins, Allan Heath, Paul King, Dawn Micklewright, John Robinson, Leo Smith, Rob Stokes, Gareth Thomas, Celia Todd, John Tucker and Tony Webb.

The images in the *Habitats* chapter have been contributed by Jim Almond, Simon Cooter, Mike Crawshaw (Natural England), John Hawkins, Stephanie Hayes (Shropshire Hills AONB), Peter and Jane Howsam, Leo Smith and Gareth Thomas. Simon Holloway and Tom Waddell contributed images for the *History* chapter.

Each photographer is acknowledged in the caption to the photograph.

Species Accounts Chapter

The status of each species was determined by Graham Walker and Leo Smith, in consultation with Geoff Holmes, in accordance with the criteria in the introduction to the *Species Accounts* (p. 61).

All maps and tetrad occupancy tables, and the Avifauna website, have been compiled by John Arnfield.

Working maps, final map distribution overlays and the maps posted on the Avifauna website were produced using Dr Alan Morton's DMAP software (www.dmap.co.uk).

To ensure consistency of judgement in the charts showing the number of occurrences each year, or in each winter period, the assessment was made by Allan Dawes, in consultation with Geoff Holmes and Graham Walker.

Bird Atlas 2007–11 (a partnership between the BTO, BirdWatch Ireland and the Scottish Ornithologists' Club) provided the impetus for the recent Shropshire Atlas, which was greatly facilitated by the data input and management systems provided by BTO for the national Atlas. Thanks are also due to the BTO for keeping open the online recording system for the two years 2012 and 2013, after data collection for the BTO national Atlas ended.

Thanks to BTO staff Dawn Balmer and Simon Gillings for advice and information on a variety of subjects, Andy Musgrove, Kate Risely and Sarah Harris for supplying BBS data; Heidi Mellan for WeBS data; Mike Toms, Kate Risely and Clare Simm, and local GBW Ambassador John Arnfield, for supplying GBW graphs and data; Rob Robinson for answering queries about the Online Ringing Reports; David Lack for responding to data requests about various BTO Atlases and survey results; and Carole Showell, BTO librarian, for sourcing various references and providing loans.

Thanks also to Natural England staff from the Telford Office (Shropshire area in the West Midlands region) who provided information and statistics for some of the species accounts, and more importantly for the *Habitats* chapter, especially about Agri-environment Schemes and SSSIs: Dave Cragg, Chris Hogarth, Denise Howe, Frances McCullagh and David Ragbourne.

Roger Riddington, editor of *British Birds*, provided copies of articles and records from back issues on several occasions.

Local residents who supplied information and comment on particular species are acknowledged in the individual accounts (*pers. comm.* or *in litt.*).

WeBS results were provided by BTO, who requested the following acknowledgement:

'Data were supplied by the Wetland Bird Survey (WeBS), a joint scheme of the British Trust for Ornithology, Royal Society for the Protection of Birds and Joint Nature Conservation Committee (the last on behalf of the statutory nature conservation bodies: Natural England, Natural Resources Wales and Scottish Natural Heritage and the Department of the Environment Northern Ireland) in association with The Wildfowl & Wetlands Trust. Although WeBS data are presented within this report, in some cases the figures may not have been fully checked and validated. Therefore, for any detailed analyses of WeBS data, enquiries should be directed to the WeBS team at the British Trust for Ornithology, The Nunnery, Thetford, IP24 2PU (webs@bto.org).

Other Chapters and Appendices

Mike Ashton (MA Design) produced Maps 1–8 in the *Habitats* chapter, and the background map for the species accounts' maps.

The Crown Copyright maps are reproduced under the terms of Crown Copyright and the Open Government Licence www.metoffice.gov.uk/about-us/legal/tandc#Use-of-Crown-Copyright and www.nationalarchives.gov.uk/doc/open-government-licence/version/3/.

Some sections of the *Habitats* chapter were largely drafted by other authors:
- John Arnfield: Climate, and Climate Change
- Sarah Gibson (SWT Communications Officer): sections dealing with SWT Reserves
- Guy Shorrock (RSPB Senior Investigations Officer): Preventing Persecution of Birds of Prey (this is an extract from a more complete article published on the Avifauna website www.shropshirebirds.com/AvifaunaSupplement)
- Prof. Ian Trueman (Wolverhampton University): Botanical Change and its Significance for Bird Populations.

Additional information and advice for the *Habitats* chapter was provided by:
- Jonathan Groom (then Shropshire Council Biodiversity Data Officer) for several sections
- Mike Kelly (Shropshire Hills AONB Natural Environment Officer) and Robert Harris (Sheffield University) on the section on Rivers and Streams
- Fiona Gomersall (Conservation Officer) and Robin Mager (Planning & Data Officer) at SWT for information about local wildlife sites
- Keith Seabridge (Farm Surveys Food and Farming Directorate, Defra, York) for providing access to government farm statistics, helping with their interpretation, and searching the archives to answer more detailed questions
- Julian Langford, leader of the ringing group at Allscott SSSI

- David Tompkins (NE Reserve Manager for Fenn's, Whixall and Bettisfield Mosses NNR), for the section on the NNR and Fenn's, Whixall, Bettisfield, Wem & Cadney Mosses SSSI
- Peter Carty (Countryside Manager, National Trust) and Andrew Perry (Ecologist, National Trust), for the section on the Long Mynd
- Tom Wall and Simon Cooter (previous and current NE Senior Reserve Manager for Stiperstones NNR), for the sections on the Stiperstones and the Hollies SSSI and the Back to Purple project.

The relevant government farm statistics can be found on the Avifauna website (see p. 73).

John Arnfield produced the maps, tetrad letter key, table of targets achieved and the summary of weather conditions in 2007–13 (the recent Atlas period), and Allan Dawes produced the charts, in Appendix 1.

John Tucker compiled the grid references in the Gazetteer (Appendix 2).

Rob Rowe supplied the scientific names for the plants and animals where these were not supplied by the relevant authors (Appendix 3).

Graham Walker compiled the list of references in a consistent format (Appendix 4).

Leo Smith and Graham Walker compiled the Shropshire List table (Appendix 6).

Proof reading of the whole book, including the species accounts, was carried out by Gay Walker.

Atlas Fieldwork and Area Co-ordinators

The fieldwork for the 2007–13 local Atlas project was organised by Allan Dawes, concurrently with the national BTO Atlas. The County's 870 tetrads were divided up into nine areas, each with a co-ordinator who helped recruit the fieldworkers, ensure coverage of every tetrad and validate records.

Area	Largest Town	No. of Tetrads	Co-ordinator
1. North West	Oswestry	126	Allan Dawes
2. North	Wem	99	David Farncombe
3. North East	Market Drayton	65	Gerry Thomas
4. North Central	Shrewsbury	100	Bob Parker
5. East	Telford	101	Glenn Bishton
6. South West	Bishop's Castle	110	Leo Smith
7. South Central	Church Stretton	100	John Arnfield
8. South East	Bridgnorth	102	Linda Munday
9. South	Ludlow	67	Jim Martin

The 1985–90 Atlas fieldwork had 10 Area Co-ordinators, but in 2007–13, Area 4 in the 1985–90 Atlas, Westbury, was divided into two, with the northern section (SJ grid squares) administered by Area 1, and the SO grid squares by Area 6.

Atlas Fieldworkers

The 650 observers who contributed winter or breeding records at the tetrad level, or who undertook TTVs, are listed below. Over 630 of these entered records themselves, and are taken from the BTO Atlas dataset, but a few submitted records to their Area Co-ordinator, and have been added to the list by them.

Mel ab Owain
C. Alder
M.G. Allderidge
Sean Allison
Jim Almond
Shirley A. Amies
Guy Anderson
Helen Anderson-Smith
Angus Andrew
Annie Andrews
Brian Andrews
John K. Andrews
Tessa Anning
Paul Anthony
Mike G. Archer
John Arnfield
John W. Arrowsmith
Paul Arrowsmith
Brian R. Ashley
Paul Ashworth
W.J. Baber
Raymond Bacciochi
Helen Backhouse
Claire S. Backshall
John C. Bacon
Ian Baggley
Mark A. Baigent
Helen Baker
James Baker
Kay Ball
Dawn E. Balmer
M. Bamber
Malcolm H. Bannister
Lee Barber
Paul Barker
Simon R.J. Barker
Dave Barnes
Dave Barrow
Jill P. Barrow
D.W. Bastin
Paul W. Bateman
Keith Bates
Vince R. Beaney
Felicity Beaumont
Kate Bedford
Marion Bell
Paul Bell
Pamela Belton
Kath Bennett
Margaret Bennett
Yvonne Benting

Bob Berry
Robin Berry
Thelma Birch
Glenn Bishton
Trevor Blackshaw
Dave Blanning
Pete Boardman
William Booth
Karen Boswarva
Holly Bowler
Hazel Bows
Simon Boyes
C. Bradley
P.G. Bradley
Christine Brandon-Lodge
Gary Branfield
Kirsty Brannan
Mark T. Breaks
Andrew C. Breed
Anne Brenchley
Emma J. Broad
Roger C. Broadbent
Jonathan Bronner
Amanda Brown
Graham Brown
Ian A.R. Brown
Mike F. Brown
Val Bruce
Sylvia M. Bryan
Anna Bunney
Michael Burman
Barbara E. Burns
David W. Burns
Jeff Butcher
John M. Butterworth
Carol Buxton
Andrew Camp
M.J. Carson
Isabel Carter
Tim Carter
Peter Carty
Dianne Cearns
Yvonne Chadwick
Jonathan Chamberlain
K.A. Chapman
Rob E. Chapman
Robert A. Chapman
David G. Chester-Master
J. Chidlow
Michael J. Chorley
J. Clarke

John E. Clarke
Graham Clarkson
Cassy Clayton
Kevin M. Clements
Peter Coffey
Alison Coggon
Steve P. Coney
Henry R.A. Cook
A.I. Cooke
Clive Cooke
David A. Cookson
Barbara Cooper
Terry Cooper
Tony A. Cooper
Simon Cooter
Paul G. Copestake
T. Corcoran
Rob A. Corfield
Julie Cowley
Stuart H. Cowper
Val Cox
Dave Cragg
Robert J. Cripps
Andrew Cristinacce
Tony Cross
Patrick Cullen
Anne C. Cummings
Mike Curtis
Rob J. Curtis
Ray Cusack
Andrew Cutts
Andrew Dale
Barbara Daniels
Jonthan D'arcy
Eric Davies
Gillian Davies
Matthew J. Davies
Phil Davies
Stephen Davies
Steven Davies
Steven Davies
Warwick Davies
Anne Davis
Cherly Davis
Jackie Davis
Allan P. Dawes
Ruth A. Dawes
Peter Dawson
Robert Dawson
Tony Dawson
John J. Day

Peter G. Deans
Bob Dennison
John S. Dennison
Jane Dibnah
David Dilks
Ian A. Dillon
Julie A. Dix
Anthony Dixon
S.G. Dodd
F.R. Dodsworth
Hamish Dolphin
Christopher R. Doughty
Vince M. Downs
Jude Duffy
Chris J. Dunkerley
Robert Eardley
Stuart Edmunds
Bill Edwards
Gareth Egarr
Ian S. Ellis
Anthony Emery
Dave Emery
Anthony D. Evans
Dave Evans
Frances Evans
John Evans
Mair Evans
A.H. Eveleigh
Brian Fahey
Michael H. Fallon
David P. Farncombe
David J. Faulkner
Lauren Fennell
Tony J. Ferris
Janet Finney
Christine Fleming
K. Fleming
Bernard Ford
Naomi Ford
Tony Ford
Sheila M. Foster
Stuart Foster
Andrew Fox
J. Fox
M. Fox
Dorcas Frame
Lorely Francis
Roger D. Franklin
Michelle Frater
Julian French
Janet Frost

Heidi Fuller
Dave W. Fulton
Roy Fussell
James Gale
Mandy Gardner
Sharon Garforth
Allan Gaunt
Martin George
Len G. Gibbons
Geoff Gibbs
Susanne Gibson
Lloyd Gifford
Gordon Gissing
Gill Glover
Andrew Godson
Peter Golborn
Kim Golden
Sheelagh M. Gooch
Liv Goode
David Goodwin
Shirley Gould
Malcolm Graham
Martin G. Grant
Vaughan Grantham
Chris Green
David B. Green
Harry Green
Nigel Green
Jeremy J.D. Greenwood
R. Greer
Stuart Greer
Scott Greeves
Frank C. Gribble
Helen J. Griffiths
James Griffiths
Richard L. Groves
James P.G. Grundy
John E. Guest
David Gurr-Gearing
Anthony C. Gutteridge
Janet Guy
Ann Hadfield
Richard Halahan
Alex Hale
Geoff C. Hall
June Hall
Stephen Hall
Trevor J. Halsey
John C. Hamlet
Peter Hammersley
Doug Hampson
John Handley
George A.G. Hann
Dale Hardgrave
Garry Hardy
Frank J. Harley

Andrew Harmer
Robert J. Harris
Robin G. Harris
Anthony J. Harrison
Michael Harrison
John Hawkins
Steve Haycox
Roger Hayes
John A. Hazell
Annie Hazlehurst
Jerry Hazzard
Sean Healey
Alan W. Heath
Bob Henning
Stephen J. Hewitt
Lisa Hickman
Brian Hicks
Anthony R. Hill
Timothy R.G. Hill
Frank R.J. Hinde
Patricia A. Holbourn-Williams
John D. Holder
Simon Holloway
Geoff E. Holmes
D. Horsley
Patrick C. House
Tessa Huggett
Estelle V. Hughes
Charles Hull
Derek Humphries
Glenn Hunt
Belinda Hunter
Moira Hurley
J.V.P. Hutchins
John Isherwood
Mel Isherwood
Andrew Jackson
Steve Jaggs
Jim Jarrett
Laura Johns
Chris Johnson
Alexander Jones
B.A. Jones
Barry Jones
Ceri M. Jones
Chris Jones
Dyfed W. Jones
Janet L. Jones
Julia Jones
Ray Jones
Richard Jones
Samuel Jones
Tim Jones
William Jones
Peter Jordan
Robin Jukes-Hughes

Jon Kean
John Kedward
Jane B. Kelsall
Heather Kidd
David B. Kightley
Sylvia Kingsbury
Ron P. Kinrade
Melvyn S. Kirby
Dan Knight
John Knowles
Gehardt Kruckow
Jo Kudlacik
Alan K. Kydd
William H. Lacey
Geoff Laight
David Lamacraft
Cath Landles
Richard P. Law
Andrew Lawrence
Peter Lawton
Miles Leach
Katrina Lear-Parkes
Thomas L. Lerwill
Lucy E. Lewis
Simon Lewis
Stephen Lewis
Alex Lickley
Michael J.S. Liley
Derek A.L. Lincoln
Jonathan D. Lingard
Allison Littlehales
John V. Lloyd
Jonathan Lloyd
Katie R. Lloyd
Malcolm Loft
Tom Lowe
Shaun Luke
John H. Luscombe
Clive Mabbutt
Hilary MacBean
Peter G.K. Maccutchan
Matthew Macfadyen
Maureen Macgregor
Donald MacKenzie
John Mackintosh
David Mapp
William Marler
Peta Marshall
Matthew Marston
Jim H. Martin
John P. Martin
Tony Mason
Denise Matthews
Dave Mayfield
Dawn McCallion
Anna McCann

Seamus McCann
Tom B. McCanna
Ian McCulloch
Sue Mcdougall
Lucy McFarlane
Alastair Mckie
Jon McLeod
David Mead
Clive J. Millard
H.J. Miller
Barrie Mills
Jill Mills
John Milner
James Milton
Jay Mitchell
Patrick G. Moore
Richard Moores
Nick Moran
Anthony Morgan
David Morris
Ray Morris
Glenn E. Morris
Jacqui Morrish
Andrew Morton
Andrew J. Morton
Christine Moss
Robert Moss
Jason Mossman
Joan Mowl
Lysbeth Muirhead
Mervin Mullard
John R. Muller
Linda Munday
John B. Murray
Neil Nash
Alison J. Nicholls
David Nicholls
Peter E. Nicholls
Steven Nichols
Peter Nickless
Benjamin Niddre-Davies
Martin Noble
Rosie Nock
Liz Norton
Stephen O'Donnell
Joe O'Hanlon
Jane Osman
Terry L. Ostler
Peter Overton
Josie Owen
Marjorie J. Owen
Derek Owens
Ian Owens
Katie Owens
Mike Paddock
Steve Paling

Gary Palmer
R.B. Palmer
Scott Pardoe
Bob Parker
Pat Parker
Patricia Parkyn
Christopher Parr
Haydn Parry
Marion Parsons
J.F. Patient
Kath Patrick
Biff Patterson
Steven Payne
Trevor Payne
Dave Pearce
Roger Pearce
Roger Peart
Martin Pennell
Christopher Penny
Gavin Peplow
Christine Phelps
Emma Phillips
N.J. Phillips
Rachel Phillips
Roy Phillips
Bob Philpot
Clare Pickering
Philip J. Pickin
Sallie A. Pittam
Richard Pitts
Nick D. Pomiankowski
Chris Poole
Clive R. Poole
S.J. Pope
Kevin Postones
Andrea Powell
Iain Prentice
Tim Preston
Catherine Price
Elizabeth Prior
Kevin T.P. Pryce
Roger Pugh
Prue M. Quayle
Ian L. Ralphs
Michael A. Rayner
Ann Reed
Rob Rees
Alan P. Reid
Anne Reid
Mary Reid
Ann Remfry
George D. Rennie
Daniel Reynolds
Anthony V. Riden
Peter L. Robberts
Andrew Roberts

Rob Roberts
David A. Rogers
Jim Rowe
Rob Rowe
Alan Rowley
Sheila Royle
J. Rushton
Paul Rutter
Elaine Sandilands
Jack Sankey
Mark Sargeant
Steve P. Satterthwaite
Steven Savage
Matt Schofield
David Scott
R.E. Scott
Dennis M. Seager
Jacob Seaward
Sid Shannon
Sue Sharpe
James Shaw
Yvonne Shaw
Anne Shepherdson
Sally P. Shipman
Graham Short
Mike Shurmer
Jack D. Shutt
Gill M. Silk
Michael Sillence
Joan Simister
John Sirrett
Jeremy Smallwood
James Smith
Justin Smith
Kevin Smith
Leo N. Smith
Margaret Smith
Michael Smith
Nick Smith
Richard N. Smith
Robin G. Smith
Susanne Smith
Steve J. Snithee
Philip A.L. Souster
Matt Southam
Derek Sparkes
Ian M. Spence
David Stafford
Jennifer Stanley
Derek Starkins
Michael Starr
Jenny Steel
Sue Stevens
Archie Stewart
Mark Stewart
Stephen P. Stocks

Peter Straughan
Tamasine A. Stretton
R.A. Stuttard
Anne Suffolk
Jane L. Sutherland
Richard J. Sutton
John H. Swift
Robert W. Swift
Richard Swindells
Sue D. Swindells
A.J. Taylor
Elisabeth C. Taylor
Lorna Taylor
Nancy Taylor
Phil Taylor
Steph Taylor-Hodge
J.E. Teare
Andrew Thomas
David H. Thomas
Gerry Thomas
Jenny Thomas
Kelly Thomas
Melanie Thomas
Roger Thomas
Ian Thompson
John Thompson
Julian P. Thompson
James H. Thomson
Kate Thorne
Celia Todd
Joy Tomkinson
Chris Toms
James Towill
Leon Towns
Chris Travis
Patricia Treves
Derrick I. Trowman
John J. Tucker
Michael R. Tucker
Michael Tupling
Louise Turner
Steve Turner
Reg E.T. Turrell
Simon Twigger
Caroline Uff
Graham Uney
Jeff Upex
Jenny Vanderhook
Rob Vaughn
Richard Vernon
Jenny Vine
John Wain
Christopher Walker
Graham Walker
Tom H. Wall
Michael J. Wall

Michael F. Wallace
Kevin W. Waterfall
Graham M. Waterman
Bill Watkins
Peter E.W. Watson
Anne Waygood
Chris Waygood
Richard Webb
Elizabeth Webster
Lynne Webster
John Wells
John N. Wells
Tom J. Wells
Dave Western
David Wheeler
Mike J. Wheeler
A. White
John R. White
Richard N. White
Matt Whittle
Janet Wilcoxon
Andrew Wilkinson
Colin E.W. Wilkinson
John Wilkinson
Brian C. Willder
Alan K. Williams
Ann Williams
B.V. Williams
David W. Williams
Heather Williams
C.S. Williamson
David J. Wilson
Stephen Wilson
Anthony J. Witts
A.J. Wood
Carol A. Wood
Hugh Wood
Malcolm Wood
G. Woodward
Ian D. Woodward
Geoff Wookey
Marian Wootton
Jean Worthington
Dan Wrench
Colin E.W. Wright
I.D. Wright
Jenny Wright
Karen M.H. Wright
Stephen D. Wright
Duncan Yapp
Phil Yates
Dorothy Young
G.A. Young
Stephen Young

British (English) Vernacular Name	IOC World Bird List International English Name	Scientific Name	Birds of Conservation Concern (4)	Shropshire Status	Breeding Season Population					Winter Population		Garden BirdWatch Rank and Reporting Rate
					Estimate from TTV Counts	Note	Unit	Rank (based on TTV counts)	Estimate in Account (if different)	Estimate from TTV Counts	Estimate in Account	
Green-winged Teal		Anas carolinensis	Green	Va								
Red-crested Pochard		Netta rufina	Green	NV1								
Pochard	Common Pochard	Aythya ferina	Red	W3 (HB)					0	70–100	<=100	
Ferruginous Duck		Aythya nyroca	Green	Va								
Ring-necked Duck		Aythya collaris	Green	P1, W1								
Tufted Duck		Aythya fuligula	Green	R4	120–150	#	P		230–380	800–920		
Scaup	Greater Scaup	Aythya marila	Red	W2								
Lesser Scaup		Aythya affinis	Green	Va								
Eider	Common Eider	Somateria mollissima	Amber	Va								
Velvet Scoter		Melanitta fusca	Red	Va								
Common Scoter		Melanitta nigra	Red	P2, W2								
Long-tailed Duck		Clangula hyemalis	Red	W1								
Goldeneye	Common Goldeneye	Bucephala clangula	Amber	W3						15–20	40–100	
Smew		Mergellus albellus	Amber	W1								
Goosander	Common Merganser	Mergus merganser	Green	R3, W4	80–90	#	P			330–370	330–370	
Red-breasted Merganser		Mergus serrator	Green	W1								
Ruddy Duck		Oxyura jamaicensis	Green	NR (eradicated)	1	#			0			
Black Grouse		Lyrurus tetrix	Red	V2 (HB)								
Red Grouse	Willow Ptarmigan	Lagopus lagopus	Amber	R3	119–128	#	P		81–102	100–110	200–400	
Red-legged Partridge		Alectoris rufa	Green	NR4 (SR)	1,000–1,500		T	50				
Grey Partridge		Perdix perdix	Red	R4 (SR)	350–400		T					
Quail	Common Quail	Coturnix coturnix	Amber	S2	13–17	#	M		5			
Pheasant	Common Pheasant	Phasianus colchicus	Green	NR7 (SR)	43,400–45,000		P					
Red-throated Diver	Red-throated Loon	Gavia stellata	Green	W1								
Black-throated Diver	Black-throated Loon	Gavia arctica	Amber	W1								
Great Northern Diver	Common Loon	Gavia immer	Amber	W1								
Storm Petrel	European Storm Petrel	Hydrobates pelagicus	Amber	Va								
Leach's Petrel	Leach's Storm Petrel	Oceanodroma leucorhoa	Amber	Va								
Fulmar	Northern Fulmar	Fulmarus glacialis	Amber	Va								
Manx Shearwater		Puffinus puffinus	Amber	Va								
Little Grebe		Tachybaptus ruficollis	Green	R4	35–70	#	P		210–350	100–110	200–400	
Red-necked Grebe		Podiceps grisegena	Red	W1								
Great Crested Grebe		Podiceps cristatus	Green	R4	40–50	#	P		110–150	80–100	30–70	
Slavonian Grebe	Horned Grebe	Podiceps auritus	Red	W1								
Black-necked Grebe		Podiceps nigricollis	Amber	P2 (HB)								
Black Stork		Ciconia nigra	Green	Va								

British (English) Vernacular Name	IOC World Bird List International English Name	Scientific Name	Birds of Conservation Concern (4)	Shropshire Status	Breeding Season Population					Winter Population		Garden BirdWatch Rank and Reporting Rate
					Estimate from TTV Counts	Note	Unit	Rank (based on TTV counts)	Estimate in Account (if different)	Estimate from TTV Counts	Estimate in Account	
White Stork		*Ciconia ciconia*	Green	Va								
Glossy Ibis		*Plegadis falcinellus*	Green	V1								
Spoonbill	Eurasian Spoonbill	*Platalea leucorodia*	Amber	P1								
Bittern	Eurasian Bittern	*Botaurus stellaris*	Amber	W1								
Little Bittern		*Ixobrychus minutus*	Green	Va								
Night-heron	Black-crowned Night Heron	*Nycticorax nycticorax*	Green	Va								
Squacco Heron		*Ardeola ralloides*	Green	NMR								
Cattle Egret	Western Cattle Egret	*Bubulcus ibis*	Green	V1								
Grey Heron		*Ardea cinerea*	Green	R4	115		P			550–580		
Purple Heron		*Ardea purpurea*	Green	Va								
Great White Egret	Great Egret	*Ardea alba*	Green	V2								
Little Egret		*Egretta garzetta*	Green	P3, W3								
Magnificent Frigatebird		*Fregata magnificens*	Green	Va								
Gannet	Northern Gannet	*Morus bassanus*	Amber	Va								
Shag	European Shag	*Phalacrocorax aristotelis*	Red	Va								
Cormorant	Great Cormorant	*Phalacrocorax carbo*	Green	W4, V3						200–240		
Osprey	Western Osprey	*Pandion haliaetus*	Amber	P2								
Honey-buzzard	European Honey Buzzard	*Pernis apivorus*	Amber	P1 (HB)								
Sparrowhawk	Eurasian Sparrowhawk	*Accipiter nisus*	Green	R4	540–570	#	P		530–1600			23 (13%)
Goshawk	Northern Goshawk	*Accipiter gentilis*	Green	NR3	9–11		P		31–50			
Marsh Harrier	Western Marsh Harrier	*Circus aeruginosus*	Amber	P2 (HB)								
Hen Harrier		*Circus cyaneus*	Red	W2, P2								
Montagu's Harrier		*Circus pygargus*	Amber	P1								
Red Kite		*Milvus milvus*	Green	R3	16–17				50			
White-tailed Eagle		*Haliaeetus albicilla*	Red	Va								
Rough-legged Buzzard		*Buteo lagopus*	Green	W1								
Buzzard	Common Buzzard	*Buteo buteo*	Green	R5	1,850–2,500		P	44	2,000–2,500			
Little Bustard		*Tetrax tetrax*	Green	NMR								
Water Rail		*Rallus aquaticus*	Green	R2, W3	2–3	#	T		2–3			
Corncrake	Corn Crake	*Crex crex*	Red	P1 (HB)		#						
Little Crake		*Porzana parva*	Green	NMR								
Spotted Crake		*Porzana porzana*	Amber	P1								
Moorhen	Common Moorhen	*Gallinula chloropus*	Green	R6	4,000–4,200		T	38		4,250–4,500	TTV estimate too low	
Coot	Eurasian Coot	*Fulica atra*	Green	W5, R4	300–340	#	P		650–1,300	1,300–1,600	1,600–2,200	
Crane	Common Crane	*Grus grus*	Amber	V2								

British (English) Vernacular Name	IOC World Bird List International English Name	Scientific Name	Birds of Conservation Concern (4)	Shropshire Status	Breeding Season Population					Winter Population		Garden BirdWatch Rank and Reporting Rate
					Estimate from TTV Counts	Note	Unit	Rank (based on TTV counts)	Estimate in Account (if different)	Estimate from TTV Counts	Estimate in Account	
Stone-curlew	Eurasian Stone-curlew	Burhinus oedicnemus	Amber	Va								
Oystercatcher	Eurasian Oystercatcher	Haematopus ostralegus	Amber	S3	18–22		P		35–40			
Black-winged Stilt		Himantopus himantopus	Green	Va								
Avocet	Pied Avocet	Recurvirostra avosetta	Amber	P1								
Lapwing	Northern Lapwing	Vanellus vanellus	Red	W6, BS4	790–860		P		800	6,200–7,200		
White-tailed Plover	White-tailed Lapwing	Vanellus leucurus	Green	Va								
Golden Plover	European Golden Plover	Pluvialis apricaria	Green	W4 (HB)						1,500–1,900		
Grey Plover		Pluvialis squatarola	Amber	P1								
Ringed Plover	Common Ringed Plover	Charadrius hiaticula	Red	P3								
Little Ringed Plover		Charadrius dubius	Green	S3	14–18		P		20–25			
Dotterel	Eurasian Dotterel	Charadrius morinellus	Red	P2								
Whimbrel		Numenius phaeopus	Red	P3								
Curlew	Eurasian Curlew	Numenius arquata	Red	S4, W3	140–160		P		160	35–40		
Bar-tailed Godwit		Limosa lapponica	Amber	P1								
Black-tailed Godwit		Limosa limosa	Red	P3, W1								
Turnstone	Ruddy Turnstone	Arenaria interpres	Amber	P2								
Knot	Red Knot	Calidris canutus	Amber	P2								
Ruff		Calidris pugnax	Red	P3, W1								
Curlew Sandpiper		Calidris ferruginea	Amber	P1								
Temminck's Stint		Calidris temminckii	Green	P1								
Sanderling		Calidris alba	Amber	P2								
Dunlin		Calidris alpina	Amber	P4, W2								
Purple Sandpiper		Calidris maritima	Amber	P1								
Little Stint		Calidris minuta	Green	P1								
White-rumped Sandpiper		Calidris fuscicollis	Green	Va								
Buff-breasted Sandpiper		Calidris subruficollis	Green	Va								
Pectoral Sandpiper		Calidris melanotos	Green	P1								
Woodcock	Eurasian Woodcock	Scolopax rusticola	Red	W6, R3		#	M		<=68	11,100–12,000	11,100–12,000	
Jack Snipe		Lymnocryptes minimus	Green	W3								
Great Snipe		Gallinago media	Green	NMR								
Snipe	Common Snipe	Gallinago gallinago	Amber	W5, BS2			P		15–20	8,800–10,000		
Red-necked Phalarope		Phalaropus lobatus	Red	P1								
Grey Phalarope	Red Phalarope	Phalaropus fulicarius	Green	P1								
Common Sandpiper		Actitis hypoleucos	Amber	P4, S2, W1	6		P		5–10			
Green Sandpiper		Tringa ochropus	Amber	P4, W3						25–30		
Lesser Yellowlegs		Tringa flavipes	Green	Va								
Redshank	Common Redshank	Tringa totanus	Amber	P3, W2, HB	2				0			

British (English) Vernacular Name	IOC World Bird List International English Name	Scientific Name	Birds of Conservation Concern (4)	Shropshire Status	Estimate from TTV Counts	Note	Unit	Rank (based on TTV counts)	Estimate in Account (if different)	Estimate from TTV Counts	Estimate in Account	Garden BirdWatch Rank and Reporting Rate
					Breeding Season Population					Winter Population		
Wood Sandpiper		Tringa glareola	Amber	P2								
Spotted Redshank		Tringa erythropus	Amber	P1								
Greenshank	Common Greenshank	Tringa nebularia	Amber	P3								
Kittiwake	Black-legged Kittiwake	Rissa tridactyla	Red	P2								
Sabine's Gull		Xema sabini	Green	Va								
Black-headed Gull		Chroicocephalus ridibundus	Amber	W6, BS4	200–270		P		65–190	11,700–12,300		
Little Gull		Hydrocoloeus minutus	Green	P2								
Mediterranean Gull		Ichthyaetus melanocephalus	Amber	P2, BS1								
Common Gull	Mew Gull	Larus canus	Amber	W4 (HB)						430–460		
Ring-billed Gull		Larus delawarensis	Green	Va								
Great Black-backed Gull		Larus marinus	Amber	W3								
Glaucous Gull		Larus hyperboreus	Amber	W2								
Iceland Gull		Larus glaucoides	Amber	W2								
Herring Gull	European Herring Gull	Larus argentatus	Red	W4						320–350		
Caspian Gull		Larus cachinnans	Amber	W2								
Yellow-legged Gull		Larus michahellis	Amber	W3								
Lesser Black-backed Gull		Larus fuscus	Amber	W5, BS2						2,300–3,000		
Gull-billed Tern		Gelochelidon nilotica	Green	Va								
Sandwich Tern		Thalasseus sandvicensis	Amber	P1								
Little Tern		Sternula albifrons	Amber	P1								
Roseate Tern		Sterna dougallii	Red	NMR								
Common Tern		Sterna hirundo	Amber	P3, BS1								
Arctic Tern		Sterna paradisaea	Amber	P2								
Whiskered Tern		Chlidonias hybrida	Green	Va								
White-winged Black Tern	White-winged Tern	Chlidonias leucopterus	Green	Va								
Black Tern		Chlidonias niger	Green	P2								
Great Skua		Stercorarius skua	Amber	Va								
Pomarine Skua	Pomarine Jaeger	Stercorarius pomarinus	Green	Va								
Arctic Skua	Parasitic Jaeger	Stercorarius parasiticus	Red	Va								
Long-tailed Skua	Long-tailed Jaeger	Stercorarius longicaudus	Green	Va								
Little Auk		Alle alle	Green	Va								
Common Guillemot	Common Murre	Uria aalge	Amber	Va								
Razorbill		Alca torda	Amber	NMR								
Puffin	Atlantic Puffin	Fratercula arctica	Red	NMR								
Pallas's Sandgrouse		Syrrhaptes paradoxus	Green	NMR								

British (English) Vernacular Name	IOC World Bird List International English Name	Scientific Name	Birds of Conservation Concern (4)	Shropshire Status	Breeding Season Population Estimate from TTV Counts	Note	Unit	Rank (based on TTV counts)	Estimate in Account (if different)	Winter Population Estimate from TTV Counts	Winter Estimate in Account	Garden BirdWatch Rank and Reporting Rate
Feral Pigeon/Rock Dove		Columba livia	Green	R5	1,500–1,600			48	1,000–3,000			30 (5%)
Stock Dove		Columba oenas	Amber	R6	7,100–7,300		T	30				
Woodpigeon	Common Wood Pigeon	Columba palumbus	Green	R7	87,000–89,000		P	5				4 (84%)
Turtle Dove	European Turtle Dove	Streptopelia turtur	Red	S2	20–22		T		<=16			
Collared Dove	Eurasian Collared Dove	Streptopelia decaocto	Green	R6	14,100–14,500		P	19	TTV estimate 'excessive'			11 (54%)
Cuckoo	Common Cuckoo	Cuculus canorus	Red	S3	90–95		P					
Barn Owl	Western Barn Owl	Tyto alba	Green	R4	175–200	#			200–220			
Tawny Owl		Strix aluco	Amber	R4	900–1,000		P	54	530			32 (4%)
Little Owl		Athene noctua	Green	NR4	50–55	#	P		150–200			
Tengmalm's Owl	Boreal Owl	Aegolius funereus	Green	NMR								
Long-eared Owl		Asio otus	Green	R2								
Short-eared Owl		Asio flammeus	Amber	W2, P2								
Nightjar	European Nightjar	Caprimulgus europaeus	Amber	S2					1–3			
Swift	Common Swift	Apus apus	Amber	S5	1,700–1,850		P	46				
Kingfisher	Common Kingfisher	Alcedo atthis	Amber	R4	80–140		P		68–170			
Bee-eater	European Bee-eater	Merops apiaster	Green	Va								
Hoopoe	Eurasian Hoopoe	Upupa epops	Green	P2								
Wryneck	Eurasian Wryneck	Jynx torquilla	Green	P1								
Lesser Spotted Woodpecker		Dryobates minor	Red	R3	30–60		P					
Great Spotted Woodpecker		Dendrocopos major	Green	R6	3,600–3,700		P	39				15 (34%)
Green Woodpecker	European Green Woodpecker	Picus viridis	Green	R4	390–400		P					
Kestrel	Common Kestrel	Falco tinnunculus	Amber	R4	570–590		P		300–350			
Red-footed Falcon		Falco vespertinus	Green	Va								
Merlin		Falco columbarius	Red	R2, W3	1		P		1–3			
Hobby	Eurasian Hobby	Falco subbuteo	Green	S3	40–45		P		<=70 in good years			
Gyr Falcon	Gyrfalcon	Falco rusticolus	Green	NMR								
Peregrine	Peregrine Falcon	Falco peregrinus	Green	R3, W3	15–17		P		20			
Ring-necked Parakeet	Rose-ringed Parakeet	Psittacula krameri	Green	NV1 (or escapee)								
Red-backed Shrike		Lanius collurio	Red	P1 (HB)								
Great Grey Shrike		Lanius excubitor	Green	W2								
Steppe Grey Shrike		Lanius meridionalis	Green	Va								
Woodchat Shrike		Lanius senator	Green	Va								
Golden Oriole	Eurasian Golden Oriole	Oriolus oriolus	Red	P1 (HB)								

British (English) Vernacular Name	IOC World Bird List International English Name	Scientific Name	Birds of Conservation Concern (4)	Shropshire Status	Breeding Season Population					Winter Population		Garden BirdWatch Rank and Reporting Rate
					Estimate from TTV Counts	Note	Unit	Rank (based on TTV counts)	Estimate in Account (if different)	Estimate from TTV Counts	Estimate in Account	
Jay	Eurasian Jay	*Garrulus glandarius*	Green	R5	2,800–2,900		T	42				28 (7%)
Magpie	Eurasian Magpie	*Pica pica*	Green	R6	11,100–11,300		T	27				10 (54%)
Nutcracker	Spotted Nutcracker	*Nucifraga caryocatactes*	Green	W1 (l)								
Chough	Red-billed Chough	*Pyrrhocorax pyrrhocorax*	Green	NMR								
Jackdaw	Western Jackdaw	*Coloeus monedula*	Green	R7	28,600–29,400		P	12				16 (30%)
Rook		*Corvus frugilegus*	Green	R7	16,100–16,750		P	18	20935			29 (7%)
Carrion Crow		*Corvus corone*	Green	R7	23,100–23,700		T	14				18 (29%)
Hooded Crow		*Corvus cornix*	Green	Va								
Raven	Northern Raven	*Corvus corax*	Green	R4	226–234		P		400			
Waxwing	Bohemian Waxwing	*Bombycilla garrulus*	Green	W2 (l)								
Coal Tit		*Periparus ater*	Green	R6	8,100–8,300		T	28				12 (52%)
Marsh Tit		*Poecile palustris*	Red	R5	1,600–1,700		T	47				
Willow Tit		*Poecile montanus*	Red	R3	70–80		P		140–200			
Blue Tit	Eurasian Blue Tit	*Cyanistes caeruleus*	Green	R7	75,300–76,500		T	7				2 (90%)
Great Tit		*Parus major*	Green	R7	51,600–52,400		T	8				6 (77%)
Bearded Tit	Bearded Reedling	*Panurus biarmicus*	Green	V1 (l)								
Woodlark		*Lullula arborea*	Green	W1 (HB)								
Skylark	Eurasian Skylark	*Alauda arvensis*	Red	R6	13,700–14,000		T	22				
Shore Lark	Horned Lark	*Eremophila alpestris*	Amber	Va								
Short-toed Lark	Greater Short-toed Lark	*Calandrella brachydactyla*	Green	NMR								
Sand Martin		*Riparia riparia*	Green	S5	380–1,200	#	N	20	2500			
Swallow	Barn Swallow	*Hirundo rustica*	Green	S6	14,000–14,200		T	20				
House Martin	Common House Martin	*Delichon urbicum*	Amber	S6	12,700–13,300		P	24				
Red-rumped Swallow		*Cecropis daurica*	Green	Va								
Cetti's Warbler		*Cettia cetti*	Green	R2								
Long-tailed Tit		*Aegithalos caudatus*	Green	R6	6,700–7,000		T	31				20 (23%)
Willow Warbler		*Phylloscopus trochilus*	Amber	S6	13,500–13,800		T	23				
Chiffchaff	Common Chiffchaff	*Phylloscopus collybita*	Green	S7, W3	27,900–28,400		T	13				
Iberian Chiffchaff		*Phylloscopus ibericus*	Green	Va								
Wood Warbler		*Phylloscopus sibilatrix*	Red	S3	55–65		M		100			
Pallas's Warbler	Pallas's Leaf Warbler	*Phylloscopus proregulus*	Green	Va								
Yellow-browed Warbler		*Phylloscopus inornatus*	Green	P1								◢
Great Reed Warbler		*Acrocephalus arundinaceus*	Green	NMR								
Sedge Warbler		*Acrocephalus schoenobaenus*	Green	S4	1,000–1,100		T	53	400–800			
Reed Warbler	Eurasian Reed Warbler	*Acrocephalus scirpaceus*	Green	S4	1,200–1350		P	51	335–670			

British (English) Vernacular Name	IOC World Bird List International English Name	Scientific Name	Birds of Conservation Concern (4)	Shropshire Status	Breeding Season Population					Winter Population		Garden BirdWatch Rank and Reporting Rate
					Estimate from TTV Counts	Note	Unit	Rank (based on TTV counts)	Estimate in Account (if different)	Estimate from TTV Counts	Estimate in Account	
Marsh Warbler		Acrocephalus palustris	Red	Va								
Icterine Warbler		Hippolais icterina	Green	NMR								
Grasshopper Warbler	Common Grasshopper Warbler	Locustella naevia	Red	S3	20–25		T		20–25			25 (9%)
Blackcap	Eurasian Blackcap	Sylvia atricapilla	Green	S7, W4	19,300–19,700		T	17				
Garden Warbler		Sylvia borin	Green	S6	4,000–4,200		T	37				
Lesser Whitethroat		Sylvia curruca	Green	S4	1,100–1,200		T	52				
Whitethroat	Common Whitethroat	Sylvia communis	Green	S6	12,000–12,350		T	25				
Dartford Warbler		Sylvia undata	Amber	Va								
Firecrest	Common Firecrest	Regulus ignicapilla	Green	P2, W2, BS1								
Goldcrest		Regulus regulus	Green	R6	5,600–5,800		T	32				34 (3%)
Wren	Eurasian Wren	Troglodytes troglodytes	Green	R7	123,500–125,000		T	1				17 (30%)
Nuthatch	Eurasian Nuthatch	Sitta europaea	Green	R6	7,900–8,300		T	29				19 (29%)
Treecreeper	Eurasian Treecreeper	Certhia familiaris	Green	R5	3,000–3,100		T	41				
Rose-coloured Starling	Rosy Starling	Pastor roseus	Green	Va								
Starling	Common Starling	Sturnus vulgaris	Red	W7, R6	11,800–12,200		P	26				14 (38%)
White's Thrush		Zoothera aurea	Green	NMR								
Ring Ouzel		Turdus torquatus	Red	P3, HB								
Blackbird	Common Blackbird	Turdus merula	Green	R7	104,000–105,500		P	4				1 (92%)
Black-throated Thrush		Turdus atrogularis	Green	Va								
Fieldfare		Turdus pilaris	Red	W6						15,800–16,400		35 (3%)
Redwing		Turdus iliacus	Red	W6						19,500–20,200		36 (2%)
Song Thrush		Turdus philomelos	Red	R7	20,000–20,350		T	16				21 (17%)
Mistle Thrush		Turdus viscivorus	Red	R5	2,400–2,500		T	43				33 (3%)
American Robin		Turdus migratorius	Green	NMR								
Spotted Flycatcher		Muscicapa striata	Red	S4	445–465		T					
Robin	European Robin	Erithacus rubecula	Green	R7	121,100–122,700		T	2				3 (88%)
Bluethroat		Luscinia svecica	Green	Va								
Nightingale	Common Nightingale	Luscinia megarhynchos	Red	S1 (HB)					0			
Pied Flycatcher	European Pied Flycatcher	Ficedula hypoleuca	Red	S4	550–650		P		500			
Black Redstart		Phoenicurus ochruros	Red	P2, W2 (HB)								
Redstart	Common Redstart	Phoenicurus phoenicurus	Amber	S5	3,100–3,300		P	40				
Whinchat		Saxicola rubetra	Red	S3	200–220		P					
Stonechat	European Stonechat	Saxicola rubicola	Green	R4	200–210		P		200			
Wheatear	Northern Wheatear	Oenanthe oenanthe	Green	P4, S3	270–280		P		90			
Desert Wheatear		Oenanthe deserti	Green	Va								
Dipper	White-throated Dipper	Cinclus cinclus	Amber	R4	45–140		P		210			

British (English) Vernacular Name	IOC World Bird List International English Name	Scientific Name	Birds of Conservation Concern (4)	Shropshire Status	Breeding Season Population					Winter Population		Garden BirdWatch Rank and Reporting Rate
					Estimate from TTV Counts	Note	Unit	Rank (based on TTV counts)	Estimate in Account (if different)	Estimate from TTV Counts	Estimate in Account	
House Sparrow		*Passer domesticus*	Red	R7	120,300–123,600		P	3				7 (76%)
Tree Sparrow	Eurasian Tree Sparrow	*Passer montanus*	Red	R5	1,850–2,000		T	45				27 (8%)
Alpine Accentor		*Prunella collaris*	Green	NMR								
Dunnock		*Prunella modularis*	Amber	R7	50,100–51,100		T	9				5 (79%)
Yellow Wagtail	Western Yellow Wagtail	*Motacilla flava*	Red	S4	200–230		T		350–500			
Grey Wagtail		*Motacilla cinerea*	Red	R4	390–410		P					
Pied Wagtail	White Wagtail	*Motacilla alba*	Green	R6	5,000–5,100		P	33				26 (9%)
Richard's Pipit		*Anthus richardi*	Green	NMR								
Meadow Pipit		*Anthus pratensis*	Amber	R5	4,350–4,450		P	34	1,500–2,000			
Tree Pipit		*Anthus trivialis*	Red	S4	700–800		P		600–900			
Water Pipit		*Anthus spinoletta*	Amber	P1, W1								
Rock Pipit	Eurasian Rock Pipit	*Anthus petrosus*	Green	P1								
Chaffinch	Common Chaffinch	*Fringilla coelebs*	Green	R7	86,400–87,700		T	6				8 (70%)
Brambling		*Fringilla montifringilla*	Green	W5 (I)		#				19,400–36,900		31 (4%)
Hawfinch		*Coccothraustes coccothraustes*	Red	W2(I), HB					0			
Bullfinch	Eurasian Bullfinch	*Pyrrhula pyrrhula*	Amber	R6	4,100–4,250		T	36				22 (15%)
Greenfinch	European Greenfinch	*Chloris chloris*	Green	R7	28,600–29,300		P	11	18,000–19,500			9 (58%)
Twite		*Linaria flavirostris*	Red	W1								
Linnet	Common Linnet	*Linaria cannabina*	Red	R6	4,100–4,300		T	35				
Common Redpoll		*Acanthis flammea*	Amber	W2								
Lesser Redpoll		*Acanthis cabaret*	Red	W5, R4	370–400	#	P		150–300			
Arctic Redpoll		*Acanthis hornemanni*	Green	Va								
Parrot Crossbill		*Loxia pytyopsittacus*	Amber	W1 (I)								
Crossbill	Red Crossbill	*Loxia curvirostra*	Green	R4, W3(I)					100–250			
Two-barred Crossbill		*Loxia leucoptera*	Green	W1 (I)								
Goldfinch	European Goldfinch	*Carduelis carduelis*	Green	R7	22,000–22,500		P	15				13 (52%)
Siskin	Eurasian Siskin	*Spinus spinus*	Green	W5, R4	690–730		P		700–750			24 (10%)
Corn Bunting		*Emberiza calandra*	Red	R4	280–320	#	T		400			
Yellowhammer		*Emberiza citrinella*	Red	R6	13,900–14,300		T	21				
Pine Bunting		*Emberiza leucocephalos*	Green	Va								
Cirl Bunting		*Emberiza cirlus*	Red	NMR								
Reed Bunting	Common Reed Bunting	*Emberiza schoeniclus*	Amber	R4	1,400–1,500		T	49				
Lapland Longspur	Lapland Longspur	*Calcarius lapponicus*	Amber	P1								
Snow Bunting		*Plectrophenax nivalis*	Amber	P1, W1								

Index to Names of Birds

Page numbers in **bold** type refer to the relevant species account. References to common and scientific names are listed separately. IOC names are only cross-referenced where they differ significantly from the British common name, i.e. they are not included when they are modified only by the addition of the prefix 'Common', 'Eurasian', 'European', 'Northern' or 'Western'. (Note that some names in everyday use are prefixed by 'Common', such as Common Gull and Common Scoter, and in these cases they are included.) This index excludes the 'Period or Date of First Record' table on pp. 12–16, and Appendix 6, *The Shropshire List* (pp. 520–30).